QUESTIONS & ANSWERS
COMPUTER PROGRAM
(VERSION 3)

The multiple-choice Questions & Answers contained in this book are available on disk (Microsoft Windows).

For the personal-use price of **$69** (*inc. GST & delivery*), users have the choice of on-screen self-testing or printing out examination papers, answer sheets and overlays of correct answers.

For a corporate price of $1600 maximum (*less for small organisations*), the program offers institutions, libraries and organisations self-testing of employees or students or on demand printing of examination papers, answer sheets and overlays of correct answers. Examination papers can be printed and marked in minutes. The questions can be randomly selected or chosen, and the order of responses can be shuffled.

To obtain a copy for personal use, you can either make a direct deposit (phone us for bank details) or send a cheque or money order for $69 to Ocean Publications. We do not have credit card facilities.

To obtain a copy for corporate use, fax, post or email your order, or contact us for a price.

UPGRADES

To upgrade your personal-use copy of a previous version, send us Disk 1 of the Program together with a payment of $45. Corporate clients, please phone.

The Australian Boating Manual

(Third Edition)

As in past editions, I have attempted to retain in the book those flavours and nuances, those sensibilities and enthusiasms that make reading and learning more enjoyable. To express my sentiment and admiration for the beauty, charm, grace, and above all, the individual personalities of vessels, I refer to them as "she", not "it".

DEDICATION

My experience of the human spirit is that it is filled with kindness and generosity.
This book is dedicated those who are less fortunate in their experience.

POINTS OF INTEREST...

1. ARE MARINE VOLUNTEERS AT RISK OF BEING SUED?
(Summarized extract from National Conference of Australian Plaintiff Lawyers' Association, 2000)

Based on over-generalised reference to US litigiousness, there is a grossly exaggerated urban myth that rescuers are constantly at risk of being sued.

Rescuers and Good Samaritans are generally encouraged and supported by the courts. Taking into account the urgency of the situation and the desirable philosophy of protecting others from harm, leniency is shown if they unwittingly do damage.

Most of the Australian emergency services legislation is concerned with granting "powers", rather than imposing duties.

2. WHICH WAY DOWN THE PLUG HOLE?

Earth spins anti-clockwise when viewed from the North Pole, and clockwise when viewed from the South Pole. You can see this difference by observing a spinning ball from the top and bottom. This is known as the Coriolis effect, and is the reason for cyclones spinning anticlockwise in the northern hemisphere and clockwise in the southern hemisphere. The Coriolis effect is zero at the Equator and maximum at the Poles, and it works only over large distances, increasing as a moving object goes faster.

The Coriolis effect has little or no influence over the spin of water draining out of a bath tub. Factors influencing the direction of its spin are mainly the shape of the bath and the waves and currents generated by filling of the bath and by the hand pulling out the plug.

In order to see the Coriolis effect in the bath tub, you will have to carry out a controlled experiment. The tub must be circular and all artificial forces must be allowed to die down by letting the water sit for many days. Such a successful experiment was reported in a 1965 edition of *Nature*, re-reported in *Helix* in the mid-90s.

The influence of the Coriolis effect is also not guaranteed in the case of storms smaller than cyclones. Dust devils can spin either way and tornadoes have been known to spin the "wrong" way.

3. SPLITTING AND WANDERING MAGNETIC POLES
(Professor Pat Quilty, Head of Antarctic Division's Research Program, quoted in the Sydney Morning Herald in 1996)

On Magnetic Poles, the magnetic lines of force run at right angles to the earth's surface. These poles not only wander but also sometimes split. In the Northern Hemisphere there are now two magnetic poles: one in Canada and the other in Siberia. The South Magnetic Pole, which is about 1500 nautical miles from the Geographical Pole and located in southern South America, is in the process of splitting into two. It has also probably shifted some 2000 miles in 400 years, which is an immense rate of movement. The nickel-iron core of the earth is surrounded by a fluid and then by a silicate mantle strewn with bits of iron. The movement of these bits of iron in the internal convection currents is the explanation given for the wandering and splitting of the magnetic poles.

(Author's Note: Another theory is that the North Magnetic Pole is moving southwards, and the South Magnetic Pole is moving northwards on the opposite sides of the earth. In a few thousand years, the two poles may come to be located over the equator and then eventually swap their current locations.)

4. FINDING DIRECTION BY THE MOON
(J.P. Jewell, Australian Volunteer Coast Guard Association)

The curve of the <u>Waxing</u> (getting larger) moon points to the <u>West</u>.   W ← ☾ WAXING

The curve of a waning moon points to the East.   WANING ☽ → E

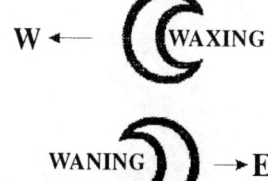

This Third Edition of the Australian Boating manual has expanded with a mind of its own by almost 50%. The author has merely helped it take shape, based on the needs that he has recognised from teaching the subject.

This fifth boating book by Dick Gandy and the Gandy's Exams Computer Program are the culmination of his decades of sea experience, teaching, expert consultations and research into hundreds of books, journals, manuals and technical papers from all over the world.

Every effort has been made to maintain accuracy of information. However, the publisher and the author disclaim any responsibility for any error and/or omissions in the work, or ordinance changes subsequent to publication.

The medical advice and procedures outlined in this book should not be taken without first consulting a doctor or a trained paramedic. At the time of writing this edition, many of the National Marine Safety Committee publications were at draft stage. References made to the NMSC publications throughout this book must therefore be treated with caution.

Copyright Dick Gandy, 1996

First Published Feb 1996
Reprinted (updated where necessary) May 1996, Feb 1997
Second Edition (Revised & Updated) 1999
Olympics Edition (incorporating the Second Edition) 2000

THIRD EDITION 2003

THIS EDITION

National Library of Australia Cataloguing-in-Publishing Data

Gandy, Dick.
    The Australian Boating Manual.

    Includes index.
    ISBN 0 9580971 0 0

    1. Boats and boating - Australia - Handbooks, manuals, etc.
    2. Seamanship - Handbooks, manuals, etc.
    3. Navigation - Handbooks - manuals, etc.
    4. Marine engineering - Handbooks, manuals, etc.
    I. Ocean Publications.
    II. Title.

623.880994

Written / Complied by
CAPT. DICK GANDY
Master Mariner, B.Sc. (Hons) Maritime Commerce (Wales), M.Sc. Marine Law (Wales), Dip.Ed. (Sydney)

Published & Distributed by
OCEAN PUBLICATIONS
(Sydney Maritime Group Pty Ltd.   ABN 71 002 249 416)
9 North Harbour St., Balgowlah, NSW 2093
Tel: (02) 9907-6744   Fax: (02) 9907-6704
e-mail: dgandy@bigpond.com

Editorial Consultants

RUSSELL VASEY
Commander RANR (Ret)
Fellow of the Australian Institute of Navigation
AYF Yacht Master Examiner
Maritime Teacher at TAFE
Sailing & Motor Cruising Trans Pacific, Trans Tasman, Australian circumnavigation and Portsmouth to Gibraltar.
Yacht Racing Sydney to Hobart five times.

ALAN BARRINGHAM
Round the world yachtsman
Commercial Skipper & Engineer
Maritime Teacher at TAFE

Cover Design
ROSS CAREY

ACKNOWLEDGEMENTS

This book reflects the Australian spirit of generosity and community effort. Back in 1995, John Anderson of the Yachting Association of NSW walked the manuscript through the Australian Yachting Federation's endorsement process. Don McKenzie, whom I had never met, phoned to tell me that he had read and 'highly recommended' it, which was endorsed by Alastair Mitchell, and remains endorsed by Phil Jones, the current CEO of AYF.

It was endorsed with equal enthusiasm by Chris Gillett, NSW Squadron Commodore of the Australian Volunteer Coast Guard Association, Keith Jenkins, FOC of the Royal Volunteer Coastal patrol, and Roy Privett, General Manager of the Boating Industry Association of NSW. My sincere thanks to Tom Reed, President of the Volunteer Marine Rescue Western Australia and Inspector Mick Lynch of NSW Water Police for their endorsement of the Third Edition.

Once again, people from most unexpected places, people with busy lives and people I have never met have spent time and effort in helping me to put this edition of the book together. To name a few: Lorraine Dodds and Bob Cowley of Volvo Penta, Darren Hood of Detroit Diesel, Alex Johnston of Muir Windlasses, Don Power of Yamaha Motors, Roger Davey and Rob Flack of Energy Power Systems, Adrian Benetti of Kinnears Pty Ltd., John Davis of Australia-wide Solar, Ian Morelly of RFI Industries, Stephen Fields of Lowrance Electronics, Russell Muller (ex-Altex Coatings), Hood Sailmakers, John Ferris of RFD Marine, and Mark Baxter of Pains-Wessex.

I remain indebted to my friends and colleagues for giving me the benefit of their vast depth of knowledge and editing skill. Russell Vasey edited the nautical section; Alan Barringham the engineering section; Ken Batt, of Bureau of Meteorology, the Meteorology chapter; John Richardson, of Pumptec, the bilges and pumps chapter; David Duff, of Chubb Fire, the fire prevention and control chapter; Des Highfield, of NSW Workcover, the ropes and lifting equipment chapters. Steve Mitchell provided the naval architectural input and Ken D'Cruz, of NSW Waterways, advised me on some of the chapters relevant to commercial certificates.

Martin Phillips, an online pharmacist from www.pharmacy2shop.com, arrived via cyberspace and rewrote the First Aid chapter. Peter Irvin of Queensland Transport helped revise the Radio chapter with the latest changes in the nick of time. I learnt a great deal from attending the excellent course on GPS & Electronic Charts conducted by Ravi Nijjer of Marine Consultancy Group of Melbourne. Ross Carey, of course, is the best cover designer in the world. Without the input of each and every one of them, the book would be incomplete.

CONTENTS

CHAPTER		PAGE
1.	**BUYING & INSURING A BOAT**	10
	Includes… Options - Stability - Buoyancy - Inspections - Technical details - Tests - Sellers - Dealers	
2.	**TRAINING SCHEMES, RECOGNITION OF DEFENCE QUALIFICATIONS & ON BOARD TRAINING**	20
	Includes… Recreation & Commercial Certificates - Competency based training - STCW95 endorsement	
3.	**NAUTICAL TERMS, FITTINGS, HULLS & RIGS**	33
	Includes… Helm orders - Points of compass - Keels - Parts of vessel - Boat construction - Sails	
4.	**ROPES, KNOTS & SPLICES**	55
	Includes… Bulldog grips - Berthing ropes - Rigging a stage - Rigging a bosun's chair - Boarding ladders	
5.	**BOAT MAINTENANCE & CARE OF SAILS**	76
	Includes… Corrosion - Osmosis - Paints - Varnishes - Anti-fouling - Slipping - Trailer maintenance	
6.	**IALA BUOYAGE SYSTEM & OTHER COLOUR ILLUSTRATIONS**	96
	Includes… International Flags and Pennants - Fire Extinguishers - Pyrotechnic Distress Signals - Symbols & Abbreviations Used on Charts - Marker buoys for safer boating - light characteristics - Morse code	
7.	**ANCHORING SYSTEMS**	114
	Includes… Anchoring Terms - Chains - Capstan - Windlass	
8.	**BOAT HANDLING & TOWING**	130
	Includes… Sailing theory - Trailers - Interaction - Bars	
9.	**LIFTING EQUIPMENT & WINCHES**	161
	Includes… Safe Working Loads - Purchases - Block systems - Cranes - Shackles - Hooks - Slings	
10.	**WATCHKEEPING DUTIES & NAUTICAL EMERGENCIES**	180
	Includes… Reporting Marine Casualties & Incidents – Salvage Claims - Planned Grounding	
11.	**INTERNATIONAL & LOCAL COLLISION PREVENTION REGULATIONS**	193
12.	**AUSTRALIAN BOATING REGULATIONS & POLLUTION CONTROL**	241
	Includes… Equipment Standards - Length Measurement - Speed Limits - Restrictions - Jet Boat Regulations - Boat Licence - Registration - National Compliance - Interstate Voyages - Overseas Voyages - NAVAREA X Navigational Warnings - Annual Notices To Mariners - USL - NMSC - Survey - Inspection - Competent Person - Operational Areas - Fast Craft - Seaworthiness - Hire & Drive - Static Electricity - Bird Droppings	

13.	**LOAD, BUOYANCY & STABILITY**		**274**
	Includes... Why Boats Float - Draft Marks - Classification Societies - Load Lines Certificates - Boat Stability		

14.	**GPS & ELECTRONIC CHARTS**	**287**
	Includes... GPS – DGPS - GLONASS - GALILEO - Electronic Charts - Plotters - RTK Machinery Guidance System - UAIS Automatic Identification System - Buying GPS	

15.	**SOUNDERS, LOGS, AUTOPILOTS & ALARMS**	**304**
	Includes... Detectors - Indicators	

16.	**RADAR, SART & OTHER RADAR TRANSPONDERS**	**324**
	Includes... Buying Radar - Horizon Distance Calculation - Anti-Collision Radar Transponders - Radar Reflective Distress Signals	

17.	17.1	**NAVIGATION – COMPASSES, BEARINGS & HEADINGS**	**358**
		Includes... Magnetic, Gyro & Fluxgate Compasses - Pelorus - Relative Bearings - Azimuth Mirror - Compass Error - Transits	
	17.2	**NAVIGATION - CHART WORK**	**379**
		Includes... Plotting Instruments - Charts - Mercator & Gnomonic Projections - Rhumb Line - Great Circle - Speed, Time & Distance Calculations - Plotting Terminology - Set, Rate & Drift - Leeway - Geographical Range - Luminous Range - Logbook - Passage Planning	
	17.3	**NAVIGATION – TIDES & CURRENTS**	**429**
		Includes... Tide Calculations - Rule Of Twelfths - Form AH130 - Tidal Streams in Torres Strait, Sydney & Broome - Surface Currents around Australia	

18.	**RADIO, SATELLITE & EPIRB COMMUNICATION**	**459**
	Includes... Operator Certificates - GMDSS - DSC - GPIRBS - LEOSAR - GEOSAR - INMARSAT - SafetyNET - Phonetic Alphabet & Numerals	

19.	19.1	**OCCUPATIONAL HEALTH & SAFETY - FIRST AID**	**504**
		Includes... Surviving Heart Attack When Alone - Marine Animal Bites & Stings - Seasickness	
	19.2	**OH&S - FIRE PREVENTION & CONTROL**	**520**
		Includes... Portable & Fixed Fire Extinguishers - Pressure Tests - Inspections - Fire Detectors	
	19.3	**OH&S - SAFETY EQUIPMENT, SEA SURVIVAL & RESCUE**	**535**
		Includes... NMSC Standard - Small Craft Particulars Form - Search Patterns - Abandoning Ship - Helicopter Rescue - Assisting Others in Distress - Safety Drills - Emergency Station List – AUSREP - REEFREP - Sydney To Hobart Yacht Race Safety Recommendations	
	19.4	**OH&S – WORK PRACTICES**	**579**
		Includes... OH&S Act (ref) - Privacy Act (ref) - Duty of Care - Risk Management – Safety Checklist - Galley Safety - Confined Spaces - Safety Symbols - Dangerous Goods - Alcohol Consumption - On Board Communication	

#	Chapter	Page
20.	**METEOROLOGY (WEATHER)** Includes… Barometer - Relative Humidity - Cyclone Contingency Plans & Safe Havens (examples)	591
21.	**HOW ENGINES WORK & THEIR PARTS**	625
22.	**ENGINE LUBRICATION & COOLING SYSTEMS**	643
23.	**INBOARD DIESEL FUEL SYSTEM** Includes… EFI - Tanks Capacity - Safety - Terminology	658
24.	**PROPULSION, GEARBOXES & PROPELLERS** Includes… Jet Drive - Conventional In-Line Drive - Surface Drive - Stern (Sail) Drive - Vee Drive - Z Drive - Gearboxes - Shaft Installation & Alignment - Selecting the Right Propeller - Thruster Units	673
25.	**ENGINE BEDS, EXHAUST, VENTILATION & SOUNDPROOFING** Includes… Engine Mounts - Vibration Isolators - Engine Inclination - Weight Distribution - Wet & Dry Exhaust Systems - Accumulation of CO - Machinery Noise	693
26.	**ENGINE OPERATION, ALARMS, CONTROLS & MAINTENANCE** Includes… Fuel Calculations - Starting Systems - Instruments - Alarms - Propulsion Control Systems - Diesel Engine & Generator Troubleshooting	708
27.	**OUTBOARD ENGINES** Includes… Two & Four Stroke - Single & Twin Cylinder - EFI - Electric - Gearbox - Remote Control - Petrol & LPG Engine Troubleshooting	729
28.	**LPG & LPG ENGINES**	776
29.	**BILGES, PUMPS, AUXILIARY POWER TAKE-OFFS, REFRIGERATION & PLUMBING** Includes… Hot & Cold Water - Cabin Heater - Refrigeration - Toilets - Drains - Holding Tanks - Colour Coding of Pipes	781
30.	**SOLAR & OTHER ELECTRICAL SYSTEMS** Includes… Electrical Installations - Batteries - Generators - Shore Power - Inverters	806
31.	**RUDDERS & STEERING SYSTEMS** Includes… Requirements for Commercial Vessels	825
32.	**INDEX**	835
33.	**CONTACTS - EMERGENCIES, WEATHER & WEBSITES**	851

Chapter 1

BUYING & INSURING A BOAT

BUYING A BOAT

IMPORTANT:
1. *The description below is not 'power' or 'sail' specific. It covers both, switching back and forth between the two types. It aims to provide an understanding of boats to help you decide your own need and make your own selection.*
2. *For second-hand boats, you may not be able to get all the information recommended below. A builder may not have provided it to the first owner or may not have known it because the boat was built without the architectural input. Nevertheless, the educational nature of this chapter aims to make you a better-informed person.*

Suitable boats are like suitable partners. Some dreadful boats have made their owners very happy, while some excellent boats have failed to give their owners the same joy. The best boat is the one that you can safely enjoy and be able to resell easily.

If you are new to boating, find someone with experience who is prepared to spend time advising you. Talk to as many knowledgeable people as possible. What should matter is not what you fancy but what suits your need. Don't buy your "dream boat" until you have hired boats a few times. Hiring may seem expensive, but it's not. The interest from the money invested instead of buying the boat and the money saved from maintaining the boat would easily pay for regular hires. Spend time on other people's boats. Become their apprentice and take your partner with you. Build up confidence in boat handling. Learn basic seamanship and the rules of the road, and if planning to go outside the heads, learn coastal navigation. You should be able to take compass bearings, correct them and plot positions on a paper chart when the GPS stops working.

People do not usually get the boat they want until about the third one they buy. The first boat usually is an all rounder and may quickly become a disappointment. In order to find your dream boat, you would need to eliminate all the different prejudices you have about boats and their equipment. Learn how different boats perform in different sea conditions. Figure out your need for comfort and speed. Establish not only your own preferences, but also of your partner.

A first time buyer shouldn't be looking for a brand new boat, just as a first time golfer wouldn't be buying the best set of golf clubs. Look for an affordable and manageable production line boat. Don't buy a one-off special or a specialized boat, such as an asymmetrical sail rigged yacht that needs a large crew to hang off the windward rail or requires water ballast. A symmetrical Masthead Rig with external ballast (heavy keel) will probably make a good first yacht. (A Masthead Rig is a yacht with one mast with her forestay reaching to the top of the mast, such as the Masthead Sloop shown in Figure 3.18 in Chapter 3.)

Most modern cruising yachts have only one mast. Boats with more than one mast (split rigging), such as a ketch, yawl or schooner, may look shippy and more secure in case one mast is dismasted. The mizzen mast can also be argued to be useful for holding up items such a wind generator, radar reflector and antennas for radar, GPS, radio and satellite communication. But such rigs are much harder to balance. Their performance lags far behind the sail-handling technology of the simpler and cheaper and yet powerful and reliable sloop rigs. The fractional rigs (the forestay extending only part of the way up the mast) too are harder to optimize for power, control and comfort.

Don't fall for the 'wrong' advertised features. It is logical for boats designed for short-distance bareboat charters (large holiday groups) to have lots of bunks, toilets and other luxury items within a given length. But a boat crammed with bunks even in and around a centre cockpit is as uncomfortable for long cruises as an asymmetrical boat that keeps veering off to one side. Forepeak bunks may feel comfortable in the harbour, but they are the first bunks to be vacated when the boat starts to pitch in a seaway. In fact they can be outright dangerous in head seas. You will then be glad that the dinette is convertible into a bunk, which usually being in the middle of the boat is perhaps the most spacious and comfortable space. Also remember that a home-style galley layout can be quite dangerous at sea, and a beautiful dining room may deprive you of the best sleeping berth.

The yacht should be of a moderate beam-to-length ratio. The maximum beam at waterline (BWL) divided by the length at waterline (LWL) should be approximately 0.3. Some ultra-light racing yachts exceed this ratio at the risk of losing seakeeping abilities and stability. The boat should have a safe cockpit from where the sheets and winches can easily and safely be reached without leaving the wheel. Centre cockpits are indeed safer and drier than the end cockpits. But they come at the cost of losing the space below deck, especially in boats less than 12 metres in length.

It is wise to stay away from laid decks. They might look classic and romantic, but if they don't already leak, sooner or later they will. Large varnished surfaces may look warm and attractive, but you are buying the constant care and work that comes with it. Similarly, stainless steel standing rigging is not a good idea. Sure, stainless steel is expensive and corrosion resistant, but it is also heavy and brutally unpredictable. It gives away without warning, often causing serious injury. The galvanized wire standing rigging, on the other hand, costs less, lasts longer and gives months of warning before catastrophic failure.

Labour-saving devices are wonderful, but be wary of relying on them 100%, in case they breakdown and you don't have skills, tools or spares to fix them. The seller may also have gone overboard on fitting a lot of expensive electronic gadgets. Beyond a VHF radio, radar detector, magnetic compass and a simple GPS, you seriously have to consider what you need and what you are paying for in the price of the boat.

For a powerboat, the cost of fuel may be a major consideration. A 4-metre aluminum runabout runs on the smell of an oily rag compared with a 6-metre fiberglass cabin cruiser. You may have to choose between frequent less comfortable boating and less frequent more comfortable boating.

WHAT SIZE BOAT & ENGINE?

There are no guidelines on safe relationship between vessel length and sea condition. Boats are built for specific activities in specific waters. A 5-metre tinnie (aluminum boat) may be suitable to go offshore but not a 20-metre ferry designed to operate in Sydney Harbour.

However, generally speaking, the bigger the boat, the more comfortable she would be at sea. The Department of Transport in Western Australia requires that all boats under 3.75 metres (12 feet) in length (including personal watercraft) stay within five nautical miles of the mainland, unless they are within the limit of a port or within one mile of any island.

The size of the boat you are looking for depends on the number of people you plan to carry and whether you intend to pot around a harbour or go cruising. For a first powerboat, a length of between 4 and 7 metre should be an adequate all rounder for a small family to learn boating and have fun. (Length-to-capacity ratio for powerboats is discussed in Chapter 12). Such a boat in good condition would hold its resale value when you decide to upgrade. You will one-day want a boat to pursue your single-minded interest, as most boaters do.

For going offshore, you not only require a suitable boat but also greater care, knowledge and experience. Many popular offshore areas are only accessible by crossing sand bars, which have contributed to many boating accidents. Bar crossing is discussed in Chapter 8. Trailable-size boats can set their comfort limit by the One and Five Rule: stay inshore if the forecast wind is over 15 knots or the sea over 1.5 metres. In summer, the typical swell height on the Australian coast is about 1.5 metres. Breakers start to appear in unusual places when it gets bigger. This is the time to stay away from the seaward edges of reefs.

The engines too have both minimum and maximum power limitations. The extra weight of an overpowered engine may unbalance the boat and lower her freeboard. A trailer boat would usually be fitted with an outboard: 4-stroke petrol or diesel, or the older 2-stroke petrol engine. A moored boat will usually have an inboard or a stern drive engine.

Normally, it is best to stick with the manufacturer's recommendations. Be warned of some dealers trying to sell you a bigger engine for extra power and better fuel economy. This may be true, but the extra weight of the engine in a small boat may dangerously unbalance the boat.

Engines without turbochargers, sometimes referred to as "unblown engines", are more reliable for intermittent use, and therefore more suitable for cruising yachts. A general rule of thumb for the "unblown" engine size is 3 to 4 kilowatts (4 to 5 horse power) per displacement tonne (weight) of vessel. Thus, a 5 tonne yacht needs a 15 to 20 kW (20 to 25 hp) engine.

WORK OUT YOUR OPTIONS

You do need to work out what type of boat you are looking for. Motorboats are a great way to get out on the water, especially to go poking around bays and inlets. You can do this on a sailing boat too, because most of them have an inboard or an outboard engine. Then there are displacement and planing hulls, and round bilge and chine hulls to consider (See Chapter 3 for explanation and larger illustration, Figure 3.5).

You need to decide whether you want a trailer-boat or a boat that sits on a mooring. Trailer-sailors offer more flexibility for weekends and holidays away from home. Being out of water most of the time they are cheaper to upkeep, and also usually free to park at home.

Larger boats with fixed keels are kept on moorings or marina berths. They offer a different sort of pleasure for a different sort of money. Not only are these boats more expensive, they bring with them quite a hefty mooring or berthing fee and a high maintenance cost because they are always in a saltwater environment. With a boat on a mooring, you will need to row out each time, and there is the security of your dinghy during the week to consider.

HULL MATERIALS & CONDITION

Most boats are built of either aluminium or fibreglass (GRP).

ALUMINIUM: Small aluminium boats are regarded tough and hardwearing. In aluminium hulls, look for corrosion. Superficial powdering is not a problem, but lots of white powder or deep pitting should be treated with suspicion. Look for cracks where rails weld to decks and sides meet the keel. In a boat with an outboard engine, look for cracks in the transom; and with an inboard engine, at the engine mounts. A painted-aluminium hull should be evenly painted. It should not have any rough welds, dents, or ripples in the hull.

GRP: Unlike metal and timber, GRP is not a uniform material. It can suffer from osmosis, delamination and gelcoat cracks. Look for signs of structural failure and cracking in areas where flat surfaces meet stiffeners, and on chines, planing strakes and bulkheads. Distinguish between cosmetic wear and tear and deep cracks. Stand back and examine the curvature of the hull to try and detect any repairs, particularly at the turn of the bilge. Osmosis is quite a common problem, and a patch of osmosis is not the end of the world, as long as the repair is of good quality. There should be no bubbles, wrinkling, or rough spots in the gelcoat. If the gelcoat has a colour pigment added, see that the colour is not blotchy. Examine the hull below the waterline. This is where many of such problems occur, and this is where osmosis is often hidden under the antifouling.

The GRP hull construction may be either "solid" or "sandwich" type. The latter is built around a lightweight core of high strength synthetic foam or timber such as balsa or western red cedar. It is thus a thicker and therefore stiffer hull, but damage to the outer layer of fibreglass can saturate the core with water. The presence of water in a sandwich hull can be checked by gently tapping the hull with a blunt instrument. It will produce a dull thud.

PLASTIC (POLYETHYLENE): Some of the modern boat tenders and sailing dinghies are constructed of high-density moulded plastic, or more accurately, polyethylene. They are one piece, without any joints or bulkheads. They are usually of sandwich construction containing a foam core, i.e., they are foam filled. Polyethylene has been used in whitewater kayaking for quite some time. Its advantage over aluminium lies not only in lightness but also in noise absorption. With the centerboard and rudder lifted up, these boats can be dragged onto a beach without causing any damage to the hull. Small skin damage can be repaired with a heat gun. However, unless UV stabilizers are used in construction, this material becomes weaker in prolonged exposure to sun.

TIMBER & PLYWOOD: There are four types of timber construction:

Carvel hulls are the most common. Their planks are butted together and fastened to frames with copper nails to form a flush hull. Larger vessels may use iron spikes or hardwood pegs called trunnels or treenails.

Clinker construction is an old method of fastening overlapping planks to frames with copper nails.

Cold Moulding consists of thin planks laid diagonally in double, triple or quadruple layers. These planks are held together with marine epoxy glue and also fastened to the frames. Not being entirely dependant on fastenings, they form a stronger hull. However, a Cold mould hull is quite complex to build and repair.

Strip planking is a more modern method of wooden boat construction. A mass of small timber strips are machined into shape, then glued and nailed together and to the frames. Such boats too are complex to repair.

You will also come across plywood boats. Due to the problem with bending plywood in more than one curve, their traditional method of construction was to build hard chine boats where the plywood panels were glued and screwed to a conventional timber frame in a manner similar to a carvel construction. The ends of such panels, relying on fastenings, are quite susceptible to water entry causing delamination and rot. The modern, cold moulded plywood boats, although expensive, are longer lasting.

Whatever the type of hull, they are all susceptible to worm infestation and rot. Check for the Teredo worm in the keel and adjacent timbers. You will need a magnifying glass. Small neat holes are their points of entry. Boats with poor antifouling are more susceptible to such infestation. Dry rot is caused by humidity inside the boat. Its telltale signs are cracks in the paintwork or mouldy and musty smell. Look for it in deck beams and ribs in areas subject to condensation and rainwater leakage. Look for mould on paintwork.

Wooden boats cannot be left out of water for any extended period of time. They require to be worked on and maintained more than any other hull material. Fastenings in hulls over 20 years old may have corroded away. The refastening option is expensive and time consuming.

FERRO-CEMENT: The occasional ferro-cement boat is cheap to buy, but hard to sell. They are usually amateur built, without any uniform rule about quality. The specially prepared mortar must be forced into the steel and wire armature in one session and then allowed to cure under controlled conditions. These boats are also difficult to survey because the steel armature is covered in concrete. An X-ray of the accessible areas can show the amount of steel used, but not the quality of the cement or the construction technique. Look for cracks and local bulging in the concrete. The latter may indicate the expansion of corroding steel.

A ferro-cement boat may be difficult to insure even when there are photographs taken during construction. The boat will be slow moving under sail or motor and cost more in fuel.

STEEL: Steel is not common in boat building. In addition to its rusting nature, steel is hot, noisy and maintenance intensive. But it is the strongest hull to have and is the easiest to fix. Look for rust due to condensation in areas behind refrigeration, shower recess and linings. Check that the longitudinal stringers have been notched to allow moisture to drain into the bilge. Examine the welding. It should be professional. A boat built over a number of years is likely to have steel of different grades and batches. Such a boat would not be electrically uniform and thus be more prone to corrosion.

Both the ferro-cement and steel hulls are magnetic. Hand-held magnetic compasses cannot be used on these boats. Even the fixed compasses have to be adjusted for the magnetic deviation caused by the hull.

STABILITY

Stability is an important consideration in all boats. As discussed in Chapter 13, all vessels are designed to heel, roll, and pitch and yaw when pushed by the forces of wind and sea. The difference between the stable and the unstable is their ability to return to upright each time. Vessels designed and operated with a low centre of gravity have a greater ability to return to upright after being heeled by an external force, such as wind and waves. (See larger illustration in Figure 13.8, Chapter 13).

The movement of a properly designed, built and loaded motorboat is like a balanced pendulum. If the pendulum has most of the weight on the bottom it will swing quickly and return to rest quickly. In a vessel, this movement is described as "stiff". A stiff vessel resists rolling and thus rolls quickly and jerkily. She is very stable but her movement is uncomfortable for those on board.

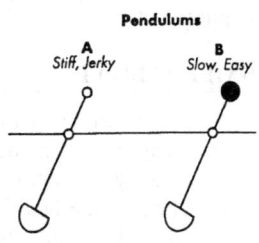

Conversely, if the pendulum is fitted with a much larger top weight, it will swing very slowly and reverse direction very slowly. In a vessel, this movement is described as "tender". A tender vessel lacks stability and runs the risk of not returning from a roll. A vessel should be designed and loaded along a well-balanced pendulum principle - neither stiff nor tender. This allows everyone on board a more comfortable motion.

Sailing boats are built "stiff" to prevent them from capsizing under pressure of wind and waves. A piece of cast iron or lead is bolted to the lower extremity of the hull, sometimes in the form of a bulb. It lowers the boat's centres of gravity. When hit by a gust of wind, the boat heels over easily at first. But the further she heels the greater is the righting moment exerted by the weight on the bottom of the keel to right her. A boat heeled even at 60° will right herself when pressure on the sails is eased.

BUOYANCY

Buoyancy is an important consideration in all vessels. It keeps the vessel afloat if she is swamped or is capsized. The weight of the engine and other fittings can make even a wooden boat sink when filled with water. A yacht with a lead keel and a dinghy with an outboard are similarly vulnerable. (See larger illustration in Figure 13.19, Chapter 13).

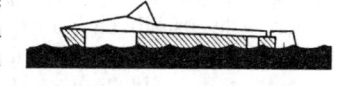

Larger, heavier, enclosed boats should be built with enough watertight bulkheads (like walls in a building) to restrict fore and aft flooding or spread of fire. Smaller, lightweight and open boats, without watertight bulkheads, need to be designed with built-in buoyancy, known as Positive Buoyancy. Generally, it means having enough foam built into the boat to compensate for the weight of the motor and to provide balance at the bow, under the seats and at the sides. Foam is a suitable buoyancy material for all lightweight construction materials, i.e., wood, aluminium and fibreglass. In the past, as a standard practice, foam was added only to aluminium boats. Now, it is also beginning to appear in GRP hulls. Usually Polyurethane (PUR) – and not Styrene - foam is used because it does not readily absorb water and does not disintegrate (melt) on coming in contact with engine fuels. However, the fumes from burning Urethane are toxic. Many aluminium boat builders use Expanded Polystyrene (EPS) instead.

Foam should be encapsulated so that it does not absorb water. It should installed either in the double skin of the boat or on the underside of her decks. In boats with outboard engines, the stern area should be made more buoyant. Most of the foam should be fitted above the waterline. Too much foam placed in the bottom of the boat will encourage her to capsize more easily, and once capsized, she will be difficult to turn upright. However, some foam should be placed below the estimated damage waterline. In case of hull damage, it will reduce the permeability of the boat, i.e., allow less water to enter. The level of flotation of a boat filled with water may be described as follows:

➢ *Not enough or no flotation*: The boat sinks quickly.
➢ *Basic flotation*: The boat remains partially afloat either by the bow or upside down.

> *Level flotation*: The boat remains afloat in a level position.
> *Additional flotation*: It helps prevent capsizing.

As discussed in Chapter 12, there is now a National Recreational Boat Compliance Plate Program, which was introduced in 2001 by the National Marine safety Committee (NMSC). Under this program, compliance plates are to be fixed to all new boats manufactured in Australia showing serial number, passenger capacity, maximum engine rating, standard of buoyancy and the manufacturer's name and model type. The new compliance plate indicates the boat's response to swamping in the form of following buoyancy standards:

> *Level Flotation*: the boat will remain floating in a level position with passengers and gear on board.
> *Basic Flotation*: the boat will remain floating with passengers clinging to her.
> *A blank space in the buoyancy section*: the boat's buoyancy standard is not specified.

As part of the compliance plate program, the manufacturers have also agreed to produce nationally-consistent owner manuals to provide operators with a broad range of safety information on vessel operation and maintenance, including explanations on passenger load and carrying capacity, stability, fuel safety and minimum onboard safety equipment requirements.

SAILING BOATS

As discussed in Chapter 8, sailing boats are fitted with a deep keel so that they can sail at an angle without capsizing. Extending up to 2 or 3 metres below the hull, the keel also provides resistance to leeway (sideways motion) resulting from wind pressure on sails. The keel in sailing dinghies is usually a retractable centerboard. The larger cruising yachts have a deep fixed keel or a combination of a sallow fixed keel and a retractable centreboard, which allows them to come into waters of depth of a metre or so. The fixed keels fall into one of three categories. (See larger illustrations in Figure 3.6, Chapter 3).

The first is the Traditional Long Keel. This full-length keel adds strength to the boat's backbone as well as provides good stability. A boat with such a keel would normally be of heavy displacement (weight). She will be seakindly and comfortable for long voyages. She would be easy to keep on course, and easy to cradle when hauled out. Under a larger sail area, she would easily point into the wind, and experience minimum leeway. However, due to a large wetted surface area, she would be slower and sluggish in performance. In some boats, the fore part of the keel is tapered off. This is known as the Cutaway Forefoot.

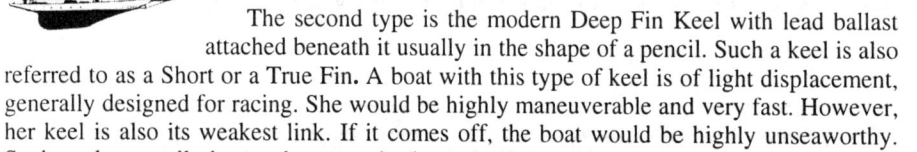

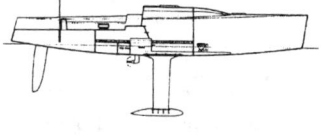

The second type is the modern Deep Fin Keel with lead ballast attached beneath it usually in the shape of a pencil. Such a keel is also referred to as a Short or a True Fin. A boat with this type of keel is of light displacement, generally designed for racing. She would be highly maneuverable and very fast. However, her keel is also its weakest link. If it comes off, the boat would be highly unseaworthy. Such yachts usually have a large spade (long slender) rudder. An ultra-light displacement yacht may need two rudders, as the one central rudder would lift too far out of water to remain effective. This is the handiest but also the least practical types of yacht. It requires constant helm to stay on course and has a large leeway She is also hard to slip. Both the rudder and the keel can easily get damaged during slipping.

The Long Fin Keel is a good marriage between the other two. It provides the boat a good backbone, but with less wetted surface area than the traditional long keel. Such boats are of moderate displacement. Their skeg-hung rudder (hanging from a metal fitting at the stern) is well supported, and if well designed, makes them quite maneuverable and easy to point. This configuration also allows the boat to be designed with good stability and minimum leeway. She is easy to slip and handy under sail. However, its disadvantage is that the propeller usually has to be enclosed in an aperture. Although protected, it is less efficient.

For your first cruising yacht, you should perhaps be looking for a long fin keel with a skeg-hung rudder. A yacht with a traditional long keel with a cutaway forefoot should be equally suitable. The short fin keel, being faster and more maneuverable, may be more fun to sail over short distances, but it may lack directional stability for longer voyages.

STERN AND BOW: A boat with the traditional Canoe Stern is probably the best option. Such a stern provides excellent seaway, particularly in heavy following seas. A stern with flat flow lines (common in many racing boats) is likely to generate a large and steep rooster tail in its wake, which can not only be dangerous, but it is also a waste of boat's energy.

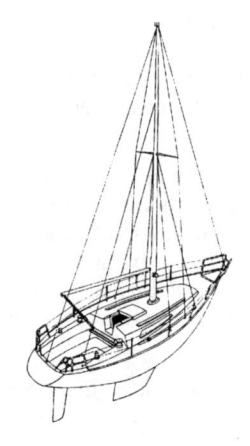

Chapter 1: Buying & Insuring a Boat

Then there is the Reverse Transom Stern (i.e., stern sloping forward, See larger illustration in Figure 3.10, Chapter 3). The builder's purpose of designing boats with such a stern is to increase their waterline length and, at the same time, reduce weight. However, such a stern can be difficult to protect because of the difficulty of reaching out with a fender to soften the blow from another boat in a busy harbour. It is also susceptible to a following sea rolling into the cockpit. Now, looking at the other end, a boat with a broad "U" shaped bow, common in racing yachts, may look spacious. But, such a bow lacks reserve buoyancy, making the boat sensitive to any additional weights forward. (Reserve buoyancy is provided by the enclosed volume above the waterline). The "U" shaped hull also causes the boat to slam when heading into the sea.

MONOHULL OR MULTIHULL? : You can cruise or race in a multihull (catamaran or trimarans) just as you would in a monohull. Some people are monohull fans, others like only the multihulls. Early cruising multihulls were almost all trimarans. But they have now been replaced almost entirely by catamarans, which are cheaper to build and offer greater volume of space in the bridgedeck between the hulls. Early trimarans can be purchased quite cheaply and make economical cruisers and floating homes.

Multihulls are faster and of shallower draft with more spacious and stable living areas. They do not roll like the monohulls, and fitted with twin propellers, they are quite maneuverable for their size. Being totally free of ballast and fitted with void spaces, they will float even when swamped or capsized. But, when the sea gets really bad, their tacked-on fin keel does not provide the sense of security that one gets from a deep-draft monohull keel. Multihulls are more expensive to buy and cost more in berthing fees. Building them requires a greater skill, therefore only the quality designs should be considered. The bridgedeck can be the fully enclosed-type or a semi-bridgedeck with a soft or hard cover over it. The latter is almost essential for cruising in the tropics. When choosing a multihull, be aware of her designed load-bearing capacity. The fatter U-shaped cruising catamaran hulls are slower but have a higher load-bearing capacity than the thin racing catamaran hulls. You can obtain more information on multihulls from multihull clubs in each State, or log on to www.insidemultihulls.com.au. (See larger illustration in Figure 8.3, Chapter 8).

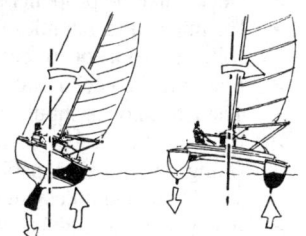

GENERATORS & FRIDGES: A generator will generally give you more power than you will need, but it is an extra engine to fuel, maintain and carry around. A fridge is a luxury we all are used to, but it will require the boat's engine or generator to run for at least an hour a day, regardless of where the boat is and regardless of your need to use the engine for propulsion.

INSPECTION

You will be lucky to find everything listed below in one boat, but the dry and wet inspection lists should make good checklists. For each boat, list all the 'for and against' points. Above all else, choose the one with the best seakeeping hull and motor combination. You can always fit out a good boat with whatever you want later. However, these days some of the factory fitted boat and motor packages are quite good value and worth a look.

DRY INSPECTION

Do not buy a boat without an inspection on the slip. Be methodical. Start at the bow and work aft. Start at the top and work your way down. Write down every question and deficiency you come across in a pocket notebook. Don't be afraid of being labeled a 'nitpicker'. Make sure you get a copy of the inventory and check that you receive all the items listed on it. The following is a list of some of the things to pay attention to.

➢ Inspect masts and spars for damage and repairs. Look for splits (shakes) in wooden masts, and corrosion in out-of-sight deck fittings, such as on masts and shrouds.
➢ In wooden boats, look for cracked ribs and rot between the planks, at the mast step and in the region of the chain plates.
➢ Look for signs of hull buckling due to tightened rigging.
➢ In a wooden hull, look for signs of rot behind fittings.
➢ Unfurl sails and check for strained seams, tears, chafe and mildew and general signs of age. Canvas rots with age while synthetics become brittle in sun's UV light.
➢ Check rigging for wear and rust.
➢ Check synthetic ropes for signs of stretching and kinks. There should be no powdery substance when you twist open strands.
➢ Cracks, splits and sprung butt ends in wooden hulls indicate loose fastenings.
➢ The boat should have good size cleats and/or bollards not only fore and aft, but on boat's shoulders and amidships. This will allow you to tie up the boat in all sorts of berths and weather and tide conditions (See larger illustration in Figure 3.4, Chapter 3).
➢ Check the condition of lifejackets (PFDs), lifebuoys, flares, dinghy and liferaft. These are discussed in Chapter 19.3.
➢ If there is water in the gearbox, or if there are signs of stain, watermarks and poor paint below decks in a wooden boat,

she might have been submerged at some stage.
- The engine should look maintained. The exhaust manifold shouldn't be flaking sheets of rust. The seller should provide a record of engine service. There should be sufficient spares on board.
- The concrete poured in the bilge might be covering a leak. And, if it is ballast, why does she need it?
- Water in the bilge may not be for a good reason or due to just laziness. (See larger illustration in Figure 29.7, Chapter 29).
- Check wooden hulls for rot and soft wood, especially at the waterline and at the ends of planks such as at the transom intersection. Do seams need re-caulking? The traditional old wooden boats were not designed to be sheathed with fibreglass. Sheaths often hide poor condition of hulls. They can work loose in time to allow rot to set in. They trap moisture, which, over a period of time, develops rot.
- Patches and repairs on the hull are usually a sign of needing more work. Patches often hide what the seller doesn't want you to see.
- Excess weed growth suggests poor maintenance and lack of antifouling.
- Examine propeller surfaces for cracks, splits and fairness. Check shaft bearings for play and for security.
- Check that the propeller shaft is not loose in its bearings.
- Examine rudder pintles and gudgeons for wear and security. Check rudder for damage. In a wooden rudder, look for splits in planks or rotten end grain.
- Examine the keel and the rudder for grounding or impact damage.
- The zinc anodes should not be unduly wasted, unless they are at the end of the maintenance cycle.
- Look at the quality of the materials and components used in her construction, inside and outside. All fixtures and fittings should be capable of withstanding the marine environment and the rough sea. All the hose clamps and screws should be of stainless steel. Look for places where electrolysis (corrosion) may occur.
- All gauges should be clearly visible to the driver, standing or seated at the wheel. Instruments, such as steering compass, GPS and sounder, should be in the driver's line of sight. All seats and bunks should be comfortable and big enough. The upholstery should not be water-sodden. It should be heavy-duty, marine grade vinyl of quality finish. In addition, the driver's seat should have good back support and allow a clear view to all points of the boat. It should be either adjustable, or mounted far enough back, to allow driving standing up.
- Sit, lie down and generally move about in the cabins to ensure that they will suit your needs. Do this both underway and at rest. Make sure all the gear has its place, both wet and dry. All hatches should close properly. Look for water leaks below portholes, underneath through-deck fittings and inside cupboards. There should be no protrusions to bump into or, or get caught on. The gunwale should be high enough for the safety of the crew and children?
- If buying the boat to go fishing, check on the number of rod holders and their position. Are there sufficient number of grab rails and handles for all the crew? In the main cockpit, make sure your toes can fit in the side pockets, so that you can brace against the gunwales when fighting a fish in a big swell. There shouldn't be a shortage of mooring cleats. They should also be strong and in the right positions. Any rocket-launcher rod holders should be fitted properly and not just tacked on.
- Lack of proper stowage space is a cause for annoyance and danger. The safety equipment box should be within easy reach and with a large lid for easy access. There should be separate activity-specific storage. A fishing boat, for example, should have storage space for fishing tackle boxes and rods. There should also be suitable storage space is for the occasionally used items, such as anchors and ropes.
- Consider the access and ease of stepping in and out of cabins. Examine the drink holders, fold-down table, courtesy lights, stereo, sink, hand shower and toilet rail. Are they functional and easy to use? Is the carpet of the easy-care clip-in variety or is it permanently laid, which may be a problem to clean?
- On boats with inboard engines, which are usually diesel, make sure the necessary engine parts are accessible for maintenance. These include belts, fuel injectors, engine and gear box oils, fuel filters, gear box inspection covers and gearbox and throttle linkages. Also check the accessibility and condition of the stern gland, sea inlet and outlet valves, and steering gear. For an engine with a starting handle, there should be adequate room for its use. On outboards, the fuel cap and oil bottle should be accessible.
- Check the wiring. All connections should be waterproofed. The wiring should be colour-coded, neat and easily accessible without being obvious. The battery should be accessible, but off the 'floor' and out of the way to avoid being soaked. It should be secure, ventilated and protected.
- In case of a trailer boat, notice how she sits on the trailer. The hull, especially along the keel, should be fully supported by rollers and bearers. Look for cracking on spray chines in way of rollers. Look at the methods used to tie the boat down and, whether hydraulic brakes are required.
- Go back and check the boat for leaks on a rainy day!

Chapter 1: Buying & Insuring a Boat

WET INSPECTION

NOTE: The following on-the-water tests may sound involved and time consuming, but you wouldn't need more than 30 minutes after launching.

- If she's a trailer boat, she should easily roll off the trailer without the need to bring the tow vehicle in the water.
- Get a feel for the boat at slow speeds, in calm water. This will help you to ascertain the throttle response and the feel of the steering. Carry out a couple of docking and turning maneuvers to see the effect of wind on her performance. In a boat over 6 metres you would want hydraulic steering. Test all on-board systems to your satisfaction. Don't take the seller's personal guarantee that everything works.
- Listen for vibrations and noises when the engine is idling and underway. A slapping noise at slow speed may indicate worn stern gland or propeller damage. Check for loose nuts on the engine bearers. The propeller shaft should be well supported by bearings and fitted with grease nipples inside the hull.
- Examine the anchors and the anchoring gear. They are discussed in Chapter 7. Is she easy to anchor? Is the anchoring arrangement through the cabin, or on the bow? Is the anchor locker big enough, including for a reef anchor? Is there a high enough bow rail to lean on or rails to grab while kneeling on the bow when anchoring? Test the anchoring gear to actually see how long she takes to anchor and retrieve. You should feel safe and secure on the bow when using all your body weight to free a snagged anchor. (See larger illustration in Figure 7.14, Chapter 7).
- Head out to sea to try full speed maneuvers. Note any smoke coming from the exhaust. As discussed under Troubleshooting in Chapter 26, blue smoke may indicate worn engine and black smoke may indicate overloading. Cruise for a while to get the real feel for the boat. Getting to know the throttle and helm response will help you to get out of trouble in heavier seas. Meanwhile check the engine room for any leaks or increased vibration. Look for water spray corrosion on tops of tanks below the air vents. Check the engine noise levels around the boat. You may get a sense of the quality and condition of the engine room insulation, shaft bearings, couplings, stern gland and engine bearers. Examine the quality of engine room ventilation. Such items are discussed in Chapters 24 and 25. (See larger illustration in Figure 25.20, Chapter 25).
- The accommodation should have an adequate flow of fresh air. Without proper ventilation, the interior can become hot and the evaporation from bilges, cooking and tank vents can increase humidity, which would increase the chances of mould, corrosion and rot. Depending on its location, the engines too will increase the cabin heat, even when running to provide refrigeration or to charge batteries. The machinery space temperature normally rises after the engine has stopped and is no longer drawing in fresh air. Where possible, even the lockers should be ventilated. Lack of ventilation would make everyone feel tired and cause headaches and nausea. Such symptoms indicate a lack of oxygen or presence of carbon monoxide, which if allowed to persist, can become dangerous.
- Depending on sea conditions, test the hull capabilities by running the waves at various angles while maintaining a healthy three-quarter throttle. You certainly don't want a boat that takes too long to 'climb out of the hole' and shows tendency to broach in a following sea. This test should also indicate if the boat will keep you dry. Wet boats range from those that draw up annoying mist into the cockpit to outright water splashes from all directions. This occurs in hulls with insufficient bow flair or poorly designed chines and spray chines.
- Test the boat under the conditions in which you will be using her. On a fishing boat, take your family or fishing mates along. Have them move around while underway to see the reaction of the hull. Have someone move quickly as if by accident to upset the boat's trim and see if the boat recovers quickly.
- Stop and check the boat's rolling behaviour at rest. Does she roll all over the place? Check her stability and buoyancy by having some of the crew move to one side, as if helping someone overboard. Does she list too far?

TECHNICAL DETAILS

- Ask for the age of the hull and the record of her last slipping and the Surveyor's report. How many past owners? Is she insured?
- If GRP, was she built in conditions of controlled temperature and humidity?
- Are there any plugs from the hull available for inspection as a result of through-hull fittings? Plugs from cutting holes in the hull are often retained to impress future buyers of the thickness and quality of the material. It is good to be able to assess the ratio of glass to resin in a GRP hull. Resin-poor hulls are prone to absorbing water.
- In addition to the Breadth/Length ratio of 0.3 discussed earlier, an expert should also ensure the following approximate values for a <u>cruising yacht</u>:
    - Ballast/Displacement ratio = 0.4 to 0.5.
      (Formula: Ballast ÷ Displacement, i.e., Weight of keel ÷ Overall weight of boat)

- o Sail Area/Displacement ratio = 14 to 17.
  (Formula: SA ÷ two-thirds of Volume of Displacement)
- o Displacement/Length ratio = 250 to 350.
  (Formula: Displacement ÷ 3(LWL x 0.01)
- o The range of positive stability in excess of 130°. It is expected that a yacht with a coach-house will have a little smaller range of positive stability than the one with a flush-deck configuration.

➢ Check the displacement and ballast (both in tonnes) from the original design specifications. An older yacht's displacement and ballast would almost certainly have changed from the original specifications. But if the boat is being slipped, her current displacement (weight) can be easily estimated from the travel-lift operator. The design office, builder or the owner should also be able to supply the vessel's hydrostatic stability data for the expert's analysis. If they are unable to supply this information, you should be wary of purchasing the boat. An old or modified boat should be subjected to a thorough expert examination. Even a small change in boat's displacement, ballast, total sail area, or hydrostatic stability can have a dramatic effect on the above ratios.

TEST & SURVEY

Do not buy a boat without testing her yourself. The published boat tests are only a guide. They are generalized, as everybody has different expectations from a boat. A boat tested for cruising on a river may or may not be able to handle offshore work. Similarly, not all good offshore boats make top bar-crossing boats. Different hull types handle varying conditions with their own idiosyncrasies.

You should also get the boat surveyed by a reputable, independent and professional boat surveyor, preferably <u>not</u> arranged by the broker. You should be able to get a list of independent surveyors from the Boating Industry Association (BIA) in your State. Obtain a written report from the surveyor. In case of a large boat, you will most likely have to pay for slipping, but it is too risky to buy a boat without such an inspection. Ask the surveyor if the boat is sound and if she would meet your need. Is the motor too big or small for what you want to do and for the hull to perform? Does she have sufficient power for casual skiing and unexpected loads like towing another boat or to get you across your local river mouth bar? Are the engine and hull in good condition? Is the equipment appropriate and sound? Does the boat have positive buoyancy? Does she have good stability? Would she float when swamped? If the boat is good, but not the motor, is it justifiable to replace the latter with a new or a good secondhand one?

If you are looking for a boat to run a business, such as charters, ferry, water taxi or mooring barge, it is best to buy one that is already in commercial 'survey'. To get a boat into survey or resurveyed (if expired) can be very expensive, time consuming and often heart-breaking. The surveyor or the inspector may have to rip the boat apart in and out of water and drill holes to assess the quality of the hull and decking.

SELLERS & DEALERS

The Trade-A-Boat monthly magazine is the easiest and the cheapest way to look for a secondhand boat. Its 600-odd pages offer everything from old hulks to million-dollar luxury boats. Reading classified sections of newspapers will also add to your knowledge. Visiting marinas to become familiar with boats can also be a pleasant way to spend a few weekends. After a while you will get a feel for the going rate for you type of boat.

If buying from a dealer, you are more likely to get a better deal from a dealer who is an agent of one or more boat manufacturers and/or marine propulsion systems. Look for a dealer who is also a member of the Boating Industry Association. When looking for a new boat, obtain written quotes.

Buying privately is riskier, but can also the way to pick up a bargain, as long as the seller is genuine. There are not only stolen boats around, but there also some sharks from 'shonky' yards who tow a boat home each night and pretend to be private sellers selling from their backyard.

Be careful not to buy a stolen boat, or a boat with money owing on it. It may look like a bargain, the seller may want to meet you in a pub, he might give dodgy reasons for selling it, his knowledge of the boat's history might be vague or too good to be true, the registration or survey papers might not be original, he may only have a mobile phone number, the list goes on.... Establish the proof of ownership, note down the seller's car driver's licence, his car registration number and address. Ask for receipts for any work done to the boat. Phone your local marine authority to get the phone number for 'Revs for Boats'. In many areas of Australia, they can tell you if the boat is stolen or if there is money owing on it. In most States, it is a legal requirement to have a Boatcode or Hull Identification Number (HIN) affixed to the transom. It provides some identification of the boat, her owner and whether or not she is encumbered.

See contact details of Boating Industry Association (BIA) in each State at the back of the book.

INSURING A BOAT

Boat insurance claims arise either from what people do to their boats or from what they do not. A large proportion of claims are due to:
- Lack of maintenance
- Poor servicing
- Inattention to existing defects
- Going to sea in conditions beyond your ability
- Navigating with insufficient knowledge of the area.

In order to judge the risk, the Insurance Company would want to know not only about your boat but also your skills. Commercial craft have to undergo surveys and their crews have to be qualified. But, there is not always such warrant of fitness for recreational boats. Therefore, when the time comes to make a claim, the difference in the interpretation of a clause can make you lose out.

For example, the "latent defect" clause may allow an insurer to avoid liability in cases such as follows:
- The fin keel of a converted trailer yacht falls off. (Claim declined due to poor fastening of the keel to the hull.)
- A fire extinguisher doesn't work when needed. (Claim declined because the policy stipulated fire extinguisher in working order.)
- Vessel is missing from her anchorage. (Claim declined because the period at anchorage was greater than the policy stipulation.)
- A yacht rigging gives way (Claim declined because the gear has gone past its constructive life span of ten years.)
- A stainless bolt on an alloy mast fails due to corrosion. (Claim declined because a protective membrane of plastic or similar not fitted between dissimilar metals.)

At the time of taking out the policy you must declare all material facts to your insurer. Ask yourself, if I were the insurer, would I want to know that? The following is a list of some of the questions you should ask about your policy:
- Does it have depreciation on any part of a claim, and if so, where?
- What if I add new gear to the boat?
- What if I replace an existing piece of equipment with that of different value?
- Is there any limit on my area of operation?
- Are there any restrictions on who may sail my boat?
- Does the policy exclude entering boat races?
- What is my third party cover?
- Can I leave the boat at a place other than my regular moorings, and for how long?

NOTES:
- If you intend to use a commercially registered vessel for both commercial and recreational purposes, make sure you cover yourself for both by obtaining it in writing from the relevant waterways authority as well as the insurance company.
- Some insurance companies may not cover vessels over a certain age or type, or may want the boat surveyed prior to giving it a cover.
- Some insurance companies require notification prior to vessel being slipped.

Chapter 2

TRAINING SCHEMES, RECOGNITION OF DEFENCE QUALIFICATIONS & ONBOARD TRAINING

RECREATION BOAT TRAINING SCHEMES

The Australian Yachting Federation (AYF) administers a diverse range of Training and Certification schemes for Powerboats as well as Sailing boats. The appropriate Training Logbook can be purchased at boating bookshops, chandlers and the AYF-credited sailing schools. The following is a summary of training and certification for the Yacht Cruising Scheme.

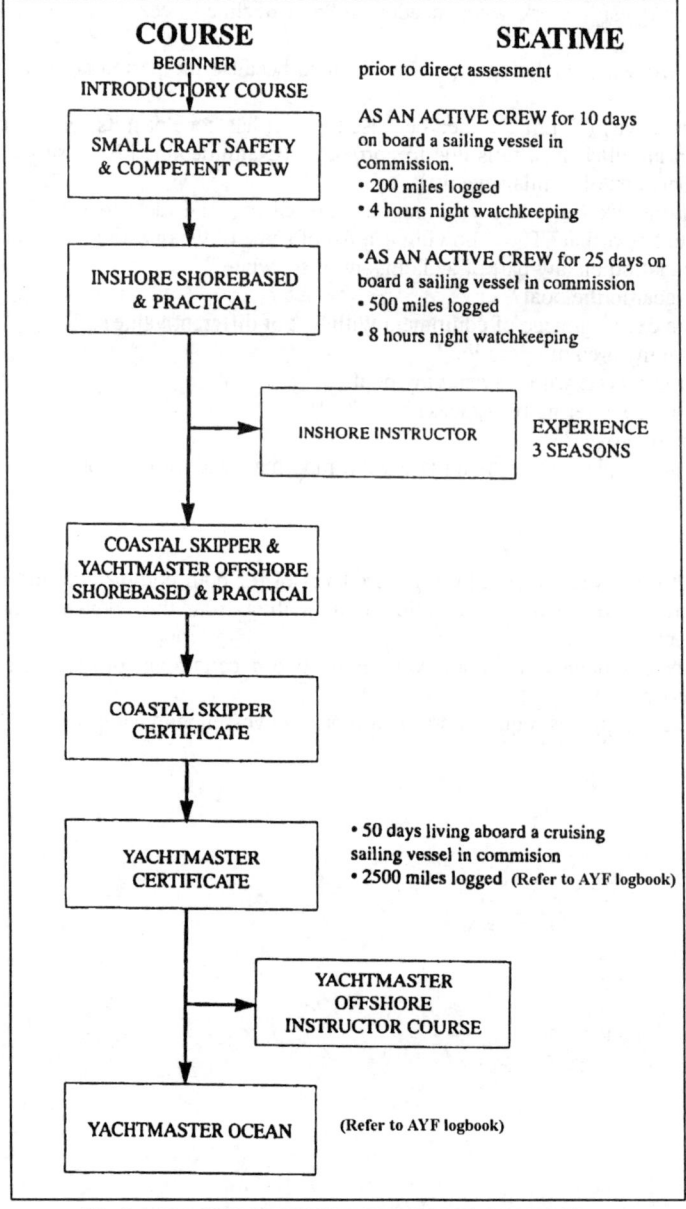

COURSE	SEATIME
BEGINNER INTRODUCTORY COURSE	prior to direct assessment
SMALL CRAFT SAFETY & COMPETENT CREW	AS AN ACTIVE CREW for 10 days on board a sailing vessel in commission. • 200 miles logged • 4 hours night watchkeeping
INSHORE SHOREBASED & PRACTICAL	• AS AN ACTIVE CREW for 25 days on board a sailing vessel in commission • 500 miles logged • 8 hours night watchkeeping
INSHORE INSTRUCTOR	EXPERIENCE 3 SEASONS
COASTAL SKIPPER & YACHTMASTER OFFSHORE SHOREBASED & PRACTICAL	
COASTAL SKIPPER CERTIFICATE	
YACHTMASTER CERTIFICATE	• 50 days living aboard a cruising sailing vessel in commision • 2500 miles logged (Refer to AYF logbook)
YACHTMASTER OFFSHORE INSTRUCTOR COURSE	
YACHTMASTER OCEAN	(Refer to AYF logbook)

Fig 2.1: YACHT CRUISING TRAINING SCHEME

Chapter 2: Training Schemes, Defence Qualifications & Onboard Training

COMMERCIAL CERTIFICATES OF COMPETENCY

NOTE: The following brief generalised summary is subject to change. Complete up-to-date details should be obtained from the relevant authorities.

In order to be the master or a crewmember on a commercial vessel, you would need to obtain one of the following certificates of competency:

CERTIFICATE	FUNCTION
General Purpose Hand or Pre-Sea Safety or Deckhand (TAFE Certificate I) or appropriate onboard safety training (Pre-requisite boating experience: nil)	Crew member
Coxswain (TAFE Certificate II) Pre-requisite experience: 12 months or 1800 hours boating experience)	Master-cum-engineer of a vessel of up to 12 metres in length, up to 15 miles from the coast (upto 30 miles in some cases).
Master Class V or Fishing Skipper Grade 3 (TAFE Certificate III) (combined with a Marine Engine Driver Grade III Certificate) (Pre-requisite experience: 30 months or 4500 hours boating experience, including experience on commercial vessels)	Master-cum-engineer of a vessel of up to 24 metres in length, up to 30 miles from the coast (up to 200 miles in some cases).
Master Class IV, and, higher certificates	

STCW-95 ENDORSEMENT

Most State-issued maritime certificate holders work within the jurisdiction of a State. There is no need for them to obtain an STCW95 endorsement on their Master Class V or higher certificate from AMSA. However, in order to work on a commercial vessel overseas, or on a vessel trading interstate in Australia, you would need to obtain the appropriate endorsement.

Complete details are available from AMSA. Briefly, however, there are numerous pre-requisites for the endorsements in terms of sea time (at least 12 months in vessels over 80 GRT for some certificates) as well as training.

For example, some of the additional training requirements for Master Class V candidate include:

- Cargo handling knowledge to Master Class IV standard.
- ARPA
- GMDSS GOC or MROCP with satellite endorsement

Similarly, a Master Class IV candidate seeking STCW95 endorsement for other than near-coastal waters would need to complete:

- Diploma of Applied Science (Master Class 3)
- Advanced Fire Fighting
- Medical First Aid (Beyond elementary first aid)
- Proficiency in Survival Craft and Rescue Boats other than Fast Rescue Boats
- GMDSS GOC

NOTES:

- The Master Class IV near-coastal waters endorsement is generally not recognised overseas.
- STCW endorsements are not given to qualifications lower than Master Class V or MED 3.

ABBREVIATIONS

AMSA: Australian Maritime Safety Authority
STCW: International Convention on Standards of Training, Certification & Watchkeeping for Seafarers.
GMDSS: Global Maritime Distress & Safety System (See Chapter 18)
GOC: General Operators Certificate (See Chapter 18)
ARPA: automatic radar plotting aids
GRT: Gross Registered Tonnes (See Chapter 13)
NMSC: National Maritime Safety Committee (See Chapter 12)

RECOGNITION OF AUSTRALIAN DEFENCE FORCE MARINE QUALIFICATIONS (Fig 2.2 – 2.8)

DEFINITIONS

- Junior Sailor is a RAN sailor of the rank of Ordinary Seaman, Able Seaman or Leading Seaman.
- Senior Sailor is a RAN sailor of the rank of Petty Officer, Chief Petty Officer or Warrant Officer.
- Officer is a RAN officer of the rank of Midshipman, Sub-Lieutenant, Lieutenant, Lieutenant Commander, Commander, Captain, Commodore or Admiral.
- National Transition Program (Maritime) Curriculum is the agreed National Curriculum for marine Certificates of Competency used by approved Training Providers.
- **ADF:** Australian Defence Force
- **ANTA:** Australian National Training Authority
- **RAN:** Royal Australian Navy
- **RPL:** Recognition of Prior Learning

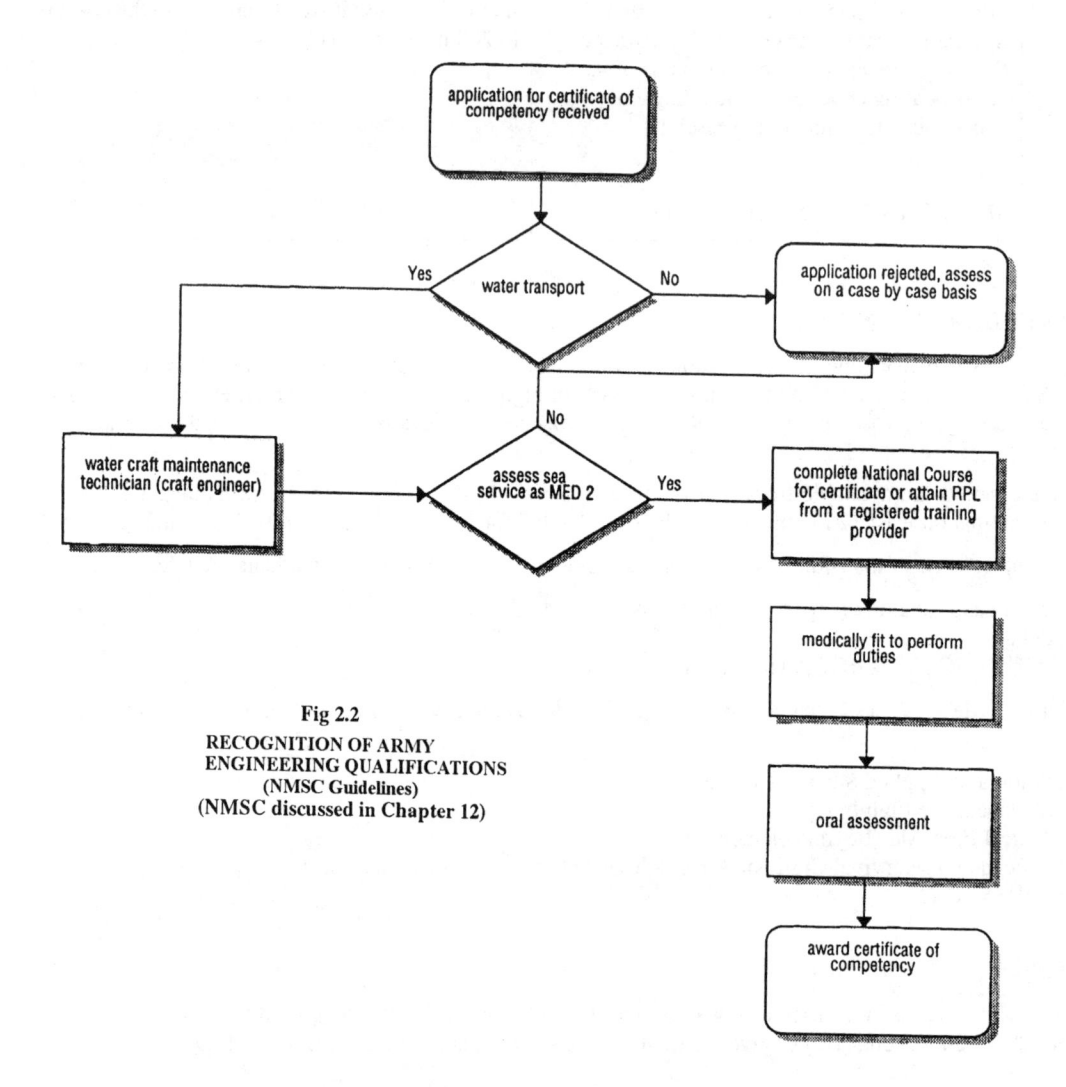

Fig 2.2
RECOGNITION OF ARMY ENGINEERING QUALIFICATIONS
(NMSC Guidelines)
(NMSC discussed in Chapter 12)

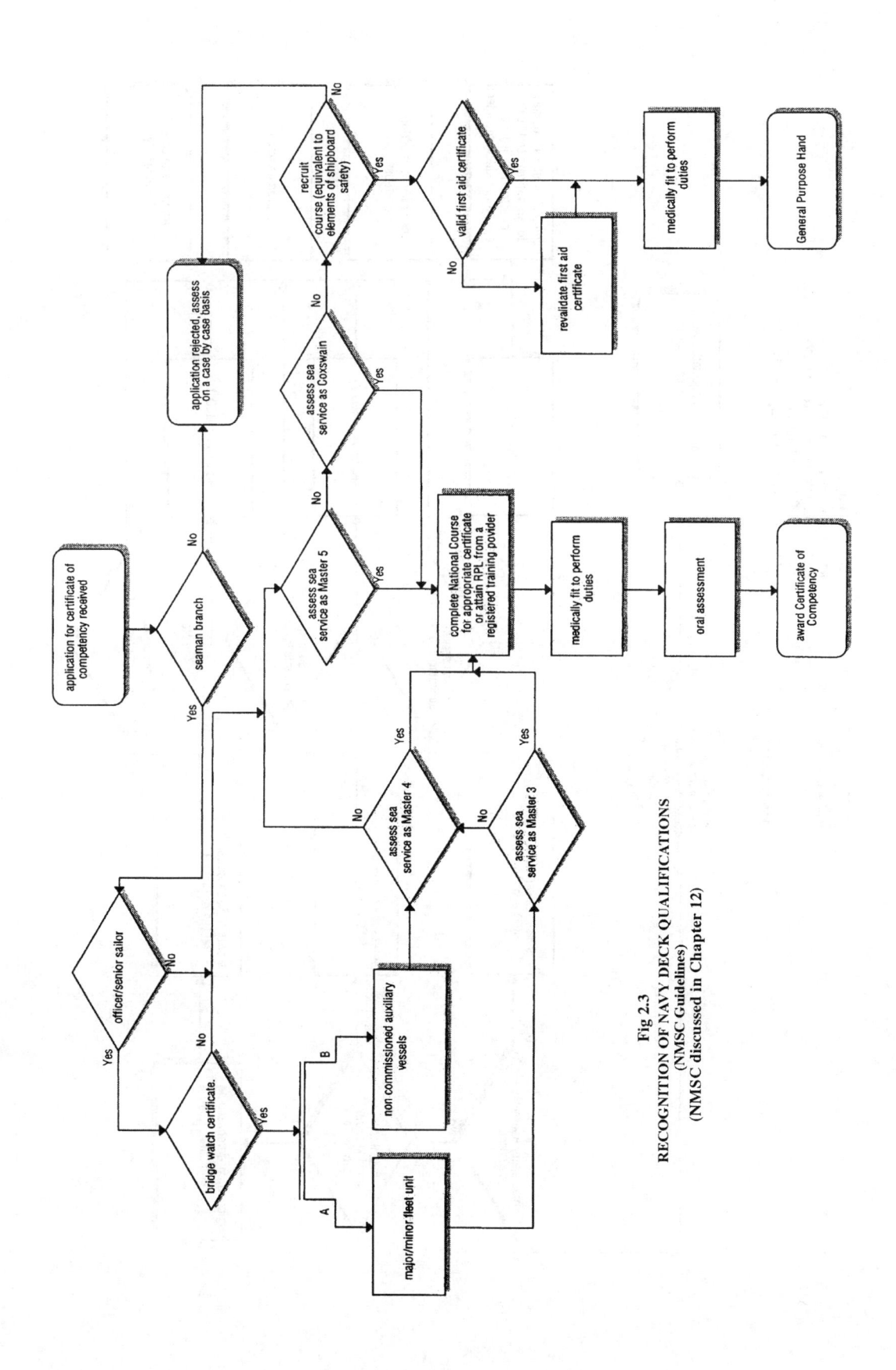

Chapter 2: Training Schemes, Defence Qualifications & Onboard Training

Fig 2.3
RECOGNITION OF NAVY DECK QUALIFICATIONS
(NMSC Guidelines)
(NMSC discussed in Chapter 12)

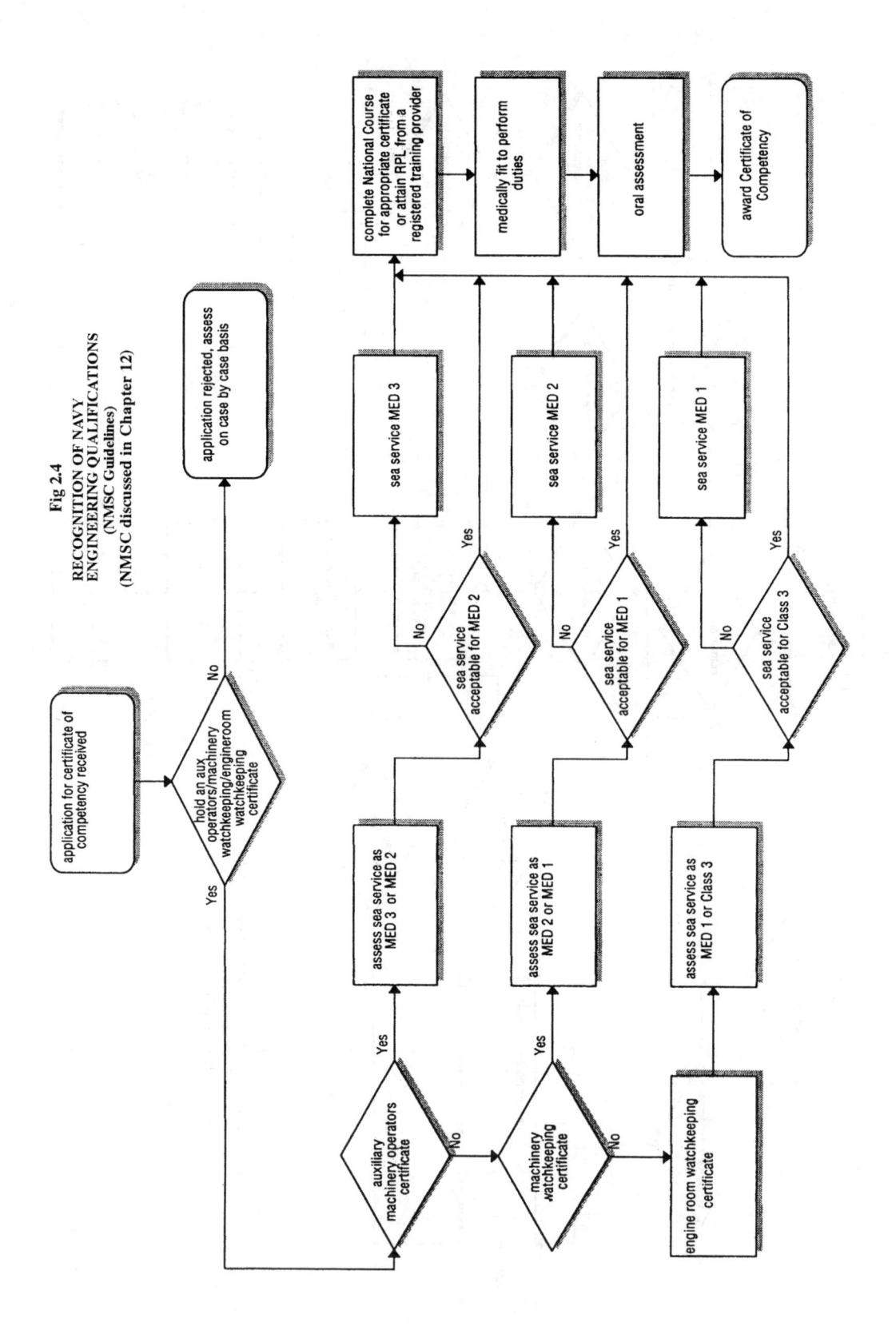

**Fig 2.4
RECOGNITION OF NAVY
ENGINEERING QUALIFICATIONS
(NMSC Guidelines)
(NMSC discussed in Chapter 12)**

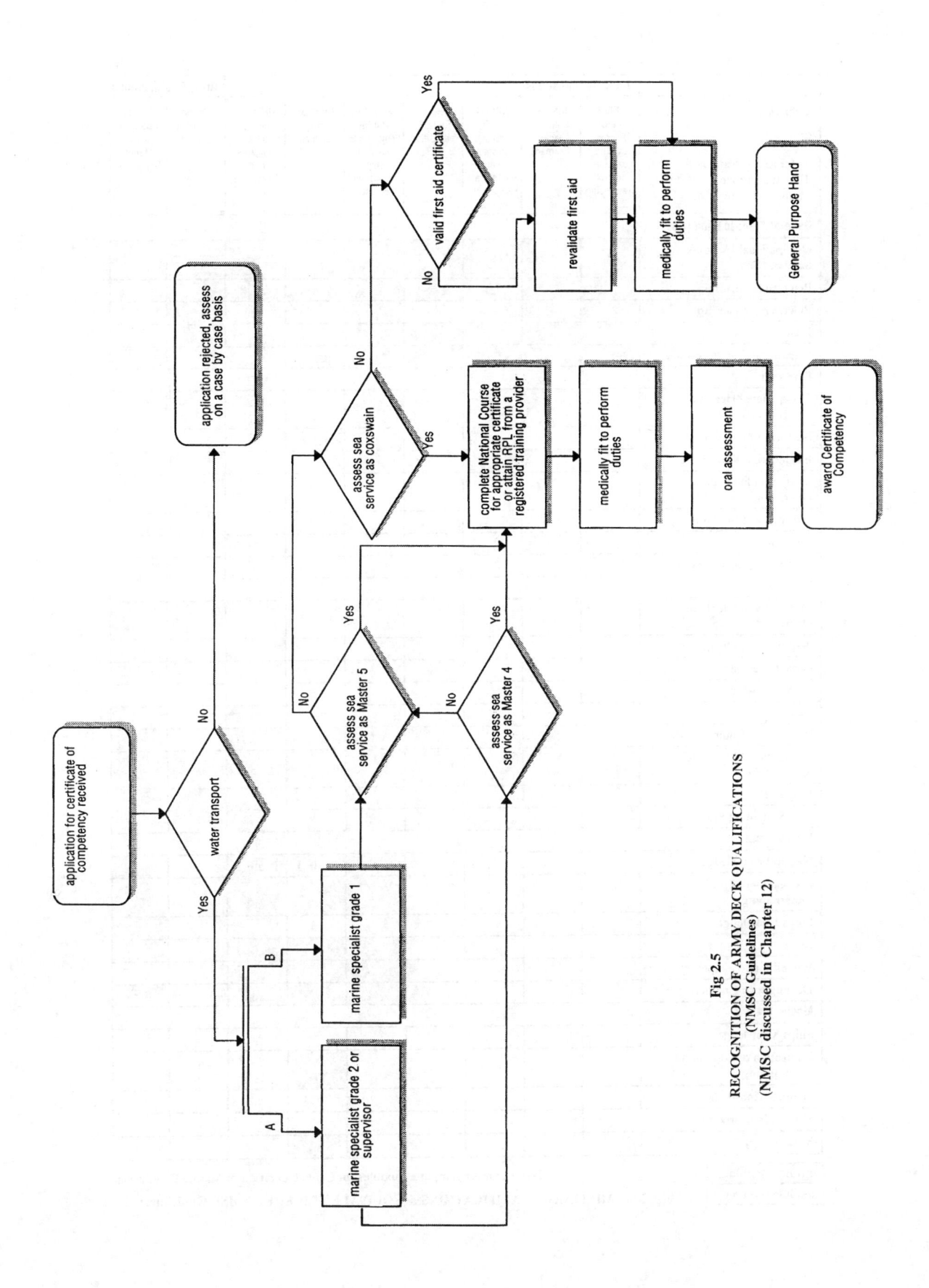

Chapter 2: Training Schemes, Defence Qualifications & Onboard Training

Fig 2.5
RECOGNITION OF ARMY DECK QUALIFICATIONS
(NMSC Guidelines)
(NMSC discussed in Chapter 12)

Fig 2.6 MILITARY QUALIFICATIONS & EQUIVALENT (DECK) (NMSC Guidelines)

Modules Refer to the National Transition Program Maritime Curriculum	Navy: JUNIOR SAILOR Recruit training damage control seamanship	Navy: JUNIOR SAILOR Leading Seaman	Navy: JUNIOR SAILOR Leading Seaman with intermediate seamanship course	Navy: SENIOR SAILOR Petty Officer	Navy: SENIOR SAILOR Petty Officer with small ship navigation course	Navy: OFFICER Bridge watch cert	Navy: OFFICER Bridge watch cert, navigation or adv. Navigation cert.	Army: Marine Specialist Grade 1	Army: Marine Specialist Grade 2 & Marine Supervisor
General Purpose Hand (NSW)									
Occupational Health & Safety at Sea	■							■	■
Practical Seamanship	■								
Practical Seamanship	■								
First Aid	■								
Life Sustaining Procedures	■								
Coxswain									
Occupational Health & Safety at Sea	■	■	■	■	■	■	■	■	■
Nautical Knowledge			■	■	■	■	■		■
Engineering Knowledge		■	■					■	■
Marine Radiotelephone Operation							■	■	
Writing Skills for Work		■	■						
First Aid			■	■	■				
Life Sustaining Procedures	■	■							
Master Class 5									
Marine Radiotelephone Operation								■	
Nautical Knowledge			■						
Coastal Navigation						■	■		
Ships Knowledge					■	■	■		
Radar						■	■		
Work Team Communication									
Occupational Health & Safety at Sea	■	■	■	■	■	■	■		■
First Aid	■			■	■				
Life Sustaining Procedures	■	■							
Master Class 4									
Nautical Knowledge					Note 1	Note 1	Note 1		■
Navigation & Position Determining					■				
Ships Knowledge					■	■	■		
Radar					■	■	■		
Wheelhouse Equipment					■	■	■		
Dealing with Conflict									■
Master Class 3									
Ships Knowledge									
Navigation & Position Determining						■			
Nautical Knowledge									
Ships Administration									
Meteorology						■			

Legend: ▒ advance standing; ☐ module required

Note 1: must complete modules for Cargo Operation & Marine Legislation

Chapter 2: Training Schemes, Defence Qualifications & Onboard Training

	Navy Qualifications			Army
	Auxiliary Machinery Operators Certificates	Machinery Watchkeeping Certificates	Engine Room Watchkeeping Certificates	Water Craft Maintenance Technician
Marine Engine Driver Grade III				
Engineering Knowledge	▓	▓	▓	▓
Occupational Health & Safety at Sea	▓	▓	▓	▓
Writing Skills for Work	▓	▓	▓	▓
Hand and Power Tools	▓	▓	▓	▓
First Aid	▓	▓	▓	▓
Marine Engine Driver Grade II				
Marine Diesel Engine Operation		▓	▓	▓
Marine Auxiliary Equipment & Systems Operation	▓	▓	▓	▓
Marine Engineering Operational Procedures and Safety	▓	▓	▓	▓
Work Team Communication	▓	▓	▓	▓
Occupational Health & Safety at Sea	▓	▓	▓	▓
First Aid	▓	▓	▓	▓
Marine Engine Driver Grade I				
Marine Engine and Propulsion Systems			▓	
Auxiliary Marine Equipment		▓		
Operational Safety and Survey Maintenance		▓	▓	
Practical Mathematics		▓		
Dealing with Conflect		▓	▓	
Occupational Health & Safety at Sea		▓		
First Aid		▓		
Marine Engineer Class III				
Auxiliary Machinery & Systems				
Marine Electrical Practice				
Marine Diesel Engines & Systems				
Engineering Mathematical Calculation				
OH & Safety for Supervisors			▓	
Advance Fire Fighting				
Writing Technical Documents			▓	
Proficiency in Survival Craft			▓	
First Aid			▓	

▓ advance standing
☐ module required

NOTE: Navy engineering qualifications are based on the 1992-93 Technical Training Plan (TTP92)

Fig 2.7 MILITARY QUALIFICATIONS & EQUIVALENT (ENGINEERING) (NMSC Guidelines)

Navy Ships	Type	Area of Operation
Major Fleet Units	Destroyers Frigates Submarines Stores Ships/Tankers	Australia and International Waters (ships are greater than 80 metres in length)
Minor Fleet Units	Patrol Boats Landing Crafts (heavy) Mine Hunters (coastal) Mine Hunters (inshore) Survey Motor Launch	Australian and International Waters (ships are between 30-60 metres in length)
Non-Commissioned Auxiliary Vessels	Minesweepers Dive Launches	Australian Coastal Waters

Army Ships	Type	Area of Operation
Auxiliary Harbour Vessels	Military Small Craft Launches and Lighters RIBs, Sharkcats Light Amphibious Ships	Restricted Offshore Operations (less than 12 metres in length)
Water Crafts	Landing Barges Light Cargo Vessels	Australian Coastal Waters (ships are between 15-30 metres in length)

Fig 2.8   MILITARY SHIP QUALIFICATIONS (NMSC Guidelines)

ONBOARD TRAINING
(COMPETENCY BASED TRAINING)

On-board learning and assessment conducted in partnership with approved registered training organisations (RTO) can lead to a reduction in the service required for qualification for a certificate of competency. The attainment of each competency should be recorded in a training logbook and signed by the candidate and the master.

Many of the main crewing competencies expected of crews on commercial vessels are listed below. The authorities encourage owners, masters and crew to use these competencies to enhance the quality of their service on their vessels. There is no compulsion on the operators or crew to complete these lists. They act as a guide to the type of questions asked by marine authorities at the oral examinations for issue of a certificate of competency.

The entries in the training log book should only be made after the competency has been considered to be properly achieved. In most cases signing should only occur after reasonable training and practice at the skill has occurred. The time taken to acquire a competency varies from person to person. Contact your State Waterways Authority or a RTO for further information on how you can participate in the competency-based training system.

See National Maritime Safety Committee's Tables below

(Fig. 2.9 – 2.12)

Chapter 2: Training Schemes, Defence Qualifications & Onboard Training

Competency Required **Occupational Health and Safety**	Competency Required **Vessel Handling and Manoeuvring** Displacement hull vessel Length – 8 to 15 metres	Competency Required **Vessel Handling and Manoeuvring** Displacement hull vessel Length 15 – 35 metres
Locate personal flotation devices on board and demonstrate how to use	Steer vessel in good conditions	Steer vessel in good conditions
Locate liferafts and demonstrate the launching procedures	Steer by a compass course in moderate winds and waves	Steer by a compass course in moderate winds and waves
Locate portable fire fighting equipment on board and demonstrate how to use	Steer a vessel into a port entrance keeping vessel on leads	Steer a vessel into a port entrance keeping vessel on leads
Receive instructions on the safety practices and precautions on board	Manoeuvre vessel in following seas without pooping/broaching	Manoeuvre vessel in following seas
Receive instructions on actions to be taken in the event of an emergency	Manoeuvre vessel safely in busy traffic adjusting speed & course	Manoeuvre vessel safely in busy traffic adjusting speed & course
Receive/demonstrate actions to take in the event of abandonment of vessel	Manoeuvre and anchor vessel	Manoeuvre and anchor vessel
Activate emergency power, lighting and pumping systems, fuel shut offs	Depart wharf/berth safely using engines, lines, crew instructions	Depart wharf/berth safely using engines, lines, crew instructions
Participate/organise emergency drills for fire/collision/damage/abandonment	Berth vessel safely using engines, lines, crew instructions	Berth vessel safely using engines, lines, crew instructions
Receive/provide instructions in operations of vessel equipment	Manoeuvre vessel safely stern first into a pen/ berth	Manoeuvre vessel safely stern first into a pen/ berth
Receive/provide instructions in precautions associated with: • Work procedures • Cargo/stores handling • Passenger control • Movement around moving vessel • Entering engine room • Heavy weather • Handling lines/nets • Confined spaces	Berth vessel safely in adverse wind/tidal conditions	Berth vessel safely in adverse wind/tidal conditions
	Manoeuvre vessel safely alongside another vessel in wind & waves	Manoeuvre vessel safely alongside another vessel in wind & waves
	Tow another vessel safely using towing bridle, correct towline	Tow another vessel safely using towing bridle, correct towline
	Manoeuvre vessel safely across a bar entrance outbound	Manoeuvre vessel safely across a bar entrance outbound
	Manoeuvre vessel safely across a bar entrance inbound	Manoeuvre vessel safely across a bar entrance inbound
Demonstrate use of personal protective and safety equipment used during work	Manoeuvre vessel to recover a person overboard (drill)	Manoeuvre vessel to recover a person overboard (drill)
Understand alcohol drug and smoking policy on board	Manoeuvre vessel in an expanding square/parallel track search pattern	Manoeuvre vessel in an expanding square/parallel track search pattern
Participate in OH&S consultation with master/owners in risk management	Manoeuvre vessel using emergency steering	Manoeuvre vessel using emergency steering
Maintain personal and vessel hygiene	Manoeuvre vessel into/from a slipway	Manoeuvre vessel into/from a slipway
Read and understand Material safety Data Sheets (MSDS) for hazardous goods	Manoeuvre a planing hull at speed in turns, crash stops	

Fig 2.9: ONBOARD TRAINING (1)
NMSC Guidelines
(NMSC discussed in Chapter 12)

Competency Required — **Navigational safety**	Competency Required — **Navigational safety**	Competency Required — **Seamanship**
Keep a lookout by day reporting all items of navigational nature to master	Switch on a marine VHF radio and a marine 27MHz radio, select channels and adjust squelch and volume controls	Tie a bowline, clove hitch and round turn and two half hitches
Keep a lookout by night reporting all items of navigational nature to master	Use marine radio correctly to obtain radio checks and weather reports	Splice an eye in synthetic rope
Determine risk of collision in head on, crossing and overtaking situations	Use marine radio correctly to contact other vessels and coast stations	Whip the end of a rope
Correctly identify lateral, cardinal, special and danger marks in daylight	Identify major chart symbols and abbreviations to identify dangers and aids to navigation	Rig a towing bridle and attach towing lines for a tow
Correctly identify lateral, cardinal, special and danger marks at night	Safely operate an auto-pilot to steady a vessel on a specified course	Rig fenders to protect vessel hull when alongside other vessels/wharves
Identify an isophase light, an occulting light and a flashing light and compare their period and rhythm with chart	Obtain tidal data for a voyage	Make mooring lines fast during berthing operations
Use a handbearing compass to take accurate navigational bearings	Obtain water depth using an echo sounder and compare with naval chart	Rig backsprings to stop vessel moving along wharf when berthed
Use a deviation card to determine the value of deviation for any heading	Identify lights and shapes of vessels constrained by their draft, restricted in their ability to manoeuvre, not under command, fishing, sailing	Rig a gangway for safe access between vessel and shore
Determine the compass variation from a naval chart for current date	Understand whistle signals sounded during manoeuvring	Turn mooring lines up on winches/bits to secure vessel
Identify all the controls on a radar set using manufacturers handbook	**Watchkeeping** on a voyage of at least 24 hours duration	Check & maintain watertight seals on hatches and doors
Switch on radar set and obtain a tuned picture to makers specifications	Plan a safe coastal passage	Prepare an anchor and tackle ready for anchoring
Use a radar on all working ranges to detect other vessels and land targets	Understudy the master in keeping a navigation watch	Let go and retrieve an anchor during anchoring
Obtain a radar bearing and correct it for ships head and compass error	Keep a navigation watch alone calling the master as required	Rig a sea anchor to control drift at sea
Obtain a radar range of a vessel target and a land target	Estimate a position at sea using course steered, speed made good and prevailing wind and sea conditions	Operate a winch safely to haul in and run out lines
Fix the vessel's position using radar ranges and radar bearings	Complete a log book with navigational details of the watch	Launch/recover dinghy/rescue boat from main vessel at sea
Use a GPS set to obtain a position and transfer it to a navigation chart	Keep an anchor watch	Transfer persons safely from boat to boat at sea
Lay off a course on a chart between two positions and calculate compass course to steer between them	Determine speed made good and course made good between fixes	Rig ropes on a small derrick boom
Lay off a course on a chart to clear a known navigational danger and calculate compass course to steer	Calculate fuel consumption for a voyage	Use small derrick boom to load/discharge heavy items
Fix a vessel's position using compass bearings and compare it to other fixes made by radar and GPS		Overhaul blocks used in a small derrick boom
Check a compass deviation when vessel is heading on a set of leads		Identify confined spaces and understand precautions necessary

Fig 2.10: ONBOARD TRAINING (2)
NMSC Guidelines
(NMSC discussed in Chapter 12)

Chapter 2: Training Schemes, Defence Qualifications & Onboard Training

Competency Required **Engineering** (inboard diesel engines)
Identify main components of engines and auxiliaries
Assist in pre-start checks on engines and auxiliaries
Assist in shut down checks on engines and auxiliaries at end of trip
Identify engine fuel system Including emergency shut-offs
Assist in checks to monitor operation of engines & auxiliaries during voyage
Read and understand gauges of operating machinery
Check, clean and replace main filters of fuel, cooling & lubrication systems
Identify main components of fuel systems
Identify fuel tanks, filling, venting and sounding arrangements
Identify & understand use of emergency fuel shut-offs
Calculate fuel capacities of tanks and calculate fuel consumption rates
Bleed a fuel system to remove air or contaminants from the system
Refuel vessel taking precautions to prevent overfilling and pollution
Identify main components of lubricating oil system
Top up and maintain lubricating oil systems for main machinery
Perform routine maintenance on main engines, and lubricating system
Identify main components of cooling system including suctions/discharges
Maintain a heat exchange cooling system – including corrosion control
Diagnose following engine faults: Fail to start, noisy engine, overheating Exhaust colour, low oil pressure
Replace faulty/defective components for repair/maintenance
Maintain gear box
Identify and check main shaft alignment, bearings, seals, glands

Competency Required **Engineering** (inboard diesel engines)
Identify components of bilge system and trace bilge lines to overboard
Operate main bilge pump to remove water from hull – without pollution
Identify and operate emergency hand bilge pump to remove water from hull
Clean bilge strainers and operate valves and manifold system
Operate L cock system to pump water onto deck
Check and replace belt drivers on engines/pumps
Identify components of a hydraulic steering system
Maintain a hydraulic oil system without causing pollution
Engage emergency steering arrangements to steer vessel
Identify components of a cable/pulley system of steering
Maintain a cable/pulley system of steering
Inspect and assist in the maintenance of a rudder
Inspect and assist in the removal/maintenance of a propeller
Inspect and assist in he maintenance of a propeller shaft and stern gland
Inspect and replace sacrificial anodes on the underwater hull
Identify and maintain hull skin fittings
Clean marine growth from hull without causing pollution
Apply environmentally friendly anti-foul to underwater sections of hull
Maintain lifting gear (derrick, goose neck, blocks, sheaves and shackles
Maintain deck machinery and associated guards
Disassemble, maintain and re-assemble a deck winch

Competency Required **Engineering** (inboard diesel engine vessels)
Connect batteries in parallel & in series and understand reasons why
Test batteries for charge and use a hydrometer to monitor condition
Explain the dangers/precautions involved when charging batteries
Maintain battery terminals and battery system in good condition
Understand the difference between ammeters and voltmeters – their use
Explain features and operation of alternators and starter motors
Check and adjust belt tension on an alternator
Identify and maintain main a low voltage electrical system
Select and use correct fuses/circuit breakers
Connect/disconnect shore power systems ensuring correct polarity
Monitor electrical systems on board for corrosion, and defects
Maintain a running and maintenance log for the machinery on board
Maintain watertight seals on hatchways, doors and windows
Maintain a timber vessel using correct preservation methods
Identify electrolytic corrosion and understand how to prevent it
Identify and understand osmosis in a GRP hull and understand treatment
Use hand tools around vessel to carry out routine maintenance
Understand the regulations applying to overboard discharge of sewage
Operate fire pump to obtain sufficient pressure for fire hose use
Identify main components of engine room fire fighting system and its use
Explain the procedures to be followed in an engine room fire

Fig 2.11: ONBOARD TRAINING (3)
NMSC Guidelines
(NMSC discussed in Chapter 12)

Competency Required
Vessel construction and stability

Inspect all main compartments and tanks of vessel including watertight bulkheads
Secure hatchways, doors and windows to a weathertight condition for rough weather conditions
Identify all exterior openings into hull and means of maintaining watertightness
Check & maintain watertight seals on hatches and doors
Test watertight closings for weathertightness
Identify and repair deteriorated or damaged fittings
Identify and maintain all hull below water fittings during slipping
Identify hull attachment points of all engines and fittings
Trace all pumping and piping systems
Operate main and emergency bilge pumping system
Operate deck wash pumping system
Identify/operate and maintain venting to all compartments
Identify fresh water tanks, fill, empty and clean out as required
Read and be familiar with stability booklet for vessel
Read and understand survey maintenance schedules
Observe annual survey inspection by marine authority surveyor
Assist in maintenance of rudder and propeller and shaft during slipping
Clean and maintain surface of hull during slipping
Identify and maintain hull surface defects (rust, electrolysis, osmosis)
Apply solvents, paints, resins etc safely to maintain vessel

Fig 2.12: ONBOARD TRAINING (4)
NMSC Guidelines
(NMSC discussed in Chapter 12)

Chapter 3

NAUTICAL TERMS, FITTINGS, HULLS & RIGS

GLOSSARY OF NAUTICAL TERMS

Abaft	behind, such as 'Abaft the Beam'.
Abeam	abreast of, or at right angles to, the fore and aft line of the vessel.
A-Cockbill	anchor lowered clear of hawse pipe (or bow roller) and hanging vertically.
Aft	towards the "stern" or rear of the vessel.
After peak	narrow space below deck in the after end of the vessel.
Amidships	in the middle section of the vessel.
Anchor aweigh	anchor has been hove out of ground and is clear of it.
Astern	behind or in the back of the vessel.
Athwartship	across the vessel; at right angle to the fore and aft direction.
Azimuth	see Bearing.
Back (Backing)	anticlockwise change in wind direction. (see Veer)
Back splice	splice made at the end of a rope to stop it from unlaying or slipping through the block.
Backing a sail	holding a sail against the wind when the wind is very light.
Bar	a shallow area formed by sand, mud, gravel or shingle, near the mouth of a river or at the approach to a harbour. A measure of atmospheric pressure.
Beam	width at the widest point of the vessel.
Bearing	the direction of an object.
Bend	to tie two ropes together.
Bight	a loop in a rope.
Bilge	the lowest inside part of the boat.
Bilge keel	small fin keels on the sides of or in lieu of the main keel.
Bitter end	the secured (tied up) end of rope.
Bitts	a pair of posts for securing (tying) mooring lines.
Block	a pulley.
Bollard	a post for securing (tying up) ropes – on vessel or jetty.
Bolt rope	rope sewn into the leading edge of a sail. It slides in the groove of a mast.
Bombora	a shallow area where waves may break.
Bosun's chair	a sling with a wooden seat for hoisting a person aloft.
Bow	the front end of a vessel.
Bow chock	see Fairlead.
Bower anchor	the anchor in the bow of boats.
Bowsprit	spar projecting out of the bow and taking one or more headsails.
Breaking out (the anchor)	to make ready and free the anchor for letting go.
Broached	vessel turned beam on to the following sea and capsized.
Brought up	vessel is riding to her anchor and cable, and the former is holding.
Bulkhead	a vertical partition in a vessel (equivalent to a wall in a building).
Cable	one-tenth of a nautical mile, 185 metres at the equator.
Camber	the transverse (sideways) curve in a deck, just as a road has camber for drainage.
Capstan	anchor winch with vertical axis. (see also Windlass) (Both types illustrated in Chapter 7)
Cardinal points	the four main points of a compass: N, E, S, W.
Careening	running a boat onto a suitable beach for repairs.
Cathead	a structure in the bow to suspend the anchor in a small vessel.
Centerboard (Centre Plate)	a board or metal plate lowered through the bottom of some sailing boats to reduce leeway or to increase the size of the keel. (see also Lee Board)

Chain stopper	a short chain temporarily holding a wire rope under load.
Chine	the angle formed where two non-curved sections of a boat's hull join. (as opposed to a round hull)
Clear anchor	see Breaking Out the Anchor. The term "Clear Anchor" is also used when weighing anchor to indicate that it is in sight and clear of turns in the cable or other obstructions.
Coaming	a fence around an opening to keep the water out, such as around a hatchway.
Cock-a-bill	see A-Cockbill.
Con	to direct a person at the helm in steering of a vessel.
Course	ship's heading.
Counter	an overhanging stern, or the underneath of that overhang.
Dan buoy	a marker buoy with a pole and a flag.
Davit	a boom for lowering a lifeboat or large liferaft.
Devil's claw	a metal claw to secure the anchor in its stored position.
Downstream	proceeding down a river or harbour out to seaward.
Draft (Draught)	the depth of boat in the water at the deepest point of the keel.
Drift	the distance traveled by the vessel due to current or tidal flow.
Drogue	see Sea Anchor.
Ebb tide	falling or running out tide.
Eddy	circular movement of water.
EPIRB	Emergency Position Indicating Radio Beacon.
Fairlead	a smooth, non-chafing channel through which mooring ropes and wires are passed.
Fairway	any navigable channel.
Fender	a tire or a padding placed between the vessel and wharf to prevent chafing.
Flare	the overhang of the ship's bow.
Fluke(s)	the sharp end(s) of the anchor that dig into the seabed.
Flood tide	the rising or run in tide.
Fore and aft	in line with the stem and stern; in line with the keel.
Forecastle (Focsle)	the compartment or the built-up deck in the bow (fore part) of the ship.
Forepeak	narrow space below deck in the forward end of vessel.
Forward	the front or towards the front of a vessel.
Foul anchor	anchor caught in an underwater cable, or it has brought old hawsers to the surface with it, or it is fouled by its own cable.
Foul hawse	both anchors are out and the cables are entwined or crossed.
Foundered	vessel filled with water and sunk.
Frames	vessel's ribs, usually running transversely (side to side)
Freeing ports	opening in the bulwark to allow water washing on deck to run off.
Gantline	a line used for hoisting through a block at the masthead.
Give way	reduce speed, stop, go astern or alter course so as to keep out of another vessel's path.
GPIRB	EPIRB with a built-in GPS (See EPIRB)
Growing	the way the anchor cable is leading from the hawse pipe, e.g., a cable is growing aft when it leads aft.
Gunwale	pronounced "gunnel", the top edge of the vessel's sides.
Gypsy	a cogged wheel on the anchor winch (windlass) over which the anchor cable runs.
Hard-a-port (or Starboard)	see "Helm Orders" below.
Hawse pipe	the tube through which the anchor cable goes from the vessel to the anchor.
Heave to	steering into the wind and sea to make minimum headway.
Helm	tiller. The steering wheel is also sometimes referred to as the helm. See "Helm Orders" below.
Hitch	tying a rope to something.
Hog	bending down of the fore and aft parts of the hull due to uneven loading or wave riding. (opposite of sag)
Hove in sight	the anchor is completely out of the water (when picking up anchor)
Jury rig	a temporary arrangement replacing a failed mechanism on board, such as a jury rudder.
Knots (speed)	one knot is a speed of one nautical mile per hour, or 1.852 km/h.
Kedge	a small anchor used for light duties.
Langs lay	the rope strands are laid in the same direction as the yarns.
Lanyard	a short rope or cord for securing or holding something.
Lazarette	a space between decks in the after part of the vessel.

Chapter 3: Nautical Terms, Fittings, Hulls & Rigs

Leads	marks used in channels and at bar entrances which when in line indicate the centre of the navigable channel.
Lee Board	vertical wooden board, pivoted in forward edge, attached to the side of some flat-bottom sailing boats and lowered into water to reduce leeway. (see also Centreboard)
Lee Shore	the shore onto which the wind blows. This term is often misunderstood. It is the shore in the lee of the vessel - not the vessel in the lee of the shore.
Lee Tide	a tidal stream setting to leeward or downwind. The water surface has a minimum of chop on it, but the combined forces of wind and tide act upon the vessel at anchor.
Leeward	downwind side.
Leeway	wind blowing the vessel sideways.
Length Overall (LOA)	the distance measured parallel to the designed (loaded) waterline from the foremost part pf the hull to the aftermost part of the hull, excluding appendages, such as outboard engines, fenders, pulpit rails, bow door or ramp, sponsons, rubbing strip, bowsprits, bumpkins, etc. Bulwarks and areas of deck that overhang beyond the extremities of the hull are included in LOA.
Let go	drop anchor in the water.
Lizard	a line or wire with an eye in the end for another line to be attached or rove through.
Lloyds Length	See Measured length.
LOA	See Length Overall.
Log	a speed and distance measuring instrument; a record keeping book.
Long stay	the anchor cable is taught and leading down to water close to the horizontal.
LWL	length at waterline.
Made fast	secured.
Making way	moving through the water (by engine, oars or sail)
Masthead	the top of the mast.
Measured Length	It is the length measured from stem to stern along the weather deck or 96% of the vessel's shoe-box length, whichever is greater.
Midships	see Amidships.
Mooring lines	ropes for securing (tying) a vessel to jetty, etc.
Monkey's fist	a heavy knot on the end of a heaving line to help it to be thrown at a distance.
Nautical mile	one minute of latitude, which is 1852 metres at the equator.
Neap tide	tide at the first and last quarters of the moon when there is smallest rise and fall in tidal level.
Nipped cable	an obstruction, such as the stem or hawse-pipe lip has caused the cable to sharply change its direction.
Overfalls	dangerous steep waves due to a current meeting an opposing wind; rippling or sudden race of water due to sudden increase of depth.
Painter	a small mooring line for a small boat or liferaft.
Panting	heaving of the bow and stern hull in and out due to variations in water pressure during pitching.
PFD	personal flotation device (lifejacket or buoyancy vest)
Poop deck	a short raised deck in the after part of the vessel.
Port side	the left hand side of a vessel when you are looking forward from the stern and the side on which a red light is displayed.
Port 10	See "Helm Orders" below.
Pulpit	guardrail around the bow of a boat.
Pushpit	guardrail around the stern.
Quarter	the part of the hull on each side of the stern.
Rake	the lean in masts, funnel, etc.
Racking	twisting of the hull during rolling.
Reef points	short lines for tying up unwanted sail.
Sag	bending of the hull in the middle due to uneven loading or wave riding. (opposite of hog)
Samson post	a vertical post used for tying mooring lines.
Scope	the ratio of the anchor cable in relation to the depth of water.
Scupper	a drain hole on deck.
Scantlings	timber or other material used in the construction of a vessel.
Sea anchor	a parachute-shaped bag tied to a liferaft or vessel to reduce her drift.
Sea cock	a valve fitted on a pipe at the hull to allow liquid in or out.
Set	the direction in which current or tidal flow will cause the vessel to travel.
Sheer	The dip in the middle of a deck, or the upward curve of deck fore and aft.

Short stay	the anchor cable is taught and leading down to water close to the vertical.
Slack water	the turn of the tide when the tide neither floods or ebbs for a short period.
Snub	to stop a rope or anchor cable running out by turning it around the bitts or braking the cable.
Sounding	measuring the depth of water, fuel its.
Sponson	a wooden or rubber fender fitted around a boat to safeguard hull against chafing and rubbing.
Spurling pipe	pipe through which anchor chain leads into the chain locker.
Spring tide	a tide of relatively large tidal range occurring near the times of a new or full moon.
Stand on	to maintain the same course and speed.
Starboard side	the right hand side of the vessel when you are looking forward from the stern, and the side on which a green light is displayed.
Starboard 10	See "Helm Orders" below.
Stay	a rope supporting a mast.
Steady	see "Helm Orders" below.
Steaming light	masthead light.
Steer	see "Helm Orders" below.
Stem	the front of the bow.
Stern	the back or rear of the vessel.
Stopper	a short rope or chain, temporarily holding a larger rope under load. (See also Chain Stopper)
Stringers	fore and aft timbers or beams joined with the frames to add rigidity to the hull.
Surge	allow cable or rope to run out or slacken while being hauled.
Tack	the direction of a sailing vessel in relation to the wind; a sailing vessel changing course.
Transit	lining up two objects. (see Leads)
Transom	the flat stern (opposite of cruiser or round stern)
Trim	the difference the fore and aft draft of a vessel.
Tumblehome	a hull built with a bulge in it between the waterline and gunwale. Vessel has greater beam on the water than on deck.
Underway	not at anchor or made fast to the shore or ground, if you are drifting you are underway.
Up-and-down	anchor cable leading vertically down to the water.
Upstream	proceeding up river or into the harbour.
Veer	let out rope or cable under control; clockwise change in wind direction. (see Backing)
Walk back	see Veer.
Warp	to move a vessel along a jetty by pulling on her mooring lines; a rope on a drum used for pulling.
Weather tide	tidal stream setting to windward or upwind. It makes the sea very choppy.
Weigh anchor	lift anchor clear of the seabed.
Windlass	anchor winch with horizontal axis. (see also Capstan) (Both types illustrated in Chapter 7)
Windage	the hull and superstructure exposed to wind which causing the vessel to experience leeway.
Windward	the side of the vessel from where the wind is blowing.

HELM, RUDDER OR STEERING ORDERS (Fig 3.1)

(Orders from the master or navigating officer to the person at the steering wheel. The helmsperson must repeat each order.)

Steady	(or "Steady As She Goes"). Steer the present course by lining up a land mark or by compass.
Hard-a-Port	Turn the wheel fully to port side. (Rudders are usually fitted with stoppers at the maximum angles of 35°.)
Starboard 10	Turn the wheel so that the rudder angle indicator reads 10° to starboard side; and keep it there. (The rudder angle indicator is fitted where it is visible from the steering position.)
Amidships	Turn the wheel to amidships position (Zero degrees on the rudder indicator.)
Steer 010	Steer a course of 010° by compass.

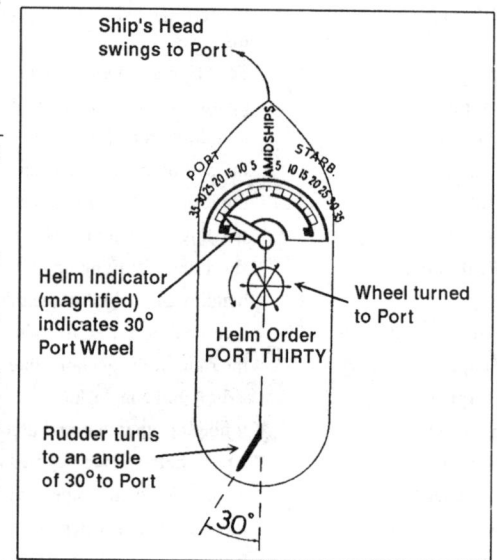

Fig. 3.1

Chapter 3: Nautical Terms, Fittings, Hulls & Rigs

POINTS OF COMPASS

A compass represents a geometrical circle of 360 degrees. In nautical terminology, a circle is divided into 32 points, known as the points of a compass, or compass points. Therefore, 1 Point = 360° ÷ 32 = 11¼°. There are 8 points in each quarter. An angle of 45° = 4 points.

The compass points provide a useful way of expressing the direction of another vessel or object relative to the boat.

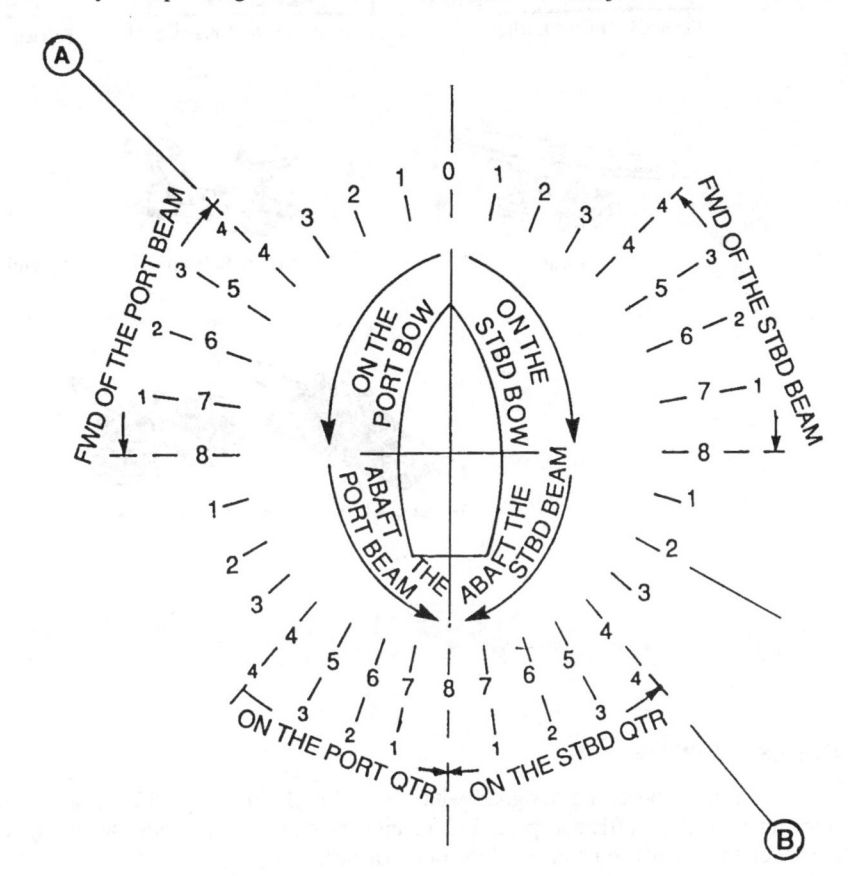

Fig 3.2
"A" is 4 points on the port bow (or 4 points forward of the port beam)
"B" is 4 points on the starboard quarter (or 4 points abaft the stbd beam)

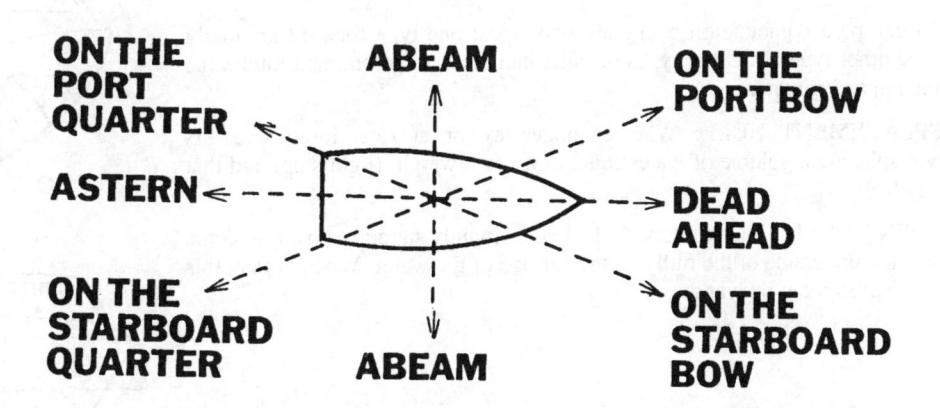

Fig 3.3

DECK FITTINGS (Fig 3.4)

Fairlead (Open Chock)    Fairlead (Single Roller Chock) 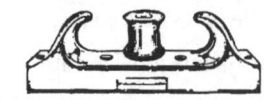   Fairlead (Double Roller Chock)

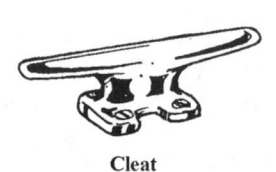

Cleat    Bitts or Bollards    Staghorn or Single Bollard

Wedging Cleats (V-cleat, clamping jaws, etc.)

HULLS –TYPES (Fig 3.5)

Cruising boats are designed with a round hull, which provides a softer ride, more internal space, efficient speed for a given engine power, and better seaworthiness. However, these hulls roll more than the other shapes.

High-speed boats are fitted with a chine (or Vee shaped) hull for planing. A chine can be "soft" (topside meets the bottom in a curve) or "hard" (topside meets the bottom in compound curves or at angle). A Soft Chine usually gives a faster ride in rough weather, but it is not very roomy inside. A Hard Chine provides more internal space. It is designed to have less 'wetted' area but can be expensive to fabricate.

Hulls can be a combination of the above shapes: one type forward gradually changing to the other type aft. Other types of hulls include flat bottom, and multihulls such as catamaran and trimaran.

DISPLACEMENT HULL: Whether underway or at rest, it achieves its buoyancy by displacing a volume of water equal to its own weight (light ship) and that of its load (deadweight).

PLANING HULL: When underway, its load is mainly supported by a dynamic action between the underside of the hull and the surface of the water. When at rest, this hull reverts to displacement buoyancy.

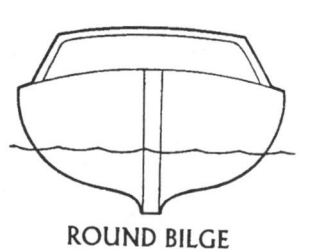

ROUND BILGE

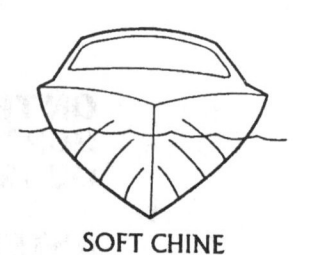

SOFT CHINE

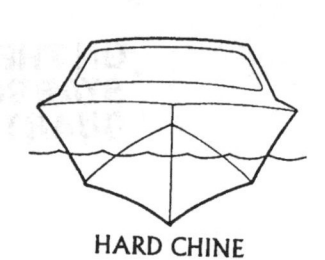

HARD CHINE

Chapter 3: Nautical Terms, Fittings, Hulls & Rigs

KEELS – TYPES (Fig 3.6)

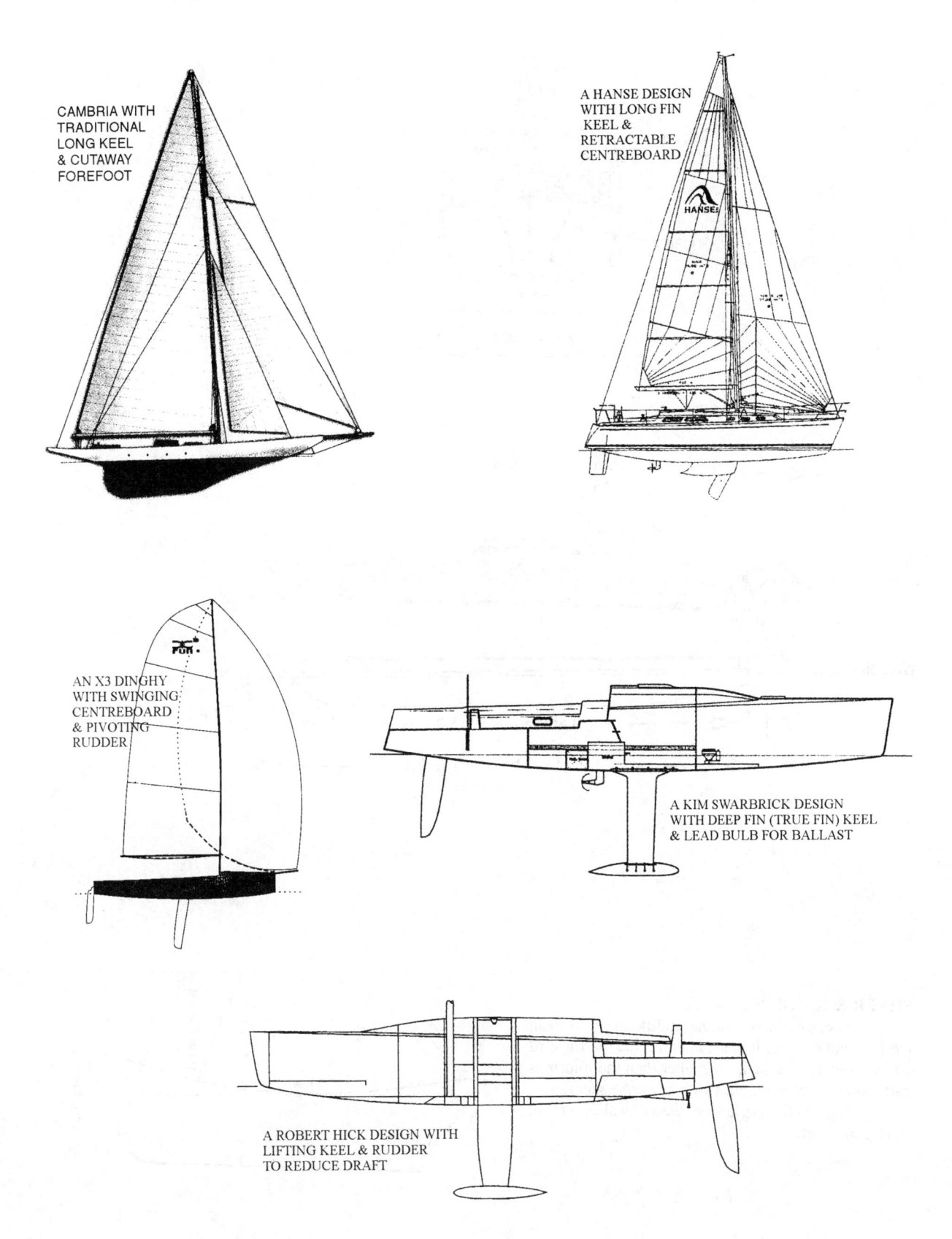

CAMBRIA WITH TRADITIONAL LONG KEEL & CUTAWAY FOREFOOT

A HANSE DESIGN WITH LONG FIN KEEL & RETRACTABLE CENTREBOARD

AN X3 DINGHY WITH SWINGING CENTREBOARD & PIVOTING RUDDER

A KIM SWARBRICK DESIGN WITH DEEP FIN (TRUE FIN) KEEL & LEAD BULB FOR BALLAST

A ROBERT HICK DESIGN WITH LIFTING KEEL & RUDDER TO REDUCE DRAFT

PARTS OF A VESSEL & TERMS RELATING TO HULL (Fig 3.7 & 3.8)

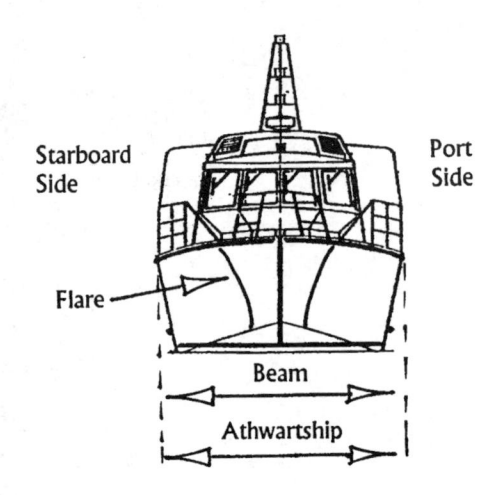

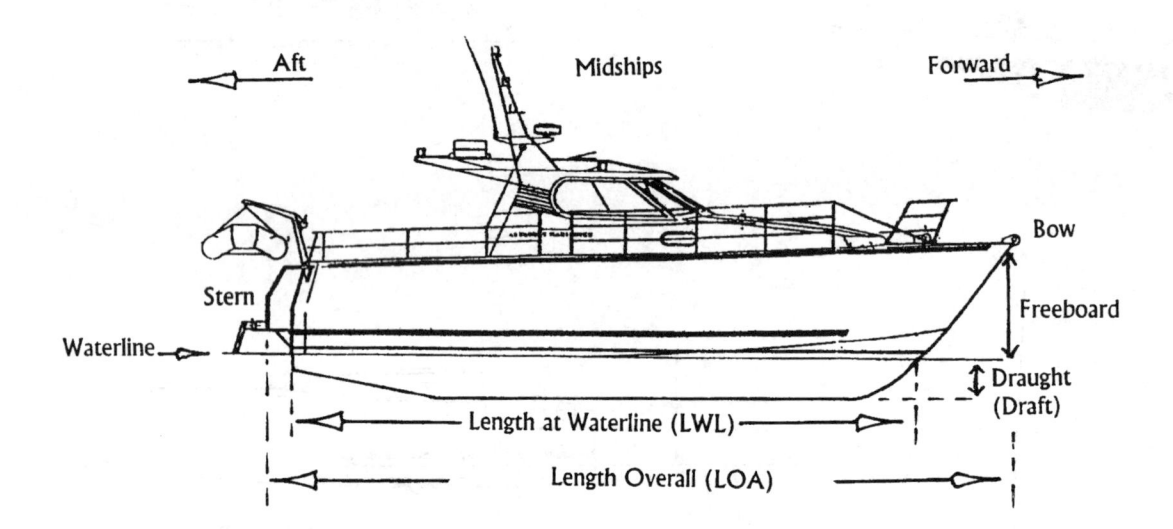

SHEER & RAKE (Fig 3.9)

Sheer is the dip in the middle of the boat in the fore and aft line. It is the amount that the forward and after ends of a deck are higher than the midships part. Some vessels are built with a reverse sheer.

Rake is the angle from the vertical of a mast, stem, funnel, etc.

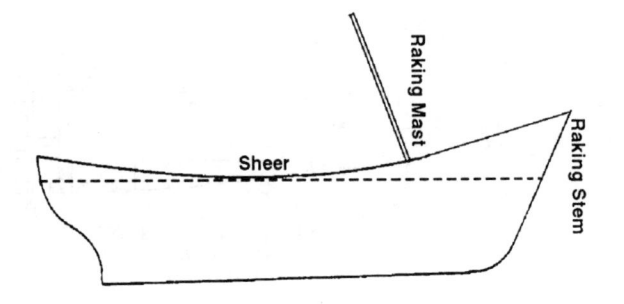

Chapter 3: Nautical Terms, Fittings, Hulls & Rigs

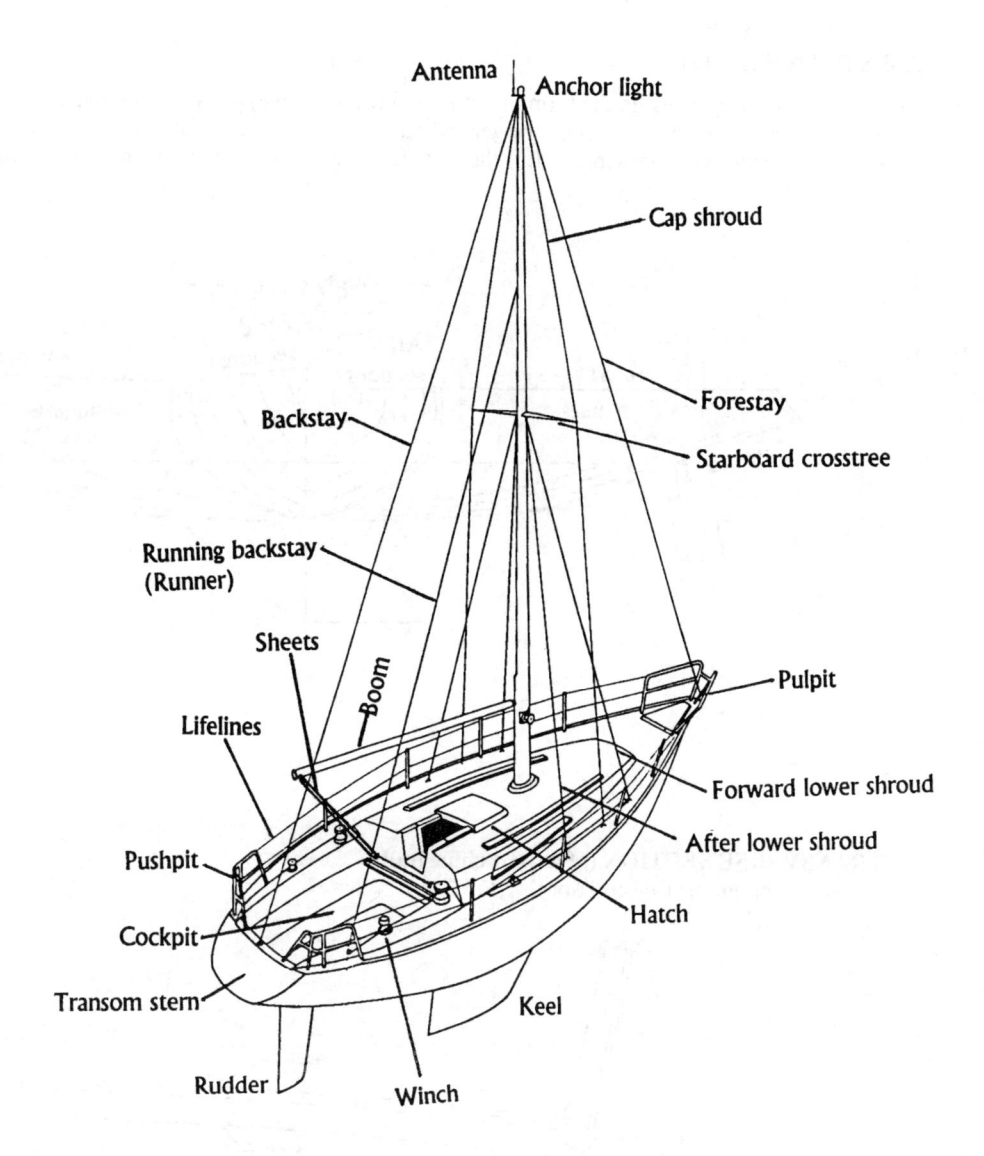

SAILING VESSEL: STANDING RIGGING & PARTS (Fig 3.10)

Labels: Antenna, Anchor light, Cap shroud, Forestay, Backstay, Starboard crosstree, Running backstay (Runner), Sheets, Boom, Pulpit, Lifelines, Forward lower shroud, After lower shroud, Pushpit, Hatch, Cockpit, Transom stern, Keel, Rudder, Winch

SAILING VESSEL: LAYOUT

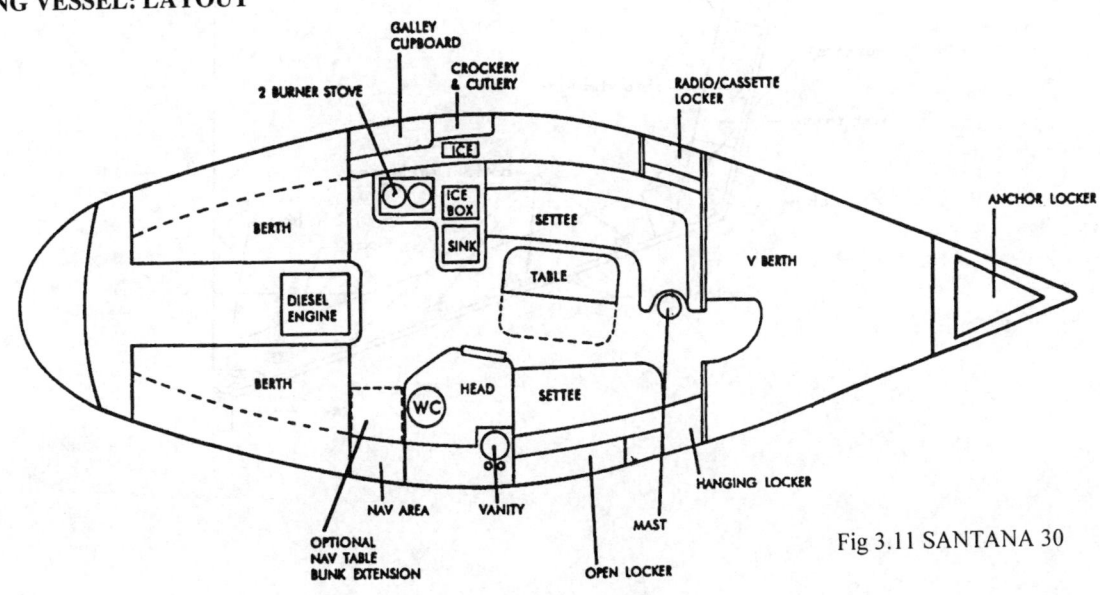

Labels: Galley Cupboard, Crockery & Cutlery, 2 Burner Stove, Radio/Cassette Locker, Ice, Ice Box, Sink, Berth, Settee, Anchor Locker, Diesel Engine, Table, V Berth, Berth, Head, WC, Settee, Nav Area, Vanity, Hanging Locker, Optional Nav Table Bunk Extension, Mast, Open Locker

Fig 3.11 SANTANA 30

LINES PLAN (Fig 3.12)

This is a scale drawing showing the form of a vessel in two (or three) planes of reference:
- Profile (or sheer plan): It gives the general outline of a boat, and her buttock lines and the designed waterline.
- Body Sections (or Body plan): It is the end elevation of a boat showing the above information.

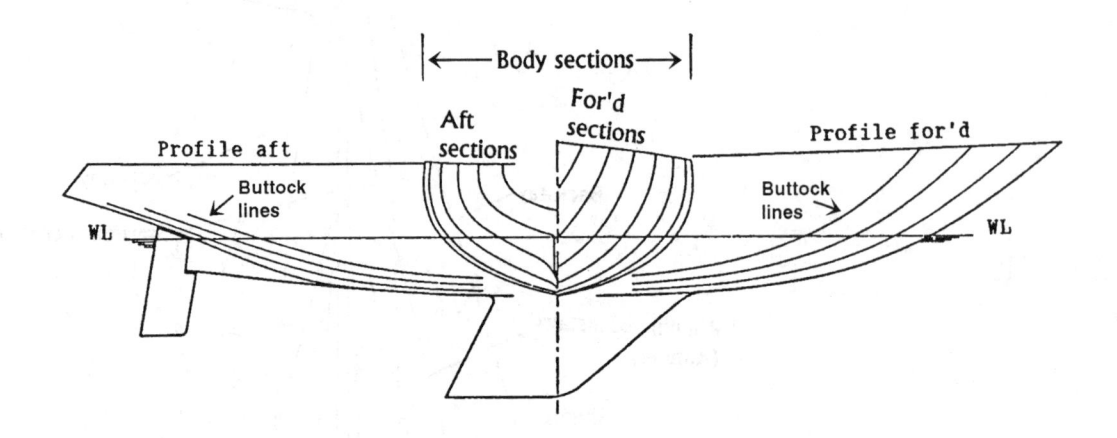

TRANSVERSE SECTION (of a hard chine hull)
(Typical Web Frame Construction)   (Fig 3.13)

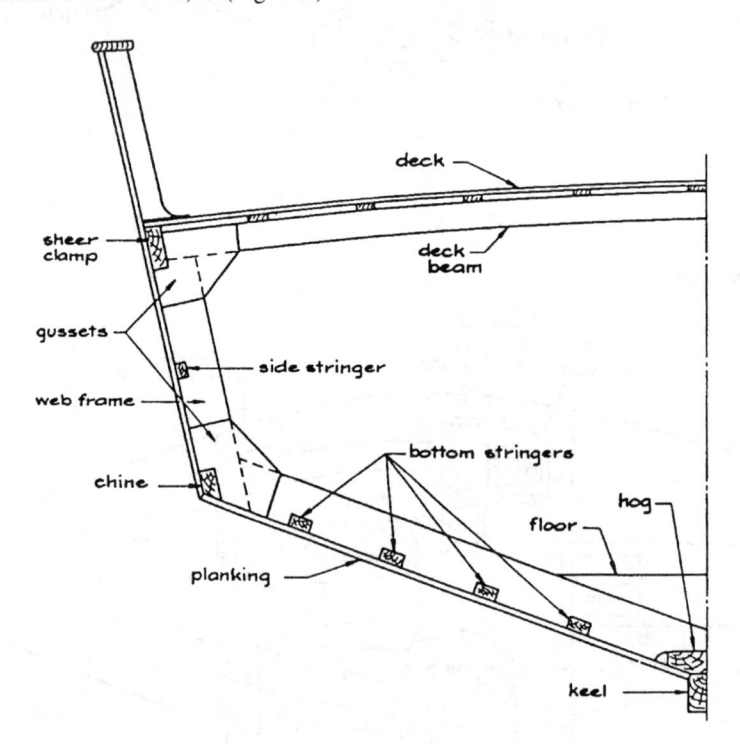

Chapter 3: Nautical Terms, Fittings, Hulls & Rigs

ISOMETRIC VIEW
Fig 3.14

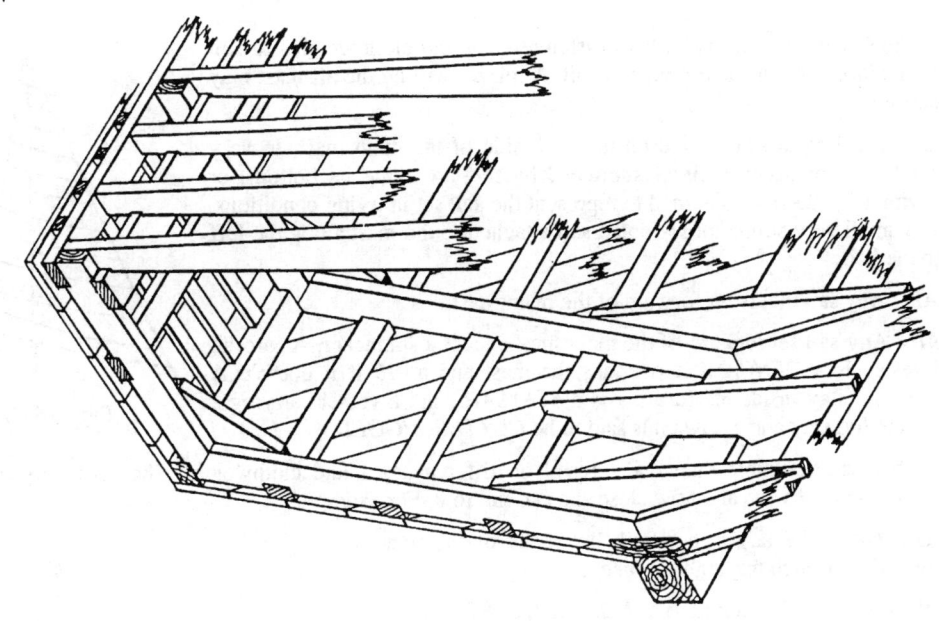

PROFILE VIEW
Fig 3.15

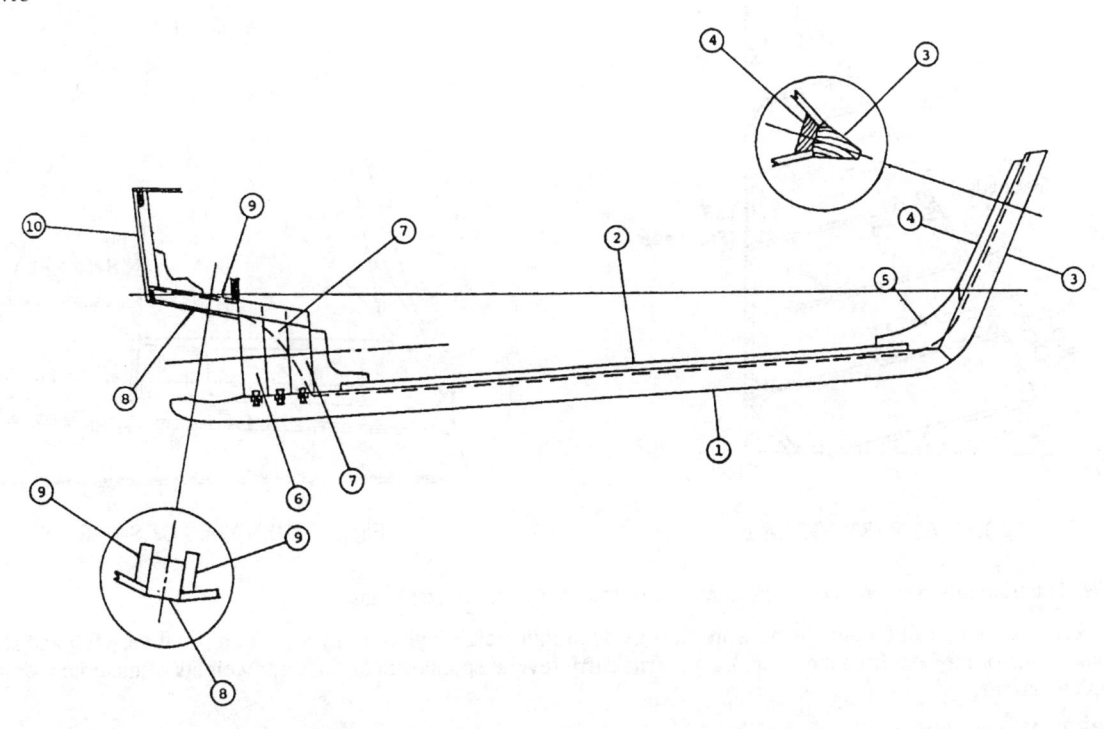

SAILS: TYPES & PARTS

A boat may carry many sails of different sizes and cloth weights to suit different conditions of wind and points of sail. A mizzen sail on the aft mast may also be carried.

MAINSAIL: The largest sail carried on the after side of the main mast. In very strong winds, the mainsail is either shortened by reefing (tying its bottom) or replaced with a *TRYSAIL* – so named to represent the sail set in trying conditions. Trysail is a small very strong loose-footed sail attached to the mast along the luff but not to the boom.

HEADSAIL: Any sail that is set forward of the main mast.

FORESAIL: Any sail set forward of the main mast except a spinnaker. A foresail on the forestay is a *GENOA* if it overlaps the mast and a *JIB* if it doesn't. A foresail set on a stay inside the forestay is a *STAYSAIL*, and a vessel carrying a foresail on the forestay and a staysail is said to be *CUTTER-RIGGED*.

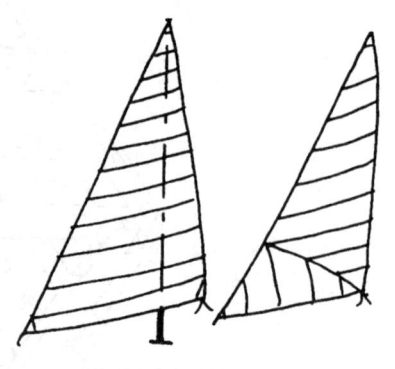

Fig 3.16 GENOA & JIB

A boat carries more than one foresail of different sizes and clothweight. She uses only one at a time to suit different conditions of wind and how close she can sail to it. For example:

- *GENOA* is the largest headsail, but made of the lightest material to catch the slightest breeze.
- *WORKING JIB* (referred to as *NO. 1 JIB*) & *HEAVY WEATHER JIB* (*NO. 2 JIB*) are in between the genoa and storm jib.
- *STORM JIB* is the smallest, and made of a heaviest cloth to withstand storm winds.

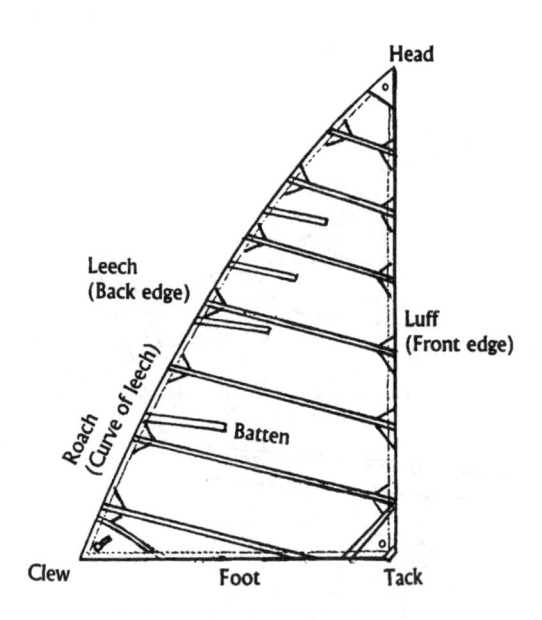

Fig 3.17 (A): PARTS OF SAILS

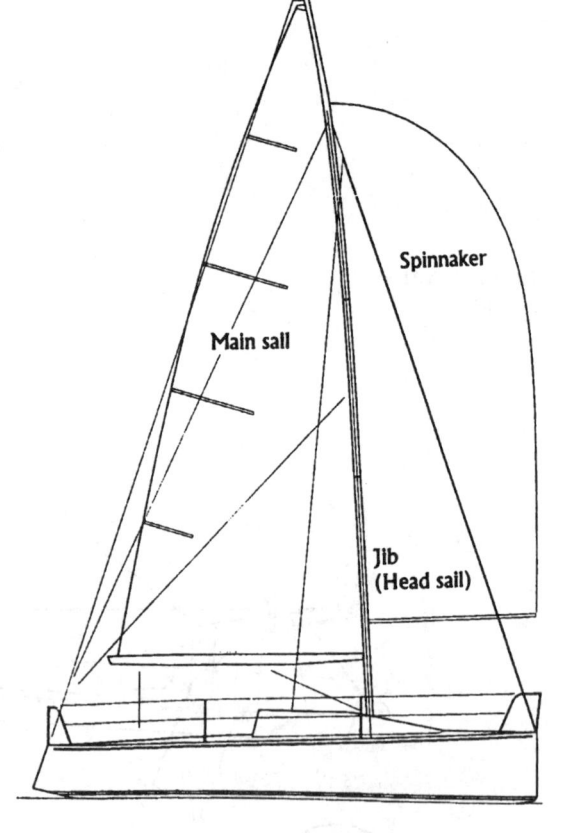

Fig 3.17 (B): TYPES OF SAILS

MIZZEN: The triangular sail hoisted on the after side of the aftermost (mizzen) mast.

SPINNAKER: A racing sail for downwind work. It is made of lightweight nylon or mylar, which is cut very full and sets on the opposite side of the boat from the main. Large boats carry several spinnakers of different weights of materials for use in different conditions.

BLOOPER: Offshore racing sail near the size of a spinnaker, but set on the opposite side and without a pole.

Chapter 3: Nautical Terms, Fittings, Hulls & Rigs

SAILING RIGS – TYPES (Fig 3.18)

BERMUDIAN RIG: A yacht with triangular mainsail and foresail. This is the modern rig.

SLOOP: A single masted yacht with a forestay. Most yachts are sloops.

MASTHEAD SLOOP: The forestay extends to the top of the mast.

FRACTIONAL SLOOP: The forestay extends only part of the way up the mast. The most common fractional rig is three-quarter, where the forestay is attached three quarters of the way up the mast.

Masthead Sloop — **Fraction Rig Sloop**

KETCH: A yacht with two masts, where the main mast is higher than the mizzen mast, and the latter is stepped forward of the rudder post.

YAWL: A yacht with two masts, where the main mast is much higher than the mizzen mast, and the latter is stepped abaft the rudder post.

Ketch (Main mast higher than Mizzen mast) — **Yawl** (Main mast much higher than Mizzen mast. The latter stepped abaft the rudder post)

GAFF-RIGGED: A yacht with 4-sided mainsail. Gaff is the pole to which the top side of the sail is attached. If the spar is pulled almost vertical, instead of at an angle of about 30°, it is referred to as the GUNTER-RIG. A yacht may be a GAFF SLOOP, GAFF KETCH or GAFF YAWL.

CUTTER: A yacht with two jibs (foresails).

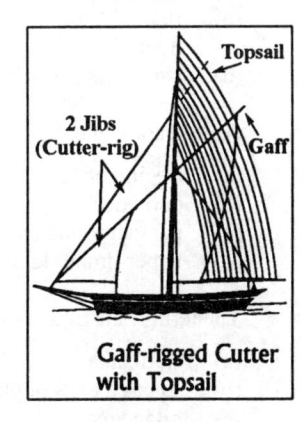

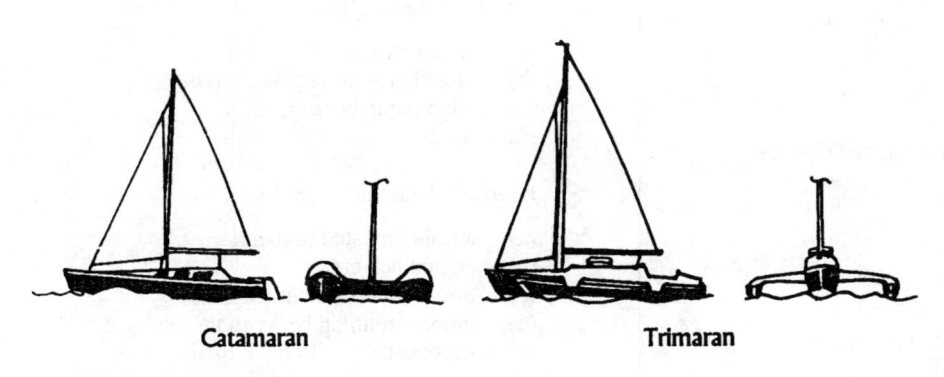

Gaff-rigged Cutter with Topsail

Catamaran — **Trimaran**

CATAMARAN: A vessel with two hulls.
TRIMARAN: A vessel with three hulls
SCHOONER: A fore-and-aft rigged vessel with two or more masts. When carrying a square fore-topsail, she may be referred to as square-rigged.
FORE-AND-AFT RIG: The sails hang in the same line as the keel.
SQUARE-RIG: The sails hang athwartship (Square sails set across the mast).

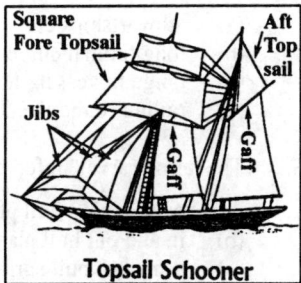

Topsail Schooner

CHAPTER 3: QUESTIONS & ANSWERS

NOTE: For some of the questions below, there is no reference in the text. The answers speak for themselves.

CHAPTER 3.1: PARTS OF BOAT & NAUTICAL TERMINOLOGY

1. An Apron is the:
   (a) strengthening timber in front of the stem.
   (b) stem assembly.
   (c) forward deadwood.
   (d) strengthening timber behind the stem.

2. The Bilge is the:
   (a) inside lower part of the hull.
   (b) outside lower part of the hull.
   (c) inside or outside lower part of the hull.
   (d) the middle of the hull.

3. The Breasthook is the:
   (a) knee holding the bow parts together.
   (b) tow hook.
   (c) hook to fasten the jib.
   (d) mooring hook.

4. A Bulkhead is a:
   (a) toilet.
   (b) floor.
   (c) berth.
   (d) partition.

5. A Bulwark is the:
   (a) freeing port.
   (b) poop deck.
   (c) parapet round the deck.
   (d) upper timber fastened to the keel.

6. The term "butt" refers to the:
   (a) stern area.
   (b) squared ends of planks joined together.
   (c) bilge area.
   (d) upper end of a yard.

7. Camber is the:
   (a) athwartships curve in a deck.
   (b) longitudinal curve in a deck.
   (c) bulge in vessel's hull.
   (d) railing round the deck.

8. The term "carvel" refers to:
   (a) overlapping hull planks.
   (b) in and out hull planks.
   (c) a smooth hull surface.
   (d) non-transom construction.

9. The term "chine" refers to the:
   (a) boat's fore and aft lines.
   (b) round bilge construction.
   (c) angle between a boat's topside and bottom.
   (d) longitudinal dip in a boat's decks.

10. A Soft Chine vessel has:
    (a) a round bilge.
    (b) only one chine.
    (c) a small chine angle.
    (d) a large chine angle.

11. A Hard Chine vessel is a vessel with:
    (a) a round bilge.
    (b) a double chine.
    (c) a single chine.
    (d) one or more chines.

12. A Cleat is a:
    (a) type of anchor.
    (b) boat hook.
    (c) hatch opening.
    (d) fitting to which lines may be fastened.

13. The term "clinker" refers to:
    (a) overlapping hull planks.
    (b) in and out hull planks.
    (c) a smooth hull surface.
    (d) non-transom construction.

14. A Coaming is a:
    (a) deck house.
    (b) hatch cover.
    (c) raised side of hatch or cockpit.
    (d) deck broom.

15. A Cutlass Bearing is a:
    (a) engine bed.
    (b) propeller shaft support bearing.
    (c) bilge pump bearing.
    (d) pallet.

16. Deadwood are the:
    (a) termite infested timbers.
    (b) engine bearers.
    (c) timbers infested with wet rot.
    (d) timbers running between the keel and sternpost.

17. A Deckhead is the:
    (a) uppermost deck.
    (b) lowermost deck.
    (c) toilet.
    (d) underside of the cabin top.

Chapter 3: Nautical Terms, Fittings, Hulls & Rigs

18. The term "Displacement" means the:
    (a) weight of vessel in salt water.
    (b) weight of vessel in fresh water.
    (c) weight of vessel.
    (d) volume of water displaced by an empty vessel.

19. A vessel's draft (draught) is:
    (a) her loading capacity.
    (b) her trim.
    (c) the vertical distance from keel to waterline.
    (d) her freeboard.

20. A Docking Strip is a:
    (a) strip of hardwood on the underside of keel.
    (b) slipway.
    (c) docking cradle.
    (d) gently shelving beach.

21. A flare in a vessel's structure:
    (a) helps seas on deck to run off quickly.
    (b) reduces seas coming on board.
    (c) is the forward flowing of the bow.
    (d) is the rake in the bow.

22. Floors are the:
    (a) floors in vessels like floors in houses.
    (b) deck beams.
    (c) vertical members connecting keel to frames.
    (d) loose planking in the bottom of a vessel.

23. Frames in a vessel are:
    (a) deck supports.
    (b) large transverse ribs.
    (c) supports for the stern frame.
    (d) rudder supports.

24. A vessel's freeboard is the:
    (a) loading capacity.
    (b) trim.
    (c) vertical distance from keel to waterline.
    (d) vertical distance from waterline to weather deck.

25. A vessel's freeing ports are the:
    (a) openings in the bulwarks.
    (b) portholes.
    (c) openings on the port side.
    (d) scuppers.

26. The term "Garboard Strake" refers to the:
    (a) garbage chute fitted in a vessel's hull.
    (b) the extra large ribs.
    (c) vessel without floors.
    (d) first plank adjacent to the keel.

27. A Gunwale (or Gunwhale) is the:
    (a) parapet round the deck.
    (b) upper timber fastened to the keel.
    (c) upper edge of the top side of a vessel.
    (d) uppermost strake.

28. A Hog is the:
    (a) upper fore & aft timber fastened to the keel.
    (b) dropping or depression in the keel.
    (c) timber keel.
    (d) floor.

29. A hogged vessel is the one with a:
    (a) dropping or depression in the keel.
    (b) droop at the bow and stern.
    (c) rake of keel.
    (d) hog timber fitted to her keel.

30. A Keelson is a:
    (a) thin keel.
    (b) large keel.
    (c) internal additional keel.
    (d) hog timber.

31. In the structure of a boat, a knee is a:
    (a) breasthook.
    (b) breastplate.
    (c) gunwale.
    (d) timber holding a deck beam to a frame.

32. Osmosis is:
    (a) dry rot.
    (b) wet rot.
    (c) blistering of fibreglass hulls.
    (d) any of the conditions stated here.

33. Panting is:
    (a) creaking noises in a hull in rough seas.
    (b) violent pitching.
    (c) screw motion of a vessel in rough seas.
    (d) the end plates of a vessel flexing in & out.

34. Racking refers to the:
    (a) stresses in a hull due to rolling & twisting.
    (b) pounding stresses in a hull.
    (c) raised edges fitted to tables & stoves.
    (d) none of the choices stated here.

35. A Sampson Post in a small vessel is a:
    (a) deck cleat with horns.
    (b) stern post.
    (c) rudder post.
    (d) a strong mooring post secured to the keel.

36. A Scupper is a:
    (a) drain hole in a deck.
    (b) freeing port.
    (c) curve in the deck to allow drainage.
    (d) fuel connection.

37. In the structure of a boat, the term "Sheer" refers to:
    (a) her camber.
    (b) her lines.
    (c) the curved line of deck edge from bow to stern.
    (d) the angle of the bow overhang.

38. A Sheer Strake is the:
    (a) bottom-most strake.
    (b) upper-most strake.
    (c) angle of deck sheer.
    (d) straight keel.

39. In the structure of a boat, a shelf is a:
    (a) beam knee.
    (b) panting beam.
    (c) rudder support.
    (d) longitudinal support for beam ends.

40. A skeg is a:
    (a) external support for rudder and shaft.
    (b) flat stern.
    (c) Transom stern.
    (d) Rake of keel.

41. A Sponson is a:
    (a) Sampson Post.
    (b) Staghorn.
    (c) Rubbing Strip protecting the topside.
    (d) any deck fitting designed for mooring.

42. Stem of a vessel is her:
    (a) stern.
    (b) after deck.
    (c) forward deck.
    (d) most forward structural member.

43. The Stringers are:
    (a) ribs.
    (b) frames.
    (c) ribs or frames.
    (d) longitudinal strength members fastened to frames.

44. A round bilge vessel has:
    (a) a single chine.
    (b) two chines.
    (c) no chine.
    (d) hard chine.

45. Gooseneck is the:
    (a) curved stem of a vessel.
    (b) boom joint to the mast.
    (c) tapering of the mast.
    (d) tapered boom.

46. Guy is:
    (a) a forestay.
    (b) a line for lateral control of a spar or boom.
    (c) another name for the sheets.
    (d) a shroud.

47. Halyard is a:
    (a) yard arm.
    (b) short mast.
    (c) wire or rope for hoisting yards, sails & flags.
    (d) mast stay.

48. Topping Lift is the:
    (a) wind in the top sail.
    (b) rope to lift or support the boom.
    (c) process of raising a top sail.
    (d) rope to lift a top sail.

49. Yard is a:
    (a) boom.
    (b) block and tackle.
    (c) spar fitted across a mast.
    (d) short stay.

50. The term 'Abaft' means:
    (a) Amidships
    (b) Abeam
    (c) Behind
    (d) Aft

51. The term 'Anchor Aweigh' means:
    (a) The anchor has been hove out of the ground.
    (b) The anchor has been hove out of the water.
    (c) The anchor has been let go.
    (d) The anchor is jammed in the hosepipe.

52. The term 'Athwartship' means:
    (a) Amidships
    (b) Across the vessel
    (c) After part of the vessel.
    (d) Fore part of the vessel.

53. Bilge is:
    (a) A pair of posts for tying mooring lines.
    (b) A Staghorn.
    (c) A Pulley
    (d) The lowest inside part of the hull.

Chapter 3: Nautical Terms, Fittings, Hulls & Rigs

54. Bitts is:
    (a) A pair of posts for tying mooring lines.
    (b) Short pieces of rope
    (c) Another name for bulkheads.
    (d) A small two-sheave pulley.

55. Bollard is:
    (a) A Bolt Rope
    (b) The keel.
    (c) A post for tying up ropes.
    (d) The front end of the vessel.

56. Staghorn is:
    (a) A post for tying up ropes.
    (b) A cleat.
    (c) The horn-shaped stern
    (d) The horn-shaped bow.

57. The term 'Breaking Out the Anchor' means:
    (a) To make ready and free the anchor for letting go.
    (b) Freeing the anchor from seabed.
    (c) Heaving the anchor above water.
    (d) Unshackling the anchor from the cable.

58. A lookout reporting sightings to navigation officer is required to:
    (a) be as accurate and descriptive as possible.
    (b) report only the relative directions of sightings.
    (c) keep lookout only ahead.
    (d) keep lookout only on starboard side on the basis of 'give way to your starboard' rule.

59. To take a sounding is to:
    (a) sound the decks for adequacy of thickness.
    (b) report any unusual sounds from the engine.
    (c) measure the depth of liquid in a tank.
    (d) measure the depth of water outside the hull or the depth of liquid in a tank.

60. The term 'Brought Up' means:
    (a) The vessel is at the jetty.
    (b) The vessel is at the anchoring position.
    (c) The vessel is riding to her anchor & cable, and the anchor is holding.
    (d) The anchor has been hove out of the ground.

61. Bulkhead is:
    (a) The deck above.
    (b) A vertical partition, equivalent to a wall in a building.
    (c) The ship's toilet
    (d) The uppermost deck.

62. Cathead is:
    (a) The deck above.
    (b) The narrow space inside the bow.
    (c) The aft cabin.
    (d) A structure in the bow to house the anchor in a small vessel.

63. The term "Clear Anchor" means:
    (a) To break out the anchor.
    (b) To wash the anchor.
    (c) To free the anchor from seabed.
    (d) To take the anchor out of the locker.

64. Hawse Pipe is:
    (a) The pipe through which the anchor cable goes into the chain locker.
    (b) A wash water pipe.
    (c) The pipe through which the anchor cable goes from the vessel to the anchor.
    (d) The pipe through which the mooring lines are secured ashore.

65. Helm is:
    (a) The Ship's Captain.
    (b) The Steering Wheel or the Tiller.
    (c) The steering Wheel
    (d) The Tiller.

66. Hitch is:
    (a) To ask for a tow.
    (b) The position plotted on chart.
    (c) To tie a rope to something.
    (d) To secure the vessel to jetty.

67. Vessel's speed is stated in:
    (a) Kilometres per hour.
    (b) Knots.
    (c) Knots per hour.
    (d) Nautical miles.

68. The term "Let Go" means:
    (a) To drop anchor in the water or to untie the vessel's mooring line(s).
    (b) To drop all the anchor chain in the water.
    (c) To drop the mooring lines in the water.
    (d) To drop both anchors in the water.

69. Leeward means:
    (a) To drift down.
    (b) The side of the vessel from where the wind is blowing.
    (c) To anchor with wind behind the vessel.
    (d) The downwind side of the vessel.

70. To Make Fast is to:
    (a) Speed Up.
    (b) Increase speed of the vessel.
    (c) Allow the anchor cable to run freely.
    (d) Secure.

71. Painter is:
    (a) A Roller Brush.
    (b) A small mooring line for a small boat or liferaft.
    (c) A Stern Anchor.
    (d) The underneath of a deck.

72. Pulpit is:
    (a) A guardrail around the stern of a boat.
    (b) A towrope.
    (c) A stern rope.
    (d) A guardrail around the bow of a boat.

73. Ship's Quarter is:
    (a) The part of the hull on each side of the stern.
    (b) One-fourth of the ship.
    (c) Ship's bow.
    (d) Right angle to the ship's fore & aft line.

74. Samson Post is:
    (a) The mast on ship's centreline.
    (b) A vertical post used for tying mooring lines.
    (c) Any vertical post.
    (d) The sternpost.

75. Sounding is to:
    (a) Measure depth outside the vessel.
    (b) Measure depth inside the vessel.
    (c) Measure depth outside or inside the vessel.
    (d) Blow the ship's whistle.

76. Sponson is:
    (a) A vertical post.
    (b) A vertical post used for tying mooring lines.
    (c) The Stern.
    (d) A wooden or rubber fender fitted around the vessel.

77. Spurling Pipe is:
    (a) The pipe through which the anchor cable goes into the chain locker.
    (b) A wash water pipe.
    (c) The pipe through which the anchor cable goes from the vessel to the anchor.
    (d) The pipe through which the mooring lines are secured ashore.

78. A Stay is:
    (a) A stay in port.
    (b) A stay at anchorage.
    (c) A rope supporting a mast.
    (d) A mooring line attached to liferaft.

79. Vessel's stem is:
    (a) The Mast.
    (b) The front of the Bow.
    (c) The Bow.
    (d) The rear of the vessel.

80. The term "Surge" means:
    (a) To pay out the cable or rope.
    (b) To stem the tide.
    (c) To stop the cable or rope from running out.
    (d) To allow the cable or rope to run out or slacken while being hauled.

81. Fairlead is:
    (a) Navigating in a channel.
    (b) Leading a rope through the chain pipe.
    (c) A fitting that ensures a rope leading in the desired direction.
    (d) A Fair wind.

82. Cleat is:
    (a) The lower corner of a square sail.
    (b) A Tackle.
    (c) The jaw end of a boom.
    (d) A fitting used for securing or controlling ropes.

83. Transom is:
    (a) The flat stern.
    (b) The stern.
    (c) To line up two objects for a bearing.
    (d) The round stern.

84. Vessel's Trim is:
    (a) The angle of the trim tabs.
    (b) The difference between the fore and aft drafts.
    (c) The difference between the port and starboard drafts.
    (d) Another name for List.

85. The term "Up-and-down" means:
    (a) The vessel is riding the waves.
    (b) The vessel is pitching.
    (c) The stem is vertical.
    (d) The anchor cable is leading vertically down to the water.

86. The term "Veer" means:

    (a) To heave in the cable or rope.
    (b) To stem the tide.
    (c) To let out rope or cable under control.
    (d) To allow the cable or rope to run out freely.

87. The term "Walk Back" means:

    (a) To heave in the cable or rope.
    (b) To go astern.
    (c) To veer the rope or cable.
    (d) To allow the cable or rope to run out freely.

88. The term "Warp" means:

    (a) To move a vessel along a jetty by pulling on her mooring lines.
    (b) To lead the anchor cable aft.
    (c) To veer the cable or rope.
    (d) To Walk Back.

89. Weigh Anchor means:

    (a) To check the weight of the anchor.
    (b) To drop anchor.
    (c) To secure anchor.
    (d) To lift anchor.

90. Windward means:

    (a) To head into wind.
    (b) The side of the vessel from where the wind is blowing.
    (c) To anchor with wind behind the vessel.
    (d) The downwind side of the vessel.

91. The helm order "Steady" means:

    (a) Wheel amidships
    (b) Maintain the wheel in its current position.
    (c) Steer the present course by lining up a landmark or by compass.
    (d) Do not allow the vessel to yaw.

92. The helm order "Hard-a-Starboard" means:

    (a) Turn the wheel fully to starboard.
    (b) Keep turning the wheel to starboard.
    (c) Do not turn the wheel to port.
    (d) Turn the vessel to starboard.

93. The helm order "Steer 025" means:

    (a) Turn the wheel to 25°.
    (b) Steer a course of 025° by compass.
    (c) Turn the wheel hard over.
    (d) Steer with rudder at an angle of 25°.

94. When compared with Displacement Hull, Chine hull:

    (a) Provides softer ride.
    (b) Is faster.
    (c) Is more seaworthy.
    (d) Is suitable for planing.

CHAPTER 3.2: SAILING RIGS & RIGGING

95. Aspect Ratio: a sail with a high aspect ratio is a:

    (a) tall and narrow sail.
    (b) short squat sail.
    (c) larger than the main sail.
    (d) square sail.

96. Bermuda Rig is the:

    (a) standard fore & aft rig of modern ocean racer.
    (b) opposite rig to Marconi rig.
    (c) square rigged vessel.
    (d) rig that includes lateen sails.

97. A Bolt Rope is:

    (a) sewn into the edges of a sail.
    (b) used for towing.
    (c) bolted to hull.
    (d) trailed behind a vessel.

98. Brigantine is a sailing vessel with:

    (i) square rig on foremast.
    (ii) fore and aft rig on aftermast.
    (iii) fore and aft rig.
    (iv) two masts.

    The correct answer is:

    (a) all except (i).
    (b) all except (ii).
    (c) all except (iii).
    (d) all except (iv).

99. Clew is:

    (a) the upper corner of a sail.
    (b) a cleat.
    (c) the lower corner of a sail.
    (d) a knot.

100. Club is a:

    (a) spar fitted to the bottom of a jib.
    (b) jib.
    (c) spar fitted to the bottom of a triangular sail.
    (d) boom.

101. Cringle is a:
    (a) loop of rope in a sail to take reefing lines.
    (b) Cunningham eye.
    (c) shortened sail.
    (d) dinghy like boat.

102. Cross Trees are:
    (a) sail supports.
    (b) floor timbers with middle on the keel.
    (c) hull frames.
    (d) shroud spreaders on a mast.

103. Crowd on sails means to:
    (a) set extra sails to press the vessel faster.
    (b) sit everyone on windward.
    (c) set all sails on one side.
    (d) store all sails together.

104. Cunningham eye is a:
    (a) loop of rope in a sail to take reefing lines.
    (b) eye in the luff of the mainsail to control its shape.
    (c) eye in the mast for securing shrouds.
    (d) eye spliced at the end of a rope.

105. Fore and Aft Rig means the vessel:
    (a) has no main sail.
    (b) has no square sails.
    (c) is sailing with wind directly behind.
    (d) is sailing with wind abeam.

106. Fore sheet is the:
    (a) foresail.
    (b) fore shroud.
    (c) rope for controlling the foresail.
    (d) leading edge of the sail.

107. Foresail is a:
    (i) spinnaker
    (ii) genoa.
    (iii) jib.
    (iv) staysail.

    The correct answer is:
    (a) (i) only.
    (b) all except (i).
    (c) all except (ii).
    (d) all except (iii) & (iv).

108. Forestay is a:
    (a) stay from the bow to the top or near to the top of the mast.
    (b) foresail.
    (c) bowsprit.
    (d) jib.

109. Fractional Rig is a yacht with:
    (a) a three-quarter rig.
    (b) forestay not extending to top of the mast.
    (c) a seven-eighths rig.
    (d) half rig.

110. Gaff is the:
    (a) top of the mast.
    (b) spar extending the top of old fashioned mainsail.
    (c) mast heel.
    (d) cross tree.

111. Genoa is a:
    (a) headsail large enough to overlap the mast.
    (b) headsail that does not overlap the mast.
    (c) staysail.
    (d) another name for a jib.

112. Head is the following part of a sail:
    (a) forward bottom corner.
    (b) upper fore corner.
    (c) aft bottom corner.
    (d) leading edge.

113. Headsail is:
    (a) any foresail.
    (b) a genoa.
    (c) a jib.
    (d) a staysail.

114. Headstay is a:
    (a) forestay.
    (b) foresail.
    (c) bowsprit.
    (d) jib.

115. Jib is:
    (a) any foresail.
    (b) any headsail.
    (c) a stay.
    (d) a foresail not overlapping the mast.

116. Ketch or a Yawl is:
    (i) fore and aft rigged.
    (ii) two masted.
    (iii) rigged with a mizzen sail.
    (iv) square rigged.

    The correct answer is:
    (a) all except (iv).
    (b) (i) only.
    (c) (ii) only.
    (d) all except (i).

Chapter 3: Nautical Terms, Fittings, Hulls & Rigs

117. Lateen is:
    (a) a triangular sail at an angle across the mast.
    (b) a square sail.
    (c) a fore and aft sail.
    (d) none of the choices stated here.

118. Leech is the:
    (a) leading edge of a sail.
    (b) lower edge of a sail.
    (c) trailing edge of a sail.
    (d) lower corner of a sail.

119. Luff is:
    (a) the trailing edge of a sail.
    (b) to turn away from the wind.
    (c) the leading edge of a sail.
    (d) sail close-hauled.

120. Main Sheet is the:
    (a) main sail.
    (b) line controlling the main sail.
    (c) main stay.
    (d) largest fore and aft sail.

121. Main Stay is the:
    (a) line controlling the main sail.
    (b) largest fore and aft sail.
    (c) for'd fore & aft line supporting the main mast.
    (d) longest shroud.

122. Marconi Rig is the:
    (a) rig without a main sail.
    (b) Bermuda rig.
    (c) square rigged vessel.
    (d) rig that includes lateen sails.

123. Masthead Rig is a yacht of the following description:
    (a) a fractional rig.
    (b) gaff sail on masthead.
    (c) forestay reaching to the top of the mast.
    (d) square rig.

124. Point refers to the following:
    (i) pointed end of a sail.
    (ii) 11.25° of a compass.
    (iii) sail as high as possible to wind.
    (iv) a boat able to sail closer to wind than others.

    The correct answer is:
    (a) all except (i).
    (b) (ii) only.
    (c) (i) only.
    (d) all of them.

125. Schooner is a sailing vessel with:
    (i) fore and aft rig.
    (ii) two masts only.
    (iii) occasionally carries square fore top-sails.
    (iv) second mast taller than the first.

    The correct answer is:
    (a) all except (ii).
    (b) all of them.
    (c) (iii) & (iv) only.
    (d) all except (iv).

126. Sloop is a sailing vessel with:
    (i) one mast.
    (ii) more than one headsail.
    (iii) fore and aft rig.
    (iv) jib stay.

    The correct answer is:
    (a) all except (ii).
    (b) all except (i).
    (c) all except (iii).
    (d) all except (iv).

127. Spinnaker is a:
    (a) full cut racing sail.
    (b) boom.
    (c) wind direction indicator.
    (d) mast.

128. The term "Tack" may be used to describe:
    (i) port tack: sailing with wind on port side.
    (ii) tacking: changing course by changing tack.
    (iii) tack: the forward bottom corner of a sail.
    (iv) starboard tack: sailing with sail on stbd side.

    The correct answer is:
    (a) all except (i).
    (b) (ii) & (iii) only.
    (c) (i) & (iv) only.
    (d) all except (iv).

129. Throat is the following part of a sail:
    (a) forward bottom corner.
    (b) head (upper fore corner).
    (c) aft bottom corner.
    (d) leading edge.

130. Top sail is:
    (a) a sail above the main sail.
    (b) the main sail.
    (c) the process of raising a sail.
    (d) the foremost sail.

131. Triatic Stay is a:
    (a) stay made up of three stays.
    (b) stay from one mast to another.
    (c) line controlling the main sail.
    (d) stay that does not share the load.

132. Yacht types: which of the following can only be a two masted vessel?
    (a) Barque: square rig except on aftermast.
    (b) Barquentine: fore & aft rig except on foremast.
    (c) Brig: square rig.
    (d) Schooner: fore & aft rig.

CHAPTER 3: ANSWERS

1 (d), 2 (c), 3 (a), 4 (d), 5 (c),
6 (b), 7 (a), 8 (c), 9 (c), 10 (b),
11 (d), 12 (d), 13 (a), 14 (c), 15 (b),
16 (d), 17 (d), 18 (c), 19 (c), 20 (a),
21 (b), 22 (c), 23 (b), 24 (d), 25 (a),
26 (d), 27 (c), 28 (a), 29 (b), 30 (c),
31 (d), 32 (c), 33 (d), 34 (a), 35 (d),
36 (a), 37 (c), 38 (b), 39 (d), 40 (a),
41 (c), 42 (d), 43 (d), 44 (c), 45 (b),
46 (b), 47 (c), 48 (b), 49 (c), 50 (c),
51 (a), 52 (b), 53 (d), 54 (a), 55 (c),
56 (a), 57 (a), 58 (a), 59 (d), 60 (c),
61 (b), 62 (d), 63 (a), 64 (c), 65 (b),
66 (c), 67 (b), 68 (a), 69 (d), 70 (d),
71 (b), 72 (d), 73 (a), 74 (b), 75 (c),
76 (d), 77 (a), 78 (c), 79 (b), 80 (d),
81 (c), 82 (d), 83 (a), 84 (b), 85 (d),
86 (c), 87 (c), 88 (a), 89 (d), 90 (b),
91 (c), 92 (a), 93 (b), 94 (d), 95 (a),
96 (a), 97 (a), 98 (c), 99 (c), 100 (c),
101 (a), 102 (d), 103 (a), 104 (b), 105 (b),
106 (c), 107 (b), 108 (a), 109 (b), 110 (b),
111 (a), 112 (b), 113 (a), 114 (a), 115 (d),
116 (a), 117 (a), 118 (c), 119 (c), 120 (b),
121 (c), 122 (b), 123 (c), 124 (a), 125 (a),
126 (a), 127 (a), 128 (d), 129 (b), 130 (a),
131 (b), 132 (c)

Chapter 4

ROPES, KNOTS & SPLICES

ROPE CONSTRUCTION

Ropes are made of fibre or wire; the latter being referred to as wire ropes. Cordage is the collective name for all fibre ropes and lines.

Simplistically speaking, ropes are manufactured by twisting fibres into yarns, yarns into strands, and strands into ropes. To prevent the rope from unlaying (untwisting), the strands are laid in the opposite direction to the yarns.

Natural fibres ropes, such as manila and coir, are made of short lengths (staple) of fibre. Therefore, they lack the *strength of synthetic ropes* in most of which each fibre (monofilament) runs the full length of the rope. Synthetic ropes are also stronger due to their ability to stretch.

Silver rope, although synthetic, is similar in construction to natural fibre ropes. It is thus weaker than other synthetic ropes. It is popular because it is 30% lighter most other ropes, does not absorb moisture and does not slip as easily as other synthetics.

THREE-STRAND ROPES

Most ropes are *right-hand laid* with three strands. They are easier to work with, especially for splicing. These so-called HAWSER LAID ropes come in three levels of STIFFNESS:
- Soft lay: silky soft but kinks easily
- Medium lay: suitable for boats
- Hard lay: difficult to work with on a boat

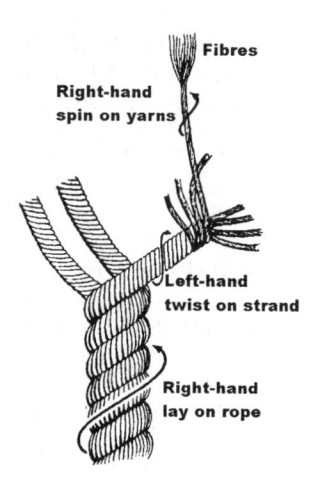

Fig 4.1: TWISTS IN A RIGHT HANDED HAWSER-LAID ROPE

RIGHT-HANDED & LEFT-HANDED ROPES

RIGHT-HAND LAID ROPE ("Z" Twist)

LEFT-HAND LAID ROPE ("S" Twist)

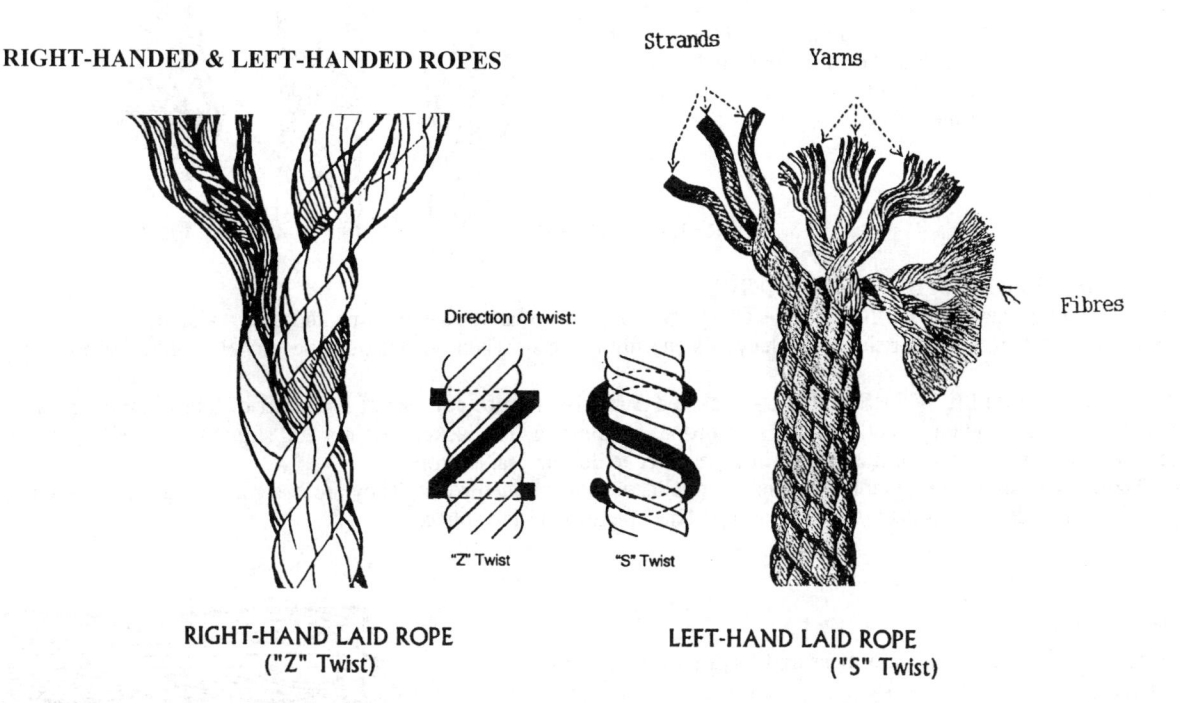

Fig 4.2: DISTINGUISHING BETWEEN RIGHT-HAND & LEFT-HAND LAYS
(The direction of twist of the strands is clock-wise or "Z" shaped in the right-hand laid rope, and anti-clockwise or "S" shaped in the left-hand laid rope. It does not alter if the page is turned around.)

TYPES OF ROPES
1. Hawser Laid
2. Shroud Laid
3. Cable Laid
4. Square
5. Braided
6. Jacketed Parallel-Core

Fig 4.3
SHROUD-LAID ROPE

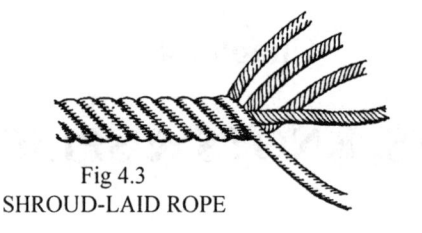

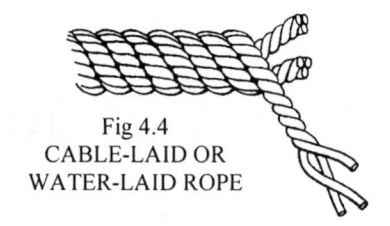

Fig 4.4
CABLE-LAID OR WATER-LAID ROPE

1. HAWSER LAID ROPES: See Three Strand Ropes above.

2. SHROUD LAID (FOUR-STRAND) ROPES: A Shroud Laid rope is made of 4 strands laid around a fibre heart. It is 11% weaker than a 3-strand rope, and not often seen these days.

3. CABLE LAID ROPES: A Cable Laid rope is made of three Hawser Laid ropes. Three ropes are twisted into one for greater elasticity for towing etc.

4. SQUARE ROPES: Plaiting 4 left-hand strands into 4 right-hand strands make a Square Rope. This tough, kink resistant and flexible rope is used for towing and berthing lines.

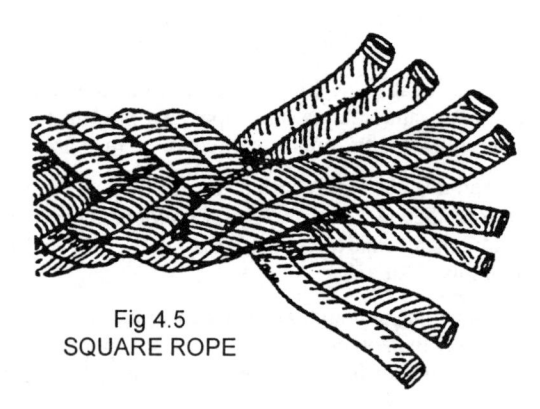

Fig 4.5
SQUARE ROPE

5. BRAIDED ROPES: These ropes can be of the following varieties:

    - SINGLE (OR SOLID) BRAIDED ROPES: These are made of 12 or 18 strands, common for flag halyards, awnings, etc.
    - PLAITED ROPES: These are common for flag halyards, awnings, etc.
    - DOUBLE BRAIDED ROPES: These have a braided core inside a braided jacket, and are commonly used on boats. But they are harder to splice. The core bears about 70% of the load, and the jacket about 30%. Their characteristics are:
        ➤ Flexibility - due to smaller strands
        ➤ Resistant to abrasions - due to wear being distributed over many strands
        ➤ Coils evenly
        ➤ Kink-free

Fig. 4.6
BRAIDED ROPES

   VARIETIES OF DOUBLE-BRAIDED ROPES:
   ➤ Nylon jacketed for berthing lines – Their advantage lies in being easy running and easy on hands.
   ➤ Polyester jacketed for sheets and halyards on sailing boats – Their advantage is lies in being of low stretch.

6. JACKETED PARALLEL-CORE ROPES: A jacketed parallel-core rope consists of a number of ropes running parallel to each other inside a thin jacket. A set of seven ropes is common. The jacket does not add strength; it merely protects the inner ropes from abrasion. It is similar in appearance to double braided rope.
    These ropes are not as suitable for dynamic applications as braided ropes. They are designed for use in long term, direct tension loads, such as deep-water moorings. They are also hard to splice.

ROPE SIZES (Fig 4.7)
Ropes, both fibre and wire, are measured by their diameter in millimetres.

MEASURING ROPE SIZE

Chapter 4: Ropes, Knots & Splices

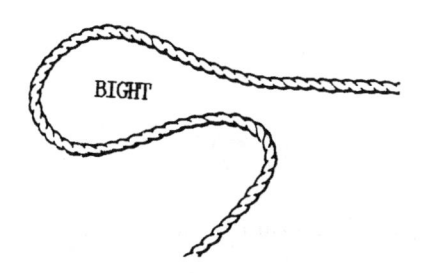

Fig 4.8: BIGHT OF A ROPE

ROPE HANDLING PRECAUTIONS

- Do not stand in the bight of a rope.
- Do not stand in way of a rope under load. A parting synthetic rope has a deadly whip.
- Wet ropes are slippery.
- When easing out (surging) a synthetic rope under load, it can momentarily weld then slip on bitts or a cleat.

COILING A ROPE

The right-handed laid ropes should be coiled clockwise and the left-handed coiled anti-clockwise. However, polyprop ropes should not always be coiled or heaved on a drum in the same direction. This practice unbalances their lay and creates kinks and hooks.

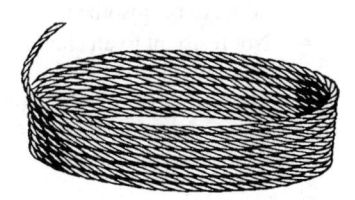

Fig 4.9
COILING A RIGHT-HANDED ROPE

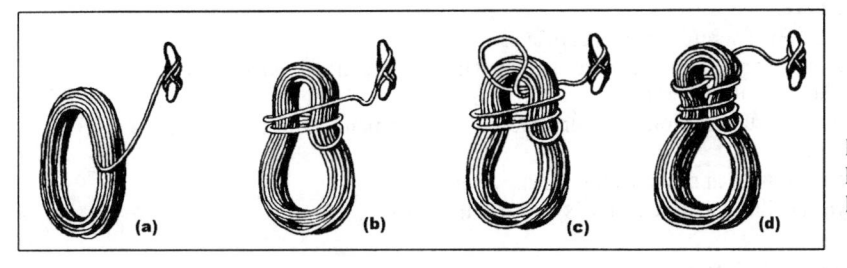

Fig 4.10
MAKING UP A LARGE COIL FOR HANGING ON A CLEAT

UNCOILING A ROPE

Improper uncoiling can lead to turns and twists in the rope. A coil of fibre rope should be uncoiled by pulling at the bottom end from the inside. The remaining coil can thus be left in its protective wrapping. A wire rope should be uncoiled from the outer end on a turntable.

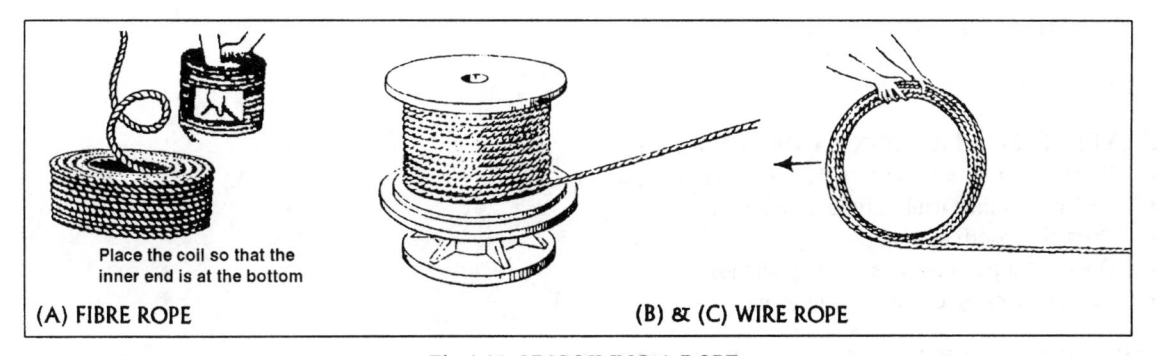

Place the coil so that the inner end is at the bottom

(A) FIBRE ROPE   (B) & (C) WIRE ROPE

Fig 4.11: UNCOILING A ROPE

ROPES WITH KNOTS & SPLICES: REDUCTION IN STRENGTH

It is usually better to use splicing for joining ropes. Knots decrease rope strength more than splicing. The approximate reduction in strength of a rope as a result of various knots is as follows:

- Fisherman's bend, Round turn & Timber hitch = 30%
- Bowline, Sheet Bend & Clove hitch = 40%
- Reef Knot   = 55%
- Hard Eye Splice = 10%
- Soft Eye Splices = 20%

CHARACTERISTICS OF SYNTHETIC FIBRE ROPES

NYLON (POLYAMIDE)
- High strength.
- High stretch (30% to 40%). Pre-stretched ropes stretch less.
- Breaking stress = 2.25 times that of natural fibre rope.
- Stretch = 4 times that of natural fibre rope.
- Its flexibility is not affected by temperature, but it shrinks when wet. It may also lose up to 15% strength when wet.
- Relative density 1.14
- Resistant to rot, mildew and alkalis.
- Good UV, weather and abrasion resistance.
- Good for shock loads but has little value for lifting gear. Recommended in applications where movement energy needs to be absorbed.
- Not resistant to all chemicals: affected by linseed oil and mineral acids such as sulphuric and hydrochloric (muriatic) acids.
- Melting point 210°C (Natural fibre rope begins to char at 150°C). Although nylon melts or fuses with excessive heat it stops smouldering when heat source is removed. It may melt with heat build up when turns are surged around warping drums.

POLYESTER (DACRON / TERYLENE)
- Lower stretch than nylon (15% to 20%). Therefore not as elastic as nylon.
- High strength. Pre-stretched: even higher strength and lower stretch. (Excellent general purpose fibre)
- Breaking stress = 2 times that of the natural fibre rope.
- Excellent UV, weather and abrasion resistance. Holds colour superbly. Recommended for repeat loading conditions and yachting rope covers.
- Its flexibility is not affected by temperature, and it does not shrink when wet.
- Relative density 1.38 (Some polyester fibres with RD of less than 1 will float)
- Melting point 260°C
- Resistance to acids, rot, mildew and heat as for nylon.
- Exposure to alkalis, such as caustic soda, can be damaging.

POLYVINYL ALCOHOL (KURALON)
- Breaking stress = 1.25 times that of the natural fibre rope.
- Features similar to Terylene.

POLYETHYLENE (TANIKLON)
- Breaking stress = 1.45 times that of the natural fibre rope.
- Features similar to Terylene.

SILVER ROPE (FLAT SPIN TANIKLON FIBRE)
- Breaking stress = 1.16 times that of the natural fibre rope.
- 30% lighter than natural fibre or nylon rope.
- Does not absorb moisture.
- Does not slip as easily as other synthetics.
- Relative density less than 1 – buoyant.

POLYPROPYLENE (LAID SHATTERED FILM TYPE)
- Breaking stress = 1.6 times that of the natural fibre rope.
- It has about 60% of the strength of nylon and polyester, and about the same stretch as polyester.
- Moderate weather and abrasion resistance.
- Absorbs very little water and does not shrink when wet.
- Relative density 0.91 (Buoyant).
- Melting point 170°C. It can lose up to 30% strength at 65°C. The low melting point makes it unsuitable for applications involving frictional forces.
- Suitable for repeat loading conditions.
- Unaffected by acids or alkalis except in a very concentrated form.

Chapter 4: Ropes, Knots & Splices

HIGH-MODULUS POLYETHYLENE (HMPE) (PLASMA, STEELITE, DYNEEMA AND SPECTRA)
- Very low stretch (3% – 5%), even lesser if pre-stretched.
- Very high strength.
- Very light weight.
- Breaking stress = 6 times that of the natural fibre rope.
- Excellent weather resistance.
- Excellent abrasion resistance.
- Relative density 0.97 (Buoyant)
- Melting point 165°C (Low resistance to heat).
- Susceptible to creep under constant high loading for long periods.
- Its flexibility is unaffected by temperature.
- Absorbs very little water and doesn't shrink when wet, remaining lightweight.

ARAMID (TWARON AND KEVLAR)
- Very stretchy, unless pre-stretched.
- Very high strength.
- Negligible creep.
- Poor UV and abrasion resistance (generally forms a core with a polyester cover).
- Poor knot strength (Susceptible to axial compression fatigue)
- Relative density 1.41
- Melting point 500°C

LIQUID CRYSTAL POLYMERS (LCP) (VECTRAN)
- Very strong with minimum stretch.
- Zero creep.
- Excellent abrasion resistance.
- Melting point 330°C
- Low resistance to UV light, so it should be used inside a cover.

WIRE ROPES
(FLEXIBLE STEEL WIRE ROPES - FSWR)

Wire ropes are made the same way as the laid fibre ropes. Wires are twisted into strands, and strands into wire ropes. Oil soaked fibre-hearts usually run between the wires and strands, which keep the wire ropes lubricated, flexible and rust free from the inside. Wire ropes used for standing rigging may have a wire strand heart, thus adding to their strength and rigidity.

Most wire ropes have six or seven strands, with number of wires in each strand ranging between 7 and 37. The more wires in a strand, the more flexible the wire rope. Standing rigging needs fewer wires per strand than running rigging.

Wire ropes are referred to as:
- 7 x 7 FSWR: Seven strands, each containing seven wires. (Common for standing rigging)
- 6 x 24 FSWR: Six strands, each containing twenty four wires.
- 6 x 37 FSWR: Six strands, each containing thirty seven wires. (Extra flexible)
- 19 x 1 FSWR: Nineteen strands, each containing only one wire. (Highly inflexible - common for standing rigging)

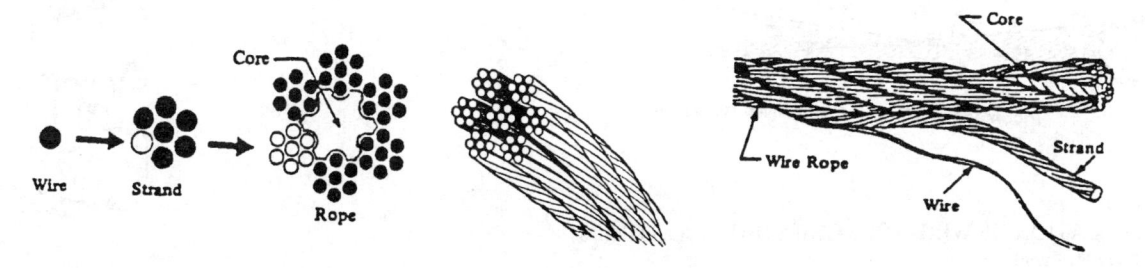

Fig 4.12:   7 x 7 WIRE ROPE

LANGS LAY

The ordinary or regular lay of a wire rope is similar to a fibre rope. For example, in a right-hand wire rope, the wires are laid left-handed and strands right-handed.

In a Langs Lay, both the wires and strands are laid in the same direction: both left-handed or both right-handed. This makes the rope more flexible and better for wear when passed through lifting blocks. However, because it is liable to rotate and unlay if a weight is hung on one end, it can only be used where both ends are anchored, such as for topping lifts for cranes. This rope is not as easy to handle as an ordinary lay.

NON-ROTATING WIRE ROPE

This is an improvement on Langs lay so that it does not rotate when a load is hung on its end. The wires and strands are made smaller and the inner strands are arranged in a non-rotational manner while maintaining the flexibility of a Langs Lay. This rope can be used for crane whips or runners.

PREFORMED ROPES (Fig 4.13)

In a preformed rope the wires do not jut out when cut, but tend to lie flat and only slightly separated. This is achieved during manufacturing by making the wires and strands lay naturally, free from internal stress.

PREFORMED ROPE 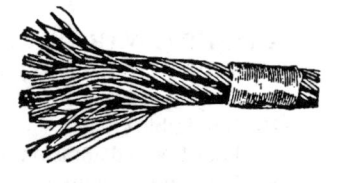 **UN-PREFORMED ROPE**

END-FOR-ENDING & CROPPING

Wear and tear at specific locations in a rope can be minimised by occasionally end-for-ending the rope or cropping a short length off its end. End-for-ending means reversing the ends, so that the outer-end becomes the inner-end and visa versa. This prolongs the rope's life.

LUBRICATION OF WIRE ROPES

Oil impregnated into wire ropes during manufacture generally does not last for the life of the ropes. To guard against corrosion wires are either galvanised or regularly coated with or soaked in a penetrating lubricant.

WIRE ROPE GRIPS (BULLDOG GRIPS) & SADDLE GRIPS

Grips are used as an alternative to splicing for making a permanent eye in a wire rope. Care must be exercised in their use. Over tightened grips would crush the rope; under tightened would slip. The correct and the incorrect way of using bulldog grips is shown in the NSW Workcover illustration below. Note that the saddle is on the live part of the rope and the U-bolt on the less heavily loaded tail of the rope. Saddle grips are safer than the bulldog grips. They have saddles on both sides.

Use at least three grips, spaced at least six rope diameters apart. The grips are suitable only for permanent fixed stays or guys. Do not use them on temporary stays or guys because of their crushing effect on the rope. Also, do not use grips on any load hoisting ropes.

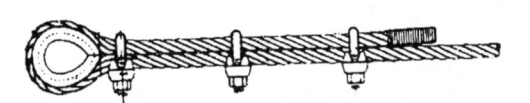

Correct method of fitting wire rope grips

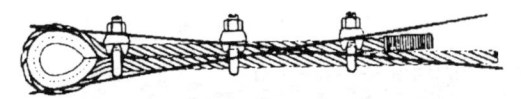

Incorrect method of fitting wire rope grips

Fig 4.14
INSTALLATION OF WIRE ROPE GRIPS (BULLDOG GRIPS)
(NSW Workcover)

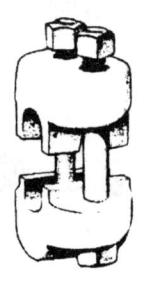

Fig 4.15 SADDLE GRIP
(NSW Workcover)

Chapter 4: Ropes, Knots & Splices

Fig 4.16
ROPE SIZES - GUIDE

Boat Size		Main Halyard	Main Sheet Traveller	Reefing Line	Main Sheet	Main Sail Outhaul	Cunningham	Genoa/Jib Halyard	Genoa/Jib Sheet	Genoa Car Control	Spinnaker Halyard	Spinnaker Sheet	After Guy/Brace	Spinnaker Pole Topping Lift	Spin Pole Fore Guy/ Tack Line	Spinnaker/Genoa Barberhauler	Running Backstay	Topmast Backstay Control	Headsail Furling Line	Lazy Jack Line	Anchor Warp	Dock Line	Dinghy Painter
2-4m	6.5-13 FT	4-5	4-8	-	-	3-4	4-5	4-5	6	4-5	5	-	-	4-5	4	-	-	-	4-5	-	-	6	-
4-6m	13-20 FT	5-6	5-8	6	6	4-5	5-6	5-6	6	5-6	6	-	-	5-6	5	-	-	-	5-6	6	-	6	-
6-8m	20-26 FT	6-8	6-8	8	8-10	5-6	6-8	6-8	8	6-8	6-8	6-8	6-8	6-8	6	6-8	5-6	4-5	6-8	6	8	10	12
8-10m	26-33 FT	8-10	8-10	10	10-12	6-8	8-10	8-10	10	8-10	8-10	8-10	8	8	6-8	10-12	6	5-6	8-10	10	10	12	12
10-12m	33-39 FT	10-12	10-12	10-12	10-12	8-10	10-12	10-12	12-14	10-12	10	10-12	8-10	10	6-8	12-14	6-8	6-8	10-12	12	12	14	12
12-14m	39-46 FT	12-14	12-14	12-14	12-14	10	12-14	12-14	14-16	10-12	10-12	12	10-12	12	8	14-16	8	10-12	12	14	14-16	14-16	12
14-16m	46-53 FT	14	12-14	14	14-16	14	14	14	14-16	14	10-12	14	12	-	8-10	14-16	8-10	12	-	16-20	16	12

Product Size (mm) — Applications

The above chart is an abbreviated version of the Southern Ocean brand of ropes application chart. The sizes indicted are an approximate guide only. Variation in the type of yacht (cruiser, cruiser racer, high tech racer) and factors such as displacement, righting moment, sail area and rig types will produce different load characteristics. As a general rule Southern Ocean's Challenge Braid and Super Braid can be the smaller of the diameters shown. You must also make sure that the line is of sufficient diameter to avoid slippage on self tailing winches, cleats and jammers.

SIZE OF MOORING (BERTHING OR DOCKING) ROPES

The handling of mooring ropes and the degree of mechanisation depends a great deal on the size and the weight of the vessel. The Ships Classification Societies (such as Lloyds Register and American Bureau of Shipping) define the minimum restrain capability that must be provided for a vessel of a given tonnage. The required size of appropriate cordage is then selected from information complied by rope manufacturers or as laid down in the Australian Standard AS 4142. Prudent boat owners always exceed the requirement by a comfortable margin. Ships normally secure to a berth using three or four head and stern lines and single springs. In severe weather they further increase the number of all mooring lines.

The preferred mooring line material for small as well as large vessels is polypropylene which floats easily. Very large ships are moored with wire ropes of high intrinsic strength and held in position by constant tension winches on which these ropes are permanently stored.

An appropriate authority or a rope manufacturer must be consulted for advice on the size of mooring lines for a vessel and her area of operation. The following is only an approximate guide to sizes of polypropylene mooring ropes for small vessels. In the absence of a suitable size, two smaller ropes may be used. The spring lines may be half the size of bow and stern lines.

SIZE OF RECREATIONAL VESSEL	SIZE OF HEAD AND STERN LINES
2 – 4 m	6 mm
6 – 8 m	10 mm
10 – 12 m	14 mm
14 – 16 m	18 mm

SIZE OF WORK BOAT OR TRAWLER	SIZE OF HEAD AND STERN LINES
6 – 8 m	20 mm
10 – 12 m	28 mm
14 – 16 m	36 mm
25 m	70 mm

At least double the above size or the number of lines for cyclone and severe weather moorings.

Consult the cyclone contingency plan for the port concerned. These are available in print as well as on the websites of marine departments of Western Australia, Northern Territory and Queensland, The website addresses are listed on the last page of this book.

Information on cyclone preparedness can be obtained from the State/Territory Emergency Services. Their phone numbers are listed on the second last page of this book. Cyclones are discussed in Chapter 20.

ANCHOR LINE SIZE

The above guideline for the bow and stern lines for recreational vessels may also be used for selecting their anchor lines. The anchor line should not be made of material that floats. Keep in mind that nylon fibre loses 10% to 20% strength when saturated with water. *(Anchor warps are discussed in Chapter 7)*

MOORING (BERTHING) LINES "LENGTH OUT" GUIDE
(As discussed in Chapter 8, use longer lines where there is a large range of tides)

- Bow and stern lines = 1½ x beam
- Spring lines = ¾ x length overall

TOW ROPES: See Chapter 8

FACTORS AFFECTING LIFE OF ROPES ARE:

- Excessive stress.
- Abrasion or cutting on sharp objects, such as burrs, rust and paint on chocks, bitts, winches and drums.
- Sharp bends (Around sharp bends only about half the rope's fibres take the load; compression makes the others ineffective.)
- Dragging over rough ground (dirt and grit picked up by ropes can work into the strands cutting the inside fibres.)
- Stowing away wet (Wet stowage of synthetic ropes is itself not damaging. The deterioration is caused if the rope

Chapter 4: Ropes, Knots & Splices

remains in the same position for long periods).
- Salt deposits (Hose down ropes regularly with freshwater).
- Inadequate ventilation.
- Exposure to chemicals, such as acids.
- Unnecessary exposure to UV light.
- Kinking.
- Movement in splices, showing clean rope.
- Passing through wrong size blocks (See matching sheave and rope size).
- Use of pulleys (sheaves) that are not free to rotate or have rough surfaces or sharp edges.
- Exposure to petrochemicals, e.g., diesel fuel, petrol, etc.
- The load capacity of mooring (berthing) lines is reduced due to factors such as their age, abrasion, knots, over-stress, temperature, and end of line configuration, i.e., knots, splices and turns.

ROPES TO BE RETIRED

Apart from rejecting a rope when obviously damaged, it is a good practice to establish a lifetime of each rope within the parameters of its intended use. If the conditions of usage remain unchanged, the rope can then be retired from that function at the end of that period.

ROPES TO BE CONDEMNED
- Reduction in diameter
- Flattened
- Frayed (Not to be confused with the furry look of a well-used synthetic rope, which may indicate a slight strength loss, but the hairy surface helps to protect the rope against further abrasion. However, this principle is only really effective on relatively large ropes (of diameter 60 mm+. Smaller ropes tend to lose too much strength through the initial abrasion process.)
- Mildew, rot or powdery substance between strands)
- Melting spots on synthetic ropes
- Kinked
- Bird-caged (strands or wires separated)

WIRE ROPES TO BE CONDEMNED (Fig 4.17)
- All the above reasons for ropes.
- More than 5% of the total wires are broken in any length of its 10 diameters. *(Ref.: AMSA Marine Order, Part 32, Para 12.6 c; and International Labour Office: Safety & Health in Dock Work, Para 9.9)*

OR
- More than 10% of the total wires are broken in any length of its 8 diameters. *(Ref: A Guide for Riggers, NSW Workcover)*

> **EXAMPLE:** A wire rope of 6 x 24 construction and 20 mm in diameter has a total of 6 x 24 = 144 wires.
> Therefore, 5% of its wires = 7.2; say 7 wires.
> Length of 10 diameters = 10 x 20 = 200 mm = 20 cm.
> Such a rope must be condemned if more than 7 wires are broken in its any length of 20 cm.

LIFTING GEAR - INSPECTION & LOG KEEPING

Cordage in constant use should be examined every 6 months and wire ropes every 3 months - more often if a broken wire or other damage is discovered. Chains up to 12 mm diameter should be inspected every 6 months, and larger chains annually. All lifting gear should be opened up and examined annually.

The fixed gear, such as cranes and derricks, must be periodically tested and maintained as per manufacturer's instructions. Removable gear, such as blocks, shackles and lifting trays usually do not require retesting as long as the original markings of SWL and certificate number are clearly visible, and there is no evidence of distortion or elongation.

Boat's crew must inspect and maintained the gear. Keeping accurate record of these inspections and maintenance may prevent an accident. Commercial vessels maintain an AMSA-approved inspection and maintenance register, commonly known as the *Chain Register*.

The Australian Boating Manual

BENDS, HITCHES, KNOTS & SPLICES
Fig 4 18 (A)

OVERHAND KNOT
(Prevents the end of a rope unlaying or passing through a block)

FIGURE OF EIGHT KNOT
(The same as the overhand knot but does not jam)

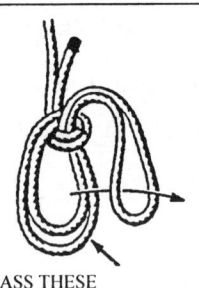

PASS THESE THROUGH THE EYE

REEF KNOT
(Joins 2 lines of equal size)

LIKE THIS

BOWLINE ON THE BIGHT
(A makeshift bosun's chair for rescue or to work on the mast)

RUNNING BOWLINE
(Pass the rope through the eye made by the bowline)

BOWLINE
(The most useful knot: makes a quick eye for mooring. Easy to undo)

Chapter 4: Ropes, Knots & Splices

Fig 4.18 (B)

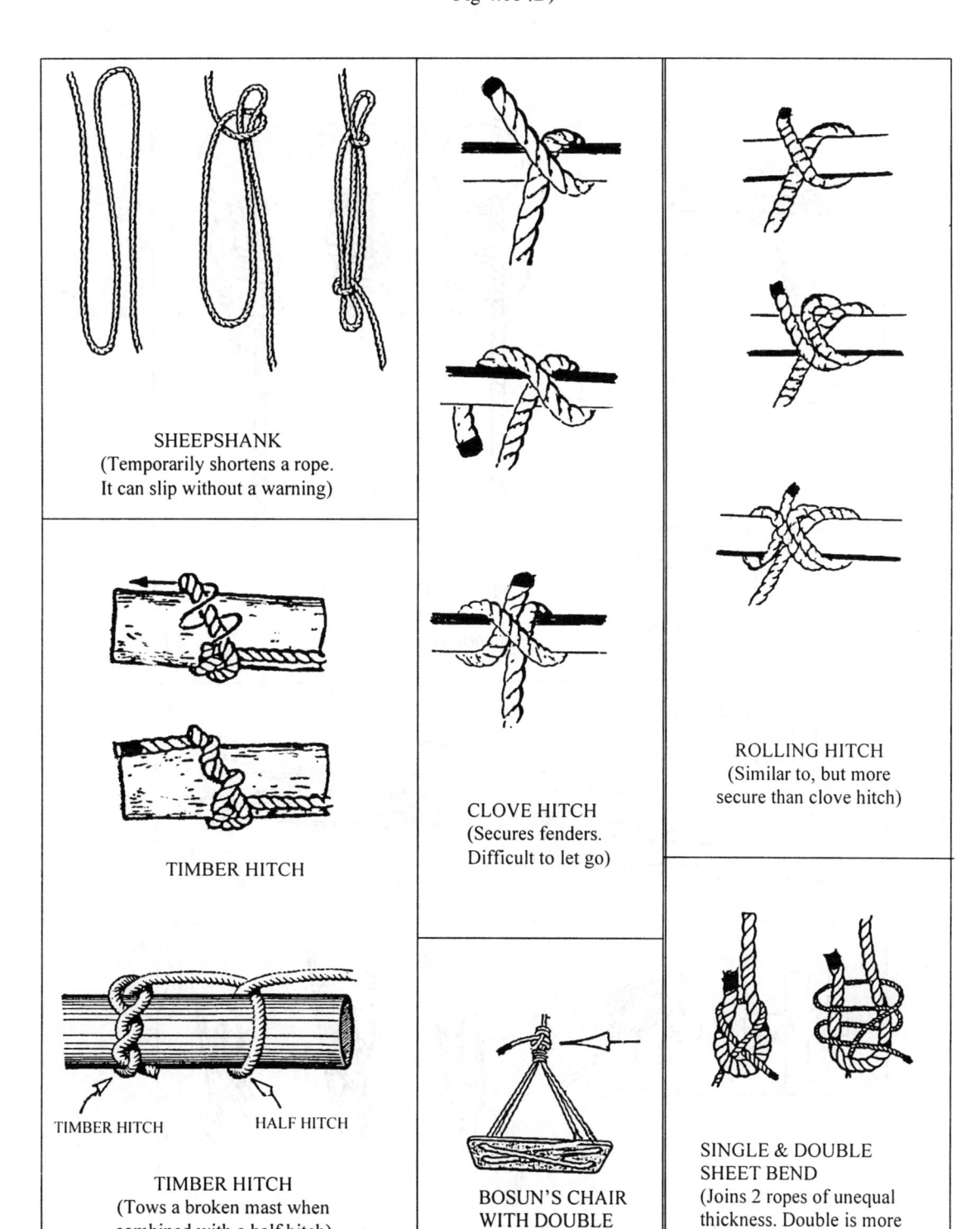

SHEEPSHANK
(Temporarily shortens a rope. It can slip without a warning)

TIMBER HITCH

TIMBER HITCH — **HALF HITCH**

TIMBER HITCH
(Tows a broken mast when combined with a half hitch)

CLOVE HITCH
(Secures fenders. Difficult to let go)

BOSUN'S CHAIR WITH DOUBLE SHEET BEND

ROLLING HITCH
(Similar to, but more secure than clove hitch)

SINGLE & DOUBLE SHEET BEND
(Joins 2 ropes of unequal thickness. Double is more secure)

66
The Australian Boating Manual

Fig 4.18 (C)

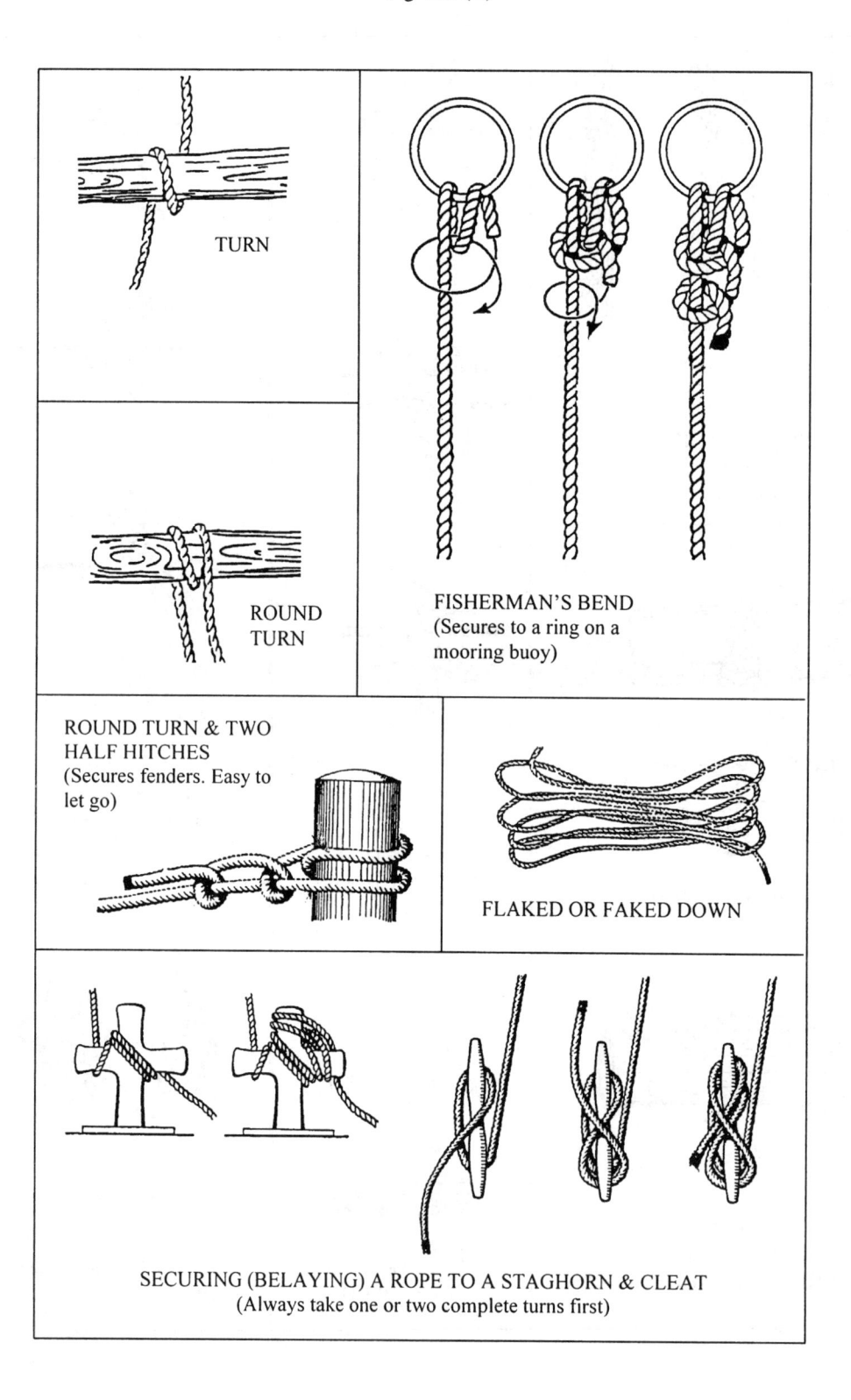

TURN

ROUND TURN

FISHERMAN'S BEND
(Secures to a ring on a mooring buoy)

ROUND TURN & TWO HALF HITCHES
(Secures fenders. Easy to let go)

FLAKED OR FAKED DOWN

SECURING (BELAYING) A ROPE TO A STAGHORN & CLEAT
(Always take one or two complete turns first)

Chapter 4: Ropes, Knots & Splices

Fig 4.18 (D)

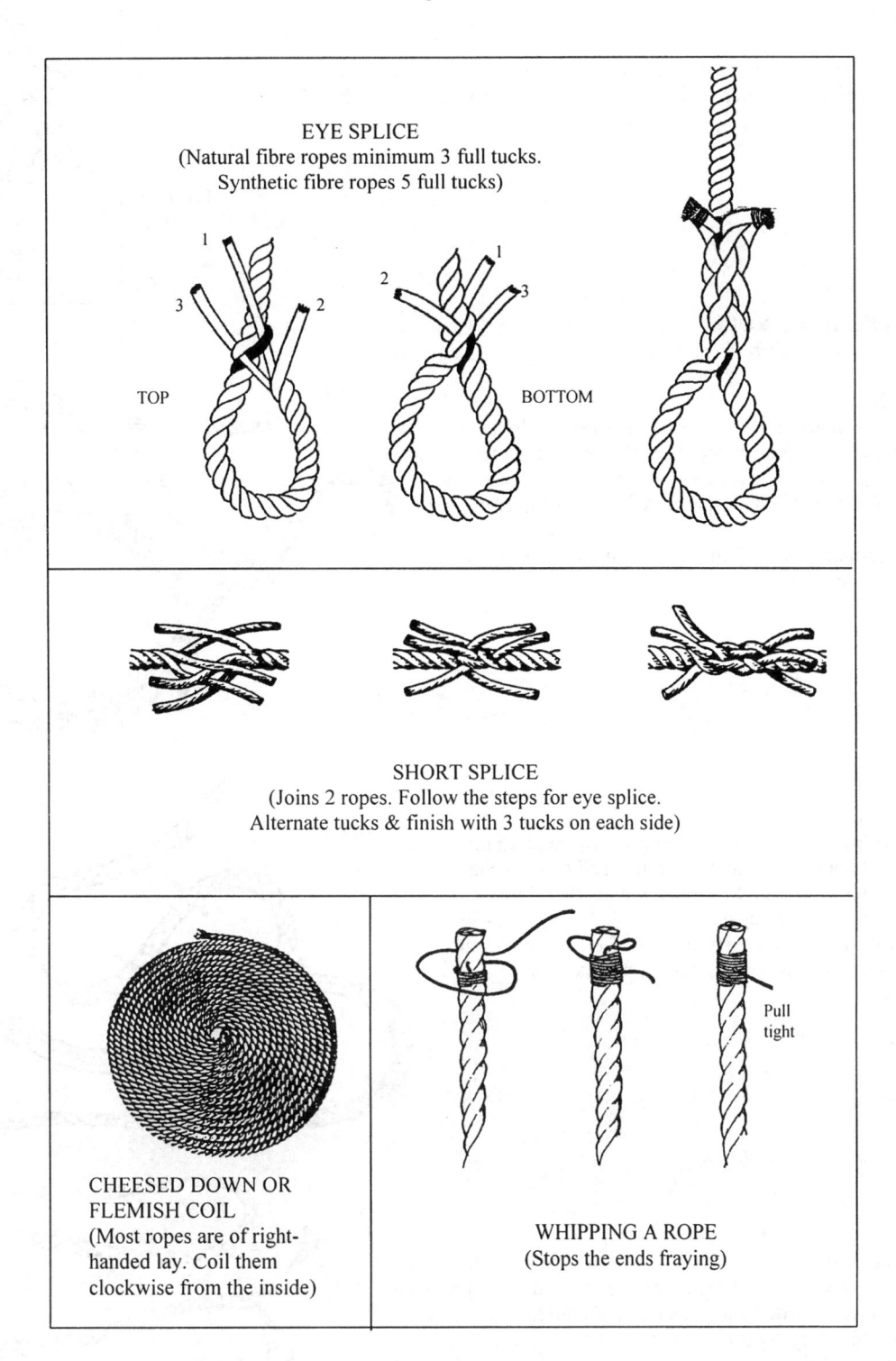

EYE SPLICE
(Natural fibre ropes minimum 3 full tucks.
Synthetic fibre ropes 5 full tucks)

TOP   BOTTOM

SHORT SPLICE
(Joins 2 ropes. Follow the steps for eye splice.
Alternate tucks & finish with 3 tucks on each side)

CHEESED DOWN OR FLEMISH COIL
(Most ropes are of right-handed lay. Coil them clockwise from the inside)

Pull tight

WHIPPING A ROPE
(Stops the ends fraying)

BACK SPLICE

It is to finish off the end of a rope that is not required to reeve through a block. Start with a *Crown Knot* (as shown) and then tuck the strands as in the Eye Splice or Short Splice.

Fig 4.19
CROWN KNOT

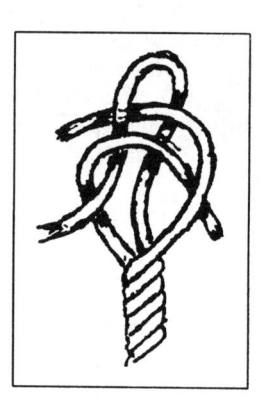

SPLICING A BRAIDED ROPE
(Example: Kinnears' Southern Ocean Racing Braid)

Step 1 (Fig 4.20 (A))
- Make two marks on the cover of the rope: the first approximately 25 diameters from the end of the rope.
- Form a loop of the required size of the eye and place the second mark made opposite mark 1. This is the crossover point.
- Tie a knot at about 1 ½ metres back up the rope to stop the core slipping.

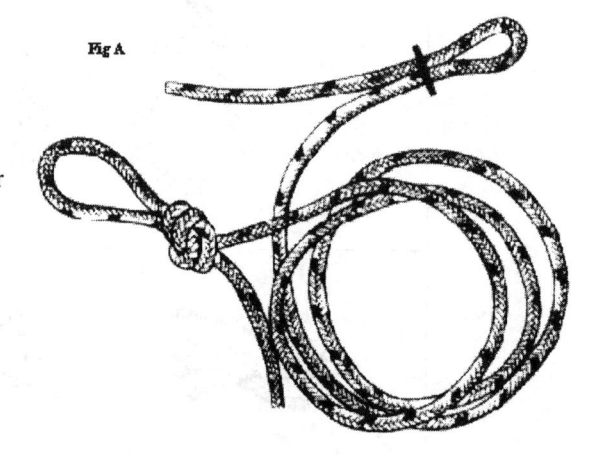

Step 2 (Fig 4.20 (B))
- Extract the core at the crossover point by bending the rope and hooking the core out with a spike. Mark the core at this point and then pull it out another 5 rope diameters and mark again.
- Taper the cover starting from about 5 cm or one colour coding towards the end by cutting the alternate strands.

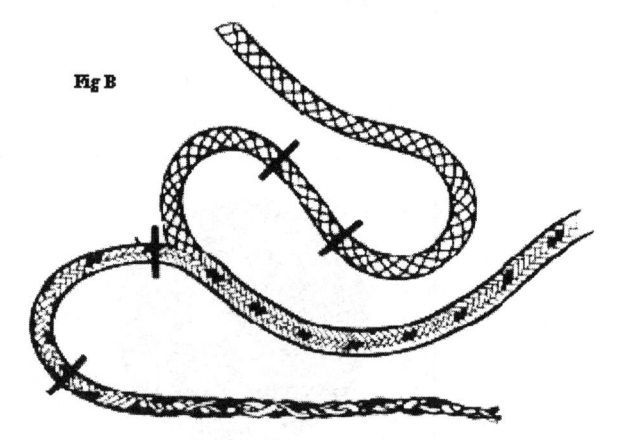

Step 3 (Fig 4.20 (C))
- Tape the tapered end of the cover to a fid, and push it into the core. Enter at mark 2 and push it in until the first cover mark reaches the entry point.
- Remove the fid from the core.

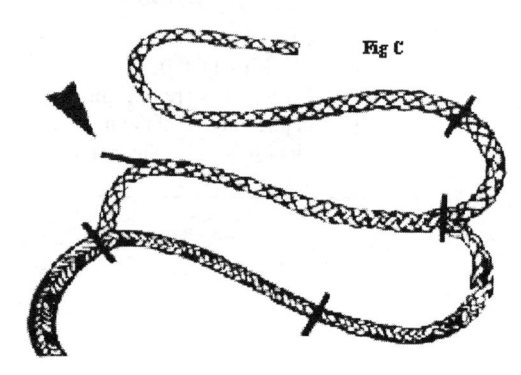

Chapter 4: Ropes, Knots & Splices

Step 4 (Fig 4.20 (D))
- Now tape the end of the core to the fid and enter it into the hollow part of the cover at mark 1 and pull it out at the crossover point or second mark on the cover.

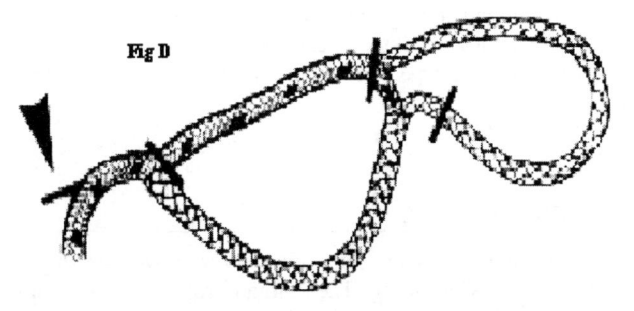

Step 5 (Fig 4.20 (E))
- When the parts of the core and cover that form the loop have been "milked" level, cut off the excess core left protruding at the crossover to make it flush with the cover. Then "milk" the cover from the 1 ½ metre knot back towards the loop.
- The core should now disappear into the cover. THE FINISHED SPLICE IS SHOWN IN FIGURE F.

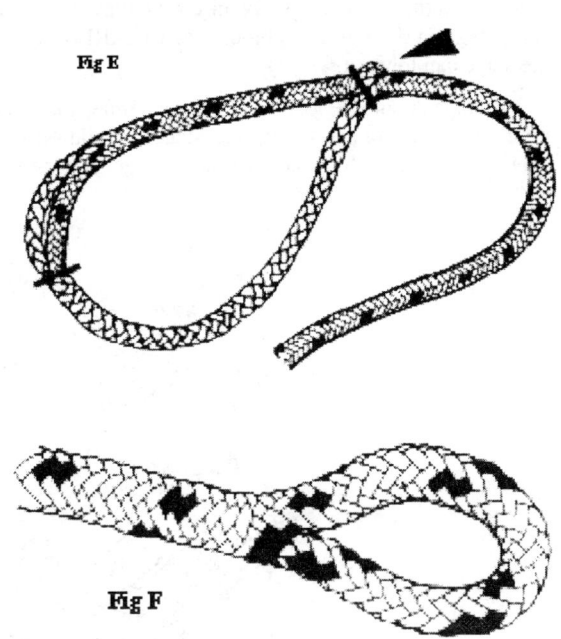

STOPPER (TEMPORARILY HOLDING A ROPE UNDER TENSION)

Such a need may arise when mooring (berthing) a large boat: when the hauling part of a mooring line is to be transferred from a winch drum to the bitts. A Rolling Hitch or a Rope Stopper may be used to temporarily hold a large rope under tension. The rolling hitch is shown above. The rope stopper is a small rope about two fathoms (4 metres) long.

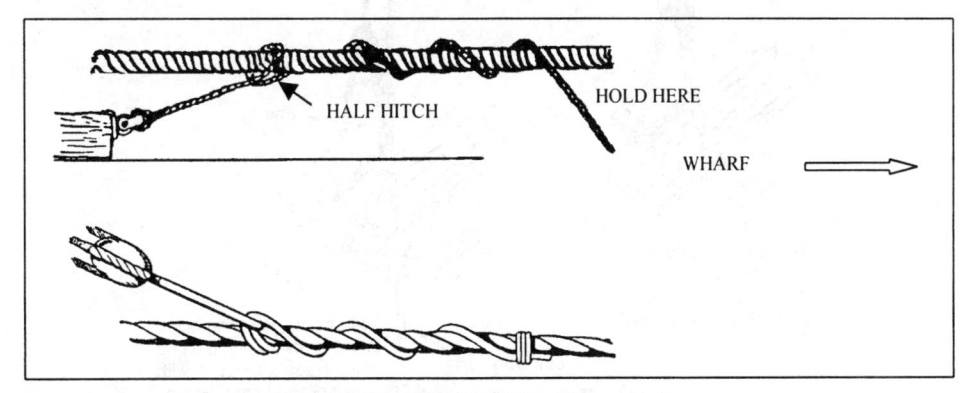

Fig 4.21: STOPPER HITCH (2 Examples)
(It holds a rope under tension. Hold or secure the end as shown.)

One end of the stopper is secured somewhere on deck and, with the other end a half hitch or two turns of a rolling hitch are passed against the lay of the mooring (berthing) line. Several additional turns are then taken around the taut line against the lay to complete the stopper hitch. One crewmember holds the end of the stopper while the other crew transfer the line from the winch to the bitts. In case of a one-person operation, the stopper end may be secure as shown.

Use synthetic rope stoppers on synthetic lines and natural fibre stoppers on natural fibre lines. To hold a wire rope, use a *Chain Stopper*.

RIGGING A STAGE

A STAGE is a plank slung by two ropes over the ship's side on which the crew can sit or stand to paint or repair the hull. It is like scaffolding. The common rope size is 20 mm, which should be long enough to reach the waterline on the bight (from their fastenings). A third rope may be used to prevent the middle of an extra long stage from sagging. The ropes can be rigged through a tail block or a LIZARD (an iron ring spliced into a rope end) or a shackle so that those working on the stage can lower it.

A stage should be load tested to four times the intended load. Everyone working on it must be provided with a lifeline long enough to trail in the water. Stages should only be rigged over water, and never whilst making way. A rope ladder should be provided for access from the stage. A person on deck must be in attendance with a lifebuoy.

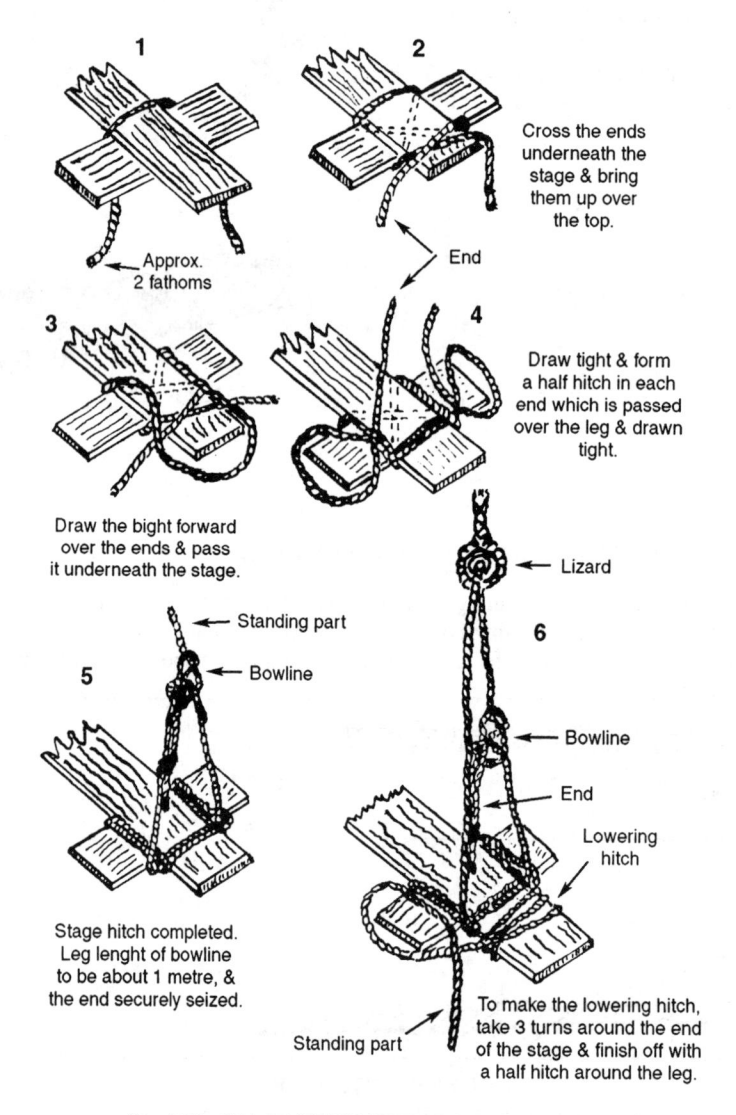

Fig 4.22: STAGE KNOT WITH LOWERING HITCH
(Illustration: C.H. Wright)

Chapter 4: Ropes, Knots & Splices

RIGGING A BOSUN'S (or BOATSWAIN'S) CHAIR

A Bosun's Chair is shown below and among the illustrations of knots earlier in this chapter. Secured to a gantline with a double sheet bend it is used for working aloft. GANTLINE is a rope used for hoisting a person or gear aloft. The gantline can be rigged through a tail block or a lizard so that the person in the bosun's chair can lower oneself with a LOWERING HITCH. For safety reason's it is better that your learn this process from someone and not just follow the steps listed below:

- ➢ Step 1: Bring the running part of the gantline through the chair bridle and grip it firmly with one hand against the standing part (or use a temporary seizing).
- ➢ Step 2: With the other hand, pull the bight of the running part through the bridle, over your head, behind the back, under the feet and up across the knees.
- ➢ Step 3: Heave the part marked "To coil" until the hitch is tight around the bridle's apex.
- ➢ Step 4: Release the grip or the seizing on the first two parts.
- ➢ Step 5: You can now lower yourself by rendering the hitch just made.

Fig 4.23
BOSUN'S CHAIR HITCH

Before using always check the bosun's chair and lines and fittings for any defects. Load test it for 4 to 5 times the intended load. Do not use a winch to haul a person aloft; do it manually. Wear a safety harness and make sure all the tools taken aloft are attached to safety lines. When riding a bosun's chair with a shackle passed around a stay, make sure the bow of the shackle (not the pin) is around the stay. Always mouse the shackle pin.

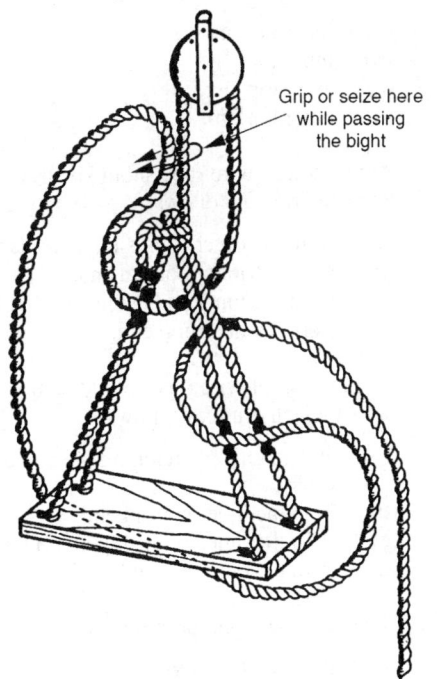

BOARDING LADDERS

Boarding ladders when fully extended should go down to about 50 cm below the water level. The steps above the water should be non-slippery. This can be achieved by fixing split rubber tubing to the steps or winding a rope around the rungs. Just in case it is accidentally dislodged, a removable metal ladder should be secured to the boat with a rope.

CHAPTER 4: QUESTIONS & ANSWERS

CHAPTER 4.1: ROPES

1. To open a new coil of a small rope, the following end should be taken out first:

    (a) inner bottom.
    (b) inner top.
    (c) outer top.
    (d) outer bottom.

2. A new coil of wire rope should be transferred onto a winch drum as follows:

    (a) from the outer end using a turn table.
    (b) from the inner bottom end.
    (c) from the inner end using a turn table.
    (d) from the outer top end.

3. A new coil of wire rope should be transferred onto a winch drum as follows:

    (a) by opening & stretching the coil on the wharf.
    (b) from the inner bottom end.
    (c) from the inner end using a turn table.
    (d) from the outer top end.

4. When coiling a polyprop rope, coil it:

    (a) always left handed.
    (b) always right handed
    (c) alternatively right & left handed.
    (d) always in the same direction.

5. In relation to use of wire rope, which of the following statements is correct:

    (a) a 6x12 wire rope is unsuited for running gear.
    (b) a 6x12 wire rope is suitable for standing gear.
    (c) a 6x24 wire rope is unsuited for running gear.
    (d) a 6x24 wire rope is unsuited for standing gear.

6. To "fake or flake" a rope is to:

    (a) cheese it.
    (b) Flemish Coil it.
    (c) undo its strands.
    (d) coil it in long lengths from side to side.

7. When a synthetic rope parts, it:

    (a) whips forward along the line of pull.
    (b) whips back along the line of pull.
    (c) curls like a wire rope parting.
    (d) uncoils its strands along the full length.

8. Synthetic ropes subjected or exposed to the following are not harmed by it:

    (a) sunlight.
    (b) kinking.
    (c) moisture.
    (d) chemicals

9. Ropes used for the following function are most easily strained:

    (a) anchoring.
    (b) heaving vessels alongside.
    (c) lifting weights.
    (d) mooring vessels to buoys.

10. The telltale signs of a worn out synthetic rope do not include:

    (a) reduction in diameter.
    (b) flattened shape.
    (c) fraying.
    (d) mildew.

11. Synthetic ropes are stronger than natural fibre ropes due to their:

    (a) length of fibres.
    (b) elasticity and ability to withstand moisture.
    (c) length of fibres and elasticity.
    (d) ability to withstand moisture.

12. The end of a manila rope can be prevented from fraying by:

    (i) heating it.
    (ii) whipping it
    (iii) tying a figure of eight knot.
    (iv) tying an overhand knot.

    The correct answer is:
    (a) all except (i).
    (b) all except (ii).
    (c) all except (iii).
    (d) all except (iv).

13. The following condition inside cordage indicates its state of wear and tear:

    (i) powdery substance.
    (ii) rot.
    (iii) mildew.
    (iv) lighter colour.

    The correct answer is:
    (a) all except (i).
    (b) all except (ii).
    (c) all except (iii).
    (d) all except (iv).

14. For condemning ropes, which of the following statements is correct:

    (a) rot in synthetic ropes.
    (b) mildew in synthetic ropes.
    (c) powdery substance inside natural fibre ropes.
    (d) 5% of broken wires in any 10 diameters of wire rope.

CHAPTER 4.2: KNOTS & SPLICES

15. When at anchor the anchor rope should be secured on board as follows:

    (a) a bowline placed over a bollard or cleat.
    (b) numerous turns on a sampson post & a hitch.
    (c) a tight and secure knot on a post.
    (d) a round turn & two half hitches.

16. The slipping end of a slip rope should be held on board as follows:

    (a) a bowline placed over a bollard or cleat.
    (b) an eye placed over a bollard or cleat.
    (c) a tight and secure knot on a post.
    (d) figure of eight turns on a bollard.

17. To join two ropes of unequal thickness the following knot should be used:

    (a) sheep shank.
    (b) reef knot.
    (c) bow line.
    (d) double sheet bend.

18. To make a temporary eye on the end of a rope the following knot should be tied:

    (a) bowline.
    (b) reef knot.
    (c) clove hitch.
    (d) figure of eight knot.

19. To secure a dinghy to a ring bolt on the jetty the following knot should be used:

    (a) sheet bend.
    (b) bowline.
    (c) a round turn and two half hitches.
    (d) clove hitch.

20. To tow a broken mast behind the vessel the following knot should be used:

    (a) two figure of eight knots.
    (b) two bowlines.
    (c) a timber hitch and a half hitch.
    (d) a round turn and two half hitches.

21. To prevent the end of an untended rope passing through a block the following knot should be used:

    (a) figure of eight.
    (b) clove hitch.
    (c) overhand knot.
    (d) sheet bend.

22. To tie a bandage round a wounded arm the following knot should be used:

    (a) sheet bend.
    (b) overhand knot.
    (c) reef knot.
    (d) clove hitch.

23. The following knot at the end of a rope can be used to improvise a bosun's chair:

    (a) bowline.
    (b) bowline on the bight.
    (c) clove hitch.
    (d) reef knot.

24. The most suitable knot to tie two lines of unequal thickness is:

    (a) sheepshank.
    (b) reef knot.
    (c) clove hitch.
    (d) sheet bend.

25. To make a temporary eye in a rope, the most suitable knot is:

    (a) figure of eight.
    (b) bowline.
    (c) bowline on the bight.
    (d) two round turns and a half hitch.

26. To make fast a line to a spar or to a larger rope or standing rigging, the most suitable knot is:

    (a) rolling hitch or clove hitch.
    (b) double sheet bend.
    (c) clove hitch.
    (d) round turn or two half hitches.

27. To tow a broken mast in a straight line behind a vessel, the most suitable hitch is the combination of a half hitch and a:

    (a) timber hitch.
    (b) sheet bend.
    (c) clove hitch.
    (d) bowline.

28. To join two ropes of equal size, the most suitable knot is:

    (a) sheepshank.
    (b) sheet bend.
    (c) reef knot.
    (d) double sheet bend.

29. The following splice is used to join two ropes without increasing their thickness:

    (a) short splice.
    (b) long splice.
    (c) back splice.
    (d) eye splice.

30. Knots and splices can reduce the strength of ropes by up to:

    (a) 10%
    (b) 20%
    (c) 40%
    (d) 60%

31. Binding the spliced area of a rope with twine, improves its:

    (a) looks.
    (b) strength.
    (c) looks and strength.
    (d) looks and durability.

32. Fibre and steel wire rope sizes are measured as follows:

    (a) Fibre rope in diameter & Wire rope in circumference.
    (b) Both types of ropes in circumference.
    (c) Both types of ropes in diameter.
    (d) Fibre rope in circumference & Wire rope in diameter.

33. Cable-laid rope is:

    (a) made of three hawser-laid ropes.
    (b) is a hawser-laid rope.
    (c) not suitable for towing.
    (d) is a 4-strand rope.

34. In relation to manufacture of ropes, which of the following statements is incorrect?

    (a) Natural fibre ropes are made of short staple of fibre.
    (b) In most ropes, the strands are laid in the opposite direction to yarns.
    (c) In Langs Lay, the wires & strands are laid in the same direction.
    (d) The term 'Cordage' includes both fibre & wire ropes.

35. In relation to ropes, which of the following statements is incorrect?

    (a) Most ropes are left-hand laid with three strands.
    (b) Square rope is tough, kink-resistant and flexible.
    (c) The core of double-braided rope bears about 70% of the load.
    (d) Braided rope is kink & abrasion-resistant and flexible.

36. In relation to ropes, which of the following statements is incorrect?

    (a) Wire ropes are internally lubricated by an oil-soaked fibre heart.
    (b) Wire ropes used for standing rigging may have a wire strand heart.
    (c) A 6 x 24 wire rope has 24 strands, each containing 6 wires.
    (d) A 10 mm wire rope containing 144 wires is more flexible than a 10 mm wire rope containing 72 wires.

37. In relation to ropes & chains, which of the following statements is incorrect?

    (a) The size of chain is the diameter of its metal rod.
    (b) Ropes are measured by their diameter in millimetres.
    (c) Synthetic ropes when surged under load can weld then slip on bitts.
    (d) The left-hand laid ropes should be coiled clockwise.

38. In relation to wire grips, which of the following statements is correct?

    (a) They should be spaced at least 6 rope-diameters apart.
    (b) The U-bolt should be fitted on the live part of the rope.
    (c) The saddle should be fitted on the less heavily loaded part of the rope.
    (d) They should be used only for making a temporary eye in a wire rope.

39. Which of the following statements is incorrect?

    (a) Chain stoppers are used on wire ropes.
    (b) Natural fibre stoppers are used on synthetic ropes.
    (c) The stopper is turned against the lay of the mooring (berthing) rope.
    (d) A rolling hitch may be used in lieu of a stopper.

40. In relation to ropes, which of the following statements is incorrect?

    (a) Polyprop rope should be coiled in the same direction every time.
    (b) A 6x24 wire rope is more flexible than a 6x12 wire rope of the same diameter.
    (c) Nylon ropes stretches more than Polyester ropes.
    (d) Nylon ropes shrink when wet.

Chapter 4: Ropes, Knots & Splices

41. In relation to ropes, which of the following statements is <u>incorrect</u>?

    (a) Langs Lay wire rope is suitable for topping lift on a crane.
    (b) Non-rotating wire rope is suitable for crane whip or runner.
    (c) Silver rope is heavier than nylon rope.
    (d) The wires in Preformed ropes do not jut out when cut.

42. In relation to ropes, which of the following statements is <u>incorrect</u>?

    (a) Nylon ropes lose up to 15% strength when saturated with water.
    (b) The spring lines can be of smaller size than the bow & stern lines.
    (c) In the absence of a suitable size mooring (berthing) rope, two smaller ropes may be used.
    (d) Nylon ropes are more buoyant than polypropylene ropes.

43. In relation to rigging a Stage, which of the following statements is correct?

    (a) The common rope size for rigging a stage is 10 mm.
    (b) The ropes should be long enough to reach the stage from their fastenings.
    (c) The stage should be load-tested to twice the intended load.
    (d) The ropes should be rigged to permit those working on the stage to lower it.

44. A bosun's chair should be load tested:

    (a) 2 – 3 times the intended load.
    (b) 3 – 4 times the intended load.
    (c) 4 – 5 times the intended load.
    (d) 5 – 6 times the intended load.

45. In relation to rigging a Bosun's chair, which of the following statements is correct?

    (a) When riding a bosun's chair with a shackle passed around a stay, the bow of the shackle should be around the stay.
    (b) When hauling a person aloft in a bosun's chair, it should be done using a winch.
    (c) The gantline should be rigged such that the person in the bosun's chair cannot lower it.
    (d) The lowering hitch is to allow a person on deck to control the bosun's chair.

46. In relation to boarding ladders, which of the following statement is correct?

    (a) When fully extended, the ladder should stay 50 cm above the water level.
    (b) Only the steps below the water level should be made non-slippery.
    (c) Steps can be made non-slippery by fixing split rubber tubing.
    (d) The ladder should not be allowed to extend below the water level.

47. In relation to ropes, which of the following statements is <u>incorrect</u>?

    (a) Natural fibre ropes make a creaking sound before parting.
    (b) Synthetic ropes dramatically reduce in diameter before parting.
    (c) Excessive heat makes the natural fibre rope dry and brittle.
    (d) The fibre core inside a steel wire rope should be dry.

48. The easiest rope to splice an eye into is:

    (a) Braided rope.
    (b) Three-strand rope.
    (c) Four-strand rope.
    (d) Plaited rope.

CHAPTER 4: ANSWERS

1 (a), 2 (a), 3 (a), 4 (c), 5 (b),
6 (d), 7 (b), 8 (c), 9 (c), 10 (d),
11 (c), 12 (a), 13 (d), 14 (d), 15 (b),
16 (d), 17 (d), 18 (a), 19 (c), 20 (c),
21 (a), 22 (c), 23 (b), 24 (d), 25 (b),
26 (a), 27 (a), 28 (c), 29 (b), 30 (d),
31 (d), 32 (c), 33 (a), 34 (d), 35 (a),
36 (c), 37 (d), 38 (a), 39 (b), 40 (a),
41 (c), 42 (d), 43 (d), 44 (c), 45 (a),
46 (c), 47 (d), 48 (b)

Chapter 5

BOAT MAINTENANCE & CARE OF SAILS

ELECTROLYTIC CORROSION

ELECTROCHEMICAL CORROSION of metals can cause serious and expensive damage to propellers, propeller shafts, rudders, keels and other equipment fitted on vessels.

The main types of corrosion are:
- Crevice and deposit corrosion
- Galvanic corrosion
- Stray current corrosion

Paint barriers are used to keep the crevice and deposit corrosion to a minimum, while the galvanic and stray corrosions are minimised by proper electrical installation and the choice of correct materials for submerged components. In addition, the underwater metal components are protected from corrosion by the use of sacrificial anodes made of zinc or magnesium.

GALVANIC CORROSION

Galvanic corrosion is the process of ELECTROLYSIS, which is also the principle on which wet-cell batteries work. When two dissimilar metals are placed in an electrolyte (acid or seawater), a current flows from the ignoble metal, i.e., the metal with the lower potential (the anode) to the noble metal, i.e., the metal with the higher potential (the cathode). The process will result in the gradual wastage of the ignoble metal.

Dissimilar metals around a vessel's stern when immersed in seawater form a large battery. Seawater acts as the electrolyte. Steel rudder and shaft corrode in the presence of a bronze propeller; or an aluminium propeller corrodes in the presence of steel.

GALVANIC SERIES

Metals can be listed in a galvanic voltage series, indicating their nobility or potential (voltage) in relation to other metals. *In the attached table, the electrolyte is seawater at a temperature range of +10°C to +26.7°C. The water flow rate is 2.4 to 4.0 metres/sec. The reference electrode is silver-silver chloride (Ag-AgCl):*

The metals higher up in the table have a greater voltage potential. They are progressively more noble than those lower down the list. The further apart the metals are in the galvanic series, the greater the current flow between them, resulting in greater corrosion (wastage) of the ignoble metal. For example, aluminium would corrode faster in the presence of lead than in the presence of steel.

One way to protect dissimilar metals from corroding in each other's vicinity is to provide them with a metal more ignoble than both of them. We place a relatively more ignoble metal (e.g., zinc or magnesium) in their vicinity. Zinc and magnesium are the most common sacrificial metals. Placed in the vicinity of steel and aluminium, zinc will then corrode rather than the other two. This is known as cathodic protection.

Metal	Voltage
Graphite	+0.19 to +0.25V
Stainless steel 18-8, 3% Mo, in a passive state *	0.00 to -0.10V
Stainless steel 18-8 in a passive state *	-0.05 to -0.10V
Nickel	-0.10 to -0.20 V
Lead	-0.19 to -0.25V
Silicon bronze (92.9% Cu, 1.50% Zn, 3% Si, 1.00% Mn, 1.60% Sn)	-0.26 to -0.29V
Manganese bronze (58.5% Cu, 39% Zn, 1% Sn, 1% Fe, 0.3% Mn)	-0.27 to -0.34V
Aluminium brass (76% Cu, 22% Zn, 2% Al)	-0.28 to -0.36V
Soft solder (50% Pb, 50% Sn)	-0.28 to -0.37V
Copper	-0.30 to -0.57V
Tin	-0.31 to -0.33V
Red brass (85% Cu, 15% Zn)	-0.30 to -0.40V
Yellow brass (65% Cu, 35% Zn)	-0.30 to -0.40V
Aluminium bronze	-0.31 to -0.42V
Stainless steel 18-8, 3% Mo, in an active state **	-0.43 to -0.54V
Stainless steel 18-8 in an active state **	-0.46 to -0.58V
Cast iron	-0.60 to -0.71V
Steel	-0.60 to -0.71V
Aluminium alloys	-0.76 to -1.00V
Galvanised iron and steel	-0.98 to -1.03V
Zinc	-0.98 to -1.03V
Magnesium and magnesium alloys consumed	-1.60 to -1.63V

* Metals are in a passive state when the metal has a thin, reaction-inhibiting coating. This coating is lacking in an active state.
** Still water.

Fig 5.1: TABLE OF GALVANIC VOLTAGE SERIES *(Volvo Penta)*

Chapter 5: Boat Maintenance & Care of Sails

CATHODIC PROTECTION

- Zinc ingots, known as *Sacrificial Anodes*, are bolted to the hull. They are also fitted inside steel hulls that are subject to seawater flooding as in the case of ballast tanks.
- Zinc anodes must never be painted or cleaned with steel wool or steel brush. This will reduce their galvanic protection. However, whenever the opportunity arises, they should be cleaned (activated) with an abrasive paper in order remove the oxide layer.
- The *outboard motor* units are usually protected against salt water corrosion as follows: Corrosion-resistant stainless steel is used for drive shafts, propeller shafts, drive pins, linkages, clips, springs, washers and screws. Moderately loaded parts and linkages are made of plastic. Motor covers are made of fibreglass. Many other parts are coated and plated with various materials to reduce corrosion.
- Lead or copper oxide paints must not be used on aluminium hulls. Such metals will corrode the parent metal of the hull.
- Engines and stern drives must never be allowed metallic contact with the keel, rudder or other metallic components below the waterline.
- Do not ground electrical equipment (such as high voltage charging equipment, navigation aids, marine radio, etc.) to the engine.
- Another way to avoid galvanic corrosion on components submerged in the water is to bond all of them to a common anode, normally made of zinc. The bonding system with its individual components should normally have no contact to the negative circuit of the boat's electrical system.

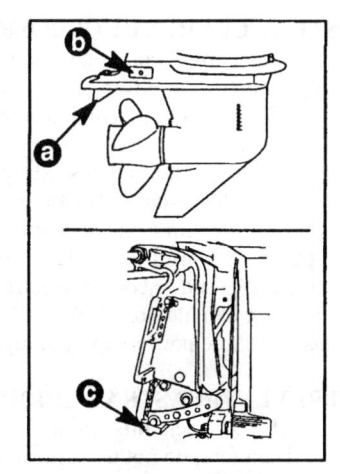

Fig 5.2: ZINC ANODES ON A MERCURY OUTBOARD
(One on the trim tab (a),
Two on the side (b),
One on the bottom of the transom bracket (c)

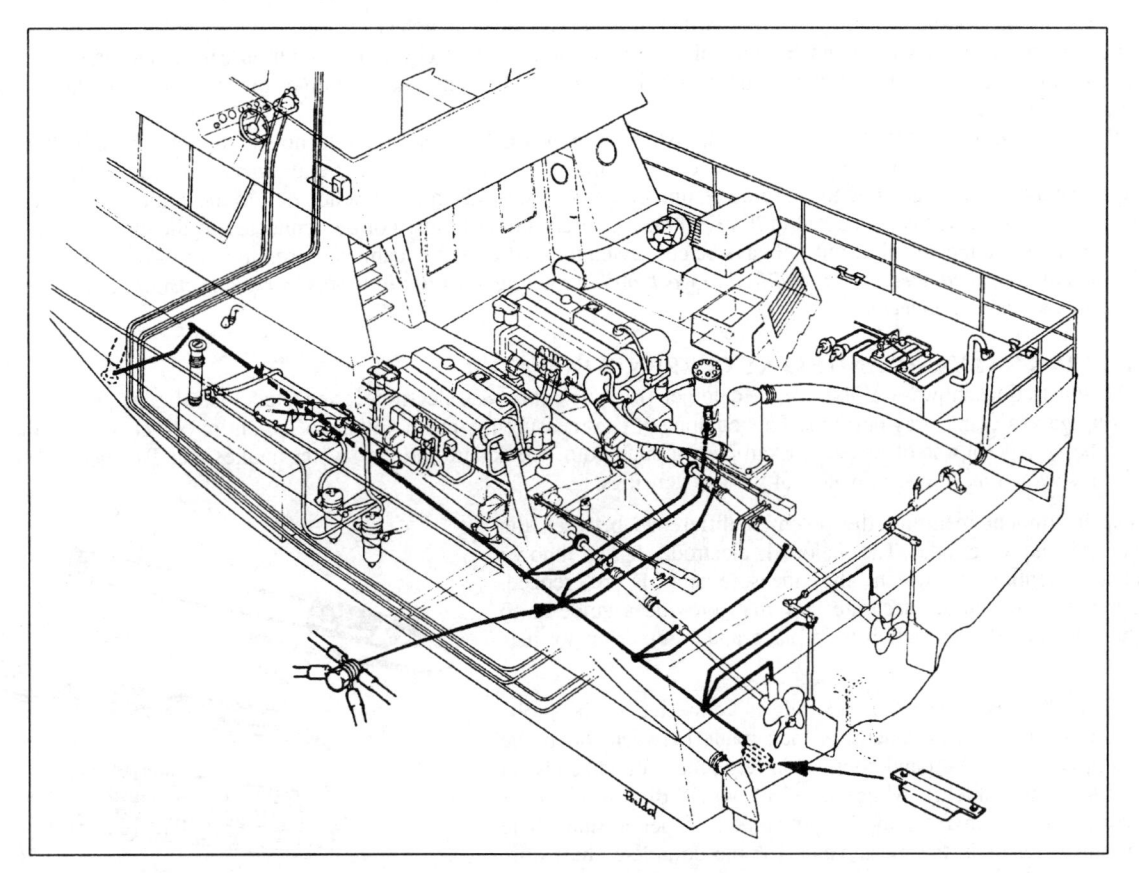

Fig 5.3: BONDING TO ANODE *(Volvo Penta)*

STRAY CURRENT CORROSION

Stray current corrosion is similar to galvanic corrosion in the way it acts but differs in the way it is caused. In galvanic corrosion, it is the potential difference in the metals that initiates corrosion.

As the name implies, stray currents cause stray current corrosion. Stray currents can arise due to faults in the vessel's electrical system. The examples are connections and splices exposed to moisture or bilge water, wear and tear in electrical equipment, and incorrect electrical installations. Stray current corrosion can also result due to stray currents from neighboring boats or shore power supply equipment.

PREVENTING STRAY CURRENTS

- All D.C. circuits must have an insulated return cable. Therefore, do not use a metal keel as a return conductor.
- Do not install electrical sockets, terminal blocks, switch panels and fuse holders where they can be exposed to moisture or bilge water.
- Route electrical cables as high as possible above the bilge water. If a cable is exposed to water, it must be housed in a watertight conduit and the connections must also be watertight.
- Cables likely to wear must be installed in self-draining conduits, sleeves, cable channels etc.
- Install a main switch for the starter battery on the (+) side. It should disconnect all equipment except equipment such as theft protection, bilge pumps and the operating switch for electrical main switches.
- If additional battery (accessories battery) is fitted, the main switch must be fitted between its (+) terminal and the fuse block for the electrical equipment. The main switch must disconnect all equipment except equipment such as theft protection, bilge pumps and the operating switch for electrical main switches.
- Engines and drivelines must not be galvanically connected to other equipment such as trim plane or bathing steps unless bonded to an anode. They must not be used as a ground for radio, navigation or other equipment where separate ground cables are used.
- All separate ground cables (ground connections for radio, navigation equipment, echo sounder etc) must be linked to a common ground point, i.e., a cable that does not normally act as a return for equipment. Do not ground electrical equipment (such as high voltage charging equipment, navigation aids, marine radio, etc.) to the engine.
- If shore-based power is connected (120V/220V), the safety ground must not be connected to the engine or any other ground point on the vessel. It must always be connected to the connection cabinet's ground terminal ashore.
- WARNING! Only an electrician qualified to work on high voltage installations may carry out installation and work on shore-connected equipment.

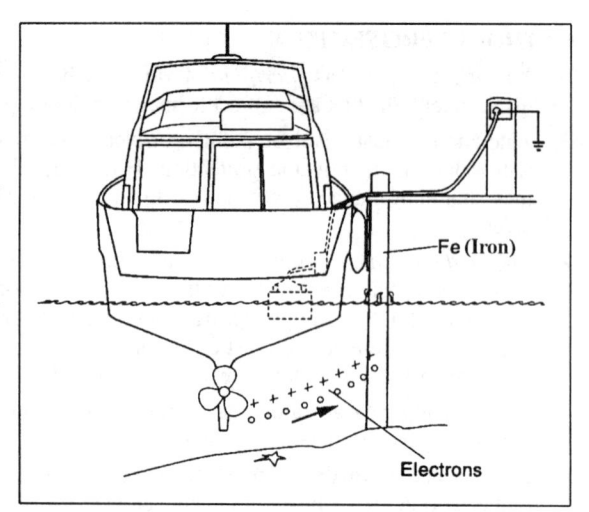

Fig 5.4: STRAY CURRENT CORROSION *(Volvo Penta)*

MEASURING GALVANIC AND STRAY CURRENTS IN WATER

The illustration shows a Volvo Penta calomel *(mercury mineral)* electrode connected to a digital probe tester for measuring galvanic and stray currents. This equipment can measure the mean voltage of an entire object, such as the shaft line, or the voltage of one of its parts. Examples of such point checks are rudders, water intakes etc. The electrode can be used in fresh water as well as in waters of various densities.

The instrument measures the potential difference between the object and the electrode. The calomel electrode has a known constant potential to which the measurement is referenced. Measurements, which are accurate to 1 mV, provide a guide as to whether the cathodic protection is in place and whether or not it is sufficient.

To use this instrument the electrode is connected to the probe tester and set to D.C. A contact is then made between the probe tester's tip and a good ground connection (e.g., the ground screw on the inside of the shield), while the electrode is dipped in water approximately 30 cm from the propeller and propeller shaft. The measurement result is the mean value for the complete shaft line. The result should lie between (minus) -900 mV and -1340 mV.

To check individual parts, move the electrode so that the tip is directed towards the surface, approximate 5 mm away from the

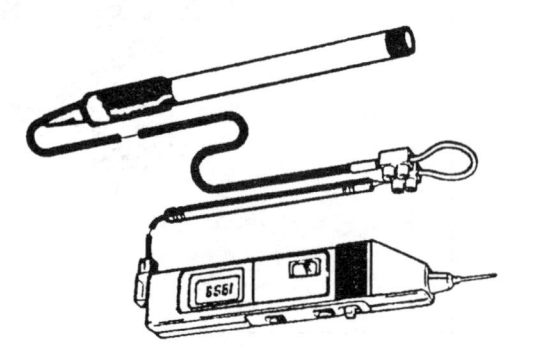

Fig 5.5: CALOMEL ELECTRODE & PROBE TESTER *(Volvo Penta)*

Chapter 5: Boat Maintenance & Care of Sails

surface where the part is fitted. The measurement result here should also lie between -900 and -1340 mV. If the result exceeds this (i.e. is a more positive value such as -800), the proportion of "noble" metals such as stainless steel, bronze etc., is excessive for the zinc anodes to overcome the corrosion current. The anodes should be increased.

The result may also be from stray currents caused by incorrect or incorrectly connected (+) cable or (+) cables exposed to bilge water.

A result of less than -1340 mV indicates overprotection. This could be due to stray currents from incorrectly connected separate ground cables for marine radio and other equipment, or due to the anodes providing too much protection current, such as from magnesium anodes in salt water.

BOAT MAINTENANCE

Whether you do it yourself or employ someone to do it, boats need regular maintenance. The hull, topside and interior need protection from weathering, the bottom needs slipping and antifouling, and yachts need rerigging. This section deals with some specific issues of hull maintenance and repairs.

OSMOSIS

GRP boats are made of fibreglass laminates held together by gelcoat and polyester. Gelcoat and polyester are semi-permeable membranes. When improperly cured they allow water to seep through and dissolve the soluble components in the laminates. This concentrated watery solution is unable to seep back out through them. Instead, it builds up 3 – 5 atmosphere "osmotic pressure", causing blistering of the laminates.

Only parts of the boat in direct contact with water run the risk of osmosis, e.g. the underwater hull. Boats berthed in fresh water run a greater risk of osmosis than those berthed in salt water. This is because the difference between the salinity/acidity of the osmosis blisters and that of water is greater in freshwater.

Osmosis blisters are difficult to detect, but when the antifouling is scraped off, they appear as small moulds, especially on the waterline. If pierced, the contents of the blister will have a sour smell and taste like vinegar. Osmosis blisters are easier to ascertain when the boat is hauled out in autumn. They reduce in size when dried in air, but the osmosis still persists.

If osmosis is allowed to spread, the fibreglass laminate may deteriorate so much that fittings such as sea valves, shaft brackets and keel bolts may begin to leak around the lead-ins

OSMOSIS & GELCOAT REPAIRS

Remove the gelcoat with an angular or eccentric grinder or by sandblasting. Carefully high-pressure hose the exposed laminate with fresh water to remove the blistering product. Allow the laminate to dry completely in correct temperature and humidity range. To check if the laminate is dry, tape a piece of plastic to the surface. If condensation occurs on the inside within 24 hours, drying is still insufficient.

As a result of sandblasting, the exposed laminate will have an uneven, porous and grayish appearance. Fill the cracks and pores with a penetrating sealer and large pores with an epoxy filler. The sealer is best applied by brush. It must not form a glossy surface. Remove any excess sealer with a rag soaked in spirit. When dry, sand it if necessary.

Paint the surface with 5-7 coats of a solvent-free epoxy primer designed for fibreglass and osmosis repairs. This should be done to a recommended thickness either with an airless spray a foam roller. Observe the recommended recoating intervals and temperature range. Temperatures above 35° C increase the risk of bad film formation. The boat should not be launched until the primer has fully cured.

FIBREGLASS LAMINATE REPAIRS

- Grind out all damaged material, making an oval or circular hole with a 12:1 bevelled edge.
- If the hole goes right through the hull, and if inside access is possible, provide a temporary backer (a firm base) for the new laminate. The backer can be made from a sheet of plastic, a pad and a brace. If inside access is not possible, make a permanent laminated backer on a sheet of plastic. Pass it through the hole and bond it to the inside using a thick glue mix.
- Cut the required number of pieces of fibreglass cloth, with increasing sizes to match the bevelled hole. The total thickness should be just under the level of the surrounding laminate.
- Accurately mix the two-pot resin with the hardener, and glue powder if recommended.
- Coat the surface with this thick resin mix, filling all voids.
- Lay the fibreglass cloth into the repair area, using the largest first, wetting each layer with the resin mix.
- Tape a sheet of plastic over the repair and press firmly to remove any excess resin or entrapped air. Remove the plastic sheet when resin is fully cured.
- Sand and overcoat the surface with a sealer or a primer.

NEW FIBREGLASS BOATS

- The surface of a new boat should be cleaned of any residues of wax and mould-releasing agents. Check with the dealer. If it has not yet been done, wash the hull thoroughly with a cleaning agent, dry it and degrease it. Make sure the cleaning agent does not contain ammonia as it can discolour gelcoat. Sand it lightly with 180-grade sandpaper, wash down thoroughly with freshwater and a sponge and dry it.
- To guard it from ultraviolet rays and to prevent it from picking up dirt, apply a marine polish and polish it straight away. A delay in polishing may allow dirt to settle on it and cause scratching when polishing later on. Repeat this process two or three time a season.
- If the boat is to be kept on a mooring or a jetty, polish only the above-waterline hull and antifoul the hull below the waterline. This process also protects the hull from the risk of osmosis by water soaking into the still porous gelcoat.
- If you have to store your boat in the open, do not store it under a plastic cover. Use a canvas cover instead. Sweating caused by plastic will ruin the gelcoat. Similarly, leaving damp suits and chamois draped over the boat to dry will draw colour from the porous gelcoat, ruining it in the process. Leaving water in the bilges and under cover during storage can cause the "hot house" effect to generate moisture which, in due course, can cause mould in the furnishings, damage the electrics, and rust out any metal.
- Always park the boat nose up with the bungs removed to keep any moisture drained (Don't forget to put them back before the next trip).
- Beaching a fibreglass boat can also damage its gelcoat. Any screw holes made in the hull, say, to add a ladder, should be competently sealed with silicon.

ALUMINUM BOATS

Aluminum is anodic to metals such as steel, lead, copper and bronze. In a marine environment it corrodes quickly in the presence of such (cathodic) metals from within the boat or boats moored in close proximity. If she is to be kept on a mooring or a jetty, paint the entire boat. She should be washed with a strong cleaning agent, rinsed with freshwater and lightly sandblasted to remove the loose oxidising layer. Where necessary, the aluminum surface can be roughened by dry rubbing with sand paper. Any irregularities should be filled with an Epoxy filler. The hull below waterline should be antifouled. Do not use alkaline detergents on aluminium.

ALUMINIUM & STAINLESS STEEL FITTINGS

There are many unpainted surfaces on board that are susceptible to galvanic corrosion or oxidation. These include pop rivets on aluminum masts and porthole frames, steel wires and other fittings. Rust is often seen at bends and welds on stainless steel. Normally it does not affect the steel but ruins the appearance. Such surfaces can be protected with a silicone-free penetrating oil or marine polish containing silicone. These coatings must be removed with a degreaser if later you decide to paint the surface.

STEEL BOATS

Steel hulls should be primed immediately after chipping, grinding or sandblasting. A new steel boat needs 4-5 coats of primer and 2-3 coats of finishing paint above the waterline. Below the waterline, apply 6-7 coats of primer followed by 2-3 coats of polishing antifouling.

FERRO HULL REPAIRS

Ferro hulls are cheaper and quicker to repair than any other type of material. The cheapest and simplest option is the cement and sand mixture. For longer lasting repairs, make up a bog of chopped strand and fibreglass. After cleaning the surface as described in surface preparation for painting, trowel the bog into the hole and build it up. This can be ground and sanded back to the original hull. Ready for mixing cement replacement products can be purchased for use as a bog.

LEAKY WOODEN BOATS

A boat that has been out of water for a long time may be too leaky to stay afloat. Thoroughly clean the hull, saturate it with one coat of primer, fill the cracks below the waterline with a sealing compound (that replaces oakum), then paint the entire boat with the primer.

WORM INFESTED BOATS

The best way to get rid of Teredo infestation in a wooden hull is to take the boat out of water. Estimates vary on how long it takes for the worms to die out of water, but two weeks seems acceptable.

TIMBER ROT

The best way to check the condition of timber is to poke it with a knife. Timber can break down for any of the following reasons:

- Fungal attack: Fungus grows on timber when its moisture content is over 20%. Its telltale signs include musty smell, fruiting growth, white threadlike or cotton wool type of growth, softening of timber, paint or varnish failure,

creaking.
- ➤ Termite (White ants) attack: Being internal, the termite damage can remain invisible to the naked eye until it is too late. The best way to safeguard against termite attack is to prevent prolonged contact between timber and the ground.
- ➤ Lyctus Borer: It attacks hardwood of high starch contact. The attack becomes evident with the appearance of tiny heaps of flour-like dust on the surface of the timber. The borer can be exterminated with propriety poisons.
- ➤ Crustacean (or Putty) borers: Their evidence is common on wooden wharf piles, typically wasted away in the shape of "hour glass". They can cause considerable damage to the unprotected underwater section of timber boats.
- ➤ Teredo (or Shipworm): The waterborne tiny larva of this mollusc bores a tiny hole in timber's surface and enters it. Once inside, it grows up to a metre in length by feeding on the inside of the timber while remaining invisible to the naked eye. Timber can be saved from such attacks by treating it with a proprietary preservative, usually containing creosote. Metal sheathing of timber hull is another solution, but the cost is usually prohibitive.

PAINTING

There is an amazing array of paints in paint shops: single pack, two pack, epoxy, polyurethane, varnish, alkyd, water-based, solvent-based. Selecting a paint system for your boat depends on factors such as condition of the current paint, type of substrate, the painting conditions and the level of finish desired. You need to establish a suitable combination of primers, undercoats and finish coats.

TERMINOLOGY
- ➤ EPOXY: A resin capable of forming tough, tight and cross-linked polymer structures. It has strong adhesion and low shrinkage qualities. It makes paints and adhesives tough and chemical resistant.
- ➤ POLYMER: Polymer means "made up of many parts of the same material". For example, plastic is a synthetic polymer, and Protein is a natural polymer.
- ➤ POLYURETHANE: It is a flexible resin. However, polyurethane paints are not as flexing as oil-based paints.
- ➤ ENAMEL: A paint that dries to a hard glossy finish.
- ➤ PRIMER: A coat of paint applied to prepare a surface for painting.
- ➤ UNDERCOAT: A coat of sealing paint applied before the top coat.

MARINE PAINTS & VARNISHES

OIL-BASED / ENAMELS	TWO-PART POLYURETHANE	TWO-POT EPOXY
IN EXISTENCE: For over 100 years.	IN EXISTENCE: Since 1970s	IN EXISTENCE: Modern
CONSTITUENTS: Old Type: Linseed oil & Shellac. Modern Type: Enamel paints containing resin (vegetable oil, glycerine & additives for gloss), solvents, agents for colour, drying, stability & UV.	CONSTITUENTS: Part A = gel. Part B = hardener.	CONSTITUENTS: Part A = gel. Part B = hardener.
APPLICATION: Final coat as well as surface preparation coats such as primers, undercoats and anti-corrosion paints.	APPLICATION: Final coat for areas above the waterline. (Not suitable for areas below the waterline.)	APPLICATION: Surface preparation coats for areas above and below the waterline.
WATERPROOFING: 2 to 5%	WATERPROOFING: 90%	WATERPROOFING: 99%
THINNING & CLEANING UP: Mineral turpentine (Water cleaning oil & plastic paints are not recommended for marine use).	THINNING & CLEANING UP: Special thinning agent (usually expensive).	THINNING & CLEANING UP: Special thinning agent (usually expensive).
ADVANTAGES: Cheaper, easier to apply & to clean up.	ADVANTAGES: Produces a hard, durable finish. Long life (5-7 years)	ADVANTAGES: Long life.

DISADVANTAGES: Short life (2-3 years) SUITABILITY: Suitable for flexing surfaces, such as timber. APPLICATION METHOD: First preference: spay painting. Second preference: short-hair roller and a wide brush for spreading. STORAGE: 1-3 years. Remove skin from the partly-used tin and then stir thoroughly.	DISADVANTAGES: Costly. More difficult to apply. SUITABILITY: Suitable for non-flexing hard surfaces, such as fibreglass, steel and aluminium. APPLICATION METHOD: Some paints can only be spray-painted. Others can be also be applied by roller or brush. Follow manufacturers' instructions. STORAGE: Part A: 2 years, opened or unopened. Part B: Unopened = 1-2 years. Opened = 1-2 months.	DISADVANTAGES: Costly. SUITABILITY: As for two-part polyurethane. However, it is also suitable for coating aluminium and steel hulls below the waterline prior to antifouling. APPLICATION METHOD: Spray painting, roller or brush. Follow manufacturers' instructions. STORAGE: As for two-part polyurethane.

GENERAL PRECAUTIONS
- Painting is best done in fair weather. Avoid early mornings and late afternoons when the condensation can cause problems. For safety and health reasons, work under shelter or out of the hot sun.
- High quality brushes and rollers will produce the best result. A high density/small cell size foam roller (as opposed to mohair and large cell foam rollers) will keep the bubble formation to a minimum. But, the paint applied would be thinner and so more coats may be required.
- Two-pack polyurethane paints, once mixed, should be applied within two hours in hot weather (25°- 30°C) and four hours in cooler weather (up to 10°C). The ideal temperature is around 15°C with low humidity. In warmer conditions, adding a little retarder thinner makes the paint easier to apply. The two-part paint should be left to dry for about 24 hours to allow bubbles generated in mixing to disperse. Do not lift the boat in a sling before the paint is fully cured, which can take a few days.
- The recommended drying time is usually for a temperature of 20°C, the relative humidity of 60-65% and a good ventilation to remove evaporating solvents. For every drop of 10° in temperature, the drying time doubles. Lack of ventilation prevents the paint solvents from evaporating. It can leave the paint uncured and blistered.
- To stop paint becoming too thick in cold weather, place the tin in lukewarm water.
- The zinc anodes on hulls must not be painted. If the boat has been out of water for a long time, the surface of the anodes would have oxidised. It must therefore be ground back to bare metal before launching.

SAFETY
As a general rule, avoid inhaling paint and solvent vapours and contact of liquid paint with skin and eyes. Provide forced ventilation when applying paint in confined spaces or stagnant air. Even when ventilation is provided, respiratory, skin and eye protection are recommended when spray painting. Also take necessary precautions against the risk of fire or explosion.

TESTING EXISTING PAINTWORK
If it was you who did the existing paintwork, the surface is in good condition and you are using the same paint as previously, your job is much more straightforward. However, if there are obvious problems with the existing paintwork or if you are not sure what paint is on the boat, one of the following tests is recommended:

The compatibility of a planned two-pack paint with the existing paintwork can be checked as follows. Soak a soft cloth or cotton-wool ball in the manufacturer-recommended epoxy thinner for the paint you have chosen. Tape the ball over the existing paintwork for approximately one hour, then remove the ball. If the thinner has dissolved, softened or washed away the existing paint then this paint must be removed prior to painting as it is not compatible with your chosen product.

A second test, suitable for both single- and two-pack systems, is to sand a small area of existing paintwork with a fine sand paper. After cleaning it with a pre-painting cleaner, apply a coat of the planned topcoat. Leave this patch for approximately 24 hours. If there are no wrinkles or lifting, it indicates that the existing paintwork is compatible with your new paint.

SURFACE PREPARATION

If the new paint system chosen is not compatible with the existing paint, you must remove all the existing paint and treat the job as if you are painting a new boat. The same applies if the current paint is badly cracked, peeling extensively, heavily chalked, flaking or if there is corrosion or damage to the substrate. If you are in any doubt about what level of preparation is needed, contact your local paint company representative for advice.

Paint can be totally stripped in one of three ways:
- Use a paint stripper
- Water-blast it
- Sand and/or grind it off.

The preferred option will depend on the boat type, surface area, cost, location and the time available.

- If it's a metal hull, check carefully for corrosion. Even the slightest blistering in the paint is usually corrosion on the brink of breaking through. Large blisters or soft spots may indicate a failure of old fairing work.
- On a fibreglass hull, blisters and soft spots may indicate osmosis. Do not paint over such a condition without first consulting an expert and repairing the affected areas.
- If the existing paint is in sound condition, remove all possible contaminants on it with a pre-painting cleaner. Then remove any wax or polish residue with a degreaser, and wash again. Allow the surface to dry these two phases.
- Before painting a new wooden boat, coat it with a wood preserver to prevent (dry) rot. Allow the preserver to completely dry, then fill any cavities, pits and uneven surfaces with an epoxy filler.
- Before painting a plywood surface, sand it and carefully remove dust. Saturate the surface, especially the edges, with a sealer. Remove the surplus to avoid a glossy layer of sealer.

SANDING & FILLING

Sand the clean surface, starting with the 100- to 200-grit sandpaper to remove all gloss from the previous finish. This is the time to feather any dents, scratches or dings. The grade of sandpaper will depend on the surface quality to be painted. If it is necessary to start with a harsher grade, then progressively work to a lighter grade of paper so that any scratch or sanding marks can be minimised.

Fill any indentations or marks with epoxy filler and sand them back. Spot prime these spots and any areas that have been sanded back to substrate level. Use either the manufacturer-recommended epoxy primer or multi-purpose primer undercoat, depending upon the selected topcoat system.

FERRO-CEMENT SURFACE PREPARATION

- Abrasive-blast or high pressure water-jet the surface to obtain a rough and firm surface free of scum and contamination. Remove dust and loose material. Alternately, the surface may be acid etched with a 5% w/w nitric of phosphoric acid solution (avoiding skin contact). Prior to etching, saturate the concrete with fresh water to prevent corrosion of the steel reinforcement. Leave the acid to react 3-4 minutes, then hose down the surface with fresh water – preferably with a 5% w/w sodium hydroxide solution - and scrub carefully.
- Dry the surface homogeneously. It should appear as an even, rough surface, but free of a loose outer layer.
- The surface must have a pH reading of between 6.5 and 8.0.
- Repeat the process if any of these conditions are not fulfilled.
- Dry the surface with good ventilation for at least 2 days at 65% relative humidity and temperature of 20° C.
- Check the pre-treatment by scraping with a strong knife. The surface should feel solid and hard, and the knife must only leave a clear scratch mark.
- Saturate the surface with a sealer. Remove surplus to avoid a glossy layer of sealer after drying.

UNDERCOATING

Undercoating fills in fine scratch marks from sanding and makes the surface adhesive for the topcoat. Two or more coats may be necessary, depending on the condition of the surface, the amount of spot priming and repairs undertaken. Strictly observe recoat times. When in doubt, refer to the product data sheet. This is particularly important at the finish coat phase. When it has dried, sand back each coat to a smooth finish with a 220-440 grit paper. On completion of undercoating, clean the surface thoroughly with the manufacturer-recommended degreasing solvent, and allow it to dry.

APPLYING THE TOP COAT

The top coat is generally a polyurethane or a marine gloss enamel paint. It can be applied in one of four ways: brush, roller or spray, or a combination of the three. Each method has its advantage and disadvantage. Spray painting generally provides the best finish, but it requires a professional to do it in a controlled environment. The do-it-yourself work is done with a brush, roller or a combination of the two. Read the manufacturer specifications closely. Closely follow all steps for surface preparation, recoat timings and the use of appropriate thinners and additives.

The topcoat should be kept dry for at least 24 hours. The presence of moisture will cause improper curing and loss of gloss. Full curing will take further 7-10 days. Only a fully cured paint will provide maximum resistance to abrasion. Do not lift the boat in a sling before the paint is fully cured.

METAL PRIMERS

There are three different mechanisms employed in primer paints to protect metal:
- Cathodic or sacrificial protection
- Inhibitive protection
- Envelope protection

Cathodic protection paints protect metals by offering themselves as sacrificial anode. They corrode in preference to the metal they are applied to. The most common of these are zinc-based or the cold-galvanising type. These primers are normally only used in industrial applications and are not commonly available on the retail market.

The inhibitive primers are the most common variety. The hydrophobic filler used in making these paints tends to repel moisture from the surface of the metal. Since corrosion results basically from the combination of water and oxygen, the paint protects the metal from corrosion by keeping it dry. The most common primer of this type is ZINC PHOSPHATE.

The envelope type primers protect metals by producing a barrier that is highly impervious to water vapour. The primer/finish type coatings are generally of this type.

NOTE: The old red lead primers were of the inhibitive type. Red lead is not metallic lead, but a salt that is highly inert. Unlike the lead metal, which is more noble than steel, they were not cathodic to steel surfaces. Lead based paints have been banned from production under legislation for quite some years.

VARNISHING

- As with painting, a good varnish finish requires good surface preparation.
- Pay particular attention to the end grains and edges of plywood. Being sensitive to moisture, they need to be sealed several times with the correct sealer before varnishing.
- To avoid a glossy surface, any excess sealer should be removed immediately with a rag soaked in spirit, followed by light sanding.
- Wood that has become dark or discoloured can be restored to its original colour with one or more coats of a restorer.
- Wash the surface thoroughly with fresh water and degrease it before varnishing. Do not wipe with a solvent after sanding. It may contaminate the surface.
- For maximum moisture resistance of timber and stability of varnish or paint, give the timber or plywood two coats of an epoxy adhesive, also known as a pre-coat application.
- Avoid direct sun or rising temperatures when coating fresh wood. Air trapped in the pores of the timber will expand and create bubbles under the epoxy coating. It is better to apply epoxy to an already warmed surface and allow it to cool down after it is coated. This will draw the epoxy down into the surface pores.
- For good penetration of varnish, dilute the first coat with thinners by 20%, followed by the second coat diluted by 10%. This should be followed by as many undiluted coats as possible. Ten or twelve coats is quite normal, with the surface lightly sanded between coats.
- To achieve a mirror-like finish and to avoid the formation of air bubbles, apply varnish with slow steady brush strokes.
- Shaking varnish immediately before use will cause air bubbles. Stir it instead.
- Teak has high oil content. Its adhesive property can be improved by degreasing it with a degreaser or roughing it with a very rough sand paper. It should be pre-coated immediately with an epoxy containing a viscosity reducing additive.
- Timbers such as Jarrah oxidize rapidly wherever they are cut. Oxidised cells are not very adhesive. Therefore, pre-coat sawn off sections immediately.

STAINING TIMBER

Use only a spirit-based stain. An oil-based or varnish stain may not allow epoxy pre-coat to stick properly. You can either apply the stain before applying the epoxy, or mix a small amount into the first coat of the epoxy. The latter method would produce a more uniform effect.

MAKING DECKS NON-SLIPPERY

Generally, standard fibreglass boats come with moulded tread marks and non-slip rubberised mats. But many older boats, especially built of ply or timber may need to make their decks non-slippery.

Follow the painting preparation procedure discussed above. The cheaper method is to sprinkle fine (silica) sand over a freshly painted enamel gloss top coat, then repaint over with several more coats. But the easier option is to use a 'Jet Dry' variety of paint that comes with pre-mixed grit.

ANTI-FOULING

Boats sitting stationary on moorings and at marinas become fouled with small marine organisms within a few weeks unless they have been treated with antifouling. The antifoulings release biocide materials that prevent the microscopic organisms from attaching themselves to the vessel and growing. The most common biocide used is copper.

(NOTE: Steel and aluminium will corrode in the presence of copper. Therefore, only the suitably designed copper-based antifoulings should be used on steel and aluminium hulls and aluminium components of propellers, Sterndrives and outdrives. There is usually no problem in using copper-containing antifouling on bronze propellers.)

All antifoulings change colour when they are immersed, because copper compounds are difficult to mask with colour pigments. Along the waterline, antifouling may look dirty or turn green. This is due to copper reacting with oxygen to form copper oxide. Some of the modern additives may reduce this problem.

There are three groups of antifouling, with different properties and suitability for different materials.

1. SOFT (ERODING) ANTIFOULING release their binder along with their biocide. In other words they are water-soluble. They are the least efficient of the three types of antifoulings, but are suitable for most wooden boats, which are not used for racing, and can also be used on fibre glass boats. They dry rather slowly (about 5 hours) but the boat should be launched before they dry out - as soon as the paint is no longer tacky. At the end of the season, scrub off as much of the old paint as possible before treating the bottom again. Apply a boat primer before reapplying antifouling.

2. HARD (CONTACT LEACHING) ANTIFOULINGS are long lasting and very resistant to abrasion and rubbing. They are ideal for fast powerboats and vessels moored in drying out mud berths of fast tidal movements. Hard antifoulings do not release their binder along with the biocides. They leave behind a porous cell structure, which needs to be wet sanded before the application of a new coat. One disadvantage of the hard antifoulings is that unless the residue is properly removed, it can build-up season after season. These antifoulings can be used on all types of hulls, except on old wooden boats where there is excessive expansion and contraction of wood. They can be sanded to a smooth finish, resulting in a 'fast" bottom. They are also fast drying (about 2 hours) and the boat can also be left ashore for up to three months or for the whole winter if protected from sun and rain.

3. POLISHING ANTIFOULINGS are the more advanced type of hard antifoulings. They are called polishing antifoulings because they get "polished" by the boat's speed through the water. They thus constantly release a fresh surface of the biocide. The boat can be launched 12 to 24 hours after applying a polishing antifouling. They are suitable for most hulls, but not for aluminium.

Only the hard or polishing antifoulings are suitable for propellers, due to their high speeds and turbulence generated wear.

APPLICATION OF ANTIFOULING ON NEW BOATS

The underwater hull of a new boat is first painted with four or five coats of an epoxy primer, followed by two or three coats of an antifouling. Fouling develops faster in sunlight. An extra coat of antifouling should therefore be applied at the waterline and on the rudder, which are the areas of the greatest wear. Do not use oil-based primers below the waterline because they are not resistant to constant water submersion.

For regular maintenance, apply two or three coats of antifouling annually on top of one or two fresh coats of a primer. One recommendation is to apply a coat of hard antifouling followed by a coat of soft antifouling.

REMOVAL OF OLD ANTIFOULING:

Soft antifoulings can be removed with a boat cleaner, but hard and polishing antifoulings have to be removed with a paint remover. Recommended personal and environmental safety precautions should be taken.

CHANGE OF ANTIFOULING

A polishing antifouling can generally be applied directly on top of a hard antifouling after wet sanding and hosing down with fresh water. But, a hard antifouling cannot be applied on a soft or a polishing antifouling. The latter must first be completely removed and a new coat of primer must be applied before painting with a hard antifouling.

ENVIRONMENTAL DAMAGE FROM ANTIFOULING

Antifouling prevents marine growth on the hull, but it is also generally damaging to the environment. Law is continually restricting the permitted leakage of the active ingredients in paints. Australia, does not allow the use of mercury-based fungicides or tin-based (TBT or tributyl tin) antifouling paints. Sweden has banned even the use of copper-based antifouling on its Baltic coastline.

Regulations for small vessels are stricter than for large vessels. The reason given is that small vessel harbours are mostly located in shallow waters or fresh waters, which are also the spawning ground for fish. Furthermore, small vessels spend most of their time tied up in harbours. This adds to the impact on the environment of these waters.

A little thoughtfulness before using an anti-fouling paint goes a long way. For example, trailer boats, which spend most of their life out of water, do not need to use them. All they need is a Teflon treatment, combined with sponge cleaning a few times during the season.

FUEL SAVING ALTERNATIVE TO ANTIFOULING

LANOLIN is made from wool grease – the water-repellent that protects sheep from harsh effects of weather. It is now commercially available in cans. When coated on hulls of vessels, it makes them slippery. It is claimed to not only make boats go faster, but also eliminate the need for anti-fouling.

NON-TBT ANTIFOULINGS (some still under trial)
1. Foulant-Release Coatings: These so called 'non-stick coatings' are silicon elastomers. Their service life is claimed to be similar to that of TBT, but are costly.
2. Ceramic-Epoxy Coatings: Suitable for fibreglass as well as metal hulls.
3. Epoxy-Copper Flake Paints: A bonded-copper system, which claims to protect boats for up to 10 years but cannot be used on aluminium hulls.
4. Electric Current Systems: They produce hyperchlorite from seawater on the surface of the hull, which is said to sterilise the surface for up to 4 years. Another similar system uses alternating current and a conductive hull coating.
5. Biological Compounds: About 50 natural substances are being trialled.

SLIPPING

TYPES OF SLIPS
There are six main types of slips or dry docks:
1. Crane or Travel lift
2. Cradle
3. Graving dock
4. Floating dock
5. Careening
6. Hard Stand

1. CRANE LIFT OR TRAVEL LIFT (Fig 5.6)
Lifting a boat out of water by a crane is a common method of slipping. Travel Lifts with slings can lift boats of considerable size.

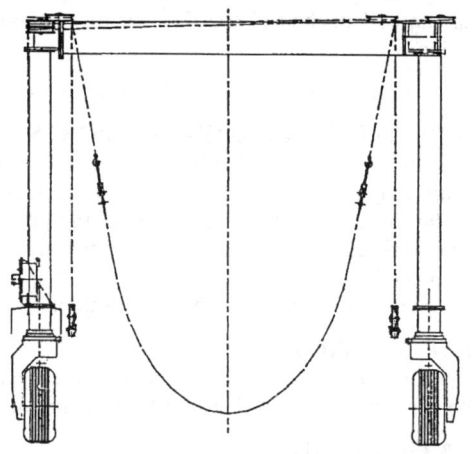

2. CRADLE (PATENT SLIP) (MARINE RAILWAY) (Fig 5.7)
Another common method is to pull the boat out of water by floating it on wheeled cradle. Once the boat is secured in place, the cradle is wheeled in on a sloping railway track with the help of a winch.

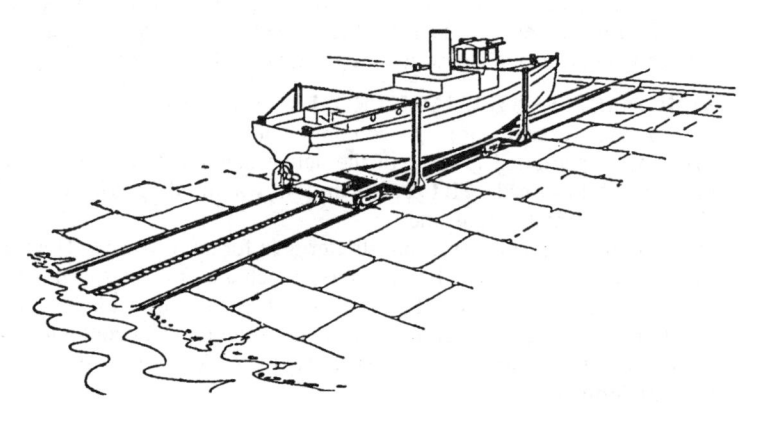

Chapter 5: Boat Maintenance & Care of Sails

3. GRAVING DOCK (Fig 5.8)

It is commonly known as a dry dock - used by ships rather than boats. It is excavated from the land and closed with a Lock Gate or Caisson. The ship is floated in and lined up in the middle, over a row of wooden blocks. The lock gate closed and water is pumped out. Ship settles on the wooden blocks and is shored on the sides.

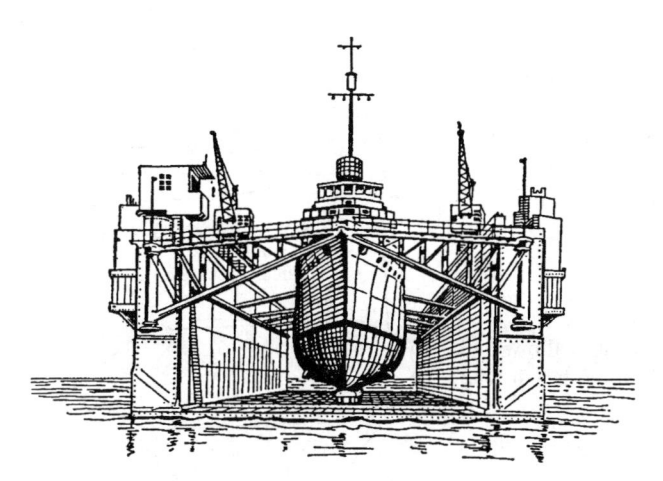

4. FLOATING DOCK (Fig 5.9)

This too is used for ships. It is a large flat pontoon tank with tall sides, which are also tanks. The whole dock forms a floating watertight structure, which can be submerged by flooding the tanks.

While the dock is submerged, the ship is floated into it and positioned above the keel blocks with the help of docking (mooring) lines. Shores are fitted to provide side support. The pontoon is then pumped out until its floor is dry.

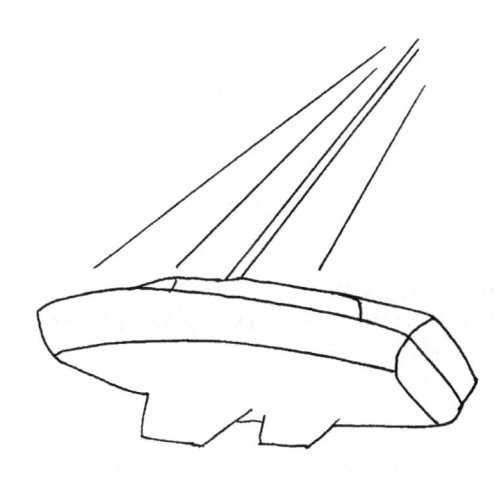

5. CAREENING (Fig 5.10)

Careening is another method of exposing a boat's bottom for cleaning or repairs when a slip is not available. It is usually done on a beach or a riverbank.

The boat in light condition is heeled at high tide. This is done by heaving her down using a tackle from one or more trees or other shore structures. Work on the underwater hull is then carried out when the tide goes down. See also Planned Grounding (Beaching) in Chapter 10.

6. HARD STAND (Fig 5.11)

A hard-standing cradle may be a permanent structure or a makeshift arrangement. It is usually cheaper and easier to slip the boat with one of the ways described above, than to construct a makeshift hard-standing cradle and pull the boat out of water. It is a skilled and arduous operation, and requires a hard-shelving ground and a large tidal range. This method of support is suitable only for vessels with a long and fairly straight keel.

(A) A PERMANENT CRADLE **(B) A MAKESHIFT CRADLE**

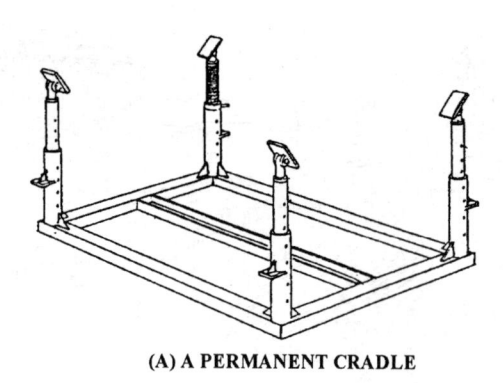

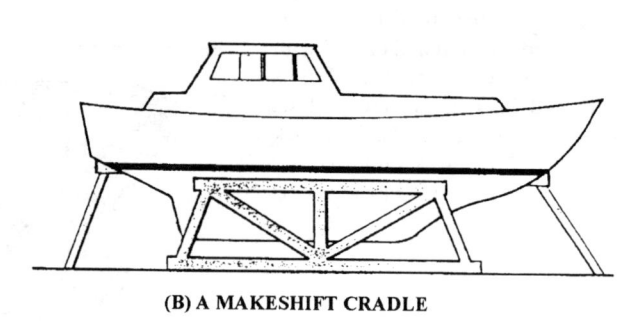

PRECAUTIONS WHEN CAREENING
(See also Planned Grounding or Beaching in Chapter 10)
- Choose a gently shelving and sheltered beach.
- Make sure there are no stones or sharp objects to damage her lower side.
- Don't heave her so far as to roll her over.
- Don't beach her at a higher high tide. Refloating may become a problem.
- Her Centre of Gravity should be made as low as possible for stability.
- Make her as light as possible by removing all removable heavy gear.
- Lower and secure any cranes and derricks.
- Ensure she will not flood on the rising tide.

USE OF BALTIC MOOR FOR CAREENING (Fig 5.12)
The anchor is dropped 3 to 4 boat lengths offshore. The boat is allowed to drift ashore until it touches the bottom. She is then secured on the beach with mooring lines. (See Boat Handling for Baltic Moor)

Secured by mooring lines on the beach (to trees, rocks, etc.)

Baltic moor offshore

PREPARATION FOR SLIPPING
- Provide the Dock Master with a copy of the Vessel's Docking Plan.
- Ensure keel and bilge blocks are properly sized and positioned.
- Make the vessel as light as possible.
- Trim her slightly by the stern.
- Make sure that the chocks and side shoring on the slipway will not rest on the vessel's outfittings, such as external keel coolers, log and echo sounders.
- Withdraw the log and other protrusions into the hull before docking.
- She should have positive stability. A top-heavy boat may fall over.
- Make a repair/job list.
- Organize the surveyor's visit - if the boat is in survey or if an inspection is required.
- Tanks should be either full or empty to minimise free surface effect. Tanks to be surveyed must be emptied and ventilated thoroughly. Drain plugs to be removed and placed in safe custody.
- Remove or secure all movable weights.
- Bilges should be pumped dry.
- Have all necessary spares and packings available.
- Before painting the underwater hull, cover transducers, anodes and other non-paintable fittings with grease and masking tape.

SLIPWAY SAFETY
- Provide a safe gangway.
- Provide safe lighting.
- No unauthorised boarding.
- Tools secured to the body with a lanyard to prevent them from falling over someone inside or under the vessel.
- Everyone to wear hardhats.
- Close off bathrooms and toilets, and plug overboard discharges.

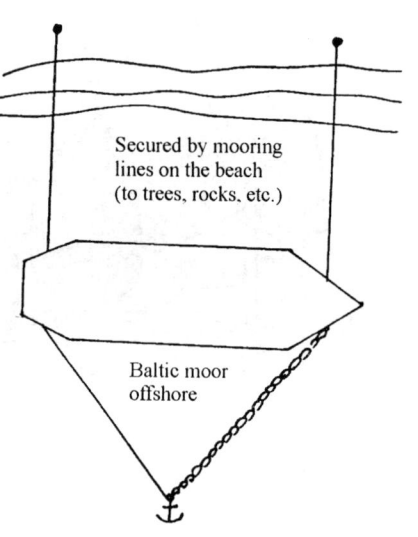

Chapter 5: Boat Maintenance & Care of Sails

HULL STRESSES WHILE ON A SLIP

A boat's hull is designed to be surrounded and supported by water. When taken out of water, it tends to bulge or bend outwards. This stress is particularly significant if the boat is loaded. Ships usually do not dry dock with cargo on board.

The bending and bulging stresses due to slipping can be minimised by supporting the hull on sides with cradle support beams and shores. These are placed on bulkheads and the frame lines. Placing them between the frames may cause the hull to buckle.

A vessel sitting on keel and bilge blocks on a slipway is also subjected to considerable local stresses. Her entire weight is supported at a few points. These stresses can be minimised by supporting her with as many blocks as possible.

COMING OFF A SLIPWAY

- Make sure the main engine is in working order.
- Maintain original stability and trim (see caution below).
- Replace all plugs in the hull and tank bottoms.
- Disconnect shore power and all other lines.
- Check the tide and wind conditions.
- Make sure the paint is dry.
- Operate all valves over full range from open to close. Use feeler gauges to check proper seating of valves with flanged covers.
- Replace all suction grids and sea strainers.
- Replace locking devices for propeller and coupling.
- Ensure rudder is free to move.

CAUTION: Vessel should be brought off the slip with the same conditions of stability and trim with which she went in. Before refloating her, all weights that may have been moved should be replaced and tanks filled or emptied as they were on slipping. This is to ensure her stability and trim are the same as for docking.

Immediately on entering water, open valves and take hull soundings to ensure that there is no leakage. Examine stern gland and adjust as necessary.

TRAILER MAINTENANCE

- Keep the trailer under cover and parked on concrete or paved area when idle.
- The trailers and wheels are normally galvanised. However, it does not make them impervious to rust. After each outing, hose down the parts subject to immersion and paint them with grease or fish oil: for example, wheel hubs and rims, axle and sub frame. Remove any corrosion with a wire brush and apply re-galvanising paint where necessary. Check high stress areas for cracks and for signs of rusty coloured fluid run offs.
- The clear view hub oilers may be better than the grease-filled axle assemblies. Their clear view lens gives an instant picture of the status of the hub and the quality and state of the oil in the hub. They keep the rear seal permanently lubricated in oil.
- Check the tow bar assembly at least quarterly for wear and tear. Spray the linkage coupling with WD 40 or similar and grease the locking assembly before each outing.
- Regularly check wheel nut security, tyre pressure and wear. Check wheel bearings by checking for play in the wheel. Keep them filled with grease.
- Remove build up of sand and road grime from joints between the springs and the main frame. Pump in grease regularly through grease nipples at these locations.
- Keep the brakes serviced. Keep the cables and rollers lubricated.
- Spray the winch with WD 40 or similar, and grease all shackles. Replace frayed cables immediately. Check the bolts that secure the winch to the trailer for tension, and the winch "stem" for cracking.
- Check the boat securing straps for rot and grease the metal tie downs.
- Do not immerse the non-submersible lights and other electrical components in the water. Check them before each outing.
- Replace deteriorated rollers, and adjust their height for even distribution of load. Grease the bolts securing them.

Fig 5.13: RUSTED AND WEAKENED TRAILER FRAME, WHEELS AND BOLTS, AND WORN ROLLER
(Photos: courtesy Powerboat Fishing)

SAILS – TAKING CARE OF THEM

BREAKING IN
The first time you use your sails should be in the weather conditions for which they were designed. Do not go out and put up your new sails in a howling gale the very first time. Learn how to sheet and trim them in a good breeze of 10 to 18 knots. Get to know the correct luff tensions and sheeting positions for various breezes. In other words, get to know your new sails before you go out racing. Knowing their sheeting positions beforehand will save you time from experimenting on the racetrack.

DURING THE SEASON
Do not put away damp or wet sails into sail bags. Synthetic sailcloth does not rot, but it does mildew. Although the cloth is treated, it will still mildew if put away when damp. If drying the sail on board is difficult, each crewmember could take one sail home and dry it. They don't need to be thoroughly washed every time, but if wet with seawater, hosing off the salt will help. The salt does not cause any damage, but sails encrusted with salt are more difficult to dry. They also tend to get hard and become difficult to pack. After the sails are dry, fold them, roll them and put them into their bags. To fold and roll a sail, lay the foot of the sail on the deck (or ground) and flake the rest of the sail on to the foot and roll it loosely.

Some boats with grooved headstay systems have 'turtle bags'. The sails are automatically flaked and put into them. The mainsail should always be folded. The easiest way to do this is to flake it over the main boom, then roll and put in the bag. If leaving the mainsail on the boom, ease the outhaul (the hauling rope), remove all the battens and cover it. Leaving it exposed to the ultraviolet sunlight will weaken the fabric and tear it more easily.

END OF SEASON MAINTENANCE
At the end of each season take your sails ashore to wash and dry them. You can wash them in a bath, on a concrete path or a lawn. Take them to a sailmaker to check and store them, if you are short of space. Many Sailmakers offer complete overhaul and reconditioning service during winter months.

STAIN REMOVAL

- Blood: Soak the stain in cold water containing one cup of ammonia per 4 litres of water. If a residual stain is still present, damp it with 1% solution of ammonia in water, allow it to stand without drying for 30 minutes, then rinse out thoroughly.

- Mildew: Remove as much of the mould as possible by scrubbing lightly with a dry stiff brush. Soak the stain for 2 hours in a 1% solution of bleach (sodium hypo-chlorite) and cold water. A proprietary brand of liquid bleach may be used according to manufacturers instructions. Wash thoroughly in water and repeat the treatment if necessary. If after final washing there is a residual smell of Chlorine this may be removed by immersing for a few minutes in a 1% solution of sodium thio-sulphate (photographers' hypo). Rinse thoroughly with water.

- Oil, Grease and Wax: Small stains of this nature can be removed by dabbing with tri-chloro-ethylene or a proprietary stain remover. For heavy stains brush on a mixture of one part strong detergent with two parts solvent, such as benzine or white spirits. After brushing it well into the fabric, leave it for about 15 minutes, and then wash off with warm water. Do this in a well-ventilated place, and remember that most solvents are flammable.
  This treatment will remove oils, greases, petroleum jelly and most lubricating mixtures, but they will not remove stains caused by the fine metallic particles in used lubricants. Such stains can be removed by the methods described below, after the oil or grease has been eliminated.

- Metallic stains: Stains caused by metals in the form of rust, verdigris (green film on copper, brass or bronze) or finally divided particles can be removed by either of the following methods:
  - Immerse the stain in a 5% solution of oxalic acid dissolved in hot water (50 grams per litre). Oxalic acid is poisonous. Thoroughly wash your hands and the fabric after using the solution.
  - Immerse the stain in a warm water solution containing 2% concentrated hydrochloric acid. Wash your hands and the fabric after using the solution.

- Pitch and Tar: Dab the stain with an organic solvent, such as per-chloro-ethylene, tri-chloro-ethylene, tri-chloro-ethane or white spirits. Do this in a well-ventilated place, and take precautions with flammable solvents.

- Varnish: Dab the stain first with tri-chloro-ethylene and then with a mixture of equal parts of acetone and amyl acetate. Shellac varnish is easily removed with alcohol or methylated spirit. Alkies-based paint strippers should not be used on these fabrics. Thoroughly wash your hands and the fabric after using the solution.

EMERGENCY HULL REPAIR KITS

A prudent skipper always carries an emergency pack of relevant material and tools for emergency hull repairs. Quick setting materials in the form of putties are available for fibreglass, timber and ferro hulls.

OTHER HULL MAINTENANCE MATTERS

The answers to some of the questions at the end of this chapter do not require an explanation.

CHAPTER 5: QUESTIONS & ANSWERS

BOAT MAINTENANCE

1. A paint barrier would keep the following type of corrosion to a minimum:

    (a) All types of corrosion.
    (b) Galvanic corrosion.
    (c) Stray current corrosion.
    (d) Crevice & deposit corrosion.

2. Galvanic corrosion is caused by:

    (a) stray current.
    (b) electrolysis.
    (c) the presence of two anodes.
    (d) the presence of two cathodes.

3. Galvanic corrosion results in:

    (a) noble metals corroding in the presence of ignoble metals.
    (b) bronze propeller corroding in the presence of steel shaft.
    (c) steel shaft corroding in the presence of bronze propeller.
    (d) two dissimilar metals corroding together.

4. To provide cathodic protection against corrosion is to:

    (a) paint the metal parts.
    (b) provide sacrificial anodes.
    (c) use stainless steel.
    (d) use more noble metals.

5. To minimise galvanic corrosion:

    (a) paint all anodes.
    (b) all metals components below the waterline must be in contact with each other.
    (c) use lead-based paints on aluminium hulls.
    (d) do not ground electrical equipment to the engine.

6. The stray current corrosion:

    (a) can be caused by faults in the vessel's electrical system.
    (b) can only result from the presence of other vessels nearby.
    (c) cannot result from using the metal keel as a return conductor.
    (d) cannot result from electrical sockets being located in moist places.

7. The function of zinc anodes on a vessel's hull is to reduce corrosion of:

    (a) metals more noble than zinc.
    (b) metals less noble than zinc.
    (c) all metals.
    (d) bronze & steel in the presence of aluminium.

8. Lead or copper oxide primer paints are not suitable for use on aluminium hulls because:

    (a) paint will act as anode.
    (b) hull will act as anode.
    (c) zinc anodes fitted to hull will act as cathode.
    (d) zinc anodes will corrode both hull & paint.

9. Antifouling paints containing more noble metals than the hull itself should only be applied to:

    (a) bare metal hull surfaces.
    (b) metal hull surfaces coated with a primer.
    (c) non-metal surfaces.
    (d) metal surfaces coated with a top coat.

10. A copper coin deposited in the bilge of an aluminium hull will result in:

    (a) the coin being corroded.
    (b) the hull being corroded.
    (c) no corrosion.
    (d) both metals corroding.

11. A zinc primer paint is:

    (a) a sealant.
    (b) porous.
    (c) porous until an electrolytic film clogs it.
    (d) cathodic on steel.

12. A chlorinated rubber or an epoxy paint is:

    (a) a sealant.
    (b) porous.
    (c) porous until an electrolytic film clogs it.
    (d) anodic on steel.

13. Weathering of a fibreglass (GRP) hull means:

    (a) the erosion of its gel coat surface.
    (b) osmosis.
    (c) cracking.
    (d) stripping off its polish or paint.

14. Dry rot does not usually occur in timber if:

    (a) its moisture content is over 20%
    (b) its moisture content is under 20%
    (c) it is wet.
    (d) it is in water.

15. Dry rot occurs only in:

    (a) dry timber.
    (b) wet timber.
    (c) timber in water.
    (d) soft wood.

16. Dry rot differs from wet rot in that:

    (a) it affects only dry timber.
    (b) it is a drier version of wet rot.
    (c) it affects both dry & wet timber.
    (d) unlike wet rot it is not a fungal attack.

Chapter 5: Boat Maintenance & Care of Sails

17. A Teredo:
    (i) cannot survive in dry timber.
    (ii) is poisoned by antifouling paint.
    (iii) remains microscopic throughout its life.
    (iv) leaves only tiny holes on timber surface.

    The correct answer is:
    (a) all except (i).
    (b) all except (ii).
    (c) all except (iii).
    (d) all except (iv).

18. In relation to Osmosis, which of the following statements is incorrect?

    (a) Only parts of the GRP vessel in direct contact with water run the risk of osmosis.
    (b) Vessels berthed in seawater run a greater risk of osmosis than those in fresh water.
    (c) Osmosis is caused by improper curing of gelcoat & polyester.
    (d) Osmosis blisters reduce in size when dried in air

19. In relation to new Fibreglass vessel, which of the following statements is incorrect?

    (a) Surface should be cleaned of any residue of wax and mould-releasing agents.
    (b) Cleaning agents containing Ammonia can discolour gelcoat.
    (c) For vessel to be kept on mooring or jetty, apply polish to the hull above-waterline & anti-foul below the waterline
    (d) For vessel to be stored in the open, store under a plastic cover.

20. In relation to Aluminium vessels, which of the following statements is incorrect?

    (a) Use only the alkaline detergents.
    (b) If vessel is to be kept on a mooring or jetty, paint the entire hull.
    (c) If vessel is to be kept on a mooring or jetty, antifoul the hull below waterline.
    (d) In marine environment, aluminium corrodes in the presence of steel, lead or bronze.

21. In relation to Steel and Ferro-cement vessels, which of the following statements is incorrect?

    (a) Ferro hulls can be repaired with mixture of cement & sand.
    (b) Ferro hulls can be repaired with a bog of chopped strand & fibreglass.
    (c) Steel should not be primed immediately after chipping, grinding or sandblasting.
    (d) Below the waterline, new steel hull needs 6-7 coats of primer followed by 2-3 coats of antifouling.

22. In relation to timber hulls, which of the following statements is incorrect?

    (a) The cure for leaky hull is to leave it out of water for two or three weeks.
    (b) Before launching a wooden hull that has been out of water for a long time, seal all cracks with sealing compound & paint it with primer.
    (c) The best way to check timber hull for rot is to poke it with knife.
    (d) The cure for Teredo infestation is to leave the vessel out of water for two or three weeks.

23. In relation to timber vessels, which of the following statements is incorrect?

    (a) Fungus grows on timber when its moisture content is over 20%
    (b) Underwater hull is susceptible to damage by Crustacean Borers.
    (c) Teredo is also known as Shipworm.
    (d) Teredo's entry points into hull are quite large & visible to naked eye.

24. In relation to paints, which of the following statements is incorrect?

    (a) Epoxy resin makes paints & adhesives tough & chemical resistant.
    (b) Polyurethane paints are more flexible than oil-based paints.
    (c) Enamel paints dry to a hard glossy surface.
    (d) The two-part Polyurethane & Epoxy paints are suitable for non-flexing hard surfaces.

25. In relation to painting, which of the following statements is incorrect?

    (a) The ideal time to paint is early mornings & late afternoons.
    (b) The ideal temperature for applying two-pack polyurethane paints is 15ºC.
    (c) Lack of ventilation can leave paint uncured, leading to blisters.
    (d) The compatibility of a proposed paint with the existing paintwork can be checked with cotton wool soaked in epoxy thinner for the chosen paint.

26. In relation to hull condition, which of the following statements is incorrect?

    (a) After painting a new timber vessel, coat it with a wood preserver.
    (b) In metal hulls, corrosion shows up as paint blisters.
    (c) In fibreglass hulls, osmosis shows up as paint blisters.
    (d) In fibreglass hulls, osmosis shows up as soft spots.

27. In relation to painting undercoats, which of the following statements is incorrect?

    (a) Degrease the surface on completion of undercoating.
    (b) Do not sand back the surface between coats.
    (c) Strictly observe recoat times.
    (d) The number of coats depends on the condition of the surface.

28. In relation to applying topcoat of paint, which of the following statements is incorrect?

    (a) It is usually polyurethane or marine gloss enamel paint.
    (b) It can be applied with brush, roller or spray or a combination of three.
    (c) Presence of moisture causes improper curing.
    (d) Full curing takes 24 hours.

29. In relation to Metal Primers, which of the following statements is incorrect?

    (a) Cathodic Primers protect metal by offering themselves as sacrificial anodes.
    (b) Cathodic Primers are the most commonly used marine primers.
    (c) Inhibitive Primers repel moisture from metal surface.
    (d) Envelope Primers provide a water impervious barrier.

30. In relation to the Inhibitive Type Metal Primers, which of the following statements is incorrect?

    (a) They are the most common variety in marine use.
    (b) Zinc phosphate is an Inhibitive Primer.
    (c) They are the least common variety in marine use.
    (d) They are hydrophobic.

31. In relation to Varnishing, which of the following statements is correct?

    (a) Do not seal the end grains and edges of plywood prior to varnishing.
    (b) Do not wash the surface prior to varnishing.
    (c) Dilute the first coat with thinners.
    (d) It is best to coat fresh wood in rising temperature.

32. In relation to Varnishing, which of the following statements is correct?

    (a) Do not degrease teak prior to varnishing.
    (b) Apply varnish with fast brush strokes.
    (c) Shake varnish immediately before use.
    (d) Pre-coat immediately the sawn off sections of Jarrah.

33. In relation to Staining timber, which of the following statements is correct?

    (a) Use only a spirit-based stain.
    (b) Use only an oil-based stain.
    (c) Use only a varnish stain.
    (d) Apply stain after applying epoxy.

34. In making decks non-slippery, which of the following statements is incorrect?

    (a) Standard fibreglass boats come with moulded tread marks.
    (b) Sprinkle fine silica over a freshly painted topcoat, and then repaint with several more coats.
    (c) Sprinkle fine silica over freshly painted topcoat.
    (d) Apply paint that comes with pre-mixed grit.

35. In relation to antifouling, which of the following statements is incorrect?

    (a) Copper is the most common biocide in antifoulings.
    (b) Copper is not used in antifoulings designed for steel or aluminium surfaces.
    (c) All antifoulings change colour when immersed.
    (d) Copper reacting with oxygen can cause antifouling along the waterline to appear dirty.

36. In relation to Soft Antifoulings, which of the following statements is correct?

    (a) They do not release their binder along with the biocide.
    (b) They are longer lasting than the hard type.
    (c) The vessel should not be launched until the antifouling is totally dry.
    (d) Apply a primer before retreating the bottom with antifouling.

37. In relation to Hard Antifoulings, which of the following statements is correct?

    (a) They are the Contact Leaching type.
    (b) They are suitable for wooden hulls.
    (c) The vessel must be launched before the antifouling dries out.
    (d) They are not suitable for fast powerboats.

38. In relation to Polishing Antifoulings, which of the following statements is correct?

    (a) They are suitable only for aluminium hulls.
    (b) They are not suitable for propellers.
    (c) They constantly release a fresh surface of the biocide.
    (d) The vessel must be launched before the antifouling dries out.

Chapter 5: Boat Maintenance & Care of Sails

39. In relation to Antifoulings, which of the following statements is correct?

    (a) Apply antifouling only to bare hull.
    (b) Use only an oil-base primer under the antifouling.
    (c) Apply an extra coat of antifouling at the waterline.
    (d) Fouling grows faster in the dark.

40. In relation to Antifoulings, which of the following statements is correct?

    (a) Polishing Antifouling can be applied on top of Hard Antifouling.
    (b) Hard Antifouling can be applied on top of Soft Antifouling.
    (c) Hard Antifouling can be applied on top of Polishing Antifouling.
    (d) Polishing Antifouling can be applied on top of Soft Antifouling.

41. When slipping a vessel, the following condition should be avoided:

    (a) a high centre of gravity.
    (b) empty tanks.
    (c) stern trim.
    (d) full tanks.

42. Before a vessel enters water following a slipping, the following precautions should be taken:

    (a) lower her centre of gravity.
    (b) raise her centre of gravity.
    (c) replace all weights as they were on slipping.
    (d) empty all tanks.

43. When Careening, the following precaution should be observed:

    (a) Beach the vessel at a Higher High Water.
    (b) Select a steep beach.
    (c) The vessel should be loaded to make her heavy.
    (d) The vessel's Centre of Gravity should be low.

CHAPTER 5: ANSWERS

1 (d),   2 (b),   3 (c),   4 (b),   5 (d),
6 (a),   7 (a),   8 (b),   9 (b),   10 (b),
11 (c), 12 (a), 13 (a), 14 (b), 15 (b),
16 (b), 17 (c), 18 (b), 19 (d), 20 (a),
21 (c), 22 (a), 23 (d), 24 (b), 25 (a),
26 (a), 27 (b), 28 (d), 29 (b), 30 (c),
31 (c), 32 (d), 33 (a), 34 (c), 35 (b),
36 (d), 37 (a), 38 (c), 39 (c), 40 (a),
41 (a), 42 (c), 43 (d)

Chapter 6

COLOUR ILLUSTRATIONS OF...

International Flags and Pennants
Marker buoys for safer boating
Fire Extinguishers
Pyrotechnic Distress Signals
Symbols & Abbreviations Used on Charts
IALA Buoyage System
Light Characteristics

A	•—	M	——	Y	—•——
B	—•••	N	—•	Z	——••
C	—•—•	O	———	1	•————
D	—••	P	•——•	2	••———
E	•	Q	——•—	3	•••——
F	••—•	R	•—•	4	••••—
G	——•	S	•••	5	•••••
H	••••	T	—	6	—••••
I	••	U	••—	7	——•••
J	•———	V	•••—	8	———••
K	—•—	W	•——	9	————•
L	•—••	X	—••—	0	—————

MORSE CODE (Fig 6.1)

Chapter 6: Buoyage System & Other Colour Illustrations

INTERNATIONAL CODE FLAGS

Fig 6.2

NUMERAL PENDANTS

SUBSTITUTES

FIRST SUBSTITUTE   SECOND SUBSTITUTE   THIRD SUBSTITUTE

All boats should be able to recognize and know the meanings of at least flags A, B, D, O and the combination NC.

The meaning previously assigned to flag "R" has been removed. However, the Morse letter "R" (1 short, 1 long, 1 short blast) may be used by vessels at anchor to warn another vessel approaching.

VICTORIA'S INLAND WATERS
Marker Buoys for Safer Boating

The marker buoy system of defining zoned water areas is now in common use on Victoria's inland waters.

RED
STOP – NO BOATS or SWIMMING
– NO BOATS: Used to mark prohibited water and swimming areas.

YELLOW
SPEED RESTRICTION: An area is set aside as a speed restriction zone because excessive speed is a risk to the operator, to the other boats or persons, or to the environment. The yellow buoys may be placed because of local or general requirements for slower speeds.

GREEN
ACCESS LANE: The waters between these buoys are unrestricted to allow skiers and boats to approach or depart from the shore at speed.

RED AND YELLOW
SPECIAL PURPOSE: These unmarked buoys are used to signify regatta areas, hazards, channels, etc.

MINI-BUOYS
Small mini-buoys of the same colour may be used in conjunction with the larger buoys to demarcate an area.

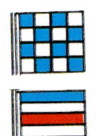

ANSWERING PENDANT

I am in distress; require immediate assistance.

You should proceed at slow speed when passing me.

See Phonetic Alphabet in the Radio Chapter

A	I have a diver down; keep well clear at slow speed.	J	I am on fire and have dangerous cargo on board; keep clear of me.
B	I am taking in, or discharging, or carrying dangerous goods.	K	I wish to communicate with you.
C	Yes.	L	You should stop your vessel instantly.
D	Keep clear of me; I am manoeuvring with difficulty.	M	My vessel is stopped and making no way through the water.
E	I am altering my course to starboard.	N	No
F	I am disabled; communicate with me.	O	Man overboard.
G	I require a pilot. *(When shown by a fishing vessel: I am hauling nets)*	P	All persons should report on board as the vessel is about to proceed to sea. *(My nets have come fast upon an obstruction)*
H	I have a pilot on board.		
I	I am altering my course to port.		
Q	My vessel is "healthy" and I request free pratique.		
R	(No meaning)		
S	My engines are going astern.		
T	*Keep clear of me; I am engaged in pair trawling.*		
U	You are running into danger.		
V	I require assistance.		
W	I require medical assistance.		
X	Stop carrying out your intentions and watch for my signals.		
Y	I am dragging my anchor.		
Z	I require a tug. *(I am shooting nets)*		

Fig 6.3 **FIRE EXTINGUISHERS**

INDICATOR	CLASS OF FIRE →		A	B	C	(E)	F	SPECIAL NOTES
POST 1995 (Fire extinguisher sign)	TYPE OF FIRE →		Ordinary combustibles (wood, paper, plastics, etc.)	Flammable and combustible liquids	Flammable gases	Fire involving energized electrical equipment	Fire involving cooking oils and fats	
PRE 1995	IDENTIFYING COLOURS	TYPE OF EXTINGUISHER	↓ EXTINGUISHER SUITABILITY ↓			↓	↓	↓
(red triangle, white)	silver/red bottles	WATER	YES	NO	NO	NO	NO	Dangerous if used on electrical fires
(red triangle, orange)	red/yellow bottles	WET CHEMICAL — Vapours can cause respiratory distress	YES	NO	NO	NO	YES	Dangerous if used on electrical fires
(red triangle, blue)	red/blue bottles	FOAM	YES	YES	NO	NO	NO	
(red triangle, white)	red bottle white band	AB(E) DRY CHEMICAL POWDER	YES	YES	YES	YES	NO	
		B(E) DRY CHEMICAL POWDER	*NO	YES	YES	YES	YES	*May be used on small surface fires.
(red triangle, black)	red bottle black band	CARBON DIOXIDE (CO₂)	*NO	NO	NO	YES	YES	*May be used on small surface fires.
(red triangle, yellow)	red bottle yellow band	VAPOURIZING LIQUID — Fumes may be dangerous in confined spaces	*YES	5kg only YES	NO	YES	NO	*Vapourizing Liquid extinguishers are not suitable for smouldering deep seated A class fires.

NOTE → CLASS **'D'** fires (involving metals e.g. magnesium) - use special purpose extinguishers only.

EXTINGUISHERS MUST BE RECHARGED AFTER USE

Copyright © 1998 NSW Fire Brigades

Chapter 6: Buoyage System & Other Colour Illustrations

Fig 6.4   **PYROTECHNIC DISTRESS SIGNALS**

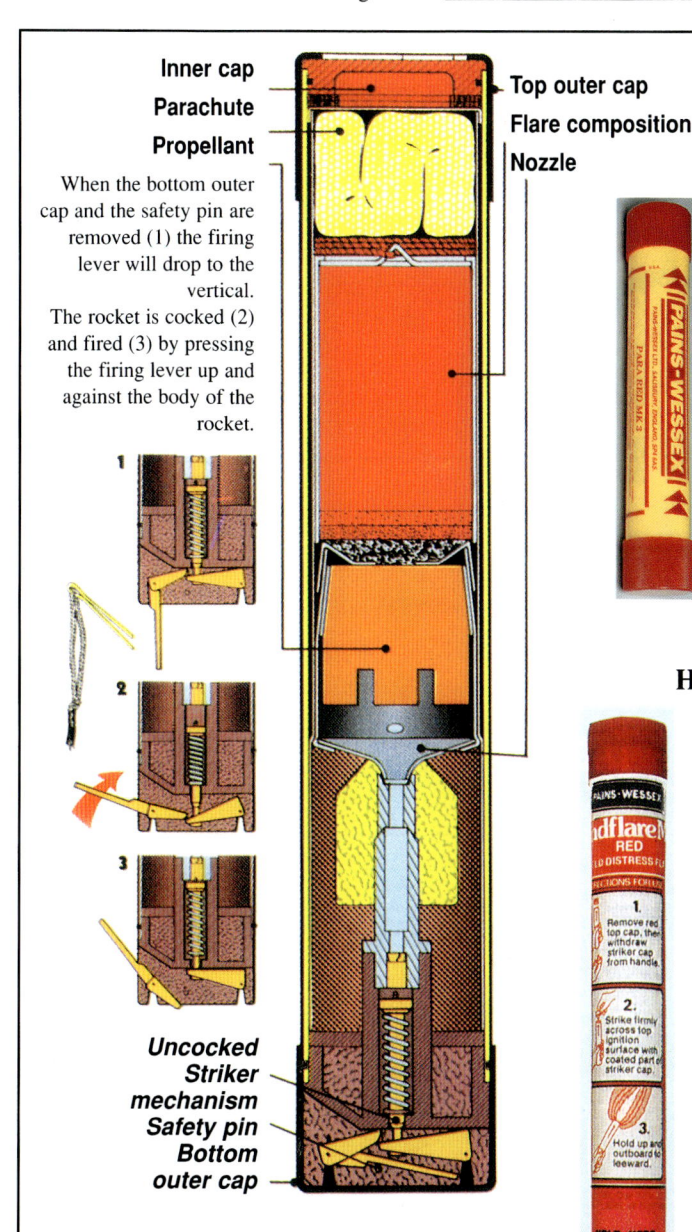

Inner cap
Parachute
Propellant
Top outer cap
Flare composition
Nozzle

When the bottom outer cap and the safety pin are removed (1) the firing lever will drop to the vertical. The rocket is cocked (2) and fired (3) by pressing the firing lever up and against the body of the rocket.

Uncocked
Striker
mechanism
Safety pin
Bottom
outer cap

PARA RED MK.3 DISTRESS ROCKET

An easy to use hand held self contained distress rocket, ejecting a parachute suspended red flare at 300 m altitude which burns for 40 secs. at a brilliant 40,000 candela, visible range 15 km. by day, 40 km. or more by night. SOLAS'83 APPROVED.

OPERATION:
Although the signal is normally fired vertically to provide maximum range of visibility, in low cloud conditions (below 300 m), it is advisable to fire the rocket at an angle of 45°. In a strong wind aim slightly down wind.

HOW THE ROCKET FUNCTIONS
On activating the trigger mechanism as diagrams 1, 2 and 3, the striker plunger activates an ignition cap which transmits a flame to ignite the propellant in the rocket motor. The inner flight tube is instantly launched from the plastic outer tube. On reaching 300 m a gunpowder charge expels and deploys the parachute suspended red flare burning at 40,000 c.d.

HAND HELD DISTRESS FLARES

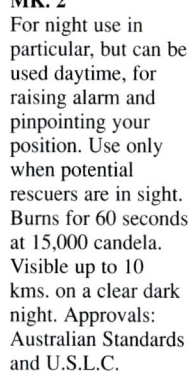

Red Handflare    Orange Handsmoke

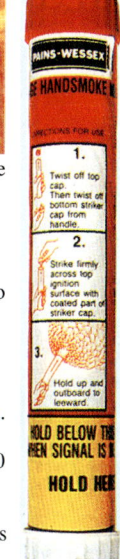

RED HANDFLARE MK. 2
For night use in particular, but can be used daytime, for raising alarm and pinpointing your position. Use only when potential rescuers are in sight. Burns for 60 seconds at 15,000 candela. Visible up to 10 kms. on a clear dark night. Approvals: Australian Standards and U.S.L.C.

ORANGE HANDSMOKE MK. 3
For day use only. To raise the alarm and pinpoint your position. Use only when potential rescuers are in sight. Produces orange smoke visible for 60 secs. up to 4 km. distant. Approvals: Australian Standards and U.S.L.C.

FLARE VISIBILITY: The potential sighting area of an approved distress signal to a VESSEL is as follows:-

SIGNAL HEIGHT	MAX SIGHTING RADIUS KM	SIGHTING AREA KM^2
2 m Orange Hand Smoke (day)	4 km	50 km^2
2 m Red Handflare (night)	10 km	314 km^2
300 m Red Para Rocket (day)	15 km	707 km^2
300 m Red Para Rocket (night)	40 km	5028 km^2

HOW TO IGNITE YOUR DISTRESS FLARES

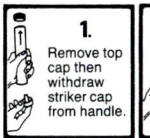

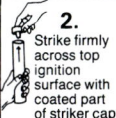

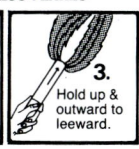

1. Remove top cap then withdraw striker cap from handle.
2. Strike firmly across top ignition surface with coated part of striker cap.
3. Hold up & outward to leeward.

The assistance of Pains-Wessex is appreciated in the preparation of this page.

Mast (1-2) Wk	On large-scale charts, wreck which does not cover, height above height datum
Mast (1-2) Wk	On large-scale charts, wreck which covers and uncovers, height above Chart Datum
5 Wk	On large-scale charts, submerged wreck, depth known
Wk	On large-scale charts, submerged wreck, depth unknown
⌐	Wreck showing any part of hull or superstructure at the level of Chart Datum
Mast (1-2) Funnel Mast (1-2) / Masts	Wreck of which the mast(s) only are visible at Chart Datum
4₆ Wk / (25) Wk	Wreck over which the depth has been obtained by sounding but not by wire sweep
4₆ Wk / (25) Wk	Wreck which has been swept by wire to the depth shown
⌀	Wreck, depth unknown, which is considered dangerous to surface navigation
‡	Wreck, depth unknown, which is not considered dangerous to surface navigation
(20) Wk	Wreck over which the exact depth is unknown, but which is considered to have a safe clearance at the depth shown
[Foul]	Remains of a wreck, or other foul area, no longer dangerous to surface navigation, but to be avoided by vessels anchoring, trawling, etc

Obstn	Obstruction or danger to navigation the exact nature of which is not specified or has not been determined, depth unknown
16₈ Obstn	Obstruction, depth known
4₆ Obstn	Obstruction which has been swept by wire to the depth shown
⊤ ⊤ ⊤ Obstn	Stumps of posts or piles, wholly submerged
#	Submerged pile, stake, snag, well or stump (with exact position)
	Fishing stakes
	Fish trap, fish weir, tunny nets
[Fish traps] / [Tunny nets]	Fish trap area, tunny nets area
🐟	Fish haven
🐟 (2₄)	Fish haven, depth known
[Shellfish Beds]	Shellfish beds
[🐟] [🐟]	Marine farm (on large-scale charts)
[🐟]	Marine farm (on small-scale charts)

SYMBOLS & ABBREVIATIONS USED ON CHARTS
(See examples on chart extracts in the Navigation Chapter)
Reproduced from Admiralty Symbols & Abbreviation (BA5011) by permission of the Controller of Her Majesty's Stationery Office & the UK Hydrographic Office.

Page 1/4: OBSTRUCTIONS, WRECKS & FOUL GROUND

Chapter 6: Buoyage System & Other Colour Illustrations

→→→→ 3kn →	Flood tide stream (with rate)
----- 3kn →	Ebb tide stream (with rate)
# →→→	Current in restricted waters
~~ (see Note)	Ocean current. Details of current strength and seasonal variations may be shown
〰 〰 〰	Overfalls, tide rips, races
⊙ ⊙ ⊙	Eddies
◇	Position of tabulated tidal stream data with designation
Co + + Co 5₈	Coral reef which is always covered
(5₈) 19 Br 18	Breakers
〰〰	Discoloured water

QUALIFYING TERMS

c	Coarse
f	Fine
h	Hard
m	Medium
sf	Soft
sy	Sticky

OTHER TERMS

ED	Existence doubtful
PA	Position approx.
PD	Position doubtful
Rep	Reported
SD	Sounding doubtful

NATURE OF SEABED

Co	Coral
Cy	Clay
G	Gravel
M	Mud
P	Pebbles
R	Rock
S	Sand
Sh	Shells
Si	Silt
S/M	Sand over Mud
St	Stones
Wd	Weed
Wk	Wreck

(1·7) (3·1)	Rock which does not cover, height above High Water
(*1₆*) *(2₇)	Rock which covers and uncovers, height above Chart Datum
#	Rock awash at the level of Chart Datum
+ ⊕	Underwater rock over which the depth is unknown, but which is considered dangerous to surface navigation
M S St G	Area of sand and mud with patches of stones or gravel
*(4₂) 1 M S	Rocky area
(*1₆) (0₉) *(2₉) 1₈ 4 0₈	Coral reef

Page 2/4 ROCKS & CORAL *(Above & above right)*
TIDAL STREAMS & CURRENTS *(Above right)*
NATURE OF SEABED *(Right)*
BRIDGE CLEARANCE *(Below)*

(8·9)	20

Vertical clearance above High Water

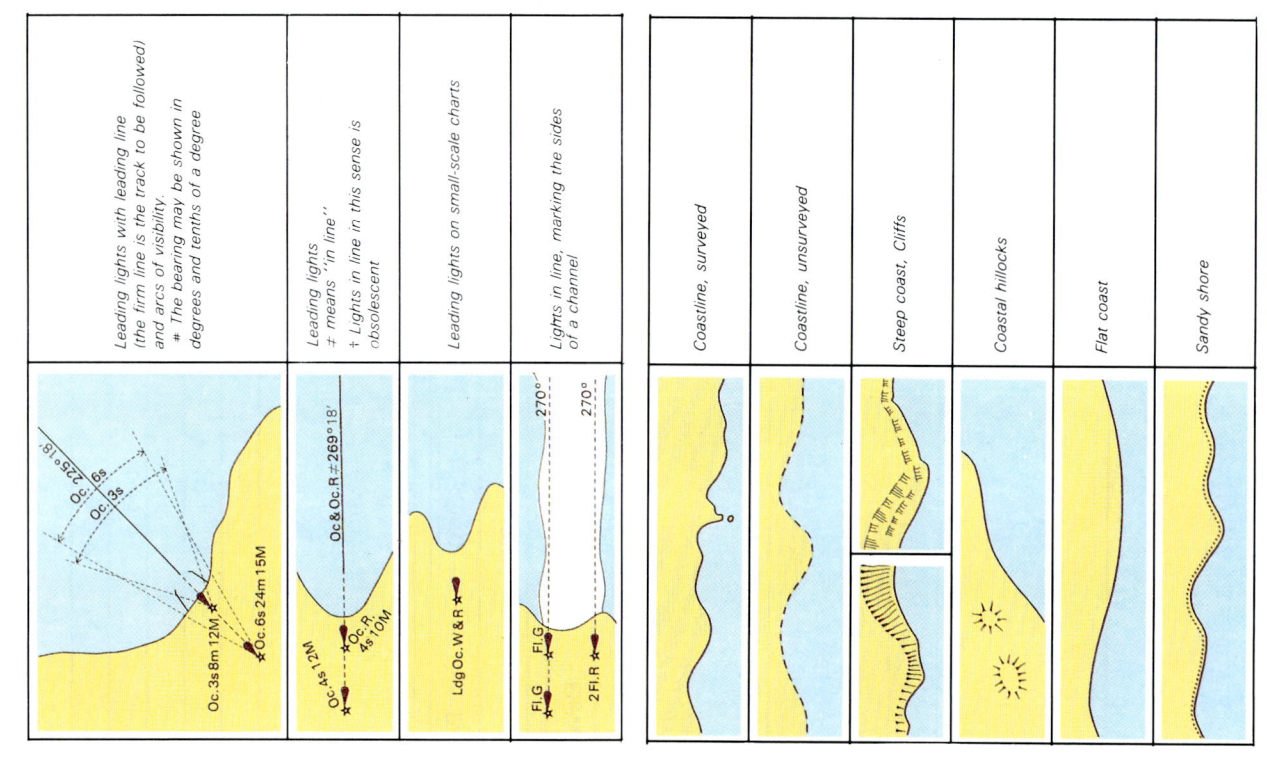

Page 3/4   **LEADING LIGHTS, TRANSITS, CLEARING LINES, ROUTES & COASTLINES**

Chapter 6: Buoyage System & Other Colour Illustrations

	↪	Recommended anchorage (no defined limits)
	Ⓐ	Anchor berths
	N53	Anchor berths with swinging circle shown
	⚓ (box)	Anchorage area in general
	No 1 ⚓	Numbered anchorage area
	Oaze ⚓	Named anchorage area
	DW ⚓	Deep water anchorage area / anchorage area for deep-draught vessels
	Tanker ⚓	Tanker anchorage area
	24h ⚓	Anchorage area for periods up to 24 hours
	⚓ (explosives)	Explosives anchorage area
	⊕ ⚓	Quarantine anchorage area
	Reserved (see Note)	Reserved anchorage area
	⚓	Seaplane landing area
	⚓ (×)	Anchorage for seaplanes
	⚓ × (box)	Anchoring prohibited
	× fish (box)	Fishing prohibited

~~	Submarine cable
⊥⊥⊥⊥	Submarine cable area
~∫~∫~	Submarine power cable
⊥⊥⊥⊥	Submarine power cable area
~~	Disused submarine cable
Oil / Gas / Chem / Water	Supply pipeline: unspecified, oil, gas, chemicals, water
Oil / Gas / Chem / Water	Supply pipeline area: unspecified, oil, gas, chemicals, water
Water / Sewer / Outfall / Intake	Discharge pipe: unspecified, water, sewer, outfall, intake
Water / Sewer / Outfall / Intake	Discharge pipe area: unspecified, water, sewer, outfall, intake
Buried 1·6m	Buried pipeline/pipe (with nominal depth to which buried)
32 Obstn	Diffuser
- - - -	Disused pipeline/pipe

Page 4/4   **SUBMARINE CABLES, PIPELINES, ANCHORAGES, PROHIBITED ANCHORING & FISHING**

The Australian Boating Manual

Fig 6.6   **IALA BUOYAGE SYSTEM "A"**

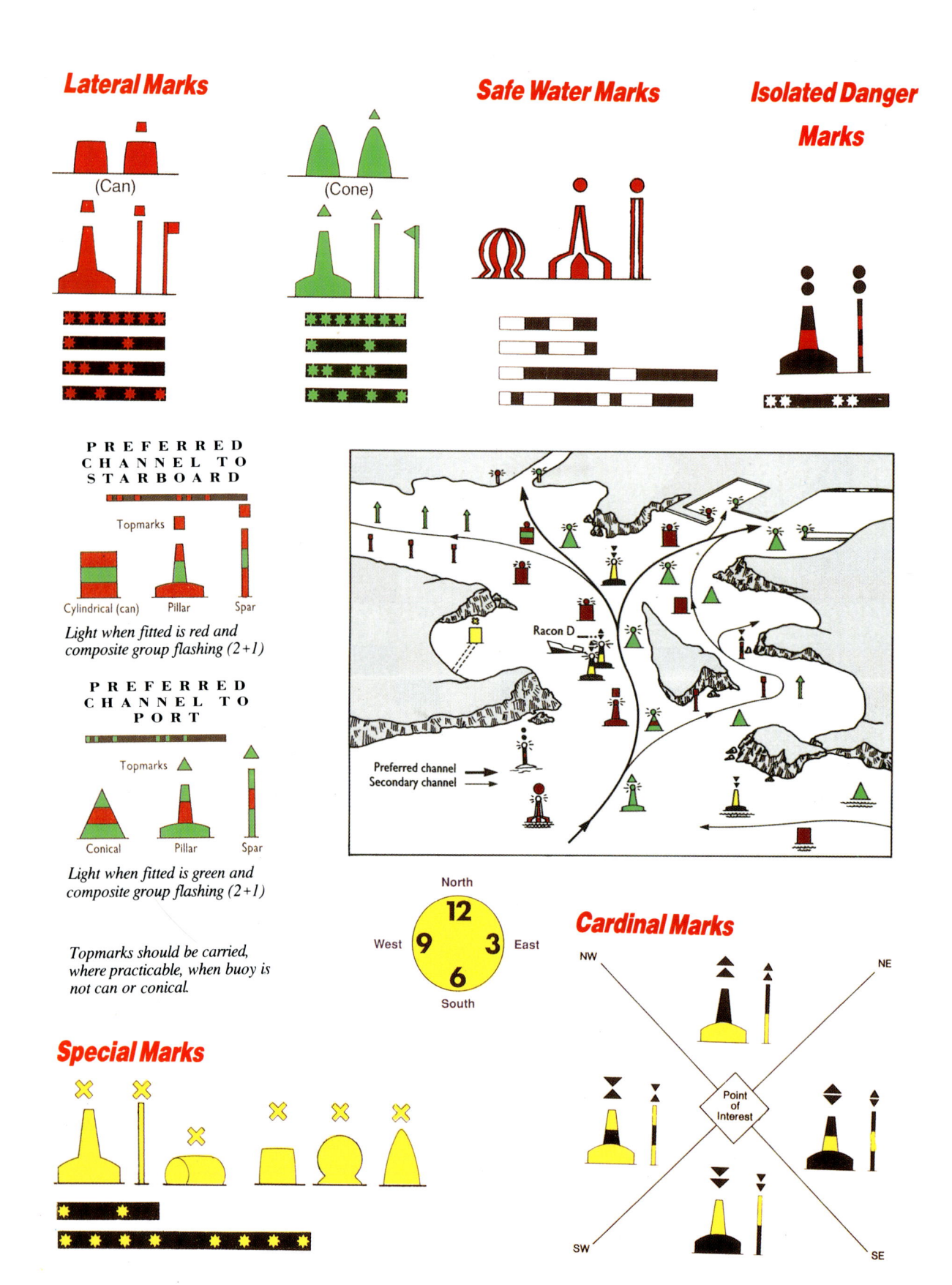

Lateral Marks

(Can)

(Cone)

Safe Water Marks

Isolated Danger Marks

PREFERRED CHANNEL TO STARBOARD

Topmarks

Cylindrical (can)   Pillar   Spar

Light when fitted is red and composite group flashing (2+1)

PREFERRED CHANNEL TO PORT

Topmarks

Conical   Pillar   Spar

Light when fitted is green and composite group flashing (2+1)

Topmarks should be carried, where practicable, when buoy is not can or conical.

Racon D

Preferred channel →
Secondary channel →

North 12
West 9   East 3
South 6

Cardinal Marks

NW   NE

Point of Interest

SW   SE

Special Marks

Chapter 6: Buoyage System & Other Colour Illustrations

LIGHT CHARACTERISTICS (Commonly used on buoys, beacons and lighthouses)

Charts indicate the colours of lights as Y, G, R, Or, etc. If colour is not indicated, then it is a white light.

DIRECTION LIGHTS: The abbreviation 'Dir' shown on charts for some harbour entrance lights indicates a direction light with narrow sector and 'narrow' course to be followed. The sector is only about 2° wide. The light could be flanked by darkness or unintensified light, or by light sectors of different character or colour. Thus a 'Dir. WRG' light is a directional white light flanked by red and green sector lights.

A range of light characteristics exist so that we can distinguish between various lighthouses, beacons and buoys. The main characteristics are shown below.

Al.WR	ALTERNATING : a light which alters in colour (e.g. white, red) in successive flashes or eclipses. 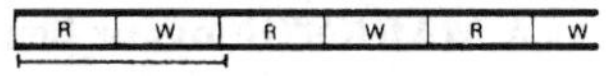
F	FIXED : a continuous steady light. For example, FR means a fixed red light.
Fl	FLASHING : single brilliant flash at regular intervals. For example, Fl 5s means one flash every 5 seconds. Duration of light always less than the period of darkness. 
FFl	FIXED AND FLASHING : a steady light and one flash at regular intervals.
Fl( )	GROUP FLASHING : two or more flashes in succession at regular intervals. For example, Fl(3)10s means 3 flashes in succession within a period of 10-seconds. 
Fl(2+1)	COMPOSITE GROUP FLASHING : e.g. two flashes in succession followed by one flash at regular intervals. 
Oc	OCCULTING : steady light with a total eclipse at regular intervals. Duration of darkness is less than that of light. 
Oc( )	GROUP OCCULTING : two or more eclipses in a group at regular intervals. 
Q	QUICK FLASHING : continuous flashing at the rate of 50 or 60 per minute. 
IQ	INTERRUPTED QUICK FLASHING : flashing in groups of 50 or 60 times per minute with a total eclipse at regular intervals. 
Q( )	GROUP QUICK FLASHING : two or more quick flashes in a group at regular intervals. For example, Q(3)10s means a group of 3 quick flashes within a 10-second period. 
VQ	VERY QUICK FLASHING : continuous flashing at the rate of 100 or 120 per minute, or in groups as for Q. 
UQ	ULTRA QUICK : continuous flashing at the rate of 240 to 300 per minute. 
IUQ	INTERRUPTED ULTRA QUICK FLASHING : UQ with a total eclipse at regular intervals. 
Iso	ISOPHASE : a light where duration of light and darkness are equal. 
Mo( )	MORSE CODE LIGHT : the characteristics are shown by the appropriate letters or figures in brackets; e.g. Mo(K).
LFl	LONG FLASHING : a flash of two or more seconds. 

Fig 6.7   Period shown ⊢────⊣

THE IALA MARITIME BUOYAGE SYSTEM "A" & "B" (See Figure 6.6)

The International Association of Lighthouse Authorities (IALA) Buoyage System is in two parts: "A" & "B". The difference is in the channel markers.

When entering a harbour or proceeding in the direction of the main stream of the flood tide, System "A" has green marks and lights to starboard and red marks and lights to port. System "B" has red marks and lights to starboard and green marks and lights to port.

System "A" applies in Europe, Australia, New Zealand, Africa, Persian Gulf and some Asian countries. System "B" applies in North, Central and South America, Japan, Korea and the Philippines.

TYPES OF IALA MARKS (Buoys or beacons)

1. LATERAL (OR CHANNEL) MARKS: They mark the port (red) and starboard (green) sides of a route to be followed, such as in a channel.
2. CARDINAL MARKS: They mark the North, South, East & West extremities around a large dangerous area. (The word "cardinal" refers to the four "main" points of a compass: i.e., North, East, South, West).
3. ISOLATED DANGER MARKS: A small area of danger, such as a reef or wreck, is marked with a single (isolated) danger buoy.
4. SAFE WATER MARKS: These are sometimes used in lieu of the channel marks to guide vessels along a safe water route, and may be passed on either side.
5. SPECIAL MARKS: These "yellow" buoys are like the Witch's Hats used on roads for directing traffic. These may be used around dredgers, land reclamation areas and construction works in the water.

COLOURS, SHAPES, LIGHTS & TOPMARKS OF IALA BUOYS & BEACONS

Lateral or Channel Marks:

When approaching a harbour entrance or any other waterway from seaward, that is when proceeding in the direction of the main stream of flood tide, you should keep:

- Green colour, conical or triangular shapes & green lights to starboard.
- Red colour, rectangular or can shapes & red lights to port.

Preferred Channel Marks:

Where a shipping channel divides into two navigable channels, the deeper or preferred channel is *sometimes* identified by Preferred Channel Marks. These are modified port or starboard hand marks shown in figure 6.6. The dominant colour and shape of these marks indicates the side on which the mark is passed by a vessel entering the preferred channel.

- If it is a *green buoy* with a *red band* in it, then it is the *starboard* hand buoy for the preferred channel and the port hand buoy for the secondary channel. Its light characteristics is distinctive *"flashing green 2+1"*.
- If it is a *red buoy* with a *green band* in it, then it is the *port* hand mark for the preferred channel and the starboard hand buoy for the secondary channel. Its light characteristics is distinctive *"flashing red 2+1"*.

Cardinal Marks:

- THE TOPMARKS ON ALL CARDINAL MARKS ARE A COMBINATION OF TWO BLACK CONES:
    - On the North mark, both cones point up.
    - On the South mark, both cones point down, the opposite to the North Cardinal Mark.
    - On the West mark, the cones point towards each other, looking like a Wine Glass.
    - On the East mark, the cones point away from each other, the opposite to the West Cardinal Mark.

- THE COLOURS ON ALL CARDINAL MARKS ARE BLACK & YELLOW HORIZONTAL BANDS. The cones point towards the black band(s) of the Mark. Therefore:
    - The North mark is Black over Yellow.
    - The South mark is Yellow over Black.
    - The West mark is Yellow Black Yellow.
    - The East mark is Black Yellow Black.

- Cardinal Marks show only white lights. The light characteristics may be a combination of Quick flashing or Very Quick flashing together with an identifying long flash for the South Cardinal Mark.

Chapter 6: Buoyage System & Other Colour Illustrations

    EXAMPLES:
- North Cardinal Mark: Continuous Q or VQ
- East Cardinal Mark :   Q or VQ (3)
- South Cardinal Mark :   Q or VQ (6) + 1 Long Flash
- West Cardinal Mark :   Q or VQ (9)

It will be seen that the number of flashes corresponds with clock face numerals, with the slight exception of the North and South Cardinal Marks.

➢ *A Very Quick (VQ) flashing light is quicker than the Quick (Q) flashing light. Therefore, its period is made shorter. The periods are as follows:*
    EAST CARDINAL MARK = VQ 5 s; Q 10 s.
    SOUTH & WEST MARKS = VQ 10 s; Q 15 s.

➢ Instead of trying to remember which side to pass a Cardinal Mark, you may find it easier to remember that YOU SHOULD PASS:
    West of the West Cardinal Mark.
    East of the East Cardinal Mark.
    North of the North Cardinal Mark.
    South of the South Cardinal Mark.

Isolated Danger Mark:
    Colour: Black & Red horizontal bands
    Shape: Pillar buoy or spar beacon
    Light: White Fl (2) [Period not defined]
    Topmark: Two *black spheres*, disposed vertically

Safe Water Mark:
    Colour: Red & White vertical stripes
    Shape: Spherical, pillar or spar
    Light: White, isophase, occulting, L.Fl.10s. or Mo "A"
    Topmark: For non-spherical marks: one red sphere

Special Marks:
As stated earlier, think of the Special Marker buoys like the Witches Hats used in directing traffic on roads. These buoys are *always yellow*. Lights, if fitted, are also yellow. Topmark, if any, is the *St. Andrews Cross*.

These buoys do however come in *different shapes* to guide the ships in the appropriate direction, if necessary. For example, a conical buoy is to be kept on the starboard side when coming into port, just like the green starboard hand buoy.

    Listed below are some of the examples where these buoys could be used:
- Channel within a channel;
- Traffic separation;
- Spoil ground;
- Military exercise zone;
- Cable or pipeline marks;
- Ocean monitoring buoy;
- Recreation zone.

NEW DANGER MARKS

A newly discovered hazard to navigation not yet shown on charts or included in sailing directions or sufficiently publicised by notices to mariners is termed a New Danger. It includes natural and man-made dangers, such as sand banks and wrecks.

A New Danger is marked by one or more Cardinal or Lateral (channel) marks. If the danger is especially grave, at least one of the marks will be duplicated until the danger has been sufficiently publicised. For example, the eastern side of the danger may be marked with two East Cardinal marks, instead of just one.

If lighted, the New Danger mark shows a quick light or very quick light of the appropriate colour: white, green or red.

HOW BUOYS ARE NUMBERED & MARKED ON CHARTS

In long channels buoys are often numbered as follows: odd numbers (1, 3, 5,...) on starboard and even (2, 4, 6,...) on port side when entering from seaward or going with the flood. [Remember it like a vessel's 1 short blast (odd number) when going to starboard, etc....]. But always check the largest scale chart available.

The buoys (on navigational charts) are NOT marked in their colours. Instead, they are marked as outline shapes with a

The Australian Boating Manual

purple (or orange) plume to indicate the ones that are lit. Their colours and light characteristics are abbreviated and written underneath and next to them. A tiny circle in the base of the shape marks their exact position on the chart. On small scale charts, they may only be marked as stars with a plume for light.

For example, letters BYB means the buoy's colours are black, yellow, black. And, Q(3) 10s. means that it carries a light: Quick Flashing 3 times every 10 seconds. The colour of the light is white unless indicated by the letter G (for green), R (for red), etc. For example: Q(3)G 10s.

DIRECTION OF THE BUOYAGE SYSTEM (Fig 6.8)

The direction of the buoyage system follows the main stream of the flood tide when entering ports, rivers and channels. In other areas, the direction is determined by authorities, usually in a clockwise direction around continental land masses. Where there may be any doubt, the direction of buoyage can be indicated on charts by an arrow symbol as shown:

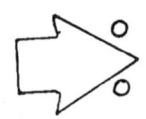

EXERCISE:

The following harbour chart displays night characteristics of various buoys and navigational lights. The direction of North is also indicated. Draw your route through the appropriate buoyed channel for leaving port to go out to sea.

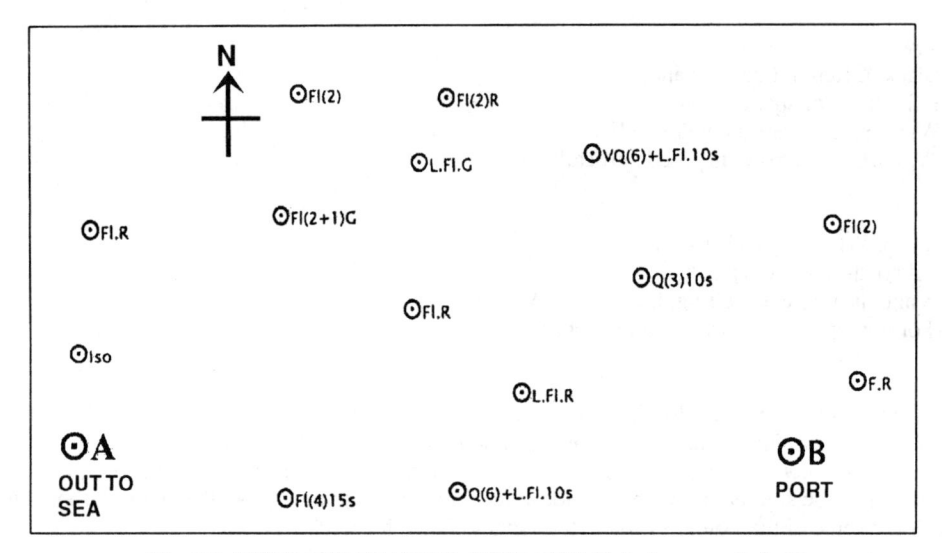

Fig. 6.9: BUOYAGE CHANNEL EXERCISE (*Solution – see below*)

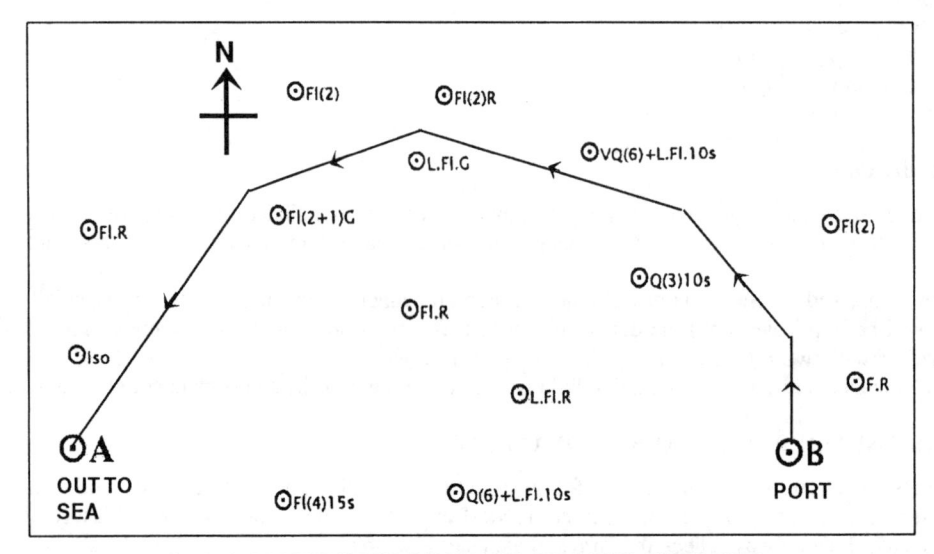

Fig 6.10: SOLUTION TO BUOYAGE CHANNEL EXERCISE

CHAPTER 6: QUESTIONS & ANSWERS

CHAPTER 6.1: INTERNATIONAL CODE FLAGS

1. A fishing vessel flying a blue flag with a white square in the middle is indicating that:

    (a) she is shooting nets.
    (b) her nets are fast on an obstruction.
    (c) she is hauling nets.
    (d) she is engaged in purse seine fishing.

2. A fishing vessel flying a flag with yellow & blue vertical stripes is indicating that:

    (a) she is shooting nets.
    (b) her nets are fast on an obstruction.
    (c) she is hauling nets.
    (d) she is engaged in purse seine fishing.

3. A fishing vessel flying the international code flag "Tango" is indicating that:

    (a) she is shooting nets.
    (b) her nets are fast on an obstruction.
    (c) she is hauling nets.
    (d) she is engaged in pair trawling.

4. A fishing vessel flying the international code flag "Zulu" is indicating that:

    (a) she is shooting nets.
    (b) her nets are fast on an obstruction.
    (c) she is hauling nets.
    (d) she is engaged in pair trawling.

5. A vessel flying a flag consisting of yellow and blue vertical stripes indicates that:

    (a) her fishing nets are caught on an obstruction.
    (b) she is shooting fishing nets.
    (c) she requires a pilot.
    (d) she requires a tow.

6. A vessel flying a flag consisting of white and blue vertical halves indicates that:

    (a) she is engaged in diving operations.
    (b) she is awaiting pilot.
    (c) she has a pilot on board.
    (d) she is hauling nets.

7. The colours of one of the international code flags in the two-flag distress signal are:

    (a) Blue & white checkers
    (b) Blue & red horizontal stripes.
    (c) Blue & yellow vertical stripes.
    (d) none of the choices stated here.

8. The colours of the flag "November" are:

    (a) Blue & white checkers.
    (b) Blue, white, red, white, blue horizontal.
    (c) Yellow & red diagonal halves.
    (d) None of the choices stated here.

9. The colours of the flag "Charlie" are:

    (a) Blue & white checkers.
    (b) Blue, white, red, white, blue horizontal.
    (c) Yellow & blue vertical bars.
    (d) White & red squares.

10. The international code flag coloured yellow, blue, yellow horizontal, means:

    (a) I require a tug.
    (b) (b) Keep clear of me; I am manoeuvring with difficulty.
    (c) Man overboard.
    (d) I am disabled; communicate with me.

11. The meaning of the all-yellow international single-letter code flag is:

    (a) The pilot is on board
    (b) I require assistance.
    (c) I require a pilot.
    (d) I request quarantine clearance.

12. The colours of the flag "Oscar" are:

    (a) Blue and yellow.
    (b) Red and yellow.
    (c) Blue and white
    (d) Yellow.

13. How do you distinguish a numeral pennant from a letter code flag?:

    (a) Shape.
    (b) Size.
    (c) Colour.
    (d) Order of hoist.

14. The international code flag meaning "I require assistance" is coloured:

    (a) White with red St. Andrews Cross.
    (b) Yellow.
    (c) Orange.
    (d) Blue with a white square.

15. The colours and meaning of the international Code Flag "A" are as follows:

    (a) Yellow and blue - Pilot required.
    (b) Yellow and blue - diver below.
    (c) Red and yellow - man overboard
    (d) White and blue - diver below.

16. The international code flag(s) signal exhibited by a vessel taking in or discharging dangerous goods consists of:

    (a) November Charlie.
    (b) Bravo.
    (c) Delta.
    (d) none of the choices stated here.

17. The signal displayed by a vessel taking-in, discharging or carrying dangerous goods is:

    (a) a black ball.
    (b) a black diamond.
    (c) the international code flag "D".
    (d) the international code flag "B".

CHAPTER 6.2: IALA BUOYAGE SYSTEM

18. A Q(3) G light on an IALA System 'A' buoy indicates that it is:

    (a) an east cardinal mark.
    (b) a west cardinal mark.
    (c) a starboard hand mark.
    (d) none of the choices stated here.

19. The colours and shapes of Special Marks in the IALA Buoyage System 'A' re:

    (a) yellow cone, can and sphere.
    (b) yellow cone.
    (c) yellow and black pillars and spars.
    (d) Black and red pillars and spars.

20. One purpose of the Special Marks in the IALA Buoyage System is to:

    (a) mark the extremities of an underwater danger.
    (b) keep unwanted vessels out of a sensitive area.
    (c) control speed in a special area.
    (d) mark new dangers.

21. When coming out of port an isolated danger mark must be passed clear:

    (a) on starboard side.
    (b) on port side.
    (c) on either side.
    (d) northwards.

22. In the IALA buoyage System 'A' South Cardinal mark must be passed as follows:

    (a) vessels keep it to south when going into port.
    (b) vessels keep south of it when going into port.
    (c) vessels keep it to south at all times.
    (d) vessels keep south of it at all times.

23. You see a buoy displaying two black cones in a vertical line pointing upwards. You should pass it as follows:

    (a) keep north of it at all times.
    (b) keep north of it when going into port.
    (c) keep north of it when going out of port.
    (d) keep it to north at all times.

24. It is not advisable to use IALA buoys for taking bearings in order to fix a vessel's position, because their:

    (a) positions are not marked on charts.
    (b) light characteristics are not marked on charts.
    (c) positions can change unnoticed.
    (d) colours are not distinguishable.

25. Buoys are marked on large scale coastal charts:

    (a) as coloured shapes.
    (b) as cone, can or sphere shapes.
    (c) by characteristics of their light.
    (d) by colour and/or characteristics of light.

26. When entering a long channel from seaward, the System 'A' IALA buoys are often painted with numbers as follows:

    (a) odd numbers on starboard, even on port buoys.
    (b) odd numbers on port buoys, even on starboard.
    (c) there is no standard system of numbering them.
    (d) none of the choices stated here apply.

27. Buoys indicating wrecks are usually:

    (a) preferred channel buoys.
    (b) special buoys.
    (c) isolated danger buoys.
    (d) cardinal buoys.

28. Proceeding into port you see ahead a red and black horizontally banded spar with a topmark consisting of two vertical black sphere. You will pass it as follows:

    (a) keep it to port.
    (b) keep south of it.
    (c) keep close to it.
    (d) keep clear on all sides.

29. Yellow buoys, with or without yellow lights at night, indicate:

    (a) a sensitive area.
    (b) a preferred channel.
    (c) a closed channel.
    (d) none of the choices stated here.

Chapter 6: Buoyage System & Other Colour Illustrations

30. The night signal for a South Cardinal Mark is a light flashing as follows:

    (a) Q(6)+1LF R 15s.
    (b) Q(6)+1LF 15s.
    (c) Q(9)+1LF R 15s.
    (d) Q(9)+1LF 15s.

31. You see ahead a spar buoy painted bottom half black and the top half yellow. It is fitted with a top-mark consisting of two black cones, pointing downwards. You would pass it as follows:

    (a) keep north of it.
    (b) keep north or south of it.
    (c) keep south of it.
    (d) keep it south of you.

32. Proceeding into port with IALA Buoyage System 'A', you see ahead a light Fl.(2)G 5s. You would pass it as follows:

    (a) keep it to port.
    (b) keep it to starboard.
    (c) keep clear on either side.
    (d) keep close on either side.

33. Steaming in a northerly direction, you see ahead three quick flashes followed by one long flash of a white light on a buoy before being engulfed by fog. It can only be a:

    (a) south cardinal mark.
    (b) east cardinal mark.
    (c) preferred channel mark.
    (d) special mark

34. Proceeding into a port, you see on your starboard bow a buoy with light Fl.(2)10s. It is:

    (a) a safe water mark.
    (b) a special mark.
    (c) an isolated danger mark.
    (d) none of the choices stated here.

35. Leaving port, you see ahead an Iso. 8s light on a buoy. You would pass it as follows:

    (a) keep well clear of it.
    (b) pass close to it.
    (c) pass north of it.
    (d) keep it to north.

36. Proceeding into a port, you see ahead a spherical buoy painted in red and white vertical stripes. Its top-mark seems to have blown away. You pass it as follows:

    (a) keep well clear of it.
    (b) pass close to it.
    (c) pass north of it.
    (d) keep it to north.

37. Sailing in the direction of the ebb, you see ahead a Q(9)15s light on a buoy. You would you pass it as follows:

    (a) keep west of it.
    (b) keep it to west.
    (c) keep clear of it.
    (d) pass close to it on either side.

38. Your course 300 degree in the approaches to a harbour. You observe a white light Q(3)10s. on a buoy bearing 15 degrees on you port bow. You would pass it as follows:

    (a) keep it on port side.
    (b) keep it on starboard side.
    (c) keep it to east.
    (d) keep east of it.

39. At sea, course 135 degrees. A spar buoy with horizontal red and black stripes appears right ahead. You would pass it as follows:

    (a) keep west of it.
    (b) keep it to west.
    (c) keep clear of it on either side.
    (d) pass close to it on either side.

40. You are steaming down river through the IALA buoyage System 'A'. A red can buoy is observed 5 degrees on your starboard bow. You would pass it as follows:

    (a) keep it on starboard side.
    (b) keep it on port side.
    (c) keep clear of it on either side.
    (d) none of the choices stated here apply.

41. The chart symbol for a light Oc. R 10s. in the IALA Buoyage System 'A' indicates:

    (a) a starboard hand mark.
    (b) a port hand mark.
    (c) an isolated danger mark.
    (d) a safe water mark.

42. The chart symbol for a light Iso. G 8s. on a buoy indicates:

    (a) a starboard hand mark.
    (b) a port hand mark.
    (c) an isolated danger mark.
    (d) a safe water mark.

43. The buoyage system in the Australian waters is:

    (a) IALA System 'A'.
    (b) IALA System 'B'.
    (c) Pre-IALA System.
    (d) none of the choices stated here.

44. Sailing in the direction of the ebb, you see a buoy painted black yellow black in 3 horizontal bands. Its top mark seems to have gone adrift. You would you pass it as follows:

    (a) keep west of it.
    (b) keep it to west.
    (c) keep clear of it.
    (d) pass close to it on either side.

45. Entering a port with the IALA Buoyage System 'A', you see an Iso. G 6s. light on a buoy. You would you pass it as follows:

    (a) keep clear of it on either side.
    (b) keep it to west.
    (c) keep it to starboard.
    (d) pass close to it on either side.

46. Leaving a port, you see a Fl. Y 4s. light on a buoy. You will take the following action:

    (a) determine its special function.
    (b) pass it on starboard side.
    (c) pass it on port side.
    (d) keep clear of it on either side.

47. Sailing out to seaward with the IALA Buoyage System 'A', you see a Fl.(2) 6s. light on a buoy. It is a:

    (a) safe water mark.
    (b) isolated danger mark.
    (c) port hand mark.
    (d) starboard hand mark.

48. A buoy with the light characteristic L.Fl. 10s is a:

    (a) safe water mark.
    (b) isolated danger mark.
    (c) special mark.
    (d) preferred channel mark.

49. A buoy with the light characteristic Q is a:

    (a) safe water mark.
    (b) isolated danger mark.
    (c) special mark.
    (d) north cardinal mark.

50. In the IALA Buoyage System 'A', a buoy with the light characteristic Fl. G 5s is a:

    (a) starboard hand mark.
    (b) safe water mark.
    (c) special mark.
    (d) port hand mark.

51. Under the IALA BUOYAGE System 'A', the direction of buoys is usually defined as:

    (a) with the ebb.
    (b) anti-clockwise.
    (c) with the ebb or clockwise.
    (d) with the flood.

52. In order to indicate the deepest water in an area, or the safe side to pass near a danger, or to draw attention to a feature in a channel such as a bend, junction, fork or end of a shoal, you are most likely to find the following buoy:

    (a) special mark.
    (b) preferred channel mark.
    (c) cardinal mark.
    (d) isolated danger mark.

53. The greatest danger of rounding a buoy too close is:

    (a) running aground.
    (b) fouling its mooring cable.
    (c) drifting out of the channel.
    (d) collision with another vessel.

54. An Occulting light consists of:

    (a) equal on and off periods.
    (b) none of the choices stated here.
    (c) longer off than on period.
    (d) longer on than off period.

55. An Isophase light consists of:

    (a) very quick flashing.
    (b) longer off than on period.
    (c) equal on and off periods.
    (d) longer on than off period.

56. A Quick Flashing light consists of:

    (a) 100 to 120 flashes per minute.
    (b) 30 to 40 flashes per minute.
    (c) 10 to 20 flashes per minute.
    (d) 50 to 60 flashes per minute.

57. Proceeding into a port, you see ahead a yellow spar with a topmark of a single yellow "X" shaped cross. You are required to pass it as follows:

    (a) at a safe navigational distance.
    (b) keep it on your port side.
    (c) pass north of it.
    (d) pass south of it.

58. An Ultra Quick Flashing light consists of:

    (a) 100 to 120 flashes per minute.
    (b) 60 to 90 flashes per minute.
    (c) 240 to 300 flashes per minute.
    (d) 300 to 400 flashes per minute.

Chapter 6: Buoyage System & Other Colour Illustrations

59. In relational to navigational marks, a Long Flash is a flash of:

    (a) about 5 seconds.
    (b) 2 or more seconds.
    (c) 5 or more seconds.
    (d) about 1 minute

60. The abbreviation 'FFl' in relation to lights marked on charts indicates:

    (a) an interrupted flashing light.
    (b) a long flash.
    (c) a steady light & a flash at regular intervals.
    (d) a fast flashing light.

61. The abbreviation 'Dir WRG' in relation to lights marked on charts indicates:

    (a) Direction white light flanked by red & green sector lights.
    (b) Directly facing white, red & green lights.
    (c) White, red & green lights in all directions.
    (d) Alternating white, red & green directional lights

62. The abbreviation 'Al.WR' in relation to lights marked on charts indicates:

    (a) flashing white & red in separate sectors.
    (b) alternating white & red in separate sectors.
    (c) Flashing white & red simultaneously.
    (d) alternating white & red in successive flashing.

63. The abbreviation 'Fl.WR' in relation to lights marked on charts indicates:

    (a) flashing white & red in separate sectors.
    (b) alternating white & red in separate sectors.
    (c) Flashing white & red simultaneously.
    (d) alternating white & red in successive flashing.

64. The topmark of a south cardinal mark is:

    (a) two cones, points together.
    (b) two cones, bases together.
    (c) Two cones, both pointing down.
    (d) Two cones, both pointing up.

65. The topmark of an east cardinal mark is:

    (a) two cones, points together.
    (b) two cones, bases together.
    (c) two cones, both pointing down.
    (d) Two cones, both pointing up.

66. Which type of navigational mark shows a red flashing light?

    (a) An isolated danger mark.
    (b) A wreck marker.
    (c) A port hand channel mark.
    (d) A safe water mark.

67. A navigational buoy is coloured yellow over black with no top mark. It is:

    (a) an isolated danger mark.
    (b) a preferred channel mark.
    (c) a north cardinal mark.
    (d) a south cardinal mark.

68. The top mark of a west cardinal mark is:

    (a) two cones, bases together.
    (b) Two cones, points together.
    (c) Two cones, both pointing down.
    (d) Two cones, both pointing up.

69. A navigational buoy is vertically striped, red and white, with no top mark. It is:

    (a) an isolated danger mark.
    (b) a preferred channel mark.
    (c) a safe water mark.
    (d) a special mark.

70. Two black balls in a vertical line on a navigational mark indicates:

    (a) an isolated danger.
    (b) a south cardinal mark.
    (c) a special mark.
    (d) a safe water mark.

CHAPTER 6: ANSWERS

1 (b), 2 (c), 3 (d), 4 (a), 5 (c),
6 (a), 7 (a), 8 (a), 9 (b), 10 (b),
11 (d), 12 (b), 13 (a), 14 (a), 15 (d),
16 (b), 17 (d), 18 (c), 19 (a), 20 (b),
21 (c), 22 (d), 23 (a), 24 (c), 25 (d),
26 (a), 27 (c), 28 (d), 29 (a), 30 (b),
31 (c), 32 (b), 33 (a), 34 (c), 35 (b),
36 (b), 37 (a), 38 (d), 39 (c), 40 (a),
41 (b), 42 (a), 43 (a), 44 (b), 45 (c),
46 (a), 47 (b), 48 (a), 49 (d), 50 (a),
51 (d), 52 (c), 53 (b), 54 (d). 55 (c),
56 (d), 57 (a), 58 (c), 59 (b), 60 (c),
61 (a), 62 (d), 63 (a), 64 (c), 65 (b),
66 (c), 67 (d), 68 (b), 69 (c), 70 (a)

Chapter 7

ANCHORING SYSTEMS

ANCHORING TERMINOLOGY

ANCHOR AWEIGH:	The anchor has been hove out of ground and is clear of it.
A-COCKBILL (or COCK-A-BILL):	Anchor lowered clear of hawse pipe (or bow roller) and hanging vertically, clear of water.
BREAKING OUT THE ANCHOR:	To make ready and free the anchor for letting go.
BROUGHT UP:	The vessel is riding to her anchor and cable, and the former is holding.
CLEAR AWAY ANCHOR (S):	See "Breaking Out the Anchor".
CLEAR ANCHOR	The term "Clear Anchor" is also used when weighing anchor to indicate that it is in sight and clear of turns in the cable or other obstructions.
COCK-A-BILL:	See A-Cockbill above
FOUL ANCHOR:	An anchor caught in an underwater cable, or it has brought old hawsers to the surface with it, or it is fouled by its own cable.
FOUL HAWSE :	Both anchors are out and the cables are entwined or crossed.
GROWING:	The way the cable is leading from the hawse pipe, e.g., a cable is growing aft when it leads aft.
HAWSE PIPE:	The tube through which the anchor cable goes from the vessel to the anchor.
HOVE IN SIGHT:	The anchor is completely out of the water (when picking up anchor).
LEE SHORE:	This term is often misunderstood. It is the shore in the lee of the vessel - not the vessel in the lee of the shore.
LEE TIDE:	A tidal stream setting to leeward or downwind. The water surface has a minimum of chop on it, but the combined forces of wind and tide act upon the vessel at anchor.
LET GO:	Drop anchor in the water.
LONG STAY:	The cable is taught and leading down to water close to the horizontal.
NIPPED CABLE:	An obstruction, such as the stem or hawse-pipe lip has caused the cable to sharply change its direction.
SHORT STAY:	The cable is taught and leading down to water close to the vertical.
SNUB CABLE:	Stop the cable running out by using the brake on the windlass.
SURGE CABLE:	Allow cable to run out or slacken while being hauled.
UNDER FOOT	Anchor dropped just below the bow with a short length of cable running nearly up and down. It is sometimes used to turn a vessel around or to hold a vessel briefly on location.
UP-AND-DOWN:	The anchor cable is leading vertically down to the water.
VEER CABLE, WALK BACK:	Pay out cable under power, i.e., using the windlass motor.
WEATHER TIDE:	A tidal stream setting to windward or upwind. It makes the sea very choppy. The two forces act in opposition on a vessel at anchor.
WEIGH ANCHOR:	Lift anchor.

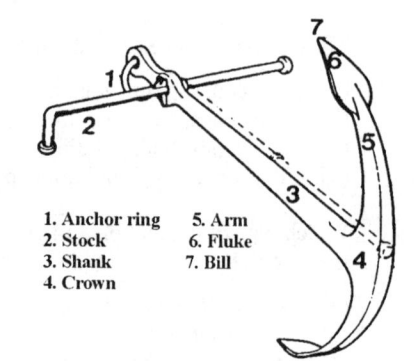

1. Anchor ring
2. Stock
3. Shank
4. Crown
5. Arm
6. Fluke
7. Bill

Fig 7.1: PARTS OF AN ANCHOR

Chapter 7: Anchoring Systems

TYPES & QUALITIES OF VARIOUS ANCHORS (Fig 7.2)

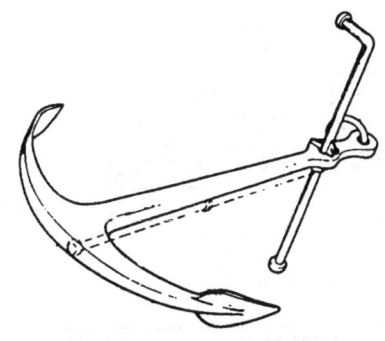

(a) KEDGE ANCHOR (ADMIRALTY OR PICK ANCHOR)
Until the invention of b & c type, this had the greatest holding power, but it is dangerous in shallow water as the vessel may accidentally sit on the one fluke that sticks out. It is now rarely used.

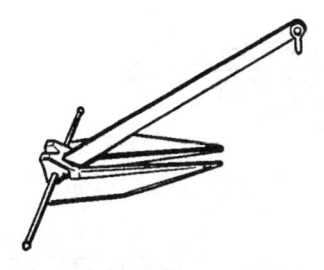

(b) DANFORTH ANCHOR (SPADE TYPE)
Excellent holding power.
Good close stowing on deck.
No disadvantage.

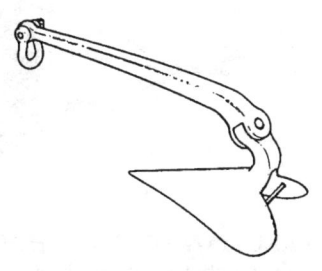

(c) CQR (PLOUGH TYPE)
Excellent holding power, but impossible to stow in a hawse pipe, and difficult to stow on deck.

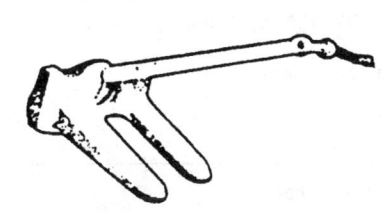

(d) STOCKLESS (PATENT)
It can be hove right into hawse pipe. But its holding power is less than type "a".

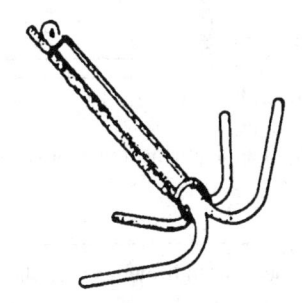

(e) GRAPNEL OR REEF ANCHOR
Practical for reef use.
It is not suitable as a general-purpose anchor.

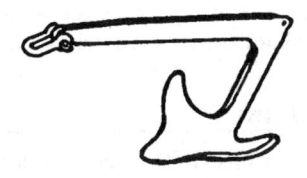

(f) BRUCE (CLAW) ANCHOR
This anchor was designed for anchoring oil rigs.
It is highly efficient, but expensive and difficult to stow on deck.

MAKING UP AN EMERGENCY ANCHOR (Fig 7.3)

Nail two pieces of hardwood with sharpened ends in the shape of a cross. Lash a stone in the middle. Pass a rope or chain around the arms of the cross and tie it over the stone to form a bridle.

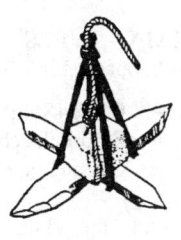

SEA ANCHOR

Also known as a floating anchor, ideally it is an elongated canvas bag with a large mouth on the upper end and a narrow mouth at the bottom. In deep waters, where it is not possible to use a "pick" anchor, sea anchor is streamed from a bridle fitted to its mouth. The bag fills up with water and helps to retard a vessel's drift and hold her head in the required direction to the wind.

When no longer required, it is pulled in from a line attached to its bottom end. Any drogue can be used for a floating anchor, such as, mooring (berthing) lines or a sail.

The length of line for a sea anchor should be such that the sea anchor and the vessel are not on the wave crest at the same time. This is to avoid the anchor being turned back on itself just when the vessel needs it to hold her against a breaking wave.

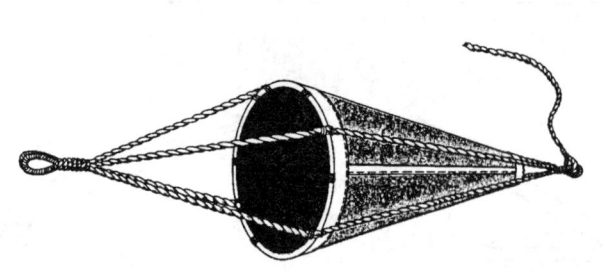

Fig 7.4: Sea Anchor

SELECTING YOUR ANCHOR

In the above list, the b & c type give a superior performance. Therefore, they are allowed a reduction in size. For example, a vessel of 10 m in length would require a stockless type anchor of 22 kg, but a spade or plough anchor of only 11 kg to give the same performance. (SEA 33).

The survey authorities determine the anchor and chain sizes for commercial vessels. For recreational vessels, the CQR anchor with hinged shank is considered quite strong and reliable. The Lewmar chart shown here is a guide only for its customers in selecting the correct size of CQR anchor. In using this chart, you need to consider the vessel's overall length and displacement – whether it is light, medium or heavy. Vessels of medium and heavy displacement should select the anchor of the next size up.

Fig 7.5: GUIDE TO SELECTING CQR ANCHOR (Lewmar Ltd.)

CHAIN

Just like ropes, chain is measured by its diameter in millimetres, i.e., the thickness of the metal.

Chains are usually made of mild steel or high *tensile* (stretchable) steel. Such chains do not require any special treatment. However, chains and other materials made of wrought iron need to be periodically *annealed* (reheated to rebind molecules) in order to prevent them from becoming brittle.

- It is easy to overlook faults in chains. They need to be carefully examined for corrosion, reduction in diameter of metal, distortion and missing studs.
- Do not use a chain that shows more than 10% wear or reduction in diameter
- Do not use any chain for lifting unless stamped with a Safe Working Load (SWL).

CHAIN – TYPES (Fig 7.6)

On board ships, 3 basic types of chains are used:

- Close (Short) link chain – used for lifting.
- Stud links chain – used for anchoring.
- Open (Long) link chain – used for cargo lashing

Stud link chain is designed for anchoring, where studs prevent cable jamming when coming in or out of chain lockers. It is not used for lifting because studs prevent it from showing elongation due to overloading or age.

Fig 7.7: CLOSE-LINK CHAIN (USED FOR LIFTING)

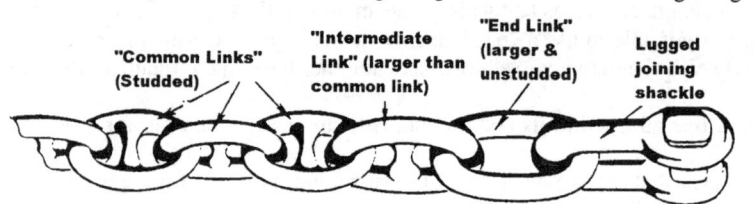

Fig 7.8: STUD LINK CHAIN (ANCHOR CHAIN)

Chapter 7: Anchoring Systems

Fit a strong **SWIVEL** between the chain and the anchor. It should be of a larger diameter than the chain. It will prevent kinks building up in the chain when the vessel swings around at anchor. Kinks make it harder to heave the chain on the gypsy and to stow it in the chain locker.

SHACKLES OR SHOTS OF CABLE

Most small boat owners can purchase the required length of cable from chandlers. For larger vessels, however, the chain is supplied in 15 or 7½ fathom lengths. Eight or nine of the 15-fathom (27.5 metre) lengths are shackled together to form an anchor cable. A vessel is thus said to have eight or nine "shackles of cable" or "shots of cable" on each anchor. A ship may pay out one shackle on her anchor for a "short stay" or four shackles for a "long stay". When in dry dock, large vessels remove the first shackle of chain from the anchor and rejoin it on the inner end. This keeps the wear in their chain as even as possible.

The joining shackles can be lugged shackles, but **Kenter shackles** are more common (See shackles in Chapter 9). A lugged shackle cannot be passed through a studded link. Therefore, an unstudded link at each end of the 15-fathom length has to be manufactured if lugged joining shackles are to be used.

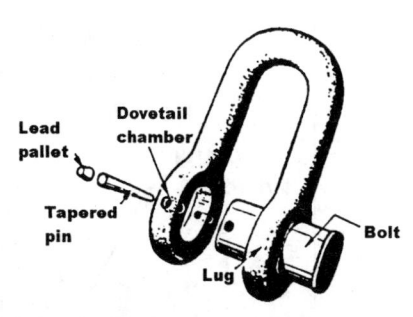

Fig 7.9: LUGGED JOINING SHACKLE

MARKING THE ANCHOR CABLE

If your anchor chain or cable is not marked, you will have to guess how much you have paid out, which can lead to problems. Many modern gypsies are fitted with a mechanical or electric revolution counter to indicate the amount of cable that is out. Still, it is important that cable be marked so that the amount paid out can always be known.

In small vessels, marking the cable at 10 metre intervals is a common practice. Seizing wire and paint marks are used on chain, and canvas and rope twine on rope cable. You can also thread the rope cable with different colours of tape or band it with paint rings (One ring for each 10 metres). Some people use electrical cable ties in a similar fashion. Intermediate 5 metre lengths can be marked with a single tie. The ties easily feed through fairleads and winches. They can be felt and counted in the dark and don't get obscured by mud.

Ships with anchor cable measured in shackles mark it in shackle lengths, as follows:

- At the end of the 1st shackle length - the first link on each side of the joining shackle is painted and marked.
- At the end of the 2nd shackle length - the second link on each side of the joining shackle is painted and marked.
- And, so on.

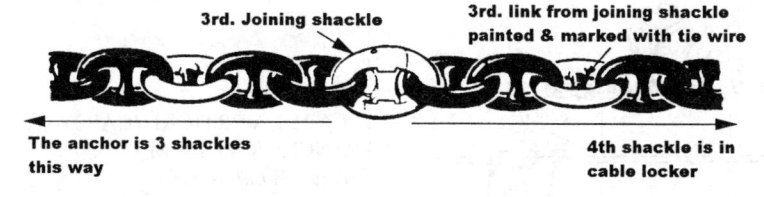

Fig 7.10 MARKING THE ANCHOR CABLE

ANCHOR CABLE - SURVEY

For survey, the cable is flaked out on deck or in a dock. All bolts and pins are driven out, cleaned and greased. The links and shackles are measured and tapped with a hammer to gauge the condition of the metal. Loose studs are renewed. The wear and tear of cable is reduced by periodically reversing it end for end. The anchor and its joining shackle are also inspected for wear.

CAPSTAN & WINDLASS

(See also Winches in Chapter 9)

Other than on small boats, the process of anchoring requires the use of either a capstan or a windlass. The Capstan is a drum that revolves around a vertical shaft, while Windlass consists of one or more drums operating on a horizontal shaft. However, the terms used in the boating industry are <u>vertical and horizontal windlasses</u>. Both types can be a simple lever ratchet arrangement or internally geared. In a capstan, the Gypsy (also known as Cable Lifter or Wildcat) is beneath the warping drum, which drives the gypsy via a ratchet and pawl. *(WARPING is to move a vessel along a wharf by heaving on her mooring lines, without the use of engines.)* A band brake controls the gypsy when the pawl is disengaged. For the choice between a horizontal and a vertical windlass, see "Cable Lockers" below.

Located on a strengthened deck on the forecastle, windlasses and capstans are either electrically or hydraulically driven, although on small boats they are often hand powered. By disengaging the cable lifter, the warping drums can be used for heaving mooring lines. Unless specifically designed for the purpose, do not secure mooring ropes on the warping drums. They are for heaving the lines; bollards are for securing the berthing lines.

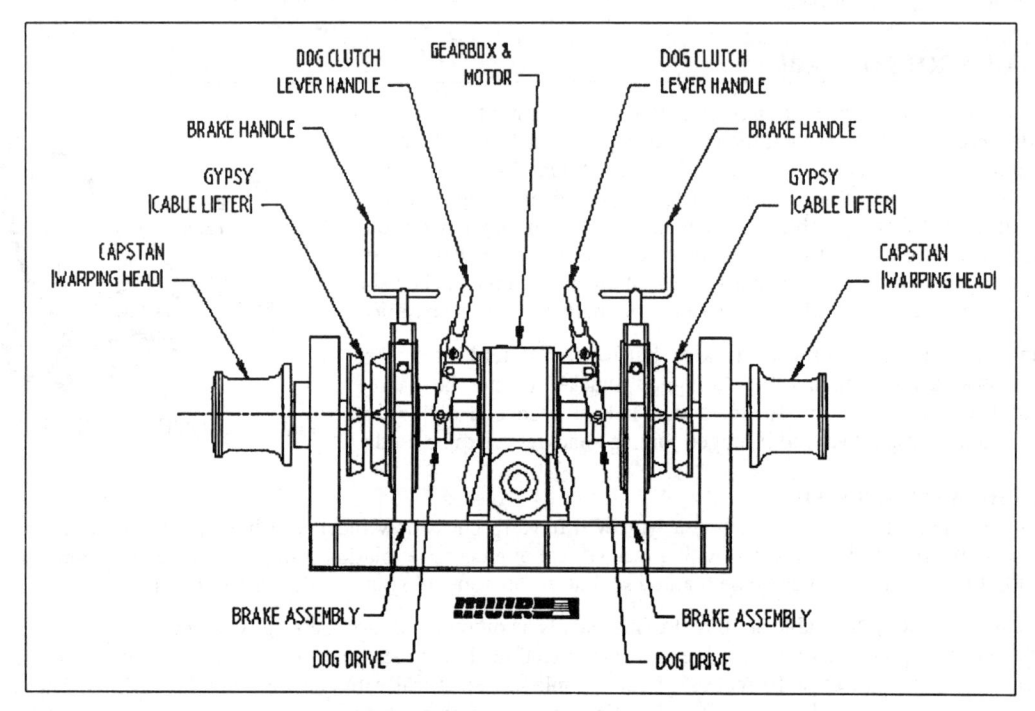

Fig 7.11: LARGE COMMERCIAL WINDLASS *(Muir Windlasses)*

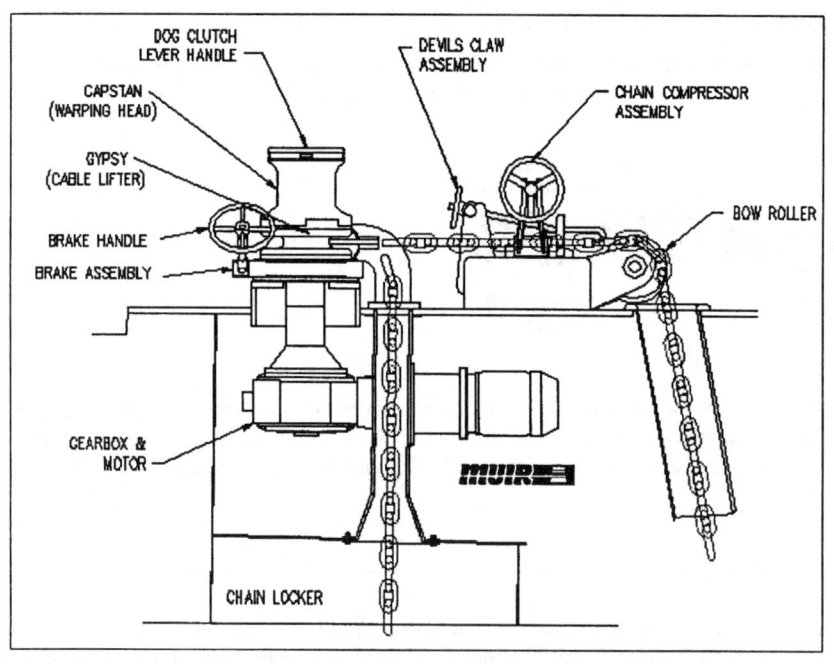

Fig 7.12
LARGE VERTICAL (CAPSTAN)
ANCHOR SYSTEM
(Muir Windlasses)

When selecting a capstan or windlass for your vessel, make sure that its PULLING POWER is at least 4 times the total weight of the rode (anchor chain or rope and chain combination) and anchor. Its CHAIN RECOVERY RATE should be around 15 metres per minute under load, which should bring the anchor home in most cases in less than 4 minutes. Electric windlasses usually speed up as the load decreases.

Ideally, you want to be able to "**LET GO**" the anchor in free fall instead of slowly "**PAYING OUT**" under power at the rate allowed by the windlass or capstan engine. This is possible only if, at the time of letting go, the anchor load is held by a brake (such as a band brake), instead of being controlled by "**self-locking**" **worm reduction gearing**. Some people may

Chapter 7: Anchoring Systems

regard the latter an advantage because it allows them to lower the anchor under electrical control from the cockpit while anchoring. But, for the anchor to get a good grip on the bottom at your chosen location, it is important to promptly lower it (drop it) and pay out sufficient chain (at least equal to twice the depth of water), as the vessel slowly moves astern. Furthermore, in windy conditions it can be difficult to hold the bow into the wind while slowly paying out cable. Unless the chain can freely run off the windlass, the bow will rapidly pay off, dragging the anchor with it. Premature or fast dragging will prevent the anchor from setting, with dangerous consequences.

Vessels wanting to handle a rope and chain rode should choose a windlass with the chain gypsy and the rope drum fitted side by side, or stacked vertically, instead of being on the opposite sides of the windlass. The base of the rope drum should be of the same diameter as the gypsy. This arrangement would allow an easy transition from rope to chain when heaving the anchor. The chain can be forced off the drum and into the gypsy by simply taking extra turns of rope around the drum.

Some gypsies are fitted with a central rope-gripping groove, similar to those on self-tailing winches. However, their transition from rope to chain can be difficult if the rope is not neatly spliced to the chain. This system is also hard on the rope when braking a free falling anchor. Such a gypsy is more suitable in a capstan with chain wrapped around it half to three-quarter of its way than with only a quarter wrap around a windlass. It would keep the rope more firmly gripped as the first links settle into the gypsy pockets. Capstans are designed with a greater wrap to prevent chain links lifting out of the vertically fitted gypsy pockets. On the horizontal windlass, a quarter wrap is sufficient because the chain is held in place by gravity.

Fig 7.13: ROPE/CHAIN COMBINATION WINDLASS *(Muir)*

A gypsy and its chain need to be compatible. Two chains of the same size and grade may require subtly different gypsy dimensions. A mismatch would cause problems when heaving or letting go. Therefore it is better to buy a standard chain and windlass or capstan.

WINDLASS MAINTENANCE
(See also winch maintenance)

Capstans and windlasses require regular inspection and maintenance as per the manufacturers instructions. Lubrication of bearings and all moving parts as well as brake maintenance is essential. Brake liners need to be inspected and renewed where fitted. Brake drums are to be kept free from sand and grit. Hydraulic pipes and cocks are to be inspected for leaks. Always test the windlass or capstan after a sea passage.

CABLE LOCKERS

Well designed cable or chain lockers are deep and narrow to ensure that the cable neither plies up in a heap nor capsize and become fouled when the vessel inclines.. In poorly designed lockers, the cable when hove in does not stow freely and automatically, i.e., it is not self-tailing. Manual assistance is often essential.

A boat whose cable locker is shallow or of unusual shape would find the horizontal windlass more suitable, especially if her rode is a rope which does not have the gravitational pull of a chain. The horizontal windlass feeds the rode

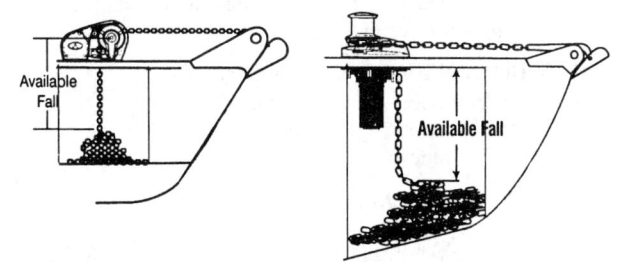

Fig 7.14: CABLE LOCKER FOR HORIZONTAL & VERTICAL WINDLASSES *(Lewmar, slightly altered)*

vertically down and directly into the locker, irrespective of its gravitational pull. On the other hand, vertical windlasses are regarded more aesthetic and more secure because the anchor rode makes a 180° wrap around the gypsy (as opposed to 90° on a horizontal windlass). However, in order to be self tailing the lockers needs to have a larger fall, i.e., it needs to be deeper to provide gravity for self tailing of the rope rode.

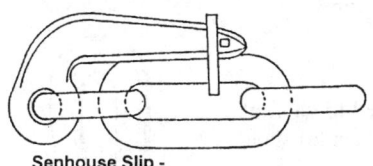

Senhouse Slip -
It goes through the link of a cable

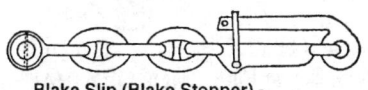

Blake Slip (Blake Stopper) -
It goes over the cable link

Fig 7.15: SENHOUSE SLIP & BLAKE SLIP

SECURING THE BITTER END

The inboard end (bitter end) of the cable is secured inside the cable locker such that it can be released in an emergency without causing injury to the person releasing it.

In large vessels the end link is usually placed through a stiffened slot cut in a bulkhead or deck surrounding the chain locker so that it projects into an adjoining compartment, where a pin is driven through the link and forelocked, or secured by a Senhouse Slip. Thus, when necessary, the cable can be slipped from outside the chain locker. On many boats it is secured to a strong point inside the chain locker via a piece of strong rope. The rope is long enough to allow the last link of chain to be retained on the gypsy, making it possible to cut the rope in an emergency.

SECURING ANCHORS FOR THE SEA

Anchors are not needed when a vessel is at sea. They must therefore be safely secured until the next port of call. Securing anchors at sea also protects the cable lifters from the forces of pounding and pitching, and the hull is protected from the risk of being punctured by anchors coming loose. Anchors are secured in place with devices such as Blake Slips, and Devil's Claws fitted with Bottle Screws, and Chain Locks and Compressors. They relieve the windlass from the weight of the anchors.

The Spurling Pipe (Cable or Chain Pipe), through which the cable enters cable locker, is covered with the supplied cover or with canvas and cement so that it does not fill up with salt water. Unless the chain locker is self-draining, it will have to be pumped out through a connection to a pumpable bilge.

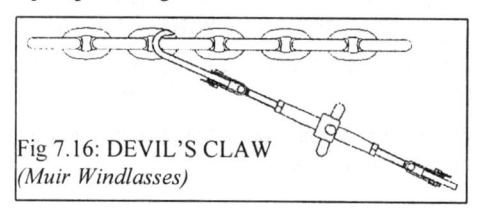

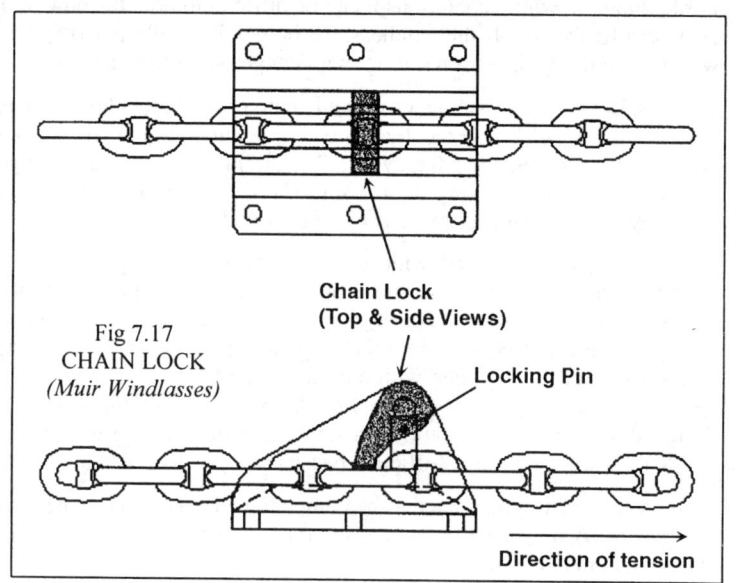

Fig 7.17 CHAIN LOCK (Muir Windlasses)

Fig 7.16: DEVIL'S CLAW (Muir Windlasses)

ANCHOR CABLE FOR SMALL VESSELS

➢ *FIRST PREFERENCE*: short link chain. The following are the recommended sizes of tested chain of at least Grade 5 (See Chapter 6 for grading)

BOAT WEIGHT	CHAIN SIZE
5-10 tonnes	8 mm
10-12 tonnes	10 mm
12-20 tonnes	12 mm

➢ *SECOND PREFERENCE*: rope + a short length of chain pennant. (Together, it is called the "**RODE**").

- *ROPE*: Nylon rope or a slightly larger diameter "silver rope" (mixed synthetics) is satisfactory. Do not use a rope that floats, such as polypropylene. It makes it difficult for the anchor to dig in, and can easily foul the propeller.

- *CHAIN PENNANT*: The chain pennant is a length of chain of approximately equal to the boat's length attached between the anchor and the rope rode. The weight of the chain will cause it to lie on the bottom and:
  ➢ Will provide a horizontal pull on the anchor, improving its holding power.
  ➢ Reduce chafe on the rope rode.
  ➢ Reduce shock load due to waves.

Fig 7.18: HARD EYE

- *ROPE & CHAIN CONNECTION:* The joining shackle should not be allowed to chafe the rope. A hard eye (eye with a thimble in it) in the rope is recommended. All shackles should be moused with non-corroding seizing wire, to prevent them coming undone due to vibrations.

Various authorities publish tables of recommended anchor sizes, anchor ropes and chains. They do vary, but only slightly. The following table appears in AMSA's Safety Education Article No. 33. For a commercial vessel, her anchoring gear forms part of the survey conducted by the State authority.

Fig 7.19 MOUSED SHACKLE

Chapter 7: Anchoring Systems

BOAT LENGTH	ANCHOR ROPE		CHAIN PENNANT	
	Size	Length	Size	Length
Up to 5 metres	8 mm	50- 70 m	6 mm	3 m
5-8 metres	10 mm	75-100 m	6 mm	6 m
8-12 metres	12 mm	100-125 m	6 mm	10 m

FOR LARGER BOATS:
- ❖ Extract rope size from Southern Ocean Ropes Chart in Chapter 6
- ❖ Determine suitable length of rope & a proportionally larger chain pennant.

NOTE: When operating in open waters more than one anchor must be carried.

SECURING THE ANCHOR RODE ON BOARD

The inboard end (bitter end) of the anchor rode should be made fast to a structurally strong point preferably inside the cable locker. It should be readily accessible and easy to cast off in an emergency.

When at anchor, the anchor rode, now taking the weight of the vessel at anchor should be made fast to the vessel's bits or centreline bollard.

If made fast to the bitts, one complete turn around one or both bitts should be taken before framing "figure eights" on the bitts.

If made fast to the centreline bollard, this should be done using the illustrated Bollard Hitch (also known as a Towing Hitch or Lighterman's Hitch). This involves placing a number of turns on the bollard before passing a bight of the rode under the tensioned rode and over the bollard in the reverse direction.

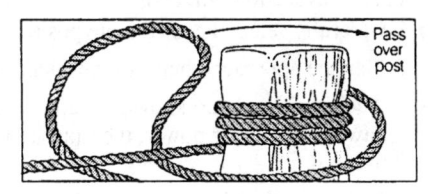

Fig 7.20

HOW MUCH SCOPE TO USE

The term Scope refers to the Ratio of the Length of the cable used to the Depth of water.

Scope = Length of cable paid out ÷ Depth

For example, if the depth of water is 10 metres and you have paid out 50 metres of cable, then the scope is 5:1.

Scope is the major factor in deciding the holding capability of an anchor. Anchors work more efficiently with the stock being pulled horizontally. Therefore, the bigger the scope the better. A large catenary (curve or dip in the cable) indicates a large scope.

RECOMMENDED SCOPE

Sea conditions	Anchor cable	Scope
Favourable	Chain	3:1
Average	Chain	5:1
Rough	Chain	7:1
Favourable	Rope (+ chain pennant)	5:1
Average	Rope	8:1
Rough	Rope	10:1

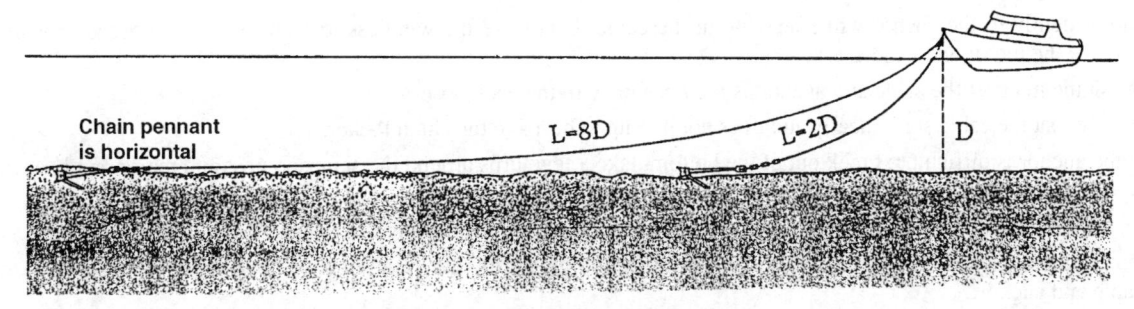

Fig 7.21:   Scope of 2:1 & 8:1 is shown

LETTING GO & PICKING UP AN ANCHOR

CHECK THE ANCHORING LOCATION
- Quality of the holding ground
- Shelter from weather
- Swinging room
- Range of tide
- Rate of tide
- Likelihood of other vessels anchoring too close

ANCHORING PROHIBITED AREAS
- Within 200 metres of submarine cables and pipelines.
- In prohibited or unsafe anchorages.
- In fairways or channels (except in emergency).
- Near leading marks.

ANCHORING PROCEDURE
- Carefully select the anchorage position having regard to the above factors.
- Plan the approach course to the anchorage position, head into wind or tide.
- Anchors have been known to jam in the hawse pipe. In getting the anchor ready for letting go, first veer it out of the hawse pipe under power, then put it on brake and disengage the motor. The letting go should be done by releasing the brake.
- Make sure the bitter (inner) end is secured on board.
- Check the depth and re-confirm the anchoring position.
- Have a slight astern movement when letting go the anchor - to prevent the cable piling up in a heap on the bottom.
- Make sure no one is standing in the way of the cable or in the bight of the cable. When letting go the anchor, protect yourself from the rust flakes or dry pieces of mud that may fly out of the cable locker.
- Let go the anchor. Allow the cable to run freely. Use only the brake to control the cable. Pay out the required length (see "scope").
- Secure the cable and wait until the vessel is "brought up", i.e., comfortably holding and riding on the anchor. She is brought up when the taut cable goes smoothly slack. If the cable shudders or tightens again the anchor could be dragging.
- After anchoring do not leave the weight of the anchor on the capstan engine. Engage the brake and disengage the gipsy from the capstan engine.
- Unlike large ships, windlasses on boats are usually not designed to hold high loads at anchor. Therefore, after anchoring, secure the anchor cable using a chain compressor or a devil's claw, taking its weight off the windlass.
- Note down two or more bearings of navigational features or landmarks, preferably on an abeam bearing. This is to check later that the anchor is holding.
- Stop vessel's engine (but only after the vessel is brought up).

HOW TO WEIGH (TAKE IN) ANCHOR
- Start vessel's engine and test all controls.
- Assign a capable person to weigh anchor.
- Engage gipsy and disengage brake.
- Motor gently to the anchor while heaving in the cable. Don't use the windlass to haul the boat to the anchor. It will damage the motor and windlass.
- Wash the mud off the cable and anchor as it comes in, with the deck hose.
- Ensure that the cable stows neatly and does not pile up in heaps in the chain locker.
- If the anchor is difficult to break out of the bottom, take a few turns around the sampson post and motor gently ahead.
- In most cases the master in the wheelhouse cannot see the cable. The person heaving in the cable should continuously report the state of the cable to the master so that the vessel can be manoeuvred to reduce the load on the windlass.
- When the anchor is up, put on brake and disengage the gipsy. In case of an anchor hove on deck, securely lash the cable and anchor.

Chapter 7: Anchoring Systems

USE OF AN ANCHOR BUOY

An anchor buoy is a small buoy or a block of wood with its pennant made fast to crown of the anchor. It is used for indicating position of anchor when on the bottom. It is particularly useful in crowded anchorages, and in the Mediterranean moor, Baltic moor, and Running and Standing (Bahamian) moor. Make sure the pennant is longer than the depth of water.

IF ANCHOR BECOMES FAST UNDER A SUBMARINE CABLE

Submarine cable and pipelines are signposted on inland waterways wherever possible. They are also marked by appropriate symbols on navigational charts. Modern submarine cables can be as small as 2 cm in diameter and are likely to carry very high voltages (up to 13,000 volts). They may be buried or simply lying on the seabed. Cutting them or handling their cut ends could prove lethal. A cable fouled on an anchor or fishing gear or brought on board could be under considerable tension. The whiplash from cutting or breaking it could cause serious injury. The weight or the tension of the cable might also upset the stability of a small vessel. The damage to a gas pipeline may release gas at high pressure. It too could cause injury, loss of life or damage to vessel. Damage to cables and pipelines might also result in disruption of services to communities. Under Commonwealth legislation serious penalties can be imposed for damaging a submarine cable or pipeline. The owner of the cable or pipeline may claim civil damages.

Therefore, take care when attempting to free your fouled gear from a submarine cable or pipeline. If you believe that your gear cannot be freed without risk or damage to the cable or pipeline, the gear should be abandoned. Under the terms of the International Conventions of 1884, 1958 and 1982 for the protection of submarine cables, which are now part of the International Law of the Sea and subsequent Australian legislation, it is obligatory that anyone hooking up a cable or pipeline must sacrifice their gear rather than damage or cut the cable or pipeline. Cable and pipeline owners must pay compensation for any gear lost under these circumstances, providing such a loss can be proven and all reasonable precautions were taken to prevent fouling or damaging the cable or pipeline.

In trying to free your anchor, try steaming up and down. If unsuccessful, seek assistance from the authorities or buoy the anchor if you need to leave in a hurry. Another way to clear the anchor from a submarine cable is to pass a slip rope under the bight of the cable that you have picked up and secure it on board. "Walk back" the anchor until it is free. You can now heave in the anchor and slip the submarine cable back into the water. However, this method could be unsafe if it is a power cable, and it could also result in damaging the cable. The best overall solution may be to sacrifice your gear. In any case inform the authorities. They would check if the cable or pipeline is damaged.

FORCES ACTING ON AN ANCHOR

1. **WIND**: Wind physically pushes the vessel with a quadratic (squared) effect. In other words, if the wind speed goes up 3 times, the force on the vessel is increased 9 times. Wind is the main cause of anchor dragging. The prospect of anchor dragging can be reduced by increasing the scope of the cable.

2. **CURRENTS**: River currents and changes in tidal flow affect the load on an anchor cable or rode. However, they have to be fairly extreme to match the force of wind. The ratio varies with vessel design, but a good rule of thumb is 1 knot of current is equal in effect to about 10 to 15 knots of wind.

3. **WAVES**: Waves cause the vessel to pitch, yaw and heave up and down. This can cause the anchor to break out. However, if the anchor line is long enough, the vessel's movement will not affect the anchor. "The anchor holds the cable, and the cable holds the ship".

ANCHOR WATCH

There is no such thing as a completely safe berth or mooring. When on anchor watch, check anchor bearings regularly to ensure that the vessel is not dragging the anchor. Also make sure that the anchor signals (day and night) are suitably displayed. Large vessels must also illuminate their decks and superstructure.

Whether in confined or open waters, an anchor watch is also essential to keep a watch that other vessels anchoring nearby maintain a sufficient swinging room for your vessel, and to guard against fire hazards.

It is particularly essential to maintain an anchor watch in an open roadstead, where a change in weather may have an adverse effect. In addition, other vessels may not move with as much care in open waters as in confined waters.

ANCHOR BEARINGS

- It is essential that when the vessel is brought up to the anchor, two or more anchor bearings or radar ranges be taken and entered in the logbook. Plot the anchor position on the chart, and take a tracing of the radar display.
- The bearings or distances should remain constant, except for a slight change due to the vessel swinging at anchor.

- Transits of prominent objects should be noted and logged.
- Note the echo sounder depth, tide and the under-keel clearance.

SIGNS OF ANCHOR DRAGGING (NOT HOLDING)
➤ Significant change in anchor bearings, radar ranges or transits.
➤ The state of the cable changing quickly from "long stay" to "short stay".
➤ Vibrations in the cable felt outboard of the fairleads.
➤ The leading forward of a "hand lead line" dropped over the side.

PREVENTING CHAFING OF HULL & ANCHOR LINE
Fairleads, roller fairleads and hawse pipes help reduce chafing. A piece of split rubber or plastic tubing can be put over the contact point between the anchor line and hull to further reduce chafing.

When anchoring for prolonged periods, it is a good practice to "freshen the nip". This means to periodically veer or haul (pay out or take in) slightly the anchor rope or chain, so that a part which has been subject to nip or chafe is moved away and a fresh part takes its place.

OTHER ANCHORING CONSIDERATIONS

ANCHORING IN TANDEM: This means anchoring with two anchors secured one after the other on one cable. The leading anchor is the working anchor. It is shackled with a short length of chain onto the crown of the second anchor, which is known as the storm anchor. This practice is recommended for stormy conditions, where the swinging room is not restricted, in preference to mooring using two anchors, each on separate cables.

ANCHORING ON A REEF OR ROCKY BOTTOM: The anchor may loose its hold on coral and come free without warning. But more importantly, it may become jammed or fouled. You may not be able to weigh it when it is time to leave. There are two ways of avoiding this problem:

1. Use a reef anchor or grapnel (see types of anchors).
2. Unshackle the anchor warp from the anchor ring. Secure it to the crown of the anchor and lead it along the shank of the anchor and lash it to the anchor ring with a small secure lashing (rope yarn). Anchor as normal.

The principle is that should the anchor fluke jam in a rock crevice, a quick hard pull will break the lashing and the anchor can be pulled out backwards from the crevice. The attached illustration appeared in AMSA's Safety Education Article No. 33.

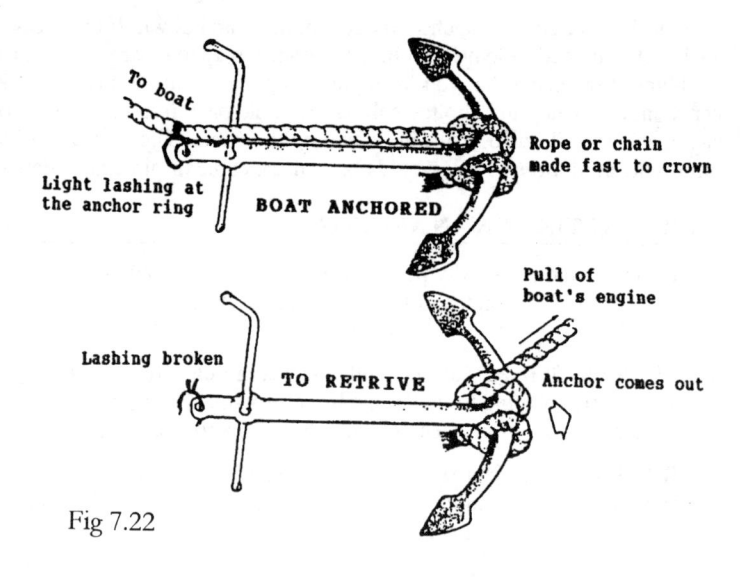

Fig 7.22

NOTE: If, at the time of anchoring, you are in doubt as to the nature of the seabed, you should buoy the anchor with a trip line attached to its crown. You will be able to heave in the anchor with this line if necessary.

ANCHORING IN DEEP WATER: If an anchor has to travel a large distance to reach the bottom, it may pick up uncontrollable speed when "let go". It should therefore be veered slowly (with windlass in gear, for example) until it is only a short distance from the bottom and then let go. Note the cable marks as it is veered.

IF ANCHOR DOESN'T HOLD: First pay out more cable. Consider letting go a second anchor. If this fails to stop the vessel dragging, rig a sea anchor, and advise a Coast Radio Station of your situation. Start up the engine and request assistance, if necessary.

RIDING OUT A GALE AT ANCHOR: Use two anchors, one with long cable to provide good holding power, and the other anchor dropped under foot (under the bow with cable only to reach the bottom) to reduce the vessel's yaw.

Alternatively, if you want to use both anchors for riding, lay them about 20 degrees apart. The risk in this method is that

Chapter 7: Anchoring Systems

if there is a large shift in the wind direction, you may end up with a turn in the cables that can be extremely cumbersome to clear.

Books also recommend that if the vessel on two anchors starts to drag, and if the sea is reasonable, you may consider slowly steaming up between the anchors to prevent dragging.

However, for very small vessels, perhaps the easiest way to ride a gale is to take her out of water if possible and secure her. You may be able to take her away somewhere safe on a trailer or run aground in a creek.

MOORING WITH TWO ANCHORS

This may be necessary where the swinging room is restricted in a tidal waterway. There are two methods:

1. ANCHORING FORE AND AFT
(ONE ANCHOR FROM BOW & ONE FROM ASTERN)

The problems with this method, if there is a wind shift, are:
- The wind or tide may drag the vessel sideways.
- The stern may be swamped by a wave from astern.

2. ANCHORING WITH TWO ANCHORS FROM THE BOW

In this method, one anchor leads forward and the other astern. The vessel sits between them and rides on one at flood tide and the other at ebb. This method is not advisable when using rope warp that may foul the propeller.

The procedure to lay these anchors is to conduct a *RUNNING MOOR* or a *STANDING MOOR*. The difference in the two terms lies in the way the two anchors are dropped:

In Running Moor, the vessel drops the first anchor while still making headway, i.e., "running" into the tide and paying out on the cable. She lets go the second anchor after she is past the proposed midway position between the two anchors. She then falls back, heaving on the first cable and slacking on the second, in order to lie midway between them.

In Standing Moor, the vessel sails past the proposed midway position without dropping an anchor. She stops (hence the term, "standing"), drops the first anchor and falls back, while paying out on the cable, to a position past the midway point between the two anchors. She then drops the second anchor. Now, paying out on this cable and heaving on the first she ends up lying midway between them.

The running moor has the advantage that it is quicker to execute and the vessel is under power, giving a better control on positioning the anchors. The Standing and Running moors are also referred to as the *BAHAMIAN MOOR*.

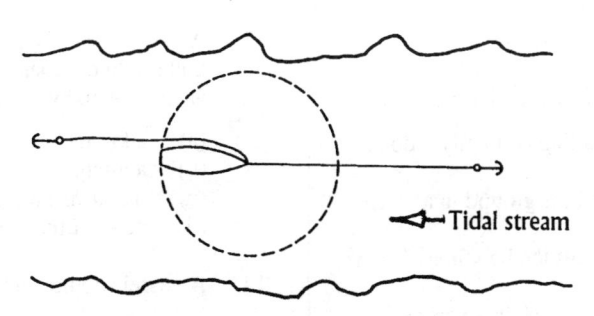

Fig 7.23: MOORING WITH TWO ANCHORS

CHAPTER 7: QUESTIONS & ANSWERS

ANCHORING SYSTEMS

1. A stud-link anchor chain is also described as:

    (a) a short-link chain.
    (b) a close-link chain.
    (c) an open-link chain.
    (d) none of the choices stated here.

2. The size of anchor chain is measured by the:

    (a) overall circumference of the link.
    (b) overall diameter of the link.
    (c) diameter of the iron of the link.
    (d) circumference of the iron of the link.

3. For small vessels, anchor chain is manufactured in the following lengths:

    (a) 15 fathoms.
    (b) 7.5 fathoms.
    (c) a continuous length.
    (d) 10 metres.

4. The term "Anchor Aweigh" means that the anchor has been:

    (a) made ready to let go.
    (b) hove out and above the water.
    (c) let go.
    (d) hove out of the ground.

5. On the anchoring command "Let Go" given to you, you would:

    (a) remove the anchor lashing.
    (b) let go the entire cable.
    (c) lower the anchor into the water.
    (d) let go the anchor into the water.

6. The anchoring tem "Up and Down" means:

    (a) the anchor cable is leading vertically down to the water.
    (b) the anchor is sitting on the ground in a vertical position.
    (c) the anchor is hanging off the bow in a vertical position.
    (d) The anchor is sitting on deck in a vertical position.

7. A 10-metre fishing vessel operating up to 15 miles offshore should be equipped with the following minimum anchor rode:

    (a) 12 mm synthetic rope with a 6 mm chain pennant.
    (b) 32 mm synthetic rope without a chain pennant.
    (c) 32 mm synthetic rope with a 12 mm chain pennant
    (d) 16 mm chain cable.

8. When anchoring a vessel, the most important aspect to consider is the:

    (a) range of tide.
    (b) rate of the tidal stream.
    (c) height of high water.
    (d) height of low water.

9. The most accurate means of checking that a vessel at anchor is holding her position is to:

    (a) feel for vibrations in the anchor cable.
    (b) check the anchor bearings.
    (c) observe a transit.
    (d) observe another anchored vessel's position.

10. Anchoring is <u>permitted</u> within 200 metres of a:

    (a) submarine cable.
    (b) submarine pipeline.
    (c) leading mark.
    (d) channel boundary.

11. A long anchor cable scope will help to prevent the anchor from dragging due to:

    (a) wind.
    (b) waves.
    (c) poor holding ground.
    (d) all of the factors stated here.

12. When anchored for the night in a depth of 7 metres in a sandy open bay, a vessel should pay out the following length of anchor rope:

    (a) 10 metres.
    (b) 20 metres.
    (c) 30 metres.
    (d) 70 metres.

13. An anchor rope secured to a sampson post or a bollard on deck should have the weight of the anchor taken by:

    (a) a knot.
    (b) a hitch.
    (c) the turns on the post.
    (d) none of the choices stated here.

14. A vessel should anchor by:

    (a) throwing it ahead as far as possible.
    (b) lowering it until it touches the bottom.
    (c) letting go while moving astern slowly.
    (d) letting go while moving ahead slowly.

15. The part of the anchor between anchor ring and crown is known as the:

    (a) stock.
    (b) fluke.
    (c) shank.
    (d) arm.

Chapter 7: Anchoring Systems

16. Among the anchors listed below, the one with the best holding power, weight for weight, and on most bottoms, is:

    (a) Kedge
    (b) Grapnel
    (c) CQR
    (d) Admiralty.

17. The holding power of an anchor with a rope warp is increased by:

    (a) securing the rope to its crown.
    (b) increasing the size of the rope.
    (c) shackling a length of chain between them.
    (d) streaming the anchor astern.

18. A submarine cable fouled in a vessel's anchor may be freed by:
    (i) cutting it.
    (ii) steaming up and down.
    (iii) steaming in a circle.
    (iv) sacrificing the anchor.

    The correct answer is:
    (a) all except (i).
    (b) all except (ii).
    (c) all except (iii).
    (d) all except (iv).

19. In case of a submarine cable becoming fouled with a vessel's anchor, the authorities must be informed:

    (a) if the submarine cable was damaged.
    (b) under all circumstances.
    (c) if a slip rope was employed to clear it.
    (d) if the anchor was sacrificed.

20. A "brought-up" vessel means her anchor is:

    (a) dragging.
    (b) dredging.
    (c) holding.
    (d) secured.

21. A catenary in an anchor cable:

    (a) reduces the anchor's holding power.
    (b) increases the anchor's holding power.
    (c) has no effect on the anchor's holding power.
    (d) helps to keep the anchor under foot.

22. A floating anchor is:

    (a) a sea anchor.
    (b) a pick anchor.
    (c) not an anchor.
    (d) a reef anchor

23. On the deck of a vessel you see two pieces of hardwood with sharpened ends, nailed together in the shape of a cross, and a stone lashed in the middle. It is an emergency:

    (a) sea anchor.
    (b) stabilizer.
    (c) pick anchor.
    (d) anti-rolling device.

24. The length of line run out on a sea anchor should permit the vessel and the anchor to be:

    (a) on wave crests at the same time.
    (b) in a trough and on a crest at the same time.
    (c) in wave troughs at the same time.
    (d) on the same wave crest at the same time.

25. In calm waters the following is the recommended scope for anchoring a vessel whose rode is a chain cable:

    (a) three times the depth.
    (b) six times the depth.
    (c) ten times the depth.
    (d) equal to the depth.

26. Anchors In Tandem means two anchors:

    (a) laid fore and aft.
    (b) with cables turned around each other.
    (c) riding one on the other.
    (d) secured one behind the other on one cable.

27. To "Freshen The Nip" is to periodically:

    (a) check anchor for dragging.
    (b) veer or haul a rope or chain slightly.
    (c) check vessel's position.
    (d) report vessel's position.

28. When anchoring in deep water, the anchor should be:

    (a) let go.
    (b) walked out all the way.
    (c) veered out slowly most of the way.
    (d) thrown overboard.

29. A vessel is repairing her engine at anchor in an open anchorage. A storm blows up and she starts to drag anchor towards the shore. She should consider:

    (a) paying out more cable.
    (b) letting go a second anchor.
    (c) requesting help.
    (d) all of the choices stated here.

30. A vessel anchoring close to a danger wants to be able to swing but restrict the swinging room. She should anchor as follows:

    (a) Anchors from bow and stern.
    (b) Two anchors from bow leading ahead.
    (c) Bahamian (Running or Standing) Moor.
    (d) One anchor from stern.

31. The seamanship term "kedging" means:

    (a) moving a vessel by laying out an anchor.
    (b) dragging an anchor.
    (c) running a vessel onto a beach.
    (d) securing to a mooring buoy.

32. The seamanship term "veer" means:

    (a) to pay out cable under control.
    (b) to stop a rope that is running out.
    (c) to stop a cable that is running out.
    (d) to allow a rope or cable to run out on its own weight.

33. The seamanship term "snub" means:

    (a) to pay out cable under control.
    (b) to stop suddenly a cable or rope that is running out.
    (c) to surge a cable.
    (d) to secure a rope.

34. The seamanship term "surge" means:

    (a) to veer a cable.
    (b) none of the options stated here.
    (c) to allow a rope or cable to run out on its own weight.
    (d) to stop a cable from running out.

35. An anchor tripping line is useful when anchoring:

    (a) on a windward shore.
    (b) on a rocky or reef bottom.
    (c) with large anchors.
    (d) on a lee shore.

36. On receiving the command "Weigh Anchor", you would:

    (a) start heaving in the anchor.
    (b) verify the weight of the anchor.
    (c) let go the anchor.
    (d) drop the anchor under foot.

37. The term "Foul Hawse" means:

    (a) the anchor is caught in an underwater cable.
    (b) both the anchors are out and the cables are entwined.
    (c) the anchor is fouled by its own cable.
    (d) the anchor has brought up old hawsers to the surface with it.

38. The stud link chain is:

    (a) made in short lengths.
    (b) designed for lifting.
    (c) designed for anchoring.
    (d) made in long lengths.

39. Marking the anchor cable in 10 metre lengths is:

    (a) common practice in small vessels.
    (b) referred to as shackles of cable.
    (c) common practice in large vessels.
    (d) an unsafe practice.

40. On a windlass the cable lifter is known as the:

    (a) Windlass drum.
    (b) Warping drum.
    (c) Gypsy.
    (d) Cable drum.

41. A Wildcat is:

    (a) a Windlass drum.
    (b) a Warping drum.
    (c) another name for a Windlass.
    (d) another name for a Gypsy

42. A self-tailing cable locker:

    (a) is shallow in construction.
    (b) is not fitted in large vessels.
    (c) is not fitted in small vessels.
    (d) allows kink-free stowage of cable.

43. The bitter-end of a cable is:

    (a) its outer end.
    (b) secured inside the cable locker.
    (c) the stud-free link at each end.
    (d) secured to the anchor.

44. A Spurling Pipe is the:

    (a) chain pipe through the hull.
    (b) anchor pipe through the hull.
    (c) chain pipe leading to chain locker.
    (d) chain pipe leading to water.

45. A shackle is moused in order to:

    (a) prevent it becoming distorted.
    (b) make a hard eye splice in it.
    (c) make a soft eye splice in it.
    (d) prevent it coming undone due to vibrations.

46. When at anchor, the knot used for making fast the anchor rode on board is the:

    (a) Bollard Hitch.
    (b) Clove Hitch.
    (c) Fisherman's Bend.
    (d) Sheet Bend.

Chapter 7: Anchoring Systems

47. The term 'Snubbing Round' means:
    (a) dropping the anchor under foot in order to turn the vessel short round.
    (b) stopping the cable from running out.
    (c) letting the cable run out freely.
    (d) paying out the cable with windlass in gear.

48. Anchoring a runabout by the stern:
    (a) is better than anchoring by the bow.
    (b) is suited when fishing.
    (c) may cause waves to break over the stern and swamp the vessel.
    (d) is suited to deep waters.

49. A vessel moored to a buoy, instead of anchoring:
    (a) will have a much smaller swinging circle.
    (b) will have a much larger swinging circle.
    (c) will be less secure.
    (d) will require more cable to be paid out.

50. The anchor chain survey involves the following steps:
    (i) renew any damaged chain links.
    (ii) remove joining shackles, bolts & pins to clean & grease.
    (iii) measure chain links for size reduction.
    (iv) Gauge metal condition by tapping with a hammer.

    The correct answer is:
    (a) all except (i).
    (b) all except (ii).
    (c) all except (iii).
    (d) all except (iv).

51. Anchor chain is periodically reversed end for end in order to:
    (a) keep it evenly worn.
    (b) prevent the inner end rusting.
    (c) reverse the direction of joining shackles.
    (d) prevent the outer end rusting.

52. A short length of chain secured between an anchor and the anchor rope is to:
    (a) help the anchor to dig deeper.
    (b) provides a horizontal pull on the anchor.
    (c) strengthen the anchor cable.
    (d) make the anchor heavier.

CHAPTER 7: ANSWERS

1 (d), 2 (c), 3 (c), 4 (d), 5 (d),
6 (a), 7 (a), 8 (b), 9 (b), 10 (d),
11 (d),12 (d),13 (c),14 (c),15 (c),
16 (c),17 (c),18 (a),19 (b),20 (c),
21 (b),22 (a),23 (c),24 (b),25 (a),
26 (d),27 (b),28 (c),29 (d),30 (c),
31 (a),32 (a),33 (b),34 (c),35 (b),
36 (a),37 (b),38 (c),39 (a),40 (c),
41 (d), 42 (d),43 (b),44 (c),45 (d),
46 (a),47 (a),48 (c),49 (a),50 (a),
51 (a),52 (b)

Chapter 8

BOAT HANDLING & TOWING

SAILING - THE THEORY
(Based on RYA's simplified explanation)

According to Bernoulli's principle, a curved surface in an airflow creates low pressure on one side and high pressure on the other side. It is for this reason that an aircraft wing is "sucked" and "pushed" up as it makes an angle to the airflow. A sail is sucked and pushed in a similar way to an aircraft wing. This theory can be easily tested by observing a spoon being "sucked" by fast water flow.

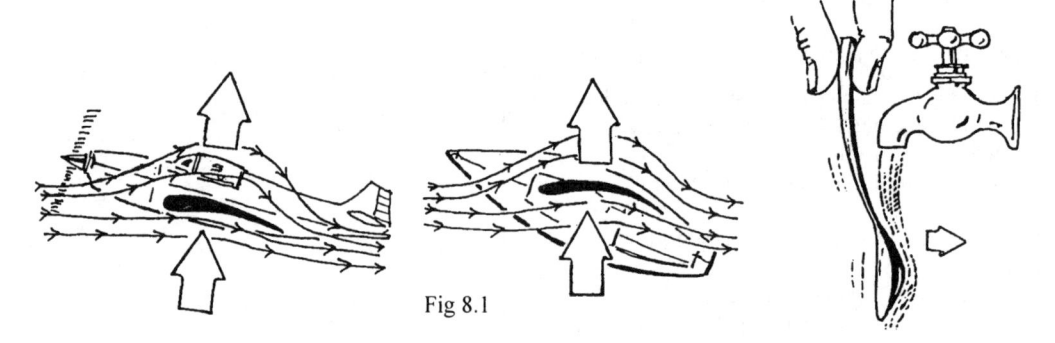

Fig 8.1

WHY DOESN'T A BOAT GO SIDEWAYS?

The sailing boats are designed with a suitable underwater shape and a keel to resist the sideways motion. They move mainly forward, with a little sideways component, known as the *LEEWAY*.

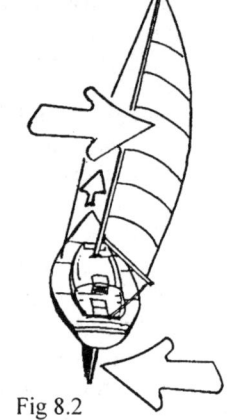

WHY DOESN'T THE BOAT BLOW OVER?

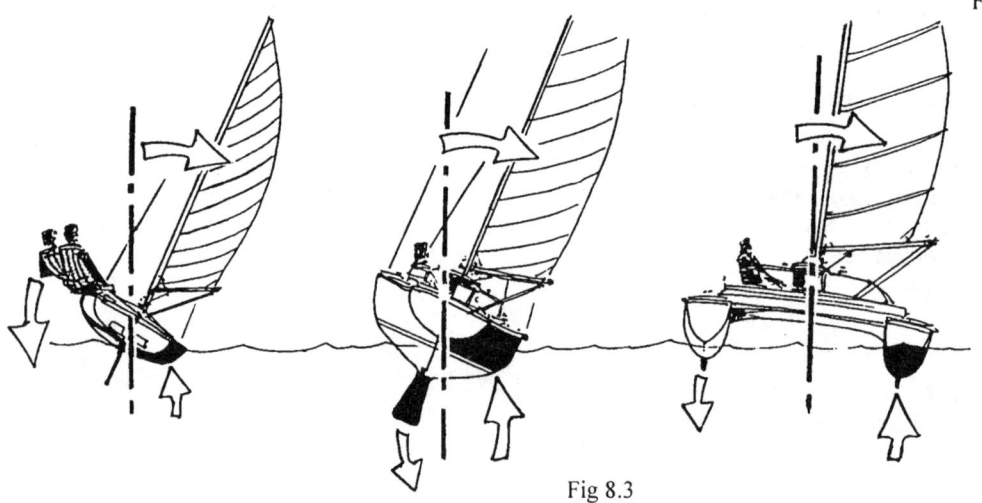

Fig 8.2

Fig 8.3

Chapter 8: Boat Handling & Towing

The combined forces on the sail and the underwater hull make the boat heel over, and sometimes turn over as sailing dinghies often do. However, most sailing boats are designed with a very low Centre of Gravity, i.e. their bottom hulls are weighed down with a heavy lead keel. They can sail at an angle without capsizing. The wind force is counter-balanced by the buoyancy of the immersed hull plus a weight, such as the weight of the crew in a dinghy, the keel of a yacht, or the hull of a catamaran.

To maximise the suction effect, the sails need to be *TRIMMED* at an optimum angle to the wind. They must be sucked, not shake or flap. This is achieved by pulling in the *SHEETS*, i.e. the line attached to the clew of a sail or the block and tackle on the end of the boom attached to a sail. When the sails are as tight as possible, the boat is said to be sailing *CLOSE HAULED*. She is then sailing as close into the wind as possible. Any effort to sail closer will result in the angle of the sail becoming wrong and losing suction. The sail will start to flap or *LUFF*. It is usually hard to sail within 30° either side of the wind direction - known as the *NO GO ZONE*.

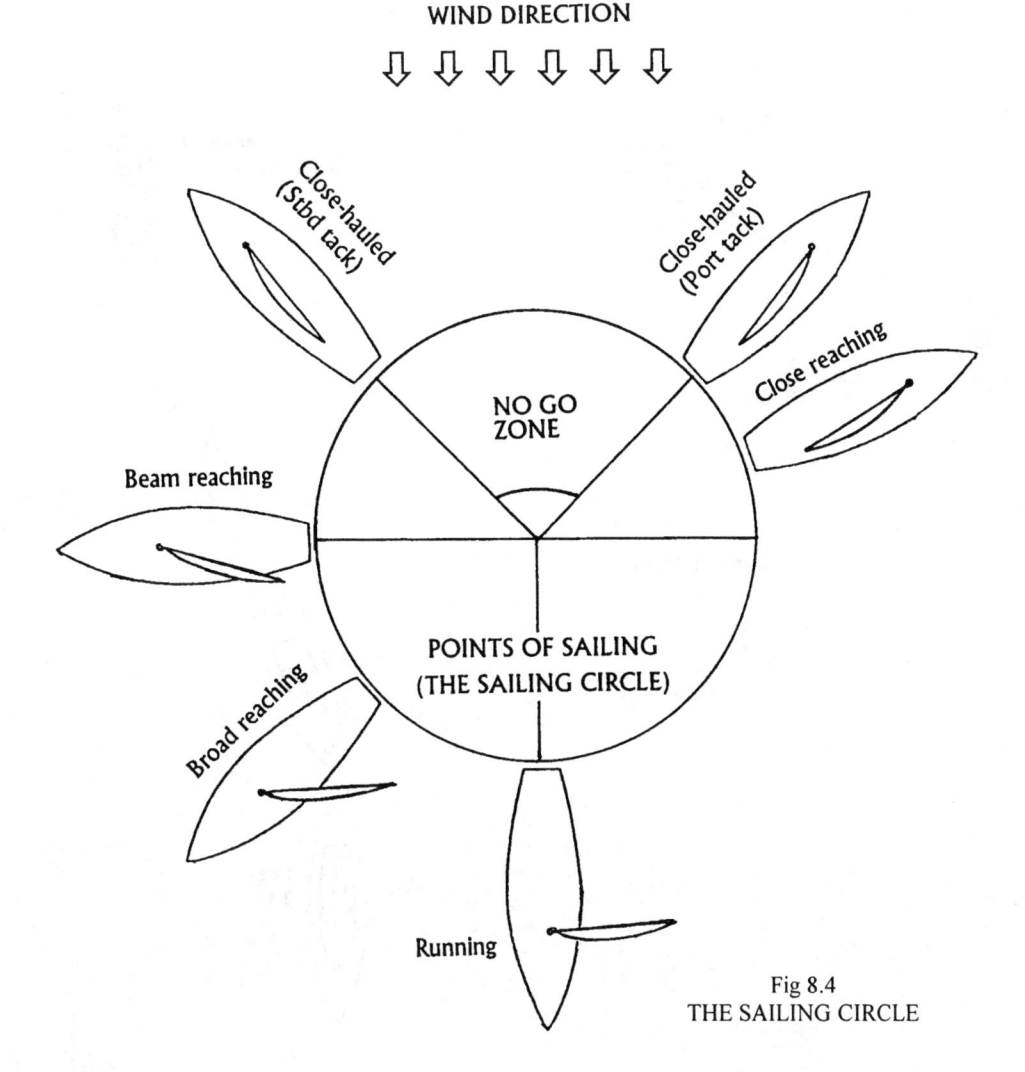

WIND DIRECTION
⇩ ⇩ ⇩ ⇩ ⇩ ⇩

Fig 8.4
THE SAILING CIRCLE

If sailing with the wind on the Port side, the boat is said to be on *PORT TACK*. If sailing with the wind on the Starboard side, she is said to be on *STARBOARD TACK*. In order to proceed on a course within the "No Go Zone", the boat is sailed on a "zigzag" path (*DOG'S-LEG COURSE*) alternating on either side of the "no go zone". She alters course by going into the wind and blowing the sails on the other side - known as **TACKING or GOING ABOUT**. Sailing in this manner is known as *WORKING TO WINDWARD* or *BEATING*. If she can sail to her destination on one tack, and back on another, she is said to have a *SOLDIER'S WIND*.

In figure 8.4, a boat *RUNNING or SAILING DOWNWIND* has set the sails to catch the most wind. If the wind is not exactly behind the boat, but a bit on one side, she is said to be on a *BROAD REACH*. If it is abeam she is *BEAM REACHING*; and if forward of the beam, but not to the extent of being close-hauled, the boat is *CLOSE REACHING*.

The Australian Boating Manual

GYBING or ***JIBING*** refers to swinging over of a fore-and-aft sail when running before the wind. This may be done purposely to alter a boat's course, in which case care should be taken that the jerk of the boom and sail is not too severe. It may also happen accidentally if, due to a yaw or a gust, the wind catches the sail on the wrong side. The sail may then fling violently across the boat, and if not *LOOSE-FOOTED*, i.e. if it has a boom attached to the foot, it may cause a serious injury to the crew. In addition, the violence of a gybe may carry away the mast and the rigging. The danger is most prevalent when sailing with the wind dead astern. ON THE PORT (OR STARBOARD) GYBE refers to a boat running free with the wind on port (or starboard) quarter.

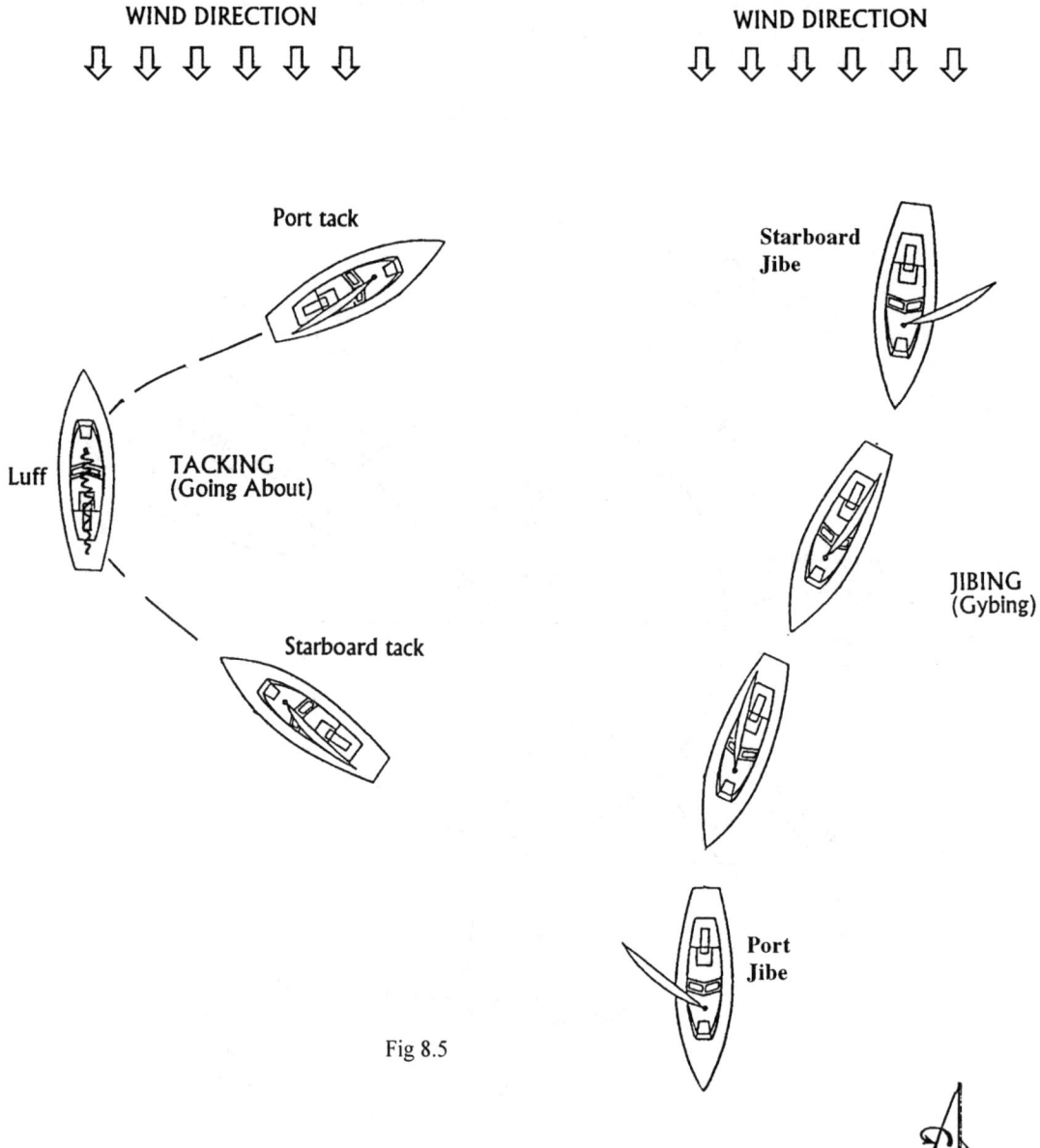

Fig 8.5

OTHER SAILING TERMS & TECHNIQUES:

In very strong winds, boats reduce the sail area by reducing the number of sails, hoisting smaller stronger sails (*STORM SAILS*) or *REEFING* the sails, i.e. rolling up the bottom or the side of the sail. A boat *LIES TO* when she has the wind nearly ahead with a small sail area hoisted. She is *HOVE-TO* when she is stopped in the water; it is achieved by backing the sails in the winds eye or by using engines. A boat is *GOOSEWINGED* when running dead before the wind with the sails set on both sides of the boat.

NO HIGHER is a term used to indicate that the boat must not be brought any closer to the wind. *NOTHING OFF* is the reverse of 'No higher'. It refers to keeping the boat's head

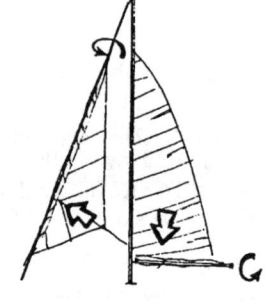

Fig 8.6
REEFING THE SAILS

Chapter 8: Boat Handling & Towing

close to wind, and not to let it fall to leeward.

IN IRONS or *IN STAYS* refers to a boat that has tried to tack but failed, and will not fall on either tack. The situation may be remedied by *BACKING THE JIB*, i.e. by holding the clew of the jib out to the original leeside.

HELM is the tiller or the wheel by which the rudder is controlled. *CARRYING LEE HELM* indicates that if the helm is left amidships the boat tends to pay off, and is likely to gybe in squally conditions. CARRYING *WEATHER HELM* indicates that if the helm is left amidships the boat tends to round up into the wind. A little weather helm is preferable so that the boat is constantly seeking wind, and it gives a "feel" to the boat. However, too much weather helm would require too much counter-acting rudder, so slowing down the boat. The position of the mainsheet traveller or the sail trim should be adjusted to balance the helm. One point on the sail surface is considered to be the *CENTRE OF EFFORT*, where all the wind force is assumed to act. One point on the lee side of the underwater hull surface is considered to be the *CENTRE OF LATERAL RESISTANCE*, usually at or near the pivot point. A couple exists between these two points, and the boat will not carry lee or weather helm when the two points are in line.

LUFF OR LUFF-UP means to steer closer to the wind.

FLAT AFT refers to a fore and aft sail, the sheets of which have been hauled as tightly as possible.

A boat is *TAKEN-ABACK* when the sails fill suddenly from the wrong side, due to a yaw or sudden change in wind direction.

HUGGING means sailing as close to the wind as possible, while *PINCHING* refers to a boat which is sailing too close to the wind with some loss of speed.

BOAT TRAILERS

Problems with trailers can take a lot of the fun out of boating. Badly adjusted, they can cause long-term damage to boats. Undersized and badly set up, they can be dangerous on the road.

SELECTION
- Choose a long trailer. The extra distance between the boat's bow and the tow hitch will place less weight on the draw bar. It will be easier to tow; and the wheels being further from the car will make it easier to reverse.
- Buy a multi-roller trailer fitted with a large number of rollers. Each roller should take the designated share of the weight of the boat; the keel rollers taking the most of the weight.
  The boat should not overhang the rearmost roller by more than a few centimetres.
- The wheel hubs should be fitted with bearing protectors to deter water entering the bearings during launching and retrieving the boat.
- The trailer should be fitted with submersible trailer lights.
- The rear of the trailer should be fitted with a bow location gadget, such as the "Boat Nabber" or "Retriever Mate". It will facilitate retrieving by locating the boat's bow.
- Tilt trailers are no longer common. However, a trailer fitted with a tilting cradle at the rear will make launching and retrieving the boat easier.
- The trailer should be fitted with heavy-duty winch and webbing straps and a chain to secure the bow, and straps to secure the rear of the boat. The boat should not squirm or bounce on the rollers.

Fig 8.7
TRAILER FITTED WITH TRAILER MATE

LOADING
If you intend to load the boat (on the trailer) with gear, make sure it is suitably distributed. Load the heavier items in the middle of the boat. Do not make the ends heavy, particularly the rear end. It will give a flywheel effect, which will exaggerate any swaying.

Do not use the outboard tilt support lever/knob when trailing a boat. The outboard could shake loose from the tilt support and fall. If the motor cannot be trailed in the "down position", use an additional support device to secure it in the "up position".

SPARES
o A correctly inflated spare wheel
o A suitable jack

The Australian Boating Manual

o   Spare wheel bearings, if working in remote regions.

LAUNCHING AT BOAT RAMPS

- Prepare the boat for launching away from the ramp, not on it.
- Attach a full bow line to the boat's bow.
- Tilt the motor and ensure the drain plug is in.
- Inspect ramp for obstacles and wind and tide effect.
- Wait your turn.
- Do not immerse the trailer's hubs or axle in water, if possible.
- Do not use the tilt action of a tilt-trailer when launching. You may damage the keel or engine by lowering at too sharp an angle. On a trailer with a tilting cradle at the rear, the boat should roll off easily.
- Secure or hold the boat and move trailer promptly away from the ramp.
- Embark extra gear and passengers after launching the boat.

RETRIEVING

- Before reversing the trailer into water, pull out the winch cable and hook it to the rear of the trailer - all ready to hook on the boat's bow.
- Make use of the rear tilting cradle or the bow location gadget if fitted.

BERTHING (OR MOORING) LINES

The function of the Head and Stern ropes (or lines) is to hold the boat at the wharf. Breast lines prevent her from stretching off the wharf. Springs control her fore and aft movement. Mooring lines can also be used to WARP (move) a boat along a jetty without using a motor. Fenders prevent hull damage from hitting or rubbing against the wharf.

In order to be able to tend to mooring lines, belay (secure) their inboard ends with turns around a fitting on deck. Place their "eye" ends over cleats or bollards on the jetty. By reducing the number of turns on the deck fitting and standing at 90° to the angle of the pull a mooring line can be slacked or heaved in without releasing a hold on the boat. Dipping of the eye is illustrated.

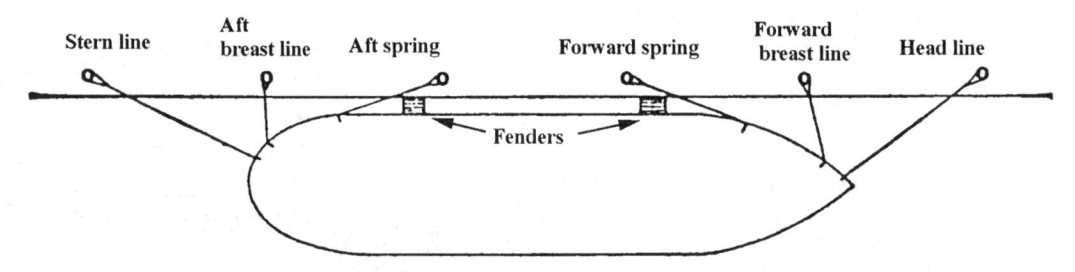

Fig 8.8: BERTHING (MOORING) LINES

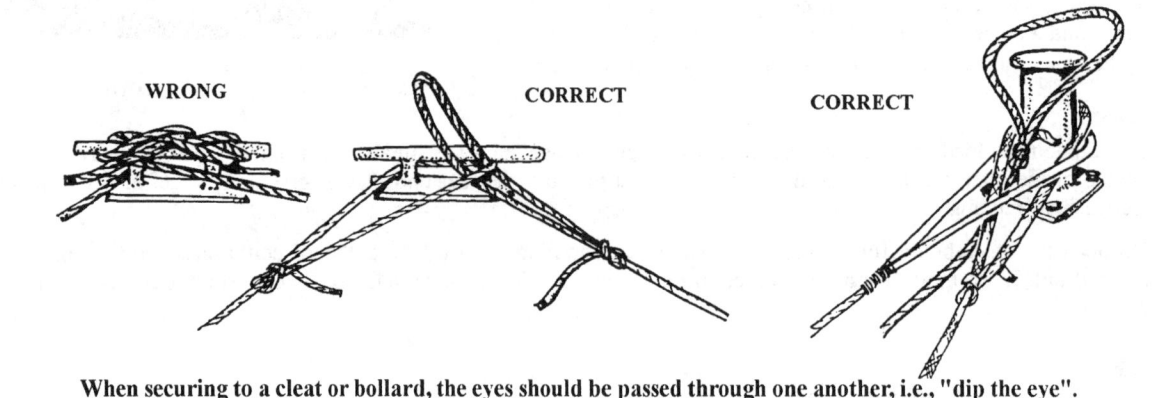

When securing to a cleat or bollard, the eyes should be passed through one another, i.e., "dip the eye".

Fig 8.9: DIPPING THE EYE PREVENTS THE LINES OF OTHER BOATS FROM BEING LOCKED IN

Chapter 8: Boat Handling & Towing

HEAVING LINE

Large vessels have large mooring (berthing) lines, which are usually too heavy to pass from ship to shore when securing alongside. Therefore, they employ a "Messenger", which is a light flexible line, to which the large mooring line is attached. A Heaving Line is a messenger, with a "Monkey's Fist" tied in one end. Together with a part of the coiled line, it is flung towards the person ashore, who gets hold of it, heaves in the heaving line, and gets to the mooring line attached to it. The Monkey's Fist is an intricate woven knot that is put in the end of the heaving line to increase its ability to fly through the air. To enclose a piece of metal inside the monkey's fist is dangerous, and should never be done.

In small vessels, the mooring lines are quite small. Properly coiled and held half a coil in each hand, they can be heaved quite a distance without a monkey's fist in their ends.

PREPARATION PRIOR TO LEAVING FOR SEA

- Batten down & secure for sea so that the boat is as watertight as possible. Many of the watertight doors and hatches and weathertight air pipes and vents can be kept closed at all times except when the opening is needed.
- Stow all mooring lines and gear below decks. Alternatively, they should be lashed on deck.
- Secure any cranes, derricks or booms on board. Also secure cranes hooks.
- Place any cargo on dunnage (timber) to allow drainage. The lashings should be provided with means of retightening them: turn buckles for wire and chain, and ratchet hand-tensioners for webbing lashings. Fibre rope lashings are hard to maintain tight. Frap and bowse them, i.e., tighten lashings with lashings.
- Test controls
- Check weather forecast
- Test motors
- Check fuel lines for leaks or fumes
- Don't overload or have excessive trim

CONTROLS TO TEST:
- Steering gear
- Main engine
- Remote controls
- Emergency trips & indicators
- Navigation lights & all other electrical circuits
- Emergency lighting
- Whistle
- Bilge system. (Pour some water in the bilge & pump it out)
- Fire fighting system
- Radio & barometer
- Winches

CHECK LIST:
- Check levels & pressures of oil, fuel, air, grease and water.
  (e.g., water level in the cooling system expansion tank, oil level in gearbox & grease level in stern tube lubrication).
- Check water inlet, outlet & cooling systems.
- Check propeller and shaft clear of obstructions; all rags and tools put away and machinery boxed up after any repairs.
- Check fuel, oil, water & provisions on board.
- Open all necessary valves.
- Check batteries and their charge.
- Check navigational charts & equipment.

MANOEUVRING DIFFICULTIES WITH LARGE VESSELS

Smaller vessels should bear in mind that they can respond to collision avoidance much quicker than the larger vessels. The size, hull form and displacement of larger vessels make them less manoeuvrable and slow to respond. Their large momentum also increases their stopping distances and turning circles. Boats should therefore avoid impeding the path of larger vessels. They should not pass so close to them so as not to allow any margin for error.

INTERACTION

Every vessel making way generates pressure waves in the water near her bow and stern, and a suction effect amidships on each side. These forces can cause a vessel to hydro-dynamically push away from, or be sucked in towards solid structures such as dock walls, piers, wharves and jetties or other vessels passing close by. In shallow water vessels also tend to get sucked towards the bottom. Such occurrences are known as interaction.

Interaction occurs whenever the normal flow of water around a hull becomes restricted. The results of interaction can vary. The person at the wheel may feel loss of control as the bow swings outward, or the stern seems intent on swinging across the channel to collide with the hull of a passing vessel. Discussed below is interaction with other vessels, the shallow water effect and the canal effect.

Factors that increase the risk of interaction are:
- High speed
- Large size of vessel
- Narrow channel (the "inner space" of rivers & docks)
- Shallow water
- Vessel's draft

(A) INTERACTION WITH LARGER VESSELS

In interaction between two vessels, the smaller vessel is at a greater risk of being sucked into the path of the larger vessel, as shown below.

EXAMPLE 1: Smaller vessel's stern is in line with the larger vessel's bow. Her stern gets pushed out by the bow wave of the larger vessel, forcing her bow to swing inwards. She risks being run over.

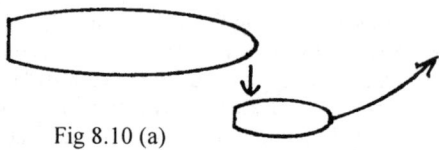

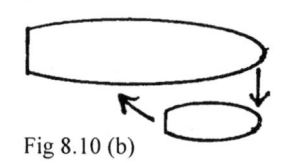

Fig 8.10 (a)

EXAMPLE 2: Smaller vessel's bow is in line with the larger vessel's bow. Her bow is pushed out and the stern pulled in. Her stern may collide with the side of the larger vessel. She may also end up colliding with another vessel or run over a bank.

Fig 8.10 (b)

EXAMPLE 3: Smaller vessel's stern is in line with the larger vessel's stern. Her stern is pushed out and the bow pulled in. Her bow may collide into the side of the larger vessel.

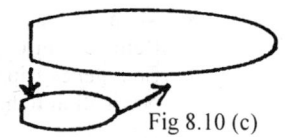

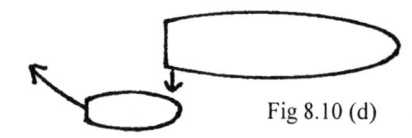

Fig 8.10 (c)

Fig 8.10 (d)

EXAMPLE 4: Smaller vessel's bow is in line with the larger vessel's stern. Her bow is pushed out and the stern swings in the wake of the larger vessel. She may end up colliding with another vessel or run over a bank.

(B) SHIPS TURN DIFFERENTLY TO MOTORCARS

Although not an interaction, the risk of being close to a turning ship should be mentioned here. The rear end of a turning motorcar does not swing in the opposite direction. But a turning ship's after-end swings out in the opposite direction, which can easily strike a nearby boat even though she is abaft the ship's beam and not in the direction of her turn. (See "Pivot Point" in this chapter).

(C) INTERACTION WITH VESSELS AT ANCHOR

When passing close to the stern of a vessel at anchor, the stern of the anchored vessel may be pulled towards the passing vessel.

(D) INTERACTION IN SHALLOW WATER (SHALLOW WATER EFFECT)

On entering shallow water, the vessel's bow and stern pressure waves move a little forward. She may also experience vibrations.

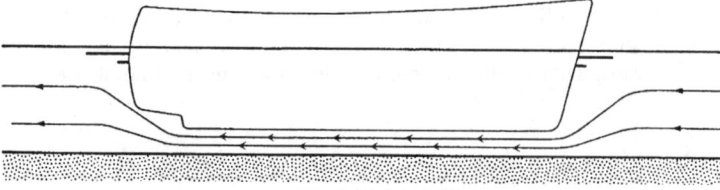

Fig 8.10 (e): SQUAT

In shallow water, the water under the hull is constrained, making it flow faster. It causes the vessel to be sucked towards the bottom, thus increasing her draft. This effect, known as Squat (or smelling the bottom), is especially pronounced in flat-bottom and deeply laden vessels. Barge shaped vessels tend to squat by the head, whereas fine-

Chapter 8: Boat Handling & Towing

hull vessels squat by the stern. Squat thus also results in a change of trim. Squat can be minimised by reducing speed.

In shallow water, there is also less water flowing past the rudder and propeller. It results in steering becoming sluggish and the output power drops. Thus in shallow water vessels take longer to turn or stop.

(E) INTERACTION IN NARROW CHANNELS (CANAL EFFECT)

In steep-banked waterways a vessel tends to find the middle of the channel where the pressure on both sides is equal. If she moves to one side to give way to another vessel, there is a risk that the cushioning by the oncoming water flow and the pressure of the bow wave against the bank would repel her bow towards the middle of the channel. The swing at the bow is exaggerated by suction between her hull and the bank – which is similar to Squat. The bow-swing is most pronounced in flat-sided vessels travelling in channels where the banks are not continuous. And, faster the vessel is moving the greater is the pressure build up. This Canal Effect is also experienced against solid structures such as dock walls, piers, wharves and jetties.

EFFECT OF USING ENGINE ALONGSIDE A WHARF

- ➢ If a vessel lying alongside a solid wharf puts her engine ahead, she will be sucked closer to the wharf. This is because the propeller will pull the water from between the boat and the wharf, creating suction.
- ➢ Putting the engine astern, on the other hand, will push her away from the wharf. In this case the propeller will push the water between the boat and the wharf.

HIGH SPEED CRAFT - CATEGORIES & RISKS

See Chapter 12

FEATURES AFFECTING VESSEL HANDLING

(This topic incorporates information from AMSA's Safety Education Article No. 19)

The three most important design features that affect the handling of any craft are:
- ➢ Underwater profile
- ➢ Rudder
- ➢ Size and efficiency of the propeller

A **FLAT-BOTTOM BARGE** with shallow displacement, smaller wetted surface area and the propeller and rudder close to the surface will be less responsive to the helm. She would also be more susceptible to being blown by wind than the two profiles shown below.

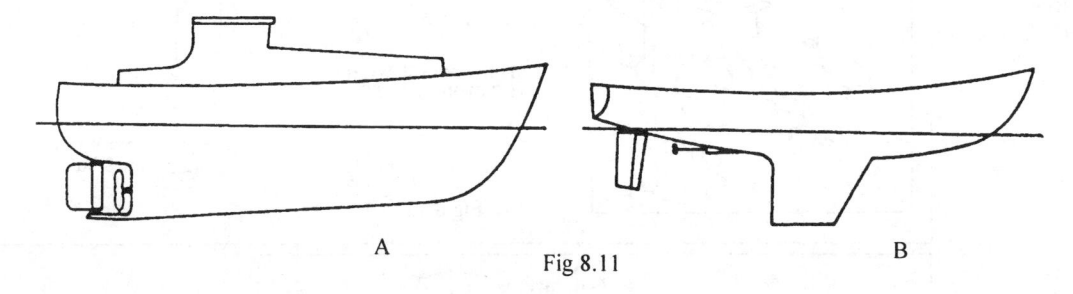

Fig 8.11

Two profiles, almost opposite extremes, are shown in figure 8.11 A & B.

FEATURES OF HULL "A"
- ➢ Long straight keel
- ➢ Heavy or full displacement
- ➢ Large wetted surface area
- ➢ Large rudder deep in the water
- ➢ Large slow-turning propeller

ITS HANDLING CHARACTERISTICS:
- Holds her course well due to large deep rudder.
- Turns slowly due to heavy displacement and long keel.
- Having large area of lateral resistance, she is less affected by wind than hull "B".
- Doesn't steer well when going astern due to large transverse thrust.

FEATURES OF HULL "B"
- Short deep fin keel
- Rudder near the waterline
- Wetted surface area concentrated midships
- Small high speed propeller

ITS HANDLING CHARACTERISTICS:
- Needs constant steering to hold course due to light displacement.
- Turns fast due to light displacement and short keel.
- Having small area of lateral resistance, she would be more affected by wind than hull "A"
- Steers well when going astern due to small transverse thrust.

TRANSVERSE THRUST & BOAT HANDLING
(Simplified explanation)

Propellers are discussed in Chapter 24. As a propeller screws its way through the water, it not only drags a boat forward on its axial thrust, it also swings the stern sideways just like a paddle wheel. In other words, a propeller gives a boat a *FORWARD THRUST* as well as a *TRANSVERSE THRUST*. The larger the propeller the bigger the transverse thrust or the **"PADDLE WHEEL EFFECT"** or the **"PROP WALK"**. The transverse thrust is most noticeable when power is first applied to a vessel stopped in the water.

Depending on the direction of rotation when viewed from astern, the propellers are described as either right-handed or left-handed. Most propellers are right-handed, i.e., they turn clockwise when going ahead. With such a propeller, the paddle wheel effect will swing the boat's stern to the right (starboard) when going ahead and the bow will swing to port.

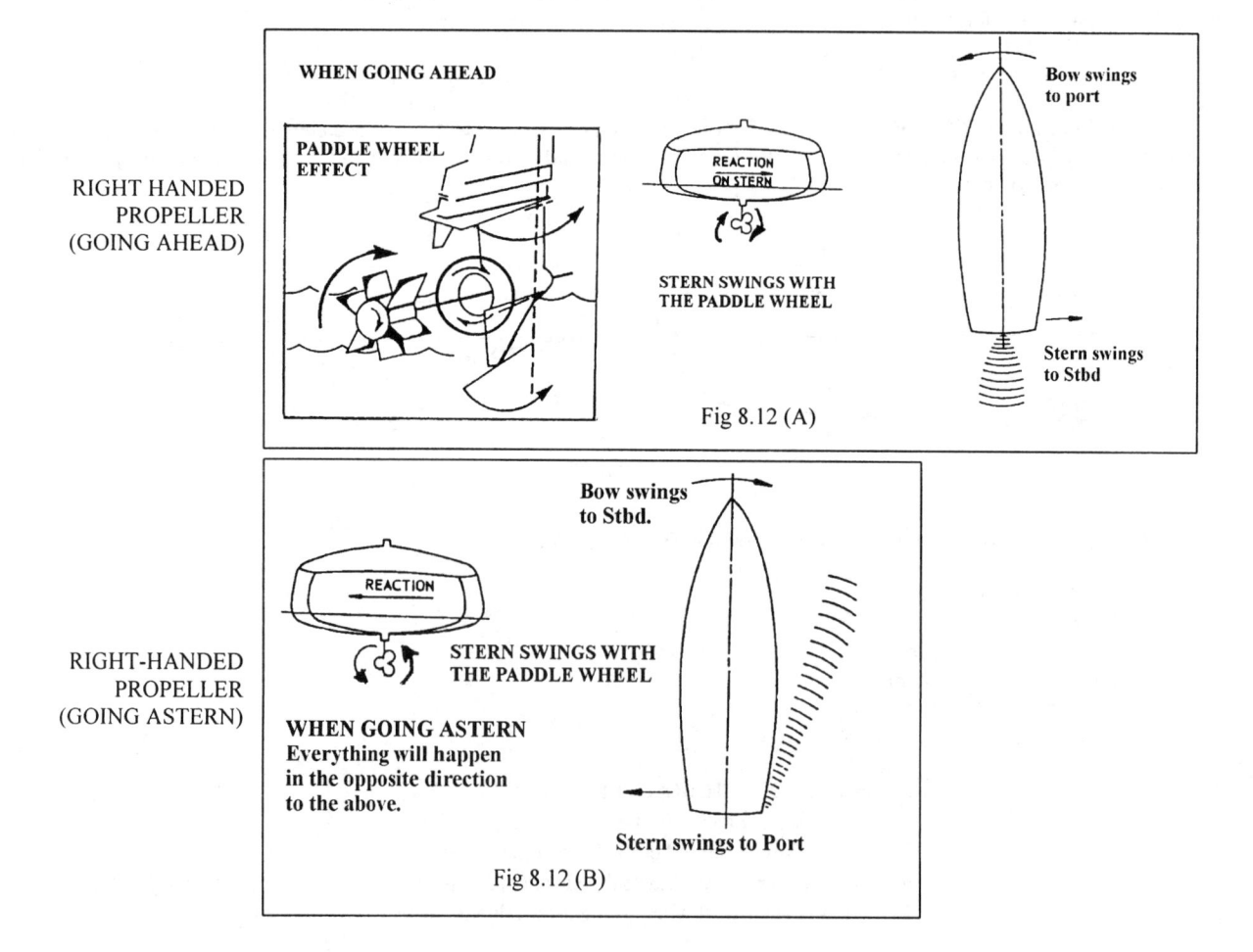

RIGHT HANDED PROPELLER (GOING AHEAD)

PADDLE WHEEL EFFECT

REACTION ON STERN

STERN SWINGS WITH THE PADDLE WHEEL

Bow swings to port

Stern swings to Stbd

Fig 8.12 (A)

RIGHT-HANDED PROPELLER (GOING ASTERN)

REACTION

STERN SWINGS WITH THE PADDLE WHEEL

WHEN GOING ASTERN
Everything will happen in the opposite direction to the above.

Bow swings to Stbd.

Stern swings to Port

Fig 8.12 (B)

Chapter 8: Boat Handling & Towing

With a **LEFT-HANDED PROPELLER,** the boat will respond in the opposite direction to those shown in figures 8.12 (A) & 8.12 (B).

In **TWIN-SCREW VESSELS** the paddle wheel effect; is cancelled by the two propellers rotating in the opposite direction - usually each turning outwards at the top.

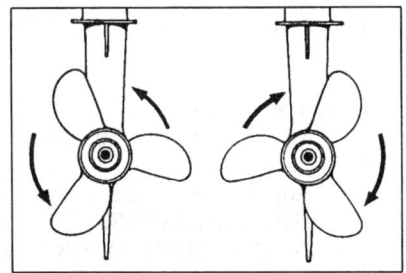

Fig 8.13: OUTWARD TURNING TWIN-SCREW

TRANSVERSE THRUST & THE RUDDER

The transverse thrust or the paddle wheel effect is not so noticeable when going ahead because the rudder readily controls the vessel. The rudder is generally placed directly astern of the propeller. As discussed in Chapter 24, it is designed to operate in the water flowing past the hull and the wash created by the propeller, both of which are effective when going ahead. But when going astern, the rudder does not have the use of the screw wash and has little effect unless fitted well below the hull. The transverse thrust is therefore more significant when going astern.

The rudder may work when going astern, but usually with unpredictable results. In any case, it is the propeller's transverse thrust that will decide which way the vessel swings. For this reason, most vessels leave the rudder amidships when going astern, that allows the paddle wheel effect of the propeller to work clearly and predictably.

USE OF TRANSVERSE THRUST
(In a single screw vessel with a right-handed propeller)

(A) TURNING SHORT AROUND:

A vessel with a single right-handed propeller has a smaller *TURNING CIRCLE* to port than to starboard. However, if it is required to turn short around in a narrow waterway, it is better to carry out this turn to starboard and gain maximum benefit from the transverse thrust when the propeller is operating astern. This manoeuvre is sometimes called a *THREE POINT TURN*. Its sequence is detailed in figure 8.14 (A) and the text below.

Figure 8.14 (B) illustrates how the transverse thrust works against the rudder effect when turning to port. This manoeuvre is not recommended.

Sequence of events:
➢ Start at the port side of the river.

➢ Wheel hard a starboard, engine slow ahead. Stop engine when on the other side of the river. The effect of the propeller wash on the rudder would have swung the boat's stern to port by about one-third of the way.

➢ Wheel amidships, engine slow or half astern. Stop engine when back on the same side of the river where the manoeuvre was commenced. THE TRANSVERSE THRUST OF THE PROPELLER WOULD HAVE SWUNG THE BOAT'S STERN TO PORT BY ANOTHER ONE-THIRD OF THE WAY.

➢ Stop engine. Wheel hard a starboard and engine slow ahead again to complete the final one-third of the turn.

Tip: You may be able to use an anchor 'under foot' to turn short round, if the 3-point turn is not possible. See 'Snubbing Round' below.

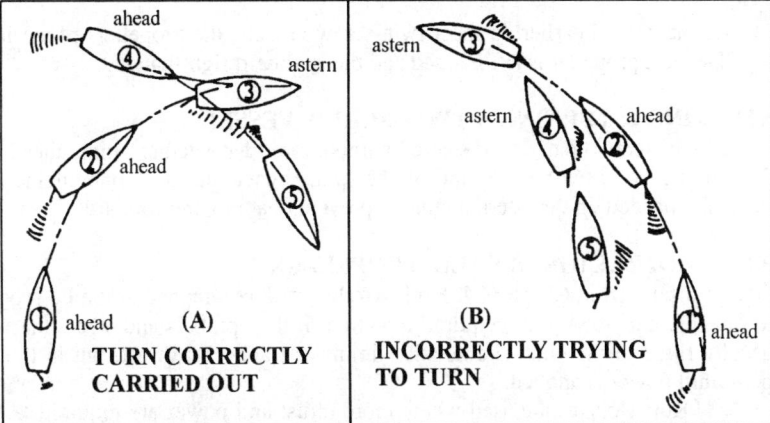

(A) **TURN CORRECTLY CARRIED OUT**

(B) **INCORRECTLY TRYING TO TURN**

Fig 8.14: TURNING SHORT AROUND

(B) COMING ALONGSIDE A WHARF:

In a single screw boat with a right-handed propeller, it is easier to come alongside a wharf PORT SIDE TO, than starboard side to.

Sequence of events:
- Approach at about 20 degrees at slow speed, until bow on the wharf.
- Stop engine, rudder amidships. Now go slow astern. THE TRANSVERSE THRUST WILL SWING THE BOAT'S STERN TO PORT - and put you alongside.

The same steps if followed for coming alongside starboard side to will swing the stern further away from the wharf, and result in collision with the wharf. (See incorrectly done steps 3 & 4).

(C) SECURING A BOAT TO A MOORING BUOY:

Sequence of events:
- Approach it on starboard bow.
- Go stern to take the way off the boat. The transverse thrust will swing the stern to port and the bow to starboard towards the buoy.

BERTHING WITH A BERTHING (MOORING) LINE
(a single screw vessel)

CORRECTLY DONE   **INCORRECTLY DONE**

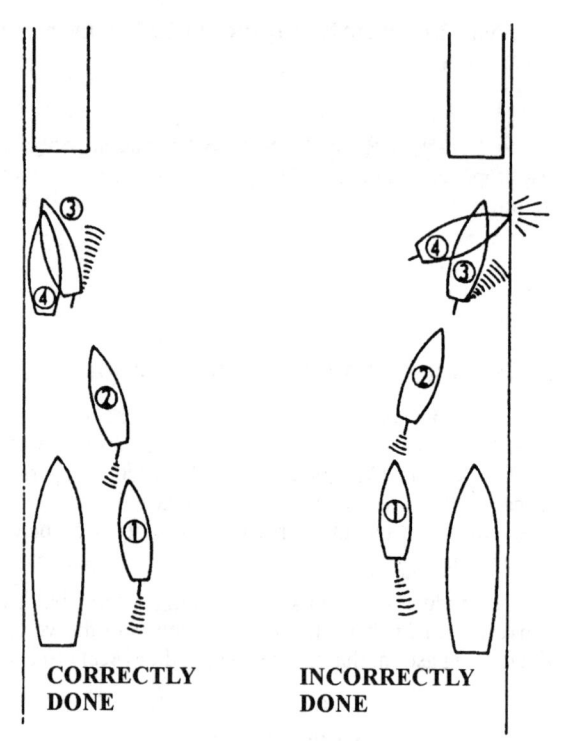

Fig 8.15: COMING ALONGSIDE A WHARF

If you are able to use a mooring line to pull the boat alongside when berthing, the effect of transverse thrust becomes irrelevant. The boat can be berthed port or starboard side to.

Sequence of events:
- Approach the jetty at about 20 degrees angle on port or starboard side with a fender on the bow or forequarter nearest to the jetty.
- Send the forward spring out as quickly as possible.
- Put the engine slow ahead and wheel hard over away from the jetty. The stern should swing alongside while the spring controls her forward motion.

BERTHING A TWIN-SCREW VESSEL

With twin screws you can either use the approach for a single screw vessel, or bring her as close as possible parallel to the wharf and then make her "walk":

For example, for going alongside port side to, go ahead on the starboard engine, astern on the port, and prevent her from turning to port with wheel to starboard. A few extra revolutions on the astern engine may be needed to prevent her from moving ahead. This will make her move sideways towards the wharf. Adjust the throttles and the wheel to keep her approach parallel to the wharf.

As mentioned earlier, in most twin-screw vessels, the propellers rotate in opposite directions, turning outwards at the top. The port propeller is left-handed and the starboard right-handed.

STEERING & TURNING A TWIN-SCREW VESSEL

A twin-screw boat can be steered without a rudder whether going ahead or astern. She can also be turned on the spot. To turn a twin-screw boat around on the spot, engage ahead on one engine and astern on the other. A few extra revs are generally needed on the astern engine to prevent boat moving forward.

KORT NOZZLE (OR DUCTED) PROPELLER

As discussed in Chapter 24, Kort Nozzle is a duct wrapped around a propeller. It may be fixed or steered like a rudder to direct water flow in the required direction. It thus propels and steers the vessel without a rudder. However, a rudder is usually fitted to the unit. The nozzle eliminates transverse thrust, but in the case of fixed nozzles the steerage is usually poor until power is applied.

Kort Nozzles are used where more thrust and power are required, e.g., commercial towing, pushing and trawling. High drag of the nozzle at high speeds makes them unsuitable for fast vessels.

Chapter 8: Boat Handling & Towing

FOLDING, FEATHERING, VARIABLE PITCH & AZIMUTH PROPELLERS
See Chapter 24.

JET DRIVES
See Chapter 24

THRUSTERS (BOW THRUSTERS & OTHERS)
See Chapter 24

BERTHING IN BAD WEATHER
Safety is paramount. If the conditions are not right for a particular manoeuvre, don't make that manoeuvre or ask for assistance. For example, if going alongside is unsafe, then consider anchoring or seek shelter elsewhere until the conditions improve.

LETTING GO YOUR OWN LINES
If there is no one available to let go your lines from ashore, replace the necessary mooring lines with *slip ropes*, also known as *pick up ropes*. These are lines passed around a bollard ashore with both ends secured on board. You can let go one end and heave away on the other.

LEAVING A BERTH (UNBERTHING)

As far as possible a boat should always leave berth by first going ahead on the forward spring and swinging out the stern. This reduces the possibility of damage to the boat and the propeller becoming fouled with rubbish floating under the jetty. However, figure 8.16 illustrates the methods recommended in AMSA's Safety Education Article No. 19, which serves well to highlight a variety of circumstances.

Fig 8.16 LEAVING A BERTH

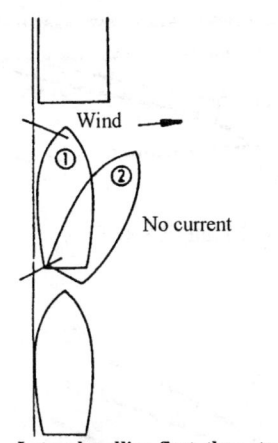

No current — Let go headline first, then stern line. When boat clear of berth, go ahead on motor.

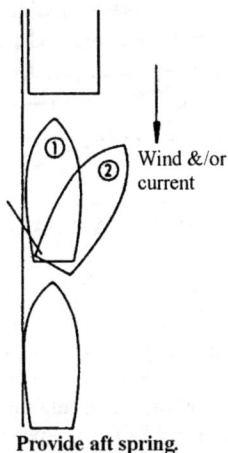

Wind &/or current — Provide aft spring. Let go lines, push bow out.

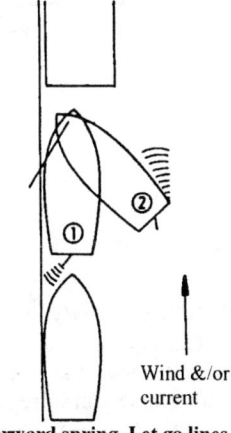

Wind &/or current — Provide forward spring. Let go lines, go slow ahead with wheel to port. When sufficiently canted, go astern.

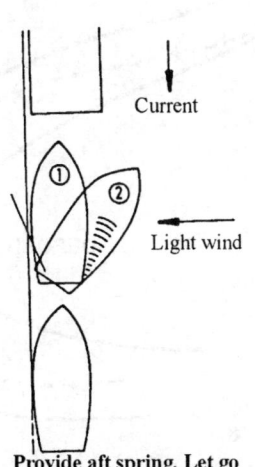

Light wind / Current — Provide aft spring. Let go forward. Go astern on motor until the bow swings out

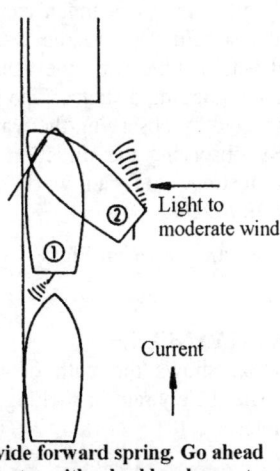

Light to moderate wind / Current — Provide forward spring. Go ahead on motor with wheel hard a-port until stern well out. Then go astern until clear of berth.

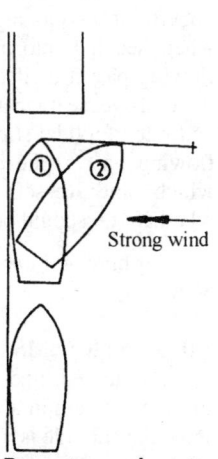

Strong wind — Run out an anchor, and pull the ahead out; watch the stern.

PRECAUTIONS WHEN BERTHING
- Fenders out
- Test controls
- Mooring lines ready
- Crew standing by
- Nothing trailing in the water
- No protruding limbs of people on board
- All extrusions (log, etc.) taken in

WHARF WITH A LARGE TIDAL RANGE

The boat's mooring lines should be secured on the jetty as far away fore and aft from the boat as possible so that they still have sufficient lead when the boat drops down on the tide. Some books recommend that the lines should be at least three times the length of the range of tide. In other words, if the difference between the high and the low tides (i.e., the range) is 5 metres, the mooring lines should be at least 15 metres long.

If you need to secure at a berth where the boat will be "high and dry" at low tide, she may need to be kept vertical as she takes the ground. Shoring or heeling against the wharf before she takes the ground can support her.

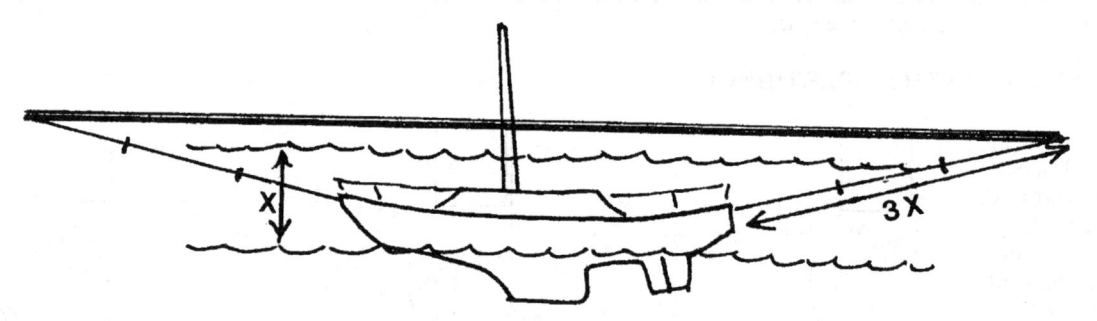

Fig 8.17: TYING AT A WHARF WITH A LARGE TIDE RANGE

MANOEUVRING IN CURRENTS & TIDES

Currents and tides do not push vessels. They cause a body of water to move like a moving walkway. The flow moves every floating object in it, regardless of its size, by the same amount in the same direction. The direction is known as *SET*, the speed is known as the RATE and the distance moved in a particular time interval is known as *DRIFT*. Set, Rate and Drift are also discussed in Chapter 17.2, which deals with Navigation.

Other than in strong winds, manoeuvring a vessel (berthing, unberthing or picking up or casting off a buoy) is much easier when heading into a current or tidal stream. Due to the water flowing past it as if the vessel itself was making way, the vessel would have steerage as well as a shorter stopping distance. The set and rate of a tidal stream can be estimated by observing the water flowing past a pile and a buoy. By observing the direction in which other vessels at anchor are resting, you can work out whether you should be *WIND OR TIDE RODE*.

Berthing and unberthing techniques are shown in Fig. 8.15 & 8.16.

TIDAL STREAMS IN RIVERS & CHANNELS

The dotted line in the illustration shows the path of the strongest stream in a winding river. After heavy rain or melting of snow the ebb tide is likely to be stronger.

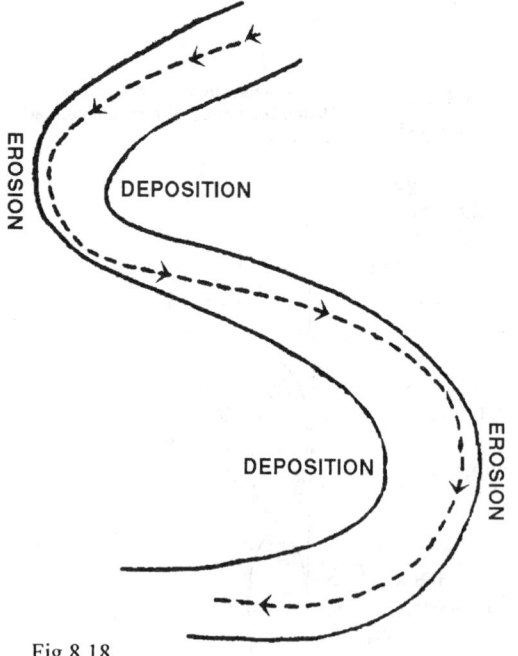

Fig 8.18

MANOEUVRING IN THE WIND

Wind "pushes" each body differently depending on its exposed hull, known as the "sail area" or WINDAGE. When steering a steady course, the sideways push by the wind is judged by observing the angle between the boat's wake and the

Chapter 8: Boat Handling & Towing

course steered. This angle is known as the LEEWAY ANGLE. If the boat's wake is on her course line, she is not experiencing any Leeway. If the angle is 5 degrees, then the Leeway is 5 degrees. Leeway is also discussed in Chapter 17.2 - Navigation.

Being aware of a boat's characteristics is helpful when manoeuvring in wind. Just like the sails on a sailing vessel, the windage of a power-driven vessel is related to the amount of hull out of water and the height and relative location of the superstructure. A vessel with a large trim by the stern offers her bow to significant windage.

A vessel of small draught (or large freeboard) with rudder partly out of water has less steerage. A propeller that is partly out of water would provide diminished forward thrust and cause greater transverse thrust.

By observation in various wind conditions, try to gain an understanding of the effect of windage on your vessel's manoeuvring characteristics. Most vessels, when going astern, will tend to "fly up into the wind". The stern tends to seek the wind.

Generally, when berthing:
- If wind is parallel to the berth, head into the wind.
- If wind is onshore to the berth, head for a point a little forward of the desired berthing position.
- If the wind is offshore to the berth, head for a point a little astern of the desired berthing position.
- When berthing alongside a larger vessel, berth in her lee.

PIVOT POINT OF A VESSEL

The point about which a vessel turns when helm is applied is known as the Pivot Point. The underwater shape of the vessel is the main determinant of the pivot point's position. Trimming her by the head would move the pivot point a little further forward, and visa versa. For most small vessels with inboard engines, when going ahead the pivot point is about one third of the length of the vessel from the bow. When going astern, it moves a little aft of amidships. (See "Ships turn differently to motor cars" in this chapter).

MOORING BETWEEN TWO BUOYS

Approach into the wind and/or tide to pick up the head mooring first, then manoeuvre astern paying out the picking up rope or bridle until the stern buoy is under the stern. Pass the stern bridle then middle the two.

When securing a boat between two buoys, do not heave in so tightly as to affect the free movement of the moorings of the buoys themselves. The pull on these moorings in tidal or windy conditions can part a boat's lines.

Ships securing to large mooring buoys need outside assistance to shackle their mooring lines or chains to the eye on top of the buoy. In the absence of outside assistance they may send one or two crewmembers to climb onto the buoy to do the job. In order to hold the ship in place while mooring lines are secured, a *pick up rope,* known as *slip rope* or *slip wire*, is first sent from the ship. This is passed through the eye of the buoy and returned to the ship where it is secured. On completion of mooring the slip rope is retained in place without any weight on it. At the time of unmooring the ship, the slip rope is tightened again to slacken the mooring lines so that they can be unshackled from the buoy. The crew on the buoy then clear away and the ship slips off the slip rope.

OTHER METHODS OF MOORING

- **BALTIC MOOR**

The Baltic Moor originated in the Baltic Sea where the prevailing wind-direction is onshore. It is useful for single screw boats. The vessel's anchor is suitably laid abeam of the vessel before she goes alongside the wharf. A hawser, previously secured to the anchor through a stern fairlead, makes it possible to control the bow and the stern while the vessel is blown down onto the wharf.

The reverse procedure is used to haul the vessel off the wharf.

This method is also used to put a vessel alongside a wharf that is not of robust construction.

- **MEDITERRANEAN MOOR**

This is used when wharf space is limited. The vessel is moored stern on to the jetty with one anchor lying ahead or both anchors, fine on each bow. It is an advantage to use anchor buoys in this manoeuvre.

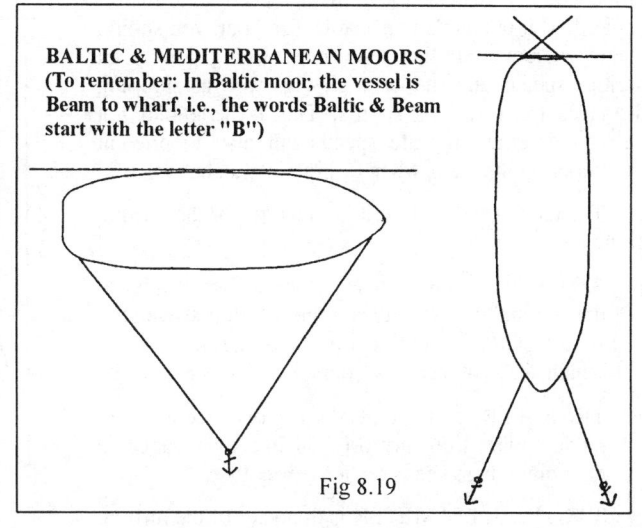

BALTIC & MEDITERRANEAN MOORS
(To remember: In Baltic moor, the vessel is Beam to wharf, i.e., the words Baltic & Beam start with the letter "B")

Fig 8.19

DREDGING DOWN (or DREDGING ANCHOR)

This manoeuvre involves dropping the weather (offshore) anchor and paying out cable as required when berthing in a strong onshore wind. This stops the bow from paying off uncontrollably. The anchor is heaved in after securing alongside the wharf or the cable could be slackened to lie on the seabed until the anchor is required to haul the vessel off the wharf.

SNUBBING ROUND

Dropping the anchor under foot in order to turn the vessel short round. The anchor is dropped after reducing the headway to minimum and having commenced the turn. It is heaved in on completion of the turn.

HULL SPEED LIMIT & FUEL CONSUMPTION

A useful rule of thumb for fuel consumption is 1 litre per hour per 5 horsepower (3.75 kilowatt). However, described below is the Hull Speed Limit, which must also be kept in mind:

1. The bow of a moving displacement hull pushes the water aside, creating a hollow around itself. The water surges back into the hollow, bouncing back off the hull. This sets of the familiar pattern of waves around a moving vessel starting with the bow waves.
2. Wavelength and wave speed are directly proportional: the faster the wave, the longer its length. A boat increasing her speed increases the speed of the bow waves, thus making them longer. The speed at which the length of the bow wave equals the length of the hull is known as HULL SPEED LIMIT.
3. Increasing boat speed beyond its Hull Speed Limit will cause her stern to drop into the trough of the bow wave, reducing propeller thrust and increasing fuel consumption. For example, the last 10% of a displacement hull's top speed consumes 27% of the fuel.

PLANING HULLS

Planing hulls skim over the surface of the water. The main resistance to their speed is their skin friction, which, for good performance, must be kept to a minimum by maintaining the hull as smooth as possible.

Planing hulls are very sensitive to boat weight. Even the turning effect of the propeller causes them to heel to one side. This problem can be corrected in three ways:

➢ The helms-person of a right-handed screw boat to sit on the starboard side. (A right-handed screw turns the boat's head to port, causing her to heel to that side. For this reason, many boats with a right-handed screw are designed for the helmsman to sit on the starboard side.)
➢ Shift weights in the boat to counteract the heel.
➢ Adjust trim tabs on the transom.

STOPPING DISTANCES & TURNING CIRCLES

A vessel's stopping distance depends on her speed, size, draught and astern power. Her turning circle is influenced by displacement, trim, speed and rudder angle. If you operate a reasonable size boat, you should make yourself familiar with her turning ability at various speeds and rudder angles, as well as stopping distances from various speeds. This is beneficial not only in determining safe speeds but also to prevent overshooting the leads when turning into a channel.

The terms used in describing features of the turning circle are:

➢ ADVANCE is the distance along a line parallel to the original course between the position at which a vessel starts to alter course and the position at which she is on her new course.

➢ TRANSFER is the perpendicular distance a vessel moves away from her original line of advance to the point where she is on her new course.

➢ TACTICAL DIAMETER is diameter of the turning circle from the original line of advance.

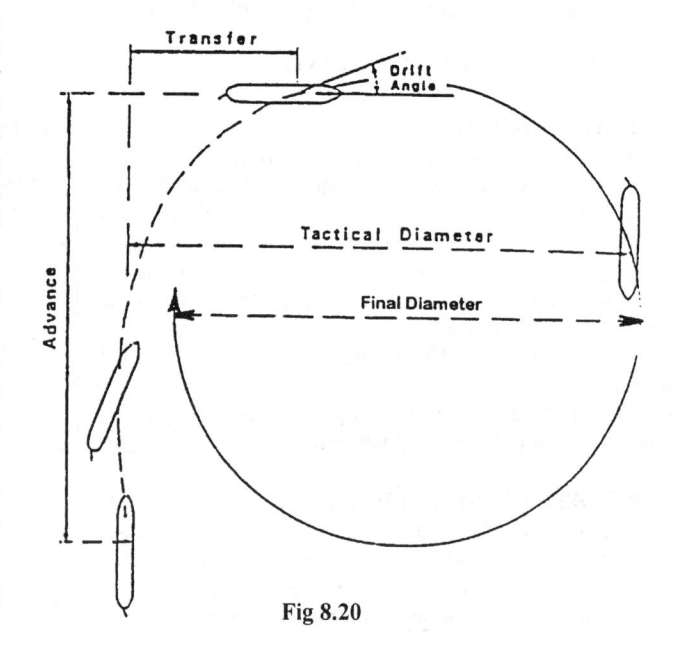

Fig 8.20

Chapter 8: Boat Handling & Towing

VESSEL STOPPING TERMINOLOGY:

- CRASH STOP: Putting the engine(s) full astern.
- INERTIA STOP: Putting the engine(s) into neutral. The distance travelled until coming to stop is known as HEAD REACH.
- FISH TAILING (or RUDDER RECYCLING): Washing off speed by yawing the vessel between 20°–40° either side of her course. On larger vessels this method is more effective than crash stopping.

LANDING BARGES

Landing craft beach themselves bow first at right angle to the beach to load and discharge passengers or cargoes through a bow door which when thrown open becomes a ramp. These vessels are used to service places where port facilities are just a ramp on a beach. Some landing barges are equipped with a stern kedge anchor, which is useful to control the approach as well as to kedge the vessel off the beach.

The manoeuvring and handling characteristics of landing barges are quite different to other twin-screw vessels. They have a flat bottom, small draft and high windage area, especially up forward. Approaching a ramp in a strong beam wind or tidal stream may require steering up to 40 degrees off course. Running short of making the ramp, the bow is likely to swing into a difficult position, risking the vessel to broaching onto the beach. This can be prevented by holding the stern with the kedge anchor by letting it go about 2 ship's lengths prior to beaching and tightening the cable after she has beached. The anchor would also prevent the vessel from being blown or pushed onto the beach further than is necessary. In rivers with strong tidal streams, beaching may only be possible at slack water.

When berthing, be aware of the tide, winds and weather. The beach gradient and tide should be such that with the bow placed on the beach the stern should be in sufficient water to manoeuvre. Firm sand is preferable to mud. It not only provides good support for loading and unloading cargoes, but the vessel is unlikely to get sucked into it as she could in mud.

Throughout her stay, the entire length of the vessel must be evenly supported by sand and water. If the tide falls below the level of hard support, she is likely to break into two at a point where the hard support ends.

For a short stay it is best to berth on a rising tide and unberth before high water. The engines in slow ahead position throughout the stay would keep the bow edging up the ramp as the tide rises. If berthing on a falling tide to load cargo, cease loading, say, every 10 minutes and ease her off the beach a little so that she does not become embedded. To prevent undue stress on the bow door hinges and skids do not run up too fast on hard concrete ramps. Don't allow the door to rest on the ramp without also the bow resting on it. In other words the vessel should not pull on the door hinges.

Prior to unberthing make sure the vessel is at right angle to the beach. Start by closing the bow door, yaw her and then go astern on both engines. The squatting effect would push the water forward, helping her to float off.

CATAMARANS

Many of the recreational vessels, harbour ferries, charter boats and inter-island transporters are catamarans. Their hulls are of three basic types: Displacement, semi-displacement and Wave-piercing.

The twin screws in catamarans are further apart than in mono hulls. As a result, they can be turned around faster and in a tighter circle. Those equipped with highly responsive engines can be accelerated and decelerated quite rapidly. All their other manoeuvring characteristics are roughly the same as of monohull twin-screw vessels.

DINGHY DOCKS

Many more dinghies (tenders) can be tied up at a dinghy dock if they are all tied up long, i.e., with a single <u>long line</u> from the bow. Short lines limit the privilege to only a few.

CROSSING BARS

WHAT IS A BAR

A bar is a shallow bank formed by the movement of sand along the coast and sediments built up at the entrances to rivers and lakes. Bars cause the waves to become steeper and, in some cases, break as they approach the bar.

KNOW THE SCEND

Scend is an abbreviation of "ascend" to mean upward rising of a vessel's bow when her stern falls into the trough of a wave. It also expresses the distance between the flat sea and the bottom of the trough. It means that with a 2-metre sea running over a bar, the depth in a trough may be 1 metre less than what it would be in a flat sea. Take this into account when calculating the depth over a bar.

WHEN ARE BARS MOST DANGEROUS TO CROSS

Whether a vessel is coming in or going out, she should avoid crossing a bar during strong onshore winds and ebb tides. Both onshore and offshore winds can cause the water to become rough. Bars are less rough when the tide is in the same direction as the incoming seas.

INFORMATION SOURCES CONCERNING BARS
- Local boating officers
- Community
- Surf life rescue centres
- Sailing Directions (Pilot book)
- Coast guard organisations
- Personal observations
- Relevant clubs

RECOGNIZING REEF & BAR OPENINGS

The openings in reefs and bars can be identified by the difference in the colour (shade) of the water. Blue colour indicates deeper water. On the windward side, an opening can also be recognized by the gap in the breaking water.

PRECAUTIONS WHEN CROSSING A BAR
- Everyone to wear a lifejacket and be seated.
- Close up the vessel.
- Vessel trim should be as designed. She should not be heavy on one end.
- Secure everything. Make sure the outboard motor is tightly secured and a chain preventer is attached.
- Test all controls to make sure boat is operating correctly.
- Move towards the breaking area and pick the line of least activity. Stay with the leads or channel markers if available.
- Watch for breakers behind you.
- Wait for a big set to roll in, position the boat on the back of a wave and stay there. Don't run down the wave face.
- Add power as needed to keep her nose up a little.
- Remember the Shallow Water Effect at excessive speed.
- Maintain a speed which ensures good directional control.

CROSSING THE BAR
- Idle towards the breaking waves. Look for a lull. Apply throttle and run through.
- If there is no lull, motor to the surf zone. Gently accelerate over the first piece of water and run to the next wave. Time this carefully. If you go too fast you may get airborne on the next wave.
- Back off power just before contact with the swell.
- As you come through or over the breaker, accelerate again and repeat the process until clear.
- Head for the lowest part of the wave (the saddle). This is the last part of the wave to break.
- Half cabin vessels can choose to proceed slowly through the entire bar system, although at some stage a wave will break over the boat. So long as the boat goes straight ahead, very little can happen to it.

IF BAR BECOMES DANGEROUS ON RETURN

Don't come in. Land somewhere safer or call for assistance.

BAR CROSSING LICENCE FOR COMMERCIAL BOATS

In most States, only licensed boats and skippers are permitted to take passengers over bars. In NSW, the licence is in the form of an endorsement for a specific bar in the vessel's Survey Record Book and the master's certificate. To obtain this endorsement, the master must gain experience and local knowledge by crossing the bar in a variety of weather conditions at least ten times in and out with a "licensed" person.

BOAT HANDLING IN HEAVY WEATHER

HULL STRESSES IN HEAVY WEATHER
- Impact of waves on the bow.
- Panting: Pitching causes the two ends of the boat to experience stress due to variation in water pressure. The bow in particular pants heavily. The boat ends are strengthened with panting beams and stringers at the time of building.
- Hogging & sagging:. These are the bending stresses on the boat when riding on waves or due to improper loading or tightening of mast stays. Members such as the keel, keelson and hog counteract hogging and sagging.

Chapter 8: Boat Handling & Towing

- Pounding: Stresses induced in the hull as the bow repeatedly lifts and bangs on the water. Landing barges are especially prone to it.
- Waves breaking on board.
- Racking: Rolling causes a transverse hull-distortion stress, known as racking. Bulkheads counteract racking stresses.

NOTE: Vessels are also subject to stress on slipways. The "dry docking" stress is counteracted by placing sufficient keel and bilge blocks and side supports.

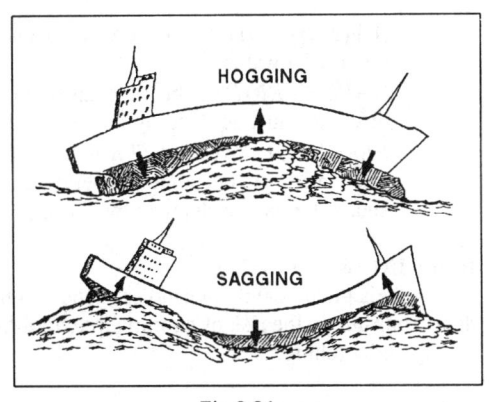

Fig 8.21

IMMEDIATE RISKS
Driving full speed into heavy seas can cause serious injuries and damage due to pounding, pitching and taking water on board. Serious hull fracture and flooding can result from the vessel colliding head-on with the seas.

REDUCING THE RISK
- Slow down.
- Reduce stern trim, if the bow is slamming. This can be done by moving passengers forward and trimming in the outboard motor leg.
- Steer 10-20 degrees off the wind and sea.
- Heave-to, if necessary, by one of the following methods:
    - Steam slowly to keep the vessel's head 10-20 degrees off the wind and sea.
    - Stream a sea anchor from the bow (see anchors).

(NOTE: Maintain a radio contact with a Maritime Coast Station (Coast Radio Station), and, don't wait too long before seeking help.)

PREPARING A VESSEL FOR HEAVY WEATHER
- Batten down and secure everything. Close all watertight doors, portholes and deadlights. Use storm boards where necessary. Turn the vessel into a watertight cocoon so that she could roll over without anything moving or being damaged.
- Clear decks of all gear. Lower and secure any booms.
- Check strum boxes (strainers) and pump bilges dry. Reduce slack tanks to minimum. Minimise free surface effect.
- Make sure all freeing ports are clear to allow the seas to run off the decks as quickly as possible.
- Take in ballast to increase stability or change trim if necessary.
- Shut all hull openings (including toilets installed below waterline) other than those required for engine operation.
- Run strips of masking tape across large window panes to avoid flying glass in case they are stove in.
- Rig lifelines or nets on the open decks.
- Make sure that lifebuoys and lifejackets are accessible. Lifejackets and safety harnesses to be worn in heavy weather
- Secure the galley and shut off gas.
- Close air pipes to fuel and fresh water tanks not in use.
- Have sea anchor ready.

PREPARING FOR A SQUALL
If close to land or an island, head for shelter. If not, secure for heavy weather and heave-to as discussed in this chapter.

PREPARING FOR A CYCLONE
Cyclones are discussed in Chapter 20.

ENCOUNTER WITH WAVES
DEFINITIONS
- *ROLLING* is the side-to-side swinging of a vessel when in a seaway.
- *PITCHING* is the downward falling of a vessel's bow as water support leaves it.
- *YAWING* is swinging to either side of an intended course.
- *PERIOD OF ROLL* (or the *Roll Period*) is time taken in seconds for a vessel to roll from one side to the other and back to the same side, i.e., one complete roll.
- *PERIOD OF PITCH* (or the *Pitch Period*) is the time taken in seconds for a vessel's bow to rise from the horizontal, fall below it and return to the horizontal.

The Australian Boating Manual

- *WAVE HEIGHT* is the vertical distance between the *TROUGH* (the lowest point) and the *CREST* (the highest point) of a wave.
- *WAVE PERIOD* is the time taken in seconds for one complete wave to pass (i.e., two successive crests). The observer must be standing on a stationary object, i.e., not on vessel underway.
- *PERIOD OF ENCOUNTER* is the time taken for one complete wave to pass, measured from a vessel underway. The period of encounter depends on the wave period and the course and speed of the vessel. The period of encounter can be changed by altering the vessel's course and/or speed.

ROLLING & PITCHING

Vessels pitch heavily in head seas and roll heavily in beam seas. Running with the sea or a QUARTERING SEA can be dangerous due to the risk of pooping, pitchpoling and broaching. Altering course to take the sea on the bow can reduce both rolling and pitching.

SYNCHRONIZATION

When a vessel's Roll Period or Pitch Period equals the Period of Encounter, it can lead to the vessel becoming synchronised with the motion of the waves, thus losing her ability to ride the waves. This may cause violent and dangerous rolling or pitching. The propeller may frequently race out of water and the vessel subjected to severe bending forces of hogging and sagging. A synchronised vessel may capsize or break up.

Synchronization can be recognized from the gradual increase in vessel's Roll Period or Pitch Period. Should such a condition start to develop, the remedy is to change the period of encounter by altering the vessel's course and/or speed.

BROACHING & POOPING

A vessel running before the sea may lose her steerage when riding on a wave crest with the stern out of water. If she now pitches forward, her bows may bury deep into the trough and the stern swing round until she is broadside to the waves. The vessel is now broached, and the waves may roll her over and capsize her.

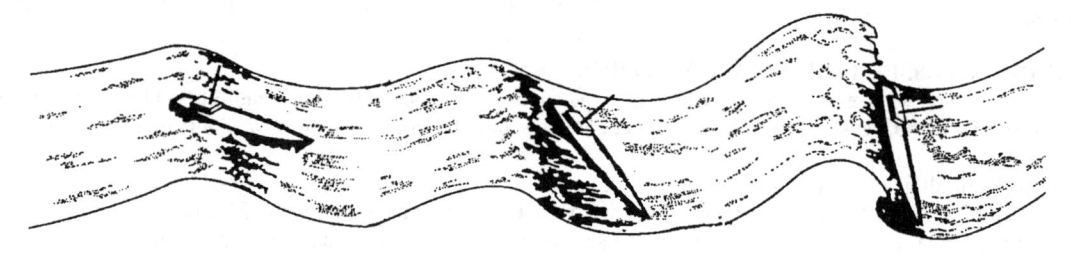

Fig 8.22: STAGES OF A BOAT BROACHING

This situation may be avoided by reducing the speed of the vessel to about half or one-third the speed of the waves. This will allow the waves to sweep under the vessel. Steering will be difficult and there is the risk of POOPING (a large wave breaking over the stern). Trimming to lighten the bow or streaming astern long warps or a drogue can aid steering.

It is worth noting that a following sea can make the vessel's speed appear slow when in fact she may be travelling quite fast.

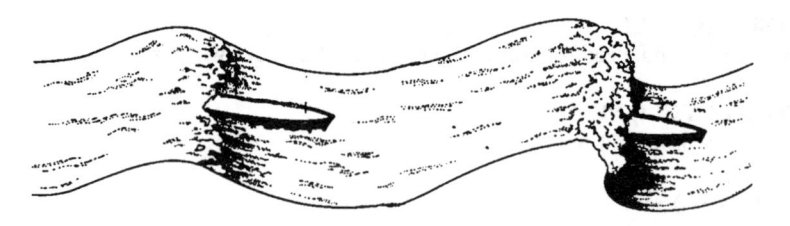

Fig 8.23: STAGES OF A BOAT BEING POOPED

ALTERING COURSE WITHOUT BROACHING

A vessel can broach when altering course in rough conditions. This risk can be avoided by watching for a lull in the pattern of the waves. This usually follows a series of large waves. He vessel should then be brought round as quickly as possible under full helm, and without using full power. Combination of full helm and full power may cause the vessel to heel excessively to leeward.

Chapter 8: Boat Handling & Towing

PITCHPOLING

A vessel surfing down a steep wave may bury her bow and slow down. This may cause her stern to lift and turn her end for end. This is known as pitchpoling. It is therefore advisable to always ride on the back of the waves in rough seas. In other words, proceed at a speed to avoid surfing.

ROLLING & PITCHING ARE AFFECTED BY LOAD DISTRIBUTION

Weights in a vessel should be evenly distributed on the longitudinal centreline and in the wings (sides), because:
- Concentration of weights in the wings makes rolling less violent.
- Concentration of weights on the centreline makes pitching less violent.

HEAVING-TO

To "heave-to" means to stop or slow down until the heavy weather abates. Three options may be considered:

1. Hold course with the sea on the bow. Steam at slow speed or stream a sea anchor from the bow. Vessels with a right-hand propeller may find it easier to keep the sea on port bow. This is the most common and the safest method.
2. Hold course with the sea on the quarter. Continue to steam slowly or stream a sea anchor from the quarter. The vessel may be hard to steer and may broach or be pooped.
3. Stop engines; secure the vessel like a cocoon and drift. She will lie beam to sea and roll heavily, and may even capsize. If course cannot be hold this may be the only option.

ROOSTER TAIL

When running before the sea, a rooster tail may be formed behind the vessel. This can gradually build up into a breaking wave, which if not checked, may eventually break over the stern. To prevent it from dangerously swamping the stern, the top of the rooster tail should be lowered as quickly as possible by streaming astern a few mooring lines or a fishing net.

TOWING

Note: Reward for towing and salvage are discussed under Claiming Salvage in Chapter 10.

TOW ROPES (TOWING HAWSERS)

The forces exerted on a towline are:
- Bollard pull and the weight of the towing vessel.
- Resistance of the tow due to its size and shape.
- Loading of the towline due to the speed of the operation.
- Friction of the towline in the water.

Rope manufacturers use only the bollard pull to calculate the required strength and size of towropes. An approximate guide to calculating **BOLLARD PULL** is 11 kg per kilowatt (or 15 kg per horsepower) of towing vessel's main engine capacity. Variables in the towing vessels propulsion systems, such as kort nozzles and twin-screws, would obviously influence this rule of thumb.

Generally wire ropes and chain cables are used in ocean towing of ships. When towing with fibre ropes do not use a rope of high stretch characteristic. At the time of parting an ordinary braided nylon rope would have stretched over 25%, resulting in a potentially lethal recoil and whiplash. Do not use such ropes for towing or in areas where they are likely to exceed their Safe Working Load (SWL).

The most suitable towing ropes are pre-stretched and made from high strength synthetic fibre, such as **HMPE** (e.g. Plasma, Steelite, Dyneema and Spectra), LCP (e.g. Vectran) and PBO (e.g. Zylon). HMPE fibre ropes are favoured because of their buoyancy and superior durability. The use of lightweight, high strength and buoyant ropes has benefits, such as:
- not heavy to handle,
- floatable to vessel in distress,
- requires a smaller winch,
- requires less deck space.

The disadvantage of their relatively smaller size is that they are more vulnerable to cuts and abrasions.

Towing of underwater vehicles is a specialised field where the depths and speeds of tow vary. Special cables are sometimes incorporated in their tow ropes.

GENERAL PROCEDURE

- The "tug's" master should take charge.

- The tow line (also referred to as the towing hawser) should be at least 20 mm in diameter and at least 3 times the length of the towing vessel.
- Towing with a bridle of an angle of less than 30 degrees will ensure good steerage and better load spreading.
- Set up a suitable communication link and the emergency alternative between the tug and the tow.
- Secure tow's rudder amidships if not manned.
- After securing the tow, gradually increase speed to the required level, keeping in steps with the tow. A sudden take off will make the tow go berserk and may part the line.
- Both skippers should closely watch as the tug gets underway.

TOWING WITH BRIDLE FOR BETTER STEERAGE & LOAD SPREADING
Bridle Angle less than 30°
Fig 8.24

TOWING IN HEAVY WEATHER

- Use a longer, stronger, heavier and pre-stretched towing hawser for good shock absorption.

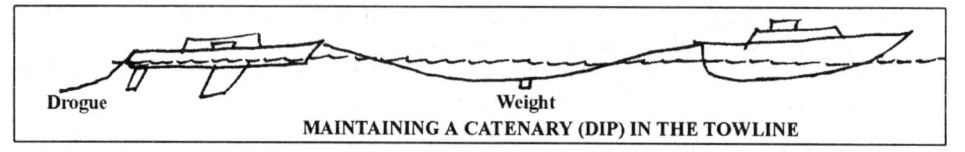

MAINTAINING A CATENARY (DIP) IN THE TOWLINE
Drogue — Weight
Fig 8.25

- Distribute the load of the tow by securing the tow line to more than one point on the tug and the tow.
- Secure the tow at the tug end in a way that permits ready and easy slipping.
- Wedge fender at chafing points.
- Display the *"restricted in ability to manoeuvre"* shapes and lights.
- Hang an anchor or a motor tyre on the towline to restrict the tow from yawing.
- Rig a stress line, as discussed below.
- Maintain a good *catenary* (i.e., a dip in the tow line) so that the tow doesn't part when riding on waves. Slow down if necessary.
- Steer the tow if possible.
- Periodically stop the vessel and "freshen the nip", i.e., pay out or take in tow line a little to prevent it chafing at the same spot.
- Be prepared to slip or cut the tow quickly if it begins to sheer badly. It can affect the towing vessel's steering and can even capsize the tug.

STRESS LINE (TOWING TELLTALE)

A synthetic tow rope when stretched to more than 25% of its original length can suffer permanent loss of strength. To show when the limit is approaching, a warning indicator can be provided by incorporating a Stress Line in the tow rope. The stress line is a loose bight of light rope spliced into the towrope. It is positioned well clear of the fittings but in clear view of the towing vessel.

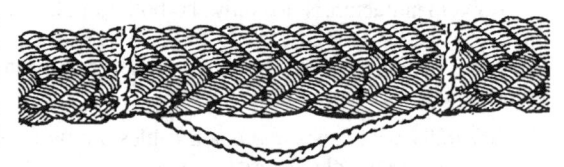

Stress Line is spliced into a Square Rope Towing Hawser & secured to it.

FIG 8.26: STRESS LINE

The Navy recommends a 4.5-metres of stress line of 10 mm polyethylene rope. Half a metre of it at each end is passed around and spliced into the strands of the towing hawser 2.8 metres apart. This creates a 3.5 metres bight in the stress line over a 2.8 metres length of the towing hawser, i.e., the bight is 25% longer.

As the tension in the towrope increases the slack in the stress line is taken up. The towrope's limit of stress is reached as soon as the stress line becomes taut. The towing vessel reduces tension in the tow before the limit is reached, preventing the towrope from weakening.

TO CONTROL THE TOW FROM YAWING & SHEERING

Yawing can cause the tow to take a big sheer, which may pull the towing vessel dangerously off course. One or more of the following methods may reduce this tendency:
- Alter course or speed.
- Shift weights to trim the tow by the stern.
- If unable to steer the tow, secure its rudder at an angle of sheer to one side.
- Stream a drogue (ropes, small sea anchor etc.) behind the tow to stop the tow shearing.

Chapter 8: Boat Handling & Towing

- Shift the point of tow to one bow.
- If two rescue vessels are available, tow the second one astern.
- Make sure that the tow can be slipped in an emergency.
- Heave-to, if all else fails.

TOWING IN CONFINED WATERS

- Shorten towline for better control. Excessive catenary in shallow waters may cause the tow line to drag along the bottom and chafe badly. It may even snag on a sunken object.

- Tow alongside (side by side) in confined waters. This allows better manoeuvring and berthing. It is also a legal requirement in most harbours. Secure tightly to maintain the "tug's" stern well astern of the tow, so that it does not lose steerage. The towlines should be secured to act as bow, spring and surge lines. Short breast lines should be avoided. The transverse pull of breast lines reduces the forward movement of the tow. Keeping the after tow line longer also reduces the transverse pull, allowing the tow greater forward movement.

Fig 8.27

- When pushing a barge in a narrow waterway, the tug's bow should be butted against the barge so tight that, for all practical purposes, the barge acquires an engine, propeller and rudder. The main connection is made up of wire ropes from the after side bitts of the tug to the outboard stern bitts of the barge on both sides. After heaving the wires tight and securing them, they can be further tightened with ratchets. This procedure locks the tug against the barge. Additional connections can be made from the tug's bow bitts to the inner stern bitts of the barge.

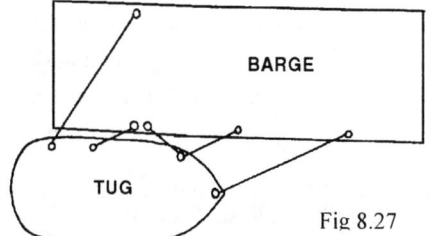

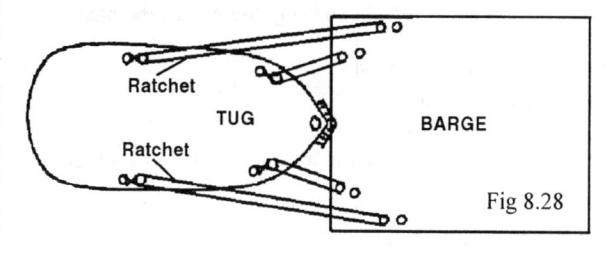

Fig 8.28

GIRTING

Girting may occur in tugs when assisting ships into ports. The Howard Smith illustration shows a tug on a quarter-line assisting the ship to turn to starboard. The tug master either underestimates the speed of the ship or the ship increases her speed. This causes the towline to run abeam of the tug. Failure to quickly release the towline results in the tug capsizing.

The risk of such an accident can be minimised by use of a GOB ROPE, which is a stopper rigged on tug's stern and passed over the tow line to prevent it from running abeam of the tug. The gob rope can be heaved in or eased out as necessary.

Fig 8.29: "LADY MOIRA" GIRTED & CAPSIZED
(Howard Smith Towage)

Fig 8.30: GIRTING
(Howard Smith Towage)

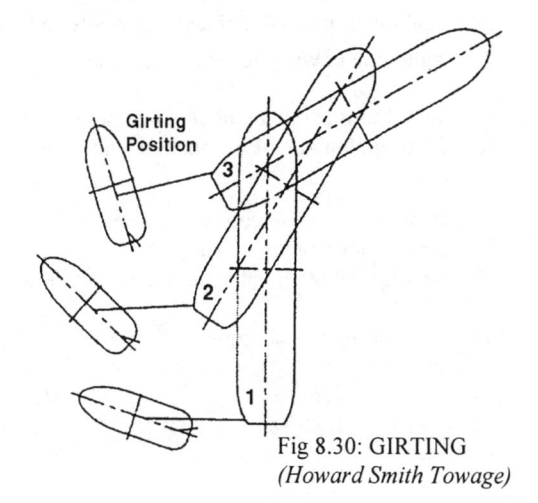

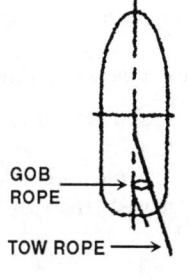

Fig 8.31: GOB ROPE

CHAPTER 8: QUESTIONS & ANSWERS

8.1: VESSEL HANDLING & TOWING

1. Steaming in calm seas at 10 knots on a dark night, you feel the vessel lifting her bow and tucking her stern down. You also hear the stern wash breaking. The most likely reason is:

    (a) the propeller has become fouled.
    (b) it is the squatting action in shallow water.
    (c) the vessel has touched bottom.
    (d) the propeller has been damaged.

2. A "starboard 10" order to a person at the wheel means turn to starboard 10°:

    (a) by compass.
    (b) on the wheel and stay there until the next order.
    (c) on the wheel.
    (d) at a time.

3. A "steer 115" order to a person at the wheel means:

    (a) steer 115° by compass.
    (b) turn vessel by 115° on the wheel.
    (c) change course by 115° by compass.
    (d) none of the choices stated here.

4. A "midships" order to a person at the wheel means:

    (a) steer the present course.
    (b) place wheel amidships and keep it there.
    (c) place wheel amidships and then steer.
    (d) do not change the present course.

5. A "steady" order to a person at the helm means:

    (a) keep the rudder amidships.
    (b) keep the rudder at its present angle.
    (c) steer the course the vessel is on now.
    (d) allow the vessel to turn in a circle.

6. In order to moor a vessel between two buoys, you should:

    (a) pick up the stern mooring first.
    (b) approach them down wind.
    (c) approach them down tide.
    (d) approach them up wind or up tide..

7. The usefulness of a slip rope is:

    (a) to slip away a vessel quietly.
    (b) to secure a vessel between two buoys.
    (c) to avoid kinks in a rope.
    (d) to cast off a vessel without outside help.

8. In a Baltic Moor a vessel lies:

    (a) stern to jetty.
    (b) with anchor abeam.
    (c) bow or stern to jetty.
    (d) with two anchors ahead.

9. In a Mediterranean Moor a vessel is moored:

    (a) stern to wharf & an anchor fine on each bow.
    (b) beam to wharf & an anchor abeam.
    (c) beam to wharf.
    (d) with both anchors abeam.

10. A Bumbora or Bombora is:

    (a) eddies.
    (b) a tidal stream.
    (c) sea breaking over rocks.
    (d) a current.

11. In minimising water entry into hull in rough weather, a vessel should:

    (i) run before the sea.
    (ii) alter course.
    (iii) alter speed.
    (iv) keep freeing ports closed.

    The correct answer is:
    (a) all except (i).
    (b) all except (ii).
    (c) all except (iii).
    (d) all except (iv).

12. In heavy weather, the best course to steer is:

    (a) head to sea.
    (b) stern to sea.
    (c) beam to sea.
    (d) 10-20 degrees off the sea.

13. In a vessel about to face heavy weather:

    (a) wearing lifejackets is obstructive.
    (b) closing of openings is suffocating.
    (c) sealing air pipe of fuel tank is not advised.
    (d) battening down hatches is dangerous.

14. For a vessel to ride out a gale in a harbour, she may consider dropping one of the anchors under foot. This will:

    (a) provide a spare anchor in readiness.
    (b) reduce her yaw.
    (c) prevent the other cable cutting across the bow.
    (d) stop her from swinging.

Chapter 8: Boat Handling & Towing

15. You are steaming in calm water close inshore at night with an onshore wind when you feel a light thump and your engine starts to slow down. The most probable cause is:

    (a) vessel is entering shallow water.
    (b) fuel blockage.
    (c) damage to cooling system.
    (d) fouled propeller.

16. The risk of vessel broaching in rough seas is minimal when:

    (a) running before the sea.
    (b) surfing.
    (c) pitching in synchronism.
    (d) riding on the back of a wave.

17. A 12 metre vessel is lying alongside a berth with engine opened up for repairs. On hearing the gale forecast for the area, the master should:

    (a) ignore it. You are in port.
    (b) box up the engine and secure the vessel.
    (c) go out to sea to prevent smashing on wharf.
    (d) wait for re-confirmation of the forecast.

18. A vessel about to leave port receives a gale warning. The master should:

    (a) close all openings prior to departure.
    (b) keep anchors in readiness.
    (c) postpone departure.
    (d) take in ballast prior to departure.

19. If a vessel is to secure to a wharf where there is a large range of tide, her berthing lines should be of the following length:

    (a) at least three times the range of tide.
    (b) at least three times the depth of water.
    (c) as short as possible.
    (d) at least three times the length of vessel.

20. You are approaching a river mouth which you know from the pilot manual is navigable by fishermen with local knowledge but the manual advises great caution. Having decided to go in unassisted, you would choose the following conditions:

    (a) offshore wind and ebb tide.
    (b) onshore wind and ebb tide.
    (c) offshore wind and flood tide.
    (d) light wind and flood tide.

21. You are to secure a deep keel vessel at a berth where she will be "high and dry" at low tide. You should:

    (a) rig strong lines for her to hang.
    (b) heel her against the wharf.
    (c) heel her away from the wharf.
    (d) keep her upright with anchors and lines.

22. Risk of propeller damage and fouling can be minimised by the vessel:

    (a) berthing bow first; unberthing stern first.
    (b) berthing stern first; unberthing stern first.
    (c) berthing bow first; unberthing bow first.
    (d) berthing stern first; unberthing bow first.

23. When slipping a berth, the berthing lines that have been 'let go' should be:

    (a) left trailing in the water for some time.
    (b) taken out of water quickly.
    (c) left uncoiled for a few hours.
    (d) stowed underdeck right away.

24. In order to slip from a berth where there is no one to let go the berthing lines from ashore, you should:

    (a) steam into the berth.
    (b) cut the lines.
    (c) let go all lines and cut the last line.
    (d) rig the necessary lines as slip ropes.

25. When secured to fore and aft mooring buoys, the vessel's mooring lines should be hove in and secured in order to affect the buoys' own moorings as follows:

    (a) stop the free movement of both the buoys.
    (b) allow the free movement of both the buoys.
    (c) stop the free movement of the upwind buoy.
    (d) stop the free movement of the uptide buoy.

26. A vessel has a right-handed screw. With her wheel amidships the effect of a short burst of astern power should be to swing her:

    (a) stern to starboard.
    (b) none of the choices stated here.
    (c) bow to port.
    (d) bow to starboard.

27. The greatest transverse thrust is produced by:

    (a) a large, slow revving propeller.
    (b) a large, fast revving propeller.
    (c) a small, slow revving propeller.
    (d) a small, fast revving propeller.

28. In a vessel going astern, the rudder action can be unpredictable. However, it can start to react in a desired manner:

    (a) after gathering sternway.
    (b) before gathering stern way.
    (c) with a port helm.
    (d) with a starboard helm.

29. In calm conditions and deep water, a vessel's drift due to a tide or current is influenced by:

    (a) her hull shape.
    (b) her size.
    (c) her draught.
    (d) none of the choices stated here.

30. A single screw vessel has a high superstructure forward. During unberthing, an offshore wind will tend to pivot her:

    (a) bow out.
    (b) stern out.
    (c) bow in.
    (d) stern in.

31. You are to turn a single screw vessel into the wind. She is at full speed ahead. Compared to a vessel with a large superstructure forward, a vessel with a large superstructure aft will:

    (a) turn slower.
    (b) turn faster.
    (c) turn at a rate independent of wind direction.
    (d) not turn.

32. A twin screw vessel is to berth starboard side to with a current running from astern and a moderate offshore wind. She should approach the jetty:

    (a) closing in parallel and then "walking" in.
    (b) closing in parallel and then "walking" out.
    (c) with one engine only.
    (d) without using the helm.

33. In order to make a three-point turn (i.e., to turn short around) with a right-handed single screw, a vessel should start with:

    (a) slow ahead on engine and rudder to starboard.
    (b) slow ahead on engine and rudder to port.
    (c) slow astern on engine and rudder to starboard.
    (d) slow astern on engine and rudder to port.

34. A twin screw vessel is proceeding at full speed. The quickest way for her to turn around in an emergency is to go ahead on one engine and astern on the other, with wheel hard over to one side as follows:

    (a) port engine astern, wheel to starboard.
    (b) starboard engine astern, wheel to port.
    (c) one engine astern, wheel to the same side.
    (d) one engine astern, wheel to the opposite side.

35. A single screw vessel is alongside port side to. The best way for her to unberth in a strong on-shore wind is to start with the following manoeuvre:

    (a) Forward spring on, slow ahead, hard a port.
    (b) Aft spring on, slow ahead, hard a port.
    (c) Stern line on, slow ahead, hard a starboard.
    (d) Head line on, slow astern, hard a starboard.

36. A single screw vessel is to berth starboard side to, in a strong off-shore wind. She should approach her berth as follows:

    (a) Forward spring on, slow ahead, hard a port.
    (b) Aft spring on, slow astern, hard a starboard.
    (c) Head line on, slow astern, hard a starboard.
    (d) Stern line on, slow ahead, hard a port.

37. A single screw vessel is alongside port side to. The best way for her to unberth in a strong wind from astern is to start with the following manoeuvre:

    (a) Aft spring on, slow ahead, hard port.
    (b) Stern line on, slow ahead, hard a starboard.
    (c) Forward spring on, slow ahead, hard a port.
    (d) Head line on, slow astern, hard a starboard.

38. Openings in a reef can be distinguished by the:

    (i) difference in the colour of the water.
    (ii) gaps in the breaking water.
    (iii) channel markers.
    (iv) speed of waves.

    The correct answer is:
    (a) all except (i).
    (b) all except (ii).
    (c) all except (iii).
    (d) all except (iv).

39. The risk of vessel pitchpoling in rough seas is minimal when:

    (a) running before the sea.
    (b) surfing.
    (c) pitching in synchronism.
    (d) riding on the back of a wave.

Chapter 8: Boat Handling & Towing

40. Vessel swinging to either side of an intended course is known as:

    (a) rolling.
    (b) pitching.
    (c) yawing.
    (d) synchronization.

41. A vessel's quarter is the:

    (a) bow.
    (b) quarter of any part of the hull.
    (c) stern.
    (d) quarter between stern and beam.

42. A vessel in the following load condition experiences long slow rolling:

    (a) Bottom heavy.
    (b) Top heavy.
    (c) Trimmed by stern.
    (d) Trimmed by head.

43. A vessel risks pitchpoling when:

    (a) surfing down the face of a steep swell.
    (b) riding on the back of a steep swell.
    (c) in a crest of a large wave.
    (d) in a trough of a large wave.

44. When turning a vessel around in a heavy sea:

    (a) use full power with full helm.
    (b) sit on the crest of a wave.
    (c) look for a lull in the waves.
    (d) sit in the trough of a wave.

45. A vessel will not lift easily, become harder to steer and take larger seas when:

    (a) trimmed by the head.
    (b) trimmed by the stern.
    (c) loaded top heavy.
    (d) loaded bottom heavy.

46. A single screw vessel is easiest to berth with tide from:

    (a) ahead.
    (b) behind.
    (c) onshore.
    (d) offshore.

47. When mooring to a buoy, it is best to approach it with wind from:

    (a) ahead.
    (b) beam.
    (c) astern.
    (d) quarter.

48. When unberthing a single screw vessel, the use of the forward spring assists in:

    (a) swinging out her stern.
    (b) swinging out her bow.
    (c) walking her out like a twin screw vessel.
    (d) none of the choices stated here.

49. To assist in turning a single screw vessel short around:

    (a) an anchor may be used.
    (b) an anchor cannot be used.
    (c) an anchor must be used.
    (d) both anchors must be used.

50. The most effective way to stop a vessel in an emergency is to:

    (a) go full astern and drop anchors.
    (b) go full astern.
    (c) put engine in neutral and drop anchors.
    (d) put engine in neutral and turn in a circle.

51. The hogging and sagging stresses in a vessel at sea are caused due to the vessel:

    (a) panting.
    (b) pounding.
    (c) riding on wave crests & in troughs.
    (d) impact of waves on the bow.

52. Synchronization means that a vessel's period of:

    (a) roll or pitch equals the wave encounter period.
    (b) roll equals the wave period.
    (c) pitch equals the wave period.
    (d) rolling has increased.

53. The safest method of heaving-to is to:

    (a) Hold course with sea on the bow.
    (b) Stop engine, secure all openings and drift.
    (c) Hold course with sea on the quarter.
    (d) Hold course with sea right astern.

54. A vessel's roll period becomes longer when the weights on board are concentrated:

    (a) fore and aft.
    (b) amidships.
    (c) in wings.
    (d) on centreline.

55. A large stern trim can lead to pooping when the vessel is:

    (a) heading into the sea.
    (b) running before the sea.
    (c) running beam to sea.
    (d) stopped at sea.

56. When running before a large sea, a vessel can avoid surfing by adjusting her speed to roughly:

    (a) equal to the speed of waves.
    (b) double the speed of waves.
    (c) half the speed of waves.
    (d) quarter the speed of waves.

57. A following sea can make a vessel's speed appear:

    (a) slower.
    (b) faster.
    (c) erratic.
    (d) unaffected.

58. When running before a sea, a rooster tail may start to appear behind a vessel. It should:

    (a) be lowered as soon as possible.
    (b) cause no concern to the vessel.
    (c) increase vessel's stability.
    (d) level out if left alone.

59. The risk of interaction is increased by:

    (i) High speed.
    (ii) Narrow channel.
    (iii) Deep water.
    (iv) Deep draught.

    The correct answer is:
    (a) all except (i).
    (b) all except (ii).
    (c) all except (iii).
    (d) all except (iv).

60. A high powered planing hull vessel with a right handed screw can be prevented from heeling at high speeds by:

    (a) reducing her load.
    (b) adjusting her trim tabs.
    (c) the helms person sitting on port side.
    (d) an even distribution of weights.

61. When towing a vessel in confined waters, such as harbours:

    (a) make the tow line as long as possible.
    (b) keep the sterns of the two vessels in line.
    (c) secure the tow alongside the tug.
    (d) keep the stern of the tow behind the tug's.

62. You are to tow a small vessel from out at sea into port. The tug should:

    (a) take charge.
    (b) pass the tow line from ahead.
    (c) pass the tow line from windward.
    (d) gradually increase speed on connecting up.

63. When launching at boat ramps, the person in-charge should:

    (a) embark passengers before launching.
    (b) immerse the trailer's hubs in water.
    (c) use the trailer's tilt action.
    (d) prepare the vessel on the ramp.

64. Lee helm means:

    (a) tiller to leeward.
    (b) tiller to leeward to point vessel upwind.
    (c) helmsman to sit on leeward side.
    (d) helm hard over to leeward.

65. Leeward means:

    (a) lee helm.
    (b) on the downwind side of the observer.
    (c) tiller to leeward to point vessel upwind.
    (d) sailing on the lee shore.

66. When crossing a sand bar in a small vessel, the following rules should be followed:

    (i) Ride on the back of the waves.
    (ii) Keep an anchor on deck, ready for letting go.
    (iii) keep the fore & aft trim as small as possible.
    (iv) Vary the engine speed.

    The correct answer is:
    (a) all except (i).
    (b) all except (ii).
    (c) all except (iii).
    (d) all except (iv).

67. A sand bar is dangerous to cross in bad weather at any time. However it is made _more_ dangerous by the following conditions:

    (a) flood tide and onshore wind.
    (b) ebb tide and strong wind.
    (c) ebb tide.
    (d) flood tide and offshore wind.

68. If a sand bar was in a dangerous condition on a vessel's return from a trip at sea, and doesn't appear to be abating, she should:

    (a) definitely wait for the weather to improve.
    (b) seek advice by radio.
    (c) heave to.
    (d) cross it seeking the line of least activity.

69. The most reliable source of the state of sea condition over a sand bar is the:

    (a) navigational chart for the area.
    (b) tide tables.
    (c) weather forecast.
    (d) local rescue organisation.

Chapter 8: Boat Handling & Towing

70. A vessel towing, notices that her tow line is losing its catenary. It indicates that the:

    (a) tow line tension is increasing progressively.
    (b) tow line has reached the correct pull.
    (c) tow is experiencing steering problem.
    (d) towing vessel is too slow.

71. The following are the suitable solutions to prevent a tow line chafing:

    (i) towing vessel should tow from her extreme aft.
    (ii) pass tow rope through fairleads.
    (iii) periodically "freshen the nip".
    (iv) pad bull bar with fender.

    The correct answer is:
    (a) all except (i).
    (b) all except (ii).
    (c) all except (iii).
    (d) all except (iv).

72. If the tow takes a big sheer on a shortened tow line, it will cause the:

    (i) tug to capsize.
    (ii) tug to be pulled astern.
    (iii) tug to be pulled off course.
    (iv) tow line to break.

    The correct answer is:
    (a) all except (i).
    (b) all except (ii).
    (c) all except (iii).
    (d) all except (iv).

73. You are towing a vessel in confined waterways on a shortened towline. You cannot reduce the risk of being pulled off course by the sudden yaw of the tow by:

    (a) altering the trim of the tow.
    (b) streaming a drogue behind the tow.
    (c) being able to quickly slip the tow.
    (d) steering by autopilot.

74. When towing in heavy weather, the tow line should not be:

    (a) heavy.
    (b) tight.
    (c) secured to more than one point.
    (d) elastic.

75. The following will not stop a towed vessel from yawing:

    (a) change of speed.
    (b) change of trim.
    (c) towing a drogue
    (d) listing the vessel.

76. On a vessel lying alongside a solid wharf, the engines are put ahead with wheel amidships. She would:

    (a) come off the wharf, bow first.
    (b) come off the wharf, stern first.
    (c) bodily move closer to the wharf.
    (d) bodily move away from the wharf.

77. The following hull shape offers the best accelerating and stopping performance:

    (a) planing hull.
    (b) small displacement hull.
    (c) large displacement hull.
    (d) fishing trawler.

78. A single screw vessel has the wheelhouse amidships. With strong wind on her starboard bow and rudder amidships, she goes ahead on her engine. She would tend to:

    (a) remain on a steady course.
    (b) fall off the wind.
    (c) head into the wind.
    (d) get blown sideways.

79. The Shallow Water Effect is that:

    (a) Vessels become more manoeuvrable.
    (b) Vessels take longer to turn or stop.
    (c) Barge shaped vessels tend to squat by the stern.
    (d) Fine-hulled vessels tend to squat by the bow.

80. Which of the following terminology is not associated with a vessel's turning circle:

    (a) Advance.
    (b) Snubbing.
    (c) Transfer.
    (d) Tactical Diameter.

81. The Turning Circle term 'Transfer' is:

    (a) the Advance.
    (b) the Tactical Diameter.
    (c) the Final Diameter.
    (d) a distance perpendicular to Advance.

82. The term "Inertia Stop" means:

    (a) putting the engine into neutral.
    (b) putting the engine full astern.
    (c) Rudder Recycling.
    (d) Fish Tailing.

83. When unberthing a Landing Barge from a beach:
    (a) first put her parallel to the beach.
    (b) do not close the door until after unberthing.
    (c) first put the engines slow ahead.
    (d) Yaw her & then go astern on both engines.

84. Throughout her stay on a beach berth, the Landing Barge should be evenly supported by:
    (a) Sand.
    (b) Water.
    (c) Sand & Water.
    (d) Anchors.

85. With her twin-screws wider apart than in a mono-hull, a Catamaran:
    (a) is less susceptible to set and drift.
    (b) turns more slowly.
    (c) has a smaller turning circle.
    (d) is more susceptible to set and drift.

86. Among the fibre ropes, the most suitable towing ropes are:
    (a) Pre-stretched.
    (b) of high stretch characteristic.
    (c) Non-buoyant
    (d) of heavy weight.

87. When towing with a fibre rope, a Stress Line is a:
    (a) Pre-stretched rope.
    (b) towing telltale.
    (c) Heavy weight rope.
    (d) Braided rope.

88. In towing, a Gob Rope:
    (a) is a Pre-stretched rope.
    (b) is a towing telltale.
    (c) prevents the tow line from running abeam of the tug.
    (d) prevents the tow from yawing.

89. To warp a vessel is to:
    (a) move her along a wharf using warping drums.
    (b) remove distortion from her decks.
    (c) heave in the fishing gear in a fishing vessel.
    (d) secure her at an angle to the wharf.

90. A Heaving Line is:
    (a) used for heaving-in the vessel.
    (b) a messenger with a monkey's fist in its end.
    (c) use for heaving in fishing nets.
    (d) Only used in small vessels.

91. A Messenger is:
    (a) a lashing rope.
    (b) not a rope.
    (c) another name for the monkey's fist.
    (d) used for passing a heavier line from one person to another.

92. Propeller in a Kort Nozzle:
    (a) causes greater transverse thrust than an open-air propeller.
    (b) causes no transverse thrust
    (c) provides less bollard pull than an open-air propeller.
    (d) causes reverse transverse thrust

93. In relation to jet drives, which statement is incorrect?
    (a) Warn passengers before crash stopping.
    (b) Do not accelerate too quickly with a skier in tow.
    (c) Jet drives respond quickly to acceleration, stopping & turning.
    (d) Do not operate in shallow water.

94. In relation to thrusters fitted to vessels, which statement is incorrect?
    (a) They are fitted to ferries to facilitate berthing & unberthing.
    (b) They help overcome crosswinds, tides and lack of space during berthing or unberthing.
    (c) They provide additional control.
    (d) They are fitted only in the bow.

95. In steep-banked waterways, when a vessel moves to one side of the channel:
    (a) she tends to stay there.
    (b) her hull tends to be repelled towards the middle of the channel.
    (c) her bow tends to be repelled towards the middle of the channel.
    (d) she experiences interaction only if proceeding at high speed.

96. The Canal Effect is experienced:
    (a) only in canals.
    (b) only in canals & such steep-banked waterways.
    (c) only where the banks of waterways are not continuous.
    (d) against all solid structures, including piers & wharves.

97. The Bank Cushioning Effect in a river can be used to advantage:

    (a) to alter course for oncoming vessels.
    (b) to negotiate turns in the river.
    (c) to reduce squat.
    (d) to stay in the middle of the channel.

98. To 'Heave-to' is to:

    (a) stop or slow down until weather abates.
    (b) hold course with the sea on the quarter.
    (c) hold course with the sea on the bow.
    (d) drift.

CHAPTER 8.2: SAILING TERMINOLOGY

99. Aback: the sailing term Aback means:

    (a) wind on the wrong side of a sail.
    (b) drifting down wind.
    (c) negative speed due to strong tide.
    (d) sailing with wind behind.

100. Bare poles means:

    (a) without sails set.
    (b) mainsail not set.
    (c) minimum sails set.
    (d) only trysail is set.

101. Beam Reach means:

    (a) sailing with wind abeam.
    (b) sailing with sea abeam.
    (c) sails abeam.
    (d) approaching a buoy from windward.

102. Beating: a sailing vessel is beating when:

    (a) its sails are flapping.
    (b) running free.
    (c) sailing close-hauled on alternate tacks.
    (d) lowering main sail.

103. Broad Reach means:

    (a) sailing with wind abeam.
    (b) sailing with sea abeam.
    (c) sailing with wind slightly abaft the beam.
    (d) approaching a buoy from windward.

104. Close-hauling a vessel is to:

    (a) close hatches.
    (b) shut her down completely.
    (c) sail as near to wind as possible.
    (d) tack immediately.

105. Close Reefed sail is one which:

    (a) has no reef points.
    (b) has been reefed as much as possible.
    (c) cannot be reefed due to windage.
    (d) has closely spaced reefs points.

106. Dead Run means:

    (a) a vessel sailing with wind directly astern.
    (b) no wind.
    (c) a vessel sailing under power and sail.
    (d) a vessel hove to.

107. Fetch means:

    (a) unobstructed distance travelled by wind or sea.
    (b) angle of a sail.
    (c) angle of the main sail.
    (d) angle of all fore and aft sails.

108. Free: a sailing vessel is said to be free when sailing:

    (a) with wind right aft.
    (b) with wind abaft the beam.
    (c) bare poles.
    (d) motoring.

109. Gybe or Jibe means to change course:

    (a) while sailing in a screw motion.
    (b) while rolling in beam seas.
    (c) while pitching in head seas.
    (d) by shifting over the boom while sailing free.

110. Luff (a boat) means to:

    (a) turn her to stop luffs from flapping.
    (b) retrimming sails to increase speed.
    (c) turn her directly into wind to make luffs flap.
    (d) unreef sails.

111. Off The Wind means:

    (a) sailing with wind abaft the beam.
    (b) close-hauled sailing.
    (c) off-shore wind.
    (d) sailing with wind abeam.

112. On The Wind means:

    (a) sailing with wind abaft the beam.
    (b) sailing close-hauled.
    (c) off-shore wind.
    (d) sailing with wind abeam.

113. Proud run means:

    (a) no wind.
    (b) a vessel sailing under power and sail.
    (c) the same as dead run.
    (d) a vessel hove to.

114. A vessel on Port Tack has the wind on:
   (a) starboard side.
   (b) port side.
   (c) port quarter.
   (d) starboard quarter.

115. Running Free: a sailing vessel is said to be running free when sailing:
   (a) with wind right aft.
   (b) with wind abaft the beam.
   (c) bare poles.
   (d) motoring.

116. Sheets on a sailing vessel refers to:
   (a) any line attached to the clew of a sail.
   (b) block & tackle attached to the boom of a sail.
   (c) any line attached to the clew or boom of a sail.
   (d) none of the choices stated here.

117. In a sailing vessel, to "Pay off" is to:
   (a) alter course away from the wind.
   (b) alter course into the wind.
   (c) ease sheets.
   (d) tighten sheets.

118. A sailing vessel 'in irons':
   (a) is constructed of steel.
   (b) is at anchor.
   (c) has accidentally jibed.
   (d) has failed to tack.

119. Carrying Lee Helm means that the sailing vessel:
   (a) tends to pay off.
   (b) is steering upwind.
   (c) is sailing as close to the wind as possible.
   (d) has failed to tack.

120. Carrying Weather Helm means that the sailing vessel:
   (a) steers downwind a little.
   (b) steers upwind a little.
   (c) is sailing as close to the wind as possible.
   (d) has failed to tack.

121. A sailing vessel Hugging means:
   (a) luffing up.
   (b) tacking on a upwind course.
   (c) sailing as close to the wind as possible.
   (d) carrying wet sails.

CHAPTER 8: ANSWERS

1 (b), 2 (b), 3 (a), 4 (b), 5 (c),
6 (d), 7 (d), 8 (b), 9 (a), 10 (c),
11 (a), 12 (d), 13 (c), 14 (b), 15 (d),
16 (d), 17 (b), 18 (c), 19 (a), 20 (d),
21 (b), 22 (a), 23 (b), 24 (d), 25 (b),
26 (d), 27 (b), 28 (a), 29 (d), 30 (a),
31 (b), 32 (a), 33 (a), 34 (c), 35 (a),
36 (a), 37 (c), 38 (d), 39 (d), 40 (c),
41 (d), 42 (b), 43 (a), 44 (c), 45 (a),
46 (a), 47 (a), 48 (a), 49 (a), 50 (a),
51 (c), 52 (a), 53 (a), 54 (c), 55 (b),
56 (c), 57 (a), 58 (a), 59 (c), 60 (b),
61 (c), 62 (d), 63 (c), 64 (b), 65 (b),
66 (b), 67 (b), 68 (b), 69 (d), 70 (a),
71 (a), 72 (b), 73 (d), 74 (b), 75 (d),
76 (c), 77 (a), 78 (c), 79 (b), 80 (b),
81 (d), 82 (a), 83 (d), 84 (c), 85 (c),
86 (a), 87 (b), 88 (c), 89 (a), 90 (b),
91 (d), 92 (b), 93 (d), 94 (d), 95 (c),
96 (d), 97 (b), 98 (a), 99 (a), 100 (a),
101 (a), 102 (c), 103 (c), 104 (c), 105 (b),
106 (a), 107 (a), 108 (b), 109 (d), 110 (c),
111 (a), 112 (b), 113 (c), 114 (b), 115 (b),
116 (c), 117 (a), 118 (d), 119 (a), 120 (b),
121 (c),

Chapter 9

LIFTING EQUIPMENT & WINCHES

SAFE WORKING LOADS OF ROPES & CHAINS

Breaking stresses of ropes and tensile strengths of chains supplied by manufacturers are usually of laboratory conditions. The strength decreases as soon as they are put into use, and continues to decrease over time. Safe Working Loads (SWL) are also difficult to recommend due to the wide range of applications of ropes, chains and other lifting equipment. Textbooks attempt to offer a variety of formulae for calculating SWL. All are approximate. The following are simplified formulae for calculating SWL of ropes and chains, as recommended in *The Guide for Riggers (published by Workcover in NSW) and supplied by manufacturers.*

SWL OF A ROPE (in Kgs) = $D^2 \times F$

D is diameter of rope in millimetres. F is the factor representing its strength.
These Factors are (IN ASCENDING ORDER):

- Manila/Silver = 1 (This applies to all Natural Fibres. The factor for Silver Rope is 1.16 but 1 is easier to remember. Silver Rope is also known as Flat Spin Taniklon)
- Polythene = 1.5 (It's 1.6 but 1.5 is easier to remember; a.k.a. Polyethylene, Polypropylene, Taniklon)
- Polyester = 2 (a.k.a. Terylene, Dacron)
- Nylon = 2.25
- Kevlar = 6 (Spectra, Kevlar, etc.)
- Wire = 8
- Stainless = 10

*NOTE: The Factors listed in some **TAFE Colleges' notes** differ from above, as follows:*
➢ Polyethylene = 1.2 (staple fibre)
➢ Polyethylene = 1.8 (monofilament fibre)
➢ Polyester = 2.5
➢ Nylon over 50 mm = 2.5
➢ Nylon under 50 mm = 3

SWL OF STEEL CHAIN BELOW 12.5 MM (in kgs) = $3D^2 \times$ Grade
(See note below about the old formula)

Depending on their material composition, welded chains below 12.5 mm in size, if tested and marked in compliance with the International Standards Association (ISO), are graded 3 upwards; i.e. 3, 4, 5.... generally up to 9. Higher numbers represent better grades. The grade numbers are usually stamped or embossed on the chain approximately a metre apart. Some manufacturers use a 2-digit marking system: G3, G4, etc. ('G' meaning grade); or A3, A4, etc. ('A' being the manufacturer's initial). Some may use a 3-digit system (200, 300, 400, etc.) or letters to indicate the grades:

➢ L = 3 (Mild steel, stress relieved)
➢ M = 4 (High tensile steel, quenched and tempered)
➢ P = 5 (ditto)
➢ S = 6 (High tensile, load binder chain)
➢ T = 8 (Higher tensile steel, quenched and tempered)
  (or CM or A)
➢ V = 10 (Very high tensile steel, quenched and tempered)

NOTE: The above formula used to be $0.3D^2 \times$ Grade, when grades were 30, 40, 50... instead of 3, 4, 5... etc.)

Unmarked chain should be treated as grade 3, i.e. the lowest grade. Lifting chain is usually upwards of grade 8. Stud link chain is not used for lifting

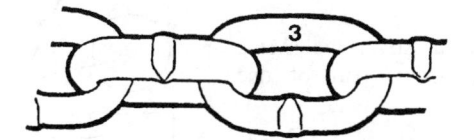

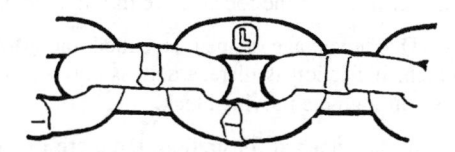

Fig 9.1: CHAIN MARKINGS (GRADE 3, 30 OR L)

because studs prevent it from showing elongation due to overloading or age. Stud link chain is designed mainly for anchoring. Studs prevent cable jamming when coming in or out of chain lockers. See anchor cable in Chapter 7.

BREAKING STRESS (& SWL) OF STEEL CHAIN 12.5 MM & OVER (in tonnes)

Chains 12.5 mm and over are graded into 3 grades, whose Breaking Stress (BS) is calculated as follows.

NOTE: The following formulae calculate BS. To calculate SWL divide the result by 4 (or 5, to comply with some TAFE notes as stated below under 'Breaking Stress'):

- Grade 1:   BS (in tonnes) = $20D^2/600$
- Grade 2:   BS (in tonnes) = $30D^2/600$
- Grade 3:   BS (in tonnes) = $43D^2/600$

BREAKING STRESS (BS)

Lifting equipment, such as ropes, chains, blocks and shackle, are tested in factories to a breaking stress of 3 to 6 times their SWL (3-4 for chain, 5-6 for ropes). (In some TAFE notes, it is stated as 5 times for chain and 6 times for ropes.) These tests are conducted immediately after manufacturing when their fibres and material are intact and in perfect condition. Legally and in practice, lifting equipment should not be subjected to stresses beyond its safe working load.

EXAMPLES OF SWL & BS CALCULATIONS

EXAMPLE 1

SWL of 15 mm nylon rope = D^2 x F
= 15 x 15 x 2.25
= 506 kgs (approx. 0.5 tonnes)

Breaking Stress = 506 x 6
= 3036 kgs (approx. 3 tonnes)

EXAMPLE 2

SWL of 15 mm wire rope = 15 x 15 x 8
= 1800 kgs (1.8 tonnes)

EXAMPLE 3

SWL of 10 mm chain of grade 3 = $3D^2$ x Grade
= 3 x 10 x 10 x 3
= 900 kgs (approx. 1 tonne)

If the same chain was of grade 9, its SWL will be 3 times higher = 3 x 10 x 10 x 9 = 2700 kgs (2.7 tonnes)

EXAMPLE 4 (*For chain greater than 12.5 mm*)

BS of 30 mm chain of grade 1 = $20D^2 \div 600$
= 20 x 30 x 30 ÷ 600
= 30 tonnes

SWL = 30 ÷ 4 = 7.5 tonnes
(or 30 ÷ 5 = 6 tonnes, as per some TAFE notes)

PURCHASES AND TACKLES

A Purchase or Tackle is a mechanical device that allows a load to be lifted with a smaller force or pull than the load. It also increases the lifting capacity of the rope above its SWL. A Tackle consists of a rope rove through two or more blocks. The reduction in the pull required (at its hauling part) depends on the number of sheaves in the blocks and the manner in which the rope is rove through them.

The advantage gained is called the Mechanical Advantage (MA), which, if friction is disregarded, is equal to the number of parts of rope rove through the moving block.

In the attached illustration, there are two parts of rope at the moving block, giving the Tackle a MA of 2. Therefore, it can lift a one tonne load using a rope of SWL of half a tonne, and with a pull of half a tonne. Determining MA for various rigs is discussed below.

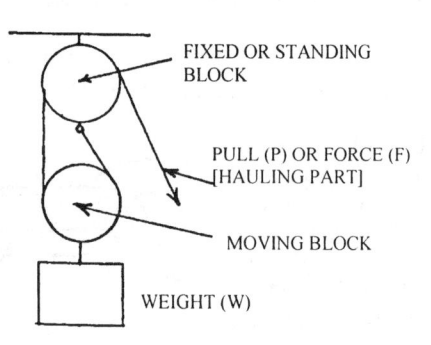

Fig 9.2: PARTS OF A PURCHASE OR TACKLE

Chapter 9: Lifting Equipment & Winches

TYPES OF PURCHASES, BLOCKS & TACKLES OR PULLEY SYSTEMS

➢ A two-fold purchase is made up of two 2-sheave blocks; a three-fold purchase of two 3-sheave blocks; and so on.
➢ A 2-fold purchase is sometimes referred to as a 'Double Luff' - not be confused with 'Luff on Luff' discussed below.
➢ A one-fold purchase is known as a Gun Tackle. It is so named because it was used for hauling out guns.
➢ Some other purchases include a Luff Tackle (or Watch Tackle, Handy Billy or Jigger) and Gyn Tackle. These consist of unequal number of sheaves in the two blocks, i.e. 2 and 1, and 3 and 2 respectively, as shown.

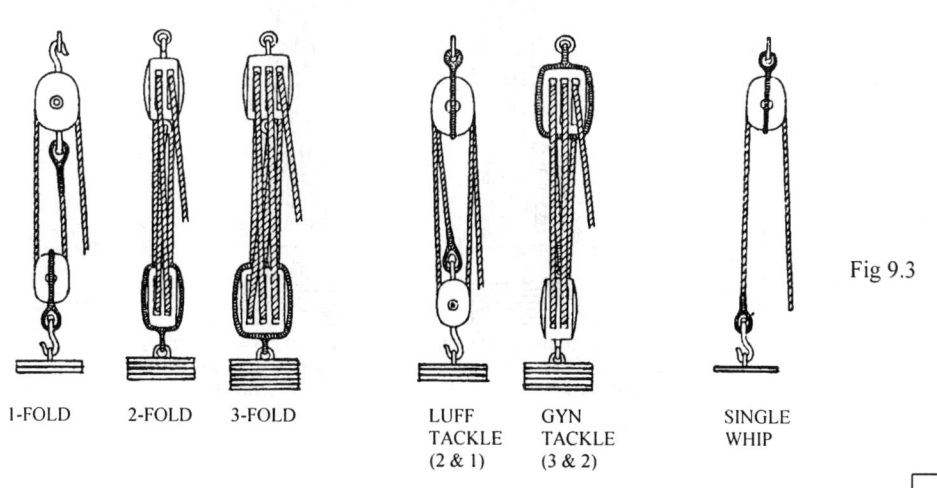

Fig 9.3

1-FOLD   2-FOLD   3-FOLD   LUFF TACKLE (2 & 1)   GYN TACKLE (3 & 2)   SINGLE WHIP

SINGLE WHIP & LEAD BLOCK

A single block used for hoisting a load does not offer a mechanical advantage. It only allows an object to be lifted above one's own height or to choose the direction of the pull. Such a rig is known as a SINGLE WHIP.

A lead block is sometimes used to lead the hauling rope from a purchase to a winch. It too does not offer a MA, and is not counted as a part of the purchase.

Fig. 9.4
LEAD BLOCK ON A PURCHASE

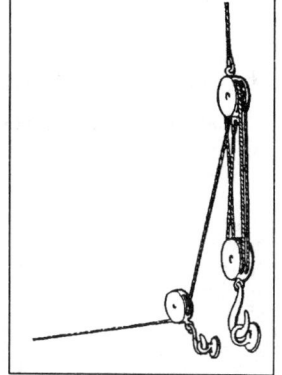

HOW TO DETERMINE MECHANICAL ADVANTAGE

To determine the M.A. of a purchase, count the number of lines (or falls) supporting the moving block. For example, in a 2-fold purchase the moving block may be supported by four lines (if the hauling line comes out of the fixed block) or five lines (if the hauling line comes out of the moving block itself).

Rig 'A' can lift 4 times the SWL of the rope, whereas rig 'B' can lift 5 times. Every purchase can be rigged in two ways: the hauling line coming out of the fixed block or the moving block. The latter will always have a M.A. of 1 more than the former, because the hauling line gets counted as a supporting line when it comes out of the moving block. These rigs are thus called "RIGGED TO DISADVANTAGE" and "RIGGED TO ADVANTAGE" respectively. To change the Mechanical Advantage of a purchase, simply turn it upside down.

As we have seen, a purchase allows us to use a smaller rope and apply a smaller pull than the load. But, remember, THE SWL OF BLOCKS, HOOKS AND SHACKLES MUST BE EQUAL TO THE ENTIRE LOAD. They do not gain from the mechanical advantage of the purchase.

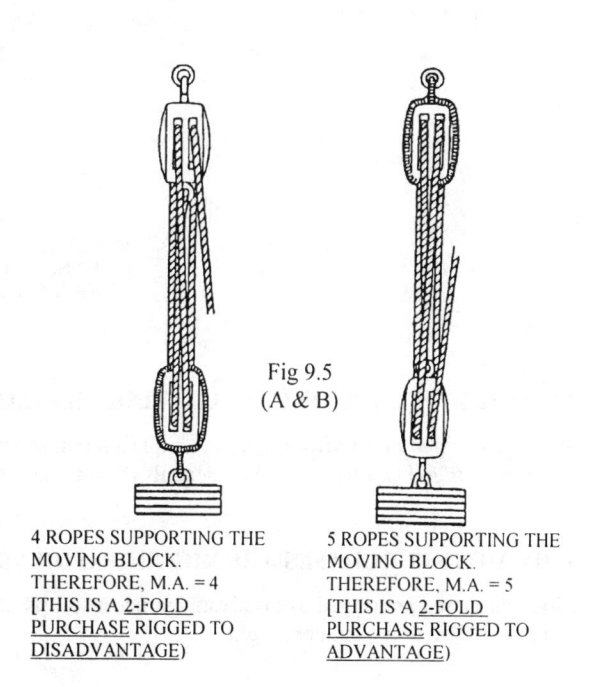

Fig 9.5 (A & B)

4 ROPES SUPPORTING THE MOVING BLOCK. THEREFORE, M.A. = 4
[THIS IS A 2-FOLD PURCHASE RIGGED TO DISADVANTAGE]

5 ROPES SUPPORTING THE MOVING BLOCK. THEREFORE, M.A. = 5
[THIS IS A 2-FOLD PURCHASE RIGGED TO ADVANTAGE]

MECHANICAL ADVANTAGE: MORE EXAMPLES

Fig 9.6
A 3-FOLD PURCHASE

RIGGED TO DISADVANTAGE
(6 ROPES SUPPORTING THE MOVING BLOCK)
M.A. = 6

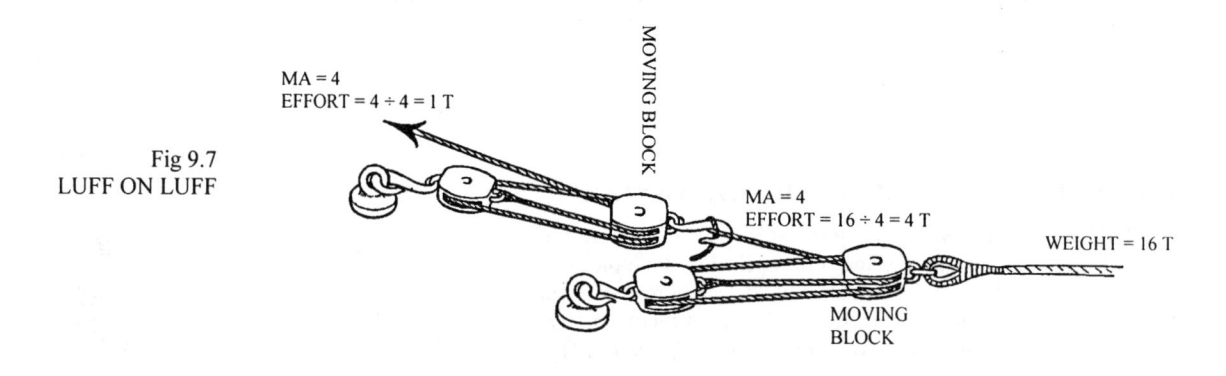

RIGGED TO ADVANTAGE
(7 ROPES SUPPORTING THE MOVING BLOCK)
M.A. = 7

LUFF ON LUFF (TACKLE ON TACKLE)

Luff on Luff is a general term for combining two tackles in which the moving block of one is secured to the hauling part of the other. In so doing, their M.A.s don't just add up, they MULTIPLY. The following figure shows the calculation.

Fig 9.7
LUFF ON LUFF

MA = 4
EFFORT = 4 ÷ 4 = 1 T

MOVING BLOCK

MA = 4
EFFORT = 16 ÷ 4 = 4 T

WEIGHT = 16 T

MOVING BLOCK

COMBINED MA = 4 x 4 = 16   (NOT 4 + 4 = 8)
(1 TONNE EFFORT IS REQUIRED TO PULL A WEIGHT OF 16 TONNES, ASSUMING AT THIS STAGE THAT THE PULLEYS ARE FREE FROM FRICTION)

ANOTHER METHOD OF CALCULATING MECHANICAL ADVANTAGE

- *When rigged to disadvantage*, M.A. equals the number of sheaves in both the blocks.
- *When rigged to advantage*, M.A. equals the number of sheaves in both the blocks + 1.

WHY MOST PURCHASES ARE RIGGED TO DISADVANTAGE

Most purchases are rigged to disadvantage because it is physically easier to pull a line downwards than upwards. The load can also be lifted to a greater height.

Chapter 9: Lifting Equipment & Winches

SNATCH BLOCK

A SNATCH BLOCK differs from a normal block in that one of its cheeks is hinged. The bight of the rope can be passed into the swallow (over the sheave) by opening its hinged cheek. A rope can thus be inserted without having to reeve its free end through the block. It is useful for changing the lead of a rope without having to reeve the end through it. The snatch block is always a single-sheave block. It should never be used for lifting.

The yachting snatch blocks usually have roller/ball bearing sheaves for low friction. They may also have a push button latch for one hand operation. Their cheeks may be made of urethane to prevent damage to the boat.

Fig 9.8: SNATCH BLOCK (YACHTING TYPE - *Lewmar*)

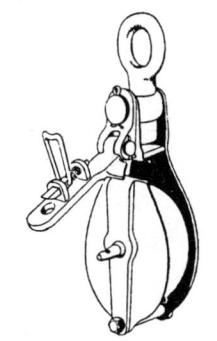

Fig 9.9: SNATCH BLOCK

PARTS OF A BLOCK (Fig 9.10)
- Blocks should be stamped with SWL.
- They should be maintained according to the manufacturers instructions.
- Use only the fibre rope blocks for fibre ropes and wire rope blocks for wire ropes.
- Lubricate their moving parts.
- The swivel eye, where fitted, should move freely.
- The axle pin should be secure.
- There should be no wear on the axle pin or the sheave bush or groove.
- Check wooden blocks for splitting.
- Painting a block may hide its markings and defects. It may also clog lubrication holes.

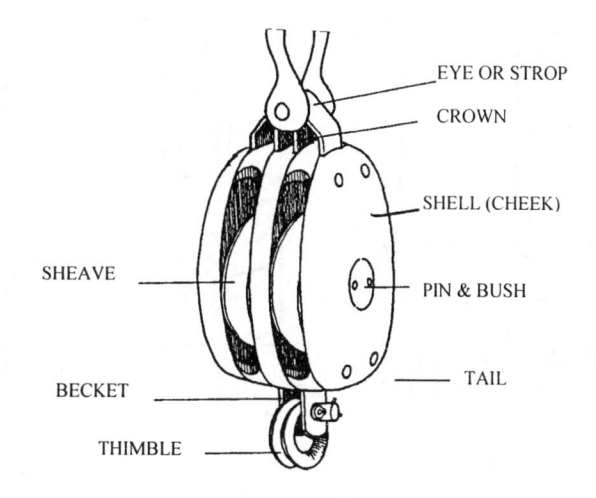

Labels: EYE OR STROP, CROWN, SHELL (CHEEK), PIN & BUSH, TAIL, SHEAVE, BECKET, THIMBLE

SOME EXAMPLES OF BLOCK SYSTEMS ON YACHTS
(Antal Marine)

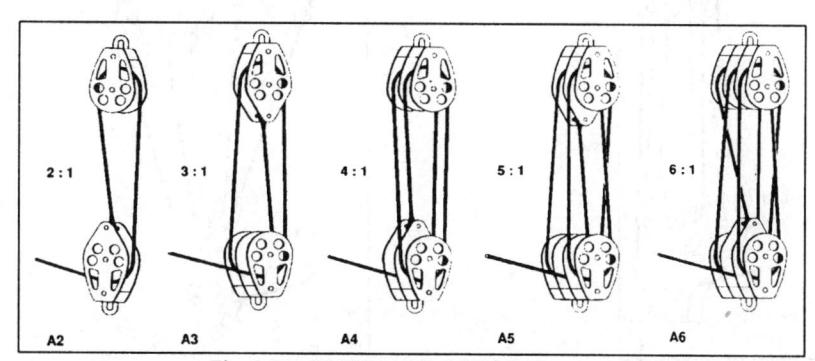

2:1 3:1 4:1 5:1 6:1
A2   A3   A4   A5   A6

Fig. 9.11: STANDARD SYSTEMS

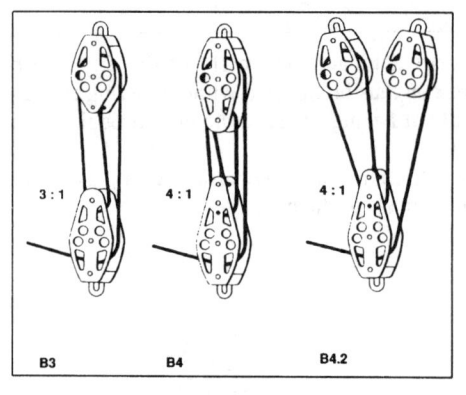

Fig 9.12: FIDDLE SYSTEMS

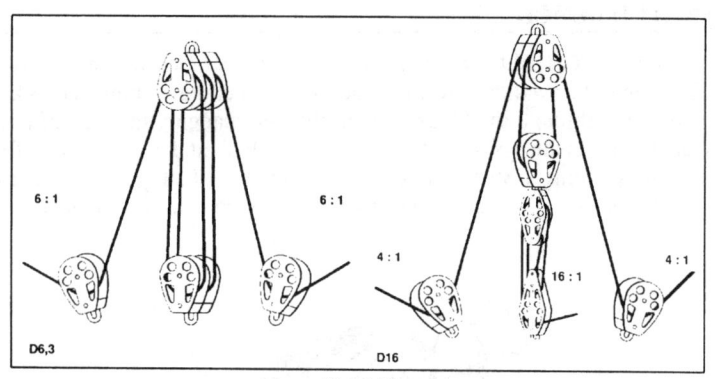

Fig 9.13
BLOCKS SYSTEMS FOR VANG AND BACKSTAY

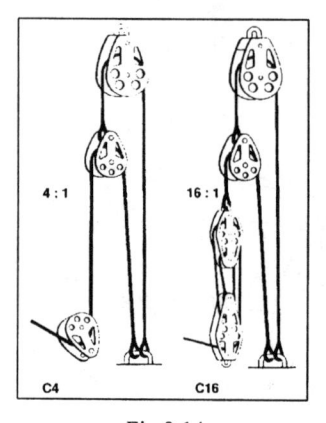

Fig 9.14
CHAIN SYSTEMS FOR VANG AND BACKSTAY

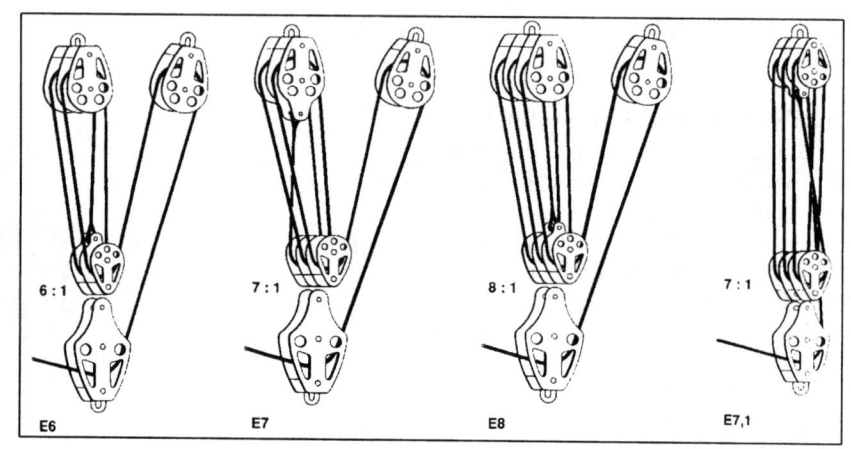

Fig 9.15: SINGLE SPEED MAIN SHEET SYSTEMS

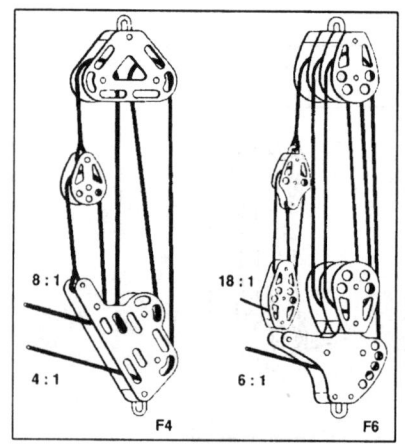

Fig 9.16: DOUBLE SPEED MAIN SHEET SYSTEMS

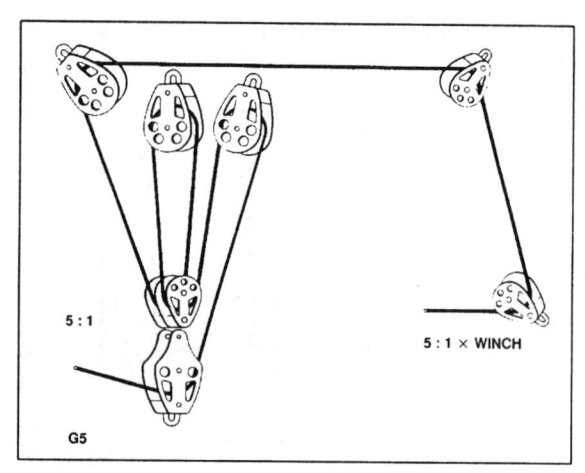

Fig 9.17: DOUBLE SPEED SYSTEM x WINCH

Chapter 9: Lifting Equipment & Winches

BLOCK LOADING depends on the angle of the rope. The following are the values for some typical angles.

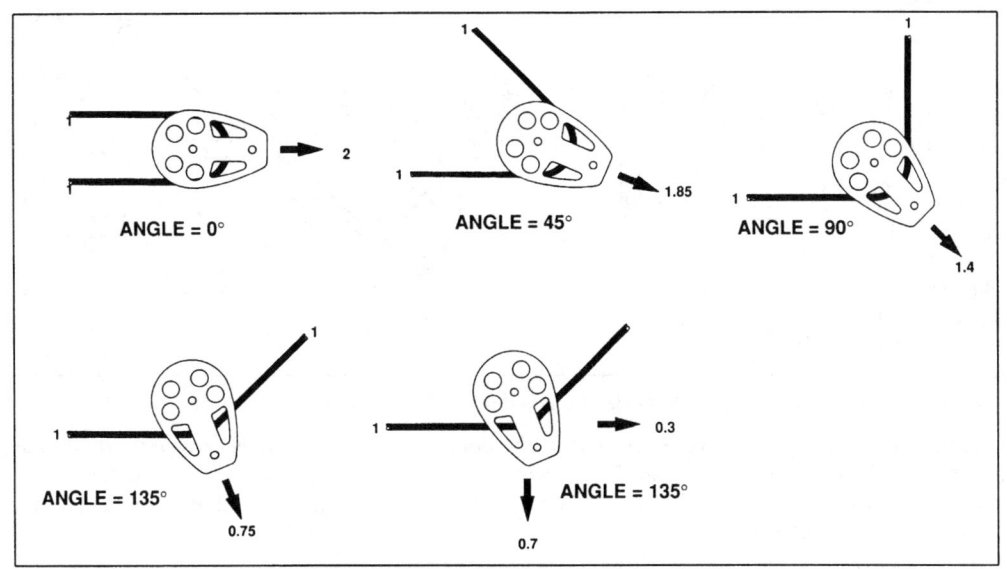

Fig 9.18: BLOCK LOADING AT DIFFERENT ANGLES (*Antal Marine*)
(Block loading depends on the angle of the line. Values for typical angles are shown here.)

FRICTION IN A PULLEY

All pulleys suffer from friction, thus increasing the load on the rope by up to 10% for each pulley. For example, in a two-fold purchase, friction may increase the pull by *4 x 10% = 40%*, and in a 3-fold purchase by *6 x 10% = 60%*. Modern blocks, using roller bearings, may have friction as little as 2.5%. But it may be safer to use 10% unless specified.

To calculate the pull required to lift a load of 2000 kgs with a two-fold purchase rigged to disadvantage, the calculation will be as follows:

Weight	= 2000 kgs
M.A.	= 4
Pull without friction	= 2000 ÷ 4 = 500 kgs [Pull or Effort = Weight ÷ M.A.]
No. of pulleys	= 4
Friction = 10% x 4	= 40% of 500 = 200 kgs
Thus, pull with friction	= 500 + 200 = 700 kgs

For lifting the same weight with a 3-fold purchase rigged to advantage, the calculation will be as follows:

Weight	= 2000 kgs
M.A.	= 7
Pull without friction	= 2000 ÷ 7 = 285.7 kgs
Friction due to 6 pulleys	= 10% x 6 = 60% of 285.7 = 171.4 kgs
Pull with friction	= 285.7 + 171.42 = 457.1 kgs

The illustration Lewmar Mainsheet System shows how a 2.5% friction would increase the trimming force (T) required at each block in a yacht's mainsheet system. The friction is low, assuming the blocks are fitted with roller bearings.

Fig 9.19
MAINSHEET SYSTEM SHOWING
LOSSES DUE TO FRICTION
(*Lewmar*)

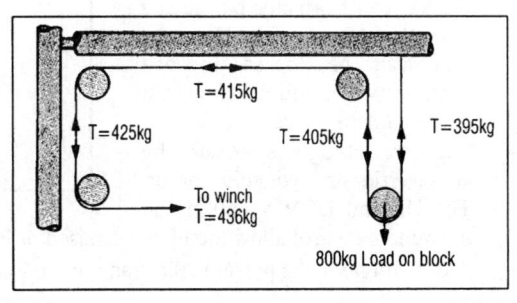

MATCHING SHEAVE (PULLEY), GROOVE & ROPE SIZE

Around sharp bends only about half the rope's fibres take the load; compression makes the others ineffective. It is therefore better to use larger size blocks. The size of a block is measured by its sheave diameter in millimetres. A sheave that is too small will damage the rope by flattening it. Small diameter sheaves also exert greater friction on their pins.

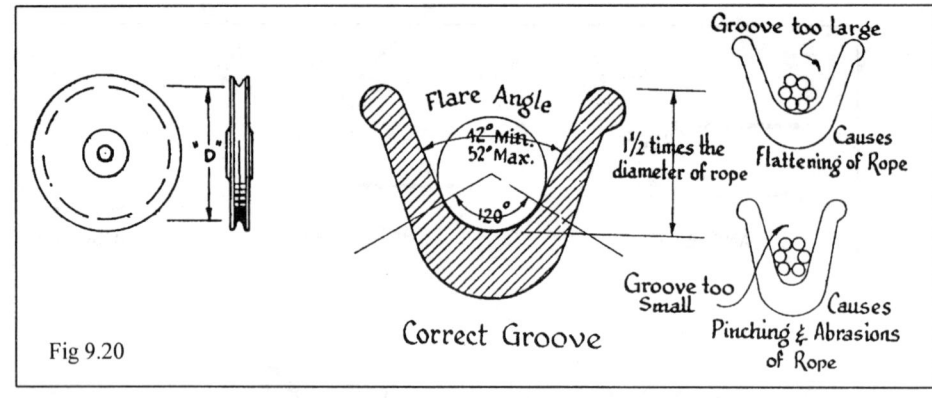

Fig 9.20

The following is recommended in the *Guide for Riggers published by NSW Workcover*.

The sheave diameter ("D" in figure above, *measured at the bottom of the groove where rope seats on the sheave*) should be 10 times the rope diameter (never less than 5). For wire ropes on power operated blocks it should be 20 times (on non-power-operated blocks it can be 10). There are also some rope materials, such as the aramid fibre, that require the sheave diameter to be 20 rope diameters.

The sheave's groove should be neither too small nor too large. It should support approximately one third of the rope's circumference as shown here.

CHAIN HOISTS (CHAIN BLOCKS)

Some boats may use chain hoists in engine rooms. The lifting hook of a chain hoist is normally its weakest part. Replace the hoist when the hook starts to spread open or show signs of wear. A distorted link is also an indication of an unsafe hoist.

CRANES (Fig 9.21)

Cranes such as MUIR, HIAB or PALFINGER are common on working boats. Fishing boats may have derricks.

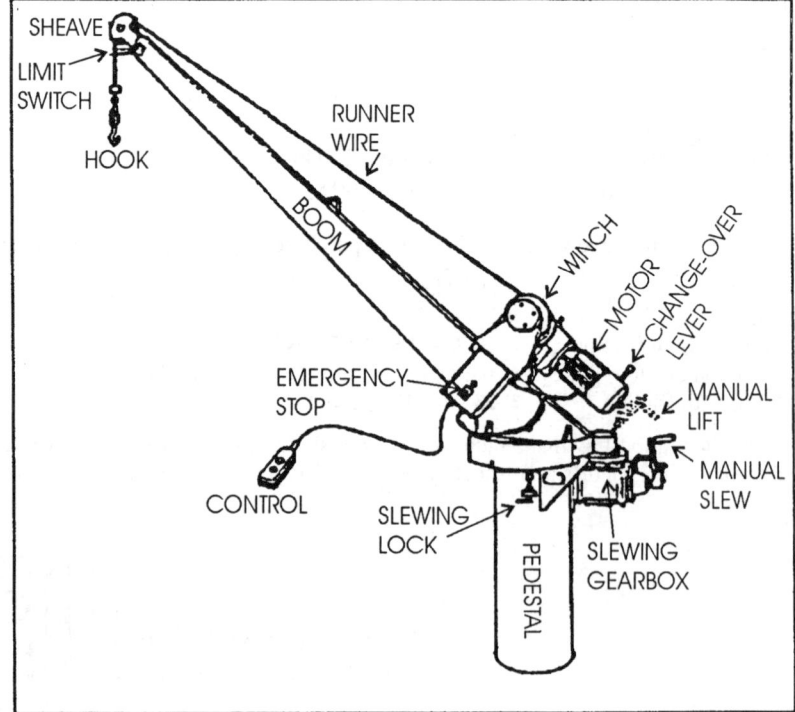

- Their SWL is stamped on them.
- The survey authorities do not usually examine cranes up to 5 tonnes SWL on vessels, and only carry out a superficial examination of cranes over 5 tonnes. Their usual requirement is that the manufacturers' operating and maintenance instructions must be complied with, and the equipment must be tested and thoroughly examined at intervals not exceeding 5 years.
- Cranes can normally be slewed through 360 degrees. However they can be locked in one position when required to hold a net or a line in place.
- The retractable jib is operated by an electric or hydraulic motor. The UP and DOWN buttons on the remote control allow the jib to be raised or lowered.
- A centrifugal brake prevents the crane or the load from moving in case of power failure or the operator letting go of the

Chapter 9: Lifting Equipment & Winches

controls.
- Only the appointed signaller should signal the crane operator, but anyone can give the STOP signal. The signaller should wear highly visible gloves or an armband to indicate his/her authority.
- The load to be lifted must be free, using only the appropriate slings. Start to lift the weight slowly so as not to exceed the SWL due to inertia. The speed of operation must be slow enough not to create jerky motion. Do not allow the load to swing, nor keep it in the air longer than necessary.
- No one should pass under a hanging load or stand where they can be crushed against a bulkhead or railing by a swinging load.
- Do not hold on to any moving part.
- When lowering the load, stop a short distance above the landing site, steady the load, recheck the suitability of the site, and then lower it.
- Do not ignore sounds indicating a fault, overload or something about to part.
- Check the lifting equipment for any damage prior to returning it to stowage.

WINCHES

In yachting, winches are used for hoisting and *trimming sails*. They are mechanically revolving drums that turn only in one direction, always clockwise. After taking a clockwise turn around the winch (i.e., wrapping a sheet (rope) around it) a handle is turned to provide the mechanical advantage through an internal gearing arrangement, making the job of heaving the rope easier. Ideally, one person grinds the winch and the other tails the line, i.e., keeps a steady pressure on the rope as it comes off the winch. However, *self-tailing winches* require only one person because the line is passed over a feeder arm and through a 'wedging' cleat (see deck fittings in Chapter 3) that prevents it from slipping back. The winch should be installed at an angle (on a base wedge) so that the rope should enter onto the drum at an angle of 5° to 10° to the base axis of the winch.

Anchor winches (windlasses) are discussed in Chapter 7. Similar winches are also used for mooring and handling of weights such as cargo and fishing gear. They can be classified as *conventional drum winches* and *traction winches*. In conventional drum winches the top layers of the rope tends to bury down into lower layers, particularly under high shock loads. The rope is thus subject to dangerously snatching out or being damaged when being freed. The modern *split-drum winches* are less prone to rope burying. Their drum consists of two compartments: one for storage of the rope not required for the job in hand; the other for the working length of the rope.

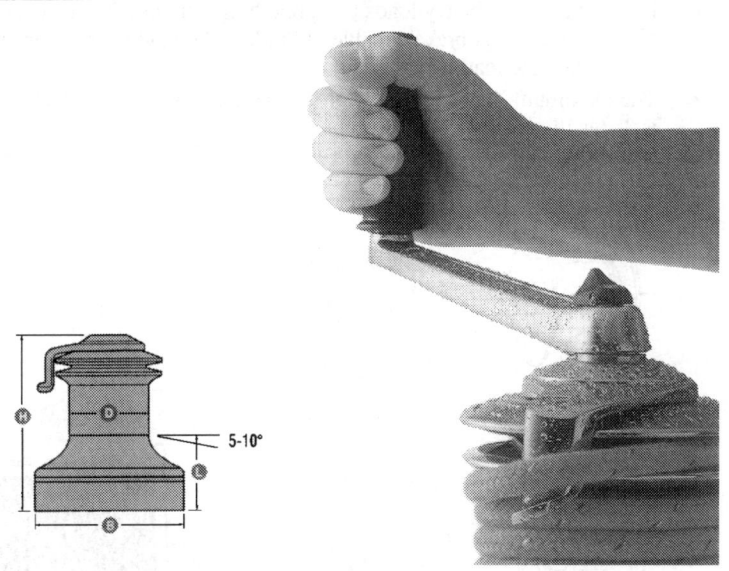

Fig 9.22 (A & B)
SELF-TAILING WINCH (CAPSTAN) (*Lewmar Winches*)

PRECAUTIONS WHEN USING WINCHES AND CAPSTANS
(See Windlass in Chapter 7)

➢ Always test a winch prior to use together with its controls, fittings, brake and pawl functions.
   (A PAWL is a pivoted lever that engages into a ratchet to prevent its reverse rotation)
➢ If in doubt, stop and check and rectify.
➢ Three turns around a warping drum is usually sufficient for heaving or slacking.
➢ Keep your hands at least half a metre clear of the drum.
➢ Stop and clear a riding turn immediately.
➢ The heat generated when surging or slipping a taught synthetic line on a rotating drum may melt or fuse it to the drum. It not only destroys the rope but the fusion suddenly coming undone or the rope parting can cause serious injury.
➢ Avoid rapid movements of taking up or stopping a load. Shock loads can stress the rope several times more than the load being lifted.
➢ Use appropriate shoes and clothing.
➢ Do not stand in the bight of a line.

- The operator must have a clear view of the operation. Never leave the winch unattended with gear running or a load suspended.
- Winches should be provided with means to prevent over-hoisting and to prevent accidental release of a load in case of power failure. They must also be provided with safety guards.
- Everyone must know the position of the emergency stop button.

WINCH DRUM WITH A WIRE ROPE SPOOLED ON IT

Drums are pulling machines that rotate, haul in and store surplus wire. Their braking system is connected to either the drum or the gearing on the drive mechanism.

- The rope must be anchored to the drum with a fixed mechanical anchorage. Make sure it is properly tightened.
- A minimum of two full turns must always remain on the drum.
- Do not overwind the drum. There must remain space equal to minimum of two full turns over the winding (as illustrated).
- Do not use your hands to guide the rope onto the moving drum.
- Do not raise very heavy loads to a great height unless the hoist brake is adjusted to take the extra torque. A brake capable of holding the load near the ground may not be able to do so when the load is high.
- The maximum fleet angle for a grooved diameter drum is 5° and for an ungrooved drum is 3° (as illustrated).
- The right-hand and left-hand laid ropes must be wound as illustrated.

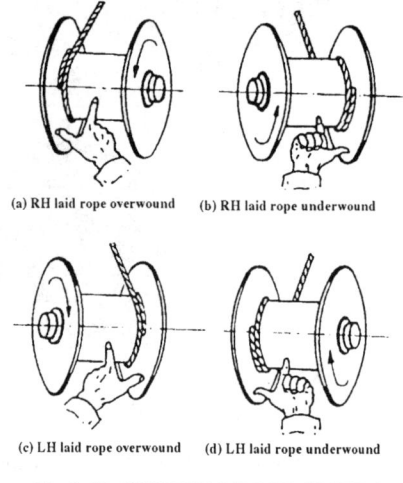

(a) RH laid rope overwound (b) RH laid rope underwound
(c) LH laid rope overwound (d) LH laid rope underwound

Fig 9.23: WINDING ROPE ONTO A DRUM *(NSW Workcover)*

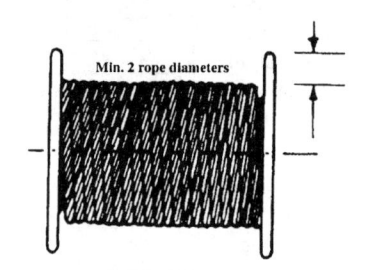

Fig 9.24 DRUM CAPACITY *(NSW Workcover)*

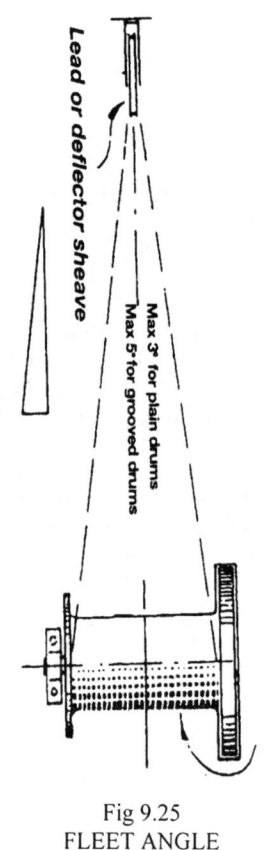

Min. 2 rope diameters

Fig 9.25 FLEET ANGLE OF A ROPE *(NSW Workcover)*

Lead or deflector sheave
Max 3° for plain drums
Max 5° for grooved drums

YACHT WINCH MAINTENANCE

(Windlass & similar winch maintenance is discussed in Chapter 7)

Because they always seem to work, winches are often neglected in vessel maintenance. Skippers are also scared that they might not be able to reassemble the gearing mechanism after opening it. The fact is that their gearing mechanism is quite simple, and in most cases only an Allen key, a winch key and a pot of grease are required.

Most winches incorporate aluminium pads, flush mounted on bronze bases with stainless steel bolts. Such a combination could easily set off electrolysis if seawater gets in. Left unattended, it will first buckle and then break the bronze base. You may not be able to find parts for older winches.

- Strip the winch to the level of the base plate.
- Take off the base plate if it is corroded or buckled.
- Cover screws & threads with lanolin grease and insulate them with plastic gaskets. This will prevent contact between above-mentioned dissimilar metals and guard against electrolysis.
- Grease all moving parts & reassemble.

SHACKLES, HOOKS & SLINGS

SHACKLES

A shackle has two parts: the body and the pin or the bolt. The body of a STANDARD or "D" SHACKLE is shaped like the letter "D", whereas an OMEGA or BOW SHACKLE is more circular. The pin (bolt) is locked into the lug of the shackle by either being threaded or fitted with a flat tapered pin known as the FORELOCK.

Forelock shackles are quite secure for permanent attachments, such as standing rigging or to withstand vibrations. If a threaded shackle is to be used for a permanent attachment it must be MOUSED against the risk of being undone due to vibrations. This is done by securing the pin to the shackle with a ceasing wire. Bow shackles are used for attachment of more than one item to it.

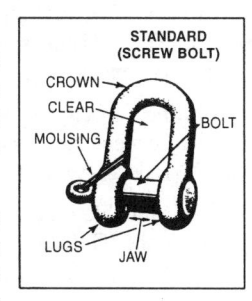

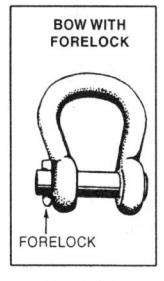

Fig 9.26

Fig 9.27

Shackles used for lifting must have SWL stamped on them. Never use a shackle with a bent pin or with a bolt not designed for it. The wear due to rubbing would be most visible on the pin or inside the crown. Discard the shackle if worn more than 10% of its original diameter. When lifting, do not place the sling on the pin of the shackle; it may roll open the pin. Place it on the crown. The pin should rest on the hook.

Shackles are designed to take vertical forces only. Diagonal forces will strain the shackle. Therefore, the crown of a lifting shackle should be pointing vertically down. If the shackle is too large for the sling and is likely to COCK-BILL OR CANT, pack its pin with washers to keep the load in the centre of the pin.

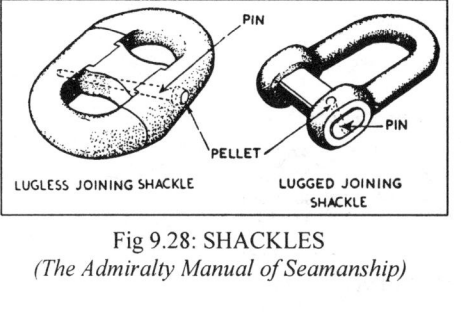

The KENTER or LUGLESS JOINING SHACKLE is used for joining lengths of anchor cable.

Fig 9.28: SHACKLES
(The Admiralty Manual of Seamanship)

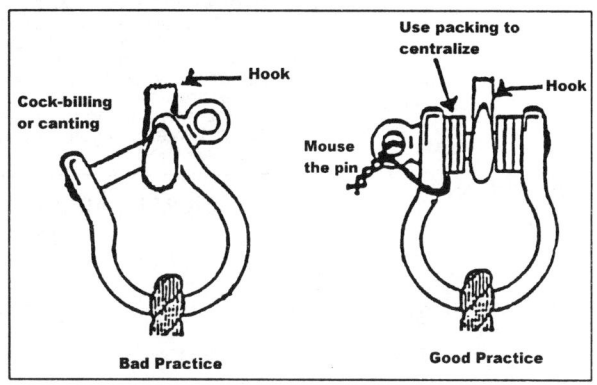

Fig 9.29: COCK-BILLING OR CANTING & PACKING THE SHACKLE PIN *(NSW Workcover)*

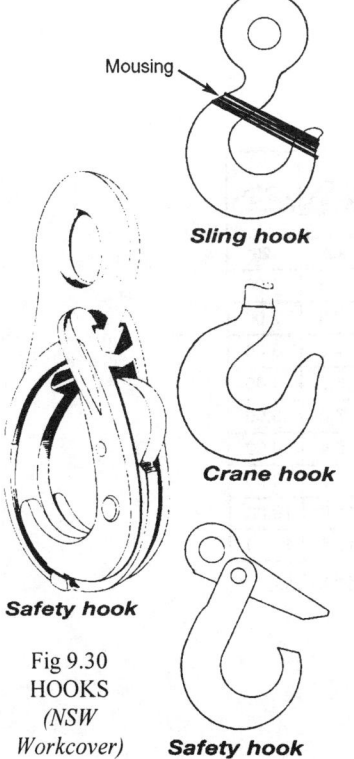

Fig 9.30 HOOKS *(NSW Workcover)*

HOOKS

Being of open construction, a hook is usually the weakest part of a lifting rig. Its throat opening must be largest enough to fit the lifting sling, shackle or ring. Do not overcrowd the hook. Do not use a hook that is damaged, distorted or bent, or whose throat opening has stretched more than 5%. Hooks must be stamped with the SWL. A load likely to twist or turn should be lifted with a swivel hook. The load can be prevented from accidentally jumping off the hook by use of a spring hook or by mousing the load to the hook.

SLINGS

Slings may be made of flat webbing, round synthetic rope, wire ropes or chain. All commercially available slings should be labelled with **WLL (Working Load Limit)**. The synthetic slings are colour coded and striped according to lifting capacity. Each stripe represents 1 tonne WLL.

The WLL is the lifting capacity of a sling for a straight lift. This figure gets altered with different slinging methods, depending on the angle between the legs of the sling and the reeve of the sling. It is then referred to as the SWL (Safe Working Load). The SWL decreases as the angle between the legs increases.

The WLL tables are available for all types of slings and angles. The following examples of label and table of stripes for synthetic slings appears in the 'Guide to Rigging' produced by NSW Workcover.

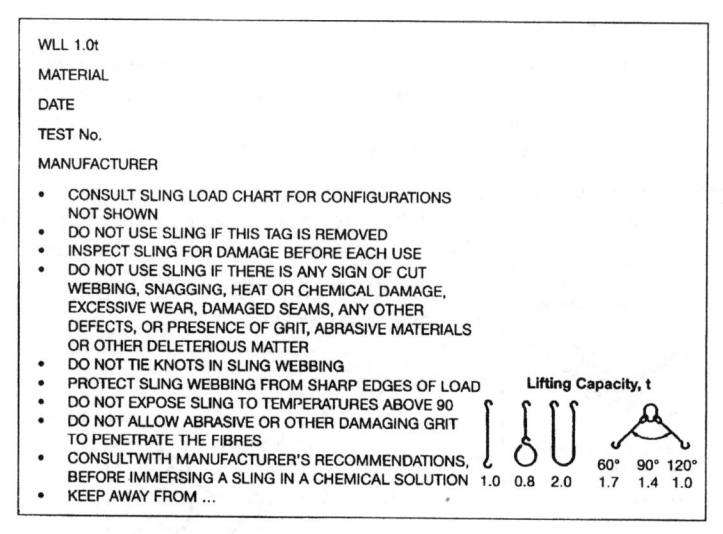

```
WLL 1.0t
MATERIAL
DATE
TEST No.
MANUFACTURER

• CONSULT SLING LOAD CHART FOR CONFIGURATIONS
  NOT SHOWN
• DO NOT USE SLING IF THIS TAG IS REMOVED
• INSPECT SLING FOR DAMAGE BEFORE EACH USE
• DO NOT USE SLING IF THERE IS ANY SIGN OF CUT
  WEBBING, SNAGGING, HEAT OR CHEMICAL DAMAGE,
  EXCESSIVE WEAR, DAMAGED SEAMS, ANY OTHER
  DEFECTS, OR PRESENCE OF GRIT, ABRASIVE MATERIALS
  OR OTHER DELETERIOUS MATTER
• DO NOT TIE KNOTS IN SLING WEBBING
• PROTECT SLING WEBBING FROM SHARP EDGES OF LOAD    Lifting Capacity, t
• DO NOT EXPOSE SLING TO TEMPERATURES ABOVE 90
• DO NOT ALLOW ABRASIVE OR OTHER DAMAGING GRIT
  TO PENETRATE THE FIBRES
• CONSULT WITH MANUFACTURER'S RECOMMENDATIONS,          60°  90° 120°
  BEFORE IMMERSING A SLING IN A CHEMICAL SOLUTION   1.0 0.8 2.0  1.7 1.4 1.0
• KEEP AWAY FROM ...
```

Fig 9.31: LABEL ON A FLAT WEBBING SLING
(NSW Workcover)

Colour No Stripes	Tonne	Vertical	Choke	Basket	30°	60°	90°	120°
Violet 1	1	1	0.8	2	1.9	1.7	1.4	1.00
Green 2	2	2	1.6	4	3.8	3.4	2.8	2.00
Yellow 3	3	3	2.4	6	5.7	5.1	4.2	3.00
Orange 4	4	4	3.2	8	7.6	6.8	5.6	4.00
Red 5	5	5	4.0	10	9.5	8.5	7.0	5.00
Brown 6	6	6	4.8	12	11.4	10.2	8.4	6.00
Blue 8	8	8	6.4	16	15.2	13.6	11.2	8.00
Olive 10	10	10	8.0	20	19.0	17.0	14.0	10.00
Grey 12	12	12	9.6	24	22.8	20.4	16.8	12.00

Fig 9.32: COLOUR CODING OF SYNTHETIC SLINGS
(NSW Workcover)

Chapter 9: Lifting Equipment & Winches

REEVE OF A SLING

A sling reeved around a square load and suspended from a single point reduces its lifting capacity to half. Its load factor (lifting capacity) is described as 0.5. The load factor (lifting capacity) of a sling reeved around a round load is 0.75.

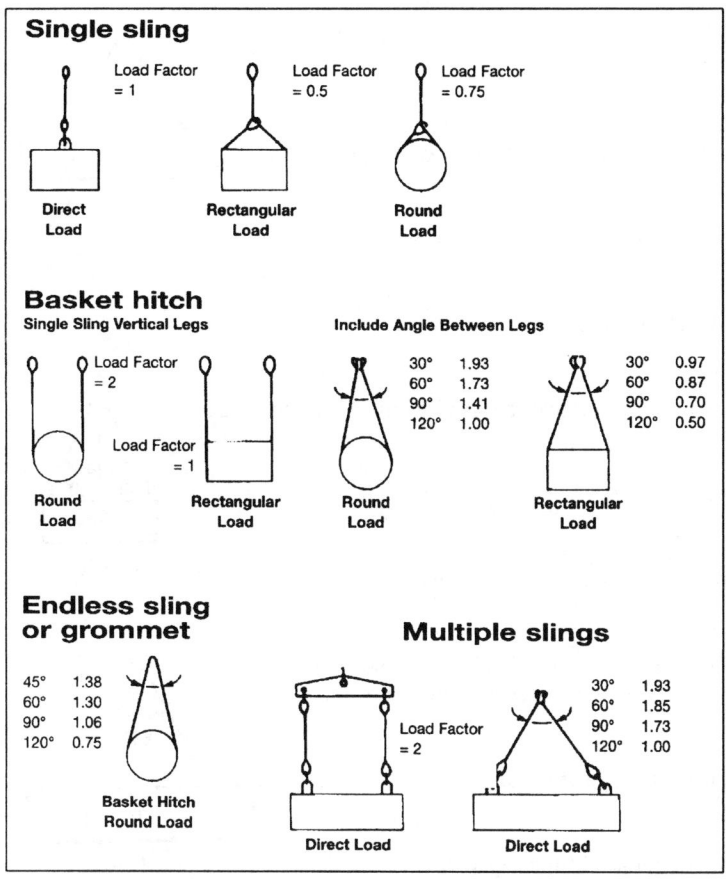

Fig 9.33: LOAD FACTORS (LIFTING CAPACITIES) OF SLINGS
(NSW Workcover)

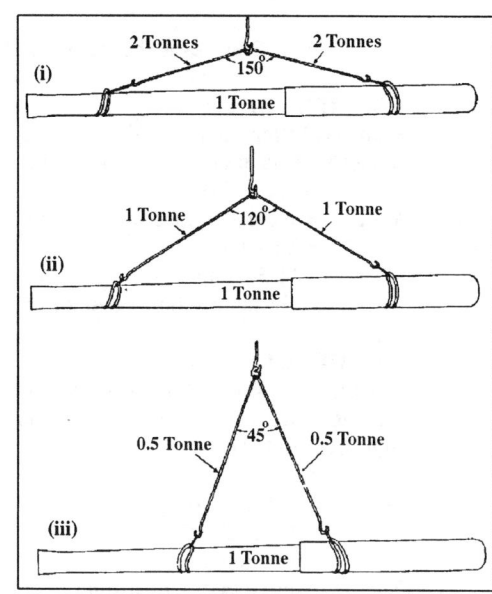

Fig 9.34
STRESSES IN THE LEGS OF A SLING
(The Admiralty Manual of Seamanship)

LOAD SLUNG BY A STROP

In illustration (i), the angles between the four legs of the strop are wide and each leg bears about 1.5 times the weight of the case. In illustration (ii) the angles are small and each leg bears about half the weight of the case. No. (i) is the wrong way.

In the case of MULTI-LEGGED SLINGS only two of the sling legs are assumed to take the entire load. Therefore, each leg has to be capable of taking half the weight of the load. The WLL of such a sling is assessed on the diagonally opposed legs that have the largest angle between them.

The Guide to Rigging suggests A ROUGH RULE OF THUMB for a good safe working angle: Make sure that the horizontal distance between the points of attachment of the load does not exceed the length of one leg of the sling. Based on the principle of equilateral triangle, this will ensure that the angle between the two diagonally opposed legs does not exceed 60°. At 60° the sling will lift = 1.73 x the WLL of one leg.

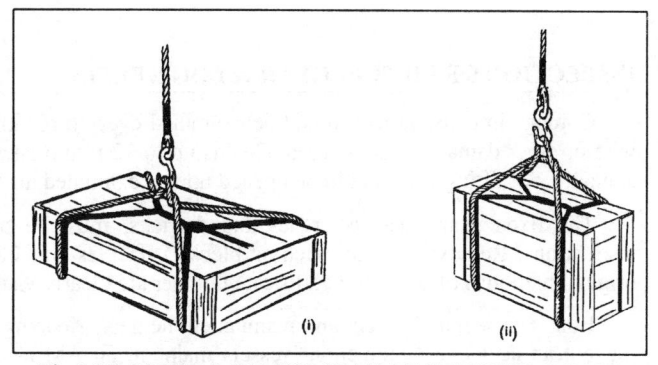

Fig 9.35: RIGHT & WRONG WAY OF SLINGING A CASE BY A STROP   (i) Wrong Way   (ii) Right Way
(The Admiralty Manual of Seamanship)

The Australian Boating Manual

SLINGS LIFTING CAPACITY CALCULATIONS:
(A Guide to Rigging, Workcover NSW)

EXAMPLE A:
Calculate the maximum load that can be lifted with a two-legged wire sling, whose WLL of each leg is 8 tonnes. The angle between the sling legs is 60°, both of which are reeved around a SQUARE LOAD.

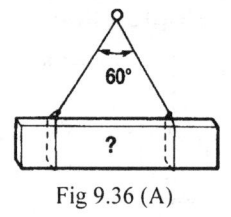

Max Load = WLL (of each sling leg) x Angle factor x Reeve factor for a square load
= 8 x 1.73 x 0.5
= 6.92 tonnes

Therefore, 6.92 tonnes is the maximum load that can be lifted.

Fig 9.36 (A)

EXAMPLE B:
Find the lifting capacity of a two-legged wire sling that can be used to lift a weight of 20 tonnes with an angle between the sling legs of 60°.

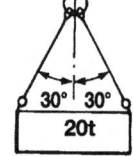

WLL = Weight ÷ Load factor (from the above table)
= 20 ÷ 1.73
= 11.56 tonnes

Therefore, the sling should be of a minimum 11.56 tonnes lifting capacity.

Fig 9.36 (B)

EXAMPLE C:
Calculate the WLL of a two-legged wire sling needed to lift a load of 4 tonnes with the angle between the sling legs of 60°, both of which are reeved around a 4 tonne SQUARE LOAD.

WLL = Weight ÷ Angle factor ÷ Reeve factor for a square load
= 4 ÷ 1.73 ÷ 0.5
= 4.62 tonnes

Therefore, use a two-legged sling of at least 4.62 tonnes lifting capacity.

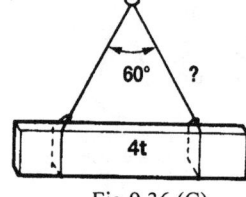

Fig 9.36 (C)

EXAMPLE (D):
Calculate the WLL of a two-legged wire sling needed to lift a 20 tonnes ROUND LOAD with the angle between the sling legs of 60°, both of which are reeved around the load.

WLL = Weight ÷ Angle factor ÷ Reeve factor for a round load
= 20 ÷ 1.73 ÷ 0.75
= 15.41 tonnes

Therefore, the sling should be of at least 15.41 tonnes lifting capacity.

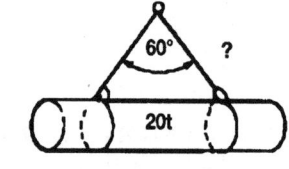

Fig 9.36 (D)

INSPECTION OF LIFTING GEAR & LOG KEEPING

Cordage in constant use should be examined every 6 months and wire ropes every 3 months - more often if a broken wire or other damage is discovered. Chains up to 12 mm diameter should be inspected every 6 months, and larger chains annually. All lifting gear should be opened up and examined annually.

The fixed gear, such as cranes and derricks, must be periodically tested and maintained as per manufacturer's instructions. Removable gear, such as blocks, shackles and lifting trays usually do not require retesting as long as the original markings of SWL and certificate number are clearly visible, and there is no evidence of distortion or elongation.

Boat's crew must inspect and maintained the gear. Keeping accurate record of these inspections and maintenance may prevent an accident. Commercial vessels maintain an AMSA-approved inspection and maintenance register, commonly known as the *Chain Register*.

Chapter 9: Lifting Equipment & Winches

CHAPTER 9: QUESTIONS & ANSWERS

CHAPTER 9.1: SAFE LIFTING - GENERAL

1. The rope and chain sizes are measured as follows:
   (a) rope in diameter & chain in circumference.
   (b) both rope & chain bar in circumference.
   (c) both rope & chain bar in diameter.
   (d) rope in circumference & chain in diameter.

2. When tested and marked, the 10 mm steel chain is graded as follows:
   (a) Grades 3, 4, 5 ..... 9.
   (b) Grades 1 & 2.
   (c) Small chain is not graded.
   (d) Small chain is not marked.

3. A steel chain marked with "9" on some of its links is of the following grade:
   (a) Superior grade.
   (b) Grade 9.
   (c) None of the choices stated here.
   (d) Inferior grade.

4. An unmarked steel chain is:
   (a) of inferior grade.
   (b) untested.
   (c) of superior grade.
   (d) grade 9.

5. In relation to Chain Hoists, which of the following statements is incorrect?
   (a) The block is normally the weakest link.
   (b) Replace the hoist when the hook starts to spread open.
   (c) Replace the hoist when the hook starts to show signs of wear.
   (d) A distorted link is an indication of an unsafe hoist.

6. The maximum number of turns of a synthetic rope around a warping drum should be:
   (a) two.
   (b) four.
   (c) eight.
   (d) ten.

7. Do not use a chain that shows reduction in diameter of more than:
   (a) 10%
   (b) 20%
   (c) 1%
   (d) 30%

8. Chains used for lifting must be:
   (a) made of mild steel.
   (b) stamped with SWL.
   (c) studded.
   (d) of Grade 1.

9. In relation to blocks, which of the following statements is incorrect?
   (a) Use only fibre rope blocks for fibre ropes.
   (b) Use only wire rope blocks for wire ropes.
   (c) Paint the blocks as often as possible.
   (d) Blocks should be stamped with the SWL.

10. In relation to snatch blocks, which of the following statements is incorrect?
    (a) The bight of the rope can be inserted into the swallow by opening the hinged cheek.
    (b) They are useful for changing the lead of a rope.
    (c) They should never be used for lifting.
    (d) They may have one or more sheaves.

11. In relation to cranes and derricks, which of the following statement is correct?
    (a) They are designed to automatically lower any load hanging in mid-air following a power failure.
    (b) They must be tested & thoroughly examined at intervals not exceeding 5 years.
    (c) Only the appointed signaller must give the 'Stop' signal to the operator.
    (d) The signaller must stand against a bulkhead or railing.

12. Holding arms crossed above the head means:
    (a) Distress.
    (b) Slow down.
    (c) Heave.
    (d) Stop.

13. In relation to winches on yachts, which of the following statements is incorrect?
    (a) They are used for hoisting & trimming sails.
    (b) Their drums turn in both directions.
    (c) The self-tailing winches can be operated by one person.
    (d) They are usually installed at an angle to the base.

14. In relation to winches & capstans, which of the following statements is correct?

    (a) One turn around a warping drum is sufficient for heaving or slacking the rope.
    (b) Rapid lifting of loads prevents shock loads.
    (c) Stop & clear a riding turn immediately.
    (d) The fusion property of synthetic ropes helps maintain stability of the hoist.

15. In relation to operating wire ropes spooled on winch drums, which of the following statements is correct?

    (a) A brake capable of holding the load near the ground is equally capable of holding it at a height.
    (b) A minimum of one full turn must always remain on the drum.
    (c) The wire rope must not be anchored to the drum.
    (d) The maximum fleet angle for a grooved diameter drum is 5°.

16. In relation to yacht winches, which of the following statements is correct?

    (a) Their gearing mechanism is quite complicated & beyond the capability of maintenance by crew.
    (b) They do not incorporate dissimilar metals in their assembly.
    (c) Their maintenance does not require them to be stripped.
    (d) They may incorporate metals such as an aluminium pad, a bronze base and stainless steel bolts.

17. In relation to shackles, which of the following statements is correct?

    (a) Forelock shackles are quite secure for permanent attachments.
    (b) The shackle pin is always locked into the lug with a tapered pin.
    (c) When lifting a weight, place the sling on the pin of the shackle.
    (d) Discard the shackle if worn more than 15% of its original diameter.

18. In relation to shackles, which of the following statements is correct?

    (a) The crown of the lifting shackle should point vertically upwards.
    (b) Shackles are designed to take diagonal forces only.
    (c) The pin of a lifting shackle should rest on the lifting hook.
    (d) A lifting shackle, if too small for the sling, is likely to cock-bill.

19. In relation to shackles and hooks, which of the following statements is correct?

    (a) Bow shackles are designed for joining lengths of anchor cable.
    (b) Kenter shackles are designed for lifting multiple items.
    (c) A hook is usually the weakest part of a lifting rig.
    (d) Discard the lifting hook whose throat opening has stretched more than 10%.

20. In relation to slings used for lifting, which of the following statements is correct?

    (a) The SWL increases as the angle between the legs increases.
    (b) Each stripe on a colour-coded sling represents 1 tonne WLL.
    (c) All wire rope slings are colour-coded.
    (d) Flat webbing slings are not used for lifting.

CHAPTER 9.2: SAFE WORKING LOAD CALCULATIONS

21. The Safe working Load of a 12 mm Nylon rope calculated with the simplified formula is:

    (a) 114 kg.
    (b) 609 kg.
    (c) 188 kg.
    (d) 324 kg.

22. The Safe working Load of a 12 mm Polyester rope calculated with the simplified formula is:

    (a) 134 kg.
    (b) 109 kg.
    (c) 288 kg.
    (d) 464 kg.

23. The Safe working Load of a 12 mm Polyethylene rope calculated with the simplified formula is:

    (a) 124 kg.
    (b) 230 kg.
    (c) 488 kg.
    (d) 524 kg.

24. The Breaking Stress of a 15 mm Nylon rope calculated with the simplified formula is:

    (a) 3036 kg.
    (b) 809 kg.
    (c) 1288 kg.
    (d) 2324 kg.

25. The Safe working Load of a 25 mm wire rope is:

    (a) 10 tonnes
    (b) 5 tonnes.
    (c) 15 tonnes
    (d) 30 tonnes

26. The Breaking Stress of a 24 mm Polyester/Terylene rope calculated with the simplified formula is:

    (a) 10.2 tonnes.
    (b) 9.5 tonnes
    (c) 4.5 tonnes.
    (d) 6.9 tonnes.

27. The Safe working Load of a 18 mm Nylon rope calculated with the simplified formula is:

    (a) 729 kg.
    (b) 1208 kg.
    (c) 480 kg.
    (d) 1324 kg.

28. The Safe working Load of a 28 mm Manila rope calculated with the simplified formula is:

    (a) 229 kg.
    (b) 784 kg.
    (c) 480 kg.
    (d) 1324 kg.

29. Calculated with the simplified formula, the size of Polythene rope needed to safely lift a weight of 300 kg is:

    (a) 20.9 mm.
    (b) 13.7 mm.
    (c) 18.6 mm
    (d) 10.5 mm.

30. The Breaking Stress of a 24 mm wire rope calculated with the simplified formula is:

    (a) 34.2 tonnes.
    (b) 19.5 tonnes
    (c) 27.6 tonnes.
    (d) 6.9 tonnes.

31. The Safe working Load of a 18 mm wire rope calculated with the simplified formula is:

    (a) 1078 kg.
    (b) 2590 kg.
    (c) 4800 kg.
    (d) 1324 kg.

32. Calculated with the simplified formula, the size of wire rope required for a bollard pull of 6 tonnes is:

    (a) 22.8 mm.
    (b) 36.2 mm
    (c) 27.4 mm
    (d) 43.5 mm.

33. The Safe Working Load of a 6 mm grade-3 steel chain is:

    (a) 324 kgs.
    (b) 450 kgs.
    (c) 1000 kgs.
    (d) 150 kgs.

34. The Safe Working Load of a 6 mm grade-4 steel chain is:

    (a) 324 kgs.
    (b) 432 kgs.
    (c) 1100 kgs.
    (d) 250 kgs.

35. The Safe Working Load of a 20 mm grade-1 chain is:

    (a) 3.33 tonnes
    (b) 13.33 tonnes
    (c) 1200 kg.
    (d) 1.8 tonnes.

36. The Breaking Stress of an 18 mm grade-3 chain is:

    (a) 4.64 tonnes
    (b) 23.22 tonnes
    (c) 14.58 tonnes
    (d) 2.9 tonnes

37. The SWL of a 15 mm grade-2 chain is:

    (a) 11.25 tonnes
    (b) 1350 kg
    (c) 2.81 tonnes
    (d) 6.75 tonnes

CHAPTER 9.3: PURCHASES & SLINGS

38. The size of a block is measured by its:

    (a) sheave diameter in millimetres.
    (b) sheave radius in millimetres.
    (c) sheave groove diameter in millimetres.
    (d) block diameter in millimetres.

39. For pulleys and lead blocks using synthetic ropes the sheave diameter should be:

    (a) 20 times the rope size.
    (b) 5 times the rope size.
    (c) 10 times the rope size.
    (d) 3 times the rope size.

40. For pulleys and lead blocks using natural fibre ropes the sheave diameter should be:

    (a) 20 times the rope size.
    (b) 5 times the rope size.
    (c) 10 times the rope size.
    (d) 3 times the rope size.

41. For manually operated pulleys and lead blocks using steel wire ropes the sheave diameter should be:

    (a) 20 times the rope size.
    (b) 5 times the rope size.
    (c) 10 times the rope size.
    (d) 3 times the rope size.

42. For power operated pulleys and lead blocks using steel wire ropes the sheave diameter should be:

    (a) 20 times the rope size.
    (b) 5 times the rope size.
    (c) 10 times the rope size.
    (d) 3 times the rope size.

43. The mechanical advantage of a two-fold purchase rigged to advantage is:

    (a) four.
    (b) five.
    (c) two.
    (d) eight.

44. The difference in the mechanical advantage of a purchase rigged to advantage and disadvantage is:

    (a) always one.
    (b) variable.
    (c) always two.
    (d) half.

45. The mechanical advantage of a gyn tackle rigged to disadvantage is:

    (a) five.
    (b) six.
    (c) four.
    (d) two.

46. The mechanical advantage of a gyn tackle rigged to advantage is:

    (a) five.
    (b) six.
    (c) four.
    (d) two.

47. A three fold purchase consists of:

    (a) two six sheave blocks.
    (b) one three sheave & one two sheave blocks.
    (c) two three sheave blocks.
    (d) three two sheave blocks.

48. A handy billy consists of:

    (a) a gyn tackle.
    (b) a two sheave & a one sheave blocks.
    (c) a two fold purchase.
    (d) a one fold purchase.

49. A one fold purchase is also known as the:

    (a) handy billy.
    (b) gyn tackle.
    (c) luff tackle.
    (d) gun tackle.

50. The mechanical advantage of a tackle can be worked out by counting the number of:

    (a) ropes supporting the standing block.
    (b) ropes supporting the moving block.
    (c) sheaves on the moving block.
    (d) sheaves on the standing block.

51. The mechanical advantage of a single whip:

    (a) is nil.
    (b) depends on the number of sheaves in it.
    (c) depends on its sheave size.
    (d) is two.

52. A lead block at the end of a purchase increases its mechanical advantage by:

    (a) one.
    (b) zero.
    (c) two.
    (d) minus one.

53. The mechanical advantage of a tackle on a tackle is the:

    (a) sum of the two.
    (b) product of the two.
    (c) M.A. of the first.
    (d) M.A. of the second.

54. The estimated friction in a tackle is based on the:

    (a) size of sheaves.
    (b) material of blocks.
    (c) number of sheaves.
    (d) number of sheaves & material of blocks.

55. The mechanical advantage of a three fold purchase rigged to advantage is:

    (a) five.
    (b) six.
    (c) seven.
    (d) eight.

Chapter 9: Lifting Equipment & Winches

56. A pulley system to lift a weight of 2 tonnes with a pull of 400 kgs (disregarding friction) will have the mechanical advantage of:

    (a) 2.
    (b) 4.
    (c) 5.
    (d) 10.

57. The mechanical advantage of a gun tackle is:

    (a) 2 or 3.
    (b) 5 or 6.
    (c) 2.
    (d) 5.

58. What is the stress on the hauling part of a 3 and 2 purchase (gyn tackle), rigged to advantage, lifting 6 tonnes (disregard friction)?

    (a) 0.5 tonnes.
    (b) 1 tonne.
    (c) 1.5 tonnes.
    (d) 2 tonnes.

59. The load factor of a sling reeved around:

    (a) a round load & suspended from a single point is 0.5.
    (b) a square load & suspended from a single point is 0.75.
    (c) a round load in a basket hitch is 2.
    (d) a rectangular load in a basket hitch is 0.97.

60. At what angle between the legs of a 2-legged sling is the stress on each leg equal to the weight being lifted?

    (a) 180°.
    (b) 120°
    (c) 60°
    (d) 45°

61. At what angle between the legs of a 2-legged sling is the stress on each leg equal to half the weight being lifted?

    (a) 180°.
    (b) 120°
    (c) 60°
    (d) 45°

62. At what angle between the legs of a 2-legged sling is the stress on each leg equal to twice the weight being lifted?

    (a) 45°
    (b) 150°
    (c) 60°
    (d) 30°

CHAPTER 9: ANSWERS

1 (c), 2 (a), 3 (b), 4 (b), 5 (a),
6 (b), 7 (a), 8 (b), 9 (c), 10 (d),
11 (b), 12 (d), 13 (b), 14 (c), 15 (d),
16 (d), 17 (a), 18 (c), 19 (c), 20 (b),
21 (d), 22 (c), 23 (b), 24 (a), 25 (b),
26 (d), 27 (a), 28 (b), 29 (b), 30 (c),
31 (b), 32 (c), 33 (a), 34 (b), 35 (a),
36 (b), 37 (c), 38 (a), 39 (c), 40 (c),
41 (c), 42 (a), 43 (b), 44 (a), 45 (a),
46 (b), 47 (c), 48 (b), 49 (d), 50 (b),
51 (a), 52 (b), 53 (b), 54 (d), 55 (c),
56 (c), 57 (a), 58 (b), 59 (c), 60 (b),
61 (d), 62 (b)

Chapter 10

WATCHKEEPING DUTIES & NAUTICAL EMERGENCIES

WATCHKEEPING

The law requires every vessel to maintain an efficient and competent watch, adequate to the prevailing circumstances and conditions. The watchkeeper should be sufficiently rested and not under the influence of alcohol or narcotics. If the watchkeeper being relieved is doubtful of the fitness of the relieving watchkeeper, the watch should not be handed over and alternative arrangements should be made. Watchkeepers need to fully understand the operation as well as the limitations of the navigational equipment on board. Prior to taking over a watch, the watchkeeper must become fully aware of the vessel's position, courses, speed as well as any dangers likely to be met during the watch. The watchkeeper must not hesitate to use the helm, engines, sails or sound signalling appliances on board.

Commercial vessels have learnt from experience that the Master's presence in the wheelhouse does not mean the watchkeeper has been relieved from his or her watchkeeping duties or from taking action to avoid collisions. Should the master wish to take over the watch, he or she would do so by expressly stating words such as "I now have the watch". Whether by day or night, the watchkeepers are required to notify the Master if in any doubt of the safety of the vessel. Masters leave written Standing Orders as well as Night Orders on ship's bridge to ensure that misunderstandings do not creep in.

WATCHKEEPING WHEN UNDERWAY:
- Do not at any time leave the bridge or cockpit unattended.
- Keep a close watch on weather and visibility - by satellite, radio, radar and visual.
- Be aware of other vessels and navigational hazards in the vicinity.
- Make effective use of the navigational aids on board.
- Do not use the automatic pilot in poor visibility, in confined waters or in heavy traffic density. Do not make large course alterations on automatic pilot.
- Frequently check the vessel's position by various methods, and cross-reference them.
- Be aware of compass, GPS and all other instruments errors.
- Be aware of tides, currents and depths in the area.
- Be aware of the state of vessel's machinery and auxiliaries.
- Keep an eye on any cargo on board.
- Maintain a fire and bilge watch.
- Maintain an appropriate radio watch.

WATCHKEEPING AT ANCHOR:
Ensure that the vessel maintains her position. Except when she is swinging due to change in the direction of tide or wind, a changing beam bearing is the best early warning of a dragging anchor. A minimum of two bearings noted down and rechecked from time to time is the best all round method to ensure that the vessel is holding her anchor.

A good lookout at anchor includes a safety watch on the passing traffic, positions of other vessels at anchor, knowledge of weather forecast and the state of the current weather and tides, display of the appropriate anchor signal, complying with pollution regulations, bilge and fire watch, engine and auxiliary machinery in readiness as per the circumstances. A radio watch must also be maintained. Masters of commercial vessels leave written Standing Orders and Night Orders on the ship's bridge, especially when anchored at a precarious location. In any case, the watchkeeper is always required to notify the Master if in any doubt of the safety of the vessel.

WATCHKEEPING IN PORT
In the interest of occupational health and safety and to minimise third party liability claims as well as in the interest of your own safety and the safety of your vessel, your vessel must be berthed or moored securely at all times in all tidal, wind and weather conditions. In places where there is a large range of tides the berthing lines must be long enough for the vessel to be able to rise and fall with the tide (as discussed in Chapter 8). The access to the vessel must be adequate and safe at all times. Appropriate signals and safety signs must be displayed when vessel is refuelling, there is a diver below or when carrying out other specific functions. You must be aware of who is on board at any given time.

REFUELLING PRECAUTIONS & POLLUTION PREVENTION: See Chapter 12.

Chapter 10: Watchkeeping Duties & Nautical Emergencies

MASTER'S INSTRUCTIONS TO WATCHKEEPERS
(STANDING ORDERS, NIGHT ORDERS, ETC.)
- Call me if you encounter restricted visibility, heavy weather or a navigation hazard or if you feel unable to deal with the traffic conditions.
- Call me if you fail to make the expected landfall or sight the navigational mark or obtain the expected sounding.
- Call me if experience difficult in the steerage or navigation or manoeuvring of the vessel, or if any of the equipment malfunctions.
- Call me if in doubt at any time.

HANDING OVER OR TAKING OVER A WATCH
Pay attention to:
- master's standing orders
- vessel's position, course speed and draft
- weather and tide conditions
- working of the navigational equipment
- compass errors
- presence and movement of other vessels
- the state of machinery and sails
- the state of any cargo
- any operational activities.

REPORTING MARINE CASUALTIES & INCIDENTS

Under the marine safety acts in all States as well as in Commonwealth, a marine casualty or a marine incident is deemed to have occurred when a vessel is in danger of serious damage or is lost, abandoned, stranded, grounded or materially damaged (whether by fire or otherwise) or has been in a collision with another vessel or with any other thing; or there is a serious danger of or an actual loss of life or injury to a person due to an accident on board.

The incidents involving danger, without actual injury or damage, should be reported in the same way as other incidents so that relevant data can be recorded and analysed.

Rules concerning responsibilities and care among those involved in marine casualties and incidents may vary slightly among States. You should consult your State's boating literature or Sailing Directions. In general, transmit a distress, urgency or safety message as appropriate, make an entry in the vessel's log book and make a full report to the appropriate authority within 48 hours. If the initial report is not made in the approved form the master must make a further report in the approved form as soon as possible. In addition, when involved in a collision with another vessel, follow the rules similar to those after a motorcar accident:
- Stop the vessel.
- Ensure the safety of own vessel.
- Give necessary assistance to the other vessel.
- Produce your certificate of competency if requested on reasonable grounds.
- Exchange names and addresses and registration details, if any.

CLAIMING SALVAGE *(See IMO website for more information – address on the last page of this book)*

Your legal duty to provide rescue at sea is limited only to the saving of lives. Saving of property is salvage unless the vessel needs to be salved in order to save lives.

No salvage claim in respect of a distressed vessel is valid unless the salvage is specifically requested by the master or the owner. The parties are free to enter a fixed price salvage agreement. However, the request for salvage is usually recorded on what is known as a "LLOYDS OPEN FORM OF SALVAGE (LOF) AGREEMENT". There is no provision to enter a monetary value on this form. It provides for the salvor to be rewarded a percentage of the property salvaged (vessel and cargo) if the salvage is successful. If unsuccessful, the salvor is not entitled to a salvage claim. Therefore, the form is also referred to as a "No Cure No Pay" agreement. (However, a reward may be payable for an unsuccessful salvage if it has averted or minimised pollution.) An independent arbitrator is appointed through a Lloyds Agent or the Law Society to assess the reward, based on the value of the property saved and the risk and cost borne by the salvor.

Most commercial vessels and salvage tugs carry this one page agreement on board. Agreement to comply with the LOF terms can also be signed on a plain sheet of paper. Even in the absence of this document, if the salvor is asked to salvage the property in distress and the salvage is successful, the salvage award is usually automatic.

Where it is necessary to tow a vessel in distress in order to save lives it is your statutory duty to do so without risking the safety of your own vessel, regardless of the owner's agreement or a clause in the insurance policy. An award for salvage of property may still apply. However, as soon as the distressed vessel is in a safe position, the towing vessel has no right to continue towing for the purpose of earning additional salvage award.

Ownership of a property found adrift or sunken at sea is not automatic, unless abandonment can be shown or the owner can not be found. However, a salvage award for rescuing such property may be successful, especially if it is found outside the territorial limit of 12 nautical miles. Such claims are dealt with through the Receiver of Wrecks (in the Australian Maritime Safety Authority (AMSA)).

Before undertaking a voluntary salvage, the salvor should take the following actions:
- In the absence of a printed form, enter an agreement headed "Lloyds Open Form of Salvage Agreement" written on a sheet of paper together with the names of the vessels and the date and signatures of the two masters.
- Obtain both owners' as well as your insurer's agreements.
- Ensure there is sufficient fuel remaining on board to complete the task.
- Obtain a weather forecast for the estimated period of salvage.
- Check to see that there is sufficient engine power for the task in rough weather.
- Assess your possible expenses (perishable cargo, etc.) in comparison with the value of the property being salved.
- Assess your prospects of success.

EMERGENCIES

GROUNDING

Grounding can be intentional or accidental. Intentional grounding, more often called beaching, means that you want to ground the vessel for reasons such as repairs or hull cleaning. More correctly, the following terms may be used:

- STRANDING: Unintentional running ashore (Strand means ocean shore).
- GROUNDING: Unintentional contact with the bottom but not on the shore line.
- BEACHING: Intentionally running up a beach
    - to save the vessel
    - to undertake repairs
    - to save the crew

STRANDING (ACCIDENTAL GROUNDING)

While some grounding or stranding incidents are nothing more than embarrassment; others have resulted in disastrous consequences. Therefore, grounding or standing, small or large, should be treated as potentially serious until the situation has been assessed.

The situations that may involve running aground are many, from running onto a mud bank in a quiet harbour, to being driven by heavy weather onto a rocky shore. In the latter case the principal concern will be the saving of life. If you run aground on a falling tide, the vessel may become damaged through hull fracture or holed by sitting on rocks or reefs.

INITIAL ACTIONS
- Stop engines and auxiliaries.
- Sound bilges and inspect voids.
- Take bearings and plot position.
- Examine chart details, survey the area and check soundings around the vessel.
- Check whether tide rising or falling. Check tidal stream.
- Obtain weather forecast.
- Don't go astern for too long or too fast. In fact avoid it. If the bottom is sand or mud, going astern may wash a quantity of this material from astern and throw it directly under the keel. This will bed the vessel down more firmly. The sand or mud will also be pumped into the engine through the water intake.
- If the bottom is rocky, going astern may cause hull damage.
- When grounded forward, a right-handed single screw vessel going astern may swing the stern to port. This may cause the hull to go broadside on shore.
- If grounded on a falling tide, work quickly to stop the hull from swinging due to the action of wind or waves. Also brace the vessel so that it will stay upright and easier to refloat.
- Move crew and passengers to lighten the grounded end of vessel.
- Jettison any weights you can.
- Control panic, check for personnel injuries.
- Check for hull damage.
- If the hull damage is evident, it may be better not to pull her off.
- Lay out an anchor to prevent her going up any further.
- Request Water Police, the State Marine Authority or the Volunteer Coastal Patrol/Coast Guard for a tow, if appropriate.

Chapter 10: Watchkeeping Duties & Nautical Emergencies

- Consider taking tow from a passer-by or ordering a commercial tug.
- Lay out anchors to prepare to pull off at the next high water, if decided. This is known as kedging. Two anchors laid with a 30° spread between them can be used to wiggle a vessel's stern, by heaving on them turn by turn.
- Hoist 'vessel aground' day shapes or lights.
- Send Pan Pan or Mayday and plan abandon ship, if necessary.

POLLUTION PREVENTION
See Chapter 12

SALVAGE CLAIM
Discussed earlier in this chapter

PREPARING TO REFLOAT
- Lay out anchors to pull her off.
- Lighten her or move weights or people as necessary.
- Pump out or flood compartments as appropriate.
- Have lifesaving appliances ready in case of a sudden need.
- Start refloating just before HW.

AFTER REFLOATING
- Check if she is taking any water and whether or not you can cope with it.
- Check propeller, rudder and engine damage, if any.
- Make for the nearest safe port and make a report to the licensing authority and owner.
- The licensing authority will decide if she needs to be slipped and checked.

IF HARD AGROUND ON A REEF: Running hard aground on a reef means the hull is likely to be holed. If the weather is calm, it may be advisable to stay there and request help by, say, a Pan Pan message on radio. Accidental refloating by rising tide should be guarded against. You could either shift or load some weight on board. Take an anchor ashore to hold her until temporary repairs can be made or help arrives. Take environmental precautions against oil or fuel leaks.

IF AGROUND ON AN UNCERTAIN SANDY LOCATION AT NIGHT: Since your position is uncertain, it is night time and it is a safe place to stay, you should seriously consider staying there until daylight.

IF AGROUND ON A MUD BANK

1. **ROUND BILGE (FLAT BOTTOM) BOAT**: Lay out the anchor with the longest available warp (line from a winch) and attempt to winch her off. Plan for rising tide (if any) and a favourable wind direction. Don't run engine. Close all openings in case she heels over and lighten the vessel as much as possible before attempting to haul off.
   Or, heel her (towards the shallower side if possible), either by manual force or with help from another vessel. This can be done by crewmembers sitting on the boom swung at right angles to the hull or the other vessel heaving on a halyard from the top of a mast. Take care not to damage the keel. Push or tow to slide the vessel off the mud bank in this position.

2. **KEEL BOAT**: Reduce the boat's draft by heeling her. Make fast a rope to the mast as high as possible, get out of the boat and pull on the rope. Or pass the line from a kedge anchor through a block at the masthead or at a high point and take it to a winch. Swing weights out on the boom. If the boat is dismasted then list her by moving weights (people) on board to one side.

PLANNED GROUNDING (BEACHING)

This may be necessary to save a vessel from fire or foundering in deep water or to carry out an underwater inspection. Consider the following when beaching:
- Selection of the site: nature of bottom, obstructions, bottom slope, exposure to wind, etc. A gently shelving, sandy, sheltered beach is the best option. Beaching on a rocky shore can be a disaster in bad weather. Both the vessel and survivors may smash against rocks. Under such circumstances, abandoning vessel into a liferaft may be a better alternative.
- Weights and trim. Keep half ballasted if possible. Load additional weight after grounding to hold her down. On refloating, increase buoyancy by discharging all ballast.
- Tidal conditions: range, tidal stream, times of high and low water. How long is the intended period of beaching.

Beach just after HW so that you can get maximum time for repairs and she will be easy to float at next HW.
- How to make the approach (90° to the beach). In order to prevent mud or sand being sucked into the cooling water intake stop engine prior to making contact and close all underwater openings.
- Assess securing arrangement for the vessel.
- Assess when and how to use anchors. Usually it is better to lay out anchors after grounding in readiness to pull her off at next HW. (See Kedging with anchors under Accidental Grounding.)
- Has the vessel a flat bottom or deep keel. A deep keel vessel might need shoring up. Refloating while lying on her side might be difficult. (See Careening and Hard Stand in Chapter 5.)
- Work out refloating procedure and the possibility of bleeding the cooling system.

COLLISIONS

If collision is imminent, take action to reduce damage to sensitive areas of both vessels. For example, go full astern on engines and turn the vessel to avoid a direct hit (a glancing blow would cause less damage). A bow to bow or bow to quarter hit would also cause less damage than one vessel cutting into the hull of the other, particularly into the engine compartment. If one vessel becomes wedged into the hull of the other, it may be safer for them not to separate until the situation has been assessed. Transferring people from one vessel into the other might also be safer while wedged.

Read procedures under Safety Drills, Hull Damage, Beaching and Reporting Marine Incidents in this Chapter and Chapter 19.3.

HULL DAMAGE/ FOUNDERING

A vessel may FOUNDER (sink) as a result of water ingress. Heeling her to the undamaged side may keep her afloat. Cushions, bedding, sails, boat hooks and dinghy paddles may be used to patch and shore up the hole. Sails and bedding can also be wrapped on the outside to stem the water flow. A water-cooled engine can be used as an additional bilge pump by connecting the water inlet to the bilges. It is wise to fit a filter over the intake. Shut off the normal intake seacock.

DAMAGE BELOW THE WATERLINE: If a vessel is holed below the waterline, one way to reduce the inflow of water in order to carry out repairs from inside is to rig a collision mat over the damaged hull section.

A COLLISION MAT, in its simplest form, is a sheet of strong canvas, with spars and rigging lines lashed to two opposite sides. After rolling it on deck and holding it with strong ropes, pass it under the hull from over the bow or stern. Move it over the hole, unroll and lash it in place. The hole can also be stuffed with a pillow or covered with a mattress, but DO NOT USE THE LIFE SAVING EQUIPMENT for this purpose.

One type of commercially available collision mat for small vessels is like an umbrella (shown here). It is inserted through the hole and opened.

A CEMENT BOX is used to repair the hole from inside the hull. Use quick setting underwater ready mix cement or mix 1-part cement with 3 parts sand. Make two wooden boxes: one larger than the other, both large enough to cover the hole. In each box make a hole on one side for a drainpipe to pass through.

Fig 10.1: COLLISION MAT

Fix the smaller box over the hole with a drainpipe for any water to run out. Now fix the larger box over the smaller box and fill it with cement. Plug the drainpipe when the cement has set.

SHORING BULKHEADS IN A DAMAGED BOAT: If the vessel is holed in the bow, the forward bulkhead may need shoring from behind to support the weight of the sea flooding forward of it. The forward bulkhead is likely to be triangular in shape. It will be best supported if you shored it roughly half way up from the base of the bulkhead to the flooding waterline.

A midships bulkhead needing support, on the other hand, is likely to be rectangular in shape. Shore it from behind, roughly one third of the way up from the base of the bulkhead to the flooding waterline.

In the case of a vessel in survey, any damage and the repairs thereof must be reported to and inspected by a licensed surveyor.

STEERING ROD PARTED WITH RUDDER JAMMED HARD OVER TO ONE SIDE

Try to pull back the rudder amidships or as near amidships as possible. Then rig a jury rudder as discussed below in "loss of steering at sea".

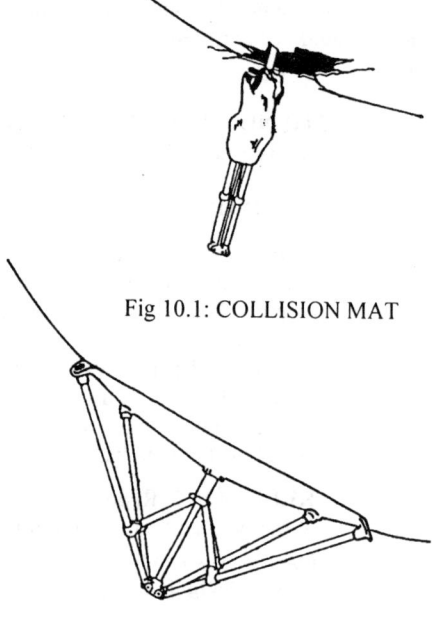

Chapter 10: Watchkeeping Duties & Nautical Emergencies

If you have to leave the rudder jammed hard over, the vessel will tend to go in circles. You should slow down to reduce the rudder effect.

In case of a twin-screw vessel, you may be able to keep her on course by using the one engine more than the other. In a single screw vessel, you will need to tow a drogue on the opposite quarter to counteract the rudder effect. By hauling it in or out, you may also be able to steer the vessel. You will need to adjust the size of the drogue and the length of the line by trial and error.

LOSS OF STEERING AT SEA

First check whether the rudder is lost or it is just the steering gear that is not working. In a non-hydraulic system, a rod (or wire) may have broken, or the chain may have slipped off the gipsy in the steering wheel assembly or off the rudder quadrant. In the hydraulic steering system, the pump may have stopped working, there could be a break in the oil line or at the tiller head, or the bearings in the steering wheel assembly may have become seized.

If the rudder is intact, and you do not have the mechanical knowledge to repair the steering gear, you should be able to steer by the emergency steering (manual tiller). It is usually required to be carried by vessels operating in open waters. If the rudder has been lost, look out for any ingress of water from the rudderpost area.

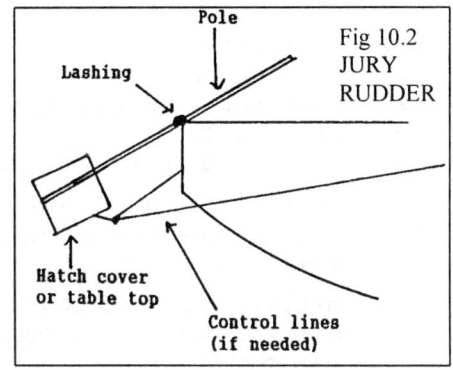

Fig 10.2 JURY RUDDER

Irrespective of the cause, make sure that the vessel is safe from running aground, collisions or being swamped or capsized by the waves.

ACTION TO BE TAKEN: Drop an anchor, if water is shallow enough. If not, use a sea anchor to hold her head into the sea to avoid capsizing. Hoist the Not Under Command day shapes or lights. Seek assistance or rig a jury rudder.

The easiest jury rudder is a drogue towed behind the vessel from a bridle secured to her quarters. (A drogue is any object offering sufficient drag.) The vessel can then be steered by hauling in the bridal line from the required quarter. Hauling on the starboard line will turn the vessel to starboard, and visa versa. She will steer straight ahead when the drogue is centred.

A cabin table or a hatch cover secured to a pole and fastened with U-bolts can also provide a substitute rudder.

A twin-screw vessel can be steered in emergency by running both engines at half speed, and then increasing or decreasing the speed on one of them to turn the vessel in the desired direction.

Record the event in vessel's logbook.

LOSS OF PROPELLER AT SEA

As discussed with loss of steering, first make sure that the vessel is safe from running aground, collisions or being swamped or capsized by the waves.

ACTION TO BE TAKEN: Drop an anchor, if water is shallow enough. If not, use a sea anchor to hold her head into the sea to avoid capsizing. Hoist the Not Under Command day shapes or lights. Seek assistance or rig a sail or row if close to the shore.

PROPELLER FOULED OR DAMAGED AT SEA

If you feel a thump & engine starts to slow down or stop without apparent reason, the propeller may have become fouled with a net or a rope and it has started to wind around the propeller shaft.

Stop the engine instantly. Make sure that the vessel is safe from running aground, collision or being swamped or capsized by the waves. Drop an anchor, if water is shallow enough. If not, use a sea anchor to hold her head into the sea to avoid capsizing. Hoist the Not Under Command day shapes or lights.

If possible, carry out a visual inspection of the propeller and shaft. You must shut down the engine without any possibility of accidentally being started before carrying out this operation. Go over the side and remove the net or rope by unwinding it from around the shaft or cutting it with a hacksaw blade or bread knife. (The synthetic ropes usually become tight and fuse around the shaft).

Check for any visible damage to the propeller blades, rudder post, propeller shaft, cutlass bearings and the spectacle frame. Also check for any damage to the gearbox.

Vibrations in the vessel indicate propeller damage as a result of striking an object. A noise from vessel's after end indicates loose or fouled propeller. It could also be rudder damage interfering with the propeller.

Once the obstruction has been removed, get the vessel slowly underway and listen for any abnormal sounds:
- A bent blade or a blade touching a bent rudderpost will go "clunk clunk".
- A bent shaft or a chipped or bent blade will cause the engine (and vessel) to vibrate.
- Leaky stern gland or chattering noise may mean that the cutlass bearing is damaged.
- Slip the vessel on returning to port for a thorough check.

CAPSIZE

Possible consequences:
- It may fill up with water and sink.
- It may continue to float upside down.
- It may right itself.
- You may be able to climb on the upturned vessel and right her.

Abandon the vessel only as a last resort. Stay close to her to improve the prospect of sighting by the rescue craft. Stay in the liferaft and don't remove your lifejackets. If you are in the water, stay together in a HUDDLE or H.E.L.P. position (see Chapter 19.3).

Don't try to swim ashore unless it is very close and suitable landing place exists. Distances can be deceptive. You may become exhausted and drown. Hypothermia is a real danger even if you are wearing a lifejacket. Keep up your spirits and maintain group morale.

Try to get the EPIRB and distress signals out of the capsized vessel and raise an alarm. Try putting up a make-up signal on the upturned vessel. Make yourselves as visible as you can to both ships and aircraft. Put on more clothes if you are able to find any or get them out of the upturned vessel.

TRANSFERRING SURVIVORS FROM STRICKEN VESSEL

Survivors and those assisting them on deck should wear lifejackets. In calm weather you may be able to go alongside the stricken vessel. Make sure all booms and other movable gear are swung inboard on both vessels. Use heavy fenders. Do not tie up to the stricken vessel in case you quickly need to get away. Maintain your position alongside using engines.

In rough weather, both vessels when unattended would lie beam to wind and sea. You should position your vessel upwind. You may be able to throw a heaving line to the stricken vessel, and then set up a heavier line between the two vessels. A raft or a dinghy travelling along the rope can then be used to transfer survivors.

Another way to send a line and an inflatable rubber raft to the stricken vessel is to secure the raft to one end of the line and let it blow downwind towards the stricken vessel. The survivors when boarded in the raft can then be hauled back with the line.

FAILURE OF MAST SUPPORTS

If a shroud or wire support has parted or slipped out of its spreader end, ease the pressure off the gear by bringing the damaged part to leeward. As an immediate measure, replace the shroud with a spare halyard while considering a better repair. A spare length of wire with hard eye in one end is always handy on a yacht.

FAILURE OF A SAIL

A sail may give way at a chafed stitching or a sharp edge may tear the cloth. An adhesive patch should provide a temporary relief, but a sewn patch is a better repair. Re-sewing the stitching with palm, needle and twine is another option. Use a double-sided tape or a single-sided tape folded in half lengthwise to hold the seams together. Pass tacking stitches at intervals along the tear or failed seam before stitching the full length of the damaged sail.

WINCH FAILURE

Rig a handy billy to take the weight of the halyard, while you decide to leave it there or move it to another winch. In the case of sheet winches, use the windward winch if the one too leeward is malfunctioning.

WINDLASS FAILURE

Haul the anchor chain with a line from another winch or with a block and tackle. Rig a stopper (see Chapter 4) on the chain while you transfer it to another winch. You may be able to bring it (in) in one haul or in short lengths by transferring the hauling line forward one haul after another. Alternatively, buoy and slip the anchor, note its position, advise the authorities and return later to retrieve it.

CHAPTER 10: QUESTIONS & ANSWERS

WATCHKEEPING & NAUTICAL EMERGENCIES

1. A navigational watchkeeper in the process of being relieved of a watch must:

    (a) make no further manoeuvres with helm, engines or sails.
    (b) hand over the watch in the presence of the master.
    (c) consider the fitness of the person taking over the watch.
    (d) stay on watch until the existing traffic is gone past & clear.

2. A person relieving someone from navigational watchkeeping must:

    (a) manoeuvre and test the engine, helm and/or sails.
    (b) not take over until the current traffic is gone past & clear
    (c) take over in the presence of the master.
    (d) become fully aware of the dangers likely to develop from the existing circumstances.

3. The master's presence on the vessel's bridge or in the cockpit during someone else's watch is an indication that:

    (a) the master is in charge of the navigational watch.
    (b) none of the statements stated here may apply.
    (c) the master is sharing the navigational watch.
    (d) only the master will take any required collision avoidance action.

4. The Master's Standing Orders are to ensure that:

    (a) misunderstandings do not arise in watchkeeping duties.
    (b) the master is not disturbed unnecessarily.
    (c) the relieving watchkeeper calls the master when taking over the watch.
    (d) the watchkeeper stays awake.

5. The Master's Night Orders:

    (a) override the Standing Orders.
    (b) allow the master to get a good night's sleep.
    (c) complement the Standing Orders.
    (d) specify the night watchkeepers' hand-over procedure.

6. When on navigational watch underway, the topmost priority is:

    (a) maintaining a proper lookout.
    (b) position fixing
    (c) proper use of GPS, autopilot & radar.
    (d) radio listening watch.

7. When on navigational watch underway, it is acceptable to:

    (a) occasionally get some sleep by engaging the vessel on autopilot & radar on watchkeeping.
    (b) leave the bridge or cockpit unattended from time to time.
    (c) make large course alterations on autopilot.
    (d) call the master when in doubt.

8. When watchkeeping at anchor:

    (a) There is no need for the master to issue Standing Orders.
    (b) It is best to leave sidelights switched on at night.
    (c) A changing beam bearing is not always an indication of a dragging anchor.
    (d) It is unnecessary to keep watch on passing traffic.

9. It is expected of a navigational watchkeeper to:

    (a) log all doubtful situations.
    (b) call the master if there is any doubt about safety of the vessel.
    (c) deal with all matters of safety with full confidence.
    (d) check the vessel's position at least every hour.

10. When watchkeeping in port:

    (a) Posting a disclaimer notice at the gangway provides little or no protection against third party injury claims.
    (b) When alongside a wharf with large tidal range, berthing lines should be kept short.
    (c) The 'diver below' signal is not necessary when diving from a vessel alongside a jetty.
    (d) It is not necessary to display flag 'B' when refuelling at a wharf.

11. Watchkeepers usually come across the following statement in the Master's Standing Orders:

    (a) Do not call me unnecessarily.
    (b) Deal with navigation hazards when you encounter them.
    (c) Fix any malfunctioning equipment immediately.
    (d) Call me if you fail to make the expected landfall.

12. Under the States, Territories & Commonwealth legislation, the following is not a Marine Casualty or Incident:

    (a) Vessel in danger of serious damage.
    (b) Postponement of sailing.
    (c) Vessel grounded.
    (d) A person injured due to an accident on board.

13. Rules concerning responsibilities and care following a Marine Incident do not require a master to:

    (a) make a log book entry.
    (b) immediately report to the authorities.
    (c) transmit Mayday.
    (d) provide name and address to any other vessel involved.

14. A person's legal duty to provide rescue at sea is as follows:

    (a) It is limited to saving lives.
    (b) The rescuer must try to save the stricken vessel.
    (c) The stricken vessel's master may demand to be towed.
    (d) The abandoned vessel must be sunk.

15. One of the characteristics of the Lloyds Open Form of Salvage is that:

    (a) it records the number of towlines used.
    (b) it records the monetary value for salvage.
    (c) the agreement can be made on a plain piece of paper.
    (d) it records the stricken vessel's choice of port of refuge.

16. In relation to salvage, which of the following statements is correct?

    (a) Salvor can commence salvaging a vessel without her consent.
    (b) Salvage must involve the stricken vessel's towline.
    (c) Salvage must involve the towing vessel's towline.
    (d) Salvor can refuse to salvage a stricken vessel.

17. In relation to salvage, which of the following statements is correct?

    (a) An incomplete salvage may still earn the tug a salvage award.
    (b) A salvage tug has the right to tow the stricken vessel to its chosen port.
    (c) The ownership of a property found adrift or sunken beyond 12-miles territorial limit is automatic.
    (d) The ownership of a property found adrift or sunken anywhere at sea is automatic.

18. The official with whom salvage claims must be registered is:

    (a) a High Court judge.
    (b) a State or Territory appointed surveyor.
    (c) the Receiver of Wrecks.
    (d) the local Water Police.

19. The master of the salving vessel should overlook the interests of the following party when undertaking salvage:

    (a) Own vessel's insurer.
    (b) Own vessel's owner.
    (c) Own vessel's cargo owner.
    (d) None of the parties stated here.

20. In a marine salvage claim, the words "No cure, No pay" means:

    (a) getting paid only if life is saved.
    (b) rescuing someone under contract.
    (c) reward for a successful salvage.
    (d) an agreed amount payable for a salvage.

Chapter 10: Watchkeeping Duties & Nautical Emergencies

21. A vessel has lost her propeller at sea. While working out an alternative means of propulsion, she should:

    (i)   stream warps.
    (ii) display a Not Under Command signal.
    (iii) hold her head into the sea.
    (iv) hold her stern into the sea.

    The correct answer is:
    (a) all except (i).
    (b) all except (ii).
    (c) all except (iii).
    (d) all except (iv).

22. A vessel's propeller's became fouled with a fishing net at sea. After clearing it when she got under way, the master heard a "clunk clunk" sound. The most likely reason is:

    (a) a blade touching the rudder post.
    (b) bent shaft.
    (c) a chipped blade.
    (d) damaged cutlass bearing.

23. A vessel becomes disabled at sea. She should:

    (i)   stream warps.
    (ii) show a Restricted in Ability to Manoeuvre signal
    (iii) show a Not Under Command signal.
    (iv) call for assistance without real emergency.

    The correct answer is:
    (a) all except (i).
    (b) all except (ii).
    (c) all except (iii).
    (d) all except (iv).

24. A small vessel becomes disabled in a shipping fairway. She should avoid:

    (a) displaying a Not Under Command signal.
    (b) anchoring in the fairway.
    (c) calling for assistance without real emergency.
    (d) securing to a navigation mark.

25. A vessel is holed in the bow. The forward triangular shape collision bulkhead is taking the weight of the sea. The master should shore it from behind at:

    (a) its centre.
    (b) one third height from its base to waterline.
    (c) its upper edge.
    (d) half way between its base and the waterline.

26. A vessel's propeller's became fouled with a fishing net at sea. After clearing it when she got under way, the vessel was vibrating. This may be due to:

    (a) a bent shaft.
    (b) a chipped blade.
    (c) bent propeller blade.
    (d) any of the reasons stated here.

27. A vessel's propeller's became fouled with a fishing net at sea. After clearing it when she got under way, the master could hear a chattering sound. This may be due to a:

    (a) damaged cutlass bearing.
    (b) chipped blade.
    (c) bent blade.
    (d) bent rudder post.

28. A vessel is holed below the waterline. The master should:

    (a) heel her to undamaged side and patch up.
    (b) abandon her.
    (c) increase engine speed to reach ashore quickly.
    (d) keep her upright.

29. A jury rudder is:

    (a) non-hydraulic steering system.
    (b) steering by emergency tiller.
    (c) a twin-screw vessel without a rudder.
    (d) a drogue towed behind a vessel from a bridle.

30. In a single screw vessel, one of your steering rods has parted and you have nothing on board to repair or replace it. The cause of the failure has also jammed the rudder hard to starboard. You wish to steer the vessel by towing a drogue. It will be most effective when towed:

    (a) on port quarter.
    (b) on starboard quarter.
    (c) amidships.
    (d) anywhere astern.

31. The weakest part of a vessel to the impact of a collision is the:

    (a) stem.
    (b) beam.
    (c) bow.
    (d) quarter.

32. A collision mat is used for:
    (a) preventing collisions.
    (b) softening the blow of a collision.
    (c) reducing the inflow of water from damaged hull.
    (d) making a life raft after a collision.

33. A cement box is used for:
    (a) temporary repair of hull damage
    (b) increasing vessel's ballast.
    (c) increasing vessel's stability.
    (d) temporary repair of lead keel.

34. A vessel has accidentally run aground. The master could:
    (i) move weights to lighten the grounded end.
    (ii) jettison weights.
    (iii) immediately go full astern on engines.
    (iv) transmit PANPAN or MAYDAY.

    The correct answer is:
    (a) all except (i).
    (b) all except (ii).
    (c) all except (iii).
    (d) all except (iv).

35. A vessel has accidentally run hard aground on a reef. She is apparently holed. The master should:
    (i) transmit PANPAN or MAYDAY.
    (ii) move weights to lighten the grounded end.
    (iii) load more weight on board.
    (iv) take an anchor ashore to hold her.

    The correct answer is:
    (a) all except (i).
    (b) all except (ii).
    (c) all except (iii).
    (d) all except (iv).

36. A vessel has run aground in an uncertain position at night. There is little tide and the bottom is sandy. The master should:
    (a) consider staying there until daylight.
    (b) move weights to lighten the grounded end.
    (c) immediately go full astern on engines.
    (d) load more weight on board.

37. One way to get a keel vessel off aground is to reduce her draught by inclining her. She may be inclined as follows:
    (i) Pulling on a rope tied to top of the mast.
    (ii) Swinging weights out on a boom.
    (iii) Loading weights on one side
    (iv) Moving weights (people) to one side.

    The correct answer is:
    (a) all except (i).
    (b) all except (ii).
    (c) all except (iii).
    (d) all except (iv).

38. Start refloating a grounded vessel:
    (a) just before high water.
    (b) at high water.
    (c) just after high water.
    (d) at a neap tide.

39. For beaching a vessel for maintenance, one of the precautions is:
    (a) not to drop forward anchor(s) while beaching.
    (b) to make the vessel top heavy.
    (c) to keep the vessel on zero trim.
    (d) to seek a steep shelving beach.

40. A soft sandy shore is not available. Beaching a vessel in distress on a rocky shore in bad weather:
    (a) will allow the vessel to jam between rocks.
    (b) would be disastrous.
    (c) is better than abandoning in a life raft.
    (d) will enable survivors to climb ashore.

41. When getting a vessel off the beach, the following should be remembered about tides:
    (a) neap lows are higher than spring lows.
    (b) spring lows are higher than neap lows.
    (c) all lows express chart datum.
    (d) neap and spring lows are of the same height.

42. When getting a vessel off the beach, the following tide should be of most assistance:
    (a) a falling high spring tide.
    (b) a rising high neap tide.
    (c) a falling high neap tide.
    (d) a rising high spring tide.

Chapter 10: Watchkeeping Duties & Nautical Emergencies

43. After refloating a grounded vessel:

    (i) Check propeller, rudder & engines for any damage.
    (ii) Check if she is taking water.
    (iii) Head for the nearest safe port.
    (iv) Trim the vessel down by the head.

    The correct answer is:
    (a) All except (i)
    (b) All except (ii)
    (c) All except (iii)
    (d) All except (iv)

44. A vessel attempting to land survivors on a rocky shore in bad weather should:

    (a) Bring the vessel ashore on the back of a wave.
    (b) Make the landing beam-on.
    (c) Land survivors by transferring them into a liferaft.
    (d) Make the landing head-on.

45. In relation to beaching a vessel, which of the following statements is incorrect?

    (a) Stop engines prior to making contact.
    (b) It is better to drop the kedge anchor(s) before the vessel takes to the beach.
    (c) Discharge of ballast prior to refloating will assist refloating.
    (d) Load half ballast prior to & half after grounding.

46. Which of the following statements is incorrect with regard to reducing damage in a collision?

    (a) Turn the vessel to avoid taking a direct hit.
    (b) A bow-to-bow hit would cause less damage than a head-on hit.
    (c) A bow-to-quarter hit would cause less damage than one vessel ramming into the other.
    (d) In a head-on collision, it is better to let go both anchors than to go full astern on the engines.

47. Which of the following methods should not be employed to reduce flooding following hull damage?

    (a) Heel the vessel towards the undamaged side.
    (b) Patch & shore up the hole using lifejackets and dinghy paddles.
    (c) Connect & pump out bilges via the engine's cooling water intake.
    (d) Rig a collision mat.

48. Following capsize and upturning of a vessel:

    (a) Stay close to the vessel.
    (b) Swim ashore if you can see the shore & are wearing a lifejacket.
    (c) Do not try to right the upturned vessel
    (d) Do not try to retrieve equipment from the upturned vessel.

49. When transferring survivors from a stricken vessel:

    (a) Remove your lifejackets for freedom of movement.
    (b) Secure your vessel alongside the stricken vessel.
    (c) Position your vessel downwind of the stricken vessel.
    (d) Set up a dinghy travelling along a rope between the vessels.

CHAPTER 10: ANSWERS

1 (c), 2 (d), 3 (b), 4 (a), 5 (c),
6 (a), 7 (d), 8 (c), 9 (b), 10 (a),
11 (d), 12 (b), 13 (b), 14 (a), 15 (c),
16 (d), 17 (a), 18 (c), 19 (d), 20 (c)
21 (d), 22 (a), 23 (b), 24 (b), 25 (d),
26 (d), 27 (a), 28 (a), 29 (d), 30 (a),
31 (b), 32 (c), 33 (a), 34 (c), 35 (b),
36 (a), 37 (c), 38 (a), 39 (a), 40 (b),
41 (a), 42 (d), 43 (d), 44 (c), 45 (b),
46 (d), 47 (b), 48 (a), 49 (d)

Chapter 11

COLLISION PREVENTION REGULATIONS
(International + Some Local)

INTERNATIONAL REGULATIONS FOR PREVENTING COLLISIONS AT SEA, 1972

(As published by the International Maritime Organisation, Australian Maritime Safety Authority and the U.S. Department of Transportation)

Part A -- General
Rule 1: Application

(a) These Rules shall apply to all vessels upon the high seas and in all waters connected therewith navigable by seagoing vessels.

(b) Nothing in these Rules shall interfere with the operation of special rules made by an appropriate authority for roadsteads, harbours, rivers, lakes or inland waterways connected with the high seas and navigable by sea-going vessels. Such special rules shall conform as closely as possible to these Rules.

(c) Nothing in these Rules shall interfere with the operation of any special rules made by the Government of any State with respect to additional station or signal lights, shapes or whistle signals for ships of war and vessels proceeding under convoy, or with respect to additional station or signal lights or shapes for fishing vessels engaged in fishing as a fleet. These additional station or signal lights, shapes or whistle signals shall, so far as possible, be such that they cannot be mistaken for any light, shape or signal authorised elsewhere under these Rules.

(d) Traffic separation schemes may be adopted by the Organization for the purpose of these Rules.

(e) Whenever the Government concerned shall have determined that a vessel of special construction or purpose cannot comply fully with the provisions of any of these Rules with respect to the number, position, range or arc of visibility of lights or shapes, as well as to the disposition and characteristics of sound-signalling appliances, such vessel shall comply with such other provisions in regard to the number, position, range or arc of visibility of lights or shapes, as well as to the disposition and characteristics of sound-signalling appliances, as her Government shall have determined to be the closest possible compliance with these Rules in respect of that vessel.

Rule 2
Responsibility

(a) Nothing in these Rules shall exonerate any vessel, or the owner, master or crew thereof, from the consequences of any neglect to comply with these Rules or of the neglect of any precaution which may be required by the ordinary practice of seamen, or by the special circumstances of the case.

(b) In construing and complying with these Rules due regard shall be had to all dangers of navigation and collision and to any special circumstances, including the limitations of the vessels involved, which may make a departure from these Rules necessary to avoid immediate danger.

Rule 3
General Definitions

For the purpose of these Rules, except where the context otherwise requires:

(a) The word "vessel" includes every description of water craft, including non-displacement craft and seaplanes, used or capable of being used as a means of transportation on water.

(b) The term "power-driven vessel" means any vessel propelled by machinery.

(c) The term "sailing vessel" means any vessel under sail provided that propelling machinery, if fitted, is not being used.

(d)   The term "vessel engaged in fishing" means any vessel fishing with nets, lines, trawls or other fishing apparatus which restrict manoeuvrability, but does not include a vessel fishing with trolling lines or other fishing apparatus which do not restrict manoeuvrability.

(e)   The word "seaplane" includes any aircraft designed to manoeuvre on the water.

(f)   The term "vessel not under command" means a vessel which through some exceptional circumstance is unable to manoeuvre as required by these Rules and is therefore unable to keep out of the way of another vessel.

(g)   The term "vessel restricted in her ability to manoeuvre" means a vessel which from the nature of her work is restricted in her ability to manoeuvre as required by these Rules and is therefore unable to keep out of the way of another vessel. The term "vessels restricted in their ability to manoeuvre" shall include but not be limited to:

    (i)   a vessel engaged in laying, servicing or picking up a navigation mark, submarine cable or pipeline;
    (ii)   a vessel engaged in dredging, surveying or underwater operations;
    (iii)   a vessel engaged in replenishment or transferring persons, provisions or cargo while underway;
    (iv)   a vessel engaged in the launching or recovery of aircraft;
    (v)   a vessel engaged in mine clearance operations;
    (vi)   a vessel engaged in a towing operation such as severely restricts the towing vessel and her tow in their ability to deviate from their course.

(h)   The term "vessel constrained by her draught" means a power-driven vessel which, because of her draught in relation to the available depth and width of navigable water, is severely restricted in her ability to deviate from the course she is following.

(i)   The word "underway" means that a vessel is not at anchor, or made fast to the shore, or aground.

(j)   The words "length" and "breadth" of a vessel mean her length overall and greatest breadth.

(k)   Vessels shall be deemed to be in sight of one another only when one can be observed visually from the other.

(l)   The term "restricted visibility" means any condition in which visibility is restricted by fog, mist, falling snow, heavy rainstorms, sandstorms or any other similar causes.

Part B -- Steering and Sailing Rules

Section I -- Conduct of vessels in any conditions of visibility
Rule 4
Application

Rules in this Section apply in any condition of visibility.

Rule 5
Look-out

Every vessel shall at all times maintain a proper look-out by sight and hearing as well as by all available means appropriate in the prevailing circumstances and conditions so as to make a full appraisal of the situation and of the risk of collision.

Rule 6
Safe Speed

Every vessel shall at all times proceed at a safe speed so that she can take proper and effective action to avoid collision and be stopped within a distance appropriate to the prevailing circumstances and conditions.

In determining a safe speed the following factors shall be among those taken into account:

(a)   By all vessels:

    (i)   the state of visibility;
    (ii)   the traffic density including concentrations of fishing vessels or any other vessels;
    (iii)   the manoeuvrability of the vessel with special reference to stopping distance and turning ability in the prevailing conditions;
    (iv)   at night the presence of background light such as from shore lights or from back scatter of her own lights;
    (v)   the state of wind, sea and current, and the proximity of navigational hazards;
    (vi)   the draught in relation to the available depth of water.

Chapter 11: Collision Prevention Regulations

(b) Additionally, by vessels with operational radar:

    (i) the characteristics, efficiency and limitations of the radar equipment;
    (ii) any constraints imposed by the radar range scale in use;
    (iii) the effect on radar detection of the sea state, weather and other sources of interference;
    (iv) the possibility that small vessels, ice and other floating objects may not be detected by radar at an adequate range;
    (v) the number, location and movement of vessels detected by radar;
    (vi) the more exact assessment of the visibility that may be possible when radar is used to determine the range of vessels or other objects in the vicinity.

Rule 7
Risk of Collision

(a) Every vessel shall use all available means appropriate to the prevailing circumstances and conditions to determine if risk of collision exists. If there is any doubt such risk shall be deemed to exist.

(b) Proper use shall be made of radar equipment if fitted and operational, including long-range scanning to obtain early warning of risk of collision and radar plotting or equivalent systematic observation of detected objects.

(c) Assumptions shall not be made on the basis of scanty information, especially scanty radar information.

(d) In determining if risk of collision exists the following considerations shall be among those taken into account:

    (i) such risk shall be deemed to exist if the compass bearing of an approaching vessel does not appreciably change;
    (ii) such risk may sometimes exist even when an appreciable bearing change is evident, particularly when approaching a very large vessel or a tow or when approaching a vessel at close range.

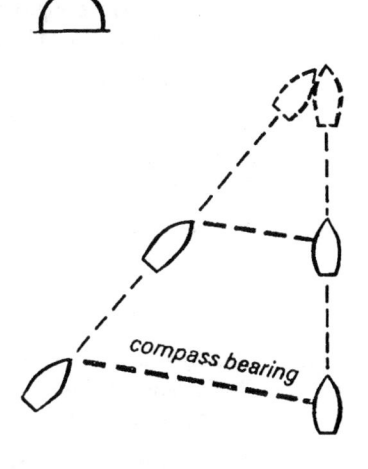

Rule 8
Action to avoid Collision

(a) Any action taken to avoid collision shall, if the circumstances of the case admit, be positive, made in ample time and with due regard to the observance of good seamanship.

(b) Any alteration of course and/or speed to avoid collision shall, if the circumstances of the case admit, be large enough to be readily apparent to another vessel observing visually or by radar; a succession of small alterations of course and/or speed should be avoided.

(c) If there is sufficient sea room, alteration of course alone may be the most effective action to avoid a close quarters situation provided that it is made in good time, is substantial and does not result in another close-quarters situation.

(d) Action taken to avoid collision with another vessel shall be such as to result in passing at a safe distance. The effectiveness of the action shall be carefully checked until the other vessel is finally past and clear.

(e) If necessary to avoid collision or allow more time to assess the situation, a vessel shall slacken her speed or take all way off by stopping or reversing her means of propulsion.

(f) (i) A vessel which, by any of these rules, is required not to impede the passage or safe passage of another vessel shall, when required by the circumstances of the case, take early action to allow sufficient sea room for the safe passage of the other vessel.

    (ii) A vessel required not to impede the passage or safe passage of another vessel is not relieved of this obligation if approaching the other vessel so as to involve risk of collision and shall, when taking action, have full regard to the action which may be required by the rules of this part.

    (iii) A vessel the passage of which is not to be impeded remains fully obliged to comply with the rules of this part when the two vessels are approaching one another so as to involve risk of collision.

Rule 9
Narrow Channels

(a) A vessel proceeding along the course of a narrow channel or fairway shall keep as near to the outer limit of the channel or fairway which lies on her starboard side as is safe and practicable.

(b) A vessel of less than 20 metres in length or a sailing vessel shall not impede the passage of a vessel which can safely navigate only within a narrow channel or fairway.

(c) A vessel engaged in fishing shall not impede the passage of any other vessel navigating within a narrow channel or fairway.

(d) A vessel shall not cross a narrow channel or fairway if such crossing impedes the passage of a vessel which can safely navigate only within such channel or fairway. The latter vessel may use the sound signal prescribed in Rule 34(d) if in doubt as to the intention of the crossing vessel.

(e) (i) In a narrow channel or fairway when overtaking can take place only if the vessel to be overtaken has to take action to permit safe passing, the vessel intending to overtake shall indicate her intention by sounding the appropriate signal prescribed in Rule 34(c)(i). The vessel to be overtaken shall, if in agreement, sound the appropriate signal prescribed in Rule 34(c)(ii) and take steps to permit safe passing. If in doubt she may sound the signals prescribed in Rule 34(d).

(ii) This Rule does not relieve the overtaking vessel of her obligation under Rule 13.

(f) A vessel nearing a bend or an area of a narrow channel or fairway where other vessels may be obscured by an intervening obstruction shall navigate with particular alertness and caution and shall sound the appropriate signal prescribed in Rule 34(e).

(g) Any vessel shall, if the circumstances of the case admit, avoid anchoring in a narrow channel.

Rule 10
Traffic Separation Schemes

(a) This Rule applies to traffic separation schemes adopted by the Organization and does not relieve any vessel of her obligation under any other rule.

(b) A vessel using a traffic separation scheme shall:

(i) proceed in the appropriate traffic lane in the general direction of traffic flow for that lane;

(ii) so far as practicable keep clear of a traffic separation line or separation zone;

(iii) normally join or leave a traffic lane at the termination of the lane, but when joining or leaving from either side shall do so at as small an angle to the general direction of traffic flow as practicable.

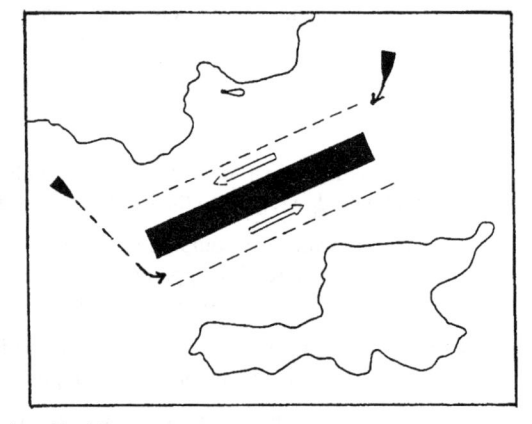

(c) A vessel shall, so far as practicable, avoid crossing traffic lanes but if obliged to do so shall cross on a heading as nearly as practicable at right angles to the general direction of traffic flow.

(d) (i) A vessel shall not use an inshore traffic zone when can safely use the appropriate traffic lane within the adjacent traffic separation scheme. However, vessels of less than 20 metres in length, sailing vessels and vessels engaged in fishing may use the inshore traffic zone.

(ii) Notwithstanding subparagraph (d)(i), a vessel may use an inshore traffic zone when en route to or from a port, offshore installation or structure, pilot station or any other place situated within the inshore traffic zone, or to avoid immediate danger.

(e) A vessel other than a crossing vessel or a vessel joining or leaving a lane shall not normally enter a separation zone or cross a separation line except:

(i) in cases of emergency to avoid immediate danger,

(ii) to engage in fishing within a separation zone.

Chapter 11: Collision Prevention Regulations

(f)   A vessel navigating in areas near the terminations of traffic separation schemes shall do so with particular caution.

(g)   A vessel shall so far as practicable avoid anchoring in a traffic separation scheme or in areas near its terminations.

(h)   A vessel not using a traffic separation scheme shall avoid it by as wide a margin as is practicable.

(i)   A vessel engaged in fishing shall not impede the passage of any vessel following a traffic lane.

(j)   A vessel of less than 20 metres in length or a sailing vessel shall not impede the safe passage of a power-driven vessel following a traffic lane.

(k)   A vessel restricted in her ability to manoeuvre when engaged in an operation for the maintenance of safety of navigation in a traffic separation scheme is exempted from complying with this Rule to the extent necessary to carry out the operation.

(i)   A vessel restricted in her ability to manoeuvre when engaged in an operation for the laying, servicing or picking up of a submarine cable, within a traffic separation scheme, is exempted from complying with this Rule to the extent necessary to carry out the operation.

Section II -- Conduct of vessels in sight of one another

Rule 11
Application

Rules in this Section apply to vessels in sight of one another.

Rule 12
Sailing Vessels

(a)   When two sailing vessels are approaching one another, so as to involve risk of collision, one of them shall keep out of the way of the other as follows:

   (i)   when each has the wind on a different side, the vessel which has the wind on the port side shall keep out of the way of the other;

   (ii)   when both have the wind on the same side, the vessel which is to windward shall keep out of the way of the vessel which is to leeward;

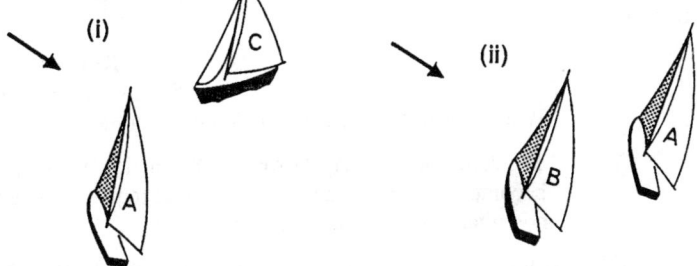

   (iii)   if a vessel with the wind on the port side sees a vessel to windward and cannot determine with certainty whether the other vessel has the wind on the port or on the starboard side, she shall keep out of the way of the other.

(b)   For the purposes of this Rule the windward side shall be deemed to be the side opposite to that on which the mainsail is carried or, in the case of a square-rigged vessel, the side opposite to that on which the largest fore-and-aft sail is carried.

Rule 13
Overtaking

(a)   Notwithstanding anything contained in the Rules of Part B, Sections I and II, any vessel overtaking any other shall keep out of the way of the vessel being overtaken.

(b)   A vessel shall be deemed to be overtaking when coming up with another vessel from a direction more than 22.5 degrees abaft her beam, that is, in such a position with reference to the vessel she is overtaking, that at night she would be able to see only the sternlight of that vessel but neither of her sidelights.

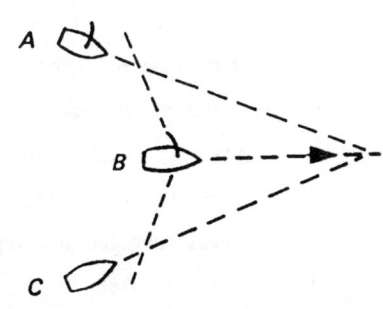

A & C must keep clear

(c) When a vessel is in any doubt as to whether she is overtaking another, she shall assume that this is the case and act accordingly.

(d) Any subsequent alteration of the bearing between the two vessels shall not make the overtaking vessel a crossing vessel within the meaning of these Rules or relieve her of the duty of keeping clear of the overtaken vessel until she is finally past and clear.

Rule 14
Head-on Situation

(a) When two power-driven vessels are meeting on reciprocal or nearly reciprocal courses so as to involve risk of collision each shall alter her course to starboard so that each shall pass on the port side of the other.

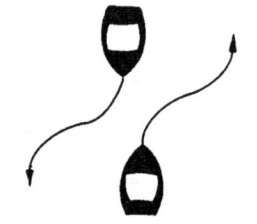

(b) Such a situation shall be deemed to exist when a vessel sees the other ahead or nearly ahead and by night she could see the masthead lights of the other in a line or nearly in a line and/or both sidelights and by day she observes the corresponding aspect of the other vessel.

(c) When a vessel is in any doubt as to whether such a situation exists she shall assume that it does exist and act accordingly.

Rule 15
Crossing Situation

When two power-driven vessels are crossing so as to involve risk of collision, the vessel which has the other on her own starboard side shall keep out of the way and shall, if the circumstances of the case admit, avoid crossing ahead of the other vessel.

Rule 16
Action by Give-way Vessel

Every vessel which is directed to keep out of the way of another vessel shall, so far as possible, take early and substantial action to keep well clear.

Rule 17
Action by Stand-on Vessel

(a) (i) Where one of two vessels is to keep out of the way the other shall keep her course and speed.

(ii) The latter vessel may however take action to avoid collision by her manoeuvre alone, as soon as it becomes apparent to her that the vessel required to keep out of the way is not taking appropriate action in compliance with these Rules.

(b) When, from any cause, the vessel required to keep her course and speed finds herself so close that collision cannot be avoided by the action of the give-way vessel alone, she shall take such action as will best aid to avoid collision.

(c) A power-driven vessel which takes action in a crossing situation in accordance with subparagraph (a)(ii) of this Rule to avoid collision with another power-driven vessel shall, if the circumstances of the case admit, not alter course to port for a vessel on her own port side.

(d) This Rule does not relieve the give-way vessel of her obligation to keep out of the way.

Rule 18
Responsibilities between Vessels

Except where Rules 9, 10 and 13 otherwise require:

(a) A power driven vessel underway shall keep out of the way of:

(i) a vessel not under command;

(ii) a vessel restricted in her ability to manoeuvre;

(iii) a vessel engaged in fishing;

(iv) a sailing vessel.

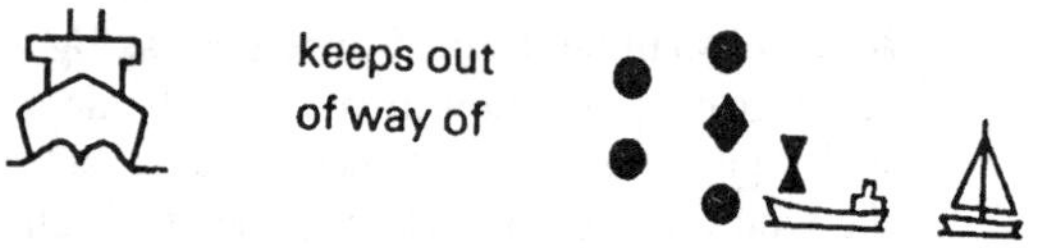

Chapter 11: Collision Prevention Regulations

(b)   A sailing vessel underway shall keep out of the way of:

       (i)   a vessel not under command;

       (ii)   a vessel restricted in her ability to manoeuvre;

       (iii)   a vessel engaged in fishing.

(c)   A vessel engaged in fishing when underway shall, so far as possible, keep out of the way of:

       (i)   a vessel not under command;

       (ii)   a vessel restricted in her ability to manoeuvre.

(d)   (i)   Any vessel other than a vessel not under command or a vessel restricted in her ability to manoeuvre shall, if the circumstances of the case admit, avoid impeding the safe passage of a vessel constrained by her draught, exhibiting the signals in Rule 28.

       (ii)   A vessel constrained by her draught shall navigate with particular caution having full regard to her special condition.

(e)   A seaplane on the water shall, in general, keep well clear of all vessels and avoid impeding their navigation. In circumstances, however, where risk of collision exists, she shall comply with the Rules of this Part.

Section III -- Conduct of vessels in restricted visibility

Rule 19
Conduct of Vessels in Restricted Visibility

(a)   This Rule applies to vessels not in sight of one another when navigating in or near an area of restricted visibility.

(b)   Every vessel shall proceed at a safe speed adapted to the prevailing circumstances and conditions of restricted visibility. A power-driven vessel shall have her engines ready for immediate manoeuvre.

(c)   Every vessel shall have due regard to the prevailing circumstances and conditions of restricted visibility when complying with the Rules of Section I of this Part.

(d)   A vessel which detects by radar alone the presence of another vessel shall determine if a close-quarters situation is developing and/or risk of collision exists. If so, she shall take avoiding action in ample time, provided that when such action consists of an alteration of course, so far as possible the following shall be avoided:

       (i)   an alteration of course to port for a vessel forward of the beam, other than for a vessel being overtaken;

       (ii)   an alteration of course towards a vessel abeam or abaft the beam.

(e)   Except where it has been determined that a risk of collision does not exist, every vessel which hears apparently forward of her beam the fog signal of another vessel, or which cannot avoid a close-quarters situation with another vessel forward of her beam, shall reduce her speed to the minimum at which she can be kept on her course. She shall if necessary take all her way off and in any event navigate with extreme caution until danger of collision is over.

Part C -- Lights and Shapes

Rule 20
Application

(a)   Rules in this Part shall be complied with in all weathers.

(b)   The Rules concerning lights shall be complied with from sunset to sunrise, and during such times no other lights shall be exhibited, except such lights as cannot be mistaken for the lights specified in these Rules or do not impair their visibility or distinctive character, or interfere with the keeping of a proper look-out.

(c)   The lights prescribed by these Rules shall, if carried, also be exhibited from sunrise to sunset in restricted visibility and may be exhibited in all other circumstances when it is deemed necessary.

(d)   The Rules concerning shapes shall be complied with by day.

(e) The lights and shapes specified in these Rules shall comply with the provision of Annex I to these Regulations.

Rule 21
Definitions

(a) "Masthead light" means a white light placed over the fore and aft centreline of the vessel showing an unbroken light over an arc of the horizon of 225 degrees and so fixed as to show the light from right ahead to 22.5 degrees abaft the beam on either side of the vessel.

(b) "Sidelights" means a green light on the starboard side and a red light on the port side each showing an unbroken light over an arc of the horizon of 112.5 degrees and so fixed as to show the light from right ahead to 22.5 degrees abaft the beam on its respective side. In a vessel of less than 20 metres in length the sidelights may be combined in one lantern carried on the fore and aft centreline of the vessel.

(c) "Sternlight" means a white light placed as nearly as practicable at the stern showing an unbroken light over an arc of the horizon of 135 degrees and so fixed as to show the light 67.5 degrees from right aft on each side of the vessel.

(d) "Towing light" means a yellow light having the same characteristics as the "sternlight" defined in paragraph (c) of this Rule.

(e) "All-round light" means a light showing an unbroken light over an arc of the horizon of 360 degrees.

(f) "Flashing light" means a light flashing at regular intervals at a frequency of 120 flashes or more per minute.

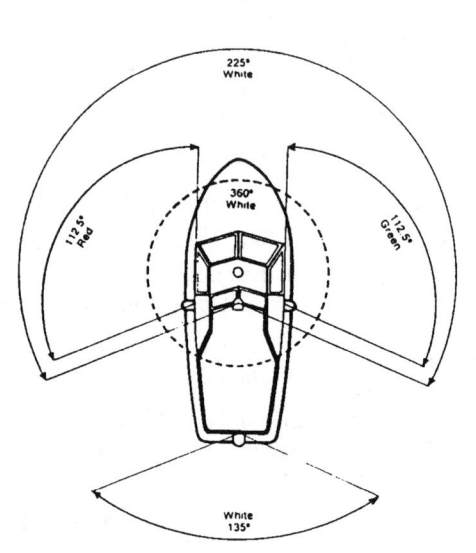

Rule 22
Visibility of Lights

The lights prescribed in these Rules shall have an intensity as specified in Section 8 of Annex I to these Regulations so as to be visible at the following minimum ranges:

(a) In vessels of 50 metres or more in length:

- a masthead light, 6 miles;
- a sidelight, 3 miles;
- a sternlight, 3 miles;
- a towing light, 3 miles;
- a white, red, green or yellow all-round light, 3 miles.

(b) In vessels of 12 metres or more in length but less than 50 metres in length:

- a masthead light, 5 miles; except that where the length of the vessel is less than 20 metres, 3 miles;
- a sidelight, 2 miles;
- a sternlight, 2 miles;
- a towing light, 2 miles;
- a white, red, green or yellow all-round light, 2 miles.

(c) In vessels of less than 12 metres in length:

- a masthead light, 2 miles;
- a sidelight, 1 mile;
- a sternlight, 2 miles;

	Less than 12m	12-50m	Greater than 50m
Masthead	2 miles	5 miles*	6 miles
Side	1 mile	2 miles	3 miles
Stern	2 miles	2 miles	3 miles
Towing	2 miles	2 miles	3 miles
All round	2 miles	2 miles	3 miles

* When length of vessel is between 12 and 20 metres the masthead light visibility is 3 miles.

Chapter 11: Collision Prevention Regulations

- a towing light, 2 miles;
- a white, red, green or yellow all-round light, 2 miles.

(d) In inconspicuous, partly submerged vessels or objects being towed:

- a white all-round light, 3 miles.

Rule 23
Power-driven Vessels underway

(a) A power-driven vessel underway shall exhibit:

(i) a masthead light forward;

(ii) a second masthead light abaft of and higher than the forward one; except that a vessel of less than 50 metres in length shall not be obliged to exhibit such light but may do so;

(iii) sidelights;

(iv) a sternlight.

(b) An air-cushion vessel when operating in the non-displacement mode shall, in addition to the lights prescribed in paragraph (a) of this Rule, exhibit an all-round flashing yellow light.

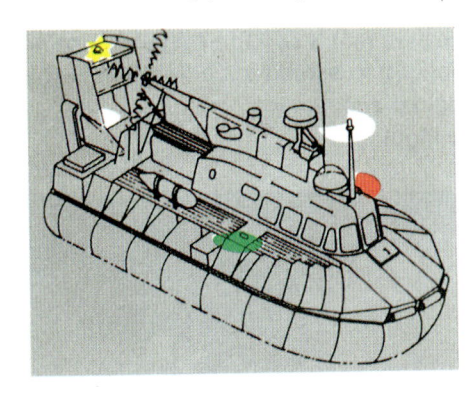

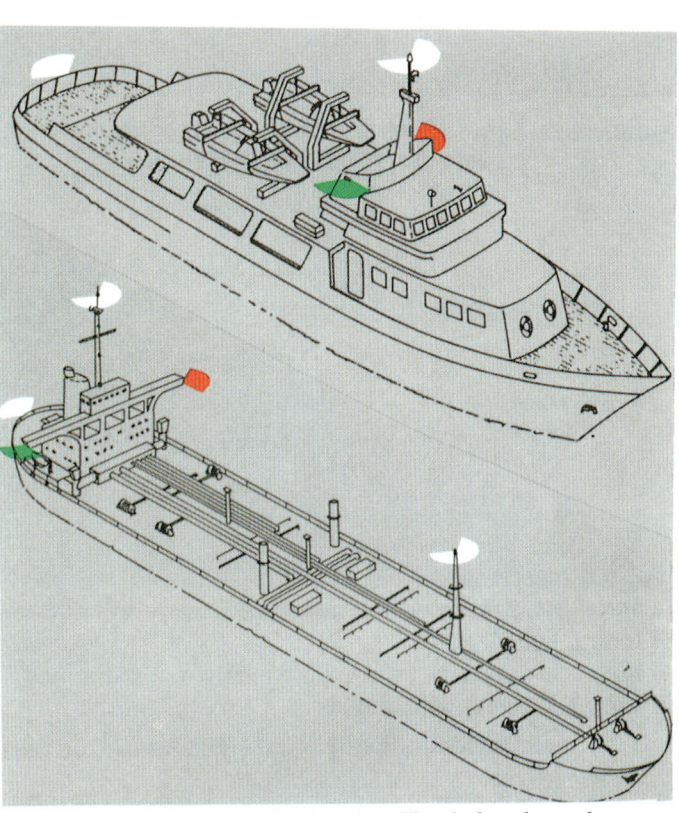

Top: Power driven vessel of less than 50 m in length - underway
Above: Power driven vessel of any length - underway

(c) (i) A power-driven vessel of less than 12 metres in length may in lieu of the lights prescribed in paragraph (a) of this Rule exhibit an all-round white light, and sidelights;

(ii) a power-driven vessel of less than 7 metres in length whose maximum speed does not exceed 7 knots may in lieu of the lights prescribed in paragraph (a) of this Rule exhibit an all-round white light and shall, if practicable, also exhibit sidelights;

Right: Power driven vessel of less than 12 m in length
Far Right: Power driven vessel of less than 7 m in length whose max speed does not exceed 7 knots

(iii) the masthead light or all-round white light on a power-driven vessel of less than 12 metres in length may be displaced from the fore and aft centreline of the vessel if centreline fitting is not practicable, provided that the sidelights are combined in one lantern which shall be carried on the fore and aft centreline of the vessel or located as nearly as practicable in the same fore and aft line as the masthead light or the all-round white light.

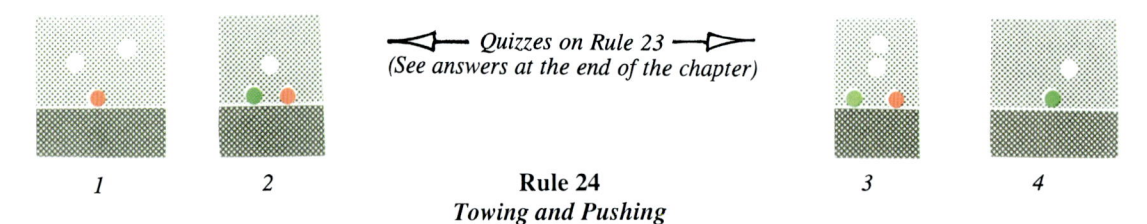

◁— *Quizzes on Rule 23* —▷
(See answers at the end of the chapter)

1   2   3   4

Rule 24
Towing and Pushing

(a) A power-driven vessel when towing shall exhibit:

(i) instead of the light prescribed in Rule 23 (a) (i) or (a) (ii), two masthead lights in a vertical line. When the length of the tow, measuring from the stern of the towing vessel to the after end of the tow exceeds 200 metres, three such lights in a vertical line;

(ii) sidelights;

(iii) a sternlight;

(iv) a towing light in a vertical line above the sternlight;

(v) when the length of the tow exceeds 200 metres, a diamond shape where it can best be seen.

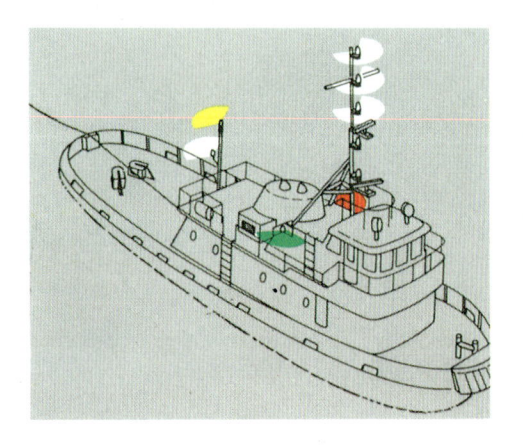

Two power driven vessels towing astern, both less than 50 m in length.
Left: Length of tow over 200 m.
Right: Length of tow 200 m or less.

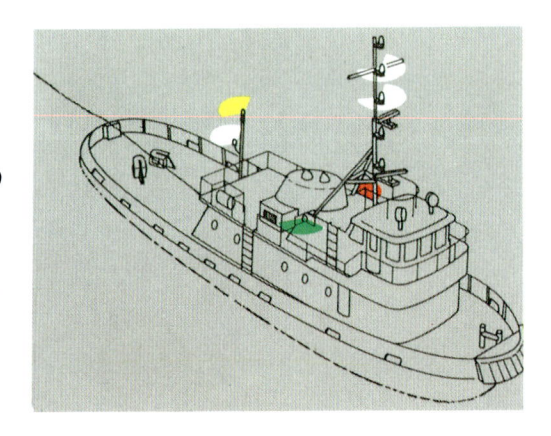

(b) When a pushing vessel and a vessel being pushed ahead are rigidly connected in a composite unit they shall be regarded as a power-driven vessel and exhibit the lights prescribed in Rule 23.

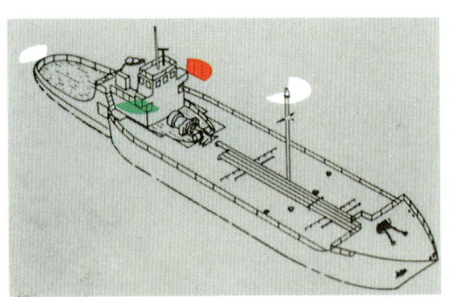

Two Composite Units under way. Left: Less than 50 m in length. Right: Any length.

Chapter 11: Collision Prevention Regulations

(c)   A power-driven vessel when pushing ahead or towing alongside, except in the case of a composite unit, shall exhibit:

   (i)   instead of the light prescribed in Rule 23 (a)(i) or (a)(ii), two masthead lights in a vertical line;

   (ii) sidelights;

   (iii) a sternlight.

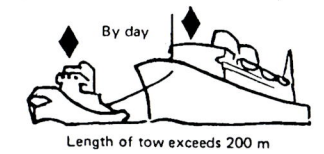

(d)   A power-driven vessel to which paragraph (a) or (c) of this Rule apply shall also comply with Rule 23(a)(ii).

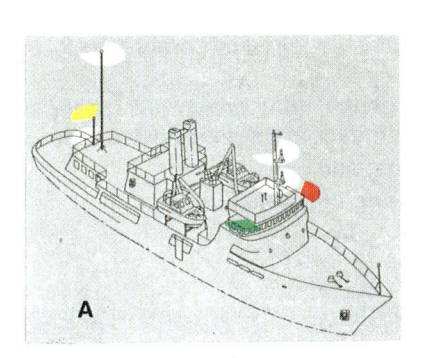

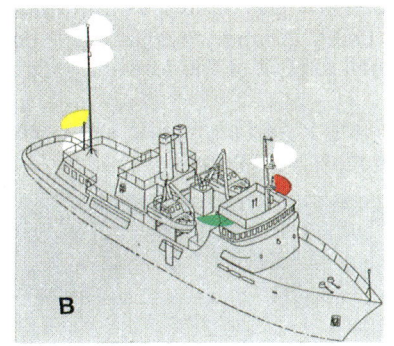

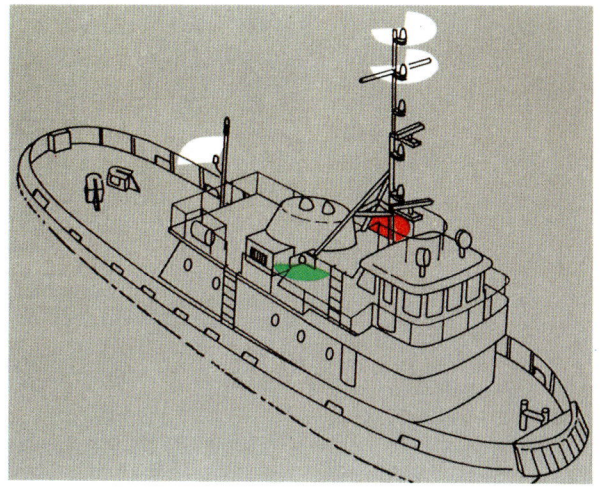

Above: P.D. vessel of less than 50 m - pushing ahead or towing alongside.
Left: Vessels A & B are towing astern. Length of tow 200 m or less. The after mast light on "A" is optional if she is less than 50 m in length. On "B" the masthead lights for towing (or pushing) are exhibited aft. A forward light is required.
Below: Vessel or object being towed.

(e)   A vessel or object being towed, other than those mentioned in paragraph (g) of this Rule, shall exhibit:

   (i)   sidelight;

   (ii) a sternlight;

   (iii) when the length of the tow exceeds 200 metres, a diamond shape where it can best be seen.

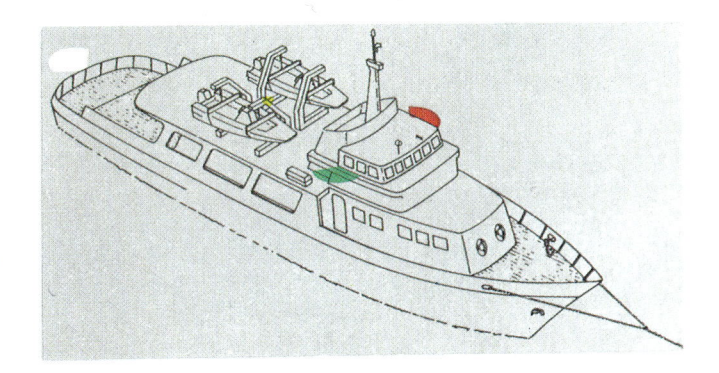

(f)   Provided that any number of vessels being towed alongside or pushed in a group shall be lighted as one vessel,

   (i)   a vessel being pushed ahead, not being part of a composite unit, shall exhibit at the forward end, sidelights;

   (ii) a vessel being towed alongside shall exhibit a sternlight and at the forward end, sidelights.

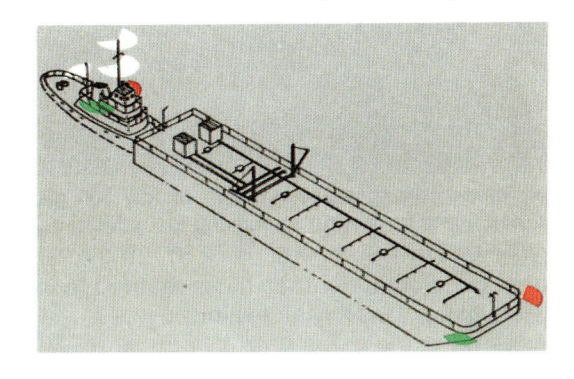

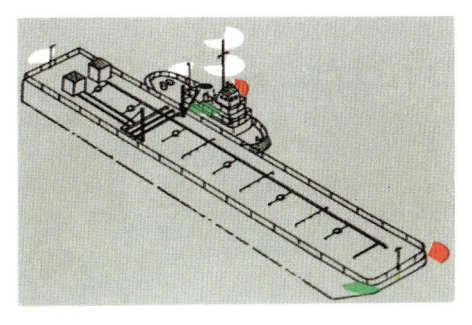

Far Left: Vessel being pushed ahead, not being part of a composite unit.
Left: Vessel being towed alongside

(g) An inconspicuous, partly submerged vessel or object, or combination of such vessels or objects being towed, shall exhibit:

   (i) if it is less than 25 metres in breadth, one all-round white light at or near the forward end and one at or near the after end except that dracones need not exhibit a light at or near the forward end;

   (ii) if it is 25 metres or more in breadth, two additional all-round white lights at or near the extremities of its breadth;

   (iii) if it exceeds 100 metres in length, additional all-round white lights between the lights prescribed in sub-paragraphs (i) and (ii) so that the distance between the lights shall not exceed 100 metres;

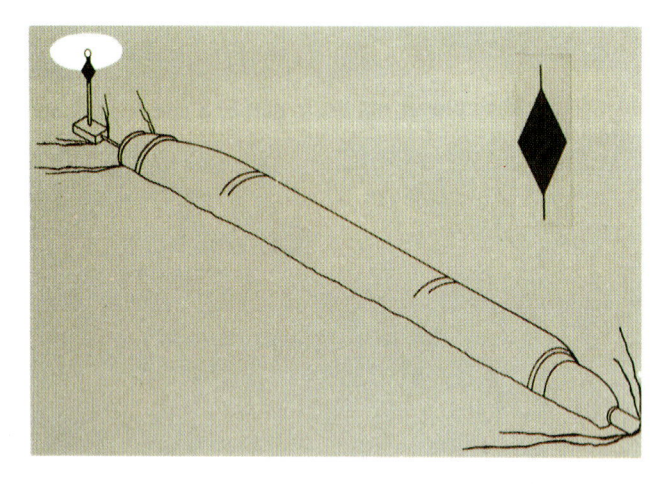

   (iv) a diamond shape at or near the aftermost extremity of the last vessel or object being towed and if the length of the tow exceeds 200 metres an additional diamond shape where it can best be seen and located as far forward as is practicable.

(h) Where from any sufficient cause it is impracticable for a vessel or object being towed to exhibit the lights or shapes prescribed in paragraph (e) or (g) of this Rule, all possible measures shall be taken to light the vessel or object towed or at least to indicate the presence of such vessel or object.

(i) Where from any sufficient cause it is impracticable for a vessel not normally engaged in towing operations to display the lights prescribed in paragraph (a) or (c) of this Rule, such vessel shall not be required to exhibit those lights when engaged in towing another vessel in distress or otherwise in need of assistance. All possible measures shall be taken to indicate the nature of the relationship between the towing vessel and the vessel being towed as authorized by Rule 36, in particular by illuminating the towline.

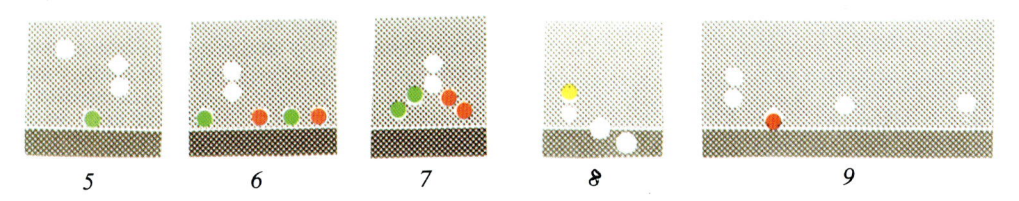

Quizzes on Rule 24
(See answers at the end of the chapter)

Rule 25
Sailing Vessels underway and Vessels under Oars

(a) A sailing vessel underway shall exhibit:

   (i) sidelights;

   (ii) a sternlight.

(b) In a sailing vessel of less than 20 metres in length the lights prescribed in paragraph (a) of this Rule may be combined in one lantern carried at or near the top of the mast where it can best be seen.

(c) A sailing vessel underway may, in addition to the lights prescribed in paragraph (a) of this Rule, exhibit at or near the top of the mast, where they can best be seen, two all-round lights in a vertical line, the upper being red and the lower green, but these lights shall not be exhibited in conjunction with the combined lantern permitted by paragraph (b) of this Rule.

Chapter 11: Collision Prevention Regulations

Sailing vessel underway

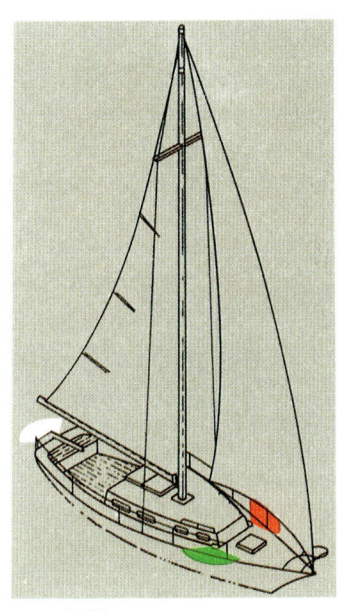

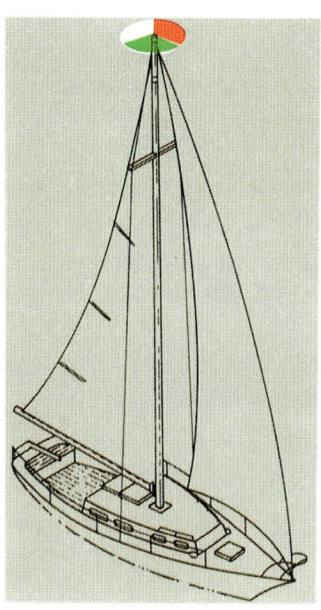

Sailing vessel underway, less than 20 m in length

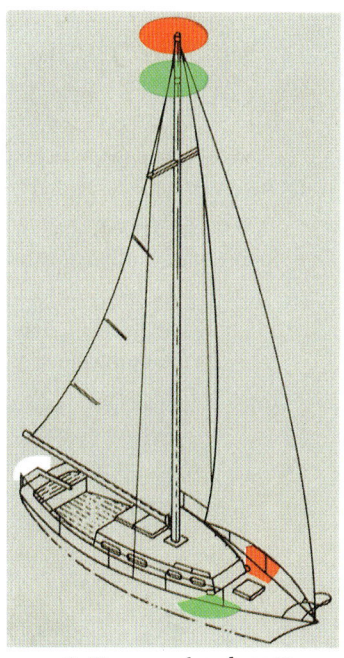

Sailing vessel underway

(d) (i) A sailing vessel of less than 7 metres in length shall, if practicable, exhibit the lights prescribed in paragraph (a) or (b) of this Rule, but if she does not, she shall have ready at hand an electric torch or lighted lantern showing a white light which shall be exhibited in sufficient time to prevent collision.

 (ii) A vessel under oars may exhibit the lights prescribed in this Rule for sailing vessels, but if she does not, she shall have ready at hand an electric torch or lighted lantern showing a white light which shall be exhibited in sufficient time to prevent collision.

(e) A vessel proceeding under sail when also being propelled by machinery shall exhibit forward where it can best be seen a conical shape, apex downwards.

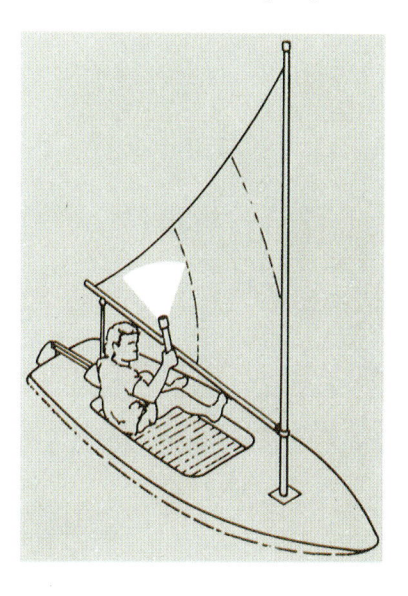

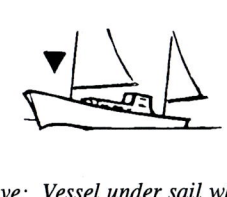

Above: Vessel under sail when also propelled by machinery
Far Left: Sailing vessel underway, less than 7 m in length
Left: Vessel under oars

Quizzes on Rule 25
(See answers at the end of the chapter)

10 *11* *12* *13*

The Australian Boating Manual

Rule 26
Fishing Vessels

(a) A vessel engaged in fishing, whether underway or at anchor, shall exhibit only the lights and shapes prescribed in this Rule.

(b) A vessel when engaged in trawling, by which is meant the dragging through the water of a dredge net or other apparatus used as a fishing appliance, shall exhibit:

  (i) two all-round lights in a vertical line, the upper being green and the lower white, or a shape consisting of two cones with their apexes together in a vertical line one above the other.

  (ii) a masthead light abaft of and higher than the all-round green light; a vessel of less than 50 metres in length shall not be obliged to exhibit such a light but may do so;

  (iii) when making way through the water, in addition to the lights prescribed in this paragraph, sidelights and a sternlight.

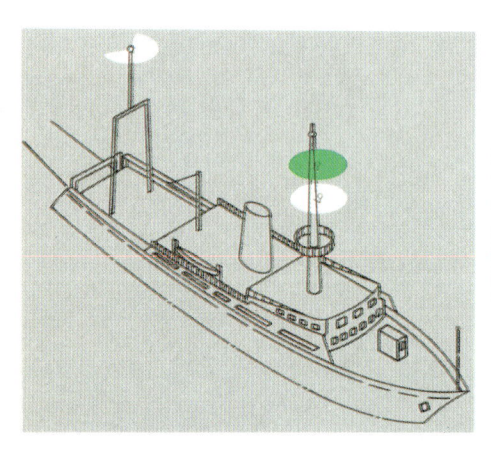

Left: Vessel of any length engaged in trawling - not making way.
Right: Vessel less than 50 m in length engaged in trawling - not making way.

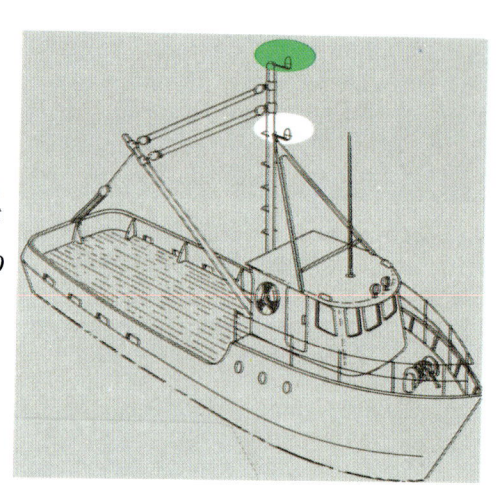

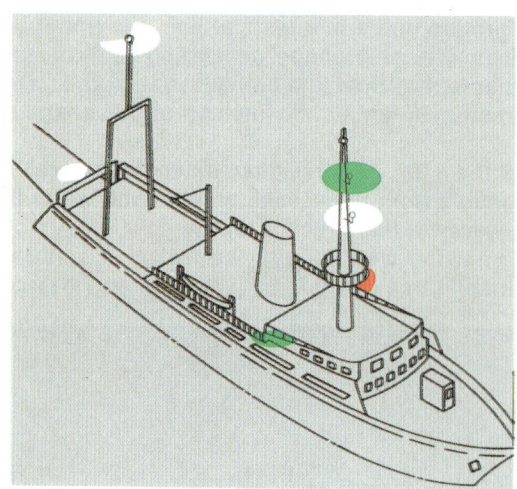

Left: Vessel of any length engaged in trawling - making way.
Right: Vessel less than 50 m in length engaged in trawling - making way.

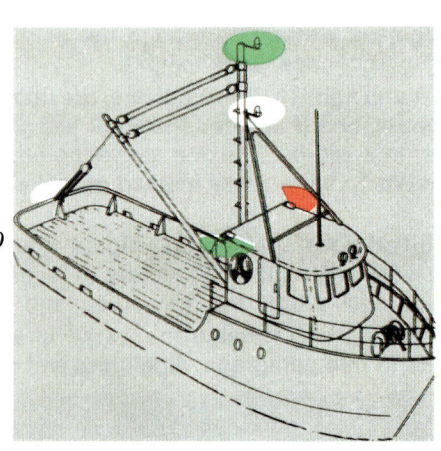

Chapter 11: Collision Prevention Regulations

(c) A vessel engaged in fishing, other than trawling, shall exhibit:

    (i)   two all-round lights in a vertical line, the upper being red and the lower white, or a shape consisting of two cones with apexes together in a vertical line one above the other.

    (ii)   when there is outlying gear extending more than 150 metres horizontally from the vessel, an all-round white light or a cone apex upwards in the direction of the gear;

    (iii)   when making way through the water, in addition to the lights prescribed in this paragraph, sidelights and a sternlight.

(d) The additional signals described in Annex II to these regulations apply to a vessel engaged in fishing in close proximity to other vessels engaged in fishing.

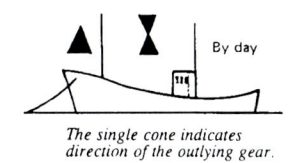

By day

The single cone indicates direction of the outlying gear.

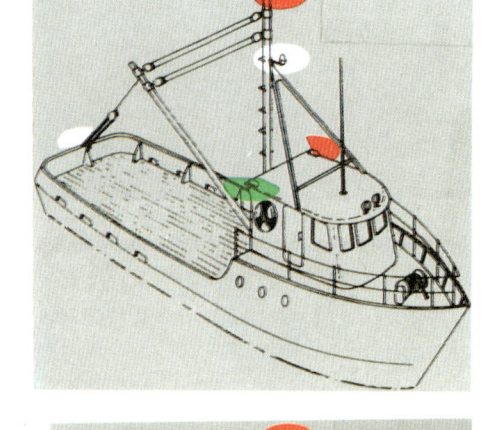

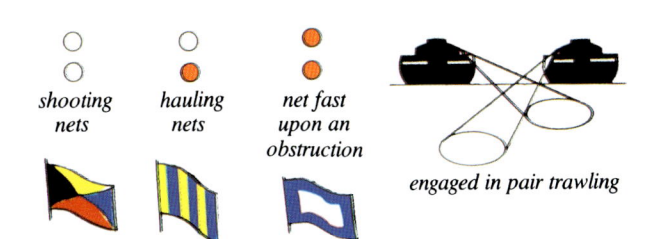

shooting nets

hauling nets

net fast upon an obstruction

engaged in pair trawling

(e) A vessel when not engaged in fishing shall not exhibit the lights or shapes prescribed in this rule, but only those prescribed for a vessel of her length.

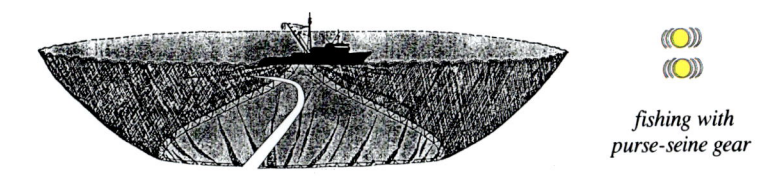

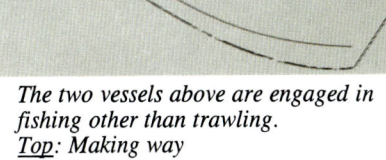

fishing with purse-seine gear

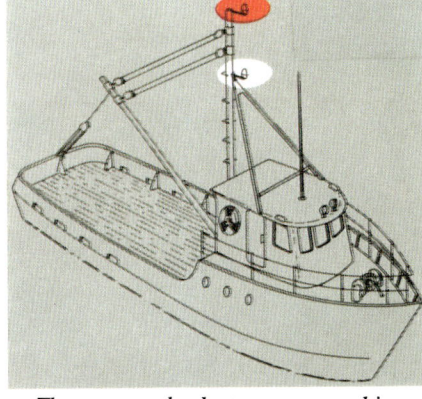

The two vessels above are engaged in fishing other than trawling.
Top: Making way
Above: Not making way.

Quizzes on Rule 26
(See answers at the end of the chapter)

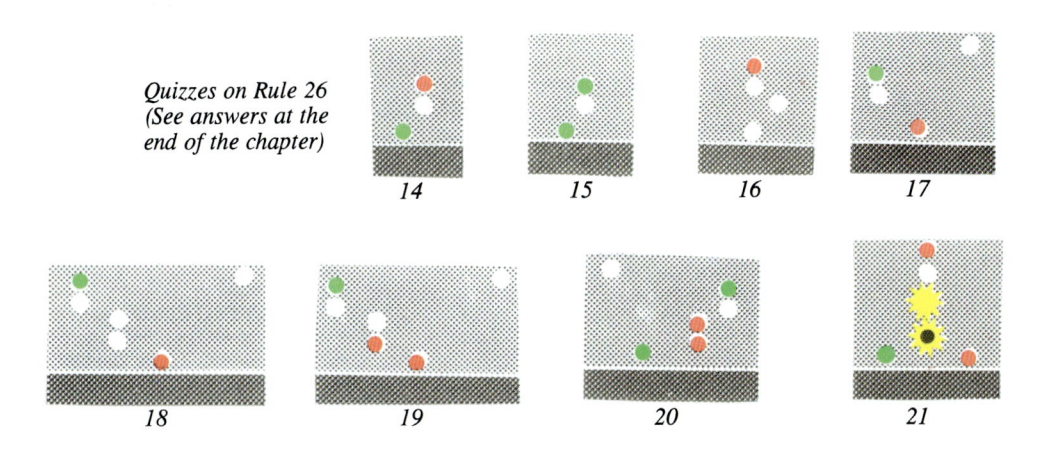

14   15   16   17

18   19   20   21

Rule 27
Vessels not under Command or Restricted in their Ability to Manoeuvre

(a) A vessel not under command shall exhibit:

    (i) two all-round red lights in a vertical line where they can best be seen;

    (ii) two balls or similar shapes in a vertical line where they can best be seen;

    (iii) when making way through the water, in addition to the lights prescribed in this paragraph, sidelights and a sternlight.

(b) A vessel restricted in her ability to manoeuvre, except a vessel engaged in mine clearance operations, shall exhibit:

    (i) three all-round lights in a vertical line where they can best be seen. The highest and lowest of these lights shall be red and the middle light shall be white;

    (ii) three shapes in a vertical line where they can best be seen. The highest and lowest of these shapes shall be balls and the middle one a diamond;

    (iii) when making way through the water, a masthead light or lights, sidelights and a sternlight, in addition to the lights prescribed in sub-paragraph (i);

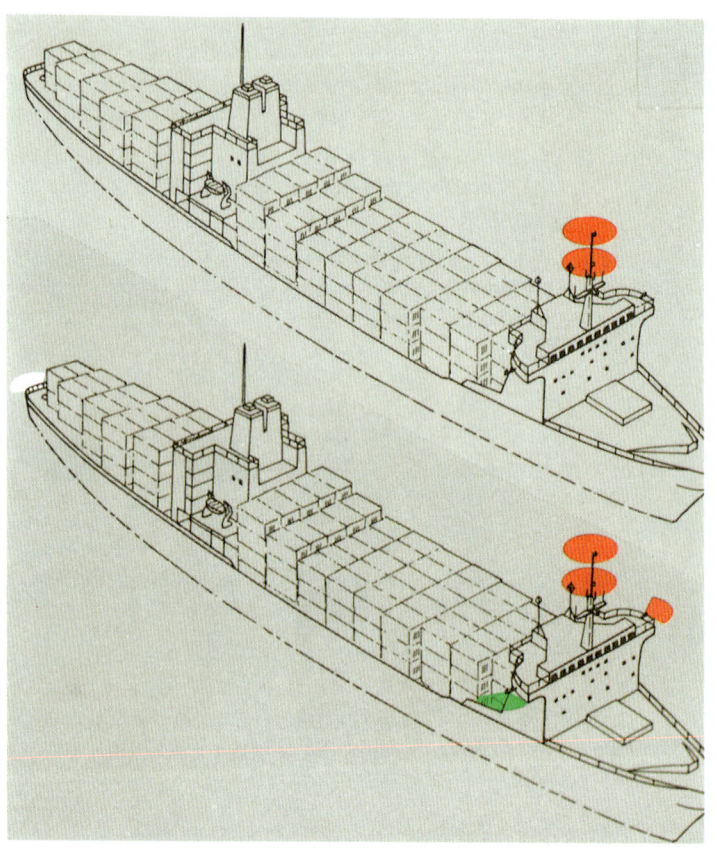

The two vessels above are Not Under Command.
Top: Not making way
Above: Making way

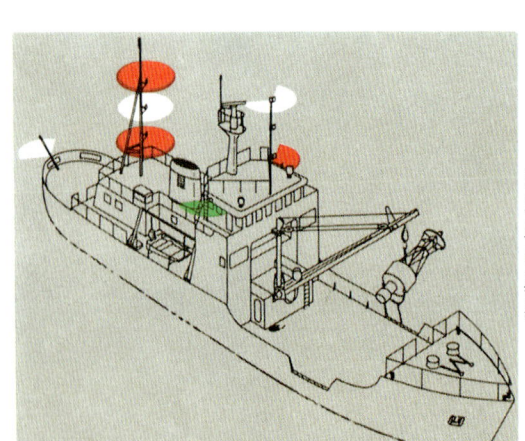

Two vessels less than 50 m in length - Restricted in Ability to Manoeuvre.
Left: Making way
Right: At anchor

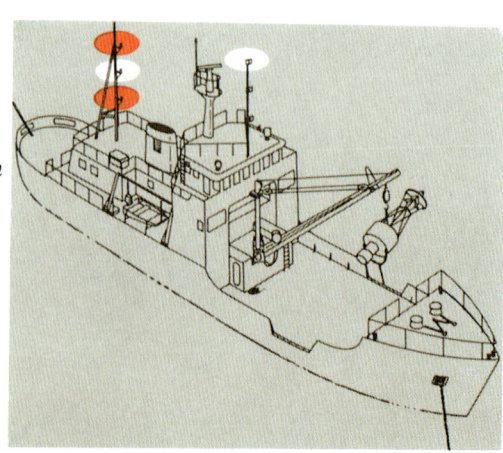

    (iv) when at anchor, in addition to the lights or shapes prescribed in sub-paragraphs (i) and (ii), the light, lights or shape prescribed in Rule 30.

Chapter 11: Collision Prevention Regulations

(c) A power-driven vessel engaged in a towing operation such as severely restricts the towing vessel and her tow in their ability to deviate from their course shall, in addition to the lights or shape prescribed in Rule 24(a), exhibit the lights or shapes prescribed in subparagraphs (b)(i) and (ii) of this Rule.

(d) A vessel engaged in dredging or underwater operations, when restricted in her ability to manoeuvre, shall exhibit the lights and shapes prescribed in sub-paragraphs (b)(i), (ii) and (iii) of this Rule and shall in addition, when an obstruction exists, exhibit:

 (i) two all-round red lights or two balls in a vertical line to indicate the side on which the obstruction exists;

 (ii) two all-round green lights or two diamonds in a vertical line to indicate the side on which another vessel may pass;

 (iii) when at anchor, the lights or shapes prescribed in this paragraph instead of the lights or shape prescribed in Rule 30.

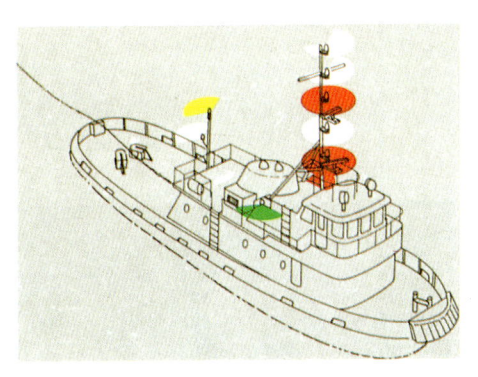

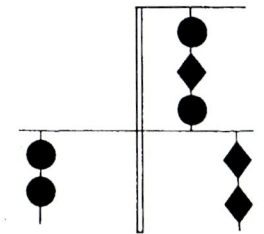

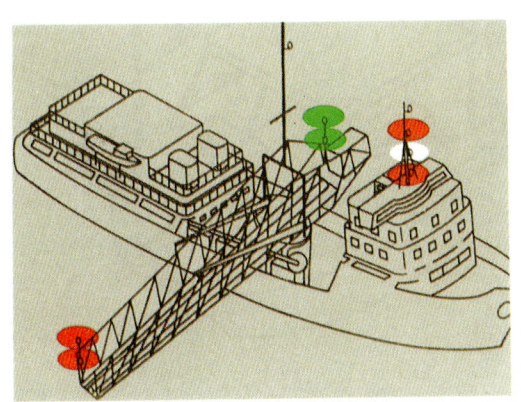

Vessels engaged in dredging or underwater operations when Restricted in ability to Manoeuvre. Obstruction on Starboard side.
Left: Not making way
Right: Making way

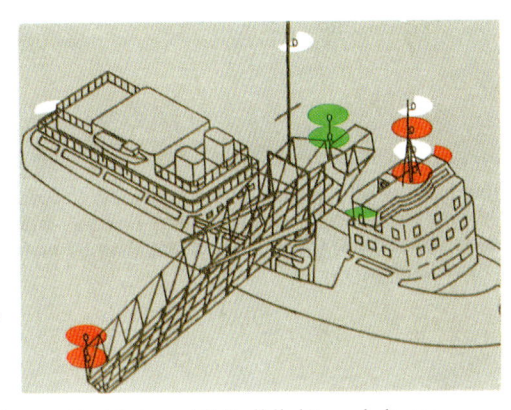

(e) Whenever the size of a vessel engaged in diving operations makes it impracticable to exhibit all lights and shapes prescribed in paragraph (d) of this Rule, the following shall be exhibited:

 (i) three all-round lights in a vertical line where they can best be seen. The highest and lowest of these lights shall be red and the middle light shall be white;

 (ii) a rigid replica of the International Code flag "A" not less than 1 metre in height. Measures shall be taken to ensure its all-round visibility.

(f) A vessel engaged in mine clearance operations shall in addition to the lights prescribed for a power-driven vessel in Rule 23 or to the lights or shape prescribed for a vessel at anchor in Rule 30 as appropriate, exhibit three all-round green lights or three balls. One of these lights or shapes shall be exhibited near the foremast head and one at each end of the fore yard. These lights or shapes indicate that it is dangerous for another vessel to approach within 1000 metres of the mine clearance vessel.

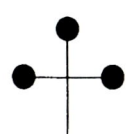

Vessels engaged in mine clearance operations.
Right: Any length
Far Right: Less than 50 m in length

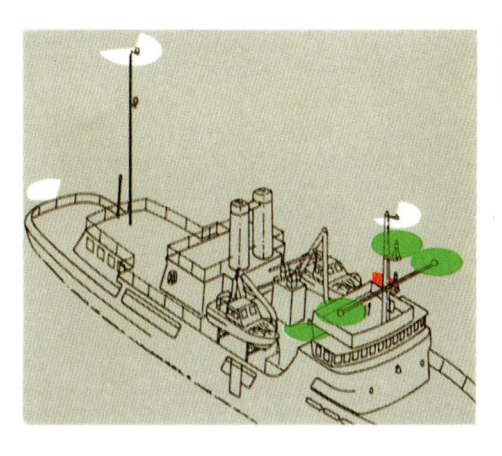

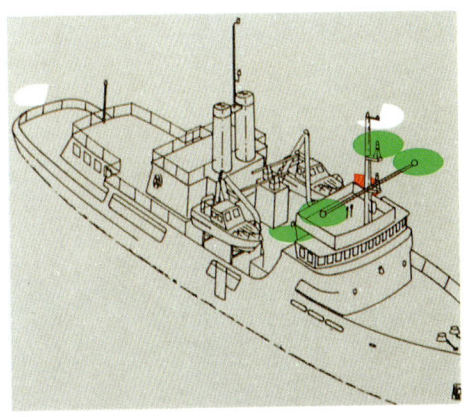

(g) Vessels of less than 12 metres in length, except those engaged in diving operations, shall not be required to exhibit the lights and shapes prescribed in this Rule.

(h) The signals prescribed in this Rule are not signals of vessels in distress and requiring assistance. Such signals are contained in Annex IV to these Regulations.

Rule 28
Vessels constrained by their Draught

A vessel constrained by her draught may, in addition to the lights prescribed for power driven vessels in Rule 23, exhibit where they can best be seen three all-round red lights in a vertical line, or a cylinder.

Rule 29
Pilot Vessels

(a) A vessel engaged on pilotage duty shall exhibit:

   (i) at or near the masthead, two all-round lights in a vertical line, the upper being white and the lower red;

   (ii) when underway, in addition, sidelights and a sternlight;

   (iii) when at anchor, in addition to the lights prescribed in sub-paragraph (i), the light, lights or shape prescribed in Rule 30 for vessels at anchor.

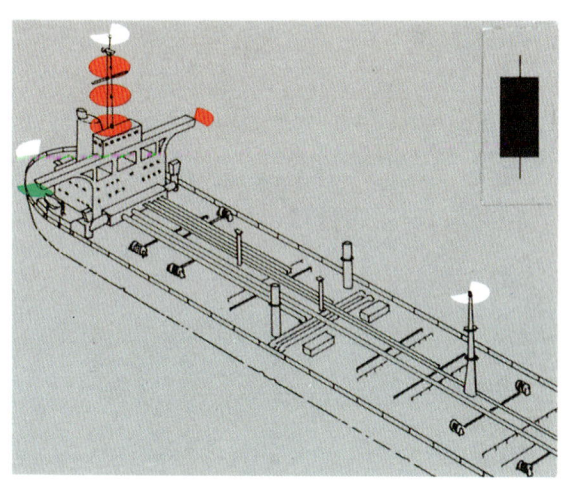

A Vessel constrained by her draught

(b) A pilot vessel when not engaged on pilotage duty shall exhibit the lights or shapes prescribed for a similar vessel of her length.

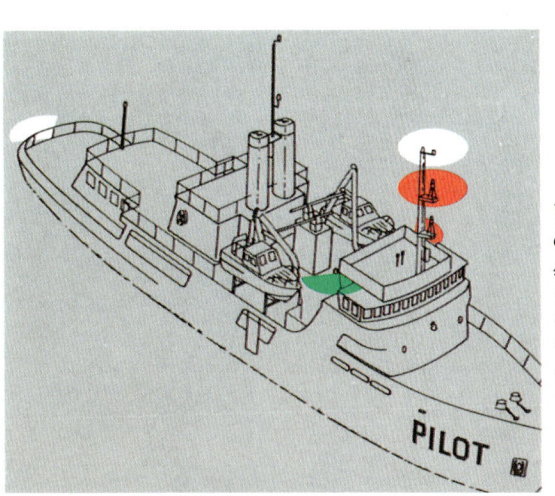

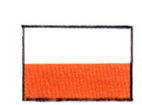

Vessels engaged on pilotage duty.
Above: Day signal - a flag.
Left: Any length - underway
Right: Less than 50 m in length - at anchor.

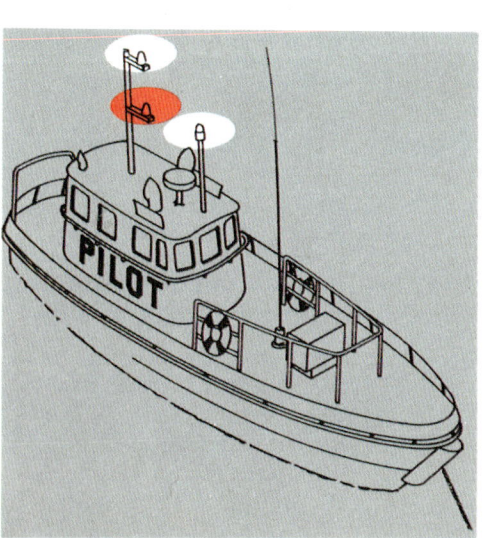

See Quizzes after Rule 30

Chapter 11: Collision Prevention Regulations

Rule 30
Anchored Vessels and Vessels aground

(a) A vessel at anchor shall exhibit where it can best be seen:

    (i) in the fore part, an all-round white light or one ball;

    (ii) at or near the stern and at a lower level than the light prescribed in sub-paragraph (i), an all-round white light.

(b) A vessel of less than 50 metres in length may exhibit an all-round white light where it can best be seen instead of the lights prescribed in paragraph (a) of this Rule.

(c) A vessel at anchor may, and a vessel of 100 metres and more in length shall, also use the available working or equivalent lights to illuminate her decks.

(d) A vessel aground shall exhibit the lights prescribed in paragraph (a) or (b) of this Rule and in addition, where they can best be seen:

    (i) two all-round red lights in a vertical line;

    (ii) three balls in a vertical line.

(e) A vessel of less than 7 metres in length, when at anchor, not in or near a narrow channel, fairway or anchorage, or where other vessels normally navigate, shall not be required to exhibit the lights or shape prescribed in paragraphs (a) and (b) of this Rule.

(f) A vessel of less than 12 metres in length, when aground, shall not be required to exhibit the lights or shapes prescribed in sub-paragraphs (d)(i) and (ii) of this Rule.

Rule 31
Seaplanes

Where it is impracticable for a seaplane to exhibit lights and shapes of the characteristics or in the positions prescribed in the Rules of this Part she shall exhibit lights and shapes as closely similar in characteristics and position as is possible.

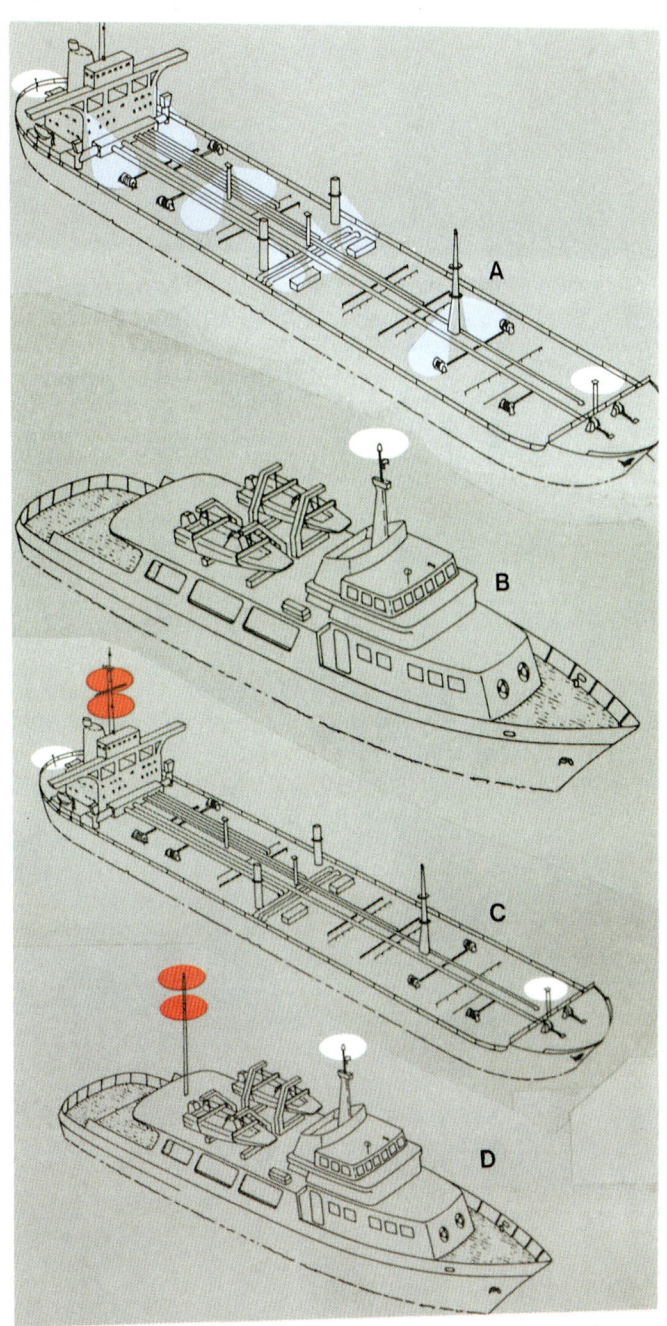

(A) Vessel of any length - at anchor with deck illumination.
(B) Vessel less than 50 m in length - at anchor.
(C) Vessel of any length - aground.
(D) Vessel less than 50 m in length - aground.

See Quizzes next page

LOCAL SIGNALS
(International Collision Regulations continue after this page)

PRIORITY OVER SAIL
In Sydney harbour, a ferry displaying an orange diamond shape has priority over sail. A sailing vessel underway must keep clear of a ferry displaying an orange diamond. No vessel may display this signal without written consent from the NSW Waterways Authority. There is no night signal for these ferries.

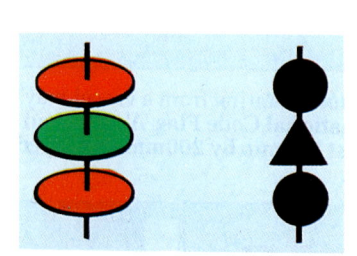

PORT CLOSED / CHANNEL BLOCKED SIGNAL
When a port is closed or channel blocked or partly blocked by a vessel or other object, it is indicated by at night by a red light over a green light over a red light and during the day by a black ball over a black cone, point upward, over a black ball.

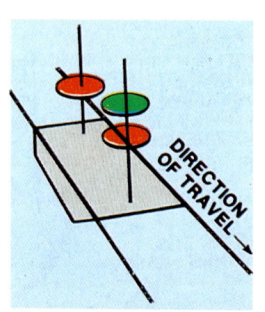

VESSELS WORKING IN WIRES & CROSS RIVER FERRIES
In NSW, vessels working in wires (Punts) show an all-round red light at each end and an all-round green light above and at the forward end, to indicate the direction in which the vessel is moving.

INTERNATIONAL & LOCAL DISTRESS SIGNALS

The following signals are internationally recognised and indicate Distress and need of assistance. Use of these Signals except for the purpose indicated is prohibited.

1. Rockets or shells, throwing red stars fired one at a time at short intervals.
2. (a) A signal made by radio or by any other signalling method consisting of the group in the Morse Code. **S.O.S.** ••• ——— •••
   (b) A signal sent by radio consisting of the spoken word – "MAYDAY".
3. A square flag having above or below it a ball or anything resembling a ball.
4. A rocket parachute flare or a hand-held flare showing a red light.
5. A smoke signal giving off orange-coloured smoke.
6. Slowly and repeatedly raising and lowering arms outstretched to each side.
7. (a) A rectangle of international orange material with a black letter V; or
   (b) A black square and circle
8. A dye marker.
9. The International Code Signal of Distress indicated by N.C.
10. Continuous sounding of sound signalling equipment – "SOS".
11. EPIRB (Emergency Position Indicating Radio Beacon.)
12. Oar with cloth on end.

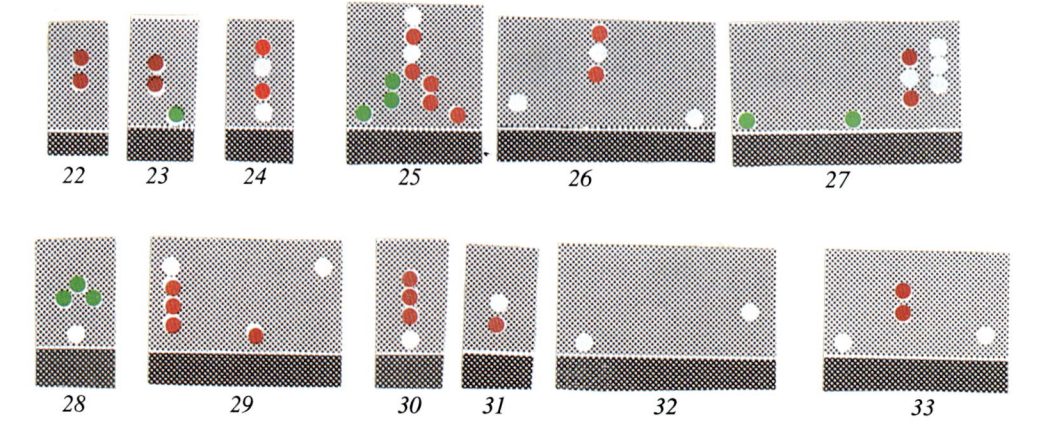

QUIZZES ON RULES 27 to 30 *(See answers at the end of the chapter)*

22 23 24 25 26 27

28 29 30 31 32 33

Chapter 11: Collision Prevention Regulations

Part D -- Sound And Light Signals

Rule 32
Definitions

(a) The word "whistle" means any sound signalling appliance capable of producing the prescribed blasts and which complies with the specifications in Annex III to these Regulations.

(b) The term "short blast" means a blast of about one second's duration.

(c) The term "prolonged blast" means a blast of from four to six seconds' duration.

Rule 33
Equipment for Sound Signals

(a) A vessel of 12 metres or more in length shall be provided with a whistle and a bell and a vessel of 100 metres or more in length shall, in addition, be provided with a gong, the tone and sound of which cannot be confused with that of the bell. The whistle, bell and gong shall comply with the specifications in Annex III to these Regulations. The bell or gong or both may be replaced by other equipment having the same respective sound characteristics, provided that manual sounding of the prescribed signals shall always be possible.

(b) A vessel of less than 12 metres in length shall not be obliged to carry the sound signalling appliances prescribed in paragraph (a) of this Rule but if she does not, she shall be provided with some other means of making an efficient sound signal.

Rule 34
Manoeuvring and Warning Signals

(a) When vessels are in sight of one another, a power-driven vessel underway, when manoeuvring as authorized or required by these Rules, shall indicate that manoeuvre by the following signals on her whistle:

- one short blast to mean "I am altering my course to starboard";
- two short blasts to mean "I am altering my course to port";
- three short blasts to mean "I am operating astern propulsion".

(b) Any vessel may supplement the whistle signals prescribed in paragraph (a) of this Rule by light signals, repeated as appropriate, whilst the manoeuvre is being carried out:

  (i) these light signals shall have the following significance:

  - one flash to mean "I am altering my course to starboard";
  - two flashes to mean "I am altering my course to port";
  - three flashes to mean "I am operating astern propulsion";

  (ii) the duration of each flash be about one second, the interval between flashes shall be about one second, and the interval between successive signals shall be not less than ten seconds;

  (iii) the light used for this signal shall, if fitted, be an all-round white light, visible at a minimum range of 5 miles, and shall comply with the provisions of Annex I to these Regulations.

(c) When in sight of one another in a narrow channel or fairway:

  (i) a vessel intending to overtake another shall in compliance with Rule 9(e)(i) indicate her intention by the following signals on her whistle:

  - two prolonged blasts followed by one short blast to mean "I intend to overtake you on your starboard side";
  - two prolonged blasts followed by two short blasts to mean "I intend to overtake you on your port side";

  (ii) the vessel about to be overtaken when acting in accordance with Rule 9(e) (i) shall indicate her agreement by the following signal on her whistle:

  - one prolonged, one short, one prolonged and one short blast, in that order.

| I intend to overtake you on your starboard side. | I intend to overtake you on your port side. | I agree to being overtaken. | I am in doubt about your signal, intentions or the safety of the manoeuvre. |

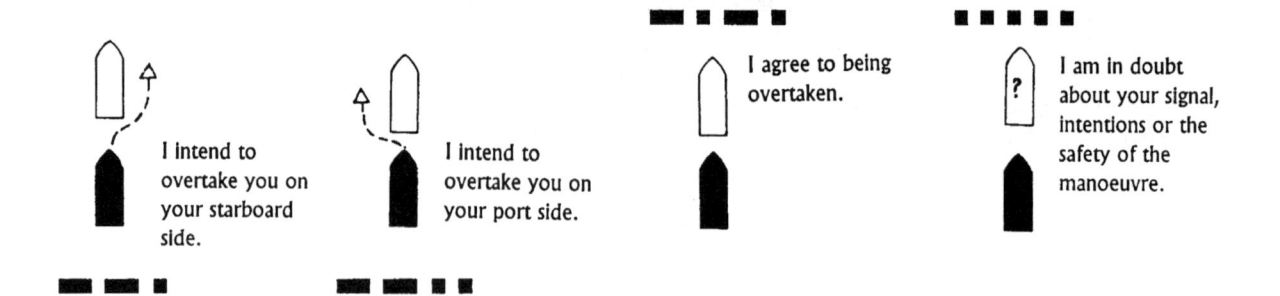

(d) When vessels in sight of one another are approaching each other and from any cause either vessel fails to understand the intentions or actions of the other, or is in doubt whether sufficient action is being taken by the other to avoid collision, the vessel in doubt shall immediately indicate such doubt by giving at least five short and rapid blasts on the whistle. Such signal may be supplemented by a light signal of at least five short and rapid flashes.

(e) A vessel nearing a bend or an area of a channel or fairway where other vessels may be obscured by an intervening obstruction shall sound one prolonged blast. Such signal shall be answered with a prolonged blast by any approaching vessel that may be within hearing around the bend or behind the intervening obstruction.

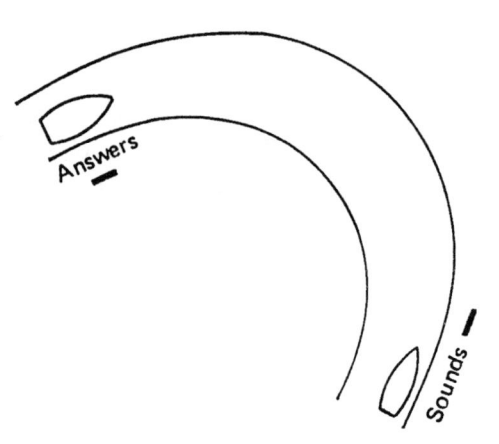

(f) If whistles are fitted on a vessel at a distance apart of more than 100 metres, one whistle only shall be used for giving manoeuvring and warning signals.

Rule 35
Sound Signals in restricted Visibility

In or near an area of restricted visibility, whether by day or night, the signals prescribed in this Rule shall be used as follows:

(a) A power-driven vessel making way through the water shall sound at intervals of not more than 2 minutes one prolonged blast.

(b) A power-driven vessel underway but stopped and making no way through the water shall sound at intervals of not more than 2 minutes two prolonged blasts in succession with an interval of about 2 seconds between them.

(c) A vessel not under command, a vessel restricted in her ability to manoeuvre, a vessel constrained by her draught, a sailing vessel, a vessel engaged in fishing and a vessel engaged in towing or pushing another vessel shall, instead of the signals prescribed in paragraphs (a) or (b) of this Rule, sound at intervals of not more than 2 minutes three blasts in succession, namely one prolonged followed by two short blasts.

(d) A vessel engaged in fishing, when at anchor, and a vessel restricted in her ability to manoeuvre when carrying out

Chapter 11: Collision Prevention Regulations

her work at anchor, shall instead of the signals prescribed in paragraph (g) of this Rule sound the signal prescribed in paragraph (c) of this Rule.

(e) A vessel towed or if more than one vessel is towed the last vessel of the tow, if manned shall at intervals of not more than 2 minutes sound four blasts in succession, namely one prolonged followed by three short blasts. When practicable, this signal shall be made immediately after the signal made by the towing vessel. ▬ ■ ■ ■

(f) When a pushing vessel and a vessel being pushed ahead are rigidly connected in a composite unit they shall be regarded as a power-driven vessel and shall give the signals prescribed in paragraphs (a) or (b) of this Rule.

(g) A vessel at anchor shall at intervals of not more than one minute ring the bell rapidly for about 5 seconds. In a vessel of 100 metres or more in length the bell shall be sounded in the forepart of the vessel and immediately after the ringing of the bell the gong shall be sounded rapidly for about 5 seconds in the after part of the vessel. A vessel at anchor may in addition sound three blasts in succession, namely one short, one prolonged and one short blast, to give warning of her position and of the possibility of collision to an approaching vessel.

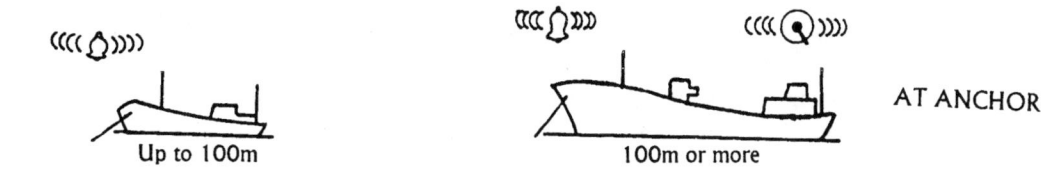

AT ANCHOR

Up to 100m   100m or more

(h) A vessel aground shall give the bell signal and if required the gong signal prescribed in paragraph (g) of this Rule and shall, in addition, give three separate and distinct strokes on the bell immediately before and after the rapid ringing of the bell. A vessel aground may in addition sound an appropriate whistle signal.

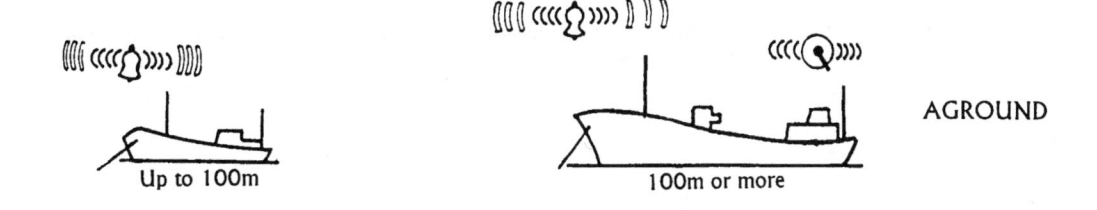

AGROUND

Up to 100m   100m or more

(i) A vessel of less than 12 metres in length shall not be obliged to give the above-mentioned signals but, if she does not, shall make some other efficient sound signal at intervals of not more than 2 minutes.

(j) A pilot vessel when engaged on pilotage duty may in addition to the signals prescribed in paragraphs (a), (b) or (g) of this Rule sound an identity signal consisting of four short blasts. ▬ ■ ■ ■ ■
                                                                                ▬ ■ ■ ■ ■

Rule 36
Signals to attract Attention

If necessary to attract the attention of another vessel any vessel may make light or sound signals that cannot be mistaken for any signal authorized elsewhere in these Rules, or may direct the beam of her searchlight in the direction of the danger, in such a way as not to embarrass any vessel.

Any light to attract the attention of another vessel shall be such that it cannot be mistaken for any aid to navigation. For the purpose of this Rule the use of high intensity intermittent or revolving lights, such as strobe lights, shall be avoided.

Rule 37
Distress Signals

When a vessel is in distress and requires assistance she shall use or exhibit the signals described in Annex IV to these Regulations.

Part E -- Exemptions

Rule 38
Exemptions

Any vessel (or class of vessels) provided that she complies with the requirements of the International Regulations for Preventing Collisions at Sea, 1960, the keel of which is laid or which is at a corresponding stage of construction before the entry into force of these Regulations may be exempted from compliance therewith as follows:

(a)   The installation of lights with ranges prescribed in Rule 22, until four years after the date of entry into force of these Regulations.

(b)   The installation of lights with colour specifications as prescribed in Section 7 of Annex I to these Regulations, until four years after the date of entry into force of these Regulations.

(c)   The repositioning of lights as a result of conversion from Imperial to metric units and rounding off measurements figures, permanent exemption.

(d)   (i)   The repositioning of masthead lights on vessels of less than 150 metres in length, resulting from the prescriptions of Section 3(a) of Annex I to these Regulations, permanent exemption.

           (ii)   The repositioning of masthead lights on vessels of 150 metres or more in length, resulting from the prescriptions of Section 3(a) of Annex I to these Regulations, until nine years after the date of entry into force of these Regulations.

(e)   The repositioning of masthead lights resulting from the prescriptions of Section 2(b) of Annex I to these Regulations, until nine years after the date of entry into force of these Regulations.

(f)   The repositioning of sidelights resulting from the prescriptions of Sections 2(g) and 3(b) of Annex I to these Regulations, until nine years after the date of entry into force of these Regulations.

(g)   The requirements for sound signal appliances prescribed in Annex III to these Regulations, until nine years after the date of entry into force of these Regulations.

(h)   The repositioning of all-round lights resulting from the prescription of Section 9(b) of Annex I to these Regulations, permanent exemption.

ANNEX I
POSITIONING AND TECHNICAL DETAILS OF LIGHTS AND SHAPES

1.   **Definition**

The term "height above the hull" means height above the uppermost continuous deck. This height shall be measured from the position vertically beneath the location of the light.

2.   **Vertical positioning and spacing of lights**

(a)   On a power-driven vessel of 20 metres a more in length the masthead lights shall be placed as follows:

           (i)   the forward masthead light, or if only one masthead light is carried, then that light, at a height above the hull of not less than 6 metres, and, if the breadth of the vessel exceeds 6 metres, then at a height above the hull not less than such breadth, so however that the light need not be placed at a greater height above the hull than 12 metres;

           (ii)   when two mast-head lights are carried the after one shall be at least 4.5 metres vertically higher than the forward one.

(b)   The vertical separation of masthead lights of power-driven vessels shall be such that in all normal conditions of trim the after light will be seen over and separate from the forward light at a distance of 1000 metres from the stem

Chapter 11: Collision Prevention Regulations

when viewed from sea level.

(c) The masthead light of a power-driven vessel of 12 metres but less than 20 metres in length shall be placed at a height above the gunwale of not less than 2.5 metres.

(d) A power-driven vessel of less than 12 metres in length may carry the uppermost light at a height of less than 2.5 metres above the gunwale. When however a masthead light is carried in addition to sidelights and a sternlight or the all-round light prescribed in rule 23(c)(i) is carried in addition to sidelights, then such masthead light or all-round light shall be carried at least 1 metre higher than the sidelights.

(e) One of the two or three masthead lights prescribed for a power-driven vessel when engaged in towing or pushing another vessel shall be placed in the same position as either the forward masthead light or the after masthead light; provided that, if carried on the aftermast, the lowest after masthead light shall be at least 4.5 metres vertically higher than the forward masthead light.

(f) (i) The masthead light or lights prescribed in Rule 23(a) shall be so placed as to be above and clear of all other lights and obstructions except as described in sub-paragraph (ii).

(ii) When it is impracticable to carry the all-round lights prescribed by Rule 27(b)(i) or Rule 28 below the masthead lights, they may be carried above the after masthead light(s) or vertically in between the forward masthead light(s) and after masthead light(s), provided that in the latter case the requirement of Section 3(c) of this Annex shall be complied with.

(g) The sidelights of a power-driven vessel shall be placed at a height above the hull not greater than three quarters of that of the forward masthead light. They shall not be so low as to be interfered with by deck lights.

(h) The sidelights, if in a combined lantern and carried on a power-driven vessel of less than 20 metres in length, shall be placed not less than 1 metre below the masthead light.

(i) When the Rules prescribe two or three lights to be carried in a vertical line, they shall be spaced as follows:

(i) on a vessel of 20 metres in length or more such lights shall be spaced not less than 2 metres apart, and the lowest of these lights shall, except where a towing light is required, be placed at a height of not less than 4 metres above the hull;

(ii) on a vessel of less than 20 metres in length such lights shall be spaced not less than 1 metre apart and the lowest of these lights shall, except where a towing light is required, be placed at a height of not less than 2 metres above the gunwale;

(iii) when three lights are carried they shall be equally spaced.

(j) The lower of the two all-round lights prescribed for a vessel when engaged in fishing shall be at a height above the sidelights not less than twice the distance between the two vertical lights.

(k) The forward anchor light prescribed in Rule 30 (a) (i), when two are carried, shall not be less than 4.5 metres above the after one. On a vessel of 50 metres or more in length this forward anchor light shall be placed at a height of not less than 6 metres above the hull.

3. Horizontal positioning and spacing of lights

(a) When two masthead lights are prescribed for a power-driven vessel, the horizontal distance between them shall not be less than one half of the length of the vessel but need not be more than 100 metres. The forward light shall be placed not more than one quarter of the length of the vessel from the stem.

(b) On a power-driven vessel of 20 metres or more in length the sidelights shall not be placed in front of the forward masthead lights. They shall be placed at or near the side of the vessel.

(c) When the lights prescribed in Rule 27 (b) (i) or Rule 28 are placed vertically between the forward masthead light(s) and the after masthead light(s) these all-round lights shall be placed at a horizontal distance of not less than 2 metres from the fore and aft centreline of the vessel in the athwartship direction.

(d) When only one masthead light is prescribed for a power driven vessel, this light shall be exhibited forward of amidships; except that a vessel of less than 20 metres in length need not exhibit this light forward of amidships but shall exhibit it as far forward as is practicable.

4. **Details of location of direction-indicating lights for fishing vessels, dredges and vessels engaged in underwater operations**

(a) The light indicating the direction of the outlying gear from a vessel engaged in fishing as prescribed in Rule 26 (c) (ii) shall be placed at a horizontal distance of not less than 2 metres and not more than 6 metres away from the two all-round red and white lights. This light shall be placed not higher than the all-round white light prescribed in Rule 26 (c) (i) and not lower than the sidelights.

(b) The lights and shapes on a vessel engaged in dredging or underwater operations to indicate the obstructed side and/or the side on which it is safe to pass, as prescribed in Rule 27 (d) (i) and (ii), shall be placed at the maximum practical horizontal distance, but in no case less than 2 metres, from the lights or shapes prescribed in Rule 27 (b) (i) and (ii). In no case shall the upper of these lights or shapes be at a greater height than the lower of the three lights or shapes prescribed in Rule 27 (b) (i) and (ii).

5. **Screens for sidelights**

The sidelights of vessels of 20 metres or more in length shall be fitted with inboard screens painted matt black, and meeting the requirements of Section 9 of this Annex. On vessels of less than 20 metres in length the sidelights, if necessary to meet the requirements of Section 9 of this Annex, shall be fitted with inboard matt black screens. With a combined lantern, using a single vertical filament and a very narrow division between the green and red sections, external screens need not be fitted.

6. **Shapes**

(a) Shapes shall be black and the following sizes:
- (i) a ball shall have a diameter of not less than 0.6 metre;
- (ii) a cone shall have a base diameter of not less than 0.6 metre and a height equal to its diameter;
- (iii) a cylinder shall have a diameter of at least 0.6 metre and a height of twice its diameter;
- (iv) a diamond shape shall consist of two cones as defined in (ii) above having a common base.

(b) The vertical distance between shapes shall be at least 1.5 metres.

(c) In a vessel of less than 20 metres in length shapes of lesser dimensions, but commensurate with the size of the vessel may be used and the distance apart may be correspondingly reduced.

7. **Colour specification of lights**

The chromaticity of all navigation lights shall conform to the following standards, which lie within the boundaries of the area of the diagram specified for each colour by the International Commission on Illumination (CIE).

The boundaries of the area for each colour are given by indicating the corner co-ordinates which are as follows:

(i) *White*

x	0.525	0.525	0.452	0.310	0.310	0.443
y	0.382	0.440	0.440	0.348	0.283	0.382

(ii) *Green*

x	0.028	0.009	0.300	0.203
y	0.385	0.723	0.511	0.356

(iii) *Red*

x	0.680	0.660	0.735	0.721
y	0.320	0.320	0.265	0.259

(iv) *Yellow*

x	0.612	0.618	0.575	0.575
y	0.382	0.382	0.425	0.406

8. **Intensity of lights**

(a) The minimum luminous intensity of lights shall be calculated by using the formula:

where: **I** is luminous intensity in candelas, under service conditions,
T is threshold factor
D is range of visibility (luminous range) of the light in nautical miles,
K is atmospheric transmissivity.

For prescribed lights the value of K shall be 0.8, corresponding to a meteorological visibility of approximately 13 nautical miles.

Chapter 11: Collision Prevention Regulations

(b) A selection of figures derived from the formula is given in the following table:

Range of visibility (luminous range) of light in nautical miles D	Luminous intensity of light in candelas for K = 0.8 I
1	0.9
2	4.3
3	12.0
4	27.0
5	52.0
6	94.0

NOTE: The maximum luminous intensity of navigation lights should be limited to avoid undue glare. This shall not be achieved by a variable control of the luminous intensity.

9. Horizontal sectors

(a) (i) In the forward direction, sidelights as fitted on the vessel shall show the minimum required intensities. The intensities shall decrease to reach practical cut-off between 1 degree and 3 degrees outside the prescribed sectors.

(ii) For sternlights and masthead lights and at 22.5 degrees abaft the beam for sidelights, the minimum required intensities shall be maintained over the arc of the horizon up to 5 degrees within the limits of the sectors prescribed in Rule 21. From 5 degrees within the prescribed sectors the intensity may decrease by 50 pr cent up to the prescribed limits; it shall decrease steadily to reach practical cut-off at not more than 5 degrees outside the prescribed sectors.

(b) (i) All-round lights shall be so located as not to be obscured by masts, topmasts or structures within angular sectors of more than 6 degrees, except anchor lights prescribed in Rule 30, which need not be placed at an impracticable height above the hull.

(ii) If it is impracticable to comply with paragraph (b) (i) of this section by exhibiting only one all-round light, two all-round lights shall be used suitably positioned or screened so that they appear, as far as practicable, as one light at a distance of one mile.

10. Vertical sectors

(a) The vertical sectors of electric lights as fitted, with the exception of lights on sailing vessels underway shall ensure that:

(i) at least the required minimum intensity is maintained at all angles from 5 degrees above to 5 degrees below the horizontal;

(ii) at least 60 per cent of the required minimum intensity is maintained from 7.5 degrees above to 7.5 degrees below the horizontal.

(b) In the case of sailing vessels underway the vertical sectors of electric lights as fitted shall ensure that:

(i) at least the required minimum intensity is maintained at all angles from 5 degrees above to 5 degrees below the horizontal;

(ii) at least 50 per cent of the required minimum intensity is maintained from 25 degrees above to 25 degrees below the horizontal.

(c) In the case of lights other than electric these specifications shall be met as closely as possible.

11. Intensity of non-electric lights

Non-electric lights shall so far as practicable comply with the minimum intensities, as specified in the Table given in Section 8 of this Annex.

12. Manoeuvring light

Notwithstanding the provisions of paragraph 2(f) of this Annex the manoeuvring light described in Rule 34(b) shall be placed in the same fore and aft vertical plane as the masthead light or lights and, where practicable, at a minimum height of 2 metres vertically above the forward masthead light, provided that it shall be carried not less than 2 metres vertically above or below the after masthead light. On a vessel where only one masthead light is carried the manoeuvring light, if fitted, shall be carried where it can best be seen, not less than 2 metres vertically apart from the masthead light.

13. High Speed Craft

The masthead light of high speed craft with a length to breadth ratio of less than 3.0 may be placed at a height related to the breadth of the craft lower than that prescribed in paragraph 2(a)(i) of this annex, provided that the base angle of the isosceles triangles formed by the sidelights and masthead light, when seen in end elevation, is not less than 27 degrees.

14. Approval

The construction of lights and shapes and the installation of lights on board the vessel shall be to the satisfaction of the appropriate authority of the State whose flag the vessel is entitled to fly.

ANNEX II
ADDITIONAL SIGNALS FOR FISHING VESSELS FISHING IN CLOSE PROXIMITY

1. General

The lights mentioned herein shall, if exhibited in pursuance of Rule 26(d), be placed where they can best be seen. They shall be at least 0.9 metre apart but at a lower level than lights prescribed in Rule 26 (b) (i) and (c) (i). The lights shall be visible all round the horizon at a distance of at least 1 mile but at a lesser distance than the lights prescribed by these Rules for fishing vessels.

2. Signals for trawlers

(a) Vessels of 20 metres or more in length when engaged in trawling, whether using demersal or pelagic gear, shall exhibit:
- (i) when shooting their nets: two white lights in a vertical line;
- (ii) when hauling their nets: one white light over one red light in a vertical line;
- (iii) when the net has come fast upon an obstruction: two red lights in a vertical line.

(b) Each vessel of 20 metres or more in length engaged in pair trawling shall exhibit:
- (i) by night, a searchlight directed forward and in the direction of the other vessel of the pair;
- (ii) when shooting or hauling their nets or when their nets have come fast upon an obstruction, the lights prescribed in 2(a) above.

(c) A vessel of less than 20 metres in length engaged in trawling, whether using demersal or pelagic gear or engaged in pair trawling, may exhibit the lights prescribed in paragraphs (a) or (b) of this section, as appropriate.

3. Signals for purse seiners

Vessels engaged in fishing with purse seine gear may exhibit two yellow lights in a vertical line. These lights shall flash alternately every second and with equal light and occultation duration. These lights may be exhibited only when the vessel is hampered by its fishing gear.

ANNEX III
TECHNICAL DETAILS OF SOUND SIGNAL APPLIANCES

1. Whistles

(a) *Frequencies and range of audibility*

The fundamental frequency of the signal shall lie within the range 7-700 Hz.

The range of audibility of the signal from a whistle shall be determined by those frequencies, which may include the fundamental and/or one or more higher frequencies, which lie within the range of 180-700 Hz (+ 1 per cent) and which provide the sound pressure levels specified in paragraph 1(c) below.

Chapter 11: Collision Prevention Regulations

(b) *Limits of fundamental frequencies*

To ensure a wide variety of whistle characteristics, the fundamental frequency of a whistle shall be between the following limits:

(i) 70-200 Hz, for a vessel 200 metres or more in length;
(ii) 130-350 Hz, for a vessel 75 metres but less than 200 metres in length;
(iii) 25-700 Hz, for a vessel less than 75 metres in length.

(c) *Sound signal intensity and range of audibility*

A whistle fitted in a vessel shall provide, in the direction of maximum intensity of the whistle and at a distance of 1 metre from it, a sound pressure level in at least one third octave band within the range of frequencies 180-700 Hz (+ 1 per cent) of not less than the appropriate figure given in the table below.

Length of vessel in metres	1/3rd-octave band level at 1 metre in dB referred to 2×10^{-5} N/m^2	Audibility range in nautical miles
200 or more	143	2.0
75 but less than 200	138	1.5
20 but less than 75	130	1.0
Less than 20	120	0.5

The range of audibility in the table above is for information and is approximately the range at which a whistle may be heard on its forward axis with 90 per cent probability in conditions of still air on board a vessel having average background noise level at the listening posts (taken to be 68 dB in the octave band centred on 250 Hz and 63 dB in the octave band centred on 500 Hz). In practice the range at which a whistle may be heard is extremely variable and depends critically on weather conditions; the values given can be regarded as typical but under conditions of strong wind or high ambient noise level at the listening post the range may be much reduced.

(d) *Directional properties*

The sound pressure level of a directional whistle shall be not more than 4 dB below the prescribed sound pressure level on the axis at any direction in the horizontal plane within +45 degrees of the axis. The sound pressure level at any other direction in the horizontal plane shall be not more than 10 dB below the prescribed sound pressure level on the axis, so that the range in any direction will be at least half the range on the forward axis. The sound pressure level shall be measured in that one-third-octave band which determines the audibility range.

(e) *Positioning of whistles*

When a directional whistle is to be used as the only whistle on a vessel, it shall be installed with its maximum intensity directed straight ahead.

A whistle shall be placed as high as practicable on a vessel, in order to reduce interception of the emitted sound by obstructions and also to minimize hearing damage risk to personnel. The sound pressure level of the vessel's own signal at listening posts shall not exceed 110 dB (A) and so far as practicable should not exceed 100 dB (A).

(f) *Fitting of more than one whistle*

If whistles are fitted at a distance apart of more than 100 metres, it shall be so arranged that they are not sounded simultaneously.

(g) *Combined whistle systems*

If due to the presence of obstructions the sound field of a single whistle or of one of the whistles referred to in paragraph 1 (f) above is likely to have a zone of greatly reduced signal level, it is recommended that a combined whistle system be fitted so as to overcome this reduction. For the purposes of the Rules a combined whistle system is to be regarded as a single whistle. The whistles of a combined system shall be located at a distance apart of not more than 100 metres and arranged to be sounded simultaneously. The frequency of any one whistle shall differ from those of the others by at least 10 Hz.

2. Bell or gong

(a) Intensity of signal

A bell or gong, or other device having similar sound characteristics shall produce a sound pressure level of not less than 110 dB at a distance of 1 metre from it.

(b) Construction

Bells and gongs shall be made of corrosion-resistant material and designed to give a clear tone. The diameter of the mouth of the bell shall be not less than 300 mm for vessels of 20 metres or more in length, and shall be not less than 200 mm for vessels of 12 metres or more but less than 20 metres in length. Where practicable, a power driven bell striker is recommended to ensure constant force but manual operation shall be possible. The mass of the striker shall be not less than 3 per cent of the mass of the bell.

3. Approval

The construction of sound signal appliances, their performance and their installation on board the vessel shall be to the satisfaction of the appropriate authority of the State whose flag the vessel is entitled to fly.

ANNEX IV
DISTRESS SIGNALS

1. The following signals, used or exhibited either together or separately, indicate distress and need of assistance:

(a) a gun or other explosive signal fired at intervals of about a minute;

(b) a continuous sounding with any fog-signalling apparatus;

(c) rockets or shells, throwing red stars fired one at a time at short intervals;

(d) a signal made by radiotelegraphy or by any other signalling method consisting of the group . . . — — — . . . (SOS) in the Morse Code;

(e) a signal sent by radiotelephony consisting of the spoken word "Mayday";

(f) the International Code Signal of distress indicated by N.C.;

(g) a signal consisting of a square flag having above or below it a ball or anything resembling a ball;

(h) flames on the vessel (as from a burning tar barrel, oil barrel, etc.);

(i) a rocket parachute flare or a hand flare showing a red light;

(j) a smoke signal giving off orange-coloured smoke;

(k) slowly and repeatedly raising and lowering arms outstretched to each side;

(l) the radiotelegraph alarm signal;

(m) the radiotelephone alarm signal;

(n) signals transmitted by emergency position-indicating radio beacons;

(o) approved signals transmitted by radiocommunication systems, including survival craft radar transponders.

2. The use or exhibition of any of the foregoing signals except for the purpose of indicating distress and need of assistance and the use of other signals which may be confused with any of the above signals is prohibited.

3. Attention is drawn to the relevant sections of the International Code of Signals, the Merchant Ship Search and Rescue Manual and the following signals:

(a) a piece of orange-coloured canvas with either a black square and circle or other appropriate symbol (for identification from the air);

(b) a dye marker.

CHAPTER 11: QUESTIONS & ANSWERS

COLLISION PREVENTION REGULATIONS

1. In a narrow channel, on which side of the centreline of the channel should you keep?:
   (a) right.
   (b) left.
   (c) either.
   (d) middle.

2. A power driven vessel should give way to a sailing vessel:
   (i) when the two are approaching head on.
   (ii) when the sailing vessel is overtaking her.
   (iii) on her starboard side.
   (iv) on her port side.

   The correct answer is:
   (a) all except (i).
   (b) all except (ii).
   (c) all except (iii).
   (d) all except (iv).

3. The duty of an overtaking vessel in open waters is to:
   (a) give a sound signal.
   (b) give a sound signal and overtake safely.
   (c) keep clear of the overtaken vessel.
   (d) overtake on starboard side

4. If two power driven vessels are meeting head-on, each shall alter course and may sound a signal as follows:
   (a) To port, two short blasts.
   (b) To starboard, one short blast.
   (c) To port, one short blast.
   (d) To starboard, two short blasts

5. You are in a power driven vessel. If another power driven vessel approaches you on your starboard side, you are required to:
   (a) alter course to port.
   (b) alter course to starboard.
   (c) stop or slow down.
   (d) give way.

6. You are in a power driven vessel. If another power driven vessel approaches you on your port side, you should:
   (a) not pass in front of her.
   (b) maintain course/speed & be prepared to give way.
   (c) alter course to starboard.
   (d) give way.

7. If a collision appears imminent, you should reduce impact by:
   (a) making sound by whistle and any other means.
   (b) going full astern and using fenders.
   (c) letting go anchors.
   (d) all of above.

8. In the international collision prevention regulations, the term "sailing vessel" means a vessel:
   (a) registered as a yacht.
   (b) under sail, whether or not also propelling.
   (c) under sail. Propeller, if fitted, not in use.
   (d) all of the above.

9. When sun's glare affects visibility, a vessel under way must:
   (a) show an all round white light on her mast.
   (b) reduce speed and increase lookout.
   (c) display a Restricted in Ability signal.
   (d) comply with all of the choices stated here.

10. The term "underway" means the vessel is:
    (a) making way.
    (b) going ahead.
    (c) not at anchor or made fast to shore, or aground
    (d) on course.

11. In the international collision prevention regulations, a power driven vessel means a vessel:
    (a) propelled by machinery alone.
    (b) propelled by machinery.
    (c) not designed to be under sail.
    (d) fitted with a motor.

12. In the international collision prevention regulations, the definition of a vessel restricted in her ability to manoeuvre includes a vessel:
    (a) engaged in mine clearance operations.
    (b) not under command.
    (c) carrying out engine room repairs.
    (d) fishing with nets, lines or trawls.

13. In the international collision prevention regulations, "safe speed" means the vessel can take proper and effective action to avoid collision and be stopped within a distance appropriate to:
    (a) half the range of visibility.
    (b) the range of visibility.
    (c) the prevailing circumstances/ conditions
    (d) her length.

14. In the international collision prevention regulations, the term "proper look-out" is stated as look-out:

    (a) ahead.
    (b) fore and aft.
    (c) all around.
    (d) by sight, hearing & all available means.

15. In the international collision prevention regulations, overtaking procedures are discussed in Rule 9 (Narrow Channels) and Rule 13 (Overtaking). The obligation of the overtaking vessel in the two rules is:

    (a) the same.
    (b) different.
    (c) to sound an overtaking signal.
    (d) to seek permission of the vessel ahead.

16. When two vessels are approaching one another on a crossing course, they are required to determine if there is a risk of collision between them, by checking each other's:

    (a) speed.
    (b) course.
    (c) bearings.
    (d) course and speed.

17. You are the stand-on vessel in a crossing situation between two power driven vessels. If the give-way vessel does not appear to take any action, you may:

    (a) alter course to port.
    (b) alter course to starboard.
    (c) sound 2 prolonged followed by 1 short blast.
    (d) sound 2 prolonged followed by 2 short blasts.

18. Between two power driven vessels crossing one another on a collision course, the "give-way" vessel is the one who has the other:

    (a) 22.5° abaft her beam.
    (b) on her port side.
    (c) on her starboard side.
    (d) right ahead.

19. Between two power driven vessels crossing on a collision course, the duties of a "stand-on" vessel include:

    (i) sounding a signal.
    (ii) slowing down or maintaining course & speed.
    (iii) alteration of course to starboard.
    (iv) alteration of course to port.

    The correct answer is:
    (a) all except (i).
    (b) all except (ii).
    (c) all except (iii).
    (d) all except (iv).

20. Between two sailing vessels crossing one another on a collision course with wind on different sides, the "give-way" vessel is the one which has:

    (a) the other on her starboard side.
    (b) the wind behind the beam.
    (c) the wind on starboard side.
    (d) the wind on port side.

21. Among the responsibilities listed in the international collision prevention regulations, Rule 18, a sailing vessel under way shall keep out of the way of a:

    (i) seaplane on the water.
    (ii) vessel Restricted in Ability to Manoeuvre.
    (iii) vessel Not Under Command.
    (iv) vessel engaged in fishing.

    The correct answer is:
    (a) all except (i).
    (b) all except (ii).
    (c) all except (iii).
    (d) all except (iv).

22. When two power driven vessels are approaching "head-on", they are required to avoid collision with each other by:

    (a) altering course to port.
    (b) altering course to starboard.
    (c) sounding at least five short blasts.
    (d) giving way to the vessel on stbd side.

23. If the give-way vessel does not give way, the stand-on vessel must:

    (a) maintain her course and speed.
    (b) continue to sound five short blasts on whistle.
    (c) alter course to starboard or slow down.
    (d) take any action to avoid collision.

24. In a traffic separation scheme, a vessel is required to avoid:

    (a) proceeding under sail.
    (b) anchoring.
    (c) laying a submarine cable.
    (d) fishing.

25. A sailing vessel becomes a power driven vessel when she is power driven:

    (a) and not under sail.
    (b) and also under sail.
    (c) whether or not under sail.
    (d) and displays a black cone on her mast.

Chapter 11: Collision Prevention Regulations

26. You are sailing but also under power, meeting a power driven vessel steaming almost head-on towards you. You are:

    (a) the stand-on vessel.
    (b) the give way vessel.
    (c) required to alter course to starboard.
    (d) required to alter course to port.

27. Your course - north. You sight the port side light of another vessel on a bearing of 030°. She is heading on a course anywhere between:

    (a) 000° and 030°
    (b) 180° and 210°
    (c) 030° and 112.5°
    (d) 210° and 322.5°

28. Your course - north. Another vessel's green light sighted on a bearing of 340°, range about 6 miles. Five minutes later the bearing is 350° and the range about 5 miles:

    (a) There is a risk of collision.
    (b) There is no risk of collision.
    (c) Risk of collision cannot be established.
    (d) A vessel with another on stbd must take action.

29. Your course 090°. You see a ship bearing 045° on an apparent collision course:

    (a) She must keep out of the way.
    (b) You must keep out of the way.
    (c) There is no give way vessel in such situations.
    (d) Both vessels are give way vessels.

30. The most suitable way to establish if another vessel is on a collision course with you, is to judge:

    (a) her speed.
    (b) your speed.
    (c) relative speed between the two vessels.
    (d) the change in her compass bearings.

31. A vessel is in the process of overtaking you, and she should pass you safely on your starboard side. She is required to observe her overtaking responsibility until:

    (a) she is forward of your beam.
    (b) she alters course to port after overtaking.
    (c) you alter course to give way to another vessel.
    (d) she is finally past and clear.

32. The Overtaking Rule 13 states that the overtaking vessel:

    (a) becomes crossing vessel by changing her course.
    (b) must overtake on starboard side.
    (c) must keep clear of the vessel being overtaken.
    (d) must give an overtaking sound signal.

33. With regard to a traffic separation scheme, a 30 metre sailing vessel:

    (a) must join it.
    (b) may stay in the inshore traffic zone.
    (c) must stay on the outside.
    (d) must proceed under power.

34. A vessel under 20 metres in length crossing a traffic lane must:

    (a) follow rules applicable to all vessels.
    (b) follow rules applicable to small vessels.
    (c) sound narrow channel signals.
    (d) cross at a small angle to the traffic flow.

35. In a traffic lane a sailing vessel must:

    (a) give way to power driven vessels.
    (b) not impede the passage of power driven vessels.
    (c) not overtake power driven vessels.
    (d) keep out of it.

36. A vessel must show her navigation lights when under way:

    (a) from sunset to sunrise and in poor visibility.
    (b) at all hours.
    (c) from sunset to sunrise.
    (d) only when motoring.

37. In the international collision prevention regulations the navigation lights required to be shown by pleasure vessels are of:

    (a) different arc from that of commercial vessels.
    (b) the same arc as for commercial vessels.
    (c) shorter range from that of commercial vessels.
    (d) longer range from that of commercial vessels.

38. In the international collision prevention regulations a vessel showing side lights and a stern light, and no masthead light, is a:

    (a) power driven vessel under 7 metres and 7 knots.
    (b) sailing vessel under 7 metres in length.
    (c) sailing vessel of any length.
    (d) none of the choices stated here.

39. You see the following lights on a vessel: A red light over a green light with another green light underneath. This is:

    (a) a sailing vessel.
    (b) one end of a cross-river ferry in chains.
    (c) a fishing vessel.
    (d) a pilot vessel.

40. At night, a power driven vessel can be distinguished from a sailing vessel because a sailing vessel carries:

    (a) a two-colour lantern.
    (b) an all-round white light on top of her mast.
    (c) side lights and a stern light only.
    (d) a mast head light only.

41. Navigation lights carried by a vessel under sail which is also under power are:

    (a) as for a sailing vessel.
    (b) as for a power driven vessel.
    (c) a red over green all-round lights on the mast.
    (d) an all-round white light on the bow.

42. A vessel at anchor at night is required to show:

    (a) one or two all-round white lights near her ends
    (b) one or two masthead lights.
    (c) side lights only.
    (d) side lights and a stern light.

43. The difference in the anchor signals of a sailing vessel and a power driven vessel is:

    (a) the positioning of the signal.
    (b) sailing vessels do not show an anchor ball.
    (c) power driven vessels show two anchor balls.
    (d) none.

44. The night signal to indicate "diver below, keep clear" consists of:

    (a) the international code flag "A".
    (b) a red light.
    (c) three all-round RWR lights in a vertical line.
    (d) no specified signal.

45. The night signal displayed on the stern of a tug to indicate that the vessel behind is in tow consists of:

    (a) only the stern light.
    (b) a 135° yellow light.
    (c) an all round yellow light.
    (d) red-white-red lights in a vertical line.

46. You observe at night a vessel displaying two red lights in a vertical line, as well as, her sidelights and a stern light. She is:

    (a) engaged in underwater operations.
    (b) engaged in towing.
    (c) constrained by draught.
    (d) not under command.

47. You observe a vessel displaying a signal consisting of ball, diamond, ball (all black in colour) in a vertical line. She is:

    (a) restricted in her ability to manoeuvre.
    (b) constrained by draft.
    (c) indicating channel blocked.
    (d) aground.

48. The night signal displayed by a dredger to indicate its side on which the obstruction exists is:

    (a) one all-round red light
    (b) two all-round red lights, rigged vertically.
    (c) one 112.5° red light.
    (d) two 112.5° red lights, rigged vertically

49. The day signal displayed by a fishing vessel not engaged in fishing is:

    (a) two black cones, apexes together
    (b) two black cones, apexes away from each other.
    (c) one black cone, apex pointing up.
    (d) none.

50. The night signal displayed by a pilot vessel berthed alongside a jetty while on pilotage duty is:

    (a) none
    (b) all-round white over red lights.
    (c) deck lights.
    (d) that of an ordinary power driven vessel.

51. Under the international collision prevention regulations, if it is not practicable for them to show the lights required by the rules, sailing vessels under 7 metres in length and vessels under oars must show:

    (a) a tri-colour lantern.
    (b) a bi-colour lantern.
    (c) an all round white light.
    (d) a torch or lantern whenever necessary.

52. A vessel about to cross astern of a tug towing another vessel astern would be warned of the danger by the:

    (a) white stern light of the tug.
    (b) masthead light(s) of the tow.
    (c) side lights of the tug.
    (d) yellow light on stern of the tug.

Chapter 11: Collision Prevention Regulations

53. A manoeuvring light is a bright all round light:
    (a) that operates in phase with a vessel's whistle.
    (b) known as the masthead light.
    (c) that identifies a tug.
    (d) carried on a tow in lieu of a masthead light.

54. In the international collision prevention regulations, the minimum lights (in a combined lantern or separately) required to be shown by a 6 metre power-boat if her maximum speed is 20 knots are:
    (a) side lights & stern light.
    (b) masthead light & stern light.
    (c) side lights, masthead light & stern light.
    (d) a lantern or a torch whenever required.

55. In the international collision prevention regulations, the minimum lights (in a combined lantern or separately) required to be shown by a 6 metre power-boat if her maximum speed is 6 knots are:
    (a) side lights & stern light.
    (b) masthead light & stern light.
    (c) side lights, masthead light & stern light.
    (d) a lantern or a torch whenever required.

56. You are overtaking a vessel over 100 metres in length at night from almost astern. The navigation lights that you should see are:
    (a) one white light.
    (b) two white lights.
    (c) three white lights.
    (d) a red, a green and a white light.

57. A vessel carrying two masthead lights, one abaft and higher than the other:
    (a) provides a definite indication of her length.
    (b) is easier to assess in her degree of course alteration.
    (c) does not need to carry a stern light.
    (d) is able to tow other vessels.

58. The day signal required to be shown by a fishing vessel indicating that her gear extends 100 metres horizontally in a particular direction is:
    (a) a black cone with apex up.
    (b) a black cone with apex down.
    (c) a black cone with apex to the fishing gear.
    (d) none.

59. The day signal for a vessel engaged in trawling is two black cones, apexes pointing:
    (a) up.
    (b) down.
    (c) away from each other.
    (d) towards each other.

60. In addition to the side lights and the stern light, the night signal for a sailing vessel which is also propelled by machinery is:
    (a) an all round white light.
    (b) the masthead light(s).
    (c) a white flood light lighting the main sail.
    (d) none at all.

61. The day signal for a sailing vessel which is also propelled by machinery is a black cone, apex pointing:
    (a) up.
    (b) down.
    (c) forward.
    (d) aft.

62. The day signal for a vessel not under command consists of the following black shape(s) hoisted in a vertical line:
    (a) a black cylinder.
    (b) ball, diamond, ball.
    (c) two black balls.
    (d) three black balls.

63. The day signal for a vessel restricted in ability to manoeuvre consists of the following black shape(s) hoisted in a vertical line:
    (a) a black cylinder.
    (b) ball, diamond, ball.
    (c) two black balls.
    (d) three black balls.

64. The day signal for a vessel constrained by her draught consists of the following black shape(s) hoisted in a vertical line:
    (a) a black cylinder.
    (b) ball, diamond, ball.
    (c) two black balls.
    (d) three black balls.

65. In addition to the navigation lights of a power driven vessel, a vessel constrained by her draught is required to show the following set of 360° lights in a vertical line:
    (a) red white red.
    (b) red green red.
    (c) two red.
    (d) three red.

66. The navigation lights of a trawler winching up her gear and not making way through the water, seen from astern, should be:
    (a) green over white; and white over red.
    (b) green over white; white; and white over red.
    (c) green over white; and red over white.
    (d) green over white; and two red.

67. You see the starboard side of a 30 metre vessel manoeuvring with difficulty. Her navigation lights should be (in or about a vertical line) as follows:

    (a) red red red green.
    (b) red white red white green.
    (c) white red red green.
    (d) white red white red green.

68. A 60 metre fishing trawler is stopped and winching up her fishing nets. The navigation lights you should see when viewed from ahead, are:

    (a) green over white; and red over red.
    (b) white; green over white; and white over red
    (c) red over white; and white over red.
    (d) none of the choices stated here.

69. You see a vessel heading towards you. She is showing four masthead lights and the side lights. She is:

    (a) very large and power driven.
    (b) over 50 metres with a tow under 200 metres.
    (c) of any length with a tow exceeding 200 metres.
    (d) none of the choices stated here.

70. A vessel overtaking <u>directly</u> from astern a large ship being towed, & unable to see the tug, should see the following navigation light(s) at night:

    (a) yellow and white.
    (b) yellow and three or more white.
    (c) two white and a yellow.
    (d) one white.

71. A towing vessel finds her tow difficult to manage. She may warn other vessels during the day by displaying a signal consisting of the following shape(s):

    (a) a black diamond.
    (b) a black diamond on each vessel.
    (c) ball, diamond, ball (black shapes)
    (d) a black cylinder.

72. A towing vessel finds her tow difficult to manage. She may warn vessels at night by displaying a signal consisting of the following light(s) in a vertical line:

    (a) two red.
    (b) three red.
    (c) red green red
    (d) red white red.

73. A small vessel is flying the international code flag A. She is required to show the following signal if continuing the indicated activity after dark:

    (a) there is no night signal for it.
    (b) two red lights.
    (c) red white red lights.
    (d) one red light.

74. A vessel is displaying a white light at her fore end and two red lights in a vertical line in the rigging. She is:

    (a) aground.
    (b) constrained by draught.
    (c) not under command.
    (d) not under command at anchor.

75. On leaving harbour in daylight you decide to pass under the bows of a ship over 100 metres in length apparently lying at anchor. Her signal to indicate that she is at anchor and not, in fact, moving slowly ahead with her anchor hanging just below the water surface, is:

    (a) a black ball on the fore end.
    (b) a black ball on each end.
    (c) a black diamond on the fore end.
    (d) a black diamond on each end.

76. You are overtaking two vessels ahead of you at night. The nearest one shows a single white light. The farthest one shows a yellow light above a white light. This indicates:

    (a) none of the choices stated here.
    (b) one is towing the other.
    (c) they are engaged in purse seine fishing.
    (d) both are at anchor, one is larger.

77. The light used to supplement the sound signals is:

    (a) a masthead light.
    (b) a 360° yellow light.
    (c) a 360° orange light.
    (d) a 360° white light.

78. The day signals for a vessel engaged in mine-sweeping is:

    (a) three red balls in a triangle.
    (b) three black balls in a triangle.
    (c) black ball diamond ball in a vertical line
    (d) black ball cone ball in a vertical line.

79. A single black cone, apex pointing upwards, is displayed by a fishing vessel:

    (a) shooting nets.
    (b) hauling nets.
    (c) with gear over 150 metres horizontally.
    (d) with gear over 200 metres horizontally.

Chapter 11: Collision Prevention Regulations

80. The additional night signal for a fishing vessel when shooting nets is two lights in a vertical line as follows:

    (a) white over red.
    (b) two red.
    (c) red over white.
    (d) two white.

81. The night signal for a channel blocked consists of the following lights displayed in a vertical line:

    (a) two red.
    (b) two green.
    (c) red green red.
    (d) red white red.

82. The additional navigation light(s) required to be exhibited by an air-cushion vessel when operating in non-displacement mode is/are:

    (a) an all-round orange flashing light.
    (b) an all-round yellow flashing light.
    (c) two all-round orange flashing lights.
    (d) two all-round yellow flashing lights.

83. Two yellow lights, flashing alternately, displayed in a vertical line on a vessel indicates a:

    (a) dredger.
    (b) small vessel not under command.
    (c) purse-seine fishing vessel.
    (d) trawler shooting nets.

84. Two trawlers in close proximity lighting the area between them with flood lights are:

    (a) indicating the safe side to pass.
    (b) pair trawling.
    (c) attracting the catch.
    (d) shooting nets.

85. In a narrow channel you hear a vessel astern of you giving a signal consisting of two prolonged blasts followed by one short blast. It means:

    (a) she wishes to overtake.
    (b) she wishes to overtake on your starboard side.
    (c) she wishes to overtake on your port side.
    (d) none of the choices stated here.

86. What signal does a vessel sound when nearing a bend in a channel? How does a vessel, if any, on the other side of the bend respond?

    (a) One and two prolonged blasts respectively.
    (b) Two and one prolonged blasts respectively.
    (c) One short blast each.
    (d) One prolonged blast each.

87. Which of the following signals indicates a vessel operating stern propulsion?

    (a) one short blast on whistle.
    (b) two short blasts on whistle.
    (c) three short blasts on whistle.
    (d) five or more rapid short blasts on whistle.

88. Which of the following describes an international distress signal:

    (a) fire on deck.
    (b) flames on vessel as from a burning oil barrel.
    (c) a national flag hoisted upside down.
    (d) a radio silence period.

89. Which of the following describes an international distress signal:

    (a) A signal transmitted by an EPIRB.
    (b) An orange sheet showing a black "V".
    (c) A hand flare showing a white light.
    (d) A gun fired every five minutes.

90. A distress signal must not be used without the authority of:

    (a) master.
    (b) master or owner.
    (c) owner.
    (d) a person with a radio operator's certificate.

91. In the international collision prevention regulations, a vessel overtaking another vessel on her starboard side, whether in a narrow channel or not, is required to:

    (a) keep clear & out of the way.
    (b) sound 1 prolonged blast every 2 minutes.
    (c) sound 2 prolonged & 1 short blast.
    (d) sound 2 prolonged & 2 short blasts.

92. Under a local rule, some vessels in Sydney harbour display an orange coloured diamond shape during the day. It means they:

    (a) do not have to give way to other vessels.
    (b) do not have to give way to sailing vessels.
    (c) do not have to give way to recreational vessels
    (d) have priority over sail.

93. The sound signal made by a power driven vessel stopped and making no way through the water in restricted visibility is:

    (a) one short blast.
    (b) two short blasts.
    (c) one prolonged blast and two short blasts.
    (d) none of the choices stated here.

94. The sound signal in restricted visibility from a vessel engaged in fishing and making no way through the water is:

    (a) one prolonged blast and two short blasts.
    (b) one prolonged blast.
    (c) two prolonged blasts.
    (d) none of the choices stated here.

95. If visibility closed in, the international collision prevention regulations require a vessel to:

    (a) run for shelter.
    (b) hoist a signal.
    (c) reduce speed.
    (d) anchor where she is.

96. In thick fog you hear the fog signal of a vessel about 30° on your starboard bow. You should:

    (a) reduce speed to minimum or stop.
    (b) alter course to starboard.
    (c) alter course to port.
    (d) continue with your course and speed.

97. The restricted visibility signal for a vessel stopped and engaged in fishing is:

    (a) one prolonged blast on whistle.
    (b) two prolonged blasts on whistle.
    (c) ringing of bell.
    (d) one prolonged and two short blasts on whistle.

98. You are fishing in thick fog when you hear dead ahead a signal consisting of one prolonged blast followed by two short blasts. You would:

    (a) reduce speed to minimum or stop.
    (b) alter course to starboard.
    (c) alter course to port.
    (d) continue with your course and speed.

99. A power driven vessel in a confined waterway, when altering course to starboard in poor visibility, should:

    (a) maintain constant speed.
    (b) sound 1 short blast.
    (c) sound 1 prolonged followed by 2 short blasts.
    (d) not sound any signal.

100. The signal for overtaking a vessel on her starboard side in a narrow channel is:

    (a) five or more short blasts.
    (b) three short blasts.
    (c) two prolonged and one short blast.
    (d) one prolonged blast.

101. The signal for agreeing to be overtaken in a narrow channel is:

    (a) one short blast.
    (b) prolonged-short-prolonged-short blasts.
    (c) two short blasts.
    (d) two prolonged and one short blasts.

102. The restricted visibility sound signal for a vessel Not Under Command which is no longer making way is:

    (a) One prolonged and two short blasts.
    (b) one prolonged blast.
    (c) ringing of bell.
    (d) none.

103. The restricted visibility sound signal for a vessel under way with divers working below is:

    (a) ringing of bell.
    (b) one prolonged blast.
    (c) two prolonged blasts.
    (d) one prolonged and two short blasts.

104. The restricted visibility sound signal for a 200 metre bulk carrier aground is the rapid ringing of a bell and a gong and:

    (a) both the signals starting and ending with 3 strokes.
    (b) the bell signal starting and ending with 3 strokes.
    (c) the gong signal starting and ending with 3 strokes.
    (d) without adding any separate strokes.

105. You should not attract the attention of another vessel by the continuous sounding of your whistle or siren, other than for the purpose of:

    (a) distress.
    (b) a warning to keep out of your way.
    (c) distress or a warning to keep out of your way.
    (d) overtaking.

106. Between two sailing vessels crossing one another on a collision course, both with wind on the same side, the "give-way" vessel is one which:

    (a) is to windward.
    (b) is to leeward.
    (c) has wind abaft the beam.
    (d) has wind forward of her beam.

Chapter 11: Collision Prevention Regulations

107. In relation to the international collision avoidance regulations for sailing vessels, the windward side is the side:

    (a) on which the mainsail is carried.
    (b) opposite to that on which the mainsail is carried.
    (c) on which the largest fore-and-aft sail is carried on a square-rigged vessel.
    (d) on which the spinnaker is carried.

108. A vessel is deemed to be overtaking when coming up with another vessel from the following direction:

    (a) more than 22.5° abaft her beam.
    (b) abaft her beam.
    (c) more than 50° abaft her beam.
    (d) right astern.

109. It is considered dangerous for another vessel to approach a mine clearance vessel within:

    (a) 1000 metres on her sides and 500 metres astern.
    (b) 1000 metres.
    (c) 500 metres.
    (d) 500 metres on her sides and 1000 metres astern.

110. In a crossing situation, the stand-on vessel shall:

    (a) slow down or stop.
    (b) alter course to starboard.
    (c) maintain her course and speed, and keep watch.
    (d) alter course to port.

111. A vessel using a traffic separation scheme shall normally join or leave a traffic lane:

    (a) at a large angle to the traffic flow.
    (b) as quickly as possible.
    (c) at right angle to the lane.
    (d) at the termination of the lane.

112. A vessel joining a traffic lane from one side of a traffic separation scheme shall do so:

    (a) at a large angle to the traffic flow.
    (b) at a small angle to the traffic flow.
    (c) as quickly as possible.
    (d) at right angle to the lane.

113. A vessel not under command but making way will display the following signal:

    (a) not under command (NUC).
    (b) restricted in ability to manoeuvre (RAM).
    (c) NUC, plus side and stern lights.
    (d) RAM, plus side & stern light.

114. The international distress signal, consisting of a square flag and a ball or anything resembling a ball, must be hoisted in the following order:

    (a) ball above the flag.
    (b) ball below the flag.
    (c) ball above or below the flag.
    (d) ball below the flag on an orange sheet.

115. One of the international distress signal consists of the following international code flags:

    (a) a chequered flag over a striped flag.
    (b) a striped flag over a chequered flag.
    (c) flags A. C.
    (d) flags R. Y.

116. One of the international distress signals consists of:

    (a) an orange V-sheet.
    (b) a hand flare throwing red stars.
    (c) a rocket parachute.
    (d) a hand flare showing a red light.

117. In the international collision prevention regulations, the dye marker distress signal is stated as follows:

    (a) dye marker.
    (b) red dye marker.
    (c) orange dye marker.
    (d) green dye marker.

118. The international distress signals may be used or exhibited as follows:

    (a) one signal at a time.
    (b) either together or separately.
    (c) in the order stated in regulations.
    (d) pyrotechnic signals before the others.

119. The restricted visibility sound signal for a 200 metre bulk carrier at anchor is the rapid ringing of a bell and a gong and:

    (a) both the signals starting and ending with 3 separate strokes.
    (b) the bell signal starting and ending with 3 separate strokes.
    (c) the gong signal starting and ending with 3 separate strokes.
    (d) without adding any separate strokes.

120. The restricted visibility sound signal for a 60-metre vessel at anchor is:

    (a) the rapid ringing of a bell and a gong.
    (b) the rapid ringing of a bell.
    (c) the rapid ringing of a bell, starting and ending with 3 separate & distinct strokes.
    (d) the rapid ringing of a gong.

121. The "prolonged blast" is a blast of the following duration:

    (a) about 1 second.
    (b) 5 to 7 seconds.
    (c) 4 to 8 seconds.
    (d) 4 to 6 seconds.

122. In the International Collision Prevention Regulations (Rule 5), the words "...all available means..." in the definition of lookout include lookout by:

    (a) Compass
    (b) Radar
    (c) Telegraph
    (d) Auto-pilot

123. In the International Collision Prevention Regulations (Rule 6), which of the following factors is not an additional determinant of safe speed for vessel fitted with operational radar?

    (a) Assessment of visibility.
    (b) Possibility of not detecting small vessels.
    (c) Presence of background lights from ashore.
    (d) Limitations of the radar equipment.

124. With regard to use of radar in the International Collision Prevention Regulations (Rules 7, 8 & 19), which of the following statements is incorrect?

    (a) A vessel fitted with radar may make a succession of small alternations of course and/or speed, instead of one large alteration.
    (b) If there is sufficient sea room, alteration of course alone may be the most effective action.
    (c) Assumptions shall not be made on the basis of scanty radar information.
    (d) A vessel detecting, by radar alone, the possibility of a close-quarters situation with another vessel, may take avoiding action in ample time.

125. In taking collision avoidance action by alteration of course based on radar information alone (Rule 19), which of the following statements is incorrect?

    (a) She shall avoid altering course towards a vessel abeam.
    (b) She shall take avoiding action in ample time.
    (c) She shall avoid altering course towards a vessel abaft the abeam.
    (d) She shall not alter course to port.

126. In restricted visibility, radar lookout is best maintained:

    (a) on a short range scale.
    (b) on a long range scale
    (c) on a medium range scale.
    (d) by varying between short & long range scales.

127. In determining safe speed in restricted visibility, a vessel fitted with radar must:

    (a) operate on a long range scale
    (b) operate on a medium range scale.
    (c) consider any constraints imposed by the radar range scale in use.
    (d) operate on a short range scale.

128. In terms of responsibility between vessels, which of the following is the 'stand-on' vessel in relation to the other three?

    (a) Engaged in trawling
    (b) Restricted in ability to manoeuvre
    (c) Sailing
    (d) Engaged in fishing

129. In relation to a power driven vessel of less than 20 metres in length in a Traffic Separation Scheme, which of the following statements is correct?

    (a) Avoid using the Inshore Traffic Lane.
    (b) Do not anchor in the Traffic Separation Scheme.
    (c) Do not impede the safe passage of a power driven vessel following a traffic lane.
    (d) Do not use the Inshore Traffic Lane.

130. In terms of arcs of the horizon of navigational lights, which of the following statements is correct?

    (a) Masthead light 225°
    (b) Sidelight 111½°
    (c) Sternlight 125°
    (d) Towing light 360°

131. In the International Collision Prevention Regulations, the term "restricted visibility" refers:

    (a) only to fog.
    (b) only to fog & rain.
    (c) to visibility of coastline
    (d) to any condition restricting visibility.

132. The Rules specify the following vessels not to impede the passage of a vessel which can safely navigate only within a narrow channel or fairway:

    (a) Vessels of less than 10 metres in length.
    (b) Sailing vessels.
    (c) Power driven vessels and sailing vessels of less than 20 meters in length.
    (d) Vessels of less than 20 metres in length & sailing vessels.

Chapter 11: Collision Prevention Regulations

133. Any action taken to avoid collision should be:

    (a) taken quickly.
    (b) Taken slowly and in small amounts.
    (c) made in ample time.
    (d) made without regard to the practice of seamanship.

134. A vessel may enter a Traffic Separation Zone:

    (a) in an emergency.
    (b) to engage in fishing within the zone.
    (c) to cross the traffic separation scheme.
    (d) for any of the reasons stated here.

135. In a narrow channel, a vessel shall

    (a) not anchor
    (b) avoid anchoring.
    (c) not engage in fishing.
    (d) avoid fishing.

136. A vessel engaged in fishing:

    (a) must keep out of the way of a vessel Not Under Command or Restricted in her Ability to Manoeuvre.
    (b) when underway, must keep out of the way a vessel Not Under Command or Restricted in her Ability to Manoeuvre.
    (c) when underway shall, so far as possible, keep out of the way of a vessel Not Under Command.
    (d) None of the choices stated here.

137. A vessel wishing to attract the attention of another vessel (Rule 36) should do so as follows:

    (a) Use a strobe light.
    (b) Use a high intensity revolving light.
    (c) Use a high intensity intermittent light.
    (d) Make a light or sound signal that cannot be mistaken for any signal authorised in the international collision avoidance regulations.

138. The term "not under command" includes the following:

    (a) Vessel with engine breakdown and at anchor.
    (b) Vessel with faulty steering, but making way.
    (c) Vessel engaged in dredging.
    (d) Vessel aground.

139. In relation to the "special rules made by an appropriate authority" (Rule 1):

    (a) The special rules shall conform as closely as possible to the international rules.
    (b) The authority may specify additional station or light signals for ships of war.
    (c) The additional signals should not be mistaken for any signal specified in the international rules.
    (d) All three statements stated here apply.

140. For a vessel approaching on a steady bearing from 3 points on the port bow at a distance of 4 miles, you should:

    (a) Commence altering course.
    (b) Commence slowing down.
    (c) Stand on and monitor the situation.
    (d) Alter course to port to put her on starboard bow.

141. As specified in Rule 1, the international collision prevention regulations apply to:

    (a) all vessels in all waters
    (b) all vessels upon the high seas.
    (c) All vessels upon the high seas and in all waters connected to the high seas.
    (d) All vessels upon the high seas and in all waters connected to the high seas that are navigable by seagoing vessels.

142. As specified in Rule 2, the international collision prevention regulations:

    (a) Exonerate those who follow these rules.
    (b) Do not exonerate those who ignore the ordinary practice of seamanship.
    (c) Do not permit a departure from the rules.
    (d) Do not permit deviation for any special circumstances.

143. The International Collision Prevention Regulations (Rule 15) state that the give-way vessel shall:

    (a) avoid crossing ahead of the other.
    (b) not cross ahead of the other.
    (c) pass port to port.
    (d) cross astern of the other.

144. Which of the following indicates a definite head-on situation?

    (a) Two masthead lights in a vertical line directly ahead.
    (b) Three masthead lights in a vertical line directly ahead.
    (c) Four masthead lights in a vertical line and/or both sidelights directly ahead.
    (d) One masthead light directly ahead.

145. It is advisable not to engage steering by autopilot:

    (a) when towing.
    (b) from sunset to sunrise.
    (c) when engaged in fishing.
    (d) in areas of restricted visibility.

146. As per Rule 31, a seaplane on the water:
    (a) must exhibit lights & shapes prescribed in the Rules.
    (b) should exhibit lights & shapes prescribed in the Rules.
    (c) exhibits navigational lights & shapes only if it can.
    (d) is not required to exhibit any navigational lights & shapes.

147. The number of masthead lights displayed by a vessel engaged in towing is:
    (a) two
    (b) three
    (c) four
    (d) two, three or four.

148. The number of masthead lights displayed by a vessel engaged in pushing another vessel, which is not a composite unit, is:
    (a) two.
    (b) Two or three.
    (c) Three.
    (d) One.

149. The number of masthead lights displayed by a composite unit is:
    (a) one or two.
    (b) two or three.
    (c) one.
    (d) two.

150. A vessel displaying the signal consisting of ball, diamond, ball (all black) in a vertical line is:
    (a) Not Under Command.
    (b) Engaged in minesweeping.
    (c) Transferring cargo at sea.
    (d) Aground

151. Unless impracticable to do so, the lights to be displayed on a 24 metres long and 24 metres wide inconspicuous partly submerged object, being towed, which is not a dracone, are as follows:
    (a) An all-round white light at each end.
    (b) An all-round white light at the after end.
    (c) An all-round white light at each end & at each extremity of its breadth.
    (d) An all-round white light at every 10 metres along its length.

152. Unless impracticable to do so, the lights to be displayed on a 24 metres long dracone, being towed, are as follows:
    (a) An all-round white light at each end.
    (b) An all-round white light at the after end.
    (c) An all-round white light at every 10 metres along its length.
    (d) An all-round white light at each end & at each extremity of its breadth.

153. Unless impracticable to do so, the day shape to be displayed on a 100-metre long inconspicuous partly submerged object, being towed, is as follows:
    (a) A diamond shape at each end.
    (b) A black diamond shape at the after end.
    (c) A black diamond shape at each end.
    (d) A diamond shape at the after end.

154. Unless impracticable to do so, the day shape to be displayed on a 210-metre long tow consisting of inconspicuous partly submerged objects is as follows:
    (a) A black diamond shape at each end.
    (b) A black diamond shape at the after end.
    (c) One diamond shape at the after end and one as far forward as practicable.
    (d) A diamond shape at the after end.

155. In a tow consisting of three towed vessels in a line, the restricted visibility sound signal for the second towed vessel is:
    (a) one prolonged followed by three short blasts.
    (b) none.
    (c) one prolonged followed by three short blasts immediately after the signal made by the towing vessel.
    (d) one prolonged followed by three short blasts at intervals of not more than 2 minutes.

156. When vessels are not in sight of one another due to restricted visibility, a vessel shall indicate her course alteration to starboard by the following signal:
    (a) One short blast.
    (b) Two short blasts.
    (c) None.
    (d) One prolonged blast.

157. When vessels are not in sight of one another due to restricted visibility, a signal consisting of at least 5 short blasts has the following meaning:
    (a) Indicate your intentions.
    (b) Indicate your presence.
    (c) I have stopped my engine.
    (d) There is no such signal for use in restricted visibility.

Chapter 11: Collision Prevention Regulations

158. The following signal indicates that the vessel is in distress:

    (a) A national flag hoisted up side down.
    (b) A large square flag.
    (c) More than 5 short blasts.
    (d) Continuous sounding of a foghorn.

159. The signal consisting of an orange diamond shape displayed by some power driven vessels in Sydney harbour indicates:

    (a) Right of way over sailing vessels.
    (b) Priority over sailing vessels.
    (c) Right of way over other vessels.
    (d) Priority over other vessels.

160. The 'channel closed' day signal consists of the following black shapes:

    (a) Ball, diamond, ball
    (b) Cylinder.
    (c) Ball, cone pointing downwards, ball
    (d) Ball, cone pointing upwards, ball

161. A 'vessel restricted in her ability to manoeuvre' is one which is:

    (a) restricted by poor visibility.
    (b) restricted by the nature of her work.
    (c) not under command.
    (d) steering with difficulty with a jury rudder.

162. You are in a sailing vessel sailing with wind on port quarter. You observe another sailing vessel approaching head-on on a steady bearing.

    (a) You should give way.
    (b) You should stand on.
    (c) Each vessel must alter course to starboard.
    (d) You should change tack.

163. The steering and sailing rules for the conduct of vessels in restricted visibility apply to vessels:

    (a) in areas of restricted visibility.
    (b) in or near an area of restricted visibility.
    (c) at all times.
    (d) other than sailing vessels.

164. A vessel is displaying two white lights in a vertical line. On one side of these, in another vertical line, are three more lights, the top and bottom of which are red and the middle one is white. This is a vessel:

    (a) Not Under Command
    (b) Constrained by her draft.
    (c) Restricted in her ability to manoeuvre
    (d) On pilotage duty.

165. A vessel restricted in her ability to manoeuvre and not making way is displaying, in a vertical line, two green lights on one side and two red lights on the other.

    (a) She is engaged in minesweeping.
    (b) One of these lights on each side is her sidelight.
    (c) She is displaying her optional sailing lights.
    (d) She is indicating her obstructed and unobstructed sides.

166. A vessel restricted in her ability to manoeuvre is displaying three green lights on one side and three red lights on the other.

    (a) You should pass her on her starboard side.
    (b) There is no such signal.
    (c) You should pass her on her port side.
    (d) She is not making way.

167. A vessel is displaying one white light, below which are three more lights in a vertical line, the top and bottom of which are red and the middle one is white. She is also displaying two green and a red light on one side and two red and a green on the other. This is a vessel:

    (a) Not Under Command
    (b) Constrained by her draft.
    (c) Restricted in her ability to manoeuvre
    (d) On pilotage duty.

168. The day signal indicating the obstructed and unobstructed sides of a vessel restricted in her ability to manoeuvre is:

    (a) Two black diamond shapes on each side.
    (b) Two black diamond shapes on the unobstructed side and two black ball shapes on the obstructed side.
    (c) Two black diamond shapes on the obstructed side and two black ball shapes on the unobstructed side.
    (d) A black diamond shape on the unobstructed side and a black ball shape on the obstructed side.

169. A vessel displaying three black balls in a vertical line is:

    (a) Aground.
    (b) Not under command.
    (c) Constrained by draft.
    (d) Restricted in ability to manoeuvre.

170. The restricted visibility sound signal of two prolonged blasts indicates:

    (a) a power driven vessel making way.
    (b) A power driven vessel that has stopped.
    (c) A power driven vessel no longer underway.
    (d) A disabled vessel.

171. As specified in Rule 35 (g), a vessel at anchor in restricted visibility may warn an approaching vessel of her position and the possibility of collision, by sounding the following <u>additional</u> sound signal:

    (a) Five or more short blasts.
    (b) One prolonged & two short blasts.
    (c) One short, one prolonged and one short blasts.
    (d) Continuous sounding of whistle.

172. Flames emerging out of a drum on the deck of a vessel, means that:

    (a) The vessel is not seeking assistance.
    (b) The crew are lighting up a BBQ.
    (c) The vessel is burning off bilge oil.
    (d) The vessel is in distress & requires immediate assistance.

173. A sailing vessel and a vessel engaged in fishing are on a crossing collision course. In this situation, the stand-on vessel is the:

    (a) Fishing vessel.
    (b) Sailing vessel.
    (c) Fishing vessel if she is on the port side of Sailing vessel.
    (d) Fishing vessel if fishing with nets.

174. A power driven vessel on a collision course with another power driven vessel, which is on her starboard bow, must:

    (a) slow down.
    (b) stop.
    (c) alter course to starboard & pass astern of her.
    (d) give way.

175. The shapes required to be displayed by the Rules, must be displayed:

    (a) at all hours.
    (b) only if there is traffic in sight.
    (c) by day.
    (d) whenever visibility permits it.

176. You see a white light over a red light, in a vertical line, on a vessel. It is:

    (a) the port side of a power driven vessel.
    (b) the port side of a power driven vessel or a vessel engaged on pilotage duty.
    (c) a vessel engaged on pilotage duty.
    (d) a vessel engaged in fishing.

177. In addition to sidelights, a manned vessel being towed, displays:

    (a) Masthead light(s)
    (b) Not under command lights
    (c) Towing lights.
    (d) Stern light.

178. The equivalent day signal, for three green lights in a triangle at night, is:

    (a) three black balls in a triangle.
    (b) three green balls in a triangle.
    (c) ball, diamond, ball, all black, in a vertical line.
    (d) a black cylinder.

179. A vessel giving a signal of 6 short and rapid flashes of light (Rule 34 (d)), means:

    (a) I am pair trawling.
    (b) I do not understand your intentions.
    (c) I am operating astern prolusion.
    (d) I am carrying out an emergency must drill.

180. A vessel giving a signal of 3 short and rapid flashes of light (Rule 34 (b)), means:

    (a) I am operating astern propulsion.
    (b) I do not understand your intentions.
    (c) I am Not Under Command.
    (d) I am carrying out an emergency muster drill.

181. In restricted visibility, a power driven vessel underway, but not making way, gives the following sound signal:

    (a) One prolonged blast.
    (b) The same signal as when making way.
    (c) Two prolonged blasts.
    (d) Two prolonged and one short blasts.

182. In restricted visibility, a vessel 'Not Under Command' underway, but not making way, gives the following sound signal:

    (a) One prolonged blast.
    (b) Two prolonged blasts.
    (c) Two prolonged and one short blasts.
    (d) The same signal as when making way.

183. A vessel showing 'Not Under Command' lights and an anchor light is a vessel:

    (a) Aground.
    (b) Not Under Command and at Anchor.
    (c) At anchor, with a 'diver below'.
    (d) Restricted in ability to manoeuvre.

184. Which of the following is 'Restricted in her Ability to Manoeuvre'?

    (a) Vessel with faulty steering system.
    (b) Vessel engaged in towing.
    (c) Vessel constrained by draft.
    (d) Vessel with a 'Diver Below.'

Chapter 11: Collision Prevention Regulations

185. When two vessels are on a crossing course, the risk of collision exists under the following circumstances:

    (a) Steady bearing and increasing range.
    (b) Changing bearing and steady range.
    (c) Steady bearing and decreasing range.
    (d) Steady bearing and steady range.

186. While making way in restricted visibility, you hear the sound signal of a vessel forward of your beam. As per Rule 19 (e), you are required to immediately:

    (a) proceed at slow speed.
    (b) reduce speed to the minimum at which your vessel can be kept on course.
    (c) proceed at safe speed.
    (d) engage astern propulsion.

187. The minimum range of sidelights of a 40-metre vessel is:

    (a) 1 mile.
    (b) 3 miles.
    (c) 5 miles.
    (d) 2 miles.

188. The minimum range of sidelights of a 10-metre vessel is:

    (a) 1 mile.
    (b) 3 miles.
    (c) 5 miles.
    (d) 2 miles.

189. The minimum range of masthead light(s) of a 30-metre vessel is:

    (a) 1 mile.
    (b) 3 miles.
    (c) 5 miles.
    (d) 2 miles.

190. The minimum range of the all-round 'red over white' lights on a 60-metre 'fishing' vessel is:

    (a) 1 mile.
    (b) 3 miles.
    (c) 5 miles.
    (d) 2 miles.

191. The minimum range of the towing lights, on vessels of 10-metres and 40-metres in length, are:

    (a) the same.
    (b) 2 & 3 miles respectively.
    (c) 3 & 5 miles respectively.
    (d) 1 & 2 miles respectively.

192. The minimum range of the all-round white light on an inconspicuous partly submerged object being towed is:

    (a) 1 mile.
    (b) 5 miles.
    (c) 2 miles.
    (d) 3 miles.

193. A towing light is:

    (a) a yellow light of the same arc as a stern light.
    (b) of the same arc as a masthead light
    (c) an all-round yellow light.
    (d) an all-round white light.

194. Sailing vessel aground displays:

    (a) Two white lights and a red light.
    (b) Masthead light only.
    (c) the same lights as a power driven vessel aground.
    (d) The anchor light(s).

195. Three lights in a vertical line, consisting of green, white & red - from top to bottom - indicate a vessel:

    (a) on pilotage duty.
    (b) engaged in trawling.
    (c) sailing.
    (d) positioned in a blocked channel.

196. Three lights in a vertical line, consisting of red, white, red, indicate:

    (a) the port side of a vessel engaged in fishing.
    (b) the 'Restricted in Ability to Manoeuvre' signal.
    (c) the port side of a vessel engaged in fishing or the 'Restricted in Ability to Manoeuvre' signal.
    (d) a vessel Not Under Command making way.

197. For a vessel to indicate that she is in distress by means of firing a gun, the firing should be at the following approximate intervals:

    (a) 1 minute.
    (b) 2 minutes.
    (c) 3 minutes
    (d) 5 minutes.

198. Which of the following is a distress signal?

    (a) Green flare
    (b) Red flare
    (c) White flare
    (d) Yellow flare

199. In relation to a vessel approaching you from 30° abaft the starboard beam:

    (a) You are the give way vessel.
    (b) She is a crossing vessel.
    (c) She is an overtaking vessel.
    (d) You should alter course or reduce speed.

200. Among the following four vessels, one vessel that should keep out of the way of the other three, is:

    (a) Engaged in fishing
    (b) Not under command
    (c) Restricted in her ability to manoeuvre
    (d) Constrained by draft

201. According to Rule 20, the rules concerning shapes shall be complied with:

    (a) when vessels are in sight of one another.
    (b) from sunrise to sunset.
    (c) in poor visibility
    (d) by day.

202. According to Rule 20, the rules concerning lights shall be complied with:

    (a) by night.
    (b) from sunset to sunrise.
    (c) from sunset to sunrise & from sunrise to sunset in restricted visibility.
    (d) When vessels are in sight of one another.

203. According to Rule 20, the rules concerning lights and shapes shall be complied with:

    (a) in all weathers.
    (b) from sunset to sunrise.
    (c) at all hours.
    (d) in restricted visibility.

204. You see a white light with two other lights below it - a green on the left and a red light on the right. It is:

    (a) A power driven vessel of less than 50 metres in length.
    (b) A power driven vessel of any length.
    (c) A power driven vessel at anchor.
    (d) A power driven vessel seen from astern.

205. The number of masthead lights, displayed by a vessel engaged in towing another vessel alongside, is:

    (a) two.
    (b) Two or three.
    (c) Three.
    (d) One.

206. A vessel is displaying a signal consisting of ball, diamond, ball - in a vertical line - on her mast. In addition, and at a lower level, she is displaying two balls in a vertical line on one side and two diamonds in a vertical line on the other side of the vessel. All these shapes are black in colour. She is:

    (a) engaged in minesweeping.
    (b) engaged in pair trawling.
    (c) Marking a blocked channel.
    (d) restricted in her ability to manoeuvre, also indicating her obstructed and unobstructed sides.

207. Someone on board a vessel raising and lowering outstretched arms is:

    (a) indicating being in distress.
    (b) indicating the direction of another vessel in distress.
    (c) undertaking physical exercise.
    (d) requesting a tow.

208. In relation to Collision Prevention Regulations, the word "breadth" means the:

    (a) Moulded breadth.
    (b) Greatest breadth.
    (c) Breadth at waterline.
    (d) Breadth amidships.

209. The Collision Prevention Regulations state that, when in doubt, assume the situation to exist and act accordingly. This is stated for the following situation(s):

    (a) Head-on
    (b) Risk of collision
    (c) All three situations stated here
    (d) Overtaking

210. In relation to Collision Prevention Regulations, when a vessel engaged in towing is in head-on situation with a power driven vessel:

    (a) Both vessels alter course to starboard and pass port to port.
    (b) The vessel engaged in towing maintains her course & speed.
    (c) The power driven vessel gives way.
    (d) The vessel engaged in towing has the right of way.

211. In Rule 21, a Flashing Light is defined as flashing at the following per minute frequency:

    (a) 60
    (b) 60 or more
    (c) 120
    (d) 120 or more

Chapter 11: Collision Prevention Regulations

212. Three red lights in a vertical line is the signal for a vessel:

    (a) Aground
    (b) Not under command
    (c) Constrained by her draft
    (d) Restricted in her ability to manoeuvre

213. In restricted visibility, you hear the sound signal consisting of one prolonged blast followed by two short blasts. It could be from the following type of vessel:

    (a) Engaged in fishing
    (b) Sailing
    (c) Not under command
    (d) Any one of the vessels stated here.

214. In restricted visibility, you hear the sound signal consisting of one prolonged blast followed by two short blasts. It could be from the following type of vessel:

    (a) Any one of the vessels stated here.
    (b) Engaged in towing
    (c) Constrained by draft
    (d) Restricted in her ability to manoeuvre

215. Ringing of bell is the restricted visibility signal for vessels:

    (a) At anchor
    (b) Aground
    (c) At anchor or aground
    (d) Of less than 12 metres in length

216. In restricted visibility, vessels of less than 12 metres in length are obliged to give the following signals:

    (a) The same signals as for larger vessels.
    (b) Any efficient sound signal at intervals of not more than 2 minutes
    (c) Bell or whistle signals
    (d) Portable horn signals

217. The restricted visibility sound signals for a pilot vessel on pilotage duty are:

    (a) as for a power driven vessel.
    (b) as for a power driven vessel, plus an optional identity signal.
    (c) as for a power driven vessel, plus an identity signal.
    (d) as for a power driven vessel, plus four short blasts.

218. Rule 36 states the following in relation to the use of signals to attract the attention of another vessel:

    (a) Only the signals authorised in the Rules must be used.
    (b) Do not direct the beam of a searchlight in the direction of the danger.
    (c) The use of strobe lights is encouraged.
    (d) Light or sound signals other than those authorised in the Rules may be used.

CHAPTER 11: ANSWERS

1 (a), 2 (b), 3 (c), 4 (b), 5 (d),
6 (b), 7 (d), 8 (c), 9 (b), 10 (c),
11 (b), 12 (a), 13 (c), 14 (d), 15 (a),
16 (c), 17 (b), 18 (c), 19 (d), 20 (d),
21 (a), 22 (b), 23 (d), 24 (b), 25 (c),
26 (c), 27 (d), 28 (b), 29 (a), 30 (d),
31 (d), 32 (c), 33 (b), 34 (a), 35 (b),
36 (a), 37 (b), 38 (c), 39 (a), 40 (c),
41 (b), 42 (a), 43 (d), 44 (c), 45 (b),
46 (d), 47 (a), 48 (b), 49 (d), 50 (b),
51 (d), 52 (d), 53 (a), 54 (c), 55 (b),
56 (a), 57 (b), 58 (d), 59 (d), 60 (b),
61 (b), 62 (c), 63 (b), 64 (a), 65 (d),
66 (a), 67 (d), 68 (b), 69 (c), 70 (d),
71 (c), 72 (d), 73 (c), 74 (a), 75 (a),
76 (b), 77 (d), 78 (b), 79 (c), 80 (d),
81 (c), 82 (b), 83 (c), 84 (b), 85 (b),
86 (d), 87 (c), 88 (b), 89 (a), 90 (a),
91 (a), 92 (d), 93 (d), 94 (a), 95 (c),
96 (a), 97 (d), 98 (a), 99 (d), 100 (c),
101 (b), 102 (a), 103 (d), 104 (b), 105 (a),
106 (a), 107 (b), 108 (a), 109 (b), 110 (c),
111 (d), 112 (b), 113 (c), 114 (c), 115 (a),
116 (d), 117 (a), 118 (b), 119 (d), 120 (a),
121 (d), 122 (b), 123 (c), 124 (a), 125 (d),
126 (d), 127 (c), 128 (b), 129 (c), 130 (a),
131 (d), 132 (d), 133 (c), 134 (d), 135 (b),
136 (c), 137 (d), 138 (b), 139 (d), 140 (c),
141 (d), 142 (b), 143 (a), 144 (c), 145 (d),
146 (b), 147 (c), 148 (b), 149 (a), 150 (c),
151 (a), 152 (b), 153 (d), 154 (c), 155 (b),
156 (c), 157 (d), 158 (d), 159 (b), 160 (d),
161 (b), 162 (a), 163 (b), 164 (c), 165 (d),
166 (a), 167 (c), 168 (b), 169 (a), 170 (b),
171 (c), 172 (d), 173 (a), 174 (d), 175 (c),
176 (b), 177 (d), 178 (a), 179 (b), 180 (a),
181 (c), 182 (d), 183 (a), 184 (d), 185 (c),
186 (b), 187 (d), 188 (a), 189 (c), 190 (b),
191 (a), 192 (d), 193 (a), 194 (c), 195 (b),
196 (c), 197 (a), 198 (b), 199 (c), 200 (a),
201 (d), 202 (c), 203 (a), 204 (a), 205 (b),
206 (d), 207 (a), 208 (b), 209 (c), 210 (a),
211 (d), 212 (c), 213 (d), 214 (a), 215 (c),
216 (b), 217 (b), 218 (d)

ANSWERS TO QUIZZES ON NAVIGATION LIGHTS

1. Power driven vessel underway - seen from port side.
2. Power driven vessel less than 50 m in length underway - seen end on.
3. Power driven vessel underway - seen end on.
4. Power driven vessel less than 50 m in length underway - seen from starboard side.
5. Vessel of any length engaged in towing - length of tow less than 200 m - seen from starboard side.
6. Vessel being towed alongside - seen end on.
7. Vessel being pushed ahead - not being part of a composite unit - seen end on.
8. Towing light and stern lights visible from the quarter.
9. Vessel under 50 m in length - towing a partly submerged object - seen from port side. The object is less than 25 m in breadth.
10. Sailing vessel underway - seen from port side.
11. Sailing vessel underway - seen from port side.
12. Sailing vessel underway - seen end on.
13. Sailing vessel less than 7 m in length or vessel under oars or a vessel seen from astern - under way.
14. Vessel engaged in fishing - other than trawling - seen from starboard side.
15. Vessel less than 50 m in length - engaged in trawling - seen from starboard side.
16. Vessel engaged in fishing - other than trawling - seen from astern - indicating the direction of her outlying gear.
17. Vessel of any length engaged in trawling - seen from port side.
18. Vessel of any length engaged in trawling - shooting nets - seen from port side.
19. Vessel of any length engaged in trawling - hauling nets - seen from port side.
20. Vessel engaged in trawling - with nets fast on an obstruction - seen from starboard side.
21. Vessel engaged in fishing with purse seine gear - seen end on.
22. Vessel Not Under Command - not making way.
23. Vessel Not Under Command - making way - seen from starboard side.
24. Vessel Restricted in Ability to Manoeuvre - underway - seen from astern.
25. Vessel Restricted in Ability to Manoeuvre - underway - indicating obstruction on port side.
26. Vessel of any length - Restricted in Ability to Manoeuvre - at anchor.
27. Vessel under 50 m in length - engaged in towing and Restricted in Ability to Manoeuvre - seen from starboard side. Length of tow exceeds 200 m.
28. Vessel engaged in mine clearance operations - underway - seen from astern.
29. Vessel constrained by her draught - underway - seen from port side.
30. Vessel constrained by her draught - underway - seen from astern.
31. Vessel engaged on pilotage duty.
32. Vessel of any length - at anchor.
33. Vessel of any length - aground.

THE FLARES THAT ARE NOT RED OR ORANGE

(Not a part of the collision REGULATIONS)

A **WHITE FLARE** is a collision-warning signal. It is not a distress signal. Small vessels, especially racing yachts carry up to 12 white flares on board. A less conspicuous small vessel may fire such a signal when she feels there is a need to draw the attention of a ship heading towards her on a collision course.

In some yacht races, white flares are also used to illuminate the starting line.

A **GREEN FLARE** is fired at night by a search aircraft (or its landing lights are switched on at 3 to 5 minutes intervals) to encourage pyrotechnic-equipped survivors to identify their location by firing a red (distress) flare. On sighting the distress flares the aircraft fires a succession of green flares (and/or switches on its landing lights). *Search and Rescue is discussed in Chapter 19.3.*

MOORING AREAS

Small vessels moored to buoys in designated mooring areas are usually not required to show an anchor signal or to be lit at night. Masters of other vessels are required to be aware of the location of such moorings.

Chapter 12

AUSTRALIAN BOATING REGULATIONS & POLLUTION CONTROL

BOATING REGULATIONS

AUSTRALIAN STANDARDS – BOATING EQUIPMENT
- AS 1799.1 Small pleasure boats code Part I — General requirements for power boats
- AS 1512 Personal flotation devices — Type 1
- AS 1499 Personal flotation devices — Type 2
- AS 2260 Personal flotation devices — Type 3
- AS 2259 PFD design
- AS 2092 Pyrotechnic, marine distress flares and signals for pleasure craft
- AS 2198 Anchors for small boats
- AS 2259 General requirements for buoyancy aids
- AS/NZS 4330 121.5 and 243.0 MHz emergency position indicating radio beacons (EPIRBS) including personal EPIRBs
- AS/NZS 4280 406 MHz satellite distress beacons
- AS 2865 Confined/enclosed spaces
- AS 4142 Ropes
- AS 1799.1 and 1851.1 Fire extinguishers
- AS 1345 Colour coding

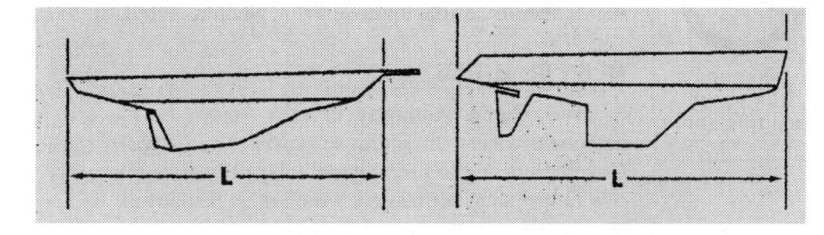

LENGTH MEASUREMENT – A GUIDE

For the purpose of vessel registration and commercial surveys, the hull length is usually measured from the point of the bow to the transom, excluding bowsprits, outboard motors and other appendages.

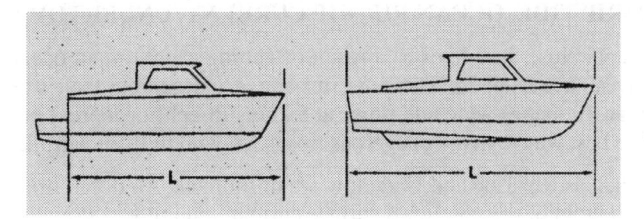

Fig 12.1: LENGTH MEASUREMENT

GENERAL BOATING

The International Regulations for Preventing Collisions at Sea, 1972, Rule 1(b) and (c), give local authorities the power to make rules and signals to suit their local needs. However, they must make sure that international shipping is not confused by such rules or signals. These rules are subject to change at short notice. The reader must consult the local boating guide, available from the marine authority in each State or Territory.

WASH is the disturbance flowing outward from the vessel which causes damage, erosion, injury or annoyance to others, created as a boat moves through the water.

Fig 12.2: A NSW WATERWAYS STICKER

Warning! It is an offence to allow a person to extend any part of their body over the bow, side or stern of a powerboat underway.

Fig 12.3

WAKE is the disturbance in the water directly astern of the vessel.

SPEED LIMITS, RESTRICTIONS, LIMITATIONS & PROHIBITIONS

When passing a DIVER signal, proceed at a safe speed and at a safe distance while maintaining proper lookout. In Western Australia the safe distance is specified as 50 metres and in Tasmania 120 metres. In New South Wales, all vessels travelling at 10 knots or more, must keep at least 30 metres from persons in the water (60 metres when towing a water skier or when driving a PWC). If the channel is too narrow to pass at this distance the vessel must reduce speed to the slowest speed while maintaining its control. Restrictions in other States include:
- *Victoria*: 5 knots within 50 metres (of a person or vessel) & 100 metres (of a diving flag).
- *South Australia*: 4 knots within 30 metres.
- *Queensland*: 6 knots within 30 metres.

Do not obstruct any FAIRWAY, CHANNEL or approach to any landing place.

The speed limit within 100 metres of a DREDGE, FLOATING PLANT, SHORE, WHARF, JETTY, etc. is 4 knots. Some State variations are as follows:
- *Tasmania*: 5 knots within 60 metres.
- *Victoria*: 5 knots within 50 metres (200 metres of shore on coastal waters).

Anchoring is prohibited near SUBMARINE CABLES. 'Anchoring Prohibited' signs are posted on enclosed waters. In some States no clearance is specified; in *Queensland* it is 50 metres.

Securing to a FLOATING OBJECT (e.g., a navigation buoy) or a STRUCTURE on land owned privately or by a public authority is not permitted without prior approval.

Securing to a PUBLIC JETTY is permitted for a limited time, without obstructing others. The vessel should be able to move away when required.

Fig. 12.4

Fig. 12.5

OPEN BEACHES

The Marine Authority in each State (see the last page of the book for websites) regulates navigable waters. In addition, in *New South Wales*, the local councils have power under the Local Government Act to prohibit the entry of vessels into areas of open waters below the high water mark. 'No Entry' signs are posted, which must be obeyed.

TIDE RIP (RIP TIDE, OCEAN RIP, RIP CURRENT, UNDERTOW)

Many people lose their lives as a result of getting caught in an ocean rip. A rip is formed by water leaving the beach after the waves have come in. There are rips at every beach but their strength and intensity varies. Typically, a dangerous rip is less than 10 metres wide and flows at 5 knots (8 kph) or more. Lengthwise, the current may extend 70 to 700 metres offshore. The best way to avoid dangerous rips is to swim only on patrolled beaches, between the red and yellow flags.

Terms such as Rip Tide and Undertow are misnomers. An ocean rip is neither a tidal movement nor a force pulling you down to the bottom of the sea. It is a surface current, pulling you straight out to sea. However, an ocean rip may knock you off your feet in shallow water, and if you thrash around and get disorientated, you may end up being pulled along the ocean bottom.. But if you relax your body, the current should keep you near the surface.

Ocean rips are terrifying because they catch you off guard. One minute you are bobbing along peacefully in surf, the next you're being dragged out to sea at top speed. They occur in all weathers and on a wide range of beaches. Unlike violent, crashing waves, you probably wouldn't notice a rip current until you are right in the middle of it.

An untrained eye is unable to track a rip. But they can be recognized from their darker, greener water, which indicates the water is deeper and in a channel. Washed up sand may give the rip a brown tinge and its water surface may have white froth, rubbish and debris moving out to sea. You may also see ripples on the surface. Waves may break on either side of a rip. Surfers often use rips to paddle out to the breakers.

Do not panic if you get caught in a rip. Do not swim against it. You will get exhausted and drown. If you are a reasonable swimmer, swim parallel to the shore until you reach breaking waves. You are then out of the current and can swim safely towards the shore. For others, the best thing to do is to lie on your back and float with the current until it loses its strength. A strong rip could carry you 200 or more metres from the shoreline, but on patrolled beaches, assistance is never far away.

KEEP OUT OF THE WAY OF

- A vessel showing a "PILOT on board" flag (flag H) or a pilotage EXEMPTION FLAG (a rectangular white flag).
- A WARSHIP displaying 4 international signal flags vertically at a yard.
- A SUBMARINE on the surface.

Chapter 12: Australian Boating Regulations & Pollution Control

- All NAVAL VESSELS (at least 100 metres).
- COMMERCIAL SHIPPING AREAS:
  Sailboarding in these areas is usually prohibited.
- WHALES: as illustrated.

DIVING & UNDERWATER OPERATIONS (DAY & NIGHT) & DIVING RESTRICTIONS

Any vessel with a diver below comes under the definition of "Restricted in Ability to Manoeuvre" (Chapter 11, Rule 3). It should exhibit the signals prescribed in Rule 27 (b) & 27 (d). If the size of the vessel makes it impracticable to exhibit the prescribed day shapes, she may exhibit the international code flag "A".

IN SUMMARY, THE FOLLOWING SIGNAL SHOULD BE SHOWN:

- *Small Vessels* = Flag "A"
- *Large Vessels* = Black ball, Black diamond, Black ball.
- *Night Signal (& in restricted visibility)* = Red, White, Red lights in a vertical line.

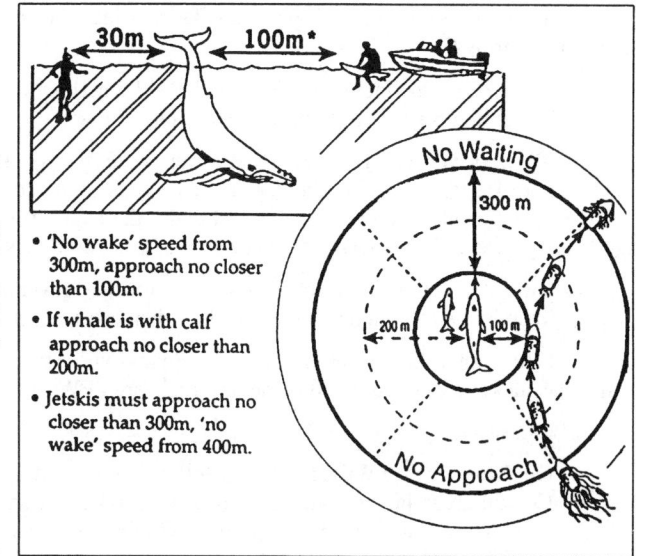

- 'No wake' speed from 300m, approach no closer than 100m.
- If whale is with calf approach no closer than 200m.
- Jetskis must approach no closer than 300m, 'no wake' speed from 400m.

Fig 12.6: LEGAL APPROACH LIMITS TO SWIMMING WHALES *(NSW Parks & Wildlife Service)*

Diving within 100 metres of a fairway or channel is not permitted without permission. *In South Australia*, diving is prohibited within 200 metres of any vessel 15 metres or greater in length navigating in a channel, or within 45 metres of a vessel 15 metres or greater in length moored at a wharf.

WATER SKIING, AQUAPLANING & PARAFLYING

- Usually maximum 2 skiers, or 3 with permission are permitted. In *Queensland*, there is no limit on the number of persons that may be safely towed. Safety is the responsibility of the owner and operator of the vessel. They must comply with the State's "General Safety Obligation". Further information is available from Queensland Transport.
- Prohibited between sunset and sunrise. In *Victoria* skiing is permitted from one hour before sunrise to one hour after sunset.
- Keep away from other persons & structures:
  - When not towing = 30 metres.
  - When towing = 60 metres.
- Keep parafliers away from overhead cables & bridges = 300 metres
- An "observer" must be carried with no sight or hearing disabilities. In *New South Wales* the observer must be at least 16 years of age or the holder of a Young Adult Licence (at least 12 years old). In Western Australia the minimum age of the observer is 14. In *Victoria and Queensland* 12, and *Tasmania* 10. In *South Australia*, the observer must be at least 16 years of age. However, a person between 12 and 15 years of age holding a motorboat operator's permit may act as an observer when accompanied by an operator of at least 18 years of age.
- See "Power Boat Licence" below for the minimum age of the boat driver.

PERSONAL WATER CRAFT (PWC) OR JET BOAT REGULATIONS

NOTE: These regulations may apply to all types of PWCs, including Jet Skis, Power Skis, Aqua Scooters, Jet Bikes, Wave Runners, Wave Jammers, etc.

In *New South Wales, Queensland, Western Australia, Victoria and most other States and Territories*, the Personal Water Craft users must observe the following Safety Code:

Always wear a PFD (lifejacket) and operate in a safe prudent manner. Be aware of the environment; obey speed limits, signs and local regulations.

Fig 12.7

Do not ride a PWC after consuming alcohol. Stay at least 60 metres away from persons in the water, small non-powered vessels, such as sailing or rowing craft under 4 metres in length, and designated surf or swimming zones *(defined below)*. (In *Queensland* you can pass a person in the water in a responsible manner at a distance of less than 60 metres at a speed of less than 6 knots). Stay away from residential areas or places where the PWC engine noise can annoy other people such as launching ramps, picnic and swimming areas.

In New South Wales, *a surf zone* is the area extending 500 metres out from the shore between surf patrol flags. A *swimming zone* is an area of water defined by signs for swimmers and extends 60 metres from shore. The only time you can enter a surf or swimming zone is when you use your PWC to rescue a person. When operating a PWC at 10 knots or more

(6 knots or more in *Queensland*) you must keep at least 30 metres from all other vessels, shore and any structure such as jetties or moorings.

In *Sydney Harbour* (including North and Middle harbours) and *Parramatta River* PWC are totally banned. In the rest of NSW, PWC riding is not permitted from sunset to sunrise.

The PWC driver and all riders must be in possession of a PWC Licence or a Young Adult PWC Licence under the supervision of a PWC Licence holder. They all must understand this code. In *Victoria*, a person over the age of 12 may operate a PWC with speed restrictions and under the supervision of a 16 years old person. In *Queensland*, a person of any age may operate a PWC under the supervision of a licensed driver.

Obey the International & Local Regulations for the Prevention of Collisions on the water. Sailing vessels have the right of way. Keep to the right in narrow channels. When crossing, the vessel on the right has the right of way.

All PWCs must be registered. Affix a "Behaviour" Sticker outlining the above rules at a prominent position on the craft. The sticker is sent out to all owners of registered PWCs.

In *Western Australia*:

➢ On Swan River, PWC Freestyling *(altering course & speed unpredictably)* is permitted only in designated areas.
➢ Certain areas in the State, including waterski areas and Swan River Marine Parks, are prohibited to PWC, PWC Freestyling and PWC Wave Jumping *(becoming airborne over a breaking wave or swell)*.
➢ PWC Wake Jumping is prohibited throughout the State.

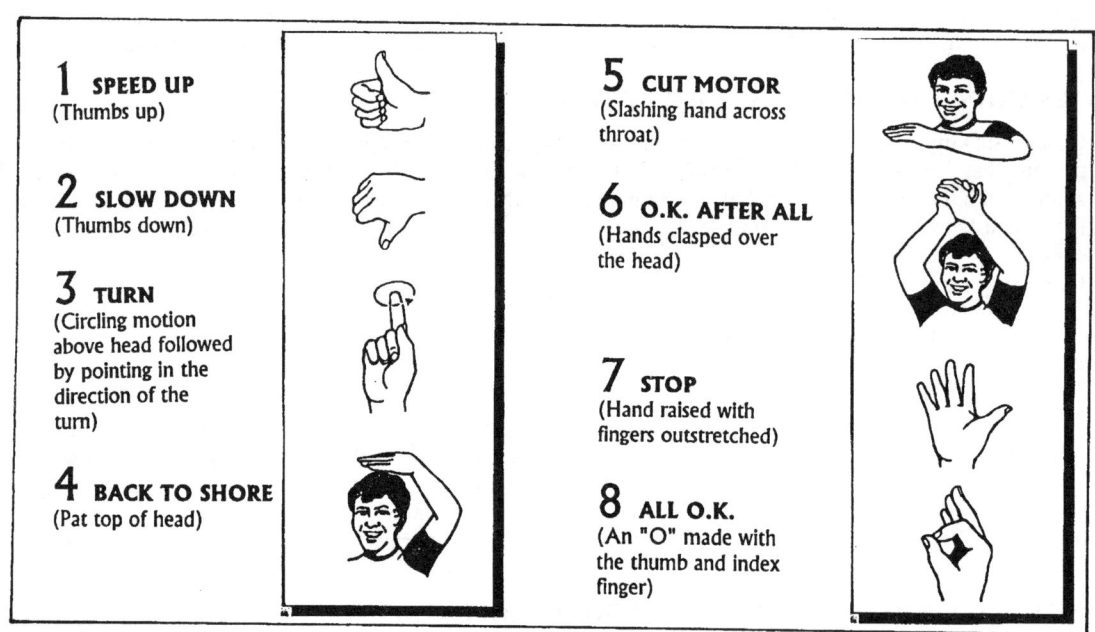

1 SPEED UP
(Thumbs up)

2 SLOW DOWN
(Thumbs down)

3 TURN
(Circling motion above head followed by pointing in the direction of the turn)

4 BACK TO SHORE
(Pat top of head)

5 CUT MOTOR
(Slashing hand across throat)

6 O.K. AFTER ALL
(Hands clasped over the head)

7 STOP
(Hand raised with fingers outstretched)

8 ALL O.K.
(An "O" made with the thumb and index finger)

Fig 12.8: Water Ski Hand Signals

PILOTAGE PORTS

In pilotage ports (e.g., *Sydney, Fremantle, Melbourne, Brisbane, Hobart, Darwin*, etc.) large vessels are navigated by a pilot or by the master or a ship's officer who has been granted a pilotage exemption certificate on passing a Local Knowledge Examination. Vessels navigated with a pilot are required to show by day the International Code Flag "H".

Small vessels are usually exempt from carrying a pilot and their masters are not required to pass the local knowledge examination. For example, in *New South Wales*, vessels of less than 30 metres in length; and in *Queensland*, vessels of less than 35 metres in length are exempt. In *Queensland*, the length exempted is proposed to be increased to 50 metres; and all vessel operating only in a pilotage area are also proposed to be exempt. You should check the regulations in your State.

Chapter 12: Australian Boating Regulations & Pollution Control

POWER BOAT LICENCE, *or in Queensland*, RECREATIONAL SHIP MASTER'S (RSM) LICENCE
(Also known as BOATING LICENCE, SPEED BOAT LICENCE or MARINE DRIVER'S LICENCE)

A person who drives a mechanically propelled vessel is required to hold a Boat Licence in most Australian States. Only in Western Australia and Northern Territory a licence is not required. Meanwhile, a national recreational boating safety system introduced by the National Marine Safety Committee (NMSC) is gradually being adopted nationwide (See below).

In New South Wales, the licence must be obtained for driving a power-driven vessel at a speed of 10 knots or more. The person must be at least 16 years of age to obtain a General (full) Licence. A Young Adult Licence may be issued to a person between the ages of 12 and under 16 to operate under the following restrictions:
- Must be accompanied by the holder of a General Licence when operating at a speed of 10 knots or more.
- Maximum speed 20 knots
- Towing of water-skiers prohibited
- Travel at under 10 knots between sunset & sunrise
- Towing of aquaplaners/parasailors limited to under 10 knots
- Must obtain the relevant Marine or Waterways Authority's permission before participating in organized events.

In South Australia, the licence is required to operate any vessel fitted with a motor, irrespective of whether the motor is used or not. A person must be at least 16 years of age to obtain the licence. A person between 12 and 15 years of age may be issued with a BOAT OPERATOR'S PERMIT. An unaccompanied permit holder is restricted to operate vessels not exceeding 4 metres in length which have a potential speed of no greater than 10 knots.

In Tasmania, it is for vessels capable of exceeding 8 knots, and the minimum age for a full licence is 17. A provisional licence may be issued to a person between the age of 12 and under 17 with restrictions similar to those in NSW.

In Queensland, the Recreational Ship Master's (RSM) Licence is required to drive a vessel by a motor exceeding 4.5 kw (6 hp) and capable of 10 knots and with a planing or non-displacement hull. The age criteria and other restrictions are similar to those in NSW.

In Northern Territory and Western Australia, the driver of a recreational vessel is not required to be licensed, and there is no restriction on the age of the boat driver, so long as an adult is in charge and directly supervises the driver. A 17-years old may be in charge independently.

In Victoria, the driver must be at least 16. A person between the age of 12 and under 16 is restricted in the same manner as in NSW.

NATIONAL STANDARD FOR BOAT LICENCES
(Summary of NMSC guidelines. NMSC is discussed below)

- A boat operator's licence to be required only for mechanically propelled boats (power boats).
- Licences to be recognised and transportable between States & Territories, just like the motor vehicle licenses.
- Persons over 16 years of age to be eligible for a full licence, and between 12 and 16 years of age for a restricted licence.
- Medical fitness standard to be similar to that required for a motor vehicle licence; minimum corrected vision 6/12 in at least one eye.
- Licence to be issued on successful completion of an approved training program from an accredited provider, or on passing the marine authority's examination.

RECOGNITION OF VISITORS' SPEED BOAT, RSM (*Qld*), AND PWC LICENCES

Pending the adaptation of National Standard for Boat Licence, temporary recognition of interstate licences currently exits. Generally speaking, the local licence is not required when:

- Your usual place of residence is in Australia but outside the State or Territory where you are a visitor.
- You have not been in the Sate or Territory continuously during the three months preceding the date on which you are driving the vessel.
- You hold a valid current licence or permit to drive a vessel of the relevant class issued under the legislation of your home State or Territory. [If you come from a State or Territory where vessel drivers are not required to be licensed, you will be required to obtain a local licence].
- You comply with any relevant terms and conditions applicable to your home State or Territory.

CANCELLATION OF A POWER BOAT LICENCE

The Authority may cancel a powerboat licence after one conviction for negligent or reckless navigation, or for driving under the influence of alcohol or drugs. It can also be cancelled after two convictions causing nuisance, annoyance or danger, water-ski offences, distance-off offences, false statement or being incapable of safe driving.

HULL IDENTIFICATION NUMBER (HIN) OR BOAT CODE

In some States the registrable vessels and vessels used as ferries and tugs are required to be affixed with a Hull Identification Number (HIN). It is a 14-character code that describes the country of origin, manufacturer, serial number of the vessel, month of production, year of production and model year. This number is used to identify each vessel.

In *NSW*, the rule applies to all new vessels manufactured in the State, and other vessels prior to transfer of registration or at new registration whichever comes first.

```
AU - SAV12345D595
```
Country Code — Manufacturers Code — Serial Number — Month Code — Model Year — Production Year

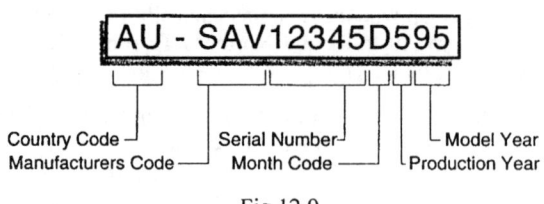

Fig 12.9

REGISTRATION OF VESSELS

NOTE: The State Registration *of recreational vessels must not be confused with the* State Survey *for Commercial Vessels. It must also not be confused with the* Commonwealth Registration *of vessels proceeding overseas (See 'Overseas Voyages' below). However, in* Queensland, *the term 'Registration' is used in relation to both the recreational and commercial vessels.*

In New South Wales, all PWCs and vessels must be registered. The exemption from this requirement applies only to vessels of less than 5.5 metres in length, powered by an engine of less than 4 kw (5 hp or less), not capable of travelling at 10 knots or more, and not subject to a mooring licence.

A typical exempt vessel is a trailer boat of less than 5.5 metres in length with less than 4 kw (5 hp) engine and not capable of travelling at 10 knots or more.

The registration number and label are required to be displayed on the hull of the vessel.

In Western Australia, all boats fitted for a motor or with a motor are required to be registered. Only a tender to a larger vessel is exempt when no more than 3.1 metres in length and not equipped with a motor bigger than 3.73 kw (5 hp).

In Victoria and South Australia all vessels fitted with an engine must be registered. In Queensland and Tasmania, the registration is compulsory for vessels fitted with a motor 3 kw and over (over 4 hp). The boat registration is not required in Northern Territory.

BOAT CAPACITY LIMITS

The safe carrying capacity of power driven recreational vessels is determined by the Australian Safety Standard AS1799, with some variations between the States and Territories:

- Length less than 3 metres   =   No. of persons   2
- 3 metres to less than 3.5 metres   =   3
- 3.5 metres to less than 4.5 metres =   4
- 4.5 metres to less than 5 metres   =   5
- 5 metres to less than 5.5 metres   =   6
- 5.5 metres to less than 6 metres   =   7

If the vessel is six (6) meters or more in length the carrying capacity is calculated as follows:

For a single decked vessel

No. of persons = $0.75 \times \text{vessel length} \times \sqrt{\text{vessel breadth}}$
(Length & breadth measured in metres)
For example, the carrying capacity of a vessel 12m length and 4m breadth = 18 persons.

For a vessel with a fly-bridge

No. of persons = $0.6 \times \text{vessel length} \times \sqrt{\text{vessel breadth}}$
(Length & breadth measured in metres)
For example, the carrying capacity of a vessel 12m length and 4m breadth = 15 persons.

The number of people on the fly bridge must not exceed 25% of the total number of people allowed on the vessel.

Notes:
(a) A child up to & including 1 year of age does not count.
(b) A child over 1 & under 12 years equals ½ adult.

(c) On recreational vessels with individual cockpits, (e.g. decked canoes or kayaks), the number of persons carried on the vessel must not exceed the number of individual cockpits, irrespective of the age of the person.

All registrable recreation vessels, except yachts with auxiliary motor, are required to have a capacity plate attached. The capacity plate(s) should be displayed near the boat's control area where the operator can see it at all times.

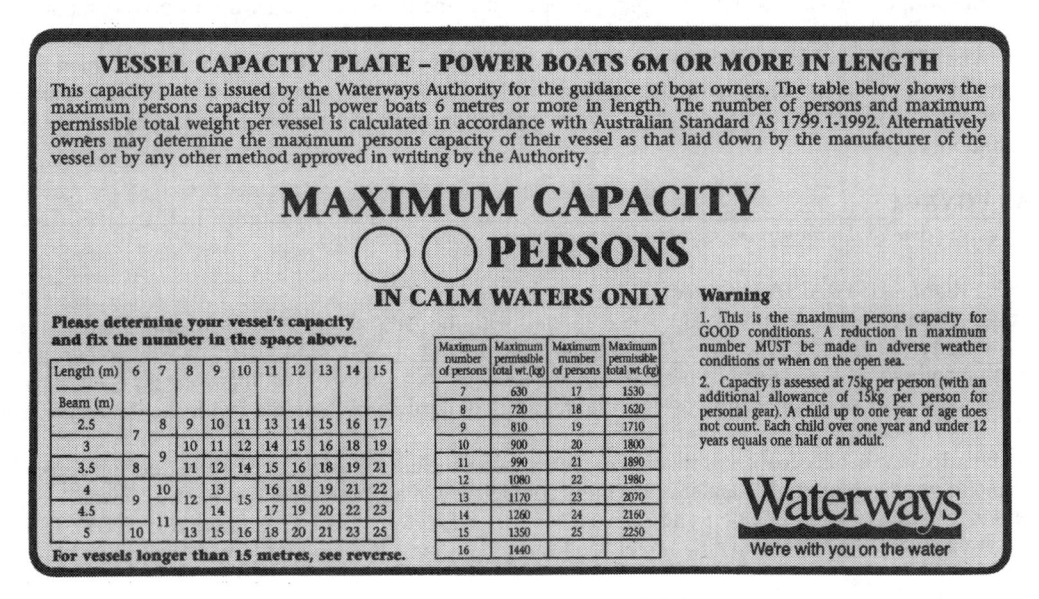

Fig 12.10: A NSW Capacity Plate

NATIONAL COMPLIANCE PLATE

A national recreational boat compliance plate program was introduced in 2001. Under this program, compliance plates are to be fixed to all new boats manufactured in Australia showing serial number, passenger capacity, maximum engine rating, standard of buoyancy and the manufacturer's name and model type.

The compliance plate will indicate the boat's response to swamping in the form of the following buoyancy standards:
- *Level Flotation*: the boat will remain floating in a level position with passengers and gear on board.
- *Basic Flotation*: the boat will remain floating with passengers clinging to it.
- *A blank space in the buoyancy section*: the boat's buoyancy standard is not specified.

As part of the compliance plate program, the manufacturers have also agreed to nationally consistent owner manuals to provide operators with a broad range of safety information on vessel operation and maintenance, including explanations on passenger load and carrying capacity, stability, fuel safety and minimum onboard safety equipment requirements.

BOAT OWNERSHIP CHANGE

When buying or selling a registrable vessel or a vessel in survey, the procedure is similar to changing ownership of a motorcar. There is printed "Change of Ownership" form for the seller and purchaser to complete and submit to the authorities.

INTERSTATE VISITORS & RECOGNITION OF VISITING VESSEL'S REGISTRATION

Temporary exemptions from registration requirements for recreational vessels are usually granted in all States and Territories for visitors from other States or Territories. Generally speaking, the local registration is not required so long as the vessel:
- Is not normally used on navigable waters in that State or Territory.
- Has not been used on navigable waters of that State or Territory for more than three months.
- Is currently registered in her home State or Territory.

LIVING ON BOARD

Living on board (which can mean staying on board for more than 48 hours) is prohibited in most parts of Australia without approval. The approval is generally based on the vessel being fitted with a suitable holding tank, and keeping record of dates, locations and quantities of pump outs. In some waterways (in parts of Gold Coast, for example) anchoring and mooring by visitors is restricted to a maximum of 24 hours in any 30-day period, unless exempt or permitted otherwise. These rules do not generally apply to vessels berthed in marinas with showers and toilet facilities.

TEMPORARY VISITOR USE OF PRIVATE MOORINGS

Under a scheme being trialled in NSW, the owners of private moorings can permit their friends and guests to use their moorings if their 'normal' vessel is absent. The purpose of the scheme is to make fuller use of private moorings. It allows mooring owners the flexibility to invite friends and guests to use their moorings while they are away overnight or for a longer cruise. The visitor use of the mooring must not exceed an aggregate of 6 months in any 12 months period. The mooring owner must not receive any payment or reward for the use of their mooring. To be eligible to participate in the scheme, the mooring owner must complete a once-a-year application form and receive a letter of approval from the NSW Waterways Authority.

OVERSEAS VOYAGES

(The following are some of the important requirements. The list is not exhaustive.)

DEPARTURE FROM AUSTRALIA
- Register the vessel, whether commercial or recreational, with the Australian Registrar of Ships (Australian Maritime Safety Authority, Canberra). The vessel would be issued with a Ship's Register, which is her passport. This identification and proof of ownership is needed for the vessel just as a person travelling overseas needs a passport.
- If vessel is to be sold or positioned overseas, the vessel may need to be 'entered' with the Australian Customs for export.
- Carry on board all available certificates relating to crew competencies and safety equipment, leaving copies back home. Be prepared to prove professional standards and crew competency in an inquiry following an accident.
- Each person must carry valid passport and relevant visa(s) or Electronic Travel Authority(s) (ETA). ETAs for some countries are available from participating travel agencies and airlines. If intending to work overseas, obtain a working visa.
- Carry a valid Vaccinations Certificate for each person, if applicable. The Yellow Fever Vaccination Certificate is usually required if the person has travelled through or landed in Tropical Zone Central Africa or South America within 6 days prior to arrival.
- Check currency restrictions for each country you intend to visit.
- Obtain health certificate for any animal on board. Check animal quarantine regulations for countries you intend to visit and re-entry to Australia.
- Obtain necessary information on Customs, Quarantine and Immigration regulations as well as Arrival Forms and Incoming Passenger Cards from the embassies/consulates of the countries you intend to visit. Check whether the vessel will be granted a Cruising Permit and if a security will be required for its temporary importation.
- If taking goods to another country with the intention of selling them, check export and import regulations with the International Chamber of Commerce.
- As discussed in GMDSS in Chapter 18, Coast Stations in some parts of the world, particularly Europe, do not provide aural (loudspeaker) watchkeeping on radiotelephony distress and calling frequencies. You would need a DSC calling facility to alert these stations to any type of calling -distress, urgency, safety or routine calling.
- Obtain an Accounting Authority Identification Code (AAIC). You would need to quote it when passing radiotelegrams and making radiotelephone calls from overseas coast stations. AAIC is available from a number of private marine communication companies, which are listed in the List of Ship Stations published by the International Telecommunication Union (ITU) or the British Admiralty. Alternatively, contact the Australian Communications Authority. The website address is listed on the last page of this book.
- For voyages longer than 24 hours, register with AusSAR RCC for AUSREP ship reporting. (AUSREP is discussed in Chapter 19.3).
- Consider joining the worldwide Automated Mutual-Assistance Vessel Rescue System (AMVER) operated by the United Stated Coast Guard (USCG). Further information is available from USCG. Its website address is listed on the last page of this book.
- Carry the relevant List of Radio Signals. The British Admiralty publishes it in numerous volumes. It contains details of International Coast Stations, Inmarsat Land Earth Stations, and Radio and Radar Beacons. It also provides details of AUSREP type ship reporting systems in various countries.
- Obtain relevant Navigational Charts, Sailing Directions (Pilot book) and International Tide Tables published by the British Admiralty or the US Hydrographer. Consult an International Chart Catalogue at a Chart Agent or a Chart Index on the UK or USA website for the publications you may need. Links can be found on the Australian Hydrographer's website address listed on the last page of this book.
- Complete the Customs and Immigration Outwards Clearance, just as it is done when departing by air. Passengers on board may be required to pay a passenger Movement Charge (Departure Tax). List major items of ship's gear on Customs Declaration Form to avoid having to pay duty on vessel's return.

Chapter 12: Australian Boating Regulations & Pollution Control

ARRIVAL OVERSEAS
- Forty-eight hours prior to arrival at first port of entry in each country, contact the Customs, Quarantine and Immigration Authorities. They can be contacted by marine radiotelephone, mobile phone or email either directly or via Harbour Control. Your first port of entry has to be where Customs, Quarantine and Immigration formalities can be completed. When contacting them, you will need to provide:
  - Vessel's name
  - Port of registry
  - Intended port of entry
  - Last port of departure
  - Number of people on board
  - Details of any illness or disease recently encountered
  - Any animals on board
  - Estimated time of arrival (ETA)
- Lodge passage details with a local boating club, search & rescue organisation, friends or relatives who may raise the alarm if you do not arrive as expected.
- Prior to entry into foreign waters, clearly display the International Pratique (quarantine clearance) flag 'Q'. See Chapter 6 for the meanings of International Code Flags. Until cleared by Quarantine, Customs and Immigration, you are required to stay on board and no other person must board the vessel. You may also be required to keep all food and animals secure until after the quarantine inspection for any infestation.
- Travel directly to the appointed boarding station or berth.
- Declare all drugs on board. This includes medications containing narcotics, hallucinogens, amphetamines, barbiturates and tranquillizers in medical kits. If using any of these drugs while in that country, record this in the vessel's logbook. A doctor's prescription may validate certain registered drugs.
- Declare all weapons and firearms. Certain weapons may be detained in safe storage for transshipment to your last port of departure. In cases where a weapon has been detained, the Master may have to contact Customs one week or more prior to departure for the return of the weapon.
- Declare all stores on board, and comply with 'duty free' limitations.
- Observe any currency restriction regulations.
- Complete relevant Arrival Forms and Incoming Passenger Cards.
- Many countries have strict quarantine regulations similar to those in Australia. They may impose penalties for illegal importation of:
- Drugs
- Animal or plant material
- Firearms, weapons or ammunition
- Protected wildlife and products made from them
- Certain food items
- Certain medicinal products, including Performance Enhancing Drugs
- Trading, especially in foodstuffs, may be prohibited.
- Contact with other vessels in port prior to clearance may be prohibited.
- Do not throw waste and foodstuffs overboard.
- The vessel may be granted a Cruising Permit or a security may be required for its temporary importation.

DEPARTURE FROM OVERSEAS
- Obtain Customs & Immigration clearance.
- Passenger Movement Charge or Departure Tax may have to be paid.

TRAVEL TO CHRISTMAS, COCOS (KEELING) & NORFOLK ISLANDS

For Immigration purposes, yachts travelling to and from Cocos (Keeling) and Christmas Islands and the mainland of Australia are deemed to have not left Australia if their trip is within 30 days of departure from the mainland Australia or these Islands. Overseas person on board these yachts must ensure that their visa covers the entire period of their stay including travel time between the mainland of Australia and these islands. Customs and Quarantine clearances are required on both arrival and departure.

Persons travelling to Norfolk Island are Immigration cleared both on arrival and departure on the mainland Australia and Norfolk Island. Therefore, overseas persons must have a multiple entry visa for return to the mainland Australia. Customs and Quarantine clearances are required on departure from and arrival to the mainland Australia.

NAVAREA X (NAVAREA 10)

Fig 12.11

Under the worldwide long-range navigational warning service, the world's oceans are divided into 16 navigational areas (NAVAREAS I to XVI). The Australian region is NAVAREA X, divided into NE, SE, SW & NW sub-regions. Australia is the area coordinator for NAVAREA X. (UK is Area I, France II, USA IV, New Zealand XIV, etc.). The NAVAREA X warnings are issued by the Australian Rescue Coordination Centre (RCC) through the AMSA Radio Network, Coast Radio Network (CRN), some Voluntary Coast Stations as well as on INMARSAT-C.

These warnings are appropriately broadcast by the relevant Radio Network for the areas close to and further from the Australian Coast. The warnings relate to navigational aids, dangers and movement of oil rig, etc. Traditionally, they have been broadcast by Coast Stations in brief text for 5 days in reverse numerical order. They are then broadcast on a numbers only basis for as long as the information remains valid or until it is promulgated by other means, such as Notices to Mariners. This schedule was due for a review at the time of writing this book.

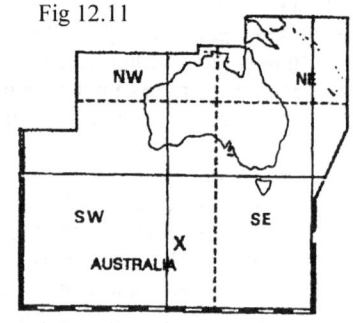

The times and frequencies for the broadcast of Maritime Safety Information by CRN for the AUSCOAST areas are listed in Chapter 18. NAVAREA X broadcast schedule is listed in the Admiralty List of Radio Signals, Volume 3. The warnings are also available from RCC Australia by direct request.

AUSTRALIAN ANNUAL NOTICES TO MARINERS

The Australian Hydrographic Office issues Fortnightly and Annual Notices to Mariners. The fortnightly notices are designed to assist mariners in keeping their charts and nautical publications corrected and up-to-date. (See Chapter 17.2). The book of Annual Notices to Mariners contains navigational safety notices of more permanent nature, such as listed below. The Notices to Mariners can be obtained direct from the Australian Hydrographic Office at 8 Station Street in Wollongong, NSW, from licensed chart agents or on the web at www.hydro.navy.gov.au.

CONTENTS OF THE ANNUAL NOTICES TO MARINERS INCLUDE:
- Requirements for reporting pollution incidents (Notice No. 31)
- Search and rescue procedures (Notice No. 4)
- Procedures for Maritime Safety Information (MSI) warnings (Notice No. 5). These include AUSCOAST warnings concerning lighthouses etc. and Safety Messages (SSM) concerning dangerous floating objects. (See Chapter 18)
- List of Weapon Practice Areas and Warnings (Notice No. 9)
- Dangers regarding submarine pipelines and cables (Notice No. 14)
- Areas dangerous due to mines and unexploded ordnance (Notice No. 12)
- Meteorological broadcasts for shipping – Analysis and Prognosis Charts as well as Schedules (Notice No. 6)
- Warnings relating to satellites-derived positions (Notice No. 27)

UNIFORM SHIPPING LAWS (USL) CODE

In 1979 the Uniform Shipping Laws Code of Australia (USL Code) was adopted as a national set of standards for survey, manning and operation of commercial vessels. Since then, it has been the basis for uniform maritime legislation in the Commonwealth, States and Northern Territory. The Australian Government publishes the Code in 18 sections, and the individual sections can be purchased separately. The sections are:

Section 1. Introduction, Definitions and General requirements
Section 2: Examinations and Certificates of Competency
Section 3: Safety Manning
Section 4: Mercantile Marine
Section 5: Construction
Section 6: Crew Accommodation
Section 7: Load Lines
Section 8: Stability
Section 9: Engineering
Section 10: Life Saving Appliances
Section 11: Fire Appliances
Section 12: Radio Equipment
Section 13: Miscellaneous Equipment
Section 14: Surveys and Certificates of Survey
Section 15: Emergency Procedures and safety of Navigation
Section 16: Collision Regulations
Section 17: River Murray Traffic Regulations
Section 18: Hire and Drive Vessels

Chapter 12: Australian Boating Regulations & Pollution Control

The Commonwealth and the State and Territory governments have enacted the USL Code with minor changes to suit individual needs. In *most States* the enactment is titled the *Marine Act*; and the *Commonwealth Act* is titled the *Navigation Act*. In NSW, it is known as the *Commercial Vessels Act*. The marine legislation in NSW is likely to be consolidated into one *Marine Safety Act* in the foreseeable future, but without changing the intent. (The Acts to be consolidated into one Marine Safety Act include the Commercial Vessels Act, NSW Navigation Act, Pilotage Act, Maritime Services Act and the Marine Safety Alcohol and Drugs Act). In *Queensland*, the Act is titled the Transport Operations (Marine Safety) Act 1994. The relevant information is usually contained in the regulations and standards pertaining to each Act.

Similar Codes exist in other countries, based on the IMO's STCW-95 Convention.
(IMO: International Maritime Organization)
(STCW: International Convention on Standards of Training, Certification & Watchkeeping for Seafarers)
Copies of the USL Code and Acts of parliament can be purchased at the relevant Government Bookshops or boat bookshops.
AUSTRALIAN MARITIME GROUP: This body represents the States and Territories for negotiations on matters of maritime concern.

NATIONAL MARINE SAFETY COMMITTEE (NMSC)

(See website address on the last page of this book)

NOTE: At the time of writing this edition, many of the NMSC publications were only at draft stage. References made to the NMSC publications throughout this book must therefore be treated with caution.

NMSC is Australia's marine safety reforms body. It promotes and facilitates uniform marine safety standards throughout Australia. Its mandate is to overcome the inadequacies and obsolesce of the USL Code resulting from its static, prescriptive and out-of-date approach.

NMSC has developed a new NATIONAL STANDARD FOR COMMERCIAL VESSELS (NSCV) which accommodates changes in technology, such as fast craft, and workplace practices including performance-based assessment. The NSCV comprises of six parts: Part A is informative, and Parts B to F are mandatory. However, all six parts are mandatory under the Occupational Health and Safety (OH&S) legislation.

The six parts of the NSCV are as follows:
Part A: Safety Obligations
Part B: General Requirements
Part C: Design and Construction
Part D: Crew Competencies
Part E: Operation
Part F: Special Craft

COMPARISON BETWEEN THE NSCV AND USL CODE

NSCV	USL CODE
Part A: Safety Obligations *(of designers, builders, suppliers, owners, employees and others)*	New
Part B: General Requirements *(in matters of safety, risk assessment, etc)*	Sections 1, 14
Part C: Design and Construction *(Engineering, Electrical and LPG Systems)*	Sections 5, 6, 7, 8, 9, 10, 11, 12, 13, 16
Part D: Crew Competencies	Sections 2, 3
Part E: Operation	Section 15
Part F: Special Craft *(Fast Craft)* *(The international term for Fast Craft is High Speed Craft or HSC)*	Section 18, new sections for Fast Craft and Unconventional Craft.

The NMSC has produced three additional documents to complement the NSCV

1. **NATIONAL MARINE GUIDANCE MANUALS** contain nationally agreed safety procedures, such as:
   - Guidelines for Onboard Safety Training
     (Due to the diverse nature of Australia's coastal marine industry, it is considered impractical to demand formal Pre-Sea training for all crew on commercial vessels. These guidelines will facilitate skippers and crew, particularly in areas outside the metropolitan areas, who can provide safety training for new arrivals.)
   - Guidelines for Accreditation of Surveyors and Marine Training Providers
   - Guidelines for Recognition of Australian Defence Force Marine Qualifications (See flowcharts in Chapter 2)
   - Guidelines for Administrative Protocol for the Mutual Recognition of Vessel Certificates of Survey
   - Guidelines for Australian Marine Pilotage Standards

2. **REGISTER FOR COMPLIANT EQUIPMENT** provides a centralised list of equipment that complies with the standards specified in the NSCV or USL Code. The register is administered by Standards Australia on behalf of NMSC and marine safety agencies. The equipment and systems for inclusion on the register undergo appraisal by independent technical experts or organisations to determine whether they meet NSCV or USL Code requirements.

   Initially, the register will cover products currently eligible for approval and inclusion under specific USL Code requirements. Products and systems such as fire safety and recreational boating equipment will be included at a later stage.

3. The **EXEMPTIONS REGISTER** provides a centralised database of generic vessel exemptions for reference purposes and application by marine authorities.

HOW TO USE THE STANDARD

The NSCV is flexible in application. It specifies performance in the form of required outcomes. While the required outcomes are mandatory, the means of achieving them are not fixed. Solutions may be either "deemed to satisfy" prescriptive solutions specified in the NSCV, or equivalent performance-based solutions proposed by the applicant, as illustrated in the flow chart below.

Chapter 12: Australian Boating Regulations & Pollution Control

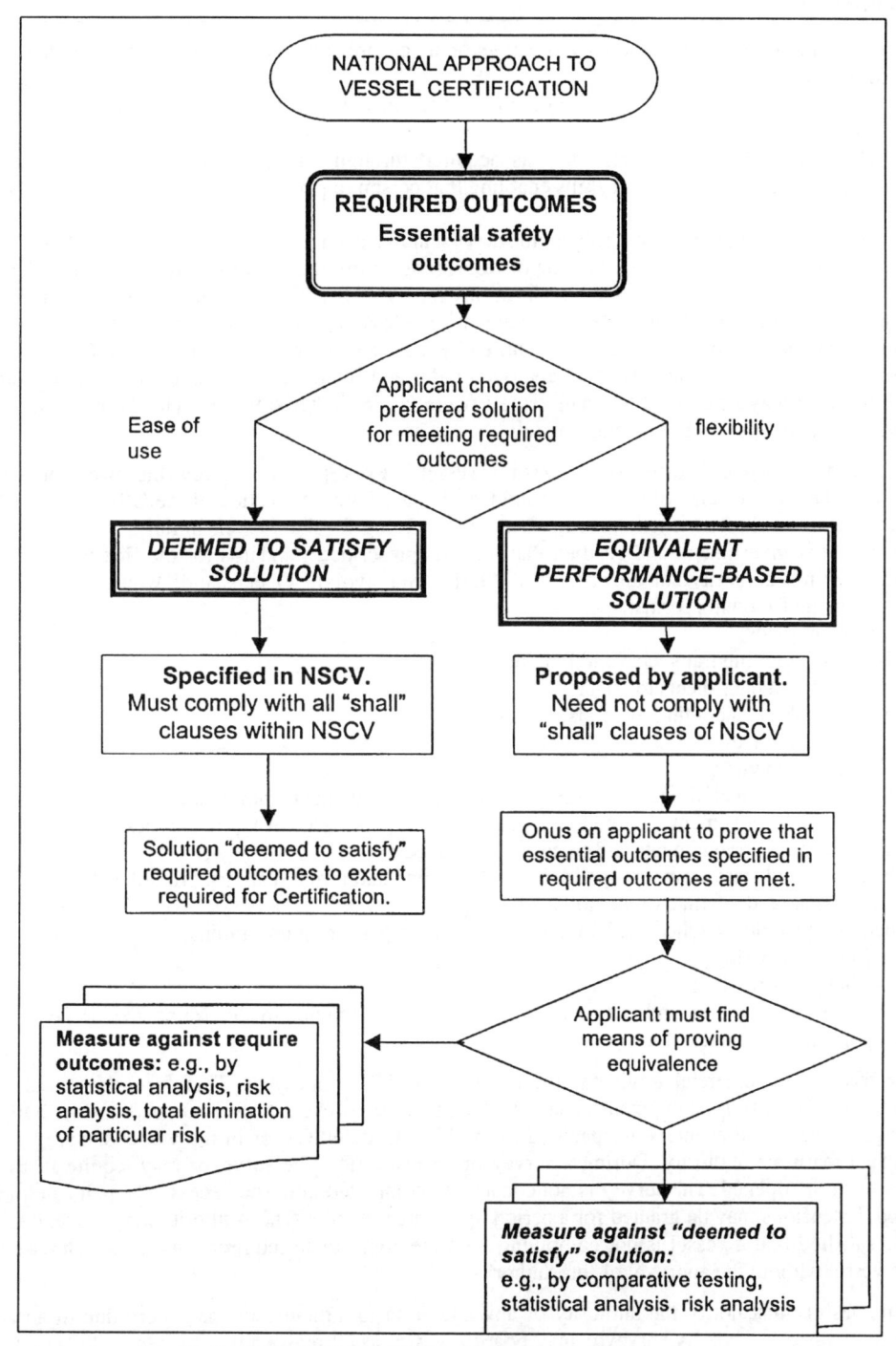

Fig 12.12: FLOWCHART FOR A NATIONAL APPROACH TO VESSEL CERTIFICATION *(NSCV Part B)*

RECREATIONAL BOATING

In the area of recreational boating, the NMSC has helped set up the following uniform programs:

- National Compliance Plate Program (discussed in this chapter)
- Principles for a National Standard for Recreational Boat Operator Licences (listed above)
- Guidelines for Recreational Boat Operator Competencies

COMMERCIAL VESSELS

> - A SURVEY is a thorough examination by a Surveyor to ensure that the vessel complies with the laws and regulations of the Authority.
> - An INSPECTION is a visual inspection by an Approved (Competent) Person to ensure that the equipment is in good order and condition.
> - A COMPETENT PERSON is a person who has acquired through training, qualification, or experience, or a combination of these, the knowledge and skills enabling that person to perform the required task. (NSCV)

Vessels wishing to operate commercially are subject to surveys and inspections by the appropriate authority. They must obtain a Certificate of Survey or a Permit or a Certificate of Registration from the local maritime authority. The Certificate is issued after the vessel has satisfactorily completed survey in respect of her hull, machinery and equipment and has a satisfactory manning level. The vessel must be provided with lifesaving and fire fighting appliances and other miscellaneous equipment including navigation lights as required by the relevant provisions of the USL Code and/or NSCV. Seagoing vessels are required to be equipped with an approved radio and fitted with a suitable compass. In some states a Survey Exemption Certificate may be granted to small vessels operating in sheltered waters. The definition of commercial vessels includes government vessels, but not defence vessels.

When a vessel is to be built under survey or an existing vessel is to be put into survey, the owner or the intending owner must comply with the requirements of the Survey and Certificate of Survey Section of the USL Code and/or NSCV, and all other sections relating to the construction, equipment and outfitting for that vessel. Initial Survey is the process by which a commercial vessel is investigated and verified that it meets prescribed standards for the first time. For the Initial Survey, the relevant maritime authority (or an accredited Marine Surveyor in Queensland) would need to assess and approve the information in the following format:

- General arrangement plans.
- Construction plans, including transverse and longitudinal sections.
- Plans, specifications and data sheets to cover:
- Scantlings of all members, including methods of fastenings.
- Details of the closing devices.
- Bilge pumping arrangements.
- Details of the fuel system, including tanks, filling and venting arrangements, piping and valves.
- Where applicable, the arrangements for the loading, carriage and discharge of liquid cargoes.
- Structural fire protection arrangements and fixed fire appliances.
- Details of the rudder and stern frame, propeller brackets, engine and thrust seatings, propeller shafting, bearing and coupling, steering gear and alternative method of steering.
- Where applicable, the welding schedule, laminating schedule or plastering programme.
- Electrical equipment or wiring.
- Preliminary stability information.
- Such further plans, information and data as the Authority may require to determine the proper construction, machinery, equipment and safety of the vessel.

Certificates of Survey or Commercial Registration are usually valid for one year. Therefore, the vessels undergo a Periodic Survey annually. The vessel's equipment is surveyed annually. However, the various parts of the hull, structure, machinery and fittings are surveyed at intervals specified in the USL Code/NSCV or in the vessel's survey record book supplied by the relevant maritime authority. During a survey or an inspection, the surveyor may require that a lining be removed or a compartment be opened. The survey is not considered completed until the necessary repairs and deficiencies have been made good. Extensions may be granted for a period up to three months. The Authority may suspend a Certificate of Survey when it is satisfied that a vessel is not complying with the appropriate requirements. In which case the owner must not operate the vessel without the approval of the Authority.

The owner of the vessel must notify the authority of any alteration to structure or machinery due to an accident or otherwise or a change in vessel's trade. A surveyor may board a vessel at all reasonable times to make an occasional or random inspection.

The commercial vessels are required to display their Permit label or Registration Certificate in a conspicuous place on board. The Survey Number is required to be painted on the hull of the vessel. In NSW, the Survey Number consists of 5 numerals. *It should not be confused with the Registration Number, which consists of letters and/or numerals, but always ends with the letter "N" for NSW.* Detailed instructions are provided in survey documents.

The trend now is to move away from regular surveys by authorities. In Queensland, for example, under the Marine Safety Act 1994, annual surveys of vessels are no longer required. The owner/master must always operate the vessel in a safe condition and manner and sign a declaration to that effect when renewing the annual registration. The authority has the right to inspect the vessel at a reasonable time.

Chapter 12: Australian Boating Regulations & Pollution Control

QUEENSLAND REGISTRATION OF COMMERCIAL/FISHING VESSELS

VESSELS UNDER 6 METRES IN LENGTH

A vessel of less than 6 metres in length (see 'measured length' in this chapter) can be commercially registered by providing a Statement Of Positive Flotation. This is the statement of swamp test issued by an accredited builder or surveyor, certifying that the vessel, if swamped with her full complement of crew/passengers, would remain afloat and upright. Alternatively, you can submit a Certificate of Compliance for Survey, as discussed below.

VESSELS 6 METRES OR MORE IN LENGTH

- You must submit the following documents with your application. These documents are issued by accredited designers, surveyors and boat builders:
- Certificate of Compliance for Design (verifies the vessel meets design standards)
- Certificate of Compliance for Survey (similar to roadworthy certificate)
- Certificate of Compliance for Stability (similar to a vehicle compliance plate)
- Certificate of Compliance for Safety Equipment (verifies that the vessel is appropriately equipped)
- For a Queensland Commercial Certificate of Registration, the vessel's carrying capacity (number of persons) is determined as follows:
    - 65 Kg per person for operation in prescribed smooth and partially smooth waters.
    - 75 Kg per person for operation beyond partially smooth waters
    - 110 Kg per person for diving operations.

SURVEY CLASSIFICATION OF COMMERCIAL VESSELS

Under the USL Code/NSCV, all commercial vessels, whether under State or Commonwealth survey, are classed in reference to their type and area of operation as follows:

VESSEL CLASSIFICATION - There are 6 types of vessels

1. CLASS 1 *(incl. Hire & Drive - Class 1F)*: a passenger vessel carrying or certified to carry more than 12 passengers.
2. CLASS 2: a non-passenger vessel (tug, workboat or small passenger charter vessel carrying 12 or less passengers).
3. CLASS 3: a registered commercial fishing or oyster vessel, which is not permitted to carry any passengers.
4. NOVEL (NOV) VESSEL: a vessel for which the hazards and risks are not adequately addressed by the requirements for conventional vessels. *(Requirements for novel vessels are contained in NSCV Part F Section 3.)*
5. SPECIAL PURPOSE (SP) VESSEL: a Class 2 vessel which by reason of its function carries on board a total of more than 12 special personnel and passengers; or a Class 3 vessel which by reason of its function carries on board more than 12 special personnel.
   Examples of Special Purpose Vessels: Ships engaged in research, expeditions and surveys; ships for training of marine personnel; fish factory ships; and ships processing other living resources of the sea. Requirements for such vessels are contained in NSCV Part F Section 4.
6. FAST CRAFT: a vessel that travels at a maximum speed of 20 knots or more (a speed and displacement relationship parameter also applies). There are two categories of Fast Craft.
    (a) F1: Vessels of length 35 metres or more in sub-Class A, B or C in Class 1 or 2 (1A, 1B, 1C & 2A, 2B, 2C).
    (b) F2: Not of Category F1 and of Class 1 (1D & 1E)

HIGH SPEED CRAFT (HSC) CODE

(It is proposed by the NMSC that the IMO International Code of Safety for High-Speed Craft (HSC Code) be applied to Category F1 vessels. NMSC does not consider it appropriate to simply apply it to all domestic fast craft.)

The HSC Code is the international standard for the construction and operation of high-speed craft. It was developed for vessels involved in international operations or operating at sea. It applies to hovercraft, hydrofoils, surface effect ships and high-speed monohulls and catamarans. It takes into account the light-weight construction that is essential for high-speed craft. It provides a set of design, construction, equipment, operational, infrastructure, survey and certification standards parallel to those of SOLAS. It also imposes strict operational conditions, including quality management in accordance with the ISM Code (discussed in Chapter 19.4) and compliance with other supporting infrastructure. The Permit to Operate is specific to a particular route or routes and contains details of operational conditions applying to the service.

The Code divides Passenger Craft into "RESCUE-ASSISTED" and "UNASSISTED" categories. Both may operate up to 4-hours voyage-time from a place of refuge. The Rescue-assisted passenger craft operate solely within waters covered by readily and rapidly available rescue services and may be certified to carry up to 450 passengers. The Unassisted passenger craft are built with the "get-home" (to a place of refuge) capability in an emergency, even with the disabled machinery space, while providing a safe refuge to its occupants. The Code also requires that all passengers be provided with a seat and that no enclosed sleeping berths are provided for passengers. CARGO VESSELS form a separate category with requirements similar to those for the 'Unassisted' passenger craft, except that they may operate up to 8-hours voyage-time from a place of refuge and the "get-home" facility is not required.

FAST CRAFT - RISKS
- ➤ Extensive penetration of the hull resulting from grounding or collision.
- ➤ Increased potential for unstable maneuvering or stability characteristics.
- ➤ Structural overloading arising from high pressures, accelerations or collision.
- ➤ Structural fatigue failure caused by high cyclic loading.
- ➤ Increased risk of personal injury due to high accelerations, operating in open seas, grounding or collision.
- ➤ Reduced time to react in emergency situations.
- ➤ Poor communications and fatigue due to high noise levels.
- ➤ Foundering in heavy seas, as craft are not designed for these conditions.
- ➤ System failure due to increased complexity.
- ➤ Operator error due to increased complexity and fatigue.

OPERATIONAL AREA – There are 5 operational areas for commercial vessels
- ➤ A: Unlimited domestic operations.
- ➤ B: Offshore operations (Up to 200 nautical miles to seaward of the coast).
(The term "offshore" in Australia has the same definition as "Near Coastal Voyages" under the STCW-95)
- ➤ C: Restricted offshore operations (Up to 30 nautical miles (50 in Queensland) to seaward of the coast or from a safe haven).
- ➤ D: Sheltered waters (Partially smooth water operations)
- ➤ E: Sheltered waters (Smooth water operations).
- ➤ (F): Class 1F vessels are Hire & Drive vessels.

LIMITS OF WATERS - The NSCV guidelines for limits of waters are as follows:

❖ INSHORE WATERS
any open stretch of water extending laterally along the coast up to and including 2 nautical miles offshore. It also includes bar entrances and waters designated as partially smooth waters or equivalent by each State/Territory marine authority.

❖ OFFSHORE WATERS
open waters more than 2 nautical miles seaward from the coast.

❖ PARTIALLY SMOOTH WATERS (See Class 'D' vessels above)
where the *significant wave height* does not exceed 1.5 metres from trough to crest for at least 90% of the time.
NOTE: Partially smooth waters are generally designated by legislation in each State or Territory. For example: Bays, such as Botany bay and Moreton Bay; ports such as Port Phillip and Port Augusta; areas of water between mainland and islands, such as Kangaroo Island and Rottnest Island; and inland waters, such as lower reaches of the river Murray.

❖ SMOOTH WATERS (See Class 'E' vessels above)
where the *significant wave height* does not exceed 0.5 metres from trough to crest at least 90% of the time.
NOTE: Smooth waters are generally designated by legislation in each State or Territory.
The SIGNIFICANT WAVE HEIGHT *(defined in NSCV Part B)* is the mean value of the highest one third of the heights measured from trough to crest recorded in a wave time history. *It is probable that* one in every 1000 waves *will have a height at least 1.86 times the significant wave height.*

Therefore, a Class 1C vessel is a passenger vessel that may operate up to 30 nautical miles (50 in Queensland) to seaward of the coast or such lesser distance to sea as the marine authority may decide.

A vessel may have more than one class. For example, she may be in survey for carrying goods and/or up to 12 passengers as Class 2C and, say, 40 passengers as Class 1E.

ADDITIONAL SEAWARD LIMITS FOR CREWING
- ➤ Australian coastal & middle water operation: 600 nautical miles to seaward.
- ➤ Inshore operations: 15 nautical miles to seaward (not applicable in Queensland).

A SPECIAL SERVICE NOTATION is added where the vessel is of a specialised type that requires special measures for the control of risk. These notations are FAST (Fast Craft), NOV (novel or unusual vessel) and SP (special purpose vessel). The vessel classification may thus read Class 1C (FAST), Class 1D (NOV), Class 3C (SP), etc.

MINIMUM LENGTH OF SEAGOING PASSENGER VESSELS:
Vessels of Classes 1A, 1B and 1C shall be a minimum of 10 metres in measured length.

Chapter 12: Australian Boating Regulations & Pollution Control

NOTE: Under the international STCW-95 Certification Standard (USL Code, Sect. 2 and NSCV Part D):
- Vessel length   35m = approx.   200 grt
- Vessel length   80m = approx. 1600 grt

[STCW = Standards of Training, Certification & Watchkeeping]
[grt or GRT = gross registered tonnes]

DOCUMENTS REQUIRED TO BE KEPT ON BOARD COMMERCIAL VESSELS

The main documents are:
- Operational manuals.
- Technical manuals.
- Maintenance & service manuals.
- Marine occupational & health manual.
- Safety management plan.
- Vessel's Survey Record Book.
- Operational records (Vessel's log book or Vessel Record Book (VRB))
- Certificate(s) of Competency (Coxswain, Master class V etc.).
- Safety equipment certificates (for liferaft, fire extinguishers etc., if applicable).
- Gas fittings certificate.
- Compass deviation card (if applicable).
- Ship Station Licence (when fitted with a marine radio).
- Radio Operator's Licence.
- Radio Operators Handbook.
- Radio Log.

VESSEL RECORD BOOKS

The requirements regarding vessel record books (VRB) for commercial vessels are contained in Section 13 of the USL Code and Part C of NSCV. Seagoing commercial vessels of over 35 metres are usually required to purchase and maintain an AMSA approved Official Log Book (OLB). Its entries include the drafts and times of arrivals and departures at each port, any death or disappearance, or illness or injury to anyone on board, times and dates of emergency drills conducted, details of any casualty to the vessel or assistance given to another vessel, and details of engine running and maintenance of machinery and equipment. Other commercial vessels operating outside smooth and partially smooth waters must maintain the same record, but may maintain it in any form of VRB.

The entries should be made as soon as possible after the occurrence; and timed and dated. In the event of a vessel being lost or abandoned, the VRB should be handed to the relevant authority. It is an offence to wilfully destroy a VRB or an entry in it, or to render an entry illegible or to make a false entry or to omit or to sign an entry knowing it to be false.

Working vessels also maintain an engine room logbook as well as a record of their lifting equipment maintenance.

Regardless of whether you operate a commercial or a recreational vessel, you must maintain some sort of a vessel record book or log book as well as a radio log book. One day, they may save your neck from a negligence claim. In general, logbooks facilitate safety of life, investigations, preventative maintenance and troubleshooting.

CREW LIST

Commercial vessels of Class 1A, 1B, 1C, 2A, 2B and 2C are required to maintain a crew list. A crew list contains the name and official number of the vessel which is found on the ship's registration document, name and address of the owner, name and address of the employer of the crew, and the name and address, date of birth, capacity and the dates of joining and leaving of every crew member.

MASTER, CAPTAIN & MASTER MARINER – WHAT'S THE DIFFERENCE?

Master is the person having command or charge of a vessel. Captain in RAN is the rank between Commander and Commodore. In the Merchant Navy the Captain is a courtesy title for the Master. A Master Mariner is a person holding a certificate entitling him or her to command a merchant vessel.

A SEAWORTHY VESSEL & INTERSTATE VOYAGES - THE LAW

Courts have defined a Seaworthy Vessel - with her master and crew - as being fit to encounter the perils of the voyage that she is about to undertake. She must be tight, staunch and strong, and properly crewed, equipped and supplied. *(Carver: Carriage by Sea; and Hedley v. Pinkney & Sons S.S. Co, 1894, etc.).*

For example, she can be considered unseaworthy for the lack of an essential chart or the shortage of crewmember or malfunctioning of a bilge pump. Just being in survey does not mean that the vessel is seaworthy at all times.

The owner, operator and the master are under obligation, before and at the beginning of every voyage, to exercise due diligence to make the vessel seaworthy. It is an offence for an owner, agent, operator or master to send or take an unseaworthy vessel to sea.

Trading vessels proceeding on interstate voyages come under the Commonwealth Navigation Act. They are not subject to the State Acts. For a one-off interstate voyage, they must obtain a single voyage permit from the Australian Maritime Safety Authority (AMSA).

Fishing vessels, when on interstate voyages remain subject to the Act of their parent state for matters of survey, licences, crew certification and welfare but are also subject to the other state's Act for matters of navigation or operational regulations. When a vessel is employed for the purpose of trading, operating or fishing in a State other than that in which she is in survey, she and her crew will be subject to the Acts of the State in which she is employed.

Under the international *SOLAS* Convention, any State or Federal marine authority may detain and deal with any vessel on matters of safety, regardless of her nationality or port of registry. *(SOLAS: Safety Of Life At Sea).*

QUEENSLAND TRANSPORT 12214

CERTIFICATE OF REGISTRATION

Ship Name	Registration No.	USL Class	Fishing Symbols
	12214QC	2C	

*Length	Breadth	Depth
5.88 m	2.30 m	1.93 m

Main Engine Details	Hull Material
OMC JOHNSON 130.55 KW	FRP

Limits and conditions of operation and maximum persons permitted to be carried

WITHIN FIFTEEN (15) NAUTICAL MILES OF LAND, OR IN SMOOTH OR PARTIALLY SMOOTH WATERS - MAXIMUM FIVE (5) PERSONS.

Name and Address of Owner

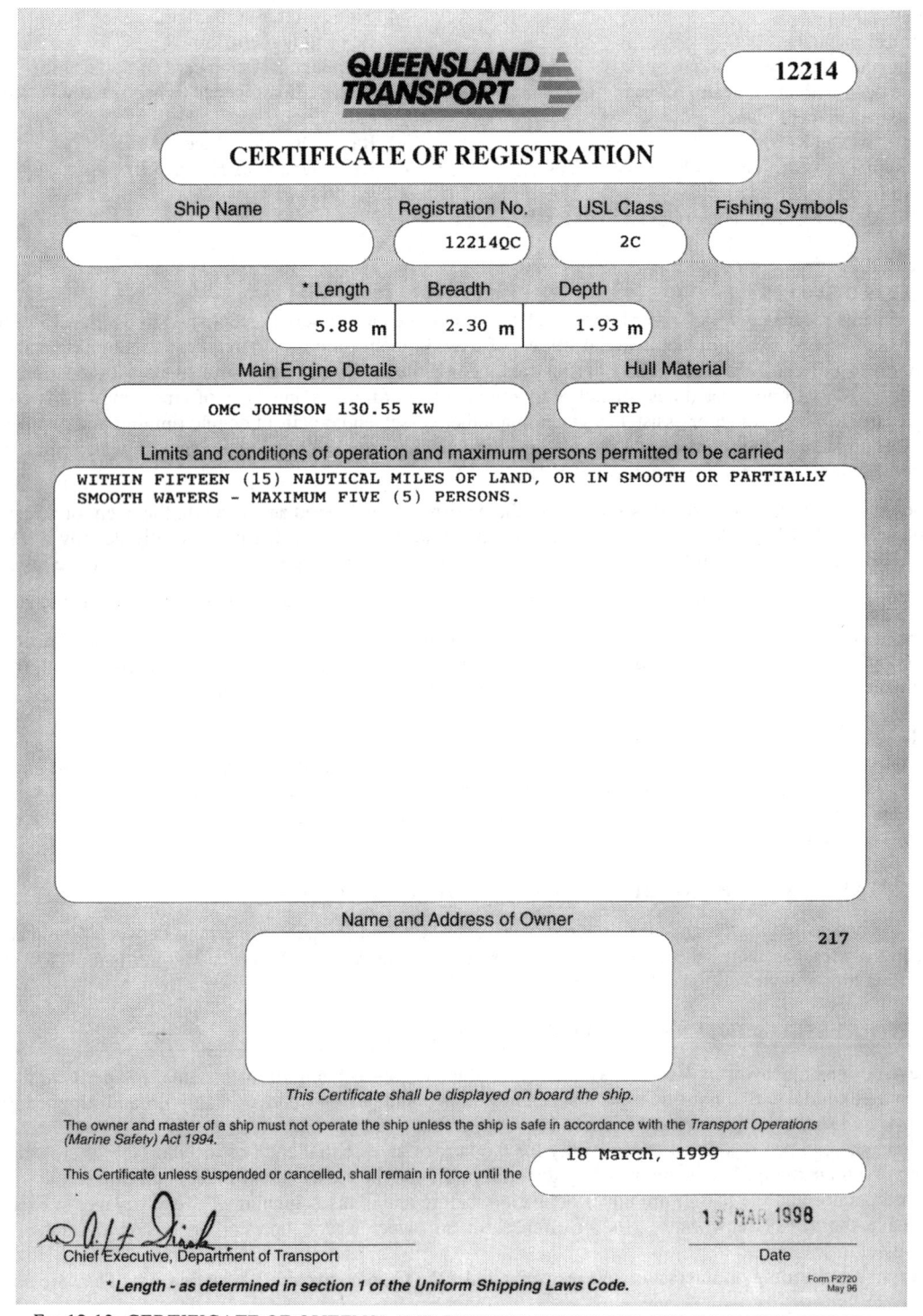

217

This Certificate shall be displayed on board the ship.

The owner and master of a ship must not operate the ship unless the ship is safe in accordance with the *Transport Operations (Marine Safety) Act 1994.*

This Certificate unless suspended or cancelled, shall remain in force until the 18 March, 1999

Chief Executive, Department of Transport   13 MAR 1998
                                                   Date

Length - as determined in section 1 of the Uniform Shipping Laws Code.   Form F2720 May 96

Fig 12.13: CERTIFICATE OF QUEENSLAND REGISTRATION FOR COMMERCIAL VESSEL

SAFETY ON HIRE & DRIVE VESSELS (CLASS 1F VESSELS)

Including Jet Boats or Personal Water Craft
NOTE: Class 1F vessels were previously referred to as Class 4

In addition to the hire & drive vessels being properly equipped and maintained as per the survey, the following is a summary extracted from Section 18 of the USL Code. The State-wise variations may apply:

- The owner of a hire & drive vessel shall maintain a register in an approved form. In it, the details of all hirings shall be shown, including:
  - the name and permit (survey) number of the vessel.
  - the full name, address and signature of each hirer.
  - the date and time of each hiring of the vessel.
  - the date and time of return of the vessel.
  - the number of persons declared by the hirer that are to be carried on the vessel.
  - the details of the "speed boat" licence if applicable.
- The register is required to be retained by the owner for a period of at least 6 months from the date of the last entry in the register, or, longer if directed by the Authority.
- The owner shall satisfy himself/herself that the hirer is competent to take charge of the vessel.
- The owner shall give the hirer clear and concise instructions on all matters of safety, correct handling of vessel and the geographical operating limits.
- On a vessel with a motor, refuelling instructions must be displayed in a position adjacent to the fuelling point and where they can best be seen. They must be of a material capable of withstanding, or of a material that is protected from, the elements of the weather.
- The instructions to hirers shall include:
  - Correct and safe handling and navigation of the vessel.
  - Correct and safe operation of machinery, fuel, gas and pumping systems and valves or openings in the hull.
  - Stowage and use of lifesaving appliances.
  - Location and use of fire appliances
  - Limits of operation of the vessel, being, if the Authority so directs, in the form of an appropriate plan.
- The owner shall obtain from the hirer a signed statement that the hirer fully understands the limits, restrictions and conditions.

MARINE POLLUTION REGULATIONS

Pollution is not just oil pollution. It includes everything and every act that harms the marine environment, marine life or affects public amenities. It includes:

- Noise pollution, such as from motors, rigging slapping against the mast all night, noisily launching boats at boat ramps where people live near by.
- Material pollution. Harmful substances, including oil, plastics, emulsifiers, degreasers, detergents, chemicals and oil dispersants.
- Visible pollution, such as food or toilet waste floating about or lying on beaches.

STATE LAW

Laws in *every State* prohibit the above polluting activities. No discharge of waste (including kitchen and toilet waste) is permitted in harbours and inland waters. "Grey water", that is, wastewater from showers, galleys and laundries, is usually not included in the Acts. Vessels are expected to bring the garbage ashore in garbage bags and may be required to fit holding tanks for toilet waste or carry portable toilets that are to be emptied into pump out facilities. Vessels may also fit an on-board waste treatment system approved by an Authority. Oily bilges must be discharged into a mobile or a shore based pump-out facility.

In addition to the public pump-out facilities in some harbours and waterways, certain marina and yacht clubs are required to install facilities for their own customers.

Many State authorities are empowered to impose "on the spot" fines for breaches of their pollution legislation. Penalties for non-compliance are high: as much as $260,000 for an individual and over $1 million for the company that owns the law-breaking vessel.

PROPOSED VESSEL SEWAGE MANAGEMENT PLAN
- No discharge of untreated sewage in harbours and inland waterways.
- No discharge of treated or untreated sewage in:

- waterways in which aquaculture, including oyster growing, occurs.
- waterway which are used for drinking water supplies
- in or near a bathing area, mooring area, marina and anchorage area.
- Class 1F commercial vessels over 6 metres (e.g. houseboats) to install holding tanks for discharge into sewage pump-out facilities.
- Class 1 commercial vessels (passenger carrying vessels) to install holding tanks for discharge into sewage pump-out facilities.
- Recreational vessel operators to suitably manage sewage from their vessels, depending on the conditions applying to waterways in which they operate, the length of the journey and the type of activity being undertaken.

COMMONWEALTH LAW

Under the International Convention for the Prevention of Pollution from Ships 1973/78 (known as MARPOL), all of the above applies. In Australia, this convention is enacted in the Protection of the Sea (Prevention of pollution from ships) Act 1983 and the Navigation Act 1912. Under the convention, the following GARBAGE DISPOSAL REGULATIONS also apply:

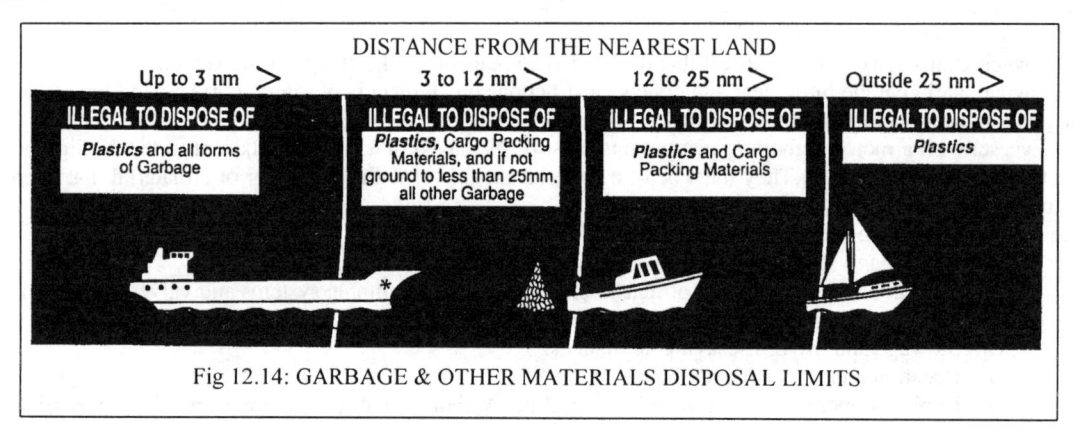

Fig 12.14: GARBAGE & OTHER MATERIALS DISPOSAL LIMITS

Up to 3 nm	3 to 12 nm	12 to 25 nm	Outside 25 nm
ILLEGAL TO DISPOSE OF *Plastics* and all forms of Garbage	ILLEGAL TO DISPOSE OF *Plastics*, Cargo Packing Materials, and if not ground to less than 25mm, all other Garbage	ILLEGAL TO DISPOSE OF *Plastics* and Cargo Packing Materials	ILLEGAL TO DISPOSE OF *Plastics*

- No discharge of plastics anywhere.
- No discharge of processed (pulverised) garbage within 3 nautical miles of the nearest land.
- No discharge of unprocessed (unpulverised) garbage within 12 nautical miles of the nearest land. (In some States, small amounts of food – including fish or fish parts – may be disposed of overboard in some cases, such as fish-feeding activities for tourist operators or in the normal course of operations for commercial fishing vessels.)
- Vessels of 12 metres and over must display placards setting out the garbage disposal requirements of MARPOL 73/78.
- Vessels of 400 gross registered tonnes (grt) and over, or certified to carry 15 persons and over, must develop and follow a shipboard waste management plan. They must also maintain a Garbage Record Book.

MARPOL contains five annexes, dealing with:
(i) Oil
(ii) Noxious Liquids
(iii) Garbage
(iv) Sewage
(v) Harmful substances in packaged form.

Under Annex IV of MARPOL, it is proposed that the discharge of sewage from ships should be controlled in all coastal areas in a manner similar to that of garbage. Australia has already decided that we should adopt the Annex. When signed and enacted, the following vessels are expected to be required to

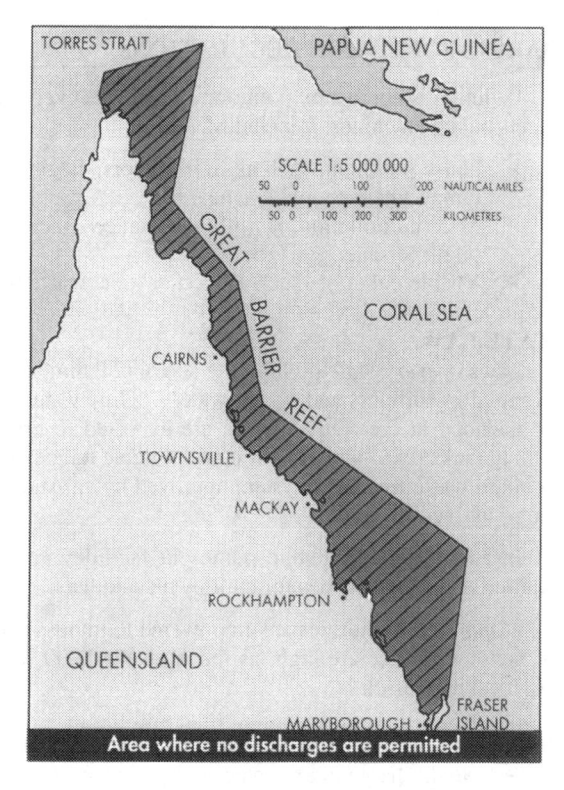

Fig 12.15

fit holding tanks and ancillary pollution control equipment:
- *New vessels of 400 gross registered tonnes and over.*
- *New vessels certified to carry more than 15 persons.*
- *Existing vessels of 400 gross registered tonnes and over (to be fitted within 10 years).*
- *Existing vessels certified to carry more than 15 persons (to be fitted within 10 years).*

GREAT BARRIER REEF
(See also REEFREP in Chapter 19.3)

Under MARPOL, no discharge of any type is permitted in the area of Great Barrier Reef. In some cases this can be as much as 150 nautical miles from the Queensland coast. Where discharges are prohibited within a certain distance from the land, these distances are measured from the outer edge of the reef.

FISHING VESSELS
Fishing vessels must make every effort to retrieve all lost or damaged fishing gear. Lost fishing gear should be reported to the Australian Rescue Co-ordination Centre (RCC) in Canberra. This can easily be done via a Coast Radio Station.

REPORTING OF POLLUTION
All pollution from your vessel must be reported to the State or Federal authorities. You are also REQUESTED to report pollution from other vessels.

DON'T CONFUSE ALGAL BLOOMS WITH OIL
Sometimes it is easy to confuse naturally accruing algal blooms with oil. Floating Sea Scum (Trichodesmium) is common in early summer. It is algal plant life in tones of red, yellow or brown. Algae decaying along the shoreline can also turn greenish or release a purple dye. It may appear in large beach slicks, often accompanied by putrid fishy, chlorine, iodine or oil smell. But, unlike oil, it will wash off in water.

Coral Spawns: Corals in places such as the Great Barrier Reef spawn for a couple of days each year between October and December. Its timing is linked to the lunar cycle. Slicks of these minute eggs can be easily mistaken for oil slicks. The spawns are of pink, orange or red colour with uniform sized particles. Many of these eggs die. They then turn whitish and clump together to form a thick, dirty and irregular scum. It too will wash off in water.

OIL POLLUTION REGULATIONS FOR "LARGE" VESSELS
[Discharge of bilges and tank cleaning, etc.]

In order for ships to operate, the following level of oil discharge is currently acceptable to the international community. The regulations are made to allow ships to pump out bilges and tankers to clean tanks at sea, without damaging the environment.

In order to abide by these regulations, ships are fitted with oily water discharge monitoring equipment, oily water separators and sludge holding tanks. They are required to record the details of all bilge pumping and tank washing operations in an Oil Record Book.

The permissible levels of discharge listed below relates to the trace of oil in the water discharged from bilges and during washing of oil tanks:

(a) MACHINERY SPACES OF ALL SHIPS OF 400 GROSS TONS (WHICH IS EQUIVALENT TO ABOUT 40 METRES IN LENGTH) AND ABOVE:
- Maximum oil content of the discharge 15 parts per million.
- Ship must be proceeding on route.
- Ship must have in operation an approved oil discharge monitoring and control system, oily water separating equipment or oil filtering equipment.

(b) TANKERS OF 150 GROSS TONS (which is equivalent to about 30 metres in length) AND ABOVE:
- Maximum discharge of oil 30 litres per mile travelled by the ship.
- Ship must be proceeding on route.
- No discharge of any oil whatsoever from cargo tanks within 50 miles of the nearest land. [In the region of the Great Barrier Reef, 50 miles is measured from the outer edge of the reef].
- Maximum quantity of oil allowed to be discharged (as a result of cleaning of tanks on a ballast passage) is 1/30,000 of the total cargo carrying capacity. This means that a for a 30,000 tonne tanker, a trace of oil amounting to 1 tonne is permissible in the discharge of water during tank cleaning at sea.

The Australian Boating Manual

DUMPING AT SEA: Under the Australian Environmental Protection (Sea Dumping) Act, licences are needed by all involved in the operation of dumping at sea. This law relates to dumping of sand, gravel, factory waste and other materials.

POLLUTION PREVENTION FROM "SMALL" VESSELS
- Observe anti-spill and fire precautions when re-fuelling (discussed below).
- Oily bilges must be discharged into a mobile or a shore based pump-out facility. Observe the above "large vessel" guidelines when discharging at sea. Bilge water can easily be cleaned by installing an oil absorbent pad or a oily water separator near the bilge pump. Bilge sponges are available from most chandlers.
- Do not discharge plastics anywhere in the water.
- Observe the discharge of garbage, toilet waste and noise pollution regulations stated above.
- Report pollution incidents.
- Decant cooking oils and fats into suitable container and take home for disposal.
- Wipe plates clean with a paper towel before washing up.
- Use minimal amounts of washing detergent.
- Engine oil must only be discharged into an oil reception barge or a shore facility.

POLLUTION MAY NOT BE AN OFFENCE WHEN it is necessary to jettison or discharge pollutants to save a vessel and her crew from grave danger.

FUEL EXPANSION IN HOT WEATHER

Fuel expands in volume about 0.5% per 1°C rise in temperature. Therefore, with a 10° rise in air temperature - a common daily fluctuation in Australia - the fuel in a tank, sitting in open air, may expand by 5%. Ignoring some expansion of the tank itself, this amounts to 5 cm rise in the sounding in a tank full of fuel, measuring 1 x 1 x 1 metre. Without sufficient ullage (space between the liquid and tank top), the fuel could overflow.

REFUELLING A VESSEL - PRECAUTIONS
- Take portable tanks out of the vessel for filling.
- [Do not carry spare fuel in plastic containers. They can rupture without being noticed.]
- Hoist flag B for refuelling internal tanks.
- Keep watch.
- No smoking.
- No fires and no motors running.
- Disconnect the battery.
- Turn off gas.
- Have a suitable fire extinguisher available near the filling station.
- Check for leaks.
- Block off deck scuppers and freeing ports to contain any spill on deck.
- Secure vessel properly alongside.
- Provide earth connection to or discharge static electricity from the fuel hose (See static electrical charge).
- Keep the fuel nozzle in contact with the filler pipe, to prevent static electricity build up.
- Make sure the fuel goes into the correct tank.
- Constantly monitor the tank being filled.
- Consider stability when filling side tanks.
- Fill slowly towards the end.
- On disconnection of fuel line, catch any spillage in a container.
- Clean up any spill immediately.
- Keep the vessel well ventilated, and close up enclosed spaces.
- Ventilate for some time before starting engine.
- *NOTE: The refuelling instructions for Hire & Drive Vessels are discussed separately.*

IN CASE OF AN OIL SPILL IN A HARBOUR
- Stop refuelling.
- Contain the spill as much as possible.
- Take fire precautions.
- Advise the authorities.
- Carry out on board clean up.
- Don't use oil dispersants in water unless advised

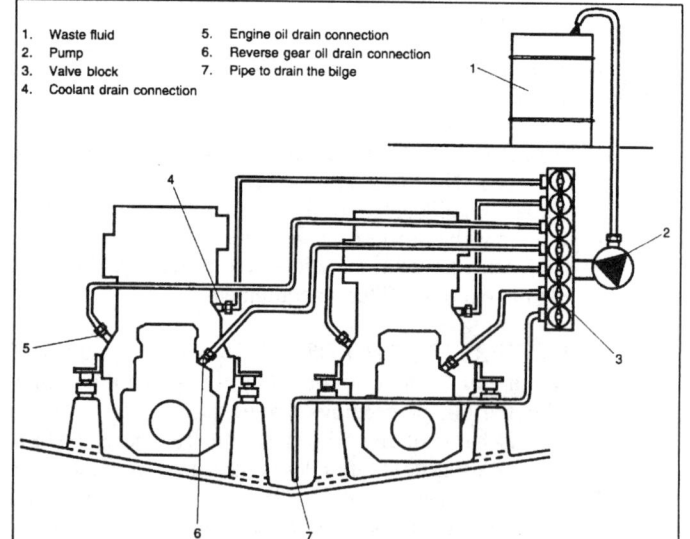

1. Waste fluid
2. Pump
3. Valve block
4. Coolant drain connection
5. Engine oil drain connection
6. Reverse gear oil drain connection
7. Pipe to drain the bilge

Fig 12.16: OIL & COOLANT DRAINAGE SYSTEM *(Volvo Penta)*

Chapter 12: Australian Boating Regulations & Pollution Control

by the local authority.

OIL AND COOLANT DRAINAGE SYSTEM

The illustration shows a central waste pump set up to extract dirty lube oil, engine coolant and bilge liquid. Once collected into a container, the liquids are disposed off ashore in an environmentally safe manner. Some engines can be drained with a hand-operated drainage pump. Installing an electric pump is another option. The pump can be run in either direction by changing the polarity. To prevent engine being accidentally drained, connect the pump hose only when changing oil.

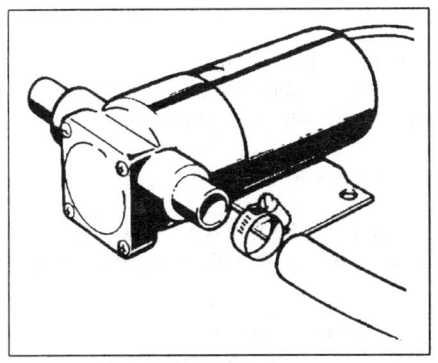

Fig 12.17
OIL DRAINAGE PUMP

POLLUTION FROM ANTI-FOULING: Anti-fouling paints prevent marine growth on hulls. But they are generally damaging to the environment. Their use and the alternative to antifouling are discussed in Chapter 5.

STATIC ELECTRICITY – A FIRE HAZARD DURING REFUELLING

Static is the electricity produced on dissimilar materials through physical contact and separation. A spark generated by it can ignite flammable vapour.

A static electrical charge can build up during refuelling when the fuel moves through a pipe. The fuel may become negatively charged and the pipe positive. The negatively charged fuel, in turn, causes the tank to become positively charged. The risk of ignition thus continues.

The risk also exists when sounding, sampling or washing a fuel tank. The positively charged sounding tape or the sampling container strikes against the negatively charged very fine particles of fuel in the tank. A positively charged water nozzle may react with the negatively charged oily residue during the washing operation.

TO SAFEGUARD AGAINST BUILD UP OF STATIC ELECTRICITY CHARGE DURING REFUELLING, the refuelling system must be bonded and grounded. It is then said to be electrically connected. The filler attachment is connected to the tank either by a direct metal-to-metal connection or by means of a copper wire (ground cable) of suitable size. The filler pipe must extend into the bottom of the tank, and the tank should be connected to the vessel's bonding system.

Foam fire fighting hoses are rubber lined to eliminate the risk of static resulting from friction with foam.

BIRD DROPPINGS

Bird droppings and perhaps even nesting often foul boats on marina berths, moorings surrounded by food sources from natural or human habitation. Bird droppings contain phosphates and ammonia and their oily nature prevents them from being washed away by rain. They can soften solvent-based paints and discolour vinyl plastic sail covers and awnings.

Not all commercial or DIY deterrents are effective with all birds in all seasons. You first need to identify the type of birds causing the problem. Swallows, cormorants and shags perch on elevated masts and spreaders, terns and seagulls congregate in large flocks on small diameter flat surfaces such as pushpits, gunwale and booms. Birds such as cockatoos, galahs and corellas just want to rip and tear through things.

ELEVATED DETERRENTS
- A plastic flag, pennant, shopping bag or the bladder from a wine cask fluttering above the masthead.
- A few widely spaced long spikes or rods. (Birds would see them as danger to their bodies and wings.)
- Cable ties with their ends pointing upwards.
- Fishing lines or thin single strand wires strung above the spreaders.

LOW LEVEL DETERRENTS
- A plastic flag, pennant, shopping bag or the bladder from a wine cask fluttering at a lower level.
- The decoy of a large predatory bird hovering in the rigging.
- A feather duster suspended in the rigging.
- A wind-pushed lightweight beam rotating on a pivot on a power boat's flat doghouse.
- Slip-on covers made of split PVC hose, studded with long nails on pulpits and pushpits.
- A net or a loose-weave shade cloth.
- A dinghy bobbing astern.

CHAPTER 12: QUESTIONS & ANSWERS

CHAPTER 12.1: BOATING REGULATIONS - ALL VESSELS

1. If the master of a vessel should be, but is not, licensed, the following person(s) are punishable:

    (a) only the master.
    (b) only the owner.
    (c) either the master or the owner.
    (d) both the master and the owner.

2. The following convictions call for cancellation or suspension of a boat driver's or RSM licence:

    (i) Negligent &/or dangerous navigation.
    (ii) Causing nuisance, annoyance or danger.

    The numbers of convictions leading to the cancellation of a licence are as follows:
    (a) either one of the above.
    (b) two of the first and one of the second.
    (c) one of the first and two of the second.
    (d) two of each.

3. In a State where the boat licence (or RSM licence) is compulsory, the following restriction is imposed on the holder of a Provisional Boat Licence:

    (a) 10 knots speed limit when towing a water skier.
    (b) 10 knots speed limit.
    (c) No towing of a water skier or an aquaplane.
    (d) May tow a water skier but not an aquaplane.

4. In NSW, the following vessels are required to be registered with the Waterways Authority:

    (a) Power driven, capable of 10 Kt. or more.
    (b) Power driven or sailing, 5.5 metres or longer.
    (c) Any vessel with a permanent wet berth.
    (d) All the vessels stated here.

5. In Victoria and South Australia, the following vessels are required to be registered with the State Waterways authority:

    (a) Only power driven, capable of 10 Kt. or more.
    (b) Only sailing vessels fitted with an engine.
    (c) Any vessel fitted with an engine.
    (d) None of the choices stated here.

6. In Queensland & Tasmania, the following vessels are required to be registered with the State Waterways authority:

    (a) Only power driven vessels.
    (b) Only sailing vessels fitted with an engine.
    (c) Vessels fitted with an engine of 3 kw & over.
    (d) None of the choices stated here.

7. In Northern Territory, the following vessels are required to be registered with the State Waterways authority:

    (a) Only power driven vessels.
    (b) Only sailing vessels fitted with an engine.
    (c) Vessels fitted with an engine of 3 kw & over.
    (d) Registration is not required.

8. In Western Australia, the following vessels are required to be registered with the State Waterways authority:

    (a) Only power driven vessels.
    (b) All motor vessels, other than a small tender.
    (c) Vessels fitted with an engine of 2.98 kw & over.
    (d) Registration is not required.

9. The following vessels are exempt from the NSW Waterways Authority registration:

    (a) personal water craft.
    (b) vessels operated as ferries or tugs.
    (c) vessels over 25 metres in length.
    (d) none of the choices stated here.

10. The following person(s) are held responsible if the registration number and the renewal label are not shown on a registered vessel:

    (a) both the master and the owner.
    (b) only the owner.
    (c) either the master or the owner.
    (d) only the master

11. On the sale of a registrable vessel, which of the following party(s) must submit to the State Waterways authority a transfer of registration form completed by both parties, together with the transfer fee:

    (a) the seller.
    (b) the buyer.
    (c) both parties.
    (d) either party.

12. In NSW, a Vessel Capacity Plate must be affixed to the following recreational vessels:

    (a) all vessels.
    (b) the mechanically propelled vessels.
    (c) all vessels over 5.5 metres.
    (d) none of the choices stated here.

13. In Victoria, vessel capacity restrictions apply to the following recreational vessels:

    (a) all vessels.
    (b) the mechanically propelled vessels.
    (c) all vessels over 5.5 metres.
    (d) none of the choices stated here.

Chapter 12: Australian Boating Regulations & Pollution Control

14. In NSW, the maximum number of persons indicated in the Vessel Capacity Plate are assessed:
    (a) for good weather conditions.
    (b) at 75kg per person, allowing for children.
    (c) at 75kg per person in good weather conditions.
    (d) none of the choices stated here.

15. The maximum number of persons for a vessel's capacity is based on the following criteria:
    (a) child up to 1 year age does not count.
    (b) child between 1 & 12 years equals ½ adult.
    (c) length of the vessel.
    (d) all the choices stated here.

16. In NSW, the Vessel Capacity Plate must be affixed to vessel such that it can be clearly seen:
    (a) on the hull.
    (b) from each steering position.
    (c) from the main steering position.
    (d) not applicable in my State.

17. In NSW, Queensland & South Australia, any vessel travelling at 10 knots or more in enclosed waters must keep from a person or other vessel or object in the water, a distance of at least:
    (a) 30 metres.
    (b) 60 metres.
    (c) 90 metres.
    (d) unspecified safe distance.

18. In Victoria, a 5 knot speed limit applies when passing a "diver below" signal within:
    (a) 30 metres.
    (b) 60 metres.
    (c) 90 metres.
    (d) unspecified safe distance.

19. In Queensland, a 6-knot speed limit applies when passing a "diver below" signal within:
    (a) 30 metres.
    (b) 60 metres.
    (c) 90 metres.
    (d) unspecified safe distance.

20. In Western Australia, a vessel is required to pass a diver signal at a distance of at least:
    (a) 30 metres.
    (b) 50 metres.
    (c) 90 metres.
    (d) unspecified safe distance.

21. When towing a water skier, the skipper must keep both the vessel and the person being towed at a distance from other vessels not engaged in water skiing or aquaplaning, of at least
    (a) 60 metres.
    (b) 30 metres.
    (c) 90 metres
    (d) unspecified safe distance.

22. When towing a water skier, the skipper must keep both the vessel and the person being towed at a distance from other persons in the water, of at least
    (a) 30 metres.
    (b) 60 metres.
    (c) 90 metres
    (d) unspecified safe distance.

23. A skipper must keep a skier using aerial equipment from any bridge, cable, wire, pipe or other structure or apparatus erected or suspended above water, at a distance of at least:
    (a) 60 metres.
    (b) 90 metres.
    (c) 300 metres.
    (d) unspecified safe distance.

24. A skipper must keep a skier using aerial equipment from any structure on the shore or any person in the water or on the water or shore, at a distance of at least:
    (a) 60 metres.
    (b) 90 metres.
    (c) 300 metres.
    (d) unspecified safe distance.

25. In NSW and SA, one of the requirements of an "observer" on a vessel engaged in water-skiing is:
    (a) to be aged 18 years or over.
    (b) to be aged 16 years or over.
    (c) to be a good swimmer.
    (d) none of the choices stated here.

26. In Queensland, one of the requirements of an "observer" on a vessel engaged in water-skiing is:
    (a) to be aged 16 years or over.
    (b) to be aged 15 years or over.
    (c) to be aged 12 years or over.
    (d) none of the choices stated here.

27. In Victoria, one of the requirements of an "observer" on a vessel engaged in water-skiing is:
    (a) to be aged 16 years or over.
    (b) to be aged 15 years or over.
    (c) to be aged 12 years or over.
    (d) none of the choices stated here.

28. In Western Australia, one of the requirements of an "observer" on a vessel engaged in water-skiing is:

    (a) to be aged 16 years or over.
    (b) to be aged 14 years or over.
    (c) to be aged 12 years or over.
    (d) none of the choices stated here.

29. Water-skiing is permitted:

    (a) at all hours.
    (b) in enclosed waters only.
    (c) in open waters only.
    (d) from sunrise to sunset only.

30. In Victoria, water-skiing is permitted:

    (a) at all hours.
    (b) in enclosed waters only.
    (c) in open waters only.
    (d) from 1 hour before sunrise to 1 hour after sunset only.

31. The maximum number of water-skiers who may be towed at one time are:

    (a) one.
    (b) two.
    (c) three.
    (d) four.

32. In Queensland, the maximum number of water-skiers who may be towed at one time are:

    (a) one.
    (b) unspecified.
    (c) two.
    (d) three.

33. "Wash" is the disturbance which causes damage, injury or annoyance to others, created as a vessel moves through:

    (a) any waters.
    (b) smooth waters.
    (c) deep waters.
    (d) shallow waters.

34. You must know who is in command of your vessel at any time because you may be:

    (a) required to identify the person.
    (b) responsible if the master is unlicensed.
    (c) responsible for damage caused by the master.
    (d) responsible for all the items stated here.

35. In certain areas, water-skiing is not permitted. This may be indicated by:

    (a) a flag.
    (b) a sign similar to a no smoking sign.
    (c) a speed limit sign.
    (d) any of the signs stated here.

36. Vessel speed restrictions apply in certain areas. This may be indicated by signs depicting:

    (a) a number with or without an arrow.
    (b) no entry.
    (c) no wash
    (d) any of the signs stated here.

37. Certain areas are "no wash zones". This may be indicated by signs depicting:

    (a) a red band through a vessel.
    (b) a red band through a vessel making wash.
    (c) a speed limit
    (d) any of the signs stated here.

38. In certain areas vessels are prohibited. This may be indicated by signs depicting:

    (a) a red band through a vessel.
    (b) "no entry".
    (c) either of the signs stated here.
    (d) neither of the signs stated here.

39. In Sydney Harbour, certain power driven vessels display an orange-coloured diamond. It indicates they have:

    (a) a right of way over sailing vessels.
    (b) a right of way over all vessels.
    (c) a priority over sailing vessels.
    (d) none of the choices stated here.

40. Power-driven cross-river ferries when underway usually exhibit the navigation lights of a:

    (a) power driven vessel.
    (b) punt.
    (c) punt, but with the top light flashing.
    (d) punt and side lights.

41. The navigation lights shown by a ferry working in chains or wires are:

    (a) those of a power driven vessel.
    (b) those of a sailing vessel.
    (c) one all round white light on each end.
    (d) none of the choices stated here.

42. When approaching a ferry working in chains or wires (a punt), every vessel is required to:

    (a) slow down to less than a specified speed.
    (b) give a sound signal.
    (c) comply with one of the rules stated here.
    (d) comply with both the rules stated here.

Chapter 12: Australian Boating Regulations & Pollution Control

43. Two vessels collide with each other, causing injury and damage. Their masters only need to be concerned with the safety of:
    (a) their respective vessels.
    (b) both vessels.
    (c) first their own, then of the other.
    (d) each other's vessels.

44. While unberthing, a vessel's superstructure is damaged by coming in contact with the wharf. She must take the following action:
    (a) notify the authorities immediately.
    (b) notify the authorities within 24 hours.
    (c) repair the damage within three months.
    (d) repair the damage before departing.

45. The courts have held that the Seaworthiness of a vessel depends on the voyage that she is about to take:
    (a) this is not true.
    (b) this is true.
    (c) this is sometimes true.
    (d) this depends on the survey schedule.

46. The following wave height is the guideline for determining partially smooth waters:
    (a) 1.5 metres in the roughest sea conditions.
    (b) 1.5 metres in the normal sea conditions.
    (c) 0.5 metres in the roughest sea conditions.
    (d) 0.5 metres in the normal sea conditions.

47. In the vicinity of submarine cables in Queensland, anchoring is prohibited within a distance of:
    (a) 200 metres.
    (b) 100 metres.
    (c) 50 metres.
    (d) a reasonable range.

48. An Unseaworthy vessel is defined as lacking in:
    (a) strong construction.
    (b) safety and navigational equipment.
    (c) proper manning & provisions.
    (d) all the essentials stated here.

49. In Australia, a master's duty to render assistance to other vessels and aircraft in distress is laid down in the:
    (i) State marine acts.
    (ii) Commonwealth marine act.
    (iii) Commonwealth radio communications Act.
    (iv) State radio communications Acts.

    The correct answer is:
    (a) all except (iii).
    (b) all except (iv).
    (c) all of them.
    (d) all except (i) & (ii).

50. For which of the following persons it is an offence to send or take an unseaworthy vessel to sea?
    (a) Master
    (b) Owner or charterer
    (c) Agent
    (d) Any of the persons stated here

51. In matters of safety, a vessel outside the State in which she is registered, is subject to the laws of the:
    (a) Commonwealth.
    (b) present port State.
    (c) State in which the vessel is registered.
    (d) any or all of the choices stated here.

52. In matters of navigation, a vessel outside the State in which she is registered, is subject to the laws of the:
    (a) Commonwealth.
    (b) present port State.
    (c) State in which the vessel is registered.
    (d) any or all of the choices stated here.

53. On applicable beaches, the 'boats entry prohibited' sign is as follows:
    (a) A red band through a black boat.
    (b) A black band through a red boat.
    (c) Words "No Entry"
    (d) Words "Keep Out"

54. When navigating near harbour entrances, especially at night:
    (a) Look for the Safe Water mark.
    (b) Stay behind the ferries.
    (c) Keep well over to starboard.
    (d) Do not use astern propulsion.

55. In relation to the National Recreational Boat Operator's Licence, which of the following is not an NMSC guideline?
    (a) The licence is required to operate a recreational boat.
    (b) The licence is transportable between States & Territories.
    (c) The licence is issued on successful completion of specified training or passing the examination.
    (d) Medical fitness standard is similar to that required for a motor vehicle licence.

56. The Hull Identification Number:
    (a) does not identify the country of origin.
    (b) is not the same as the Boat Code.
    (c) is a 9-digit number.
    (d) is a 14-character code.

57. In relation to the national recreational boat "Compliance Plate" program, which of the following is incorrect?

    (a) It is to be fixed to all new and existing boats.
    (b) It is to be fixed to all boats operating in Australia.
    (c) It is to include an indication of the boat's response to swamping.
    (d) It is to include the boat's passenger capacity.

58. Prior to leaving for an overseas voyage, the requirement to obtain a Ship's Register applies to:

    (a) commercial vessels only.
    (b) recreational vessels only.
    (c) all vessels.
    (d) motor vessels only.

59. Under the worldwide long-range navigational warning service:

    (a) the world's oceans are divided in 10 navigational areas.
    (b) the Australian region is NAVAREA X.
    (c) The Australian region is divided into 6 sub-regions.
    (d) The Navigational Warnings are issued by IMO.

60. A Race Control Boat carries the following sign:

    (a) Sign reading 'No Entry'
    (b) Sign reading 'Race Control'
    (c) A white flag
    (d) Two yellow flags

61. A PWC Rider does not need to:

    (a) carry the PWC licence.
    (b) wear a PFD
    (c) observe 'safe distances' applicable to power boats.
    (d) ride alone.

62. The Annual Notices to Mariners do not include:

    (a) the year's Chart Corrections.
    (b) the requirements for reporting pollution incidents.
    (c) Search & Rescue procedures.
    (d) the procedures for MSI warnings.

63. The information on the use of helicopters for search & rescue can be found in the:

    (a) relevant State Marine Act
    (b) USL Code.
    (c) Annual Notices to Mariners
    (d) NSCV

64. In relation to the National Marine Guidance Manuals produced by NMSC, which of the following statements is incorrect? The Manuals contain guidelines for:

    (a) onboard safety training.
    (b) the recognition of Australian Defence Force marine qualifications.
    (c) calculating ship stability.
    (d) the accreditation of Surveyors and Marine Training Providers

65. In relation to the Register for Compliant Equipment produced by NMSC, which of the following statements is incorrect?

    (a) It provides a centralised list of the compliant equipment for vessels.
    (b) It is one of the National Marine Guidance Manuals.
    (c) It is administered by Standards Australia.
    (d) It complements NSCV.

66. On a powerboat underway, it is not an offence to:

    (a) sit with legs hanging over the side.
    (b) sit with legs hanging over the bow.
    (c) sit with legs hanging over the stern.
    (d) sit on the bow.

67. In measuring the vessel length for the purpose of registration or commercial survey, which of the following statement is incorrect?

    (a) The length is measured from the point of the bow to the transom.
    (b) Bowsprits are included.
    (c) Outboard motors are excluded.
    (d) All appendages are excluded.

68. In relation to Fast Craft, which of the following statements is incorrect?

    (a) Excessive penetration of the hull following grounding or collision.
    (b) Greater risk of operator error.
    (c) Better communication & less personnel fatigue.
    (d) Increased potential for unstable manoeuvres.

69. The following entry is not required to be made in the Vessel Record Book:

    (a) Times & dates of emergency drills conducted.
    (b) A death or disappearance.
    (c) Details of any casualty to the vessel or assistance given to another vessel.
    (d) Master's arrival & departure times from the vessel.

CHAPTER 12.2: COMMERCIAL VESSELS

70. The purpose of the NSCV Exemptions Register is to provide a centralised database of:

    (a) the marine equipment exempt from entry in the Register for Compliant Equipment
    (b) the crew certification exemptions
    (c) the generic vessel exemptions.
    (d) none of the choices stated here.

71. An in-survey vessel documentation consists of:

    (a) a permit plate.
    (b) a permit and a permit plate.
    (c) a survey schedule.
    (d) all of the documents stated here.

72. Commercial vessels exempt from government survey are:

    (a) Class 1C carrying less than 12 passengers.
    (b) Classes 2C and 3C
    (c) all of the classes stated here.
    (d) none of the classes stated here.

73. Unless otherwise endorsed, a coxswain's certificate permits the holder to operate vessels limited to the following length and seaward radius from a port:

    (a) 12 metres, 15 miles.
    (b) 15 metres, 12 miles.
    (c) 12 metres, 12 miles.
    (d) none of the choices stated here.

74. In addition to displaying the commercial vessel certificate inside the vessel, a vessel in "in-survey" must also display on her hull:

    (a) the registration number.
    (b) the permit number.
    (c) two of the numbers stated here.
    (d) a special category number.

75. A vessel operating as a LFB (Licensed Fishing Boat) is issued with:

    (a) a LFB number.
    (b) a "survey-exempt" permit number.
    (c) two of the numbers stated here.
    (d) a special category number.

76. The State registered commercials vessels in Australia are regulated by the:

    (a) Uniform Shipping Laws (USL) Code.
    (b) State Act of Parliament.
    (c) National Standard for Commercial Vessels.
    (d) USL Code, NSCV and State Act of Parliament.

77. The in-survey commercial vessels are surveyed:

    (a) when due for a periodic survey.
    (b) after a structural damage or alteration.
    (c) in two of the cases stated here.
    (d) annually.

78. You have been appointed master of a commercial vessel. Her seaworthy condition should be evident from the inspection of her:

    (a) survey certificate.
    (b) hull.
    (c) hull, equipment, supplies and manning.
    (d) sea trials.

79. A Class 1C commercial vessel (except in Queensland) may operate within:

    (a) 50 miles to seaward of the coast.
    (b) 10 miles to seaward of the coast.
    (c) 30 miles to seaward of the coast.
    (d) smooth or partially smooth waters.

80. A Class 1F commercial vessel is a:

    (a) passenger vessel.
    (b) non-passenger vessel.
    (c) hire and drive vessel.
    (d) fishing vessel.

81. In the USL Code/NSCV, the term Sheltered Waters means:

    (a) smooth waters.
    (b) partially smooth waters.
    (c) smooth or partially smooth waters.
    (d) none of the choices stated here.

82. A commercial vessel may:

    (a) carry more than one survey classification.
    (b) not carry more than one survey classification.
    (c) not carry passengers unless survey classed 1.
    (d) not carry passengers unless survey classed 2.

83. On board a commercial vessel outside the State in which she is registered, the crew disputes of non-criminal nature are dealt with under the laws of the:

    (a) Commonwealth.
    (b) present port State.
    (c) State in which the vessel is registered.
    (d) any or all of the choices stated here.

84. A vessel in State survey, when trading on an interstate voyage, must obtain an additional permit from:

    (a) her home State authority.
    (b) the State she intends to trade with.
    (c) both the State authorities.
    (d) AMSA.

85. Following any repairs of structural damage, a commercial vessel is not permitted to operate without obtaining a:

    (a) sea trial certificate.
    (b) work completion certificate.
    (c) authority's inspection certificate.
    (d) gas free certificate.

86. Under exceptional circumstances, a commercial vessel may be granted an extension of her certificate of survey, for a period up to:

    (a) 1 month.
    (b) 2 months.
    (c) 3 months.
    (d) 6 months.

87. Class 3 vessels are:

    (a) Fishing vessels, including game fishing.
    (b) Registered commercial fishing vessels.
    (c) Hire & Drive vessels.
    (d) Landing barges.

88. A commercial vessel's permit plate or certificate of registration should be:

    (a) held in safe custody in the owner's office.
    (b) prominently displayed on board at all times.
    (c) filed among vessel's survey documents.
    (d) displayed in owner's registered office.

89. The master of a commercial vessel in breach of survey requirements may be prosecuted under the:

    (a) USL Code
    (b) NSCV
    (c) SOLAS Convention
    (d) relevant Marine Act

90. The operating limit for class "A" commercial vessels is:

    (a) Unlimited seagoing
    (b) Smooth or partially smooth waters.
    (c) Inshore waters.
    (d) Australian & middle waters.

91. The operating limit for class "E" commercial vessels is:

    (a) Smooth or partially smooth waters.
    (b) Inshore waters.
    (c) Australian & middle waters.
    (d) Smooth waters.

92. The operating limit for class "D" commercial vessels is"

    (a) Inshore waters.
    (b) Partially smooth waters.
    (c) Up to 30 miles offshore.
    (d) Up to 200 miles offshore.

93. In relation to the USL Code, which of the following statements is incorrect?

    (a) It is a national set of standards for survey, manning & operation of commercial vessels.
    (b) It is the law in every State & Territory and the Commonwealth of Australia.
    (c) It includes a section on Stability.
    (d) It includes a section on Emergency Procedures and Safety of Navigation.

94. In relation to the NSCV, which of the following statements is incorrect?

    (a) It comprises of 6 parts.
    (b) All its parts are mandatory under the OH&S legislation.
    (c) Three additional documents complement the NSCV.
    (d) None of its parts deal with crew competencies.

95. In the USL Code/NSCV, a Survey is defined as:

    (a) a thorough examination by a competent person.
    (b) a visual inspection by a surveyor.
    (c) a thorough examination by a surveyor.
    (d) a visual inspection by an approved person.

96. In the USL Code/NSCV, an Inspection is defined as:

    (a) a thorough examination by a ship's crewmember.
    (b) a visual inspection by a surveyor.
    (c) a thorough examination by a surveyor.
    (d) a visual inspection by an approved (competent) person.

97. In the USL Code/NSCV, a Competent Person is someone who has:

    (a) the knowledge & skills to perform the task.
    (b) received formal training.
    (c) received formal qualification.
    (d) the experience.

98. In the Initial Survey, a commercial vessel:

    (a) is investigated to ensure that it meets the prescribed standards.
    (b) is investigated and verified that it meets the prescribed standards for the first time.
    (c) is verified that it meets the prescribed standards.
    (d) is investigated and verified that it meets the prescribed standards.

Chapter 12: Australian Boating Regulations & Pollution Control

99. Which of the following items is not inspected at the time of the Initial Survey of a commercial vessel?

    (a) Bilge Pumping Arrangement.
    (b) Electrical Equipment.
    (c) Periodic Survey Certificate
    (d) Stability Information.

100. In relation to Periodic Surveys of commercial vessels, which of the following statements is incorrect?

    (a) Extensions cannot be granted.
    (b) Once a year, depending on the State or Territory, the vessel's equipment is surveyed or declared by the owner or master to be in working order.
    (c) The vessel's Survey Record Book contains the list of items that do not require to be surveyed annually.
    (d) The survey is not complete until the necessary deficiencies are made good.

101. In relation to vessels in Survey, which of the following statements is incorrect?

    (a) The Survey Number must be painted on the hull.
    (b) Structural alterations can only be made at the time of periodic surveys.
    (c) The Survey Certificate must be displayed on board.
    (d) In some States & Territories of Australia, commercial vessels are not surveyed by the Government Authorities.

102. For a Commercial Certificate of Registration for a Dive Vessel in Queensland, the carrying capacity is determined at:

    (a) 65 kg per person.
    (b) 75 kg per person.
    (c) 125 kg per person.
    (d) 110 kg per person

103. In NSCV, a Novel Vessel is defined as a vessel:

    (a) that travels at a speed of 20 knots or more.
    (b) designed for a special purpose.
    (c) for which the hazards & risks are not adequately addressed by the USL/NSCV requirements for conventional vessels.
    (d) to which the High-Speed Craft Code applies.

104. In NSCV, which of the following is not a "Special Purpose Vessel"?

    (a) A class 2 or 3 vessel carrying more than 12 passengers.
    (b) A vessel engaged in research, expedition or survey.
    (c) A vessel engaged in training of marine personnel.
    (d) A fish factory ship.

105. In NSCV, which of the following statements is incorrect in relation to Fast Craft?

    (a) These are vessels that travel at a maximum speed of 20 knots or more.
    (b) These are only Class 1 fast vessels.
    (c) The IMO standard for the construction & operation of high-speed craft is the HSC Code.
    (d) There are two categories of Fast Craft.

106. In NSCV, which of the following is not a Special Service Notation?

    (a) HSC
    (b) FAST.
    (c) NOV.
    (d) SP.

107. Only the following Classes of vessels are required to maintain a Crew List:

    (a) Seagoing Class 1 vessels.
    (b) Class 1 & 2 vessels.
    (c) Class 1 A, 1B & 1C vessels.
    (d) Class 1A, 1B, 1C, 2A, 2B & 2C vessels.

108. Which of the following entries is not a requirement in a statutory Crew List?

    (a) Official Number of the vessel.
    (b) Length of the vessel.
    (c) The address, DOB, capacity & dates of joining & leaving of each crew member.
    (d) Name & address of the owner.

CHAPTER 12.3: HIRE & DRIVE VESSELS

109. Operators of Hire and Drive vessels do not need to:

    (a) supply qualified masters to hirers.
    (b) satisfy themselves that the hirers are competent.
    (c) maintain a register of hirings.
    (d) instruct hirers on matters of safety and operating limits.

110. In the register of hirings for Hire and Drive vessels, the following entry is not required to be made:

    (a) Signature of each hirer
    (b) Names and addresses of everyone on board.
    (c) Survey number of the vessel.
    (d) Date and time of return of the vessel.

CHAPTER 12.4: MARINE POLLUTION

111. The pollution regulations permit vessels to discharge:

    (a) toilet waste in tidal waters.
    (b) food scraps within 3 miles of land.
    (c) oily bilges outside 12 miles of nearest land.
    (d) metal cans outside 12 miles of nearest land.

112. Marine pollution is not an offence when it is:

    (a) essential to save lives.
    (b) accidental.
    (c) outside 12 miles of nearest land.
    (d) in the form of fishing nets.

113. Under the pollution control regulations, the following vessels may discharge toilet waste in harbours and inland waters:

    (a) less than 6 metres in length.
    (b) with no sleeping accommodation.
    (c) with no toilet.
    (d) none of the choices stated here.

114. Oily bilges can be pumped overboard when the vessel meets the following condition:

    (a) over 50 miles from nearest land.
    (b) over 12 miles from nearest land.
    (c) outside the harbour areas.
    (d) none of the choices stated here.

115. The following action should be taken after an accidental oil spill in the water:

    (a) clean with oil dispersants.
    (b) clean with oil dispersant and advise authority.
    (c) contain spill and advise authority.
    (d) dilute and break up the spill.

116. In relation to MARPOL, which of the following statements is incorrect?

    (a) It is the International Convention for the Prevention of Pollution from Ships 1973/78.
    (b) It is the basis for Commonwealth Protection of the Sea (Prevention of pollution from ships) Act 1983.
    (c) It is the basis for the Commonwealth Garbage Disposal Regulations.
    (d) Its Annex IV deals with pollution from oil.

117. In relation to MARPOL, which of the following statements is incorrect?

    (a) Vessels of 12 metres & over must display placards setting out the MARPOL Garbage Disposal Requirements.
    (b) Vessel certified to carry 15 persons & over must develop & follow a shipboard Waste Management Plan.
    (c) MARPOL contains 20 annexes.
    (d) No discharge of any type is permitted in the area of Great Barrier Reef.

118. Under MARPOL, ships of 400 GRT and above may pump out the engine room bilges as follows:

    (a) The ship must be stopped.
    (b) Maximum oil content 15 ppm.
    (c) No discharge within 50 miles of land.
    (d) Maximum discharge of oil 30 litres per mile travelled by the ship.

119. Under MARPOL, tankers of 150 GRT and above may pump out their tank washing as follows:

    (a) The ship must be stopped.
    (b) Maximum oil content 30 ppm.
    (c) No discharge within 12 miles of land.
    (d) Maximum discharge of oil 30 litres per mile travelled by the ship.

120. Under MARPOL, the maximum quantity of oil from tank cleaning allowed to be discharged by tankers is the following proportion of the total cargo carrying capacity:

    (a) 1/30,000.
    (b) 1/20,000.
    (c) 1/60,000.
    (d) 1/10,000.

121. In relation to the Australian Environment Protection (Sea Dumping) Act, which of the following statements is correct?

    (a) No dumping, whatsoever, is permitted.
    (b) It relates to throwing garbage overboard.
    (c) It relates to pumping out oily bilges.
    (d) It relates to dumping of sand, gravel & factory waste.

122. POLREP is the pollution report made to AMSA following an incident involving a discharge of:

    (a) sewage.
    (b) oil
    (c) garbage
    (d) plastic

Chapter 12: Australian Boating Regulations & Pollution Control

123. In preventing build up of static charge in fuel, which of the following statements is <u>incorrect</u>?

    (a) The filler pipe should extend to the bottom of the tank
    (b) A static charge can build up during refuelling.
    (c) The vessel's refuelling system must not be connected to her bonding system.
    (d) A static charge can build up when sounding, sampling or washing a fuel tank.

124. In relation to environmental damage from antifouling, which of the following statements is correct?

    (a) The use of tin-based antifoulings is prohibited in Australia.
    (b) The use of copper-based antifoulings is prohibited in Australia.
    (c) Stricter regulations are applied to larger vessels.
    (d) Teflon treatment of trailer boats is as damaging to the environment as biocide-releasing antifoulings.

12.5: HSC CODE – ADDITIONAL QUESTIONS

125. The HSC Code imposes 4-hours voyage-time restriction on the following categories of high-speed craft:

    (a) Only the 'rescue-assisted' passenger craft.
    (b) Only the 'unassisted' passenger craft.
    (c) All categories of high-speed craft.
    (d) Both categories of passenger craft.

126. The HSC Code imposes 8-hours voyage-time restriction on the following categories of high-speed craft:

    (a) 'Rescue-assisted' passenger craft.
    (b) 'Unassisted' passenger craft.
    (c) Cargo craft.
    (d) All categories of high-speed craft.

127. The HSC Code imposes the "get-home" capability on the following categories of high-speed craft:

    (a) All passenger craft.
    (b) Only the 'unassisted' passenger craft.
    (c) Only the 'unassisted' cargo craft.
    (d) All 'unassisted' craft.

128. The HSC Code permits the following:

    (a) None of the choices stated here.
    (b) Standing room for 25% of passengers.
    (c) Enclosed sleeping berths for passengers.
    (d) Enclosed standing room for passengers.

CHAPTER 12: ANSWERS

1 (d), 2 (c), 3 (b), 4 (d), 5 (c),
6 (c), 7 (d), 8 (b), 9 (b), 10 (a),
11(b), 12 (b), 13(a), 14(c), 15(d),
16 (b), 17(a), 18(c), 19(a), 20 (b),
21(a), 22 (b), 23 (c), 24 (a), 25 (b),
26 (b), 27 (c), 28 (b), 29 (d), 30 (d),
31(c), 32 (b), 33 (a), 34 (d), 35 (b),
36 (a), 37 (b), 38 (c), 39 (c), 40 (c),
41(d), 42 (d), 43 (c), 44 (b), 45 (b),
46 (b), 47 (c), 48(d), 49 (b), 50 (d),
51 (d), 52 (d), 53 (a), 54 (c), 55 (a),
56 (d), 57 (a), 58 (c), 59 (b), 60 (b),
61 (d), 62 (a), 63 (c), 64 (c), 65 (b),
66 (d), 67 (b), 68 (c), 69 (d), 70 (c),
71 (d), 72 (d), 73 (a), 74 (b), 75 (a),
76 (b), 77 (c), 78 (c), 79(c), 80 (c),
81 (c), 82 (a), 83 (c), 84 (d), 85 (c)
86 (c), 87 (b), 88 (b), 89 (d), 90 (a),
91 (d), 92 (b), 93 (b), 94 (d), 95 (c),
96 (d), 97 (a), 98 (b), 99 (c), 100 (a),
101 (b), 102 (d), 103 (c), 104 (a), 105 (b),
106 (a), 107 (d), 108 (b), 109 (a), 110 (b),
111 (d), 112 (a), 113 (d), 114 (d), 115(c),
116 (d), 117 (c), 118 (b), 119 (d), 120 (a),
121 (d), 122 (b), 123 (c), 124 (a), 125 (d),
126 (c), 127 (b), 128 (a)

Chapter 13

LOAD, BUOYANCY & STABILITY

LOADING OF VESSELS

DENSITY (RELATIVE DENSITY or R.D.):
Relative density (or Density) of a material is its weight divided by the weight of an equal volume of freshwater. For ease of understanding, look at it as follows:
- One cubic metre of freshwater weighs one tonne (or 1 cubic metre = 1000 litres = 1000 kg = 1 tonne). In other words, the space to weight ratio of freshwater is 1. Therefore, the density of freshwater is 1.
- One cubic metre of iron weighs 7.8 tonnes, meaning it is 7.8 times heavier than freshwater, or its space to weigh ratio is 7.8. Therefore, the density of iron is 7.8 tonnes/cubic metre (stated simply as 7.8).
- One cubic metre of seawater weighs 1.025 tonnes (or 1025 kg), which means it is 25 kg heavier than one cubic metre of freshwater. Therefore, its density is 1.025 tonnes (or 1025 kg) / cubic metre (stated simply as 1.025 or 1025).
- Oils are less dense than water. Therefore, their density is always less than one, say 0.8.

ARCHIMEDES PRINCIPLE (as applied to ships):
For an object to float, it must displace its own weight of liquid in which it floats.

WHY DO VESSELS FLOAT?
If we immerse the above-mentioned one cubic metre iron into a pool full of freshwater, only one cubic metre (i.e., one tonne) of freshwater will overflow and the iron will sink. The iron sank because its weight of 7.8 tonnes was greater than the one tonne of water it displaced.

Now, let us mould this 7.8 tonnes of iron into a shape equal to 7.8 cubic metres. This time, when immersed, 7.8 cubic metres (or 7.8 tonnes) of water will overflow, and the iron will float just below the surface of the water. The iron shape has displaced water equal to its own weight. The hole in water can now accommodate the weight of the iron; therefore it floats. It floats by displacing water equal to its own weight.

As the next step, let us mould this 7.8 tonnes of iron into a shape of 10 cubic metres. What do you think will happen? It will float 7.8 cubic metres underwater and the rest (2.2 cubic metres) above water. The height of the underwater section is called the DRAFT and the height of the above-water section is the FREEBOARD.

DRAFT (OR DRAUGHT) MARKS

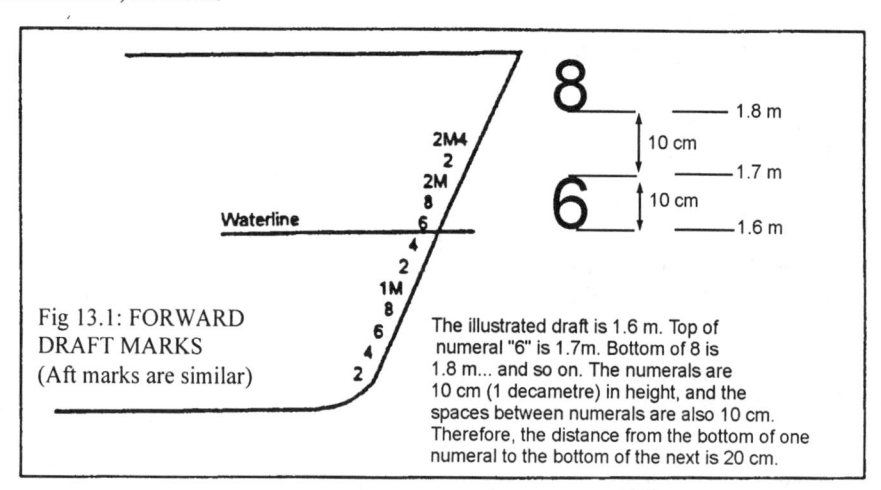

Fig 13.1: FORWARD DRAFT MARKS
(Aft marks are similar)

The illustrated draft is 1.6 m. Top of numeral "6" is 1.7m. Bottom of 8 is 1.8 m... and so on. The numerals are 10 cm (1 decametre) in height, and the spaces between numerals are also 10 cm. Therefore, the distance from the bottom of one numeral to the bottom of the next is 20 cm.

BLOCK COEFFICIENT (COEFFICIENT OF FINENESS) (Cb)
Imagine shaping a boat out of a rectangular block of wood. If the final boat is 80% of the volume of the original block, her Cb is said to be 0.8. If it is 60%, then her Cb is 0.6. More precisely, this ratio is applied only to her underwater hull. Thus, a vessel's Cb is the ratio between her underwater shape and a rectangular block of the same waterline length, breadth and depth.

Chapter 13: Load, Buoyancy & Stability

Planing hulls and yachts needing high speed are designed with Cb as small as 0.30, while barges designed for maximum carrying capacity have Cb as high as 0.95.

DISPLACEMENT

We have seen above that vessels displace water equal their own weight. Therefore a vessel's weight is also referred to as her displacement or displacement tonnes. LIGHT DISPLACEMENT is the weight of an empty ship (the only external weight on board is the lubricating oil in her machinery) and LOAD DISPLACEMENT is the weight of a loaded ship.

A vessel displacing one cubic metre of freshwater weighs one tonne. If she were a box-shaped vessel of 2 metres length and 1 metre width, her freshwater *draft* would be 0.5 metres (Freshwater displacement = L x B x draft = 2 x 1 x 0.5 = 1 cubic metre or 1 tonne). If she is now moved into seawater, her weight (or displacement tonnes of seawater) would not change, but she would have a smaller draft. This is because seawater is denser (1.025 tonnes per cubic metre), needing a smaller volume of displacement than freshwater.

We can calculate the displacement of a box-shaped vessel in any liquid by multiplying the above formula by the relative density of the liquid.

Displacement = L x B x draft x density
OR
Draft = Displacement ÷ (L x B x density)
Therefore, draft in seawater = 1 ÷ (2 x 1 x 1.025)
 = 0.49 metres.

For a vessel that in not box-shaped, the displacement calculation must also take into account her Block coefficient. The approximate displacement of a boat can thus be calculated as follows:

DISPLACEMENT = L x B x draft x density x Cb.

Let us shape the above-mentioned box-shaped vessel into a boat-shape of Cb of 0.7. Calculating her displacement in freshwater or seawater would give the same result provided we use the appropriate draft and density:

Displacement in freshwater = 2 x 1 x 0.5 x 1 x 0.7
 = 0.7 tonne
(Freshwater draft = 0.5; Freshwater density = 1)
OR
Displacement in seawater = 2 x 1 x 0.49 x 1.025 x 0.7
 = 0.7 tonne
(Seawater draft = 0.49; Seawater density = 1.025)

Barges etc Cb = 0.75 - 0.95

Tugs and coasters Cb = 0.65 - 0.75

Trawler and workboats Cb - 0.55 - 0.65

Fishing boats Cb = 0.50 - 0.55

Displacement cruisers Cb = 0.45 - 0.55

Semi displacement patrol boats etc Cb = 0.40 - 0.45

High speed planing boats Cb = 0.35 - 0.40

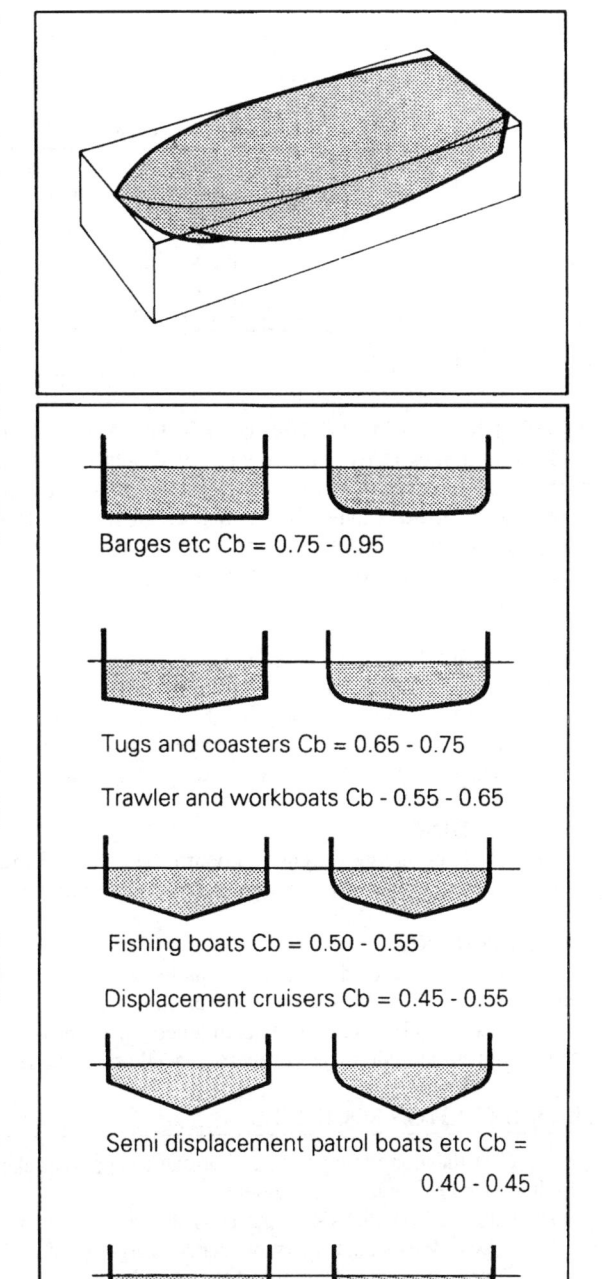

Fig. 13.2: BLOCK COEFFICIENT (Cb)
(*Lister Petter Marine*)

Naval architects calibrate and supply commercial ships with a Draft and Displacement Table. The ship's master can check the ship's displacement at any stage of loading by reading the ship's mean draft and then referring to the Table.

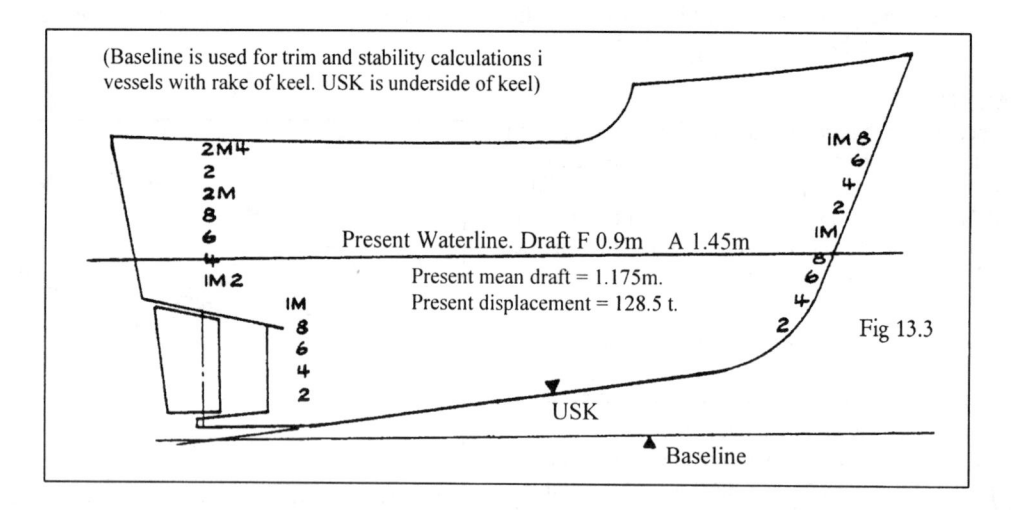

(Baseline is used for trim and stability calculations i vessels with rake of keel. USK is underside of keel)

Present Waterline. Draft F 0.9m   A 1.45m
Present mean draft = 1.175m.
Present displacement = 128.5 t.

Fig 13.3

DEADWEIGHT (DWT) & DEADWEIGHT CAPACITY

Deadweight capacity is the amount of weight (cargo, fuel, water, stores, etc) that a vessel can legally carry. Deadweight is the amount of weight on board at any given time, which may be less than her deadweight capacity. Sizes of cargo ships are normally expressed in deadweight capacity. A 20,000 tonnes ship means she can carry 20,000 tonnes of cargo.

GROSS & NET REGISTERED TONNES (GRT & NRT)

These misleading terms do not represent the weight (displacement) of a vessel. They are the cubic measurements of her internal gross space and net (cargo carrying) space. Converted into an approximate carrying capacity in tonnes, they are sometimes used for calculating taxes and charges levied by port and canal authorities.

SAMPLE DISPLACEMENT TABLE

S.W. MEAN DRAFT (in Metres)	DISPLACEMENT (in Tonnes)
1.0	105
1.1	118
1.2	132
1.3	166

Fig 13.4

DECK LOADING

Before taking heavy weights on board, be aware of the load bearing capacity of the deck on which the loads will be placed. Decks supported from underneath can generally take heavier loads than the unsupported decks.

POINT LOADING

Unlike the distributed weight of a palletised load, a shipping container exerts its weight through its four feet. A 20-tonne container would thus subject a deck to point-loading of 5-tonnes at four places. Such loads must therefore be placed only where the deck is suitably strengthened for point loading. The weight of point loading can be distributed to a larger area through suitable dunnage (timbers) placed under the feet of the load.

CLASSIFICATION SOCIETIES

Classification of ships makes it easier to apply similar and comparable rules and regulations to them. Ships are classed according to three operational conditions:
- Climatic Conditions -tropical, coastal, ocean or ice conditions.
- Type of load – passengers, oil, general cargo or freezer.
- Type of manning – unmanned or manned machinery spaces.

Classification of ships and their machinery improves safety and creates better documentation for insurance and other commercial transactions, such as mortgages and sale and purchase of vessels. Classification Societies are commercial organisations with surveyors stationed in ports around the world. Almost all large commercial vessels and some small commercial and recreational vessels choose to be CLASSED with a Classification Society of their choice. They thus carry a CLASSIFICATION CERTIFICATE. Governments in most countries authorise Classification Societies to enforce classification rules on commercial vessels classed with them. The un-classed commercial vessels are inspected by government surveyors.

The world's oldest Classification Society, the Lloyd's Register of Shipping, was founded in 1760. Its initials "LR" appear on the sides of many ships. Today, there are about 12 major Classification Societies, including:
- Lloyds Register of Shipping (LR)
- Det Norske Veritas (DnV)

Chapter 13: Load, Buoyancy & Stability

- Bureau Veritas (BV)
- American Bureau of Shipping (AB)
- Germanischer Lloyd (GL)
- Nippon Kaiji Kyokai (NK)

LOAD LINES CERTIFICATE

The maritime world is divided into loadline zones based on the expected weather conditions in different latitudes. Most of the Australian waters fall into the (permanent) Summer Zone. In addition to draft marks, the loadlines-assigned commercial vessels are marked with loadlines for various zones. They thus carry a survey certificate as well as a load line certificate. Certificates for a classed vessel are issued by her Classification Society.

The characteristics of loadlines marks are:

- Plimsoll mark is the circle with a horizontal line through it. The top of this line represents the Summer Loadline.
- The Plimsoll mark and the appropriate loadlines are painted on each side of the ship at a specified distance below the uppermost watertight deck.
- A deckline mark is painted to indicate the position of the uppermost watertight deck.
- All lines are 25 mm in thickness.
- The Tropical Loadline, marked "T", is the highest and the Winter Loadline (W) is the lowest. This is so because, in calmer tropical waters, vessels are permitted to carry greater loads than in the rougher waters of the middle or high latitudes (known as the Summer and Winter Zones).
- Summer Freshwater loadline (F) and Tropical Freshwater loadline (TF) are also marked to indicate the additional sinkage a vessel would experience when loading upriver in these waters.
- The distance between S and F (or T & TF) is known as the vessel's **FRESH WATER ALLOWANCE.**
- Vessels must not load beyond the top of the appropriate loadline.

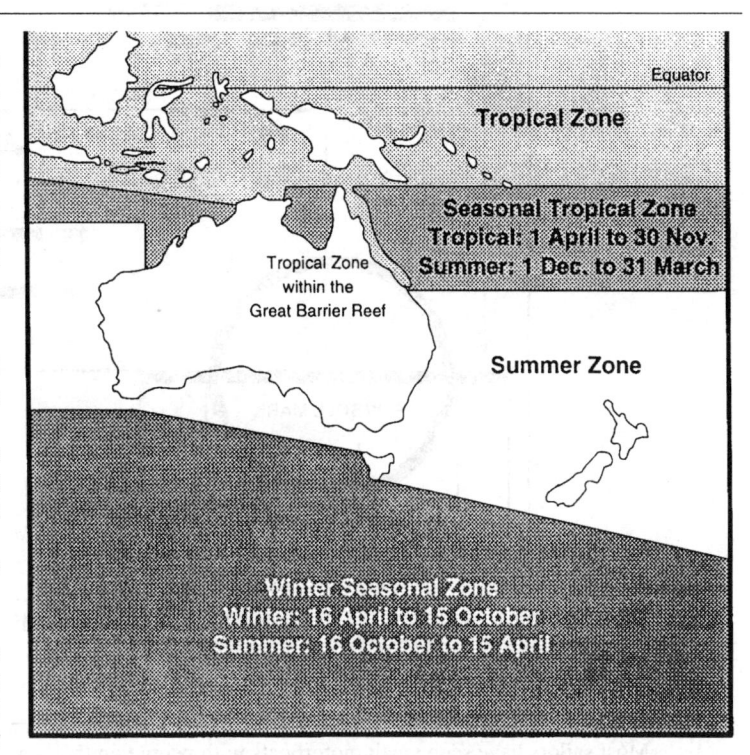

Fig 13.5: LOADLINE ZONES: AUSTRALIAN REGION

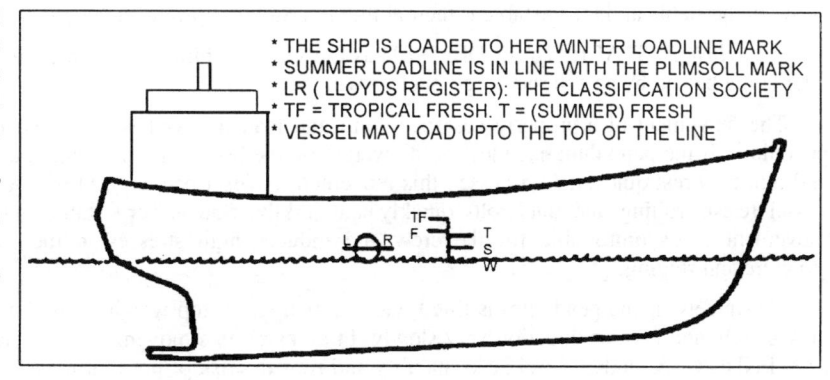

* THE SHIP IS LOADED TO HER WINTER LOADLINE MARK
* SUMMER LOADLINE IS IN LINE WITH THE PLIMSOLL MARK
* LR (LLOYDS REGISTER): THE CLASSIFICATION SOCIETY
* TF = TROPICAL FRESH. T = (SUMMER) FRESH
* VESSEL MAY LOAD UPTO THE TOP OF THE LINE

Fig 13.6: LOADLINES MARKS WELDED & PAINTED ON A SHIP

THE FOLLOWING VESSELS ARE EXEMPT from showing loadlines because under the international loadline convention, the term "vessel" does not include:

- A vessel of less than 24 metres measured length
- A commercial fishing vessel *(Author's note: Loadlines are impossible to comply with when loading in open seas and swell.)*
- A smooth or partially smooth water vessel operating solely on the basis of the maximum number of passengers permitted to be carried.

A naval architect calculates the vessel's hydrostatic particulars during the design phase, and then verifies them by forcibly inclining the vessel on completion of construction (This is known as the INCLINING EXPERIMENT). The load lines are assigned based on meeting the CONDITIONS OF ASSIGNMENT with regard to vessel's watertight integrity and reserve buoyancy. A Load Line Certificate is then issued, which is valid for 5 years. However, load line surveys are conducted annually to ensure that vessel's openings can be closed watertight, her structural strength and watertight integrity are intact and that the load line marks have not been fraudulently moved higher. The surveyor also makes sure that the vessel's stability information booklet is retained on board.

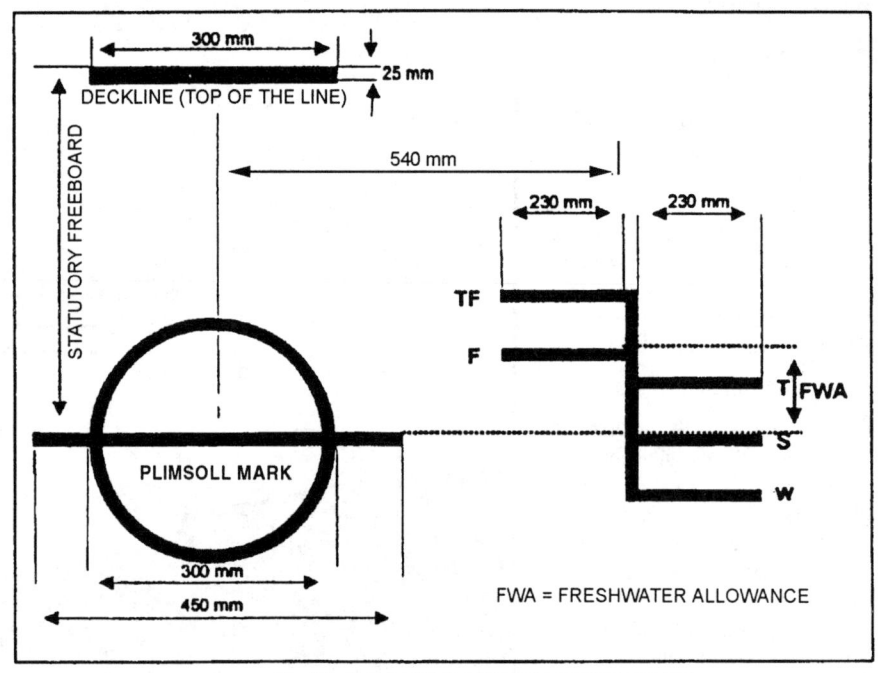

Fig 13.7: LOADLINES MEASUREMENTS

VESSEL STABILITY

Most sailors have seen small motorboats with people on their cabin tops and yachts heeled to an angle of deck edge immersion, and wondered if they'll make it. Well, some do and some don't!

All vessels are designed to heel, roll, and pitch and yaw when pushed by the forces of wind and sea. The difference between the stable and the unstable is their ability to return to upright each time.

HENCE, STABILITY IS DEFINED as the vessel's ability to return to upright after being heeled by an external force, such as wind and waves.

The movement of a properly designed, built and loaded vessel is like a balanced pendulum. If the pendulum has most of the weight on the bottom it will swing quickly and return to rest quickly. In a vessel, this movement is described as "STIFF". A stiff vessel resists rolling and thus rolls quickly and jerkily. She is very stable but her movement is uncomfortable for the crew and induces high stresses in the ship's structure and rigging.

Conversely, if the pendulum is fitted with a much larger top weight, it will swing very slowly and reverse direction very slowly. In a vessel, this movement is described as "TENDER". A tender vessel lacks stability and runs the risk of not returning from a roll.

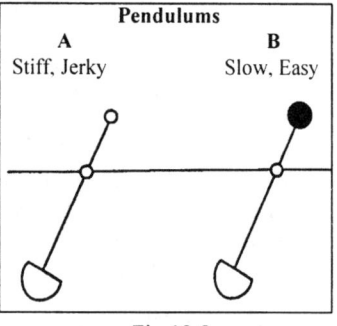

Fig 13.8

Most commercial craft are designed and loaded along a well-balanced pendulum principle - neither stiff nor tender. This allows the passengers, crew and cargoes a more comfortable motion. A sailing vessel, on the other hand, is designed to be stiff - as discussed below.

CENTRE OF GRAVITY

The stability of a vessel is largely determined by the position of her Centre of Gravity (G) at any given moment. The lower the centres of gravity, the more stable the vessel.

The Centre of Gravity (G) is the centre of mass. It is the point from which an object can be suspended in balance. It is the point from which the whole weight of a body (the vessel) and everything in it acts vertically downwards. If a weight in a vessel is moved, her G would move

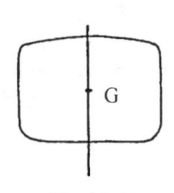

Fig 13.9

proportionally.

As fuel and water from the bottom tanks are consumed, vessel's G gradually rises. It also rises when seas break on her deck. In other words, removing a weight from the bottom of a vessel has the same effect on her G as loading a weight higher up. If, on the other hand, a large weight from the upper deck is jettisoned or shifted to the lower hold, her G will shift downwards, a distance depending on the mass of the weight and the distance it has been moved vertically downwards.

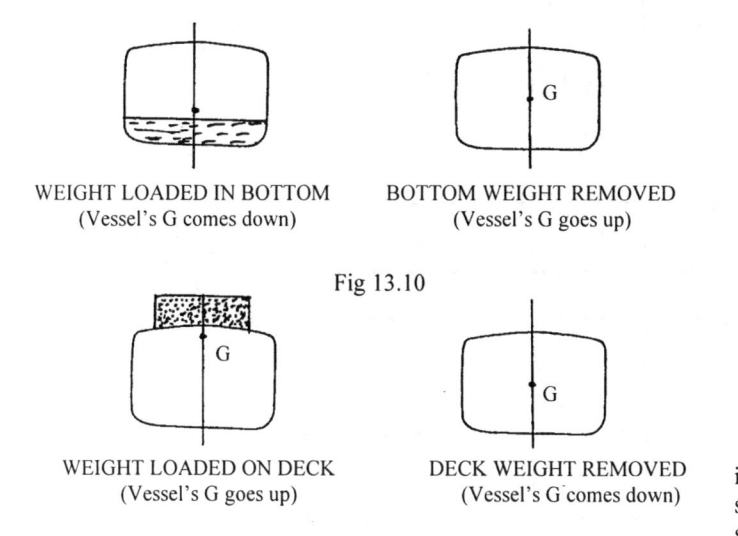

WEIGHT LOADED IN BOTTOM
(Vessel's G comes down)

BOTTOM WEIGHT REMOVED
(Vessel's G goes up)

Fig 13.10

WEIGHT LOADED ON DECK
(Vessel's G goes up)

DECK WEIGHT REMOVED
(Vessel's G comes down)

A weight suspended on a derrick is considered to act at the derrick head

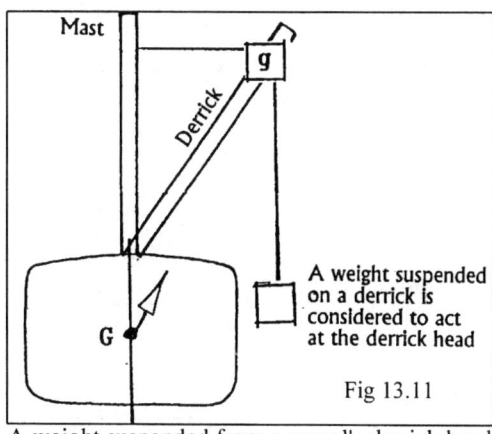

Fig 13.11

A weight suspended from a vessel's derrick head is equivalent to being positioned at the point of suspension at the top of the derrick head. It will significantly raise the vessel's G.

Ideally a vessel's G should be neither too low nor too high. Early action should be taken to move weights in the vessel to correct for excessive "stiffness" or "tenderness".

CENTRE OF BUOYANCY

Just as the weight of a vessel pushes her vertically downwards through her Centre of Gravity, the buoyancy of water lifts her vertically upwards through her Centre of Buoyancy (B).

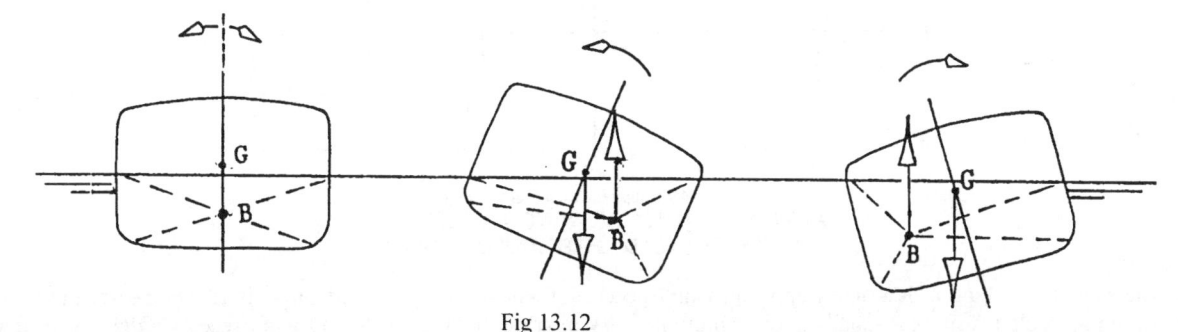

Fig 13.12

The Centre of Buoyancy is the geometric centre of the vessel's underwater section. It moves as the underwater shape of the hull changes during rolling, pitching, loading, discharging or shifting of weights. The G & B of a vessel floating upright (i.e., neither <u>listed</u> to one side <u>by an internal force</u> nor <u>heeled by an external force</u>) is on her centreline: G acting vertically downward and B acting vertically upward.

SAILING VESSEL STABILITY

Heeling is to be expected in all monohull sailing vessels. However, it must be kept under control. A vessel needs to be "stiff" to prevent her from capsizing under pressure of wind and waves. A solid weight is therefore fitted to most sailing vessels, low in the hull, in order to lower their centres of gravity. A piece of cast iron or lead is bolted to the lower extremity of the keel, sometimes in the form of a bulb. When hit by a gust of wind, these vessels heel over easily at first. But the further a vessel heels the greater is the righting moment exerted by the weight on the bottom of the keel to right the vessel. A vessel heeled even at 45 degrees will right herself when pressure on the sails is eased.

SAILING VESSEL STABILITY

Sailing vessels engaged in commercial activities are supplied with a set of Maximum Steady Heel Angle Curves. In figure 13.13, the recommended maximum heel angle is 27°. Operating within the limit would ensure that the critical downflooding openings on the vessel would not be immersed if her heel angle increased when struck by a gust.

The figure also shows the curves of maximum steady heel angle beyond which the vessel will suffer downflooding in the event of a squall. Operation of the vessel in conditions where severe squalls are imminent requires the recommended maximum steady heel angle to be reduced, depending on the apparent wind speed in accordance with the curves.

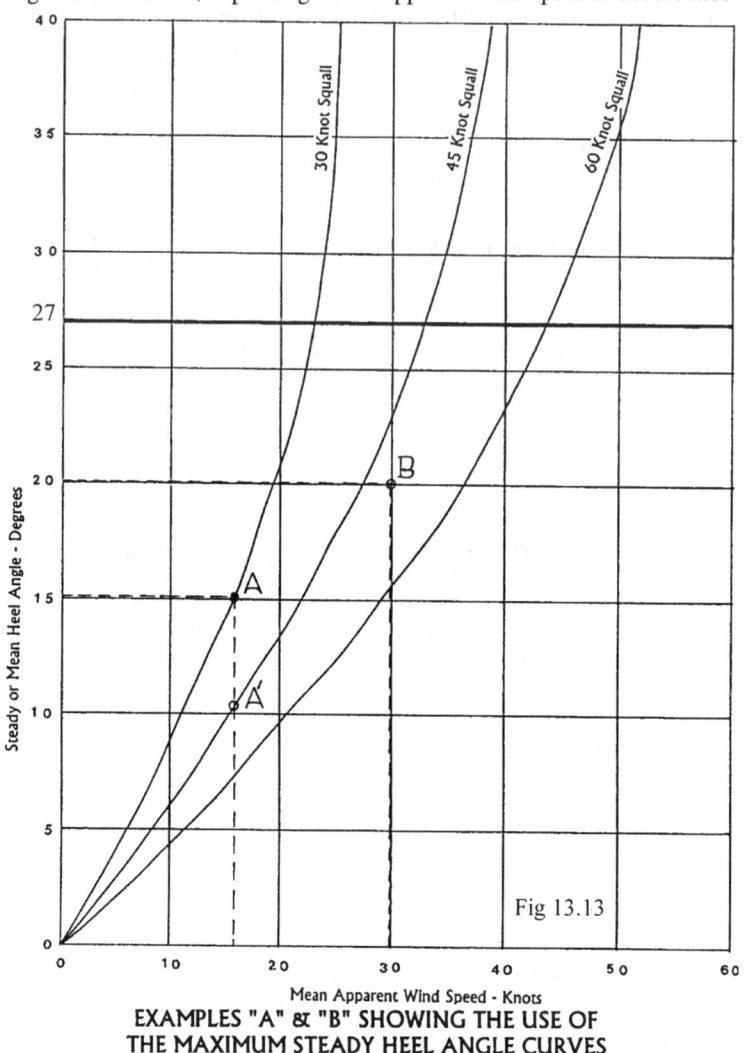

EXAMPLES "A" & "B" SHOWING THE USE OF THE MAXIMUM STEADY HEEL ANGLE CURVES

Fig 13.13

EXAMPLE A

The yacht is reaching with a steady apparent wind speed of 16 knots. The mean heel angle is 15°. Forecasts and visible Cumulo-nimbus clouds suggest squalls may be imminent. By plotting the heel angle and wind speed (POINT A in figure 13.13) the indication is that the vessel will be in danger of heeling to the angle of downflooding in squalls of 30 knots. In order to increase safety from downflooding, say, to withstand squalls of up to 45 knots, sails should be handed or reefed to reduce the mean heel angle to 10.5° or less.

EXAMPLE B

The yacht is beating in gusty conditions with a mean apparent wind speed of 30 knots. The mean heel angle is 20°. No squalls are expected. The heel angle is significantly less than 27°, the maximum recommended steady heel angle. Therefore, there is a good safety margin against downflooding in a strong gust. Plotting these values of wind speed and heel angle (POINT B in figure 13.13) also indicates that the vessel would not be vulnerable to downflooding in a squall of wind speed less than 50 knots.

CATAMARAN STABILITY

Catamaran hulls are usually unballasted, i.e., they do not have lead or steel ballast attached to the bottom of their keels. A cat's stability lies in her load-spreading shape. She is extremely stable (in fact, stiff) up to about 12° angle of heel. After that her stability diminishes rapidly. She would capsize at a much smaller angle than a ballasted monohull. Once upturned, she would tend to remain upturned. An upturned monohull, on the other hand, is top heavy. She would readily right herself when force is applied. This is the principle used in designing SELF-RIGHTING boats.

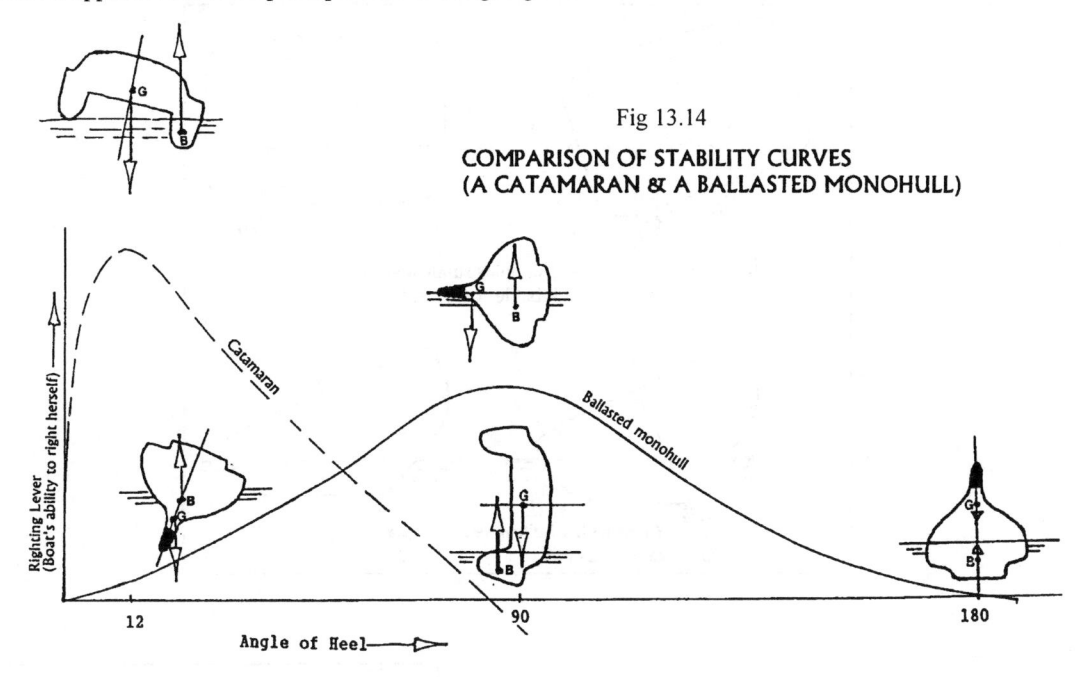

Fig 13.14

**COMPARISON OF STABILITY CURVES
(A CATAMARAN & A BALLASTED MONOHULL)**

FREE SURFACE EFFECT (FSE)

All fluids (and fluid-like materials, such as fish, grain, etc.) on board that are free to move with the motion of the vessel will act like a pendulum and reduce her stability. This is known as the Free Surface Effect (FSE). Partially filled tanks, unpumped bilges and water on deck cause free surface effect. During rolling such liquids run across from side to side. The energy of their movement builds up with rolling, which further increases the vessel's angle of roll and make her less stable.

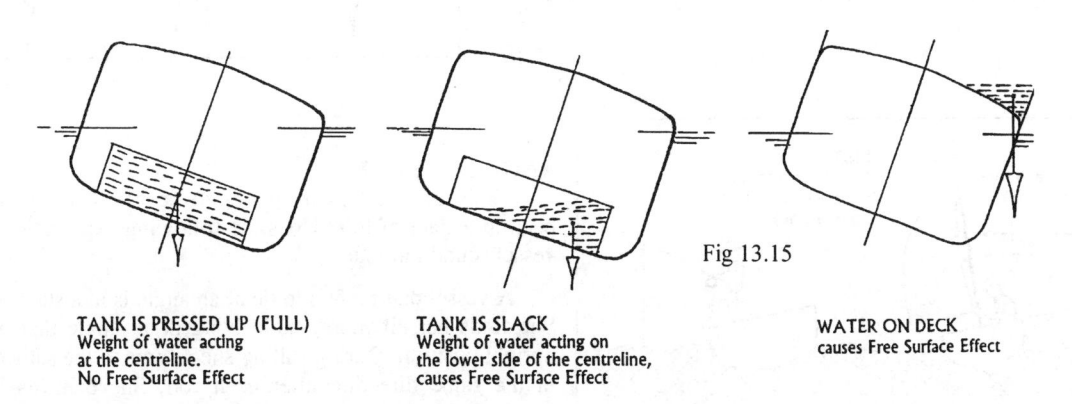

Fig 13.15

TANK IS PRESSED UP (FULL)
Weight of water acting
at the centreline.
No Free Surface Effect

TANK IS SLACK
Weight of water acting on
the lower side of the centreline,
causes Free Surface Effect

WATER ON DECK
causes Free Surface Effect

HOW TO MINIMISE FREE SURFACE EFFECT

You cannot avoid FSE altogether. You can only minimise it.

- Keep bilges free of any liquid.
- Keep deck scuppers and freeing ports open at sea so that seas breaking on deck can run off. Water on deck not only creates a large free surface but its height and weight also raises vessel's G.
- Consumption of water and fuel should be carefully managed to ensure that tanks are emptied before drawing from others, so that the least number of tanks are partially filled at any one time.
- Tanks and fish holds, etc. should be subdivided with longitudinal bulkheads or baffles.
- Yachts may be designed with a hollow keel.

The FSE of a liquid depends upon its free-surface area, NOT its quantity. Larger the quantity of liquid, lesser the empty space available for it to move inside the compartment. The effect of the larger quantity is therefore cancelled by the reduction in distance of its movement.

Fig 13.16 Tanks & fish holds subdivided or fitted with baffles cause less FSE

Yacht with a hollow keel causes less FSE

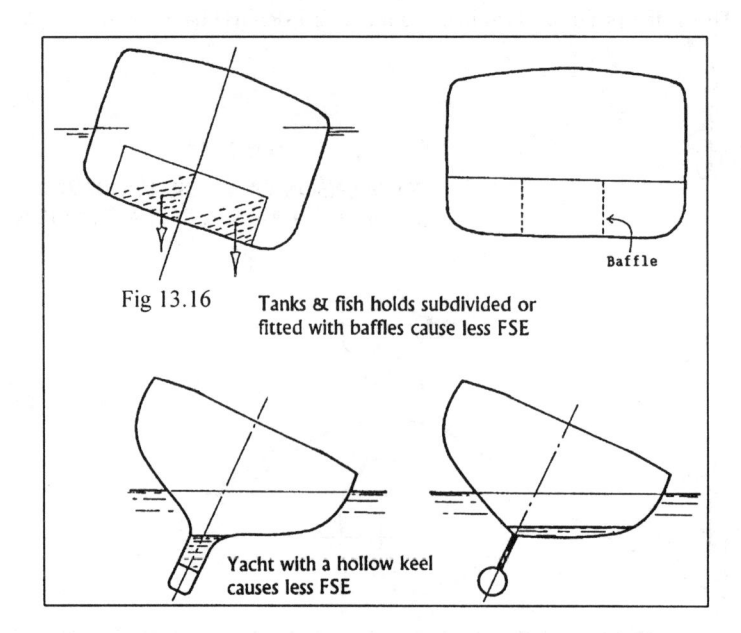

In the attached illustration, the liquid in vessel "B" is smaller in quantity than in vessel "A". However, its centre of gravity is at a greater distance from the vessel's centreline than that of "A". Both cause similar heeling or listing moments (weight x distance), and therefore similar FSE.

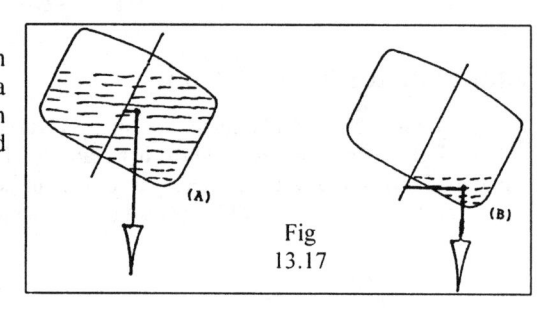

Fig 13.17

HEEL (due to an external force, e.g., wind)

LIST (due to an internal force, e.g., a weight on one side)

ANGLE OF LOLL

Free Surface Effect (Liquid, fish, etc. in the vessel)

Fig 13.18

ANGLE OF LOLL

Think of "loll" as it applies to people. Have you ever been in a state of loll? Does the following explanation for a vessel sound familiar?

A vessel that prefers to lie at an angle is in a state of loll. She finds herself more stable lying at an angle than in the upright position. During rolling she returns to the same angle in the same direction after each roll. She behaves like a drunk. If you try to straighten her up from one side she is likely to flop to the same angle on the other side.

DIFFERENCE BETWEEN HEEL, LIST & LOLL

List should not be confused with loll. A vessel lists because she has more weight on one side than on the other. She cannot be made to lie at the same angle on the lighter side. A list may not have anything to do with a lack of stability.

A vessel in a state of loll, on the other hand, is unstable or has very little stability in the upright position. She prefers to lie at an angle of loll because she lacks stability in the upright position.

LOLL OR LIST - HOW TO TELL
A vessel lying at an angle could be listing or may be at an angle of loll. It is important to know whether it is a list or a loll, in order to take proper correcting action.

IT IS A LIST, IF:
- The heavier weights are stowed low in the hull.
- She has negligible Free Surface Effect.
- She has more weights on one side than on the other.

IT IS LOLL, IF:
- The heavier weights are stowed high in the hull, and she is "tender" during rolling.
- She has a large Free Surface Effect.
- Her load on port and starboard sides are even.

CORRECTING A LIST
A list can be corrected simply by balancing the weights on two sides. For example:
- Shift weights from low to high side. Don't use derricks and cranes in rough weather to shift weights. Weights can become dangerously uncontrollable.
- Add weight or pump in ballast in the higher (lighter) side.
- Discharge weight or pump out ballast from the low side.

CORRECTING LOLL
Correcting loll is not simply a matter of balancing weights like correcting a list. The vessel is unstable. A sudden movement during loading, discharging or shifting of weights, or the introduction of FSE while filling or emptying a tank can cause her to capsize.

The object is to lower G gently, keeping the FSE to a minimum. Any or all of the following actions can be taken:
- Gently and gradually remove or reduce the Free Surface Effect causing the problem. Drain high and press-up low compartments, one by one. Allow the vessel to become upright, but gently and gradually. Don't risk her returning to the upright so fast that she may continue going and flop over to the other side.
- Drain top tanks into the lower ones - one at a time, starting with the smallest one. This is to keep the FSE to a minimum. Transfer tank contents from centreline to centreline or the first tank on the low side followed by a tank on the high side. Once again, this is to avoid returning her to the upright position too fast. Loading on the low side may sound illogical, but it too lowers her G. Draining tanks in this manner will have a two-fold effect on lowering of G. First G is lowered because the weights are removed from the top, and, then it lowered because they are added to the bottom.
- Shift other weights, taking precautions as above.
- Fill up bottom tanks, taking precautions as above.
- Jettison top weights and drain top tanks, taking precautions as above.

DANGER SIGNS OF LACK OF STABILITY

- Vessel becoming tender.
  (Rolling becomes longer and she is slower to return to upright).
- Vessel developing an angle of loll.
  (She wants to lie at an angle in spite of all the weights being evenly distributed).
- Her gunwale is beginning to go under water during rolling.
  (She has reached her "angle of deck-edge immersion" which is a critical point in stability).

VESSEL'S ROLL PERIOD
All vessels have a natural roll period. It is determined by their length, beam and hydrostatic feature. One roll period is the time the vessel takes to roll from her upright position to one side, then the other, and back to the upright position. Larger vessels usually have loner roll periods. The roll period of masted monohull yachts is usually between 4 and 6 seconds, whereas the roll period of a fishing trawler will be around 10 seconds.

You can measure your boat's natural roll period by forcing her to roll in calm waters alongside a jetty (with mooring lines all loose) or at an anchorage. She can be set to roll by a person on the jetty pulling on a rope tied to the top of her mast. At anchor she can be set to roll by rhythmically stepping on the gunwale. For this experiment she should not have excessive free surface effect, and she should be as light (unloaded) as possible. You need to take an average of 8 or 10 rolls.

Knowing your vessel's natural roll period is useful. Depending on the sea conditions, the magnitude of roll (the angle of rolling) will change, but not the period. A vessel rolling in a seaway with period above her natural period indicates that her centre of gravity is higher than usual and visa versa, i.e., she is noticeably 'tender' or 'stiff', respectively. You should expect small changes in roll periods with changes in loading conditions. A noticeable increase in the period, however, is usually a sign of a rise in her centre of gravity to an unacceptable level, which may capsize her. In a yacht it could indicate that her external ballast has fallen off. In a fishing vessel it could be due to too much loading on deck. In all vessels it could indicate a large free surface effect due to hull flooding, or in freezing conditions, the unacceptable level of ice build up on rigging.

The STABILISERS fitted in some vessels, such as hydroplanes, anti-roll tanks, bilge keels and paravanes, do not reduce the vessel's roll period, but the magnitude (the angle) of the roll. Almost all these remedies work only on power driven vessels. In any case, the need for such devices is greater in power driven vessels than in sailing vessels. Sailing hulls have the benefit of their inbuilt anti-roll features. Their magnitude of roll is reduced by at least three factors:
- Ballasted keel.
- Asymmetrical forces of wind in the sails. They oppose rolling to windward more than to leeward, thus reducing the roll magnitude.
- Roll damping by sails at a large lever arm.

HOW TO MAKE VESSELS MORE STABLE

DESIGN (STRUCTURAL) FEATURES
- LARGER BEAM: However, vessels with excessive beam can be uncomfortable in heavy seas. They are also harder to propel.
- HIGH COAMING ON OPEN CRAFT: Barges etc. are provided with high coamings to keep the water out due to waves or rolling. But once water floods over the gunwale, the vessel ceases to have positive stability, and will sink if she has insufficient built-in buoyancy.
- LARGER FREEBOARD: Freeboard increases vessel's range of stability and helps to keep the water off the deck. However, too much freeboard will make the vessel cranky; i.e., she will heel easily with movement of passengers or weights. She will also be difficult to manoeuvre in strong winds.

OPERATIONAL FEATURES
- KEEP HER BOTTOM HEAVY: However, as discussed above, don't make her too "stiff".
- KEEP HATCHES & DOORS CLOSED: In decked vessels, hatches and doors leading to lower spaces should be kept closed when not in use and in bad weather. The cockpit drains and freeing ports should be kept open at all times. Keeping a vessel watertight helps to maintain her stability.
- MINIMISE FREE SURFACE EFFECT
- REDUCE TRIM & LIST: The stability of a vessel is drastically reduced when she has a list or is not trimmed correctly. She should be kept upright and loaded at her designed fore and aft trim.
- ALTER COURSE FOR EASIER RIDING: Vessels pitch excessively in head seas and roll excessively in beam seas. Running before the sea can be dangerous due to the risk of pooping, pitchpoling and broaching. Altering course to take the sea on the bow can reduce both rolling and pitching.

HOW TO INCREASE BOAT'S BUOYANCY
- WOODEN HULL: Wood is buoyant. But the weight of the engine and other fittings can even make a wooden boat sink when filled with water. A yacht with a lead keel and a dinghy with an outboard are similarly vulnerable.
- WATERTIGHT BULKHEADS: Vessels are designed to remain afloat after an accident by either fitting a number of watertight bulkheads or putting buoyancy foam somewhere in their hull. Watertight bulkheads prevent the whole of the vessel from flooding following damage in one area. They also provide strength (and restrict the spread of fire in a vessel).
- FOAM BUOYANCY: Foam is a suitable buoyancy material only in the lightweight construction materials, such as wood, aluminium and fibreglass. Usually polyurethane (not styrene) foam is used because it does not readily absorb water and does not disintegrate (melt) on coming in contact with engine fuels. However, the fumes from burning urethane are toxic.

Foam is installed either in the double skin of the vessel or on the underside of her decks. In vessels with outboard engines, the stern area is made more buoyant. Foam is not placed solely in the bottom of the vessel. This will encourage her to capsize more easily, and once capsized, she will be difficult to turn upright. Foam is also placed below the estimated damage waterline. In case of hull damage, it will reduce the permeability of the compartment, i.e., allow less water to enter.

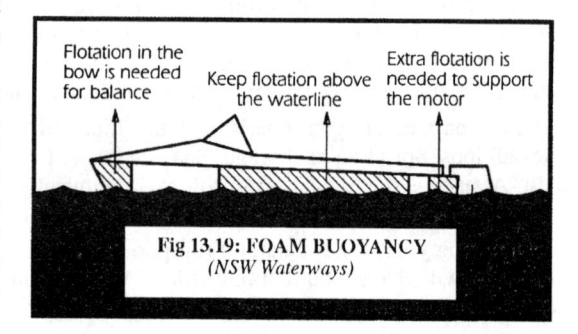

Fig 13.19: FOAM BUOYANCY
(NSW Waterways)

CHAPTER 13: QUESTIONS & ANSWERS

LOAD, BUOYANCY & STABILITY

1. Vessel stability is defined as her ability to:

   (a) stay upright.
   (b) return to upright.
   (c) withstand heavy sea conditions.
   (d) stay watertight.

2. The presence of liquid in a vessel's bilge:

   (a) does not affect her stability.
   (b) can cause a list.
   (c) increases her stability.
   (d) decreases her stability.

3. Polyurethane foam is a suitable buoyancy material in the lightweight vessels constructed of timber, aluminium and fibreglass. However, foam should not be installed solely in the bottom of the vessel, because it:

   (a) readily absorbs water.
   (b) disintegrates on contact with fuel.
   (c) may cause the vessel to capsize more easily.
   (d) makes the vessel stiff.

4. The following items contribute to reducing the Free Surface Effect in vessels:

   (i) Freeing ports.
   (ii) Tank sub-divisions.
   (iii) Bilge pump.
   (iv) Guard rails.

   The correct answer is:
   (a) all except (i).
   (b) all except (ii).
   (c) all except (iii).
   (d) all except (iv).

5. A vessel's large trim:

   (a) affects her stability.
   (b) affects her buoyancy.
   (c) improves her seaworthiness.
   (d) reduces her drift.

6. The period of a vessel's roll is the time a vessel takes to roll from:

   (a) one side to the other.
   (b) one side to the other and back again.
   (c) upright to one side and then to the other.
   (d) none of the choices stated here.

7. A period of wave encounter is:

   (a) the wave period observed from a vessel.
   (b) the period between a wave crest and a trough.
   (c) the period of synchronization.
   (d) none of the choices stated here.

8. Concentration of weights in a vessel's wings will make her:

   (a) roll period shorter and pitch period longer.
   (b) roll period longer and pitch period shorter.
   (c) roll period longer.
   (d) roll and pitch periods shorter.

9. When a weight is removed from the bottom of a vessel:

   (a) the vessel's stability increases.
   (b) the vessel's stability decreases.
   (c) the vessel's stability remains unaffected.
   (d) the vessel becomes more stiff.

10. A vessel is in danger of becoming unstable, if she:

    (i) has a long roll period.
    (ii) has a short roll period.
    (iii) lies at an angle of loll on one side.
    (iv) lies at an angle of loll on either side.

    The correct answer is:
    (a) all except (i).
    (b) all except (ii).
    (c) all except (iii).
    (d) all except (iv).

11. A vessel's stability can be improved by:

    (i) moving passengers to upper deck.
    (ii) pumping out bilges.
    (iii) reducing the number of slack tanks.
    (iv) filling up bottom tanks.

    The correct answer is:
    (a) all except (i).
    (b) all except (ii).
    (c) all except (iii).
    (d) all except (iv).

12. A 10 metre patrol-boat is carrying some weights on deck. She has 50% fuel on board equally divided between two unbaffled tanks in the bottom of the vessel. She is rolling dangerously. The following action will reduce her chances of capsizing:

    (i) Transfer all fuel into one tank.
    (ii) Jettison the contents of one fuel tank.
    (iii) Shift weights from deck to down below.
    (iv) Alter course and slow down.

    The correct answer is:
    (a) all except (i).
    (b) all except (ii).
    (c) all except (iii).
    (d) all except (iv).

13. Before leaving port, a master can make sure that the vessel has a positive stability to cope with bad weather:

    (a) by taking maximum fuel.
    (b) by taking minimum fuel.
    (c) by storing more weights below decks.
    (d) by storing more weights above decks.

14. A reduction in the stability of a vessel makes:

    (a) her roll period shorter.
    (b) her roll period longer.
    (c) the period of wave encounter shorter.
    (d) the period of wave encounter longer.

15. A vessel's list is:

    (a) also known as the heel.
    (b) caused by an internal force.
    (c) caused by an external force.
    (d) caused by an internal or external force.

16. A sailing vessel sailing in a heeled position, does not capsize because:

    (a) she is listed by an internal force.
    (b) she has a high centre of gravity.
    (c) she has a low centre of gravity.
    (d) she is listed by an external force.

17. Factors which contribute to dangerous rolling include:

    (i) a low centre of gravity.
    (ii) Free Surface Effect.
    (iii) Synchronism.
    (iv) improper weight distribution.

    The correct answer is:
    (a) all except (i).
    (b) all except (ii).
    (c) all except (iii).
    (d) all except (iv).

18. A stiff vessel has:

    (a) a shorter roll period.
    (b) a longer roll period.
    (c) no influence on the roll period.
    (d) alternative long and short roll periods.

19. The problem associated with a vessel being tender can be remedied as follows:

    (a) Raise weights within the vessel.
    (b) Lower weights within the vessel.
    (c) Discharge weights from the lower part of the vessel.
    (d) Load weights in the higher part of the vessel.

20. Which of the following statements is incorrect? Vessels are classed by Classification Societies:

    (a) only if they are commercial vessels.
    (b) according to climatic conditions
    (c) according to type of cargoes.
    (d) according to manning.

21. In relation to Classification Societies, which of the following statements is incorrect?

    (a) Classification of vessels creates better documentation for commercial transactions.
    (b) Classification Societies are commercial organisations.
    (c) Governments authorise Classification Societies to enforce classification rules on all vessels.
    (d) Ships choose to be classed with a Classification Society of their choice.

22. In relation to the International Load Lines Convention, which of the following statements is incorrect?

    (a) The Tropical Load Line mark is higher than the Summer Load Line mark.
    (b) Commercial fishing vessels are exempt from the International Load Lines Convention.
    (c) Vessels of less than 24 metres in length are exempt from the International Load Lines Convention.
    (d) The Winter Load Line mark is higher than the Tropical Load Line mark.

23. In relation to Draft Marks, 1.2 metres is marked as:

    (a) 1.2M.
    (b) 1M2
    (c) 120CM
    (d) 1.2

24. In relation to Draft Marks, the top of numeral 4 below the 1M Mark is:

    (a) 0.5 metres.
    (b) 0.4 metres.
    (c) 0.45 metres.
    (d) There is no such mark.

25. In relation to Draft marks, the bottom of numeral 6 between 1M & 2M Marks is:

    (a) 1.5 metres.
    (b) 0.6 metres.
    (c) 1.6 metres
    (d) 1.55 metres.

CHAPTER 13: ANSWERS

1 (b), 2 (d), 3 (c), 4 (d), 5 (a),
6 (b), 7 (a), 8 (c), 9 (b), 10 (b),
11 (a), 12 (b), 13 (c), 14 (b), 15 (b),
16 (c), 17 (a), 18 (a), 19 (b), 20 (a),
21 (c), 22 (d), 23 (b), 24 (a), 25 (c)

Chapter 14

GPS & ELECTRONIC CHARTS

GLOBAL NAVIGATIONAL SATELLITE SYSTEMS (GNSS)

GNSS is the name given to the satellite based radionavigation systems, of which there are three:
- *GPS* is owned and operated by the US military.
- *GLONASS* is owned and operated by the Russia military.
- *GALILEO* is owned and to be commercially operated by the international community. It is the result of the European community's initiative. Its development phase is 2002-2005.

Combined GPS/Galileo/GLONASS receivers should become available by the year 2005. With almost three times the number of satellites at their disposal, they will be better at minimising and eliminating many of the inaccuracies that exist in the single system receivers, particularly the error due to HDOP/PDOP. These inaccuracies are discussed below.

GPS (GLOBAL POSITIONING SYSTEM)

IMPORTANT: Regular crosschecks of GPS positions by radar and visual bearings on paper charts are essential. Piloting in channels and ports by GPS alone is not yet recommended.

HOW GPS WORKS

GPS is a continuously transmitting 3-dimensional radio navigation system, consisting of 24 earth-orbitting satellites located some 11,000 miles above the earth. The system was developed and is controlled by the US Department of Defense. The GPS receiver tracks these satellites and obtains its own position, time and velocity by measuring simultaneous ranges (distances) from at least 3 satellites for a 2-D fix or 4 satellites for a 3-D fix. To achieve this, most GPS receivers have 12 parallel channels. Three of these channels lock on to three satellites for triangulation to obtain latitude and longitude. Another channel locks on to a fourth satellite for altitude in a 3-D fix for aircraft. These four channels continuously and simultaneously track the four satellites in the best geometrical positions. The additional eight channels track all other visible satellites and add this data to the data from the original four satellites. The additional channels also ensure reliable, continuous and uninterrupted navigation even in adverse locations such as valleys or dense woods.

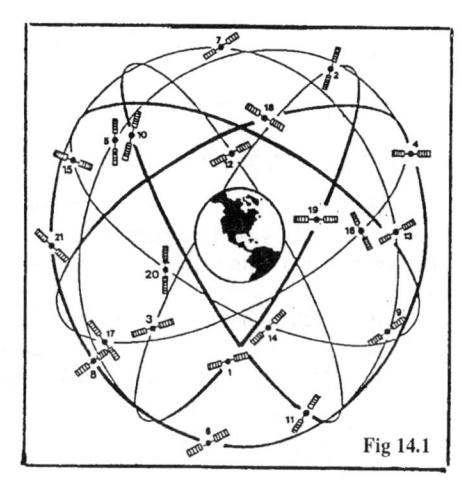

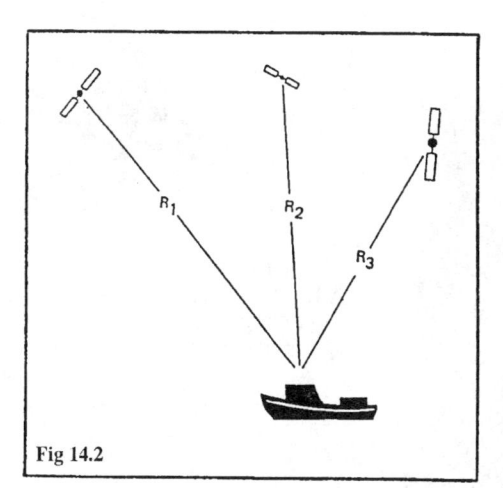

Fig 14.1

Accurate clocks inside the satellites and receivers measure the time taken between transmission and reception of radio signals. Signals are coded to prevent interference. Time is measured in ns, nanoseconds (billionths of a second). An error of 1 ns creates an error of 0.3 metres in range. The ranges are calculated by the GPS receiver as: Distance (range) = Speed of radio signal x Time of travel. The speed of radio signals is almost constant at approximately 300,000,000 metres per second.

GPS is used in all forms of position fixing and navigation. As discussed below, it can also provide vessel's course and speed over the ground, and by calculation, set and drift.

A marine in-dash GPS receiver unit consists of a mushroom shaped Antenna and a Control Head (or Display Unit). Typical GPS antenna is a combination of a QUADRIFILAR spiral and a PATCH microstrip. The former is a length of coiled wire designed to receive transmissions from low altitude satellites, and the latter is a flat bar, which is better at detecting satellites that are directly overhead. An active antenna includes a low-noise amplifier (with one or more associated filters) to boost the weak signals. This type of antenna is used if the antenna and receiver are separated by some distance. A detachable antenna is always better so that it can be positioned to receive clear signals.

Fig 14.2

The receiver's operation is controlled by a microprocessor. The more powerful the microprocessor, the more tasks it can perform, such as waypoint navigation and conversion to various chart datums. Instructions for running the receiver are embedded in the memory of the microprocessor or in auxiliary integrated circuits within the receiver. The antenna and the display unit communicate with each other via a NMEA (National Marine Electronics Association) data cable, which is similar to a telephone cable.

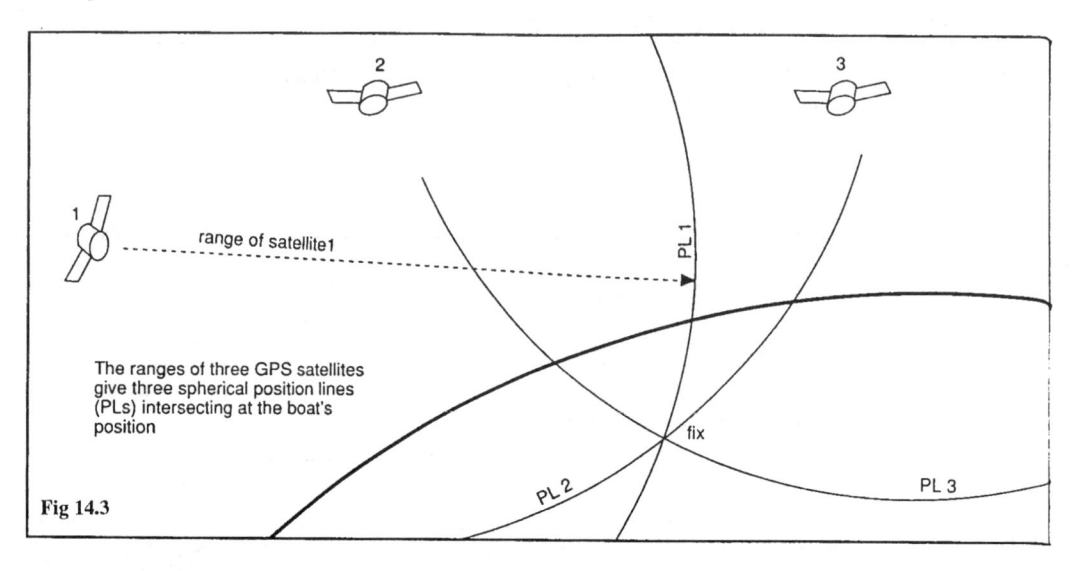

Fig 14.3 The ranges of three GPS satellites give three spherical position lines (PLs) intersecting at the boat's position

SITING THE ANTENNA

GPS signals are subject to the Ground Error and the Multipath Error. This is due to the signals being reflected by the ground and the vessel's superstructure before being received by the antenna. The ground error is not considered serious in the marine environment. Only the shore-based GPS antennae are shielded against it. They are mounted on a flat round disk of about 30 cm diameter.

The multipath error can be avoided by carefully siting the antenna on board a vessel. It should be sited where it has clear visibility, and as far away from metal masts as possible, and outside the radar scanner beam. Enclosing the antenna even inside a fibreglass structure will reduce its signal strength.

Fig 14.4
A RAYTHEON GPS
CONTROL HEAD
& ANTENNA

Chapter 14: GPS & Electronic Charts

TYPES OF GPS DISPLAY SCREENS

There are two types of display screens:
- Cathode ray tube and LCD. A Cathode Ray Tube (CRT) display - as in most TV sets - has the advantages of superior resolution and a larger screen. However, it is bulkier, requires more power (approximately 40-50 Watts, 4 Amp), and is more expensive.
- A Liquid Crystal Display (LCD) needs less power (approximately 4-5 Watts, 0.5 Amp) and is more portable. Its disadvantages are: poorer resolution and the possibility of the liquid bleeding. Colour LCD displays offer better resolution but a shorter battery life in hand-held units. Active colour displays can be viewed from larger angles than the passive colour displays.

Whether CRT or LCD, a larger screen is easier to read.

WAYPOINT NAVIGATION

Many boat owners buy a GPS for waypoint navigation. Keeping in mind some of the errors discussed below, GPS can be programmed with a number of waypoints. It can then provide track control data to an autopilot, enabling it to steer the boat from waypoint to waypoint.

A waypoint is any chosen position. For a sea angler or diver it might be the position of a wreck. For a yacht on passage, it might be a position where the boat will be turned onto a different heading. For all mariners, a particularly useful waypoint is the position of the home port which can be called up at the end of the day.

A waypoint can be entered in two ways: either by a 'save' function, or by entering the latitude and longitude. The 'save' function simply memorises the current position as a waypoint. This is particularly useful to anglers and divers, for if an interesting fishing mark or wreck appears on the fishfinder screen the position can be saved in the memory for later recall by pressing a single button.

Once entered, the waypoint is automatically given a reference number in the sequence. Upon recall, the waypoint position is displayed with its bearing and distance from the present position.

For a passage using a number of waypoints, several waypoints can be pre-programmed before starting the trip. As the boat arrives at each waypoint - signalled by an alarm - the next waypoint is automatically sequenced, and its details are displayed on the screen.

Fig 14.5: WAYPOINT NAVIGATION ON A LOWRANCE GPS

DTG = Distance to go to the waypoint *BRG = Compass bearing of the waypoint* *COG = Course over the ground*
XTE = Cross-track error *TTG = Time to go* *WPT NAME = Waypoint name*

SPEED MEASUREMENT

GPS can calculate a vessel's speed to an accuracy of 0.1 knot by calculating the Doppler shift in the frequency of radio signals received from a satellite. (See Doppler Log for the meaning of Doppler effect). However, measurement of Doppler shift requires an additional circuit in the receiver, which is often missing in cheaper receivers. These calculate speed by measuring the carrier wave's time difference, which is not as accurate. Most receivers also give readout of the distance travelled.

TWO TYPES OF GPS TRANSMISSIONS

GPS satellites transmit radio signals on two L band frequencies: L1 (1575.42 MHz) and L2 (1227.60 MHz). Most civilian receivers are designed to receive only the L1 frequency. Although multi-channel, they are single frequency receivers. These are known as the SPS (Standard Positioning Service) code receivers or Coarse Acquisition (C/A) receivers.

The military receivers and those used for surveying make measurements from both L1 and L2 frequencies. They are known as PPS (Precise Positioning Service) code receivers or, more commonly, the *Phase Carrier Receivers*. Radio signals from satellites at an altitude of 11,000 miles are refracted as they pass through the earth's atmosphere. This refraction increases their path length and induces errors called the IONOSPHERIC DELAY. Phase Carrier Receivers use the information received on the second frequency to correct errors due to Ionospheric Delay. Receivers using only the SPS code employ a filter, which is a mathematical model, to minimise the Ionospheric Delay.

D.R. INTERFACE

Many GPS sets are fitted with log and compass input capability, which allows them to extrapolate and provide Dead Reckoning positions when the GPS input is off-line.

PERSON OVERBOARD BUTTON ON GPS

Most GPS sets are fitted with a **MOB (man overboard) button** which, like the 'waypoint save' function, memorises the boat's current position when it is pressed, and changes the screen to display it, plus some additional information such as the time of pressing the button, and the distance and bearing to the initial MOB position. It then continually gives a course to steer back to that position.

Allowance must be made for any set and drift due to the delay in pressing the MOB button. For example, if the tide is running at 4 knots, the victim will be 100 metres away from the position indicated after only one minute. The best approach may be to return to the initial MOB position as quickly as possible and then head in the same direction as the tidal stream or the current is running. This information may have to be obtained from an external source, such as the chart or tidal stream atlas.

THE PLOTTER

There are two types of GPS plotters: chart plotters and chartless track plotters. The chart plotters are discussed below in the Electronic Chart Systems. A chartless track plotter does not hold any chart information. Its screen may be bordered by the four compass points or a grid showing the lines of latitude and longitude. The unit shown here is a combined GPS and fishfinder. It gives a bird's eye view of the boat's track, which can be very helpful when, for example, trying to locate a wreck. The completed search pattern can be seen at a glance. On some models, the screen can be split to display both the plotter and echo sounder pictures together, so that the seabed profile can be monitored during the search. To be truly useful for this work, the plotter should have a minimum range of a quarter of a mile or less.

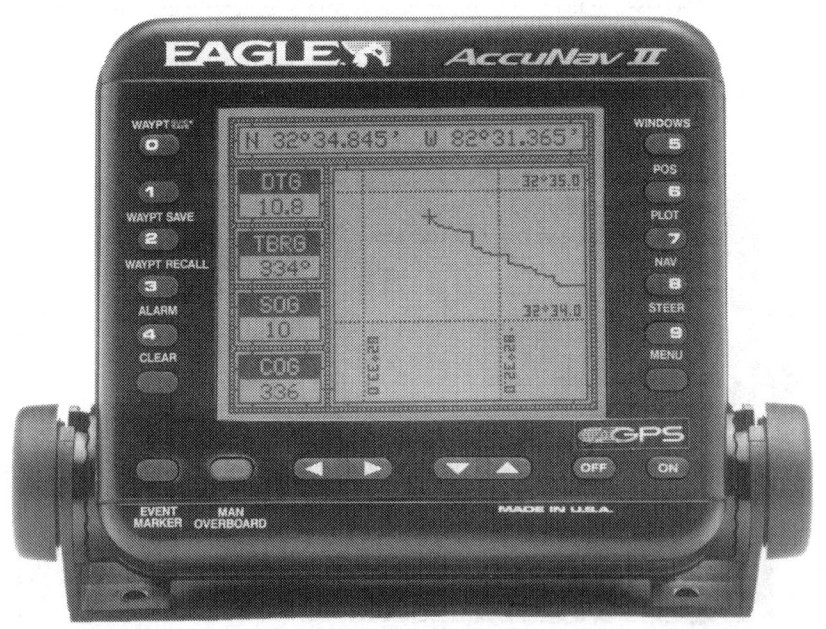

Fig 14.6: AN EAGLE CHARTLESS TRACK PLOTTER

The plotter can also be used to give a visual display of track as a straight line between a start position and a waypoint. As the boat moves towards the waypoint, its actual course is plotted, enabling the operator to identify any cross-

Chapter 14: GPS & Electronic Charts

track error immediately.

The screens of some of the models can be customised to show a combination of the plotter, steering and maybe echo sounder displays at the same time. These are known as 'window displays', and can be real help when searching for a wreck: all the relevant search information is shown on the screen at the same time, so there is no need to constantly switch from one display to the other.

ELECTRONIC CHART SYSTEMS

In the interest of international shipping, the requirements for electronic charts are supervised by IMO (International maritime Organisation). Trials are underway for totally interactive charts for the ECDIS system for use of ships. Meanwhile, small craft are using ECS. These charts are produced by government as well as private manufacturers under licence for use of government charts.

TERMINOLOGY:
- **ECDIS**: Electronic Chart and Display Information System.
- **ENC**: Electronic Navigation Charts are <u>digitised</u> charts for use with ECDIS. However, they are not yet widely available. An ENC is the most versatile chart. In fact it is not a chart but a database of all the chart information. The software draws the chart on the fly based on the user's requirement. It produces the customised display and raises the necessary alarms.
- **RNC**: Raster Navigation Charts are <u>scanned</u> charts for use with ECDIS. These are photocopies or fixed images of paper charts in pixels, like a non-digital Television screen. However, these charts are geo-referenced, which allows the computer to relate to every point on the chart and sound the necessary alarm.
- **ECS**: Electronic Chart System is the generic name for all Non-ECDIS charts. Like RNCs they are still mostly scanned.

Most GPS packages include an Electronic Chart System (ECS), which can be run on a laptop or a dedicated plotter. The plotters range from tiny mono units to large colour displays. The system may consist of only the chart information or an extensive navigation system, which also displays vessel's position, compass heading, log speed and soundings imported from inbuilt or remote sensors. Although the equipment is liable to stop working at any time, the ECS has the following advantages over paper charts:
- Constant real time position and track information.
- Position fixing in poor visibility.
- Regular contractual (via dealers) updates are available for ECS charts
- ECS chart portfolio extension available by phone or email from anywhere in the world. (Vector charts only, discussed below)

Charts may be SCANNED (known as Raster Charts, or more correctly, Digital Raster data or Raster Navigation Charts or RNC) or DIGITISED (known as Vector Charts, or Vector data or Electronic Navigation Charts or ENC).

RASTER CHARTS

Raster charts are made by scanning paper charts. Their quality depends on the quality of scanning. The information cannot be corrected or added to. The suppliers offer the updating and correcting facility in the form of replacement CDs or cartridges. The system usually incorporates one or more layers of port and tide information which can be called up on the screen. The Australian Hydrographic Office manufactures Australian raster charts under the name of Seafarer Charts. The UK Hydrographic Office manufactures ARCS (Admiralty Raster Chart Service) which covers the globe. NOAA (National Oceanographic and Atmospheric Administration) does the same in USA.

Raster charts are available on CDs, which require a suitable software program to run them. Such programs include SeaPro 2000, TQM C-Plot and Endeavour Navigator. The benefit of Raster Charts is their familiarity. They appear just like paper charts, and are also cheaper than Vector Charts.

VECTOR CHARTS

These charts are manufactured by digitising paper charts. The information is then stored on cartridges or CDs in a form of layers (vectors). There might be one layer for lights, one for soundings, another one for the coastal features, and so on. All details can be called up at once, or just one or two layers displaying, for instance, a channel with a 2-metre depth contour. For safety reasons there are some layers that cannot be removed. This system also incorporate layers of port and tide information which can be called up on the screen.

Manufacturers producing Vector charts include C-MAP, Transas Dataco and Navionics. The operating software manufacturers include TQM and Winchart. However, cartridges used in conjunction with a plotter do not usually require additional software.

A GOOD ECS should offer:
- A continuous display of the vessel's position.
- Greater flexibility of manoeuvre in blind pilotage conditions.
- Ability to operate within the safe depths selected by the mariner.
- Real-time visual indication and audio warning of any breach of navigational integrity.
- Facility for automatic chart updating.
- "Black box" recording of navigational and vessel handling parameters.

ZOOMING-IN ON ELECTRONIC CHARTS

The accuracy and precision of a chart is determined by its survey data and its original scale. Zooming-in on a chart does not improve its data or accuracy. It merely enlarges the view. For example, on scale of 1:10,000, 1 mm = 10 metres; and on 1:150,000, 1 mm = 150 metres. A pencil thickness on the former would be more than 2.5 metres wide but on the latter it is more than 36.5 metres wide. This fact would not alter by zooming-in on either chart. In view of the precision required when docking ships, the docking electronic charts are being produced on scale of 1:1000; i.e., 1 mm = 1 metre.

PLOTTERS AND LAPTOPS

Dedicated plotters are weatherproof and have fewer moving parts, but they are less versatile than laptops. The disadvantage of laptops is that they are not weatherproof like the plotters. They are thus more vulnerable in marine environment. You might have to stow away the laptop when it is most needed in the cockpit in bad weather. The advantages of laptops is their additional ability to handle HF email, Inmarsat C, weather fax, etc. The best option is to carry both, so that you are not relying on a single instrument for navigation as well as outside contact.

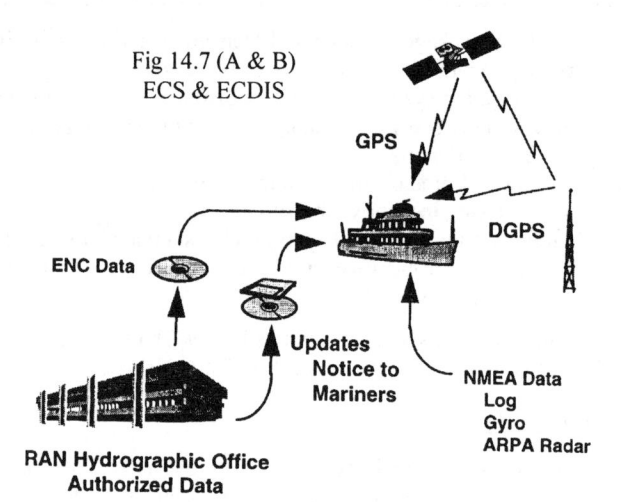

Fig 14.7 (A & B) ECS & ECDIS

Shown here are two examples of the electronic chart technology: an ECS developed by Navionics and ECDIS (Electronic Chart and Display Information System) developed by the Hydrographic Sciences of Australia. For ECDIS, the Hydrographic Office supplies the Electronic Navigational Charts (ENC) as well as regular updates. The navigator can add local knowledge, temporary and provisional notices and other relevant material.

PAPER OR ELECTRONIC CHARTS?

Safe navigation requires that vessels fitted with an ECS system must still carry paper charts for back up. However, as per SOLAS V Regulation 19.1.2 (of the international Safety Of Life At Sea convention) that enters into force on 1 July 2002, ships may carry paper nautical charts or ECDIS. Being 'SOLAS V' compliant means:
- approved equipment
- authorised charts
- adequate back-up support (for safe changeover without resulting in critical incident)
- approved training.

TYPES OF GPS RECEIVERS

As stated earlier, civilian GPS receivers are single frequency SPS receivers. They receive satellite signals on the L1 frequency. However, they are of two types: SEQUENCING CHANNEL and PARALLEL-CHANNEL. The former are cheaper. They have only one or two channels that pick up one or two satellite signals at a time, cycling through a few

Chapter 14: GPS & Electronic Charts

satellites. They are more susceptible to multipath errors causing them to lose contact with satellites. It makes them less accurate.

Most modern receivers are Parallel-channel receivers, which have 12 or more channels that lock onto at least 4 satellites at the same time. They don't easily lose contact with satellites, thus are inherently more accurate. A MULTIPLEXING CHANNEL is a variation of sequencing channel, with a faster ability to sequence through satellites.

CAUSES OF GPS ERRORS & MALFUNCTIONS

A mariner should always carry manual position fixing aids and a copy of the paper navigation chart. A GPS can malfunction or breakdown due to the following reasons:

- Radio and other RF interference can cause range measurement errors. See 'Buying a GPS' above.
- Electrical power surges, especially when starting engine in a vessel. It is more common with outboard motors. This can be avoided by switching off GPS before starting engine.
- Problems in the vessel's alternator.
- Poor weather conditions, such as lightning
- Vibrations in the vessel.
- Salt in the marine environment.
- Computation error in the receiver.
- Mismatching of satellite signals in the receiver.
- Satellite clock error: Satellite clocks are highly accurate instruments. Nevertheless, they can drift from time to time, causing an error in fixes.
- Receiver clock error: The quartz-crystal clocks in receivers are not as accurate as those in satellites. An error in their time measurement causes an error in ranges. An error of 1 billionth of a second causes an error of 0.3 metres in range. In most cases, the range error is common to all ranges. Therefore the receiver eliminates the error in position by automatically resolving the resulting 'cocked hat'. (See Chapter 17.2 for cocked hat). However, whenever the clock error is not common, it results in an error in the fix.
- Satellite orbit information error: An error in the position information of a satellite creates an error in its range. This so-called EPHEMERIS (precise orbital data) error is usually kept to a minimum by the Master Control Station and monitor stations in USA.
- Multipath error: as discussed in "Siting the Antenna" above.
- Ionospheric delay: This atmospheric propagation error is discussed above. The delay in receiving refracted satellite signals results in incorrect range. Satellites do transmit known refraction data to the receivers. However, unpredictable refraction can cause range errors. Satellites with high elevation suffer less ionospheric delay.
- Vessel movement creates only a minor Ranging Error in the receiver. For some reason, the error at slow speed (less than about 5 knots) is greater than at normal speed. The author understands that manufacturers set speed limits in GPS receivers for specific markets so that a ground receiver cannot be used to navigate an aircraft.
- Wet sails: Wet sails interfere with and partially block the high frequency GPS signals. The operator is made aware of the downgrading of signals by the HDOP value going up to, say, 8 or more. (HDOP is discussed below). Heavy rainfall may be argued to have a similar effect. However, there are sufficient openings in rainfall for the signals to penetrate at an acceptable level.
- Cross Track error: This is not a GPS error. It is an off-course error due to Set and Leeway. The GPS computes it when programmed with a course between two waypoints.
- GPS satellites orbit the earth at an *inclination* of about 55 degrees. Therefore, in *tropics and high latitudes*, the signal strength can vary from extremely good to extremely poor. During periods of poor signal strength, some receivers automatically increase their gain. But, when suddenly, there is an exceptionally good strength, the receiver with gain turned up too high may simply lock up. It would then need to be switched off and on again.

ACCURACY OF GPS POSITIONS

The accuracy and reliability of GPS positions may be affected by the following factors:

1. Above listed errors.
2. Since May 2000 the US Department of Defense has stopped deliberately reducing the accuracy of signals for civilian users. Switching off this *SELECTIVE AVAILABILITY (SA)* or *DITHERING* of signals has improved the official accuracy of GPS for civilian users from 100 metres to 20 metres at 95% confidence level. Although SA is not expected to be reintroduced in the near future, its application is a matter of US policy.
3. Undetected breakdowns in satellite transmissions.
4. The GPS system may be turned off or degraded without notice in a national or international emergency.

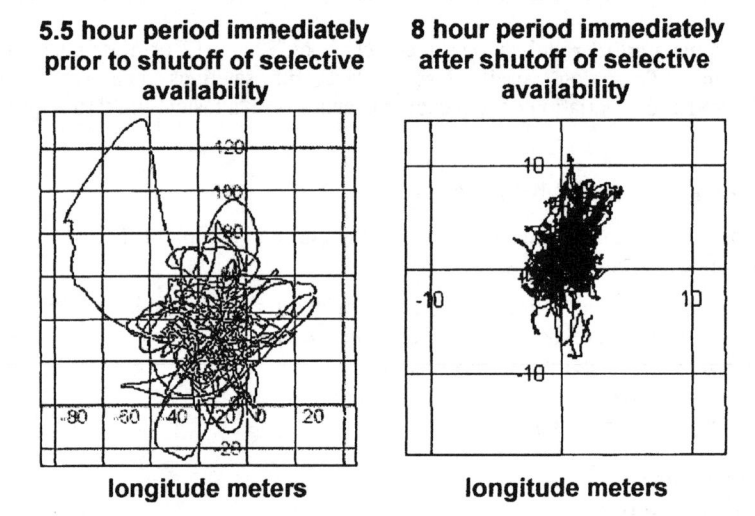

Fig. 14.8: GPS POSITION ACCURACY OF A STATIONARY RECEIVER BEFORE AND AFTER SHUTOFF OF SELECTIVE AVAILABILITY (*ASTRON*)

5. HORIZONTAL DILUTION OF POSITION (HDOP)

Azimuth spread (geometrical configuration) of satellites can affect the precision of a fix. Ideally, one satellite should be overhead and three on the horizon 120 degrees apart. The value of HDOP or PDOP (Precise Dilution of Position) in relation to the configuration of satellites is displayed on most receivers. Typically, good values are between 2 and 4 - the lower the better. Values in excess of 6 are known as OUTAGES. These are periods of poor accuracy when weak geometry of satellites makes GPS fixes unsuitable. A value of about 20 represents a fix of incalculable inaccuracy. An outage can last up to 30 minutes, and repeated every 12 hours. Some receiver can indicate outage in advance if queried. On some GPS receivers, positions resulting from poor HDOP can be warned against by setting a pre-defined limit.

Clocks in receiver are never as accurate as those in satellites, which makes them unable to accurately replicate the information they receive from satellites. The ranges (distances) measured by them are thus biased and known as Pseudoranges. This ranging error is known as the User Equivalent Range Error or "UERE", which in stand-alone SPS receivers is approximately 22.5 metres. (Prior to May 2000, during periods of Selective Availability UERE was up to 62.5 metres.)

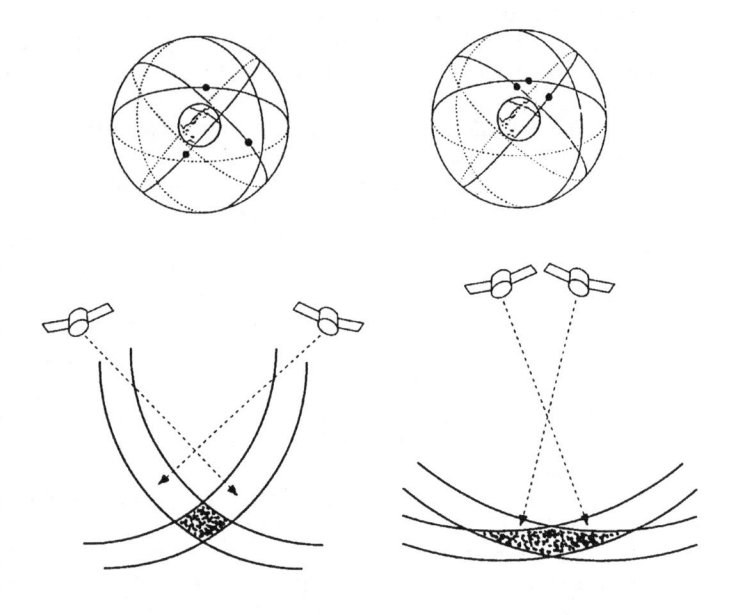

Fig 14.9: HDOP

Error caused by pseudoranges can be calculated by multiplying the maximum UERE value with HDOP at the time of reading the position. If, for example, the GPS receiver displays HDOP of 2.0, then the position accuracy due to range errors is 22.5 x 2.0 = 45 metres. In other words, it is a 95% probability that the error caused by pseudoranges in the GPS position is within 45 metres.

6. RECEIVER AUTONOMOUS INTEGRITY MONITORING (RAIM)

As discussed above, a high HDOP reading indicates poor satellite geometry, which can be excluded for position fixing on some receivers by pre-setting an acceptable HDOP limit. RAIM is available in some receivers. It continuously monitors every satellite and eliminates the faulty ones from position fixing. For RAIM to start working as soon as the

Chapter 14: GPS & Electronic Charts

receiver is switched on, there must exit a separate battery powered memory in the receiver. It is similar to the lithium battery in a desktop computer that maintains its calendar when it is switched off. Without it, the GPS would take 12.5 minutes to reacquire the satellite almanac when switched on. HDOP limit and RAIM equipped receivers give timely visual and audible warnings to the user. Obviously RAIM is of greater use to aviators than mariners.

7. DISCREPANCY BETWEEN GPS & CHARTED POSITIONS:

GPS positions relate to the World Geodetic System 1984 (WGS84), which is based on earth's true centre. The true centre is mathematically defined because the earth is not a perfect sphere. Charts, on the other hand, have been traditionally based on "off-centred" surveys of different countries, including the Australian Geodetic Datum 1966 (AGD66). Due to these discrepancies, navigators often have to adjust GPS positions before plotting them on charts because the relative features on charts are usually more reliable for navigation than the absolute GPS positions.

All new charts are being produced on the WGS84 datum, but it is going to take some time before all old datum charts disappear. Meanwhile, the amount of adjustment needed to GPS positions is being indicated on the reprinted charts. Navigators must take note of this adjustment. It can be quite considerable, depending on its location. In Australia the *datum shift* between AGD and WGS can be around 200 metres. In Chile it can be as much as 800 metres. Figure 14.11 shows the variations in the position of Sydney Observatory when compared between AGD66, Clarke 1858 Spheroid datum and the WGS84 datum.

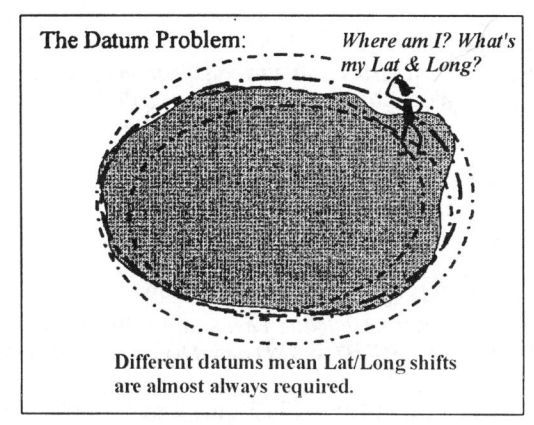

Fig 14.10 (A & B): THE DATUM PROBLEM *(AHS)*

HORIZONTAL DATUM DIFFERENCES - SYDNEY OBSERVATORY

The chart user is advised to refer to the note in the Chart Title under the heading "*Positions*", and in particular, the notes appearing on the chart under the heading "*Satellite-Derived Positions*". These notes inform the mariner to either plot the GPS position directly onto the chart, if the chart is on the WGS 84 datum, or else give corrections that must be applied before plotting. A vessel's GPS receiver may have the ability to offer a wide range of datums. The user should be aware of the datum in use, and guard against applying the datum shift twice or in the wrong direction.

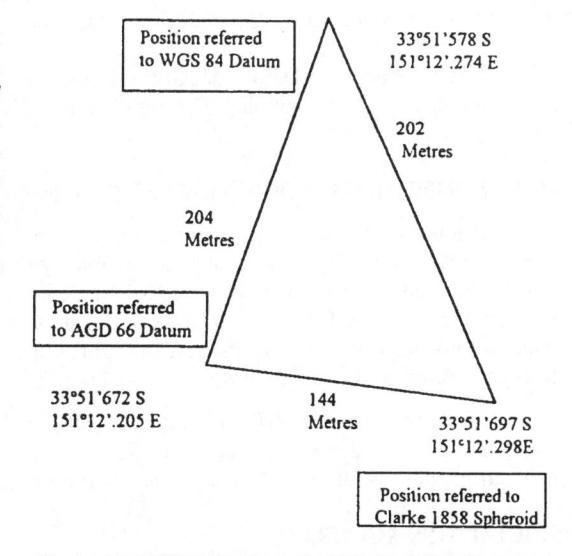

Fig 14.11: HORIZONTAL DATUM DIFFERENCES – SYDNEY OBSERVATORY

STEPS TO CONSIDER BEFORE PLOTTING A GPS POSITION ON AN AUSTRALIAN CHART

The following examples of chart datums and notes appearing on the Australian charts at present, together with a worked example, are based on the assumption that the GPS receiver is receiving its position relative to the WGS 84 datum and the user has not pre-selected another datum. The GPS user manual should be consulted as regards setting datum selection. If the user has a GPS receiver with the facility to convert WGS datum to other datums on charts, positional discrepancies can still occur due to the inadequacy of the original source survey data. All GPS users should therefore exercise caution. Navigation in inshore or shoal areas should be supplemented by visual observations.

EXAMPLE 1
CHART AUS 21 MELVILLE ISLAND - SNAKE BAY
Positions are related to the World Geodetic System 1984.

SATELLITE-DERIVED POSITIONS
Positions obtained from satellite navigation systems are referred to the WGS Datum and can be plotted directly onto this chart.

Fig 14.12: STEPS TO CONSIDER BEFORE PLOTTING A GPS POSITION ON AUSTRALIAN CHART

EXAMPLE 2
CHART AUS 830 RUSSEL ISLAND TO LOW ISLETS
Positions are related to the Australian Geodetic Datum 1966.

SATELLITE-DERIVED POSITIONS
Positions obtained from satellite navigation systems are normally referred to the WGS Datum; such positions should be moved 0.09 minutes SOUTHWARD and 0.05 minutes WESTWARDS to agree with the chart.

In the case of chart AUS 830, the correction would be as follows:

GPS Position	17° 00.00' S	146° 05.00' E
Correction	+0.09'	-0.05'
Chart Position	17° 00.09' S	146° 04.95' E

On some charts the notes state that the "adjustments" cannot be determined. In such cases, it should not be assumed that the difference is negligible. On such charts, only the relative positioning method may be reliable.

NOTE: Geocentric Datum of Australia (GDA) 94 is a revised version of WGS 84 for a few centimetres greater accuracy for surveying and mapping. For maritime use WGS 84 and GDA 94 are regarded one and the same.

AMSA MARINE NOTICE 9/1995 ON USE OF GPS: AN EXCERPT

As a result of making navigational decisions based on GPS data alone, a number of vessels have risked stranding. Although GPS is normally an accurate and reliable system, it is not a substitute for sound watchkeeping and navigational practice. The accuracy and reliability of a position determined by GPS depends to some extent on the quality of the receiver employed. It is possible for a GPS satellite to transmit an erroneous signal for some hours before users can be warned by messages transmitted by GPS satellites. This can result in large position errors on GPS receivers. There are also problems with chart datums, as discussed above.

For these reasons, a GPS should be used only in conjunction with other aids such as compasses, radar, depth sounder, log and navigational publications. This is especially important when navigating in confined or shallow waters, dense traffic, poor visibility or in strong winds or tidal streams.

OFFICIAL GPS ACCURACY

The official accuracy of marine GPS receivers *(i.e., the SPS-type)* is within 20 metres about 95% of the time.

Chapter 14: GPS & Electronic Charts

(Prior to switching off Selective Availability, it was 100 metres 95% of the time. SA is discussed below). This level of accuracy is predicted on a 24-satellite constellation (additional satellites are a bonus), a 5-degree satellite elevation mask, no obstruction to incoming signals, and at least 4 satellites in view with a PDOP of no greater than 6.

DIFFERENTIAL GPS (DGPS)

DGPS is a method of minimising the GPS error due to Selective Availability and other natural causes. Although, as discussed above, Selective Availability is no longer a US policy, the DGPS system is still with us for now. It consists of Reference Stations (Differential Beacons) at known coastal locations. These stations independently check the integrity of GPS signals on site and measure their errors. They estimate the error by comparing their GPS position with the accurate knowledge of their own position. They then broadcast the corrections to DGPS receivers in the vicinity, which automatically apply them to their own GPS positions.

A Reference Station can extend the period of a faulty satellite's use by transmitting accurate corrections, or it may notify users of its unsuitability, by actuating an alarm.

The most common form of DGPS used internationally for maritime navigation operates in the LF/MF band (285-325 kHz) and conforms to the RTCM Recommended Standards for Differential Navstar GPS Service for the transmission of data.

The Australian Maritime Safety Authority (AMSA) has established 16 DGPS broadcasting stations along Australia's coastline, providing better than 10 metres (95% probability) position fixing accuracy within each station's coverage area. The following map shows coverage in 2002. The latest information can be obtained from AMSA website listed on the last page pf this book. Based on acceptable reception levels, the Sydney station covers coastal waters between Newcastle and Jervis Bay. The Mackay station covers the Hydrographers Passage beyond its eastern channel. Each of the Queensland stations has a coverage of at least 130 nautical miles.

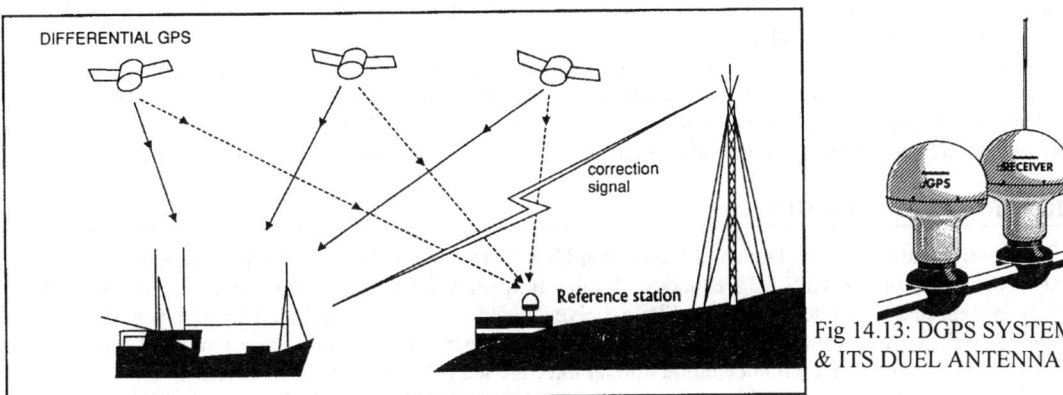

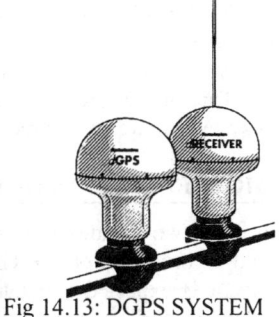

Fig 14.13: DGPS SYSTEM & ITS DUEL ANTENNA

The DGPS radiobeacons are remotely controlled and monitored 24 hours a day. Any DGPS radiobeacon in out-of-tolerance condition is reported to users via AUSCOAST and SafetyNET Warnings.

To utilise AMSA's DGPS service a vessel needs to have a DGPS receiver. Alternately it may use a *"Differential-ready"* GPS receiver combined with a DGPS Beacon Receiver. Most GPS receivers are DGPS ready. The difference between DGPS and DGPS-ready GPS receivers, is that the former have a built-in DGPS Beacon Receiver, where as the latter are ready to be connected to one.

The DGPS service is free as the fee is included in the navigation levy paid by commercial vessels calling at Australian ports.

Although selective availability has been discontinued, the DGPS system is still regarded a worthwhile integrity monitor. It is possible for an unhealthy GPS satellite to transmit an erroneous signal for some hours before a GPS receiver can be warned by a message transmitted by the satellite. A Reference Station quickly detects an unhealthy satellite, and transmits a message alerting DGPS receivers not to use that satellite. It thus reduces the GPS integrity check interval from several hours to several seconds.

The DGPS system is not totally error-free. A DGPS Receiver <u>cannot resolve or correct</u>:
- HDOP errors.
- Multipath errors.
- Datum discrepancy between WGS & non-WGS positions.
- Range errors.
- DGPS signals can also be lost during thunder and lightning.

DGPS is limited to a range of approximately 50 miles from a beacon. Within this range, the error varies with the distance from the beacon. The greater the distance, the larger the error. The position accuracy offered by the DGPS is in metres, not hundreds, of metres. Outside the differential beacon's range, the DGPS receiver works as a GPS.

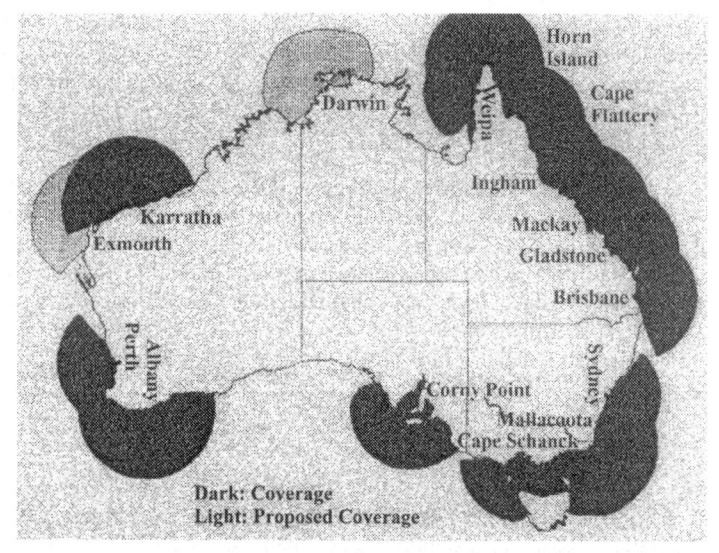

Fig 14.14: AMSA'S THEORETICAL DGPS COVERAGE IN 2002 (assuming 95% availability)

REAL TIME KINEMATICS (RTK) MACHINERY GUIDANCE SYSTEM

This system provides for real-time DGPS-style processing of both code and phase signals for highly accurate measurement of position and velocity. Using three frequencies, it provides three applications: GPS at sea, DGPS in coastal waters and RTK for precision docking of ships.

It claims to provide accuracies of 10 cm for automatic machinery guidance in applications such as docking, dredging, surveying and farming. As discussed earlier, the electronic docking charts are of scale 1:1000 (i.e., 1 mm = 1 metre). This Three-in-One receiver, combined with other ECDIS components, allows ships to navigate blindly from dock to dock. This system is already in practice in Canada where poor visibility is a problem for most of the year.

GPS II - MODERNIZATION OF GPS

In 1998, two decades after the launch of the current GPS system, the US government announced that a second civil signal would be broadcast on the GPS L2 frequency. A third frequency L5 was also announced for the aeronautical use. Accordingly, it is expected that duel-frequency SPS receivers would become operational by the year 2008. This three frequency "wide lanes" program would be achieved without compromising the spectral separation of civil and military GPS services. In fact, it would significantly reduce unintentional interference in the overall GPS services.

The additional L2 frequency would reduce the ionospheric error from typically 7 metres to 0.01 metres. On stand-alone receivers, the current accuracy of approximately 22.5 metres (95%) will be reduced to 8.5 metres. A technique known as the Accuracy Improvement Initiative (AII) is further expected to reduce the GPS clock and ephemeris contribution to UERE (see HDOP above) to approximately 1.25 metres. These actions should improve the stand-alone GPS horizontal accuracy for all users from 100 metres to 6 metres or better.

The system would allow more accurate navigation around sandbars and other hazards. Fishing vessels would be able to pinpoint their favourite spots in inland waters as well as at sea. Lobster traps would become easier and quicker to locate and recover.

Furthermore, in order to improve the response time on emergency calls, it will soon be required that all new mobile (cellular) phones be equipped with more accurate GPS technology. Mobile phones could thus become the preferred method for making emergency calls ashore and in coastal areas.

Finally, the additional signals on L2 and L5 frequencies would also improve the performance of high-precision (RTK) applications, such as berthing, dredging and surveying. In the non-marine field, it would improve applications such as aircraft precision landing, precision farming and machine control.

UNIVERSAL AUTOMATIC IDENTIFICATION SYSTEMS (UAIS or AIS)

The AIS Transponder Units exchange GPS, Gyrocompass, Log and other navigational information between vessels, or between vessels and shore.
- ➢ They assist vessels with collision avoidance in poor visibility by exchanging their location, speed and heading information without the use of radar.
- ➢ They automatically transmit vessels' position, speed and heading information to AIS Base Stations ashore, as part of the Ship Reporting System.

Chapter 14: GPS & Electronic Charts

- ➤ They allow vessels to access marine data, such as tide and wind information, from sensors installed at critical locations en route or during docking.

BUYING A GPS

- Quality brands are more expensive.
- High-energy interference from a nearby transmitter (such as TV or microwave tower) or harmonic interference from one of your other systems is possible. Some military exercises in USA have been known to have affected GPS receivers up to 300 nautical miles away. An interference may affect one brand but not another. Try it out in your vessel and in the area you will use it most. Ask your friends who may use the same model.
- Check the receiver's dynamic accuracy. Take it for a drive in a motorcar and subject it to sudden take offs, stops and accelerations. Interference from a poorly suppressed communication device in a nearby car or a nearby factory or laboratory should disappear once you move beyond that area.
- Civilian GPS receivers are single frequency SPS receivers. They receive satellite signals on the L1 frequency. However, they are of two types: Sequencing (or Multiplexing) Channel and Parallel Channel.
- Typical GPS antenna is a combination of a QUADRIFILAR spiral and a PATCH microstrip.
- Know the difference between GPS, DGPS and RTK receivers. Most GPS receivers are advertised as DGPS-ready. The difference between DGPS and DGPS-ready GPS receivers is that the former have a built-in DGPS beacon receiver, where as the latter are ready to be connected to one.
- Be aware that there are three separate GNSS (Global Navigation Satellite Systems): GPS, Galileo and GLONASS. Combined receivers should become available by the year 2005.
- Most GPS receivers have internal DC power supplies, some of which are in the form of rechargeable batteries. Some hand-held receivers may use penlight batteries. Find out their life span. The latest receivers draw very little current, thus extending battery life and recharging periods. Hand-held penlight receivers that can also accept external power are obviously better. In-dash (fixed) receivers usually cannot be used as separate mobile units.
- GPS display can be CRT, LCD colour or LCD monochrome.
- It is better to use the receiver with its default WGS 84 chart datum. However, if you wish to use other datums, make sure the receiver recognises them.
- Some receivers have a data port to allow interchange of information with computers.
- All GPS receivers indicate positions in latitude and longitude and allow some waypoint navigation. Some others are chartless track plotters and some are chart plotters. These features are discussed in this chapter. Chart plotters may have some uncorrectable built-in charts or they may allow input from chart cartridges or from a downloading facility such as a CD, computer or INMARSAT. See Chapter 18 for Inmarsat.
- GPS signals are extremely weak. They are below the ambient (surrounding) RF noise level. They can be easily jammed or interfered with by microwaves in general and objects such as radar, EPIRB and motors. GPS signals can also be masked by ships and buildings in coastal waterways and harbours. The receiver could then be displaying DR positions without indicating. Better quality receivers indicate whenever they switch over from satellite positioning to the DR positioning mode.
- Make sure the receiver can measure distance in nautical miles.
- Some receivers have rotatable screens, rotating from a vertical to a horizontal position. These might suit some users.
- Receivers with user-changeable fields allow users to select only the information they want to look at.
- Not all GPS receivers are sealed to make them completely waterproof. Many are merely water-resistant.
- The receiver should be equipped to indicate (audio and visual) when something may be causing inaccurate positioning due to poor satellite reception, signal interference or receiver malfunctioning.
- Check the quality of the receiver's manual. It should not omit vital operational information, such as receiver's errors and limitations, its rate of position updates and multipath rejection ability. It should also state whether the receiver calculates vessel's speed by measuring Doppler shift or the carrier wave's time difference. Doppler shift measurement requires an additional circuit and is more accurate.
- Consider the receiver's ease of use.
- Some receivers are fitted with RAIM facility and a separate lithium battery for it to start working immediately.
- Compare various receivers for their the ability to track poor signals.
- Receivers able to place an elevation mask to cut out low altitude satellites provide more accurate positions.
- Check the manual for the receiver's operating temperature limits. They do vary.
- Check that the receiver is capable of indicating outage in advance.
- Do you want a stand-alone GPS/DGPS or an integrated navigation unit incorporating GPS, Plotter, Echo Sounder, Radar and AIS, etc.?
- Consider the manufacturer and dealership size, after sale service and availability of parts.

CHAPTER 14: QUESTIONS & ANSWERS

CHAPTER 14.1: GPS & ELECTRONIC CHARTS

1. A GPS receiver obtains its position by monitoring radio signals from at least three satellites for their:

   (a) doppler shift.
   (b) ranges.
   (c) ionospheric delay.
   (d) selective availability.

2. A GPS receiver can calculate a vessel's speed over the ground by calculating the:

   (a) doppler shift of the transmitted signals.
   (b) ranges.
   (c) ionospheric delay.
   (d) selective availability.

3. In which of the following phenomenon does the Doppler Effect occur?:

   (a) Sound
   (b) Light
   (c) Radio
   (d) All three

4. A GPS receiver calculates doppler effect by measuring the apparent change in the radio frequencies received from satellites as its relative distance:

   (a) changes.
   (b) increases.
   (c) decreases.
   (d) remains constant.

5. Most civilian GPS receivers are:

   (a) SPS code receivers.
   (b) PPS code receivers.
   (c) single channel receivers.
   (d) switching receivers.

6. The error caused by the refraction of radio signals from satellites is known as the:

   (a) computation error.
   (b) multipath error.
   (c) ionospheric delay.
   (d) satellite orbit information error.

7. A continuous tracking GPS receiver has:

   (a) numerous receiver channels.
   (b) only one receiver channel.
   (c) one tracking and one monitoring channel.
   (d) one tracking and two monitoring channels.

8. The following GPS error can be avoided by switching off the GPS receiver before starting the engine:

   (a) electrical noise.
   (b) computation error.
   (c) receiver clock error.
   (d) voltage fluctuation error.

9. The following GPS error can be avoided by carefully citing the antenna:

   (a) mis-matching of satellite signals.
   (b) satellite clock error.
   (c) satellite orbit information error.
   (d) multipath error.

10. The following GPS error is usually automatically resolved by the receiver:

    (a) receiver clock error.
    (b) satellite clock error.
    (c) error due to vessel movement.
    (d) computation error.

11. Which of the following GPS errors changes due to variations in the atmospheric propagation?:

    (a) Satellite orbit information error.
    (b) Error due to mismatching of satellite signals.
    (c) Error due to ionospheric delay.
    (d) Multipath error.

12. The accuracy of a GPS derived position is least affected by the following:

    (a) radio & RF interference.
    (b) azimuth spread of satellites.
    (c) difference between chart datums and WGS.
    (d) vessel's movement.

13. The discrepancy between chart datums and WGS means that the:

    (a) positions of coastlines are less accurate.
    (b) GPS-derived positions are less accurate.
    (c) GPS receives mis-match satellite signals.
    (d) GPS system is occasionally turned off.

14. The following HDOP values in a GPS receiver indicate a good azimuth spread of satellites:

    (a) above 20.
    (b) as low as possible.
    (c) between 2 and 6.
    (d) below 20.

Chapter 14: GPS & Electronic Charts

15. In GPS receivers an outage can last for periods up to
    (a) 10 minutes.
    (b) 30 minutes.
    (c) 2 hours.
    (d) 4 hours.

16. In GPS receivers HDOP values in excess of 6 are known as:
    (a) outages.
    (b) dithering.
    (c) selective availability.
    (d) precise position service.

17. The discontinuation of "dithering" of GPS signals has eliminated errors due to:
    (a) HDOP.
    (b) difference between chart datums and WGS.
    (c) selective availability.
    (d) ionospheric delay.

18. The necessary adjustments to GPS derived positions due to discrepancy between chart datums and WGS is notified:
    (a) on some charts.
    (b) in marine notices.
    (c) in GPS handbooks.
    (d) on all charts.

19. The term DGPS stands for:
    (a) Digital GPS.
    (b) Differential GPS.
    (c) Direct GPS.
    (d) Dithering GPS.

20. The DGPS minimises the following GPS error(s):
    (a) outages only.
    (b) error due to vessel movement only.
    (c) dithering only.
    (d) SA (if reintroduced) & several other natural errors.

21. In terms of interference to GPS signals in harbours & coastal waters, which of the following statements is incorrect?
    (a) Interference from nearby TV & Microwave towers is possible.
    (b) Interference from military exercises 200 miles away is possible.
    (c) Interference from other equipment within the vessel is possible.
    (d) All brands of GPS Receivers in the same area & in the same vessel suffer from the same level of interference.

22. All GPS Receiver:
    (a) allow input from INMARSAT.
    (b) are chart plotters.
    (c) allow some waypoint navigation.
    (d) allow input from chart cartridges.

23. In relation to GPS signals, which of the following statements is incorrect?
    (a) GPS signals are generally weaker than the surrounding RF noise levels in and around the vessel.
    (b) On its signals being interfered with, a GPS Receiver will indicate that it is displaying a DR position.
    (c) Radar, EPIRB or a motor operating in a vessel can interfere with GPS signals.
    (d) In coastal waters and harbours, other ships and buildings can mask GPS signals.

24. In relation to GPS, which of the following statements is incorrect?
    (a) The GPS network consists of 24 or more satellites.
    (b) The Receiver measures simultaneous ranges from at least 3 satellites for a 2-D fix & 4 satellites for a 3-D fix.
    (c) Most GPS Receivers have 12 or more parallel channels.
    (d) The clocks in the Receivers are generally more accurate than those in the Satellites.

25. An "Active" GPS Antenna means that it:
    (a) contains a 'low-noise' amplifier & one or more filters.
    (b) is mushroom shaped.
    (c) contains a Patch microstrip.
    (d) contains a Quadrifilar spiral.

26. An 'Active' GPS Antenna is used:
    (a) in all GPS Receivers.
    (b) where the Antenna and Receiver are separated by some distance.
    (c) in hand-held Receivers.
    (d) in 'in-dash' Receivers.

27. Which of the following terminology is not associated with the Electronic Chart Systems?
    (a) ECDIS.
    (b) ENC.
    (c) SEC.
    (d) RNC.

28. In relation to Electronic Chart Systems, which of the following statements is incorrect?

    (a) RNC are scanned charts.
    (b) ENC are digitised charts.
    (c) RNC are more versatile.
    (d) ENC are in the form of a database.

29. In relation to Electronic Chart Systems, which of the following statements is incorrect?

    (a) RNC are also known as Raster or Digital Raster Charts.
    (b) ENC are also known as Digitised or Vector Charts.
    (c) ECS is the generic name for all Non-ECDIS charts.
    (d) Raster charts can be automatically corrected through Internet downloads.

30. In terms of the advantages of Electronic Chart Systems, which of the following statements is incorrect?

    (a) Raster charts can be automatically corrected through Internet downloads.
    (b) ECS provide real time position fixing & track information.
    (c) ECS allow position fixing in poor visibility.
    (d) Vector charts can be automatically corrected through Internet downloads.

31. In relation to Electronic Chart Systems, which of the following statements is incorrect?

    (a) A 'SOLAS V' compliant vessel may carry either paper nautical charts or ECDIS.
    (b) Zooming-in on an electronic chart improves its accuracy.
    (c) A good ECS provides the ability to operate within the safe depth selected by the mariner.
    (d) 'Black Box' recording of navigational information is possible with ECS.

32. In being a 'SOLAS V' compliant vessel, which of the following statements is incorrect?

    (a) The crew must undergo approved training.
    (b) The vessel must carry authorised equipment.
    (c) The vessel must carry paper nautical charts for back up.
    (d) The vessel must have adequate back-up support.

33. In relation to GPS Receivers, which of the following statements is incorrect?

    (a) Some Receivers can be pre-set to an acceptable HDOP limit.
    (b) HDOP values in excess of 6 are known as Outages.
    (c) A HDOP value between 2 & 4 is acceptable.
    (d) All Receivers are equipped with RAIM, which eliminates faulty satellites from position fixing.

34. In relation to GPS positions, which of the following statements is incorrect?

    (a) WGS84 is based on the earth having numerous centres.
    (b) All new paper charts are being produced to WGS84 standard.
    (c) On non-WGS84 charts, the datum-shift is written under the heading "Satellite Derived Positions".
    (d) Positions obtained from GPS Receivers relate to WGS84, unless otherwise set.

35. In relation to DGPS, which of the following statements is incorrect?

    (a) In order to utilize DGPS service, a vessel must be equipped with a DGPS Receiver or a DGPS-ready GPS Receiver.
    (b) In order to utilize DGPS service via a DGPS-ready GPS Receiver, it must be combined with a DGPS Beacon Receiver.
    (c) Most GPS Receivers are not DGPS-ready.
    (d) An Australian DGPS radiobeacon in the 'out of tolerance' condition is reported to users via AUSCOAST & SafetyNET Warnings.

36. DGPS Receivers:

    (a) can resolve HDOP & multipath errors.
    (b) can resolve datum discrepancy between WGS & non-WGS positions.
    (c) can correct Range Errors.
    (d) have a range of approximately 50 miles from the Beacon.

37. In relation to 'GPS II', which is expected to become operational by the year 2008, which of the following statements is incorrect?

    (a) It will be a duel-frequency SPS receiver.
    (b) The typical ionospheric error will be 7 metres.
    (c) The typical position accuracy will be 8.5 metres.
    (d) It will improve the performance of RTK applications.

Chapter 14: GPS & Electronic Charts

38. In relation to GPS Accuracy Improvement Initiative (AII), which of the following statements is <u>incorrect</u>?

    (a) It will reduce the GPS clock error to 4 metres.
    (b) The response time on emergency calls will be improved.
    (c) The typical horizontal accuracy of GPS will be 6 metres.
    (d) All new mobile (cellular) phones are expected to be compulsorily equipped with GPS for the emergency calls.

39. In relation to Global Navigational Satellite Systems, which of the following statements is <u>incorrect</u>?

    (a) GPS is owned & operated by the US military.
    (b) GLONASS is owned & operated by the Russian military.
    (c) GALILEO is owned & operated by IMO.
    (d) Combined GPS/Galileo/GLONASS Receivers will provide greater HDOP accuracy.

40. The accuracy of a GPS fix can be found with 95% reliability by:

    (a) multiplying HDOP & UERE values.
    (b) dividing HDOP value by UERE value.
    (c) dividing UERE value by HDOP value.
    (d) adding HDOP & UERE values.

41. In Australia, the DGPS service is provided for the following areas:

    (a) Great barrier Reef
    (b) Complete coastline
    (c) Pilotage ports
    (d) Critical areas

42. The DGPS provides position fixes with 95% probability of being accurate to within:

    (a) 100 metres.
    (b) 10 metres.
    (c) 95 metres.
    (d) 1 metre.

43. Switching off the selective availability of signals has improved the official accuracy of SPS-code GPS positions (at 95% confidence level) from:

    (a) 20 metres to 1 metre.
    (b) 100 metres to 50 metres.
    (c) 100 metres to 20 metres.
    (d) 1000 metres to 20 metres.

44. The total error in GPS range measurement is referred to as:

    (a) DOP
    (b) Total Error
    (c) HDOP
    (d) UERE

45. In relation to the function of a DGPS Reference Station, which of the following statements is <u>incorrect</u>?

    (a) It may extend the period of use of a faulty satellite by transmitting corrections to DGPS receivers.
    (b) It may notify DGPS users of unsuitability of satellite positions by actuating an alarm.
    (c) It may advise DGPS users to disregard information from a faulty satellite.
    (d) It may switch off affected DGPS receivers.

46. When using GPS for coastal navigation, the GPS positions must be:

    (a) verified by some other means of navigation.
    (b) plotted every 10 minutes.
    (c) plotted only on WGS charts.
    (d) considered inaccurate.

CHAPTER 14.2: AIS UNITS

47. In relation to AIS Units, which of the following statement is <u>incorrect</u>?

    (a) They are Radar Transponders.
    (b) They assist vessels with collision avoidance.
    (c) They can access tidal and wind data from sensors.
    (d) They facilitate Ship Reporting Systems.

CHAPTER 14: ANSWERS

1 (b), 2 (a), 3 (d), 4 (a), 5 (b),
6 (c), 7 (a), 8 (d), 9 (d), 10 (a),
11 (c), 12 (d), 13 (a), 14 (b), 15 (b),
16 (a), 17 (c), 18 (a), 19 (b), 20 (d),
21 (d), 22 (c), 23 (b), 24 (d), 25 (a),
26 (b), 27 (c), 28 (c), 29 (d), 30 (a),
31 (b), 32 (c), 33 (d), 34 (a), 35 (c),
36 (d), 37 (b), 38 (a), 39 (c), 40 (a),
41 (d), 42 (b), 43 (c), 44 (d), 45 (d),
46 (a), 47 (b)

Chapter 15

SOUNDERS, LOGS, AUTOPILOTS & ALARMS

ECHO SOUNDERS

ECHO SOUNDERS AND FISH FINDERS
A fish finder and an echo sounder (also known as depth sounder) are one and the same instrument. Only the sweep (cone or beamwidth) in the water is different. An echo sounder has two basic components: a transducer and a display.

SONAR
It is an acronym for Sound Navigation And Ranging. It is an echo sounder that determines not only the depth of an underwater object but also its relative bearing and range. On its screen it displays the depth as well as bearings and ranges of echoes.

TRANSDUCER
It consists of a ceramic element or oscillator embedded in a plastic casing. Some through-hull transducers have bronze casing. Transducers come in many shapes and sizes. Some protrude from the hull, others are installed flush with the hull.

Triggered by an electrical signal, the transducer vibrates to transmit sound pulses through the water. It then receives the vibrations from the returning echo and converts them to an electrical signal, which is passed on to the instrument control box. The instrument measures the time taken by each pulse to return from the seabed. Since the speed of sound in water is nearly constant (1500 metres/sec), it can calculate depth (Depth = speed of sound x half the time taken by echo to return). The returned signal, being weak, is amplified before appearing on a screen or recorded on paper. The components of an echo sounder are: POWER SUPPLY, OSCILLATOR, TRANSDUCER, AMPLIFIER, AND DISPLAY/RECORDER.

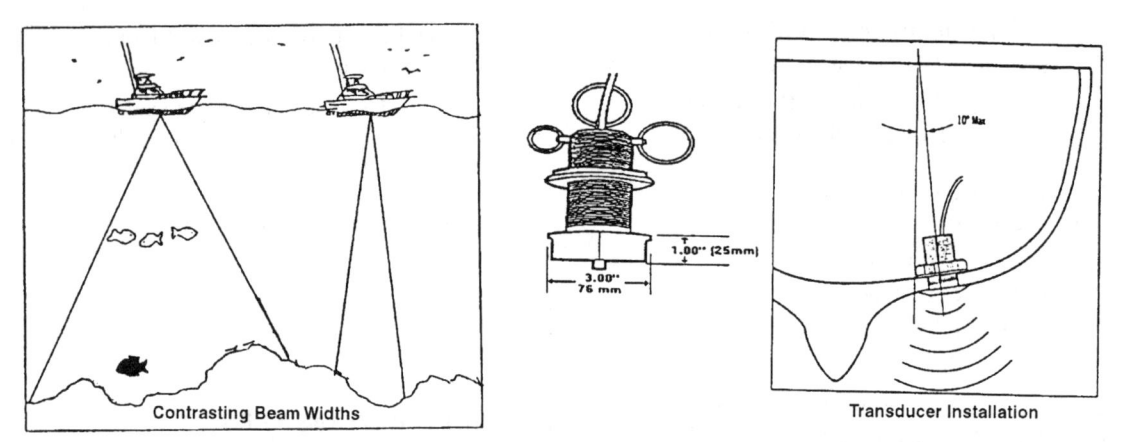

Fig 15.1: ECHO SOUNDER TRANSDUCER

POWER OUTPUT & BEAM WIDTH
The power output of echo sounders ranges between 200 and 3000 watts. The higher the power output, the greater the depth penetration; hence a greater ability to see smaller fish at depth. For fishing at sea in up to 60 metres of water, a sounder of at least 600 watts of power, and a transducer of not less than 20 degrees beamwidth should be sought. In deeper waters, a narrower 8-degree beamwidth is better.

The really high-power sounders can examine the seabed for fish at depths of more than 200 metres with a wide-beam transducer, and more than 400 metres with a narrow-beam transducer. A vessel intending to navigate and fish in all depths could fit both a narrow-and-wide-beam transducer.

A sounder's performance is always better in fresh water, because there is less suspended material to attenuate the pulse.

Chapter 15: Sounders, Logs, Autopilots & Alarms

FREQUENCY, PULSE LENGTH & RESOLUTION
Commercial fishing vessels, interested only in large schools of fish in deep water, generally use low-frequency sounders. These sounders transmit their pulses at about 50 kHz (or 50,000 cycles per second). Some transmit as few as 10 pulses per minute. The pulses being longer, suffer less attenuation on their journey through the water. They penetrate deeper, providing soundings at greater depths and a better ground discrimination. They have a low resolution because the long pulses of energy cannot resolve targets close to each other.

Such sounders are less suitable for anglers and other sports users. For example, if two fish were swimming close together, a low frequency sounder will display them as one echo on the display, whereas a high-frequency unit (operating on a shorter pulse length) will be able to resolve them into two separate echoes. Similarly, the echo of fish swimming close to seabed will merge into the echo of the seabed on a low-frequency unit. Most sports fishing sounders operate at a frequency of around 200 kHz (or 200,000 cycles per second, transmitting up to 600 pulses per minute). They are known as high-frequency sounders, which are better able to discriminate smaller echoes in shallower waters.

Some sounders are designed to switch automatically to an appropriate frequency when selecting different ranges. A few sounders allow the operator to manually change the pulse length to short, medium or long.

FISH WITHOUT SWIM BLADDERS
Some older low-frequency sounders were only able to pick up fish with swim bladders. They detected the difference in the density between the air in the bladder and the water. It was often thought that fish such as mackerel, which have no swim bladder, couldn't be detected. However, most modern sounders work on picking up the sound pulse reflected from the body of the fish. Their echoes appear as arches on the display.

DISPLAYS - three types:
1. Liquid Crystal Graph (LCG or LCD) display.
2. Video display.
3. Record-keeping display.

All can be changed to a digital mode, and may provide visual and audible warnings at selected depths.

1. LIQUID CRYSTAL SOUNDERS
(Note: TFT (thin-film transistor) is an LCD display)

These sounders currently dominate the boating market. They are lightweight, compact and sealed against the elements. Many are pressurised with an inert gas, making them even more waterproof, and the screen are condensation-free so they can be mounted in exposed outside positions. The LCD screens have very little metal in their construction and can usually be positioned within a few centimetres of a magnetic compass without affecting it. However, these displays are relatively smaller and the liquid inside them may leak.

The LCD screens can be in colour or monochrome or switchable to either. A colour sounder uses a different colour for any of up to 16 different signal strengths. Strong echoes show up as red and the weak ones as green or blue. Monochrome sounders use different levels of grey to show the different levels of signal strength. Strong echoes such as a hard bottom show up very dark and weak echoes such as debris as cloudy grey.

In most cases the LCD viewability increases in direct sunlight. And the contrast increases with back or front lighting, which is also necessary for night use. But it lacks the level of detail of a CRT screen. LCD sounders run on low power - some models using only 0.25 amps.

Fig 15.2: SOME LCD DISPLAYS
(Digital, LCG and 3-D)

2. VIDEO SOUNDERS

These sounders present the picture on a cathode ray tube (CRT) similar to a TV set. This type of display offers high contrast in normal or low light conditions, making the echoes appear bright and clear. But the screen fades away in bright sunlight and is not usually suitable for outdoors. The colour video models are more popular with commercial fishing vessels. Different colours represent different strengths of echo. The strongest signals - such as the sea bed - are normally indicated by red, while the weaker signals are displayed progressively as orange, yellow, green, and shades of blue. Their high-resolution large screens can be seen from across a wheelhouse, and the different colours provide quick and easy interpretation of the signals. Most Video models run on a 12-volt as well as a 24-volt battery.

The disadvantages of a video sounder are:

- They are bulky.
- Their picture fades in sunlight and they are not waterproof. Therefore they need to be installed in a sun-shaded and waterproof location.
- They consume a significant amount of power, normally between 1-2 amps.
- Incorrect adjustment of the sensitivity control can produce anomalous colouring which gives very small echoes the appearance of large targets.
- They need to be installed at least a metre away from a magnetic compass.

3. RECORD-KEEPING SOUNDERS

Such sounders are used mainly for survey work and in large vessels requiring permanent record of soundings.

They can be the older recording-paper type or the modern computerised record-keeping sounders. In the paper version, which is now rarely seen, a stylus fitted on a revolving belt inscribes a roll of paper first at the start of a pulse and then on its return. The greater the depth, the greater the distances between the markings. The paper is moved from one roller to the other by a clockwork mechanism, thus producing a timed permanent record.

In the computerised version, a high-resolution black-and-white LCD screen displays a picture looking similar to the paper version. Some have built-in selectable dual frequency of 50 kHz or 200 kHz, or can be used simultaneously with split screen displays. A GPS can also be connected to it. Programmable greyscale values can be set to match the user's preferences in game fish, baitfish, bottom contour and structure identification, etc. The data can be recorded, stored, reviewed, edited and transported on digital multi-media cards.

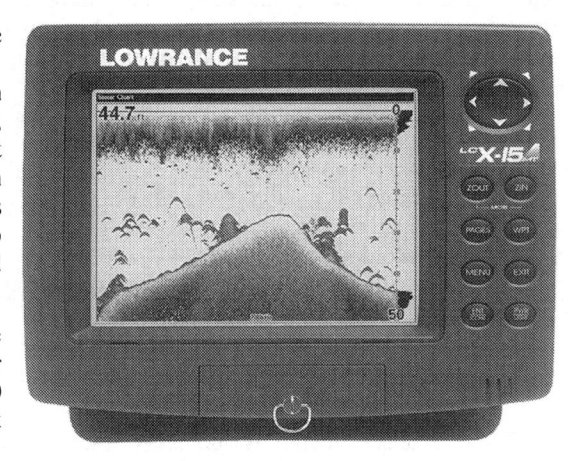

Fig 15.3: COMPUTERISED RECORD-KEEPING SOUNDER

ECHO SOUNDER CONTROLS

- *GAIN* or *SENSITIVITY* controls the signal amplification. Adjust it to obtain the sharpest signal. (See below)
- *RANGE* control enables the change of operating range.
- *PHASE* control enables selection of upper and lower depth limits.
- *WHITE LINE* enables discrimination of bottom echoes from the seabed. (See below)
- *TIME VARIED GAIN (TVG)* is similar to radar's sea clutter control. It suppresses weak echoes at very close range.

Chapter 15: Sounders, Logs, Autopilots & Alarms

DISPLAY INTERPRETATION, Fig 15.4 (A)

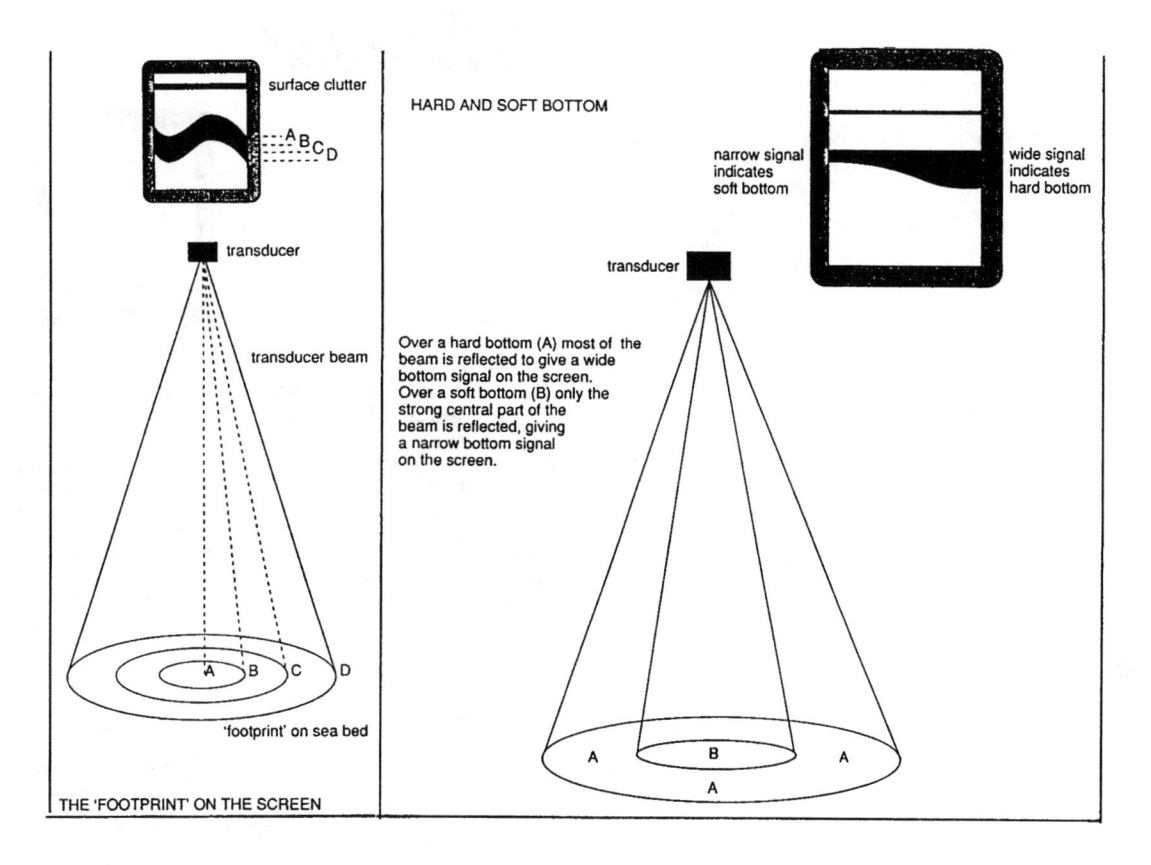

DISPLAY INTERPRETATION, Fig 15.4 (B)

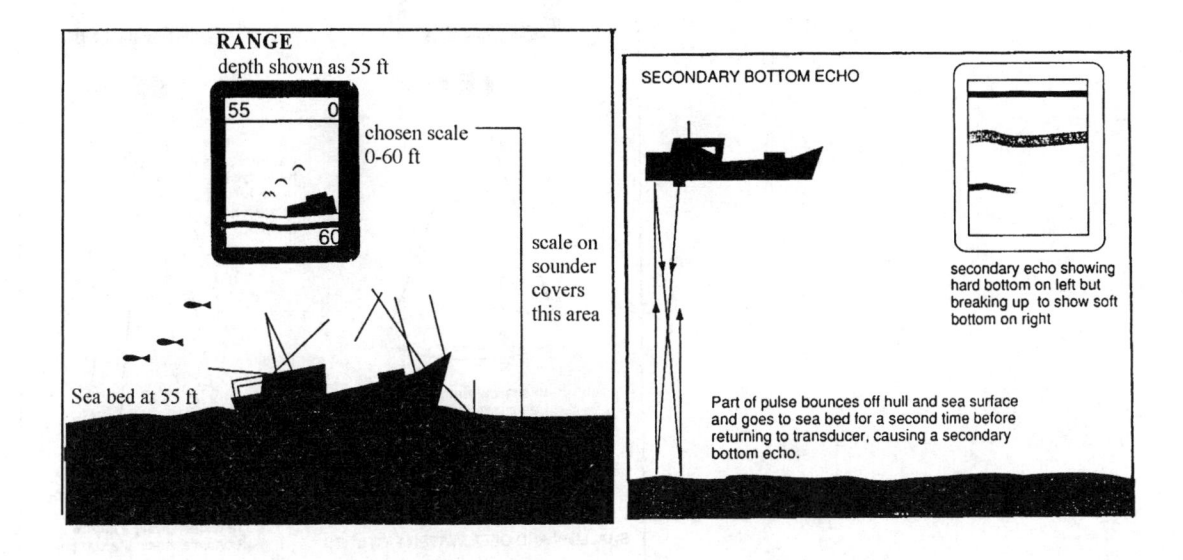

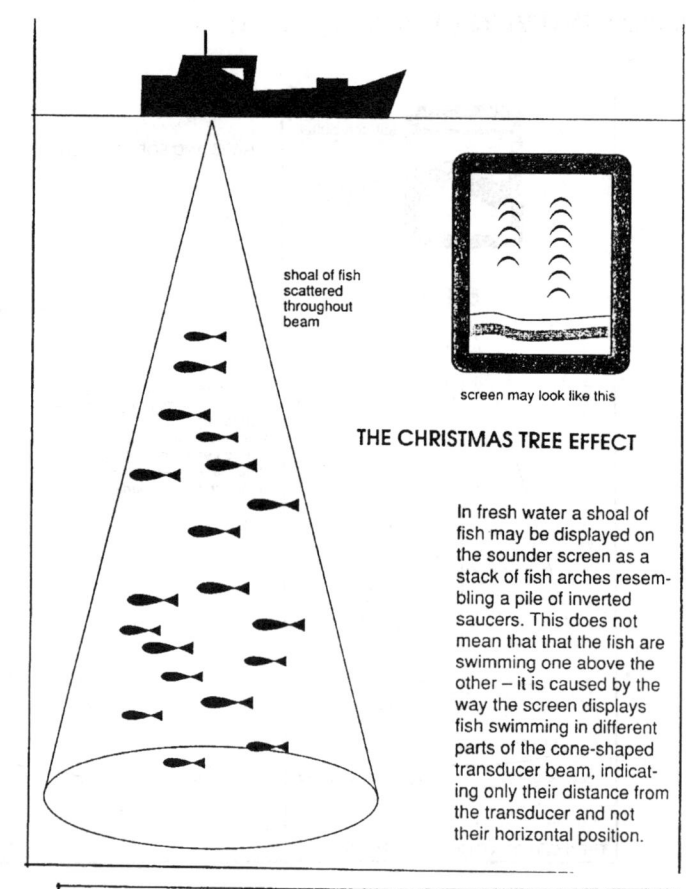

DISPLAY INTERPRETATION
Fig 15.4 (C)

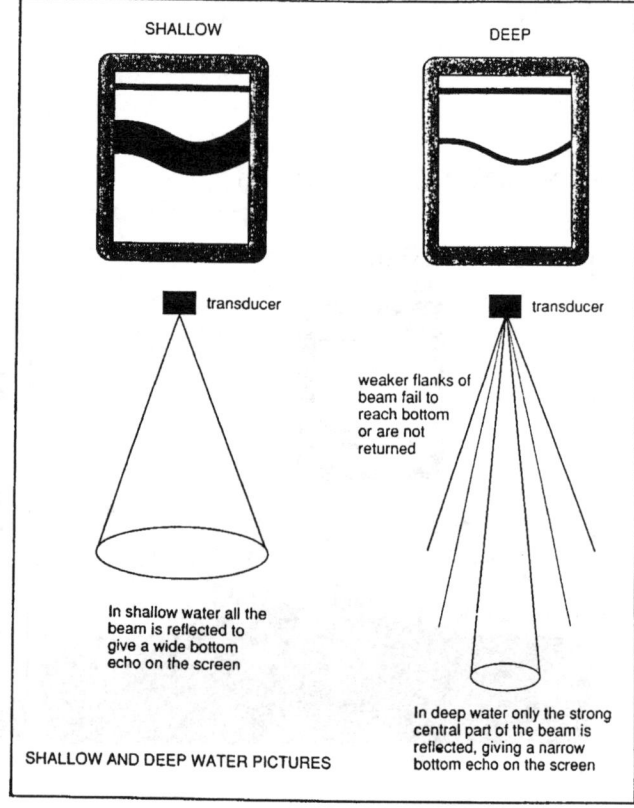

shoal of fish scattered throughout beam

screen may look like this

THE CHRISTMAS TREE EFFECT

In fresh water a shoal of fish may be displayed on the sounder screen as a stack of fish arches resembling a pile of inverted saucers. This does not mean that that the fish are swimming one above the other — it is caused by the way the screen displays fish swimming in different parts of the cone-shaped transducer beam, indicating only their distance from the transducer and not their horizontal position.

DISPLAY INTERPRETATION
Fig 15.4 (D)

SHALLOW

DEEP

transducer

transducer

weaker flanks of beam fail to reach bottom or are not returned

In shallow water all the beam is reflected to give a wide bottom echo on the screen

In deep water only the strong central part of the beam is reflected, giving a narrow bottom echo on the screen

SHALLOW AND DEEP WATER PICTURES

Chapter 15: Sounders, Logs, Autopilots & Alarms

SIDE LOBES AND GHOST ECHOES

Side lobe reflection causes ghost echo to appear on screen

DISPLAY INTERPRETATION, Fig 15.4 (E)

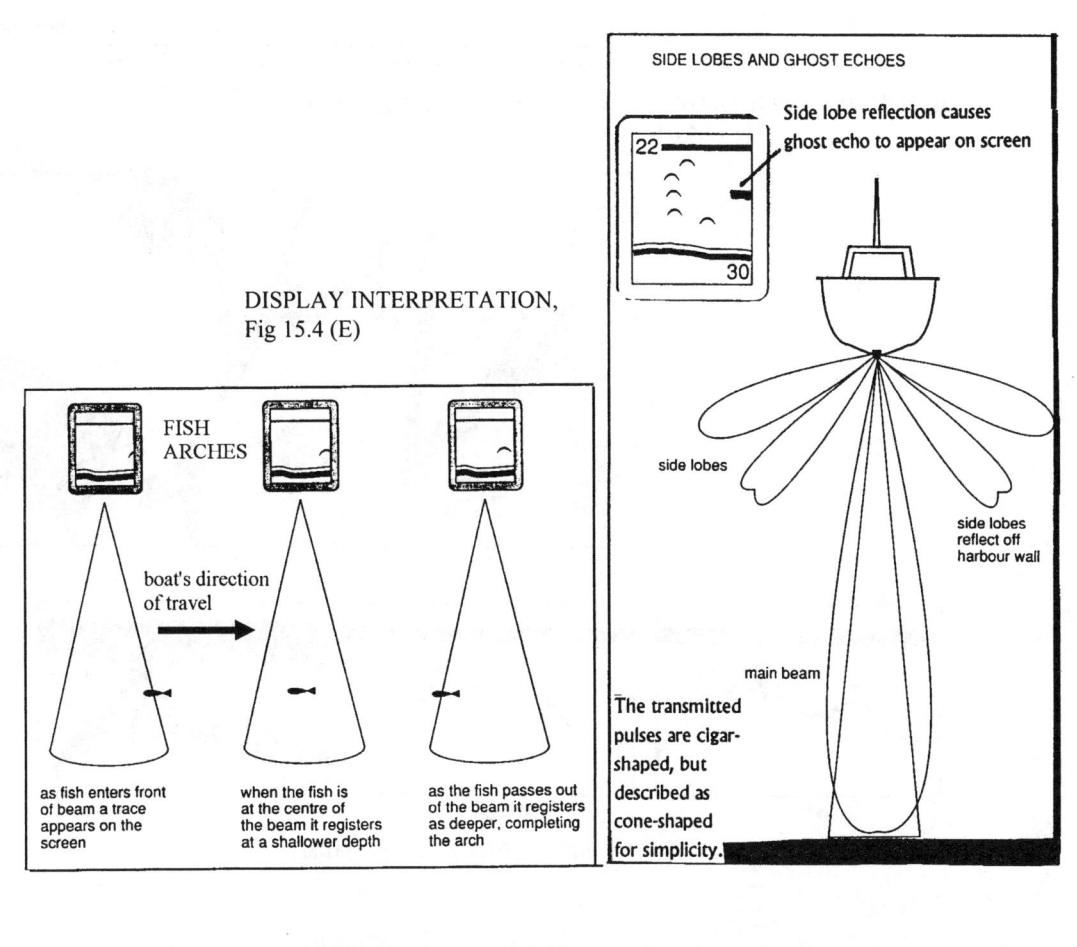

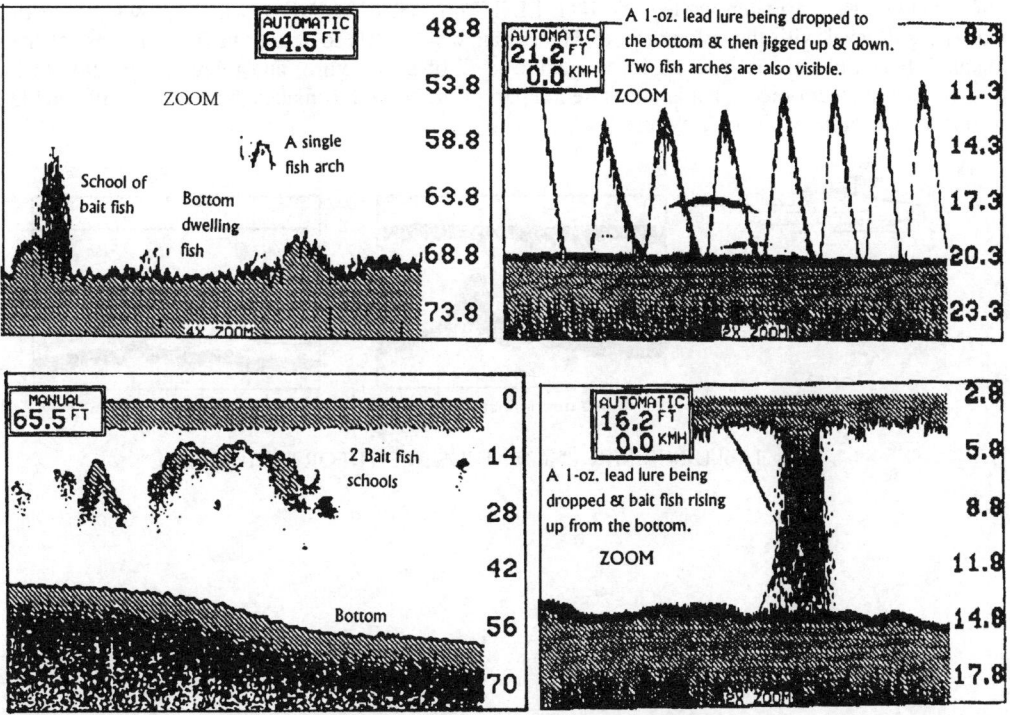

FISH ARCHES

boat's direction of travel

as fish enters front of beam a trace appears on the screen

when the fish is at the centre of the beam it registers at a shallower depth

as the fish passes out of the beam it registers as deeper, completing the arch

side lobes

side lobes reflect off harbour wall

main beam

The transmitted pulses are cigar-shaped, but described as cone-shaped for simplicity.

AUTOMATIC 64.5 FT

ZOOM

School of bait fish

Bottom dwelling fish

A single fish arch

A 1-oz. lead lure being dropped to the bottom & then jigged up & down. Two fish arches are also visible.

AUTOMATIC 21.2 FT 0.0 KMH

ZOOM

MANUAL 65.5 FT

2 Bait fish schools

Bottom

A 1-oz. lead lure being dropped & bait fish rising up from the bottom.

AUTOMATIC 16.2 FT 0.0 KMH

ZOOM

DISPLAY INTERPRETATION, Fig 15.4 (F) & (G)

DISPLAY INTERPRETATION,
Fig 15.4 (H)

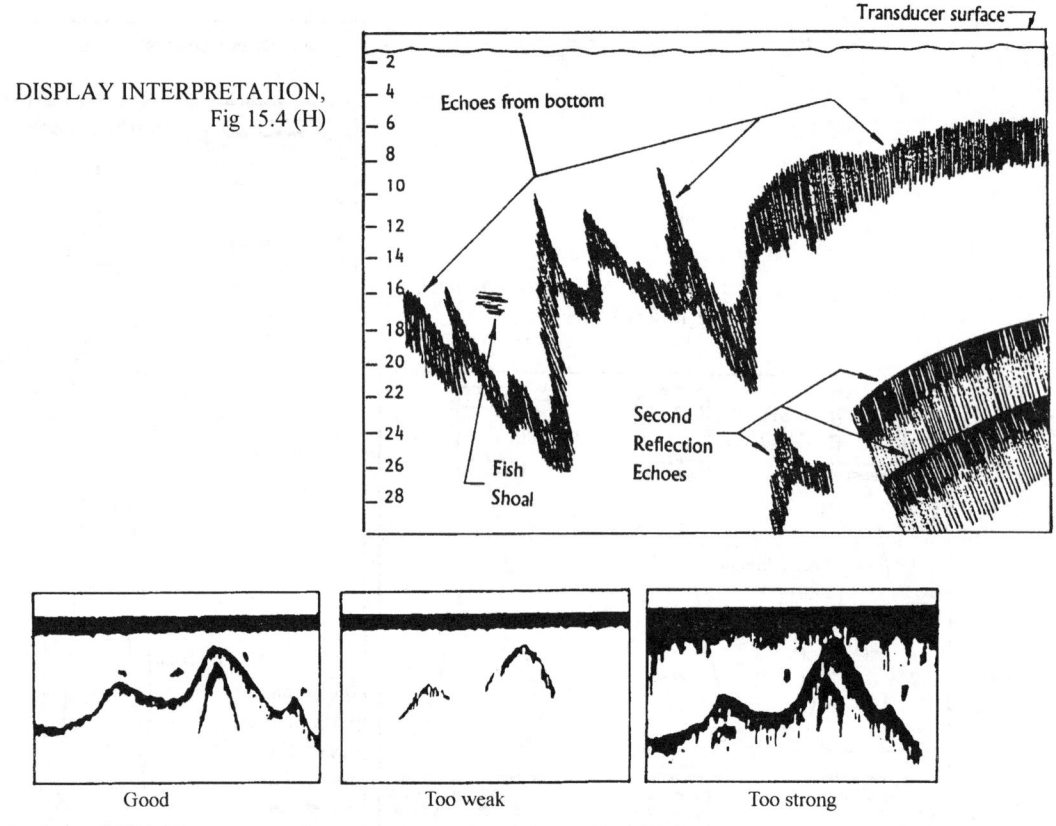

Fig 15.5: SENSITIVITY ADJUSTMENT

Some sounders have a feature called "WHITE LINE" (or "GREY LINE"), which allows the operator to replace any colour or strength level with the colour white. The colour white tends to stand out from the rest of the colours on a sounder's display. It is a useful tool to look for a specific level of echo return, and more specifically, to locate bottom-feeding fish, wrecks and other objects that are not integral part of the seabed. It makes them identifiable on the bottom since they appear to sit on top of the white or grey line.

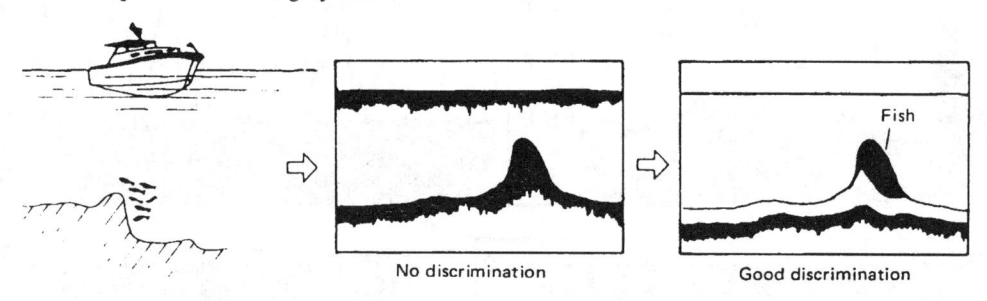

Fig 15.6: DISCRIMINATION ADJUSTMENT FOR "WHITE LINE"

Chapter 15: Sounders, Logs, Autopilots & Alarms

TRANSDUCER INSTALLATION

The Echo Sounder transducer should be installed to bring the waterflow hard up against its face, shielded from air bubbles and propeller noise. As shown, the cable from the transducer to the display should be installed well away from the boat's electrical and radio installations, particularly the engine electrics.

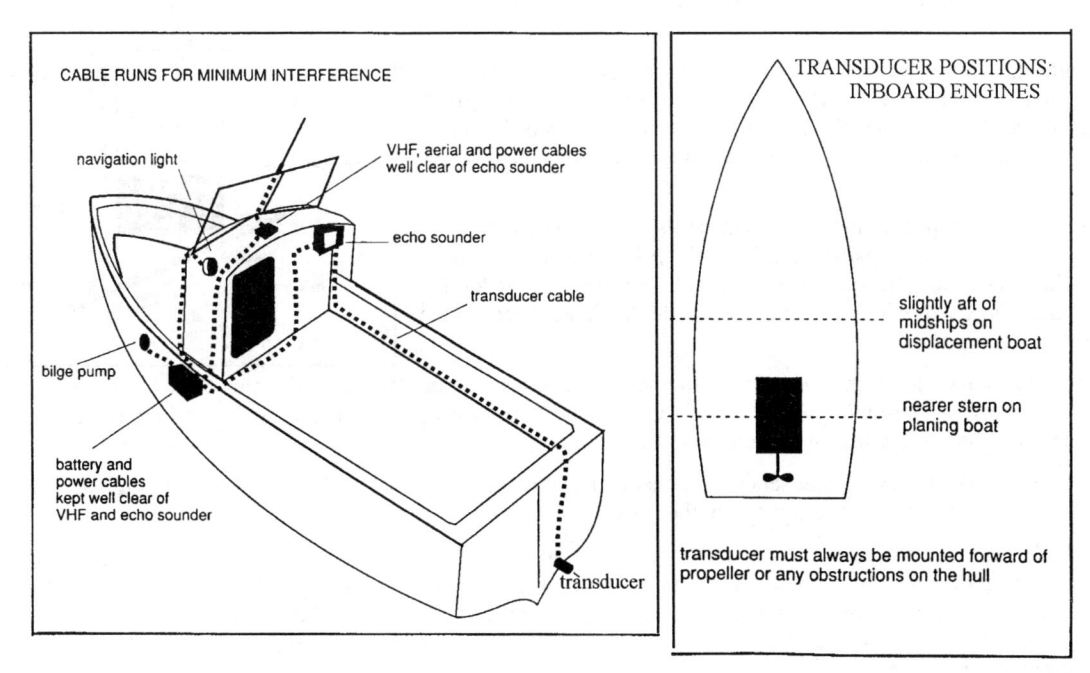

Fig 15.7: TRANSDUCER INSTALLATION

TRANSOM MOUNTING OF TRANSDUCER

Many liquid crystal sounders are supplied with a transom-mounted transducer as standard equipment. These transducers are ideal for stern-drives and outboards and boats with planing and cathedral hulls where the transom is often the only part of the boat in contact with a good solid lump of water.

A transom-mounted transducer must be installed below the waterline with face pointing straight down or canting slightly forward. The face of the transducer should be in level with or protrude slightly below the bottom of the boat. If it is positioned in front of the outboard or stern-drive propeller it will not be affected by propeller turbulence, except when the boat is going astern.

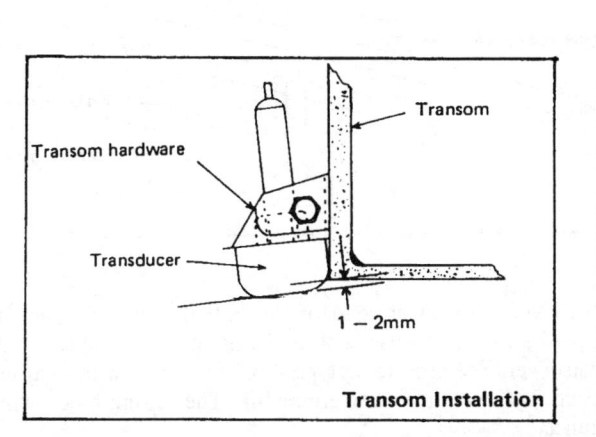

Fig 15.8

IN-HULL MOUNTING

In boats with inboard engines where the propellers are forward of the transom, it is not possible to transom-mount the transducer because the disturbed water passing under it would disrupt the signal. In such cases and for a better appearance, a different mounting method needs to be considered.

In single-skin fibreglass (GRP) boats, a transducer can be mounted inside the hull - provided the GRP is no more than, say, 20 mm (0.75 inches) thick. Most in-hull transducers come in the form of a small circular pod, which is stuck to the inside of the hull using a recommended epoxy resin.

Once again, it is important that the transducer is pointing straight down, and that it is located over a part of the hull which has a smooth flow of water moving under it. The quality of the hull construction under the transducer must be good, with no air bubbles in the laminate, because the air will distort the signal.

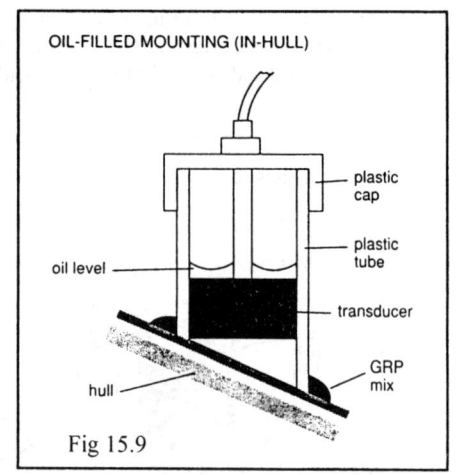

Fig 15.9

One way to find a good in-hull location is to anchor the boat in about 20 metres of water, switch on the display and, with a dab of grease or vaseline on the face of the transducer, push it up against the hull in various places to find the location that gives the best picture on the screen.

If the hull is of double-skin foam sandwich construction the transducer must be installed on the outer skin, since it is not able to transmit through the air gap between the skins.

If the hull is steeply angled, try mounting the transducer on a small horizontal platform of epoxy or similar material so that it points straight down. An alternative is to cut the base of a plastic tube to an angle that matches the deadrise of the hull, and glue it in place. Insert the transducer, then fill the tube below the face with a liquid. Any liquid will do the job, but in cold climates castrol oil is often preferred because it has a lower freezing temperature than water, and is not so likely to generate air bubbles in rough weather.

Whichever method is used, it is important that the transducer is facing straight down, and that there is a solid or liquid interface between the transducer and the hull. The presence of any air will severely degrade the sounder's performance.

THROUGH-HULL MOUNTING

The main advantage of mounting the transducer inside the hull is obvious: it requires no holes beneath the waterline, and the job can be done while the boat is on the water. However, some attenuation of the signal is inevitable; and the thicker the hull, the worse the problem. So the in-hull mounting is not recommended for those aiming to go fish-finding in very deep waters, or if the boat has a thick hull. It also cannot be used on wooden, aluminium or steel hulled boats, and the only option - assuming transom mounting is not possible - is to mount it through the hull.

Fig 15.10

As before, the transducer should be positioned vertically in an area of the hull that is free from turbulence, normally slightly aft of midships. Some transducers can be mounted flush with a chamfered head, allowing them to be embedded in the hull so that there is no projection whatsoever. The protruding type should be fitted with a fairing block to protect them from being damaged by submerged flotsam and in the event of grounding. The fairing block will also help smooth out the water flow and so reduce water interference.

Chapter 15: Sounders, Logs, Autopilots & Alarms

TRANSDUCER IN DEEP KEEL BOATS

In deep keel boats, allowance should be made for keel offset in measuring depth below the keel, as shown below. Also allow for the heeling error as shown.

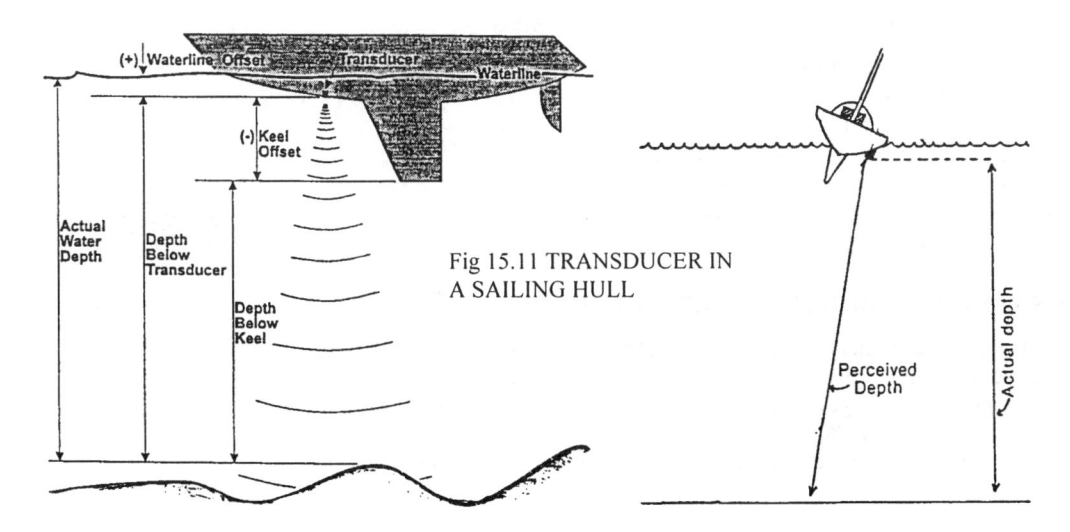

Fig 15.11 TRANSDUCER IN A SAILING HULL

TAKING CARE OF TRANSDUCER

Once fitted, the transducer, although hidden from view, must not be forgotten. It must be kept free of weed. Through-hull and transom-mounted transducers should not be painted over with anything other than a light coat of antifoul. They must not be subjected to the pressure of chocks on a slipway.

SHALLOW WATER SOUNDINGS

In shallow water an echo sounder may display twice or three times the real depth. This is due to the phenomenon known as DOUBLE ECHOES. In shallow water echo sounders tend to return double (or triple or more) echoes. This is especially the case when the sea floor is hard and the sensitivity is set too high. These are echoes reflected twice between the sea floor and boat's bottom before being received by the instrument. The instrument thus indicates a depth equal to twice the real depth. Reducing sensitivity would help, but a hand lead line is more reliable in such circumstances. See hand lead line in Chapter 19.3. Always compare the sounder readings with depths indicated on charts, but don't forget to make allowance for the tide.

DEEP WATER SOUNDINGS

In deep water an echo sounder may indicate shallower depth than it really is. This is caused by echoes having to travel too far and returning one or two cycles after their transmission. They thus appear at a shorter range than they really have travelled. This phenomenon is known as SECOND TRACE RETURN. Changing the range scale or using another echo sounder with a different pulse rate usually helps.

COMPARING SOUNDER INFORMATION WITH CHARTED DEPTHS

For a true comparison, an allowance must be made for the state of the tide and the vessel's draft.

ECHO SOUNDER MAY GIVE POOR RESULT DUE TO:

- Poor reflecting surfaces such as:
  - Mud or kelp beds
- False echoes caused by:
  - Rolling & Pitching
  - Sand suspended in water (say, over a bar)
  - A school of fish
  - Strong underwater currents
  - Echo returning from a great depth

- Temperature or salinity layers
➢ Equipment faults and maladjustment such as:
   - Gain or sensitivity too high - cross noise
   - Gain or sensitivity too low - weak echo display
   - Marine growth on transducer face

➢ Interference due to "noise" (air bubbles):
   - Rough sea
   - Excessive head trim
   - Boat going astern or turning sharply
   - Boat in the wake of another boat
   - Hull with excessive marine growth.
   - Bad siting of transducer.

ECHO SOUNDERS MAY INDICATE QUALITY OF SEA BED

➢ A smooth trace indicates mud or sand. Firm sand gives quite a clear return.
➢ A trace with steep changes in depth is probably rock or coral.
➢ Rock gives the strongest return. It may even give a double echo, the second trace being weaker.
➢ Mud over rock may show a light trace above a hard one.
➢ Soft mud, especially if covered by weed, gives the weakest return.

BUYING AN ECHO SOUNDER

- Consider your need for a sounder: navigation or fishing or both?
- Determine the type of display you would need: LCD, CRT, Colour or Monochrome?
- Knowing the approximate water depth in your area of operation will determine the choice of power output level of the sounder.
- Do you need a wide or narrow beamwidth transducer?
- How will the transducer be fitted in the vessel: through-hull, transom-mount or in-hull?
- Do you want to be able to recall the sounder's record?
- Do you want a stand-alone echo sounder or an integrated navigation unit incorporating Echo Sounder, GPS/DGPS, Plotter, Radar, AIS, etc?
- Quality brands are more expensive.
- Know the difference between a sounder and a sonar.
- Consider the sounder's ease of use.
- Check the power supply suitability.
- Try it out. Ask your friends who may use the same model.
- Some sounders have rotatable screens. These might suit some users.
- Receivers with user-changeable fields allow users to select only the information they want to look at.
- Not all sounders are sealed to make them completely waterproof. Many are merely water-resistant.
- The sounder may be equipped with shallow-water and deep-water danger alarms.
- The quality of a sounder's manual is often an indication of the sounder's quality. The manual should not omit vital operational information, such as sounder errors and limitations.
- Consider the manufacturer and dealership size, after sale service and availability of parts.

Chapter 15: Sounders, Logs, Autopilots & Alarms

LOGS

Logs (or speed logs) measure distance, which is converted by time into speed. Most logs measure the distance travelled relative to the water flowing past the Transducer. Only the Doppler log and GPS measure the actual vessel speed over the ground (GPS is not really a log). Most transducers are self-sealing, moulded from glass-filled polyester. For timber hulls they are usually made of bronze.

Distance registered by a log is affected by the following factors:

- Sea conditions
- Vessel's draught
- Vessel's trim
- Marine growth on hull

Distance obtained from a log can be used to calculate boat's DR position. A comparison between the DR position and a Fix allows us to calculate the set and drift of any existing current.

Fig 15.12
AN AUTOHELM LOG
READOUT & TRANSDUCER

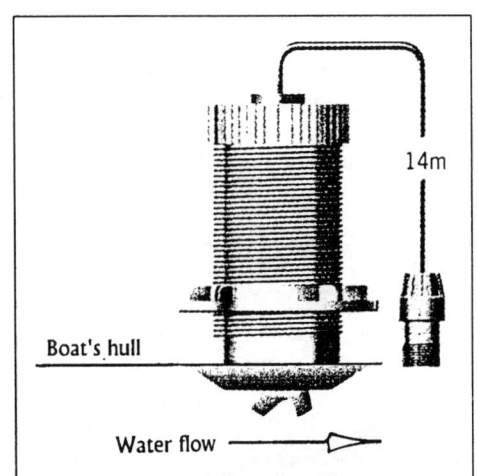

TYPES OF LOGS

1. **SPEED CALCULATION FROM RPM:** In the absence of a log or GPS, speed can be calculated from a calibration table of the propeller shaft revolutions (RPM). However, such a table can only apply to a given draught, trim, bottom cleanliness and weather.

2. **STREAMED LOG:** A metal rotor is towed astern on a plaited log line and its rotation is recorded on a counter. This is now an outdated technology. The rotor is easily lost at sea, it is inaccurate in following seas, and has to be taken in before going astern.

3. **IMPELLER LOG:** This works by measuring the rotation of an impeller or a paddle wheel in the transducer due to the travelling of the boat. It is a simple log. It registers even at low speeds, but the impeller can become fouled or damaged. Registration in rough seas can become inaccurate. It is sometimes referred to as the Chernikeef Log.

4. **PRESSURE LOG:** This works by measuring the pressure built up in a small tube in the transducer due to the water flow past the boat. Its advantage is that it has no moving or projecting parts to get fouled or damaged, but it does not register speeds under 2 knots. It is sometimes referred to as the Pitometer Log.

5. **ELECTROMAGNETIC LOG:** An electromagnetic log consists of two electrodes in a probe in the transducer. A voltage is induced in them when water flows between the probes. This is measured by an electronic circuit and converted to speed and distance on the indicator. This log too has no moving parts, but it can affect the boat's compass. It is very accurate, but expensive.

6. **DOPPLER LOG:** *The pitch of a police siren is higher when it travels towards you than when it is moving away. This variation in frequency of sound waves emitted by a travelling object is known as the DOPPLER EFFECT. It is a highly accurate method of measuring speed and distance.*

    The Doppler logs are designed on the principle of the Doppler Effect. The transducer is fitted in the fore part of the underwater hull, facing ahead and inclined downwards at an angle between 45° and 60°. It transmits and receives sound waves and then measures the difference in the frequency of the transmitted and received sounds caused by the vessel's movement through the water. Speed through the water is thus measured. When fitted with a transverse axis beam, it can also measure the sideways movement of a vessel due to wind (Leeway) or berthing speed.

    Some Doppler logs have a second beam pointing aft. These dual-beam instruments (**JANUS CONFIGURATION**) are more accurate.

    In depths up to 200 meters a Doppler Log is capable of measuring a vessel's true speed, i.e. the speed over the ground. This is because its sound waves reflect off the seabed, and it is referred to being in BOTTOM LOCK or

BOTTOM TRACK mode. In depths above 200 metres it works by the sound reflecting from a deep layer of water. In this WATER LOCK or WATER TRACK mode it is more likely to read speed through the water. Any difference in speed in bottom lock and water lock will indicate the presence of a current.

The log works best with about 1.5 metre water under the keel. Ideally, it is operated in bottom lock in depths up to 30 metres and in water lock in depths greater than 200 metres. For in-between depths the operator may choose either mode. Doppler logs are expensive and not generally fitted in small craft.

CAUSES OF ERRORS IN DOPPLER LOGS:
Doppler logs have no moving parts or hull projections but can become inaccurate due to aeration caused by rough weather or misalignment of the transducer from the fore and aft line. Vessel's trim or pitching motion can result in misalignment in the vertical plane. Other sources of errors are variations in water temperature, and to a lesser extent, variations in salinity. Some logs are compensated for errors caused by changes in sea temperature. However, none of these errors are very large. Only rolling will cause a significant error in a CROSS TRACK LOG.

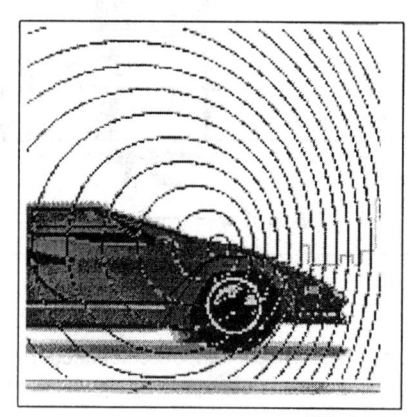

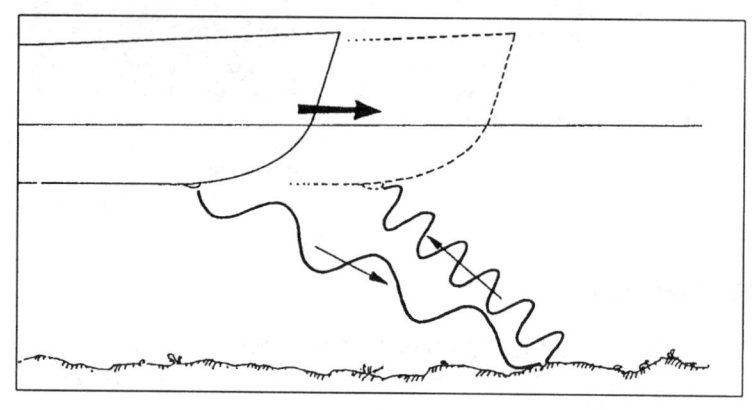

Fig 15.13: THE DOPPLER EFFECT & DOPPLER

7. **SPEED CALCULATION BY GPS:** GPS can calculate a vessel's motion (speed) to an accuracy of 0.1 knot.

8. **DUTCHMAN'S LOG:** This log is only worth mentioning in passing. A larger ship may drop a float overside from the bow and measure the time it takes to reach the stern. Speed can be measured provided the ship's length is known.

CALIBRATION OF LOGS
Logs are calibrated over a measured distance, marked on some charts inside and near major harbours. You can also use transit bearings at two places, a mile or so apart, to use as a measured distance, for log calibration.

PRECAUTIONS WITH LOGS (Transducers, Impellers, Heads)
With logs that project beyond the hull, care must be taken when coming into shallow water and when slipping the boat. If it is retractable, it must be retracted and the valve closed. The responsibility for lowering and retracting the log must be clearly assigned and the procedure displayed at suitable locations.

CAN LOGS MEASURE CURRENT?
Yes, but only when the vessel is at anchor or secured to a buoy or jetty.

POSSIBLE REASONS FOR A BOTTOM LOG NOT WORKING
- The log may be withdrawn.
- Impeller fouled or damaged.
- Faulty electrical contacts.

Chapter 15: Sounders, Logs, Autopilots & Alarms

SELF-STEERING & AUTOPILOT

WINDVANES

Sailing vessels often use a wind controlled self-steering device. A small sail or vane maintains the boat's heading at a set angle to the wind. The boat will steer the required course, provided the wind does not shift.

The rudder trim-tab turning the rudder is about one-fifth of the rudder area. Fitted to a metal rod, it turns on its own axis. The vane is made of plywood or aluminium. After bringing the vessel onto the desired course, the vane is set and locked to point into the wind. Should the boat wander, the wind bears on one side or the other of the vane and moves it, which turns the tab, which turns the rudder. However, the system is risky. If left unattended, a wind shift may bring the sails aback or take the boat the wrong way.

AUTOPILOTS

Electric autopilots range from the very basic to quite sophisticated units that can be interfaced with GPS, log, radar, chart plotter, and the wind instrument. A compass-sensing element activates a steering motor to turn the rudder whenever the boat goes off course. While large ships use gyrocompasses, any type of compass can be used with autopilots. Many boats use fluxgate compasses. The commands are transmitted to the wheel by chain drive, belt drive or hydraulic drive.

For small vessels, autopilots are sold in three types of units:

- Tiller-mounted autopilots
- Above-deck wheel autopilots
- Below-deck heavy-duty autopilots

TILLER-MOUNTED AUTOPILOTS have an extension arm, one end of which is attached to the tiller and the other to a cockpit seat or coaming. Earlier models had a compass on top, which had to be rotated to the ship's course before pressing the autopilot button. On the newer models, pressing the autopilot button automatically locks the unit to the compass course being steered. There are additional buttons to make adjustments for sea conditions and to change tack on a sailing boat. Tiller-mounted autopilots are not very strong. On sailing vessels, in strong winds, excessive weather helm needs to be avoided by keeping the sails properly trimmed and early reefing, or by reduction of sail area.

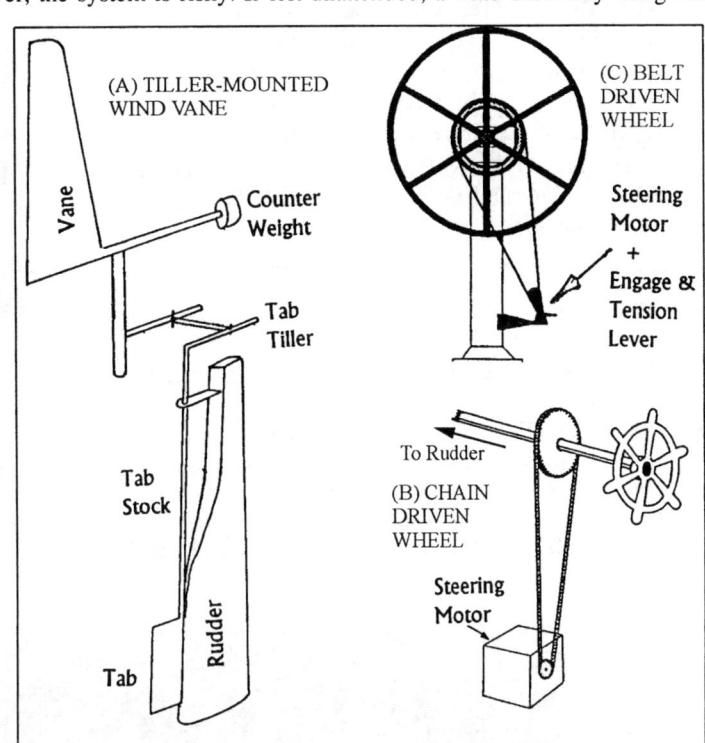

Fig 15.14 (i & ii) SELF-STEERING DEVICES

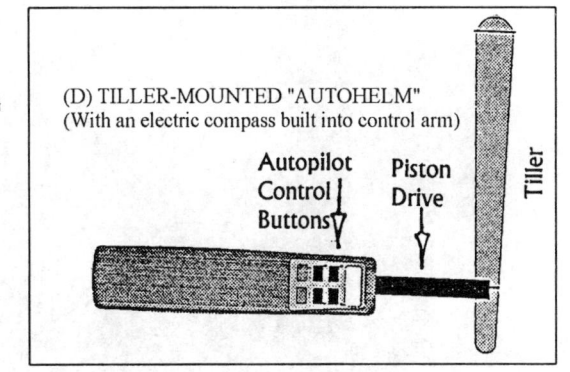

ABOVE-DECK WHEEL AUTOPILOTS are positioned somewhere near the wheel in the cockpit where they can be attached to the wheel via a chain or a belt-drive. The motor in the belt-drive units is quite small. Its power is enhanced by its gearing arrangement. Pressing a button locks the autopilot to the boat's course. The chain or the belt-drive must always be kept tight. The unit and any electrical connection on deck must be waterproof.

BELOW-DECK AUTOPILOTS are usually heavy-duty computerised units consuming 3 to 15 amps of power depending on whether the drive unit is mechanical or hydraulic, as shown below. Mechanical and hydraulic steering systems are discussed in Chapter 31.

AUTOPILOT CONTROLS

In computerised units, such as Raypilot 650, the following controls are available. Extra control units can be added, such as the hand-held remote control and a joystick.

RAYPILOT CONTROLS
(Illustrated IN Fig 15.16)

- AUTO: Engage autopilot
- RESUME: Change over to manual
- STANDBY: Temporary manual steering
- LIGHTS: Switch on autopilot light
- RESPONSE: Make autopilot use more rudder
- NAV: Display GPS information
- COURSE: Change course on autopilot

(In some units, RESPONSE setting may be referred to as RUDDER, RATIO or HELM. The term "HELM" applies to rudder, not the tiller.)

Other autopilots may carry settings labelled:

- WEATHER: to desensitise off-course setting in rough weather
- RUDDER: to decide on the size of rudder angles for steering
- COUNTER RUDDER: to decide on the size of rudder angle whenever vessel overshoots her course
- *BIAS is another control that may be found on some units. As the term suggests, this setting gives more helm on one side than on the other: when motoring on one engine in a twin-screw boat or with trawl gear on one side. In a sailing vessel, more helm to windward may be needed in order to maintain the course. This setting is also referred to as STANDING HELM, WEATHER HELM PERMANENT HELM or TRIM.*

Fig 15.15: AUTOHELM "SPORTPILOT"

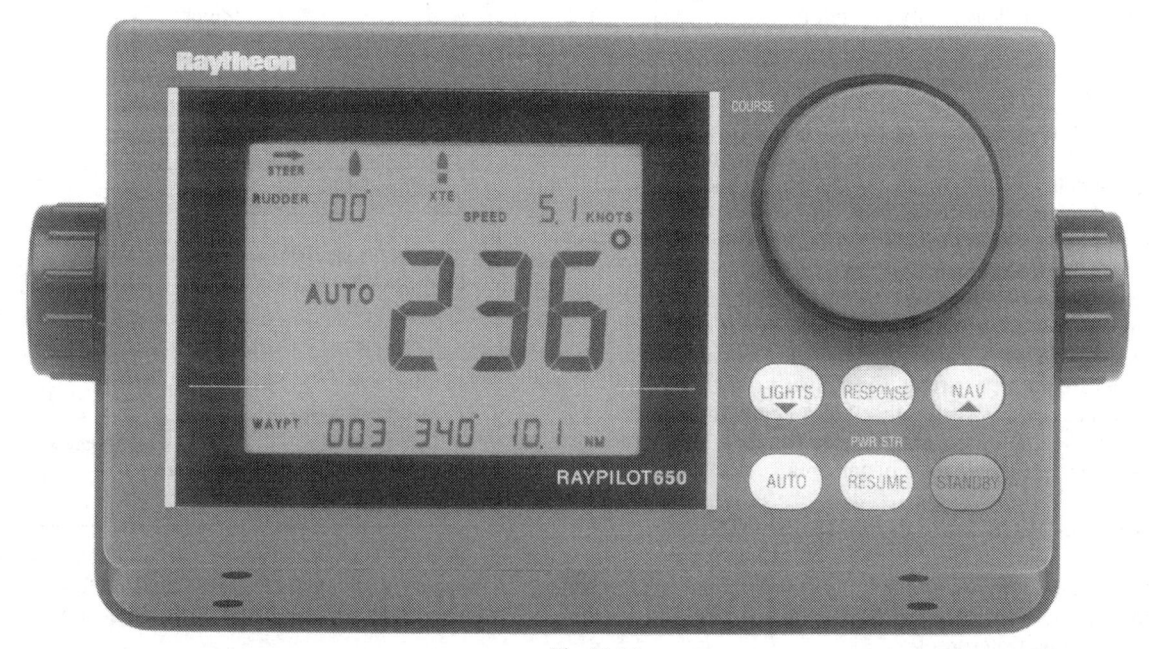

Fig 15.16

NAVIGATIONAL STEERING

An autopilot interfaced with GPS can be set to steer for a waypoint or to counteract a known cross track error (XTE). It would then continually correct the course to keep to its target. XTE is illustrated under GPS below.

AUTOPILOT UNIT

The "Course Indicator-cum-Control Panel" is fitted with a fluxgate or gyrocompass. It provides steering information to the hydraulic (or mechanical) drive unit connected to the rudder. (Steering systems are discussed in Chapter 31.)

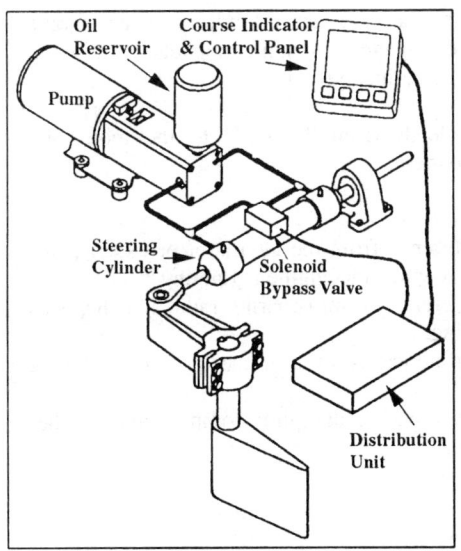

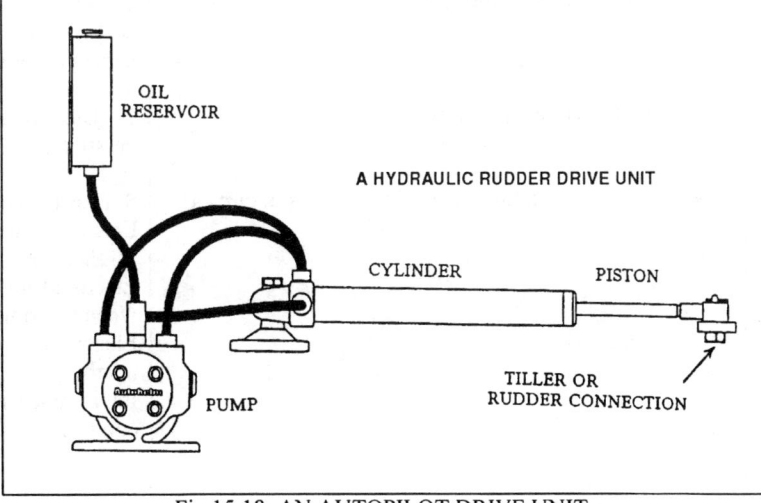

Fig 15.17: AN AUTOPILOT UNIT
(Silva Nexus)

Fig 15.18: AN AUTOPILOT DRIVE UNIT
(Autohelm)

PRECAUTIONS WHEN ON AUTOPILOT

- Always maintain a proper lookout.
- For large course alterations, the autopilot should be disengaged, the vessel brought on to the new course manually and then the autopilot re-engaged. An alteration of more than 180 degrees should never be applied by autopilot.
- Many autopilots use magnetic sensing elements. They are thus subject to magnetic influences of nearby wiring, radios, food cans and tools - just like a magnetic compass.
- Test manual steering when entering confined waters after prolonged use of autopilot.

WHEN NOT TO USE AUTOPILOT

- In very rough seas. Autopilots cannot anticipate and react to large waves to minimise deviating from course.
- In poor visibility. Emergency course alterations must be handled manually.
- In confined waterways.
- In close proximity of dangers.
- In areas of high traffic density

TROUBLESHOOTING AUTOPILOT

SYMPTOM	CAUSE
1. Vessel veers off to one side.	1. Insufficient counter-rudder. Increase the counter-rudder setting, which may also be marked as response, rudder, helm or ratio.
2. Vessel over-steers.	2. Excessive counter-rudder. Decrease the counter-rudder setting.
3. Autopilot reacts too late.	3. The sea-state or deadband setting is too high, which may or may not be adjustable. Also, make sure the steering gear itself is not defective, e.g.; slack cables, loose linkages or air in the hydraulic system.

4. The vessel veers off to leeward.	4. Increase bias setting, if fitted. It may also be marked as standing helm, weather helm or trim
5. Autopilot shows a lack of power in a sailing vessel.	5. Balance the sails, possibly sacrificing performance.
6. Autopilot operates sluggishly.	6. Low voltage at the drive motor; wheel brake accidentally left engaged; or something else is binding some part of the steering system.
7. Autopilot operates backwards.	7. The power leads from the CPU to the motor are crossed; reverse them.
8. Vessel turns to a different heading or turns in circles.	8. Possible causes: Electrical interference from the alternator or its regulator if it happens only when the engine is running. Electrical interference from a marine radio if it happens when transmitting. Electrical interference from a transistor radio playing nearby. Interference between the autopilot compass and another compass nearby.
9. Auto-pilot CPU trips off.	9. Possible causes: Voltage spike from the alternator due to poor condition of batteries. Solution: Connect the CPU leads to their own battery. Voltage-drop when starting the engine or using a winch. Solution: Increase the size of the leads to prevent voltage drop. Poor wiring connection, terminals and plugs.
10. Auto-pilot stops working.	10. CPU has developed an internal problem.

ALARMS, DETECTORS & INDICATORS
(For engine alarms, see Chapter 26; for fire alarms, Chapter 19.2)

Listed below are some of the alarms, detectors and indicators commonly fitted in vessels.

- SMOKE DETECTORS are discussed in Chapter 19.2.
- BILGE ALARM is set off by a given level of liquid in the engine room bilge. See illustration in Chapter 29.
- OFF-COURSE ALARM warns the watchkeeper in case autopilot veers beyond the set level of yawing. A similar alarm may also be fitted for keeping an eye on the helmsperson.
- WATCHKEEPING ALARM operates at regular preset intervals unless it is manually cancelled prior to its sounding. It is to ensure that a proper watch is being maintained.
- REFRIGERATION SPACE ALARM fitted outside but activated and cancelled only from within, in case someone accidentally gets locked in the refrigeration space.
- EQUIPMENT FAILURE ALARMS to indicate faults or failures in machinery and navigational equipment.
- NAVIGATION LIGHTS ALARM. Seagoing vessels in survey are required to carry two sets of navigation lights: main electric and emergency electric or battery-powered. The main lights are wired into visual and audible alarms to indicate any light failure. The watchkeeper can then immediately switch over to emergency lighting until the globe is changed or fault rectified.
- ENGINE ALARMS. See Chapter 26.
- LOUD VOICE ALARM INDICATORS: These systems use voice alarms to announce "Fire! Fire!", "Warning! Carbon monoxide", etc.

CHAPTER 15: QUESTIONS & ANSWERS

CHAPTER 15.1: AUTOPILOT

1. An Autopilot should not be used:

    (a) for making small alterations of courses.
    (b) in poor visibility.
    (c) for long coastal passages.
    (d) for ocean passages.

2. The term "Helm" on a vessel's autopilot control box may also be marked as:

    (i) Standing helm.
    (ii) Rudder.
    (iii) Ratio.
    (iv) Response.

    The correct answer is:
    (a) all except (i).
    (b) (ii) only.
    (c) (i) & (ii) only.
    (d) (ii) & (iv) only.

3. The term "Bias" on a vessel's autopilot control box may also be marked as:

    (i) Standing helm.
    (ii) Weather helm.
    (iii) Trim.
    (iv) Ratio.

    The correct answer is:
    (a) all except (i).
    (b) all except (ii).
    (c) all except (iii).
    (d) all except (iv).

4. In calm sea, when steaming on one engine in a twin-screw vessel, or with wind abeam in a sailing vessel, the following auto pilot setting will be inappropriate:

    (a) Helm - a high setting.
    (b) Yaw - a low setting.
    (c) Bias - a high setting.
    (d) Counter rudder - a low setting.

5. In moderate following seas, the following will be appropriate settings on a vessel's autopilot:

    (i) Helm - a high setting.
    (ii) Yaw - a low setting.
    (iii) Bias - a low setting.
    (iv) Counter rudder - a higher setting.

    The correct answer is:
    (a) all except (ii).
    (b) all of them.
    (c) all except (iv).
    (d) all except (iii).

6. It is considered appropriate to use a vessel's automatic steering device (auto pilot) in order to:

    (a) reduced the need for lookout.
    (b) make small course alterations.
    (c) improve steering in tide rips and overfalls.
    (d) improve steering in heavy seas.

7. A vessel's auto pilot steering device is best suited for use in:

    (a) confined waterways.
    (b) the watchkeeper's need to take a nap.
    (c) poor visibility.
    (d) open seas.

CHAPTER 15.2: LOGS & ECHO SOUNDERS

8. Speed calculation from a vessel's RPM table:

    (a) is as good as that from a bottom log.
    (b) applies only to a given draft & trim.
    (c) is independent of weather.
    (d) is speed over the ground.

9. At 1900 hours a vessel's log read 13497.8. At 1924 she altered course and the log read 13503.2. She travelled at the following speed:

    (a) 13.5 knots.
    (b) 12.5 knots.
    (c) 10.2 knots.
    (d) 2.16 knots.

10. The advantage of an electromagnetic log is that it:

    (a) measures speed over the ground.
    (b) does not affect vessel's compass.
    (c) measures vessel's drift.
    (d) has no moving parts exposed to the sea.

11. The log that is able to measure speed over the ground is a:

    (a) bottom log.
    (b) electromagnetic log.
    (c) doppler log.
    (d) pressure log.

12. The RPM reading on a vessel's tachometer works out to a speed of 8 knots. The vessel estimates that the following sea is giving her half a knot while the bottom growth is costing her about a knot. The chart shows a tidal stream of 2 knots working against her. The vessel's estimated speed over the ground is:

    (a) 6.5 knots.
    (b) 5.5 knots.
    (c) 8 knots.
    (d) 10 knots.

13. At 0800 hours a vessel sets a course of 130 degrees, log reading 124.8, speed 10 knots. At 0900 hours her log reads 136.2. She has covered the following distance at the following speed:

    (a) 11.4 miles, 11.4 knots.
    (b) 11.0 miles, 11.4 knots.
    (c) 11.4 miles, 11.0 knots.
    (d) 11.0 miles, 11.0 knots.

14. In a vessel with planing hull, the transducer of an echo-sounder should be fitted:

    (a) one third the hull length from the bow.
    (b) away from the keel centreline.
    (c) inside the hull.
    (d) near the stern.

15. An echo sounder tends to return multiple echoes of the seabed when operating in:

    (a) shallow waters.
    (b) deep waters.
    (c) muddy waters.
    (d) rough seas.

16. The following factor does not contribute to false echoes in an echo sounder:

    (a) An underwater current.
    (b) An echo returning from a great depth.
    (c) Excessive marine growth on hull.
    (d) Sand suspended in water.

17. A strong return in an echo sounder trace indicates the following type of seabed:

    (a) rock.
    (b) firm sand.
    (c) soft mud.
    (d) weed.

18. You switch on the echo sounder at sea. The depth selection is on 0-30 metres. It is operating correctly but no bottom trace is showing. You should:

    (a) try it later.
    (b) switch to another scale and try it.
    (c) reduce vessel's speed and try it.
    (d) alter vessel's trim and try it.

19. Your sounder is giving a good bottom trace but is "painting" a dark interference over the rest of the paper. This is due to:

    (a) vessel's speed.
    (b) bad siting of the transducer.
    (c) marine grown on the transducer's face.
    (d) excessive gain setting.

20. A vessel draws 4 metres. The transducer is fitted on the lowest part of the keel. The echo sounder shows a depth of 10 metres. Its zero point has not been adjusted for draught. The depth of water is as follows:

    (a) 6 metres.
    (b) 10 metres.
    (c) 14 metres.
    (d) cannot be estimated without tidal information.

21. The navigational usage of an echo sounder include:

    (i) combining a sounding with a bearing.
    (ii) combining a sounding with a transit.
    (iii) combining two soundings.
    (iv) using a line of soundings.

    The correct answer is:
    (a) all except (i).
    (b) all except (ii).
    (c) all except (iii)
    (d) all except (iv).

22. At your vessel's assumed position the chart reads a depth of 10 metres. But the echo sounder does not show a sounding on the 0-30 metre scale. The reasons are:

    (i) the receiver gain is too low.
    (ii) the depth of water is greater than 30 metres.
    (iii) the depth is extremely shallow.

    The correct answer is:
    (a) all of them.
    (b) none of them.
    (c) all except (iii).
    (d) all except (ii).

23. Rough sea makes the echo sounder trace:

    (i) broken.
    (ii) weaker.
    (iii) uneven.
    (iv) appear with tails.

    The correct answer is:
    (a) all except (i).
    (b) all except (ii).
    (c) all except (iii).
    (d) all except (iv).

24. Tails below a depth sounder's trace become larger when:

    (a) sea is rough.
    (b) vessel is over a hard seabed.
    (c) vessel is over seaweed.
    (d) vessel is in shallow water.

25. An extra echo line on a depth sounder's trace indicates:

    (a) a second trace echo.
    (b) soft sea bed.
    (c) deep water.
    (d) rough sea.

26. The second-trace returns on the echo sounder display are caused by the previous pulse returning from:

    (a) a deep seabed.
    (b) a shallow depth.
    (c) the transducer facing forward.
    (d) the transducer facing aft.

27. The ceramic transducer in echo sounders transmits:

    (a) DC generated radio waves.
    (b) AC generated electrical waves.
    (c) DC generated sound waves.
    (d) AC generated sound waves.

28. The rotating impeller of the impellor log generates voltage in direct proportion to:

    (a) vessel's speed over the ground.
    (b) distance between the impeller & recorder.
    (c) the speed of sound.
    (d) the speed of the water.

29. The problem of double echoes appearing on an echo sounder can be resolved by:

    (a) adjusting Gain control.
    (b) changing the scale.
    (c) Removing 'White Line'.
    (d) Using 'Phase range'.

CHAPTER 15.3: ALARMS

30. The specification for the audible alarm indicating failure of a navigation light is to:

    (a) be audible throughout the vessel.
    (b) be of the two-tone type.
    (c) be of one minute duration.
    (d) continue sounding until cancelled by the watchkeeper.

31. On hearing the navigation light failure alarm, the watchkeeper's first action should be to:

    (a) switch over to the emergency power supply.
    (b) check the fuse.
    (c) replace the light bulb.
    (d) switch on the NUC lights.

CHAPTER 15: ANSWERS

1 (b),   2 (a),   3 (d),   4 (a),   5 (a),
6 (b),   7 (d),   8 (b),   9 (a),   10 (d),
11 (c), 12 (b), 13 (a), 14 (d), 15 (a),
16 (c), 17 (a), 18 (b), 19 (d), 20 (c),
21 (c), 22 (a), 23 (d), 24 (b), 25 (a),
26 (b), 27 (c), 28 (d), 29 (a), 30 (d),
31 (a),

Chapter 16

RADAR, SART & OTHER RADAR TRANSPONDERS

RADAR

Scanner Unit (Contains Pulse Generator, Transmitter, T/R Switch, Wave Guide

Display

Power supply, normally 12V, 24V

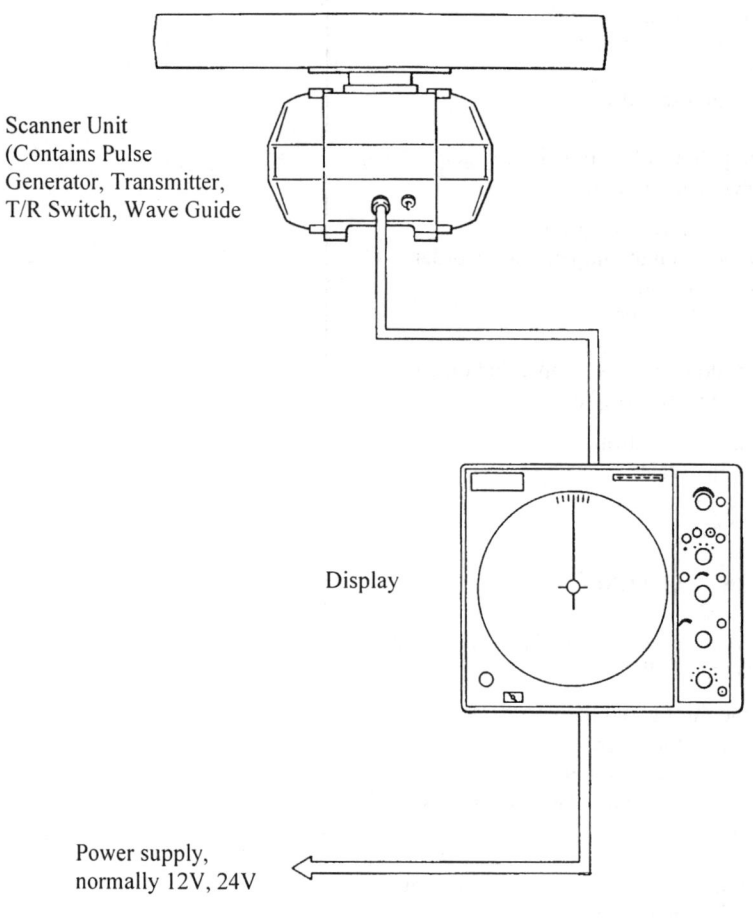

Fig 16.1

Chapter 16: Radar, SART & other Radar Transponders

RADAR = RAdio Direction And Ranging.

A Radar set is a transmitter and receiver of radio energy, alternating between the two functions at intervals of microseconds.

It sends out short pulses of high frequency radio waves, and receives them when they return after being reflected from distant radar-reflective objects. It measures the direction and distance of the objects, by relating to the antenna bearing, and the time taken by the pulses to travel to and from the reflecting object. It thus provides the operator with a bird's eye view of where the targets are in relation to the vessel, even when they are invisible to the naked eye due to darkness or poor visibility.

PARTS OF RADAR (See Block Diagram)

➢ **T/R SWITCH**: A Transmit/Receive Switch, also known as the T/R Cell or the Duplexer, blocks off the receiver during transmission. It then opens the receiver for incoming echoes. The T/R cell is fitted in the scanner unit at the junction of the two wave-guides - from the transmitter and receiver.

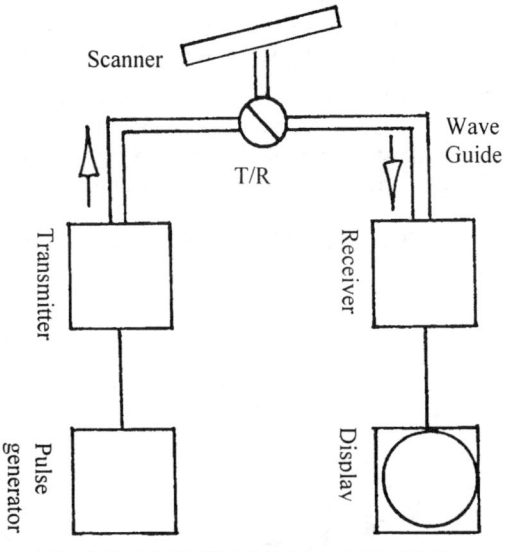

Fig 16.2: BLOCK DIAGRAM OF RADAR

➢ **WAVEGUIDE**: A radar wave-guide is a rectangular tube through which the pulses of radio frequency are piped from the transmitter to the scanner, and the reflected pulses from the scanner to the receiver. The T/R switch is fitted at the junction of the two tubes at the scanner. In modern radar sets for small vessels, the transmitter and receiver are usually inside the scanner unit. The wave-guide is therefore very short and hidden.

TYPES OF RADARS (FREQUENCY RANGE)

Most marine radars are of the following two types:

1. X-BAND or I-BAND RADAR (also known as the 3 cm radar):

Frequency:	10,000 MHz range: Wavelength 3 cm. (Often called the anti-collision frequency).
Advantages:	Better picture at shorter ranges.
	Needs shorter scanner.
Disadvantages:	Poorer performance in rain. [10 cm wavelength suffers less attenuation (loss in rain, etc.)].
	Produces more sidelobes than the 10 cm wavelength.
Suitability:	Suitable for small vessels & for better definition on shorter ranges.

2. S-BAND or EF-BAND RADAR (also known as the 10 cm radar):

Frequency:	3000 MHz range: Wavelength 10 cm (Long range surveillance frequency).
Advantages:	Better picture at longer ranges.
	Better performance in rain.
Disadvantages:	Uses greater power.
	Uses a very long scanner.
Suitability:	Suitable for large, more stable vessels & for better definition on longer ranges.

TYPES OF RADAR SCANNERS AND BEAMWIDTH

When focused into a narrow beam, light travels a longer distance and is brighter. Similarly, a radio pulse travels further when transmitted in a narrow beam. Inside the fibreglass cover of a radar scanner is a slotted wave-guide designed to focus the radar transmission into a beam with a **HORIZONTAL BEAMWIDTH** of approximately 2° and a **VERTICAL BEAMWIDTH** of 20° to 30°. The vertical beamwidth is greater so that the targets are not lost during boat's rolling.

Longer scanners have more slots than the shorter ones. Their horizontal beamwidth is smaller (narrower) and more directional. They can thus radiate energy much further and provide a greater operating range. Horizontal beamwidth is also a measurement of the antenna's target discrimination. The smaller it is, the better the radar is able to separate close targets and provide more accurate bearings, i.e., better **BEARING DISCRIMINATION**. See also bearing errors. The horizontal beamwidth of a 60 cm radome is about 7°, of a 80 cm open array scanner it is about 2.5°, and of a 120 cm scanner it is 1.2°.

Wider beamwidth will make objects appear distorted (wider). While this may help in detecting buoys and small vessels, entrances to harbours and rivers would be more difficult to locate. Beamwidth distortion can be reduced by reducing the gain, as long as you remember to turn it up again.

RADOME SCANNERS: Unlike powerboats, the scanners in sailing vessels are usually covered in a plastic dome, known as RADOME. It eliminates the possibility of the rigging and sails fouling the rotating scanner. The fibreglass dome also keeps moisture and salt out and away from critical components. But, the scanners inside are usually of shorter length than the open array scanners. A shorter scanner means larger beamwidth, and thus shorter range and less target detail.

PATCH SCANNERS: This, more modern scanner, employs an array of copper pads on a printed circuit to focus the beam.

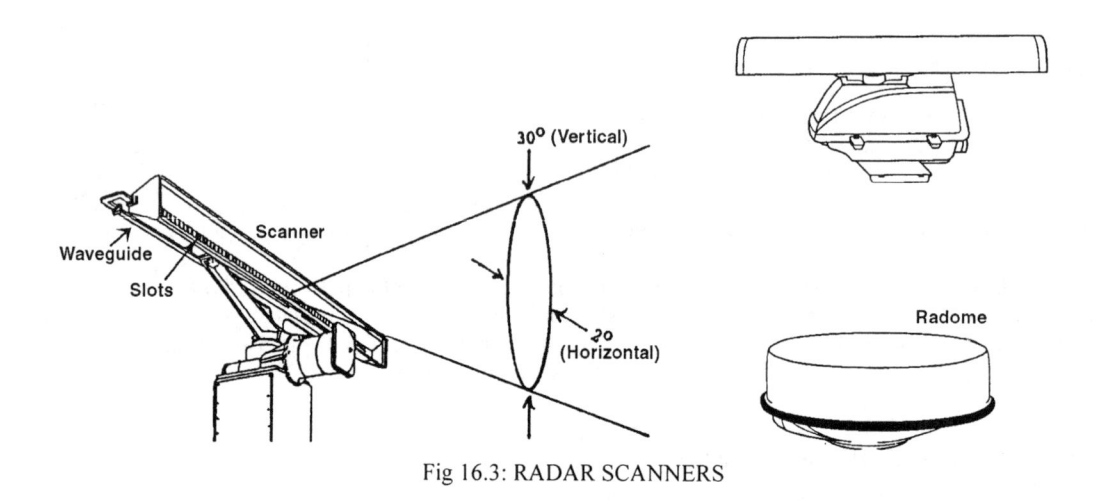

Fig 16.3: RADAR SCANNERS

TYPES OF RADAR DISPLAYS

Fig 16.4: TYPES OF RADAR DISPLAYS

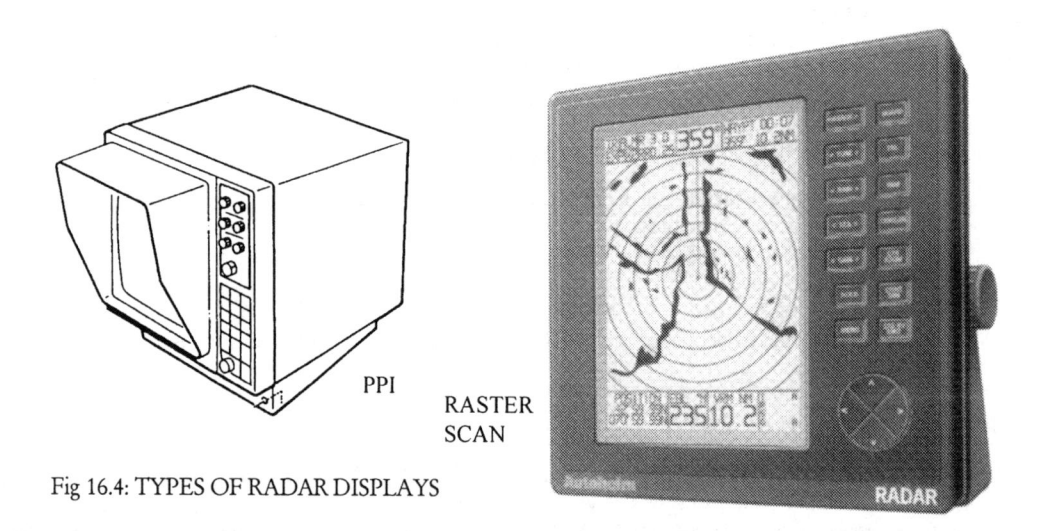

1. PLAN POSITION INDICATOR (PPI)

This is the conventional circular screen with a bright rotating radar beam (known as the sweep or trace) synchronised with the scanner. It consists of a cathode ray tube (CRT), similar to a TV set. This type of display offers high contrast in normal or low light conditions, making the targets appear bright and clear. But the screen fades away in bright sunlight and is not usually suitable for outdoors.

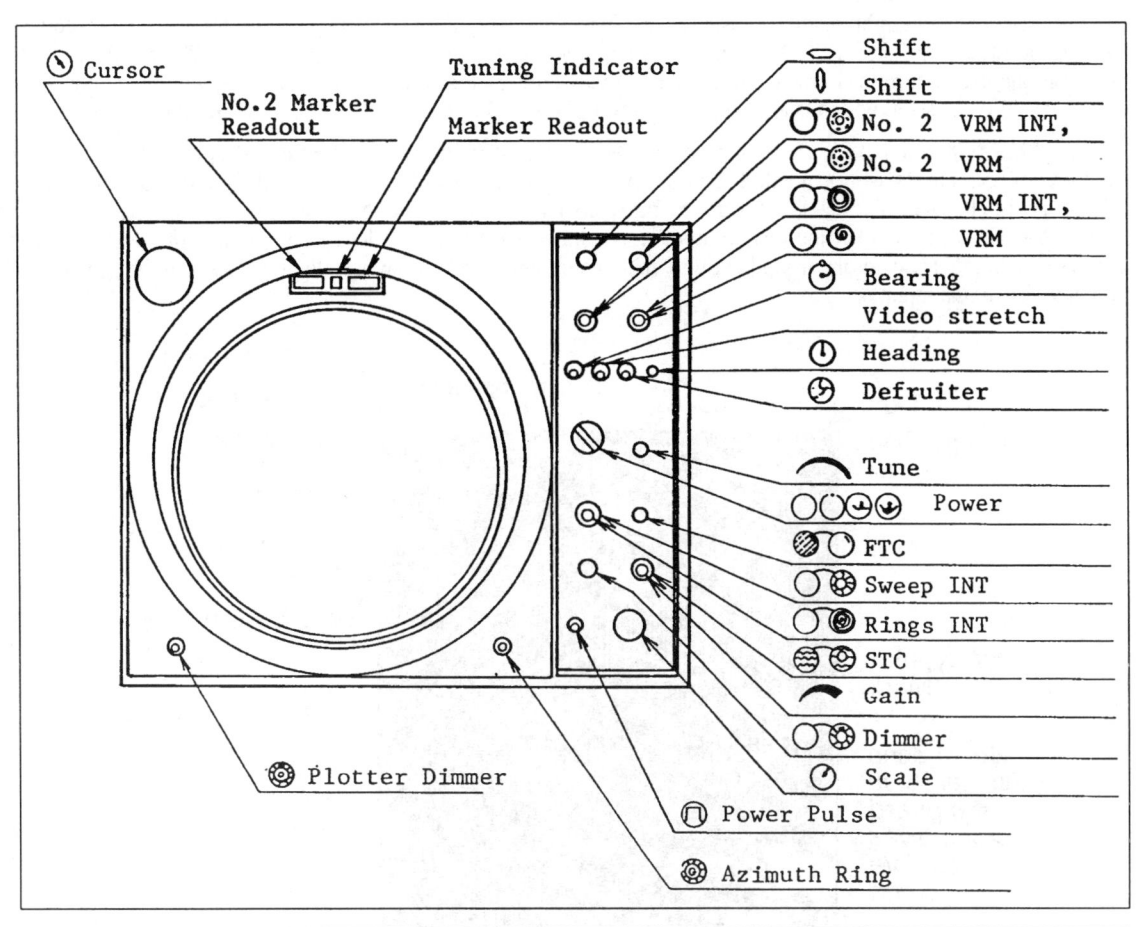

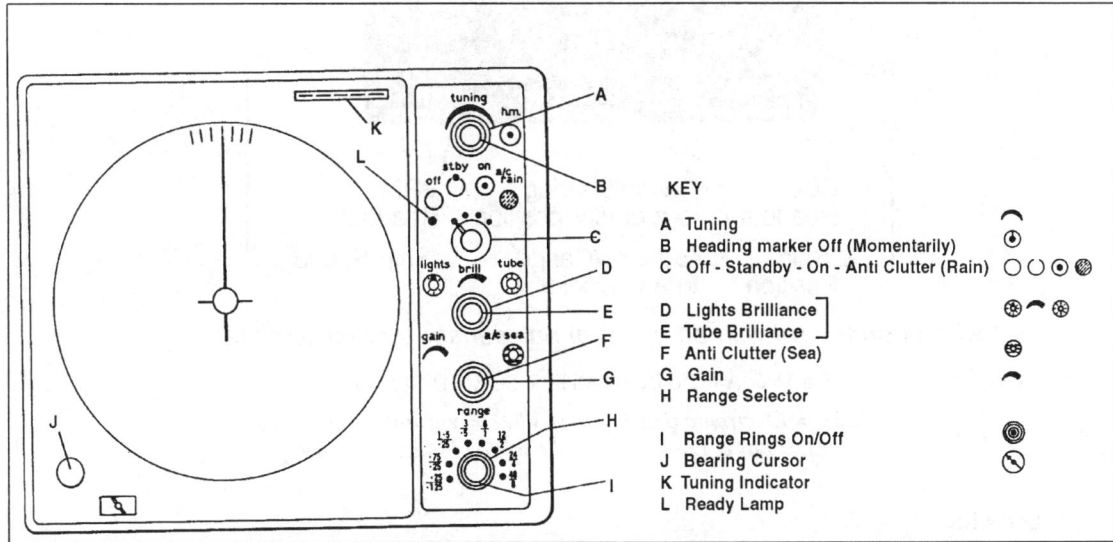

Chapter 16: Radar, SART & other Radar Transponders

Fig 16.5: TWO EXAMPLES OF TRADITIONAL PPI DISPLAYS

2. **RASTER SCAN DISPLAY (RSD)**

This is a rectangular screen, which looks like a TV screen. Instead of a rotating scanner "refreshing" the presentation on every sweep, the echoes are digitally processed by integrated circuits and the information appears on the screen much like a TV set. The presentation is "refreshed" more often and is easier to use. It can also be temporarily frozen for more accurate measurement of range and bearings.

The so-called "daylight reading" raster-scan radar uses a liquid crystal display (LCD). It is more compact and usually waterproof. Being flat, an LCD screen offers larger viewable area than its equivalent CRT screen. It is also more economical in the use of current. In most cases the LCD viewability increases in direct sunlight. And the contrast increases when it is back-lit or front-lit, which is also necessary for night use. But it lacks the level of detail of a PPI screen.

The AMSA Marine Notice 19/1994 reported two problems with some Raster Scan Displays (RSD): loss of video input and loss of azimuth (bearing) signal. The loss of video input can result in freezing of the picture, an effect not noticed until the range is changed. The cause appears to be related to the fact that the screen image is generated by a video processor and if the signal is lost, the display does not redraw or refresh. The loss of azimuth signal leads to rotation of targets or targets being depicted on wrong bearings. In some cases, the RSD may not display an alarm, or indicate that there is a problem with signal input.

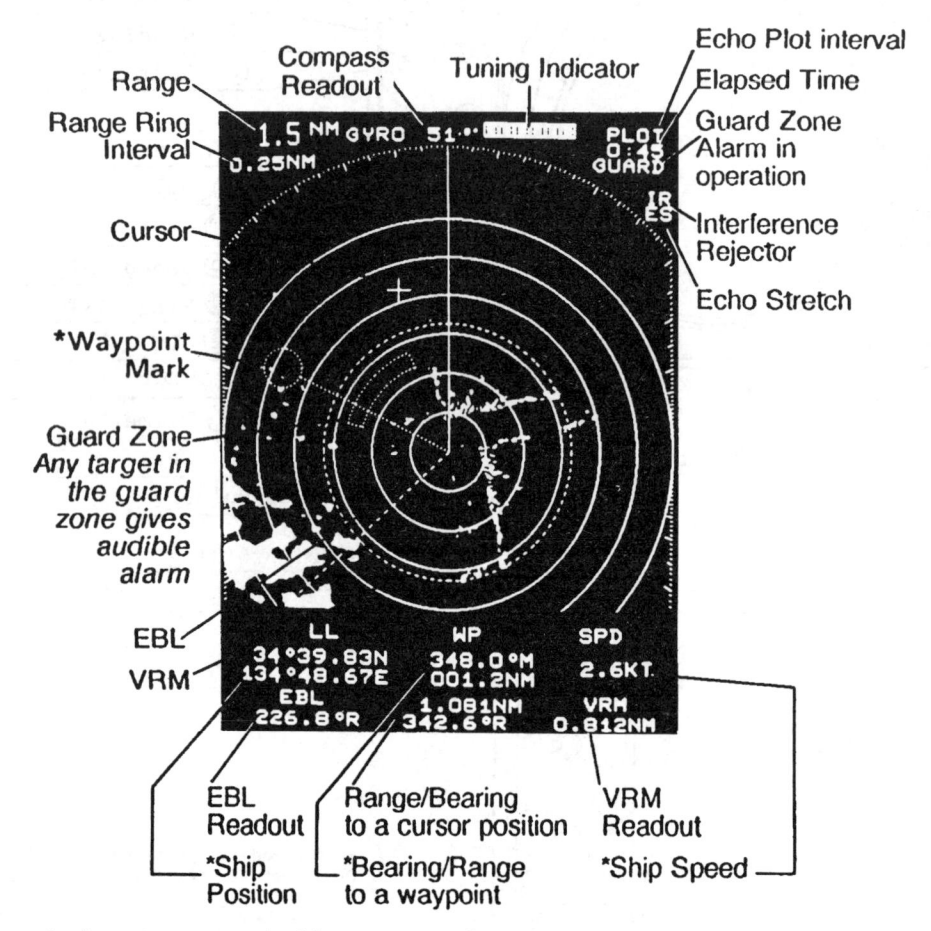

when connected with an external nav sensor outputting data

Fig 16.6: A FURUNO RASTER-SCAN DISPLAY

(EBL = Electronic Bearing Line   VRM = Variable Range Marker
LL = Lat & Long   WP = Way Point   SP = Speed)

COLOUR RADARS

In colour radars, different colours are assigned to echoes of various strengths - thus distinguishing various types of targets. Sometimes colours are also assigned according to the functions. For example, range rings may be white, targets yellow, and heading marker blue. The operator can change them to suit day or night viewing. Blue background for daytime

Chapter 16: Radar, SART & other Radar Transponders

and black for night time is common.

The safety risk associated with colour radar is that the observer may tend to pay more attention to red echoes without considering the possibility of risk of collision with, say, a yellow echo.

GUARD ALARM (OR GUARD ZONE ALARM)

A guard zone alarm alerts the observer when the radar detects a target entering a designated area or when own vessel approaches a dangerous area. The guard zone can be set for a specified distance in a given direction or all around the vessel. The observer needs to be aware that the radar will only raise an alarm if it is able to detect the target. Therefore, reliance on a guard zone alarm can lead to a collision with an undetected target.

RADAR INSTALLATION

- Place scanner at a practical height and where there are minimum shadow and blind sectors. It should be high enough to give a good radar horizon, but not too high to cause stability problems on a boat. You should also be able to take it down for servicing.
- Install away from compass and the autopilot sensor. The scanner & transceiver should be at least 2 metres away & the display at least 1 metre away from a fixed compass. The spare or old magnetron should not be stored near these units. Study the changes in compass error with the radar switched on and off.
- People should stay at least 1 metre away from the low powered scanners and 3 metres from the high-powered scanners, when the Radar is operating. There is a microwave radiation hazard, and the scanner should not be fitted where people can look directly into it. Microwave radiation can damage eyes seriously because of their high liquid content.
- Radar circuits develop thousands of volts of electrical potential. The unit should be properly protected, marked and opened only by qualified technicians.
- Place the display under cover & out of direct sunlight. Protect it from spray & rain. Install it near steering position, facing aft, such that the observer faces forward when viewing it. Light issuing from the display unit should not interfere with the night lookout. With the visor hood removed (on a CRT display), two observers should be able to view the display simultaneously.
- In order to avoid electrical interference, connect the radar set with separate cables direct from batteries. Lead these cables well clear of radio aerials and other electrics. Obtain the right lengths of custom cables; do not cut them to size. Place the alternator where its noise and vibration will not cause inconvenience.
- After installation, check the range and heading marker accuracy as per the manufacturer's instructions. Make a chart of any shadow or blind sectors, and display it near the radar.

Some of the indications of radar interference on other equipment are:

- Spurious flashes of light on echo sounder.
- A herringbone pattern on the echo sounder recording.
- Whining noise on radio.
- Fluctuations in compass error.

CARE & MAINTENANCE

Other than keeping the screen clean with a soft cloth and anti-static solution, all the on-board maintenance is restricted to the scanner & radome, as follows:

- Salt and dirt deposits can reduce scanner efficiency. It should be washed three or four times a year, using a light soap solution and a soft cloth.
- Check scanner and radome for cracks. Seal them as a temporary repair. Consult a technician as soon as possible.
- Consult the Operators' Manual regarding greasing the direct gear drive or tightening or replacing the belt drive (whichever is fitted) on the scanner motor.
- Check and renew the scanner motor brushes, if necessary.
- Paint the scanner if required. But don't paint the scanner aperture or radome.

IMPORTANT: Radar contains high-tension circuits. Switch off main switch and remove fuses before carrying out any of the above maintenance.

PERFORMANCE MONITOR

There are no targets on the radar screen. Is it because there are none around or is the radar not detecting them? The efficiency of radar can be checked at any time with an external or a built-in performance monitor. In its simplest form it is a metal echo box mounted behind the scanner. When switched on, the external type shows a narrow plume towards its direction on the screen. The length of the plume is compared with what it was at the time of installation of the set. A shorter than the original plume indicates a deterioration or fault in transmitted power, receiver sensitivity and/or tuning. The built-in type produces a circular pattern instead of a plume. But, it is used in the same manner.

OPERATING MODES

1. RELATIVE MOTION UNSTABILISED DISPLAY

Most radars on small vessels are very basic Relative Motion Unstabilised Displays. A Relative Motion radar shows the picture and not the vessel in motion. The Unstabilised Display is a radar screen without signals from a compass. It does not know the direction in which the vessel is heading. Its heading marker is therefore set permanently to shows 000 degrees relative. When vessel's course is altered, the heading marker does not alter. Instead, all the echoes rotate in the opposite direction by the amount of the course alteration.

2. RELATIVE MOTION STABILIZED DISPLAY

More expensive radars provide for compass heading input. The heading marker of such a radar therefore indicates the vessel's compass heading. The radar is then called **STABILIZED**. Since it is the picture and not the vessel that is seen to move, the display is still Relative Motion. However, the navigator can choose to operate Stabilized radar in the **HEAD-UP** or **NORTH-UP** mode, as shown below.

In the Head-Up mode, the picture is orientated so that the heading marker appears at the top of the screen. This mode is most suitable for navigating in congested waters and narrow channels. In the North-Up mode the radar picture is stabilized so that North is at the top of the screen, and the heading marker wanders according to the orientation of the vessel's heading.

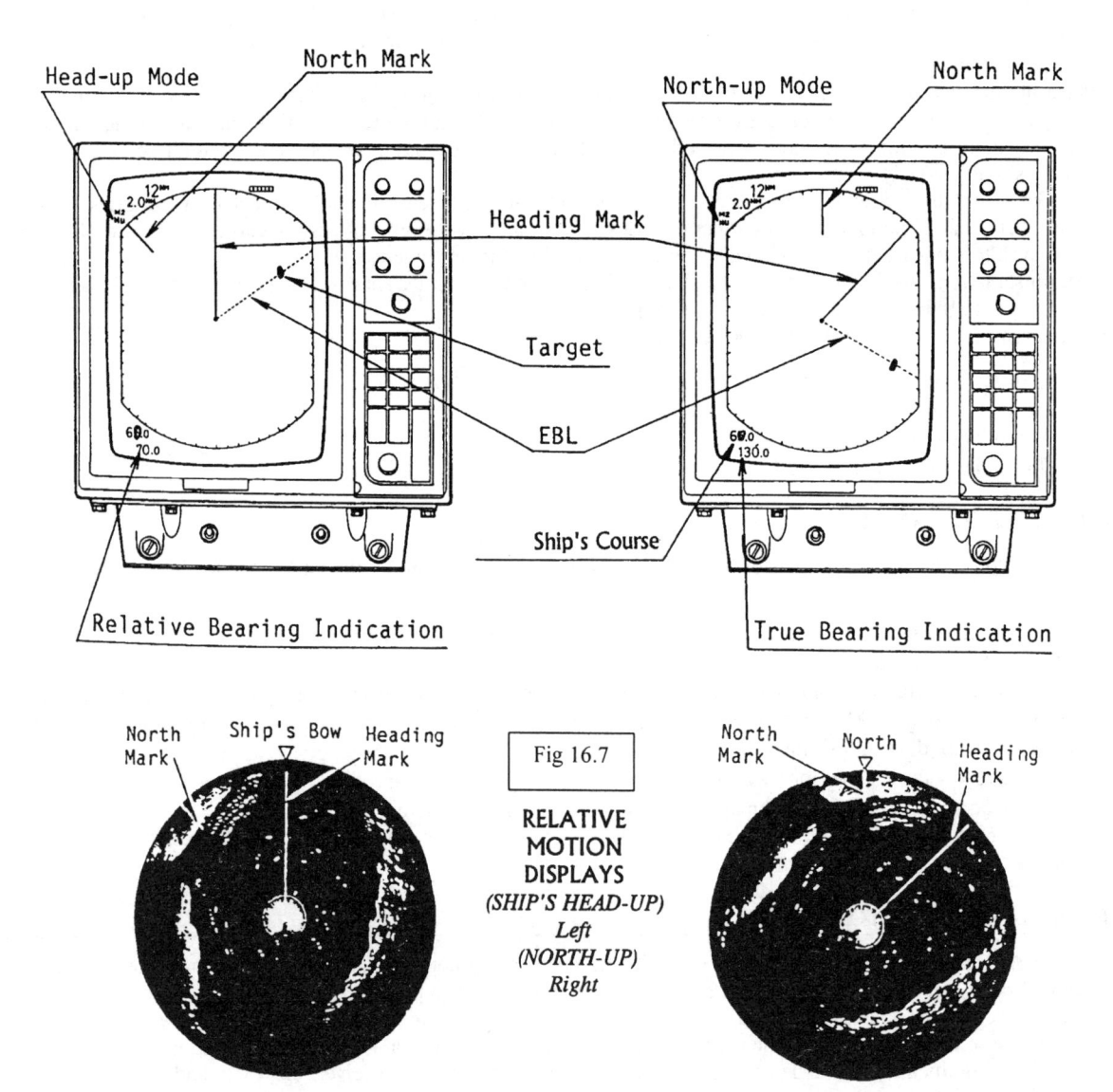

Fig 16.7

RELATIVE MOTION DISPLAYS
(SHIP'S HEAD-UP)
Left
(NORTH-UP)
Right

Chapter 16: Radar, SART & other Radar Transponders

3. TRUE MOTION STABILIZED DISPLAY

The more expensive stabilized displays can also be switched to a True Motion mode, so that the vessel is no longer a fixed spot in the centre of the screen. The spot can be made to start at any point on the screen - usually at the bottom. It then moves on the screen in TRUE MOTION. The vessel is then seen passing the coastline, instead of the coastline going past it. This display requires an input from a compass, and from the boat's log.

True motion radars can be "*Sea Stabilized*" or "*Ground Stabilized*". The latter is fed with the course and speed, and the direction and rate of tide.

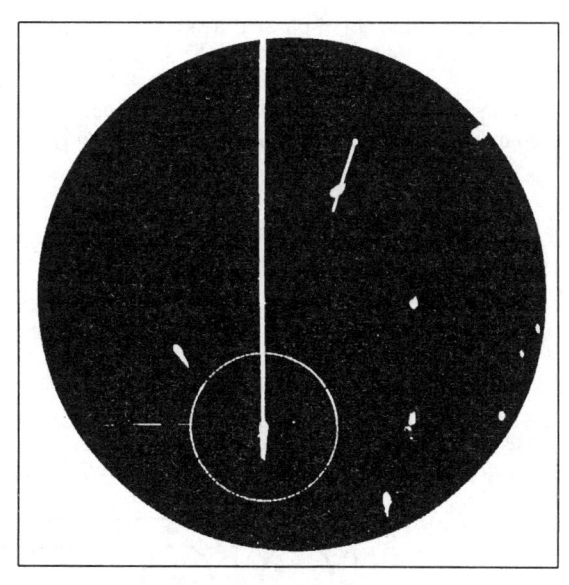

Fig 16.8: A TRUE MOTION DISPLAY

RADAR OPERATION & ITS LIMITATIONS

A target's ability to appear on a radar screen depends on four factors:

1. Type of target
2. Weather conditions
3. Settings of radar controls
4. Siting of the radar antenna.

1. THE TARGET

The size and strength of a target's echo depends on its:

(A) Size
(B) Shape
(C) Composition
(D) Aspect

(A) SIZE
Radar's ability to 'see' is no different to that of the human eye. The target that presents a large surface area and is high is detected more easily and at longer range. A low-lying small object may not be detected at all or only when it is close by.

(B) SHAPE
Objects with rough surfaces, such as rocky outcrops, are better reflectors of radar energy than those with smooth surfaces, such as ships' hulls.

(C) COMPOSITION

Metal objects are better radar reflectors than wooden objects. GRP objects reflect little or no radar energy. GRP is almost transparent to radar signals.

GRP vessels are 'picked up' by radar mainly due to the reflective ability of their masts, booms, engines, winches and other metal reflectors. These numerous but small reflectors do not act as one large reflector. On the contrary, their close proximity can subject the signal to multiple reflections, making its return out of phase with the scanner. The echo of such a vessel on the radar screen may not be constant, and a very small change in the vessel's distance from the radar antenna may make the difference between it being 'OUT OF PHASE' or 'IN PHASE'.

A radar signal from a small vessel can also become out of phase if it skips or bounces off the sea surface.

(D) ASPECT

A target that is beam-on to the radar transmission is likely to generate a stronger echo than a target lying at an angle of 45° to the transmission. Therefore, a radar pulse reflecting off a ship's beam is likely to return a stronger echo than from a ship's bow.

2. WEATHER CONDITIONS

Waves and rain showers generate unwanted radar echoes. The resulting Sea and Rain Clutter on the radar screen often hide the weak echoes of small vessels. Sea Clutter, Rain Clutter and other meteorological phenomenon are discussed below.

3. RADAR CONTROLS

(Some radar manufacturers do not use these IMO defined control symbols)
The controls that determine picture quality are: Brilliance, Gain, Range, Tuning, Sea Clutter and Rain Clutter

1. Radar Off
2. Radar On
3. Radar Stand-by
4. Scanner rotating
5. North-up Presentation
   (Radars with compass input)
6. Head-up Presentation
7. Heading Marker Alignment
8. Range selector
9. Short Pulse
10. Long Pulse
11. Tuning
12. Gain
13. Anti-clutter Rain (MIN)
14. Anti-clutter Rain (MAX)
15. Anti-clutter Sea (MIN)
16. Anti-clutter Sea (MAX)
17. Scale Illumination
18. Display Brilliance
19. Range Rings Brilliance
20. Variable Range Marker
21. Bearing Marker
22. Transmit Power Monitor
23. Transmit/Receive Monitor

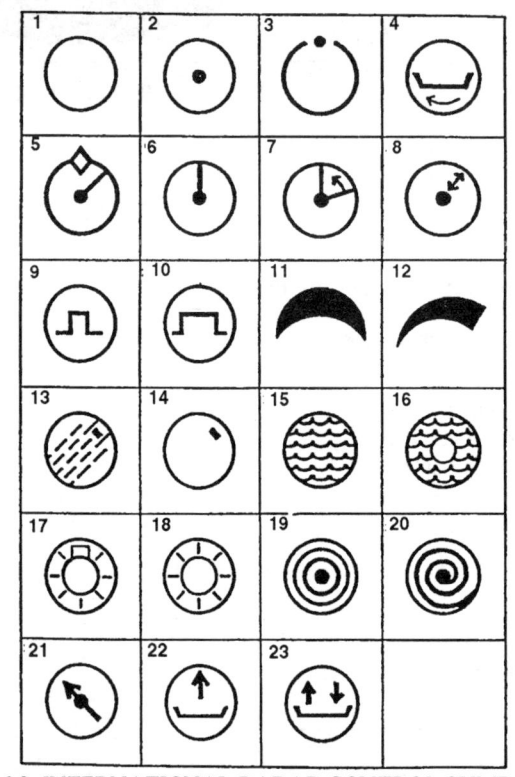

Fig 16.9: INTERNATIONAL RADAR CONTROL SYMBOLS

Chapter 16: Radar, SART & other Radar Transponders

BRILLIANCE CONTROL (Fig 16.10)

This is similar to the *Brilliance* or *Brightness* Control on a TV set. It allows the adjustment of brightness of the display, providing a *contrast* between the picture and its background. In practice, a low level brightness is better. Excessive brightness obscures echoes of small targets.

When switching on radar, turn the Brilliance up until the heading marker and the rotating trace are just visible. With a Raster-scan set, turn up brilliance until a bright picture is obtained. It can later be adjusted to suit the target contrast from time to time.

Brilliance should be turned off before switching the set off or on. This avoids screen damage when switching on the set.

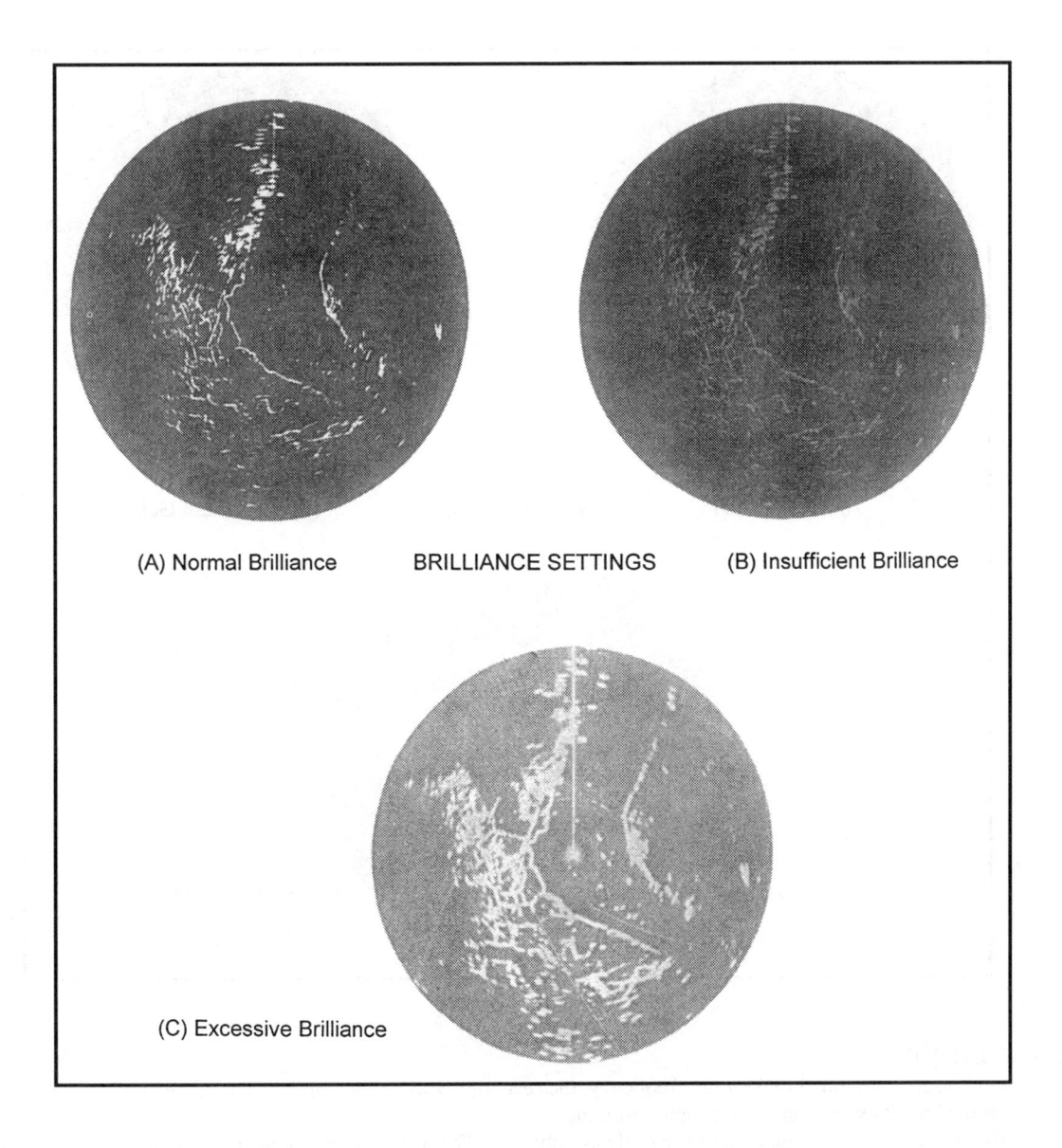

(A) Normal Brilliance   BRILLIANCE SETTINGS   (B) Insufficient Brilliance

(C) Excessive Brilliance

GAIN CONTROL (Fig 16.11)

This is similar to the *Volume* control on a radio receiver. If set too high, the reception will be "noisy". If too low, it will not be audible. Full Gain on radar will cover the screen with receiver "noise". It will turn it almost white with speckles. Zero gain will leave it black and heading marker faint. In either case the echoes will not be seen.

The correct Gain setting on a raster-scan may be obtained by turning up the Gain control until speckles ("noise") start to appear on the screen. Then, turn it back to where it's clear again. With a PPI, some speckle should be left on the screen.

A radar operating at a lower range scale usually needs less gain than on a higher scale. Accordingly, the Gain control may need to be adjusted whenever the set is switched from one range to another. Gain may also need to be reduced for clearer definition when taking bearings of islands, etc. If so, it should be restored to the higher setting after taking the bearings. Gain should be turned off before switching the set off or on. This avoids screen damage when switching on the set.

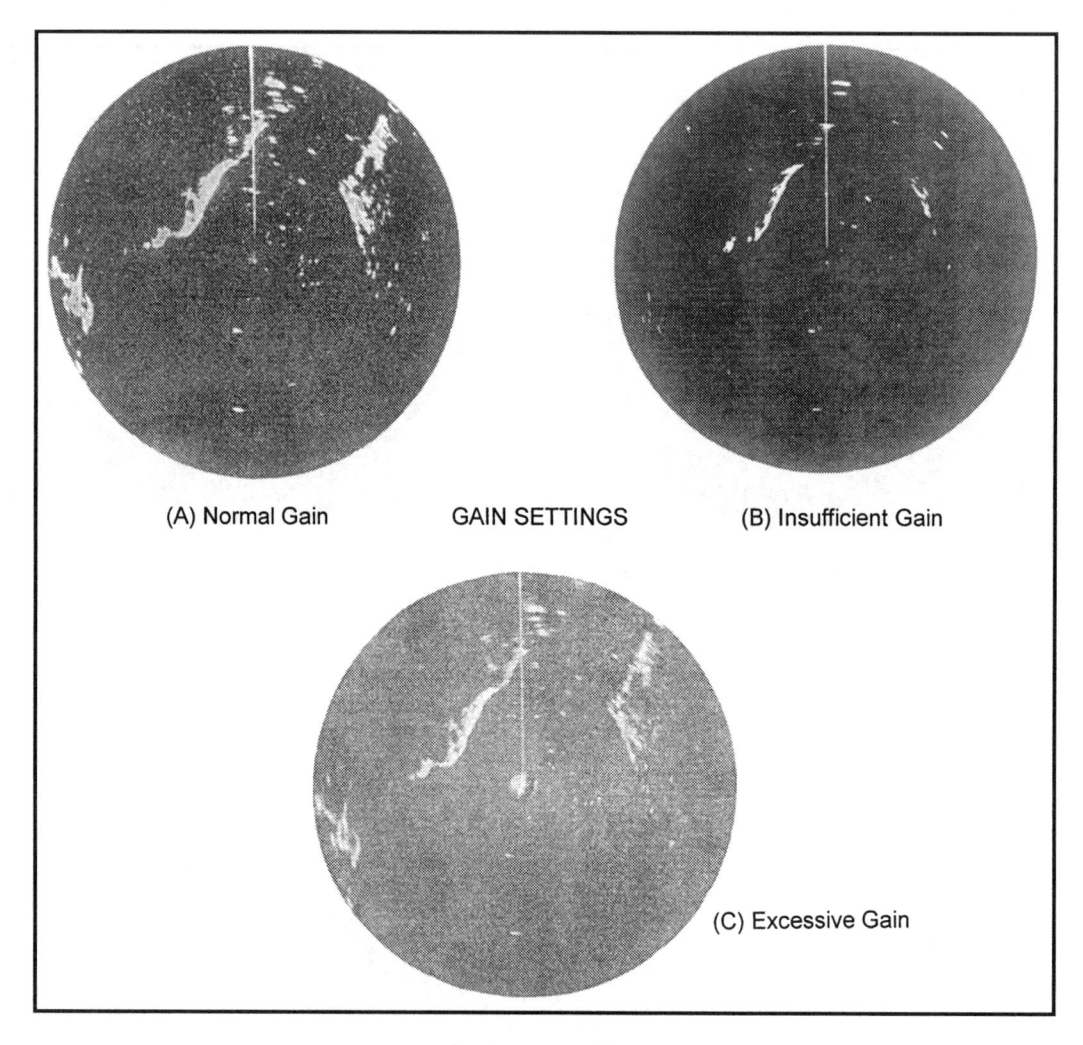

(A) Normal Gain   GAIN SETTINGS   (B) Insufficient Gain

(C) Excessive Gain

TUNING CONTROL

If a radio set is not tuned to the right frequency, the reception will be incomplete. The same applies to radar. Once the set has warmed up, it usually requires a slight re-tuning.

Unlike a radio set, out-of-tune radar is not always obvious. The picture remains clear, but some of the weak echoes are lost. The best way to check that the set is properly tuned is to slightly turn the control back and forth until the intensity of a distant echo is at its maximum.

When first switching the radar, it should be tuned to get the maximum reading on the tuning indicator and the maximum sea clutter. In the absence of other targets, the Sea Clutter is the best guide for tuning the set. It should be set to maximum at a short range and then switched to long range, if necessary. Sea Clutter is discussed below.

On modern radar sets, there is a tuning meter. The set should be tuned for the maximum reading on the meter. Some radars

Chapter 16: Radar, SART & other Radar Transponders

have automatic tuning.

SEA CLUTTER CONTROL (Fig 16.12 & 16.13)

Millions of radar echoes of the waves and seas around the vessel create a milky patch in the centre of the screen. This is known as Sea Clutter. The rougher the sea, the larger and denser the clutter. It may hide targets on radar for about 4 miles around the vessel.

An Anti-Clutter Control switch is provided to suppress these unwanted echoes. It reduces the receiver sensitivity (GAIN) of echoes from nearby targets without affecting the Gain of echoes of the more distant objects.

This control should not be used indiscriminately. If set too high, it will also suppress the weak echoes of nearby targets. The control cannot differentiate between wave echoes and target echoes. It will eliminate every echo of the same strength, particularly the weak echoes of small boats and buoys without radar reflectors. The aim should be not to eliminate the clutter altogether, only reduce its density.

Fig 16.12: SEA CLUTTER
(More clutter on the windward side)

The clutter occurs in an unsymmetrical pattern around the vessel. The Anti-clutter setting should be adjusted while searching each area around the centre of the screen. Further adjustments are necessary if the weather changes or if the vessel moves into calmer waters.

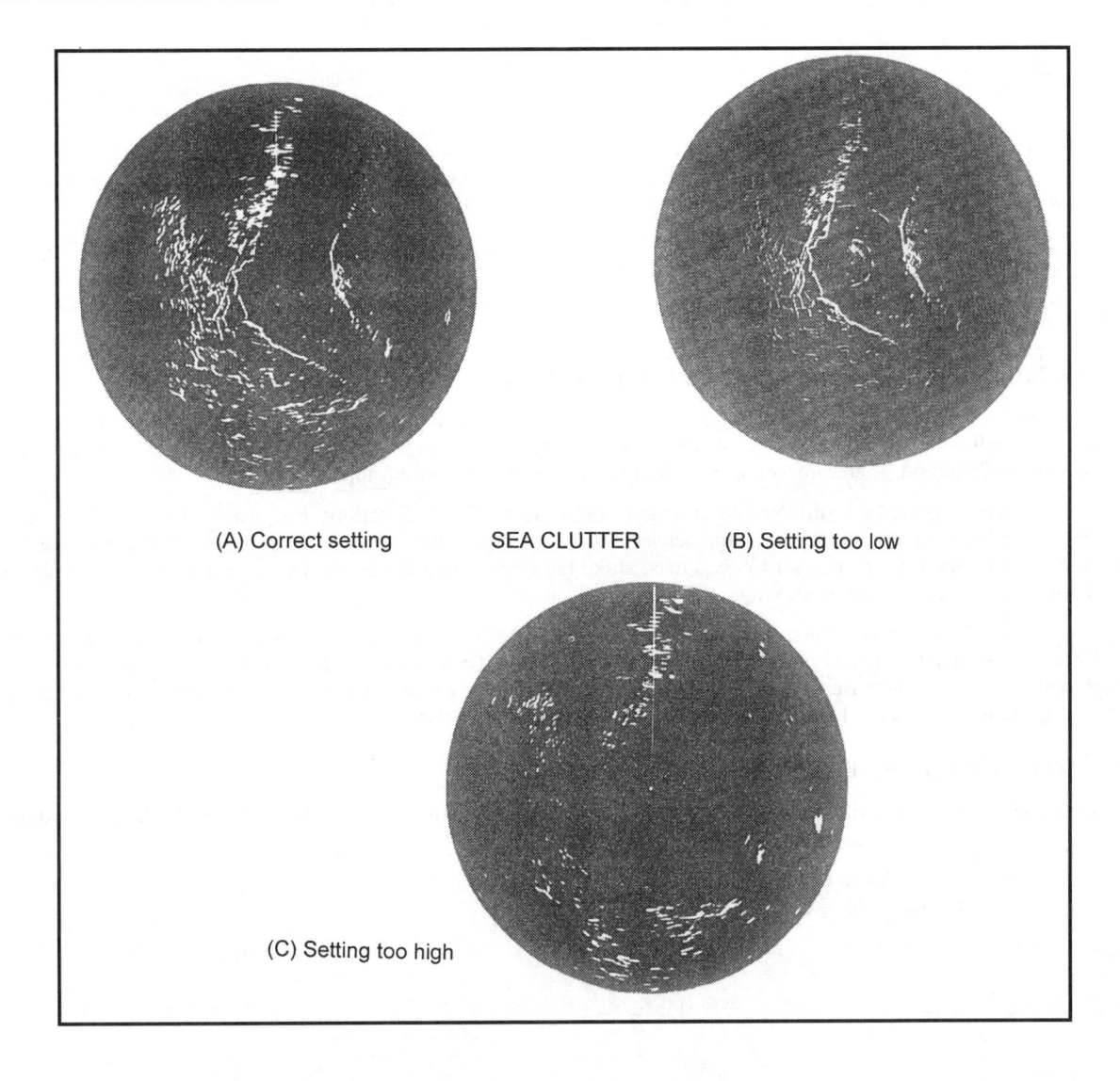

(A) Correct setting   SEA CLUTTER   (B) Setting too low

(C) Setting too high

RAIN CLUTTER CONTROL

Rain shows up on radar for up to 25 miles away. Intense rainstorms show up at even greater range. Rain reduces the ability of radar to detect targets in two ways:

❖ Rain produces a strong echo of its own, which often blots out echoes of targets in the rain. It can cover any part of or the whole screen in a cotton wool looking mass.
❖ Rain absorbs and weakens the radar energy passing through it. Consequently, the targets beyond the rain receive weakened signals and return weaker echoes. The absorption of radar energy by atmospheric gasses and moisture (including rain) is known as ATTENUATION. Targets at a closer range than the rain are not affected.

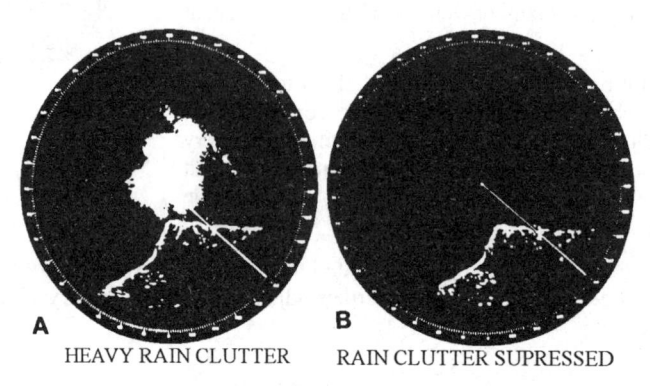

A HEAVY RAIN CLUTTER **B** RAIN CLUTTER SUPRESSED

Fig 16.14

A Rain Clutter Control differentiates between the nearest edge of the rain and its overall mass. It reduces the strength of the mass of rain but not of its leading edge, which shows up on the screen as a bright line. This allows the targets in the rain to show up but only those that are stronger than the suppressed rain echoes.

The weak echoes beyond the rain do not benefit in any way from the Rain Clutter Control. On the contrary, some will be suppressed altogether. The others will lose their mass, showing only the nearest edge on the screen. It is therefore important to increase Gain from time to time to detect these echoes. Some radars have circuits which automatically eliminate the Sea and Rain Clutter.

RANGE DISCRIMINATION & PULSE LENGTH CONTROL

Radar does not transmit radio frequency in continuous waves. It transmits in pulses, up to a few hundred metres long, every millionth of a second (micro-second) or so. A long pulse gives larger range because of more power in it, but it results in less definition. Therefore two objects close behind one another would appear as one.

Different radars have different pulse lengths and different Pulse Repetition Frequencies (PRF or the number of pulses transmitted per second). These range between 500 and 6000 pulses per second. In order to get a reasonable echo from a weak target, approximately 10 pulses must strike the target while the beam is sweeping across it. The higher the PRF, the greater the detection probability.

A high PRF increases the target definition but shortens the pulse length, which reduces the range. The two must be balanced for good detection and range. It is common to have two pulse lengths in radar sets: short pulse for short range and high definition, and long pulse for long-range detection. Some radars change automatically to long pulse at about the six mile range scale while others have a selection switch for short or long pulse.

CONTROLS TO TURN OFF WHEN NOT NEEDED

The following controls are likely to obscure echoes of small targets, and therefore should be turned off when not required.

❖ Range rings
❖ Variable Range Marker (VRM)
❖ Electronic Bearing Marker (EBM)

Chapter 16: Radar, SART & other Radar Transponders

RANGE (DISTANCE) & BEARING MEASUREMENT BY RADAR

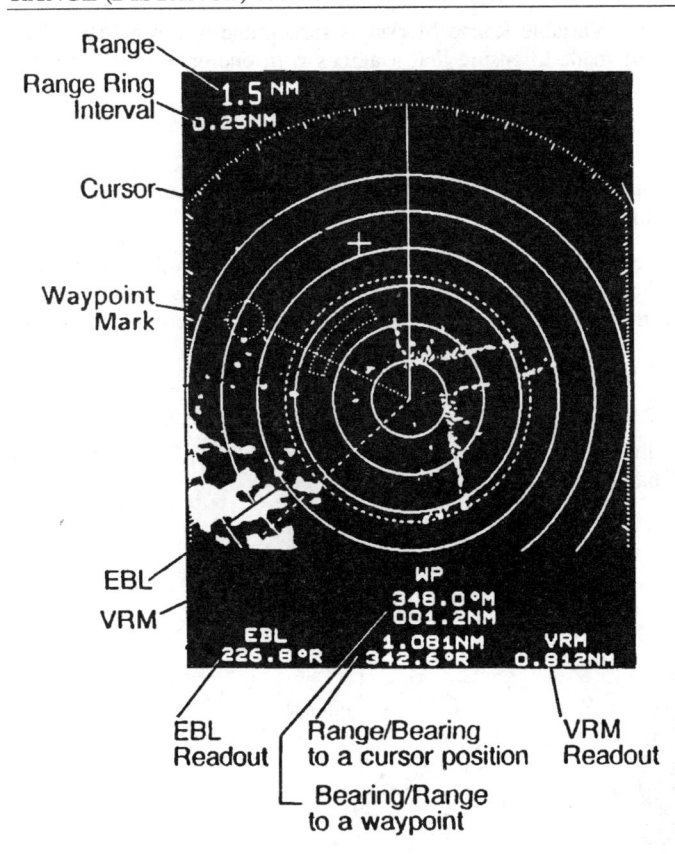

The range and bearing of a target is measured by moving the electronic cursor (in the shape of a cross) to the target and reading the digital readouts on the screen. They can also be measured by counting the range rings and measuring the angle relative to ship's head. The ranges and bearings should be measured to the middle of small targets, such as ships and small islands, and to the edge of large targets. Radar bearings are not as reliable as radar ranges. Typically, the latter are accurate to better than 1% of the range scale in use.

Labels on upper figure:
- Range: 1.5 NM
- Range Ring Interval: 0.25NM
- Cursor
- Waypoint Mark
- EBL
- VRM
- EBL Readout: 226.8°R
- Range/Bearing to a cursor position: 1.081NM 342.6°R
- Bearing/Range to a waypoint: WP 348.0°M 001.2NM
- VRM Readout: 0.812NM

Labels on lower figure:
- SIX ELECTRONIC RANGE RINGS ACCURATELY SPACED AT 0.5 n.m. INTERVALS ON THE 3 n.m. RANGE SCALE OR 2 n.m. INTERVALS ON THE 12 n.m. RANGE SCALE
- TARGET ECHO
- ELECTRONIC HEADING MARKER (SHOWS OWN HEADING)
- BEARING CURSOR

Fig 16.15 (A & B) MEASURING RANGE & BEARING BY RADAR

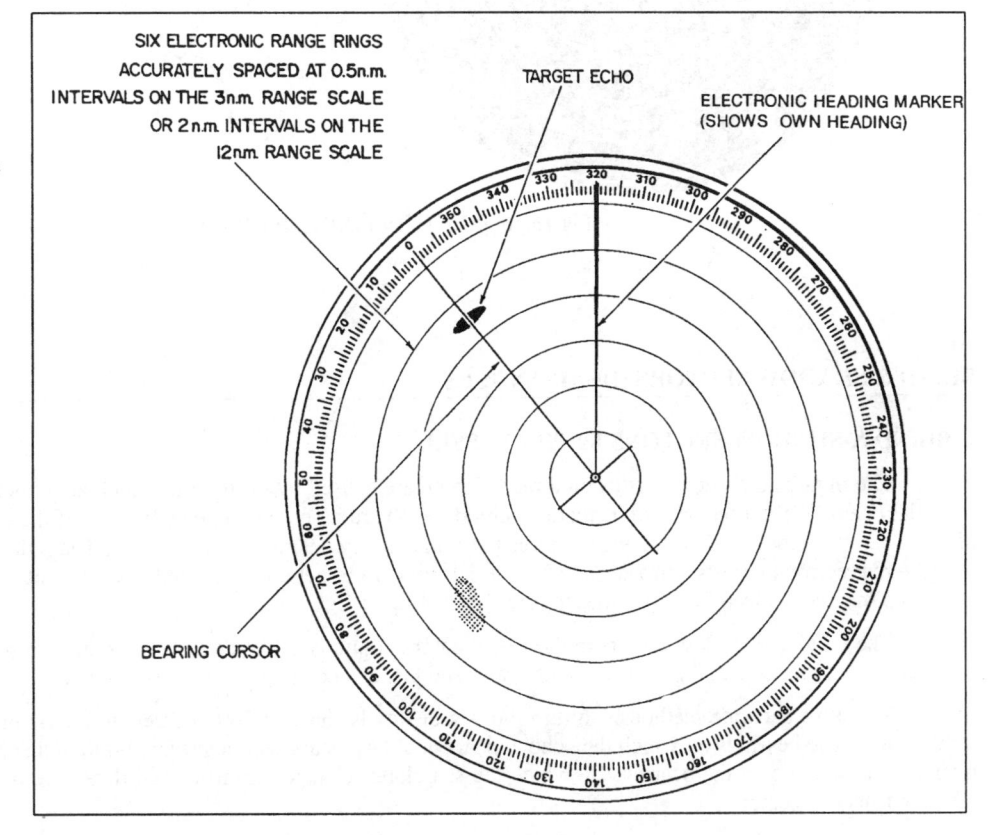

RANGE MARKER ERROR

While the Fixed Range Rings are quite accurate, the Variable Range Marker is susceptible to errors. Therefore, prior to measuring ranges with the VRM, a check should be made to ensure that it agrees with one or more fixed range rings.

BEARING ERRORS

YAW ERROR: In a yawing vessel, this error can be overcome by watching the heading marker at the time of reading the bearing. This error does not arise on stabilised displays.

BEAMWIDTH ERROR: The width of a target's echo is usually equal to the scanner's beamwidth. A scanner of 6° beamwidth would display a 6° wide echo. The correct bearing is in the middle of the echo. In other words the bearing error is equal to half the beamwidth. Therefore to make the bearing of a headland more accurate, adjust it by half the beamwidth towards the land (not towards the sea).

POSITION FIXING BY RADAR

A vessel's position can be fixed by taking the bearing and range of a navigational mark, or better, by taking two or more ranges as shown. (See RELATIVE BEARINGS in Chapter 17.1)

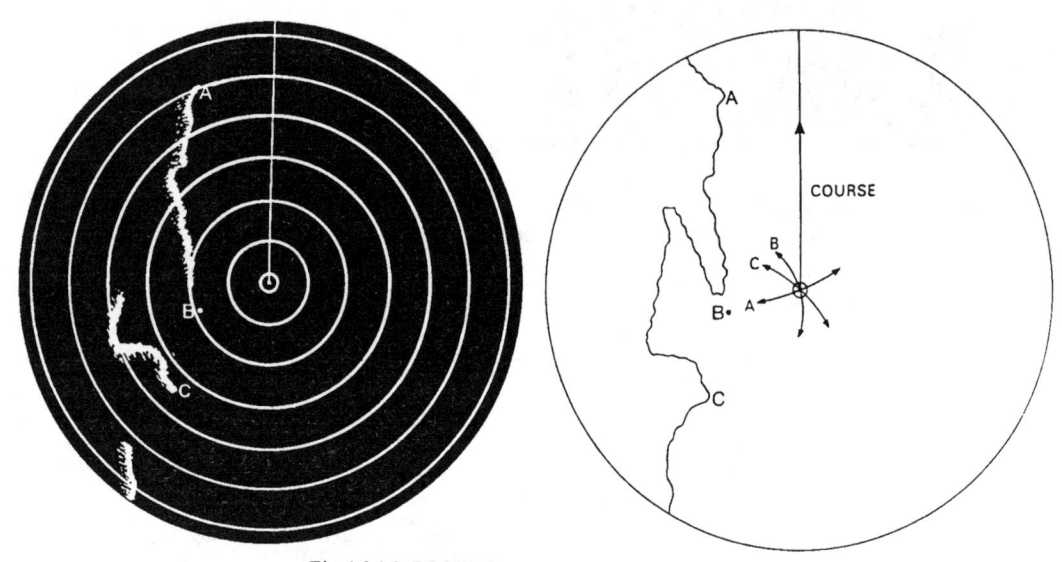

Fig 16.16: POSITION FIXING BY RADAR

BLIND & SHADOW SECTORS (DEAD ANGLES)

1. BLIND & SHADOW SECTORS IN THE VESSEL

The blind and shadow sectors on a radar screen are caused by obstructions such as funnel and masts in the way of the radar beam. A blind sector is immediately behind the obstruction that completely cuts off the radar beam. No targets are picked up in this area. Shadow sectors are areas not immediately behind the obstruction. The radar beam can bend (diffract) to some degree round masts and ridges, where it can pick up targets but with reduced intensity. The detection range of a large ship may be reduced from 10 miles to, say, 5 miles.

The blind and shadow sectors in the vessel can be minimised by carefully selecting the scanner location. They can be identified and recorded in the form of a diagram, and this can be done in one of two ways:

The most accurate method is to turn the vessel slowly through 360 degrees in the vicinity of a buoy or a similar small object. The bearings on which the echo appears and disappears will determine blind and shadow sectors. The second method is to note down the bearings on which the sea clutter changes from light to dark colour. The dark sectors are the blind and shadow sectors.

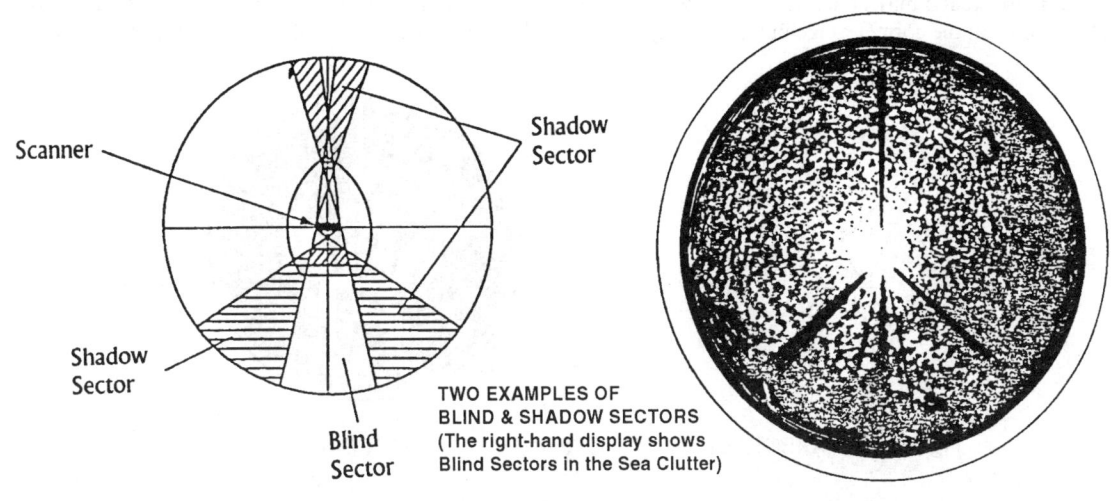

Chapter 16: Radar, SART & other Radar Transponders

Fig 16.17: BLIND & SHADOW SECTORS IN THE VESSEL

TWO EXAMPLES OF BLIND & SHADOW SECTORS
(The right-hand display shows Blind Sectors in the Sea Clutter)

2. BLIND & SHADOW SECTORS ON THE COAST

The blind and shadow areas also exist on land. One hill may obstruct another from being picked up by radar; or the earth's curvature may shadow the lower part of the coastline from being picked up by radar. It is the blind and shadow areas on land that cause discrepancy between the information on the chart and the radar picture of land. They make it difficult for the observer to identify the coastal features on radar. The shadow areas distort the picture, and continue to alter it as the vessel moves along the coast. Bays and inlets may not show on radar and high promontories may appear as islands.

There is nothing one can do about this problem, other than to exercise caution when plotting positions. Fixing positions with ranges taken from sloping coastline should be avoided. It would be difficult to be sure if the radar has picked up the nearest part of the land. Steep cliffs should be used where possible.

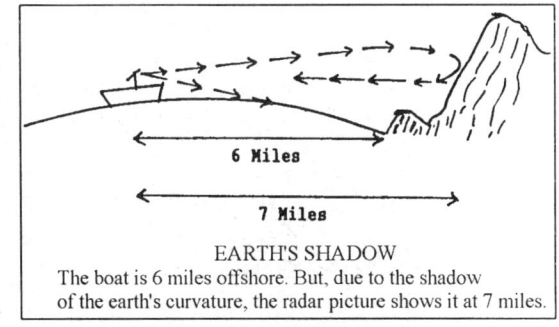

EARTH'S SHADOW
The boat is 6 miles offshore. But, due to the shadow of the earth's curvature, the radar picture shows it at 7 miles.

Fig 16.18
BLIND & SHADOW SECTORS ON THE COAST

FALSE ECHOES

False echoes are the result of:

1. Indirect (mirror) echoes
2. Side lobe echoes
3. Multiple echoes
4. Second trace returns
5. Effects of meteorological phenomenon
6. Spoking and other radar interference
7. Effect of external magnetic field.

1. INDIRECT (MIRROR) ECHOES OR VIRTUAL IMAGE

The objects that cause blind sectors also sometimes cause Indirect or Mirror Echoes. A target close to your vessel may appear at two locations on the screen. One of these is the true echo reflected directly by the target and the other is the false echo resulting from the radar beam entering the scanner after being reflected by a close-by

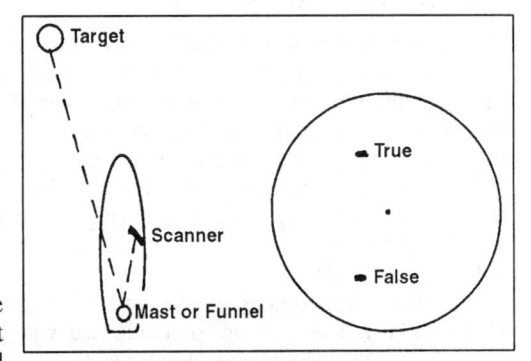

Fig 16.19: INDIRECT ECHO REFLECTED FROM A MAST OR FUNNEL

object. The indirect echo may be caused by a mast or funnel or an object close to the vessel, such as a metallic bridge. The false echo appears at the correct range but the wrong bearing, and always appear in the shadow (or blind) sectors.

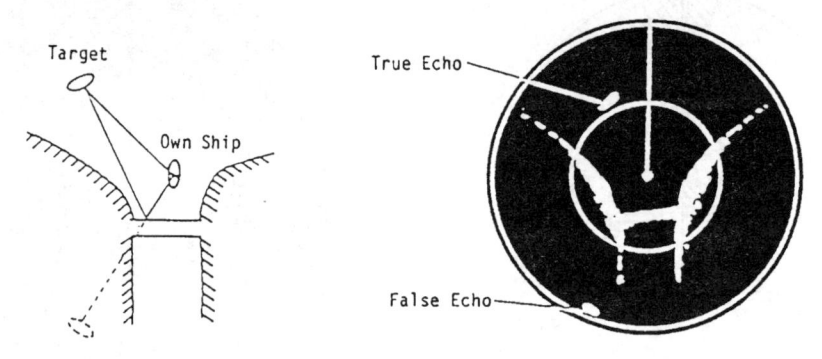

Fig 16.20: INDIRECT ECHO REFLECTED FROM AN IRON BRIDGE

2. SIDE LOBE ECHOES

The radar energy is meant to be transmitted by the scanner in one single beam. This should create one echo per target. However, some scanners also radiate energy on either side of the main beam. This Side Lobes or Side Echoes Effect may create three echoes per target. The primary echo is in the middle and a weaker echo on each side of it. Since all three echoes are at the same range, they often appear to form an arc of a circle.

In most cases, the presence of side echoes on radar screen is not really a problem. If necessary, they can be eliminated by temporarily reducing gain. But it may cause the loss or reduction in size of a small real echo.

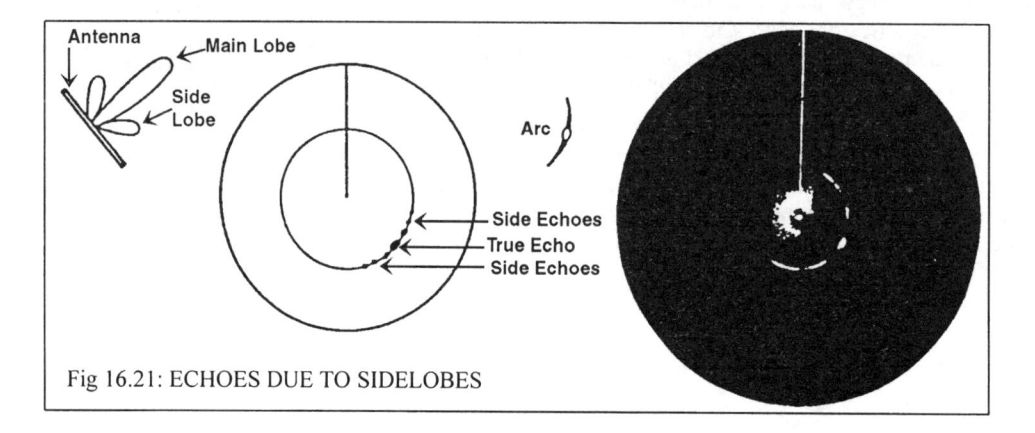

Fig 16.21: ECHOES DUE TO SIDELOBES

3. MULTIPLE ECHOES

Sometimes, when a radar pulse is reflected back by a nearby target (ship) abeam, not all of it returns to the scanner. Some of it strikes own ship's hull and bounces back and forth between the two hulls. Each time a smaller part of it returns to the scanner. This produces a row of echoes on the same bearing - all from a single target - reducing in size and strength with range. The closest echo is the strongest and the actual target.

These false echoes are usually not a problem. They can be left as they are or minimised with the Anti-Clutter Control or by reducing the Gain.

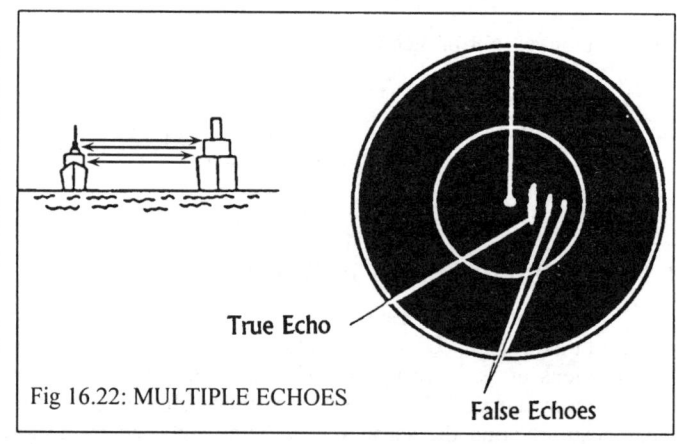

Fig 16.22: MULTIPLE ECHOES

Chapter 16: Radar, SART & other Radar Transponders

4. SECOND TRACE RETURNS (NON-STANDARD REFRACTION)

Echoes are sometimes received from targets hundreds of miles away on radar operating on perhaps a 48 miles range. These echoes are the result of radar pulses transmitted on a previous trace. They have returned after being reflected by a target a long distance away and picked up on perhaps the second or third trace. Such echoes are known as the Second Trace Returns.

This usually occurs under an atmospheric condition known as SUPER REFRACTION. A warm dry air blowing over a cold sea causes the radar (radio) waves to bend (*refract*) more than usual and travel a greater distance around the curvature of the earth.

This phenomenon occurs near very warm land masses such as the West Coast of Australia, Bass Strait with NNW winds, Persian Gulf, Mediterranean Sea and Red Sea. In all these places the warm air from land often flows on to the cooler sea.

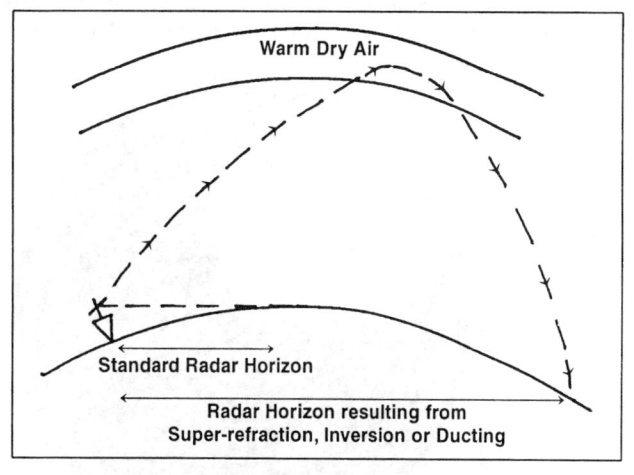

Fig 16.23: NON-STANDARD REFRACTION

In meteorology, the condition causing Super Refraction is known as *TEMPERATURE INVERSION*. It is so-named because the normal change in TEMPERATURE with height has been reversed. Instead of getting cooler further up from the ground, it gets warmer. The visual indication of temperature inversion is the *Mirage Effect*.

The Temperature Inversion is particularly strong when the weather is calm and there is no turbulence. This causes a layer of warm air to stay in the upper atmosphere, refracting radar (radio) waves to greater distances and picking up targets hundreds of miles away. This channelling effect is also known as DUCTING.

Ducting that takes place between a layer of warm air and the ground is known as *Surface Ducting*. If it takes place between two layers of air above the earth's surface, it is known as *Elevated Ducting*. The latter, however, is extremely rare.

All these false "second-trace" echoes not only appear at a wrong range but also distort the shape of the reflected coastline in the middle. Therefore, they do not offer any navigational value. Changing the range scale will make these echoes to either disappear or appear at a different range.

The opposite of "Super Refraction" is "SUB REFRACTION". This occurs when cold moist air blows over a warm sea. It creates high humidity in the upper atmosphere. Under such conditions, the radar (radio) waves bend (refract) less than usual and do not pick targets even when operating on a suitable range scale.

Sub-refraction is experienced near cold landmasses, e.g., East Coast of USA, Japan, Canada and Siberia. Super-Refraction is more common than Sub-Refraction.

The Super and Sub Refraction conditions are also known as:

- Non-standard Refraction
- Anomalous Propagation
- Freak Propagation

5. EFFECTS OF OTHER METEOROLOGICAL PHENOMENON

- *DUST STORMS* cause a general reduction in radar detection, similar to fog.
- *FOG* usually does not reflect or absorb radar energy. It does not produce echoes on the screen. However, dense fog banks, such as in Polar Regions, may show up as a faint haze on the screen. This can reduce radar visibility up to 40%.
- *CLOUDS* do not reflect radio (radar) energy, unless there is precipitation inside them. Ice in a cloud shows up as a bright line. Precipitation in cold fronts also shows up as a line. Echoes from a warm front can vary.
- *HAIL STONES* may be larger in size than raindrops, but they are not as dense. Their effect, although similar to rain, is not of the same strength.

6. SPOKING & MUTUAL RADAR INTERFERENCE

Radial lines, like the spokes of a wheel, appearing on a part of or the complete radar screen are caused by dirty or faulty contacts on the rotating parts of the radar. Only a technician should fix the problem.

Spoking should not be confused with interference, which may appear on the screen, from another radar operating in the vicinity. This shows up as bright speckles scattered over the screen, randomly or in a radial pattern but not in straight lines like spoking. Such "Mutual Radar Interference" is more common on longer-range scales.

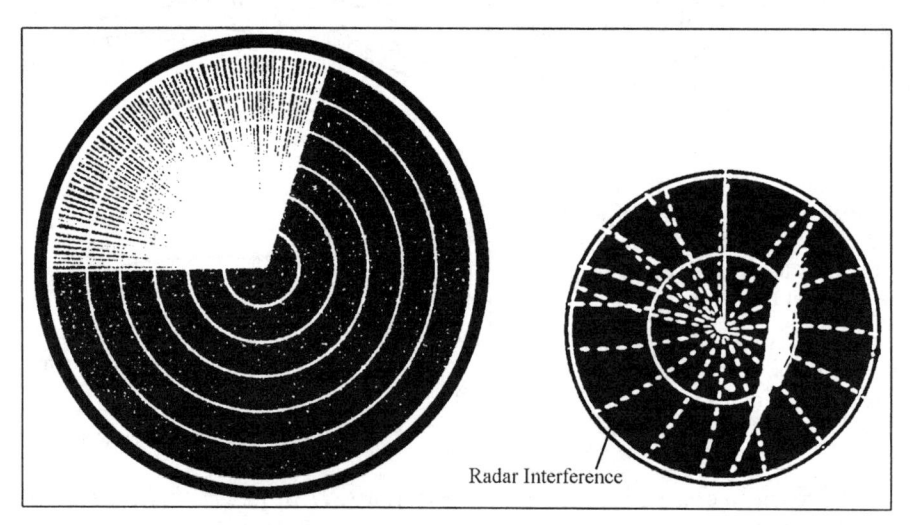

Fig 16.24: SPOKING (TWO EXAMPLES)

7. EFFECT OF EXTERNAL MAGNETIC FIELD

As discussed in 'Location & Safety' below, radar needs to be installed away from on-board magnetic influences. Radar is not usually affected by the earth's magnetic field such as magnetic variation. However, a bearing error is likely to arise when there is a large change in the earth's magnetism.

RADAR ERRORS

1. HEADING MARKER ERROR

The Heading Marker Alignment Control on the radar set allows the operator to vary the heading marker position. However, if the contacts inside the scanner housing are not precisely aligned with the fore and aft line of the vessel, the heading marker will be permanently out of alignment with the vessel's heading. It will cause an error in all the bearings by an amount equal to the heading marker error.

The error can be easily confirmed by heading the vessel towards an isolated object at some distance. If there is no error, its echo will lie on the heading marker.

Some Operators' Manuals explain how to adjust the position of the heading marker contacts. If not, a technician will be required.

2. CENTRING ERROR

This is caused by the centre of the radar picture not being directly under the centre of the Bearing Cursor. A "Centre Spot Control" is provided to correct the situation. This adjustment is sometimes necessary on the rotating trace radars, but not on the raster-scan displays.

3. PARALLAX ERROR

Lines drawn on two surfaces, which are not hard up against each other, will appear to move in relation to one another when viewed from different angles. It will be difficult to decide if a line on one surface is exactly above a line on the other. This is known as Parallax Error.

The screen of the Bearing Cursor is not hard up against the radar screen. Therefore, bearings taken with the cursor can be subject to a Parallax Error. It can be minimised by standing in such a way as to read the bearings with the centre of the cursor passing through the centre of the picture.

4. ERROR DUE TO YAWING

A vessel's head yawing from side to side will make the radar picture continually shift if the radar is operating in the ship's head up mode. This makes the task of taking accurate bearings difficult.

Chapter 16: Radar, SART & other Radar Transponders

RADAR RANGE CALCULATION (MAXIMUM)

Radar's maximum range depends on the internal characteristics of the radar set and the external limitations of the scanner height and the height of the target. The internal characteristics are Peak Power, Pulse Length and Receiver Noise. Some receivers are better at cutting out the noise received with the echoes, which can mask weak echoes.

However, from an operational point of view, the maximum range is calculated by calculating the Horizon Distances of the scanner and the target by using the formula:
Horizon Distance = $2.21\sqrt{H}$
("H" is the height of the scanner or target above the water in metres. *For heights in feet, the formula is $1.15\sqrt{H}$*)

EXAMPLE 1: The Horizon Distance of a radar set with the scanner height of 9 metres is calculated as follows:
= $2.21\sqrt{9}$
= 2.21 x 3 = 6.63 miles.

EXAMPLE 2: The geographical range at which a scanner 4 metres above the water will pick up the peak of an island of 144 metres height is calculated as follows:
= Distances $a_1 + a_2$

= $2.21\sqrt{h} + 2.21\sqrt{H}$
= $2.21\sqrt{4} + 2.21\sqrt{144}$
= 2.21 x 2 + 2.21 x 12
= 4.42 + 26.52
= 30.94 nm.

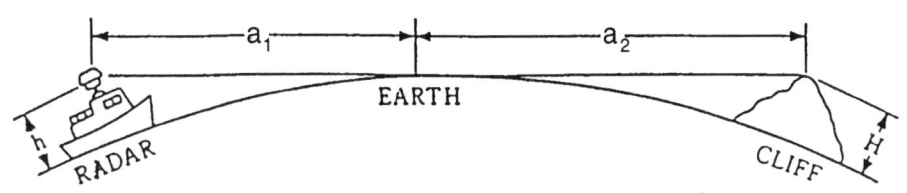

Fig 16.25: VISIBLE RANGE (DISTANCE) CALCULATION

HORIZON DISTANCE CALCULATION: The above formula can also be used to calculate the approximate geographical distance of horizon, a ship or a landfall in normal visibility.

EXAMPLE 3: The visible horizon distance of a person, whose height of eye is 4 metres, is calculated as follows:
= $2.21\sqrt{H}$
= $2.21\sqrt{4}$
= 2.21 x 2
= 4.42 nautical miles.

EXAMPLE 4: A person with the height of eye of 5 metres will see a ship, whose superstructure is 25 metres above the water, at a distance of:
= $2.21\sqrt{H} + 2.21\sqrt{H}$
= $2.21\sqrt{5} + 2.21\sqrt{25}$
= 4.94 + 11.05
= 15.99 nm.

EXAMPLE 5: A person with the height of eye of 3 metres will see the first visible flash (not the loom) of a 20 metres high lighthouse at a distance of:
= $2.21\sqrt{H} + 2.21\sqrt{H}$
= $2.21\sqrt{3} + 2.21\sqrt{20}$
= 3.83 + 9.88 = 13.71 nm.
(For greater accuracy, you also need to take into account the intensity of the light, as discussed in the Luminous & Geographical Range Tables in Chapter 17.2)

PARALLEL INDEXING

In addition to position fixing with terrestrial bearings and ranges, radar can be used for piloting a vessel along a pre-determined track by a method known as Parallel Indexing or **CROSS-INDEX RANGING**.

Plot the course on the chart. Select prominent charted features that would make good radar targets (steep-to headlands, large pile structures etc.) and measure their distance off the track at right angles to the course, i.e., their planned CPA (Closest Point of Approach).

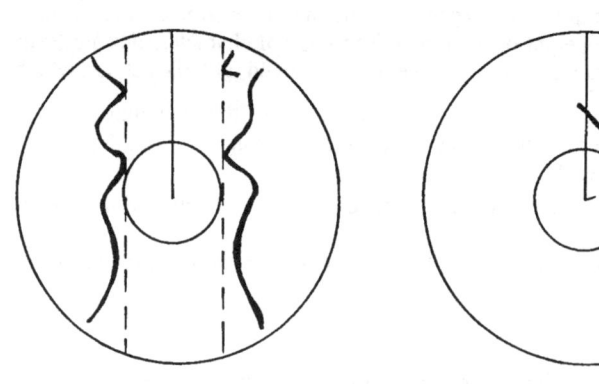

Fig 16.26: PARALLEL INDEXING

Set the variable range marker (VRM) at that range and align the bearing cursor on the planned course. Using a chinagraph pencil, draw a line on the screen (or reflection plotter) parallel to the proposed course and tangential to the variable range ring. The distance can also be set by a range ring, VRM or a line engraved on the mechanical bearing cursor. Then adjust course as necessary to ensure the radar target of the charted feature tracks down the line you have drawn on the PPI or reflection plotter. Do this on both sides if necessary.

To plan a course alteration position, mark the alter course point on the chart. Find the distance from a suitable object to this point. This is known as **DEAD RANGE** (i.e., you'll be dead if you go past this range). Set the VRM to this range or draw it on the screen around the centre. Alter course as soon as the echo of the selected object reaches the range, making allowance for the vessel's turning circle.

RADAR AS A NAVIGATIONAL AND COLLISION AVOIDANCE TOOL

Radar is a useful navigational and watchkeeping tool in any state of weather. Its use can improve the efficiency of position fixing and collision avoidance, provided the lookout by sight and sound and the need for visual fixing are not compromised. It can show targets outside the visual range and help to gauge the state of visibility by checking the radar distance of a target that has just become visible. See International Collision Prevention Rules 5, 6, 7 and 19 in Chapter 11.

COLLISION AVOIDANCE

The guidelines for the use of radar in collision avoidance are stated in Chapter 11 in Rules 5, 7 and 19 of the International Regulations for the Prevention of Collisions at Sea. See also rules 6, 8 and 35.

Chapter 16: Radar, SART & other Radar Transponders

RADAR PLOTTING (RELATIVE MOTION)

When on passage in poor visibility at sea, another vessel's time and distance to Closest Point of Approach (CPA) can be calculated, provided she is at sufficient range to permit it, and neither vessel alters course or speed during plotting. The plotting is done either using a chinagraph pencil and a REFLECTIVE PLOTTER that can be fitted on the radar screen or on a plotting sheet shown here.

In the following plot, each range ring is two miles. The other ship's echo was observed and plotted as follows:

1000 hours:	45° on stbd bow, range 7 miles
1005 hours:	41° on stbd bow, range 6 miles
1010 hours:	35° on stbd bow, range 5 miles

The plot shows that the ship will cross ahead and at the closest distance of 3 miles, at 1027 hours.

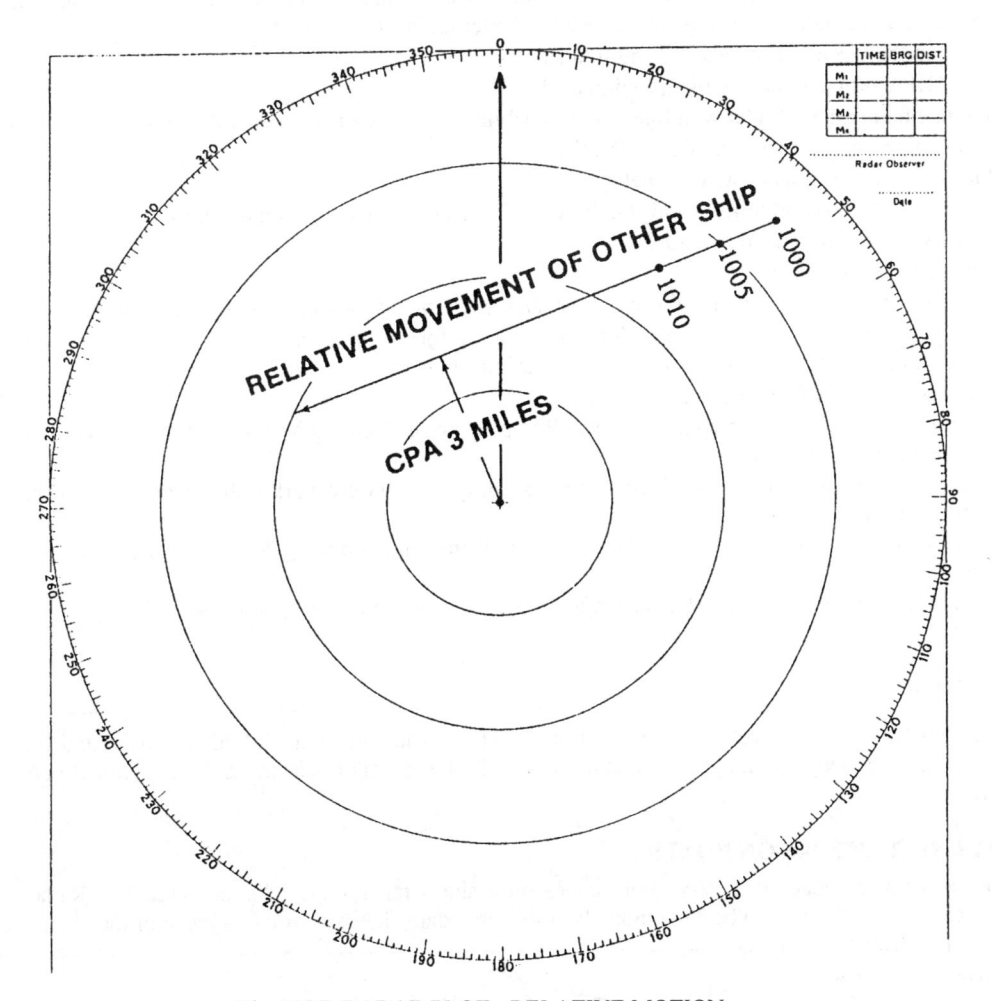

Fig 16.27: RADAR PLOT - RELATIVE MOTION

AUTOMATIC RADAR PLOTTING AID (ARPA)

This is a computerised target-plotting device. By extracting from radar the bearings and ranges of a number of targets, the computer provides the operator their closest points and times of approach and their courses and speeds. It can acquire the targets automatically from the guard zone or they can be manually selected by the operator. It then determines safe avoiding manoeuvres for the vessel, while continuously updating the target information. By way of trails, ARPA can also respond to what the consequences would be of altering course in a particular direction.

BUYING A RADAR

- Longer scanners have narrower beamwidth, which provides greater range and better discrimination between targets close to each other. The radar bearings are thus more accurate. Wider beamwidth makes objects appear distorted (wider). While this may help in detecting buoys and small vessels, entrances to harbours and rivers would be more difficult to locate. See also bearing errors.
- Scanners can be open or radome.
- Consider your boat size.
- Consider power requirements (AC/DC).
- Output power. If you want to use radar for long-range views you will want a unit with greater power output. Units with more power also perform better in adverse conditions, such as rain and fog.
- PPI (rotating) or raster-scan display. The former is also known as the CRT display and the latter an LCD display. If the unit is to be mounted where it will be exposed to direct sunlight, an LCD unit may be your best choice. A backlit, high resolution LCD screen is easy to see in most lighting conditions. For inside mounting, however, a CRT display will offer a sharper, deeper and more detailed picture as well as better nighttime visibility.
- Know the difference between X-band and S-band frequencies.
- Do you require ARPA – an automatic target plotting aid?
- Minimum and maximum range (Radar with long range capability is wasted on a vessel with a limited scanner height).
- Consider the operating range scales (e.g., 0.25 - 48 miles).
- Radar on which pulse length can be altered is better.
- Know the difference between unstabilised and stabilised, and true and relative motion displays.
- Is the radar fitted with a performance monitor?
- Quality brands are more expensive.
- Try it out in a vessel and in the area you will use it most. Ask your friends who may use the same model.
- Check the quality of the radar manual. It should not omit vital information, such as how to check range and heading marker accuracy. A good tutorial on the limitations of radar indicates customer care.
- Compare the ease of use of different models.
- Electronic features such as: Variable Range Marker (VRM), Electronic Bearing Marker or Line (EBM / EBL) and the Target guard zone are now common.
- Guard alarms are good, but don't fall into a false sense of security. It will only work if the radar has been able to detect the target entering the guard zone.
- Do you want a stand-alone radar or an integrated navigation unit incorporating Radar, Plotter, Echo Sounder and GPS/DGPS, etc.?
- Consider the manufacturer and the dealership size, after sale service and availability of parts.

RADAR REFLECTORS

There are two types of radar reflectors: Passive and Active. The former are fitted on boats, and the latter on lighthouses in areas that experience prolonged periods of poor visibility. Smaller beacons and navigational buoys may be fitted with either.

PASSIVE RADAR REFLECTORS ON BOATS

Fibreglass and timber boats give very poor echoes on a ship's radar, especially in rough sea. Radar reflectors mounted on masts can help small craft to become more "visible" on radar. They are simply a piece of metal so arranged to return radar waves. They can be purchased from boat chandlers or homemade.

Most radar reflectors on boats consist of three diamond shaped plates fitted at right angles to each other. This 8-corner reflector is known as the *"octahedral cluster"*. The plates are either metal or covered with tin foil. The unit is approximately one metre across and is placed as high as possible in the boat.

A radar reflector hoisted on a halyard may not provide a consistent return in windy conditions. Therefore, it

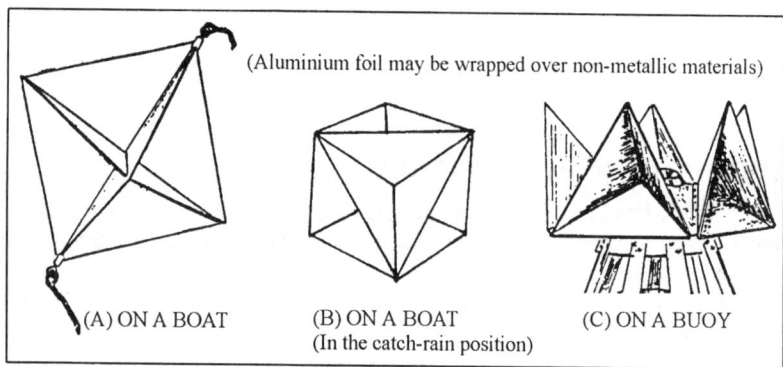

(Aluminium foil may be wrapped over non-metallic materials)

(A) ON A BOAT (B) ON A BOAT (C) ON A BUOY
(In the catch-rain position)

Fig 16.28: PASSIVE RADAR REFLECTORS

Chapter 16: Radar, SART & other Radar Transponders

should be attached to the mast. To ensure that no one plane is horizontal, it should be fixed at an angle of 45° to the mast with one of its openings (cups) facing upwards, i.e., in the "Catch Rain" position. It may not however work correctly on certain angles of heel in a sailing boat.

PASSIVE RADAR REFLECTORS ON NAVIGATIONAL BUOYS

Spherical, cylindrical and conical objects, such as navigational buoys and beacons, do not return as much reflected energy as objects with flat surfaces facing the radar. Most of it "slips and slides off" their surfaces. Only a small portion of it returns to the radar.

The radar detection range for buoys is about 2 or 3 miles. Their echoes may also appear to twinkle because of their to-and-fro motion in the sea. They are therefore usually fitted with Pentagonal Cluster (5 corner) reflectors.

ACTIVE RADAR REFLECTORS

These are small transmitters mounted in beacons, lighthouses and buoys on coastlines that are featureless or where visibility tends to be poor. When triggered by an operating radar, they radiate an identifying signal which appears uniquely coded on radar displays. Radars pick up these signals upto 20 miles away. Racon is the most common active reflector. A less common one is Ramark, which no longer exists in Australia.

The Ramark is a continuously transmitting beacon. It is suitable for taking a bearing only. Its echo appears on the radar screen as a bright radial line on the bearing of the beacon. The line extends from the centre to the edge of the screen. The most common method of providing its identification is by breaking up the radial line into dots and dashes on the radar screen. Its major disadvantage is that it might mask other important echoes on the screen.

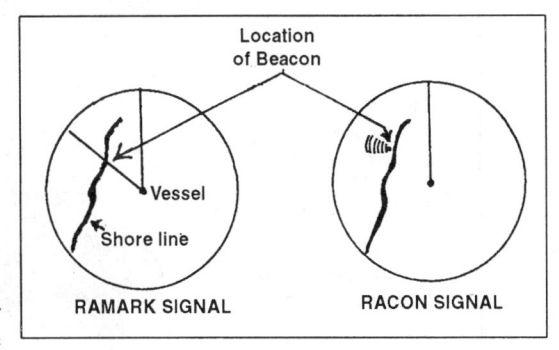

Fig 16.29: ACTIVE RADAR REFLECTORS

The Racon (RAdar beaCON) transmits its signal in pulses. It thus provides a bearing as well as range. Its echo appears on the radar screen as segments of concentric arcs or a morse code signal (dots and dashes). The number of segments and their distance apart identifies the beacon. The transmission being non-continuous, there is a slight delay in the reception of pulses. The signal therefore appears on the screen at a slightly greater range than of the beacon in which it is located. The echo of the beacon appears in front of the identifying signal when it is in range. The signal does however appear at the correct bearing.

Both Ramark and Racon are subject to giving false echoes, and are likely to mask targets close to them.

SEARCH AND RESCUE RADAR TRANSPONDERS (SARTs)
(also known as the Survival Craft Radar Transponders)

A SART is a small, buoyant, battery-powered, omni-directional radar detector and position-indicating unit. On receipt of a radar signal from a nearby aircraft or ship, it responds by transmitting a signal which shows up on the radar screen of the aircraft or ship as a series of 12 blips extending approximately 8 miles outwards from the SART's position along its line of bearing. This unique radar signal is easily recognised and allows the rescue vessel or aircraft to locate the vessel in distress or the survival craft. A ship with a radar scanner at a height of 15 metres will detect it at a range of at least 5 miles and an aircraft flying at a height of 1000 metres will detect it at a range of at least 30 miles. In order to increase its range of detection, it is advisable to mount the SART on a distress vessel or survival craft as high as possible. Like radar it transmits in the line of sight.

On activation, SARTs indicate their correct operation by a visible and/or audible signal. The batteries fitted to SARTs allow operation in the stand-by mode for at least 96 hours and in the transmitting mode for further 8 hours. A small speaker fitted in the unit sounds a beep signal when in transmitting mode. They thus indicate when being interrogated a radar signal from an aircraft or ship.

SARTs operate in the 10 GHz or 10,000 MHz band. [9.3 to 9.5 GHz, to be precise]. Therefore they respond only to radars operating on those frequencies. These "X band" or 3 cm wavelength radars are designed for short-range clarity of picture. SARTS do not respond to 3 GHz or 3000 MHz radars. Also known as the "S band" or 10 cm wavelength radars, these are designed for long-range operation.

The spacing between SART's blips is about 0.6 miles. Its accurate measurement will lead to making positive identification of a SART. Therefore, IMO recommends that searching ships operate their radars on short pulse and on 6 or 12 miles range, so that distances of 0.6 miles would be easier to measure. On becoming locked onto a radar, as the SART switches over from receive to transmit mode, there is a slight range delay of approximate 150 metres. Therefore a ship approaching a SART may find it 150 metres out of position on its radar.

Like EPIRB, SART is a distress signal transmitter. It forms a part of the GMDSS equipment.

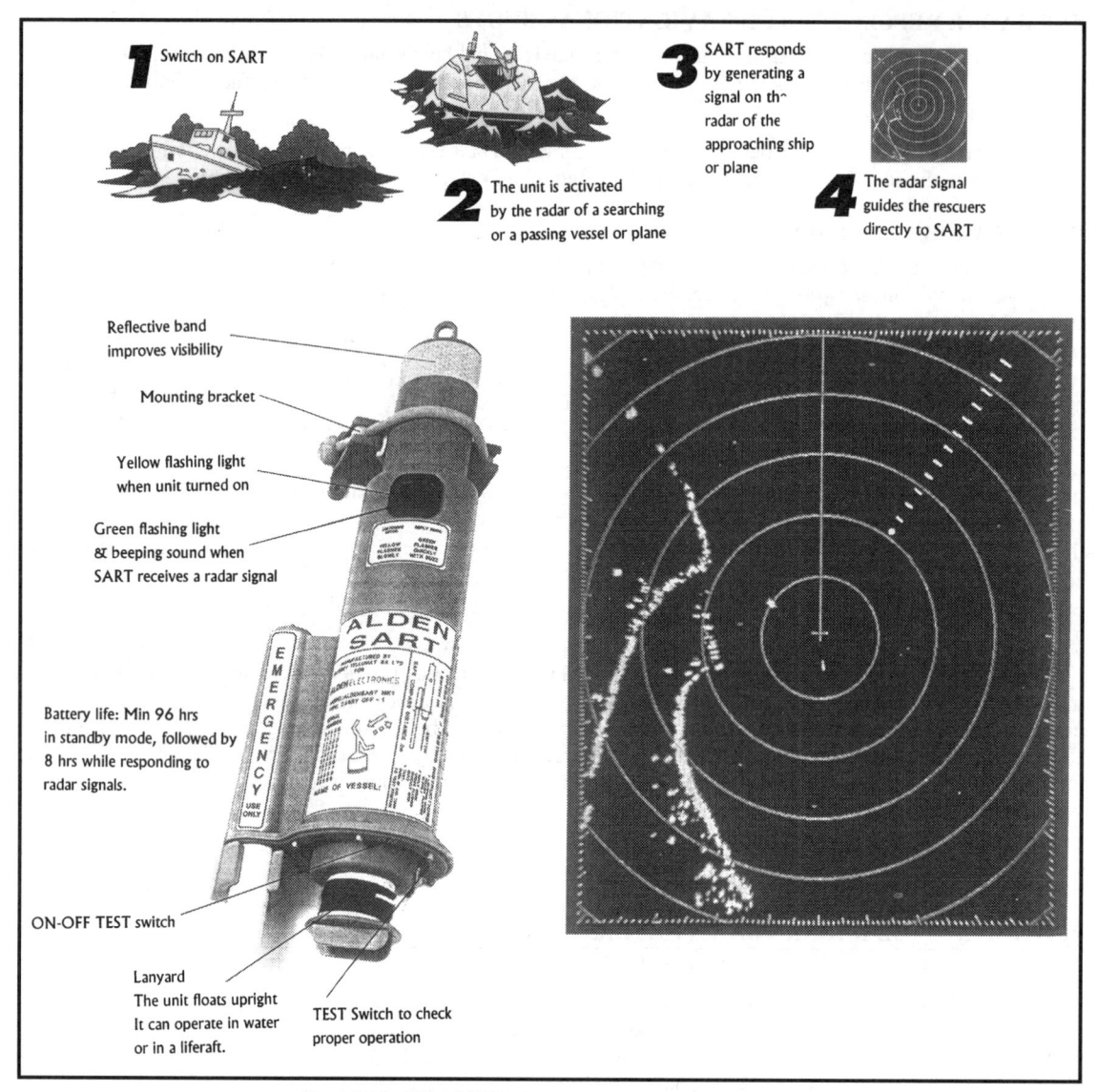

Fig 16.30: AN ALDEN SART & HOW IT WORKS
(The SART signal is seen on the rescuer's radar as a series of 12 highly visible lines leading directly to its location. The rescuer pilots the vessel or plane in the indicated direction to find the SART.)

ANTI-COLLISION RADAR TRANSPONDERS

When fitted on a small vessel, it assists other vessels to detect the faint radar echo of the small vessel. Furthermore, by a visual or audible warning, it alerts the small vessel that a radar (i.e., a vessel) is in the vicinity. An anti-collision radar transponder produces a line of 5 blips on the interrogating vessel's radar screen, extending outwards for approximately 1 mile from the transponder's position along its line of bearing.

RADAR REFLECTIVE DISTRESS SIGNALS

Radar reflective rocket distress signals are available. They eject a cloud of radar reflective material that produces a distinctive echo on a ship or aircraft radar for 10 to 20 minutes and are detectable from a considerable range. They can be used during the day or night.

CHAPTER 16: QUESTIONS & ANSWERS

CHAPTER 16.1: RADAR & RADAR REFLECTORS

1. Temperature inversion:
   (a) may reduce radar transmission range.
   (b) may bend radar (radio) waves.
   (c) does not affect radar transmission.
   (d) is a trapped layer of cold air.

2. High PRF radar operation:
   (a) increases target definition, decreases range.
   (b) increases range, decreases target definition.
   (c) increases range and target definition.
   (d) decreases range and target definition.

3. Long pulse radar operation:
   (a) increases target definition, decreases target definition.
   (b) increases range, decreases target definition.
   (c) increases range and target definition.
   (d) decreases range and target definition.

4. In relation to radar operation, which of the following statements is correct?
   (a) On a long pulse, two objects close behind each other are likely to appear as two.
   (b) An alteration of operating range does not alter the pulse length.
   (c) Radars are not fitted with a pulse length selection switch.
   (d) On a long pulse, two objects close behind each other are likely to appear as one.

5. Range discrimination is the radar's ability to distinguish two targets:
   (a) on the same bearing but slightly different ranges.
   (b) on the same bearing and range.
   (c) on the same range but slightly different bearings.
   (d) of different composition

6. Cotton-wool type patches on radar screen are usually caused by:
   (a) incorrect Tuning.
   (b) incorrect Gain.
   (c) rain
   (d) large ships.

7. Radar bearings of targets are more accurate when:
   (a) the pulse length is smaller.
   (b) the horizontal beamwidth is smaller.
   (c) the vertical beamwidth is smaller.
   (d) the horizontal beamwidth is larger.

8. The effect of super-refraction is to:
   (a) refract the radar rays upwards and decrease radar range.
   (b) refract the radar rays downwards and decrease radar range.
   (c) refract the radar rays upwards and increase radar range
   (d) refract the radar rays downwards and increase radar range.

9. The effect of sub-refraction is to:
   (a) refract the radar rays upwards and decrease radar range.
   (b) refract the radar rays downwards and decrease radar range.
   (c) refract the radar rays upwards and increase radar range
   (d) refract the radar rays downwards and increase radar range.

10. The authenticity of an echo that appears in the radar scanner's shadow sector can be checked by:
    (a) plotting it.
    (b) altering the range scale.
    (c) altering course.
    (d) increasing Gain.

11. Vessel's heading is 175°T. The relative radar bearing of target is 105°. Therefore, the true bearing of the target is:
    (a) 170°
    (b) 105°
    (c) 280°
    (d) none of the choices stated here.

12. Vessel's heading is 295°T. The relative radar bearing of target is 170°. Therefore, the true bearing of the target is:
    (a) 465°
    (b) 105°
    (c) 125°
    (d) 065°

13. Vessel's heading is 180°T. The radar bearing of a target is R55°. Therefore, its true bearing is:
    (a) 235°
    (b) 055°
    (c) 125°
    (d) 005°

14. Vessel's heading is 000°T. The radar bearing of a target is G175°. Therefore, its true bearing is:
    (a) 175°
    (b) 185°
    (c) 125°
    (d) 005°

15. The following type of land is most suitable for position fixing by radar:

    (a) Gently shelving.
    (b) Steep.
    (c) Clear of trees.
    (d) Forested.

16. In restricted visibility, radar lookout is best maintained:

    (a) on a short range scale.
    (b) on a long range scale
    (c) on a medium range scale.
    (d) by varying between short & long range scales.

17. In determining safe speed in restricted visibility, a vessel fitted with radar must:

    (a) operate on a long range scale
    (b) operate on a medium range scale.
    (c) consider any constraints imposed by the radar range scale in use.
    (d) operate on a short range scale.

18. Prior to measuring ranges with the Variable Range Marker:

    (a) its accuracy should be checked against the fixed range rings.
    (b) it should be turned to maximum, then minimum.
    (c) it should be turned to minimum, then maximum.
    (d) it should be set to zero.

19. The International Collision Prevention Rule 19, regarding conduct of vessels in restricted visibility, applies as follows:

    (a) Only in restricted visibility.
    (b) To vessels not in sight of one another.
    (c) To vessels not in sight of one another when navigating in or near an area of restricted visibility.
    (d) To all vessels at all times.

20. Radar transmits and receives:

    (a) continuous radio waves
    (b) radio waves in pulses
    (c) T/R pulses
    (d) echoes

21. Wave Guide in radar is a tube:

    (a) that generates radio pulses.
    (b) through which radio pulses travel between the receiver & the display.
    (c) through which radio waves travel from the transmitter to the scanner, and from the scanner to the receiver.
    (d) through which radio waves travel from the transmitter to the scanner.

22. X-Band radar:

    (a) operates on 3 cm. Wavelength.
    (b) works in the 3000 MHz frequency range.
    (c) needs longer scanner than the S-Band radar.
    (d) produces less sidelobes than the S-band radar.

23. S-Band radar

    (a) is more suitable for small vessels than the X-Band radar.
    (b) works in the 10,000 MHz frequency range.
    (c) needs shorter scanner than the X-Band radar.
    (d) needs greater power than the X-band radar.

24. Radar pulse:

    (a) travels further when transmitted in a narrower beam.
    (b) travels further when transmitted in a wider beam.
    (c) travels the same distance regardless of the beamwidth.
    (d) does not travel in a beam.

25. Radome radar scanners seen on sailing vessels:

    (a) do not have a slotted waveguide.
    (b) transmit unguided radar pulses.
    (c) have a beamwidth of about 10°.
    (d) have a beamwidth just like any other radar scanner.

26. The approximate beamwidth of 80-cm. radar scanners is:

    (a) horizontal 2.5°; vertical 30°
    (b) horizontal 30°; vertical 2°
    (c) horizontal 10°; vertical 2°
    (d) horizontal 2°; vertical 2°

27. Longer radar scanners:

    (a) provide poorer bearing discrimination.
    (b) have lesser number of slots.
    (c) are more directional.
    (d) provide shorter operating range.

28. Shorter radar scanners:

    (a) provide better target discrimination.
    (b) provide a shorter operating range.
    (c) have a greater number of slots.
    (d) are more directional.

Chapter 16: Radar, SART & other Radar Transponders

29. The two types of radar displays are:
    (a) PPI & CRT
    (b) Raster Scan & LCD
    (c) RSD & LCD
    (d) PPI & Raster Scan

30. On Raster Scan radar display:
    (a) a rotating trace is displayed.
    (b) the presentation is refreshed on every sweep.
    (c) the echoes are digitally processed & displayed much like a TV screen.
    (d) A CRT displays the information.

31. Wide beamwidth radar scanners:
    (a) provide greater range.
    (b) provide better target discrimination.
    (c) make harbour & river entrances easier to locate.
    (d) cause echoes of buoys & small vessels to appear distorted.

32. LCD radar display:
    (a) is also known as CRT display.
    (b) is more difficult to read in daylight than the PPI.
    (c) is easier to read in daylight than the PPI display.
    (d) when viewed in the dark, offers a sharper, deeper & more detailed picture than PPI.

33. Loss of video image can freeze radar picture which the observer may not notice until the range is changed. This can happen in:
    (a) in all types of radars.
    (b) only in Raster Scan radars.
    (c) only on CRT screens.
    (d) only on PPI screens.

34. Loss of azimuth signal in radar leads to targets being displayed on wrong bearings. This can happen:
    (a) in all types of radars.
    (b) on CRT screens.
    (c) in Raster Scan radars.
    (d) on PPI screens.

35. Safety risk with colour radar is that:
    (a) the observer may pay more attention to red echoes.
    (b) the observer may pay more attention to vessels with smaller CPA.
    (c) targets may appear yellow.
    (d) range rings may appear white.

36. Guard Zone Alarm in radar:
    (a) allows the navigator to attend to duties other than watchkeeping.
    (b) detects all targets inside the guard zone and sounds an alarm.
    (c) overcomes tuning deficiency.
    (d) may not detect all targets inside the Guard Zone.

37. Radar scanner should be installed:
    (a) alongside the steering compass.
    (b) where there are minimum shadow and blind sectors.
    (c) in the navigator's direct line of sight.
    (d) close to the autopilot sensor.

38. Electrical interference in radar can be minimised by:
    (a) connecting it with separate power cables direct from the battery.
    (b) cutting the power cables to the shortest length required.
    (c) positioning the alternator so that it is below the level of the radar.
    (d) minimising shadow & blind sectors.

39. Which of the following is not an indication of radar interference on other equipment:
    (a) Spurious flashes of light on echo sounder.
    (b) Whining noise on radio.
    (c) Autopilot's failure to engage.
    (d) Fluctuations in compass error

40. Which of the following does not constitute care and maintenance of radar.
    (a) Wash down scanner 3 or 4 times a year.
    (b) Repair any visible cracks in the scanner or radome.
    (c) Check & renew scanner motor brushes.
    (d) Paint scanner aperture and radome every 2 years.

41. In the absence of targets, radar performance can be checked as follows:
    (a) Only with an external performance monitor.
    (b) Only with a built-in performance monitor
    (c) With an external or built-in performance monitor.
    (d) None of the choices stated here.

42. With regard to Radar Performance Monitor, which of the following statements is incorrect?

    (a) The external type is a plastic box mounted behind the scanner.
    (b) It can be the external or the built-in type.
    (c) The external type shows a narrow plume on radar screen.
    (d) The built-in type produces a circular pattern on radar screen.

43. With regard to Relative Motion Unstabilised radar display, which of the following statements is correct?

    (a) Own vessel is seen in motion.
    (b) Coastline is seen as stationary.
    (c) Heading marker indicates vessel's compass heading.
    (d) When vessel's course is altered, the echoes on the screen rotate in the opposite direction to the alteration.

44. With regard to Relative Motion Stabilised radar display, which of the following statements is correct?

    (a) Own vessel is seen in motion.
    (b) Coastline is seen as stationary.
    (c) Heading marker indicates vessel's compass heading.
    (d) It can be operated only in North-UP mode.

45. With regard to radar display, which of the following statements is correct?

    (a) In the Head-Up mode, the Heading Marker wanders according to the orientation of radar.
    (b) The Head-Up mode is more suitable for navigating in narrow channels.
    (c) In the North-Up mode, the North Marker wanders according to the orientation of radar.
    (d) Stabilized radar can only be operated in the North-Up mode.

46. With regard to True Motion Stabilized radar display, which of the following statements is correct?

    (a) Vessel's position can be made to start at any point on the screen.
    (b) Coastline is seen in motion.
    (c) Own vessel is seen as stationary.
    (d) The display does not require compass input.

47. A target's ability to appear on a radar screen depends on the:

    (a) type of target
    (b) weather conditions
    (c) settings of radar controls
    (d) all the factors stated here.

48. The size and strength of a target's echo depends on its:

    (a) shape & size
    (b) none of the factors stated here.
    (c) composition & aspect
    (d) all four factors stated here.

49. The radar echo of a 'beam-on' ship is:

    (a) larger than the 'bow-on' echo.
    (b) smaller than the 'bow-on' echo.
    (c) of the same size as of 'bow-on' echo.
    (d) difficult to see.

50. In terms of reflecting radar energy, objects with rough surfaces reflect:

    (a) less than those with smooth surfaces.
    (b) more than those with smooth surfaces.
    (c) the same as those with smooth surfaces.
    (d) little or no energy.

51. In terms of reflecting radar energy, which of the following statements is correct?

    (a) GRP is a better reflector than wood.
    (b) Wood is a better reflector than steel.
    (c) Steel is a better reflector than wood or GRP.
    (d) Steel is a poorer reflector than wood or GRP.

52. GRP hulls fitted with engines, masts and other metal fittings:

    (a) are good radar reflectors.
    (b) generate a constant echo.
    (c) remain out of phase.
    (d) may or may not remain in phase.

53. Which of the following radar settings does not determine picture quality?

    (a) Gain
    (b) Brilliance
    (c) Tuning
    (d) Scale illumination

54. The "Radar Off" IMO symbol is a:

    (a) circle
    (b) circle with a dot in the middle.
    (c) circle with a dot on the circumference
    (d) circle with a line inside it.

55. The "North-Up Presentation" IMO radar symbol is:

    (a) a circle with a heading marker pointing upwards
    (b) a circle with a diamond on the circumference & the heading marker pointing to the right of it.
    (c) a spiral inside a circle.
    (d) a circle with a dot on the circumference.

Chapter 16: Radar, SART & other Radar Transponders

56. The "Variable Ranger Marker" IMO radar symbol is:

    (a) a series of concentric circles.
    (b) a crescent shaped symbol.
    (c) a spiral inside a circle.
    (d) two heading markers with an arrow indicating rotation.

57. The "Gain Control" IMO radar symbol is a:

    (a) half crescent.
    (b) crescent.
    (c) circle with a light bulb symbol inside it.
    (d) spiral inside a circle.

58. The "Bearing Marker" IMO radar symbol is:

    (a) two radial lines in a circle with an arrow in the middle.
    (b) a circle with a dot in the middle and a radial line with an arrow on each end.
    (c) a circle with a dish-shaped symbol and an arrow inside.
    (d) a circle with a dot in the middle and a line with an arrow through the dot.

59. With regard to Brilliance Control on radar, which of the following statements is correct?

    (a) Brilliance can never be excessive.
    (b) Only a radar technician should adjust brilliance.
    (c) Brilliance should be left on when turning the set on or off.
    (d) Brilliance is adjusted to provide a suitable contrast between the picture & its background.

60. With regard to Gain Control on radar, which of the following statements is correct?

    (a) There is no correlation between the radar Gain control and the Volume control on radio receivers.
    (b) Excessive Gain setting will cover the screen with the receiver noise.
    (c) Radar operating at a lower range needs a higher gain setting.
    (d) To see all the echoes, maximum Gain setting is essential.

61. With regard to Tuning Control on radar, which of the following statements is correct?

    (a) Just like an out-of-tune radio receiver, out-of-tune radar is obvious.
    (b) There is no way to check if radar is properly tuned.
    (c) Radar tuning can be verified by tuning for the best reception of a distant echo.
    (d) Echoes appear distorted on an out-of-tune radar.

62. With regard to the effect of waves and rain on the reception of other radar echoes, which of the following statements is correct?

    (a) Echoes of waves and rain can hide echoes of vessels.
    (b) Waves do not generate echoes.
    (c) Rain does not generate echoes.
    (d) Echoes of targets beyond the rain are not affected.

63. With regard to Sea Clutter Control on radar, which of the following statements is correct?

    (a) It is best to set the Sea Clutter Control as high as possible.
    (b) The Sea Clutter Control cannot differentiate between wave echoes & other target echoes.
    (c) The Sea Clutter pattern is symmetrical around the vessel.
    (d) Sea Clutter is not determined by the sea state.

64. With regard to Rain Clutter Control on radar, which of the following statements is correct?

    (a) Rain shows up as a milky patch in the centre of the screen.
    (b) Rain does not affect the strength of echoes of other targets.
    (c) Rain Clutter Control does not differentiate between the nearest edge of the rain and its overall mass.
    (d) Some radar sets have a circuit to automatically eliminate rain Clutter.

65. With regard to Pulse Length Control on radar, which of the following statements is correct?

    (a) Use of longer pulse length increases target definition but reduces radar's range capability.
    (b) All radar sets work on the same pulse length.
    (c) Use of shorter pulse length increases target definition but reduces radar's range capability.
    (d) Radars are not capable of automatically changing pulse length when the range is changed.

66. Which of the following controls is likely to obscure echoes of small targets, but should have only a spring loaded turn-off switch?

    (a) Range Rings
    (b) VRM
    (c) EBM
    (d) Heading Marker

67. With regard to radar, which of the following statements is incorrect?

    (a) The range & bearing of a target should be measured only to the edge of the target.

    (b) The range & bearing of a target can be measured by moving the cross-shaped electronic curser to target's position.

    (c) The range & bearing of a target can be measured by counting the range rings & measuring its angle relative to ship's head.

    (d) The range & bearing should be measured to the middle of small targets, and to the edge of large targets.

68. With regard to radar operation, which of the following statements is incorrect?

    (a) The bearing error due to vessel yawing does not arise on Stabilized displays.

    (b) To make the bearing of a headland more accurate, adjust the bearing by half the beamwidth towards the sea (not towards the land).

    (c) A scanner of 6° beamwidth would display a 6° wide echo.

    (d) The bearing error is equal to half the beamwidth.

69. With regard to radar operation, which of the following statements is correct?

    (a) Radar bearings are more reliable than radar ranges.

    (b) Fixing a vessel's position by taking two or more ranges is likely to be more accurate than by taking a bearing & a range.

    (c) Shadow sectors are areas immediately behind the obstruction to radar beam.

    (d) Land does not cause Blind & Shadow Sectors.

70. With regard to fixing vessel's position by radar, which of the following statements is incorrect?

    (a) Blind & Shadow Sectors of land can cause discrepancies between information on the chart & the radar picture of the land.

    (b) Bays & inlets may not appear on radar.

    (c) Lower parts of coastline may not appear on radar screen.

    (d) Fixing positions with ranges from sloping coastline is more accurate than from steep cliffs.

71. In relation to Indirect Echoes on radar, which of the following statements is incorrect?

    (a) They appear at correct bearings but the wrong ranges.

    (b) They may be caused by vessel's own masts and funnels or metallic bridges ashore.

    (c) They are also referred to as Mirror Echoes

    (d) They are also referred to as Virtual Images.

72. In relation to False Echoes on radar, which of the following statements is incorrect?

    (a) The Side Lobe Effect creates two additional echoes per target.

    (b) The error known as Multiple Echoes creates a row of echoes at the same range

    (c) The Side Lobe Effect can be eliminated by temporarily reducing Gain.

    (d) The Multiple Echoes can be eliminated by temporarily reducing Gain.

73. In relation to Second Trace Returns on radar, which of the following statements is incorrect?

    (a) Surface Ducting takes place between a layer of warm air & the ground.

    (b) Super Refraction is caused by warm dry air blowing over a cold sea.

    (c) Temperature Inversion is particularly strong when there is air turbulence.

    (d) The shape of the coastline reflected in a Second Trace Echo is usually distorted in the middle.

74. In relation to Non-standard Refraction in radar, which of the following statements is correct?

    (a) Super-refraction causes radio waves to travel shorter distances.

    (b) Sub-refraction causes radio waves to travel longer distances.

    (c) Super-refraction occurs near cold landmasses.

    (d) Super-refraction occurs near warm landmasses.

75. Which of the following terminology is not associated with Super-Refraction in Radar?

    (a) Virtual Image.
    (b) Non-standard Refraction.
    (c) Anomalous Propagation.
    (d) Freak Propagation.

Chapter 16: Radar, SART & other Radar Transponders

76. In relation to radar reception, which of the following statements is correct?

    (a) Precipitation in cold fronts does not show up.
    (b) Clouds do not show up unless there is precipitation inside them.
    (c) Fog and rain produce similar echoes.
    (d) Hailstones produce stronger echoes than rain.

77. In relation to unwanted appearances on radar screen, which of the following statements is correct?

    (a) Radial lines, like the spokes of a wheel, are caused by interference from another radar operating nearby.
    (b) Bright speckles scattered randomly are caused by dirty or faulty contacts on the rotating parts of radar.
    (c) Radar is not affected by on-board magnetic influences.
    (d) Mutual Radar Interference is more common on longer-range scales.

78. In relation to radar Heading Marker, which of the following statements is correct?

    (a) Radar is not provided with a control switch to allow operators to vary the heading marker position.
    (b) An out-of-alignment heading marker will not cause an error in radar bearings.
    (c) A permanently out-of-alignment heading marker indicates misalignment of contacts inside the scanner housing.
    (d) The operator cannot verify the heading marker error by heading towards a distant isolated object.

79. In relation to the radar Centre Adjustment by the operator, which of the following statements is correct?

    (a) It is sometimes necessary, but only on raster-scan displays.
    (b) It is sometimes necessary, but only on rotating trace displays.
    (c) It is sometimes necessary on all types of displays
    (d) There is no control switch provided for the operator to correct the Centring Error.

80. In relation to Parallax Error in bearings taken by radar, which of the following statements is correct?

    (a) It occurs in all bearings whether taken with mechanical curser or EBM.
    (b) It occurs only in bearings taken with EBM.
    (c) The error does not vary between bearings.
    (d) It can be minimised by standing at a suitable position above the radar.

81. Which of the following is not a factor in determining radar's maximum range?

    (a) Pulse length
    (b) Peak Power
    (c) Receiver Noise
    (d) Screen size

82. Which of the following is not a factor in determining radar's maximum range?

    (a) Height of scanner
    (b) Height of target
    (c) Meteorological visibility
    (d) Pulse length

83. The height of a radar scanner is 4 metres above the water. At what geographical range will it pick up the peak of an island 144 metres high?

    (a) 30.94 nm.
    (b) 26.52 nm.
    (c) 4.42 nm.
    (d) 22.10 nm.

84. The height of a radar scanner is 6 metres above the water. At what geographical range will it pick up a ship 20 metres above the water?

    (a) 17.41 nm.
    (b) 9.88 nm.
    (c) 15.29 nm.
    (d) 21.2 nm.

85. The height of a radar scanner is 5 metres above the water. At what geographical range will it pick up an island 100 metres high?

    (a) 22.1 nm.
    (b) 35 nm.
    (c) 11.7 nm.
    (d) 27.04 nm.

86. The height of a radar scanner is 7 metres above the water. At what geographical range will it pick up a peak of 210 metres height?

    (a) 39.77 nm.
    (b) 37.87 nm.
    (c) 32.02 nm.
    (d) 21.07 nm.

87. Which of the following is improper use of radar?

    (a) Position fixing.
    (b) Collision avoidance.
    (c) Gauging state of visibility.
    (d) Replacing visual lookout.

88. In relation to use of radar, which of the following terms is not related to Parallel Indexing?

    (a) Second trace Returns.
    (b) Cross-Index Ranging.
    (c) Dead Range.
    (d) Piloting a vessel along a pre-determined track.

89. In collision avoidance, which of the following may constitute improper use of radar in poor visibility?

    (a) Long range scanning.
    (b) Radar plotting.
    (c) Immediate course alteration for an echo that has suddenly appeared close on starboard bow.
    (d) Making systematic observations of detected objects.

90. In the definition of Lookout in the International Collision Prevention Rule 5, the words "...by all available means..." include a reference to:

    (a) Compass
    (b) Radar
    (c) Telegraph
    (d) Auto-pilot

91. In the International Collision Prevention Rule 6, for vessels fitted with operational radar, one of the additional factors in determining Safe Speed is the:

    (a) State of visibility.
    (b) Possibility of not detecting small vessels.
    (c) Presence of background lights ashore.
    (d) Size of own vessel.

92. With regard to use of radar in the International Collision Prevention Rules 7, 8 & 19, which of the following statements is incorrect?

    (a) A vessel fitted with radar should make a succession of small alternations of course and/or speed, instead of one large alteration.
    (b) If there is sufficient sea room, alteration of course alone may be the most effective action.
    (c) Assumptions shall not be made on the basis of scanty radar information.
    (d) A vessel detecting the possibility of a close-quarters situation with another vessel in ample time may take avoiding action based on radar information alone.

93. In taking collision avoidance action by alteration of course based on radar information alone (Rule 19), which of the following statements is incorrect?

    (a) She shall avoid altering course towards a vessel abeam.
    (b) She shall take avoiding action in ample time.
    (c) She shall avoid altering course towards a vessel abaft the abeam.
    (d) She shall not alter course to port.

94. Which of the following statements is not true about ARPA?

    (a) It is a computerised target-plotting device.
    (b) It extracts the bearings & ranges of targets from radar and calculates their closest points & times of approach.
    (c) It can only acquire targets from the guard zone.
    (d) It determines collision-avoiding manoeuvres.

95. Which of the following statements is true about Passive Radar Reflectors?

    (a) They are not suitable for small vessels.
    (b) They are usually made of fibreglass.
    (c) None of the "cups" should be in the "catch rain" position.
    (d) They should not be hoisted on halyards.

96. In relation to Radar Reflectors, which of the following statements is incorrect?

    (a) Radar detection range of navigational buoys not fitted with radar reflectors is about 2 or 3 miles.
    (b) Racon appears on radar screen as a bright radial line
    (c) Active radar reflectors transmit coded identifying signals to radars operating upto 20 miles away.
    (d) Racon provides a bearing as well as range.

97. With regard to RACON, which of the following statements is incorrect?

    (a) It appears on radar screen slightly closer than its true range.
    (b) It appears on radar screen as segments of concentric arcs or a Morse code signal.
    (c) It is subject to giving false echoes.
    (d) It is likely to mask targets close to it.

98. Which of the following surfaces would reflect the most radar energy back to radar?

    (a) Cylindrical
    (b) Spherical
    (c) Conical
    (d) Plane

99. Which of the following surfaces would reflect the most radar energy back to radar?

    (a) Cylinder
    (b) Sphere
    (c) Cone
    (d) Ball

CHAPTER 16.2: SARTS, RADAR REFLECTIVE DISTRESS SIGNALS & ANTI-COLLISION RADAR TRANSPONDERS

100. In relation to Anti-Collision Radar Transponders, which of the following statement is <u>incorrect</u>?

    (a) They produce a line of 10 blips on the interrogating vessel's radar.
    (b) The blips on the interrogating vessel's radar extend outward for approximately 1 mile from the Transponder's position.
    (c) They alert the host vessel that another vessel with operating radar is in the vicinity.
    (d) They are fitted on small vessels, which may otherwise be difficult to detect by radar.

101. Which of the following statements is true about Radar Reflective Distress Signals?

    (a) They eject a radar reflective flare.
    (b) They do not show up on aircraft radar
    (c) Their echo is visible on radar screen for 10 to 20 minutes.
    (d) They can be used during daytime only.

102. A SART:

    (a) is the search & rescue radar transponder.
    (b) detects operating radars of ships only.
    (c) is an anti-collision radar transponder.
    (d) detects operating radars of aircraft only.

103. In relation to SARTS, which if the following statement is <u>incorrect</u>?

    (a) It responds to radar signals from nearby aircraft or ship.
    (b) It responds only to 10 cm wavelength radar.
    (c) It appears on radar screen as a series of 12 blips.
    (d) A ship with radar scanner at a height of 15 metres will detect it at a range of approximately 8.5 miles.

104. In relation to SARTS, which if the following statement is <u>incorrect</u>?

    (a) The spacing between SART's blips is 1.2 miles.
    (b) IMO recommends searching ships operate their radars on short pulse & on 6 to 12 miles range.
    (c) A ship approaching a SART may find it 150 metres out of position on its radar.
    (d) The range of SART's transmission can be increased by mounting it at a height.

CHAPTER 16: ANSWERS

1 (b),   2 (a),   3 (b),   4 (d),   5 (a),
6 (c),   7 (b),   8 (d),   9 (a),   10 (c),
11 (c), 12 (b), 13 (c), 14 (a), 15 (b),
16 (d), 17 (c), 18 (a), 19 (c), 20 (b),
21 (c), 22 (a), 23 (d), 24 (a), 25 (d),
26 (a), 27 (c), 28 (b), 29 (d), 30 (c),
31 (d), 32 (c), 33 (b), 34 (c), 35 (a),
36 (d), 37 (b), 38 (a), 39 (c), 40 (d),
41 (c), 42 (a), 43 (d), 44 (c), 45 (b),
46 (a), 47 (d), 48 (d), 49 (a), 50 (b),
51 (c), 52 (d), 53 (d), 54 (a), 55 (b),
56 (c), 57 (a), 58 (d), 59 (d), 60 (b),
61 (c), 62 (a), 63 (b)   64 (d), 65 (c),
66 (d), 67 (a), 68 (b), 69 (b), 70 (d),
71 (a), 72 (b), 73 (c), 74 (d), 75 (a),
76 (b), 77 (d), 78 (c), 79 (b), 80 (d),
81 (d), 82 (c), 83 (a), 84 (c), 85 (d),
86 (b), 87 (d), 88 (a), 89 (c), 90 (b),
91 (b), 92 (a), 93 (d), 94 (c), 95 (d),
96 (b), 97 (a), 98 (d), 99 (a), 100 (a),
101 (c), 102 (a), 103 (b), 104 (a)

Chapter 17.1

NAVIGATION

COMPASSES, BEARINGS & HEADINGS

COURSES AND BEARINGS

Courses or Headings must not be confused with Bearings. As illustrated, a course is the direction in which the boat is heading. Bearing is the direction of another object, such as an island, lighthouse or another vessel. A compass is used to measure these directions.

True (T) course or True bearing implies that it was measured with a compass which was free of errors, or for which the errors were known and applied. Courses and bearings read from a navigational chart are always True because charts are printed in the True north-south direction. The compass circles (*Compass Rose*) printed on charts are used to measure true directions. (See Fig 17.27)

A True Course is therefore the angle between true north and the vessel's heading. And, the True Bearing of an object is the angle at the vessel between true north and the object

Course and Bearings measured with a compass whose error has not been applied are written with the letter "C" (for Compass), not "T" (for True).

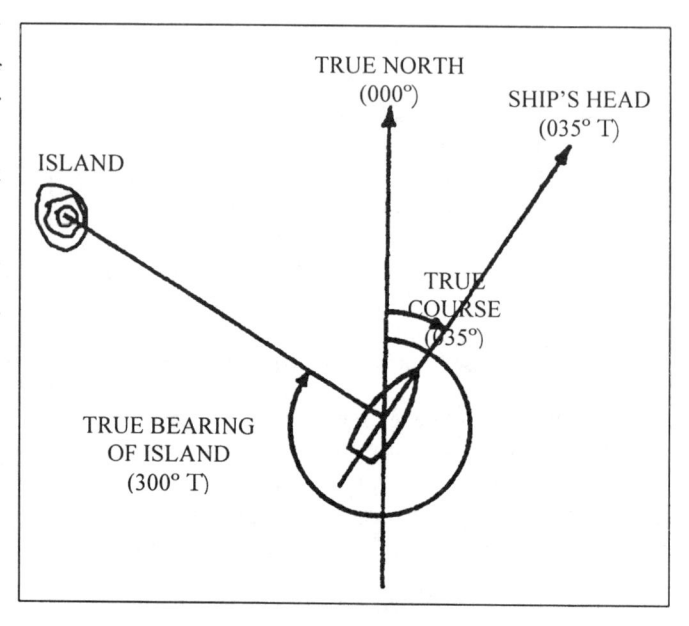

Fig 17.4
Vessel is steering a course of 035°(T). The island bears 300°(T).
(T is for True)

RELATIVE BEARINGS

Unless your radar incorporates a compass, bearings measured by radar are ***relative bearings***. A Relative bearing (Red, Green or Rel.) is an angle of an object from the vessel's heading. The Green and Red relative bearings are measured to starboard or port of the vessel's heading (ship's head). However, the Rel. (Relative) bearings are always measured clockwise from the ship's head. *(Measuring relative bearings with a pelorus is discussed below)*.

To convert a Rel. (Relative) bearing into a True bearing, add to it the True Ship's Head, as shown:

- Bearing 040° Rel + Ship's Head 070° T = Bearing 110° T
- Bearing 260° Rel + Ship's Head 070° T = Bearing 330° T
- Bearing 310° Rel + Ship's head 070° T = Bearing 380° T (Since the compass is of 360°, the bearing is 380 - 360 = 020° T)

To convert the Green and Red relative bearings to true bearings – add the Green or subtract the Red from the ship's head. The above Green and Red relative bearings can be converted to True bearings, as follows:

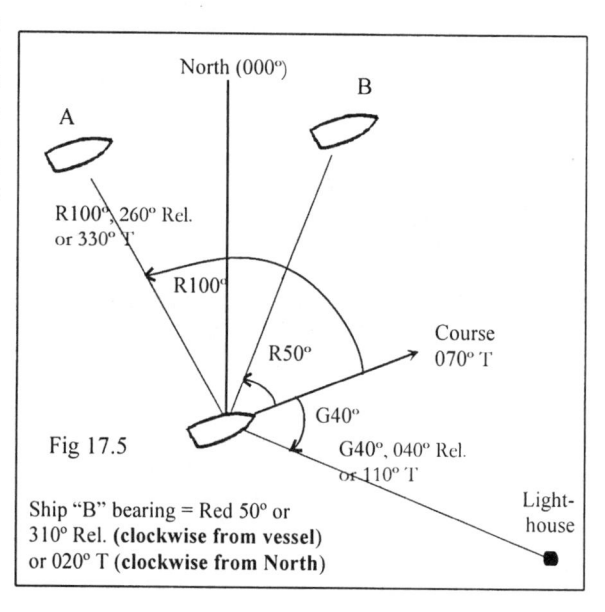

Fig 17.5
Ship "B" bearing = Red 50° or 310° Rel. **(clockwise from vessel)** or 020° T **(clockwise from North)**

Chapter 17.1: Navigation - Compasses, Bearings & Headings

- Ship's Head 070° T + G40° = Brg 110° T
- Ship's Head 070° T - R50° = Brg 020° T
- Ship's Head 070° T - R100° = Brg -30°T
  Add   +360°
  True Brg = 300°

COMPASSES

There are three types of compasses: Magnetic, Gyro and Fluxgate.

MAGNETIC COMPASSES

COMPASS CONSTRUCTION:

The earth is magnetic. Therefore, a freely suspended bar magnet or magnetic needle aligns itself parallel to the earth's magnetic lines of force, and thus establishes a DIRECTION. This is the principle of magnetic compass.

A typical compass contains a number of bar magnets or magnetic needles, the directional force of which acts together. The compass card, made of non-magnetic material, is mounted on a wire frame. Under the frame is a cap containing a BEARING, also called the JEWEL, which rides on a hard sharp point called the PIVOT. The magnets are attached to the frame and aligned with the N-S markings of the card.

A hollow float is sometimes attached under the frame to partially suspend the card and magnets, and thus reduce friction and wear on the bearing and pivot.

The complete mechanism is fitted inside a bowl, with a transparent sealed cover. Most magnetic compass bowls are filled with liquid through a pluggable hole, to DAMPEN the swing of the compass card. The liquid is mainly distilled water with a little alcohol added to prevent it from freezing in cold climates. Some compasses are filled with mineral oil instead of water. An expansion bellows is fitted in the bowl to allow for expansion and contraction of liquid with changes in temperature without bubbles being formed.

An adequately dampened and steady compass is also referred to as a DEAD BEAT COMPASS. Such a compass has short but very strong needles (magnets) and a comparatively large bowl.

The compass card may be of the Back Reading or the Front Reading type. Compasses are usually gimballed for a 45° tilting angle fore and aft and unlimited tilting angle athwartships. Sailboat compasses are also marked with 45° lubber lines for tacking references. Some also come with a built-in Clinometer.

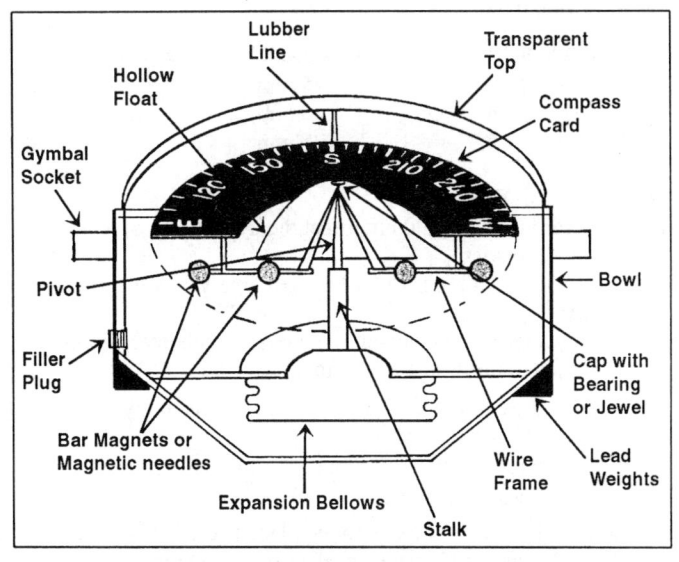

Fig 17.6: COMPASS CONSTRUCTION

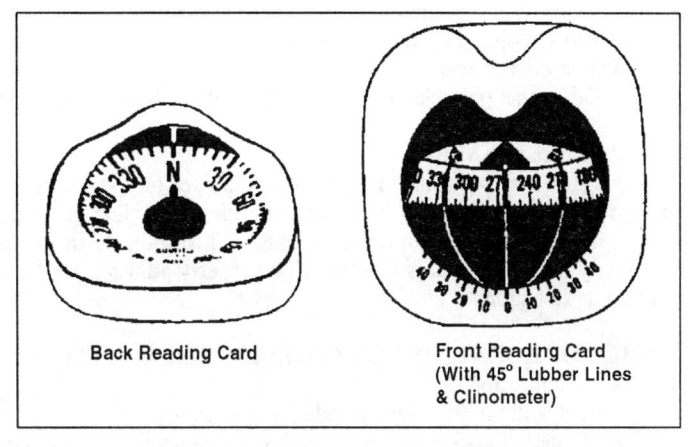

Fig 17.7: SUUNTO FLUSH MOUNT & SAILBOAT COMPASSES

RULES FOR INSTALLING COMPASSES

The USL Code Section 13 Appendix B and NSCV Part C list the rules for installation of compasses on small vessels (Class C & D). These are summarised as follows:

STEERING COMPASSES

- To be located so that the view of the horizon from the compass position shall be uninterrupted for a minimum arc of

115° from right ahead on either side of the vessel.
- To be located forward of the steering position in such a manner that it can be easily read from the normal steering position.
- To be fitted with an efficient means of illumination together with a device for dimming the illumination.
- To be suspended by gimbals so that the bowl remains horizontal when the binnacle is tilted 40 degrees in any direction.
- To be located in such a position as to permit proper adjustment.
- To be provided with a compass card according to the following table:

Vessel Length	Diameter of Compass Card
Less than 10m	75 mm
Less than 20m	100 mm
20 & over	125 mm

Fig 17.8: A SILVA MAGNETIC COMPASS

MAGNETIC COMPASSES
- A power failure should not affect the availability of a suitable compass.
- Be provided with at least 24 hours of emergency electrical supply.
- For new commercial vessels, a general arrangement plans submitted to the Authority, to include details of:
    - the equipment or magnetic materials likely to affect the compass.
    - the items that will impair the visibility of the horizon from a compass.
- Electrical equipment must not be placed near a compass closer than the "safe distance" determined by test or recommended by the manufacturer.
- All electrical equipment close to a compass should be checked - in operating and non-operating mode - for any effect on the compass.
- Do not operate portable electrical equipment, such as microphones, telephone and transistor radios, close to a compass.

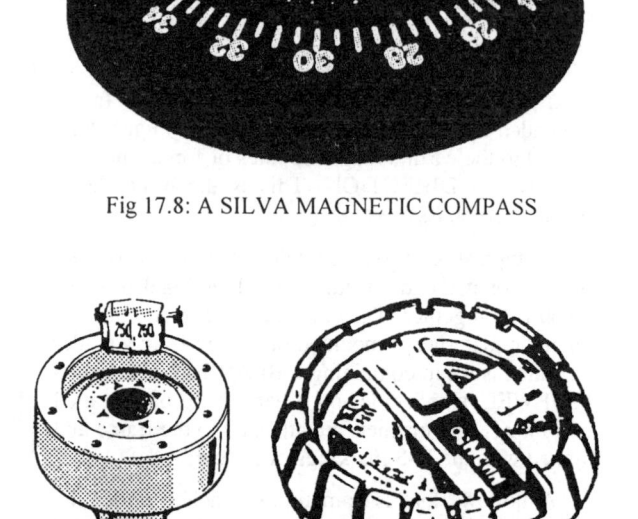

Fig 17.9: HAND-BEARING COMPASSES
(Two Types - Both Magnetic)
Batteries

COMPASS ALIGNMENT
It is often not practical to mount the compasses on the centreline of the vessel. This is not a problem as long as a line through the centre of the compass card and the lubber line is parallel to the centreline of the vessel. The lubber line is a vertical line on the inside of the compass bowl to indicate the vessel's heading. This alignment is important, because an improperly aligned compass can never be properly adjusted for Deviation. It will have a constant error on all headings in addition to deviation.

CHOICE OF LOCATION FOR COMPASS INSTALLATION
- Choose a location of least vibration.
- Locate it suitably for steering and taking bearings.
- The compass location should be protected from sea and salt sprays, and the compass should be protected from the sun when not in use.
- Choose a location of minimum Deviation as detailed below.

HOW TO MINIMISE COMPASS DEVIATION
As discussed below, the magnetic compass error is made up of two components:
- *Variation*: due to the earth's magnetic field
- *Deviation*: due to the vessel's magnetic field

All vessels using magnetic compasses have to cope with Magnetic Variation, but not necessarily deviation. A vessel without any magnetic materials in its hull or equipment will not have any deviation in its compass. Such a vessel is rare.

Nevertheless, in non-ferrous hulls a carefully chosen location for a compass can mean little or no error due to magnetic

Chapter 17.1: Navigation - Compasses, Bearings & Headings

deviation. The following precautions should be observed:

1. ZERO-IN THE INTERNAL MAGNETS
Compasses are normally sold with internal compensatory magnets set to zero, but it is a good practice to double check before installation. The slots of the screws for N-S and E-W compensators on the back of the compass should be horizontal. This is necessary to eliminate the effect of these magnets on the compass Deviation. Any Deviation will then come only from external sources.

2. KEEP AWAY FROM MAGNETIC INFLUENCES
Whether a mounted or a hand-bearing compass, it should not be used in the vicinity of electrical wiring, vessel's radio and ferrous materials such as engines and winches. Do not install a compass immediately above the engine, or above the tool cupboard. Do not build a shelf in the compass surrounds where magnetic influences, such as metal tea mugs, tools and portable radios, may be placed. Do not use a hand-bearing compass in a steel hull.

3. CORRECT FOR EXTERNAL DEVIATION
In steel hulls, deviation can be as large as 50 degrees. However, it can be reduced to almost zero by inserting magnets in the compass binnacle and fitting soft iron correctors, known as the Quadrantal Correctors, on the sides and a soft iron Flinder's Bar in front of the binnacle. A Deviation Card is then made to record any remaining deviation in the adjusted compass. Commercial vessels are required to have their compasses adjusted at least every three years. A qualified Compass Adjuster should be employed to do this work and provide a certified Deviation Card for the compass. (See Navigation and Chartwork). The Deviation Card must be displayed in the wheelhouse.

4. CHECK THE CARD FOR DEFLECTION
- Connect the compass light to battery with a "return" wire, i.e., both terminals connected directly to the battery.
- Check the card for deflection due to electromagnetic field by switching lights and other equipment on and off. Have the compass re-adjusted after completing a new electrical installation.

Fig 17.10: COMPASS BINNACLE
(A Saura Compass)

COMPASS MAINTENANCE & REGULAR CHECKS
- Check compass error regularly. You may be able to compare the compass heading of the boat with the jetty's true direction, provided the vessel is berthed securely parallel to the jetty. After allowing for magnetic Variation, the two should read the same (assuming no Deviation). As illustrated here, and discussed below, transit bearings of navigational features and landmarks offer the same facility.
- Make sure no one has disturbed the magnets inserted by the compass adjuster in the binnacle.
- There should be no air bubbles in the liquid in the compass bowl.
- Compass card should rotate freely and gimbals free to move.
- Compass light should be in working order.
- Compass should be clean of dust and salt, etc.

IF COMPASS BEHAVES ERRATICALLY
The magnetic compass card is supported only in the centre on a pivot. Its erratic behaviour is an indication that either the pivot point is worn, there is a large air bubble in the liquid or the rotation of the compass card is being obstructed for some other reason.

Fig 17.11: TRANSIT BEARING

A NEW ERROR IN THE COMPASS
A sudden and unexplained error in a magnetic compass could be due to a significant metallic or electrical alteration in the vessel or interference from a nearby portable object such as a metal tea mug, pair of pliers or a transistor radio.

HOW TO REMOVE AIR BUBBLE IN A COMPASS BOWL
An air bubble in a liquid filled compass bowl will make the compass inaccurate and unstable. To remove it, turn the compass bowl on its side. Take out the screw plug on the side of the bowl. Work the bubble under the hole. Using an eyedropper, insert a few drops of distilled water until it overflows. Screw back the plug. *Note: the bubbles should only be*

removed by a qualified compass technician.

Some compasses are fitted with a "bubble trap" in the base of the bowl. Slow inversion and subsequent erection will trap the bubble in the bowl base.

GYRO COMPASSES

A gyrocompass is non-magnetic. It operates on the principle of a fast spinning wheel maintaining its direction. By holding its orientation in space, it is able to sense changes in the direction of the vessel.

This compass is not affected by the errors that affect magnetic compasses, i.e., magnetic deviation and variation. Provided it gets a steady power supply, its error is usually negligible. Voltage fluctuations may cause it to run a degree or two High or Low, like a clock running a little fast or slow.

Modern gyro compass units, such as the Furuno shown in figure 17.12, can be temporarily run with a 24 Volt battery in case of the mains power failure. It draws about 2.2 Amps current in the settled condition.

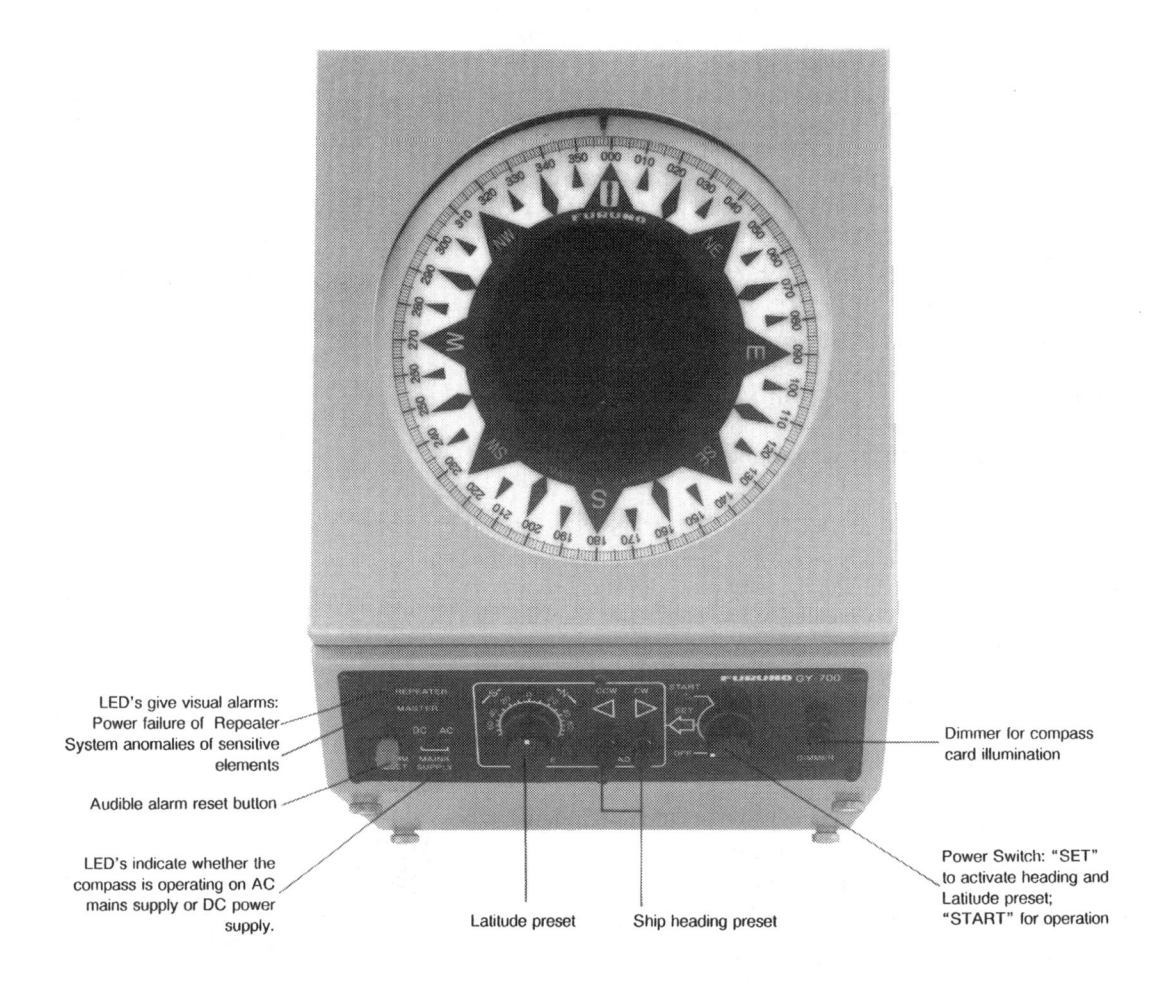

Fig 17.12: FURUNO MASTER GYRO COMPASS

ADVANTAGES OF A GYRO COMPASS:
- It is Deviation and Variation free.
- Numerous repeaters (slaves) can be run from one master compass.
- The course information can be easily fed to GPS, autopilot, radar, etc.

DISADVANTAGE:
- The compass information depends on the power supply.

Chapter 17.1: Navigation - Compasses, Bearings & Headings

Fig 17.13
A GYRO COMPASS
REPEATER OR SLAVE

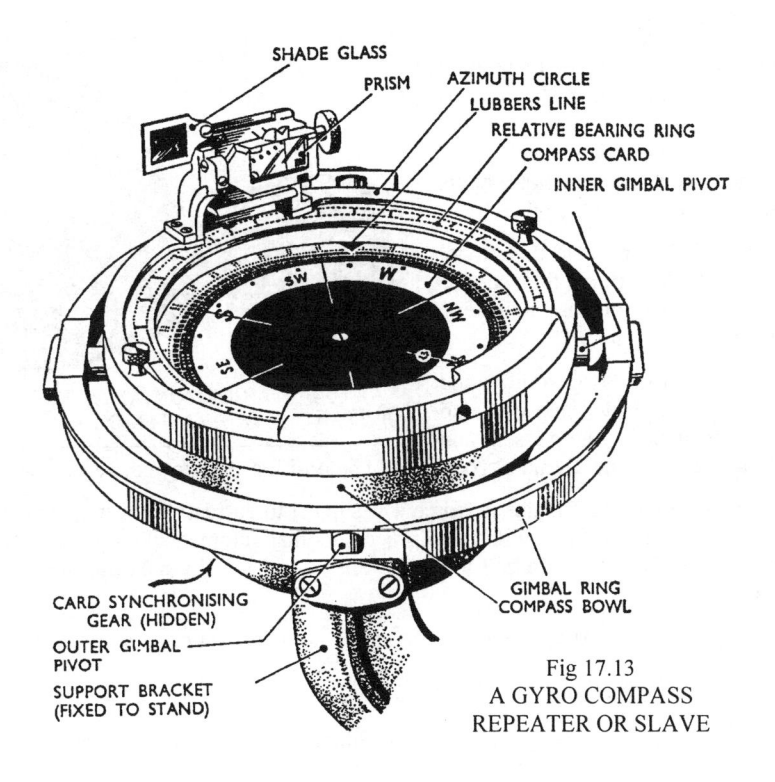

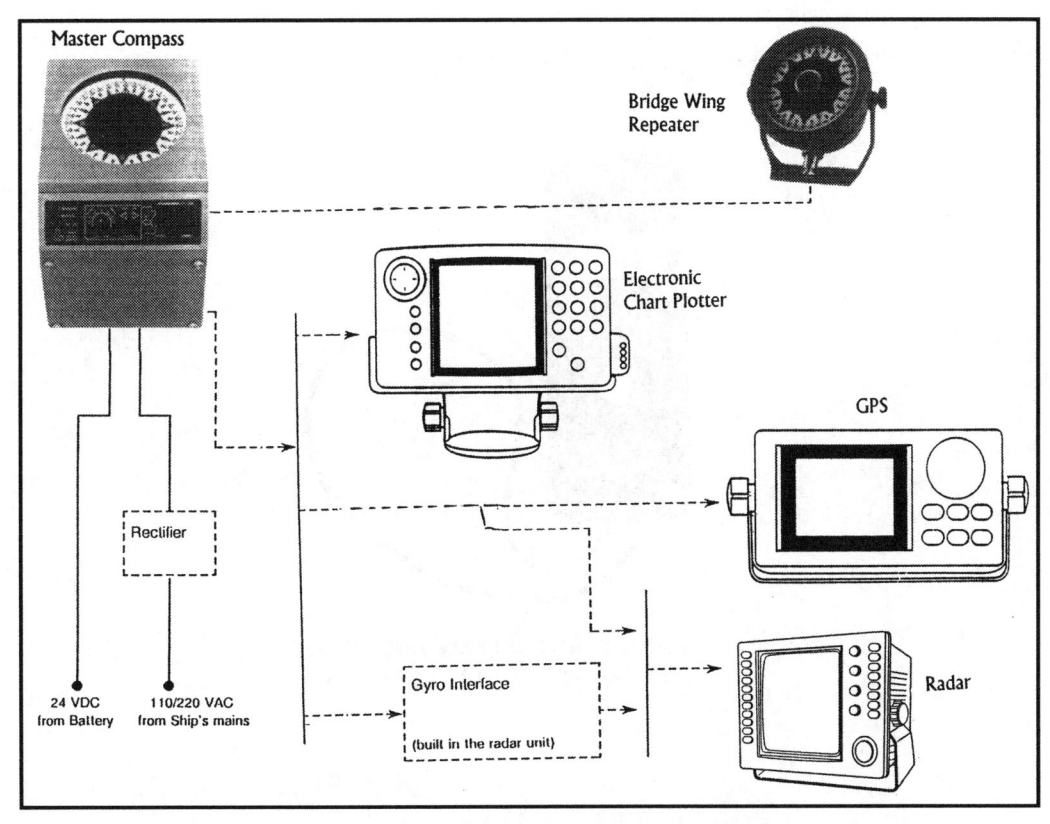

Fig 17.14: GYRO COMPASS & OFFSHOOTS

FLUXGATE COMPASSES

The fluxgate compass provides direction by reading the earth's magnetic field. It consists of two coils carrying an electrical current. One coil is mounted horizontally and the other vertically. Any alteration of the vessel's heading (in the earth's magnetic field) alters the flow of current in one coil relative to the other.

These signals are amplified to provide directional information. Being electrical, the information can be easily used for heading input to compass repeaters, autopilots, radars, GPSs and chart plotters.

Fluxgate compasses are sensitive to rolling and pitching. They should be well gimballed and mounted in a position of least movement in the vessel, and aligned with the vessel's fore and aft line. Remote repeaters can be positioned in any convenient location.

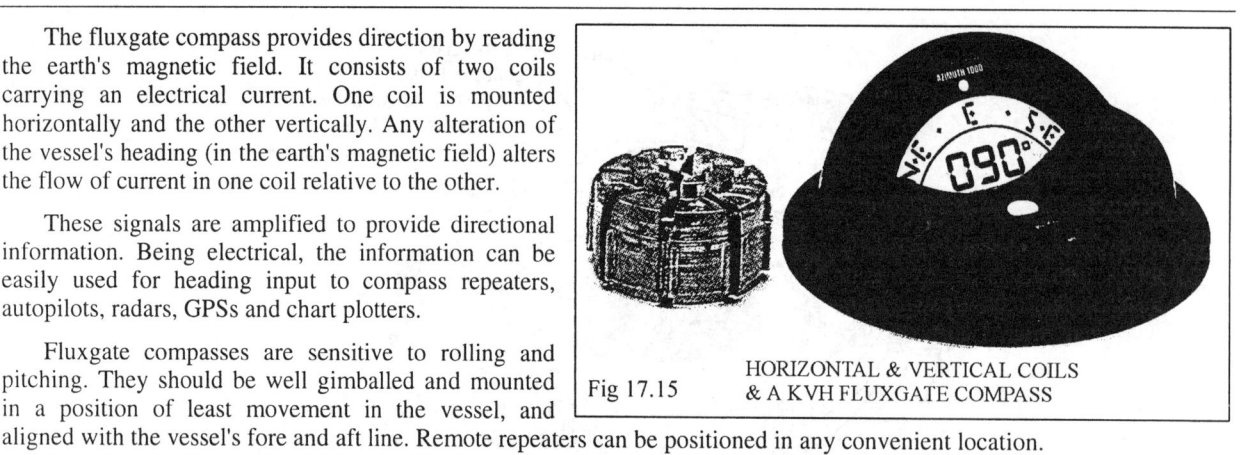

Fig 17.15 HORIZONTAL & VERTICAL COILS & A KVH FLUXGATE COMPASS

Since this compass relies on reading a magnetic force it is subject to correction for magnetic Variation and Deviation. However, many fluxgate compasses are fitted with microprocessors that automatically compensate for both Deviation and Variation and provide a true heading to within half a degree. They automatically measure the Deviation but the Variation has to be put into their memory.

Fluxgate compasses require a power supply but far less than required for a gyrocompass.

HAND-HELD FLUXGATE COMPASSES

Fluxgate Hand Bearing Compasses, being sensitive to heeling, must be held absolutely horizontally when recording bearings.

To overcome the possibility of out-of-level errors these compasses allow for taking a series of bearings, which are memory-stored and computer-averaged on recall.

The bearing sights and the LCD digital display are illuminated for night use.

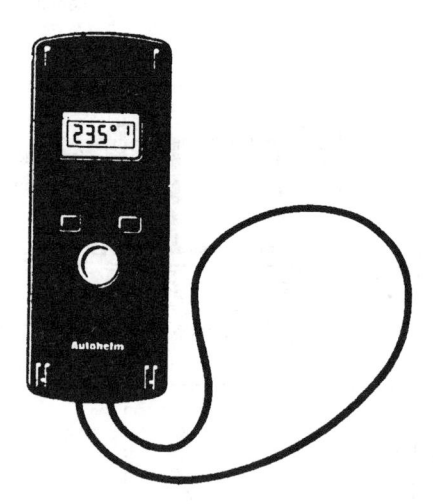

Fig 17.16: AUTOHELM HAND-HELD FLUXGATE DIGITAL COMPASS

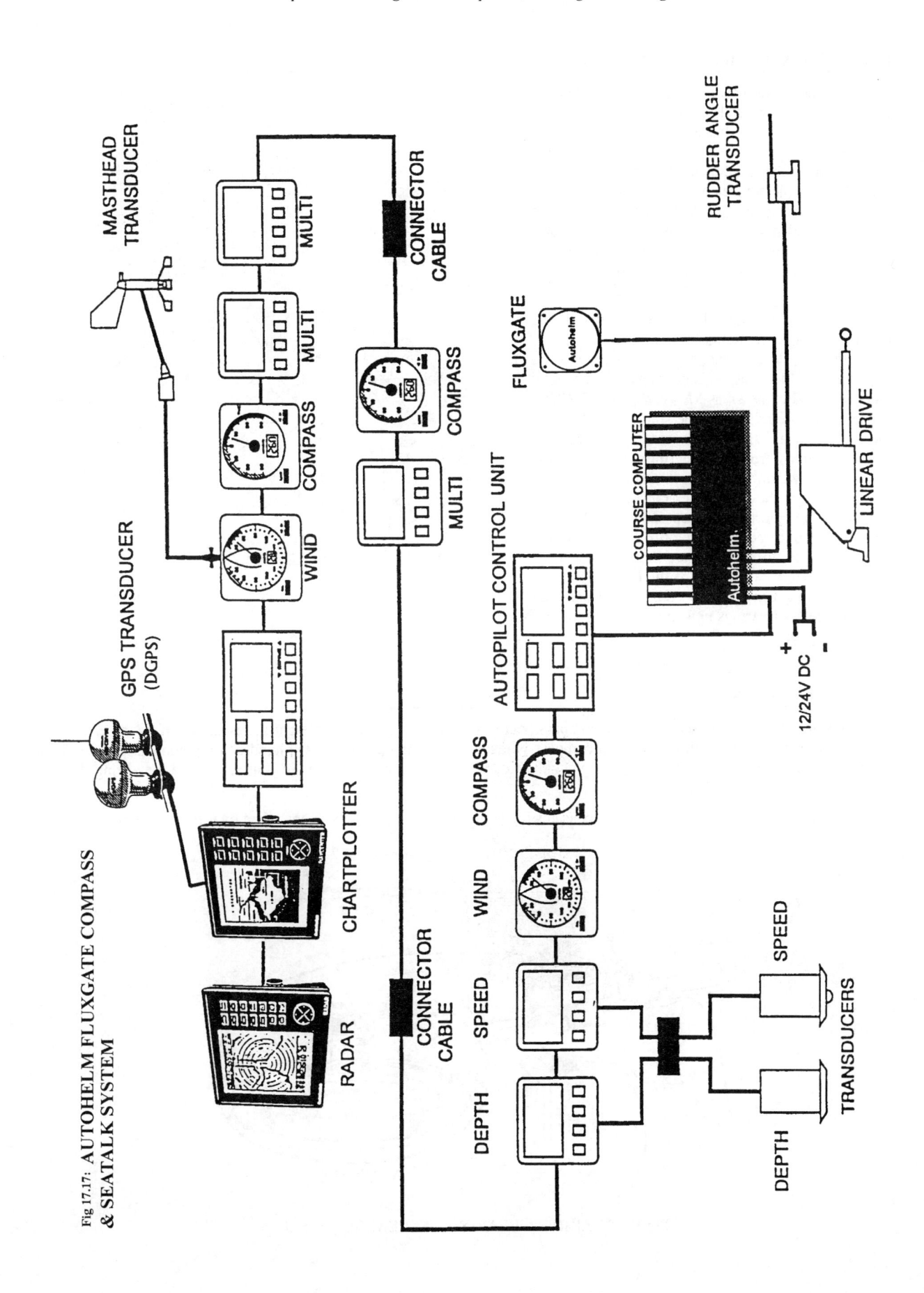

Chapter 17.1: Navigation - Compasses, Bearings & Headings

Fig 17.17: AUTOHELM FLUXGATE COMPASS & SEATALK SYSTEM

PELORUS – A DUMB COMPASS

Hand held compasses are common in boats. However, as discussed earlier, such compasses cannot be used in ferrous hulls. In a steel hull if there is a need to take a bearing from a location obscured from the fixed compass, a pelorus may be used. This simple device consists of a set of sighting vanes mounted over a circular card or a plate marked in 360 degrees - just like a compass card, but without a magnetic needle. It can be temporarily positioned anywhere in the vessel but must be lined up with the fore and aft line of the vessel.

To take a bearing, clamp the pelorus in the desired position. Rotate the disc so that the scale is set with 000° dead ahead, over or parallel to the centreline of the boat. The bearings that are now taken with it are known as *RELATIVE BEARINGS* - because they are relative to the ship's head. Convert these to Compass Bearings by adding the vessel's heading as read from the steering compass at the time of observation. The observer should call "stop" to the person at the Steering Compass at the instant of reading the Pelorus. As discussed under 'Relative bearings' above, add the Pelorus Bearing to the Ship's Course to obtain the Compass Bearing. (Subtract 360° from the sum of the Pelorus Bearing and the Ship's Course if the sum is greater than 360°). Compass Bearings must be converted to True before plotting.

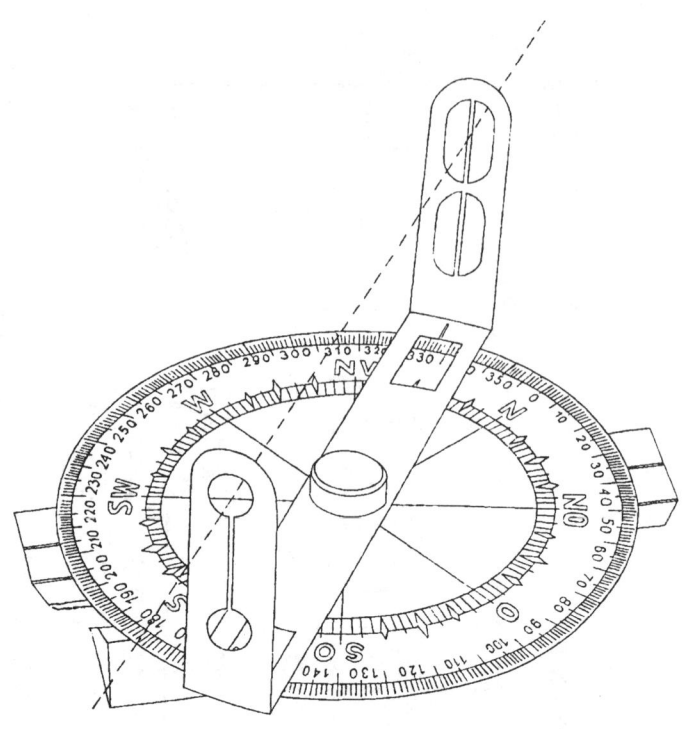

Fig 17.18: PELORUS

AZIMUTH MIRROR

An Azimuth Mirror is used for taking compass bearings (azimuth), just like the sight vanes shown in figure 17.18. It is basically a reflecting prism through which compass is read while taking a bearing. On hand-bearing compasses, it is fitted to the rim (See Figure 17.9). On larger vessels with fixed bearing compasses, the prism is attached to a metal ring that sits and rotates on the compass rim.

The curve on the face of the prism compensates for minor misalignment of bearings taken over the compass card.

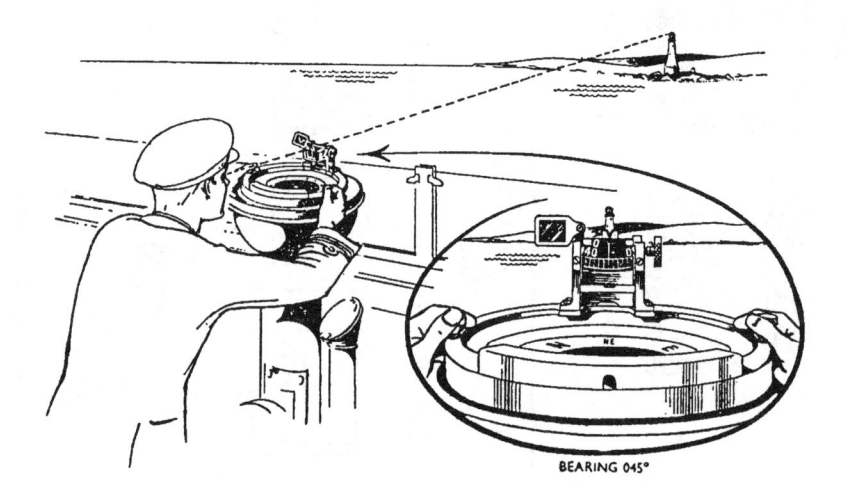

Fig 17.19 TAKING A BEARING ON A LARGE VESSEL

Chapter 17.1: Navigation - Compasses, Bearings & Headings

COMPASS ERROR

As discussed above, among the three types of compasses (gyro, magnetic and fluxgate), only the gyrocompass is free of magnetic errors.

GYRO COMPASS ERROR

A gyrocompass is not affected by the errors that affect magnetic compasses, i.e., magnetic deviation and variation. However, voltage fluctuations in its power supply and its mechanical wear and tear can cause it to run a degree or two fast or slow.

The error is referred to as HIGH (H) when the gyro is running fast, and LOW (L) when it is running slow. It is best remembered and applied like the error in a clock. The gyro bearings and courses are shown followed by the letter 'G'

For example, if the gyro error were 2°H, a bearing of 150°G would be 148°T; and a course of 000°T would be steered as 002°G.

Vessels usually have only one gyrocompass. Gyro repeaters (or slaves) are fitted in suitable locations in the vessel and incorporated into the desired equipment, such as radar and autopilot. They all therefore reflect the gyro error.

MAGNETIC COMPASS ERROR

The Magnetic Compass Error consists of two components: Variation & Deviation.

VARIATION

The magnetic compass points to the north magnetic pole. The positions of the North and South magnetic poles are not the same as that of the North and South geographic or True poles. Consequently, except for special regions, magnetic North differs from True North. This difference is referred to as the Magnetic Variation and is measured east or west of True North.

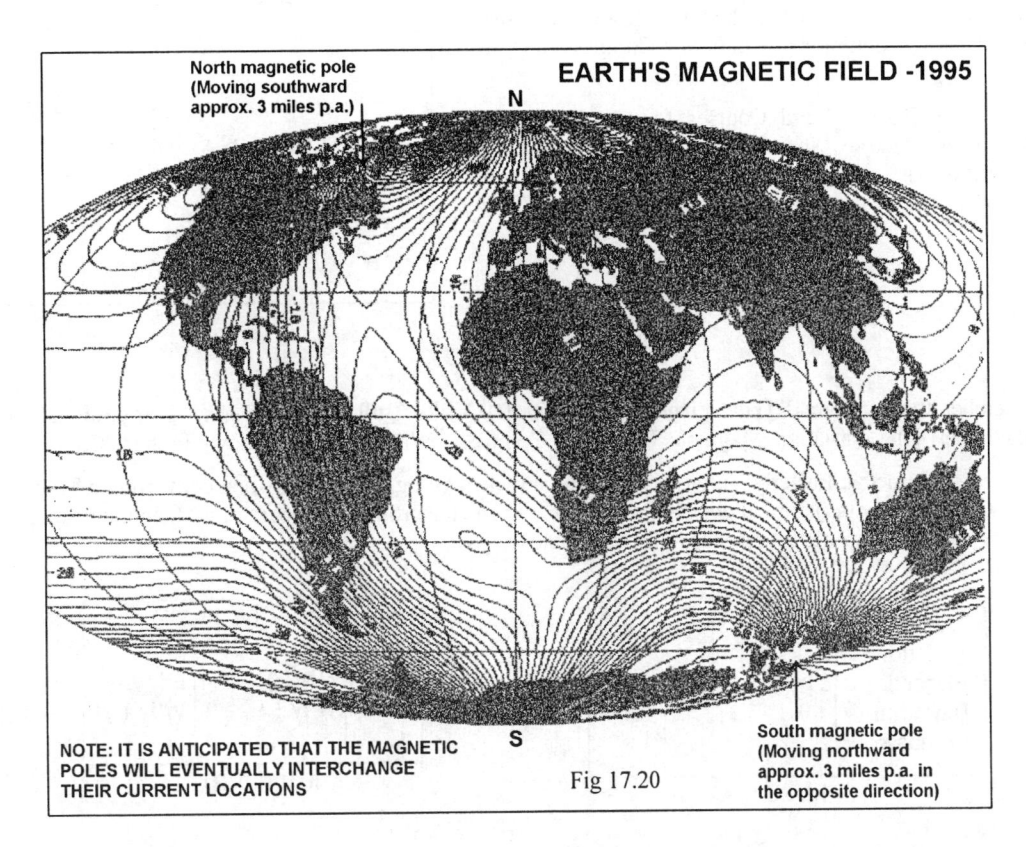

Fig 17.20 — EARTH'S MAGNETIC FIELD - 1995

North magnetic pole (Moving southward approx. 3 miles p.a.)

South magnetic pole (Moving northward approx. 3 miles p.a. in the opposite direction)

NOTE: IT IS ANTICIPATED THAT THE MAGNETIC POLES WILL EVENTUALLY INTERCHANGE THEIR CURRENT LOCATIONS

Fig 17.21 MAGNETIC VARIATIONS AUSTRALIA - 1995

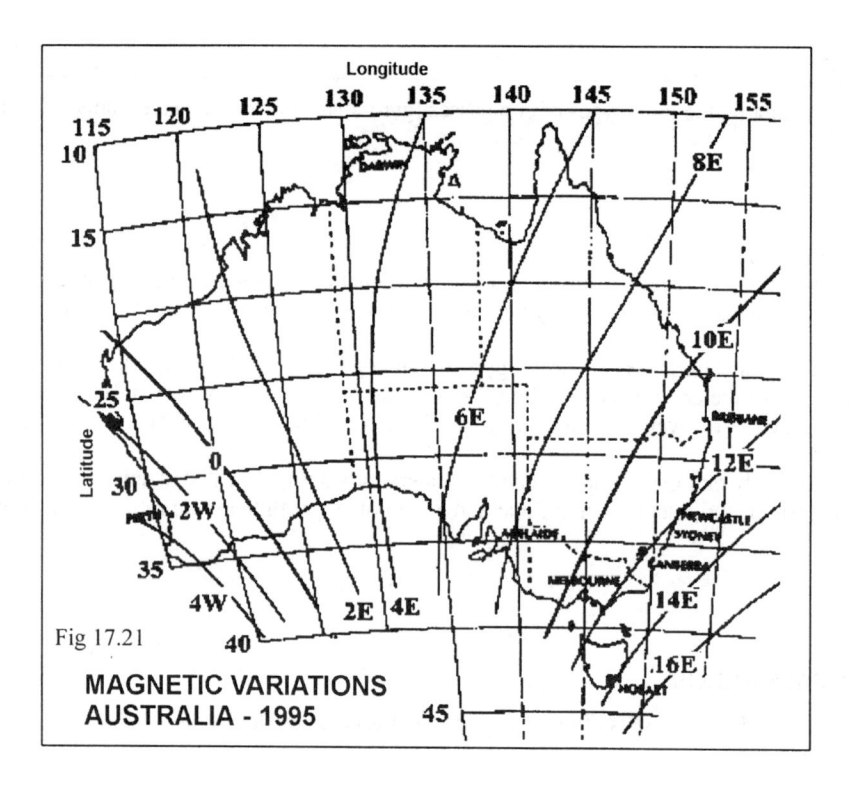

Magnetic Variation is different in direction (east or west), and in amount and rate of change throughout the world. The direction, amount and annual rate of change are shown on Compass Roses on charts. The attached illustration is a section of a Compass Rose from the Whitsunday's (Queensland) chart, AUS252.

To obtain True Bearings and Courses for plotting, Easterly Variation is always added to Magnetic Bearings and Courses; and Westerly Variation is subtracted. *(The meaning of Compass Bearings and Magnetic Bearings is explained in "Deviation" below).*

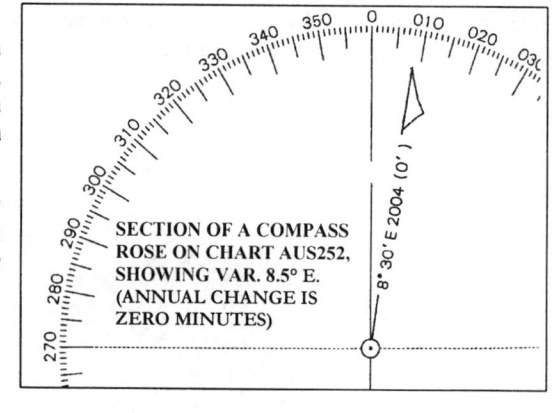

SECTION OF A COMPASS ROSE ON CHART AUS252, SHOWING VAR. 8.5° E. (ANNUAL CHANGE IS ZERO MINUTES)

RULE 1:

Remember the acronym **"True Virgins** *(male/female)* **Make Dull** *(or Delightful)* **Companions"** in order to remember the order "**T**rue, **V**ariation, **M**agnetic, **D**eviation, **C**ompass". Write it down as shown in the table.

You must subtract an Easterly Variation or Deviation when going down (i.e., when converting a True reading into a Magnetic or Compass reading); or add when going up (i.e., when converting a Compass or a Magnetic reading into a True reading). For WESTERLY values reverse the 'plus' and 'minus' signs.

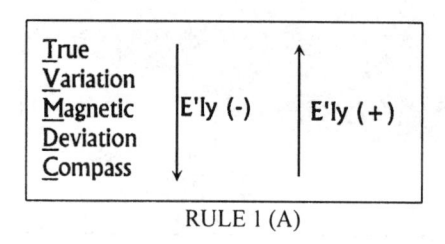

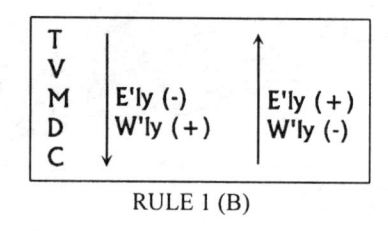

RULE 1 (A): True, Variation, Magnetic, Deviation, Compass — E'ly (-) going down, E'ly (+) going up

RULE 1 (B): T V M D C — E'ly (-) W'ly (+) going down, E'ly (+) W'ly (-) going up

EXAMPLE A:
A magnetic bearing *(i.e., the bearing is obtained with a hand-held compass)* reads 200°. Find the true bearing if Variation was 10° E

ANSWER (in the adjacent table) = 210° T

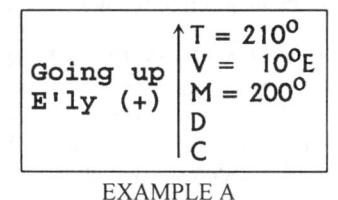

```
            ↑ T = 210°
Going up    | V = 10°E
E'ly  (+)   | M = 200°
            | D
            | C
```
EXAMPLE A

EXAMPLE B:
A magnetic bearing *(i.e., the bearing is obtained with a hand-held compass)* reads 200°. Find the true bearing if Variation was 10° W

ANSWER (in the adjacent table) = 190° T

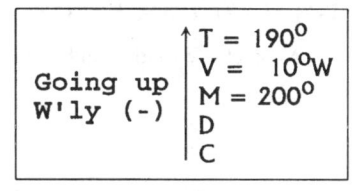

```
            ↑ T = 190°
Going up    | V = 10°W
W'ly  (-)   | M = 200°
            | D
            | C
```
EXAMPLE B

ALTERNATELY, THE ABOVE CAN BE REMEMBERED AS FOLLOWS:

RULE 2:   ERROR EAST COMPASS LEAST
          ERROR WEST COMPASS BEST

This means that when total compass error (or variation or deviation) is east, the compass bearing is smaller than the true bearing. When error is west, it is larger than the true bearing. (This is discussed in more details below).

DEVIATION

Most vessels have, to a greater or lesser degree, their own magnetic field. This may arise from the material of construction (in the case of iron and steel ships), DC electrical circuits and the presence of other magnetic material in the vicinity of the compass.

When a magnetic compass is influenced by the vessel's magnetic field the compass will be deflected east or west of the magnetic North. This magnetic deflection is called DEVIATION, and the amount and direction of the deviation changes with changes in the vessel's heading.

Deviation may be reduced by fitting to the compass binnacle, soft iron spherical correctors (known as the Quadrantal Correctors or Thompson's Balls) on either side and a vertical soft iron bar (Flinder's Bar) on the forward side. Deviation may be further reduced by a Compass Adjuster who will position compensatory magnets inside the binnacle. The residual deviation for each 10 degrees of change of the vessel's heading is observed and tabulated in the form of a Deviation Card.

The *DEVIATION CARD* is signed and dated by the Compass adjuster and is valid for up to three years (USL Code, Section 13, Para 6; and NSCV Part C). See Deviation Card (1) for an adjusted compass and Deviation Card (2) for an unadjusted compass (Figures 17.21 & 17.22).

The compass should be re-adjusted every two years or immediately after the vessel has undergone any alterations, additions or repairs and if the compass has suffered any interference or damage.

HAND BEARING COMPASSES, being portable do not have a constant relationship with the vessel's magnetic field and consequently Deviation Cards cannot be drawn up for such compasses. However, Deviation can be ignored when using a Hand Bearing Compass provided the vessel is not constructed of steel or ferro-cement, and provided the compass is used as far away as possible from the engine and other magnetic fittings.

Bearings and courses which have been corrected for Deviation or which do not need to be corrected for Deviation, as in the case of Hand Bearing Compasses in "non-magnetic" vessels, are known as *MAGNETIC BEARINGS AND COURSES*. These need to be corrected for *VARIATION ONLY* in converting them to TRUE for plotting.

In all types of hulls, including steel and ferro-cement hulls, hand-held compasses may be used for taking bearings for the purpose of finding **HORIZONTAL ANGLES BETWEEN LAND MARKS**. If the bearings are taken from the same position in the vessel, an equal amount of deviation will affect all of them. The angles between these "equally incorrect" bearings will not be affected by this deviation error.

For example, the angle between the bearings of 120° and 200° is 80°. If both bearings are incorrect by, say, 10°, i.e., 130° and 210°, the angle between them is still 80°.

Navigators often use the two horizontal angles between three landmarks for fixing a vessel's position. It is a very accurate position fixing method, not affected by unknown values of Variation or Deviation.

FLUXGATE COMPASS ERROR

These compasses are subject to both DEVIATION and VARIATION, and must be maintained in the horizontal plane to obtain accurate readings. Free gymbol mounting in an area of least movement in the vessel are two important aspects of Fluxgate compass installation. Some fluxgate compasses are fitted with microprocessors, which automatically compensate for Deviation and Variation.

TOTAL COMPASS ERROR

Magnetic Variation and Deviation make up the Total Compass Error.

DEVIATION CARDS

DEVIATION CARD 1
*(Small deviation values as a result of compass having been adjusted..
No interpolation necessary for values between tabulated courses.)*

Compass Heading	Dev	Magnetic Heading
000°	2½° E	002½°
020°	2° E	022°
040°	1° E	041°
060°	½° E	060½°
080°	0	080°
100°	0	100°
120°	½° W	119½°
140°	1° W	139°
160°	2° W	158°
180°	2½° W	177½°
200°	2° W	198°
220°	1° W	219°
240°	½° W	239½°
260°	0	260°
280°	0	280°
300°	½° E	300½°
320°	1° E	321°
340°	2° E	342°
360°	2½° E	002½°

Fig 17.21

Chapter 17.1: Navigation - Compasses, Bearings & Headings

DEVIATION CARD 2
(Large deviation values due to compass un-adjusted.
Interpolation may be necessary for values between tabulated courses.)
For example, Deviation for Compass Heading of around 010° is 9°E.
Similarly, Deviation for Magnetic Heading of around 270° is 0°.
Interpolation to the nearest half degree Deviation is sufficient.

Compass Heading	Dev	Magnetic Heading
000°	10° E	010°
020°	8° E	028°
040°	6° E	046°
060°	4° E	064°
080°	2° E	082°
100°	2° W	098°
120°	4° W	116°
140°	6° W	134°
160°	8° W	152°
180°	10° W	170°
200°	8° W	192°
220°	6° W	214°
240°	4° W	236°
260°	2° W	258°
280°	2° E	282°
300°	4° E	304°
320°	6° E	326°
340°	8° E	348°
360°	10° E	010°

Fig 17.22

CHECKING DEVIATION BY TRANSIT (Ø)

(FINDING DEVIATION WHEN DEVIATION CARD IS UNRELIABLE OR MISPLACED, OR JUST TO CHECK ITS ACCURACY)

(Ø is the symbol for transit)

When two landmarks (beacons, buildings, cliff edges, chimneys, etc.) are seen in a line, they are said to be in Transit. If they are marked on the chart, their transit bearing can be used to find the true line of position.

By looking through a compass along a transit, the compass bearing can be obtained. The difference between the True bearing and the Compass bearing is the Total Compass Error for the vessel's heading at the time. By applying the correction for Variation, the Deviation on that heading can be obtained.

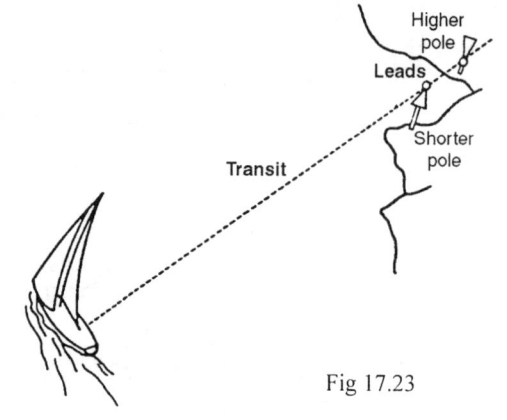

Fig 17.23

EXAMPLES

FINDING COMPASS DEVIATION: (EXERCISE 1)

A transit bearing on chart AUS252 gave the total compass error of 1°W. What is the Deviation on the vessel's heading?

SOLUTION:

Setting out the information in the following manner may make the answer more obvious:

Var = 8½° E *(Printed on chart)*
Dev = ?

Err = 1° W

Answer: The Deviation must be 9½°W to give the Total Error of 1°W.

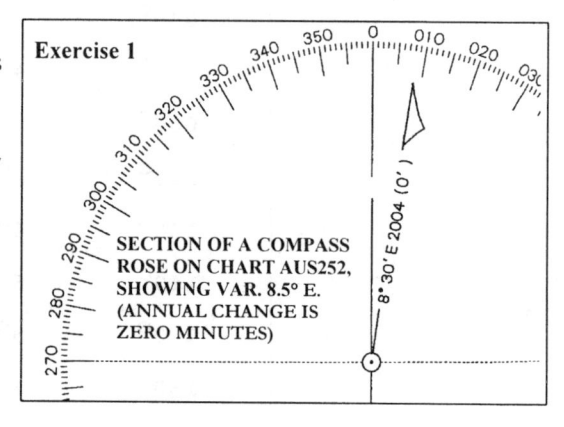

Exercise 1

SECTION OF A COMPASS ROSE ON CHART AUS252, SHOWING VAR. 8.5° E. (ANNUAL CHANGE IS ZERO MINUTES)

FINDING COMPASS DEVIATION & TOTAL COMPASS ERROR (EXERCISE 2)

On chart AUS 802 (figure 17.32), a transit is shown to read 196° T. If it reads 190° by compass, find the Total Compass Error and deviation on that heading.

As shown in the table, Deviation on that heading = 7° W

Total Error (on the same heading) is the combined effect of Variation and Deviation:

Variation = 13° E *(The compass is deflected 13° E)*
Deviation = 7° W *(The compass is deflected 7° W)*

Total Error = 6° E *(The resultant deflection is 6° E)*

```
            | T = 196°   (Given in the question)
            | V = 13°E   (Given on the chart, 12°55'E)
E'ly (-)    | M = 183°   (Subtracted 13 from 196)
W'ly (+)    | D = 7°W    (7 must be added to 183 to obtain 190)
            ↓ C = 190°   (Given in the question)
```

EXERCISE 2

Chapter 17.1: Navigation - Compasses, Bearings & Headings

EXERCISE 2A: Complete the following tables:

T	105°		009°
V	15°E		
M			357°
D	5°W	5°E	
C		215°	
Error		14°E	10°E

C		223°	071°	
D	4°W	5°E		
M				351°
V	13°E		11°W	
T	114°			003°
Error		11°E	6°W	9°E

EXERCISE 2A
(The right hand table is inverted merely for practice. It can either be rewritten in the form of the left-hand table or the arrows discussed earlier can be reversed.

ANSWER:			
T	105°	229°	009°
V	15°E	9°E	12°E
M	090°	220°	357°
D	5°W	5°E	2°W
C	095°	215°	359°
Error	10°E	14°E	10°E

C	105°	223°	071°	354°
D	4°W	5°E	5°E	3°W
M	101°	228°	076°	351°
V	13°E	6°E	11°W	12°E
T	114°	234°	065°	003°
Error	9°E	11°E	6°W	9°E

EXERCISE 2A - ANSWERS

CHAPTER 17.1: QUESTIONS & ANSWERS

CHAPTER 17.1.1: COMPASSES

1. One point of a compass equals:

    (a) one degree.
    (b) 11¼ degrees.
    (c) 11½ degrees.
    (d) 45 degrees.

2. The USL Code/NSCV lists the rules for installing magnetic steering compasses in vessels. Two of these rules for vessels less than 10 metres in length relate to the compass card diameter and a minimum arc of uninterrupted view from the compass position. These are:

    (a) 100 mm; 115 degrees.
    (b) 75 mm; 115 degrees.
    (c) 75 mm; 115 degrees on each side.
    (d) 100 mm; 115 degrees on each side.

3. The corrector magnets in a compass may be located:

    (a) only underneath the compass.
    (b) only on the sides of the compass.
    (c) only in front of the compass.
    (d) underneath & on 3 sides of the compass.

4. Your wooden vessel is fitted with an engine & steel fittings. She has two magnetic compasses: One is a fixed steering compass and the other a hand bearing compass. When using the Hand Bearing Compass, you should:

    (a) use deviation card of the steering compass.
    (b) disregard deviation under all circumstances.
    (c) obtain transit bearings whenever possible.
    (d) not observe transit bearings.

5. The USL Code/NSCV states the following requirement regarding compasses in vessels:

    (i) provide min 24 hrs of emergency electric supply
    (ii) check electrics for an effect on the compass.
    (iii) a vessel's compass may be electrically operated
    (iv) do not operate portable electrics near compass.

    The correct answer is:
    (a) all except (i).
    (b) all except (ii).
    (c) all except (iii).
    (d) all except (iv).

6. Unless exempt or otherwise stated by the Authority, a magnetic compass is required to be swung for deviation at intervals not exceeding:

    (a) one year.
    (b) two years.
    (c) three years.
    (d) four years.

7. The reasons for an erratic behaviour of a magnetic compass are:

    (i) worn out pivot point of the compass card.
    (ii) a large air bubble in the liquid.
    (iii) an obstruction inside the bowl.
    (iv) the need for a new deviation card.

    The correct answer is:
    (a) all except (i).
    (b) all except (ii).
    (c) all except (iii).
    (d) all except (iv).

8. Your compass has developed a sudden and unexplained error. It may be due to:

    (a) an air bubble in the compass liquid.
    (b) the need for a new deviation card.
    (c) a transistor radio standing nearby.
    (d) a recently replaced lifebuoy.

9. After installing wiring or other electrical equipment, check for any deflection in the vessel's compass card by switching the electrical equipment on and off. If there is deflection in the card, it means that:

    (a) the compass is in good working order.
    (b) the compass deviation has become unpredictable.
    (c) the vessel should be re-swung for deviation.
    (d) the compass now has larger deviation.

10. You have installed a new compass, carefully aligned it and then had it adjusted and swung by a licensed compass adjuster before setting out on a voyage. All goes well during the afternoon but after dark your compass shows a large and obvious error. This is because the compass is being affected by:

    (a) a change in magnetic deviation after sunset.
    (b) a change in magnetic variation after sunset.
    (c) the electrical supply to vessel's lights.
    (d) the rolling and pitching of the vessel.

11. A gyro compass differs from a magnetic compass in operation and errors as follows:

    (a) it is error free.
    (b) it is affected by variation only.
    (c) it is affected by deviation only.
    (d) it is not affected by deviation or variation.

Chapter 17.1: Navigation - Compasses, Bearings & Headings

12. A liquid magnetic compass is preferable to a dry compass card, mainly because liquid:

    (a) helps to dampen the card.
    (b) prevents freezing.
    (c) provides lubrication.
    (d) minimises magnetic deviation.

13. The most necessary fluid in a compass-bowl in vessels operating in Australia is:

    (a) alcohol.
    (b) mixture of distilled water and alcohol.
    (c) distilled water.
    (d) oil.

14. A compass-bowl is designed to cope with the expansion of its fluid by:

    (a) adding alcohol to it.
    (b) fitting expansion chamber in it.
    (c) filling it with only distilled water.
    (d) maintaining an air bubble in it.

15. An air bubble in a compass-bowl:

    (a) allows its fluid to expand in summer.
    (b) should only be displaced with alcohol.
    (c) cannot be removed.
    (d) makes the compass inaccurate and unstable.

16. A fluxgate compass is subject to:

    (a) all magnetic errors.
    (b) error due to Variation only.
    (c) error due to Deviation only.
    (d) error due to fluctuations in power supply only.

17. A compass-bowl is designed to remain level at sea by:

    (a) its liquid.
    (b) fore and aft gimbals.
    (c) athwartship gimbals.
    (d) fore and aft and athwartship gimbals.

18. The north seeking end of a bar magnet or a magnetic needle is painted:

    (a) red or marked 'north'.
    (b) blue.
    (c) red.
    (d) blue or marked 'north'.

19. A soft iron bar commonly attached to the front of large compasses is known as the:

    (a) Flinder's bar.
    (b) vertical corrector.
    (c) compass corrector.
    (d) deviation adjuster

20. There are two soft iron correctors attached to some compass binnacles. The master may:

    (a) move them closer to reduce compass deviation.
    (b) move them apart to reduce compass deviation.
    (c) not tamper with their positioning.
    (d) remove them if necessary.

21. Positioned inside the compass binnacle of large compasses, there are:

    (a) deviation compensatory magnets.
    (b) Flinder's bars.
    (c) spare magnets.
    (d) compass adjusting tools.

22. The electrical circuit in a vessel may affect the:

    (a) Variation in her magnetic compass.
    (b) Deviation in her fluxgate compass.
    (c) Error in her gyro compass.
    (d) Variation & Deviation in her magnetic compass.

23. A lubber line is the:

    (a) fore and aft mark on the compass bowl.
    (b) vessel's steering position.
    (c) north-south line on the compass bowl.
    (d) mark on a vessel to assist helmsman to steer.

24. The south seeking end of a bar magnet or a magnetic needle is painted:

    (a) red.
    (b) blue.
    (c) red or marked 'south'.
    (d) blue or marked 'south' or not marked at all.

25. There are deviation compensatory magnets inside some compass binnacles. The master may:

    (a) remove them if they are not effective.
    (b) re-position them if they are not effective.
    (c) not tamper with them.
    (d) take out only half of them if necessary.

26. An azimuth mirror is an instrument for measuring:

    (a) terrestrial bearings.
    (b) celestial bearings.
    (c) terrestrial & celestial bearings.
    (d) horizontal angles.

27. A pelorus is a:

    (a) dumb compass dial.
    (b) hand held compass.
    (c) azimuth mirror.
    (d) magnetic compass.

28. The power supply to the gyro compass alarm circuit:
    (a) is independent of the power supply to the gyro compass.
    (b) is common to the power supply to the gyro compass.
    (c) is fitted with a fuse.
    (d) is fitted with a loop.

29. Which of the following equipment does not need to be placed or installed at a safe distance from a magnetic compass:
    (a) Portable radio
    (b) Radar
    (c) Speakers
    (d) Aluminium coffee mug

30. When observing bearings with a hand-bearing compass, the observer does not need to take into account the proximity of the following personal items:
    (a) Steel-framed spectacles.
    (b) Steel wristwatch
    (c) Gold chain
    (d) Steel pen.

31. The electromagnetic influence on compass of the compass light wiring can be minimised by:
    (a) running the wires parallel to each other.
    (b) using light gauge wires.
    (c) twisting the wires together.
    (d) installing an isolating switch.

32. The magnetic compasses of a vessel laid up for an extended period of time will experience a:
    (a) permanent change in magnetic variation.
    (b) permanent change in magnetic deviation.
    (c) temporary change in magnetic variation & deviation.
    (d) temporary change in magnetic deviation.

33. The gyro compass error can be corrected by:
    (a) stopping & restarting the gyro.
    (b) Adjusting its lubber line.
    (c) Adjusting the slew control.
    (d) Slowing it down.

34. Some gyro compasses need to be periodically adjusted for changes in:
    (a) Speed & Distance.
    (b) Course & Speed.
    (c) Latitude & Course.
    (d) Latitude & Speed.

35. Vessels with gyro compasses usually install:
    (a) a gyro repeater at the steering location only.
    (b) Gyro repeaters at the steering location as well as at the bearing locations.
    (c) the master compass at the bearing location.
    (d) one gyro repeater in the engine room only

36. An out of alignment gyro repeater can be realigned by:
    (a) switching off the master gyro.
    (b) switching off the repeater.
    (c) adjusting it with the re-set button.
    (d) switching it off & on again

CHAPTER 17.1.2: COMPASS ERROR, COURSES & BEARINGS – THE CONCEPT

37. In the list of pre-departure checks relating to magnetic compasses, the master should have a:
    (a) deviation card for the steering compass.
    (b) deviation card for each hand-held compass.
    (c) common deviation card for all compasses.
    (d) deviation card no more than 5 years old.

38. A hand bearing magnetic compass in a steel hulled vessel:
    (a) cannot be used at all.
    (b) can be used for taking bearings only.
    (c) can be used for taking horizontal angles only.
    (d) can be used for all purposes.

39. A vessel should be re-swung for compass deviation on the following occasions:
    (a) Change in waterway area of operation.
    (b) Annual survey.
    (c) Repairs after structural damage to vessel.
    (d) Change of ownership.

40. You were caught in a storm out of sight of land. The deviation card for the steering compass on your steel hull vessel has been destroyed. You should:
    (a) take a copy from a sister vessel in the area.
    (b) use your hand-held magnetic compass instead.
    (c) swing the vessel to make an interim card.
    (d) check log book for deviations used previously.

41. Your vessel has two fixed magnetic compasses. A deviation card:
    (a) is necessary for only one of them.
    (b) common to both compasses is sufficient.
    (c) is not required if you have two compasses.
    (d) is required for each of them.

Chapter 17.1: Navigation - Compasses, Bearings & Headings

42. It is pointless to swing a vessel to make a deviation card for a hand bearing compass, because such a compass:

    (a) is not subject to magnetic deviation.
    (b) is not fixed at one location.
    (c) is not used in a steel hull vessel.
    (d) can be battery powered.

43. The values of magnetic variation:

    (a) are supplied in the deviation card.
    (b) alter with change of vessel heading.
    (c) must be checked after re-wiring a vessel.
    (d) are the same regardless of vessel size.

44. Magnetic deviation that needs to be applied to a magnetic compass is supplied in a deviation card for:

    (a) a standard heading.
    (b) a range of headings.
    (c) a range of bearings.
    (d) a range of headings and bearings.

45. The term azimuth means:

    (a) course.
    (b) bearing
    (c) position.
    (d) angle above the horizon.

46. Compass course is the course:

    (a) plotted on the chart.
    (b) corrected for deviation and variation.
    (c) steered by compass.
    (d) known as the magnetic course.

47. One transit bearing is sufficient to:

    (a) check a compass error.
    (b) fix a vessel's position.
    (c) print a measured distance on a chart.
    (d) plot a course.

48. A deviation card reads deviation 5°E for compass course 090°, and 3°W for course 200°. The vessel's course by compass is 090°, and a bearing of a lighthouse taken with the same compass is 200°. The chart reads variation 10°E. The true bearing of the lighthouse is:

    (a) 185°
    (b) 193°
    (c) 207°
    (d) 215°

49. A vessel has two compasses, each with its own deviation card. Having misplaced one card she must solve the problem as follows:

    (i) compare two compasses to check error.
    (ii) use transit bearings to check error.
    (iii) use one deviation card for both compasses.
    (iv) swing the vessel to check deviation.

    The correct answer is:
    (a) all except (i).
    (b) all except (ii).
    (c) all except (iii)
    (d) all except (iv).

50. A transit bearing indicates a compass error of 1° East. The chart shows a Magnetic Variation of 7° West. Therefore the Magnetic Deviation is:

    (a) 8° West.
    (b) 8° East.
    (c) 6° West.
    (d) 9° East.

51. A transit bearing indicates no compass error. The chart shows a Magnetic Variation of 10° East. Therefore the Magnetic Deviation is:

    (a) 20° West.
    (b) 10° East.
    (c) 10° West.
    (d) zero.

52. Magnetic Variation is 8° West. Magnetic Deviation is 4° East. Therefore the compass error is:

    (a) 4° West.
    (b) 12° West.
    (c) 12° East.
    (d) 8° East.

53. A transit bearing indicates 10° West compass error. The chart shows a Magnetic Variation of 10° East. Therefore the Magnetic Deviation is:

    (a) zero.
    (b) 20° East.
    (c) 20° West.
    (d) 10° West.

54. A transit bearing provides a:

    (a) true bearing without using a compass.
    (b) magnetic bearing.
    (c) fix.
    (d) magnetic position line.

55. You are in the wheelhouse of a vessel. The compass is set forward of the wheel and well below the sill of the wheelhouse windows. You wish to take an accurate bearing of a lighthouse 3 miles away at about 35° on your starboard bow. It should be best achieved as follows:

    (a) turn the vessel's head directly towards it.
    (b) turn the vessel's stern directly towards it.
    (c) turn the vessel's head at right angle to it.
    (d) line it up relative to a window sill.

56. If you had to choose a vessel's position from a large cocked hat, and if it was not possible to get another fix, you should assume her to be in the following position:

    (a) in the middle of the cocked hat.
    (b) on the cut with the largest angle.
    (c) on the cut closest to the danger.
    (d) anywhere outside the cocked hat.

57. When fixing a vessel's position with compass bearings, which of the following reasons would cause a cocked hat?

    (i) Deviation card for hand held compass too old.
    (ii) A metallic object left near the compass.
    (iii) Incorrect identification of bearings.
    (iv) incorrect compass error used.

    The correct answer is:
    (a) (i) only.
    (b) all except (i).
    (c) (iv) only.
    (d) all of them.

58. When fixing a vessel's position, the following bearings should be avoided:

    (a) within 30° of each other.
    (b) a 60° cut between three bearings.
    (c) a 90° cut between two bearings.
    (d) angles between 30° and 150°.

59. A vessel's true course is 150°. The magnetic deviation is 5° West and variation 5° East. Her compass course is:

    (a) 160°
    (b) 155°
    (c) 145°
    (d) 150°

60. A vessel's compass course is 170°. The magnetic deviation is 8° East and variation 4° West. Her true course is:

    (a) 158°
    (b) 174°
    (c) 166°
    (d) 182°

61. A vessel is steering a compass course of 090°. She takes two bearings with the same compass, which is fitted forward of her steering position. Bearing "A" reads 150°, and bearing "B" 180° by compass. The chart indicates a magnetic variation of 12° East. The deviation card reads as follows:

    Compass heading 090°: Dev 5°E
    Compass heading 150°: Dev 0
    Compass heading 180°: Dev 5°W

Therefore the corrected bearings are:
(a) 167° & 197°
(b) 162° & 187°
(c) 138° & 173°
(d) 133° & 163°

62. A vessel is steering a compass course of 212°. She takes two bearings with the same compass, which is fitted forward of her steering position. Bearing "A" reads 212°, and bearing "B" 265° by compass. The chart indicates a magnetic variation of 12° West. The deviation card reads as follows:

    Compass heading 210°: Dev 3°W
    Compass heading 220°: Dev 4°W
    Compass heading 260°: Dev 5°E
    Compass heading 270°: Dev 7°E

    Therefore the corrected bearings are:
    (a) 227° & 271°
    (b) 197° & 259°
    (c) 227° & 280°
    (d) 197° & 250°

63. A vessel is steering a course of 359° by a compass fitted forward of her steering position. She takes two bearings with another compass, which is a hand held compass. Bearing "A" reads 270°, and bearing "B" 310°. The chart indicates a magnetic variation of 8° East. The deviation card of the steering compass reads as follows:

    Compass heading 270°: Dev 8°W
    Compass heading 310°: Dev 4°W
    Compass heading 000°: Dev 5°E

    Therefore the corrected bearings are:
    (a) 283° & 323°
    (b) 278° & 318°
    (c) 257° & 297°
    (d) 270° & 314°

CHAPTER 17.1: ANSWERS

1 (b), 2 (c), 3 (d), 4 (c), 5 (c),
6 (c), 7 (d), 8 (c), 9 (b), 10 (c),
11 (d), 12 (a), 13 (c), 14 (b), 15 (d),
16 (a), 17 (d), 18 (a), 19 (a), 20 (c),
21 (a), 22 (b), 23 (a), 24 (d), 25 (c),
26 (c), 27 (a), 28 (a), 29 (d), 30 (c),
31 (a), 32 (b), 33 (c), 34 (d), 35 (b),
36 (c), 37 (a), 38 (c), 39 (c), 40 (d),
41 (d), 42 (b), 43 (d), 44 (b), 45 (b),
46 (c), 47 (a), 48 (d), 49 (c), 50 (b),
51 (c), 52 (a), 53 (c), 54 (a), 55 (a),
56 (c), 57 (b), 58 (a), 59 (d), 60 (b),
61 (a), 62 (d), 63 (b)

Chapter 17.2

NAVIGATION

CHART WORK

YOU WILL NEED CHARTS AUS197 (Approaches to Port Jackson in NSW), AUS252 (Whitsunday Passage in Queensland) and AUS802 (Cape Liptrap to Cliffy Island in Victoria) to plot and answer many of the questions at the end of this chapter. You will also need:

CHART PLOTTING INSTRUMENTS & AIDS

- Pencil: 2B
- Eraser: soft white rubber
- Drawing compass: long legged
- Dividers: long legged
- Protractor
- Rolling Rule or Parallel Ruler

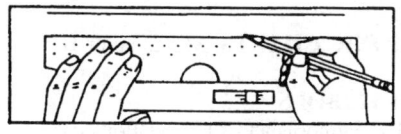

Fig 17.24: ROLLING RULER

A wide variety of navigational rulers are available. The tradition Parallel Ruler opens and shuts on swinging lattice arms, which enables a navigator to transfer a bearing or a course from a compass rose to the required point on the chart or visa versa by a crab-walk. It is not as suitable on a small vessel as a plastic Rolling Ruler, which is also the cheapest option. Heavy brass Rolling Rulers are dangerous in a small vessel. There are many other types of navigational rulers and protractors, which a navigator may find more suitable after becoming familiar with chart plotting.

TAKING & PLOTTING BEARINGS

(a) Take bearings of at least two landmarks, and apply compass error, as discussed in Chapter 17.1 and below.
(b) Place ruler on a compass rose on the chart on one of the corrected bearings. The outer circle represents true bearings.
(c) Transfer bearing from compass rose to the landmark. Repeat procedure for each bearing. Boat's position is where the bearings intersect.

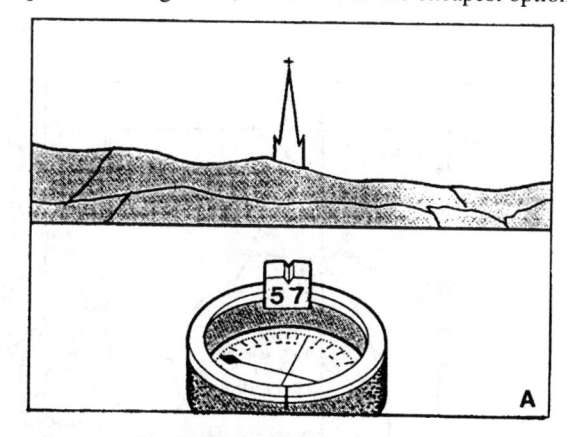

Fig 17.25: TAKING A BEARING

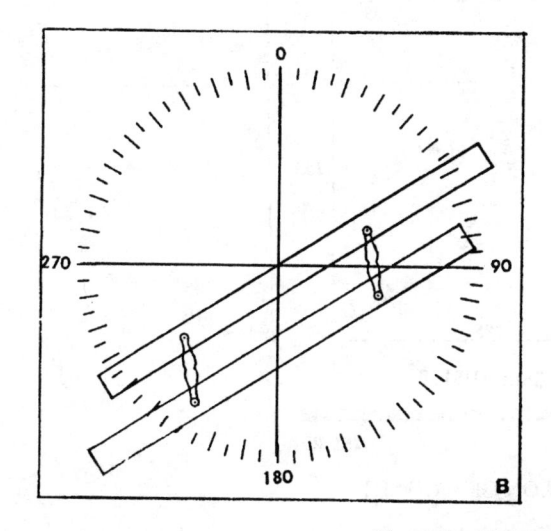

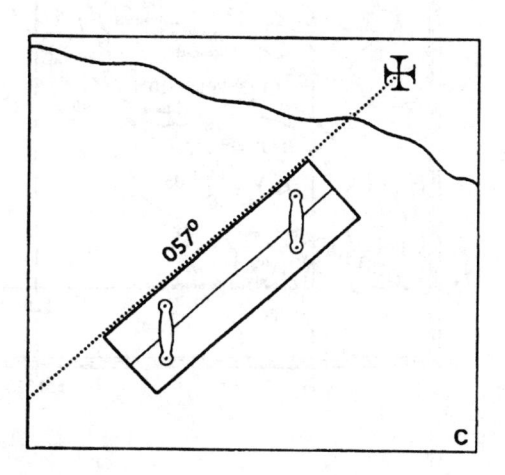

Fig 17.26: PLOTTING THE BEARING

NAVIGATIONAL CHARTS

CHART NUMBERS

Charts are produced by a country's Hydrographic Office (Hydrographer of the Navy). Charts commonly used in the Australasian region are AUS (Australian), NZ (new Zealand), BA (British Admiralty) and HO (Hydrographic Office of USA). Charts are identified by a alpha-numeric catalogue number, e.g., AUS 201, NZ 520, BA 3325, HO 2121, etc. Among the English speaking countries, only Britain and USA publish charts for the entire world. There is now a move to avoid duplication among countries, and to publish International ocean charts (pre-fixed 'INT').

Charts needed for a passage can be selected from a Chart Catalogue or Chart Index, which is available at the licensed chart agents and the sub-agents. A chart catalogue is like an atlas. The Australian Chart Index (Index of Australian Charts) is contained in two chart-sized sheets, numbered AUS65000 (Northern portion) and AUS65001 (Southern portion). For an interactive chart index on Australian Hydrographer's website, see the address on the last page of this book.

Figures 17.27, 17.31 & 17.32 show chart extracts **AUS198**, **AUS252** and **AUS802** respectively.

ELECTRONIC CHARTS: See Chapter 14

CHART SYMBOLS

The Symbols and Abbreviations used on Charts are common throughout the world. These are published in the form of a booklet (Catalogue No. BA5011) by the British Admiralty. The New Zealand and the US Hydrographic Offices publish similar information.

See Chapter 6 for an extract of the Chart Symbols & Abbreviations. Some are also explained in figures 17.31 & 17.32 in this chapter.

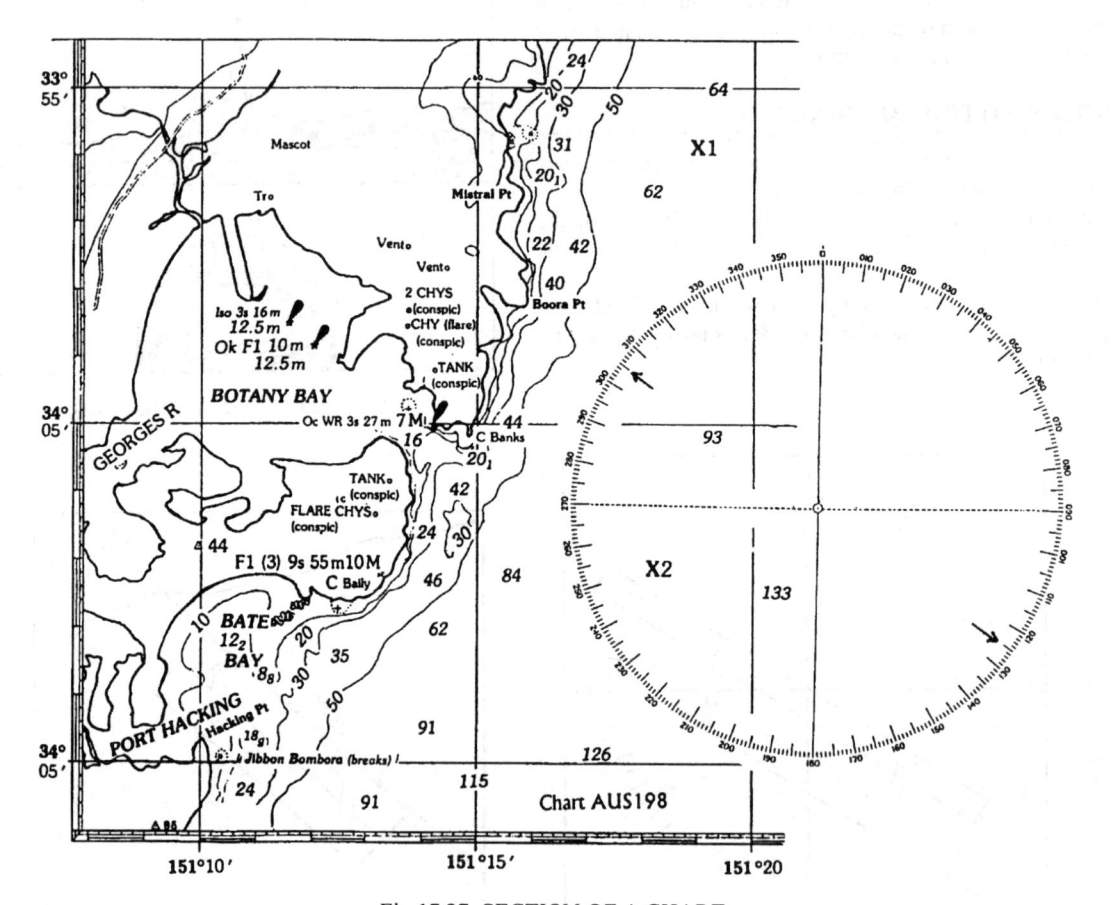

Fig 17.27: SECTION OF A CHART

Chapter 17.2: Navigation - Chartwork

CHART CORRECTION (& CORRECTIONS FOR OTHER NAUTICAL PUBLICATIONS)

State marine authorities issue *MARINE NOTICES* or *STATE NOTICES TO MARINERS* to advise public of any changes in the waterways. These are published in the Public Notices of newspapers, and can also be obtained from the relevant authority via mailing list or from their website. Urgent information is immediately broadcast by the coast radio stations and harbour control towers. These sources are quite sufficient for the local boat operators.

The information is supplied to the interstate and international shipping via fortnightly published national *NOTICES TO MARINERS* for charts published in Australia and New Zealand, and weekly for the charts published in UK and USA. These notices are published by the Hydrographer of the Navy in each country. It is also broadcast as Maritime Safety Information (MSI) by radio and Inmarsat Satellites. *(Inmarsat is discussed in Chapter 18)*. The Notices to Mariners can be obtained direct from the Australian Hydrographic Office at 8 Station Street in Wollongong, from licensed chart agents or on the web at www.hydro.navy.gov.au.

Australian charts are available through chart agents registered as "A Class" and "B Class". "A Class" chart agents provide charts corrected up-to-date with recent Notices to Mariners. "B Class" agents provide uncorrected charts together with copies of appropriate Notices to Mariners to permit up-to-date corrections to be made by the purchaser. Some "A Class" agents provide facilities for complete chart folios of ships to be corrected (at a cost). They can also order and supply charts published overseas.

In most cases, charts and relevant publications only need to be corrected with a chart-correcting pen. For example, a wreck symbol or a new buoy may need to be entered on a chart. For more involved corrections, replacement pages, tracings or cut-outs (block) are supplied.

Each correction is identified with a number (Notices to Mariners Number). Every time a chart is corrected, the relevant NM number is noted outside the margin at the bottom left hand corner of the chart together with its year of publication. The list is headed Notices to Mariners (or Small Corrections on older charts). It may thus appear as: Notices to Mariners 2001 – 285, 2002 – 23, 46, 203. The corrections for other nautical publications are similarly noted on the inside front cover of the publication.

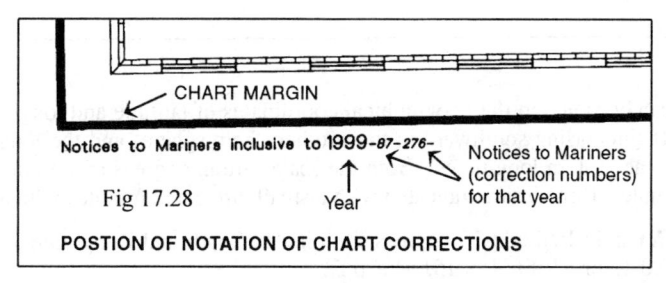

Fig 17.28 POSTION OF NOTATION OF CHART CORRECTIONS

Some Notices to Mariners are prefixed (T) for Temporary Notices or (P) for Preliminary Notices. These should be noted on charts in pencil to permit subsequent confirmation or withdrawal.

A chart or other nautical publication may have numerous corrections in a year or none at all. The noted last correction does not therefore indicate the currency of the publication. Charts and publications should be up-to-date when purchased from a "Class A" chart agent. Thereafter, it is the navigator's responsibility to maintain their currency. Navigators responsible for maintaining a large folio of charts and publications keep a log or catalogue of corrections.

READING A CHART

RELIABILITY OF CHARTS

Most Australian charts are printed with a Zone of Confidence (ZOC) diagram to indicate the quality of their survey and in particular the reliability of depths printed on them. The older Reliability Diagrams have been replaced with the ZOC system to provide conformity between digital and paper charts. The five ZOC quality categories are A1, A2, B, C and D, with a sixth category (ZOC U) to indicate that the data has not been assessed. Most modern surveys of critical areas are ZOC A2. The category ZOC A1 is for surveys under exceptionally stringent conditions for very special reasons.

Fig 17.29

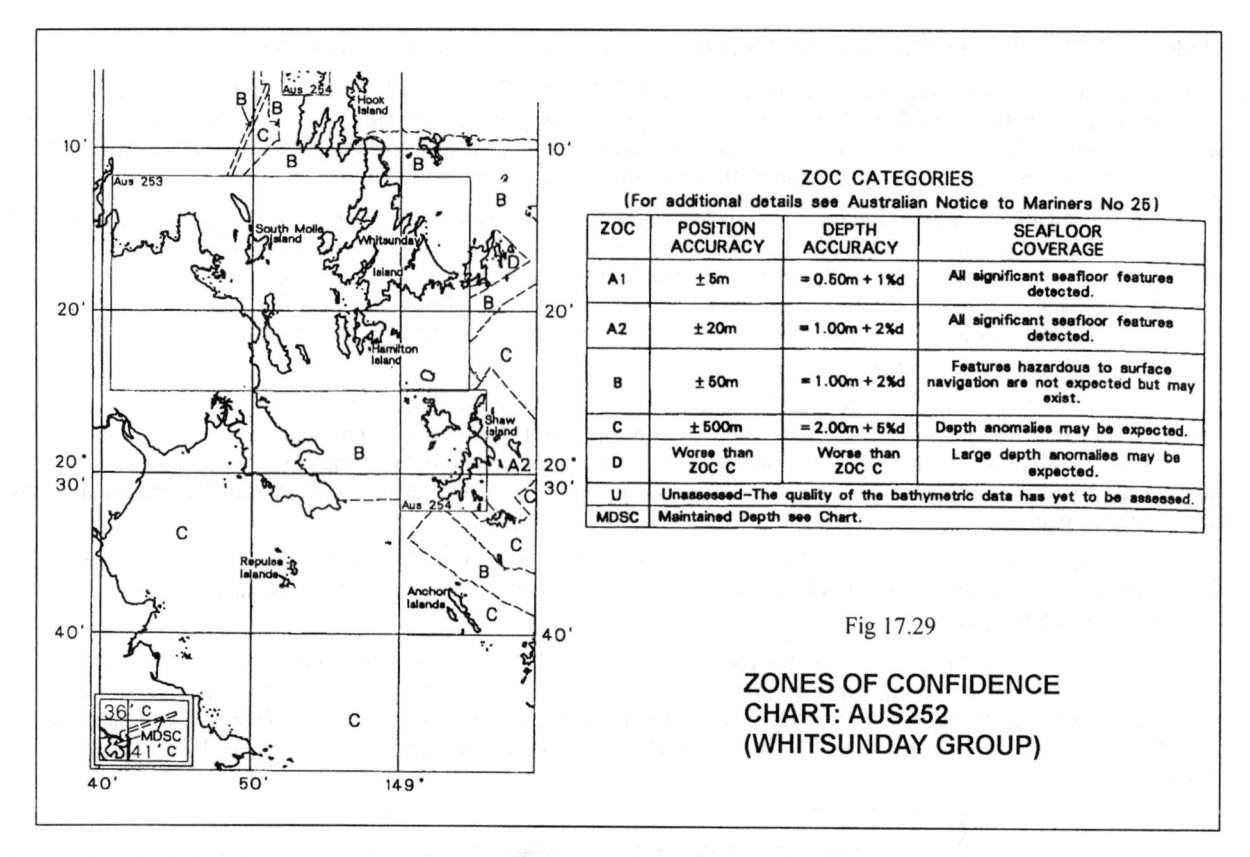

**ZONES OF CONFIDENCE
CHART: AUS252
(WHITSUNDAY GROUP)**

ZOC	POSITION ACCURACY	DEPTH ACCURACY	SEAFLOOR COVERAGE
A1	± 5m	= 0.50m + 1%d	All significant seafloor features detected.
A2	± 20m	= 1.00m + 2%d	All significant seafloor features detected.
B	± 50m	= 1.00m + 2%d	Features hazardous to surface navigation are not expected but may exist.
C	± 500m	= 2.00m + 5%d	Depth anomalies may be expected.
D	Worse than ZOC C	Worse than ZOC C	Large depth anomalies may be expected.
U	Unassessed—The quality of the bathymetric data has yet to be assessed.		
MDSC	Maintained Depth see Chart.		

(For additional details see Australian Notice to Mariners No 25)

LATITUDE & LONGITUDE

A vessel's position on a chart can be stated in the geographical coordinates of latitude and longitude. The latitude scale is marked on the sides of the charts (increasing southward in the southern hemisphere) and the longitude scale on the top and bottom (increasing eastward in the eastern longitudes). Both the scales are in degrees and minutes and, on larger scale charts, in tenths or twentieths of minutes. Degrees are indicated with a small circle and minutes with a dash.

In figure 17.27, **Position X1** is in Latitude 33 degrees 56 minutes South and Longitude 151 degrees 19 minutes East (written as 33° 56.0' S 151° 19.0' E, or 33° 56'.0 S 151° 19'.0 E).

NAUTICAL MILE & CABLE

The nautical mile is the basic measure of distance in all aspects of navigation, and a CABLE is one-tenth of a nautical mile.

One nautical mile equals one minute of latitude and there are 60 nautical miles in one degree of latitude. All distance measurement is therefore taken from the latitude scales at the sides of the chart closest to where the distance is measured. Being expressed in minutes of latitude, the distance can also be indicated with a dash. Use a pair of dividers to transfer distance to and from the latitude scale as shown in figure 17.30. Kilometres are never used in navigation even on metric charts.

A vessel's position can also be expressed as a bearing and distance of a prominent landmark. In figure 17.27, **Position X2** is Cape Banks bearing 305 degrees distance 3.4 miles (written as Cape Banks 305° x 3.4'). Sometimes the bearing is expressed from landward, in which case Position X2 will be 125° x 3.4' from Cape Banks. However, in navigation the bearings are always expressed from seaward, i.e., as observed from the vessel.

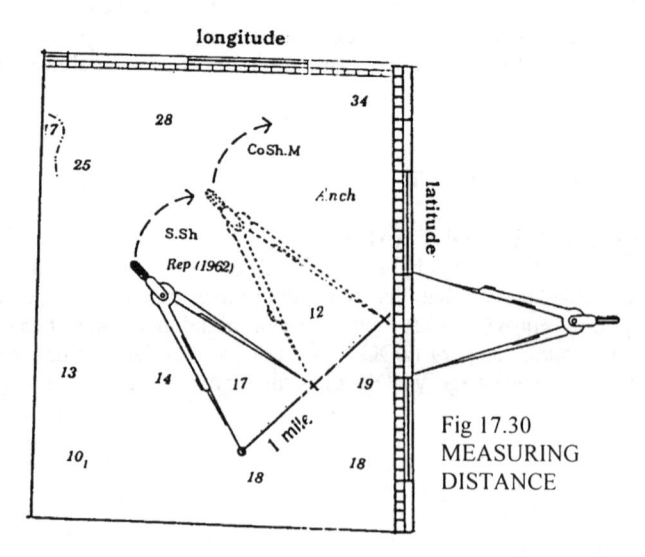

Fig 17.30 MEASURING DISTANCE

Chapter 17.2: Navigation - Chartwork

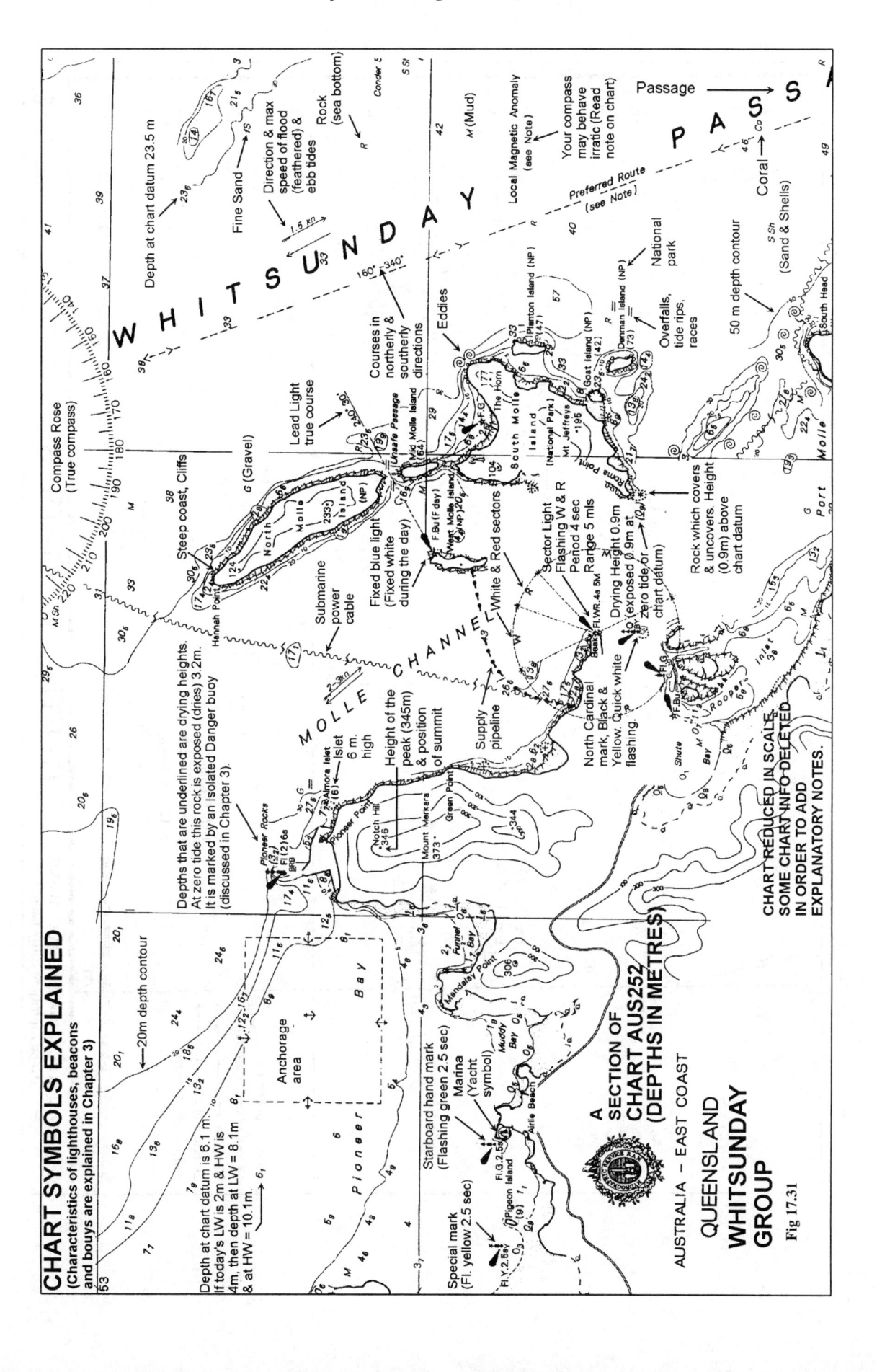

Fig 17.31

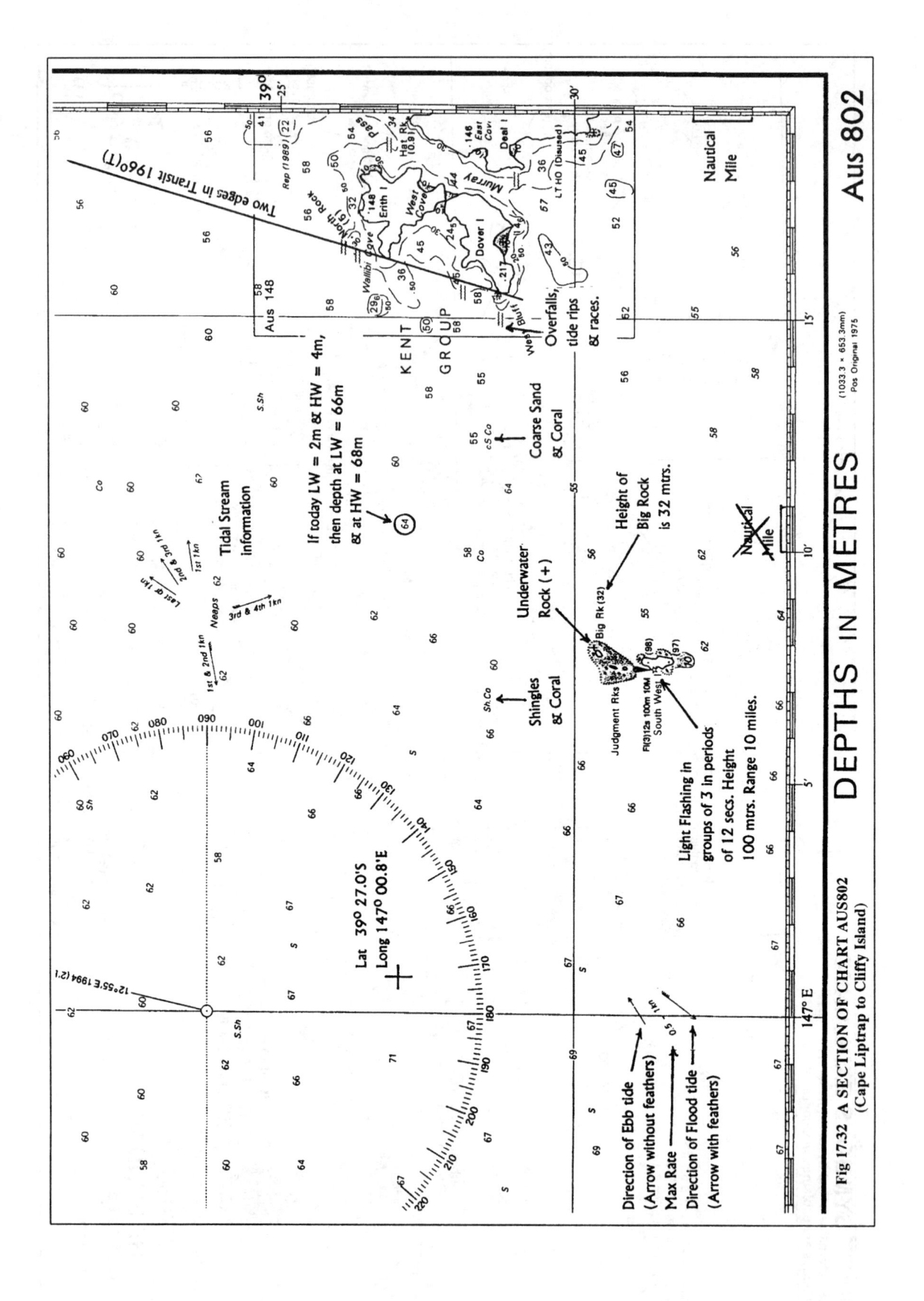

Fig 17.32 A SECTION OF CHART AUS802
(Cape Liptrap to Cliffy Island)

DEPTHS PRINTED ON CHARTS

The depths printed on charts are *SOUNDINGS* that have been reduced to a common level known as *CHART DATUM*. This datum is set on or close to the lowest low water spring tide or the Lowest Astronomical Tide (LAT). <u>The charted depths therefore represent the depth of water at zero tide and are consequently minimum depths</u>. The height of the tide must be added to the chartered depth to obtain the actual total depth at that time. Depths on "older" charts are in fathoms and on modern charts in metres. (See chart extract above.)

On metric charts, a depth marked as 2_4 is 2.4 metres (equivalent to 7.84 feet). The same depth on a fathom chart is 2 fathoms 4 feet (equivalent to 16 feet - more than twice the depth). One *FATHOM* = 6 feet.

<u>Underlined depths or depths written as "Dries ....m or ft" or "Dr. .....m or ft" are *DRYING HEIGHTS* of underwater rocks</u> (see Figures 17.31 & 17.32). An underlined depth of 2 metres means the rock in that location will be exposed 2 metres at zero tide. If the height of tide was 5 metres, then there will be 3 metres water over it. On fathom charts, the drying heights are in feet, not fathoms.

CLEARANCES UNDER BRIDGES

The heights of objects such as bridges, lighthouses and other landmarks are expressed on charts as being above Mean High Water Springs (*MHWS*) or Mean Higher High Water (*MHHW*). This is the mean of all the high waters at spring tides. The tide rarely rises above this level. Therefore, only on rare occasions will the clearance under the bridge be less than that marked on a chart. Where there is no tide, the height is expressed above the Mean Sea Level (*MSL*). See Chart Symbols in Chapter 6.

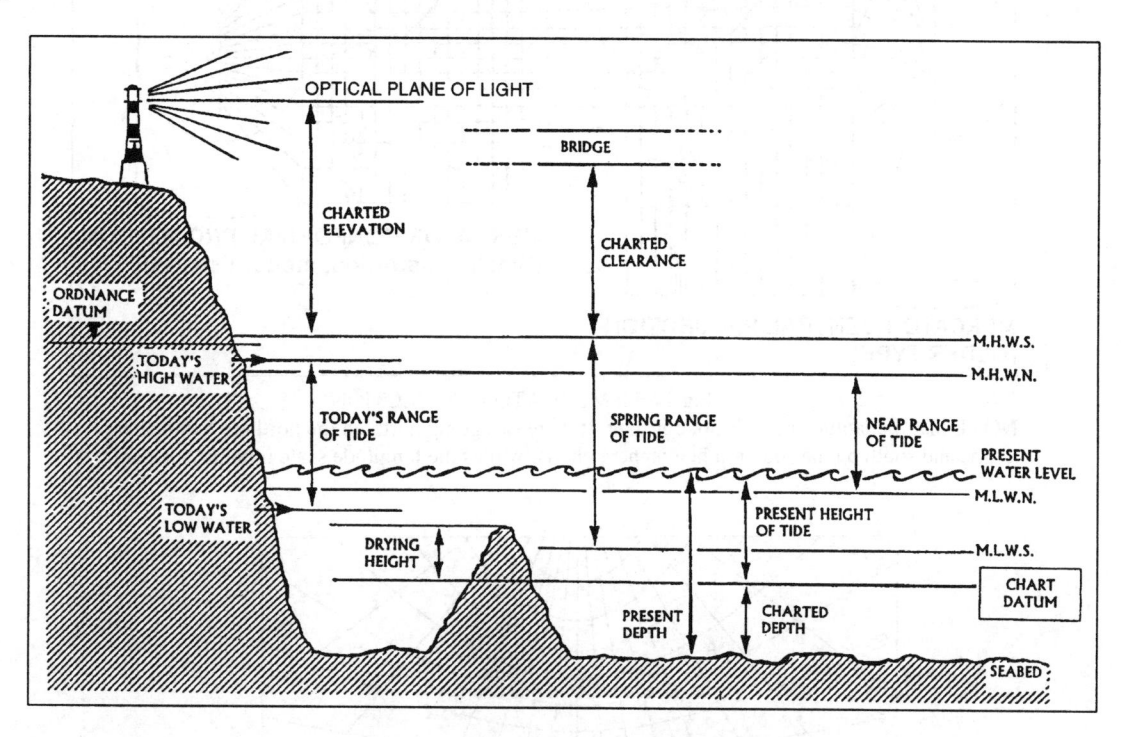

Fig 17.33: SEA LEVELS REFERENCE CHART
(Tidal Levels and Chartered Data or Tide chart)

LEAD LIGHTS & LIGHTHOUSES

Lead lights are unidirectional, relatively narrow beam "flashing" or "fixed" lights installed in pairs on extensions of the channel into a harbour. As illustrated in figure 17.11 (showing a Transit), the rear light is higher than the front light and, by keeping the lights vertically aligned, a vessel can follow the central line of the channel safely into the waterway. During the day, the vessel can keep the leads vertically aligned to follow the centre line of the channel.

Lighthouses, on the other hand, generally show an all-round "flashing" white light and have a range of 10 to 20 miles. The flash rate varies with the location. The frequency of the flash, and occasionally the colour of the lights, varies along the coast. The details of the light characteristics are printed on the relevant navigational charts and in the publication titled, Admiralty List of Lights.

TYPES OF CHARTS (CHART PROJECTIONS)

There are numerous ways of representing the spherical earth on a flat chart. The most common is the MERCATOR PROJECTION. Mercator charts are made by projecting the globe on to a cylinder, then opening it into a flat chart. On Mercator charts the circles of latitude and longitude appear as straight lines

Most charts - including the ones used for plotting exercises in this book - are Mercator charts. These are used everywhere except near the earth's poles (usually, in latitudes greater than 60°), where their picture becomes extremely distorted. There, the GNOMONIC PROJECTIONS or Gnomonic charts are used, on which the latitudes appear as curves and the longitudes as straight lines. ('G' in 'gnomonic' is silent).

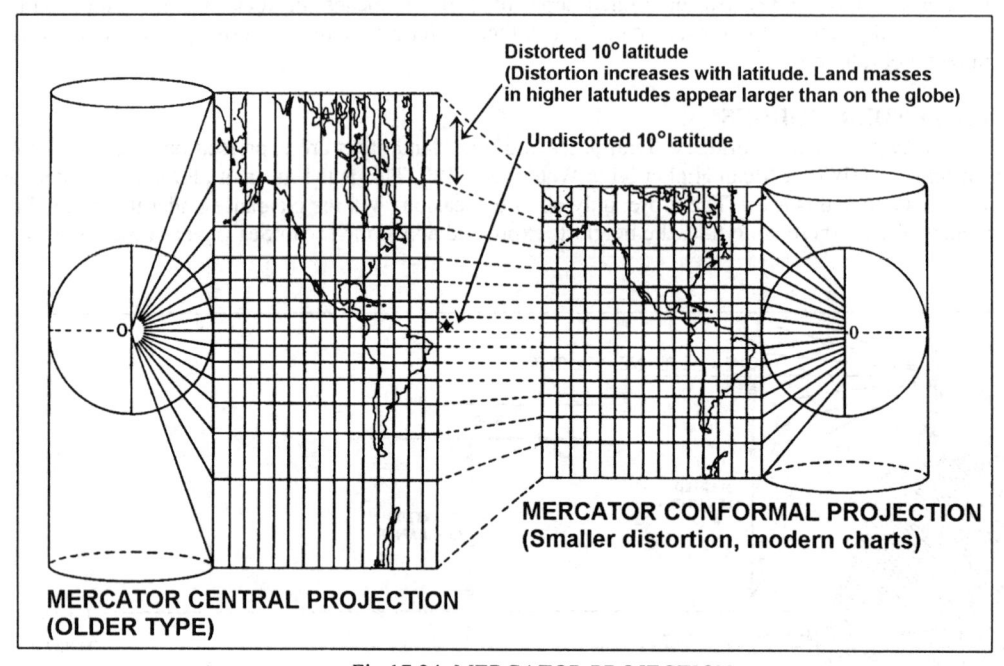

Fig 17.34: MERCATOR PROJECTION

NOTE that the latitude scale is not constant. It increases going north on the northern hemisphere charts and south on the southern hemisphere charts, whiles the longitude scale remains constant.

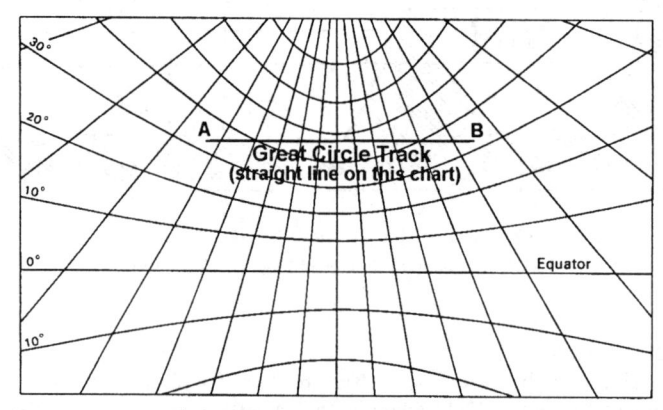

Fig 17.35: GNOMONIC CHART

A straight line drawn on a Mercator chart is a **RHUMB (or RHOMB) LINE**. A straight line drawn on a Gnomonic chart is a **GREAT CIRCLE**. The latter is the shortest distance between two points on the earth's surface.

To cover a large distance in high latitudes, ships usually sail along a GC track. Such **GREAT CIRCLE NAVIGATION** saves them a considerable amount of distance. They first draw the course on a Gnomonic chart, and then transfer it to a Mercator chart for ease of navigation. However, being the arc of a circle, a GC track cannot be transferred to

Chapter 17.2: Navigation - Chartwork

a Mercator chart in one piece. It is transferred in sections of short straight lines by latitude and longitude, usually in coordinates of 5° longitude.

Note that the rhumb line track cuts each meridian (longitude line) at the same angle, where as the GC track cuts them at varying angles.

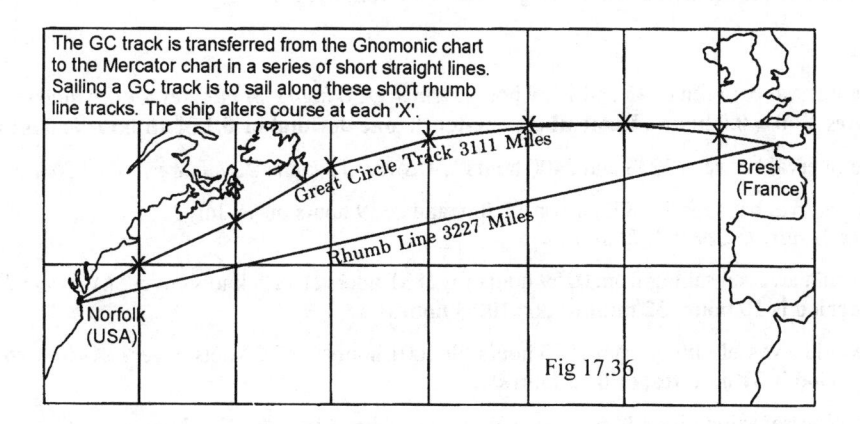

Fig 17.36

SPEED, TIME & DISTANCE (STD) CALCULATIONS

(All figures rounded to 2 decimal places)

If you are not good at these calculations you may want to examine the following STD exercises. The most common error is overlooking that there are 60 (not 100) minutes in an hour. This can be avoided by writing down as shown here in subtracting 0140 hours from 0727 hours.

```
      7h   27m
  =  6h   87m
    -1h   40m
    ---------
     5h   47m
```

STD EXERCISES:

1. What is the time interval between 0845 and 1539 hours? *(Answer: 6 hours 54 minutes, or 6.9 hours)*
   (Note: 54 minutes ÷ 60 = 0.9 hours. Remember, one-tenth, one decimal or 0.1 of an hour is 6 minutes.)

2. What is the time interval between 1238 and 2400 hours? *(Answer: 11 hours 22 minutes, or 11.36 hours)*

3. What is the time interval between 1818 hours on 17 July and 0339 hours on 18 July?
   (Answer: 9 hours 21 minutes, or 9.35 hours)

4. What distance would a vessel sailing from 0759 hours till 1831 hours at 12.5 knots cover? *(Answer: 131.6 miles)*
   (Note: Sailing period is 10 hours 32 minutes, or 10.53 hours)

5. What distance would a vessel sailing from 1530 hours till 1601 hours at 13.2 knots cover? *(Answer: 6.86 miles)*
   (Note: Sailing period is 31 minutes, or 0.52 hours)

6. How long would a vessel sailing for 132.5 miles at 10.8 knots take? *(Answer: 12.27 hours, or 12 hours 16 minutes)*
   (Note: 0.27 hours x 60 = 16 minutes)

7. Departing at 0735 hours on a voyage of 65.4 miles, what would be the vessel's ETA at 9.6 knots?
   (Answer: 1423 hours) **(Note: The vessel would need to sail for 65.4 ÷ 9.6 = 6.81 hours, or 6 hours 48 minutes. Therefore ETA = 07h 35m + 6 h 48m = 14h 23m)**

8. A vessels sails a distance of 4.2 miles in 49 minutes. What is her speed? *(Answer: 5.1 knots)*
   (Note: 49 minutes = 0.82 hours)

9. Departing at 2045 hours on 15 October on a voyage of 126 miles, what would be the vessel's ETA at 8.8 knots?
   (Answer: 1104 hours on 16 Oct.) **(Note: The vessel would need to sail for 126 ÷ 8.8 = 14.32 hours, or 14 hours 19 minutes. Therefore ETA = 20h 45m + 14h 19m = 35h 04m = 16 Oct. 11h 04m)**

The "Speed, Time and Distance Triangle" is a useful tool for those struggling with the above calculations. In STD exercises 4 to 8 above, Speed, Time and Distance are three pieces of information of which two are known in each case, and you need to find the third.

Rules:
- **Always place Distance at top** of the triangle.
- Draw a line under the word "Distance".
- Place Speed and Time on the bottom in any order.
- **Time must be in hours**. For example, if it is 2 hours 12 minutes, convert 12 minutes into (0.2) hours by dividing it by 60, then write it as 2.2 hours.

Fig 17.37

Using the line under "Distance" as the "dividing in" line, the triangle offers us three equations, only one of which is needed for a given problem:

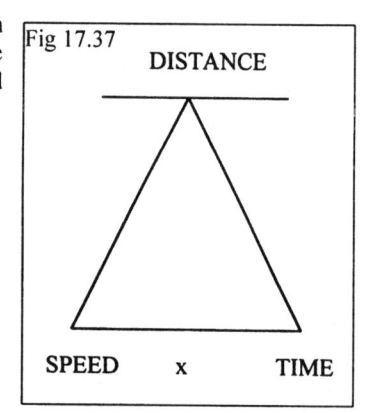

> $\text{Speed} = \dfrac{\text{Distance}}{\text{Time}}$ *(as in STD exercise 8 above)*

> $\text{Time} = \dfrac{\text{Distance}}{\text{Speed}}$ *(as in STD exercises 6 & 7)*

> Distance = Speed x Time   *(as in STD exercises 4 & 5)*

USING COMPASS & CHART

IMPORTANT: An understanding of Compasses (Chapter 17.1) is essential prior to studying the following explanation.

FINDING THE COMPASS COURSE TO STEER

All Courses and Bearings plotted on a chart are True courses and True bearings. To find the course to steer by Compass, the Variation and Deviation must be applied as per the "Rule".

For example: If the true course is 075°, Variation 10°E and Deviation 0°, then the compass course to steer is 065°(C), as shown in the table.

```
              ↓ T = 075°    (From the chart)
E'ly  (−)       V = 10°E    (From the chart)
W'ly  (+)       M = 065°
                D =   0°    (From Deviation Card 1)
              ↓ C = 065°
```

COMBINING VARIATION AND DEVIATION

As discussed in Chapter 17.1, a magnetic compass suffers from two errors, together known as the **TOTAL COMPASS ERROR**:
   ➢ VARIATION is available from the chart.
   ➢ DEVIATION is available from the deviation card.

These two components of the total error can be applied individually as shown above or they may be combined as follows, and applied collectively by the same Rule, i.e., *"Subtract an E'ly Error when going down from True to Compass and add when going up"*.

Variation and Deviation swing the compass card in the easterly or westerly direction. The total error is the result of their collective pull. Some examples:

Var	=	10° E	8° W	0°	7° W
Dev	=	10° W	4° E	5° W	8° E
Err	=	0	4° W	5° W	1° E

FINDING TRUE COURSE: (EXERCISE 3) *(Continued from Ex 1 & 2 in Chapter 17.1)*

Using Deviation Card 1 in Chapter 17.1, find the true course of a vessel that is steering 180° by Compass on Chart AUS802.

```
              ↑ T = 190°
E'ly  (+)       V = 13°E    (From Chart AUS802)
W'ly  (−)       M = 177°
                D =  3°W    (From Deviation Card 1)
                C = 180°    (Given in the question)
```

Answer: True Course = 190° (T)

EXERCISE 3

FINDING TRUE BEARINGS: (EXERCISE 4)

A bearing taken with a hand-held compass reads 247°. What bearing will you plot on chart AUS 252?

```
              ↑ T = 255½°
                V = 8½°E    (From chart AUS 252)
E'ly  (+)       M = 247°    (Given in the question)
                D
                C
```

Answer: the bearing to plot = 255½° (T)

EXERCISE 4

MAKING A "COURSE & BEARING CORRECTION BOX"

THE DEVIATION CARD MUST BE ENTERED USING ONLY THE SHIP'S HEADING to find the appropriate Deviation, whether correcting courses or bearings.

A Deviation for a given course is the Deviation to be applied to all bearings while on that course. If a person went around the world without changing the vessel's course, the Deviation will be the same for every bearing taken during the voyage.

One way to avoid making the mistake is to ALWAYS make a box as shown below. In it, the course is corrected first, and the Deviation Card is not re-entered for bearings. The use of the box is shown in the following examples.

	COURSE	BEARINGS (1)	(2)
T			
V			
M		— Same →	
D			
C		— Same →	

FINDING TRUE BEARINGS (EXERCISE 5)

On chart AUS802, a vessel heading 180°C observed two landmarks bearing 260°C and 309°C. The vessel's Deviation Card is No. 1 (in Chapter 17.1). Find the true course and bearings.

		COURSE	BEARINGS (1)	(2)
	T	190½	270½	319½
GOING UP	V	(+) 13E	— Same →	
E'ly (+)	M	177½	257½	306½
W'ly (−)	D	(−) 2½W	— Same →	
	C	180	260	309

EXERCISE 5

Answer: Course = 190½°(T); and Bearings = 270½°(T) and 319½°(T)

Chapter 17.2: Navigation - Chartwork

PLOTTING POSITION & COURSE ON CHART (EXERCISE 6)

In figure 17.38 (Chart AUS802), a vessel is heading 180°C in the vicinity of Kent group of islands (approx. Lat 39° 27.4'S). At 1100 hours she observed the following compass bearings:
- Erith Island peak (height 148) = 072°C
- Dover Island peak (height 217) = 119°C

Her deviation card is No. 2 (in Chapter 17.1).

(a) Give the vessel's position at 1100 hours.
(b) She now wishes to alter course to pass Big Rock (32) (approx. Lat 39° 30.4'S) at a distance of 1 mile. What is her next compass course?
(c) If her log speed is 10 knots. At what time will she be abeam of the Big Rock? *(ABEAM means at right angle to the vessel's fore & aft line)*

SOLUTION:

(a) The true bearings are as calculated as shown in the table. They are then plotted to obtain the position at 1100 hours, which can be stated in one of two ways:

*Answer (a): 39° 27.4' S 147° 12.9' E **OR** Dover Island peak bearing 122° T, distance 2.6 miles*
(REMEMBER: THE BEARINGS ARE TAKEN FROM THE VESSEL, NOT FROM THE LAND)

		COURSE	BEARINGS (1)	BEARINGS (2)
	T	183	075	122
GOING UP	V	(+) 13E	——Same——>	
E'ly (+)	M	170	062	109
W'ly (−)	D	(−) 10W	——Same——>	
	C	180	072	119

EXERCISE 6 (See plotting on chart below)

(b) With a drawing compass, an arc of 1 mile is drawn around Big Rock. Then a course is drawn from the above position as a tangent to this arc. This course reads 245° T. The deviation for this course is 3½° W (say, 4° W). APPLYING THE "TVMDC" RULE, THE COURSE IS CONVERTED TO 232°M and 236°C.
Answer (b): The next compass course to steer = 236° C

(c) As stated above, ABEAM means at a right angle to the vessel's fore & aft line. The distance between the 1100 hours position and the right angle mark is measured. It is 4.9 miles (say, 5 miles).
Answer (c): At a speed of 10 knots, she would be abeam of the Big Rock at 1130 hours.

PLOTTING POSITION BY RADAR DISTANCES (EXERCISE 6-A)

In Exercise 6, instead of taking bearings, the navigator of a radar-equipped vessel could have the 1100 hours position by measuring the radar ranges (distances) as follows. The advantages of ranges over bearings are twofold: Ranges are free of compass errors and the radar range discrimination is highly accurate.

- West Bluff (western edge of Dover island) = 2.4 miles
- Wallibi Bluff (western edge of Erith island) = 2.6 miles.

With a drawing compass the navigator would draw arcs for the above distances. The vessel's position would be at the intersection of the arcs. It would be the same position as in Exercise 6. Some interpretation may be required where the arcs intersect at two places and a third arc is not available to confirm the correct position.

NOTE: The bearings are referred to as POSITION LINES and the arcs as POSITION CIRCLES. The vessel's position can also be found by intersecting one or more position lines with one or more position circles.

Fig 17.38: PLOT FOR EXERCISE 6 ABOVE

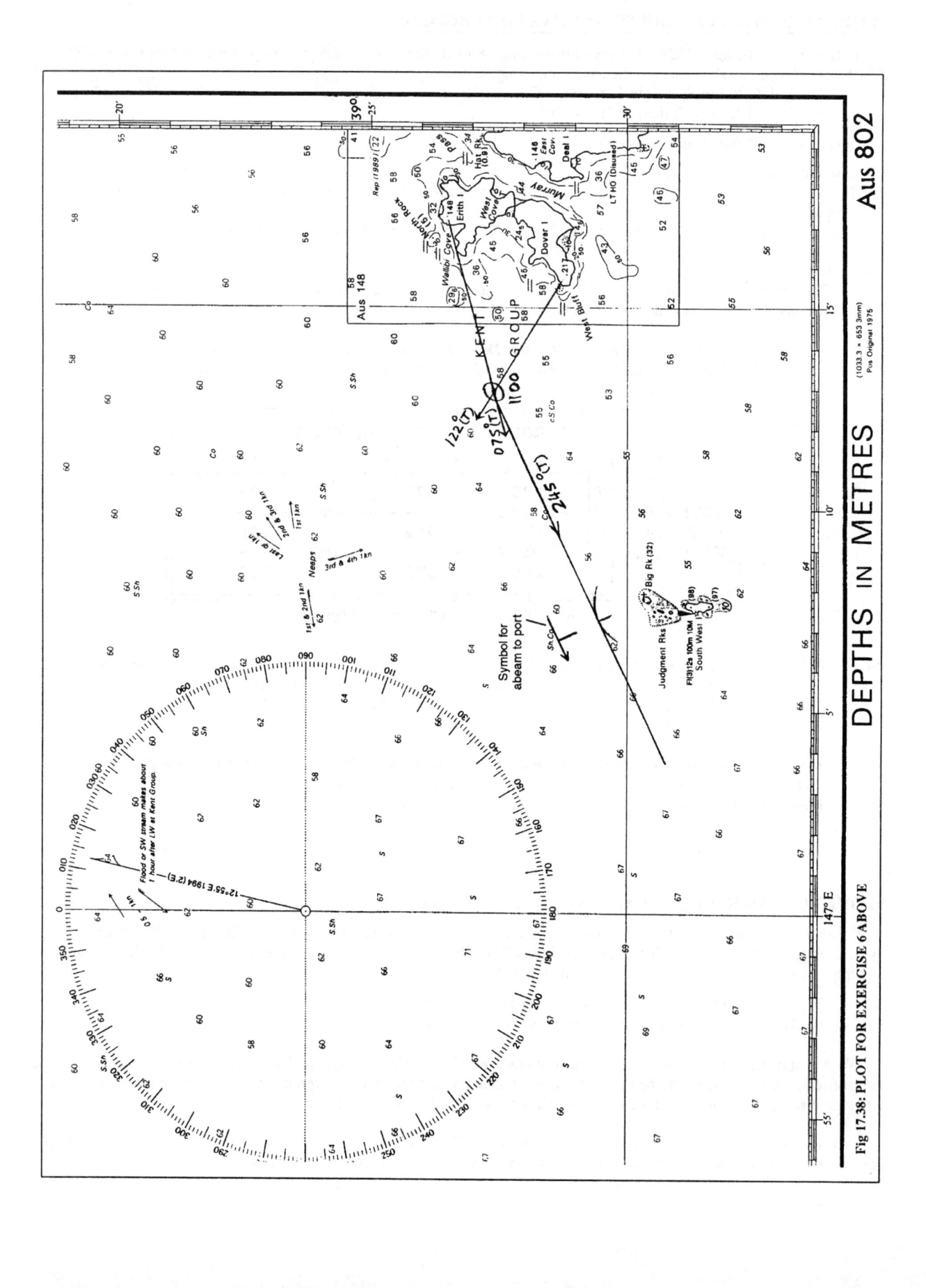

Chapter 17.2: Navigation - Chartwork

CHART PLOTTING TERMINOLOGY, LABELLING & SYMBOLS
Fig 17.39

DR (+): The expected position (dead reckoning) after sailing a certain course for a certain distance. *(Time is written next to the position)*	 —+—
EP (△): Estimated position after allowing for set and drift. *(Time is written next to the position)*	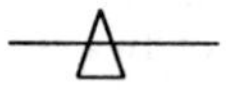 —△—
Fix (O): The definite and confirmed position. *(Time is written next to the position)*	 ⊗
Transit (Ø): Two landmarks in line.	Ø
CTS: Course to Steer (or Course Steered): Course through water: Line with one arrow in the middle.	 —►—
CMG: Course Made Good (or Course to make Good): Course over the ground. Line with two arrows in the middle.	 —►►—
SMG: Speed made Good (Speed over the ground).	
SET & DRIFT: Line with three arrows in the middle.	 ⊢►►►⊣
BEARING or **POSITION LINE** (arrow is on the end away from the landmark.)	——►
Positions lines obtained through celestial navigation or by means other than terrestrial bearings have an arrow at each end.	◄——►
POSITION CIRCLE – the above position line symbol is in the shape of an arc.	
ABEAM TO STARBOARD	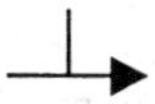 ↓→
ABEAM TO PORT	 ←↑
TPL (Transfer Position Line) (two arrows at one end).	——►►
(Also shown with two arrow at each end – more common in celestial navigation)	
TRANSFER POSITION CIRCLE – The above TPL symbol is in the shape of an arc.	

COCKED HAT

A cocked hat is a triangle on a chart formed by three position lines or (radar) ranges that do not intersect at any one point. It indicates an error in fixing the position (due to unsteady observation, faulty compass, unaccounted error, wrong reading of bearings or ranges, wrongly identifying targets, excessive time lag between bearings or ranges, etc.). If the error remains unknown and a fresh observation is not possible then the intersection of two bearings or ranges closest to a danger should be assumed to be the fix.

When fixing a vessel's position, always try to observe more than two bearings or ranges. Two will intersect even if they are wrong. The intersection of the third will confirm the fix.

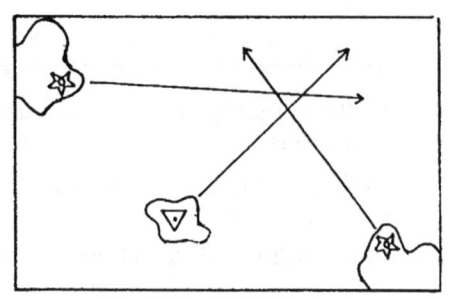

Fig 17.40: COCKED HAT

BEARINGS TO AVOID

Avoid fixing position by bearings with small angles between them, as illustrated.

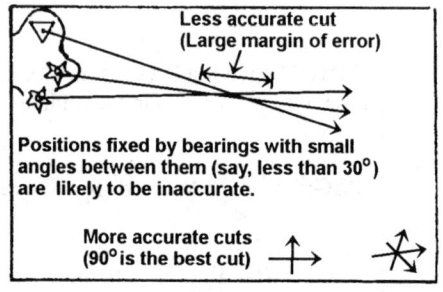

Fig 17.41

PILOTING BY TRANSITS (Ø) & SECTOR LIGHTS

Figures 17.11 (Chapter 17.1) and 17.32 illustrate the use of Transits to obtain accurate *POSITION LINES*. Two transits obtained simultaneously will provide a *FIX*, i.e., the vessel's position.

Transits and Sector Lights are also used to guide vessels into and out of ports. In the accompanying diagram, the vessel enters port by staying in the white sector of light "A" until she reaches the white sector of light "B", when she turns to port.

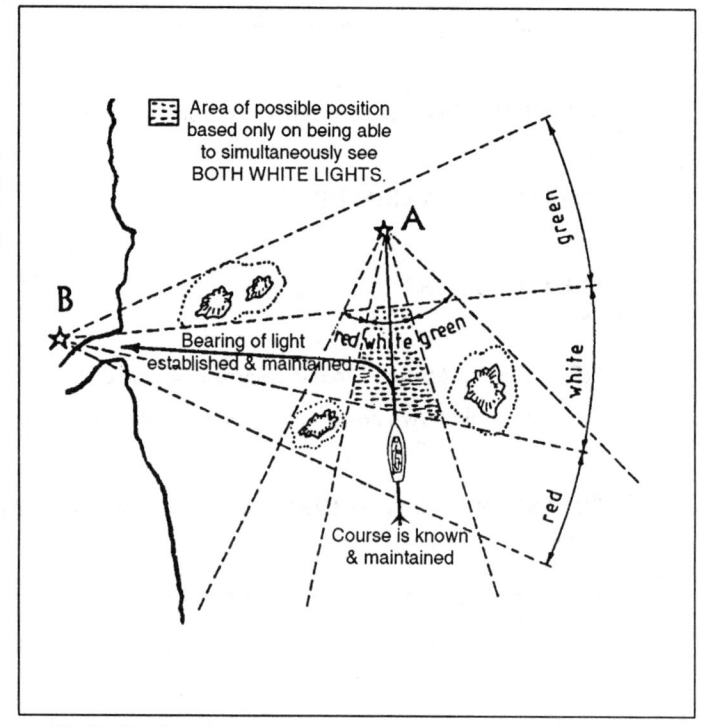

Fig 17.42: ENTERING PORT BY USING SECTOR LIGHTS

Chapter 17.2: Navigation - Chartwork

CLEARING OR DANGER BEARING

In the absence of proper fixes, transits or sector lights, the clearing or danger bearing method can be used to keep clear of a dangerous area. The clearing bearing is established between two landmarks or a landmark and the charted boundary of the danger. The selected landmark must meet the following 3 conditions:

- It must remain visible while passing the danger area.
- It must be marked on the chart.
- Its true bearing from the danger area should be in the same general direction as the vessel's course when she sails past the area.

The value of this method decreases as the angle between the course and the clearing bearing increases.

In the attached example, the vessel would keep clear of the danger as long as the bearings of the lighthouse remain above 15°T. However, for a danger to the left of the course the safe bearings would have to be below the clearing bearing.

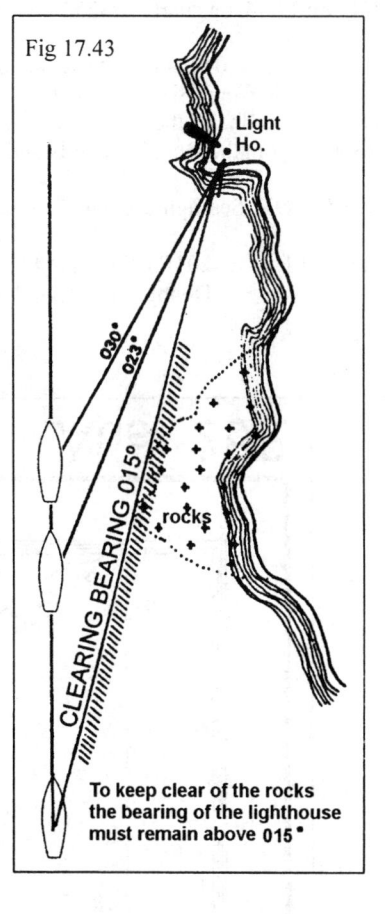

Fig 17.43

To keep clear of the rocks the bearing of the lighthouse must remain above 015°

PLOTTING POSITIONS ON CHARTS (EXERCISE 7)
(Figure 17.44, on chart AUS252)

At 0900 hours, while sailing on the "preferred" southeasterly route, a vessel observed that the northern edges of the two Double Cone Islands (44) and (104) (approx. Lat 20° 06.0'S) were in transit. At the same time the radar range of the closest coastline of the island was 1.3 miles. Give the vessel's position in latitude and longitude, and in transit bearing and distance from the Double Cone Islands.

SOLUTION: See figure 17.44. This exercise does not require the use of either Deviation or Variation.

ANSWER: 20° 05.1' S 148° 44.3' E, or
Double Cone Island Northern Edge (44) Ø Northern Edge (104) x Double Cone Island 1.3'

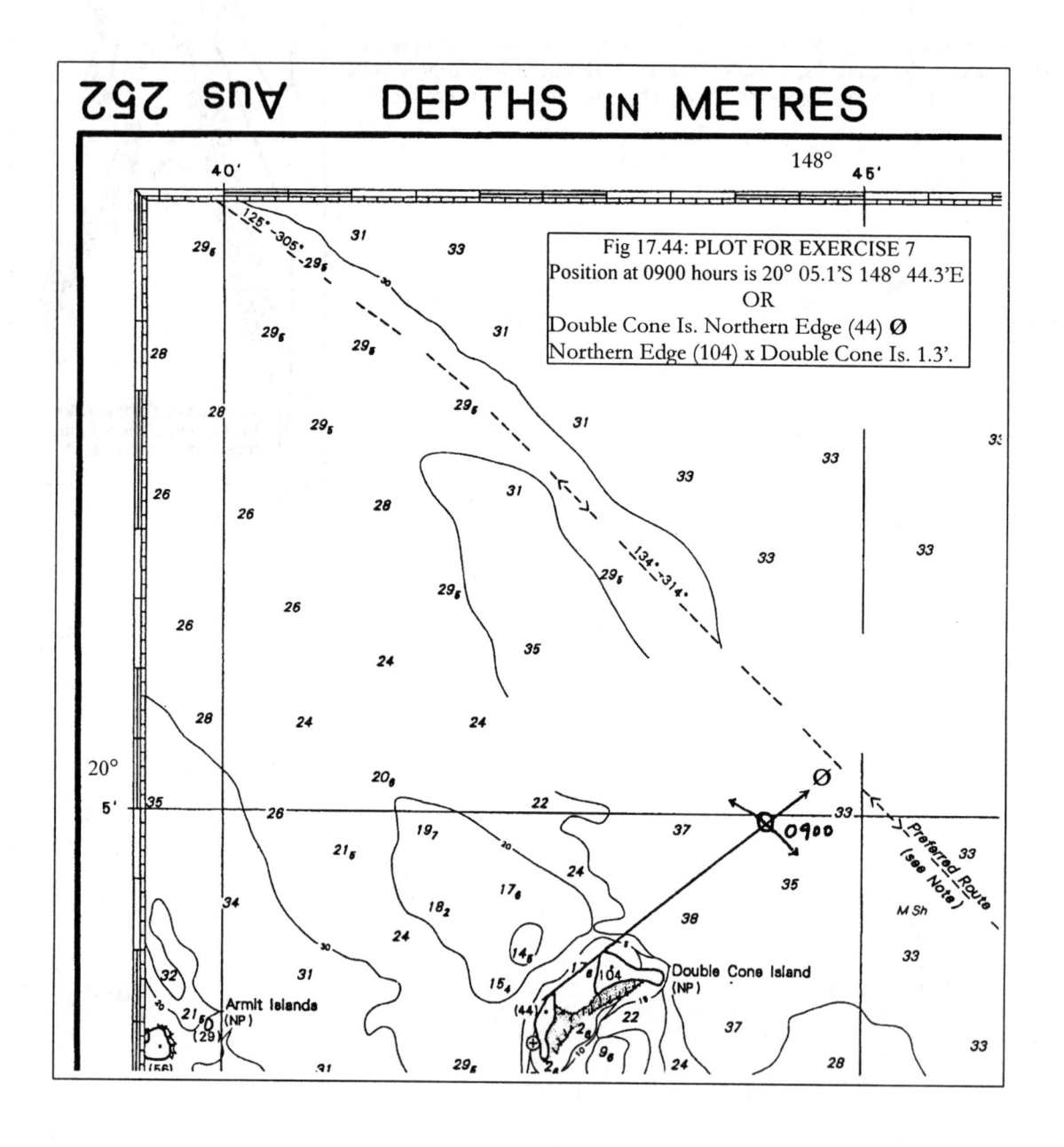

Fig 17.44: PLOT FOR EXERCISE 7
Position at 0900 hours is 20° 05.1'S 148° 44.3'E
OR
Double Cone Is. Northern Edge (44) Ø
Northern Edge (104) x Double Cone Is. 1.3'.

Chapter 17.2: Navigation - Chartwork

SET, RATE AND DRIFT

As discussed under boat handling in Chapter 8, the Set of a tidal or ocean current is the direction in which the water is moving. Everything in that water will move with it at the speed of the water. <u>The water will set a super tanker and a dinghy at the same rate, just as an escalator in a shopping centre will move an adult and a child at the same rate</u>. The <u>Rate</u> is the speed of the current measured in knots. The <u>Drift</u> is the distance moved by the current in a particular time interval.

A vessel sailing across a fast flowing river will reach the opposite bank some distance downstream. The direction in which she has been displaced is the <u>Set</u> and the distance displaced is the <u>Drift</u>. The speed of the river flow is the Rate (in knots).

Information on the direction and strength of tidal streams for critical areas is published in tide tables, tide atlases, charts and Sailing Directions (Pilot Books). Similarly, the information on currents is given on relevant navigational charts and Sailing Directions. Special charts and atlases showing ocean currents are also published. On relevant navigational charts, flood tide arrows are marked with feathers and ebb tide without feathers. Maximum rate is also indicated. See figures 17.31 & 17.32.

FINDING ESTIMATED POSITION (EXERCISE 8):

(Vessel steers a known course at a known speed for a known interval of time.)

In figure 17.45, a vessel sails from "A" towards "B" on a course of 105° T at 4 knots for one hour, through a current setting 170° at 1 knot. Obviously, she will not arrive at "B". The plot shows the Estimated Position (EP) after one hour, and course made good (CMG) and speed made good (SMG).

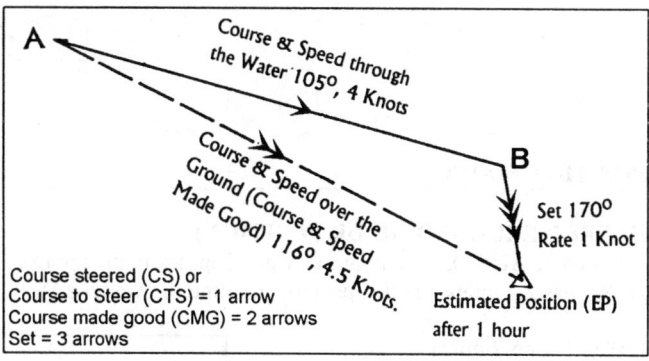

Fig 17.45: FINDING EP, CMG & SMG
(Or observing a vessel being set)

FINDING COURSE TO STEER (EXERCISE 9)

In figure 17.46, the vessel wishes to proceed from "A" to "B", i.e., she wishes to make good the track A - B, through a current setting 170° T at 1 knot. Her speed is 4 knots. What course would she need to steer and what will be her speed made good?

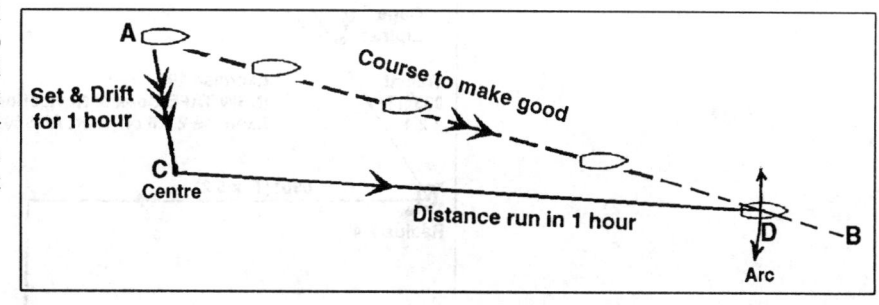

Fig 17.46: APPLYING SET & DRIFT

PROCEDURE:
- To find the course to steer to arrive at "B", we must plot the Set and Drift of the current at the beginning of the plot. Only then can we counteract the effect of the current.
- From "A" plot AC = 170° T, 1 mile
- From "C" as centre and vessel's run for 1 hour (4 miles), describe an arc to cut AB. It cuts at D.
- Join CD
- CD is the course to steer from "A", which will take the vessel along track A – B. This course is 094° T. *(Note that the vessel does not move down to "C" and then sail to "D". She sails on a course of 094° from "A".)*
- CMG & SMG are measured along A – D (105° T, 4.7 knots)
- ETA at B can be calculated by dividing AB by SMG (4.7 knots).

LEEWAY

As discussed under boat handling in Chapter 8, Leeway is the amount a vessel is blown off course by wind. Do not confuse Leeway with Set and drift. Wind causes Leeway. Water movement causes Set and Drift.

Leeway can be estimated by looking astern and observing the angle between the vessel's course and the direction of the wake. To counter the effect of Leeway, the course should be altered upwind by the amount of that angle. In figure 17.47, the vessel's CMG (plotted on chart) is 052°, and her Leeway Track is 047°. Unlike set & drift, the leeway is not plotted on chart. It is simply added to or subtracted from the course.

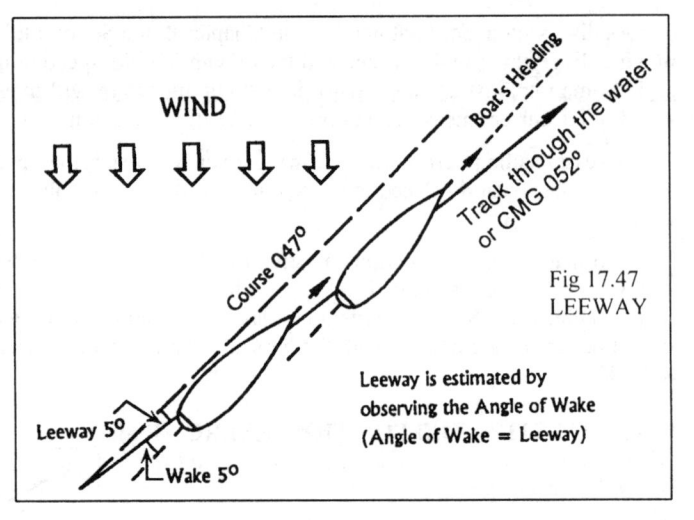

Fig 17.47 LEEWAY

POSITION FIXING

EXERCISE 10 (FINDING DR POSITION)

On chart AUS802, the range of Cape Liptrap light (approx. Lat 38° 54.6'S), when it was in transit with Liptrap Hill (168), was 2.4 miles. Fix this position and find the vessel's D.R. position after sailing the following courses:

090°T for 5.2 miles
180°T for 4.4 miles
123°T for 3.1 miles
019°T for 5.9 miles

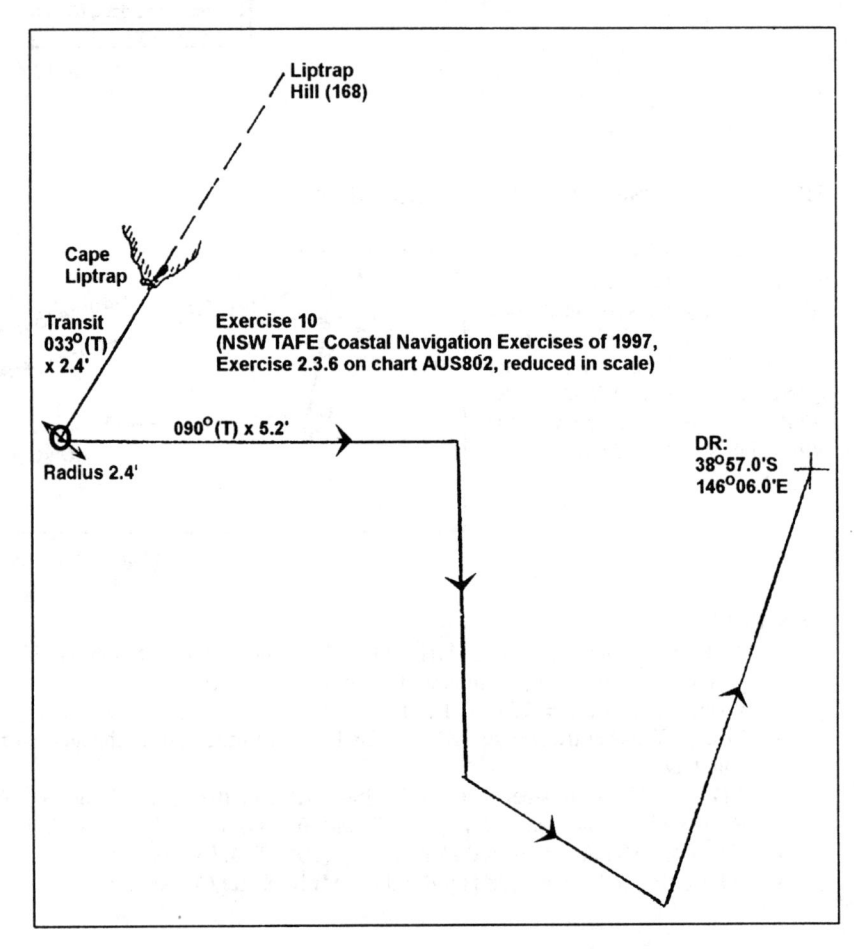

Exercise 10
(NSW TAFE Coastal Navigation Exercises of 1997, Exercise 2.3.6 on chart AUS802, reduced in scale)

DR: 38°57.0'S 146°06.0'E

Chapter 17.2: Navigation - Chartwork

EXERCISE 11 (FINDING EP BY APPLYING LEEWAY & CURRENT)

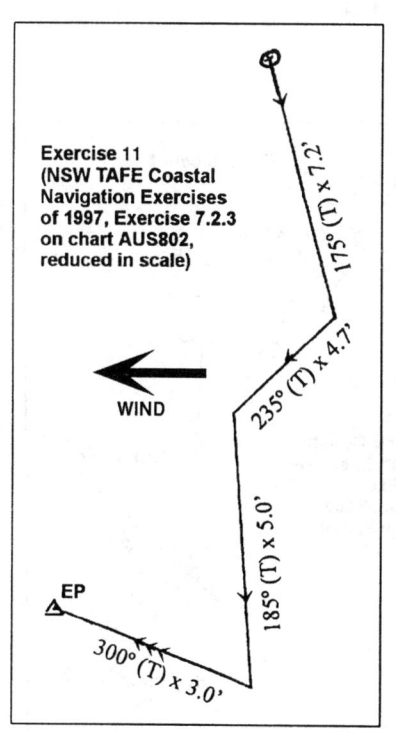

On chart AUS802, from a position where the SW tip of Norman Island (approx. Lat 39° 01.4'S) bore 081°T at a range of 2.6 miles, a yacht sailed the following tacks in an E'ly wind:

165°T for 7.2 miles leeway 10°
230°T for 4.7 miles leeway 5°
175°T for 5.0 miles leeway 10°

The current during the period was estimated to be setting 300°T with a drift of 3.0 miles. Find the E.P.

Exercise 11
(NSW TAFE Coastal Navigation Exercises of 1997, Exercise 7.2.3 on chart AUS802, reduced in scale)

SOLUTION
The vessel has already steered the courses. We have merely watched her set downwind & down current. Therefore, the Leeway due to an easterly wind is added to each course.

"Current during the period": Just like a person being moved on a moving walkway, the water on which the vessel has been sailing moved 3 miles in the direction of 300°. The vessel's final position is shifted accordingly by plotting the current as the last leg.

ANSWER: Starting position: 39° 02.2'S 146° 11.0'E;
EP: 39° 15.7S 146° 03.3'E

EXERCISE 12

(FINDING CMG & SMG BY APPLYING LEEWAY & CURRENT)

On chart AUS802, after Waterloo Point Light (approx. Lat 39° 05.4'S) bore 000°T and South East Point (approx. Lat 39° 08.0'S) bore 300°T, a vessel sailed 042°T at 10.0 knots. If a N'ly wind caused 10° leeway and a current set 090°T at 3.0 knots, find the course made good (CMG), and vessel's E.P. after 2 hours.

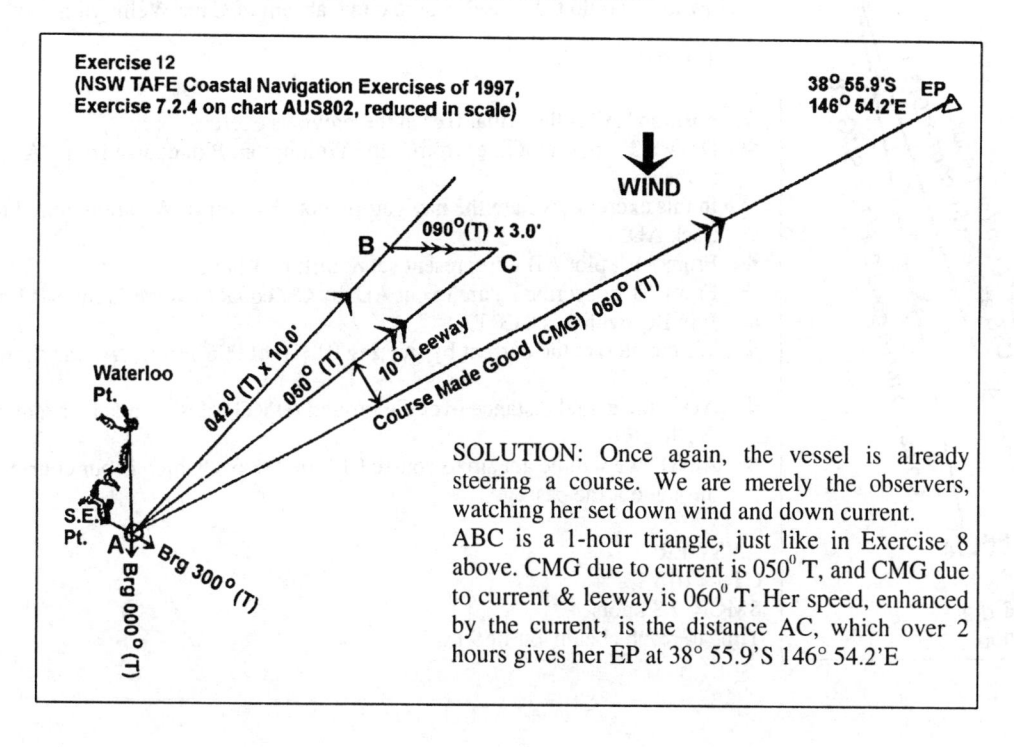

Exercise 12
(NSW TAFE Coastal Navigation Exercises of 1997, Exercise 7.2.4 on chart AUS802, reduced in scale)

SOLUTION: Once again, the vessel is already steering a course. We are merely the observers, watching her set down wind and down current.
ABC is a 1-hour triangle, just like in Exercise 8 above. CMG due to current is 050° T, and CMG due to current & leeway is 060° T. Her speed, enhanced by the current, is the distance AC, which over 2 hours gives her EP at 38° 55.9'S 146° 54.2'E

**REMEMBER: WHEN FINDING CMG, APPLY CURRENT FIRST, LEEWAY LAST.
WHEN FINDING CTS, APPLY LEEWAY FIRST, CURRENT LAST.**

EXERCISE 13 (FINDING SET & DRIFT)

On chart AUS802, from a position with Sugarloaf (94) (approx. Lat 39° 31.4'S) bearing 135°T and Curtis Island (335) (approx. Lat 39° 28.4'S) bearing 042°T, a vessel steers a course of 350°T at 7.5 knots. Two hours later W. Moncoeur Island (97) (approx. Lat 39° 13.8'S) bore 326°T and E. Moncoeur Island (101) bore 034°T. What were the set, rate and drift of the current experienced?

SOLUTION
Plot the given course from the first fix. Mark DR position at 15'. Plot the 2nd. fix.
The difference between the 2nd. Fix & DR is due to set & drift.

ANSWER:
Initial fix: 39° 29.8'S 146° 37.2'E
DR: 39° 15.2'S 146° 33.8'E
Final fix 39° 14.8'S 146° 31.4'E
Set = 277°T, Drift (2 hrs) = 1.85 miles, Rate = 0.9 knots

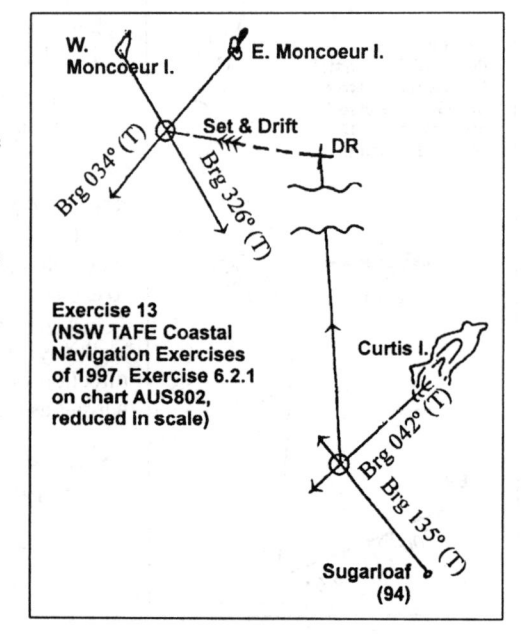

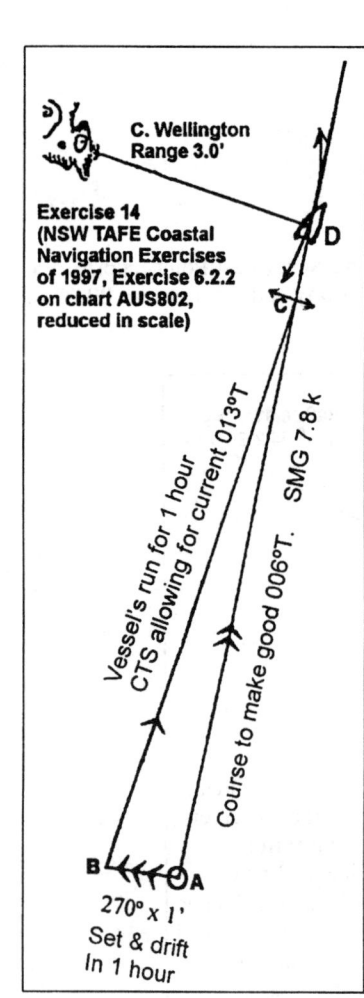

EXERCISE 14 (COUNTERACTING A CURRENT)

On chart AUS802, from the final position found in the previous exercise, lay off a course to pass Cape Wellington (approx. Lat 39° 04.0'S) at a distance of 3.0 miles to counteract a current setting 270°T at 1.0 knot. Vessel's speed is 8.0 knots. What are the C.T.S. and the S.M.G?
How long would the vessel take to come abeam of Cape Wellington?

SOLUTION

- Position "A" is the "final fix" in the previous exercise.
- Draw a 3' arc (<u>not a line</u>) from Cape Wellington. Plot course from "A" at tangent to the arc.
- In this exercise, we are the navigators (not observers). We are required to stay on track AD.
- From "A", plot AB to represent set & drift of 1 hour.
- From "B", describe an arc to cut AD at "C", equal to vessel's run for 1 hour (8').
- Join BC, which is 013°T.
- We counteract the current by steering 013° (not 006°) from "A" (as shown in Ex. 6).
- AD is the actual distance over the ground (DMG) of 9' to reach abeam of Wellington.
- At "D" we will be abeam to course 013° (not 006°), which is our course throughout the passage.

ANSWER:
CTS = 013°T
SMG = 7.8 knots
Time to reach abeam = 1 hr 9 m.

Chapter 17.2: Navigation - Chartwork

RUNNING FIX
(TRANSFERRED POSITION LINE (TPL) OR TRANSFERRED POSITION CIRCLE)

As we have seen above, fixing a vessel's position by visual means requires at least two <u>simultaneous</u> bearings (position lines), two ranges (distances) or a bearing and a range. But, sometimes only one bearing or one range is available. In such cases, positions can fixed by the Running Fix or TPL method, provided there is no unknown current. For a known current, an allowance can be made.

TPL: Shown in Exercises 15 and 16 are two bearings at two different times. They could be from the same or different landmarks. The object is to "artificially" transfer the first bearing to the time of the second, so that their intersection would give us a fix. This is done by moving the first bearing forward by the amount equal to the distance sailed by the vessel between the bearings. The intersection of the re-drawn first bearing (TPL) and the second bearing is the vessel's position at the time of the second bearing.

EXERCISE 15

Chart AUS802. At 1600 hours E. Moncoeur Island Light (approx. Lat 39° 13.8'S) bore 244°T. At 1630 hours, after steering a course of 315°T at 6.0 knots, it bore 185°T. Find the boat's position at 1630 hours.

- Boat sailed 3' between bearings.
- Her course was 315°, which can be AB, CD, EF or any other line parallel to these.
- But, only CD is 3' long (not AB or EF). Therefore, the boat must have sailed on CD.
- But, not knowing it, pick any point on the first bearing as the starting point. Let's pick E.
- Mark EG = 3' (equal to the distance sailed between bearings)
- At G, draw a line parallel to the first bearing. This line (with arrow at each end) is known as TPL. It intersects the 2nd. bearing at D, which gives us the actual Position.
- We would have achieved the same result by starting at A and cutting 3' to H. TPL at H also cuts the 2nd. bearing at D.
- Clearly, the boat was first at C and then at D. The answer is confirmed because CD measures 3'.

ANSWER
1630 Position: 39° 10.5'S 146° 32.7'E

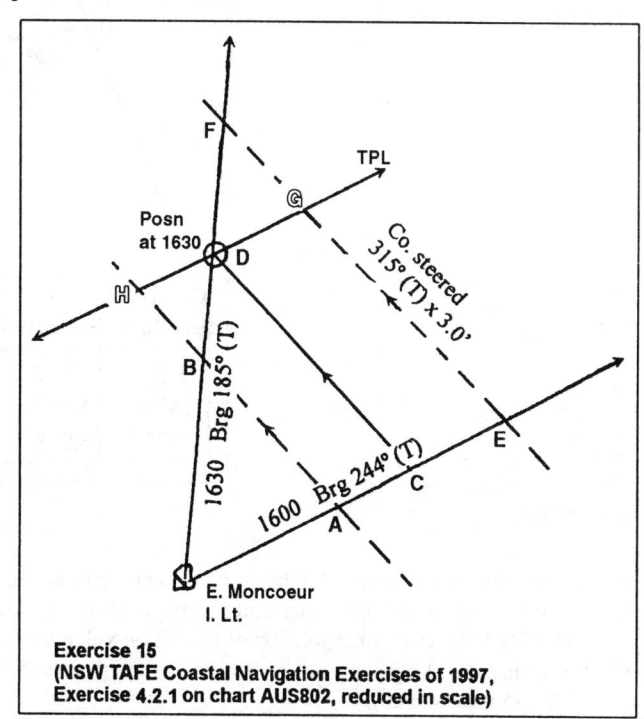

Exercise 15
(NSW TAFE Coastal Navigation Exercises of 1997, Exercise 4.2.1 on chart AUS802, reduced in scale)

EXERCISE 16

Chart AUS802. At 1215 a yacht steering 326°C, logging 5.0 knots. She observed Hogan Island Light (approx. Lat 39° 13.6'S) bearing 011°M. At 1300 the same light bore 059°M. The current was estimated to be setting 280°T at 2.0 knots. An extract of her deviation card is attached. Find the position at 1300.

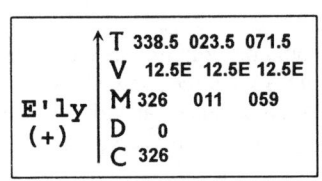

```
         T  338.5  023.5  071.5
         V  12.5E  12.5E  12.5E
 E'ly    M  326    011    059
 (+)     D  0
         C  326
```

The bearings are Magnetic;
Deviation does not apply.

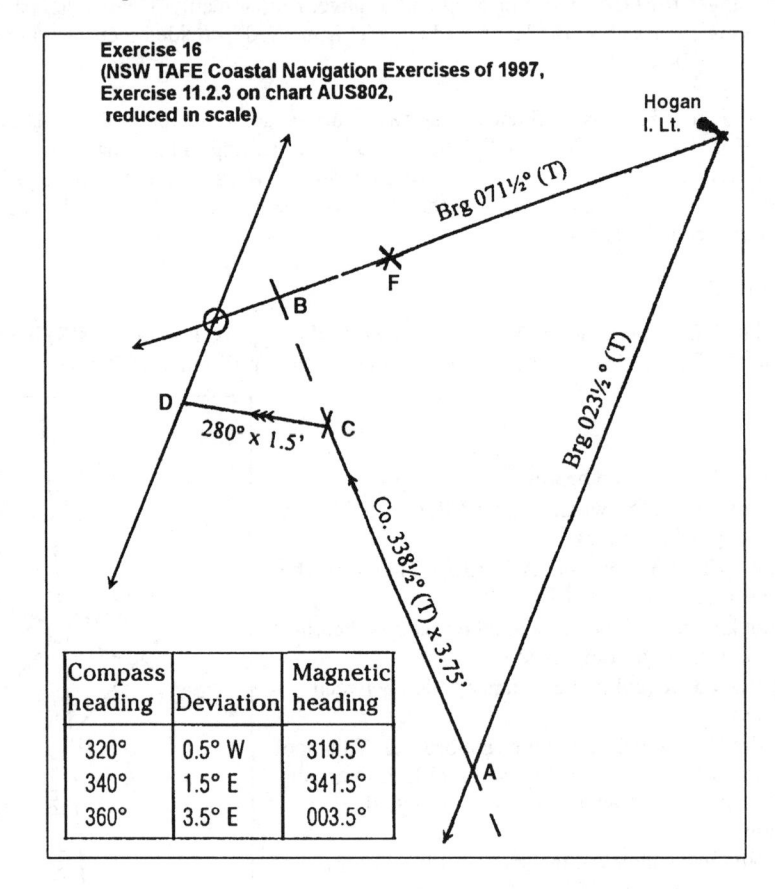

Exercise 16
(NSW TAFE Coastal Navigation Exercises of 1997,
Exercise 11.2.3 on chart AUS802,
reduced in scale)

Compass heading	Deviation	Magnetic heading
320°	0.5° W	319.5°
340°	1.5° E	341.5°
360°	3.5° E	003.5°

SOLUTION

- In this running fix, we need to **take the current into account**.
- We also must correct the course and bearings before plotting. Variation on chart AUS802 is 12.5°E. Make sure you understand the tricky interpolation of Deviation value for application to the course.
- We'll start the plot just as we did in exercise 15. Plot the course at A, which is anywhere on the 1st. bearing. Cut AC = 3.75' (45 minutes run).
- If there were no current, we would have placed TPL at C, which would have intersected the 2nd. bearing at F; and that would have been the boat's position.
- But, here, we will first move C to D, which is equal to 45 minutes drift, then draw TPL at D. The final position is thus at E. (Do not apply drift after plotting TPL.)

ANSWER
Posn. at 1300: 39° 15.0'S 146° 52.1'E

Chapter 17.2: Navigation - Chartwork

TRANSFER POSITION CIRCLES

A position circle (i.e., a range) is a circular position line. Just as with TPL, a running fix can be obtained by advancing one position circle to intersect with another at a later time. A circle is advanced by advancing its centre, as shown in Exercise 17.

EXERCISE 17

On chart AUS802, the range of Citadel Island Light (approx. Lat 39° 07.0'S) was 1.6 miles. After sailing on a course of 191°C for 2.1 miles the range was then 3.5 miles and the light was bearing <u>approximately</u> north. Fix the position. (In the vessel's Deviation Card, deviation for compass heading 180° is 9°W, and for 200° is 3°E. Magnetic variation on chart AUS802 reads 12.5°E)

SOLUTION

- The "approximate north" statement is not a bearing. It merely indicates that we are somewhere on the southern side of the island.
- Unlike TPL, the course in transfer circle exercises must only be plotted from the centre of the initial circle.
- Don't forget to correct the compass course before plotting. Interpolated deviation for 190°C is 3°W. (Tricky interpolation, as you learnt in Ex. 16: Check to make sure you come to the same conclusion).

ANSWER (POSN.): 39° 10.5'S 146° 13.8'E

NOTE: exercises involving ½° increments emphasise the need for accuracy. However, the prospect of achieving such accuracy in small vessels is quite remote.

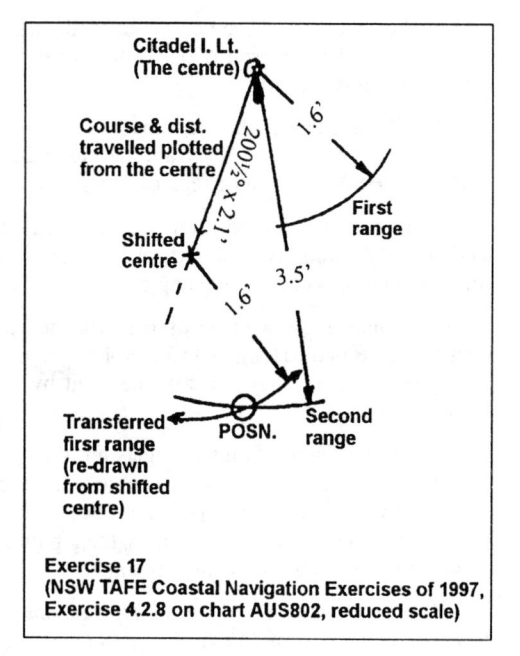

Exercise 17
(NSW TAFE Coastal Navigation Exercises of 1997, Exercise 4.2.8 on chart AUS802, reduced scale)

EXERCISE 18

Chart AUS802: At 1320 hours E Moncoeur Island (101) (approx. Lat 39° 13.8'S) was 2.0 miles off bearing approximately north. At 1400 hours Rodondo Island (350) (approx. Lat 39° 14.0'S) bore 228°M. Course steered was 274.5°C at 10.5 knots, with a current estimated to be setting 060°T at 2.7 knots.
What was the final position at 1400 hours? Use Dev 5°W and Var 12°E. For this question you may ignore the rules for Separation Schemes shown on the chart.

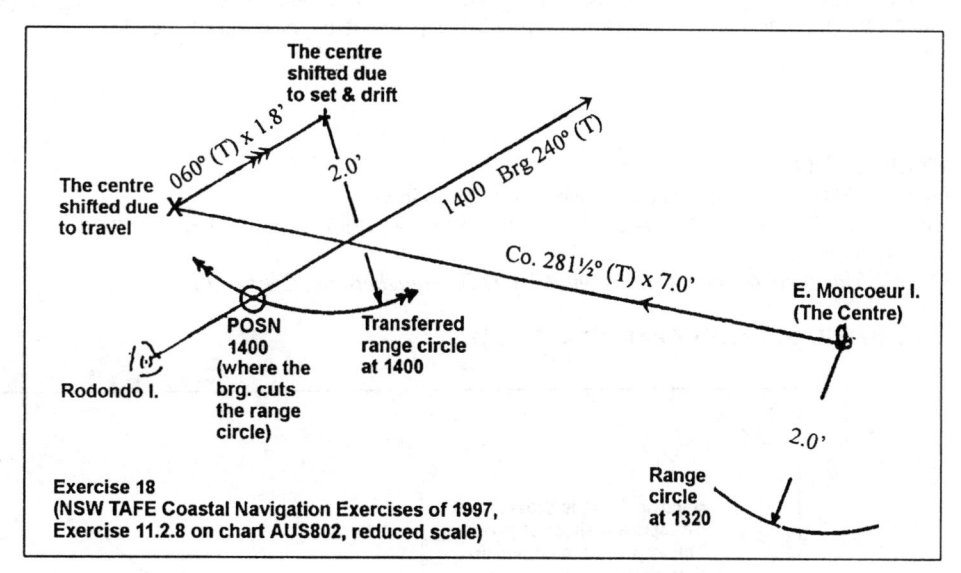

Exercise 18
(NSW TAFE Coastal Navigation Exercises of 1997, Exercise 11.2.8 on chart AUS802, reduced scale)

SOLUTION

- This running fix is a combination of transfer position circle and position line.
- The "approximate north" statement is not a bearing. It merely indicates that we are somewhere on the southern side of the island.
- The course and bearing have been corrected. Deviation and variation are supplied in the question.
- The centre is shifted first for the distance run, and then for Set and Drift before plotting the 2' arc. The principle is the same as that shown in exercise 16.

ANSWER (POSN.): 39° 13.3'S 146° 24.5'E

POSITION BY HORIZONTAL ANGLES (HA OR HSA)

The simultaneous bearings taken with a compass, whose error is unknown, cannot be plotted as bearings. However, the angles between the bearings (known as horizontal angles) can be plotted to produce an accurate fix. As long as the vessel's heading is not altered between taking bearings, whatever the error, it will be common to all the bearings. Therefore, compass error will not affect the accuracy of the angles.

For example, if the bearing of landmark "A" is 100° and of "B" is 150°, then the angle between them is 50°. Now, if the compass had an error of 10°E, the corrected bearings would be 110° and 160 °respectively. As we can see, the angle between them is still 50°.

The angles between landmarks can also be measured with a sextant (if you happen to have one). They are then referred to HORIZONTAL SEXTANT ANGLES (HSA) and can be read to a high degree of accuracy.

In order to plot the position, we need to measure two angles: one between landmarks "A" and "B" and another between "B" and "C". Note that "B" is the common line between the two angles.

Horizontal angles can be used to fix the vessel's position using a Station Pointer (illustrated), Douglas Protractor (Station Pointer without the pointers) or Transparent paper. They can also be used by geometrically constructing their position circles.

To use the Station Pointer, angles are set up in accordance with the printed instrument instructions. The instrument is then placed on the chart and manoeuvred until the three pointers pass precisely through the three charted landmarks. The vessel's position is then plotted by pencil through the hole provided in the centre of the instrument.

With the Douglas Protractor or Transparent paper, the appropriate angles are drawn on either side of a vertical line. As with the Station Pointer, the Protractor or Transparent paper is then placed over the chart and adjusted until the plotted lines pass precisely through the three charted landmarks. The vessel's position is plotted by pressing the tip of a pencil through the point of intersection of the three lines.

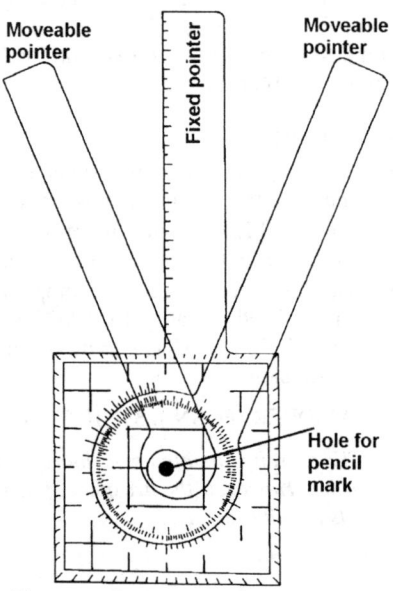

Three-arm Protractor or Station Pointer

EXERCISE 19
On chart AUS802, horizontal angles were taken as follows:
Citadel Island Light 70° Cleft Island (113) 56° Forty Foot Rocks (approx. Lat 39° 12.2'S)
Find the vessel's position.
(NOTE: The angle between Citadel and Cleft is 70°, and between Cleft and Forty Foot is 56°.)

(A) THE 'TRANSPARENT PAPER' METHOD

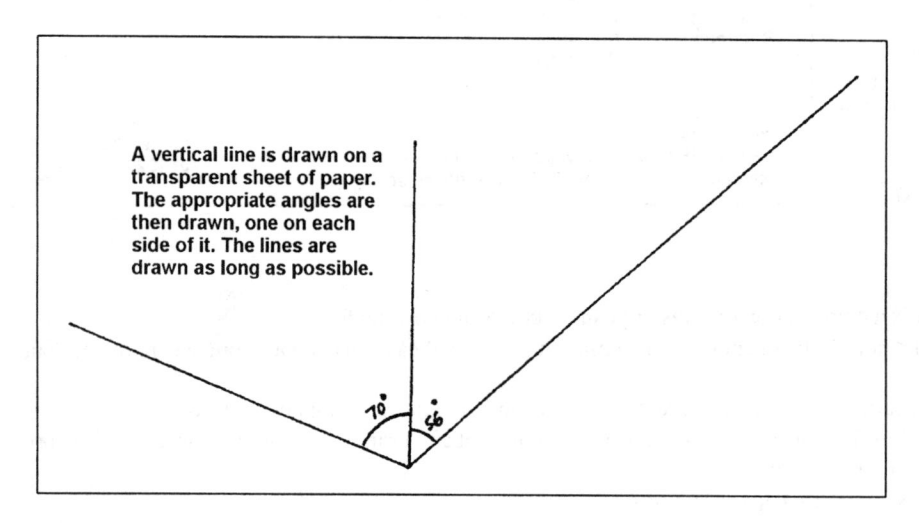

A vertical line is drawn on a transparent sheet of paper. The appropriate angles are then drawn, one on each side of it. The lines are drawn as long as possible.

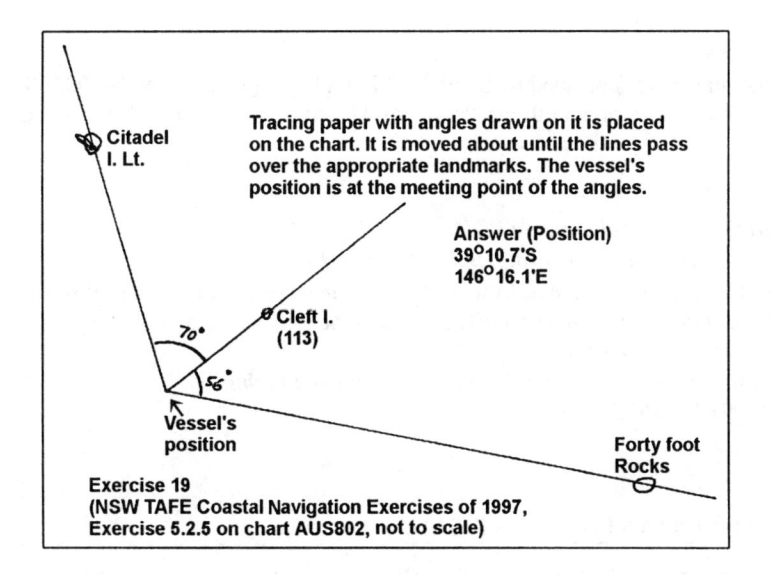

Chapter 17.2: Navigation - Chartwork

> Tracing paper with angles drawn on it is placed on the chart. It is moved about until the lines pass over the appropriate landmarks. The vessel's position is at the meeting point of the angles.
>
> **Answer (Position)**
> 39°10.7'S
> 146°16.1'E
>
> Citadel I. Lt.
> Cleft I. (113)
> 70°
> 56°
> Vessel's position
> Forty foot Rocks
>
> **Exercise 19**
> (NSW TAFE Coastal Navigation Exercises of 1997, Exercise 5.2.5 on chart AUS802, not to scale)

(B) THE 'GEOMETRIC' METHOD (AN ALTERNATIVE)

The geometric solution to Exercise 19 is based on the following theorems:

- Theorem 38: "The angle at the centre of a circle is double the angle at the circumference, standing on the same chord."
- Theorem 39: "All angles subtended by a chord at the circumference are equal."
- Theorem 16: "The sum of the interior angles of a triangle equals 180°."
- A chord of a circle is a line joining any two points on the circumference.
- Theorem 32: "One circle and one circle only can pass through three points, not in the same straight line."

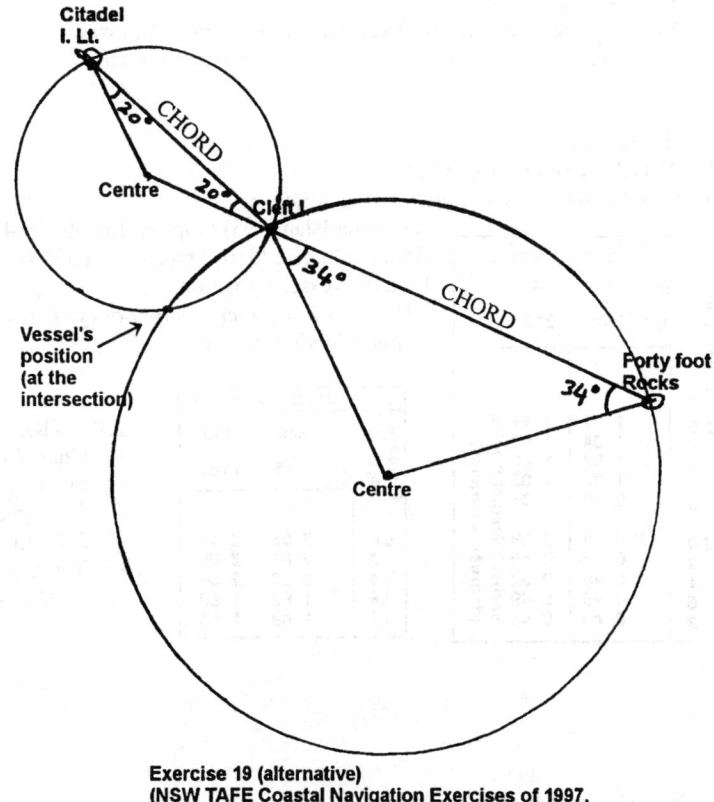

PROCEDURE:

- Join the landmarks with lines. These are the two chords.
- The given HA between Citadel and Cleft is 70°. Subtract it from 90°, and draw the difference (20°) at each end of the chord.
- The third angle in the triangle thus created is 140°, which is twice the given HA, and is thus the centre of the circle on which our HA of 70° is located.
- The circle drawn from this centre would pass through the two landmarks, and the vessel's position is somewhere on this circle (according to Theorem 32 above)
- Similarly, the vessel's position is somewhere on the second circle. It is therefore at the intersection of the two circles. The answer is given above.

Exercise 19 (alternative)
(NSW TAFE Coastal Navigation Exercises of 1997, Exercise 5.2.5 on chart AUS802, not to scale)

NOTE: For a HA greater than 90°, subtract 90° from it and draw the angle of difference on the other side of the chord to find the centre of the circle.

EXERCISE 20
(NSW TAFE Exercise No. 5.2.8)
On chart AUS802, the horizontal angle subtended by South East Point Light (approx. Lat 39° 08.0'S) and Cape Wellington (approx. Lat 39° 04.0'S), which was bearing north-westerly, was 44°. At the same time W. Moncoeur Island (97) (approx. Lat 39° 13.8'S) was bearing 190°M. Find the position.

HINTS
- *"Bearing north-westerly" is a general direction*
- *In this exercise, there is one HA and one bearing, instead of two HAs.*
- *Draw the <u>corrected</u> bearing on the chart, and draw the lines of the angle on a transparent sheet.*
- *Place the sheet on the chart and move it about until the lines of the angle pass over the respective landmarks and the tip of the angle touches the line of bearing.*
- *The vessel's position is where the tip of the angle touches the line of bearing.*
- *Answer (Position): 39° 09.6'S 146° 32.8'E*

POSITION BY VERTICAL SEXTANT ANGLES (VSA)

If you had a sextant and were close to a landmark of known height, you could find out your distance off the landmark by measuring the Vertical Sextant Angle subtended by it.

The distance can be measured by one of two ways:
1. By the formula:
   Tan • = Height ÷ Distance.
2. By use of the "**Distance by Vertical Angle Table**" (*An extract and an abridged version of the table are shown in Exercise 21.*)

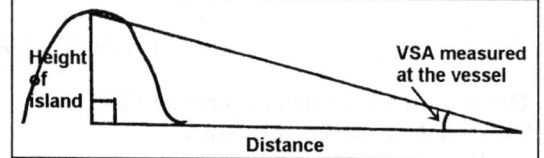

EXERCISE 21
(NSW TAFE Exercise No. 5.2.2)
On chart AUS802, VSAs were taken as follows:
Kanowna Island (95) (approx. Lat 39° 09.4'S) = 1° 01'
Rodondo Island (350) (approx. Lat 39° 14.0'S) = 1° 55'
Find the vessel's position.
NOTE: The observer's height of eye is ignored; i.e., a correction for the height of the ship's bridge is not necessary.

Distance mls	Height of object 90 m / 295 ft	Height of object 95 m / 312 ft
2.5	1° 07'	1° 11'
.6	1 04	1 08
.7	1 02	1 05
.8	1 00	1 03
.9	0 58	1 01
3.0	0 56	0 59
.2	0 52	0 55
.4	0 49	0 52
.6	0 46	0 49
.8	0 44	0 46

(Vertical Sextant Angles — degrees & minutes)

Distance mls	Height of object 340 m / 1115 ft	Height of object 360 m / 1181 ft
5.0	2° 06'	2° 14'
.2	2 01	2 08
.4	1 57	2 04
.6	1 53	1 59
.8	1 49	1 55

SOLUTION BY USE OF VSA TABLE:
➤ Chart AUS802 being metric, the heights of islands are in metres
➤ The table shows that Kanowna Island is at a distance of 2.9', and Rodondo Island at 5.6'.
➤ The arcs of these ranges will intersect at position 39° 11.8'S 146° 16.4'E, which is the vessel's position.

DANGER ANGLE (VDA)

The principle of VSA can also be used to set a danger angle to pass a danger at a specified distance.

EXERCISE 21 (A)
What will be the danger angle to set on the sextant to pass a lighthouse of 90 metres height at a range of not less than 2.7 miles?

ANSWER (from the VSA Table in exercise 21) = 1° 02'

Chapter 17.2: Navigation - Chartwork

DISTANCE BY VERTICAL ANGLE

Distance mls	\multicolumn{12}{c	}{Height of Object}
	m 7	8.5	10	11.5	13	14.5	16	17.5	19	20.5	22	23.5	25	26.5
	ft 23	28	33	38	43	48	52	57	62	67	72	77	82	87
	° ′	° ′	° ′	° ′	\multicolumn{4}{c}{Vertical Angles in degrees & minutes}	° ′	° ′	° ′	° ′	° ′	° ′
0.1	2 10	2 38	3 05	3 33	4 01	4 12	4 56	5 24	5 51	6 19	6 46	7 14	7 41	8 09
.2	1 05	1 19	1 33	1 47	2 01	2 14	2 28	2 44	2 56	3 10	3 24	3 38	3 52	4 06
.3	0 43	0 53	1 02	1 11	1 20	1 17	1 39	1 48	1 58	2 07	2 16	2 25	2 35	2 44
.4	0 33	0 39	0 46	0 53	1 03	1 07	1 14	1 21	1 28	1 35	1 42	1 49	1 56	2 03
0.5	0 26	0 32	0 37	0 43	0 48	0 54	0 59	1 05	1 11	1 16	1 22	1 27	1 33	1 38
.6	0 22	0 26	0 31	0 36	0 40	0 45	0 49	0 54	0 59	1 03	1 08	1 13	1 17	1 22
.7	0 19	0 23	0 27	0 30	0 34	0 38	0 42	0 46	0 50	0 54	0 58	1 02	1 06	1 10
.8	0 16	0 20	0 23	0 27	0 30	0 34	0 37	0 41	0 44	0 47	0 51	0 55	0 58	1 01
.9	0 14	0 18	0 21	0 24	0 27	0 30	0 33	0 36	0 39	0 42	0 45	0 48	0 52	0 55

Distance mls	\multicolumn{12}{c	}{Height of Object}
	m 26.5	28	29.5	31	32.5	34	35.5	37	38.5	40	41.5	43	44.5	46
	ft 87	92	97	102	107	112	116	121	126	131	136	141	146	151
	° ′	° ′	° ′	° ′	\multicolumn{4}{c}{Vertical Angles in degrees & minutes}	° ′	° ′	° ′	° ′	° ′	° ′
0.1	8 09	8 36	9 03	9 30	9 57	10 24	10 51	11 18	11 45	12 11	12 38	13 04	13 31	13 57
.2	4 06	4 19	4 33	4 47	5 09	5 15	5 28	5 42	5 56	6 10	6 24	6 37	6 51	7 05
.3	2 44	2 53	3 02	3 12	3 21	3 30	3 39	3 49	3 58	4 07	4 16	4 26	4 35	4 44
.4	2 03	2 10	2 17	2 24	2 31	2 38	2 45	2 52	2 59	3 05	3 12	3 19	3 26	3 33
0.5	1 38	1 44	1 49	1 55	2 01	2 06	2 12	2 17	2 23	2 28	2 34	2 40	2 45	2 51
.6	1 22	1 27	1 31	1 36	1 41	1 45	1 50	1 54	1 59	2 04	2 08	2 13	2 18	2 22
.7	1 10	1 14	1 18	1 22	1 26	1 30	1 34	1 38	1 42	1 46	1 50	1 54	1 58	2 02
.8	1 01	1 05	1 08	1 12	1 15	1 19	1 22	1 26	1 29	1 33	1 36	1 40	1 43	1 47
.9	0 55	0 58	1 01	1 04	1 07	1 10	1 13	1 16	1 19	1 22	1 26	1 29	1 32	1 35
1.0	0 49	0 52	0 55	0 58	1 00	1 03	1 06	1 09	1 11	1 14	1 17	1 20	1 23	1 25
.1	0 45	0 47	0 50	0 52	0 55	0 57	1 00	1 02	1 05	1 07	1 10	1 13	1 15	1 18
.2	0 41	0 43	0 46	0 48	0 50	0 53	0 55	0 57	1 00	1 02	1 04	1 07	1 09	1 11
.3	0 38	0 40	0 42	0 44	0 46	0 49	0 51	0 53	0 55	0 57	0 59	1 01	1 04	1 06
.4	0 35	0 37	0 39	0 41	0 43	0 45	0 47	0 49	0 51	0 53	0 55	0 57	0 59	1 01

Distance mls	\multicolumn{12}{c	}{Height of Object}
	m 46	47.5	49	50.5	52	53.5	55	56.5	58	59.5	61	62.5	64	65.5
	ft 151	156	161	166	171	176	180	185	190	195	200	205	210	215
	° ′	° ′	° ′	° ′	\multicolumn{4}{c}{Vertical Angles in degrees & minutes}	° ′	° ′	° ′	° ′	° ′	° ′
0.1	13 57	14 23	14 49	15 15	15 41	16 07	16 32	16 58	17 23	17 49	18 14	18 39	19 04	19 29
.2	7 05	7 18	7 32	7 45	7 59	8 13	8 27	8 40	8 54	9 08	9 21	9 35	9 48	10 02
.3	4 44	4 53	5 02	5 12	5 21	5 30	5 39	5 48	5 58	6 07	6 16	6 25	6 34	6 43
.4	3 33	3 40	3 47	3 54	4 01	4 08	4 15	4 22	4 29	4 36	4 42	4 49	5 49	5 03
0.5	2 51	2 56	3 02	3 07	3 13	3 18	3 24	3 29	3 35	3 41	3 46	3 52	3 57	4 03
.6	2 22	2 27	2 31	2 36	2 41	2 45	2 50	2 55	2 59	3 04	3 09	3 13	3 18	3 22
.7	2 02	2 06	2 10	2 14	2 18	2 22	2 26	2 30	2 34	2 38	2 42	2 46	2 50	2 54
.8	1 47	1 51	1 54	1 57	2 01	2 04	2 08	2 11	2 15	2 18	2 21	2 25	2 28	2 32
.9	1 35	1 38	1 41	1 44	1 47	1 50	1 53	1 56	2 00	2 03	2 06	2 09	2 12	2 15
1.0	1 25	1 28	1 31	1 34	1 36	1 39	1 42	1 45	1 48	1 50	1 53	1 56	1 59	2 02
.1	1 18	1 20	1 23	1 25	1 28	1 30	1 33	1 35	1 38	1 40	1 43	1 45	1 48	1 50
.2	1 11	1 13	1 16	1 18	1 20	1 23	1 25	1 27	1 30	1 32	1 34	1 37	1 39	1 41
.3	1 06	1 08	1 10	1 12	1 14	1 16	1 18	1 21	1 23	1 24	1 27	1 29	1 31	1 34
.4	1 01	1 03	1 05	1 07	1 09	1 11	1 13	1 15	1 17	1 19	1 21	1 23	1 25	1 27
1.5	0 57	0 59	1 01	1 02	1 04	1 06	1 08	1 10	1 12	1 14	1 15	1 17	1 19	1 21
.6	0 53	0 55	0 57	0 59	1 00	1 02	1 04	1 06	1 08	1 09	1 11	1 12	1 14	1 16
.7	0 50	0 52	0 53	0 55	0 57	0 58	1 00	1 02	1 03	1 05	1 07	1 08	1 10	1 12
.8	0 47	0 49	0 50	0 52	0 54	0 55	0 56	0 58	1 00	1 01	1 03	1 04	1 06	1 08
.9	0 45	0 46	0 48	0 49	0 51	0 52	0 54	0 55	0 57	0 58	1 00	1 01	1 03	1 04

DISTANCE BY VERTICAL ANGLE

Distance mls	m 65.5 ft 215	67 220	68.5 225	70 230	75 246	80 262	85 279	90 295	95 312	100 328	105 344	110 361	120 394	130 427
	° ′	° ′	° ′	° ′	Vertical Angles in degrees & minutes					° ′	° ′	° ′	° ′	° ′
0.5	4 03	4 08	4 14	4 19	4 38	4 56	5 15	5 33	5 51	6 10	6 28	6 46	7 23	7 59
.6	3 22	3 27	3 32	3 36	3 52	4 07	4 22	4 38	4 53	5 08	5 24	5 39	6 10	6 40
.7	2 54	2 58	3 01	3 05	3 19	3 32	3 45	3 58	4 11	4 25	4 38	4 51	5 17	5 44
.8	2 32	2 35	2 39	2 42	2 54	3 05	3 17	3 29	3 40	3 52	4 03	4 15	4 38	5 01
.9	2 15	2 18	2 21	2 24	2 35	2 45	2 55	3 05	3 16	3 26	3 36	3 47	4 07	4 28
1.0	2 02	2 04	2 07	2 10	2 19	2 28	2 38	2 47	2 56	3 05	3 15	3 24	3 42	4 01
.1	1 50	1 53	1 56	1 58	2 07	2 15	2 23	2 32	2 40	2 49	2 57	3 05	3 22	3 49
.2	1 41	1 44	1 46	1 48	1 56	2 04	2 11	2 19	2 27	2 35	2 42	2 50	3 05	3 21
.3	1 34	1 36	1 38	1 40	1 47	1 54	2 01	2 08	2 16	2 23	2 30	2 37	2 51	3 05
.4	1 27	1 29	1 31	1 33	1 39	1 46	1 53	1 59	2 06	2 13	2 19	2 26	2 39	2 52
1.5	1 21	1 23	1 25	1 27	1 33	1 39	1 45	1 51	1 58	2 04	2 10	2 16	2 28	2 41
.6	1 16	1 18	1 19	1 21	1 27	1 33	1 39	1 44	1 50	1 56	2 02	2 08	2 19	2 31
.7	1 12	1 13	1 15	1 16	1 22	1 27	1 33	1 38	1 44	1 49	1 55	2 00	2 11	2 22
.8	1 08	1 09	1 11	1 12	1 17	1 22	1 28	1 33	1 38	1 43	1 48	1 53	2 04	2 14
.9	1 04	1 05	1 07	1 08	1 13	1 18	1 23	1 28	1 33	1 38	1 42	1 47	1 57	2 07

Distance mls	m 130 ft 427	140 459	150 492	160 525	170 558	180 591	190 623	200 656	210 689	220 722	230 755	240 787	250 820	260 853
	° ′	° ′	° ′	° ′	Vertical Angles in degrees & minutes					° ′	° ′	° ′	° ′	° ′
1.5	2 41	2 53	3 05	3 18	3 30	3 42	3 55	4 07	4 19	4 32	4 44	4 56	5 09	5 21
.6	2 31	2 42	2 54	3 05	3 17	3 29	3 40	3 52	4 03	4 15	4 26	4 38	4 49	5 01
.7	2 22	2 33	2 44	2 55	3 05	3 16	3 27	3 38	3 49	4 00	4 11	4 22	4 32	4 43
.8	2 14	2 24	2 35	2 45	2 55	3 05	3 16	3 26	3 36	3 47	3 57	4 07	4 17	4 28
.9	2 07	2 17	2 26	2 36	2 46	2 56	3 05	3 15	3 25	3 35	3 44	3 54	4 04	4 14
2.0	2 01	2 10	2 19	2 28	2 38	2 47	2 56	3 05	3 15	3 24	3 33	3 42	3 52	4 01
.1	1 55	2 04	2 13	2 21	2 30	2 39	2 48	2 57	3 05	3 14	3 23	3 32	3 41	3 49
.2	1 50	1 58	2 07	2 15	2 23	2 32	2 40	2 49	2 57	3 05	3 14	3 22	3 31	3 39
.3	1 45	1 53	2 01	2 09	2 17	2 25	2 33	2 41	2 49	2 57	3 05	3 13	3 22	3 30
.4	1 41	1 48	1 56	2 04	2 11	2 19	2 27	2 35	2 42	2 50	2 58	3 05	3 13	3 21
2.5	1 36	1 44	1 51	1 59	2 06	2 14	2 21	2 28	2 36	2 43	2 51	2 58	3 05	3 13
.6	1 33	1 40	1 47	1 54	2 01	2 08	2 16	2 23	2 30	2 37	2 44	2 51	2 58	3 05
.7	1 29	1 36	1 43	1 50	1 57	2 04	2 11	2 17	2 24	2 31	2 38	2 45	2 52	2 59
.8	1 26	1 33	1 39	1 46	1 53	1 59	2 06	2 13	2 19	2 26	2 32	2 39	2 46	2 52
.9	1 23	1 30	1 36	1 42	1 49	1 55	2 02	2 08	2 14	2 21	2 27	2 34	2 40	2 46

Distance mls	m 260 ft 853	270 886	280 919	290 951	300 984	320 1050	340 1115	360 1181	380 1247	400 1312	450 1476	500 1640	550 1805	600 1969
	° ′	° ′	° ′	° ′	Vertical Angles in degrees & minutes					° ′	° ′	° ′	° ′	° ′
1.5	5 21	5 33	5 45	5 58	6 10	6 34	6 59	7 23	7 47	8 12	9 12	10 12	11 12	12 11
.6	5 01	5 12	5 24	5 35	5 47	6 10	6 33	6 56	7 18	7 41	8 38	9 35	10 31	11 27
.7	4 43	4 54	5 05	5 16	5 27	5 48	6 10	6 31	6 53	7 14	8 08	9 01	9 55	10 47
.8	4 28	4 38	4 48	4 58	5 09	5 29	5 49	6 10	6 30	6 51	7 41	8 32	9 22	10 12
.9	4 14	4 23	4 33	4 43	4 52	5 12	5 31	5 50	6 10	6 29	7 17	8 05	8 53	9 41
2.0	4 01	4 10	4 19	4 29	4 38	4 56	5 15	5 33	5 51	6 10	6 56	7 41	8 27	9 12
.1	3 49	3 58	4 07	4 16	4 25	4 42	5 00	5 17	5 35	5 52	6 36	7 20	8 03	8 46
.2	3 39	3 47	3 56	4 04	4 13	4 29	4 46	5 03	5 20	5 36	6 18	7 00	7 41	8 23
.3	3 30	3 38	3 46	3 54	4 02	4 18	4 34	4 50	5 06	5 22	6 02	6 42	7 21	8 01
.4	3 21	3 29	3 36	3 44	3 52	4 07	4 22	4 38	4 53	5 09	5 47	6 25	7 03	7 41
2.5	3 13	3 20	3 28	3 35	3 42	3 57	4 12	4 27	4 42	4 56	5 33	6 10	6 46	7 23
.6	3 05	3 13	3 20	3 27	3 34	3 48	4 02	4 17	4 31	4 45	5 20	5 56	6 31	7 06
.7	2 59	3 05	3 12	3 19	3 26	3 40	3 53	4 07	4 21	4 34	5 09	5 43	6 17	6 51
.8	2 52	2 59	3 05	3 12	3 19	3 32	3 45	3 58	4 12	4 25	4 58	5 30	6 03	6 36
.9	2 46	2 53	2 59	3 05	3 12	3 25	3 37	3 50	4 03	4 16	4 47	5 19	5 51	6 22

Chapter 17.2: Navigation - Chartwork

POSITION BY DOUBLING THE ANGLE ON THE BOW

EXERCISE 22
On a certain chart, a vessel steering 300°T at 3.0 knots, in still water with no leeway, observed an island bearing 45° on the starboard bow. One hour later, the island bore abeam. How far off the island was the vessel when abeam?
NOTE: The 45°-angle is also referred to as 4-points.

PROCEDURE
In this exercise, the observations were made when the island was 45° and 90° off the bow. In other words, the angle on the bow doubled.

You do not need to plot such exercises on charts, nor do you need to draw them as illustrated. The answer becomes clear when seen in the light of two geometric principles:

1. The sum of three angles of a triangle is 180°.
2. Equal angles carry equal sides.

EXPLANATION
➢ Such position fixing is accurate only when there is no set or leeway.
➢ When vessel was at A, the bearing of the island must have been 015° for the angle on the bow to be 45° (015° − 330° = 45°).
➢ Since the outside angle at B is 90°, the inside angle is also 90°.
➢ Having established angle A = 45° and B = 90°, angle C must be 45°.
➢ Angles A & C being equal, sides AB & BC must be equal. Since AB = 3', BC must also be 3'.
➢ ANSWER: the vessel was 3' off when abeam of the island.

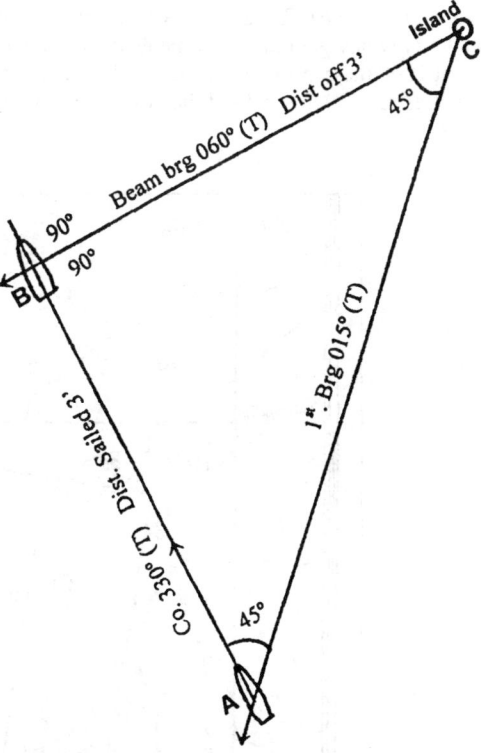

Exercise 22
NSW TAFE Coastal Navigation Exercises of 1997,
Exercise 8.1.3 on chart AUS252,
but PLOTTING NOT REQUIRED)

The concept of doubling the angles applies to any combination of angles: 30° & 60°, 35° & 70°, etc. **The distance travelled between the angles is always equal to the distance off at the second angle.**

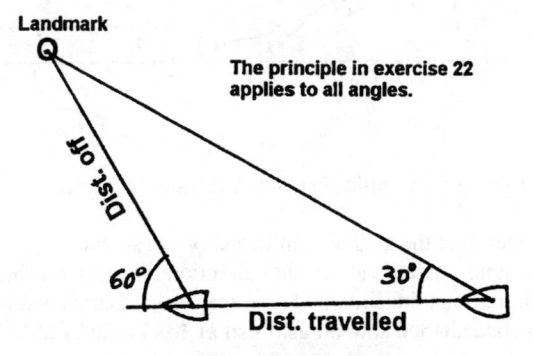

The principle in exercise 22 applies to all angles.

**The angle at the bow has doubled.
Therefore, distance travelled = distance off.**

POSITION BY A LINE OF SOUNDINGS

EXERCISE 22 (A)

On chart AUS252, at 1300 hours, a vessel is about 3 miles north of Pioneer Bay anchorage (approx. Lat 20° 14.0'S). On a course of 170°T, logging 4 knots, she is trying to reach the anchorage. The visibility is almost zero due to heavy rain, and the only navigational instruments in working order are the log and the echo sounder.

At 1300 the echo sounder recorded the corrected under-keel sounding of 6.9 metres, followed by 15 metres at 1322 and 10 metres at 1330 hours. Find the vessel's position at 1330 hours.

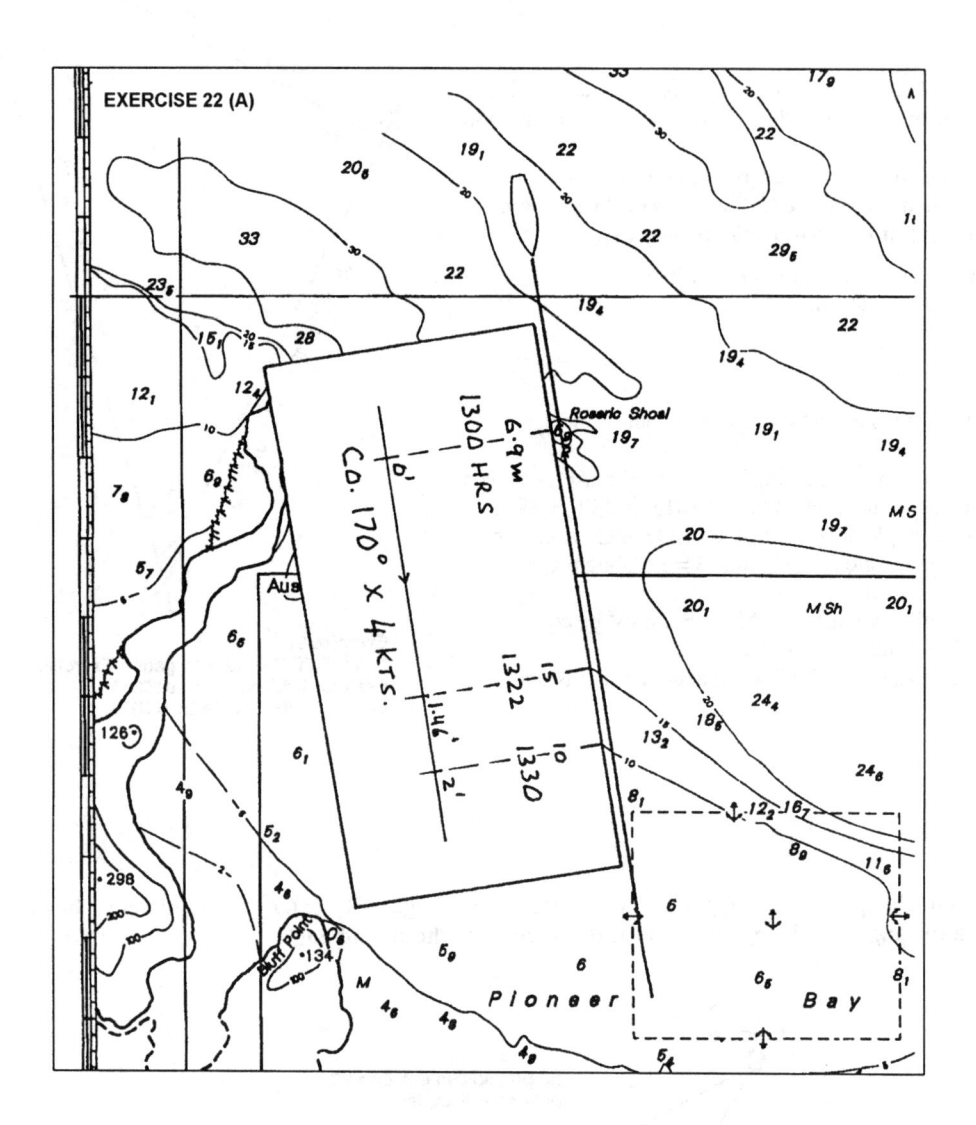

PROCEDURE:
- At 4 knots, the vessel has sailed 1.46 and 0.53 miles between the times indicated. (Being log readings, these are only approximate measurements.)
- On the edge of a plain sheet of paper, plot the available information, as shown.
- Place the paper on the chart in the approximate vicinity and direction (course) indicated. Without changing its direction (170°), manoeuvre the paper until the soundings marked on it match the charted soundings. The vessel's approximate positions are therefore on the matched charted soundings. **Posn at 1330 = 20° 12.9'S 148° 42.9'E**

CAUTION
Positions obtained by this method should be treated with caution. The tide levels may vary the soundings, or there may be more than one track where a good match can be found.

Chapter 17.2: Navigation - Chartwork

FINDING CMG BY RATIOS (THE THREE-BEARINGS METHOD)

EXERCISE 23
On chart AUS802, at 0340 hours Citadel Island Light (approx. Lat 39° 07.0'S) bore 000°M. 15 minutes later it bore 037°M. After a further 20 minutes it bore 068°M. Find the Course Made Good. The chart reads magnetic variation 12°E.

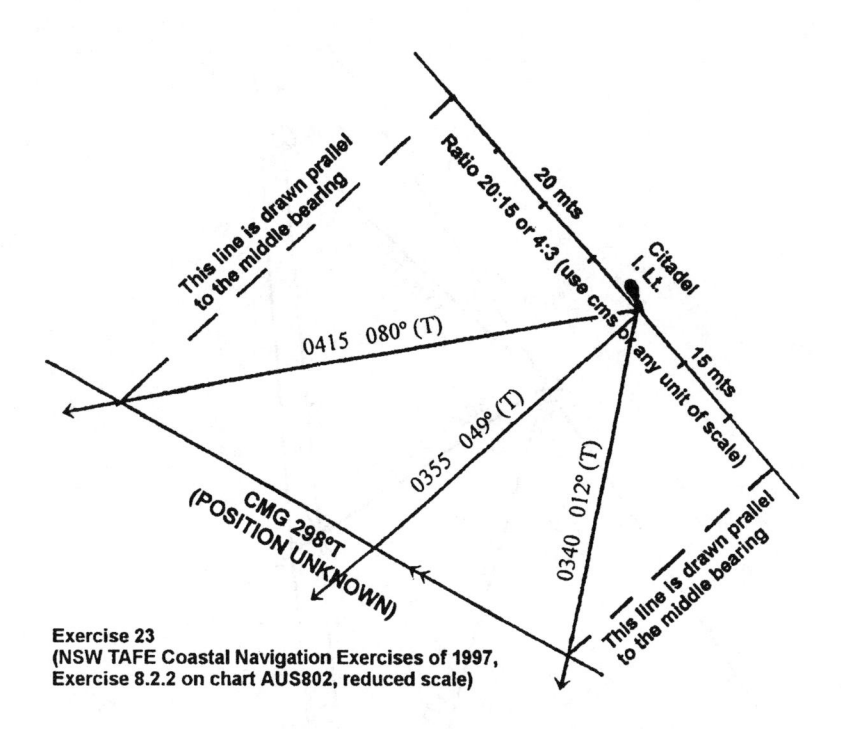

Exercise 23
(NSW TAFE Coastal Navigation Exercises of 1997, Exercise 8.2.2 on chart AUS802, reduced scale)

CMG CAN BE FOUND BY THIS METHOD <u>EVEN WHERE THE CURRENT IS UNKNOWN</u>.

PROCEDURE:
- Correct & plot the three bearings.
- Draw a line <u>in any direction</u> through the landmark. (The line drawn at right angles to the middle bearing will provide greater plotting accuracy.)
- Mark the ratio of time (or log) travelled between the 1st. & 2nd. and 2nd. & 3rd. bearings.
- Draw lines, representing the ratio, parallel to the middle bearing.
- Draw a line where the above lines cut the 1st. & 3rd. bearings. This line represents the above ratio (of 4:3). It is thus the course made good.

NOTE: The line marked CMG is merely the direction of travel, <u>not the course over the ground</u>. Any line drawn parallel to it would carry the same ratio. The course over the ground (position fixing) is shown in the next exercise.

EXERCISE 24

On chart AUS802, a vessel heading towards Bass Strait observed Cliffy Island Lt bearing 215°M at 1115. Twenty minutes later it bore 290°M at a range of 2.8 miles. After a further 25 minutes it bore 331°M. Find (a) CMG (b) vessel's position at 1115 and (c) vessel's position at 1200. The chart reads magnetic variation 12°E.

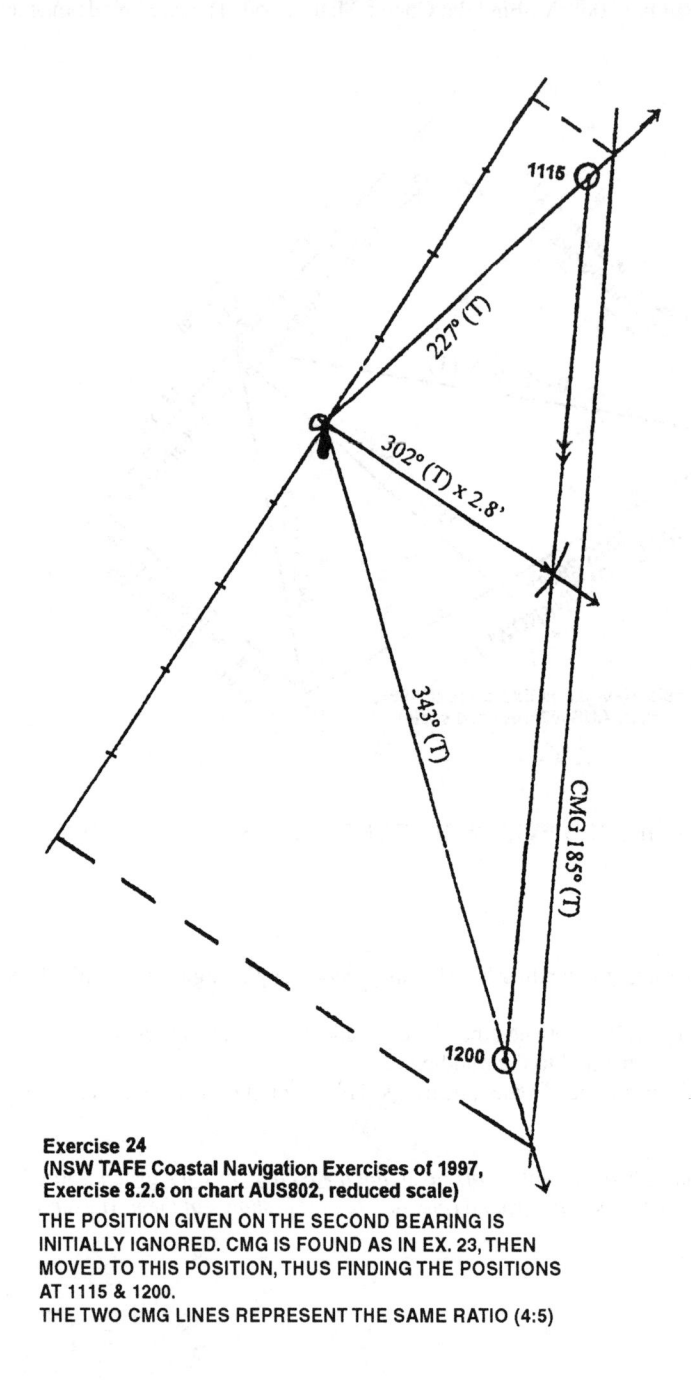

Exercise 24
(NSW TAFE Coastal Navigation Exercises of 1997, Exercise 8.2.6 on chart AUS802, reduced scale)
THE POSITION GIVEN ON THE SECOND BEARING IS INITIALLY IGNORED. CMG IS FOUND AS IN EX. 23, THEN MOVED TO THIS POSITION, THUS FINDING THE POSITIONS AT 1115 & 1200.
THE TWO CMG LINES REPRESENT THE SAME RATIO (4:5)

ANSWER
CMG = 185°T
Posn at 1115 = 38° 54.6'S 146° 45.8'E
Posn at 1200 = 39° 03.6'S 146° 44.8'E

Chapter 17.2: Navigation - Chartwork

CTS TO INTERCEPT (RENDEZVOUS) ANOTHER VESSEL
(Another application of ratios)

EXERCISE 25

You wish to intercept a vessel that is sailing due west at a speed of 3 knots. The positions of the two vessels are shown. Your speed is 6 knots. What course would you steer and at what time would you intercept her?

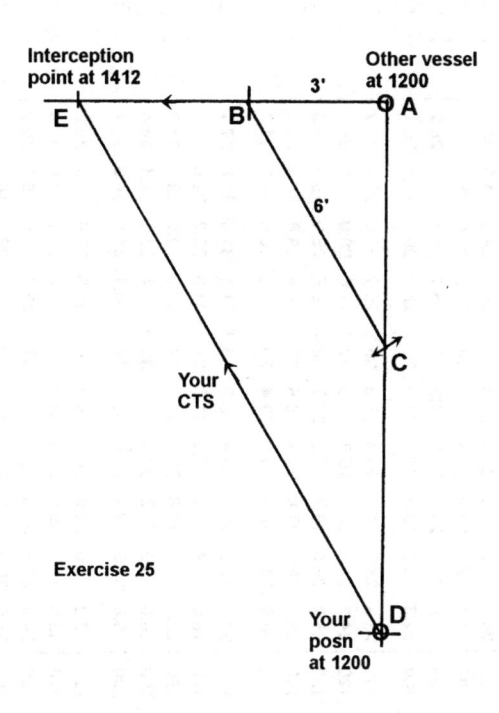

PROCEDURE:
- Join the initial positions of both vessels – line AD
- Plot the course & then the position of the other vessel after 1 hour's run at 3 knots – point B.
- With B as centre and your distance run in 1 hour at 6 knots (6') as radius, describe an arc to cut AD at C.
- Join CB. If you were at C, this would be your track to intercept the other vessel at B in 1 hour.
- But, since you are at D, plot DE parallel to BC. This is your course and distance to the point of interception. The ratio AB:BC is the same as AE:DE.

ANSWER: your course will be = 328° and, the time of interception = 1412 hours.

CAUTION: BC (not AC) is 6 miles.

GEOGRAPHICAL RANGE TABLE (Fig 17.48)

If the earth were flat, we would be able to see coastlines, lighthouses and ships at sea as far as we can see objects in space with powerful telescopes. But, the "hump" of the round earth limits the sighting distance or GEOGRAPHICAL RANGE of terrestrial objects. The greater the height of the object and/or our height of eye, the greater the geographical range.

The formula for calculating geographical range is shown under Radar in Chapter 16. However, most nautical tables and navigation exercise books contain a GEOGRAPHICAL RANGE TABLE, a sample of which is illustrated here. It shows, for example, that from a height of eye (above water) of 2 metres, we should be able to see an object of 100 metres high at a distance of 23.2 miles.

Elevation		Height of Eye of Observer in feet/metres									
ft	m	3 / 1	7 / 2	10 / 3	13 / 4	16 / 5	20 / 6	23 / 7	26 / 8	30 / 9	33 / 10
		Range in Sea Miles									
180	55	17.1	17.9	18.6	19.1	19.6	20.0	20.4	20.8	21.2	21.5
197	60	17.8	18.6	19.3	19.8	20.3	20.7	21.1	21.5	21.8	22.2
213	65	18.4	19.2	19.9	20.4	20.9	21.4	21.7	22.1	22.5	22.8
230	70	19.0	19.9	20.5	21.1	21.5	22.0	22.4	22.7	23.1	23.4
246	75	19.6	20.5	21.1	21.7	22.1	22.6	23.0	23.3	23.7	24.0
262	80	20.2	21.0	21.7	22.2	22.7	23.1	23.5	23.9	24.3	24.6
279	85	20.8	21.6	22.2	22.8	23.3	23.7	24.1	24.5	24.8	25.1
295	90	21.3	22.1	22.8	23.3	23.8	24.2	24.6	25.0	25.4	25.7
312	95	21.8	22.7	23.3	23.9	24.3	24.8	25.2	25.5	25.9	26.2
328	100	22.3	23.2	23.8	24.4	24.9	25.3	25.7	26.1	26.4	26.7
361	110	23.3	24.2	24.8	25.4	25.8	26.3	26.7	27.0	27.4	27.7
394	120	24.3	25.1	25.8	26.3	26.8	27.2	27.6	28.0	28.3	28.7
427	130	25.2	26.0	26.7	27.2	27.7	28.1	28.5	28.9	29.2	29.6
459	140	26.1	26.9	27.6	28.1	28.6	29.0	29.4	29.8	30.1	30.5
492	150	26.9	27.7	28.4	28.9	29.4	29.9	30.2	30.6	31.0	31.3
525	160	27.7	28.6	29.2	29.8	30.2	30.7	31.1	31.4	31.8	32.1
558	170	28.5	29.4	30.0	30.5	31.0	31.5	31.9	32.2	32.6	32.9
591	180	29.3	30.1	30.8	31.3	31.8	32.2	32.6	33.0	33.3	33.7
623	190	30.0	30.9	31.5	32.1	32.5	33.0	33.4	33.7	34.1	34.4
656	200	30.8	31.6	32.2	32.8	33.3	33.7	34.1	34.5	34.8	35.1
722	220	32.2	33.0	33.6	34.2	34.7	35.1	35.5	35.9	36.2	36.5
787	240	33.5	34.3	35.0	35.5	36.0	36.4	36.8	37.2	37.6	37.9
853	260	34.8	35.6	36.3	36.8	37.3	37.7	38.1	38.5	38.8	39.2
919	280	36.0	36.9	37.5	38.0	38.5	39.0	39.4	39.7	40.1	40.4
984	300	37.2	38.1	38.7	39.2	39.7	40.2	40.6	40.9	41.3	41.6
1050	320	38.4	39.2	39.9	40.4	40.9	41.3	41.7	42.1	42.4	42.8
1115	340	39.5	40.3	41.0	41.5	42.0	42.4	42.8	43.2	43.5	43.9
1181	360	40.6	41.4	42.1	42.6	43.1	43.5	43.9	44.3	44.6	45.0
1247	380	41.6	42.5	43.1	43.7	44.1	44.6	45.0	45.3	45.7	46.0
1312	400	42.7	43.5	44.1	44.7	45.2	45.6	46.0	46.4	46.7	47.0

Elevation		Height of Eye of Observer in feet/metres									
ft	m	3 / 1	7 / 2	10 / 3	13 / 4	16 / 5	20 / 6	23 / 7	26 / 8	30 / 9	33 / 10
		Range in Sea Miles									
0	0	2.0	2.9	3.5	4.1	4.5	5.0	5.4	5.7	6.1	6.4
3	1	4.1	4.9	5.5	6.1	6.6	7.0	7.4	7.8	8.1	8.5
7	2	4.9	5.7	6.4	6.9	7.4	7.8	8.2	8.6	9.0	9.3
10	3	5.5	6.4	7.0	7.6	8.1	8.5	8.9	9.3	9.6	9.9
13	4	6.1	6.9	7.6	8.1	8.6	9.0	9.4	9.8	10.2	10.5
16	5	6.6	7.4	8.1	8.6	9.1	9.5	9.9	10.3	10.6	11.0
20	6	7.0	7.8	8.5	9.0	9.5	9.9	10.3	10.7	11.1	11.4
23	7	7.4	8.2	8.9	9.4	9.9	10.3	10.7	11.1	11.5	11.8
26	8	7.8	8.6	9.3	9.8	10.3	10.7	11.1	11.5	11.8	12.2
30	9	8.1	9.0	9.6	10.2	10.6	11.1	11.5	11.8	12.2	12.5
33	10	8.5	9.3	9.9	10.5	11.0	11.4	11.8	12.2	12.5	12.8
36	11	8.8	9.6	10.3	10.8	11.3	11.7	12.1	12.5	12.8	13.2
39	12	9.1	9.9	10.6	11.1	11.6	12.0	12.4	12.8	13.1	13.5
43	13	9.4	10.2	10.8	11.4	11.9	12.3	12.7	13.1	13.4	13.7
46	14	9.6	10.5	11.1	11.7	12.1	12.6	13.0	13.3	13.7	14.0
49	15	9.9	10.7	11.4	11.9	12.4	12.8	13.2	13.6	14.0	14.3
52	16	10.2	11.0	11.6	12.2	12.7	13.1	13.5	13.9	14.2	14.5
56	17	10.4	11.2	11.9	12.4	12.9	13.3	13.7	14.1	14.5	14.8
59	18	10.6	11.5	12.1	12.7	13.2	13.6	14.0	14.4	14.7	15.0
62	19	10.9	11.7	12.4	12.9	13.4	13.8	14.2	14.6	14.9	15.3
66	20	11.1	12.0	12.6	13.1	13.6	14.1	14.5	14.8	15.2	15.5
72	22	11.6	12.4	13.0	13.6	14.1	14.5	14.9	15.3	15.8	15.9
79	24	12.0	12.8	13.5	14.0	14.5	14.9	15.3	15.7	16.0	16.4
85	26	12.4	13.2	13.9	14.4	14.9	15.3	15.7	16.1	16.4	16.8
92	28	12.8	13.6	14.3	14.8	15.3	15.7	16.1	16.5	16.8	17.2
98	30	13.2	14.0	14.6	15.2	15.7	16.1	16.5	16.9	17.2	17.5
115	35	14.0	14.9	15.5	16.1	16.6	17.0	17.4	17.8	18.1	18.4
131	40	14.9	15.7	16.4	16.9	17.4	17.8	18.2	18.6	18.9	19.3
148	45	15.7	16.5	17.1	17.7	18.2	18.6	19.0	19.4	19.7	20.0
164	50	16.4	17.2	17.9	18.4	18.9	19.3	19.7	20.1	20.5	20.8

USING THE LUMINOUS RANGE DIAGRAM (Fig 17.49)

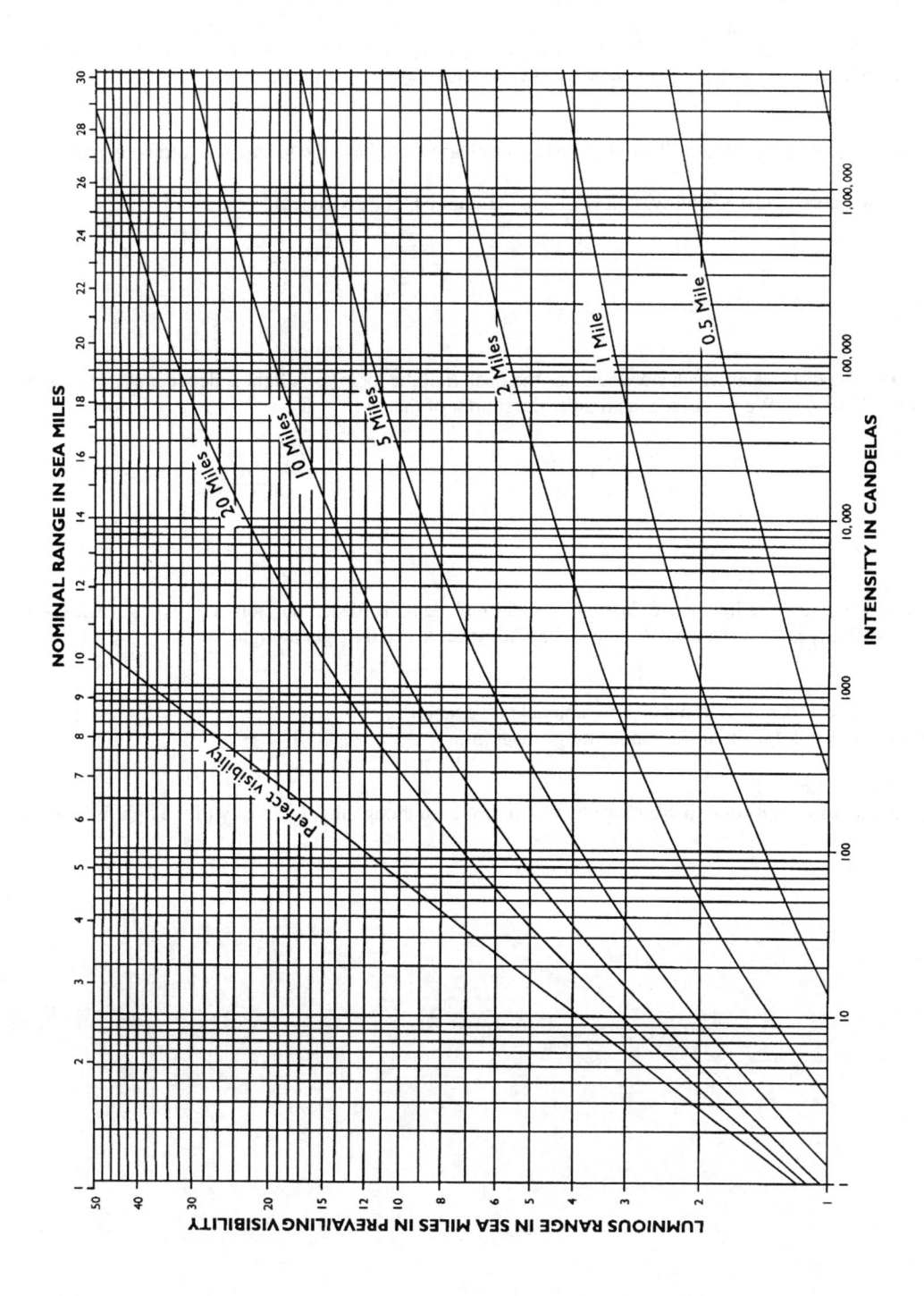

The ranges of lights printed on navigational charts are their NOMINAL RANGES, which indicates their intensity in candelas, as shown in the above diagram. For example, the intensity of a 14-mile light is 10,000 candelas.

The diagram also provides the LUMINOUS RANGE of a light for a given state of visibility. This is the range at which the light will be seen (if the earth were flat) for a given meteorological visibility. It should be remembered that the STANDARD METEOROLOGICAL VISIBILITY (for an unlit object) is 10 miles. Therefore, on the 10-mile visibility curve, the Nominal Range of any light is the same as its Luminous Range.

EXAMPLE: A lighthouse indicating 15-miles range on a chart would have the following Luminous Ranges:

Meteorological Visibility (miles)	Luminous Range (miles)
20	24.2
10	15
7.5	12
1	2.7

CONCLUSION: The distance at which we would see the flash (not the loom) of a light is dependent on two ranges: geographical and luminous. We would see the light at the shorter of the two.

EXERCISE 26

The light characteristics of a lighthouse shown on a chart include 55m 16M. Your height of eye is 1 metre and the meteorological visibility is 15M. At what distance will you expect to see the light at night?

PROCEDURE
Geographical Range = 17.1 miles (from the geographical range table)
Luminous Range = 21 miles (from the luminous range diagram)

ANSWER = 17.1 MILES
(The "hump" of the earth limits our visible range to 17.1 miles, even though the luminosity of the light is 21 miles)

Chapter 17.2: Navigation - Chartwork

BRIDGE LOGBOOK

A BRIDGE WATCHKEEPING LOG is a diary of events, and a record of navigational information in sufficient detail to permit complete reconstruction of a vessel's voyage on charts. A selection of commercially printed logbooks are available. However, many navigators prefer to make their own from a hard cover exercise book. This enables them to match their ideas with the entry headings.

Cruising vessels may supplement the logbook with a hardback pocket notebook. The watchkeeper can record all bearings, fixes, courses steered, speed and sail changes and other data in it while the facts are still fresh in his/her mind. Twice daily, the entries can be transferred into the fair logbook.

Passage from .... To ....

Date/Time	Posn	Log	True Co.	Var	Dev	Comp Co.	L'way	Co. Steered	Remarks

Fig 17.50: SAMPLE LOG BOOK PAGE

PASSAGE PLANNING

1. CHARTS: Select the charts needed for the passage from a Chart Catalogue or Chart Index, which is available at the licensed chart agents and the sub-agents. ("A" and "B" Class chart agents have been discussed earlier under Chart Correction). A chart catalogue is like an atlas. The Australian chart index (Index of Australian Charts) is contained in two chart-sized sheets, numbered AUS65000 and AUS65001. For an interactive chart index see the Australian Hydrographer's website address on the last page of this book.

    Obtain charts of varying scales: small scale charts to look at the bigger picture, large scale charts for sailing close to land, and plans (port charts) for the ports you intend to call or may need to call in an emergency en route. Make sure they are all corrected and up-to-date.

    Mark off dangers and plot the intended courses on the charts prior to departure. These courses will probably be amended many times during the passage. However, plotting them prior to departure will provide a passage summary, highlight any dangers en route and allow you to calculate distances. You can then estimate the fuel and other requirements. Through the process of plotting, you will also know if any chart is missing prior to departure.

2. WEATHER INFORMATION: Obtain necessary weather information, and the equipment and sources from which to receive weather information en route.

3. Consult the following publications, most of which need to be kept up-to-date through weekly (or fortnightly) Notices to Mariners. The digital versions are updated by periodic replacement of the CDs.

    a. TIDE TABLES

    b. Appropriate SAILING DIRECTIONS (PILOT BOOKS). These books, published in numerous volumes, are the "written down charts". Used in conjunction with charts, their description of landmarks, channels, currents and tides is invaluable, especially when making ports. The British Admiralty, US Hydrographic Office as well as some local waterways authorities publish them. The yachting versions, written by independent authors, are also available. They can be purchased from chart agents.

    c. Admiralty LIST OF LIGHTS. These books, also published in eleven volumes, give full details of all the lighthouse and beacons throughout the world. The description in these books is much more detailed than that found on charts. Each light is identified by a serial number and its geographical position.

    d. OCEAN PASSAGES OF THE WORLD: This large manual, published by the British Admiralty, contains descriptions of recommended ocean routes for both power and sailing ships for different months of the year, together with a series of chartlets.

    e. ROUTING CHARTS: These are the American equivalent of the Ocean Passages of the World. A monthly chart for each ocean gives information on the directions and strengths of currents and winds.

    f. ANNUAL NOTICES TO MARINERS are discussed elsewhere in this book.

    g. Make arrangements to obtain the WEEKLY (OR FORTNIGHTLY) NOTICES TO MARINERS, either from the Australian Hydrographic Office, local chart agents en route or on the web. These will allow you to keep the charts and most of the above nautical publications up-to-date.

    h. MARINE NOTICES (or STATE NOTICES TO MARINERS): Do not confuse the Marine Notices issued by AMSA with those issued by the State waterways authorities. The AMSA Notices deal with matters of onboard safety. They are usually issued following an accident or investigation of a safety matter on board ships.

    The Marine Notices (or State Notices to Mariners) issued by the States, on the other hand, are the precursor or forerunner to the national Notices to Mariners. They advise the public as well as the Hydrographer of the changes in the waterways. The Hydrographer announces these changes in the Notices to Mariners. Local boaters often make arrangements to receive the Marine Notices in order to keep their charts and nautical publications up-to-date.

CHAPTER 17.2: NAVIGATION - CHARTWORK

(Qns. 1–63 are in Chapter 17.1)

CHAPTER 17.2.1: READING CHART & OTHER PUBLICATIONS

64. The nautical publication, in addition to a navigational chart, which can help you pilot your vessel in unknown waterways is:

    (a) Tide Tables.
    (b) Ocean Passages of the World.
    (c) Notices to Mariners.
    (d) Sailing Directions.

65. For practical purpose of a mariner, the depths printed on navigational charts are the:

    (a) maximum depths.
    (b) depths at low tide.
    (c) depths at zero tide.
    (d) depths at high tide.

66. The clearances under bridges and overhead cables marked on navigational charts are the clearances at:

    (a) LAT (lowest astronomical tides).
    (b) LW (low water).
    (c) HW (high water).
    (d) MHWS (mean high water springs)

67. The rule of thumb for estimating leeway direction is to observe and estimate the:

    (a) direction of current.
    (b) wind strength.
    (c) angle between the vessel's heading and her wake.
    (d) angle of wind direction.

68. The chart symbols for a wreck which is dangerous to surface navigation consists of:

    (a) 3 lines intersecting a 4th inside dotted border
    (b) 3 lines intersecting a 4th.
    (c) the word "Foul".
    (d) the word "Wreck".

69. The chart symbol for chart datum depth 3.2 metres is:

    (a) 3.2
    (b) 3
    (c) 3 with subscript 2.
    (d) none of the choices stated here.

70. The chart symbol for rock which is dangerous to surface navigation is:

    (a) the sign "+".
    (b) the letter R.
    (c) the letters Rk.
    (d) none of the choices stated here.

71. The chart symbol for overfalls and tide-rips is:

    (a) one or more long purple wavy lines.
    (b) spirals.
    (c) dotted semi-circles.
    (d) groups of short black wavy lines.

72. The chart symbols for drying height 0.3m is:

    (a) 0.3 underlined.
    (b) 0 underlined and 3 subscript.
    (c) – 0.3
    (d) 0.3

73. The chart symbol "(5)" printed next to a small island on a metric chart indicates:

    (a) the shallowest depth of 5 metres.
    (b) 5 metres depth.
    (c) height of the island in metres.
    (d) drying height.

74. The chart symbol for eddies is:

    (a) groups of short black wavy lines.
    (b) spirals.
    (c) one or more long purple wavy lines.
    (d) dotted semi-circles.

75. The chart symbol for rock which uncovers 1 metre at chart datum is:

    (a) Dr.1m.
    (b) the symbol +.
    (c) the symbol Rk.
    (d) – 1.

76. The chart symbol for imperfectly known coast or unsurveyed shoreline is:

    (a) sand dunes on shore line.
    (b) sandy shore line.
    (c) straight shore line.
    (d) broken shore line.

77. The chart symbol for stones and gravel on foreshore is:

    (a) G and St.
    (b) series of tiny circles.
    (c) short lines inside the coastline.
    (d) series of "+" symbols on coastline.

78. The chart symbol for rocky seabed is:

    (a) Rk.
    (b) R.
    (c) Rock.
    (d) a tiny dotted circle.

79. The chart symbol for rock awash at chart datum is:

    (a) the sign "+".
    (b) R.
    (c) the sign "+" with dots between lines.
    (d) Rk.

80. The chart symbol for 5 metre sounding line (depth contour) is:

    (a) a line with "5 metres" written on it.
    (b) a line with "5" written on it.
    (c) a line made up of dots..
    (d) a line made up of dashes.

81. On a navigational chart, one nautical mile equals:

    (a) one minute of latitude.
    (b) one minute of longitude.
    (c) one-tenth of one minute of latitude.
    (d) one-tenth of one minute of longitude.

82. On a chart, a measured distance is printed:

    (a) on a course into harbour entrance.
    (b) between the two ends of harbour entrance.
    (c) on a course in and out of harbour entrance.
    (d) on a course between two transits.

83. On a navigational chart, the latitude scale is used for measuring distances in:

    (a) only the north south direction.
    (b) only the east west direction.
    (c) all directions.
    (d) none of the choices stated here.

84. On a navigational chart, distance is measured with:

    (a) only the latitude scale.
    (b) only the longitude scale.
    (c) both the latitude and longitude scales.
    (d) none of the choices stated here.

85. In one degree of latitude, there are:

    (a) 600 miles.
    (b) 10 miles.
    (c) 5 or 10 miles, depending on the scale of chart.
    (d) 60 miles.

86. The official means of keeping the boating public informed of safety matters within your State is through the:

    (a) Notices to Mariners.
    (b) State Marine Notices or State Notices to Mariners.
    (c) Sailing Directions.
    (d) Commonwealth Marine Notices.

87. The State Marine Notices or State Notices to Mariners, announcing activities or changes in harbours and waterways, are published:

    (a) weekly.
    (b) in newspapers.
    (c) on relevant websites & newspapers.
    (d) on relevant websites.

88. When navigating in a buoyed narrow channel, the vessel should:

    (i) take fixes by compass bearings.
    (ii) use radar.
    (iii) post extra lookout.
    (iv) hand steer the vessel.

    The correct answer is:
    (a) all except (i).
    (b) all except (ii).
    (c) all except (iii).
    (d) all except (iv).

89. On Mercator charts, the circles of latitude and longitude appear as:

    (a) straight & parallel lines
    (b) straight & equidistant lines
    (c) curved lines
    (d) straight lines.

90. On Gnomonic charts, the circles of latitude appear as:

    (a) straight lines.
    (b) Curved lines.
    (c) Curved & concentric lines.
    (d) Straight & parallel lines.

91. The shortest distance between two points on the earth's surface is a:

    (a) Rhumb line
    (b) Straight line on a Mercator chart.
    (c) Curved line on a Gnomonic chart.
    (d) Great Circle.

92. A straight line drawn on a Mercator chart is a:

    (a) Great Circle (GC) track.
    (b) Neither a GC or a RL track.
    (c) Rhumb Line (RL) track..
    (d) Both a GC or a RL track.

93. A straight line drawn on a Gnomonic chart is a:

    (a) Great Circle (GC) track.
    (b) Rhumb Line (RL) track.
    (c) Neither a GC nor a RL track.
    (d) Both a GC & a RL track.

94. The Zone of Confidence diagram, that appears on some navigational charts, indicates the reliability of the:

    (a) scale of the chart.
    (b) depths printed on the chart.
    (c) Chart's compliance with GPS data.
    (d) Chart's compliance with WGS.

95. On a navigational chart, a cable equals:

    (a) one-tenth of a nautical mile.
    (b) One tenth of a degree.
    (c) One mile.
    (d) One minute of latitude.

96. The commonwealth Notices to Mariners are published:

    (a) on State Authorities' websites.
    (b) on Hydrographer's website.
    (c) On AMSA's website.
    (d) in newspapers.

97. The list of corrections appearing on the bottom left hand corner of a navigational chart indicates the:

    (a) currency of the chart.
    (b) small corrections made to the chart.
    (c) age of the chart.
    (d) reliability of the chart.

98. The compatibility of a navigational chart with the GPS-obtained positions is stated on the chart under the heading:

    (a) GPS Positions.
    (b) WGS Positions
    (c) Omissions from Chart
    (d) Satellite Derived Positions.

99. Leeway is:

    (a) vessel's set and drift.
    (b) vessel's course over the ground.
    (c) the effect of wind on vessel's course made good.
    (d) loss of vessel's speed due to wind.

CHAPTER 17.2.2: PLOTTING TRANSITS, SIMULTANEOUS BEARINGS & COURSES

CHART AUS197

100. Chart AUS197. Macquarie Light (approx. Lat 33° 51.4'S) bears 270°T while Port Jackson leading lights are in transit. Therefore the vessel's position is:

     (a) 33° 50.3' S   151° 21.0' E
     (b) 33° 51.9' S   151° 20.5' E
     (c) 33° 50.4' S   151° 21.0' E
     (d) 33° 51.3' S   151° 21.5' E

101. Chart AUS197. Macquarie Light (approx. Lat 33° 51.4'S) is in transit with the Water Tr. (128). At the same time a sounding of 82 metres is obtained. Therefore the vessel's position is:

     (a) 33° 44.2' S   151° 24.0' E
     (b) 33° 45.1' S   151° 24.6' E
     (c) 33° 46.0' S   151° 23.5' E
     (d) 33° 45.1' S   151° 25.6' E

102. Chart AUS197. Temple (196) (approx. Lat 33° 41.3'S) is in transit with Mona Vale Hospital (approx. Lat 33° 41.3'S). At the same time a sounding of 100 metres is obtained. Therefore the vessel's position is:

     (a) 33° 41.9' S   151° 29.6' E
     (b) 33° 41.2' S   151° 30.5' E
     (c) 33° 40.9' S   151° 29.6' E
     (d) 33° 42.7' S   151° 31.2' E

103. Chart AUS197. Temple (196) (approx. Lat 33° 41.3'S) is in transit with Mona Vale Hospital (approx. Lat 33° 41.3'S). The compass bearing of this transit is 255°. Therefore the compass error is:

     (a) 15°E
     (b) 10°E
     (c) 14°w
     (d) 2°E

104. Chart AUS197. A vessel sails into Port Jackson holding exactly on the lead lights. Her compass course reads 280° and Macquarie Light (approx. Lat 33° 51.4'S) bears 230°C by the same compass. Therefore her position is:

     (a) 33° 51.5' S   151° 19.4' E
     (b) 33° 50.0' S   151° 18.0' E
     (c) 33° 50.5' S   151° 19.2' E
     (d) 33° 49.4' S   151° 18.9' E

105. Chart AUS197. A fibreglass vessel observed following bearings with a hand held compass:
Broken Hd (north-eastern edge of the land) (approx. Lat 33° 27.0'S) = 297½°
Tudibaring Head Trig Point (approx. Lat 33° 29.4'S) = 270°
Barrenjoey Lt (approx. Lat 33° 34.9'S) = 234°
Her position is:

   (a)   33° 31.4' S   151° 32.0' E
   (b)   33° 30.7' S   151° 31.1' E
   (c)   33° 29.9' S   151° 32.9' E
   (d)   33° 30.4' S   151° 32.0' E

106. Chart AUS197. At 1200 hours, Port Jackson lead lights were in transit and corrected sounding of 100 metres was obtained. From this position find the compass course to steer to pass 2.5 miles off Barrenjoey Light (approx. Lat 33° 34.9'S). At what time would the Long Point (behind Long Reef) (approx. Lat 33° 44.6'S) be abeam if the vessel's speed was 8 knots? (Deviation is 5°W).

   (a)   355° C; 1255 hours
   (b)   343° C; 1235 hours
   (c)   350° C; 1257 hours.
   (d)   001° C; 1245 hours

107. Chart AUS197. At 1130 hours, Botany Bay lead lights were in line and C. Baily Light (approx. Lat 34° 02.2'S) bore 267°T. Find the vessel's position. From this position she steered a course of 015°C, deviation 2½° E. At what time will the Port Jackson lead lights be in line if the vessel's speed was 8 knots?

   (a)   34° 02.1'S 151° 16.3'E; 1257 hours
   (b)   34° 03.5'S 151° 16.5'E; 1300 hours
   (c)   34° 01.2'S 151° 15.2'E; 1243 hours
   (d)   34° 04.8'S 151° 17.8'E; 1305 hours

108. Chart AUS197. At 0530 hours a vessel, approaching from <u>approximately</u> southward direction, obtained a bearing of C. Baily Light (approx. Lat 34° 02.2'S) of 000°C and radar distance of C. Banks (approx. Lat 34° 00.2'S) of 4.8'. Her course was 041°C and speed 12 knots. What is the distance to sail to arrive abeam of the southern tip of C. Banks? How far abeam would she pass it and at what time? (The Deviation Card reads 3°W for ship's head of 000°C and 1°E for 041°C).

   (a)   3.8 miles; 1.5 miles; 0548 hours
   (b)   4.2 miles; 3.5 miles; 0540 hours
   (c)   5.3 miles; 3.3 miles; 0611 hours
   (d)   4.2 miles; 2.5 miles; 0551 hours

109. Chart AUS197. At 0830 hours, C. Baily Light (approx. Lat 34° 02.2'S) bore 258°C and Mistral Point (approx. Lat 33° 56.6'S) bore 297°C, vessel's course 030°C. From this position the vessel set a course for position 33° 51'S 151° 25'E. Find the compass course to steer and her ETA at a speed of 10 knots. (Vessel's deviation card reads as follows: Dev for Ship's Head 260°C = 5°E; SH 300°C = 3°W; SH 030°C = 1°E).

   (a)   340°C; 1000½ hours
   (b)   338°C; 0945½ hours
   (c)   351°C; 0938½ hours
   (d)   002°C; 0920½ hours

CHART AUS252

110. Chart AUS252. A vessel steering 162°G with a gyrocompass, that reads 2°L, she observed the following bearings:
Hayman Island (247) (approx Lat 20° 03.1'S) = 039°G
Whitsunday Cairn (approx Lat 20° 10.2'S) = 102°G
North Molle Island (233) (approx Lat 20° 14.0'S) = 173°G
Give the vessel's position and then the gyro course to steer to pass Planton Island 1 mile off (Planton Island is east of South Molle Island).

   (a)   20° 09.2' S 148° 48.8' E; 150°G
   (b)   20° 08.3' S 148° 48.8' E; 159°G
   (c)   20° 07.2' S 148° 47.7' E; 154°G
   (d)   20° 08.1' S 148° 48.9' E; 155°G

111. Chart AUS252. A vessel, steering 148°C through Whitsunday Passage, takes the following radar bearings:
Cape Conway (approx Lat 20° 32.2'S) = 042° Rel
SW tip of Shaw Island (approx Lat 20° 28.3'S) = 339° Rel
Cole Island (approx Lat 20° 25.5'S) = 265° Rel.

Fix and state her position in latitude & longitude. What compass course should she steer to pass 2 miles clear of Hammer Island (approx Lat 20° 38.8'S)? (Deviation is 1°E on the first course, and 0° on the second.)

   (a)   20° 25.2'S 148° 57.8'E; 159°C
   (b)   20° 26.7'S 148° 57.9'E; 157°C
   (c)   20° 26.5'S 148° 57.2'E; 152°C
   (d)   20° 27.9'S 148° 59.1'E; 157°C

Chapter 17.2: Navigation - Chartwork

112. Chart AUS252. At 1410 hours, a vessel steering 330°C at 7 knots, observed North Molle Island (233) (approx Lat 20° 14.0'S) bearing 250°C. Her Deviation Card entries read as follows:

HEADING (C)	Dev	HEADING (M)
250°	6½°W	243½°
260°	6½°W	253½°
320°	½°W	319½°
330°	½°E	330½°

    At what time could she alter course to enter Pioneer Bay to clear Hannah Point (approx Lat 20° 12.9'S) by 0.65 miles and Pioneer Rocks (approx Lat 20° 13.6'S) by 0.8 miles and what would be the compass course to steer?

    (a) 1401 hours; 257°C
    (b) 1435 hours; 274°C
    (c) 1428 hours; 244°C
    (d) 1427 hours; 257°C

113. Chart AUS252. You are in <u>approximate</u> position 20° 26.7'S 148° 57.6'E. You do not have a reliable deviation card. But, you observed that the northern-most tips of Cole Island (approx Lat 20° 25.5'S) and Little Lindeman Island (approx Lat 20° 25.5'S) were in transit. If the compass bearing of the transit was 075°C, what is the compass error and deviation for the course the vessel is steering?

    (a) Error 10°E; Dev 1½°W
    (b) Error 10°W; Dev 1½°W
    (c) Error 10°E; Dev 1½°E
    (d) Error 10°W; Dev 1½°E

114. Chart AUS252. At 1300 hours, a vessel is <u>about</u> 3 miles north of Pioneer Bay anchorage (approx Lat 20° 14.0'S). On a course of 170°T, logging 4 knots, she is trying to reach the anchorage. The visibility is almost zero due to heavy rain, and the only navigational instruments in working order are the log and the echo sounder.
    At 1300 the echo sounder recorded the charted depth to be 6.9 metres, followed by 15 metres at 1322 and 10 metres at 1330 hours. Find the vessel's positions at 1330 hours

    (a) 20° 12.9'S 148° 42.9'E
    (b) 20° 13.8'S 148° 43.4'E
    (c) 20° 10.9'S 148° 42.6'E
    (d) 20° 13.9'S 148° 44.0'E

115. Chart AUS252. The following radar ranges were obtained:
    Nicolson Island Summit (approx Lat 20° 18.4'S) = 4.8 miles
    Pentecost Island Summit (approx Lat 20° 23.8'S) = 8.1 miles
    At the same time, Whitsunday Craig Summit (approx Lat 20° 18.4'S) was observed bearing 282°C.
    Find the vessel's position and the deviation for the course steered at the time.

    (a) 20° 20.0'S 149° 11.4'E, Dev 5°W
    (b) 20° 20.9'S 149° 10.2'E, Dev 5°W
    (c) 20° 20.9'S 149° 10.2'E, Dev 1°W
    (d) 20° 20.7'S 149° 10.3'E, Dev 4°E

116. Chart AUS252. On a course of 150°C, you are entering the Whitsunday Passage. Your deviation card is unreliable, but Hanna Point (approx Lat 20° 12.9'S) is observed in transit with Bluff Point (20° 13.8'S) bearing 249°C. What are the compass error and the true course being steered?

    (a) Error 14°W; Course 136°T
    (b) Error 10°E; Course 160°T
    (c) Error 12°W; Course 164°T
    (d) Error 14°E; Course 164°T

117. You observed the following radar ranges.
    Southern tip of Triangle island (approx Lat 20° 29.5'S) = 2.4'
    Eastern edge of Keyser Island (approx Lat 20° 31.4'S) = 3.1'
    Southern tip of Mansell Island (approx Lat 20° 28.3'S) = 2.3'
    What is your vessel's position in latitude and longitude?

    (a) 20° 31.2'S 149° 08.8'E
    (b) 20° 31.3'S 149° 07.6'E
    (c) 20° 30.9'S 149° 08.0'E
    (d) 20° 30.5'S 149° 06.7'E

118. Chart AUS252. A vessel roughly east of Craig Point (approx Lat 20° 19.4'S), observed the southernmost peak (211) of Whitsunday Island (approx Lat 20° 18.9'S) in transit with the southern edge of Craig Point. At the same time the echo sounder gave a charted depth of 4.9 metres. Therefore the vessel's position is:

    (a) 20° 20.5'S 149° 03.5'E
    (b) 20° 19.4'S 149° 02.0'E
    (c) 20° 19.3'S 149° 04.8'E
    (d) 20° 21.0'S 149° 02.5'E

119. Chart AUS252. A vessel steering 148°C through Whitsunday Passage observed the following <u>relative</u> bearings on her radar:
    Edge of Cape Conway (approx Lat 20° 32.2'S) = 042°
    Edge of Shaw Is. near Platypus Rk. (approx Lat 20° 28.3'S) = 339°
    Cole Island Summit (32) (approx Lat 20° 25.5'S) = 265°
    (The magnetic deviation on this heading is 1° E.)
    Her position is:

    (a) 20° 26.0'S 148° 52.4'E
    (b) 20° 24.7'S 148° 56.4'E
    (c) 20° 26.8'S 148° 57.9'E
    (d) 20° 25.2'S 149° 00.4'E

120. Chart AUS252. A vessel fixed her position with the following radar ranges:
Hannah Pt (approx Lat 20° 12.9'S) = 5.05'
Closest edge of Double Cone Island (approx Lat 20° 06.2'S) = 4.3'
Closest edge of Langford Island (approx Lat 20° 04.8'S) = 5.35'
Her position is:

(a) 20° 07.9' S   148° 47.7' E
(b) 20° 08.5' S   148° 48.3' E
(c) 20° 04.8' S   148° 45.6' E
(d) 20° 06.3' S   148° 49.1' E

121. A vessel observed two peaks in transit, which on the chart read 085° and by her compass 075°. The chart also gave magnetic variation 8°E. Therefore her compass error and magnetic deviation are:

(a) 10° E;   2° W
(b) 10° W;   2° E
(c) 10° W;   2° W
(d) 10° E;   2° E

122. Chart AUS252. A vessel observed the following radar ranges:
Closest edge of Dungurra Island (approx Lat 20° 21.5'S) = 1.5'
Surprise Rock (approx Lat 20° 21.4'S) = 1.5'
Closest edge of Pentecost Island (approx Lat 20° 23.9'S) = 1.1'
Her position is:

(a) 20° 22.7' S   149° 01.0' E
(b) 20° 21.5' S   149° 02.6' E
(c) 20° 20.8' S   149° 02.1' E
(d) 20° 22.0' S   149° 03.4' E

CHART AUS802

123. Chart AUS802. A vessel is <u>approximately</u> 5 miles southeast of Devils Tower (approx. Latitude 39° 23'S), steering 240°C. By radar, she observed Devils Tower bearing 045° Rel. at a range of 5.5 miles. What is the vessel's position? (Use deviation 8°E and variation 12°E)

(a) 39° 25.8'S 146° 50.4'E
(b) 39° 26.5'S 146° 40.1'E
(c) 39° 20.4'S 146° 39.3'E
(d) 39° 27.3'S 146° 45.7'E

124. Chart AUS802. At 0800 hours, Cliffy Island Light (approx. Latitude 38° 57'S) bore 175°T, and Seal Island (approx. Latitude 38° 55.6'S) at a radar range of 3.3 miles. Find the compass course to steer to pass 1.5 miles clear of Cliffy Island on its eastern side. (Deviation 3°E, Variation 12°E).

(a) 172°C
(b) 152°C
(c) 150°C
(d) 142°C

CHAPTER 17.2.3: RUNNING FIXES

CHART AUS197

125. Chart AUS197. At 1500 hours Barrenjoey Light (approx Lat 33° 34.9'S) bore 260°T. Forty-five minutes later it bore 300°T. The course steered was 200°T at a speed of 9 knots. Find the vessel's position at 1545 hours by Transfer Position Line method.

(a) 33° 39.3' S 151° 29.0' E
(b) 33° 38.6' S 151° 28.4' E
(c) 33° 41.0' S 151° 26.2' E
(d) 33° 40.8' S 151° 31.5' E

126. Chart AUS197. At 1600 hours C. Baily Light (approx Lat 34° 02.2'S) bore 276°T and thirty minutes later it bore 242°T. If the vessel steered 030°T at a speed of 4 knots between the bearings, what was the vessel's position at 1630 hours?

(a) 34° 59.9' S 151° 16.6' E
(b) 34° 00.7' S 151° 16.7' E
(c) 34° 01.2' S 151° 15.3' E
(d) 34° 00.2' S 151° 18.5' E

127. Chart AUS197. A vessel on a course of 172½°C at a speed of 4.5 knots observed Barrenjoey Light (approx Lat 33° 34.9'S) bearing 247½°C at 1127 hours. At 1327 hours she observed Mona vale Hospital (approx Lat 33° 41.3'S) bearing 292½°C. What is her position at 1327 hours? (Deviation for 172½°C is zero.)

(a) 33° 42.9' S 151° 22.8' E
(b) 33° 41.6' S 151° 20.4' E
(c) 33° 43.3' S 151° 21.9' E
(d) 33° 40.8' S 151° 22.5' E

CHART AUS252

128. Chart AUS252. At 1100 hours Hammer Island (131) (approx. Lat 20° 38.8'S) bore 025°T. At 1130 Blacksmith Island (157) (approx. Lat 20° 38.1'S) bore 076°T. The vessel was steering 349°T at 11 knots. Find her position at 1130 hours by the running fix method.

(a) 20° 39.9'S 148° 58.2'E
(b) 20° 40.7'S 149° 01.6'E
(c) 20° 37.8'S 149° 00.3'E
(d) 20° 39.1'S 148° 59.7'E

129. Chart AUS252. You have departed from an anchorage south of Defiance Island (in Repulse Bay) on a course of 104°C at 12 knots. At 1010 hours, East Repulse Island (62) (approx. Lat 20° 36.1'S) bears 150°C. What is your position at 1034 hours when the same peak bears 252°C?

(a) 20° 34.8'S 148° 56.6'E
(b) 20° 35.9'S 148° 56.6'E

(c) 20° 35.9'S 148° 57.9'E
(d) 20° 36.7'S 148° 55.1'E

130. Chart AUS252. A vessel is steering 000°C at a speed of 11.5 knots. At 0112 hours, she observed the easternmost edge of Cape Conway (approx. Lat 20° 32.2'S) at 7.5 miles radar range on her port bow. At 0148, the same landmark was found to be at 3.1 miles. Fix the vessel's position at 0148 hours, stating it in relation to Cape Conway. (Deviation 3.5°E)

    (a) C. Conway brg. 280° x 3.1'
    (b) C. Conway brg. 260° x 3.1'
    (c) C. Conway brg. 100° x 4.2'
    (d) C. Conway brg. 280° x 7.5'

131. Chart AUS252. A vessel is steering 165°C at a speed of 9.5 knots. At 0800 Hayman Island Summit (247) (approx. Lat 20° 03.1'S) bore 068°C. At 0848 Whitsunday Peak Summit (approx. Lat 20° 16.0'S) bore 123°C. Find her position at 0848 hours.

    (a) 20° 11.5'S 148° 50.6'E
    (b) 20° 10.6'S 148° 52.2'E
    (c) 20° 11.1'S 148° 50.2'E
    (d) 20° 11.4'S 148° 51.5'E

132. Chart AUS252. Your vessel is steering 358°C at a speed of 10 knots. At 2112 the radar range of the easternmost edge of Cape Conway (approx. Lat 20° 32.2'S) was 6.9 miles. At 2148 the range had decreased to 4.1 miles.
    Find the vessel's position at 2148 and determine the compass course to steer to enter Kennedy Sound, keeping 0.7 miles clear of Seaforth Island (approx. Lat 20° 28.4'S). (Deviation is 3.5°E on the given course, and 4.5°E on the course to find.)

    (a) 20° 32.3'S 149° 00.4'E; 047°C
    (b) 20° 31.5'S 148° 59.6'E; 035°C
    (c) 20° 32.4'S 149° 00.3'E; 021°C
    (d) 20° 32.4'S 149° 59.0'E; 022°C

CHART AUS 802

133. Chart AUS802. While steering 260°C at a speed of 8 knots, a vessel observed Hogan Island Light (approx. Latitude 39° 14'S) bearing 319°C. One hour later, the same light bore 038°C. Find her position at the time of the second bearing. (Her deviation card read as follows: Deviation 6°W for SH 040°C; 6°E for SH 260°C; and 1°W for SH 320°C. The chart read magnetic variation of 12°E).

    (a) 39° 15.3'S 146° 50.1'E
    (b) 39° 17.2'S 146° 51.4'E
    (c) 39° 19.2'S 146° 51.3'E
    (d) 39° 14.5'S 146° 49.7'E

CHAPTER 17.2.4: DOUBLING THE ANGLE ON THE BOW

134. Chart AUS252. A vessel sailing due west at 8 knots observed Pinnacle Point Light (approx. Lat 20° 03.8'S) bearing 225°T. After sailing 3.5 miles the same light bore 180°T. Give the vessel's distance from the light at the time of taking the second bearing.

    (a) 3.5 miles
    (b) 7.0 miles
    (c) 8 miles
    (d) 0.5 miles

135. Chart AUS252. A vessel sailing 334°T at a speed of 12 knots saw Dent Island Light (on west coast of Dent Island) (approx. Lat 20° 21.1'S) bearing 354°T. After sailing 2.6 miles, the same light bore 014°T. Find the vessel's position at the second bearing.

    (a) Dent Island brg. 334°T x 2.6'
    (b) Dent Island brg. 354°T x 1.6'
    (c) Dent Island brg. 014°T x 5.2'
    (d) Dent Island brg. 014°T x 2.6'

136. Chart AUS 252. Your vessel is on a course of 154°T at 4.4 knots. She observed Dent Island Light (on the west coast of Dent Island) (approx. Lat 20° 21.1'S) at an angle of 30° on the port bow. Half an hour later the same light was observed at 60° on the port bow. Find the vessel's position at the second observation.

    (a) Dent Is. Lt. brg. 124°T x 2.2'
    (b) Dent Is. Lt. brg. 094°T x 2.2'
    (c) Dent Is. Lt. brg. 094°T x 4.4'
    (d) Dent Is. Lt. brg. 124°T x 4.0'

137. Chart AUS252. Vessel's course 090°T, speed 6 knots. At 1030, Pinnacle Point Light (approx. Lat 20° 03.8'S) was 4 points on her starboard bow, which at 1050 became 8 points on the same side. What was her position at 1050 hours?

    (a) Pinnacle Pt. Lt. brg 180°T x 2'
    (b) Pinnacle Pt. Lt. brg 135°T x 2'
    (c) Pinnacle Pt. Lt. brg 090°T x 4'
    (d) Pinnacle Pt. Lt. brg 180°T x 3'

CHAPTER 17.2.5: VERTICAL & HORIZONTAL ANGLES

138. Chart AUS252. Find the vessel's position from the following horizontal angles: Pentecost Island (288) 90° Nicolson Island (108) 32° Workington Island (91). (The approximate latitudes of the islands are: Pentecost 20° 23.9'S; Nicolson 20° 18.4'S; Workington 20° 16.3'S)

    (a) 20° 19.2'S 149° 08.4'E
    (b) 20° 20.8'S 149° 07.2'E
    (c) 20° 20.7'S 149° 09.2'E
    (d) 20° 21.6'S 149° 07.3'E

139. Chart AUS252. A vessel observed Lindeman Island Light (approx Latitude 20° 27.7'S) bearing 086°M while subtending a VSA of 0° 59.5'. Course was then set to pass Platypus Rock (approx Latitude 20° 31.4'S) 5 cables abeam to port. Her deviation card entries read as follows:

COURSE (C)	Dev	COURSE (M)
140°	3°E	143°
160°	4°E	164°
180°	6°E	186°

    What is the Vertical Danger Angle on Mt Arthur peak on Shaw Island (250) (approx Latitude 20° 28.3'S), and what is the compass course to steer? (Use variation 8°E).

    (a) VDA 3° 26'; CTS 158°C
    (b) VDA 4° 24'; CTS 162°C
    (c) VDA 4° 24'; CTS 158°C
    (d) VDA 3° 26'; CTS 170°C

140. Chart AUS252. Find the vessel's position from the following horizontal angles: Teague Island (87) 53° Pentecost Island (288) 74° Mansell Island (170). (Their respective approximate latitudes are 20° 18.1'S, 20° 23.9'S and 20° 28.3'S)

    (a) 20° 20.1'S 149° 06.8'S
    (b) 20° 23.3'S 149° 07.7'S
    (c) 20° 22.1'S 149° 05.4'S
    (d) 20° 23.2'S 149° 08.8'S

141. Chart AUS252. What is the magnetic bearing and distance of Ladysmith Island (121) from a position with Midge island (53) and South Repulse Island (60) subtending a HA of 46° when the corrected chart sounding was 7.6 metres? (The approximate latitudes of the islands are Ladysmith 20° 39.3'S; Midge 20° 41.5'S; South Repulse 20° 36.8'S)

    (a) 258°M; Distance 8.8'
    (b) 078°M; Distance 7.8'
    (c) 087°M; Distance 3.5'
    (d) 102°M; Distance 10.9'

142. What is the Danger Angle to set on a sextant to pass a lighthouse at a distance of 1.2 miles? The charted height of the lighthouse is 35 metres.

    (a) 5° 25'
    (b) 1° 06'
    (c) 0° 54'
    (d) 54°

CHAPTER 17.2.6: FINDING CMG BY RATIOS (THREE-BEARINGS METHOD)

143. Chart AUS252. The following bearings were taken of Double Cone Island (104) (approx. Lat 20° 06.2'S).
    1330: bearing 310°M, range 3.3'
    1400: bearing 265°M
    1420: bearing 237°M
    Find the CMG and position of the vessel on the last bearing.

    (a) 012°T; 20° 04.8'S 148° 46.3'E
    (b) 192°T; 20° 04.8'S 148° 46.3'E
    (c) 021°T; 20° 05.6'S 148° 47.5'E
    (d) 201°T; 20° 06.2'S 148° 46.9'E

CHAPTER 17.2.7: INTERCEPTING (RENDEZVOUS WITH) ANOTHER VESSEL

144. Chart AUS252. At 1300 you take the following bearings:
    South Repulse Island (60) (approx. Lat 20° 36.8'S) = 331°M
    Middle of Midge Island (53) (approx. Lat 20° 41.5'S) = 255°M
    Stewart Peninsula (117) (approx. Lat 20° 46.9'S) = 208°M
    You receive a call for assistance from another vessel in position 20° 50.0'S 149° 10.0'E *(Sometimes large distances need to be examined from positions on the edges of charts, as in this exercise.)* She is steering 320°T at 8 knots. You decided to go to her assistance at your best possible speed of 10 knots. What compass course must you steer to intercept the other vessel and at what time will the interception take place? (Use deviation 0° and variation 8°E)

    (a) 108°C; 1505 hours
    (b) 116°C; 1400 hours
    (c) 100°C; 1402 hours
    (d) 280°C; 1433 hours

CHAPTER 17.2.8: PLOTTING SET, DRIFT & LEEWAY

CHART AUS802

145. Chart AUS802. A vessel's deviation card is unreliable. At 1000, while on a course of 020°C at 10 knots, she observed Deal island Light (approx Latitude 39° 30'S) in transit with the South West Island Light (approx Latitude 39° 31'S) on a compass bearing of 070°C. At the same time the radar range of the South West Island was 5.0 miles. At 1100 hours, she again fixed her position at 39° 24.6'S 147° 09.8'E
Find the (i) true course steered, (ii) set and rate of drift experienced between 1000 and 1100 hours.

    (a) 010°T; 146°T x 1.8 knots
    (b) 030°T; 125°T x 1.9 knots
    (c) 041°T; 106°T x 2.1 knots
    (d) 020°T; 110°T x 1.2 knots

146. Chart AUS802. A vessel fixed her position with Curtis Island (335) (approx Latitude 39° 28.4'S) bearing 202°M and the radar range of Devils Tower (111) (approx Latitude 39° 22.7'S) of 2.8 miles. Find her compass course to steer to pass southward of Crocodile Rock (1) (approx Latitude 39° 21.5'S) at a distance of 1 mile, counteracting a current setting 240°T at 3 knots and 6° leeway due to northerly wind. Vessel's speed is 8 knots. (Use deviation 3°W and variation 12°E)

    (a) 296°C
    (b) 314°C
    (c) 296°C
    (d) 290°C

CHART AUS252

147. Chart AUS252. What is the compass course to steer from a position with Dent Island Light (approx. Latitude 20° 22.2'S) bearing 082°M at a range of 0.8 miles, to follow the "preferred route" in a northerly direction, allowing for a current setting 140°T at 1.7 knots and leeway of 8° from a strong westerly wind? Vessel's speed through the water is 10 knots. (Deviation 1°W)

    (a) 347°C
    (b) 339°C
    (c) 321°C
    (d) 354°C

148. Chart AUS252. A vessel is 2.2 miles due east of South Repulse Island (60) (approximate Latitude 20° 36.8'S). She wishes to pass 1.5 miles off Hammer Island (131) (approximate Latitude 20° 38.8'S), counteracting a current setting 150°T at 2 knots and a leeway of 5° from a southerly wind. Her log speed is 6 knots. Find her compass course to steer and speed made good. (Use Deviation Card 2 in the Australian Boating Manual and Variation of 8°E)

    (a) CTS 102°C; SMG 7.5 knots
    (b) CTS 114°C; SMG 7.4 knots
    (c) CTS 108°C; SMG 5.6 knots
    (d) CTS 111°C; SMG 9.1 knots

149. Chart AUS252. From a position with South Repulse Island (60) (approximate Latitude 20° 36.8'S) bearing 024°M when Gould Island and Midge Point (approximate Latitude 20° 39.8'S) were in transit, a vessel steered 107°C for 13 miles when Bellows Island bore 061°M, range 2.4 miles. What are the CMG and set and drift of the current experienced? (Use deviation 25°W and variation 8°E).

    (a) CMG 085°T; Set 122°T; Drift 0.7'
    (b) CMG 176°T; Set 090°T; Drift 2.4'
    (c) CMG 086°T; Set 302°T; Drift 1.5'
    (d) CMG 226°T; Set 301°T; Drift 1.4'

150. Chart AUS252. A vessel was steering 156°C at 7 knots. At 0900, she observed Double Cone Island (104) (approximate Latitude 20° 06.2'S) bearing 211°M and at 0945 it bore 292°M. What was the vessel's position at 0945 if a current set 000°T at 3.5 knots and a leeway of 5° from a NE'ly wind? (Use deviation 19°W and variation 8°E)

    (a) 20° 08.7'S 148° 46.2'E
    (b) 20° 05.6'S 148° 45.3'E
    (c) 20° 07.8'S 148° 48.9'E
    (d) 20° 07.8'S 148° 46.1'E

151. Chart AUS252. A vessel in position 20° 10.0'S 149° 10.0'E *(Note: this position being slightly off the main chart area is OK)* was steering 290°C at 5.5 knots. The current was estimated to be setting 230°T at 2.5 knots. Find her estimated CMG and SMG. (Use deviation 23°E and variation 8°E)

    (a) 302°T; 5.8 knots
    (b) 297°T; 6 knots
    (c) 245°T; 5 knots
    (d) 273°T; 7.2 knots

152. Chart AUS252. A vessel steering 092°C at 7.5 knots observed Pinnacle Point Light bearing 119°M at a range of 2.7 miles. 35 minutes later, it bore 218°M at a range of 2.2 mile. What were the CMG and the set and rate of current experienced? (Use deviation 0° and variation 8°E)

(a) 091°T; Set 318°T x 0.9 knots
(b) 100°T; Set 138°T x 0.9 knots
(c) 091°T; Set 318°T x 1.5 knots
(d) 092°T; Set 090°T x 2.1 knots

CHAPTER 17.2.9: GEOGRAPHICAL & LUMINOUS RANGE CALCULATION

153. What is the greatest range at which you would expect to sight a navigational light, whose charted height is 57 metres and nominal range 30 miles. Your height of eye is 3 metres and the existing meteorological visibility 18 miles.

(a) 30.8 miles
(b) 30 miles
(c) 18.9 miles
(d) 17.3 miles

154. What is the greatest range at which you would expect to sight a navigational light, whose charted height is 150 feet and nominal range 20 miles. Your height of eye is 10 feet and the existing meteorological visibility 18 miles.

(a) 20 miles
(b) 17.1 miles
(c) 20.8 miles
(d) 18 miles

155. Chart AUS252. At what maximum range would a yacht sight the Pinnacle Point Light (height 21 metres) (approximate Latitude 20° 03.8'S) in meteorological visibility of 2 miles? The observer's height of eye is 1 metre.

(a) 5 miles
(b) 10 miles
(c) 11.35 miles
(d) 3.3 miles

CHAPTER 17.2: ANSWERS
(Answers 1-63 are in Chapter 17.1)

64 (d), 65 (c),
66 (d), 67 (c), 68 (a), 69 (c), 70 (a),
71 (d), 72 (b), 73 (c), 74 (b), 75 (a),
76 (d), 77 (a), 78 (b), 79 (c), 80 (b),
81 (a), 82 (d), 83 (c), 84 (a), 85 (d),
86 (b), 87 (c), 88 (a), 89 (a), 90 (b),
91 (d), 92 (c), 93 (a), 94 (b), 95 (a),
96 (b), 97 (b), 98 (d), 99 (c), 100 (d),
101 (b), 102 (b), 103 (a), 104 (c), 105 (d),
106 (c), 107 (a), 108 (d), 109 (c), 110 (d),
111 (b), 112 (d), 113 (c), 114 (a), 115 (b),
116 (d), 117 (a), 118 (b), 119 (c), 120 (a),
121 (d), 122 (a), 123 (a), 124 (d), 125 (a),
126 (b), 127 (c), 128 (d), 129 (b), 130 (a),
131 (d), 132 (c), 133 (b), 134 (a), 135 (d),
136 (b), 137 (a), 138 (b), 139 (c), 140 (d),
141 (b), 142 (c), 143 (a), 144 (c), 145 (b),
146 (a), 147 (c), 148 (a), 149 (c), 150 (d),
151 (b), 152 (c), 153 (c), 154 (b), 155 (d)

Chapter 17.3

NAVIGATION

TIDES & CURRENTS

TIDES

Tides are the rise and fall of water due to the gravitational pull of moon and sun. Moon being closer to earth exerts a greater pull. The Hydrographic Office publishes tide tables based on this astronomical and some other non-astronomical data. Local tides can generally be predicted with a high degree of accuracy from analysis of long-term tidal records. However, one thing that tide tables cannot predict is the effect on tides of the day-to-day weather changes and wind directions and velocities.

Roughly twice a month the moon is in line with the sun, i.e., on new moon and full moon. Their combined gravitational pull creates bigger tides. The high water is higher and low water is lower. These are known as the SPRING TIDES. Similarly, twice a month, at half moon, the sun and moon are at right angle to each other. Here, they work in different directions, producing smaller tides. The high tides are not so high and low tides not so low. These are known as the NEAP TIDES.

The difference in the height between a consecutive HW and LW is the RANGE OF TIDE. The tide range on the Australian coastline varies from approximately 12 metres in the Kimberleys to about 0.7 metres in places such as Sydney and Fremantle. The range of spring tides is always greater than that of neap tides.

A rising tide, i.e., a tide flowing into harbours and rivers, is known as the FLOOD TIDE. A falling tide, i.e., a tide going out, is known as the EBB TIDE. The horizontal flow of water when tide is flooding or ebbing is known as the TIDAL STREAM or TIDAL CURRENT. The Tidal Currents are strongest halfway between the times of HW and LW. See figure 17.33 for additional terminology on tides.

The UK Hydrographic Office offers free tidal prediction service for some 4000 ports around the world on its website (see address on the last page of this book). This 'EasyTide' service provides tidal information for up to seven days from the date of logging on.

THE EFFECT OF WEATHER ON TIDES

The effects of weather conditions on tides are not taken into account in Tide Tables. For example, a high atmospheric pressure does not allow a tide to rise to its full extent, while a low pressure allows it to rise higher than normal. Wind direction may also affect the height of tide. A strong onshore wind will tend to push more water onshore, causing a higher tide than predicted. An offshore wind will result in a lower tide. Consequently, the tidal heights do not always correspond exactly with the published predictions.

Coastal flooding caused by the onshore wind and low atmospheric pressure in a storm is known as a STORM SURGE. Slow moving storms of large diameter create higher storm surges. A storm surge may raise the sea level by over a metre. (See also Storm Tide in Chapter 20)

DEPTHS PRINTED ON CHARTS & CLEARANCE UNDER BRIDGES
See Chapter 17.2

CALCULATING HEIGHT OF TIDE BY RULE OF TWELFTHS

Tides change approximately every six hours in most places in the world. The heights of High Water and Low Water are published in tide tables and newspapers, and broadcast on radio and television. There are numerous methods of calculating the height of tide at any time between a HW and a LW.

One reasonably accurate method is the rule of twelfths. It is based on the assumption that tides rise and fall with simple harmonic motion. That is, between one Slack Water (the time of change of tide at HW or LW) and the next, the tide starts rising or falling slowly, runs strongly in the middle of the period and then slows down when approaching the next slack water.

The rule assumes that the tide rises or falls each hour approximately in the proportions of 1 - 2 - 3 - 3 - 2 - 1. In other words, the hourly rise or fall of tide is as follows:

In the	1st hour	= 1/12th of the tide range
	2nd hour	= 2/12ths
	3rd hour	= 3/12ths
	4th hour	= 3/12ths
	5th hour	= 2/12ths
	6th hour	= 1/12ths

EXERCISE 27

At a certain port on a certain day, HW is 1.8 m at 0600 hours and LW is 0.6 m at 1200 hours. Find the height of tide for every hour between HW and LW by the rule of twelfths

SOLUTION

HW at 0600 hours = 1.8 m
LW at 1200 hours = 0.6 m

Range = 1.2 m

1/12th of 1.2 m = 0.1 m; thus	height of tide at 0700 hours	= 1.7 m
2/12ths of 1.2 m = 0.2 m; thus	height of tide at 0800 hours	= 1.5 m
3/12ths = 0.3;	height of tide at 0900 hours	= 1.2 m
3/12ths again ...	at 1000 hours	= 0.9 m
2/12ths	at 1100 hours	= 0.7 m
1/12ths	at 1200 hours	= 0.6 m

CALCULATING HEIGHT OF TIDE BY FORM AH 130 (THE L-SHAPED GRAPH)

Instead of the Rule of Twelfths, the following graph, supplied in the Australian Tide Tables (published by the Australian Hydrographic Office) can be used as shown. It is obviously a more accurate method, does not involve calculations and can be retained as a record or for subsequent reuse.

EXERCISE 28

The tides at Cooktown this afternoon are as follows. Find the height of tide at 1800 hours by use of Form AH 130.

LW	1425 hrs	0.8 m
HW	1930 hrs	1.9 m

STEPS TO PLOTTING ON FORM AH 130 (Instructions for use printed on the form)

1. Mark the time of LW and HW on the time scale (Note: each division is 20 minutes)
2. Mark the height of LW & HW on the (There are two height scales. Use the larger one whenever possible)
3. Join the two marks on the time scale with a line
4. Join the two marks on the height scale with a line
5. On the time scale, project a vertical line (up or down) from 1800 hours, to intersect the time line that you have drawn.
6. From the above point of intersection, project lines to the tide curve, height line and the height scale, as shown. **The required height of tide shows up as 1.65 metres**. *(Warning: When reading the height of tide, be careful to read the right scale. There are two scales on the height section of the graph.)*

The above example applies equally in reverse. In other words, to find out the time when the height of tide would be 1.65 metres, reverse the order in which the lines with the arrows are drawn.

431

Chapter 17.3: Navigation - Tides & Currents

Fig 17.51: TIDE GRAPH (FORM AH 130)

INSTRUCTIONS

1. Plot the times and (using the appropriate scale) the heights of high and low water.
2. Join these points to obtain a time-line and height-line.
3. To obtain an intermediate height for a given time, plot the time on the time axis, project up to the time-line, across to the curve, down to the height-line and across to the height axis to read off the result. To find the time for a given height simply use the reverse procedure.

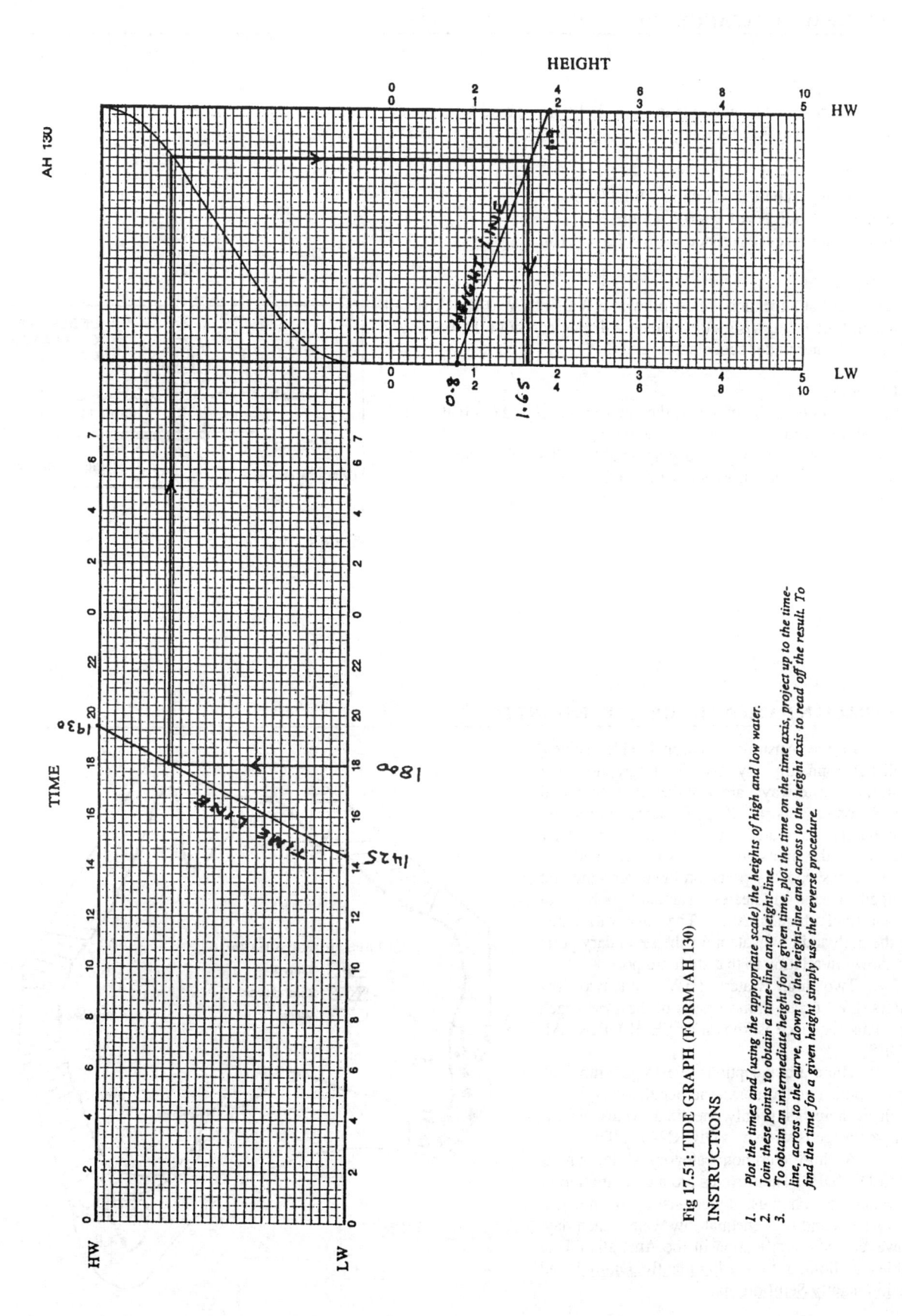

UNDER KEEL CLEARANCE

EXERCISE 29

The charted depth at the entrance to a port is 1.0 m. The tides for the day are as follows: HW 2.2m. LW 0.6m. What will be the under keel clearance for a yacht, drawing 1.5 metres at low water?

SOLUTION:
Depth of water = charted sounding + tidal level.
Here, the depth at LW = 1.0 + 0.6 = 1.6m
Therefore, under keel clearance = 1.6 – 1.5 = 0.1m.

EXERCISE 30

Today's tides at a location are as follows: HW 2.3m LW 0.7m. A vessel, with a draft of 1.2m, wishes to sail over a drying height of 0.8m at HW. What will be her under keel clearance?

SOLUTION:
The drying height of 0.8m means the outcrop is exposed 0.8m at zero tide. In other words, it is a negative sounding.
At HW, the sounding over the drying height = 2.3 – 0.8 = 1.5m.
Therefore, under keel clearance = 1.5 – 1.2 = 0.3m.

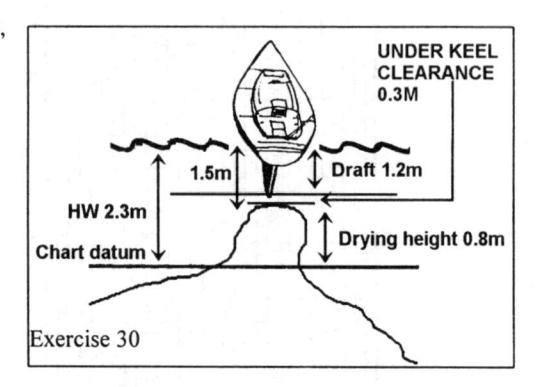

Exercise 30

AUSTRALIAN NATIONAL TIDE TABLES (ANTT)

In the Australian national Tide Tables, published annually by the Hydrographer, the Australian waterways are divided into Standard and Secondary ports. The Standard ports are individually tabulated with their own daily calendar of the times and heights of high and low waters. The secondary ports, on the other hand, are grouped around a nearest standard port with similar tidal characteristics. The time difference for the high and low waters for the secondary ports are shown in relation to that standard port.

Two high waters (HW) and two low waters (LW) occur in most parts of the world each day. This sequence is known as SEMI-DIURNAL TIDES.

There are exceptions due to geographical, topographical and atmospheric peculiarities, which result in there being only one tidal cycle per day. This sequence is known as DIURNAL TIDES.

A third common category is known as MIXED TIDES. Their cycles are a combination of the other two. They usually consist of two unequal high waters and two unequal low waters each day. However, the classification in the Australian Tide Tables is limited to <u>Predominantly Diurnal</u> and <u>Predominantly Semi-diurnal</u>.

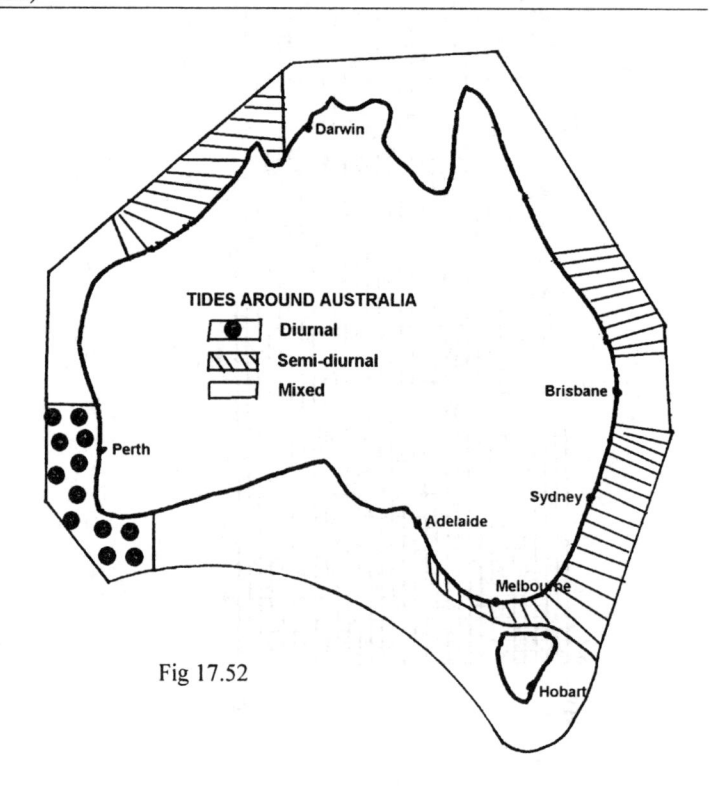

Fig 17.52

Chapter 17.3: Navigation - Tides & Currents

An extract of
PREDOMINANTLY SEMI-DIURNAL STANDARD PORT OF SYDNEY

Time difference from UTC (Greenwich Time in UK) You may ignore it.

The exact position of tide

Local Standard Time (add daylight saving where applicable)

AUSTRALIA, EAST COAST - SYDNEY (FORT DENISON)

LAT 33°51′ S   LONG 151°14′ E

TIME ZONE -1000   TIMES AND HEIGHTS OF HIGH AND LOW WATERS   YEAR XXXX

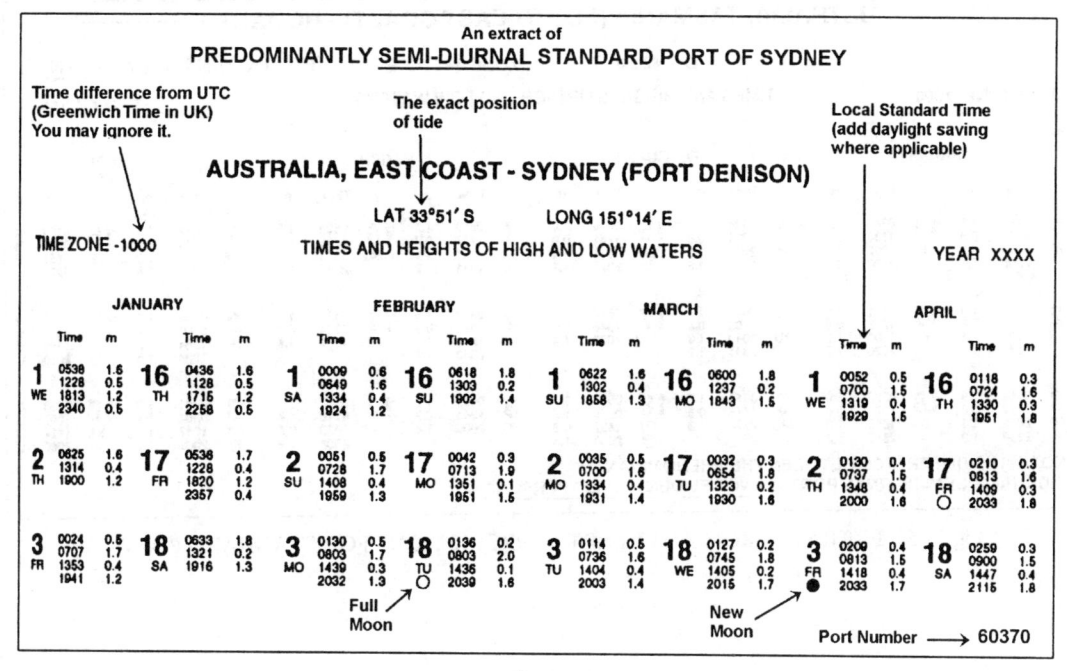

JANUARY		FEBRUARY		MARCH		APRIL	
Time m	Time m	Time m	Time m	Time m	Time m	Time m	Time m
1 WE 0538 1.6 / 1228 0.5 / 1813 1.2 / 2340 0.5	**16** TH 0436 1.6 / 1128 0.5 / 1715 1.2 / 2258 0.5	**1** SA 0009 0.6 / 0649 1.6 / 1334 0.4 / 1924 1.2	**16** SU 0618 1.8 / 1303 0.2 / 1902 1.4	**1** SU 0622 1.6 / 1302 0.4 / 1858 1.3	**16** MO 0600 1.8 / 1237 0.2 / 1843 1.5	**1** WE 0052 0.5 / 0700 1.5 / 1319 0.4 / 1929 1.5	**16** TH 0118 0.3 / 0724 1.6 / 1330 0.3 / 1951 1.8
2 TH 0625 1.6 / 1314 0.4 / 1900 1.2	**17** FR 0536 1.7 / 1228 0.4 / 1820 0.4 / 2357 0.4	**2** SU 0051 0.5 / 0728 1.7 / 1408 0.4 / 1959 1.3	**17** MO 0042 0.3 / 0713 1.9 / 1351 0.1 / 1951 1.5	**2** MO 0035 0.5 / 0700 1.6 / 1334 0.4 / 1931 1.4	**17** TU 0032 0.3 / 0654 1.8 / 1323 0.2 / 1930 1.6	**2** TH 0130 0.4 / 0737 1.5 / 1348 0.4 / 2000 1.6	**17** FR 0210 0.3 / 0813 1.5 / 1409 0.3 / 2033 1.8 ○
3 FR 0024 0.5 / 0707 1.7 / 1353 0.4 / 1941 1.2	**18** SA 0633 1.8 / 1321 0.2 / 1916 1.3	**3** MO 0130 0.5 / 0803 1.7 / 1439 0.3 / 2032 1.3	**18** TU 0136 0.2 / 0803 2.0 / 1436 0.1 / 2039 1.6 ○ Full Moon	**3** TU 0114 0.5 / 0736 1.6 / 1404 0.4 / 2003 1.4	**18** WE 0127 0.2 / 0745 1.8 / 1405 0.2 / 2015 1.7	**3** FR 0209 0.4 / 0813 1.5 / 1418 0.4 / 2033 1.7 ● New Moon	**18** SA 0259 0.3 / 0900 1.5 / 1447 0.4 / 2115 1.8

Port Number ——→ 60370

Fig 17.53

Some of the
PREDOMINANTLY SEMI-DIURNAL SECONDARY PORTS GROUPED AROUND SYDNEY

Note the difference in these headings from the predominantly diurnal group of ports below. (It is relevant only when selecting the right conversion form below)

PORT No.	PORT NAME	TIME DIFFERENCES MHW / MLW	TIDAL LEVELS (in metres, related to LAT) MHWS / MHWN / MSL / MLWN / MLWS
		(Zone -1000)	
60370	SYDNEY	(standard port)	1.5 / 1.3 / 0.9 / 0.5 / 0.3
60320	GOSFORD		0.8 / 0.5 / 0.4 / 0.3 / -0.1
60325	ETTALONG		0.9 / 0.8 / 0.5 / 0.2 / 0.1
60330	LITTLE PATONGA		1.6 / 1.3 / 0.9 / 0.5 / 0.3
60340	PITTWATER	+0016 / +0015	1.6 / 1.3 / 0.9 / 0.4 / 0.2
60360	CAMP COVE	-0003 / -0003	1.6 / 1.3 / 0.9 / 0.5 / 0.3
60390	BOTANY BAY	+0022 / +0023	1.5 / 1.3 / 0.9 / 0.5 / 0.3

HW & LW time difference from Sydney in hrs & min, e.g., Botany Bay HW occurs 22 min & LW 23 min after Sydney (Blank spaces indicate time not calculated owing to insufficient observations)

LAT = Lowest Astronomical Tide
MHSW = Mean HW Spring
MHWN = Mean HW Neap
MSL = Mean Sea Level
MLWN = Mean LW Neap
MHW = Mean HW
MLW = Mean LW
MLWS = Mean LW Spring

Fig 17.54

AUSTRALIA, TASMANIA (SOUTH-EAST COAST) - HOBART

LAT 42°53' S   LONG 147°20' E   (PREDOMINANTLY DIURNAL)

TIME ZONE -1000   TIMES AND HEIGHTS OF HIGH AND LOW WATERS   YEAR xxxx

JANUARY

Day	Time	m	Day	Time	m
1 WE	0447 / 1247 / 1914 / 2205	1.8 / 0.8 / 1.3 / 1.3	16 TH	0345 / 1133 / 1802 / 2130	1.8 / 0.7 / 1.3 / 1.2
2 TH	0523 / 1331 / 2006 / 2227	1.8 / 0.8 / 1.3 / 1.3	17 FR	0431 / 1226 / 1900 / 2220	1.9 / 0.6 / 1.3 / 1.2
3 FR	0559 / 1412 / 2056 / 2258	1.8 / 0.8 / 1.3 / 1.3	18 SA	0526 / 1321 / 1953 / 2315	1.9 / 0.6 / 1.3 / 1.2

FEBRUARY

Day	Time	m	Day	Time	m
1 SA	0534 / 1331 / 2001 / 2301	1.7 / 0.8 / 1.3 / 1.3	16 SU	0515 / 1251 / 1917 / 2322	1.9 / 0.6 / 1.4 / 1.1
2 SU	0615 / 1409 / 2040 / 2341	1.7 / 0.8 / 1.3 / 1.3	17 MO	0617 / 1345 / 2003	1.9 / 0.6 / 1.4
3 MO	0654 / 1445 / 2120	1.7 / 0.8 / 1.3	18 TU	0035 / 0718 / 1437 / 2053	1.1 / 1.8 / 0.6 / 1.4

MARCH

Day	Time	m	Day	Time	m
1 SU	0510 / 1237 / 1911 / 2311	1.7 / 0.9 / 1.4 / 1.2	16 MO	0508 / 1215 / 1841 / 2345	1.8 / 0.7 / 1.5 / 1.0
2 MO	0552 / 1312 / 1940 / 2353	1.6 / 0.9 / 1.4 / 1.2	17 TU	0612 / 1304 / 1923	1.7 / 0.7 / 1.5
3 TU	0633 / 1345 / 2009	1.6 / 0.9 / 1.4	18 WE	0100 / 0716 / 1353 / 2008	1.0 / 1.7 / 0.8 / 1.6

APRIL

Day	Time	m	Day	Time	m
1 WE	0012 / 0614 / 1230 / 1907	1.1 / 1.5 / 1.0 / 1.5	16 TH	0120 / 0723 / 1259 / 1927	0.8 / 1.5 / 1.0 / 1.7
2 TH	0105 / 0703 / 1254 / 1934	1.1 / 1.4 / 1.0 / 1.5	17 FR	0231 / 0840 / 1340 / 2010	0.8 / 1.4 / 1.1 / 1.7
3 FR	0204 / 0802 / 1317 / 2004	1.0 / 1.4 / 1.1 / 1.5	18 SA	0338 / 1000 / 1424 / 2055	0.7 / 1.4 / 1.2 / 1.7

NOTE: THE DIFFERENCE BETWEEN THE PREDOMINANTLY DIURNAL & SEMI-DIURNAL PORTS IS NOT EVIDENT TO LAYPERSONS

Fig 17.55: EXTRACT FROM ANTT PART I – PREDICTIONS FOR STANDARD PORTS

Some of the PREDOMINANTLY DIURNAL SECONDARY PORTS GROUPED AROUND HOBART

Note the difference in terminology from the table above (It helps in selecting the correct form below)

PORT No.	PORT NAME	TIME DIFFERENCES MHW / MLW	MHHW	MLHW	MSL	MHLW	MLLW
		(Zone -1000)					
61220	HOBART	(standard port)	1.5	1.0	0.8	0.7	0.2
61110	SWAN I	+0043 / +0043	1.4	1.3	0.8	0.3	0.2
61120	EDDYSTONE POINT	-0009 / +0005	1.3	0.8	0.6	0.5	-0.0
61170	SPRING BAY	-0011 / +0005	1.3	0.8	0.7	0.7	0.2

MHLW = Mean Higher LW   MHHW = Mean Higher HW
MLLW = Mean Lower LW   MLHW = Mean Lower HW

Fig 17.56: EXTRACT FROM ANTT PART II –
SECONDARY PORTS TIME DIFFERENCES & TIDE LEVELS

In order to deduce the times and heights of tides at a secondary port, one of the following forms is used. One form is for semi-diurnal and the other for diurnal ports. Matching the terminology on the form with that of the secondary port identifies the correct form.

The procedure for making entries in the two forms is identical. Only some of the terminology is different. The boxes or entries in both forms are similarly numbered or itemised.

Chapter 17.3: Navigation - Tides & Currents

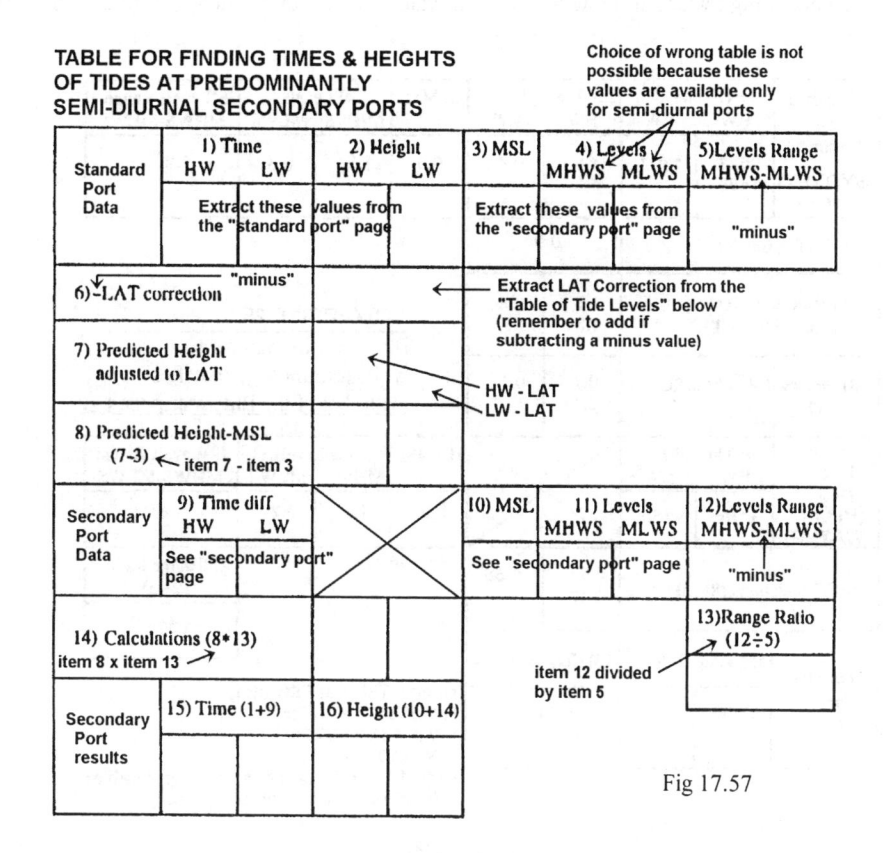

TABLE FOR FINDING TIMES & HEIGHTS OF TIDES AT PREDOMINANTLY SEMI-DIURNAL SECONDARY PORTS

Choice of wrong table is not possible because these values are available only for semi-diurnal ports

Standard Port Data	1) Time HW LW	2) Height HW LW	3) MSL	4) Levels MHWS MLWS	5) Levels Range MHWS-MLWS
	Extract these values from the "standard port" page		Extract these values from the "secondary port" page		"minus"
6) -LAT correction	"minus"		← Extract LAT Correction from the "Table of Tide Levels" below (remember to add if subtracting a minus value)		
7) Predicted Height adjusted to LAT			← HW - LAT ← LW - LAT		
8) Predicted Height-MSL (7-3) ← item 7 - item 3					
Secondary Port Data	9) Time diff HW LW See "secondary port" page	✕	10) MSL See "secondary port" page	11) Levels MHWS MLWS	12) Levels Range MHWS-MLWS "minus"
14) Calculations (8∗13) item 8 x item 13 →					13) Range Ratio (12÷5) item 12 divided by item 5
Secondary Port results	15) Time (1+9)	16) Height (10+14)			

Fig 17.57

TABLE FOR FINDING TIMES & HEIGHTS OF TIDES AT PREDOMINANTLY DIURNAL SECONDARY PORTS

Different terminology from the above form

Standard Port Data	1) Time HW LW	2) Height HW LW	3) MSL	4) Levels MHHW MLLW	5) Levels Range MHHW-MLLW
6) -LAT correction					
7) Predicted Height adjusted to LAT					
8) Predicted Height-MSL (7-3)					
Secondary Port Data	9) Time diff HW LW	✕	10) MSL	11) Levels MHHW MLLW	12) Levels Range MHHW-MLLW
14) Calculations (8∗13)					13) Range Ratio (12÷5)
Secondary Port results	15) Time (1+9)	16) Height (10+14)			

Fig 17.58

EXERCISE 29

Find the times and heights of high water and low water at Pittwater (60340) on Monday, 16 March, year xxxx.

Standard Port Data (SYDNEY) 16 March	1) Time HW / LW	2) Height HW / LW	3) MSL	4) Levels MHWS / MLWS	5) Levels Range MHWS-MLWS
	0622 / 1302 / 1858	1.6 / 1.3 / 0.4	0.9	1.5 / 0.3	1.2
6) –LAT correction		0.0			
7) Predicted Height adjusted to LAT		1.6 / 1.3 / 0.4			
8) Predicted Height-MSL (7-3)		0.7 / 0.4 / -0.5			

EXERCISE 29
The semi-diurnal form is used as its terminology matches with that of the Pittwater page.

Secondary Port Data (PITTWATER)	9) Time diff HW / LW		10) MSL	11) Levels MHWS / MLWS	12) Levels Range MHWS-MLWS
	+16m / +15m		0.9	1.6 / 0.2	1.4
14) Calculations (8*13)		0.82 / 0.47 / -0.58			13) Range Ratio (12÷5) 1.17
Secondary Port results	15) Time (1+9)	16) Height (10+14)			
	0638 / 1317 / 1914	1.72 / 1.37 / 0.32			

Hence, Tides at Pittwater:
HW at 0638 = 1.72m
LW at 1317 = 0.32m
HW at 1914 = 1.37m
(Add 1 hour if daylight saving applicable)
(There are only 3 tides on this day)

Fig 17.59

The following table, which is included in the Tide Tables, is required only for the LAT values. All other entries in the above form are extracted from either the Standard Port or the Secondary Port page.

TABLE I - TIDAL LEVELS AT STANDARD PORTS
(A sample)
PART 1: PREDOMINANTLY DIURNAL TIDES

PORT	HAT	MHHW	MLHW	MSL	MHLW	MLLW	LAT
Hobart	2.1	1.9	1.4	1.3	1.2	0.7	0.5
Honiara	1.0	0.8	0.7	0.4	0.1	0.0	-0.2
Ince Point	3.7	2.9	2.2	1.8	1.3	0.6	0.0
Karumba	4.2	3.2	2.8	1.5	0.3	-0.1	-0.5

TABLE I - TIDAL LEVELS AT STANDARD PORTS
(A sample)
PART 2: PREDOMINANTLY SEMI-DIURNAL TIDES

PORT	HAT	MHWS	MHWN	MSL	MLWN	MLWS	LAT
Sydney	2.1	1.6	1.3	1.0	0.6	0.3	0.0
Thevenard I.	2.8	2.4	1.8	1.5	1.2	0.6	0.0
Townsville	3.8	2.8	1.9	1.6	1.3	0.5	-0.2
Westernport	3.3	2.9	2.4	1.7	1.0	0.6	0.0

Fig 17.60 & 17.61: EXTRACT FROM ANTT
SUPPLEMENTARY TABLES – TABLE I

Chapter 17.3: Navigation - Tides & Currents

TIDAL STREAMS

The direction and strength of tidal streams is included in Tide Tables where necessary. Vessels in Torres Strait, for example, may be set by tidal streams of up to 8 knots. The following is an extract of the tidal streams information contained in the Australian National Tide Tables.

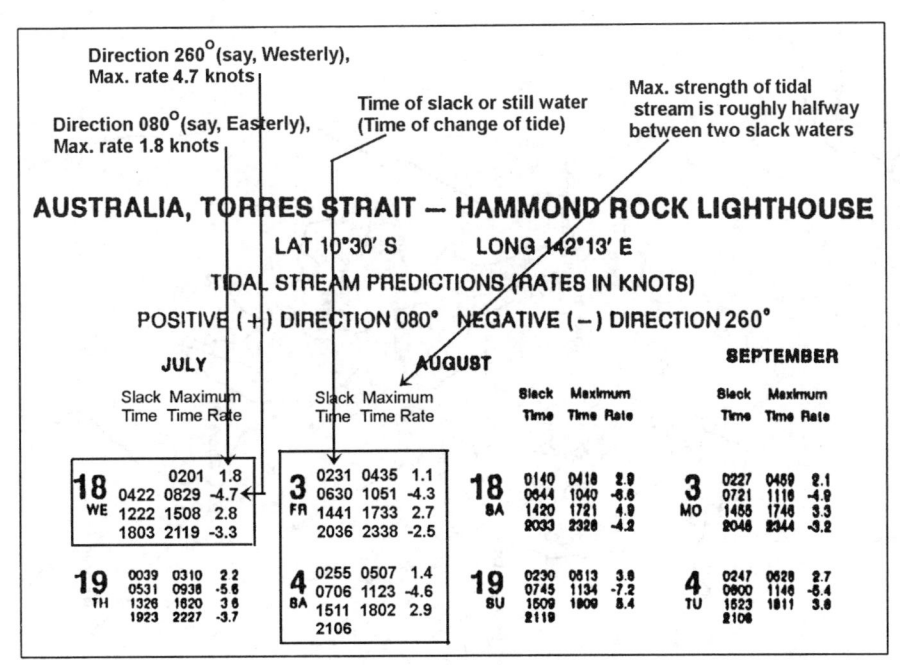

Fig 17.62

TORRES STRAIT

From the tidal point of view, Torres Strait is probably the most complex area in Australia. Its narrow and shallow channels connect two oceans with different mean sea levels caused by the general oceanic circulation patterns. This difference introduces a westward equalising current. In addition, tidal regimes on both sides of the Strait are completely different - diurnal tides to the west, and semi-diurnal to the east.

The contrast in regimes is caused by the difference of semi-diurnal components of tide at either entrance, diurnal part being generally uniform in the area. At some phases of the moon it can be high water at one entrance when it is low water at the other. In consequence, marked differences between the levels at the entrances occur resulting in strong tidal streams. While the tides may have a large diurnal component (especially at the western entrance), the tidal streams are **predominantly semi-diurnal.**

Throughout the Prince of Wales Channel and its approaches, from Twin Island in the east to a few miles west of Goods Island, the streams flow at the times predicted for Hammond Rock (see extract). The rates diminish as the channel becomes less restricted and at its western entrance are only about 30% of those predicted at Hammond Rock. At Booby Island the rates are comparatively weak and the streams are of different character.

In the vicinity of Harvey Rock and Saddle Island the streams commence and reach their maximum rates about 30 minutes earlier than Hammond Rock, but in these more open waters the rates are comparatively weak. In Endeavour Strait the streams commence and reach their maximum rates about 40 minutes later than at Hammond Rock and, except for the more restricted parts of the strait, their rates do not exceed 30% of those at Hammond Rock.

It has to be remembered however, that tidal stream predictions for Hammond Rock do not include any non-tidal flows, like the equalising current mentioned above or currents caused by meteorological influences. In addition, the El Nino Southern Ocean Oscillation can cause a drop of sea level of about 0.5m on the eastern side of the Strait in a very short time. The resultant changes to the water levels in the Strait, and to the current and stream direction and rates are impossible to predict.

SYDNEY & BROOME
(Tidal streams for these ports are shown in a graphical form, and the rates are empirical rather than predicted.)

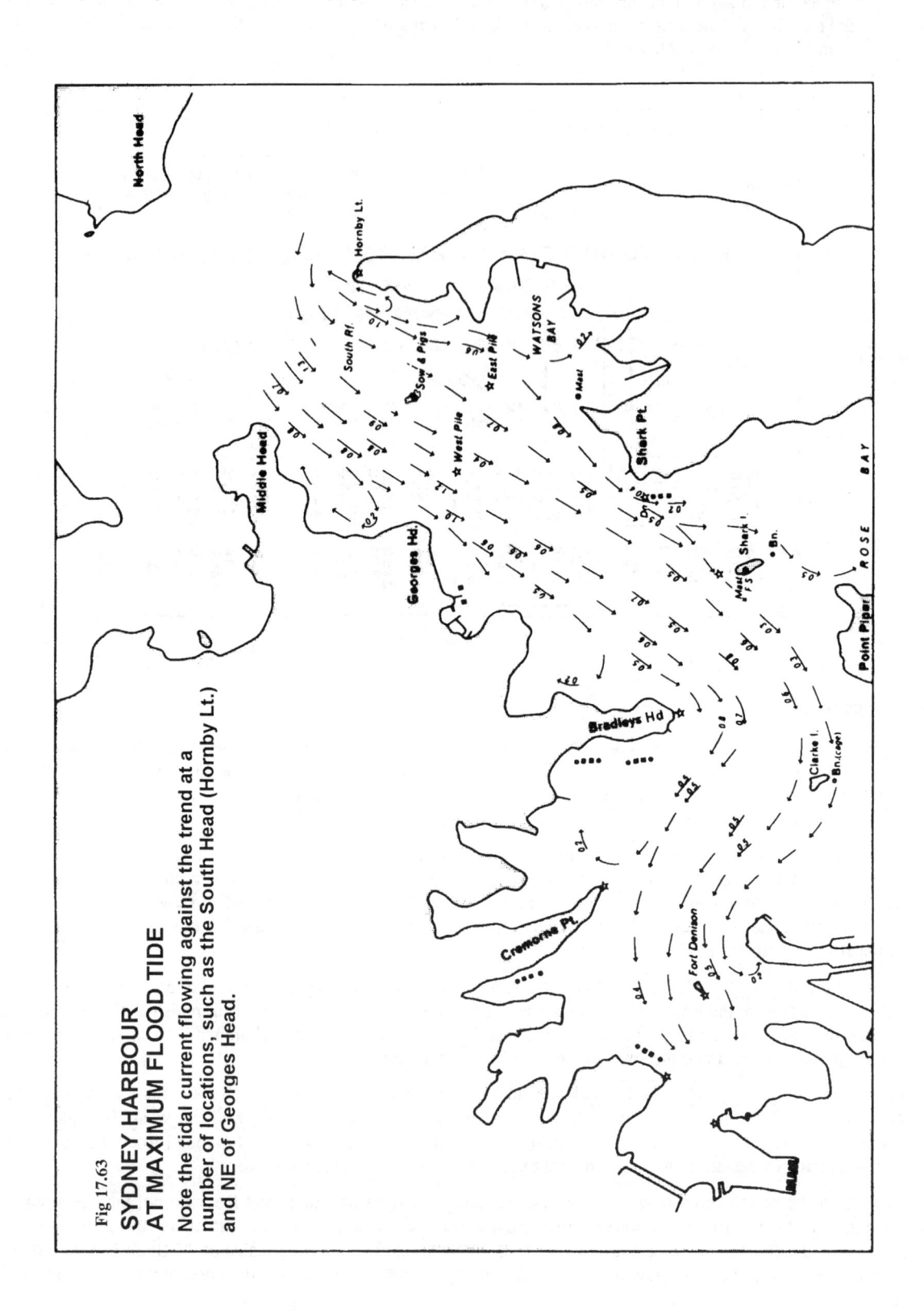

Fig 17.63
SYDNEY HARBOUR AT MAXIMUM FLOOD TIDE
Note the tidal current flowing against the trend at a number of locations, such as the South Head (Hornby Lt.) and NE of Georges Head.

Chapter 17.3: Navigation - Tides & Currents

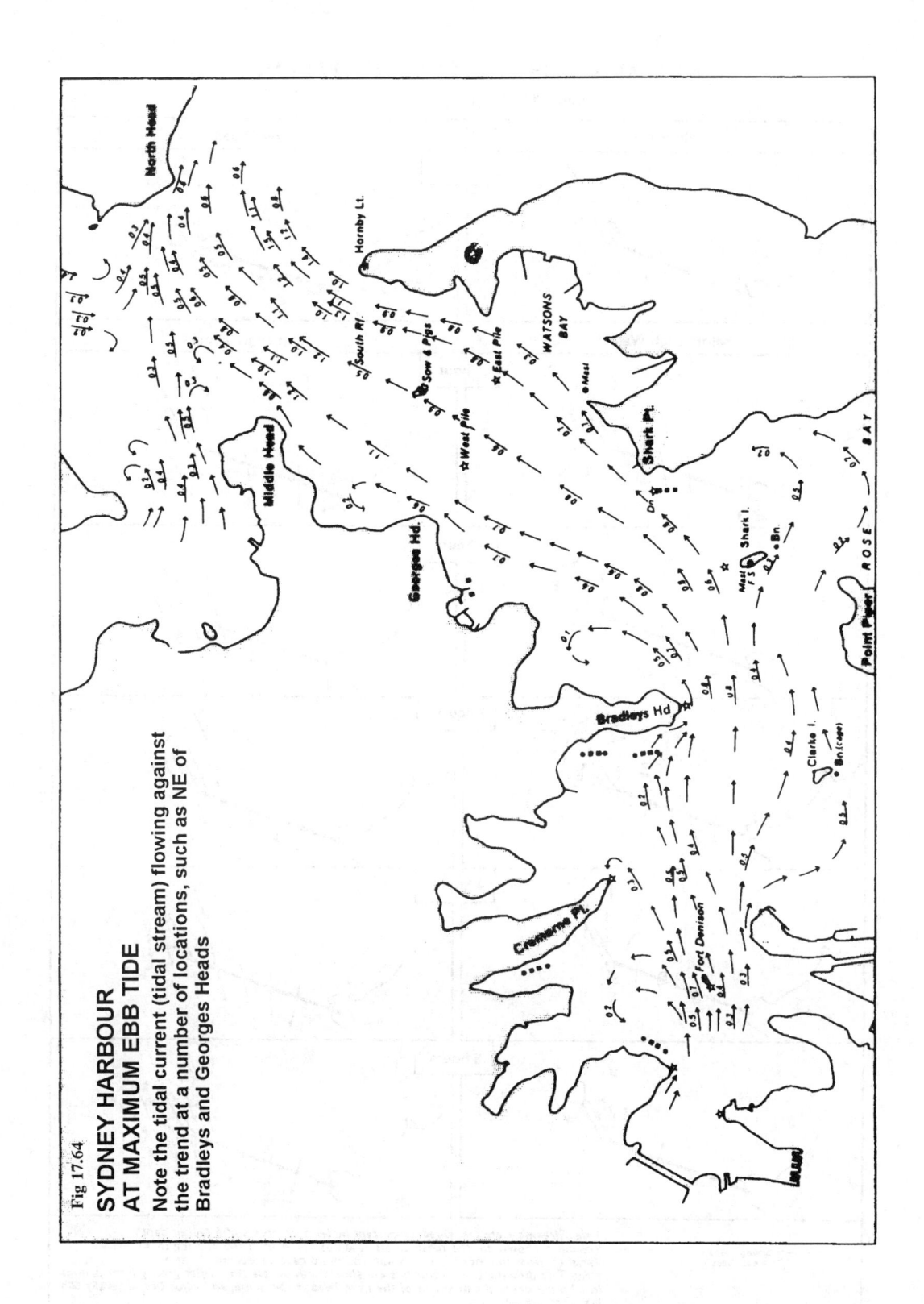

Fig 17.64
SYDNEY HARBOUR AT MAXIMUM EBB TIDE
Note the tidal current (tidal stream) flowing against the trend at a number of locations, such as NE of Bradleys and Georges Heads

AUSTRALIA—NORTH WEST COAST—BROOME.
Lat. 18°00' S. Long. 122°13' E.

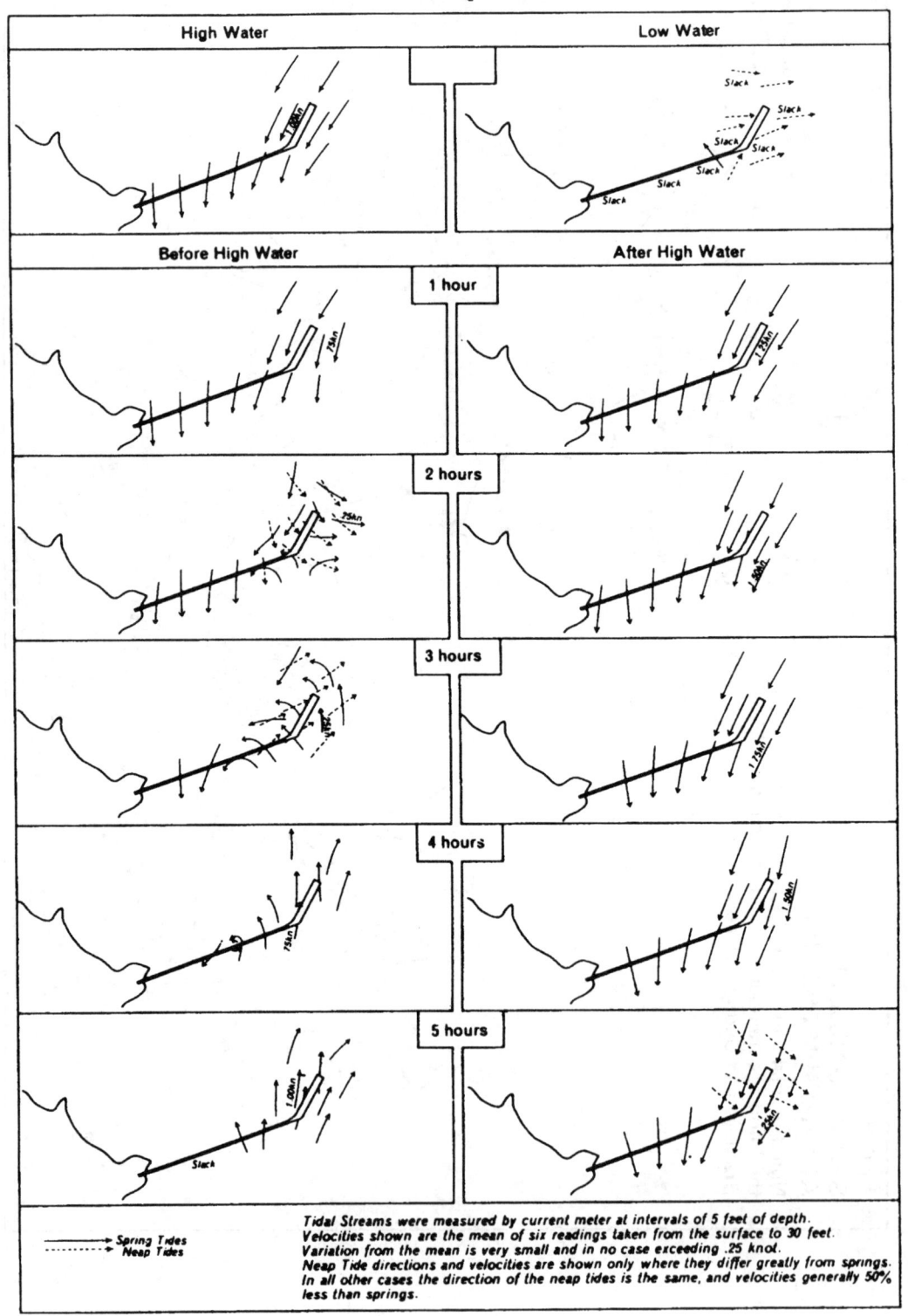

Fig 17.65

Chapter 17.3: Navigation - Tides & Currents

OCEAN CURRENTS

Ocean currents should not be confused with tidal streams. They are large-scale movement of water in oceans, resulting from a combination of the rotation of the earth, wind, geography of landmasses and differing water salinities and temperatures. The direction and strength of ocean currents are seasonal in nature, like the Trade Winds.

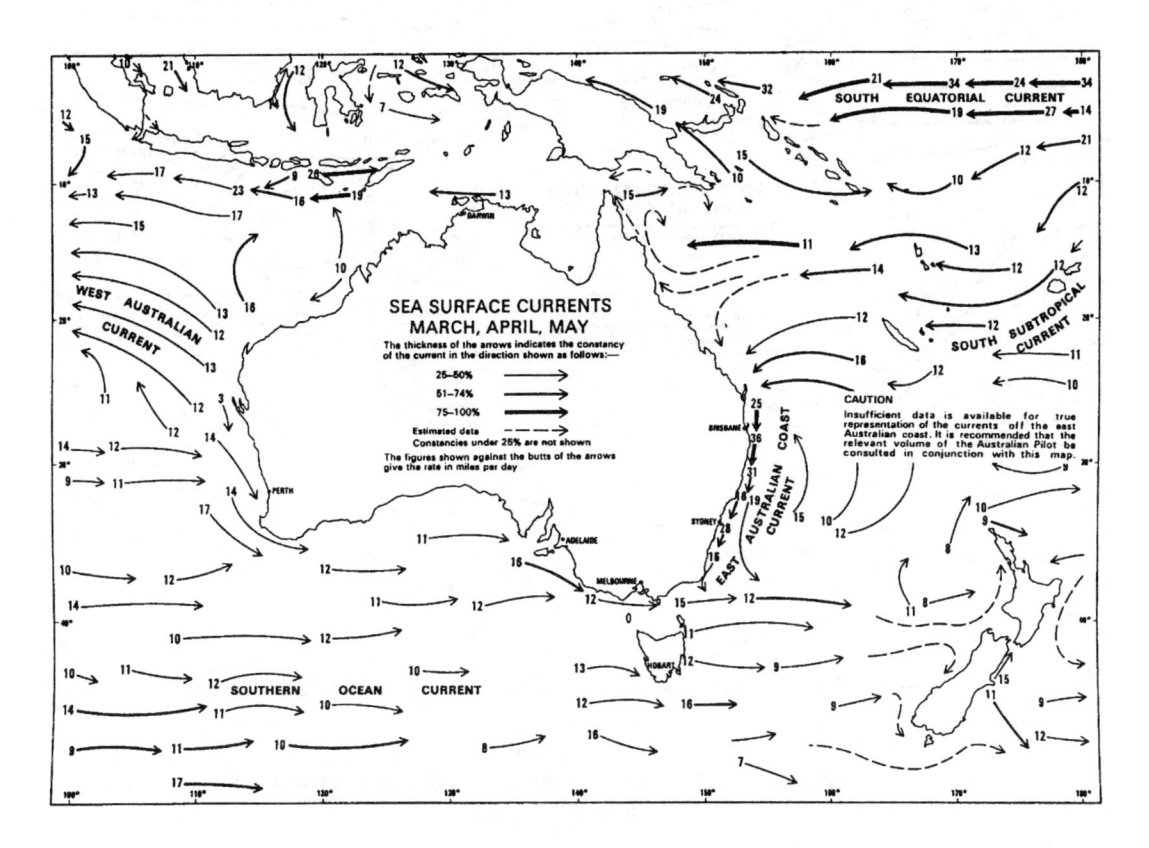

Fig 17.66
SEA SURFACE CURRENTS IN AUSTRALIAN WATERS
(March April May)

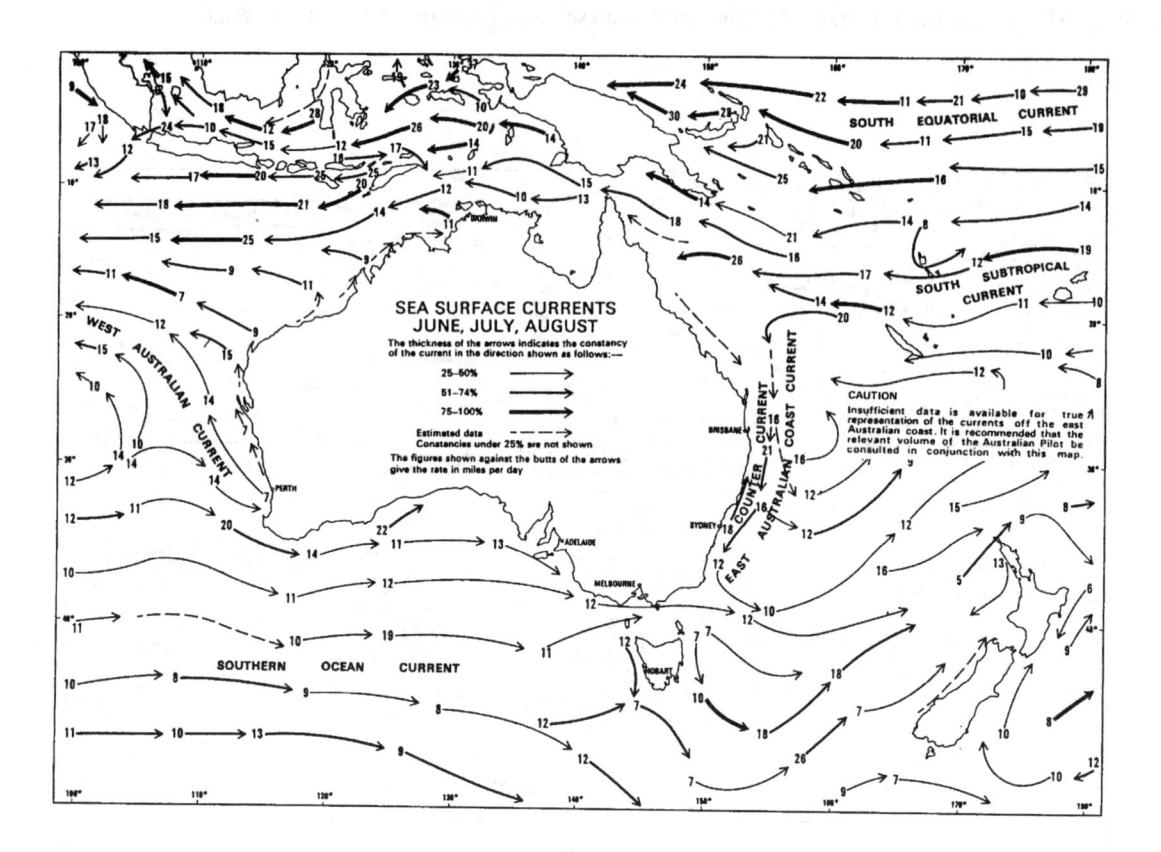

Fig 17.67
SEA SURFACE CURRENTS IN AUSTRALIAN WATERS
(June July August)

Chapter 17.3: Navigation - Tides & Currents

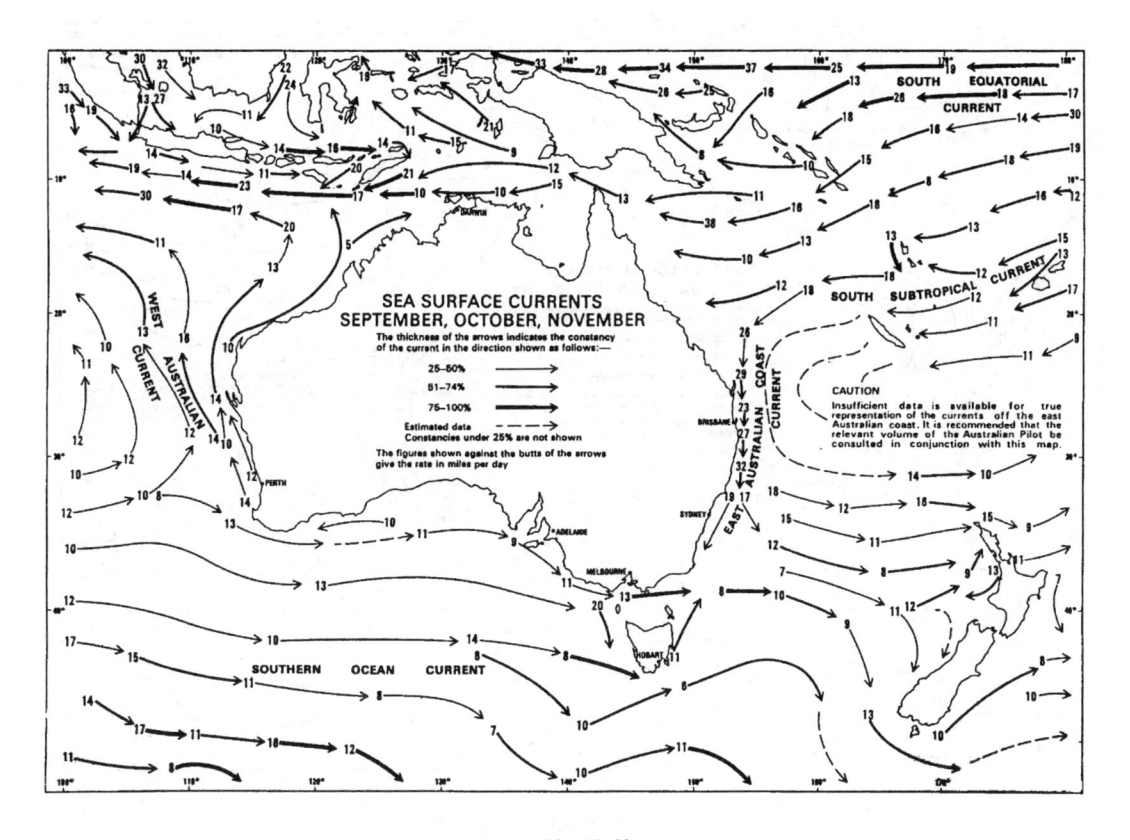

Fig 17.68
SEA SURFACE CURRENTS IN AUSTRALIAN WATERS
(September October November)

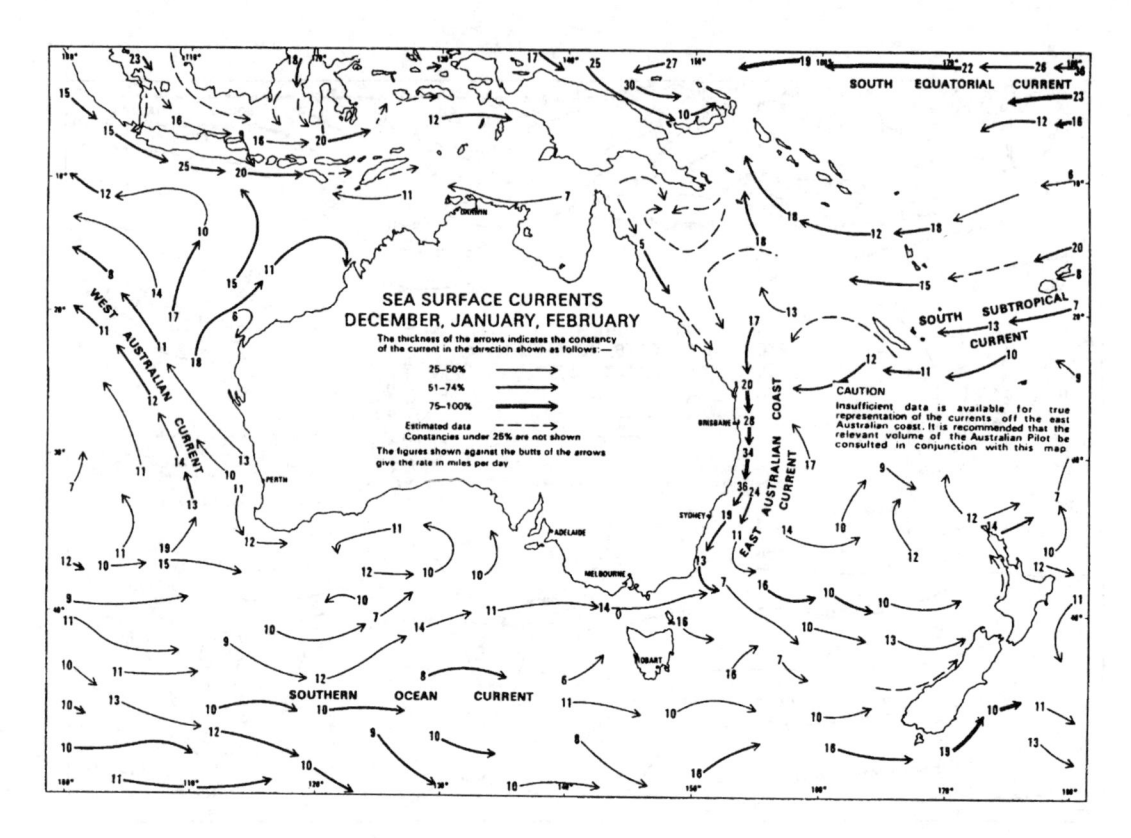

Fig 17.69
SEA SURFACE CURRENTS IN AUSTRALIAN WATERS
(December January February)

Chapter 17.3: Navigation - Tides & Currents

TIDE TABLES EXTRACT
(The following tables and forms are required for answering questions on tide calculations at the end of this chapter)

TABLE 1 - TIDAL LEVELS AT STANDARD PORTS
(AUTHOR'S NOTE: This table is required only for extracting "LAT" values)

PART 1: PREDOMINANTLY DIURNAL TIDES

PORT	HAT	MHHW	MLHW	MSL	MHLW	MLLW	LAT
Albany	1.4	1.1	0.8	0.8	0.7	0.5	0.1
Alotau	1.3	1.1	✹	0.7	✹	0.3	0.0
Anewa Bay	1.8	1.5	✹	0.8	✹	0.2	-0.1
Booby Island	4.4	4.4	2.9	2.5	2.1	0.6	0.0
Bunbury	1.3	0.9	0.7	0.7	0.7	0.4	0.1
Cairns	3.3	2.5	1.6	1.5	1.4	0.5	-0.2
Carnarvon	1.8	1.4	1.1	0.9	0.7	0.4	-0.1
Cocos Island	1.4	1.2	0.8	0.8	0.7	0.3	0.3
Denham	1.6	1.3	1.0	0.9	0.8	0.5	0.1
Dreger Harbour	1.9	1.4	✹	1.2	✹	1.0	0.4
Eden	1.9	1.6	1.0	0.8	0.6	0.1	-0.2
Esperance	1.5	1.1	0.8	0.8	0.6	0.5	0.2
Fremantle	1.3	0.9	0.7	0.7	0.7	0.5	0.1
Geelong	1.1	0.9	0.6	0.5	0.4	0.0	-0.1
Geraldton	1.3	1.0	0.9	0.6	0.4	0.3	0.0
Goods Island	4.0	3.8	2.8	2.2	1.6	0.6	0.0
Hobart	2.1	1.9	1.4	1.3	1.2	0.7	0.5
Honiara	1.0	0.8	0.7	0.4	0.1	0.0	0.0
Ince Point	3.7	2.9	2.2	1.8	1.3	0.6	-0.2
Karumba	4.2	3.2	2.8	1.5	0.3	-0.1	-0.5
Lae	1.8	1.6	1.5	1.2	0.9	0.7	0.6
Madang	1.3	1.3	1.1	0.8	0.5	0.3	0.0
Melbourne (Williamstown)	1.0	0.9	0.6	0.5	0.5	0.1	0.0
Milner Bay	2.3	1.7	1.6	1.1	0.5	0.5	0.0
Mourilyan	3.0	2.3	1.4	1.3	1.2	0.3	-0.4
Port Lincoln	2.1	1.7	1.2	1.1	0.9	0.4	0.2
Port Moresby	3.1	2.6	1.7	1.7	1.6	0.8	0.2
Port Pirie	3.4	2.9	1.9	1.7	1.4	0.5	-0.1
Portland	1.2	1.0	0.7	0.5	0.3	0.1	-0.1
Rabaul	1.2	1.1	1.0	0.7	0.4	0.3	0.0
Seeadler Hr	1.2	1.0	✹	0.5	✹	0.0	0.0
Thevenard	2.2	1.6	1.4	1.0	0.6	0.4	-0.2
Thursday Island	3.8	3.0	2.3	1.8	1.3	0.6	0.0
Turtle Head (Hammond Island)	3.7	3.2	2.5	1.9	1.2	0.6	0.0
Twin Island	3.8	2.9	1.9	1.7	1.5	0.5	0.0
Wallaroo	1.9	1.6	1.2	0.9	0.5	0.1	-0.2
Weipa	3.2	2.9	2.2	1.8	1.5	0.7	0.0
Wewak	1.4	1.5	1.0	0.9	0.7	0.3	0.0
Whyalla	3.2	2.7	1.8	1.6	1.3	0.4	-0.2

PART 2: PREDOMINANTLY SEMI-DIURNAL TIDES

PORT	HAT	MHWS	MHWN	MSL	MLWN	MLWS	LAT
Barrow I. (W.L.)	3.7	3.2	2.2	1.8	1.5	0.6	0.1
Barrow I. (T.M.)	4.7	4.1	2.7	2.3	1.9	0.6	-0.1
Brisbane Bar	2.7	2.1	1.8	1.2	0.7	0.3	0.1
Broome	9.6	8.5	5.6	4.5	3.5	0.3	-0.9
Bugatti Reef	3.4	2.5	2.0	1.5	1.0	0.5	0.0
Bundaberg	3.2	2.5	1.9	1.3	0.8	0.2	-0.3
Burnie	3.6	3.2	2.9	1.9	0.9	0.6	0.0
Cape Domett	8.0	6.9	5.1	4.0	3.0	1.3	-0.1
Cape Voltaire	7.7	6.4	4.3	3.7	3.0	0.9	0.0
Coffs Harbour	2.0	1.5	1.2	0.8	0.4	0.2	-0.1
Dampier	5.2	4.5	3.2	2.7	2.2	0.9	0.1
Darwin	7.9	6.8	4.9	4.0	3.1	1.2	-0.1
Derby	10.8	10.0	7.5	5.2	2.7	0.6	0.3
Devonport	3.4	2.9	2.7	1.7	0.8	0.5	-0.2
Georgetown	3.6	3.3	3.0	2.0	1.1	0.8	0.2
Gladstone	4.8	3.9	3.1	2.3	1.5	0.7	0.0
Gove	3.9	3.1	2.6	2.1	1.5	1.0	0.2
Hay Point	7.1	5.8	4.5	3.3	2.2	0.9	0.0
Learmonth	3.0	2.6	1.8	1.5	1.2	0.5	0.0
Lucinda	3.6	2.7	1.9	1.6	1.3	0.5	0.0
Mackay	6.6	5.3	4.0	3.0	1.9	0.7	-0.2
Newcastle	2.1	1.6	1.4	1.0	0.6	0.4	-0.1
Norfolk Island	1.9	1.6	1.4	0.9	0.4	0.2	0.1
Onslow	2.6	2.1	1.4	1.1	0.8	0.2	0.0
Point Murat	2.5	2.0	1.5	1.2	1.0	0.4	-0.4
Port Adelaide (Inner Harbor)	3.2	2.6	1.6	1.6	1.6	0.4	0.0
Port Adelaide (Outer Harbor)	3.1	2.6	1.6	1.6	1.6	0.6	0.2
Port Hedland	7.8	6.9	4.8	4.1	3.4	1.1	0.2
Port Kembla	2.1	1.6	1.3	0.9	0.6	0.3	0.0
Port Phillip Hds.	1.8	1.5	1.2	0.9	0.6	0.3	0.0
Port Walcott	5.8	5.1	3.4	2.8	2.3	0.4	-0.4
Stanley	3.9	3.5	3.2	2.2	1.2	0.9	0.2
Sydney	2.1	1.6	1.3	1.0	0.6	0.3	0.0
Thevenard I.	2.8	2.4	1.8	1.5	1.2	0.6	0.0
Townsville	3.8	2.8	1.9	1.6	1.3	0.5	-0.2
Westernport (Stony Point)	3.3	2.9	2.4	1.7	1.0	0.6	0.0
Wyndham	8.1	7.5	5.7	4.2	2.6	0.9	-0.3
Yampi Sound (Koolan Island)	11.1	10.1	7.0	5.7	4.4	1.2	0.2

✹ Tide is usually diurnal

Fig 17.70

446

The Australian Boating Manual

Fig 17.71

Chapter 17.3: Navigation - Tides & Currents

EXTRA TIDES

Month	Day	Time	Ht(m)
TWIN ISLAND			
Feb	22	1927	1.2
	23	2151	1.5
July	21	1735	1.4
		2352	2.2
Sep	2	2022	2.2
	30	2252	1.4
INCE POINT			
Feb	7	2034	1.6
		2350	1.8
Mar	22	1949	1.8
	10	1744	1.0
		2308	2.0
July	20	1730	1.3
		2340	2.2
Aug	17	1718	1.3
		2341	2.2
Sep	1	1724	1.9
		1946	2.1
	30	1853	2.2
THURSDAY ISLAND			
Feb	7	2316	1.7
	22	1708	1.9
		1942	2.1
Mar	21	1850	1.8
	22	1629	1.4
		2148	2.0
Aug	17	1329	1.7
		1647	1.4
Sep	13	1842	1.5
		2345	2.3
	14	1921	1.7
	29	2327	2.2
		1755	2.2
		2312	1.4

Month	Day	Time	Ht(m)
FREMANTLE			
Feb	21	2230	0.6
Aug	3	1215	0.7
		1715	0.6
GERALDTON			
Feb	7	2138	0.7
Oct	1	1505	0.3
		2225	0.8
	29	2204	0.8
	30	2159	0.8
Nov	17	2046	0.7
MILNER BAY			
Feb	24	1726	1.0
Aug	20	1945	0.1

Fig 17.72

AUSTRALIA, EAST COAST - SYDNEY (FORT DENISON)

LAT 33°51′ S LONG 151°14′ E

TIME ZONE -1000 TIMES AND HEIGHTS OF HIGH AND LOW WATERS YEAR XXXX

JANUARY

Day	Time	m	Day	Time	m
1 WE	0538 1228 1813 2340	1.6 0.5 1.2 0.5	16 TH	0436 1128 1715 2258	1.6 0.5 1.2 0.5
2 TH	0625 1314 1900	1.6 0.4 1.2	17 FR	0536 1228 1820 2357	1.7 0.4 1.2 0.4
3 FR	0024 0707 1353 1941	0.5 1.7 0.4 1.2	18 SA	0633 1321 1916	1.8 0.2 1.3
4 SA	0105 0746 1429 2018	0.5 1.7 0.4 1.3	19 SU	0051 0728 1411 2009	0.4 2.0 0.1 1.4
11 SA	0001 0548 1152 1827	1.3 0.6 1.4 0.4	26 SU	0113 0730 1323 1929	1.5 0.5 1.3 0.5
12 SU	0049 0641 1238 1907	1.3 0.6 1.3 0.5	27 MO ◐	0209 0841 1425 2022	1.5 0.6 1.1 0.6
13 MO ◑	0141 0741 1334 1954	1.4 0.7 1.2 0.5	28 TU	0309 0956 1538 2123	1.5 0.6 1.1 0.6

FEBRUARY

Day	Time	m	Day	Time	m
1 SA	0009 0649 1334 1924	0.6 1.6 0.4 1.2	16 SU	0618 1303 1902	1.8 0.2 1.4
2 SU	0051 0728 1408 1959	0.5 1.7 0.4 1.3	17 MO	0042 0713 1351 1951	0.3 1.9 0.1 1.5
3 MO	0130 0803 1439 2032	0.5 1.7 0.3 1.3	18 TU ○	0136 0803 1436 2039	0.2 2.0 0.1 1.6
4 TU ●	0207 0837 1508 2104	0.4 1.7 0.3 1.4	19 WE	0230 0853 1519 2125	0.2 1.9 0.1 1.6
11 TU	0057 0715 1306 1910	1.4 0.6 1.2 0.6	26 WE	0227 0923 1509 2044	1.5 0.6 1.1 0.7
12 WE ◑	0157 0830 1420 2012	1.5 0.6 1.1 0.6	27 TH	0334 1035 1633 2157	1.4 0.6 1.1 0.7
13 TH	0304 0958 1545 2129	1.5 0.6 1.1 0.6	28 FR	0443 1138 1736 2300	1.5 0.6 1.1 0.7

MARCH

Day	Time	m	Day	Time	m
1 SU	0622 1302 1858	1.6 0.4 1.3	16 MO	0600 1237 1843	1.8 0.2 1.5
2 MO	0035 0700 1334 1931	0.5 1.6 0.4 1.4	17 TU	0032 0654 1323 1930	0.3 1.8 0.2 1.6
3 TU	0114 0736 1404 2003	0.5 1.6 0.4 1.4	18 WE	0127 0745 1405 2015	0.2 1.8 0.2 1.7
4 WE ●	0150 0810 1431 2034	0.4 1.6 0.3 1.5	19 TH ○	0220 0833 1446 2100	0.2 1.8 0.2 1.7
11 WE	0026 0704 1259 1844	1.5 0.5 1.2 0.6	26 TH ◐	0140 0844 1435 2000	1.5 0.6 1.1 0.8
12 TH ◑	0129 0820 1415 1953	1.5 0.6 1.1 0.7	27 FR	0245 0947 1552 2119	1.4 0.6 1.1 0.8
13 FR	0241 0941 1539 2116	1.5 0.5 1.2 0.6	28 SA	0354 1046 1655 2229	1.4 0.6 1.2 0.7

APRIL

Day	Time	m	Day	Time	m
1 WE	0052 0700 1319 1929	0.5 1.5 0.4 1.5	16 TH	0118 0724 1330 1951	0.3 1.6 0.3 1.8
2 TH	0130 0737 1348 2000	0.4 1.5 0.4 1.6	17 FR ○	0210 0813 1409 2033	0.3 1.6 0.3 1.8
3 FR ●	0209 0813 1418 2033	0.4 1.5 0.4 1.7	18 SA	0259 0900 1447 2115	0.3 1.5 0.4 1.8
4 SA	0247 0851 1450 2109	0.4 1.5 0.4 1.7	19 SU	0348 0946 1526 2155	0.3 1.4 0.5 1.8
11 SA	0224 0921 1529 2106	1.6 0.5 1.3 0.7	26 SU	0258 0947 1601 2143	1.4 0.6 1.3 0.8
12 SU	0335 1022 1634 2220	1.6 0.4 1.4 0.6	27 MO	0357 1034 1652 2246	1.4 0.6 1.3 0.7
13 MO	0440 1115 1730 2325	1.7 0.4 1.5 0.5	28 TU	0449 1116 1735 2339	1.4 0.5 1.4 0.6

Fig 17.73

Chapter 17.3: Navigation - Tides & Currents

AUSTRALIA, TASMANIA (SOUTH-EAST COAST) - HOBART

LAT 42°53' S   LONG 147°20' E

TIMES AND HEIGHTS OF HIGH AND LOW WATERS

TIME ZONE -1000   YEAR XXXX

Fig 17.74

Fig 17.75

Fig 17.76

SECONDARY PORTS
TIME DIFFERENCES & TIDAL LEVELS

PORT No.	PORT NAME	TIME DIFFERENCES MHW	MLW	TIDAL LEVELS (in metres, related to LAT) MHWS	MHWN	MSL	MLWN	MLWS
		(Zone -1000)						
60370	SYDNEY	(standard port)		1.5	1.3	0.9	0.5	0.3
60320	GOSFORD			0.8	0.5	0.4	0.3	-0.1
60325	ETTALONG			0.9	0.8	0.5	0.2	0.1
60330	LITTLE PATONGA			1.6	1.3	0.9	0.5	0.3
60340	PITTWATER	+0016	+0015	1.6	1.3	0.9	0.4	0.2
60360	CAMP COVE	-0003	-0003	1.6	1.3	0.9	0.5	0.3
60390	BOTANY BAY	+0022	+0023	1.5	1.3	0.9	0.5	0.3
60400	PORT HACKING			1.5	1.3	0.9	0.5	0.3
60440	JERVIS BAY			1.5	1.3	0.9	0.6	0.3
60460	ULLADULLA HARBOUR			1.4	1.2	0.8	0.4	0.2
60470	BATEMANS BAY			1.5	1.2	0.9	0.5	0.3
60480	MORUYA			1.3	1.1	0.8	0.5	0.3
60930	DEVONPORT	(standard port)		3.2	2.9	2.0	1.0	0.8
60630	WINTER COVE	-0020	-0020					
60650	GREAT GLENNIE I.	-0010	-0010	2.2	2.0	1.2	0.4	0.2
60830	SURPRISE BAY	-0015	-0014	1.5	0.9	0.8	0.7	0.1
60840	GRASSY	-0003	-0017	1.5	0.9	0.8	0.8	0.1
60870	STACK I.	+0109	+0102	2.0	1.8	1.0	0.3	0.1
60980	WATERHOUSE I.	-0025	-0025					
61010	PRESERVATION I.	-0040	-0040					
61020	GOOSE I.	-0025	-0025					
61030	BIG RIVER COVE	-0027	-0027	2.7	2.4	1.6	0.8	0.6
61060	ROYDON I.	-0025	-0025					
61090	LADY BARRON Hr.	-0055	-0055	1.6	1.4	0.9	0.5	0.3
60950	GEORGETOWN	(standard port)		3.1	2.9	1.9	0.9	0.7
60970	LAUNCESTON	+0100	+0104	3.9	3.7	2.4	1.1	0.9

Fig 17.77: AUSTRALIAN NATIONAL TIDE TABLES – PART III

Chapter 17.3: Navigation - Tides & Currents

PORT No.	PORT NAME	TIME DIFFERENCES MHW	MLW	TIDAL LEVELS (in metres, related to LAT) MHHW	MLHW	MSL	MHLW	MLLW
		(Zone -1000)						
61220	HOBART	(standard port)		1.5	1.0	0.8	0.7	0.2
61110	SWAN I.	+0043	+0043	1.4	1.3	0.8	0.3	0.2
61120	EDDYSTONE POINT	-0009	+0005	1.3	0.8	0.6	0.5	-0.0
61170	SPRING BAY	-0011	+0005	1.3	0.8	0.7	0.7	0.2
61180	PIRATES BAY	-0016	-0005	1.1	0.5	0.5	0.5	-0.0
61200	PARSONS BAY	+0010	+0001	1.2	0.8	0.6	0.5	0.0
61210	IMPRESSION BAY	+0007	+0005	1.3	0.8	0.6	0.5	-0.0
61270	MAATSUYKER I.	+0031	+0026	1.2	0.7	0.6	0.5	-0.0
61280	BRAMBLE COVE			0.8	0.7	0.5	0.3	0.2
61300	CAPE SORELL, PILOT B			1.0	0.8	0.6	0.3	0.1
61320	PIEMAN R.			1.1	0.8	0.6	0.4	0.2
61410	PORTLAND	(standard port)		1.0	0.8	0.6	0.4	0.2
61360	PORT CAMPBELL	+0012	-0003	1.1	0.8	0.6	0.5	0.2
61380	WARRNAMBOOL	-0007	-0005	0.9	0.5	0.5	0.5	0.1
		(Zone -0930)						
61600	ADELAIDE OUTER Hr.	(standard port)		2.4	1.3	1.3	1.3	0.3
61520	EMU BAY	-0028	-0034	1.2	0.7	0.6	0.5	0.0
61530	KINGSCOTE	-0056	-0059	1.4	0.8	0.7	0.6	0.0
61540	AMERICAN R.	-0020	-0028	1.4	0.9	0.8	0.6	0.2
61550	HOG BAY	-0047	-0046	1.4	1.0	0.8	0.6	0.2
61561	CAPE JERVIS	-0028	-0032	1.1	0.7	0.6	0.4	-0.0
61570	SECOND VALLEY	-0008	-0005	1.7	1.0	1.0	0.9	0.3
61580	PORT NOARLUNGA	-0010	-0010	1.9	1.1	1.1	1.1	0.3
61590	BRIGHTON	-0005	0000	2.0	1.2	1.1	1.1	0.3
61650	ARDROSSAN	-0002	-0003	2.9	1.6	1.6	1.6	0.4
61670	PORT VINCENT	-0030	-0030	2.2	1.2	1.1	1.0	0.0
61680	WOOL BAY			2.7	1.9	1.9	1.9	1.1
61690	EDITHBURGH	-0030	-0028	2.0	1.1	1.1	1.1	0.3
61610	ADELAIDE INNER Hr.	(standard port)		2.4	1.4	1.4	1.4	0.3
61780	WALLAROO	(standard port)		1.7	1.4	1.0	0.6	0.3
61740	PONDALOWIE BAY			1.0	0.7	0.5	0.4	0.1
61770	CAPE ELIZABETH			1.5	1.1	0.8	0.6	0.2
61790	PORT BROUGHTON			1.6	1.3	1.0	0.7	0.4
61860	ARNO BAY	-0014	-0029	1.6	1.1	0.9	0.6	0.1

Fig 17.78 AUSTRALIAN NATIONAL TIDE TABLES, PART III –
TIME DIFFERENCE & TIDE LEVELS

The Australian Boating Manual

Standard Port Data	1) Time HW LW	2) Height HW LW	3) MSL	4) Levels MHWS MLWS	5) Levels Range MHWS-MLWS
6) -LAT correction					
7) Predicted Height adjusted to LAT					
8) Predicted Height-MSL (7-3)					
Secondary Port Data	9) Time diff HW LW	✕	10) MSL	11) Levels MHWS MLWS	12) Levels Range MHWS-MLWS
14) Calculations (8∗13)					13) Range Ratio (12÷5)
Secondary Port results	15) Time (1+9)	16) Height (10+14)			

Standard Port Data	1) Time HW LW	2) Height HW LW	3) MSL	4) Levels MHHW MLLW	5) Levels Range MHHW-MLLW
6) -LAT correction					
7) Predicted Height adjusted to LAT					
8) Predicted Height-MSL (7-3)					
Secondary Port Data	9) Time diff HW LW	✕	10) MSL	11) Levels MHHW MLLW	12) Levels Range MHHW-MLLW
14) Calculations (8∗13)					13) Range Ratio (12÷5)
Secondary Port results	15) Time (1+9)	16) Height (10+14)			

Fig 17.79 (A & B)

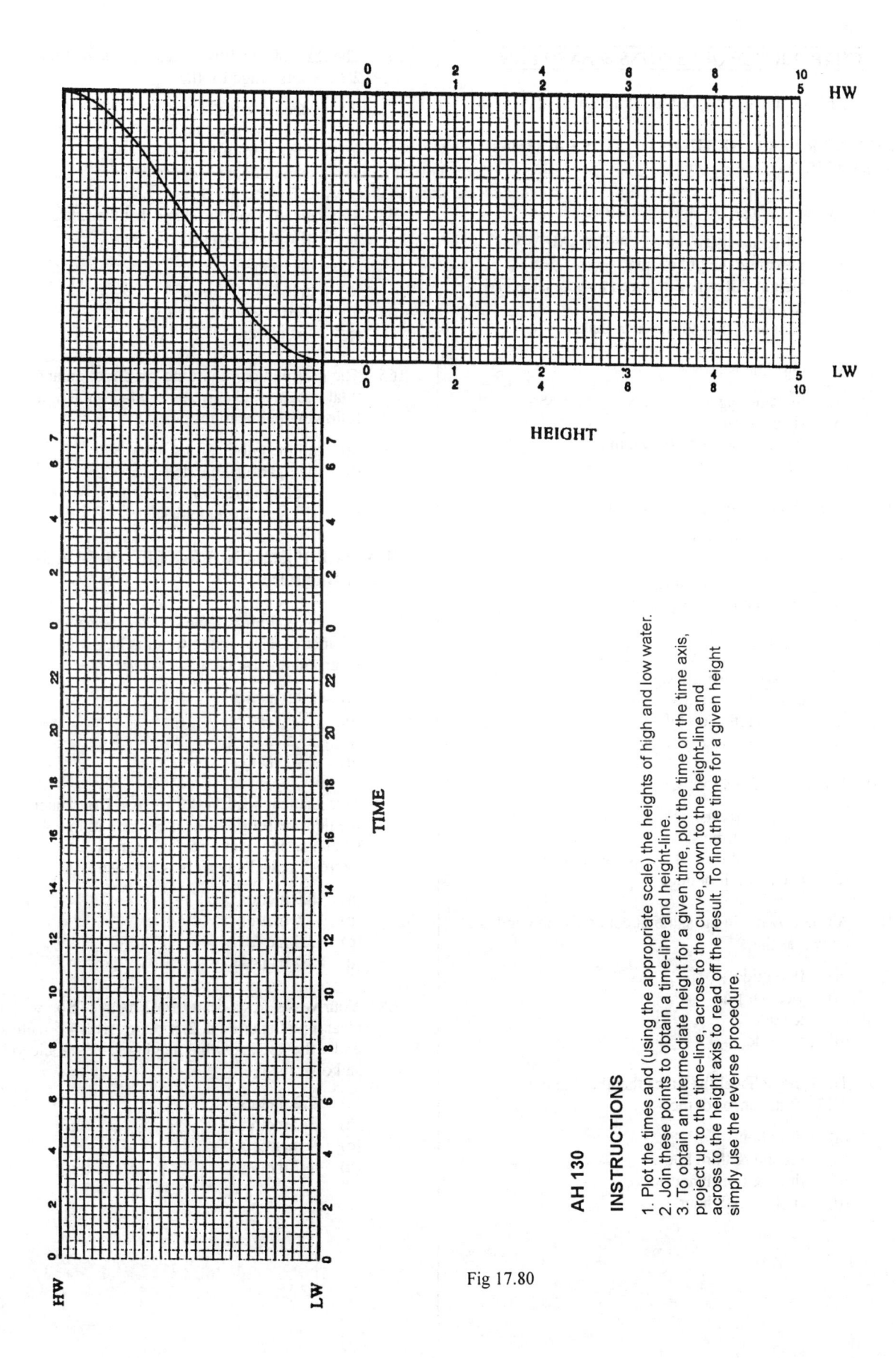

Chapter 17.3: Navigation - Tides & Currents

HEIGHT

TIME

HW

LW

AH 130

INSTRUCTIONS

1. Plot the times and (using the appropriate scale) the heights of high and low water.
2. Join these points to obtain a time-line and height-line.
3. To obtain an intermediate height for a given time, plot the time on the time axis, project up to the time-line, across to the curve, down to the height-line and across to the height axis to read off the result. To find the time for a given height simply use the reverse procedure.

Fig 17.80

CHAPTER 17.3: QUESTIONS & ANSWERS

(Qns. 1-63 are in Ch. 17.1, & 64-155 in Ch. 17.2)

CHAPTER 17.3.1: HEIGHTS OF TIDES & TIDAL STREAMS

156. The term "range" of tide means the:
    (a) difference between MHWS and MLWS.
    (b) maximum height of tide.
    (c) Height difference between a successive HW & LW.
    (d) Height of tide above chart datum.

157. A Spring Tide is a:
    (a) tide coming in.
    (b) slack water.
    (c) tide at new moon or full moon.
    (d) Christmas tide.

158. A high atmospheric pressure:
    (a) causes higher tides.
    (b) causes lower tides.
    (c) has no effect on tides.
    (d) affects only spring tides.

159. An onshore wind:
    (a) causes higher tides.
    (b) causes lower tides.
    (c) has no effect on tides.
    (d) affects only neap tides.

160. The tidal streams are the:
    (a) ocean currents.
    (b) coastal currents.
    (c) horizontal flow of water due to tides.
    (d) rise and fall of tides.

161. A tide flowing in the same direction as the wind is known as the:
    (a) flood tide.
    (b) weather tide.
    (c) lee tide.
    (d) ebb tide.

162. The Rule of Twelfths states that the tide moves $3/12^{th}$ of its range for the:
    (a) third hour.
    (b) second & third hour.
    (c) third & fourth hour.
    (d) first & sixth hour.

163. The Rule of Twelfths states that the tide moves $2/12^{th}$ of its range for the:
    (a) second & fifth hour.
    (b) second & third hour.
    (c) third & fourth hour.
    (d) second hour.

164. The Rule of Twelfths states that the tide moves $1/12^{th}$ of its range in the:
    (a) first hour.
    (b) first & sixth hour.
    (c) second & third hour.
    (d) sixth hour.

165. The information regarding the direction and rate of tidal streams and currents can be found in the following publications:
    (a) all the publications stated here.
    (b) navigational charts.
    (c) Sailing Directions.
    (d) tide and current atlases.

CHAPTER 17.3.2: UNDER-KEEL CLEARANCE CALCULATIONS

166. A vessel's draft is 2 metres and the height of tide is 1.5 metres. How much water is below her keel when chart datum reads 24 metres?
    (a) 20.5 metres
    (b) 27.5 metres
    (c) 23.5 metres
    (d) 24.5 metres

167. Tide Table reads LW 1.7 metres. How much water would a vessel find under her keel at low water when passing over a reef of charted depth of 3.7 metres? Vessel's draft is is 1.9 metres.
    (a) 3.9 metres
    (b) 3.5 metres
    (c) 7.3 metres
    (d) 5.2 metres

168. Your vessel has a draft of 3.2 metres. You wish to clear a reef charted at 2.5 metres with 1.5 metres under the keel. At what height of tide should you sail over the reef?
    (a) 2.2 metres
    (b) 3.2 metres
    (c) 0 metres
    (d) 1.7 metres

Chapter 17.3: Navigation - Tides & Currents

169. A shoal is charted as 3.0 metres and the height of tide is found to be 1.7 metres. How much clearance would a vessel have if her draft were 2.5 metres?

    (a) 3.9 metres
    (b) 3.5 metres
    (c) 4.7 metres
    (d) 2.2 metres

170. Tide Table reads LW 3.1 metres. How much water would a vessel find under her keel at low water when passing over a reef of charted depth of 2.0 metres? Vessel's draft is 1.5 metres.

    (a) 3.6 metres
    (b) 4.6 metres
    (c) 0.5 metres
    (d) 6.6 metres

CHAPTER 17.3.3: HEIGHT OF TIDE CALCULATIONS

171. Find the time and height of the evening high water at Geraldton on Thursday, 1 October year XXXX.

    (a) 2225 hours = 0.8 m
    (b) There is no evening HW
    (c) 1505 hours = 0.3 m
    (d) 0028 hours = 0.9 m

172. Find the times and heights of HW and LW at Geraldton on Monday, 2 November, year XXXX.

    (a) 1153 = 0.4 m; 1800 = 0.6 m
    (b) 1800 = 0.6 m; only one tide
    (c) 1220 = 0.4 m; only one tide
    (d) 1153 = 0.4 m; only one tide

173. Find the local (daylight saving) times and heights of LW at Fort Denison on Saturday, 1 February, year XXXX.

    (a) 0749 = 1.6 m; 2024 = 1.2 m
    (b) 0109 = 0.6 m; 1434 = 0.4 m
    (c) 0009 = 0.6 m; 1334 = 0.4 m
    (d) 2309 = 0.6 m; 1234 = 0.4 m

174. Find the local (daylight saving) time of AM HW at Sydney on Monday, 16 March, year XXXX.

    (a) There is no AM HW
    (b) 0400 hours
    (c) 0500 hours
    (d) 0600 hours

175. Find the direction of tidal stream and its approximate rate of flow at Hammond Rock Lighthouse at 0800 hours on Friday, 3 August, year XXXX.

    (a) 260°; 2 knots
    (b) Nil, it is slack water
    (c) 260°; 4.3 knots
    (d) 080°; 4.3 knots

176. What are the direction of tidal stream and its rate of flow at Hammond Rock Lighthouse at 1511 hours on Saturday, 4 August, year XXXX?

    (a) 260°; 2.9 knots
    (b) 080°; 2.9 knots
    (c) Nil, it is slack water
    (d) Slack water; 2.9 knots

177. Find the direction of tidal stream and the maximum rate of flow near Harvey Rock between 0131 and 0352 hours on Wednesday, 18 July, year XXXX.

    (a) 260°; 1.8 knots
    (b) 080°; 1.8 knots
    (c) 080°; 2.8 knots
    (d) 260°; 4.7 knots

178. Find the times and heights of the morning tides at Pittwater (60340) on Friday, 27 March, year XXXX.

    (a) 0301 = 1.48 m; 1002 = 0.55 m
    (b) 0245 = 1.4 m; 0947 = 0.6 m
    (c) 0301 = 1.4 m; 1002 = 0.4 m
    (d) 0229 = 1.48 m; 0947 = 0.55 m

179. Find the times and heights of the two high waters at Spring bay (61170) on Friday, 27 March, year XXXX.

    (a) 0223 = 1.05 m; 1655 = 0.85 m
    (b) 0201 = 0.95 m; 2118 = 0.7 m
    (c) 0201 = 0.95 m; 1645 = 0.79 m
    (d) 0956 = 1.05 m; 2118 = 0.85 m

180. Find the time and height of the pm HW at Pieman River (61320) on Friday, 27 March, year XXXX.

    (a) 1656 = 0.67 m
    (b) 0.67 m; Time = no observations
    (c) 1656 = 1.4 metres
    (d) Observations not available

181. Find the time and height of the pm LW at Waterhouse Island (60980) on Friday, 27 March, year XXXX.

    (a) 2338; Height = no observations
    (b) No pm LW
    (c) No observations for time or height
    (d) 1118 = 0.3 m

182. Find the time and height of the am HW at Gosford (60320) on Friday, 27 March, year XXXX.

    (a) 0.55 m; Time = no observations
    (b) 0.77 m; Time = no observations
    (c) 0245 = 0.77 m
    (d) 0947 = 0.6 m

183. A vessel enters Lady Barren Harbour (61090) at 2354 hours on Monday, 6 January, year XXXX. What will be the height of tide at that time?

   (a) 1.1 m
   (b) 2.6 m
   (c) 1.33 m
   (d) No observation available

184. What is the height of tide at 0600 hours <u>local</u> (daylight saving) time at Sydney on Tuesday 18 February, year XXXX? (Use form AH130)

   (a) 0.5 m
   (b) 2.2 m
   (c) 2.1 m
   (d) 1.12 m

185. At what time will the height of tide be 2.0 metres at Devonport on a falling tide in the afternoon of Monday, 6 January, year XXXX? Give your answer in standard time.

   (a) 1209 hours
   (b) 1407 hours
   (c) 1310 hours
   (d) 1353 hours

186. Find the height of tide at Pittwater (60340) at 0700 hours on Friday, 27 March, year XXXX.

   (a) 0.88 m
   (b) 1.2 m
   (c) 1.8 m
   (d) 1.1.76 m

187. At what time will the height of tide be 0.5 metres at Spring Bay (61170) on the morning of Friday, 27 March, year XXXX?

   (a) 0800 hours
   (b) 0655 hours
   (c) 0650 hours
   (d) 0740 hours

CHAPTER 17.3: ANSWERS

(Answers 1-63 are in Ch. 17.1, & 64-155 in Ch. 17.2)

156 (c) 157 (c), 158 (b), 159 (a), 160 (c),
161 (c), 162 (c), 163 (a), 164 (b), 165 (a),
166 (c), 167 (b), 168 (a), 169 (d), 170 (a),
171 (a), 172 (c), 173 (b), 174 (d), 175 (a),
176 (c), 177 (b), 178 (a), 179 (c), 180 (b),
181 (a), 182 (b), 183 (c), 184 (d), 185 (b),
186 (a), 187 (c)

Chapter 18

RADIO, SATELLITE & EPIRB COMMUNICATION

IMPORTANT: You must also study battery maintenance (in Chapters 30) for the Marine Radio examination.

WHY A MOBILE PHONE (CELL PHONE) ISN'T SUFFICIENT

Marine Radio is superior to the Cell Phone because it broadcasts emergencies to all listeners, and can be located with detection equipment in a search & rescue situation. The VHF (*very high frequency*) radio, in particular, provides a very clear line of sight communication with a typical range of 30 miles (ship-to-shore). It is immune to all but the most severe electrical interference. The HF radio is designed for global communication. However, it is susceptible to electrical interference from machinery close by and from thunderstorms. This can also cause problems receiving weak signals. The VHF and MF/HF (*medium/high frequency*) radios equipped with the digital selective calling device are more efficient, as discussed below.

SOLAS *(Safety of Life at Sea)* SEA AREAS CLASSIFICATION

SEA AREA A1: Area within the VHF radio range
SEA AREA A2: Area excluding sea area A1, within the MF radio range (2 MHz)
SEA AREA A3: Area excluding sea areas A1 & A2, within the INMARSAT satellite communication range
SEA AREA A4: Area outside sea areas A1, A2 & A3

TYPES OF MARINE RADIOS

There are three types of marine radio transceivers:

- 27 MHz
- VHF (Very High Frequency) - Operating at around 150 MHz
- MF/HF (Medium & High Frequency) - referred to as HF - Operating between 2 - 16 MHz

The first two are used for short range, i.e., for line of sight communication. Most boats are fitted with a VHF radio. The MF/HF radio is necessary on sea-going vessels for long-range communication. *The INMARSAT Satellite telecommunication system is discussed later in this chapter.*

27 MHz RADIO

In spite of its widespread use among small vessels, the 27-MHz radiotelephone is regarded as a "hobby" marine radio. Its use in emergency is restricted to volunteer coast stations and some small vessels.

AMSA Radio network, Coast Radio Network and commercial vessels are usually not fitted with this type of radio. They are therefore unable to hear calls broadcast on 27 MHz.

Among the three types of marine radios, this one is most affected by induced noise such as from engines and thunderstorms.

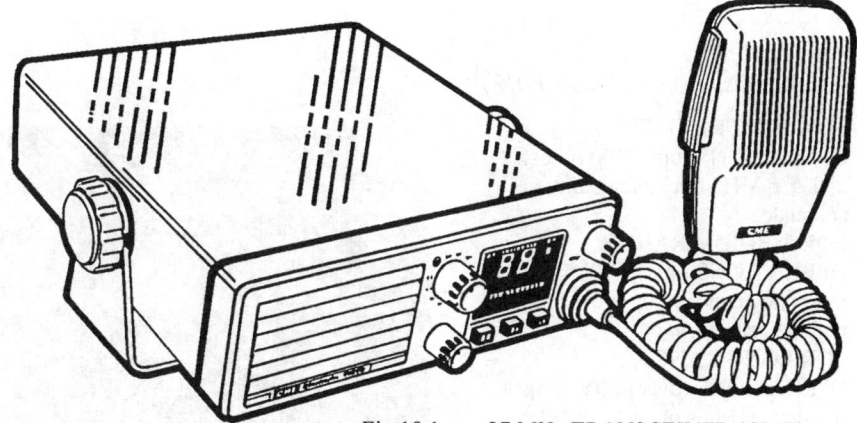

Fig 18.1: 27 MHz TRANSCEIVER (GME)

The distress and calling frequency on this radio is 27.88 MHz (known as channel 88), and the supplementary distress and calling frequency is 27.86 MHz (known as channel 86).

VHF RADIO

This is the preferred and the most commonly used marine radio in coastal waters. Among the three types of marine radios, this one is least affected by induced noise such as from engines and thunderstorms. Vessels can communicate among themselves or call a coast station when they are in the vicinity of the station. Its range is about 10 miles for communication between small vessels with low antennas and about 35 miles with high antennas. In some areas, the VHF range is extended to 100 or so miles by the use of VHF marine repeaters positioned at high altitude.

Channel 16 is the VHF calling and distress radiotelephony channel. The supplementary calling and distress channel is 67. Harbour control authorities usually monitor movements of ships entering and leaving ports on Channel 12 or 13. Channel 70 is reserved for Digital Selective Calling (*discussed later in this chapter*).

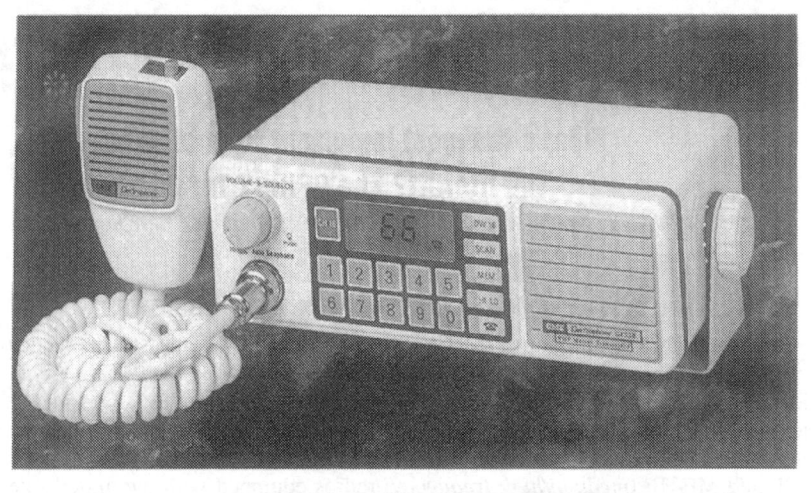

Fig 18.2: VHF TRANSCEIVER [GME]

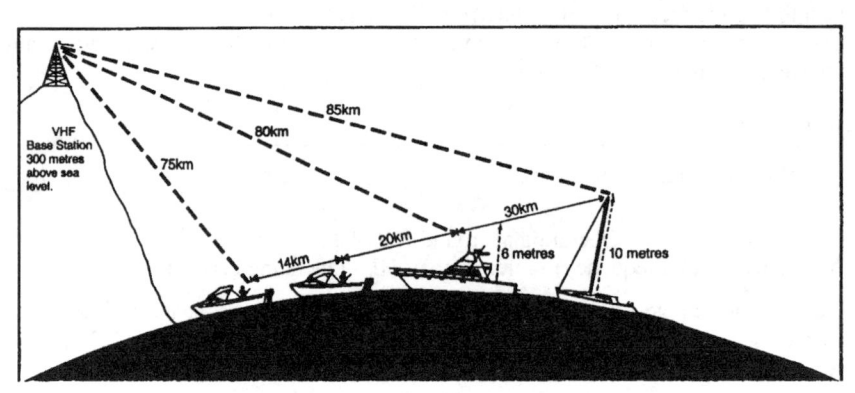

Fig 18.3: TYPICAL EFFECTIVE VHF RADIO RANGES (Telstra)

HF RADIO *(Also referred to as MF/HF)*

- ❖ FOR SHORT RANGE OPERATIONS (up to 300 miles): 2, 4 & 6 MHz frequency bands are adequate.
- ❖ FOR MEDIUM RANGE OPERATIONS (up to 600 miles): frequency bands of up to 8 MHz may be necessary.
- ❖ FOR LONG RANGE OPERATIONS (over 600 miles), frequency bands of up to 16 MHz may be necessary.

On HF radio, it is advisable to use low frequencies for communication where possible. However, during the day a higher frequency may have to be

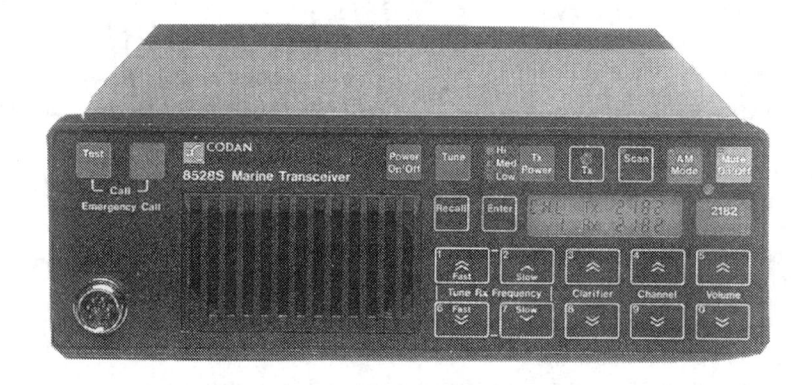

Fig 18.4: HF TRANSCEIVER [CODEN]

Chapter 18: Radio, Satellite & Epirb Communication

used in order to cope with the sun altering the ionospheric characteristics. A rule of thumb is that the higher the sun, the higher the frequency to use.

These days all HF radios transmit in the SINGLE SIDE BAND (SSB) mode. It means the carrier waves carrying the transmitted signals pick up less atmospheric interference. SSB is an option offered on some 27 MHz radios, but it should not be used on Channel 88.

IMPORTANT RADIOTELEPHONY FREQUENCIES

VHF   Channel 16   - For Distress & Safety Traffic - *On-scene communication*
    [Additional Supplementary Channel 67, for broadcasting maritime safety information]

HF   2182 kHz (or 2.182 MHz)   - For Distress & Safety Traffic - *On-scene communication*
    4125 kHz (or 4.125 MHz)   - For Distress & Safety Traffic - *Medium to long range communication*
    6215 kHz (or 6.215 MHz)   -   ditto
    8291 kHz (or 8.291 MHz)   -   ditto
    [Additional Supplementary frequency 8176, for broadcasting maritime safety information]
    12290 kHz (or 12.290 MHz) - For Distress & Safety Traffic - *Long-range communication*
    16420 kHz (or 16.420 MHz) -   ditto

RE-DESIGNATION OF HF FREQUENCIES

The above frequencies have been internationally re-designated from "Distress & Calling" to "Distress & Traffic". This is to discourage their use for calling for the purpose of establishing routine communication. Supplementary frequencies and channels should be used for the latter as well as for traffic reports and radio checks.

SUPPLEMENTARY FREQUENCIES

In addition to the Supplementary frequencies mentioned above on the VHF and 8 MHz band, the Supplementary frequencies of 2201, 4426, 6507 & 12365 kHz on the 2, 4, 6 & 12 MHz bands are also assigned for broadcast of weather and maritime safety information and to establish routine calls.

MARINE RADIO OPERATOR'S CERTIFICATE

Operators of VHF or MF/HF radio are required to hold one of the following licences:

- Third Class Commercial Marine Radio Operators Certificate
- Restricted Radiotelephone Operators Certificate of Proficiency (**RROCP**) for Radiotelephony, issued prior to 1 February 1999.
- Marine Radio Operators Certificate of Proficiency (**MROCP**), introduced in 1999 to include the use of Digital Selective Calling *(DSC is discussed in this chapter)*.
- Marine Radio Operators VHF Certificate of Proficiency (**MROVCP**), introduced in 1999 to include the use of DSC. (Operators on vessels fitted with VHF only may choose to obtain this limited certificate).

Operators of 27 MHz radio are not required to be qualified. However, a MROVCP is recommended.

Operators of the satellite telecommunication equipment INMARSAT A, B or C require one of the above certificates, plus the Marine Satellite Communication Certificate of Endorsement. However, the operators of INMARSAT-C equipment capable of being used <u>only</u> to support a Vessel Monitoring System (VMS) are exempt from obtaining this qualification. *(INMARSAT and VMS are discussed in this chapter)*. Where the peripheral equipment is fitted allowing operator control over priority, address and content of transmitted messages, then the station must be controlled by a person holding the Satellite Endorsement. The Endorsement qualification is not required for operators using the INMARSAT-M equipment.

While holders of the pre-1999 RROCP and the earlier 3rd Class Commercial Operators Certificates are considered qualified (for life) to operate small vessel marine radio equipment, including that with DSC facilities, the Australian Communications Authority (ACA) recommends that they re-qualify for one of the new certificates (MROCP or MROVCP).

The Marine Radio Operator's Examinations are conducted and certificates are issued by the Launceston-based Australian Maritime College on behalf of ACA. Training providers and examination centres are spread throughout Australia at TAFE colleges, private boating colleges, yacht clubs & volunteer marine rescue groups. Examinations are based on the knowledge contained in the Marine Radio Operators Handbook (for small vessels). It is published by the Australian Maritime College and is available from boat bookshops and training providers. The handbook is required to be retained on board at all times. However, the information in this chapter forms the bulk of the examination knowledge.

RADIO EQUIPMENT LICENCE & CALLSIGN

The ACA's Ship Station Licence is required only for the MF/HF equipment. Vessels intending to use such equipment must apply for a Class B Ship Station Licence. It assigns the vessel a unique callsign (like a car number plate), which together with her name, becomes her unique *STATION IDENTIFICATION*. ACA can be contacted on the website listed on the last page of this book. Callsigns for VHF radio can be requested from ACA.

RADIO PROPAGATION

Radio energy travels through space in two different ways: by Ground Waves and Sky Waves. Under normal circumstances, VHF and 27 MHz communication is by Ground waves only, which is over a short distance, approximately equal to the combined line of sight distance of each station. However, abnormal circumstances, known as DUCTING, particularly during summer months, may temporarily allow VHF communication over many hundreds or thousands of miles. Ducting is discussed under Radar in chapter 16. Abnormal atmospheric conditions may also cause 27MHz radio energy to travel via Sky waves, which too would temporarily allow long-range communication. This situation is popularly referred to as SKIP. Skipping is discussed later in this chapter.

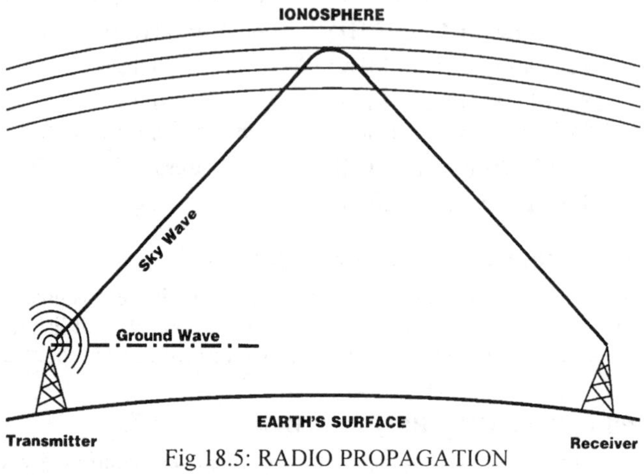

Fig 18.5: RADIO PROPAGATION

HF radio propagation is always by Ground as well as Sky waves, making it suitable for short and long-range communication.

COAST RADIO STATIONS (MARITIME COMMUNICATION STATIONS)

There are three types of Maritime Communication Stations:

1. AMSA RADIO NETWORK
2. COAST RADIO NETWORK (CRN)
3. VOLUNTEER (LIMITED) COAST STATIONS – Volunteer Marine Rescue (VMR), etc.

1. AMSA RADIO NETWORK

It consists of <u>two stations</u>, located at Charleville (Qld) and Wiluna (WA). These Commonwealth-run stations meet Australia's international Safety of Life at Sea <u>(SOLAS) obligation (as Sea Area A3) for ships over "300 gross registered tonnes"</u>. They provide the following automated long-range services:

➤ HF DSC and INMARSAT Satellite communication (They do not maintain aural listening watch. Radiotelephone communication can be conducted only after DSC alert. DSC & Inmarsat are discussed below)
➤ Bureau of Meteorology High Seas and Coastal Weather forecasts.

The inland sites of these stations are regarded strategic and electronically quiet. The stations are controlled from a Network Control Centre located within AusSAR (Australian Search & Rescue Centre) in Canberra. The network is designed to sense the power of a transmission and respond from the appropriate station. One MMSI and callsign can be used to access the network from anywhere within the Australian search and rescue area of responsibility.

The network callsigns are:
'RCC Australia' (RCC – Rescue Coordination Centre at AusSAR)
VIC
MMSI: 005030001 (Maritime Mobile Service Identity – discussed below)

WEATHER BROADCAST FROM AMSA NETWORK

See the last two pages of this book

Chapter 18: Radio, Satellite & Epirb Communication

2. COAST RADIO NETWORK (CRN)

It consists of <u>nine</u> State-run HF Coast Radio Stations, located at Perth, Port Hedland, Darwin, Cairns, Gladstone, Sydney, Melbourne, Hobart and Adelaide. Some of these operate from the Port and Harbour Control Towers and some from the Water Police Coordination Centres.

They provide the following service:

- 24-hour, 7-day monitoring of International Distress & Safety Radiotelephony Traffic on frequencies of 4125, 6215 & 8291 kHz for distress and emergency calls from vessels. They do not monitor the frequency of 2182 kHz.
- VHF Channel 16 optionally monitored by some stations.
- HF DSC and Inmarsat Service optionally provided by some stations.
- Safety (SECURITÉ) broadcasts as necessary on 4125, 6215 & 8291 kHz, and optionally on VHF Channel 16/67.
- Maritime Safety Information broadcasts on 8176 kHz twice a day for the Station's AUSCOAST area and the adjacent areas.
- Local weather broadcasts on VHF channel 16/67 (optional, as per locally published data)
- The Station ID of all 9 Stations is "COAST RADIO" – Coast Radio Sydney, Coast Radio Perth, etc.

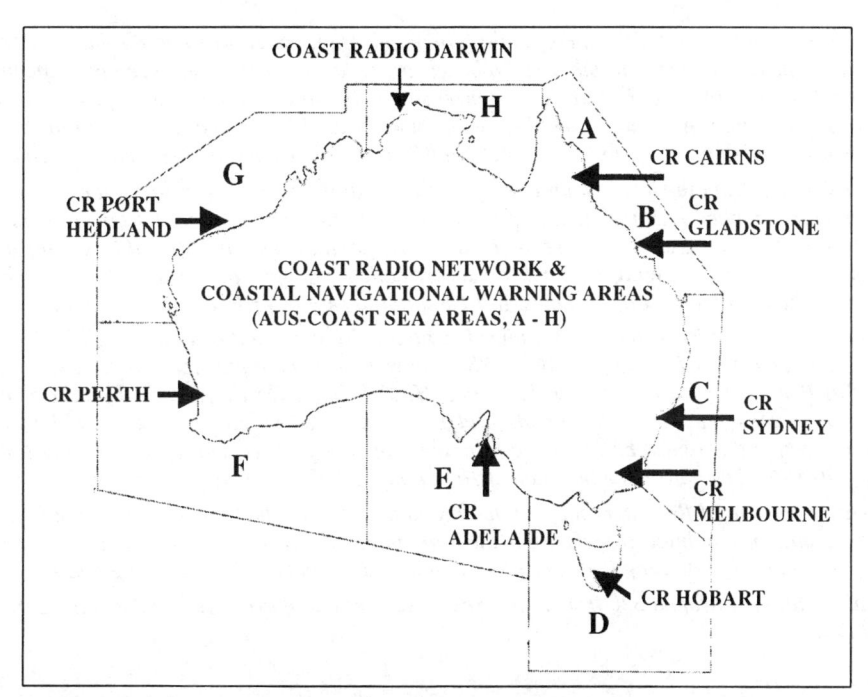

Fig 18.6: MARITIME SAFETY INFORMATION (MSI) BROADCAST AREAS

MSI BROADCAST SCHEDULE BY CRN ON 8176 KHZ
(See also NAVAREA X & its MAP in Chapter 12)

STATION	BROADCAST 1	BROADCAST 2	AUSCOAST AREA
Adelaide	0357 UTC	0757 UTC	D, E, F
Cairns	2357 UTC	1257 UTC	H, A, B
Darwin	0157 UTC	0957 UTC	G, H, A
Gladstone	2257 UTC	1157 UTC	A, B, C
Hobart	0557 UTC		C, D, E
Melbourne	0257 UTC	2157 UTC	C, D, E
Perth	0657 UTC	1057 UTC	E, F, G
Port Hedland	0457 UTC	0857 UTC	F, G, H
Sydney	0057 UTC	1357 UTC	B, C, D

It is estimated that a vessel within 200 miles of the Australian coast making an emergency call on 4125, 6215 or 8291 kHz should be heard by at least two Coast Radio Network stations.

3. **VOLUNTEER COAST STATIONS (LIMITED COAST STATIONS)**

    These stations are situated all over coastal Australia and usually provide:

    - Distress and emergency radiotelephony services on HF, VHF and/or 27 MHz.
    - Weather forecasts
    - Position Reporting and Radio Checks (Vessels are encouraged to use VHF REPEATER CHANNELS 21, 22, 80, 81 & 82 *(see below)* and HF Supplementary Channels for this purpose. The Primary Distress & Safety Traffic Channels should not be used for these services.)

    <u>*VHF REPEATER TOWERS or STATIONS*</u>, *operating on one of the "INTERNATIONAL DUPLEX" VHF CHANNELS (21, 22, 80, 81 or 82), are generally installed and maintained by volunteer rescue groups, such as AVCGA, VMR & RVCP. These channels being duplex, the towers are able to simultaneously receive and transmit (pass-on) messages between ship and shore or ship to ship. Duplex channels are designed to receive and transmit on different frequencies. Channel 16, on the other hand, is a Simplex channel. It receives and transmits on the same frequency.*

    The Repeaters increase the VHF 'line-of-sight' communication range to some 80 miles. Many of the towers are networked, which when combined with other radio bases, cover large areas. There are nine repeater channels between Sydney and the Queensland border, and 30 in Queensland, operated by over 20 repeater towers. The switching-repeater network in South Australia, together with other radio bases, covers the area from Ceduna to Port MacDonnell, Kangaroo Island and all places in between. Tasmanian waters too are well covered by Repeaters.

    You should find out from the local volunteer group the position of the closest Repeater to your operating area, the appropriate channel number and the hours of monitoring. Some are monitored from 7am to 7pm with seasonal variability. When making a passage, find out the location at which the channel should be changed to switch over from one station to the next. On changing channel, do a test call, reset the squelch and have the change logged by the station. To ensure the use of duplex frequencies, it is essential to set the VHF unit to "International".

    Two primary uses of the Repeaters are weather broadcasts by shore stations and logging of aural Position Reports by boats. (Vessels fitted with HF DSC or INMARSAT may report their positions to AUSREP via the AMSA Radio Network. AUSREP is discussed in Chapter 19.3, and HF DSC & INMARSAT are discussed below.) The Repeater towers are often located on remote headlands and powered by solar batteries. To avoid the drain on batteries, communication through them must be restricted to matters of safety, movement of vessels and test calls. The Distress and Urgency calls should also be made on channel 16 to notify other vessels in the vicinity.

    To discourage the use of Repeaters for ship-to-ship chit chat, they have a 30-second time-out facility This means that after 30 seconds of continual transmission they automatically switch off for a couple of seconds. They can be reactivated by momentarily releasing the transmit key on the radio and then transmitting again.

    An annual donation of around $20 from repeater users is greatly appreciated by the volunteers who maintain and man the facilities.

VHF SERVICE - SUMMARY

The VHF radiotelephone has been chosen as the primary marine radio for small vessels in Australia. The use of 27MHz is discouraged. The Australia's populated coastline has been divided into VHF-capable districts, covering sea areas adjacent to Townsville, Fraser Island to Tweed Coast, Newcastle to Nowra, Melbourne/Port Phillip Bay, Perth and Darwin. The *VHF network does not cover remote regions, such as the Kimberley and the western side of the Gulf of Carpentaria. Boats wanting to communicate from these regions have to use a Satellite phone.*

Each district is covered by a chain of limited coast stations, which is either automated or manned by volunteers, and backed up an automated "on-line" 24-hour VHF base station. Thus, a limited coast station, choosing to close its service at the end of the day, merely switches the radio over to its 24-hour district station, which may be, say, 150 miles up or down the coast with its antenna or a repeater (an unmanned automated relay base) located high on a coastal hill.

The 24-hour base station in effect becomes a Coast Station-cum-Emergency Centre. Its operators can call on State-run Search and Rescue organisations (Water Police, Volunteer Services, etc) to deal with maritime emergencies within their district. If necessary, they are also able to access the full range of safety and distress resources of the Canberra-based search and rescue body, AusSAR.

The local knowledge of the Base Stations thus saves Canberra-based AusSAR from having to respond to calls arising due to false alarms, engine breakdowns, flat batteries, running out of fuel or local medical emergencies.

Chapter 18: Radio, Satellite & Epirb Communication

NOTES

(A) The Australian **RADPHONE** (HF Phone Calls) service is now extinct, and **SEAPHONE** (VHF Phone Calls) service may or may not continue. The need for public correspondence with vessels is increasingly being met by Satellite Phones. This has eliminated the need for broadcasts of Traffic Lists by coast stations, which consisted of periodic announcement of names of vessels for whom there were messages or phone calls.

(B) As a result of international discussion, it is expected that VHF aural watchkeeping by GMDSS ships will continue indefinitely, and not end in 2005, as reported earlier. This is considered necessary because small vessels in developing countries may not be able to transmit Digital Selective Calling (DSC) distress alerts in the foreseeable future. The aural VHF distress and emergency calls from small vessels in Australia should therefore continue being heard by passing ships.

RADIO EQUIPMENT REQUIREMENT FOR NON-SOLAS VESSELS *("under 300 T") (AMSA Marine Order 27)*

They must carry (a) two independent means of transmitting distress (e.g., radio & Epirb), (b) the means to carry out follow-up communication, (c) the means to transmit locating signal and (d) the means to receive maritime safety information.

RADIO COMMUNICATION RULES & PROCEDURES

RADIO WATCH

Vessels are required to keep their radiotelephones switched on from the time of leaving "port" until arriving back. They should report the opening and closing of the ship station on an appropriate calling channel to the nearest volunteer coast station. The watch must be kept on the appropriate calling/distress frequencies, as follows:

- Channel 88 on 27 MHz radio.
- Channel 16 on VHF radio.
- 2182, 4125, 6215 & 8291 kHz (scanned) on HF radio

You may be excused from radiotelephony watchkeeping if it interferes with safe navigation or when communicating on another frequency.

WHEN NOT TO USE SHIP'S RADIO
- During the silence periods. *(See Fig 18.7)*
- When a distress, urgency or a safety message is being transmitted.
- Vessels at anchor or in a port should not use the ship's radio for making routine calls unless access to a telephone is not possible.
- Due to the risk of explosion, do not make radio transmissions when loading fuel or when loading or discharging flammable cargo.

CONTROL OF WORKING FREQUENCIES:
- Coast Stations control the working frequencies when communicating with Ship Stations.
- A vessel in distress controls the working frequencies when communicating with another station.
- The Ship Station called controls the working frequencies when communicating with another Ship Station.

MINIMISE RADIO INTERFERENCE
- Listen before transmitting. Make sure the channel is clear before using it.
- Use minimum transmitting power. (See Transceiver Controls, discussed below)
- Use the frequency nominated for the purpose.
- Keep test signals to a minimum.
- Observe internationally nominated silence periods.

MINIMISE ELECTRICAL INTERFERENCE ON RADIO
- Use engine suppressors. It is mainly the HF radios that are susceptible to electrical interference from close by machinery.
- Use insulators.
- Solder all connections. Connections made only by twisting wires corrode.
- Keep cables leading to radio away from other electrics.
- Turn off fluorescent lights.

OBSERVE SILENCE PERIODS: With the exception of distress calls and messages, all transmissions on all distress and calling frequencies must cease for 3 minutes every half hour. This is to give a vessel in distress with a weak signal or in a busy period a chance to be heard.

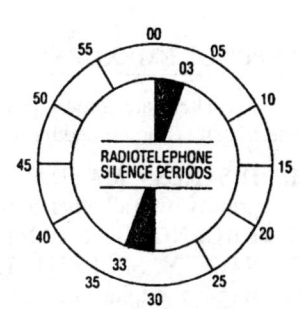

Fig 18.7

HOW TO MAKE A ROUTINE CALL

EXAMPLE:
(Call preferably on a Repeater Channel, or VHF Channel 16, HF 2182 kHz etc.)
COAST GUARD SYDNEY (spoken 3 times)
THIS IS
VANITY VVL 7232 (spoken 3 times)
CALLING ON CHANNEL 16 (or...)
OVER

Coast Guard Sydney will reply:
VANITY THIS IS COAST GUARD SYDNEY - GO TO CHANNEL .....
(Both stations switch to the nominated working frequency.)
Vanity initiates traffic on the working frequency by:
COAST GUARD SYDNEY THIS IS VANITY ON CHANNEL ......
(Traffic follows)

THE PHONETIC ALPHABET: When it is necessary to spell out words, the phonetic pronunciation shown in figure 18.13 is recommended.

CONFIDENTIALITY OF MESSAGES: Under the International Radio Regulations, the operator must preserve secrecy of communication in messages handled on behalf of other persons. This restriction does not apply to distress, urgency or safety messages.

AUSTRALIAN SHIP REPORTING (AUSREP) SYSTEM: See Chapter 19.3.

RADIO DOCUMENTS REQUIRED TO BE KEPT ON BOARD AT ALL TIMES

1. Radio operator's certificate
2. Ship station licence (the radio equipment licence, if applicable)
3. Radio Operators Handbook (available from Australian Communications Authority)
4. Radio log book (a suitable note book)

RADIO LOG BOOK: The legal requirement for log keeping extends only to distress messages. The operator must record all distress messages received and the help given by your vessel or another station, whether physical or by relay of such messages. Naturally, to comply with the above requirement your ship station's opening & closing times must be logged.

Log keeping on operational matters is at the operator's discretion, except where required under the Commonwealth or State legislation. The following format for a radio log book is suggested in ACA's handbook:

Date and Time	Station/MMSI from	Station/MMSI to	Details of Calls, Signals & Distress Working	Frequency/Channel

Fig 18.8: SUGGESTED FORMAT FOR RADIOTELEPHONE LOG BOOK PAGE

TYPES OF EMERGENCY TRANSMISSIONS

There are three types of maritime emergencies. It is essential that skippers are fully conversant with the meaning and transmission procedure for each:

1. **DISTRESS**: MAYDAY indicates grave and imminent danger to vessel (i.e., lives on vessel). It commands absolute priority over all other communications.
2. **URGENCY**: PAN PAN indicates a threat to the safety of a vessel or a person.
3. **SAFETY**: SECURITÉ The signal indicates that the station transmitting has an important navigational or weather warning to transmit.

Chapter 18: Radio, Satellite & Epirb Communication

SHOULD MAYDAY OR PAN PAN ... BE USED?

This question is answered by the AusSAR as follows:

"In both cases we will assume that a risk to life exists, but the key to the answer is whether the ship is in serious and imminent danger. If it is NOW then make a Mayday call; if it is LATER then make it a pan-pan call. Perhaps a couple of examples will assist:

(a) You are drifting helplessly in bad weather a couple of miles off a rocky lee shore ... MAYDAY.
(b) You are drifting helplessly in bad weather about 20 miles off a rocky lee shore ... PAN-PAN.

In other words, are you going to be dashed on the rocks within the hour or in about ten hours time? NOW or LATER?

In many cases brought to our attention the circumstances did not warrant the use of either Mayday or Pan-Pan; a routine call would have been sufficient. For example, the skipper of a craft which has broken down in good weather in a sheltered area should seriously consider making a routine call for assistance and anchoring if necessary, and not immediately start calling Mayday. Don't forget that a Mayday call indicates that you are threatened by grave and imminent danger and the scenario above hardly meets this criteria."

RADIOTELEPHONY ALARM SIGNAL: Some HF transceivers are fitted with an alarm unit to transmit a distinctive 2-tone alarm signal on 2182 kHz. When activated, it un-mutes the 2182-watchkeeping receivers on other vessels in the vicinity fitted with such alarm receivers. Its purpose is to attract the attention of ships and coast stations to the distress message that is to follow. Some Coast stations also use this alarm signal before transmitting distress calls and messages, and at times, before sending a safety message concerning an urgent cyclone warning. *(NOTE: On GMDSS vessels the radiotelephony alarm signal has been replaced by Digital Selective Calling (discussed below).*

MASTER'S AUTHORITY: Radios on vessels (i.e., ship stations) operate under the authority of the master or the person responsible for the safety of the vessel. Distress calls and messages can only be sent on the authority of the master of the vessel.

LEGAL DUTY OF MASTER WITH REGARD TO DISTRESS: It is the statutory duty of every vessel to:

- Maintain radio watch on a distress frequency.
- Assist a vessel in distress. The vessel may not be in a position to give physical assistance, but she may be able to assist by radio as follows:
  - Communicate with the vessel in distress.
  - Relay the distress message.

DISTRESS COMMUNICATION

DISTRESS CALL - PROCEDURE

MAKE SURE YOU ARE ON A DISTRESS CHANNEL. (Listed above)
ACTIVATE RADIO TELEPHONE ALARM SIGNAL IF FITTED.

Distress signal & call:
    MAYDAY MAYDAY MAYDAY
    THIS IS ...   *(Your vessel's name & call sign spoken 3 times)*
    MAYDAY ... *(Your vessel's name & call sign spoken once)*

Distress message:
    GIVE YOUR POSITION *(Range & bearing or Latitude & Longitude).*
    DESCRIBE THE NATURE OF DISTRESS & SPECIFY THE HELP REQUIRED.
    SUPPLY OTHER INFORMATION TO ASSIST YOUR RESCUE.

Example of distress message:
    15 MILES NORTHWEST OF ROTTENEST ISLAND LIGHT.
    ON FIRE; 4 PERSONS ABANDONING SHIP INTO A LIFERAFT; ALL WEARING LIFEJACKETS;
    EPIRB/GPIRB ACTIVATED; HAVE HAND FLARES.
    REQUIRE IMMEDIATE ASSISTANCE.

Say: OVER

Then: LISTEN ON THE SAME FREQUENCY FOR A REPLY. (If a reply not received, repeat the signal and message. Try during a silence period and/or switch to another channel.)

DISTRESS CALL FROM A LIFERAFT OR LIFEBOAT: When transmitting a distress call from a lifeboat or a liferaft, say "Number 22" after the vessel's call sign. This indicates that you have abandoned the vessel and taken to the raft or boat.

ACTION ON RECEIPT OF A DISTRESS OR URGENCY MESSAGE

1. Immediately acknowledge it if you are in the immediate vicinity.
2. Defer acknowledgement for a short interval if reliable communication between the distressed vessel and a Maritime Communication Station is evident. This is to allow the MCS to acknowledge it.
3. Follow step 2 if distress ship is a long distance away. This is to allow acknowledgement by a ship closer to her.
4. Acknowledge a distress call that has not been acknowledged by any other station. Relay the message if you can't provide assistance.

DISTRESS MESSAGE ACKNOWLEDGEMENT: Allow a few seconds for a Maritime Communication Station to acknowledge, and then acknowledge. Write down the message as received. Consider your plan of action if waiting for another station to acknowledge.

Form of acknowledgement:

MAYDAY
STELLA VL3245 STELLA VL3245 STELLA VL3245
THIS IS
TURNPIKE VM4608 (spoken 3 times)
RECEIVED MAYDAY
IN POSITION ....
PROCEEDING AT 10 KNOTS
ESTIMATE AT YOUR POSITION IN ONE HOUR THIRTY MINUTES
OVER

IF DISTRESS MESSAGE IS HEARD & ANOTHER VESSEL IS ASSISTING: Log the details and continue to monitor the situation.

MAYDAY RELAY

(Transmission of distress message by a station not itself in distress)

Vessel "A", not herself in distress, may transmit a distress message for vessel "B" which is in distress, if:

- "B" is unable to transmit.
- "A" believes that "B" needs further assistance.
- "A" has heard the distress message of "B" that has not been acknowledged and "A" herself is unable to assist.

DISTRESS RELAY - PROCEDURE

MAKE SURE YOU ARE ON A DISTRESS CHANNEL.

ACTIVATE RADIO TELEPHONE ALARM SIGNAL IF FITTED.

Say: MAYDAY RELAY MAYDAY RELAY MAYDAY RELAY
THIS IS ... (Vessel A's name & call sign spoken 3 times)
MAYDAY... (Vessel B's name & call sign spoken once)

Then: GIVE VESSEL B's POSITION.
DESCRIBE THE NATURE OF B's DISTRESS AND HELP REQUIRED BY B.
SUPPLY OTHER INFORMATION TO ASSIST B's RESCUE.

Say: OVER

NOTE: A ship should not acknowledge receipt of a Mayday Relay message transmitted by a coast station unless definitely in a position to provide assistance.

RADIO INTERFERENCE FROM OTHER VESSELS DURING DISTRESS

The ship in distress or the station in control can impose radio silence on any or all stations interfering with distress traffic by sending the instruction **SEELONCE MAYDAY**.

"Seelonce" means silence. This instruction must not be used by any station other than the ship in distress or the station controlling the distress traffic. If another station near the distressed ship believes that silence is necessary it should use the instruction **SEELONCE DISTRESS** followed by its own name and call sign.

RADIO SIGNAL TO INDICATE THAT MAYDAY IS OVER

MAYDAY
HELLO ALL STATIONS (or CQ spoken as CHARLIE QUEBEC) (spoken 3 times)
THIS IS
NAME & CALL SIGN OF THE STATION SENDING THE MESSAGE.
DISTRESS MESSAGE OF ... (time of origin)
FOR ...   (name & call sign of ship in distress)
SEELONCE FEENEE
NOTES:
  (a) Feenee means finished.
  (b) The international code letters CQ (the radiotelegraphy general call, Charlie Quebec) are some times used (spoken 3 times) instead of "Hello All Ships".
  (c) If the vessel in distress is not yet finally out of danger, but help is at hand and complete radio silence is no longer necessary, then the word PRUDONCE should be used instead of SEELONCE FEENEE.

URGENCY SIGNAL (PAN PAN) & MESSAGE

(It may be addressed to all stations or to a particular station)

Urgency signal & call (on a distress frequency):

  PAN PAN, PAN PAN, PAN PAN
  HELLO ALL STATIONS (spoken 3 times)
  THIS IS ... (Your vessel's name & call sign spoken 3 times)

Urgency message:

  (There is no need to change to a working frequency unless it is a lengthy message or it concerns an urgent medical case. Don't forget to announce the frequency you are changing to):
  GIVE YOUR URGENT MESSAGE

Say: OVER
Then: LISTEN FOR A REPLY.

MEDICAL EMERGENCIES:

Medical emergencies are usually communicated as PANPAN messages. It may appear appropriate to send a MAYDAY, but it is unnecessary and perhaps unwise to shout "help" to everyone for a specific medical advice. However, if unable to raise a Maritime Communication Station and when surrounded by people who do not understand PANPAN, the use of MAYDAY followed by a request for medical assistance may be the only choice.

Where necessary, Coast Radio Stations are able to arrange for a vessel to speak directly with a doctor, usually from the Royal Flying Doctor Service.

SAFETY SIGNAL & MESSAGE
(SECURITÉ pronounced SA-CURE-E-TAY)

(It is always addressed to all stations.)

Safety signal & call (on a distress frequency):

  SA-CURE-E-TAY, SA-CURE-E-TAY, SA-CURE-E-TAY
  HELLO ALL STATIONS (spoken 3 times)
  THIS IS ...   (Your vessel's name & call sign spoken 3 times)
  NAVIGATION WARNING, LISTEN ON 8176 kHz (or VHF Ch. 67)

Safety message (after changing channel):

  SECURITÉ (spoken 3 times)
  HELLO ALL STATIONS (spoken 3 times)
  THIS IS
  SHIP'S NAME & CALL SIGN
  A LARGE TREE TRUNK SIGHTED FLOATING IN POSITION...
  OUT

NOTE: The coast radio stations transmit safety messages (if any) at the end of the Silence Periods.

GMDSS & DSC

GLOBAL MARITIME DISTRESS AND SAFETY SYSTEM (GMDSS) & DIGITAL SELECTIVE CALLING (DSC)

The Global Maritime Distress and Safety System (GMDSS) is designed to ensure the quickest and widest global coverage of Distress and Urgency alerts and the reception of Maritime Safety Information (MSI), i.e., navigational and meteorological information. It can also be used for commercial communication.

GMDSS commenced in 1992 and came into full effect worldwide in February 1999. It uses modern satellite and terrestrial technology in the form of EPIRBs/GPIRBs, SARTs, INMARSAT and digital selective calling (DSC) techniques on HF and VHF bands. DSC-fitted VHF and HF sets are available.

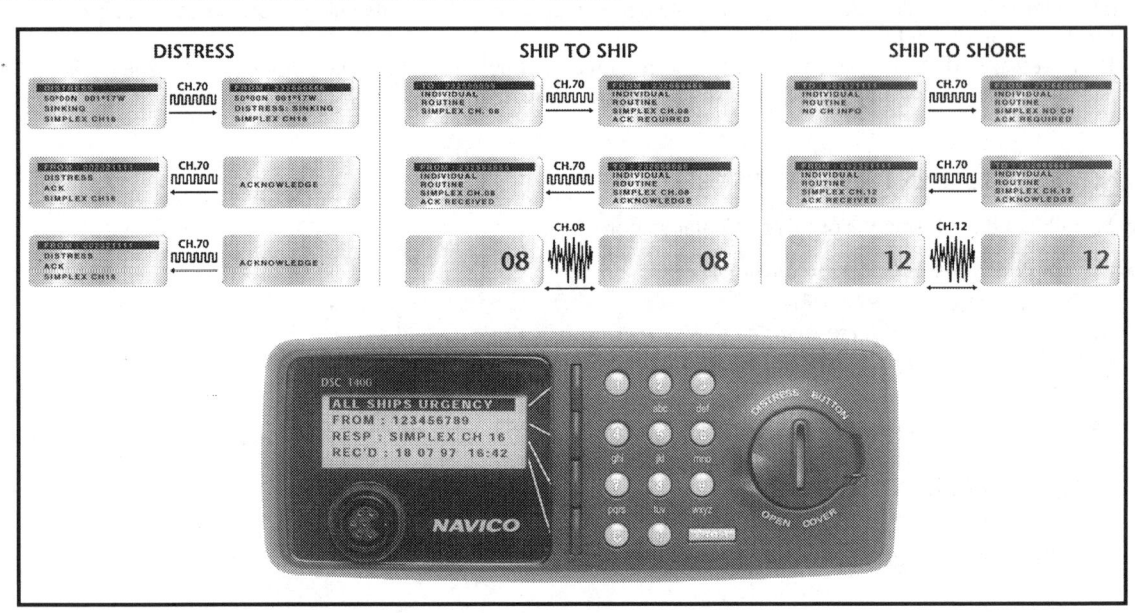

Fig 18.9: EXAMPLES OF TYPICAL DSC MESSAGES (NAVICO DSC-VHF)

The full GMDSS regulations apply only to vessels subject to the international SOLAS (Safety of Life at Sea) Convention. These are usually over 300 Gross Registered Tonnes. AMSA's obligation as the GMDSS Area 3 or "A3" has been discussed above. The State and Voluntary Radio Networks for non-SOLAS vessels have also been discussed.

DIGITAL SELECTIVE CALLING (DSC)

The DSC equipment is a HF or VHF transceiver with a display screen and a keypad on which messages can be composed and transmitted via a built-in modem. It is not necessary for vessels to keep a listening watch on their transceivers because an audio alarm, such as a beep, will sound on the vessel(s) being called.

The system works like the paging system used by doctors in hospitals or the dispatch systems used by taxis, couriers and tow trucks. It is similar to the mobile phone text messages. It is designed to establish initial contact between stations. A DSC message is a brief burst (typically 0.5 seconds on VHF and 7 seconds on HF) of digitised information transmitted from one station to another for alerting or responding to an alert. Due to its greater resistance to noise and fading, <u>a DSC signal has a greater range then a radiotelephone transmission</u>. DSC controllers (modems) have provision for interfacing to GPS for updating of position and time information.

A marine DSC signal contains the following information:

- Identity of the calling station (MMSI)
- Priority of the call - distress, urgency, safety or routine
- Station being called (a specific station or all stations)
- Frequency or channel for subsequent communication (optional)
- Vessel's position and time - by manual entry or GPS interface (for distress alerts)

The major benefits of DSC are:

- It is far less subject to interference and fading than voice transmissions. It thus provides a greater transmission range

Chapter 18: Radio, Satellite & Epirb Communication

than voice transmissions.

- It allows for automatic and semi-automatic transmission and reception of initial distress, urgency and safety calls. A single dedicated button pushed by an operator can initiate such a call.
- It removes the need for the traditional loudspeaker watchkeeping on distress and calling frequencies.
- It reduces the problem of hoax distress calls because it is impossible to transmit a DSC alert without automatic inclusion of the sender's I/D code.

The VHF DSC system operates at far greater speed than HF system. VHF operates at 1200 bps (bits of information per second) and HF at 100 bps. Information transmitted by DSC is generally known as DSC alert. The operator (sender) is able to encode manually the DSC transmission for "single station" or "all station" alerting. While every station listening on the DSC frequency will receive the alert, only the station(s) selected by the sender will hear the audible alarm and decode the message on the equipment's display screen. However, the DSC distress alerts are received and decoded by all stations. Once the contact has been established between stations, normal voice procedures are used for subsequent communication.

DSC alerts are transmitted on internationally agreed and specifically reserved frequencies, listed below. Voice transmissions are prohibited on these frequencies. For each DSC frequency there is an associated radiotelephony frequency nominated for voice communication subsequent to the DSC alert. Due to their relatively low transmission speeds (of 100 baud), the HF DSC frequencies are reserved exclusively for distress, urgency and safety alerts. If too many calls were permitted on HF channels, they would quickly become overloaded to the point where a distress call may be blocked. Therefore only the VHF channel 70 may additionally be used for routine station-to-station DSC alert. *(NOTE: DSC is not used in the 27 MHz marine band).*

DSC FREQUENCY		ASSOCIATED RADIOTELEPHONY FREQUENCY
VHF	Channel 70	Channel 16
MF/HF	2187.5 kHz	2182 kHz (2.182 MHz)
	4207.5 kHz	4125 kHz
	6312 kHz	6215 kHz
	8414.5 kHz	8291 kHz
	12577 kHz	12290 kHz
	16804.5 kHz	16420 kHz

RADIO FREQUENCIES TO MEMORISE

Candidates for radio examination are not expected to memorise every radio frequency and channel number. They must however know the significance of and remember the VHF Channels 16 and 67, and HF frequencies 2182, 4124, 6215 & 8291, as well as be able to recognize the above listed additional frequencies.

DSC IDENTIFICATION (MMSI)

Each DSC-equipped radio must be permanently programmed with a unique 9-digit identification number known as its Maritime Mobile Service Identity (MMSI). If the manufacturer has programmed MMSI into the transceiver, the operator must register it with the authorities (usually ACA). If the set has been purchased without a pre-programmed MMSI, the operator must obtain an MMSI from the authorities and programme it into the set.

MMSI is the electronic equivalent of the station's callsign, which is automatically included in all DSC transmissions. It automatically identifies the sender.

The MMSI also acts as an electronic filter on receipt of DSC alerts. It ensures that the routine DSC alerts not intended for the station, are not decoded by it. The Urgency and Safety DSC alerts may be similarly addressed to a particular station or to all stations. On receipt of a distress DSC alert, the filter is deactivated on all receiving stations.

➢ MMSI for vessels: First 3 digits: Maritime Identification Digits (MIDs) indicate the country of registration of a vessel (Australia's MID = 503). Remaining 6 digits: unique identity of the ship station.   [For example, 503123456]

➢ MMSI for coast & limited coast stations: First 2 digits: 00. Next 3 digits: country's MID (Australia's MID is 503). Remaining 4 digits: unique identity of the coast station. [For example, 005031234]

A full list of MIDs appears in Appendix 43 of the Radio Regulations published by the International Telecommunication Union (ITU).

ROUTINE DSC ALERTS FOR PUBLIC CORRESPONDENCE

Routine ship-to-ship DSC alerts are permitted on VHF DSC distress, urgency and safety channel 70, but not on MF/HF DSC distress, urgency and safety frequencies. On MF/HF, some other frequencies are specifically set aside for DSC commercial calls. The sender needs to know and program the MMSI of the desired vessel to make routine calls.

In countries where commercial DSC facility is available, routine ship to shore calls can be made via VHF or MF/HF by programming in the details of a coast station's MMSI followed by the desired telephone number.

WATCHKEEPING ON DSC FREQUENCIES

Large trading ships (GMDSS-fitted ships) are not required to maintain HF aural watchkeeping, but only the DSC watch. They are thus not able to receive HF aural distress calls. They will however continue to receive VHF aural calls.

DSC CALL OPTIONS

1. **DISTRESS ALERT**: *This is addressed to all stations. It contains vessel's MMSI, position, and possibly the nature of distress. As with the INMARSAT-C system, the user is allowed the option of either sending a pre-programmed distress message by operating a single button or composing a message with the equipment.*
   The ship's position and UTC (GMT) time are generally automatically included from a GPS interface. Under GMDSS regulations, the accuracy of this information should be verified on the DSC controller's display at least once every watch. If the DSC controller is not connected to a GPS, the vessel's position and time must be manually entered at least every two hours. A menu of nature of distress generally allows a selection such as "on fire", "collision" etc. DSC controllers without such a menu will transmit the default setting of "UNSPECIFIED".
   Once selected and initiated, a DSC distress alert continues to be automatically and randomly repeated 5 times on a single frequency unless manually stopped by the operator, or when it is acknowledged by another station and decoded by the distressed vessel.

2. **DISTRESS ALERT ACKNOWLEDGEMENT**: *It is normally sent only by* Maritime Communication Stations. *May be used by ship stations under certain circumstances.*

3. **DISTRESS RELAY ALERT**: *As per Nos. 1 & 2 above.*

4. **ALL STATIONS ALERT**: *Used prior to broadcasting an Urgency or Safety message. All stations include coast and limited coast stations.*

5. **SINGLE STATION ALERT**: *As per No. 4 above. Also used to alert a single station to a routine call, in which case the MMSI of the called station must be entered into the transceiver.*

DSC DISTRESS ALERT PROCEDURE

➢ Switch to the DSC Distress channel (VHF Channel 70, HF 2187.5 kHz etc. The HF set automatically tunes its antenna).

➢ If time permits, key in or select the nature of distress and vessel's position.

➢ Initiate a distress alert (usually by pushing an accident-proof button).

➢ Do not wait for a DSC Distress Acknowledgement from other stations. A brief pause is sufficient to allow them to receive the DSC alert.

➢ Switch to the appropriate radiotelephony channel/frequency (see list above).

➢ Transmit the standard radiotelephone distress call and message (see procedure in the earlier part of the chapter, including the two-tone radiotelephony alarm on 2182 kHz, if fitted.)

ACKNOWLEDGEMENT OF RECEIPT OF A DSC DISTRESS ALERT
(on 2187.5 kHz or VHF channel 70)

- Take note of the content of the DSC distress alert. (Do not activate DSC Acknowledgement because it will terminate transmission of the DSC distress alert from the vessel in distress).

- Immediately switch to the associated radiotelephony frequency (2182 kHz, VHF Channel 16 etc.) and listen for the MAYDAY voice message that should follow.

- If you receive the MAYDAY voice message and are able to provide assistance, then send a voice acknowledgment (RECEIVED MAYDAY) to the vessel in distress. Also advise an appropriate Maritime Communication Station.

- If you are not able to provide assistance, and you hear other stations indicating involvement in the distress situation, then do not send any acknowledgement.

Chapter 18: Radio, Satellite & Epirb Communication

- If you have received a DSC distress alert but no MAYDAY message has been heard within 5 minutes, and no other station is heard communicating with the vessel in distress, and the DSC distress alert continues to be received, then you should send a voice acknowledgment (RECEIVED MAYDAY) to the vessel in distress. Substitute her MMSI with her name and call sign when calling her. You must also contact an appropriate Maritime Communication Station and fully advise the situation.

- If, on the other hand, you have received a DSC distress alert but no MAYDAY message has been heard within 5 minutes, and no other station is heard communicating with the vessel in distress, and the DSC distress alert is NOT continuing, then you should not send an acknowledgment. You should immediately also contact an appropriate Maritime Communication Station and fully advise the situation.

- In all cases, enter details in radio log.

ACTIONS BY SHIPS UPON RECEPTION OF VHF / MF DSC DISTRESS ALERT

COMSAR/Circ.21

AMSA — Australian Maritime Safety Authority

In no case is a ship permitted to transmit a DSC distress relay call on receipt of a DSC distress alert on either VHF or MF channels.

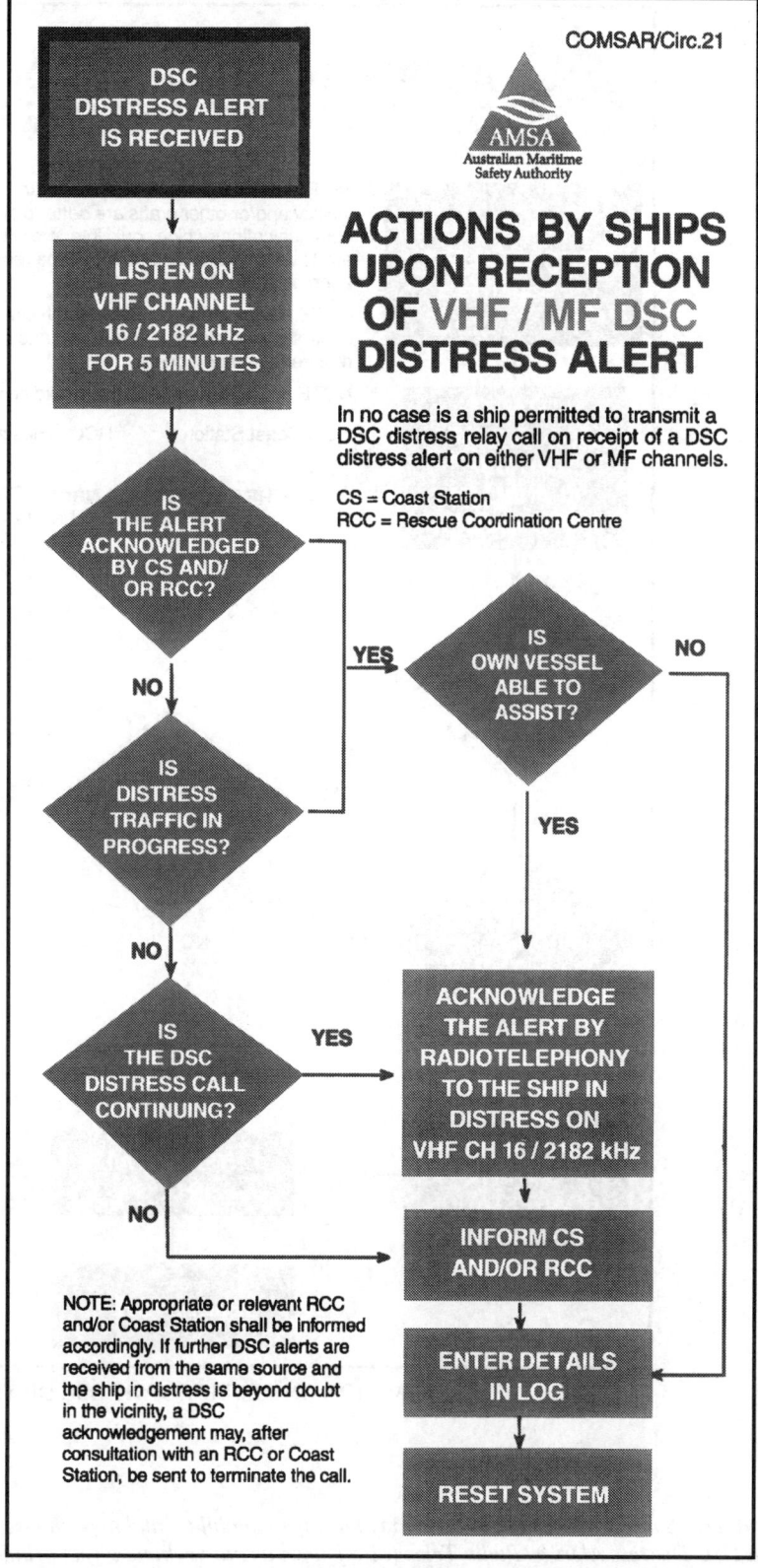

CS = Coast Station
RCC = Rescue Coordination Centre

Flowchart:
- DSC DISTRESS ALERT IS RECEIVED
- LISTEN ON VHF CHANNEL 16 / 2182 kHz FOR 5 MINUTES
- IS THE ALERT ACKNOWLEDGED BY CS AND/OR RCC?
  - YES → IS OWN VESSEL ABLE TO ASSIST?
    - NO → (end)
    - YES → ACKNOWLEDGE THE ALERT BY RADIOTELEPHONY TO THE SHIP IN DISTRESS ON VHF CH 16 / 2182 kHz
  - NO → IS DISTRESS TRAFFIC IN PROGRESS?
    - NO → IS THE DSC DISTRESS CALL CONTINUING?
      - YES → ACKNOWLEDGE THE ALERT BY RADIOTELEPHONY TO THE SHIP IN DISTRESS ON VHF CH 16 / 2182 kHz
      - NO → INFORM CS AND/OR RCC
- INFORM CS AND/OR RCC → ENTER DETAILS IN LOG → RESET SYSTEM

NOTE: Appropriate or relevant RCC and/or Coast Station shall be informed accordingly. If further DSC alerts are received from the same source and the ship in distress is beyond doubt in the vicinity, a DSC acknowledgement may, after consultation with an RCC or Coast Station, be sent to terminate the call.

Fig. 18.10: ACTION ON RECEIPT OF VHF DSC DISTRESS ALERT

ACTIONS BY SHIPS UPON RECEPTION OF HF DSC DISTRESS ALERT

COMSAR/Circ.21

AMSA Australian Maritime Safety Authority

HF DSC DISTRESS ALERT IS RECEIVED

NOTE 1: If it is clear the ship or persons in distress are not in the vicinity and/or other crafts are better placed to assist, superflous communications which could interfere with search and rescue activities are to be avoided. Details should be recorded in the appropriate logbook.

NOTE 2: The ship should establish communications with the station controlling the distress as directed and render such assistance as required and appropriate.

NOTE 3: Distress relay calls should be initiated manually.

CS = Coast Station   RCC = Rescue Coordination Centre

↓

LISTEN ON ASSOCIATED RTF OR NBDP CHANNEL(S) FOR 5 MINUTES

HF DSC RTF AND NBDP CHANNELS (kHz)

DSC	RTF	NBDP
4207.5	4125	4177.5
6312.0	6215	6268
8414.5	8291	8376.5
12577.0	12290	12520
16804.5	16420	16695

↓

IS THE ALERT ACKNOWLEDGED OR RELAYED BY CS AND/OR RCC?

— YES → **IS OWN VESSEL ABLE TO ASSIST?** — YES → **CONTACT RCC VIA MOST EFFICIENT MEDIUM TO OFFER ASSISTANCE**

NO ↓   NO ↓

IS DISTRESS COMMUNICATION IN PROGRESS ON ASSOCIATED RTF CHANNELS?

NO ↓

TRANSMIT DISTRESS RELAY ON HF TO COAST STATION AND INFORM RCC → **ENTER DETAILS IN LOG**

↓

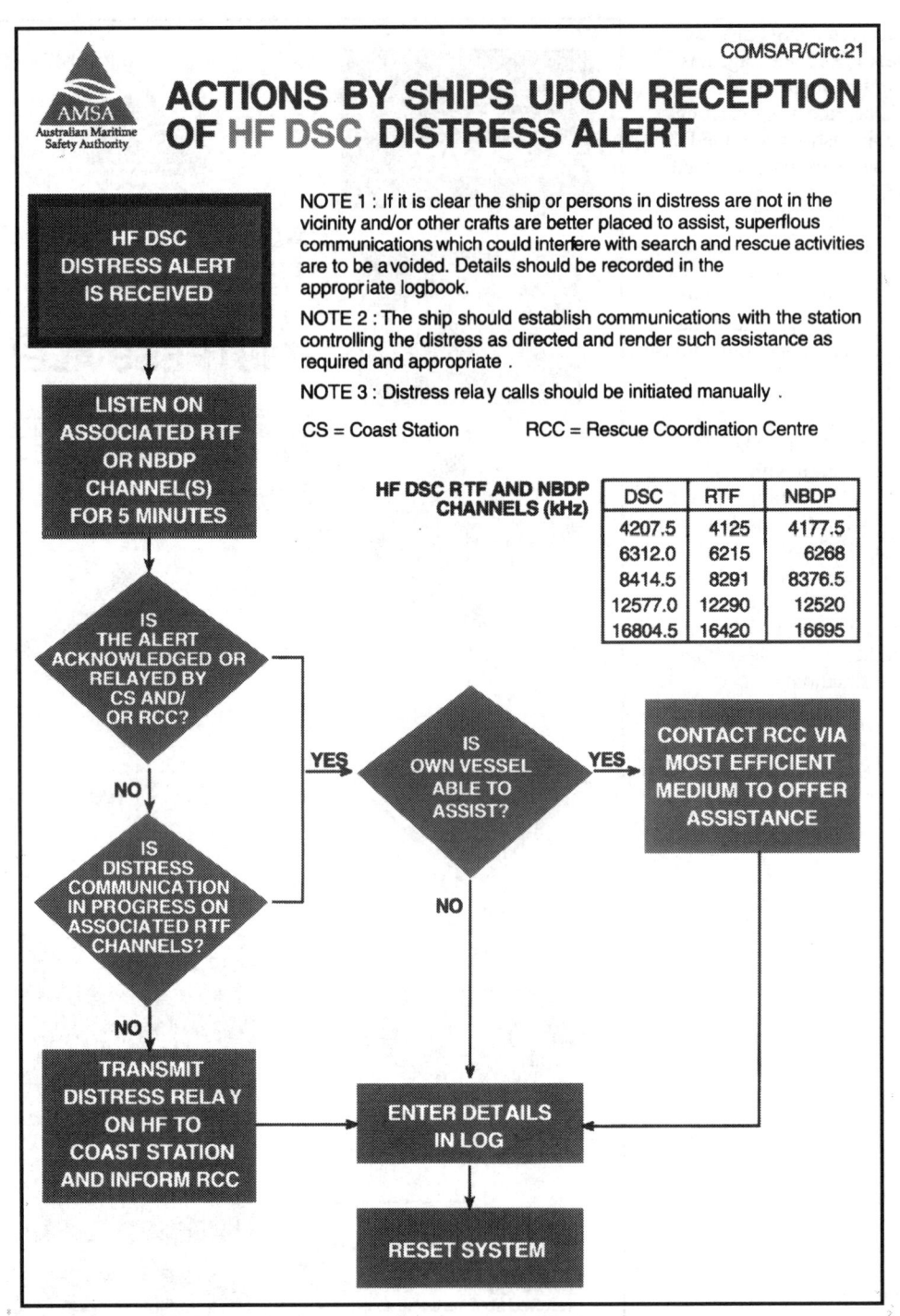

RESET SYSTEM

Fig 18.11: ACTION ON RECEIPT OF HF DSC DISTRESS ALERT

NOTE: Some large trading vessels may have the capability, and may choose, to conduct HF communication subsequent to a DSC Distress Alert by Radio Telex (also known as narrowband direct printing or NBDP) on a frequency dedicated to this use. This will be made apparent to you by a reference in the final piece of DSC information received and displayed. If it reads "J3E" then the vessel will be using radiotelephony for subsequent traffic. If it reads "F1B" then the vessel will be using radio telex. NBDP is not used on VHF.

Chapter 18: Radio, Satellite & Epirb Communication

DSC DISTRESS ALERT RELAY PROCEDURE
(Normally transmitted by a Maritime Communication Station after receiving and acknowledging a DSC Distress Alert)

- Transmission of a distress alert relay should be considered only when a DSC distress alert has been received on an MF/HF frequency of 4 MHz or above and no other station is heard communicating with the vessel in distress on the associated radiotelephony (voice) channel. The distress alert relay must be addressed to an appropriate coast or limited coast station. The distress alert relay MUST NOT BE ADDRESSED TO "ALL SHIPS".

- If you receive a DSC distress alert on either 2187.5 kHz or VHF Channel 70, you should not transmit a distress alert relay. Instead, make a radiotelephony acknowledgement to the vessel in distress on 2182 kHz or VHF Channel 16. The nearest Maritime Communication Station should also be informed.

- You may transmit a DSC distress alert relay if you have learnt that another vessel in distress is unable to transmit the distress alert and you consider that further help is necessary. In this case the DSC distress alert relay may be in the "all ships" format or, PREFERABLY, addressed to an appropriate coast or limited coast station.

- If your small vessel transceiver does not have the DSC distress alert relay facility, you may send a MAYDAY RELAY message on the associated voice channel. You must also advise to a Maritime Communication Station. In cases where a vessel's name and call sign are not known, a MMSI may be used.

NOTE: You should exercise careful judgement in relaying DSC distress alerts received on the higher frequencies as these could be received from and by vessels at distances of thousands of miles. Indiscriminate relaying will merely increase the area that stations are alerted without performing any useful function.

ACKNOWLEDGEMENT OF A DSC DISTRESS RELAY ALERT

Follow the procedure laid down for the acknowledgement of a DSC Distress Alert. *A ship should not acknowledge receipt of a Mayday Relay message transmitted by a Maritime Communication Station unless definitely in a position to provide assistance.*

CANCELLATION OF AN INADVERTENT DSC DISTRESS ALERT

- Immediately switch off the radio in question *(to cancel the automatic repeats of the DSC distress alert)*.
- Switch the radio back on and broadcast an "all station" voice message on the radiotelephony frequency/channel associated with the DSC frequency/channel on which the inadvertent alert was transmitted. Broadcast the vessel's name, call sign, MMSI and the cancellation of the distress alert with the approximate time of the inadvertent transmission.
- If, for some reason, the above procedure is not possible, then use some other means to advise authorities that the alert was accidental. *(A ship station operator is not penalised for reporting an inadvertent distress alert).*

TRANSMISSION OF DSC URGENCY ALERT

- Tune the radio to the appropriate DSC frequency or channel (2187.5 kHz, VHF channel 70, etc.).
- Select the "all ships" call format (or the "single ship" format, if desired, and key in its MMSI).
- Select the "Urgency" priority.
- Operate the appropriate transmission button.
- Tune to the associated radiotelephony frequency/channel.
- Transmit the Urgency PAN PAN signal and the message using the standard voice procedure detailed earlier in the chapter.

NOTE: Stations receiving a DSC Urgency Alert should not acknowledge it. Instead, they should tune their radio to the associated radiotelephony frequency/channel and await voice message.

TRANSMISSION OF A DSC SAFETY ALERT

- Tune the radio to the appropriate DSC frequency or channel (2187.5 kHz, VHF channel 70, etc.).
- Select the "all ships" call format (or the "single ship" format, if desired, and key in its MMSI).
- Select the "Safety" priority.

- Operate the appropriate transmission button.
- Tune to the associated radiotelephony frequency/channel.
- Transmit the Safety SECURITÉ signal by voice and announce the working frequency/channel for the transmission of the message.
- Change to the working frequency/channel and announce the message using the standard voice procedure detailed earlier in the chapter.

NOTE: Stations receiving a DSC Safety Alert should not acknowledge it. Instead, they should tune their radio to the associated radiotelephony frequency/channel and await voice message.

DSC TEST

Routine and test calls are permitted on VHF DSC Channel 70. On HF DSC sets, a "TEST" call protocol is incorporated for system verification and operator familiarisation. These calls, in the form of safety priority messages, when generated by a vessel, are automatically acknowledged by a return DSC message from a suitably equipped coast station. There is normally no subsequent radiotelephone or NBDP communication. Under GMDSS regulations such tests should be carried out at least once a month.

MODERNISATION OF DSC (DSC2 & DSC3)

Many HF receivers are designed to scan all six DSC channels in rapid sequence (2 seconds each). Following the reception of the alert, they also automatically change to the associated voice frequency or to the frequency included in the alert. On others, the operator needs to select the associated frequency indicated above. Some are combined HF and VHF sets with touch screen control panels. They may include features commonly found on mobile phones: memory dialling, log of calls, etc. Eventually, direct dialling telephone calls to anywhere in the world could become a common feature of VHF DSC.

RADIO EQUIPMENT

BATTERIES & BATTERY MAINTENANCE: See Chapter 30.

RADIO SPARES KIT *[Ref: USL Code, Section 12 and NSCV Part C]:*

- Fuses (four of each type) and fuse wire.
- If not fitted with reserve antenna, a spare antenna of similar electrical dimensions and design to the fitted antenna and capable of being rapidly assembled and erected (a spare long line aerial is easy to store).
- Spare globe for electric light illuminating system.
- Pliers.
- Screw driver
- Hydrometer or voltmeter.
- Spare microphone.
- A length of electric cable.
- Etc.

ANTENNA & EARTH:

The VHF and MF/HF whip-type antennas are constructed of either fibreglass or aluminium. Exposure to sun will cause fibreglass to deteriorate over time to the point where moisture can enter the antenna. This will affect the signal radiating efficiency of the antenna, usually making it useless. These antennas should be regularly checked for signs of damage and loose mountings.

Any insulators used with the whip-type antennas or the HF wire antennas should be regularly washed down with fresh water and kept clean. A build up of salt and soot will seriously degrade their insulating properties, and allow most of the signal to be lost to the ground, rather than being radiated.

An EARTH CONNECTION is required for MF/HF transceivers, but not for VHF or 27MHz. In non-ferrous hulls, a heavy wire or, preferably, a copper strip, from the radio is connected to a metal plate (also known as dyna plate) on the underside of the hull. In steel hulls, the earth wire may be connected to any bare metal part of the superstructure. In fibreglass hulls, earth is often connected to a metal plate within the layers of the hull.

Chapter 18: Radio, Satellite & Epirb Communication

The earth and antenna connections should be soldered, because twisted wire connections corrode. It is important to keep the antenna and the insulators clean from build up of salt, dirt and soot.

TRANSCEIVER CONTROLS

1. ANTENNA TUNING UNIT (ATU): For optimum performance, an MF/HF transceiver needs the following lengths of aerial for the frequencies shown:

        32 metres for 2 MHz
        17 metres for 4 MHz
        12 metres for 6 MHz

Such lengths are obviously not possible on small vessels. As an alternative, an aerial length of 4 or 8 metres (i.e., a factor of 32) is rigged together with a "gearbox" arrangement, known as the Antenna Tuning Unit. The ATU is a device fitted to the radio that can be manually or automatically tuned to match the antenna to the various frequency bands. Correctly tuned for a particular frequency, it ensures maximum transfer of transmitted power to the antenna on that frequency.

The operator needs to learn to tune both the radio and the ATU for the best transmitter performance, unless it is an automatic unit.

Posted next to the radio, there should be directions on how to tune the radio and ATU, and possibly a tabulation (chart) on the approximate settings of the controls for best performance for different frequencies. [NOTE: Usually the receiver and transmitter perform the best on the same ATU settings].

An ATU is not needed for the 27 MHz and VHF sets.

2. SQUELCH OR MUTE CONTROL: It allows the operator to stop the constant and annoying background roar from the receiver in the absence of an incoming signal. On VHF and 27MHz transceivers, it is usually an adjustable control. The correct setting is so that the roar <u>cannot just</u> be heard. Further muting with the control will desensitise the receiver, which will prevent the reception of weak signals. If provided on MF/HF transceivers, the level of muting is pre-set and can only be turned on or off.

3. AM/SSB (or H3E/J3E) **CONTROL**: It is found on most MF/HF transceivers and on those 27MHz transceivers with Single Side Band (SSB) option. It allows the selection of SSB or AM mode of transmission and reception, which has been discussed earlier in this chapter.

4. RF GAIN CONTROL: This control is found only on some 27MHz and MF/HF transceivers. It is similar to, and additional to, volume control. It is usually kept turned up to the maximum, and turned down only when receiving unusually strong signals. Only the volume control is used to adjust the strength of the received signal.

5. NOISE LIMITER (NOISE BLANKER): This control minimises the effect of loud static or ignition interference. It should be used with care as it may also desensitise the receiver to wanted signals.

6. POWER SELECTOR: This switch allows the selection between High and Low power for the transmitted signal. On VHF transceiver, it may be marked '25W/1W' (25 watts or 1 watt) or 'High/Low'. As listed in Radio Communication Rules above, the use of more power than necessary is a breach of rules. It not only causes interference to others, but also drains the battery at a faster rate.

7. DUAL WATCH (DW): Found on VHF transceivers, this switch permits keeping a listening watch on two channels simultaneously, usually channel 16 and one other.

8. CLARIFIER: This control is found on most MF/HF transceivers and those 27MHz transceivers fitted with SSB option. It allows fine-tuning of incoming SSB signals, when they sound distorted or 'off station'. It has no effect on outgoing signals.

9. RADIOTELEPHONY ALARM SIGNAL GENERATING DEVICE (ASGD): It has been discussed earlier in this chapter.

10. INTERNATIONAL/USA CONTROL: Found on some VHF transceivers, it permits communication with those stations in USA that do not conform to the international VHF channel plan. It should be set to 'international' at all times unless in the coastal waters of USA.

SKIPPING (SKIP ZONE)

Some antennas on certain frequencies may cause 27MHz and MF/HF radio signals to "skip" or miss nearby stations while reaching more distant stations. This phenomenon is known as skipping. The problem can be usually overcome by changing to a lower frequency or by trying later when ionospheric conditions are different.

DUPLEX SHIP STATIONS

Duplex channels are paired channels, used for simultaneous transmission and reception of signals, just as through a telephone handset. Most of the working channels in the MF/HF and VHF marine bands are duplex. They are pre-programmed into transceivers and selected automatically by use of the channel select control. However, transceivers on most small vessels are not set up for duplex operation, which requires fitting of duplex filter units and widely separated two antennas, one for transmission and another for reception. Vessels fitted with the 'duplex' facility do not need to use the transmission button on the microphone, nor do they need to use the word 'over' between transmission and reception, when on a duplex working channel. They can communicate just as they would via a telephone ashore. VHF Channel 16 is a Simplex channel. (*Duplex Repeater Stations are discussed above*).

RADIO TROUBLESHOOTING

Most radio operating problems are easily solved by making sure that the following components are in working order:

1. Radio switched on
2. Microphone connected
3. Squelch control (or mute control) turned just below static noise.
4. Battery fully charged and clean
5. Battery isolating switch turned on
6. Aerial connected
7. Fuse good
8. Battery lead connected to radio
9. Earth connected

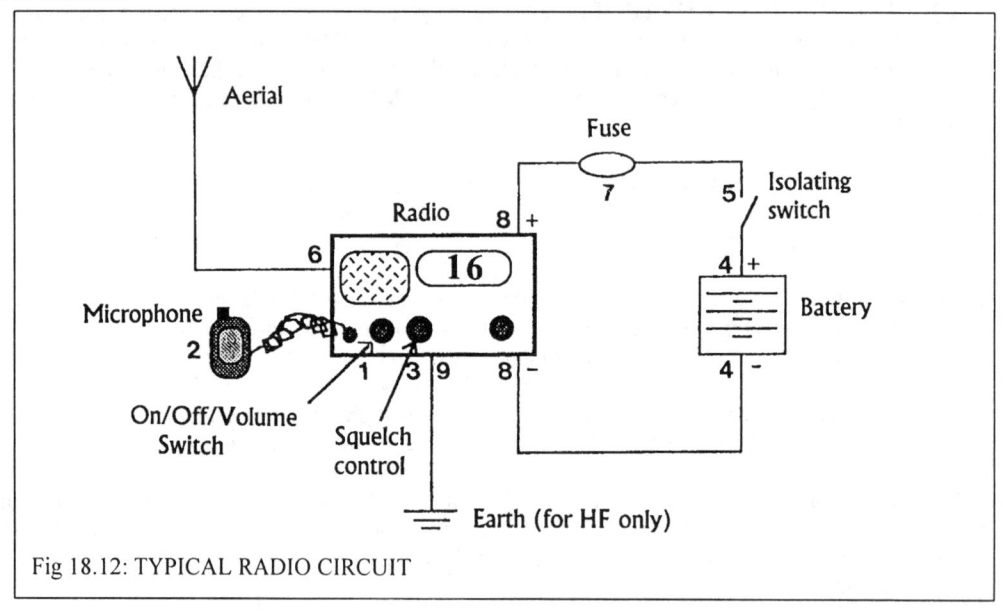

Fig 18.12: TYPICAL RADIO CIRCUIT

COMMON PROBLEMS AND THEIR REASONS

PROBLEM: Noise on squelch (mute) off position, and drop in communication
REASON: Faulty antenna

PROBLEM: Transmitter & receiver dead
REASON: Blown fuse or no power

PROBLEM: Sharp crackling noise during rolling. The transmitter output indicator not consistent when the button is pressed
REASON: Intermittent break or shorting in the aerial system

Chapter 18: Radio, Satellite & Epirb Communication

PROBLEM: Receiver working OK, but transmitter & its dial light die when button is pressed
REASON: Battery almost flat, or its terminals are corroded with white-green powder

PROBLEM: Dial lights flicker when radio switched on & intermittent noise on squelch off position
REASON: Loose electrical connection
PROBLEM: MF/HF cabinet gives a shock or sharp burning sensation when touched
REASON: Faulty or non-existent earth connection

PROBLEM: Receiver does not receive and makes no noise when squelch (mute) is switched off. Dial lights are on & the case is hotter to touch
REASON: Transmitter is stuck in the "on" position

PROBLEM: Calls only from vessels close by are received
REASON: Mute (squelch) control is set too high

PROBLEM: The ATU cannot tune the HF set, or it can tune only on some channels, or its settings have changed
REASON: The aerial system is faulty: it could be broken, shorting to earth, internally fractured, or the insulators are broken or covered in salt

PROBLEM: MF/HF transmitter goes out of tuning when hand is taken off the set
REASON: The earth system is faulty: the wire could be broken or the dyna plate detached from the hull

PHONETIC ALPHABET & FIGURE CODES
Fig 18.13

When it is necessary to spell out call signs and words the following letter spelling table should be used:

Letter to be transmitted	Code word to be used	Spoken as*
A	Alfa	**AL** FAH
B	Bravo	**BRAH** VOH
C	Charlie	**CHAR** LEE or **SHAR** LEE
D	Delta	**DELL** TAH
E	Echo	**ECK** OH
F	Foxtrot	**FOKS** TROT
G	Golf	GOLF
H	Hotel	HOH **TELL**
I	India	**IN** DEE AH
J	Juliett	**JEW** LEE ETT
K	Kilo	**KEY** LOH
L	Lima	**LEE** MAH
M	Mike	MIKE
N	November	NO **VEM** BER
O	Oscar	**OSS** CAH
P	Papa	PAH **PAH**
Q	Quebec	KEH **BECK**
R	Romeo	**ROW** ME OH
S	Sierra	SEE **AIR** RAH
T	Tango	**TANG** GO
U	Uniform	**YOU** NEE FORM or **OO** NEE FORM
V	Victor	**VIK** TAH
W	Whiskey	**WISS** KEY
X	X-ray	**ECKS** RAY
Y	Yankee	**YANG** KEY
Z	Zulu	**ZOO** LOO

* The syllables to be emphasised are underlined.

FIGURE CODE

When it is necessary to spell out figures or marks, the following table should be used:

Figure or mark to be transmitted	Code word to be used	Spoken as*
0	Nadazero	NAH-DAH-ZAY-ROH
1	Unaone	OO-NAH-WUN
2	Bissotwo	BEES-SOH-TOO
3	Terrathree	TAY-RAH-TREE
4	Kartefour	KAR-TAY-FOWER
5	Pantafive	PAN-TAH-FIVE
6	Soxisix	SOK-SEE-SIX
7	Setteseven	SAY-TAH-SEVEN
8	Oktoeight	OK-TOH-AIT
9	Novenine	NO-VAY-NINER
Decimal point	Decimal	DAY-SEE-MAL
Full stop	Stop	STOP

* Each syllable should be equally emphasised.

Chapter 18: Radio, Satellite & Epirb Communication

E.P.I.R.B. (EPIRB) & GPIRBS

As mentioned in Chapter 19.3, every vessel venturing more than 2 miles offshore around Australia's coastline is required to carry an EPIRB (Emergency Position Indicating Radio Beacon). It is a small, self-contained, battery-operated radio transmitter that is both watertight and buoyant. As the name suggests, its primary purpose is to indicate the position of survivors in an emergency. When switched on, it transmits a signal of continuous series of descending tones for a minimum of 48 hours, which is picked up by a COSPAS-SARSAT satellite or an aircraft flying-overhead. *[SARSAT: Search & Rescue Satellite. COSPAS is Russian for SARSAT].*

Operating licences are not required for operating EPIRBs. They work best from the water plane. The steps to activate an EPIRB are:

- Erect aerial
- Switch on
- Unroll and secure the tether line to the vessel or liferaft.
- Throw EPIRB overboard.

Once activated, an EPIRB should not be switched off until rescue is completed. However, once the incident is over, it must be deactivated. The Emergency Locating Beacons (ELBs) designed for aircraft use are not suitable for shipboard use.

There are two main types of EPIRBS:

➤ 121.5/243 MHz Epirb (in the process of being phased out)
➤ 406 MHz Epirb

However, as discussed below, they come in four different varieties, which is in addition to the VHF DSC and INMARSAT Epirbs.

FIG 18.14

COSPAS-SARSAT SATELLITE SYSTEM(S)

The COSPAS-SARSAT network is an international search and rescue satellite system. In operation, a satellite picks up the EPIRB signal and relays it to a ground receiving station, known as the Local User Terminal (LUT). LUTs have been set up in carefully selected locations around the world. Australia's two LUTs are located at Albany (WA) and Bundaberg (Qld), and these are linked directly by landline to the AusSAR Rescue Coordination Centre (RCC) in Canberra. A third LUT for the Southwest Pacific region is located at Wellington, New Zealand. This is also monitored by Australia.

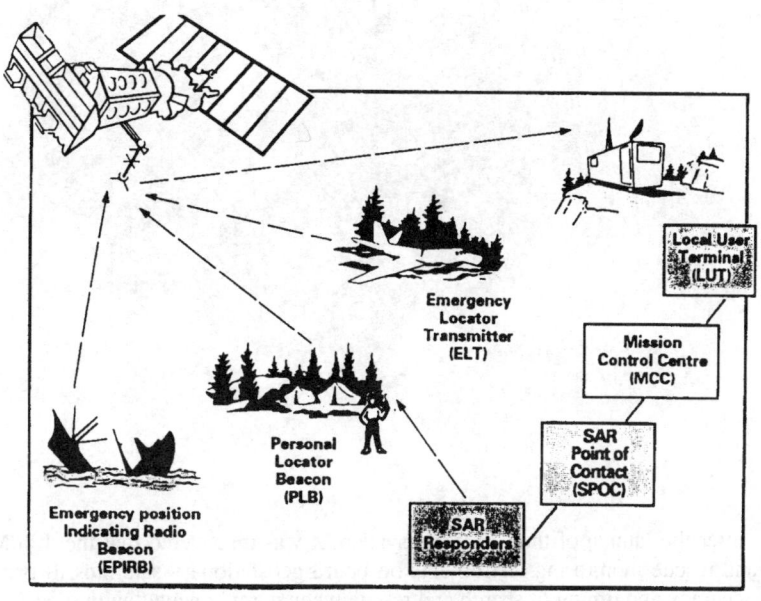

Fig 18.15: COSPAS-SARSAT SATELLITE SYSTEM

THE TWO TYPES OF SYSTEMS

Initially, there were only the LEOSAR satellites and LEOLUTs. Now there are two systems:

1. **LEOSAR** (LOW EARTH ORBIT SEARCH AND RESCUE) SYSTEM
2. **GEOSAR** (GEOSTATIONARY SEARCH AND RESCUE) SYSTEM

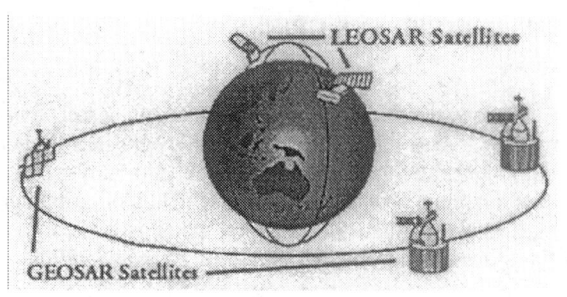

FIG. 18.16: LEOSAR & GEOSAR SYSTEMS
(Cospas-Sarsat Organisation)

1. LEOSAR

This is the system we have initially known as the COSPAS-SARSAT system. It is a network of four low-altitude polar-orbiting satellites that orbit the earth approximately every 100 minutes. The plane of their orbit is fixed in space, while the earth rotates inside the fixed orbits. The LUT stations on earth are called LEOLUTS. Being at a low altitude of around 900 km., the footprint of these satellites is rather small, i.e., they see only a small portion of the earth at a time. Although the typical waiting time in mid-latitudes is less than an hour, there can be a delay of up to 6 hours between an EPIRB being switched on and a satellite spotting it (see illustration).

The advantages of LEOSAR system are:

a) Being in orbit, the satellite can calculate the EPIRB's position through doppler effect on the received frequency, even though these positions can be inaccurate by up to 3 miles for 406 MHz and 11 miles for 121.5/243 MHz Epirbs.
b) Only the LEOSAR system provides coverage for the Polar Regions, which are beyond the coverage of GEOSAR system.
c) Since the satellite is continuously moving with respect to the Epirb, the LEOSAR system is less susceptible to obstructions, which may block an Epirb signal in a given direction.

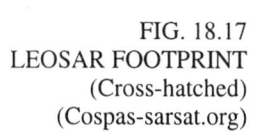

FIG. 18.17
LEOSAR FOOTPRINT
(Cross-hatched)
(Cospas-sarsat.org)

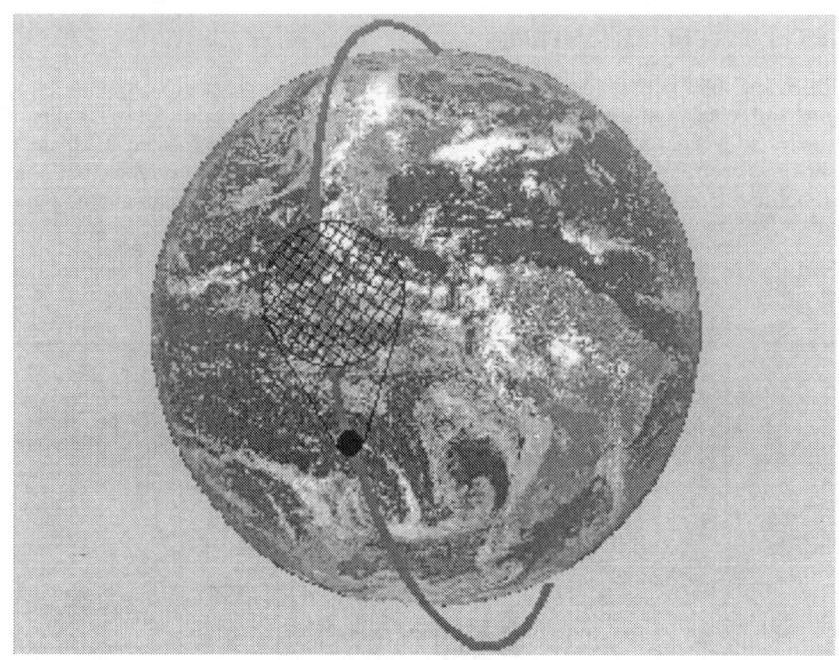

2. GEOSAR

About ten years after the launch of the LEOSAR system, it was discovered that the 406 MHz Epirbs can also be detected by search and rescue monitoring instruments on board geostationary satellites. It proved to be a tremendous advancement in the accuracy and timing of search and rescue when complemented with the LEOSAR system.

Chapter 18: Radio, Satellite & Epirb Communication

The GEOSAR system consists of 406 MHz repeaters carried on board various geostationary satellites and the associated ground facilities called GEOLUTs to process the satellite signal. These satellites do not orbit the earth. They remain fixed relative to the earth over the equator. Being stationary, they do not experience the Doppler effect. They are thus unable to calculate an EPIRB's position. However, being located at a very high altitude over the equator, their footprint is enormous, making them able to spot an activated EPIRB almost instantaneously anywhere in the world other than in the Polar Regions. Modern EPIRBs are thus being fitted with GPS engines to calculate their own position and transmit it to GEOSAR satellites. They are called **GPIRBs**, and have the advantage of indicating their position to within tens of metres.

While still capable of being located by LEOSAR satellites on their periodic sightings, GPIRBs in most parts of the globe can now also be located instantly by GEOSAR satellites. Currently there are four satellites in the GEOSAR network, providing 95% coverage of the temperate waters of the world. Additional satellite launches are planned.

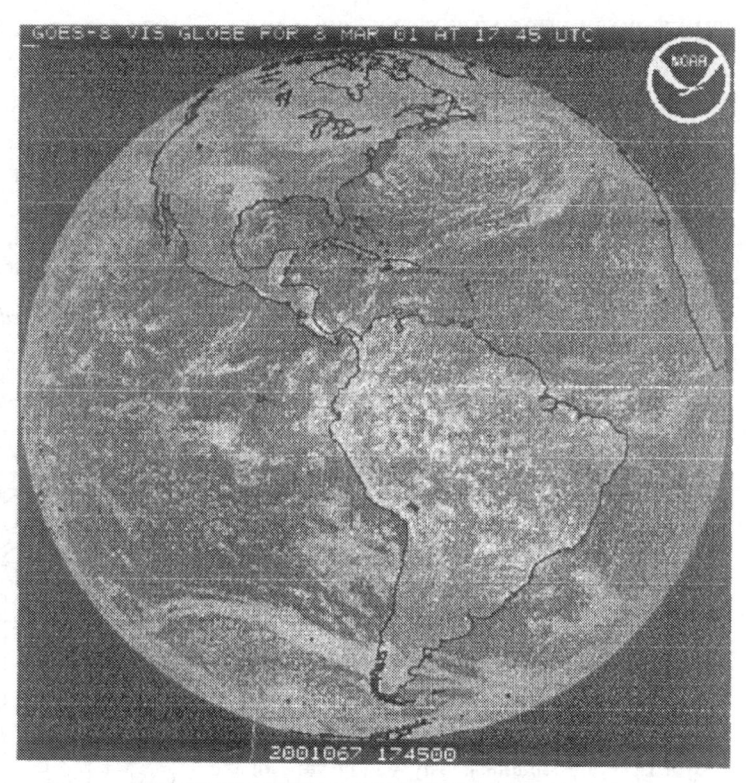

FIG. 18.18: GEOSAR FOOTPRINT (NOAA)

VARIETIES OF EPIRBS & GPIRBS

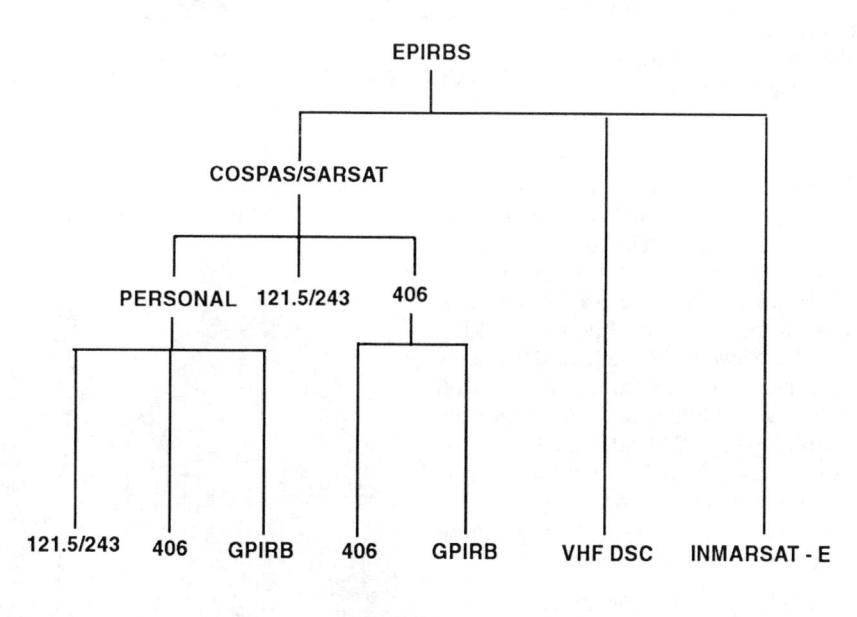

Fig 18.19

1. 121.5/243 MHz EPIRBs transmit an analogue signal through the LEOSAR network. They will be discontinued in 2009.
2. 406 MHz EPIRBs transmit a digital signal through LEOSAR network. They can be programmed to transmit their identification number and the type of emergency.

3. GPIRBs are 406 MHz Epirbs equipped with a GPS engine to calculate and broadcast their own position through LEOSAR as well as GEOSAR networks.
4. Personal EPIRBs or Personal Locator Beacons (PLBs) can be of the EPIRB or GPIRB type.
5. VHF DSC EPIRBs do not operate via the COSPAS-SARSAT system. They alert authorities on VHF DSC channel 70.
6. INMARSAT EPIRBs also do not operate via the COSPAS-SARSAT system. The INMARSAT system is discussed later in this chapter.

(1) 121.5/243 MHz EPIRB

These beacons will be discontinued in 2009. The decision is based on the IMO and the International Civil Aviation Organisation guidelines, as well as the fact that 121.5 MHz false alerts inundate search and rescue authorities. While the 406 MHz beacons cost more, they transmit a more reliable and complete information of the emergency. The implication of the decision is that all users should eventually switch to the 406 MHz beacons; and those planning to purchase a new Epirb should take the announcement into account.

The 121.5/243 MHz EPIRB (illustrated above) is small and inexpensive, and operates on the aircraft VHF distress frequencies of 121.5 and 243 MHz. Once activated, it is capable of being detected and located by aircraft (typical range 200 nautical miles at 10,000 metres altitude) and the LEOSAR system.

Signals radiated from a 121.5/243 MHz EPIRB are detected and relayed by a satellite to a search and rescue authority only if the activated EPIRB and a LUT are simultaneously within view of the satellite. Its detection and location is therefore limited to particular geographical areas surrounding a LUT. The system can locate a 121.5/243 MHz EPIRB to an accuracy of within 11 nautical miles. The figure shows the approximate geographical limits and median detection time for 121.5/243 MHz EPIRBS using the combined resources of the LUTs at Albany, Bundaberg and Wellington.

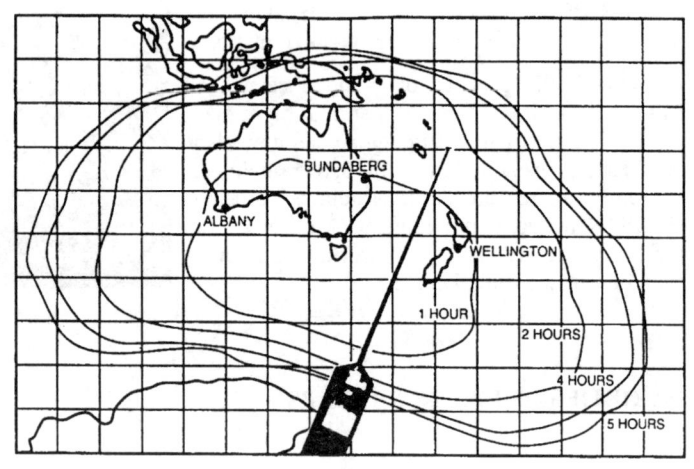

Fig 18.20:
AVERAGE TIME TO DETECT 121.5/243 MHz EPIRB

(2) 406 MHz EPIRB

The 406 MHz EPIRB is more expensive and sophisticated. It provides a global coverage because the satellites are able to store information received from a 406 MHz beacon. It also operates on 121.5 MHz for aircraft homing. Signals radiated from this EPIRB are detected and relayed to earth by COSPAS-SARSAT system in a similar fashion to the 121.5/243 MHz model. However, these signals are stored in the satellite's memory when the activated EPIRB and a LUT are not simultaneously within view of the satellite. Because of the satellite's ability to memorise signals from a (digital) 406 MHz EPIRB, its detection and location does not suffer the geographical limitations of the 121.5/243 MHz model. As the satellite's path brings it into view of a LUT, the information is retrieved from its memory and relayed down to the LUT. A 406 MHz EPIRB can be detected and located at any place on the Earth's surface to accuracy better than 3 nautical miles.

Every 406 MHz EPIRB has a unique identity code, which is transmitted as part of its signal. It is programmed into the beacon by the supplier before it is offered for purchase. It therefore identifies the vessel in distress, as well as her country of origin. The purchaser of a 406 MHz EPIRB is expected to complete the registration form provided by the supplier and mail it to the AusSAR RCC in Canberra, in order to register its identity code.

Fig 18.21
A 406 EPIRB & ITS REMOTE ACTIVATION UNIT

Chapter 18: Radio, Satellite & Epirb Communication

The 406 MHz beacons are available in two types:

- those that require manual activation; and
- those that will float-free and activate automatically should the vessel sink. They can also be manually activated.

The manual activation type may offer an electronic menu of distress situations. Selection by an operator prior to activation will provide the AusSAR Rescue Coordination Centre with an indication of the vessel's type of distress, as well as its identity and country of origin. Vessels compulsorily fitted with 406 MHz EPIRBs under Commonwealth legislation must carry the float-free type.

(3) PERSONAL EPIRB

Personal EPIRBS (or Personal Locator Beacons or Pocket EPIRBs) may be of the 121.5/ 243 MHz or 406 MHz type. Designed to be worn by individuals, they can be as small as a cigarette packet. There are two types of PLBs. In one type, when activated by a person overboard, an onboard receiver raises the alarm and indicates its direction. The other type is detected by a satellite just like any other COSPAS-SARSAT EPIRB. The wearer of a PLB must make sure that its antenna is extended and kept pointing vertically.

(4) GPIRB (GPS EPIRB)

As discussed above, a GPIRB is a smart 406 MHz Epirb that makes use of the LEOSAR as well as GEOSAR satellites. When activated, its onboard GPS engine locates itself and then loads its position into the transmission "sentence" for broadcast to the satellites. After that, the GPS shuts down to conserve power. It restarts every 20 minutes to refresh and reload its position information, and then shuts down again. The continuous updating of the Gpirb position not only reduces the uncertainty of false alarms, but the rescue centre can calculate the vessel's course and speed or the Gpirb's drift.

(5) VHF EPIRBS (VHF DSC EPIRBS)

Instead of transmitting to a satellite, these EPIRBS transmit the distress alert on the VHF DSC channel 70. In order for their signal to be located by searching ships and aircraft, they are also capable of transmitting on the X band radar – just like SARTS discussed below.

DESIGN FEATURES OF EPIRBS/GPIRBS

- They are orange or yellow in colour.
- Most brands weigh around one 2 kilograms. Personal units weigh around 200 grams.
- Small personal units are available for attachment to lifejackets.
- An orange coloured, rot proof, chafe resistant and buoyant tether line of 20 metres is permanently attached to every unit, and a shorter line (about 2 metres) to the personal models.
- Some personal unit are fitted with a strobe light.
- Their battery operating life is at least 48 hours with shelf life of at least five years. The special lithium battery is designed for long life low power consumption.
- Global coverage is provided by the Cospas-Sarsat Epirbs.
- They are fitted with transmitter and fault indicators. Some may also have an audible transmission signal.
- GPIRBs are fitted with GPS position acquired and GPS position transmitted indicators.

TESTING & CARE OF EPIRBS/GPIRBS

- Periodically examine them for water tightness and battery expiry date.
- Periodically test their transmission capability with the test switch provided. Never activate them for testing. In an emergency you can reassure yourself that it is working by tuning to 121.5 MHz on VHF and listening for the EPIRB/GPIRB signal. Signal presence can also be detected on an FM radio tuned to 99.5 MHz or an AM radio tuned to any vacant frequency and located close to an Epirbs. The 406 MHz models also transmit on 121.5 MHz to facilitate homing for search and rescue.
- 406 MHz models use a special lithium battery designed for long life low power consumption. It should be replaced only by an authorised dealer.
- You must register your 406MHz unit. Unregistered 406MHz units cannot be identified, and their signal my not be picked up by satellite.

PREVENTING INADVERTENT ACTIVATION OF EPIRBS/GPIRBS

- Do not store Epirbs/Gpirbs in lockers with equipment that may activate them by movement or pressure.
- Do not store them where they may lie in water, and become activated by water entering into the circuitry through damaged seal.

- Ensure they are completely destroyed and battery removed before disposing into garbage.
- Ensure shock-free transportation to and from the vessel.
- Educate everyone on board on the use and capability of the equipment.
- The float-free Epirbs/Gpirbs are usually "armed", i.e., they will activate as soon as removed from their cradle. Switch them off before removing for transportation.
- Advise authorities if an Epirb/Gpirb is accidentally activated.
- Never activate more than one Epirb/Gpirb at a time.

UNIVERSAL AUTOMATIC IDENTIFICATION SYSTEMS (UAIS or AIS)
See Chapter 14

SART & ANTI-COLLISION RADAR TRANSPONDERS
See Chapter 16.

THE FOLLOWING SECTION IS FOR THE SATELLITE COMMUNICATIONS ENDORSEMENT (INMARSAT) OF A MROCP CERTIFICATE

EXEMPTION: In Australia, the fisheries management authorities require commercial fishing vessels to be equipped with INMARSAT-C to enable them to participate in the VESSEL MONITORING SYSTEM (VMS). It allows the authorities to interrogate a vessel's position, course and speed, as well as the interchange of fish data. Operators of INMARSAT-C for VMS only are exempt from the satellite endorsement qualification.

INMARSAT (SATCOM) SATELLITE COMMUNICATION SYSTEM

Run by the London-based **INTERNATIONAL MOBILE SATELLITE ORGANIZATION,** Inmarsat consists of four communication satellites to facilitate mobile communications on land, at sea and in the air. Initially (in 1979), Inmarsat was set up to provide global communications for shipping via satellites on a commercial basis and was then known as the International Maritime Satellite Organization (Inmarsat). It then changed its name to the International Mobile Satellite Organization when it was extended to include aeronautical and land-mobile communications in 1985 and 1988 respectively, but retained its acronym. Inmarsat provides data transfer such as direct-dial telephone, telex, fax and email. It also provides automatic position and status reporting for aircraft and land transport, and distress and safety services, including GMDSS.

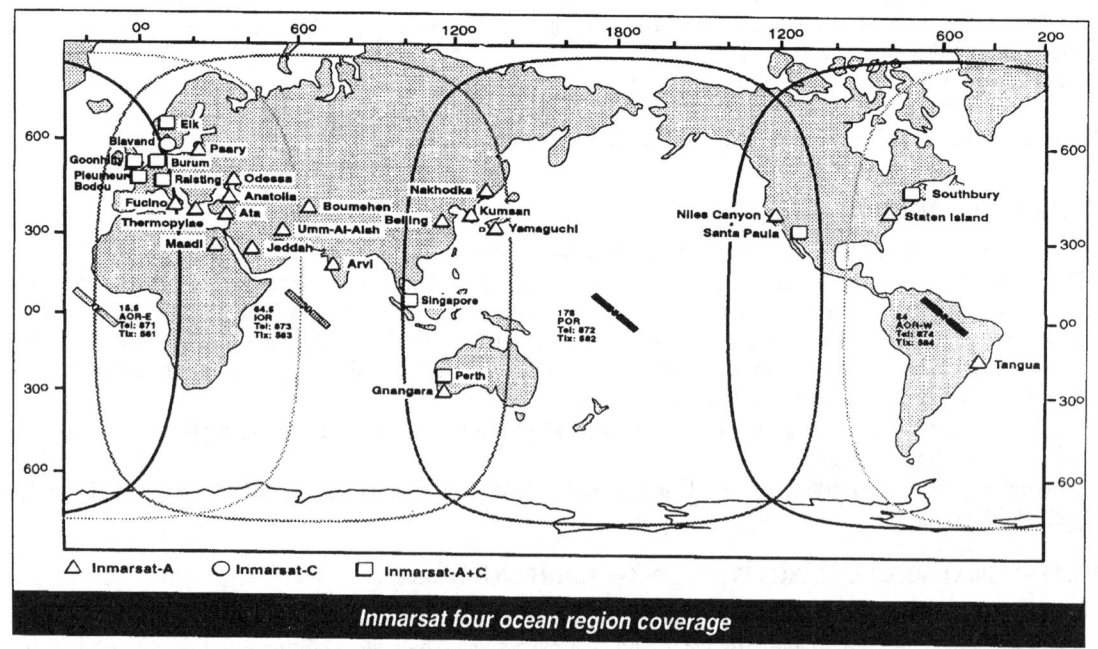

Fig 18.22: INMARSAT SATELLITE PLACEMENT AND COVERAGE

Chapter 18: Radio, Satellite & Epirb Communication

The satellites are in geostationary orbits above the equator, over the Atlantic, Indian and Pacific Oceans. Being geostationary, each satellite rotates with the earth, and thus remains in the same position over the earth. Each satellite has its own footprint or area of coverage.

Operating at super high frequency (SHF) in the 1.5 to 1.6 GHz and 4 to 6 GHz bands, these solar powered satellites provide continuous high quality telecommunication services to virtually the entire earth's surface.

The chart shows the four Inmarsat satellites and their coverage areas (or ocean regions), which are designated as follows:

- the Pacific Ocean Region (POR)
- the Indian Ocean Region (IOR)
- the Atlantic Ocean East Region (AOR East)
- the Atlantic Ocean West Region (AOR West)

The INMARSAT service provides:

1. Telephone, fax and email services (in both 'real time', and 'store and forward' modes)
2. Telex in both 'real time', and 'store and forward' modes
3. Distress and safety communication. "Distress priority" transmission is automatically and immediately routed to an appropriate Maritime Rescue Coordination Centre (MRCC).
4. Most Inmarsat equipment incorporate the Enhanced Group Calling (EGC) facility. As discussed below, this feature allows vessels to automatically receive Maritime Safety Information (MSI) (also known as SafetyNET) and FleetNET. The equipment provides audible and/or visual alarms on receipt of an EGC message.

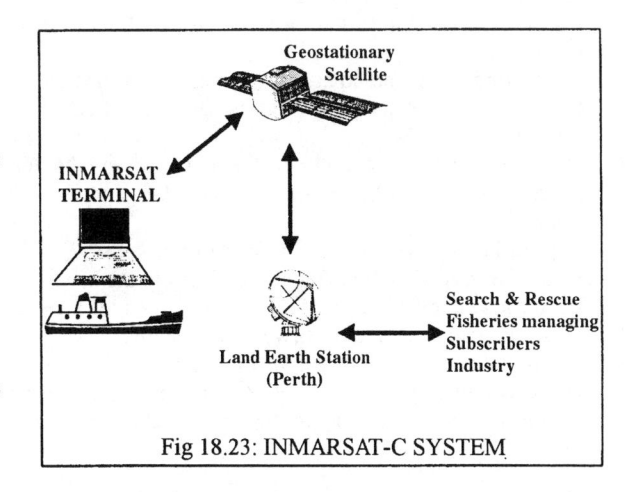

Fig 18.23: INMARSAT-C SYSTEM

INMARSAT TERMINOLOGY

> **SHIP EARTH STATIONS (SES)** are the Inmarsat units (fixed or portable) aboard vessels.

> **LAND EARTH STATIONS (LES)** connect the ship to shore telephones. They are fully automated communication interface installations on land. Each LES has an associated MRCC. The Australian LES, located at Perth, serves both the Indian (IOR) and Pacific (POR) ocean regions. Its associated MRCC is located in Canberra, which is operated by AusSAR *(discussed in Chapter 19.3)*. Details of LES, their ID numbers and charges for commercial communications are published by ITU and British Admiralty.

> **NETWORK COORDINATION STATIONS (NCS)** are responsible for overall management of regional systems.

TYPES OF SES TERMINALS

There are basically two types of SES terminals: Inmarsat-A (or B) and Inmarsat-C. In Australia, Telstra provides the Inmarsat service under the trade name of SATCOM. They are thus also referred to as Satcom-A (or B) and Satcom-C. Satcom-M, not approved for GMDSS, is mentioned below.

INMARSAT-A (or B)

(Inmarsat-B is the digital version of Inmarsat-A and will eventually replace it. However, at present the two systems coexist. Inmarsat-B is recommended for high level users, such as the offshore exploration and cruise ship industries.)

1. Top of the range real time communication system. *(Real Time communication means a live telephone or telex hook-up)*.
2. A multi-channel communication facility, incorporating telephone, fax, email and telex.
3. Consists of a dish aerial (about 80 cm), a terminal (computer and monitor), keyboard, telephone, fax machine and call alarms.
4. Designed for large vessels where its dish antenna, stabilised and enclosed in a fibreglass domes, can be installed.
5. On first switching-on, the operator enters the vessel's position and course into the terminal, which locks it on to the satellite. (The dish antenna may require repositioning after a power failure on board.)
6. A distress alert can be sent by either pressing a pre-programmed button or by input of a brief keyboard code. It assures an instant and automatic direct connection to a MRCC. There is no need to know the address of a LES.

7. Distress messages can be sent by telex or telephone.
8. The procedure for telex or the telephone mode is the same: Initiate Distress Alert – receive MRCC Answerback – transmit Distress Situation.

SITING THE (INMARSAT-A or B) DISH ANTENNA

- Site the dish antenna where the ship's superstructure will cause minimum obstruction and shadows on its "view".
- Protect against radiation hazard by avoiding prolonged human exposure to within 7 metres at just below, just above and at the level of the dish.

INMARSAT-C

1. A lightweight, portable non-voice, and non-real-time, communication unit.
2. Requires relatively low power.
3. Consists of a small, unstabilised, omni-directional (all round) antenna, a terminal (computer and monitor), keyboard and printer.
4. No moving parts, and no need for antenna stabilization.
5. Operates just like an email or a fax sent from a computer, i.e., it transmits and receives packets (or bursts) of data at 600 bits/sec.
6. On first switching on, either the operator or the terminal automatically logs-in with the regional NCS. This synchronises the SES's terminal to the NCS common channel and informs the NCS that the SES is in an operational status.
7. Log-off before switching off for an extended period. If this is not done, the LES will keep trying to deliver messages to your SES instead of advising non-delivery to the sender. You may also get charged for repeated attempts.
8. Used for sending and receiving telex, fax or email messages.
9. Messages can be stored or forwarded instantly.
10. The routine delivery time is 2 to 7 minutes.
11. On successful delivery of a message, the sender is notified.
12. Received messages can be printed or viewed on monitor.

DISTRESS & PRIORITY COMMUNICATION BY INMARSAT-C

1. The Inmarsat-C equipment contains a distress alert generator. When you press two control buttons simultaneously, it generates a default distress alert containing the following information:
    a. The identity of the SES.
    b. Unspecified maritime distress (unless you have selected a type of distress from the menu).
    c. Vessel's position, course & speed (from GPS or manually fed).
2. A distress alert can be transmitted even if the SES is not logged-in.
3. You do not need to nominate a LES or a MRCC when transmitting a distress alert. The equipment software and the NCS automatically route the alert to the right LES for forwarding it to the associated MRCC.
4. Alternatively, you may type a distress alert message on the keyboard, select the "distress priority" option and transmit it. The Inmarsat-C equipment is equipped with the "distress", "urgency" and "safety" message selection modes.
5. The equipment will indicate to you that the distress alert is being transmitted. It will also indicate the acknowledgement of its receipt by a LES.
6. Repeat the distress alert if acknowledgement is not received from a LES as well as from a MRCC within 5 minutes.
7. Conduct communication after the initial distress alert by keyboard and selection of "distress priority".
8. You can read the received messages on the display monitor and/or print them out.
9. Commence urgency messages with the words "PAN PAN" and safety messages with the word "SECURITÉ".
10. There is a set of Inmarsat 2-digit codes for automatic routing of messages to the appropriate authority. The complete list of 10 or so of these codes is published in the relevant AMSA Marine Notice (available on AMSA's website). However, the two important codes are:
    a. Code 32: seeking medical advice
    b. Code 42: reporting a navigational hazard
11. You will receive audible and/or visual alarms on receiving a distress or urgency alert via EGC. Such alarms usually require manual cancellation/resetting. Ship-to-shore messages will appropriately commence with the word

Chapter 18: Radio, Satellite & Epirb Communication

"MAYDAY", "PAN PAN" or "SECURITÉ".

12. An unintentional distress alert be must be cancelled immediately by sending a message with "distress priority" to the appropriate MRCC. Provide your vessel name, call sign and Inmarsat identity.

13. Inmarsat-C terminals are usually interfaced with or have an inbuilt GPS receiver. It provides up-to-date vessel's position in the event of a distress transmission. It also ensures that the terminal's EGC receiver would respond to messages relevant to the vessel's position. If not interfaced with a GPS, vessel's position, course and speed must be manually entered at intervals not exceeding 2 hours.

PERFORMANCE VERIFICATION TEST

Whenever you feel concerned about the condition of your Inmarsat-C equipment, you may initiate a performance verification test (PVT). The test may take up to 20 minutes to complete, depending on the level of Inmarsat traffic congestion.

SITING THE INMARSAT-C ANTENNA

The antenna is of a compact size and easy to locate. However, it should be kept at least 1 metre away from superstructure and other large objects that cause a shadow sector of more than 2°.

Unlike the dish antenna of Inmarsat-A or B, this antenna does not transmit a concentrated form of radio energy. Therefore, the radiation hazard is minimum. However, the terminal should be shut down if a person is likely to spend time within 1 metre of an Inmarsat-C antenna.

SOME NON-GMDSS INMARSAT SYSTEMS

INMARSAT-M is another type of lightweight portable communication unit. It uses "spot beams" on Inmarsat satellites to boost signal strengths. However its dome antenna, although quite small, requires stabilisation on vessels. Commercially known as **MINISAT or MINI-M**, it provides a low quality telephone and data (email, etc.) service in real time mode. Basic "briefcase" systems require manual pointing towards the satellite. They provide only voice communication using a handset at a stationary base. When connected to a computer, some models also provide data, fax and internet communications at 2400 bits/sec. Inmarsat-M is not approved for GMDSS.

INMARSAT D+ is the size of a personal CD player. It offers two-way data communication. With a built-in GPS, it is designed for supervisory systems such as tracking road transport, rental cars and container movements. It too is not approved for GMDSS.

INMARSAT ENHANCED GROUP CALLING (EGC) RECEIVERS

EGC is the GMDSS solution to distributing weather and safety information to geographically targeted ships. Most Inmarsat sets are fitted with EGC receivers. The EGC service broadcasts text messages to selected Ship Earth Stations (SESs) in an ocean region. This information includes Maritime Safety Information (MSI), which includes distress alerts, navigational warnings, meteorological warnings and weather forecasts, and other important safety information for vessels.

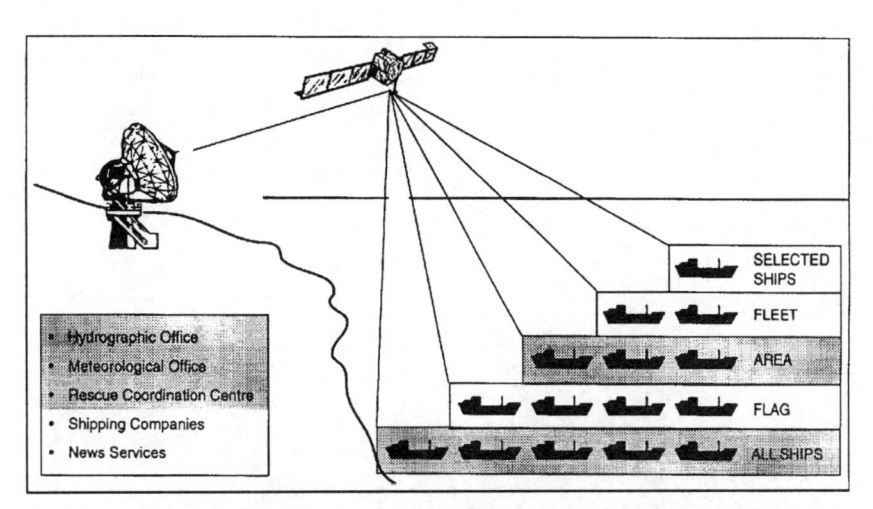

Fig 18.24: ENHANCED GROUP CALLING (EGC) SYSTEM
(The shaded area indicates the SafetyNET service)

There are two types of EGC messages: SafetyNET and FleetNET (both names are registered trademarks of Inmarsat).
- SafetyNET allows authorised organisations to broadcast MSI. The authorised organisations include:
    - Hydrographic offices
    - Meteorological offices
    - MRCCs
- FleetNET allows authorised organisations to broadcast information to selected SESs. The selected SESs may belong to a particular fleet, flag or service. The authorised organisations include:
    - ship owners, broadcasting company information to their fleet of ships
    - governments, broadcasting information to their national fleet
    - news subscription services, broadcasting news bulletins

Only the ships for which the message is intended will receive it. All other EGC receivers will reject it.

BROADCAST OF SafetyNET INFORMATION

If not interfaced with a GPS, the EGC receiver's positional information must be manually updated at intervals not exceeding 2 hours. On most Inmarsat-C equipment, the position routinely entered into the distress alert generator also updates the EGC receiver. The receiver must know its correct position at all times so that it accepts only its area specific MSI messages. If the receiver is not updated within 12 hours of switching on, it will accept all MSI messages with priorities higher than "routine" for the entire ocean region.

Reception of distress alerts and messages carrying "urgent priority" received from ashore will set off the audible and/or visual alarms on your vessel to attract your attention.

INMARSAT EPIRBS (INMARSAT-E)

This type of Epirbs are not approved for use on Australian vessels. However, a brief knowledge is required for the satellite endorsement of your MROCP certificate.

The signal from an Inmarsat Epirb (also known as the "L" band Epirb) is picked up by an Inmarsat satellite and passed on to a MRCC. However, since the Inmarsat satellites are not able to indicate the location of the Epirb, the Epirb provides its location to the satellite from a GPS receiver incorporated in its body.

The advantage of the Inmarsat Epirb over the Cospas-Sarsat Epirb is that its detection is not subject to the delay of a satellite passing overhead.

CHAPTER 18: QUESTIONS & ANSWERS

CHAPTER 18.1: RADIOTELEPHONY

1. The AMSA AUSREP system is available to small vessels:

    (a) on voyages under 200 miles.
    (b) when equipped with HF DSC or Inmarsat.
    (c) on voyages over 24 hours duration.
    (d) only through a Coast Radio Station.

2. In Australian waters, it is normal practice to observe silence periods as follows:

    (a) twice every half hour.
    (b) 15 and 45 minutes past each hour.
    (c) on the 2 MHz distress & calling frequency only.
    (d) on all distress frequencies.

3. The correct phonetic spelling of the vessel name "ROCKS" is:

    (a) Roger Oscar Charlie Kilo Sugar
    (b) Roger October Charlie Kilo Sugar
    (c) Romeo Oscar Charlie Kilo Sam
    (d) Romeo Oscar Charlie Kilo Sierra

4. On receiving a radio call which the vessel believes is intended for her but is not certain, she should:

    (a) reply & ask if she was being called.
    (b) wait for the station to call again.
    (c) make an "all stations" call.
    (d) identify herself.

5. The radio frequency 2182 kHz is permitted to be used for:

    (i) distress communication.
    (ii) broadcasting a Safety Message.
    (iii) broadcasting an Urgency Message.
    (iv) alerting to a Safety Message broadcast.

    The correct answer is:
    (a) all except (i).
    (b) all except (ii).
    (c) all except (iii).
    (d) all except (iv).

6. The radiotelephone silence periods are for three minutes:

    (a) commencing each quarter & three quarter hour.
    (b) before each hour & half hour.
    (c) each hour on the hour.
    (d) commencing each hour & half hour.

7. When a ship and a coast station are communicating, the following station has the right to choose frequencies, transmission time, etc.:

    (a) Coast station.
    (b) Ship station.
    (c) Arranged by mutual agreement.
    (d) Either of the two.

8. Phonetic words "WHISKY INDIA JULIA FOXTROT" are:

    (a) all incorrect
    (b) three correct.
    (c) all correct.
    (d) two correct.

9. On hearing a call, which you were not certain, was meant for you, you should:

    (a) call "all stations" & ask who is calling.
    (b) call & check with the calling station.
    (c) wait for a second call.
    (d) ignore it.

10. A Limited Coast Station is:

    (a) permitted to charge for phone connections.
    (b) not permitted to charge for phone connections.
    (c) permitted to charge for outgoing phone calls.
    (d) not permitted to accept phone connections.

11. The correct example of calling another ship station is:

    (a) Paluma to Merdeka do you read me?
    (b) VL5432 Merdeka calling VL4321 Paluma.
    (c) VL4321 Paluma this is VL5432 Merdeka.
    (d) Paluma VL4321 this is Merdeka VL5432.

12. Before going to sea, a vessel should switch on the radio transceiver and check that it is working, as follows:

    (a) Check the receiver noise.
    (b) Wait to hear a station calling.
    (c) Make a test call.
    (d) Listen to the receiver & the transmitter noise.

13. The approximate VHF communication range between two small vessels is:

    (a) 120 miles.
    (b) 1 mile.
    (c) 10 miles.
    (d) 35 miles.

14. By day a vessel that is 100 miles from a coast station has been unable to get a reply to calls on 2182 kHz. She should next try calling on:

    (a) 8291 kHz
    (b) 4125 kHz
    (c) 6215 kHz
    (d) 4428 kHz

15. Another vessel's 'person overboard' radio call has not been responded to by anyone. You should immediately:

    (a) proceed to her assistance.
    (b) advise her that you are on your way.
    (c) relay her message to a Coast Station.
    (d) advise her to re-transmit her message.

16. You receive a radio distress call from a vessel. You should:

    (a) reply immediately.
    (b) not reply immediately.
    (c) relay it immediately.
    (d) call a Maritime Coast Station immediately.

17. If you sighted a shipping container afloat at sea, you should:

    (a) advise a Maritime Coast Station.
    (b) transmit a SECURITÉ message.
    (c) sink it.
    (d) tow it ashore.

18. A vessel has picked up a survivor from an aircraft casualty. The master should:

    (a) make her comfortable and keep her quiet.
    (b) question her and advise a Coast Station.
    (c) warm her body with a glass of brandy.
    (d) rush her to a hospital.

19. A vessel has lost her propeller in bad weather 10 miles offshore, and she is fast drifting towards a reef. She should send the following radio signal:

    (a) MAYDAY.
    (b) PANPAN.
    (c) SECURITÉ.
    (d) PANPAN or SECURITÉ.

20. A vessel needs to contact a doctor by radio. She should:

    (a) transmit MAYDAY.
    (b) transmit PANPAN.
    (c) contact a Maritime Coast Station.
    (d) transmit SECURITÉ.

21. A distress call and message by radiotelephony should include:

    (a) the end of message with "over and out".
    (b) vessel name & call sign spoken 3 times.
    (c) vessel position in Lat. & Long.
    (d) distress signal spoken 4 times.

22. You hear a radio signal consisting of two tones, one high, one low, transmitted alternately which goes on for about a minute. It indicates the:

    (a) start of a distress call.
    (b) start of a weather forecast.
    (c) start of a silence period.
    (d) end of a silence period.

23. A single low tone lasting ten seconds after the two tone radiotelephony alarm signal identifies:

    (a) the start of a safety message.
    (b) its use by a coast station.
    (c) its use by a ship station.
    (d) the start of an urgency message.

24. One of your crew has been badly injured and needs urgent medical attention. The most appropriate radio signal to receive priority attention is:

    (a) PANPAN.
    (b) MAYDAY.
    (c) SECURITÉ.
    (d) none of the choices stated here.

25. The signal SEELONCE MAYDAY is for use of a Station to indicate that silence is necessary for:

    (a) an urgent medical message to follow.
    (b) a distress working.
    (c) its control of distress traffic.
    (d) none of the choices stated here.

26. In the following list, the international distress and calling channel is:

    (a) 27 MHz Channel 88
    (b) VHF Channel 67
    (c) 4134 kHz
    (d) 6215 kHz

27. A distress call and message may only be transmitted on the authority of:

    (a) the person in charge of the vessel.
    (b) the vessel's master or owner.
    (c) the vessel's owner.
    (d) a person holding a radio operator's licence.

Chapter 18: Radio, Satellite & Epirb Communication

28. The radiotelephony alarm signal may be used before transmitting:

    (a) a request for medical assistance
    (b) an important navigational warning.
    (c) a distress or urgency call or cyclone warning.
    (d) distress calls only.

29. The correct example of contacting Coastal Patrol Adelaide is:

    (a) Coastal Patrol Adelaide (x3), this is Rocks PLC 223 (x3).
    (b) Coastal Patrol Adelaide, this is Rocks PLC 223.
    (c) This is Rocks PLC 223 calling Coastal Patrol Adelaide.
    (d) Rocks to Coastal Patrol Adelaide, come in please.

30. The Urgency Signal has priority over all signals except:

    (a) the Safety signal.
    (b) a Cyclone warning.
    (c) a Distress signal.
    (d) an urgent navigational warning.

31. A message requesting urgent medical assistance or advice may be preceded by the following priority signal:

    (a) PAN PAN
    (b) RADIOMEDICAL
    (c) SEELONCE RADIOMEDICAL
    (d) PRU-DONCE

32. In the interest of Safety of Life at Sea, ship stations equipped with MF/HF transceiver should keep the best possible watch on:

    (a) 4215 KHz
    (b) 2281 KHz
    (c) 6125 KHz
    (d) none of the choices stated here.

33. The frequency of 8291 kHz is intended for

    (a) distress & safety traffic.
    (b) distress & calling.
    (c) routine calling.
    (d) radiotelephone calls on 8 MHz.

34. The signal to advise all stations that distress traffic has finished and normal working may be resumed, is:

    (a) PRU-DONCE
    (b) SEELONCE FEENEE
    (c) SEELONCE MAYDAY
    (d) MAYDAY FEENEE

35. The main and supplementary VHF channels used for distress and calling in Australia are:

    (a) 16 & 12
    (b) 16 & 67
    (c) 12 & 16
    (d) 16 & 72

36. The station normally in control of distress working is:

    (a) the nearest coast station.
    (b) the nearest rescue vessel.
    (c) the vessel in distress, unless control delegated.
    (d) the organisation responsible for rescue.

37. The procedure for a MAYDAY RELAY signal includes broadcasting the name of the vessel in distress:

    (a) once.
    (b) three times.
    (c) not at all.
    (d) as many times as possible.

38. A PAN PAN message:

    (a) must be addressed to a coast station.
    (b) must be addressed to all stations.
    (c) must be broadcast on a distress frequency.
    (d) none of the choices stated here apply.

39. An urgency message is sent if:

    (a) the vessel is broken down & is being swamped.
    (b) a floating log is sighted nearby.
    (c) an airplane is in distress.
    (d) someone on board the vessel is injured.

40. The international signal "SAYCURETAY" (SECURITÉ) means:

    (a) An urgent signal follows.
    (b) a distress message has been received.
    (c) a weather or navigational warning follows.
    (d) a distress message follows soon.

41. MAYDAY MAYDAY MAYDAY
    THIS IS
    VC2234 GUMTREE   VC2234 GUMTREE
    VC2234 GUMTREE   ON BEHALF OF VL4454
    POINTER   OUT OF FUEL 10 MILES
    SOUTHEAST OF BRAMPTON ISLAND AND
    WITH FOUR PEOPLE ABOARD.

    The message written above is:
    (a) a distress message.
    (b) a safety message.
    (c) an urgency message.
    (d) incorrectly written.

42. Which of the following is classed as Distress and Safety Traffic frequency:

    (a) 2201 kHz
    (b) 2638 kHz
    (c) 2182 kHz
    (d) None of the choices stated here

43. The following station has the absolute right to control distress traffic:

    (a) The nearest official Coast Station.
    (b) The nearest Limited Coast Station.
    (c) The vessel in distress.
    (d) The rescuing vessel.

44. MAYDAY MAYDAY MAYDAY
    THIS IS
    SEABIRD VL2865 SEABIRD VL2865 SEABIRD VL2865
    MAYDAY THIS IS SEABIRD VL2865
    POSITION 25 MILES SOUTHEAST OF BRAMPTON ISLAND - ON FIRE - REQUIRE IMMEDIATE ASSISTANCE - FOUR PERSONS ABOARD - HAVE A LIFERAFT - EPIRB ACTIVATED.

    The radio transmission written above is a:
    (a) safety message.
    (b) urgency message.
    (c) distress message.
    (d) distress call & message.

45. Seelonce Mayday means:

    (a) distress message follows.
    (b) keep silent, distress working in progress.
    (c) distressed station to be quiet.
    (d) distress is finished.

46. PANPAN is a call to announce:

    (a) distress traffic after the initial Mayday call.
    (b) urgent message to follow.
    (c) navigational warning to follow.
    (d) cancellation of a navigational warning.

47. A distress message is sent if a vessel:

    (a) loses the propeller.
    (b) catches fire.
    (c) has an engine breakdown.
    (d) is holed. Pumps aboard can cope with the water.

48. A distress message may be sent:

    (a) only on 2182 kHz during a silence period.
    (b) on any frequency at any time.
    (c) only on 2182, 27.88 & Channel 16.
    (d) only during a silence period on any frequency.

49. The purpose of the two tone radiotelephony alarm is to alert ships to a:

    (a) coast station closing.
    (b) urgency message.
    (c) navigational warning.
    (d) distress call or urgent cyclone warning.

50. In the following list, the on-scene distress & safety traffic frequency or channel is:

    (a) 8291 kHz.
    (b) 4125 kHz.
    (c) 6215 kHz.
    (d) Channel 16 (VHF).

51. The initial radiotelephony call to establish communication with Coast Guard Sydney on 4125 kHz is as follows:

    (a) This is Vanity VVL 7232 (x3) calling Coast Guard Sydney (x3)
    (b) Coast Guard Sydney, this is Vanity VVL 7232
    (c) Vanity to Coast Guard Sydney, do you read me?
    (d) Coast Guard Sydney (x3), this is Vanity VVL 7232 (x3).

52. A radio logbook is required to be carried in order to log:

    (a) all incoming & outgoing calls.
    (b) any distress messages acknowledged.
    (c) any distress messages heard.
    (d) all distress, urgency & safety messages.

53. The documents required to be carried on board a vessel under the Radiocommunications Act are:

    (i) radio log book.
    (ii) ship station licence, where applicable.
    (iii) radio operator's licence.
    (iv) radio repair manual.

    The correct answer is:
    (a) all except (i).
    (b) all except (ii).
    (c) all except (iii).
    (d) all except (iv).

54. A small vessel is on a voyage lasting several days. She should pass her position information to a coast station or limited coast station at following intervals:

    (a) At the commencement of the voyage, each day, and on completion of the voyage.
    (b) At the commencement and the completion of the voyage.
    (c) Only at the commencement of the voyage.
    (d) Daily, but only during bad weather conditions.

Chapter 18: Radio, Satellite & Epirb Communication

55. A vessel wishing to get a HF Radio Check should call:

    (a) a Coast Radio Station on 4125 kHz
    (b) a Volunteer Coast Station on 4125 kHz
    (c) a Coast Radio Station on 2182 kHz
    (d) a Volunteer Coast Station on 2524 kHz

56. A vessel wishing to make a telephone call ashore should call:

    (a) a Coast Radio
    (b) a Volunteer Coast Station
    (c) none of the choices stated here.
    (d) a AMSA Network Station

57. The following network of stations does <u>not</u> broadcast routine weather forecasts on HF radio:

    (a) Coast Radio Network
    (b) Volunteer Coast Stations Network
    (c) AMSA Radio Network
    (d) None of the choices stated here.

58. The radiotelephony alarm signal is:

    (a) the spoken word MAYDAY.
    (b) an uninterrupted single audible alarm.
    (c) two tones, transmitted alternately to produce a warbling audio sound.
    (d) An interrupted single audible alarm.

59. In relation to an antenna-tuning unit (ATU), which of the following statements is correct?

    (a) It is employed on MF/HF & VHF transceivers.
    (b) It fine-tunes SSB signals.
    (c) It varies the physical length of the antenna.
    (d) It ensures maximum transfer of transmitted power to the antenna on different frequency bands.

60. Which of the following is the correct use of 8291 kHz frequency?

    (a) Establish routine communication with a coast station.
    (b) Establish routine communication with another vessel.
    (c) Broadcast the contents of a routine weather message.
    (d) Distress & Safety Traffic communication.

61. The priority radiotelephony signal preceding a Strong Wind Warning advice from a coast station is:

    (a) none.
    (b) "Priority Wind Warning"
    (c) the Safety Signal
    (d) PANPAN

62. Which of the following rules is correct in relation to MF/HF radio communication:

    (a) the lower the frequency, the greater the communication range.
    (b) The higher the frequency, the greater the communication range.
    (c) The higher the frequency, the lower the communication range.
    (d) It should not be used during the hours of darkness.

63. The function of the Clarifier control on a MF/HF transceiver is to

    (a) fine-tune SSB reception.
    (b) cut out the annoying receiver roar.
    (c) fine-tune all types of reception.
    (d) vary the received signal strength.

64. The search and rescue coordination authority for waters beyond 200 miles from the coast is:

    (a) State & Territory Water Police
    (b) Volunteer Marine Rescue Organisations
    (c) State & Territory Marine Authorities.
    (d) AusSAR RCC

65. A vessel, equipped with MF/HF radio, is on fire. Which of the following items should form a part of her Mayday procedure?

    (a) The two-tone radiotelephony alarm in lieu of the DSC alert.
    (b) Mayday (x3), This is (x1), Ship's name & Callsign (x3).
    (c) Mayday (x3), This is (x1), Ship's name & Callsign (x1).
    (d) Mayday (x3), This is (x3), I am on fire (x3)

66. MV MISTY (VLW3456) wishes to broadcast a Mayday Relay call and message on behalf of MV SEADOG (VNW6789). Which of the following items should form a part of her Mayday Relay procedure?

    (a) Radiotelephony Alarm Signal, followed by the spoken words 'Mayday Relay' (x1)
    (b) This is Misty VLW3456 (x1)
    (c) Mayday Seadog VNW6789 (x1)
    (d) Misty's position

67. A vessel, equipped with HF radio, wishes to broadcast a warning of a floating container. Which of the following items should form a part of the transmission procedure?

    (a) PanPan (x1) (on 4125 kHz).
    (b) PanPan (x3) (on 2182kHz)
    (c) Hello All Stations (x1) (on 4125 kHz)
    (d) Shipping container floating (on 4426 kHz)

68. A vessel has received a HF distress message from another vessel. Which of the following items should form a part of her response?

    (a) Mayday (x3)
    (b) Name & callsign of vessel in distress (x1)
    (c) Own vessel name & callsign (x1)
    (d) Received Mayday (x1)

69. A vessel wishes to seek urgent medical assistance via radio. Which of the following items of her call and message is incorrectly stated?

    (a) PanPan (x3)
    (b) Hello Coast Radio Sydney (x3)
    (c) PanPan (x1)
    (d) I require medical assistance

70. For distress and safety purposes, Australia has been declared a GMDSS Area 3 or "A3". This means the AMSA Coast Stations must provide:

    (a) VHF listening service.
    (b) HF listening service.
    (c) Satellite watchkeeping service.
    (d) HF DSC watchkeeping service.

71. The international radiotelephony Distress & Safety Traffic frequency/channel is:

    (a) 2218 kHz
    (b) VHF channel 67
    (c) 6215 kHz
    (d) 4215 kHz

72. The radiotelephony signal SEELONCE DISTRESS may be used by the following station(s):

    (a) Only the station in distress or the station in control of the distress traffic.
    (b) Any station near a distressed vessel, which believes that silence is necessary due to distress working.
    (c) Any station to indicate that the distress working has finished.
    (d) Only the station in distress to indicate that the distress working has finished.

73. Normally, the VHF radio communication range is as follows:

    (a) Short range, approximately equal to the combined 'line of sight' distance of the stations involved.
    (b) Long range, due to radio energy's travel via sky waves.
    (c) Long & short range, due to radio energy's travel via sky & ground waves.
    (d) Long range, due to ionospheric 'ducting'.

74. Normally, the 27MHz radio communication range is as follows:

    (a) Long range, due to radio energy's travel via sky waves.
    (b) Long & short range, due to radio energy's travel via sky & ground waves.
    (c) Short range, approximately equal to the combined 'line of sight' distance of the stations involved.
    (d) Long range, due to ionospheric 'ducting'

75. Normally, the MF/HF radio communication range is as follows:

    (a) Short range, approximately equal to the combined 'line of sight' distance of the stations involved.
    (b) Long & short range, due to radio energy's travel via sky & ground waves.
    (c) Long range, due to radio energy's travel via sky waves.
    (d) Long range, due to ionospheric 'ducting'.

76. The following radio equipment would provide the best overall safety for an international voyage:

    (a) 27 MHz, fitted with all available channels.
    (b) 121.5/243MHz EPIRB
    (c) VHF, fitted with DSC & international channels.
    (d) MF/HF, fitted with DSC & all international channels/frequencies.

77. The Mute or Squelch control on VHF transceiver is used to:

    (a) increase transmitter performance.
    (b) remove distortion in the received signal.
    (c) stop annoying background roar from the receiver in the absence of an incoming signal.
    (d) adjust the 'electrical length' of the antenna.

78. In relation to Marine Radiotelephony service in Australia, which of the following statements is correct?

    (a) A CRN Station cannot connect a vessel to speak directly with a doctor.
    (b) VHF Channel 16 is a duplex channel.
    (c) VHF Repeater Channels include Channel 16.
    (d) VHF Repeater Channels are suitable for Position Reports & Radio Checks.

79. The following vessels should send their Position Reports to the State-run CRN:

    (a) None of the choices stated here
    (b) DSC/INMARSAT equipped vessels
    (c) SOLAS vessels
    (d) Non-SOLAS vessels

CHAPTER 18.2: EPIRBS
(IMPORTANT: Also study questions on SARTS in Chapter 16)

80. An EPIRB is a:
    (a) voice transmitter & receiver.
    (b) voice & direction finding transmitter.
    (c) direction finding transmitter & receiver.
    (d) transmitter.

81. A 406MHz EPIRB/GPIRB allows rescue authorities to:
    (a) communicate with the vessel in distress.
    (b) detect & locate it within 500 miles of coast.
    (c) detect & locate it anywhere on the earth.
    (d) reply to a distress call.

82. The purpose of an EPIRB/GPIRB is to:
    (a) communicate with a rescue vessel.
    (b) communicate with a rescue aircraft.
    (c) indicate distress and location.
    (d) indicate that the vessel has been abandoned.

83. An EPIRB/GPIRB works best when:
    (a) tied to the highest point on the vessel.
    (b) tied to a mast.
    (c) placed on water surface.
    (d) placed next to vessel's radio antenna.

84. An EPIRB/GPIRB should only be tested:
    (a) below decks.
    (b) in accordance with the manufacturer's instructions.
    (c) prior to taking it on board.
    (d) when requested by a rescue organisation

85. An "armed" float-free EPIRB/GPIRB:
    (a) will activate only when switched on.
    (b) will activate as soon as removed from its cradle.
    (c) should be stored in a locker.
    (d) must be carried by vessels operating offshore.

86. In addition to being received by satellites, an Epirb/Gpirb signal may also be heard by:
    (a) Coast radio stations
    (b) Water police
    (c) Large commercial ships
    (d) Aircraft

87. The EPIRB/GPIRB battery has:
    (a) shelf life of at least 48 hours.
    (b) operating life of at least 96 hours.
    (c) shelf life of at least 5 years.
    (d) shelf life of at least 12 months.

88. In relation to EPIRBS/GPIRBS, which of the following statements is correct?
    (a) The 406MHz units are for coastal vessels.
    (b) The 121.5/243MHz units are for seagoing vessels.
    (c) The 406MHz units are programmable.
    (d) Only the 406MHz units are satellite compatible.

89. In relation to EPIRBS/GPIRBS, which of the following statements is incorrect?
    (a) LEOSAR satellites are unable to calculate the unit's position.
    (b) Only the LEOSAR system provides coverage for Polar Regions.
    (c) 406MHz EPIRB positions can be inaccurate by up to 3 miles.
    (d) 121.5/243MHz EPIRB positions can be inaccurate by up to 11 miles.

90. In relation to GPIRBS, which of the following statements is incorrect?
    (a) GEOSAR satellites are geostationary & carry 406MHz repeaters.
    (b) GPIRB positions are accurate to within tens of metres.
    (c) GEOSAR satellites have large footprint than LEOSAR satellites.
    (d) GEOSAR satellites calculate the GPIRB positions.

91. In relation to LEOSAR & GEOSAR systems, which of the following statements is incorrect?
    (a) LEOSAR satellites are polar orbiting.
    (b) GEOSAR satellites are geostationary over the equator.
    (c) LEOSAR satellites have the larger footprint.
    (d) GEOSAR satellites themselves do not calculate the EPIRB or GPIRB positions.

92. In relation to EPIRBS and GPIRBS, which of the following statements is incorrect?
    (a) GPIRBS calculate & transmit their own position.
    (b) GEOSAR satellites calculate positions of EPIRBs.
    (c) GPIRBS are capable of being located by both the LEOSAR & GEOSAR systems.
    (d) The 121.5/243MHz EPIRBS will be discontinued in 2009.

93. In relation to EPIRBs and GPIRBs, which of the following statements is incorrect?

    (a) GPIRBs are 406MHz EPIRBs equipped with GPS engines.
    (b) Personal EPIRBs (PLBs) may be of the EPIRB or GPIRB type.
    (c) VHF DSC EPIRBs operate via the COSPAS-SARSAT system.
    (d) The 121.5/243MHz EPIRBs transmit an analogue signal.

94. In relation to EPIRBs/GPIRBs, which of the following statements is incorrect?

    (a) A satellite receiving a 121.5/243MHz signal must have a LUT simultaneously in view for the signal to be relayed.
    (b) A 406MHz unit must always be manually activated for transmission.
    (c) A 406MHz unit automatically transits its unique identity code as part of its signal.
    (d) A 406MHz unit can be programmed for the vessel's distress situation.

95. In relation to EPIRBs/GPIRBs, which of the following statements is incorrect?

    (a) They are orange or yellow in colour.
    (b) Personal units can be attached to lifejackets.
    (c) The length of the tether line on the personal unit is 2 metres.
    (d) They are never fitted with a strobe light.

96. In relation to EPIRBs/GPIRBs, which of the following statements is incorrect?

    (a) They are fitted with transmitter and fault indicators.
    (b) They should be periodically tested by activating them for short periods.
    (c) Some units are fitted with an audible transmission signal.
    (d) GPIRBs are fitted with GPS position acquired & GPS position transmitted indicators.

97. In relation to EPIRBs/GPIRBs, which of the following statements is correct?

    (a) The 406MHz units also transmit on 121.5 MHz.
    (b) Do not store them where they may lie in water.
    (c) Do not switch off "armed" units when removing them from their cradle.
    (d) A vessel carrying more than one unit should activate them simultaneously.

98. In relation to EPIRBs/GPIRBs, which of the following statements is incorrect?

    (a) The unit's test transmission can be heard on VHF, FM or AM radio located close to the unit.
    (b) A 406MHz unit must be registered with AMSA.
    (c) The accidental activation of a 406MHz unit cannot be mistaken for a distress signal.
    (d) Only an authorised dealer should replace the 406MHz unit's battery.

99. In relation to EPIRBs/GPIRBs, which of the following is unnecessary?

    (a) Educate everyone on board in the use and capability of the unit.
    (b) Periodically test the unit's internal circuitry with a test meter.
    (c) Ensure shock-free transportation to and from the vessel.
    (d) Advise authorities of an accidentally activated unit.

100. An EPIRB is required to be carried by:

     (a) commercial vessels only.
     (b) all vessels.
     (c) vessels operating more than 2 miles offshore.
     (d) commercial vessels operating in open waters.

101. Under SOLAS regulations, the radio equipment requirement for non-SOLAS vessels includes:

     (a) Two independent means of transmitting distress
     (b) Two independent means of receiving MSI
     (c) One DSc capable HF radio
     (d) One VHF radio

CHAPTER 18.3: MAINTENANCE & FAULT FINDING
(IMPORTANT: Also study questions on BATTERIES & ELECTRICAL SYSTEMS in Chapter 30)

102. In looking after a vessel's radio equipment, the following statement is correct:

     (a) aerials should be connected to earth.
     (b) batteries should be run flat twice a year.
     (c) a fractured whip aerial should be cut short.
     (d) earth should be connected to the sea.

103. A noise on a radio transceiver in the squelch off position combined with a drop in communication indicates:

     (a) faulty antenna.
     (b) faulty earth.
     (c) squelch setting too low.
     (d) transmitter stuck in the "on" position.

Chapter 18: Radio, Satellite & Epirb Communication

104. As a result of a blown fuse, a radio transceiver will:
    (a) not transmit.
    (b) not receive.
    (c) not receive or transmit.
    (d) give shock when touched.

105. Sharp cracking noise in a radio transceiver during rolling indicates:
    (a) loose electrical connection.
    (b) loose contact in the microphone.
    (c) intermittent break in the aerial system.
    (d) faulty earth.

106. A radio transceiver works, but the transmitter and its dial light die when button is pressed. The reason is the:
    (a) almost flat battery.
    (b) shorting aerial.
    (c) shorting earth.
    (d) shorting transceiver cabinet.

107. The dial light flickers when the radio transceiver is switched on. There is also intermittent noise in the squelch off position. The reason is:
    (a) the faulty aerial.
    (b) a loose electrical connection.
    (c) almost flat battery.
    (d) the faulty earth.

108. When transmitting, the MF/HF transceiver gives a sharp burning sensation when touched. The reason is the:
    (a) overcharged battery.
    (b) faulty aerial.
    (c) faulty or non-existent earth.
    (d) frayed power supply cable.

109. The radio receiver does not receive and makes no noise when the squelch is switched off. Dial lights are on and the case is hot to touch. The reason is:
    (a) The transmitter is stuck in the "on" position.
    (b) a loose electrical connection.
    (c) the faulty receiver.
    (d) the fuse wire is too heavy.

110. The radio transceiver receives calls only from the vessels close by. The reason is the:
    (a) faulty earth.
    (b) almost flat battery.
    (c) mute control is set too low.
    (d) mute control is set too high.

111. Faults in the aerial system, such as a broken antenna, an internal fracture, shorting to earth or broken or salt encrusted insulators are indicated by the following symptom in the HF radio transceiver:
    (a) The ATU cannot be tuned.
    (b) The radio works only when touched.
    (c) The dial light flickers.
    (d) The radio cabinet gives a shock when touched.

112. With the muting (squelch) off, there is a noise on the VHF receiver. A test call is not heard by the two nearby vessels. The most likely cause is a fault in the:
    (a) earthing.
    (b) aerial connection.
    (c) transmitter.
    (d) receiver.

113. The radio band most affected by ignition noise:
    (a) 27 MHz
    (b) VHF
    (c) MF/HF SSB
    (d) they are all equally affected.

114. The radio band least affected by ignition noise and thunderstorms is:
    (a) 27 MHz
    (b) VHF
    (c) MF/HF SSB
    (d) they are all equally affected.

115. A blown fuse in the radio power leads should be replaced by:
    (a) a fuse of a slightly higher rating.
    (b) a fuse of manufacturers recommended rating.
    (c) a fuse of a slightly lower rating.
    (d) a thin piece of silver paper.

116. The most likely cause of broken reception and transmission, together with flickering dial lights, is the faulty connection between the transceiver and:
    (a) Earth.
    (b) Antenna
    (c) Microphone.
    (d) Battery.

CHAPTER 18.4: DSC ALERTS

117. The contents of a DSC distress alert do not include:
    (a) MMSI of the calling station.
    (b) call priority.
    (c) MMSI of the called station.
    (d) a radiotelephony frequency/channel.

118. A DSC routine alert can be made on:
    (a) VHF Channel 70.
    (b) 2187.5 kHz.
    (c) VHF Channel 70 & 2187.5 kHz.
    (d) none of the options listed here

119. The DSC distress, urgency and safety alert frequencies are <u>not</u> exclusively reserved for this task on:
    (a) VHF
    (b) MF/HF
    (c) MF & VHF
    (d) HF

120. The MMSI of own station must be programmed into the DSC controller:
    (a) on acquiring the set.
    (b) prior to transmitting distress alerts only.
    (c) prior to transmitting all alerts.
    (d) prior to transmitting routine alerts only.

121. The unique identity of a coast station in its MMSI is represented by the:
    (a) first two digits.
    (b) first five digits.
    (c) last four digits.
    (d) third to fifth digits.

122. A vessel receiving a DSC distress alert via VHF or 2 MHz frequency, must:
    (a) acknowledge it via DSC.
    (b) transmit a distress alert relay addressed to "all ships".
    (c) acknowledge it on the associated radiotelephone frequency, and inform the nearest coast or limited coast station.
    (d) transmit a distress alert relay addressed to the nearest coast or limited coast station.

123. After transmitting a DSC distress alert:
    (a) stay on the DSC frequency until the alert is acknowledged.
    (b) do not transmit the distress call by voice until the DSC alert is acknowledged.
    (c) continue to broadcast the distress call & message by voice until a coast station acknowledges it on a DSC frequency.
    (d) transmit the distress call and message on the radiotelephony frequency.

124. A vessel receiving a DSC distress alert must:
    (a) acknowledge it via DSC.
    (b) stay tuned to the DSC frequency until a coast station has acknowledged it.
    (c) acknowledge it via DSC, then switch to the associated radiotelephony frequency for more details.
    (d) take note of the contents & switch to the associated radiotelephony frequency for more details.

125. A vessel receiving a DSC distress alert on 6312 kHz should immediately:
    (a) acknowledge it.
    (b) relay it.
    (c) make note of it, then switch to the associated radiotelephony channel for more details.
    (d) acknowledge it, then relay it.

126. A vessel receiving a DSC distress relay alert from a vessel should:
    (a) acknowledge it via DSC.
    (b) stay tuned to the DSC frequency until a coast station has acknowledged it.
    (c) acknowledge it via DSC, then switch to the associated radiotelephony frequency for more details.
    (d) switch to the associated radiotelephony frequency for more details.

127. A vessel receiving a DSC distress relay alert from a coast station should:
    (a) stay tuned to the DSC frequency until another vessel has acknowledged it.
    (b) acknowledge it via DSC.
    (c) acknowledge it via DSC only if definitely in a position to provide assistance.
    (d) acknowledge it via DSC, then switch to the associated radiotelephony frequency for more details.

128. To cancel an inadvertent DSC distress alert, the recommended first step for a vessel is to:
    (a) immediately switch off the radio in question.
    (b) advise all stations by voice.
    (c) advise a coast station.
    (d) wait for the message to be acknowledged.

129. A DSC urgency alert is always transmitted:
    (a) to all stations.
    (b) to all stations or to a single station.
    (c) on a distress priority mode.
    (d) on a non-distress frequency.

130. A DSC safety alert:
    (a) should not be acknowledged.
    (b) should be transmitted on a non-distress frequency.
    (c) may only be transmitted to all stations.
    (d) should be transmitted on urgency priority.

Chapter 18: Radio, Satellite & Epirb Communication

131. The MF/HF DSC operation is verified for correct operation as follows:
    (a) ask a station for a DSC radio check
    (b) make a routine call.
    (c) make a test call to a coast station.
    (d) activate the "Test" mode.

132. Under GMDSS regulations, the DSC equipment tests:
    (a) are not permitted for operator familiarization.
    (b) must be carried out at least once a month.
    (c) must not be carried out more than once a month.
    (d) are designed to generate routine calls.

133. In the following list, the correct pair of DSC and its associated radiotelephony frequency is:
    (a) VHF Channel 70 & 67
    (b) 2287.5 & 2182 kHz.
    (c) 4207.5 & 4125 kHz.
    (d) 6312 & 6125 kHz.

134. In the following list, the correct pair of DSC and its associated radiotelephony frequency is:
    (a) 6312 & 6215 kHz
    (b) 2278.5 & 2182 kHz.
    (c) 4207.5 & 4215 kHz.
    (d) 8415.5 & 8291 kHz.

135. After receiving a DSC alert via a 4MHz frequency, a vessel does not hear a distress message nor hear any other station communicating with the vessel in distress for 5 minutes. She should:
    (a) transmit a DSC distress relay to an appropriate coast station or a limited coast station.
    (b) contact the vessel in distress.
    (c) acknowledge the distress alert by DSC.
    (d) transmit a DSC distress relay to all stations.

136. A DSC distress alert relay:
    (a) must only be transmitted on receipt of a distress alert.
    (b) may be transmitted on behalf of another vessel in distress which is unable to transmit distress alert.
    (c) must only be transmitted to all ships.
    (d) must only be transmitted to a coast or limited coast station.

137. A DSC distress alert relay, when transmitted on behalf of another vessel in distress, which is unable to transmit distress alert, must be addressed to:
    (a) all stations.
    (b) a coast station.
    (c) a coast or limited coast station
    (d) all stations or a coast or limited coast station.

138. Transmission of a DSC distress alert or radiotelephony distress call and message must be on the authority of:
    (a) the vessel owner.
    (b) the master or the mate.
    (c) the master.
    (d) any adult on board.

139. Communication after the transmission of a DSC alert by a small vessel is normally carried out using the:
    (a) Inmarsat satellite system.
    (b) Narrow band direct printing.
    (c) Standard radiotelephone procedure.
    (d) DSC keyboard.

140. Vessel 'ABC' has sighted vessel 'XYZ' totally on fire, and believes that vessel 'XYZ' is unable to transmit a DSC distress alert. Vessel 'ABC' should take the following action:
    (a) Transmit a DSC urgency alert.
    (b) Transmit a DSC distress alert.
    (c) Transmit a DSC safety alert.
    (d) Transmit a DSC distress alert relay.

141. Radiotelephony communication after a DSC distress alert made on VHF is <u>normally</u> carried out on:
    (a) Channel 70.
    (b) DSC channel.
    (c) Channel 16
    (d) Channel 67.

142. Which stations are normally permitted to use DSC allocated channels and frequencies for radiotelephony transmissions?
    (a) None
    (b) Coastal stations
    (c) All stations
    (d) Coast stations and limited coast stations

143. In the following list, the international DSC frequency is:
    (a) 2182 kHz
    (b) 4125 kHz
    (c) 6215 kHz
    (d) 4207.5 kHz

144. Listed below are some steps for a vessel's transmission of a DSC Safety Alert. Which is the <u>incorrect</u> step?
    (a) Step 1 = Select DSC frequency
    (b) Step 2 = Select "all ships" call format
    (c) Step 3 = Select "Urgency" priority
    (d) Step 4 = Initiate DSC Safety Alert

145. The DSC distress alerts are received and decoded by the following stations within the telecommunication range:

    (a) Only the stations to which the alert is addressed.
    (b) All stations keeping a DSC calling watch.
    (c) All stations keeping a radiotelephony watch.
    (d) Only the coast stations.

146. Listed below are some steps for a vessel's transmission of a DSC Urgency Alert. Which is the incorrect step?

    (a) Step 1 = Select DSC frequency
    (b) Step 2 = Select "all ships" call format
    (c) Step 3 = Select "Urgency" priority
    (d) Step 4 = Initiate Distress Alert

CHAPTER 18.5: SATELLITE ENDORSEMENT

147. Which of the following is not an Inmarsat service?

    (a) HF DSC
    (b) Direct-dial telephone
    (c) Email
    (d) Fax

148. In relation to Inmarsat, which of the following statements is incorrect?

    (a) It consists of four communication satellites over four oceans.
    (b) Positioned above the equator, the satellites rotate with the earth.
    (c) It is an exclusive maritime communication system.
    (d) It is a continuous SHF global telecommunication system.

149. Which of the following statements is incorrect? The Inmarsat service provides:

    (a) Email service in 'store & forward' mode only.
    (b) 'Distress Priority' transmission service.
    (c) EGC service.
    (d) Telephone service in both 'real time', and 'store and forward' modes.

150. In relation to Inmarsat-C, which of the following statements is incorrect?

    (a) It requires a dish antenna.
    (b) SES is synchronised with NCS either automatically or by the operator.
    (c) Switching off without first logging off can cause message delivery problems.
    (d) Senders are notified of successful delivery of messages.

151. In relation to Inmarsat-C, which of the following statements is correct?

    (a) On pressing a single control buttons it generates a distress alert.
    (b) The default distress alert does not include vessel's position.
    (c) A distress alert cannot be transmitted if the SES is not logged-on.
    (d) The equipment will indicate acknowledgement of receipt of distress alert by a LES.

152. In relation to Inmarsat-C, which of the following statements is correct?

    (a) It cannot store messages. They must be transmitted immediately.
    (b) The routine messages delivery time is 30 seconds.
    (c) It operates just like an email or fax sent from a computer.
    (d) Received messages can only be viewed on a monitor. They cannot be printed.

153. In relation to Inmarsat-C, which of the following statements is correct?

    (a) A LES or MRCC must be nominated before transmitting a distress alert.
    (b) Distress Alert transmission messages cannot be typed on the keyboard.
    (c) Distress Alert acknowledgement must be received from a LES as well as from a MRCC.
    (d) Communication following the initial distress alert is conducted by keyboard.

154. In relation to Inmarsat-C, which of the following statements is correct?

    (a) The 2-digit Inmarsat code for automatic routing of "seeking medical advice" is 42.
    (b) Receipt of Distress or Urgency Alert via EGC is indicated by audible and/or visual alarm.
    (c) The 2-digit Inmarsat code for automatic routing of "navigational hazard report" is 32.
    (d) An unintentional Distress Alert must be cancelled by sending a message to all MRCCs.

Chapter 18: Radio, Satellite & Epirb Communication

155. In relation to Inmarsat-C, which of the following statements is <u>incorrect</u>?

    (a) Performance Verification Test (PVT) can be performed whenever the operator feels concerned about the equipment
    (b) PVT may take up to 20 minutes depending on the level of Inmarsat traffic congestion.
    (c) The antenna's concentrated field of radiation lies in a radius of 7 metres around it.
    (d) The antenna should be kept at least 1 metre away from objects causing shadow sector of more than 2°.

156. Which of the following is a GMDSS–approved Inmarsat?

    (a) Inmarsat-C
    (b) Inmarsat-M
    (c) MINISAT
    (d) Inmarsat-D+

157. The purpose of EGC is to distribute:

    (a) Weather information.
    (b) Safety information.
    (c) Weather & safety information
    (d) Weather & Safety information to geographically targeted ships.

158. In relation to EGC, which of the following statements is <u>incorrect</u>?

    (a) Safety NET is an EGC message.
    (b) FleetNET is an EGC message.
    (c) MSI is broadcast via FleetNET.
    (d) EGC Receivers reject messages not intended for them.

159. Which of the following is <u>not</u> a SafetyNET Authorised Organisation?

    (a) Hydrographic Office.
    (b) Ship Owner.
    (c) Meteorological Office.
    (d) MRCC

160. Which of the following is <u>not</u> a MSI message?

    (a) Fleet information broadcast by a Ship Owner.
    (b) Navigational Warning.
    (c) Distress Alert.
    (d) Weather Forecast.

161. The consequence of not updating EGC Receiver's positional information is as follows:

    (a) It will not receive MSI messages.
    (b) It will accept all MSI messages not intended for it.
    (c) It will not set off alarm on receipt of distress alerts.
    (d) It will accept all MSI messages with priority higher than "routine" for the entire ocean.

162. In relation to Inmarsat Epirb, which of the following statements is <u>incorrect</u>?

    (a) It is known as Inmarsat-E.
    (b) It is an "L" band Epirb.
    (c) Inmarsat satellites indicate Epirb's position without needing GPS input.
    (d) Its detection is not subject to the delay of a satellite passing overhead.

CHAPTER 18: ANSWERS

1 (b), 2 (d), 3 (d), 4 (b), 5 (b),
6 (d), 7 (a), 8 (b), 9 (c), 10 (d),
11 (d), 12 (c), 13 (c), 14 (b), 15 (c),
16 (b), 17 (a), 18 (b), 19 (a), 20 (c),
21 (b), 22 (a), 23 (b), 24 (a), 25 (c),
26 (d), 27 (a), 28 (c), 29 (a), 30 (c),
31 (a), 32 (d), 33 (a), 34 (b), 35 (b),
36 (c), 37 (a), 38 (d), 39 (d), 40 (c),
41 (d), 42 (c), 43 (c), 44 (d), 45 (b),
46 (b), 47 (b), 48 (b), 49 (d), 50 (d),
51 (d), 52 (c), 53 (d), 54 (a), 55 (d),
56 (c), 57 (a), 58 (c), 59 (d), 60 (d),
61 (c), 62 (b), 63 (a), 64 (d), 65 (b),
66 (c), 67 (d), 68 (d), 69 (c), 70 (c),
71 (c), 72 (b), 73 (a), 74 (c), 75 (b),
76 (d), 77 (c), 78 (d), 79 (a) 80 (d),
81 (c), 82 (c), 83 (c), 84 (b), 85 (b),
86 (d), 87 (c), 88 (c), 89 (a), 90 (d),
91 (c), 92 (b), 93 (c), 94 (b), 95 (d),
96 (b), 97 (a), 98 (c), 99 (b), 100 (c),
101 (a), 102 (d), 103 (a), 104 (c), 105 (c),
106 (a), 107 (b), 108 (c), 109 (a), 110 (d),
111 (a), 112 (b), 113 (a), 114 (b), 115 (b),
116 (d), 117 (c), 118 (a), 119 (a), 120 (a),
121 (c), 122 (c), 123 (d), 124 (d), 125 (c),
126 (d), 127 (c), 128 (a), 129 (b), 130 (a),
131 (d), 132 (b), 133 (c), 134 (a), 135 (a),
136 (b), 137 (d), 138 (c), 139 (c), 140 (d),
141 (c), 142 (a), 143 (d), 144 (c), 145 (b),
146 (d), 147 (a), 148 (c), 149 (a), 150 (a),
151 (d), 152 (c), 153 (d), 154 (b), 155 (c),
156 (a), 157 (d), 158 (c), 159 (b), 160 (a),
161 (d), 162 (c)

Chapter 19.1

OCCUPATIONAL HEALTH & SAFETY

FIRST AID

In the first edition of the book, this chapter was based on the AMSA's Safety Education Articles on First Aid and the AYF's recommended medical kit. Over the years, the chapter has gone through major revisions and additions. The main contributor to the current information is Martin Phillips, an online pharmacist at www.pharmacy2shop.com.

WARNING: An apparently straightforward medical situation may in fact be complicated. A casualty may unknowingly be allergic to a medicine or may react negatively to a treatment. Masters of vessels have a duty of care to use all available means at their disposal to assist (but not harm) the casualty until trained medical help arrives.
State and Territory Poisons Acts vary in relation to procurement of drugs and their storage requirements. Masters are normally required to take charge of supplementary items and keep under lock and key.

For these reasons, it is recommended that masters …
1. SEEK AND FOLLOW RADIO MEDICAL ADVICE
2. ADMINISTER RESTRICTED DRUGS ONLY UNDER MEDICAL DIRECTION
3. RECORD ALL COMMUNICATIONS & ACTIONS IN THE SHIP'S LOG

O.H. & S. STANDARD
Under the Occupational Health and Safety standards, it is the responsibility of the owners and masters of vessels to equip their vessels of three things:
- Knowledge of first aid
- A well equipped first aid kit
- An instruction book on administering first aid.

DRABC (Doctor ABC) ACTION PLAN

Follow the five DRABC steps in every first aid situation:
- D   Check for <u>DANGER</u> to you and the victim.
- R   Check the casualty's <u>RESPONSE</u> to your questions.
- A   Clear and then open the casualty's <u>AIRWAY</u>.
- B   Help a casualty, who can't breathe, to <u>BREATHE</u>.
- C   <u>CIRCULATE</u> the blood around the casualty's body if there is no sign of circulation, such as movement, coughing or pulse.

After following the DRABC plan, deal with any other injuries or illnesses, usually (but not strictly) in the following order:
- Bleeding
- Other wounds, such as burns
- Shock
- Fractures

IF SOMEONE HAS COLLAPSED

ASSESSMENT
- Ensure safety for the rescuer and the collapsed person.
- Stay with the collapsed person and call for help.
- Commence appropriate treatment, caring for the unconscious persons first and follow the sequence in the Basic Life Support Flow Chart in Figure 19.7.

DEFINITION OF CONSCIOUS
A conscious person is able to respond to the spoken word and obey a shouted command. To check if a person is conscious, carry out the "TOUCH & TALK TEST": Touch the person on the shoulder (do not shake), ask for name, and give a simple command: "Open your eyes". A conscious person will respond.

DEFINITION OF UNCONSCIOUS
An unconscious person is unable to respond to the spoken word and to obey a shouted command.

CARE OF COLLAPSED PERSON WHO IS CONSCIOUS
- Reassure the person.
- Allow the person to assume a comfortable position.
- Keep airway, breathing and circulation under close observation.
- Handle gently avoiding unnecessary movement.
- Treat injuries according to priorities:
    - Control bleeding
    - Splint fractures
- Elevate the legs if possible but keep head level with heart.
- Moisten lips but do not give anything by mouth.
- Protect from cold and prevent from shivering but do not overheat.
- Arrange for further care of the person.

CARE FOR COLLAPSED PERSON WHO IS UNCONSCIOUS
- Airway
- Breathing
- Circulation (and control any bleeding)
- Avoid lifting, bending or twisting the head or neck. If possible, have an assistant hold the head when moving the person.
- An unconscious person who is breathing should remain in the recovery (lateral) position.

CAUSES OF UNCONSCIOUSNESS INCLUDE
- Fainting
- Head injury
- Fits
- Alcohol ingestion or drug overdose
- Disease: stroke, diabetes, etc.
- Heat stroke
- Lack of oxygen: near drowning, cardiac arrest, etc.

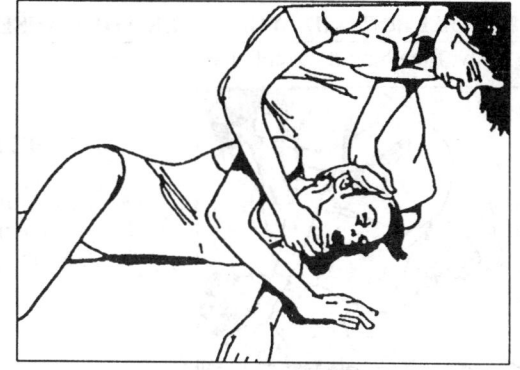

Fig 19.5: Recovery (Lateral) Position

IF A PERSON IS IN SHOCK

Any seriously injured person will suffer from shock, which can be serious. Injuries can be made worse by unnecessary or rough movement, so do not move an injured person unless:
- There is danger from fire, road traffic, hot road surfaces, surf conditions, shark.
- It is necessary to provide life saving treatment:
    - to establish a clear airway
    - to begin expired air resuscitation

Always handle an injured or collapsed person gently, and except in life threatening situations, splint fractures before moving an injured person.

Persons affected by shock are likely to have cold, clammy skin and a pulse that may be slow at first especially in children and young adults, but later becomes rapid and feeble. The breathing may be rapid and shallow - a condition sometimes called "AIR HUNGER". If blood loss is severe, the level of consciousness may be altered and the patient may become unconscious. A careful check must be kept on:
- The state of consciousness
- Airway
- Breathing
- Circulation

TREATMENT
- If unconscious, turn on side and care for airway, breathing and circulation.
- Stop any bleeding.
- Elevate legs if possible but keep the head level with the heart.
- Protect from extremes of temperature.
- Moisten lips but do not give anything by mouth.

➤ Seek medical help.

EXPIRED AIR RESUSCITATION (E.A.R.)

When breathing stops, the oxygen supply to the brain is interrupted. If it is not restored quickly, death or irreversible brain damage will occur.

The air we exhale contains enough oxygen to sustain the life of a person who has stopped breathing provided expired air resuscitation (E.A.R.) is commenced promptly. Expired air resuscitation - mouth to mouth or mouth to nose technique - is used whenever breathing has stopped in cases such as:
➤ near drowning
➤ drug overdose
➤ smothering smoke inhalation.

Therefore, whatever the cause of collapse and stoppage of breathing, follow the following "A B C" OF FIRST AID. If applying E.A.R. in the water, do mouth-to-nose, not mouth-to-mouth.

A B C OF FIRST AID: WHATEVER THE CAUSE OF COLLAPSE (Fig. 19.6)
(Airway, Breathing & Circulation)

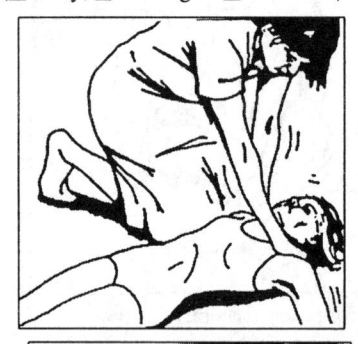

1. IS THE PERSON CONSCIOUS OR UNCONSCIOUS?

➤ Assess the response to "touch & talk" (no shaking)
➤ Ask the person's name
➤ Give the person a command: "open your eyes"

2. If the person does not respond, CLEAR THE AIRWAY

➤ Quickly turn on side to clear airway of loose foreign material
➤ Leave firmly fitting dentures in place
➤ Put head in backward tilt
➤ Support jaw

3. CHECK FOR BREATHING

➤ Watch for the movement of lower chest & abdomen
➤ Listen & feel for escape of air from nose & mouth for 10 seconds only

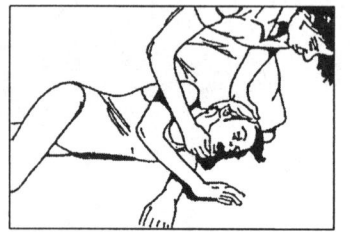

4. IF BREATHING: Leave lying on side in lateral (recovery) position with head in backward tilt, jaw supported & face pointing slightly downward. Observe airway, breathing & circulation constantly.

Chapter 19.1: OH&S - First Aid

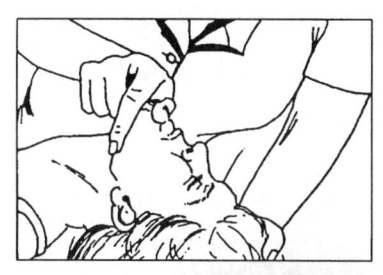

5. **IF NOT BREATHING,** quickly turn victim on his/her back.

   Kneel beside the person. Support the chin with a "pistol grip" & tilt head back. The chin is supported with the knuckle of middle finger, with little & ring finger clear of neck & jaw.

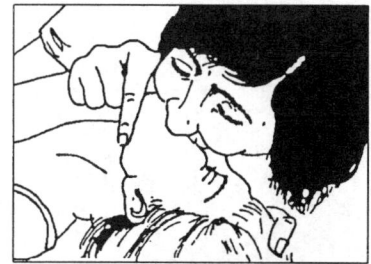

6. **BLOW:** Take a deep breath. Place widely open mouth over the victim's slightly open mouth - sealing nose with rescuer's cheek. Blow till chest rises. Give 5 full breaths within 10 seconds. If chest does not rise, check for obstruction, leaks and head tilt (adults), or blow harder.

7. **LOOK FOR SIGNS SUCH AS MOVEMENT, COUGHING OR PULSE (CAROTID)** in the neck for 10 seconds only. If absent: start CPR, checking for any of the above signs & breathing every minute & then at least every 2 minutes. If any of the above signs is present…

8. **RECOVERY:** Repeat E.A.R. 15 times per minute, i.e., once every 4 seconds for adults.

   When the victim begins to breathe, turn him/her on his/her side to the lateral (recovery) position & keep a constant check on airway, breathing & circulation.

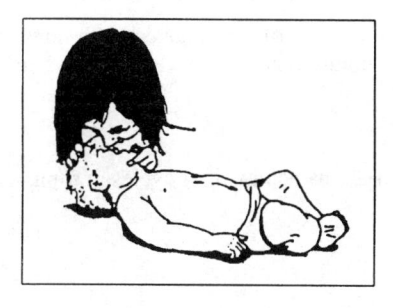

9. **FOR INFANTS & SMALL CHILDREN:** After clearing the airway, support the jaw without tilting the head backwards. Cover the mouth & nose before puffing gently into the lungs at a rate of 20 per minute, i.e., once every 3 seconds.

PROBLEMS & SOLUTIONS WITH E.A.R.

- Signs of airway leaks:
    - Bubbling round mouth / nose
    - Failure of chest to rise
- Correction of airway leak:
    - Rescuer's mouth widely open
    - Victim's nose adequately sealed
- Failure of chest to rise is due to:

- Blocked airway
- Insufficient head tilt (adults)
- Poor air seal
- Not blowing hard enough
- If chest does not rise:
  - Check airway
  - Check air seal
  - Blow a little harder
- Stretching of the stomach is due to:
  - Blowing too hard
  - Blocked airway
  - Insufficient head tilt *(not applicable to babies. Their head tilt must be neutral)*
- To prevent stomach stretching:
  - Do not compress stomach
  - Check the airway
  - Do not blow too hard
  - When chest rises stop blowing

CARDIOPULMONARY RESUSCITATION (C.P.R.)

C.P.R. = E.A.R. + E.C.C.

If you are trained to do CPR, after the first five breaths (in 10 seconds), feel for the carotid pulse by sliding two or three fingers into the groove in the neck between the large muscle and the Adam's apple. Feel with the pulps, not the tips of the fingers.

- If the carotid pulse is present:
  - Continue E.A.R. at a rate of one breath every 4 seconds (15/min) for adults.
  - One breath every 3 seconds (20/min) for infants and small children.
- If the carotid pulse is absent, and the patient is:
  - UNCONSCIOUS and NOT BREATHING, then the patient has suffered a CARDIAC ARREST.

The treatment for cardiac arrest is C.P.R. (i.e., E.A.R. + E.C.C.)

EXTERNAL CARDIAC COMPRESSION (E.C.C.)

E.C.C. involves compressing the heart against the backbone by pushing the breast bone or sternum downwards in a rhythmical manner so as to artificially stimulate the pumping action of the heart and maintain the circulation of blood.

When two rescuers are present, one performs E.C.C. and the other E.A.R. in the ratio of 5 cardiac compressions to 1 lung inflation, achieving 12 cycles per minute. There must be no pause in the compressions, the ventilations being interrupted between the 5th compression of one cycle and the first of the next.

For children there should be 20 cycles per minute.

When only one rescuer is present, the ratio is 15 compressions to 2 lung inflations achieving 4 cycles per minute for adults and 6 for children.

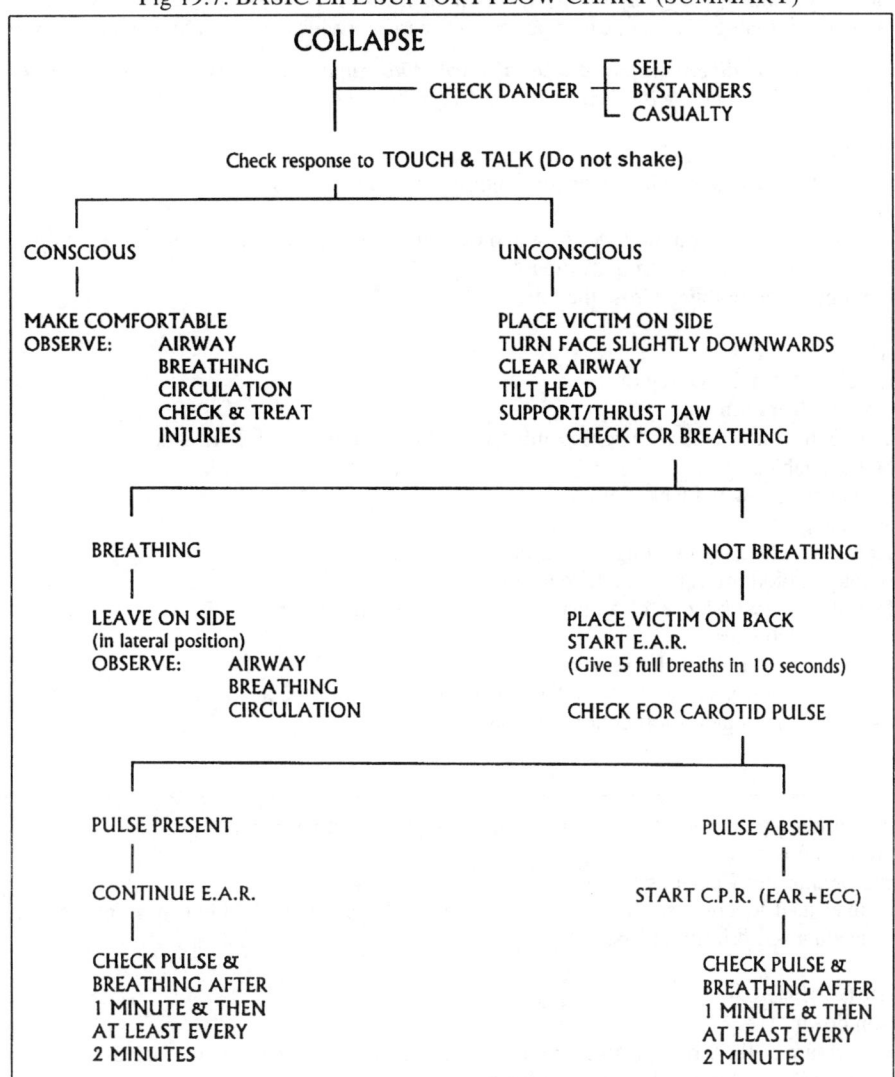

Fig 19.7: BASIC LIFE SUPPORT FLOW CHART (SUMMARY)

```
                            COLLAPSE
                               |
                               |— CHECK DANGER —┬ SELF
                               |                ├ BYSTANDERS
                               |                └ CASUALTY
                   Check response to TOUCH & TALK (Do not shake)
          ┌────────────────────┴─────────────────────┐
      CONSCIOUS                                  UNCONSCIOUS
          |                                          |
   MAKE COMFORTABLE                         PLACE VICTIM ON SIDE
   OBSERVE:   AIRWAY                        TURN FACE SLIGHTLY DOWNWARDS
              BREATHING                     CLEAR AIRWAY
              CIRCULATION                   TILT HEAD
              CHECK & TREAT                 SUPPORT/THRUST JAW
              INJURIES                      CHECK FOR BREATHING
                                   ┌───────────┴────────────┐
                               BREATHING                NOT BREATHING
                                   |                         |
                             LEAVE ON SIDE            PLACE VICTIM ON BACK
                             (in lateral position)    START E.A.R.
                             OBSERVE:  AIRWAY         (Give 5 full breaths in 10 seconds)
                                       BREATHING
                                       CIRCULATION   CHECK FOR CAROTID PULSE
                                                ┌────────────┴────────────┐
                                          PULSE PRESENT              PULSE ABSENT
                                                |                         |
                                          CONTINUE E.A.R.           START C.P.R. (EAR+ECC)
                                                |                         |
                                          CHECK PULSE &             CHECK PULSE &
                                          BREATHING AFTER           BREATHING AFTER
                                          1 MINUTE & THEN           1 MINUTE & THEN
                                          AT LEAST EVERY            AT LEAST EVERY
                                          2 MINUTES                 2 MINUTES
```

SURVIVING A HEART ATTACK WHEN ALONE

One is often alone on a boat. If your heart stops beating properly and you begin to feel faint, you have only about 10 seconds before losing consciousness. However, you can help yourself by coughing repeatedly and vigorously.

Take a deep breath before each cough, and make the cough deep and prolonged, as if producing sputum from the bottom of the lungs. Repeat the breath and cough pattern about every two seconds. Do not stop until help arrives or until you feel the heart beating normally again.

Deep breaths inject oxygen into the lungs and coughing squeezes the heart and keeps the blood circulating. The squeezing pressure on the heart also helps to restore its normal rhythm.

This procedure will allow you to dial a phone and/or call for help between breaths.

BURNS

A burn is the damage caused by heat to skin and at times to deeper tissues. The heat could be from flames, scald, electricity, chemical agents or solar radiation.

First aid treatment consists of flooding the burnt area with cool water for a minimum of 20 minutes, to relieve pain and lessen damage to tissues.

Burns are not only painful; they may be life threatening from severe loss of body fluids, and tissue destruction. Even when the burnt area is not large, they can cause permanent scarring or loss of function, such as of joints.

A first aider should not differentiate the type of burns into Superficial, Intermediate and Deep. Looks can be deceptive. Home medication should not be applied to burns. All burns should be flooded with cool water.

TREATMENT OF SCALDS
- Immediately cool the burnt area with water for a minimum of 20 minutes.
- If water is not available:
  - Remove all clothing (but not over face and/or airway). Clothing soaked with hot fluids retains heat.
  - Mop hot liquid from natural body creases.
  - Do not pull wet clothing across the face.

TREATMENT OF FLAME BURNS
- Remove the person from the source of flame.
- Prevent the person from running.
- Immediately cool the burnt area with cool running water for a minimum of 20 minutes.
- If water is not available:
  - Smother flames with blanket or coat.
  - Force the person to lie on the floor.
  - Remove smouldering clothing if not stuck to the skin.
  - Do not pull clothing across the face if possible.
  - Cover the area with loose, clean dry cloth to prevent contamination.
  - Do not break blisters.
  - Do not peel clothing stuck to body.
  - Do not apply lotions or ointments. They make it difficult for the hospital to assess the burns.
  - Cover burns with damp soft cotton material.

BLEEDING

Bleeding or haemorrhage is loss of blood from blood vessels. Bleeding may be:
- External - from an obvious wound.
- Internal - into organs, body cavities or tissues.

Severe bleeding may lead to collapse or death, depending on the amount of blood loss and how quickly it is lost. Therefore, stopping bleeding is a priority in first aid.

TREATMENT OF BLEEDING
- Lie the person down.
- Apply direct pressure to the wound with a firm pad or "shell" dressing - or fingers if necessary.
- Elevate the wound if on a limb and there is no fracture.
- If bleeding continues, place another pad over the first one, bandage firmer than the first time.
- If bleeding is profuse, or a firm dressing is not available, grasp the sides of the wound and press them firmly together.
- When bleeding is controlled, apply a standard dressing to the wound. Do not touch the part of the dressing in contact with the wound.

MARINE ANIMAL BITES & STINGS

Blue Ringed Octopus
Distribution: Indo-Pacific Ocean, extending from Japan to Southern tip of Australia.
Envenomation: Contrary to popular belief the octopus does not carry poison in its iridescent blue rings. It injects toxin into human skin by biting with its beak.
Signs and symptoms:
- The initial bite is usually minor and relatively painless, but may develop into a small red blister.
- Facial numbness, swelling, headache, nausea, vomiting, and visual disturbances with progressive breathing difficulties may occur within 10 to 15 minutes.
- Muscular paralysis, respiratory and cardiac arrests may occur in severe cases.

First Aid:
- DRABC checklist.
- Immobilise limb and apply ice to reduce circulation, which would decrease the spread of venom.
- Rest and reassurance.

Chapter 19.1: OH&S - First Aid

- ➢ EAR / CPR as needed. EAR may need to be performed for hours.
- ➢ Seek medical aid.

Cone Shell
Distribution: Widely. Especially Indo-Pacific, Red Sea, Caribbean areas.
Envenomation, signs and symptoms, first aid: as above (per octopus).

Sea Snake.
Distribution: Widely, except Atlantic Ocean.
Envenomation signs and symptoms, first aid: as above (per octopus).

Spine-Bearing.
Distribution: Widely, dependent on species. Includes various fish, rays, urchins, and starfish.
Injury/envenomation: Physical puncture, which may introduce secondary infection into wound. Spine fragments may break off and remain lodged deep in wound.
Signs and symptoms:
- ➢ Pain, swelling and bleeding at puncture site.
- ➢ May spread if toxin-producing species involved.
- ➢ Nausea, vomiting and shock (rarely progressing further).

First Aid:
- ➢ DRABC checklist.
- ➢ Control bleeding only if severe, then remove spine (if present, and only if loose).
- ➢ Irrigate away contamination.
- ➢ Immerse area in hot water. (If hot water therapy does not provide relief, try icepack).
- ➢ Inspect, clean, disinfect and dress wound.
- ➢ Analgesics if directed.
- ➢ Seek medical aid.

Box jellyfish
Also relevant for: Hydroids and Anemones found in both coastal and reef systems. *Coral:* see below.
Distribution: Widely. Tropical coastal regions (part of breeding cycle). Action of wind and waves breaks up the delicate animals, which usually do not survive beyond inshore limits.
Envenomation: Tentacle shafts inject toxin into skin.
Signs and symptoms:
- ➢ Sharp pain on contact.
- ➢ Raised weals/rash; blisters if severe.
- ➢ Lymph gland and general aching nerve pain if toxins spread.
- ➢ Respiratory difficulty, shock, unconsciousness, possible death.

First Aid:
- ➢ DRABC checklist.
- ➢ Irrigate with vinegar.
- ➢ Glove up and remove tentacles.
- ➢ Do not rub skin as this may promote toxin delivery.
- ➢ Wrapped icepacks may ease pain.
- ➢ Immobilisation bandages may also assist recovery.
- ➢ Seek medical aid (including stronger pain relief options).

Coral cuts:
- ➢ Irrigate with sterile water and clean wound.
- ➢ Apply antiseptic to reduce risk of secondary infection.
- ➢ Monitor site (may require antibiotics if sepsis confirmed).

Blue Bottle (Portuguese Man-Of-War)
Distribution: Widely. Risk of contact worse at ocean beaches exposed to onshore winds.
Envenomation: Barb-covered tentacle shafts inject toxins into the skin.
Signs and symptoms:
- ➢ Sharp pain on contact.
- ➢ Raised weals/rash; blisters if severe.
- ➢ General aching nerve pain.
- ➢ Respiratory difficulty, shock, unconsciousness, possible death.

- NB: Scarring may result at points of contact.

First Aid:
- DRABC checklist.
- Glove up and remove visible tentacles then irrigate with clean saltwater.
- Do NOT use water, alcohol or vinegar as these may stimulate further toxin release from the barb cell pumps.
- Wrapped icepacks may ease pain and swelling, and reduce circulating toxins.
- Seek medical aid.

PREVENTION FROM ALL ANIMAL STINGS
- Prevention is better than cure.
- Know your aquatic environment, respect it, and keep a look out.
- Before entering water, suit up or use rash vests.
- Maintain buoyancy.
- Petroleum jelly may help prevent tentacles from attaching to the skin.

SEASICKNESS (MOTION SICKNESS)
Motion sickness prevention is more than just popping a pill...

Motion sickness is said to occur when, over a period of time, the brain receives conflicting sensory signals from the eyes, inner ears, muscles and joints. This confusion results in loss of balance, with stomach reaction (nausea, and sometimes vomiting). The trigger stimuli may even come from watching fellow travellers vomiting.

WHO IS AT RISK
Anyone can suffer from motion sickness, but the type of vessel or motion that affects one person may not affect another. However the motion of a liferaft makes almost everyone seasick.
Stress, diet and lifestyle play a role in seasickness. These are discussed under OH&S – Work Practices in Chapter 19.4.

WARNING SIGNS
- Drowsiness
- Faintness
- Skin pallor
- Salivating.

HOW TO REDUCE THE RISK
It is helpful to acclimatise oneself by sleeping on board the night before departure. Alternatively, on boarding, a brief nap in your berth may also assist getting your 'sea legs'.

ANTI-MOTION SEASICKNESS MEDICATION
Ideally, preventative medication needs to be taken before embarking the vessel, or well prior to experiencing open water swell.

OVER-THE-COUNTER MEDICATIONS include:
- *Dramamine* tablets (dimenhydrinate 50mg)
- *Kwells* chewable tablets (hyoscine hydrobromide 0.3mg)
- *Travacalm* tablets (dimenhydrinate 50mg, hyoscine 0.2mg, caffeine 20mg)
- *Phenergan* tablets (promethazine: child 10mg and adult 25mg)

Dosage: 2 tablets with water at least 30 minutes prior departure; repeat at 4 hourly intervals. The *Travacalm* tablet tends to be stronger and preferred for those who are prone to seasickness. *Dramamine* also comes in two junior forms for children (from 2 years of age). *Phenergan* tablets are longer-acting sedating preparations (8-12 hours), normally taken at bedtime the night before. Caution: these may cause hangover effect.
Side effects: commonly include secretion drying (eyes, nose, mouth), and occasionally drowsiness. Rarely may cause agitation, blurred vision, urine flow inhibition, dizziness, etc.
Medication that may be obtained from the UK (not currently available in Australia):
- *Scop* skin patches (hyoscine) are placed behind the ear 6 hours before travel, and last up to 3 days before a new patch is needed. Suitable for long cruises.
- *Stugeron* tablets (cinnarizine 30mg, then 15mg, 6-8 hourly is the European drug of choice).

PRESCRIPTION MEDICATIONS commonly include:
- *Stemetil* tablets (prochlorperazine 5mg) - for greater gut and vestibular control.

Chapter 19.1: OH&S - First Aid

- *Maxolon* tablets (metoclopramide 10mg) - used less often.
  Caution: Both cause drowsiness.

NATURAL ALTERNATIVES
Natural, non-drowsy, therapy is effective and works directly on the stomach and gut. It is in common use in Europe and Australasia.
- Blended formulae, with ginger as the base, appear the best, eg: *Travelaid* capsules (ginger zingiber officinale 500mg, cardamom 1.5mcg, slippery elm 2.6mcg) dosed as 2-3 capsules with water at least 30 minutes prior to departure, then two more every 2-3 hours, if necessary. Ginger may also be taken as cookies or non-alcoholic ginger beer. It may be combined with any over-the-counter or prescription tablets for added effect. It may cause heartburn. Consult doctor if pregnant or suffer from gallstones.
- *Sea-Band* acupressure wristbands work in some cases, but may not be practical on the wrist for crew or divers.
- For a DIY acupressure, find a point on the underside of your arm about 4 cm above the wrist crease between two tendons. Press in there fairly hard, but not so hard that you get a bruise.

HELPFUL HINTS WHEN FEELING SEASICK
If your feel seasick, try and remove as many unwanted stimuli as possible...

- Alert the skipper, who may be able to alter course or speed.
- Find fresh air on deck - away from enclosed spaces that trap paint fumes, fuel vapours, engine exhaust emissions, and hot food odour.
- Keep your head up - avoid reading or putting your head down.
- Watch a fixed point on the horizon – a land point if possible. Avoid focussing on moving objects.
- If you want to rest, lie down mid-ships (fore-aft axis) with eyes closed.
- If you feel up to it, sip water and nibble on a dry biscuit, barley sugar or dried fruit.
- And a tip if you are about to vomit... Make sure you are standing (if safe to do so) at the leeward railing, and not on the windward side

SAFETY NOTE FOR MEDICINES

Unless previously used, always trial any planned medication regime prior to actual use. Drug interactions may occur with other medications. Knowledge of many newer drugs in relation to underwater activities is limited. Brand names mentioned are the registered trademarks of their respective companies, and no endorsement is given on products or procedures. Individuals should see the above information as of a general nature only and not rely on it in part or whole, but seek personal medical or pharmacy advice, as relevant.

BASIC MEDICAL KIT with AUDIT CHECKLIST*

Last checked: / / by _____.

Store kit within a waterproof container, with this inventory checklist affixed inside lid.

Apparent Medical Condition	Possible Treatment Option	Suggested Kit contents	No. in kit	Exp date	Notes & Warnings#
Abrasion, minor cut	Strips-fabric & w/proof *Band-aid, Elastoplast*	1 Box of each type of use			Assorted sizes/widths. May require irrigation and antiseptic.
Bites, stings & skin allergies -mild / general	Toxin neutraliser – *Stingose* spray	100ml			Use – insect bites and minor stings.
	Anaesthetic in an antiseptic base - *Paxyl* spray (lignocaine 2.5%) - *Medicreme* cream (lignocaine 2%)	125ml 50g tube			(P) Symptomatic relief of minor irritations.
- moderate	Hydrocortisone 1% topical: *Derm-Aid* crm	30g			(P) Add to therapy if no relief with above.
- mod / severe (non-drowsy)	Antihistamine: Telfast 180mg tabs (Fexofenadine)	1 Box 10			(P) Use ASAP if: medical history risk, or skin weals/wheeziness.

- severe (drowsy risk)	Phenergan 25mg tabs (promethazine)	1 Box 50			Dose: Telfast 1 daily or Phenergan 1-2 per day.
Bites & stings - Marine animals	Vinegar wash; Gloves to remove tentacles	500ml			+ Box jellyfish. Use pressure bandage.
	Salt water wash; gloves to remove tentacles; ice pack				+ Bluebottle (Portuguese Man-of-War)
	Hot (45°C) water for at least ½ to 1 hour, after cleaning puncture & flushing bleed area.				+ Fish, ray & urchin spines. NB: tetanus vaccine.
	Ice pack; Pressure bandage				+ Blue ringed octopus, cone shell & sea snakes
Burn – mild / intact	Cool 15 mins water; First-aid spray optional	See entry below			+ Alternative: Aloe Vera gel
Burn – moderate/severe	Non-stick wound pad- *Cutilin, Melolin, Telfa* Non-stick paraffin pad- *Bactigras, Jelonet* W/proof clear seal- *Op-Site, Tegaderm* Burn cream: *Silvazine*	1 Box of each type of use (5-10 per box) 100g tube			+ Cool area. Some ointments and sprays may warm or irritate major burns. (Rx) *Silvazine* Cream
Congestion – Ears	*Aqua Ear* drops	35ml			(P) Dries & cleans
- Sinuses	*Sudafed* 60mg tabs (Pseudoephedrine)	1 Box 30			(H) (P) Dose 1 tab every six hrs
	Vicks nasal spray	15ml			(P) Alternate: *Vaporub*
Dehydration	*Gastrolyte, Repalyte*	1 Box			(P) Oral rehydration
Diarrhoea - frequency	*Imodium* 2mg tab (Loperamide)	1 Box 8			(P) Non-drowsy. Dose 2 at onset, then 1 every 2-3 hrs (max 8/day)
- frequency with cramps	*Lomotil* 2.5mg/25mcg (Diphenoxylate/ Atropine)	1 Box 8			(D) (P) Dose 2 every six hrs (max 8/day)
Eye injury/foreign body	Saline (if appropriate), then: *Alcaine* eye drops (proxymetacaine 0.5%)	15ml			(Rx) (S) Anaesthetic pain relief. Dose: 2 drops, as needed.
Infection - eye/ear	*Chlorsig* eye drops (chloramphenicol 0.5% drops, or ointment 1%)	10ml or 4g tube			(Rx) (S) Can use in eye or ear. Dose 2 drops every 2 hrs for 3 days.
Infection -tooth/abscess	*Tetrex* 250mg caps (tetracycline)	1 Pack 25			(Rx) (S) Dose 2 to start, then 1 six hourly with water before food
Infection - wound/ burn	*Neosporin* ointment (Polymyxin, Bacitracin & Neomycin mixture)	4g or 15g tube			(Rx) (S) Can use for eye / ear & skin. Dose: apply thinly 3-4x/day
	Diclocil 500mg caps (dicloxacillin) = 'penicillin' antibiotic.	1 Box 24			(Rx) (S) Skin sepsis. Dose 1 six hourly with water before food.
Laceration / open cut	*Leukostrip, Steri-strip, J&J Butterfly clip* Wound pads as above.	10 strips			(P) Non-invasive, prefer over needle and suture
Pain relief - mild (Head, back, tooth, etc)	*Panadol* 500mg tabs (paracetamol)	1 Box 24			Dose 2 tabs every four to six hrs (max 8/day)
	Aspro 300mg tabs (aspirin)	1 Box 24			(A) Dose 2 tabs every six hrs with fluid /food
Pain relief - moderate	*Panadeine* tabs (paracetamol 500mg + codeine 8mg);	1 Box 24			(P) (S) Dose 2 tabs every four to six hrs (max 8/day)

Chapter 19.1: OH&S - First Aid

Pain relief - severe	*Panadeine Forte* tabs (paracetamol 500mg + codeine 30mg)	Box of 20 Qty as needed		(D) (Rx) (S) Dose 2 tabs every four to six hrs (max 8/day)
- non-codeine alternative	*Di-Gesic* tabs (Paracetamol 325mg + dpropoxyphene 32.5mg	Box of 20 Qty as needed		(D) (Rx) (S) Dose 2 tabs every four to six hrs (max 8/day)
- injection	*Morphine* (10mg amps)	1 Box 5		(D) (Rx) (S)
Pain relief - inflamed	*Nurofen* 200mg tab (Ibuprofen)	1 Box 24		(A) (D-rare) (P) Dose 2 tabs with each meal (max 6/day)
Sea-sickness - nausea	*Travacalm* (strong), or *Kwells*, or *Dramamine* (both moderate) tabs	Box 10 Qty as needed		+ (D) (P) Initial dose 2 tabs prior embarking or 30 mins prior onset sea swell: repeat 4 hourly
- nausea, dizziness, (vertigo)	*Stemetil* 5mg tabs (Prochlorperazine)	1 Box 25		+ (D) (Rx) (S) Dose 1-2 tabs three times a day
General Items:				
Antiseptic - general - hospital strength	*Savlon* (chlorhex.) crm, *Betadine* lotion (Povidone-Iodine)	30g tube, 15ml bottle		(P) Povidone preferred
Bandages	Med elastic crepe 10cm Heavy crepe 10-15cm	Min 2 of each size		Medium – general use; Heavy – pressure use.
	Triangular bandage	2		Each comes with a clip. (S) Increase as needed.
Basin	Enamel with drop sheet	1		Use - preparation area
Cotton Buds (or balls)	Sterile pack of 50	1		Cleaner / applicator
Eye Bath	Surgi eye bath	1		Use – irrigation
Forceps	S/Steel 100mm	1		Bayonet point
Gauze	WOW 50mm-100mm	Min 2 ea		Use - bandage / pad.
	Sterile pack of 5 swabs	2		(S) Increase as needed.
Gloves - disposable	*Ansell Handy*	1 Box 24		Use – infection control
Ice pack	Hot / Cold packs	1 ea type		Gel & instant chemical
Irrigation solution	Saline poly-amp 10ml	5		(P) Irrigate wound or
	Sterile water poly-amp	5		eye after general flush
Reference text- first aid	Current edition	1		Work place accredited
Safety pins & clips	Assorted	1 Box		Various sizes
Scissors	Surgical s/steel 125mm	1		Sharp / Blunt tip
Swabs - sterilising	Alcohol wipes	5		Use – infection control
Sun block SPF30+	Water resistant rating	200ml		Apply before exposure
Syringe / needle	3ml syr / 25G tip	5		(P) (S) Sterile injection
Tape - sensitive area - high stick	*Micropore* paper tape	1 roll		(P) Tape width 2.5cm minimum
	Leukotape (w/proof)	1 roll		
Thermometer	Clinical / digital	1		In protective case
Tweezers	S/Steel	1		Flat tip
Other items (fill in):				

* Disclaimer: Table and contents not exhaustive. Suggestions based on common first aid treatment scenarios.

\# Notes (A) Aspirin-type allergy risk / may upset stomach ulcers or asthma.
    (D) Potential drowsy or incoordination warning.
    (H) Not suitable with high blood pressure / heart conditions / anti-depressants.
    (P) Pharmacy or non-script item; (Rx) available only via Doctor prescription.
    (S) Supplementary item primarily for extended / isolated voyages beyond seven days; Increase quantity of continual dosage tablets and other consumables accordingly.

\+ Additional topic information is available within this chapter, (viz: bleeding, burns, bites & stings, seasickness).

CHAPTER 19.1: QUESTIONS & ANSWERS

OH&S: FIRST AID

1. The main task of the First Aider in treating a casualty is to:

    (a) send for an ambulance.
    (b) preserve life.
    (c) prevent the injury from becoming worse.
    (d) promote recovery.

2. Pain, nausea and headache are examples of:

    (a) signs.
    (b) symptoms.
    (c) history.
    (d) diagnosis.

3. A casualty with head injuries and in unconscious state should be placed in the following position:

    (a) Lying flat.
    (b) Sitting up.
    (c) Stable side position.
    (d) None of the choices stated here.

4. Absence of oxygen to the brain may result in irreversible damage after:

    (a) one minute.
    (b) four to six minutes.
    (c) three to four minutes.
    (d) fifteen minutes.

5. An unconscious casualty is bleeding from a minor head wound and is breathing. You should first:

    (a) lie the casualty on the back.
    (b) sit the casualty up.
    (c) apply a dressing to the wound.
    (d) place the casualty in the stable side position.

6. A person has collapsed and has no pulse. You commence resuscitation. When you compress the sternum you hear and feel a rib crack. You would:

    (a) re-check your hand position and continue resuscitation.
    (b) stop resuscitation immediately.
    (c) continue resuscitation immediately.
    (d) only assist breathing to avoid further damage.

7. With internal bleeding due to an injury to an internal organ, you should:

    (a) give no food or drink.
    (b) give frequent small drinks.
    (c) apply a tight bandage around the abdomen.
    (d) apply a cold compress over the injured area.

8. When approaching an injured person, and after checking for danger, you should first:

    (a) check the pulse and breathing.
    (b) establish the level of consciousness.
    (c) complete a thorough examination of the casualty.
    (d) place the casualty in the stable side position.

9. When treating a person who has fainted and collapsed in a busy shop and is now not responding to voice and touch, you should:

    (a) sit the casualty down comfortably to rest.
    (b) establish the level of consciousness.
    (c) lay the casualty down and raise the legs.
    (d) place the casualty in the stable side position for safety.

10. The following problem has the most urgent priority:

    (a) A casualty who has taken poison, but is conscious.
    (b) An unconscious casualty breathing noisily.
    (c) A severely burned casualty who is conscious and in pain.
    (d) A conscious casualty with severe bleeding.

11. Signs of a casualty's injuries are determined by:

    (a) listening to the casualty's story.
    (b) examining the casualty.
    (c) getting the story of how it happened from onlookers.
    (d) checking the surroundings.

12. The main objective in the management of a wound is to:

    (a) ease pain.
    (b) control bleeding.
    (c) prevent swelling.
    (d) prevent infection.

13. The most urgent treatment for a sucking wound to the chest is to:

    (a) lie the casualty on the left side.
    (b) place the casualty in a comfortable position.
    (c) make an airtight seal.
    (d) lie the casualty on the right side.

14. The following treatment is recommended to stop bleeding caused by a foreign body embedded in the wound:

    (a) Remove the foreign body and apply direct pressure.
    (b) Probe the wound to determine the depth of the foreign body.
    (c) Apply direct pressure and a dressing.
    (d) Apply a ring pad bandage.

Chapter 19.1: OH&S - First Aid

15. A casualty suffering from shock should not be overheated in order to maintain the casualty's:

    (a) blood pressure.
    (b) pulse rate.
    (c) blood flow to the vital organs.
    (d) resistance to fever.

16. When required to help an injured person the first thing you do is:

    (a) ensure that air can get into the lungs.
    (b) ask someone to get help.
    (c) control bleeding.
    (d) check for danger to yourself.

17. The treatment for a nosebleed is to:

    (a) lie the person down.
    (b) apply ice to the person's nose.
    (c) sit the person up with the head tilted back.
    (d) apply pressure to the flap of the nostril for at least 10 minutes with head tilted forward.

18. When performing C.P.R. with two operators, you should apply:

    (a) fifteen compressions and two breaths.
    (b) five compressions and two breaths.
    (c) five compressions and one breath.
    (d) fifteen compressions and one breath.

19. A person has collapsed into unconsciousness beside you. You should immediately:

    (a) ensure a clear and open airway.
    (b) count the pulse rate.
    (c) count the breathing rate.
    (d) see he has plenty of fresh air.

20. When attempting mouth-to-mouth resuscitation, if the chest does not fill you should:

    (a) place the casualty in the stable side position.
    (b) breath into the mouth as hard as you can.
    (c) re-check the airway.
    (d) commence mouth to nose resuscitation.

21. A person has swallowed a corrosive poison and is conscious, you should first:

    (a) contact poison information centre.
    (b) make the person vomit.
    (c) give milk to dilute the poison.
    (d) check the airway.

22. Noisy breathing is usually a sign of:

    (a) unconsciousness.
    (b) too much alcohol.
    (c) partially blocked airway.
    (d) hysteria.

23. A casualty with a chest injury and breathing difficulties is best positioned:

    (a) with head low and legs raised.
    (b) lying flat with the head raised.
    (c) half sitting and inclined to the injured side.
    (d) half sitting and inclined to the uninjured side.

24. The treatment for an amputated part is to put it in a bag:

    (a) with ice.
    (b) with ice and water.
    (c) that is dry.
    (d) seal the bag and put the bag in ice and water.

25. The most important thing happening in the lungs is:

    (a) movement of air into small air spaces.
    (b) blood flow into the capillaries.
    (c) adding of oxygen to the blood.
    (d) change in colour of red blood cells.

26. If dressings and bandage do not control bleeding, you should:

    (a) apply a bigger bandage over the first one and bandage firmly.
    (b) reassure the casualty and send for urgent medical aid.
    (c) apply constant pressure over the dressing with your hand.
    (d) remove the dressing and apply a bigger one.

27. An unconscious casualty with suspected spinal injuries should be placed in the following position:

    (a) In the stable side position.
    (b) On the back with support under the natural curves of the spine.
    (c) On the back with the head elevated.
    (d) On the back with the feet elevated for shock.

28. When about to resuscitate an elderly person, you notice dentures are present. You should:

    (a) leave them in place to assist in making a good seal around the mouth.
    (b) remove them before beginning resuscitation.
    (c) remove them as they would prevent an airtight seal around the mouth.
    (d) leave them in, in case they get lost.

29. A person has been burned and is conscious. After checking for danger, your immediate treatment would be to:

    (a) apply butter to the burnt area.
    (b) examine the casualty for other injuries.
    (c) cool the burnt area for at least 20 minutes.
    (d) send for an ambulance.

30. Your immediate treatment for kerosene poisoning should be to:

    (a) induce vomiting at once.
    (b) give nothing by mouth and call a poison information centre on 131126.
    (c) give milk to dilute the poison.
    (d) give sips of water.

31. If a casualty has a blister from any cause, it should not be broken because:

    (a) intact skin helps prevent infection.
    (b) loss of fluid may cause shock.
    (c) bleeding may result.
    (d) pain may cause shock.

32. In addition to controlling external bleeding by direct pressure, you should:

    (a) apply warmth to the casualty.
    (b) elevate the part above heart level.
    (c) apply a cold compress to the wound.
    (d) place the casualty at rest.

33. In case of a suspected internal injury, the casualty requires:

    (a) careful observation of pulse and breathing rates.
    (b) gentle massage to the affected area.
    (c) treatment for shock and urgent medication.
    (d) careful observation of pulse & breathing rate, treatment for shock and urgent medical attention.

34. Burnt areas are prone to infection if the skin is broken. The infection should be prevented by:

    (a) covering the burnt area with a clean dressing.
    (b) covering the burnt area with a burn cream.
    (c) covering the burnt area with butter.
    (d) washing the burnt area with antiseptic.

35. The pulse felt in the neck, beside the Adam's apple, is the:

    (a) Radial pulse.
    (b) Carotid pulse.
    (c) Temporal pulse.
    (d) Femoral pulse.

36. If a casualty is trapped in a crashed vehicle and is unconscious, you should immediately:

    (a) remove the casualty disregarding any injuries.
    (b) leave the casualty alone.
    (c) tilt head with jaw support.
    (d) check for signs of spinal injury.

37. Pale, clammy skin, rapid and weak pulse, rapid breathing and restlessness in a casualty following an accident probably indicates:

    (a) diabetes.
    (b) severe bleeding.
    (c) asphyxia.
    (d) concussion.

38. If a First Aider is in doubt as to whether a casualty is dead, the recommended action is to

    (a) begin treatment.
    (b) send for a doctor.
    (c) stay with the casualty in case the casualty is alive.
    (d) move the casualty.

39. With injuries to the head, neck, chest or abdomen, you should first:

    (a) control bleeding to the best of your ability.
    (b) prevent further contamination of the injured part.
    (c) immobilise the injured part to prevent further damage.
    (d) see that the casualty has an open airway and is breathing.

40. The circulatory system of the body consists of:

    (a) heart, blood vessels and blood.
    (b) heart, brain and lungs.
    (c) heart, lungs and blood vessels.
    (d) heart and lungs.

41. The most reliable signs of a cardiac arrest are:

    (a) dilated pupils, blue lips, no response to stimuli.
    (b) absent respiration, shivering, clammy skin.
    (c) unconsciousness, absent breathing, absent pulse.
    (d) pale skin, sweating, noisy breathing.

42. All poisons, sprays and chemicals are a hazard to children. The safest way to avoid an accident is to:

    (a) label all containers clearly.
    (b) warn your children of the dangers.
    (c) never transfer chemicals into soft drink bottles.
    (d) securely lock away all harmful substances with a childproof lock.

43. One of the greatest dangers to a casualty in a motor vehicle accident is:

    (a) carbon monoxide poisoning.
    (b) blocked airway.
    (c) broken bones.
    (d) explosions.

Chapter 19.1: OH&S - First Aid

44. When you count the breathing and pulse of a casualty, you should count over a period of:

    (a) one half of a minute and multiply by two.
    (b) one minute for breathing and one minute for pulse.
    (c) one quarter of a minute and multiply by four.
    (d) two minutes to make sure you check the rhythm.

45. To make a traffic accident scene safe, you should ensure safety from:

    (a) petrol or L.P. gas.
    (b) electric power lines.
    (c) other oncoming traffic.
    (d) all of the choices stated here.

46. When making an assessment of the casualty, you should first:

    (a) find out what happened.
    (b) ask the casualty if there is any pain.
    (c) look for a medical warning bracelet.
    (d) look for swelling and deformity.

47. You noticed an unconscious victim of an accident bleeding from the ear canal. You should take the following action:

    (a) Place a small cotton wool plug in the bleeding ear.
    (b) Place the casualty in the stable side position with the injury downwards after placing a small cotton wool plug in the ear.
    (c) Place the casualty in the stable side position with the injured ear upwards.
    (d) Place the casualty in the stable side position with the injured ear downwards, with a gauze pad covering the ear.

48. An adult is choking from a foreign object. You should first:

    (a) tell the casualty to swallow.
    (b) seek urgent medical aid.
    (c) give the casualty several sharp blows between the shoulder blades, with the casualty's head low.
    (d) slap the casualty's chest three or four times.

49. When performing mouth-to-mouth resuscitation to an adult, you should administer:

    (a) twelve to fifteen breaths per minute.
    (b) fifteen breaths per minute.
    (c) ten breaths per minute.
    (d) seventeen breaths per minute.

50. The main danger to a severely burnt casualty is:

    (a) pain.
    (b) shock.
    (c) loss of body fluids.
    (d) infection.

51. A person suffering from shock should be:

    (a) sat up and reassured.
    (b) laid down with the head raised.
    (c) given a drink to increase blood volume.
    (d) laid down with the head low and legs raised.

52. The main purpose of placing the unconscious casualty in the stable side position is to:

    (a) prevent vomit blocking the airway and to allow drainage from the mouth.
    (b) encourage blood flow to the brain.
    (c) enable lungs to work more efficiently.
    (d) permit observation of the casualty.

53. When performing C.P.R. on an adult, the correct hand position on the breastbone is:

    (a) the top half.
    (b) the middle.
    (c) the bottom half.
    (d) location is not important.

54. The cause of shock could be:

    (a) severe bleeding.
    (b) pain.
    (c) burns.
    (d) all of the choices stated here.

55. The stable side position is used for:

    (a) an unconscious non-breathing casualty.
    (b) a casualty suffering from shock.
    (c) an unconscious breathing casualty.
    (d) a casualty complaining of nausea.

56. When performing C.P.R. with one operator, you should apply:

    (a) 12 compressions and 2 breaths.
    (b) 15 compressions and 1 breath.
    (c) 10 compressions and 2 breaths.
    (d) 15 compressions and 2 breaths.

CHAPTER 19.1: ANSWERS

1 (b), 2 (b), 3 (c), 4 (c), 5 (d),
6 (a), 7 (a), 8 (b), 9 (d), 10 (b),
11 (b), 12 (b), 13 (c), 14 (d), 15 (c),
16 (d), 17 (d), 18 (c), 19 (a), 20 (c),
21 (a), 22 (c), 23 (c), 24 (d), 25 (c),
26 (a), 27 (a), 28 (a), 29 (c), 30 (b),
31 (a), 32 (b), 33 (d), 34 (a), 35 (b),
36 (c), 37 (b), 38 (a), 39 (d), 40 (a),
41 (c), 42 (d), 43 (b), 44 (b), 45 (d),
46 (a), 47 (d), 48 (c), 49 (b), 50 (b),
51 (d), 52 (a), 53 (c), 54 (d), 55 (c),
56 (d)

Chapter 19.2

OCCUPATIONAL HEALTH & SAFETY

FIRE PREVENTION & CONTROL

FIRE EXTINGUISHERS IN COLOUR: *See Chapter 6*

FIRE TRIANGLE & TETRAHEDRON

Fire needs three things to exist: fuel (material), heat and oxygen. Combined, they are called a Fire Triangle. The rule for fighting a fire is to break this triangle by removing one of its sides, as follows:

- Fuel: Fires can be starved of fuel by removing the burning material.
- Heat: Heat is removed by cooling, e.g. water.
- Oxygen: Fires can be starved of oxygen by a blanket, foam or CO_2. Shutting down the ventilation of a compartment, which is on fire, has the same effect.
  The Dry Chemical extinguisher works a little differently. It excludes oxygen by producing large quantities of CO_2 while breaking up the chain reaction of the fire, as shown in the FIRE TETRAHEDRON.

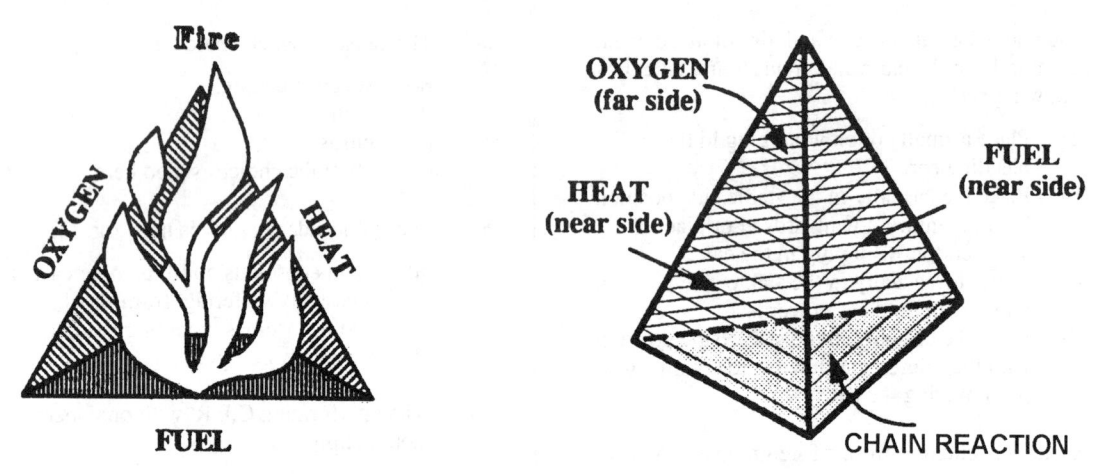

Fig 19.8: FIRE TRIANGLE (*Left*) & FIRE TETRAHEDRON (*Right*)

CLASSIFICATION OF FIRES

Class A: Solids: Fires in flammable materials that are solid in nature. For example, wood, paper, cloth and rubber

Class B: Liquids: Fires in flammable and combustible liquids and greases. Best tackled after shutting off the fuel.
NOTE: Due to their melting characteristic, plastics on fire are regarded a Class B fire.

Class C: Gases: Fires in combustible gases. Best tacked after shutting off the gas.

Class D: Metals: Fires in metal filings and in combustible metals, such as phosphorus and magnesium.

Class (E): Electrical Fires - see below.

Class F: Fat fires: Fires in cooking oils and fat.

➢ Class "A" is the most common fire. Most Class "B" fires generate a great deal of thick black smoke. Class "C" fires are usually explosive in nature. Most fires in vessels are engine or galley fuel related.

➢ **ELECTRICAL FIRES:** Any class of fire can be started by electricity. Consequently, electrical fires are not separately classed. Water type fire extinguishers (including foam) must not be used on electrical fires unless the power has been turned off or disconnected. Water conducts electricity, which may electrocute the fire fighter. For safety, the fire extinguishers that are safe to use on electrical fires are marked with the bold capital letter "E" or "SUITABLE FOR ELECTRICAL FIRES". See fire extinguishers colour illustration in Chapter 6.

PORTABLE FIRE EXTINGUISHERS

(See COLOURED illustration in Chapter 6)

The primary colour for fire extinguishers is red (See Chapter 6 for the colour chart and Chapter 29 for the Australian Standard Colour Coding System). In addition, there are five secondary colours: Oatmeal (Beige), Blue, White, Black and Yellow. The secondary colours identify their contents. In the past, fire extinguishers containing Wet Chemical or Foam were painted in their secondary colours or Red with the appropriate secondary colour band. They are now painted red with the correct colour band only. Most types of portable fire extinguishers last between 45 and 60 seconds in continuous use.

1. **WATER**
   Grey with a red band (or the older type, completely Red)
   It extinguishes fire by cooling it. Most suitable for Class A fires.

2. **FOAM**
   Red with a blue band (or the older type, completely Blue)
   Foam extinguishes fire by blanketing and starving it of oxygen. Most suitable for Class B fires.
   The most common type is the synthetic *"A Triple-F" (AFFF) or Aqueous Film Forming Foam..* For alcohol fires there is a special alcohol resistant AFFF foam. The ordinary AFFF foam is not suitable because its water content mixes with alcohol, causing the foam blanket to breakdown.

3. **DRY POWDER**
   Red with a white band
   Also known as Dry Chemical, it extinguishes fire by interrupting its chain reaction. It is a general-purpose fire extinguisher, suitable for almost all types of fires. There are two types of Dry Powder extinguishers: AB(E) and B(E). The latter is not very suitable for Class A fires.

4. **WET CHEMICAL**
   Red with an oatmeal band (or the older type, completely oatmeal colour)
   It is specifically designed to extinguish cooking fat fires, which it does by emulsifying it on surface. In other words, it turns cooking fat into soap.

5. **CO_2**
   Red with a black band
   It starves fires of oxygen. It is suitable for all types of fires, but not very effective on a Class A or Class C fire.

6. **VAPORISING LIQUID**
   Red with a yellow band
   Also known as NAF P-III. See explanation for NAF S-III]. Like the dry powder type, it extinguishes fire by interrupting its chain reaction. It is suitable for most types of fires, including electrical fires. However, it is not very effective on a deep seated Class A fire or on a Class C or F fire.

NOTE: Due to their ozone depletion potential, the yellow coloured Halon (BCF) fire extinguishers ceased to be manufactured and serviced for general use in Australia on 31.12.1995. They are now manufactured only for specific use where there is no substitute available. The thermal breakdown products of Halon are extremely toxic.

HOW TO USE
Remember the acronym **PASS**   Pull the pin (and test the extinguisher)
                                 Aim the extinguisher
                                 Squeeze the handle
                                 Sweep the fire

AFTER USING A FIRE EXTINGUISHER
Lay down a used or partly used portable fire extinguisher on its side, which indicates that it requires recharging.

SAFETY PRECAUTIONS
- Water, Foam & Wet Chemical type are electrically conductive.
- CO_2 & Vaporising Liquid: asphyxiating in confined spaces. CO2 may cause frostbite due to its extremely low

temperature of sublimation (passing from solid to gas).
- Dry Chemical type is safe in all environments.
- Water: Aim at seat of class "A" fire. If used on class "B" fires, use it in the form of a spray.
- CO_2, Dry Chemical, Wet Chemical & Vaporizing Liquid: Aim at seat of fire in a sweeping motion.
- Foam: Aim at a bulkhead nearby. The foam should slide down and form a blanket over the class "B" fire. Use judgement in other fires.

PRESSURE TEST OF PORTABLE EXTINGUISHERS

Most fire extinguishers work by storing extinguishing agents under pressure. They must therefore pass a periodic pressure test. To comply with the Australian Standards AS 1851-1 all fire extinguishers need to be emptied and pressure tested every 6 years by an authorised person. For extinguishers in an aggressive environment, of which salt air could be one, testing has to be carried out every 3 years. However, the USL Code has set the pressure-testing interval for on-board extinguishers at 5 years, except for the CO_2 extinguishers, which may be tested at intervals of 10 years for the first and second tests.

PERIODIC INSPECTIONS OF PORTABLE EXTINGUISHERS

All extinguishers, other than CO_2, have a pressure gauge indicating their state of charge, and a security seal on the squeezer. They should be recharged if the seal is broken or the pressure gauge is not in the centre of the green sector of the scale. Tap the gauge lightly to make sure that the needle is not stuck. The dry chemical fire extinguisher should be regularly inverted and shaken to prevent compacting of the powder at the bottom of the cylinder. It should be recharged if the powder has become compacted on the bottom. CO_2 fire extinguishers must be checked by weight. If the loss of weight is 10% or greater of the net weight of its content, it must be recharged.

Every extinguisher must be recharged after use. Never keep an empty or partially discharged extinguisher in service or put back in its assigned location. Prior to each sailing visually inspect all extinguishers for their charge and locations. In addition inspect them as follows to comply with the Australian Standards AS 1851-1:

6-MONTHLY INSPECTIONS: Check accessibility, anti-tamper seal, safety pin, support hook, condition of printed operating instructions, damage, corrosion, discharge nozzle, outlet hose, content or pressure indicator, operating lead, actuating device without discharge.

ANNUAL INSPECTION: Six-monthly program, plus the following:
Empty the extinguishers whose contents are likely to break down over time, such as foam and wet chemical. Check their internal components and condition and then recharge.

3-YEARLY INSPECTION: Six-monthly & annual programme, plus the following:
Empty the water type fire extinguisher and check its internal components and condition.

6-YEARLY INSPECTION (3-YEARLY IN AGGRESSIVE ENVIRONMENT & 5-YEARLY UNDER USL CODE):
Six-monthly, annual & 3-yearly program, plus the following:
Empty and pressure test all extinguishers.

MAINTENANCE RECORD: The dates of recharge and inspection of a fire extinguisher must be stamped on its body or on a metal tag attached to it. The tag is designed like a calendar. A numeric number is punched into a month and a year on it, as follows:
"1" for 6-monthly inspections
"2" for annual inspections
"3" for 3-yearly inspections
"4" for 6-yearly inspections
"5" for a service after use

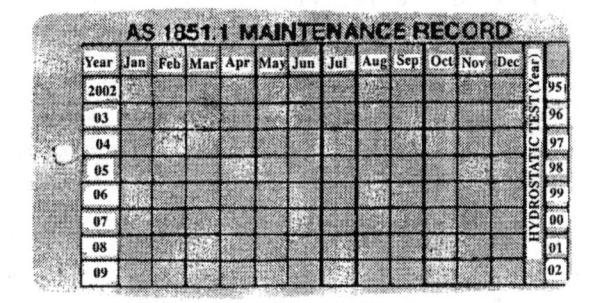

Fig 19.9 FIRE EXTINGUISHER MAINTENANCE RECORD TAG

FIXED FIRE EXTINGUISHING INSTALLATION

Commercial vessels of classes 1 (A-E), 2 (A-C) and 3 (A-C) of 12.5 metres and over in length are required to be fitted with a fixed fire extinguishing or "bulk" system in their machinery space. It is to comply with the requirements of Section 11 of the USL Code and Part C of the NSCV.

Chapter 19.2: OH&S - Fire Prevention & Control

Such a system would incorporate alarm signals to warn people to evacuate machinery spaces. The fuel shut off valve must be operable from outside the engine room, as must be the controls for the extinguishing agent. The master and every crewmember must make themselves thoroughly familiar with the system on board, in case there is only person left to operate it. Some of the fixed systems, whose safety levels must be checked with the manufacturers, are as follows.

1. NAF S-III

It is an instantaneous VAPORIZING LIQUID designed for SYSTEMS. It is a blend of hydro-chloro-fluorocarbons (HCFCs) and an ingredient added to decrease the production of breakdown products such as hydrogen fluoride (HF). NAF S-III is stored as a liquid, which on releasing, instantaneously vaporises into gas to flood a compartment. It does not leave a residue. It is also colourless and non-corrosive. The cylinders may be installed inside or outside the machinery space, but must be discharged manually only, from outside the machinery space. A detection system and a pre-gas-release warning system must be installed, and operating instructions displayed.

Note: **NAF P-III is a VAPORIZING LIQUID designed for PORTABLE** fire extinguishers. It does not vaporise instantly when expelled. It is released as a stream of liquid to aim at a fire and then vaporise.

2. CO_2

It is an asphyxiating gas. Therefore, the installation of a CO_2 system requires close scrutiny and monitoring, such as effective sealing of the machinery space and ensuring that the personnel are evacuated from the space before introducing the gas. The storage of CO_2 bottles is critical and requires well-ventilated cabinets or lockers at or above the main deck. A detection system and a pre-gas-release warning system must be installed, and operating instructions displayed. The CO_2 system is generally bulky as a larger number of bottles are required compared to other gas systems.

3. HI-FOG

Hi-Fog is a system that combines standard hydraulic pressure and a small amount of fresh water. It works by propelling very small droplets of water with high momentum to penetrate hot flue gases and reach the combustion source. It absorbs a fire's energy and cools the surrounding hot air and gases. The steam generated serves to smother the fire. The cylinders must be located outside the machinery space and operated manually. A detection system is required and operating instructions must be displayed.

4. FM 200

It is similar to NAF S-111. It is a trade name for the chemical HFC-227ea, which extracts heat from flames.

5. INERGEN

It is a compressed gas made from nitrogen, argon and CO_2. It works by depleting the oxygen concentration to a point below the level necessary to support combustion. CO_2 is incorporated to increase the breathing rate of individuals exposed to the discharge, with a view to enhancing their survivability in the depleted oxygen concentration. In other words CO_2 stimulates the human body to breathe more deeply and rapidly to compensate for the lower oxygen content in the atmosphere. However, it can cause heavy breathing and reduced physically capability.

6. PYROGEN

It is an aerosol fire suppressant, which stored as a solid chemical in non-pressurised canisters. The aerosol is released when the chemical is ignited electrically (manually) or thermally (automatically). The hot aerosol, chemically cooled upon discharge, is a mixture of tiny dry chemical particles and gases, but it leaves no residue. The dry chemical particles are mainly potassium carbonates, and the gasses are mainly CO_2, nitrogen and water vapour. Pyrogen behaves like a gas, attacking the fire chemically and physically. It does not deplete oxygen to suppress the fire. Instead, it absorbs heat to decompose itself.

THE INSTALLATION AND CARE OF A FIXED FIRE EXTINGUISHING SYSTEM:

- It must conform to the Authority's rules.
- It must be carried out by an approved company.
- It must be serviced annually.
- Alarms and detectors must be tested regularly.
- Operating instructions must be posted next to the control panel.
- The control panel must be kept free from obstructions.

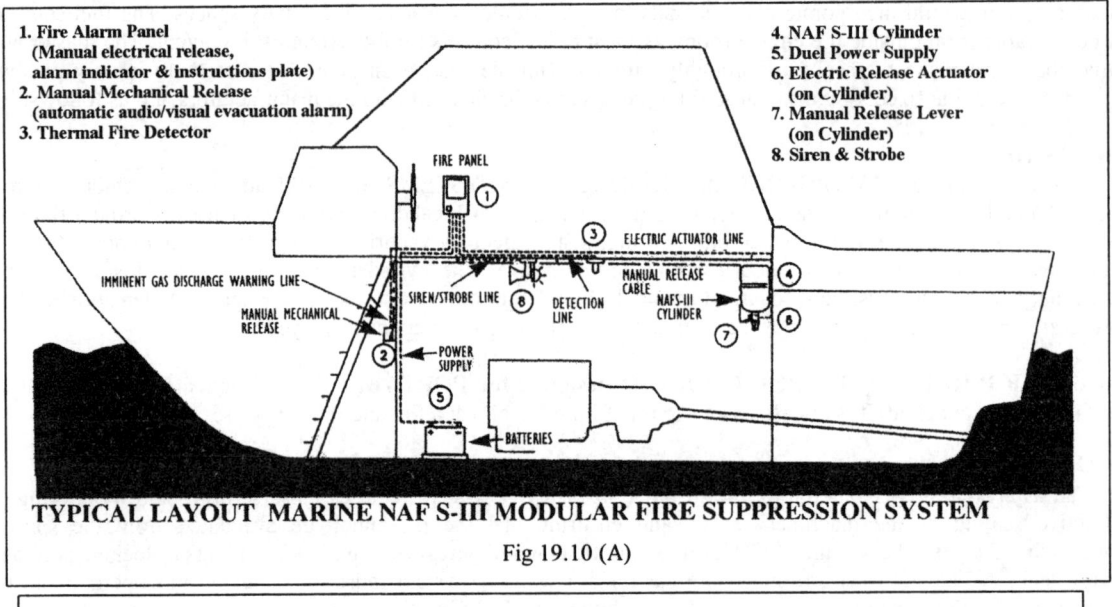

TYPICAL LAYOUT MARINE NAF S-III MODULAR FIRE SUPPRESSION SYSTEM

Fig 19.10 (A)

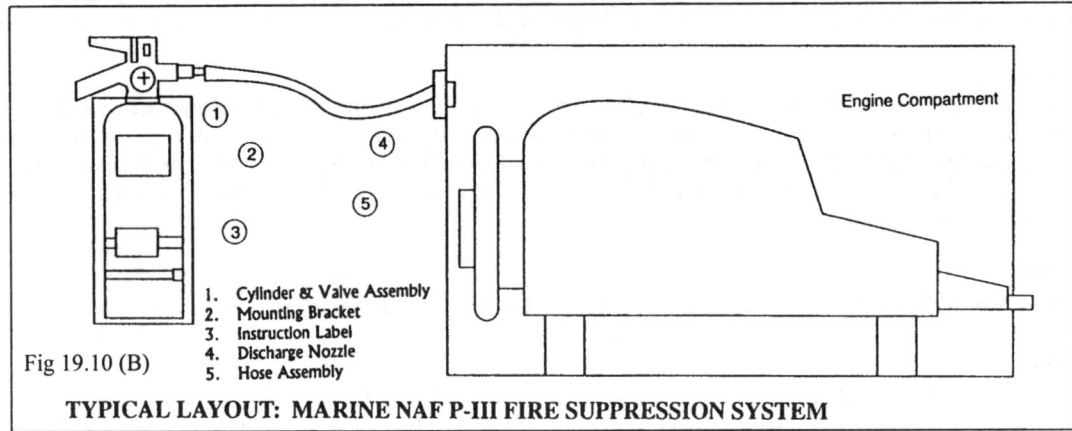

1. Fire Alarm Panel
   (Manual electrical release, alarm indicator & instructions plate)
2. Manual Mechanical Release
   (automatic audio/visual evacuation alarm)
3. Thermal Fire Detector
4. NAF S-III Cylinder
5. Duel Power Supply
6. Electric Release Actuator (on Cylinder)
7. Manual Release Lever (on Cylinder)
8. Siren & Strobe

Fig 19.10 (B)

1. Cylinder & Valve Assembly
2. Mounting Bracket
3. Instruction Label
4. Discharge Nozzle
5. Hose Assembly

TYPICAL LAYOUT: MARINE NAF P-III FIRE SUPPRESSION SYSTEM

DETECTORS: SMOKE, HEAT & RADIATION

Listed below are four types of smoke, heat and radiation detectors. They consist of two basic parts: a sensor and an electronic alarm. The alarm may be a high-pitched beeper, bell or loud voice announcing "Fire! Fire!", "Warning! Carbon monoxide", etc.

1. IONISATION SMOKE DETECTOR

An ion is an electrically charged atom that causes an electrical current flow. Inside the ionisation sensor is a small amount of Americium (radioactive element), which ionises the air in the chamber, thus producing a small electrical current flow. Smoke entering the ionisation chamber disrupts this current flow, setting off the alarm.

This is the most common type of smoke detector because it is inexpensive and better at detecting smaller amounts of smoke produced by combustion. The detector reacts to visible as well as invisible products of combustion. Its response sensitivity can be adjusted in three stages by means of a switch. One detector can normally monitor an area of 50 – 70 square metres.

(Note: The amount of radiation in a smoke detector is extremely small. It is also predominantly alpha radiation. Alpha radiation cannot penetrate a sheet of paper, and it is blocked by several centimetres of air. The Americium in the smoke detector would be dangerous only if you inhaled it by playing with it, poking at it, or disturbing it in any way to make it airborne.)

2. OPTICAL (or PHOTOELECTRICAL) SMOKE DETECTOR

An optical smoke detector senses the lack of light due to smoke and triggers an alarm. It reacts to the visible smoke produced in the early stage of smouldering type of fires, such as from mattresses and electrical installations. Combined with an ionisation smoke detector, it is particularly suitable for spaces with electrical installations. One detector can normally monitor an area of 50 – 70 square metres.

3. **RADIATION DETECTOR**
It responds to infra-red energy radiated by flames, making it suitable for fires accompanied by flames. Its sensitivity can be adjusted in four stages by means of a switch. A single unit can monitor an open area of up to 1000 square metres, as long as there are no obstacles obstructing its direct line of sight.

4. **HEAT (THERMAL) DETECTOR**
It is employed in areas where an outbreak of fire is likely to produce a rapid rise in temperature. It is the least sensitive of the four types stated here. One detector can monitor an area of between 10 – 25 square metres.

MISCELLANEOUS

HOW FIRES SPREAD: Fires are spread through transfer of heat in three different ways:
- *Radiation*: heat transferred in all directions, as around an open fire, a room heater or the sun. One object on fire, radiating heat, may spread the fire in all directions.
- *Convection*: heat transferred by air currents (or liquid currents). Since hot air rises, convection movement tends to spread the fire upwards.
- *Conduction*: heat travelling through solids. Heat conducted through a steel beam or bulkhead will transfer the heat (and fire) to another part of the vessel.

RATE OF BURNING: The rate of burning is the speed at which a fuel (solid, liquid or gas) is consumed by fire. The rate of burning depends on:
- the rate of vaporisation of fuel. Every form of fuel burns through vaporisation. Wood shavings vaporise more easily than a solid block of wood, they thus have a higher rate of burning than solid timber. Heated petrol vaporises so fast that it may explode.
- the rate at which oxygen is supplied to the fire. For example, opening the door of a cabin in which there is fire may cause a flare up.

SPONTANEOUS COMBUSTION: Scrunched up oily rags left lying in galleys and engine rooms can build up internal heat over a period of time and spontaneously catch fire. Metal filings are equally susceptible to spontaneous combustion, especially when left lying among oily rags.

BILGES: Fuel and oil leaks in bilges are a fire hazard. Keep the bilges pumped out and dry.

OXIDATION (AUTO COMBUSTION): Most people are familiar with heat generated inside garden compost heaps. Similarly, a pile of greasy rags or a pile of metal filings can generate enough heat to burst into flames. This process is known as oxidation. Build-up of grease in the galley air extractors can also lead to fire due to oxidation.

SMOKE IS THE MAIN CAUSE OF DEATHS: Smoke can contain many dangerous gases, which may stop lungs from functioning and irritate throat and eyes. Smoke also blocks light, making it difficult to fight fires or find an escape route. To escape smoke stay as low on floor or deck as possible. The real killer in smoke is carbon monoxide (CO), which sends sleeping persons into deeper sleep and ultimately death.

FIRE BLANKETS: Usually made of woven glass cloth, they are spread over the fire to exclude the 'oxygen' side of the fire triangle. Removing the blanket too early from a partially smothered fire may re-ignite it. Wait until the fire has cooled down.

SAND: Sand is no longer a commonly used fire extinguisher on vessels. But it does smother small fires and contain spilled flammable liquids.

FIRE BUCKETS: These too are rarely seen on vessels. They are metal buckets, painted red, with a long enough lanyard fitted to reach the water. They have a round bottom so that they do not sit upright when misused (used for any other purpose). They are stowed upright only in their cradle.

FIRE HOSES: A vessel with a dedicated fire hose would normally have large enough crew to warrant practical fire drills. The crew need to practise:

- Proper use of hoses to fight fires
- maintaining water pressure at critical times
- adjusting the nozzle to produce jet, spray and fog of water
- providing the attack team heat protection by building a water wall with a second hose
- keeping hose kink free
- use of hoses for boundary cooling.

EMERGENCY FIRE PUMP: In the event of fire in the engine room, the main fire pump will become inaccessible and out of action. Commercial vessels of certain categories are thus required to carry an emergency fire pump outside the machinery space. Depending on the classification and size of the vessel, the emergency pump may be a:

- fixed, independently-driven, power-operated pump, permanently connected to the fire main;
- portable independently-driven power-operated pump; or
- manually operated

An emergency generator or another source of power outside the engine room supplies power to the emergency fire pump. With the emergency pump running, an isolating valve fitted in the fire main system prevents back-flooding of the machinery space. At the time of fire drills, the emergency fire pump should be tested by operating it with the main fire pump stopped. A fixed emergency pump sucks water through the fire main suction, whereas the suction hose of the portable pump is thrown overboard, keeping the pump as close to the fire as possible.

FIRE SAFETY ON BOARD

- Regularly check the fire extinguishers pressure gauges.
- Regularly look inside the fire extinguishers nozzles for chewing gum blockage or insect infestation.
- Keep an eye on any fire hose boxes.
- Fix or report any fire hazards: overloaded electrical sockets, unattended oily rags, uncovered or leaking flammable liquids or gases.
- Keep the vessel (and the engine room) clean and tidy.
- Maintain bilges as dry as possible.
- Clean all spills immediately.
- Have fire extinguishers always properly charged and available.
- Exercise regular fire drills.
- Follow all refuelling precautions.
- Install sniffer(s), especially in the bilges.

- Keep battery area ventilated.
- Maintain and don't overload the wiring system.
- Dispose of or stow rags and rubbish, safely.
- Turn off all fuel supplies when not in use.
- Follow strict smoking rules.
- Ensure vessel is vapour free before starting motors.
- Maintain engine room ventilation.
- Maintain fuel lines in good condition.
- Keep pipe joints and valve glands tight.
- Use drip trays, where necessary.
- Maintain wooden floors free of oil.
- Stow flammables in spaces designated for them.
- Secure & fasten stoves permanently.
- Prevent cooking grease build up in galley.
- Secure galley gear and stores.
- Do not overload power points and electrical circuits.

PAY ATTENTION TO THE INSTALLATION OF:
[These would normally have been approved for vessels in survey]

- Fuel tanks
- Fuel lines
- Fuelling arrangements
- Electrical systems
- LPG or Compressed Natural (CN) Gas system
- Engine ventilation system
- Galley ventilation system

FUEL TANK LEAKS - ACTION TO BE TAKEN

- Switch off motor
- Switch off all electrical appliances
- Disconnect battery
- Remove ignition key
- Repair or replace the fuel line
- Clean up thoroughly
- Ventilate the area, or the complete vessel if necessary
- Re-connect fuel lines and check for leaks
- Start motor and check for leaks

REFRIGERANT PLANT - SAFETY PRECAUTIONS

- Some refrigerants are toxic. Guard against leakage in a confined space.
- If a section of the plant is accidentally isolated (due to incorrect operation), the entrapped hot refrigerant can rise in pressure and explode or rupture the pipe work.
- Do not carry out welding or cutting on the liquid receivers and relief devices. They both are pressure vessels and may explode in spite of being fitted with relief valves.

BASIC PRINCIPLES OF FIRE FIGHTING

- Raise alarm immediately - even if you can easily fight it single-handed.
- Treat all fires seriously, no matter how small.
- Fight them when they are still small.
- Follow the principle of fire triangle.
- Don't allow yourself or others to be asphyxiated or burnt.

IF FIRE BREAKS OUT ON A VESSEL IN HARBOUR

- Raise an alarm on board.
- Call the Coast Radio Station or Harbour Control.
- Seek assistance from fire brigade.
- Avoid drifting on to other vessels.
- Drop anchor or go alongside a wharf.
- Keep the fire end downwind.
- Organize passengers to avoid panic.
- Fight fire / bring it under control.

IF FIRE BREAKS OUT IN THE ENGINE ROOM AT SEA

- Sound alarm
- Stop engine
- Cut off fuel supply
- Shut ventilators and hatches to the engine room
- Release the extinguishing agent
- Send PANPAN or MAYDAY as appropriate
- Feel bulkheads to establish if fire is still burning. Plan re-opening with care. Fire can easily re-kindle with fresh supply of air.
- Prepare to abandon vessel in case the need arises.

TO ENSURE THAT THE EXTINGUISHED FIRE DOES NOT RE-IGNITE

- Turn over and wet all rubbish from the fire.
- Do not admit air into the smothered compartment until it has cooled down.
- Do not open gas or fuel valve to the compartment until it has cooled down.

RECORD THE INCIDENT

- to prevent it happening again,
- to debrief the crew, and
- to facilitate any enquiries.

FIRE FIGHTING - DANGER OF ACCUMULATED WATER:

As discussed in Vessel Stability in Chapter 13, accumulated water can cause a large Free Surface Effect. Water trapped in the upper decks of a vessel raises her centre of gravity. Both factors reduce stability and can cause the vessel to capsize. Therefore, you should keep scuppers clear of debris to ensure adequate drainage.

CHAPTER 19.2: QUESTIONS & ANSWERS

(Qns. 1-56 are in Chapter 19.1)

OH&S - FIRE PREVENTION & CONTROL

57. Vessels that are required to carry one or more fire extinguishers are:

    (a) all vessels.
    (b) power driven vessels.
    (c) vessels carrying fuel of any kind.
    (d) commercial vessels.

58. Under the Standards Australia regulations, modern portable fire extinguishers in a <u>non-aggressive</u> environment are required to be emptied, tested and recharged at least once every:

    (a) 12 months.
    (b) 5 years.
    (c) 6 years.
    (d) 10 years.

59. Under the Standards Australia regulations, modern portable fire extinguishers in an <u>aggressive</u> environment are required to be emptied, tested and recharged at least once every:

    (a) 6 months.
    (b) 3 years.
    (c) 5 years.
    (d) 6 years

60. Under the USL Code, modern portable fire extinguishers are required to be emptied, tested and recharged at least once every:

    (a) 12 months.
    (b) 3 years.
    (c) 4 years.
    (d) 5 years

61. The modern portable fire extinguishers are required to be externally inspected at least once every:

    (a) 12 months.
    (b) 6 months.
    (c) month.
    (d) 2 years.

62. A portable fire extinguisher, whose contents are <u>not</u> likely to breakdown over time, is required to be recharged under the following circumstances:

    (i) 12 months have elapsed since last recharge.
    (ii) The seal is broken.
    (iii) The pressure gauge does not indicate full.
    (iv) The last recharge date is not known.

    The correct answer is:
    (a) all except (i).
    (b) all except (ii).
    (c) all except (iii).
    (d) all except (iv).

63. A portable fire extinguisher, whose contents are likely to breakdown over time, is required to be recharged under the following circumstances:

    (i) 12 months have elapsed since last recharge.
    (ii) The seal is broken.
    (iii) The pressure gauge does not indicate full.
    (iv) 6 months have elapsed since last recharge.

    The correct answer is:
    (a) all except (i).
    (b) all except (ii).
    (c) all except (iii).
    (d) all except (iv).

64. Portable fire extinguishers whose contents are likely to breakdown after a 12 months period include:

    (a) foam and wet chemical.
    (b) foam and dry chemical.
    (c) water and foam.
    (d) CO2 and vaporising liquid.

65. A used or partly used portable fire extinguisher should be stored as follows:

    (a) Laid down on its side
    (b) Hung back on its bracket
    (c) Stood up in a cupboard
    (d) Hung back on its bracket and marked "Discharged".

66. "Check accessibility, anti-tamper seal, safety pin, support hook, condition of printed operating instructions, damage, corrosion, discharge nozzle, outlet hose, content or pressure indicator, operating lead, actuating device without discharge."

    The above portable fire extinguishers inspection list is for the following periodic inspection:

    (a) 3 yearly.
    (b) 2 yearly.
    (c) Annual.
    (d) 6 monthly.

Chapter 19.2: OH&S - Fire Prevention & Control

67. Which of the following is the passive means of limiting the risk & spread of fire?

    (a) Bulkheads.
    (b) Sprinklers.
    (c) Portable fire extinguishers
    (d) Fixed fire fighting systems

68. Which of the following offers a structural means of limiting the risk & spread of fire?

    (a) Sprinklers
    (b) Portable fire extinguishers
    (c) Bulkheads
    (d) Fixed fire fighting systems

69. Radiation causes fires to spread:

    (a) in all directions.
    (b) upwards only.
    (c) through contact between materials.
    (d) downwards only.

70. Convection causes fires to spread:

    (a) in all directions.
    (b) upwards only.
    (c) through contact between materials.
    (d) downwards only.

71. Convection causes fires to spread by means of:

    (a) motion of heated matter (air or liquid currents).
    (b) downwards motion of heat.
    (c) contact between materials.
    (d) infra-red radiation.

72. Conduction causes fires to spread:

    (a) in all directions.
    (b) upwards only.
    (c) through contact between materials.
    (d) downwards only.

73. Scrunched up oily rags and metal filings can cause a fire due to:

    (a) radiation.
    (b) convection.
    (c) conduction.
    (d) spontaneous combustion.

74. The heat generated by welding activity is spread mostly through the process of:

    (a) Radiation
    (b) Conduction
    (c) Convection
    (d) Liquid currents

75. A radiation fire detector works by reacting with:

    (a) visible flames.
    (b) visible smoke particles.
    (c) invisible smoke particles.
    (d) rise in temperature

76. Portable fire extinguishers for marine use are manufactured under the standards set down by the:

    (a) State Marine Authorities
    (b) AMSA
    (c) SOLAS
    (d) Standards Australia

77. The spread of fire depends on the rate at which:

    (a) fuel vaporises and heat escapes.
    (b) fuel vaporizes & oxygen is supplied.
    (c) visible smoke particles increase.
    (d) smoke escapes.

78. The following type of smoke detector reacts with the visible smoke.

    (a) Thermal
    (b) Photoelectric
    (c) Ionisation
    (d) Radiation

79. An optical smoke detector works by sensing the:

    (a) flames.
    (b) lack of light caused by visible smoke.
    (c) material on fire.
    (d) rise in temperature

80. Combustion in a fire causes visible as well as invisible products. An ionisation type of smoke detector reacts with:

    (a) Visible products only.
    (b) Invisible products only.
    (c) Both visible & invisible products.
    (d) Neither visible nor invisible products.

81. The following extinguishing agent will cool down a hot bulkhead in the shortest time.

    (a) Water fog
    (b) Dry chemical
    (c) Foam
    (d) Carbon dioxide

82. The silent killer in atmosphere ahead of a fire or in smoke is:

    (a) Carbon monoxide
    (b) Carbon dioxide
    (c) Heat
    (d) Nitrogen

83. A burning fire is sustained by:
    (a) Fuel, oxygen & heat.
    (b) Fuel, oxygen, heat & chain reaction.
    (c) Carbon monoxide, fuel, heat & chain reaction.
    (d) Fuel, heat & chain reaction.

84. Combustion in magnesium is classed as
    (a) A Class fire
    (b) B Class fire
    (c) C Class fire
    (d) D Class fire

85. Fire in cooking oil is classed as
    (a) D Class
    (b) B Class
    (c) F Class
    (d) A Class

86. A fuel tank is most likely to explode when it is:
    (a) full.
    (b) half full and has normal vapour content.
    (c) 20% full and has high vapour content.
    (d) 25% full and has low vapour content.

87. Kinks in electrical wiring can result in:
    (a) drop in voltage
    (b) drop in current
    (c) reduction in power supply
    (d) a fire

88. The greatest risk of fire during refuelling is presented by:
    (a) smoking
    (b) loose mooring lines
    (c) filling half filled tanks
    (d) filling tanks while vessel is listed.

89. In relation to fire safety, which of the following statements is incorrect?
    (a) Use non-sparking tools when working on electrical installations.
    (b) Gas vapour tend to accumulate in the lowest spaces in vessels.
    (c) Fuel vapour tend to accumulate in the lowest spaces in vessels.
    (d) LPG cylinders should be installed in a cool space below decks.

90. The fire detection system that responds to rise in temperature is known as the:
    (a) Thermal detector
    (b) Ionisation detector
    (c) Optical detector
    (d) Radiation detector

91. The ideal location for the control panel of a fire detection system is the:
    (a) Engine space
    (b) Wheelhouse
    (c) Sleeping space
    (d) Dining room

92. Paint lockers in vessels of less than 25 metres are separated from adjacent compartments by:
    (a) Steel divisions.
    (b) Insulated divisions
    (c) Non-combustible divisions
    (d) Flame retarding divisions

93. Fire Retardant Paints:
    (a) limit the spread of fire.
    (b) are a substitute for fire detectors.
    (c) are suitable only for outdoors.
    (d) are not permitted for marine use.

94. Carbon dioxide is
    (a) non-conductor of electricity.
    (b) lighter than air.
    (c) non-toxic
    (d) non-asphyxiating.

95. When confronted by smoke from a fire:
    (a) walk upright towards an exit.
    (b) cover your nose and mouth with a handkerchief.
    (c) crawl along the floor towards an exit.
    (d) head towards the seat of fire.

96. The fire extinguisher least likely to prevent re-ignition of fire is:
    (a) Water
    (b) Foam
    (c) Wet Chemical
    (d) Carbon dioxide

97. When fighting a fire with a portable CO_2 extinguisher, its horn should be directed at:
    (a) a nearby bulkhead.
    (b) the flames.
    (c) the seat of the fire.
    (d) above the seat of the fire.

98. When fighting a fire with a portable Foam extinguisher, its nozzle should be directed at:
    (a) a nearby bulkhead.
    (b) the flames.
    (c) the seat of the fire.
    (d) above the seat of the fire.

Chapter 19.2: OH&S - Fire Prevention & Control

99. When fighting a fire with a portable Water extinguisher, its nozzle should be directed at:

    (a) a nearby bulkhead.
    (b) the flames.
    (c) the seat of the fire.
    (d) above the seat of the fire.

100. You discover a small fire in the galley, where there is also a portable fire extinguisher close by. You would:

    (a) extinguish it, then raise an alarm.
    (b) Raise an alarm, then start extinguishing it.
    (c) extinguish it, then advise the master.
    (d) Raise an alarm so that a fire team can extinguish it

101. The flash point of a fuel is the lowest temperature at which:

    (a) it will continue to burn.
    (b) it will continue to burn provided an external source of ignition remains present.
    (c) it will stop burning.
    (d) it will explode.

102. The storage of following substances together in a cupboard increases the risk of fire:

    (a) Poisonous & flammable.
    (b) Toxic & corrosive.
    (c) Corrosive & flammable.
    (d) Corrosive & rags.

103. The ideal location for a fire blanket is:

    (a) Wheelhouse
    (b) Galley
    (c) Engine room
    (d) Paint locker

104. At Crew Fire Drills, a practical demonstration of the use of portable fire extinguishers shall be given by expending the charge of at least:

    (a) one extinguisher.
    (b) one extinguisher of each type on board.
    (c) one extinguisher from the galley and one from engine room.
    (d) one fire extinguisher from the inside & one from outside the engine room.

105. Which of the following statements is true with regard to fires:

    (a) Class B fires generate thick black smoke.
    (b) Class C fires are in flammable solid materials.
    (c) Class A fires usually lead to an explosion.
    (d) Class D are electrical fires.

106. The most suitable extinguisher for fighting cooking fat fires is:

    (a) Dry chemical.
    (b) Wet Chemical.
    (c) CO2
    (d) AFFF.

107. The following fire extinguisher may be used on an electrical fire:

    (a) CO2.
    (b) AFFF.
    (c) water.
    (d) foam.

108. A fire has started in the galley, which you feel you can easily fight single-handed. You should:

    (a) extinguish it.
    (b) extinguish it and then advise the crew.
    (c) extinguish it before raising alarm.
    (d) raise alarm before attempting to extinguish it.

109. The distinctive colour(s) of a portable dry chemical fire extinguisher are:

    (a) red.
    (b) red with a black band.
    (c) red with a white band.
    (d) none of the choices stated here.

110. The distinctive colour(s) of a wet chemical fire extinguisher are:

    (a) red with a white band.
    (b) oatmeal (or red with an oatmeal band).
    (c) blue (or red with a blue band).
    (d) red (or grey with a red band).

111. A vessel with an engine and cooking facilities would be required to carry:

    (a) at least two fire extinguishers.
    (b) at least one fire extinguisher.
    (c) at least three fire extinguishers
    (d) none of the choices stated here.

112. The most common fires in a small vessel are of:

    (a) Class A.
    (b) Class B.
    (c) Class C.
    (d) Class D.

113. On observing a leak in a vessel's fuel line, while under way, the master should:

    (i) replace the fuel line.
    (ii) ventilate the vessel.
    (iii) Isolate the batteries.
    (iv) speed up to return to berth quickly.

    The correct answer is:
    (a) all except (i).
    (b) all except (ii).
    (c) all except (iii).
    (d) all except (iv).

114. Regulations require all portable fire extinguishers on board a vessel to have a uniform method of operation. This means all the extinguishers in a vessel must be:

    (a) either water, foam, CO2 or dry chemical type.
    (b) either liquid or dry content type.
    (c) upright operating type.
    (d) of the same size.

115. The following is a fire hazard in a vessel:

    (a) unventilated batteries.
    (b) dry bilges.
    (c) fuel sniffers.
    (d) tight fuel pipe joints.

116. In case of fire in a vessel, the master should:

    (a) avoid anchoring.
    (b) keep the fire end downwind.
    (c) not ask passengers for assistance.
    (d) avoid berthing.

117. The most suitable fire extinguisher for blanketing a Class B fire is of the following colour:

    (a) Blue (or red with a blue band).
    (b) Red (or grey with a red band).
    (c) Red with a white band.
    (d) Red with a black band.

118. The contents of the following coloured fire extinguisher is not electrically conductive:

    (a) Red (or grey with a red band).
    (b) Blue (or red with a blue band).
    (c) red with a white band.
    (d) oatmeal (or red with an oatmeal band).

119. The contents of the following coloured fire extinguisher is asphyxiating:

    (a) Red (or grey with a red band).
    (b) Blue (or red with a blue band).
    (c) Red with a white band.
    (d) Red with a black band.

120. The following coloured fire extinguisher should not be aimed directly at a Class B fire:

    (a) Oatmeal (or red with an oatmeal band).
    (b) Blue (or red with a blue band).
    (c) Red with a white band.
    (d) Red with a black band.

121. The safety precautions when refuelling a vessel include:

    (i) unblocking scuppers.
    (ii) isolating batteries.
    (iii) turning off gas.
    (iv) keeping fuel nozzle & filler pipe in contact.

    The correct answer is:
    (a) all except (i).
    (b) all except (ii).
    (c) all except (iii).
    (d) all except (iv).

122. The basic method of fighting a fire is to:

    (a) attack all sides of the fire triangle.
    (b) attack one side of the fire triangle.
    (c) cool it.
    (d) starve it of oxygen.

123. To check if a CO2 fire-extinguisher is fully charged, you would:

    (a) get it checked professionally.
    (b) read its dial.
    (c) weigh it.
    (d) read its inspection plate.

124. In the absence of a suitable fire extinguisher, water can be used on an oil fire in the form of a:

    (a) jet.
    (b) spray.
    (c) fire wall.
    (d) none of the choices stated here.

125. A blue coloured fire extinguisher is of the following type:

    (a) water
    (b) wet chemical
    (c) dry chemical
    (d) foam

126. The following types of fire extinguishers are electrically conductive:

    (a) water only.
    (b) water and foam.
    (c) water and wet chemical.
    (d) water, foam and wet chemical.

Chapter 19.2: OH&S - Fire Prevention & Control

127. A CO2 fire extinguisher is of the following colour:
    (a) red.
    (b) red with a white band.
    (c) red with a black band.
    (d) black with a red band.

128. Fire-buckets are identified as follows:
    (a) Painted red
    (b) Painted blue.
    (c) White background with the word "Fire" in red.
    (d) Any colour background colour with the word "Fire" in red.

129. The following fire extinguisher is designed to act like a fire blanket:
    (a) water.
    (b) foam.
    (c) dry chemical.
    (d) none of the choices stated here.

130. When fighting a fire in a vessel with a water hose, there is a risk of her capsizing due to the increased:
    (a) free surface effect.
    (b) buoyancy.
    (c) freeboard.
    (d) stability.

131. Shutting down ventilation for fire below decks:
    (a) stops smoke from escaping.
    (b) stops heat from escaping.
    (c) starves it of oxygen.
    (d) cools it down.

132. The following type of fire extinguisher is designed to be used on all classes of fire on board a vessel:
    (a) water
    (b) foam.
    (c) wet chemical.
    (d) dry chemical

133. A foam fire extinguisher extinguishes a fire by:
    (a) starving it of oxygen & cooing it.
    (b) cooling it.
    (c) chemically reacting with flames.
    (d) changing its chemical composition.

134. The most suitable portable fire extinguisher for a vessel's engine room is:
    (a) dry chemical.
    (b) water.
    (c) wet chemical.
    (d) foam.

135. The following types(s) of portable fire extinguishers are fitted with a pressure gauge:
    (a) all types.
    (b) all except the CO2 type.
    (c) all except the CO2 & dry chemical type.
    (d) all except the foam type.

136. You see a portable fire extinguisher on a bracket in a vessel. It is a red container with a thick white band. It is of the following type:
    (a) water.
    (b) foam.
    (c) dry chemical.
    (d) CO2

137. The fire buckets in a vessel must be fitted with a lanyard of the following length:
    (a) long enough to reach water from the deck.
    (b) two metres.
    (c) two fathoms.
    (d) five metres.

138. A small hole should be drilled in the fire buckets on vessels to ensure they do not collect corrosive salt water; and they must be constructed of the following material:
    (a) steel.
    (b) plastic.
    (c) metal.
    (d) aluminium.

139. The portable fire extinguishers whose primary colour is red are:
    (i) Foam
    (ii) Water
    (iii) CO2
    (iv) Dry chemical

    The correct answer is:
    (a) all of these.
    (b) all except (ii).
    (c) all except (iii).
    (d) all except (iv).

140. You are on passage with wind astern. The after part of the vessel's accommodation is on fire, on which you have discharged your last fire extinguisher without success. In order to fight the fire while preventing it from spreading, you should:
    (a) speed up.
    (b) slow down.
    (c) turn head to wind.
    (d) turn 90 degrees.

141. Your vessel is at anchor in a moderate wind. There is a fire in the forward part of your vessel, which you have not been able to extinguish with a fire-extinguisher. In order to fight the fire while preventing it from spreading, you should:

  (a) slip the cable & hold the stern into the wind.
  (b) hold the bow into the wind.
  (c) pick up anchor and sail downwind.
  (d) pick up anchor and sail up wind.

142. While on watch at sea, you hear a muffled explosion from the engine room and fierce flames leap out from the engine room hatch on the main deck. You should:

  (a) cut off the fuel supply.
  (b) keep the engine room hatch open.
  (c) ventilate the engine room.
  (d) investigate before sounding the fire alarm.

143. In case of fire on board a vessel, shore assistance should be sought:

  (a) immediately.
  (b) only if the fire becomes out of control.
  (c) only if the vessel has no fire extinguishers.
  (d) only if the vessel is alongside a wharf.

144. Your fuel tank is now empty. To minimise the risk of fire from flammable vapour in it, you should:

  (a) rinse it with water.
  (b) leave its cap off.
  (c) treat it like a full tank.
  (d) gas free it.

145. The term inflammable means the substance:

  (a) is not flammable.
  (b) is flammable.
  (c) has a high flash point.
  (d) has a low flash point.

CHAPTER 19.2: ANSWERS

(Answers 1-56 are in Chapter 19.1)

57 (c), 58 (c), 59 (b), 60 (d),
61 (b), 62 (a), 63 (d), 64 (a), 65 (a),
66 (d), 67 (a), 68 (c), 69 (a), 70 (b),
71 (a), 72 (c), 73 (d), 74 (b), 75 (a),
76 (d), 77 (b), 78 (b), 79 (b), 80 (c),
81 (a), 82 (a), 83 (b), 84 (d), 85 (c),
86 (c), 87 (d), 88 (a), 89 (d), 90 (a),
91 (b), 92 (c), 93 (a), 94 (a), 95 (c),
96 (d), 97 (c), 98 (a), 99 (c), 100 (b),
101 (a), 102 (c), 103 (b), 104 (b), 105 (a),
106 (b), 107 (a), 108 (d), 109 (c), 110 (b),
111 (a), 112 (b), 113 (d), 114 (c), 115 (a),
116 (b), 117 (a), 118 (c), 119 (d), 120 (b),
121 (a), 122 (b), 123 (c), 124 (b), 125 (d),
126 (d), 127 (c), 128 (a), 129 (b), 130 (a),
131 (c), 132 (d), 133 (a), 134 (d), 135 (b),
136 (c), 137 (a), 138 (c), 139 (a), 140 (c),
141 (a), 142 (a), 143 (a), 144 (c), 145 (b)

Chapter 19.3

OCCUPATIONAL HEALTH & SAFETY

SAFETY EQUIPMENT, SEA SURVIVAL & RESCUE

FOR PYROTECHNIC DISTRESS SIGNALS IN COLOUR: *See Chapter 6*

Item	Quantity	Area of operation		
		Sheltered	Inshore	Offshore
Fire bucket	1	✓	✓	✓
Bilge pump or bailer (B1)	(B2)	✓	✓	✓
PFD 1	(P)	—	✓	✓
PFD 1, 2 or 3	(P)	✓	—	—
Anchor & chain and/ or line	1	✓	✓	✓
Waterproof/buoyant torch	1	✓	✓	✓
Fire extinguisher (F1)	(F2)	✓	✓	✓
Flares—orange	2	—	✓	✓
Flares—red (hand held)	2	—	✓	✓
Flares—parachute	2	—	—	✓
Navigation lights (N1)	(N2)	✓	✓	✓
EPIRB	1	—	—	✓
V-sheet	1	—	—	✓
Marine radio	1	—	—	✓
Compass	1	—	—	✓

KEY

✓ Required
— Not required

(B1) Bilge pump (electric or manual) shall be provided on boats with closed under-floor compartments or covered bilges. For other boats, a bailer shall be carried.

(B2) Bilge pumps shall be capable of draining each compartment of the vessel. This may require more than one bilge pump to be fitted.

(P) A PFD shall be carried for each person on board the vessel.

(F1) Fire extinguishers shall be provided on all boats with an electric start motor, gas installation or fuel stove.

(F2) The number of fire extinguishers shall be appropriate for the number and size of potential sources of fire.

(N1) Navigation lights are required from sunset to sunrise and in restricted visibility.

(N2) Quantity and type of Navigation lights fitted are to be in accordance with the Regulations for the Prevention of Collision at Sea (as amended).

Fig 19.11: "REQUIRED" SAFETY-EQUIPMENT FOR RECREATIONAL BOATS (*NMSC*)

MINIMUM SAFETY EQUIPMENT FOR RECREATIONAL VESSELS

At the time of writing of this book, the safety equipment requirements for recreational vessels varied between States and Territories. The National Marine Safety Committee (NMSC) has proposed the above table for a national standard. See Chapter 12 for NMSC definitions of Inshore, Offshore, etc.

COMMERCIAL VESSELS: They are supplied with individual lists in their survey record book. However, they must be equipped with two methods of distress alerting and one method of receiving weather and Maritime Safety Information on a 24-hour 7 days week basis. The detailed requirements for life-saving appliances and fire-fighting appliances are listed in the USL Code, Sections 10 & 11, and the NSCV Part C, and in the State Boating Safety Equipment Regulations.

ADDITIONAL EQUIPMENT

NMSC recommends that operators of recreational vessels should undertake a safety assessment of their particular vessel and her intended operation. In addition to the safety-equipment carried in accordance with this standard, vessels should carry any other additional safety-equipment that may be appropriate to control risks to acceptable levels. The following table provides a list of additional safety-equipment that should be considered. This list is informative and should not be considered to be exhaustive.

Fig 19.12 **Additional safety-equipment for recreational boats**
(NMSC)

Item
• Fire blanket
• Upgrade to PFD1
• Inflatable PFD – refer to (a) below
• First-Aid Kit
• Personal strobe light
• Signalling mirror
• Lifebuoy and line
• Boarding ladder
• Liferaft and / or dinghy (b)
• Tool kit
• Oars, paddles or other alternative means of propulsion
• Tow rope
• Appropriate chart
• GPS and plotter

(a) An inflatable PFD may encourage wearing of the device in potentially hazardous situations of normal operation.

(b) A dinghy that is relied upon for the purposes of safety equipment should be arranged with arrangements to support the weight of persons on board in the event of swamping.

NMSC STANDARD FOR BASIC ITEMS OF RECREATIONAL BOATING SAFETY-EQUIPMENT

All safety equipment carried in accordance with this standard shall be located so as to be readily accessible in time of need; and maintained in accordance with the manufacturer's instructions. Where the equipment carries a manufacturer's expiry date, it shall not exceed the prescribed expiry date.

ANCHOR WITH CHAIN &/OR LINE: The anchor and chain and/or line shall be suitable for the purpose of securing the vessel given the vessel's size, weight and the area of operation. The chain and/or line shall be of sufficient strength and durability for the purpose and is to be securely attached to both the anchor and the vessel. Where applicable, the anchor should comply with AS 2198. Anchors are discussed in Chapter 7.

Chapter 19.3: OH&S - Safety Equipment, Sea Survival & Rescue

FIRE EXTINGUISHER: Fire extinguishers carried shall be of a type suitable for the type(s) of fuel carried on board the vessel, as specified in AS 1799.1. They shall be designed and manufactured in accordance with an Australian Standard specification for portable fire extinguishers. Extinguishers shall be stowed, so as to be readily accessible in the case of fire. Fire extinguishers are discussed below. For colour illustrations see Chapter 6.

FIRE BUCKET WITH LANYARD: The bucket shall be suitable for collecting water for use in case of fire of solid combustibles. The bucket shall be manufactured from waterproof and robust material, and shall be designed so as not to collapse, distort or lose the handle when full of water. The bucket shall not to be used for any other purpose, apart from being used as a bailer, and shall be readily available at all times.

The bucket shall have a lanyard (rope) attached, which is of sufficient length to allow the bucket to be cast over the side and retrieved full of water.

BAILER & LANYARD: A bailer shall be suitable for bailing water from the vessel and shall have a lanyard (rope) securely attached to prevent loss from the boat. The bailer shall be readily accessible and shall not used for any other purpose.

A fire bucket carried in accordance with this standard, may double as a bailer provided it satisfies the above requirements.

PFD: A personal flotation device (PFD) Type 1, Type 2, or Type 3 is to be carried for each person on board the vessel and is to be the correct size for the intended wearer. The type of PFD carried shall be appropriate for the purpose for which it is used. In selecting a PFD due consideration shall be given to the purpose for which it was designed, as set out in AS 1512, AS 1499 and AS 2260, and the time of day, type of activity, area of operation and likelihood of rescue. The personal flotation devices shall be designed and manufactured in accordance with the latest AS 2259 standard. PFDs are discussed below.

WATERPROOF TORCH: A water resistant, floating type torch in operational order that is capable of being used to signal.

BILGE PUMP: Required for vessels with closed under-floor compartments or covered bilges. The pump or pumps shall be capable of draining each compartment of the vessel. They may be either manual or power operated, and shall have a strainer fitted to the suction pipe. The strainer shall be of a sufficiently small mesh size to prevent choking of the pump. Bilge pumps are discussed in Chapter 29.

V-SHEET (DISTRESS SIGNAL): A fluorescent orange-red coloured sheet of dimensions not less that 1.8 m x 1.2 m with a black V superimposed on the sheet in the centre. The letter V on the sheet shall be of a width no less than 150 mm.

COMPASS: Liquid damped with rotating card showing the cardinal points. Compasses are discussed in Chapter 17.1.

Fig 19.13

DISTRESS FLARES (*See Colour Illustrations in Chapter 6*): Distress flares shall be designed and manufactured in accordance with the provisions of AS 2092 as they relate to red hand-held distress flares and orange hand-held smoke signals. The flares shall not exceed the manufacturer's expiry date.

STOWAGE OF PYROTECHNICS: Distress signals are designed to show a brilliant continuous red light or emit dense orange smoke. All pyrotechnic distress signals are fitted with a self-contained means of ignition. Although they can be operated when wet and they burn underwater, it is recommended to store them in their protective seals and weathertight containers. Moisture build-up and temperature changes can cause problems. The manufacturers and marine authorities recommend that they be replaced every three years. Do not use them all at once.

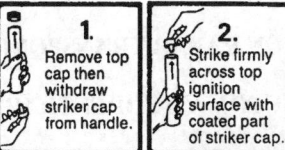

1. Remove top cap then withdraw striker cap from handle.
2. Strike firmly across top ignition surface with coated part of striker cap.
3. Hold up & outward to leeward.

RED HAND-HELD FLARES: (Fig 19.14 (A)) Depending on the brand, these are ignited either by turning the handle to arm the flare and then pushing it in to strike; or as shown below. It is visible to flying aircraft at a range of about 10 kms (5 nautical miles) on a clear dark night and 5 kms (2.5 nm) during the day for a period of 60 seconds. The burning flare should be held down wind to prevent sparks falling in the vessel.

ORANGE SMOKE SIGNALS: (Fig 19.14 (B)) It is designed for day use (with light

Fig 19.14 (A & B)

winds), and is visible for approximately 4 kms (2 nm) on a clear day. Smoke signals are of two types: hand-held flare-type and buoyant smoke canisters. The former is ignited like the hand flares, emitting orange smoke for 60 seconds. The latter is operated by method similar to pulling the tab off a soft drink can. It emits dense orange smoke for 3 minutes. The purpose of smoke signals is to attract an aircraft's attention. However, they are also attached to lifebuoys to help rescue a person overboard.

PARACHUTE FLARES: These are usually fitted with a firing level underneath the bottom cap. The rocket is cocked and fired by pressing the firing lever up and against the body of the rocket. It ejects a parachute suspended red flare to a height of 300 metres, which burns for at least 40 seconds in its descent. It provides the most suitable means of attracting attention of a ship on the horizon. They are visible for about 40 kms (25 miles) at night and less than 15 kms (8 miles) during the day. Parachute rockets are normally fired vertically slightly upwind. However, if there is a risk of disappearance into low clouds, they should be fired at an angle of 45°.

DYE MARKER: This distress signal is not orange in colour. It releases in the water a green dye, which an aircraft, flying overhead, may see but only during the day. On hearing an aircraft, the sachet containing the dye should be emptied on the lee side.

EPIRB/GPIRB: An Emergency Position Indicating Radio Beacon (EPIRB) suitable for marine use, that can transmit on either 121.5 or 406MHz and conforms to Australian Standard AS/NZS 4330 and AS/NZS 4280. Any 406MHz unit shall be properly registered with the Australian Maritime Safety Authority (AMSA). EPIRBS and GPIRBS are discussed in detail in Chapter 18.

MARINE RADIO: A HF (27 MHz) or VHF marine radio transceiver approved by the Australian Communications Authority (ACA). Marine radios are discussed in Chapter 18.

NAVIGATION LIGHTS: Navigation lights are to be positioned and perform in accordance with the provisions in the Annexes to the International Collision Regulations. (See Chapter 11.)

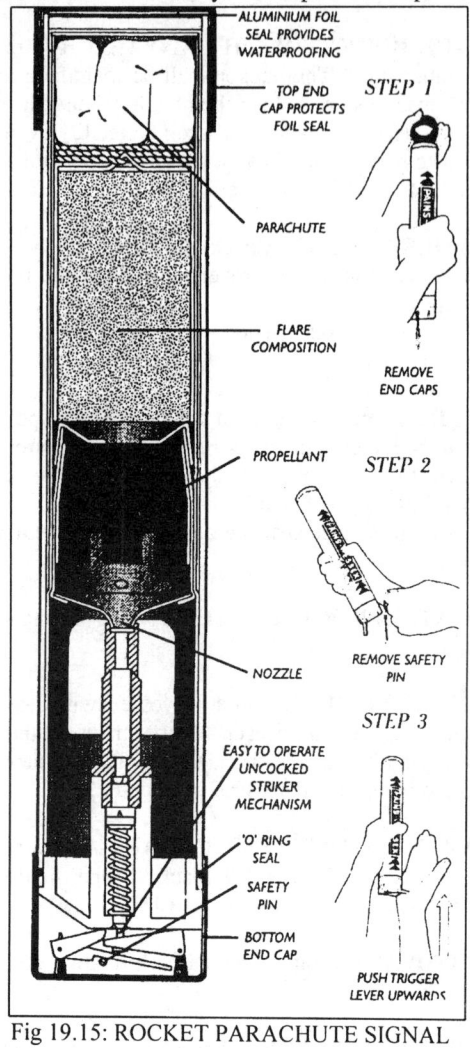

Fig 19.15: ROCKET PARACHUTE SIGNAL

OTHER SAFETY EQUIPMENT

SART: *See Chapter 16.*

ANTI-COLLISION RADAR TRANSPONDER: *See Chapter 16.*

UNIVERSAL AUTOMATIC IDENTIFICATION SYSTEMS (UAIS or AIS): *See Chapter 14.*

HAND LEAD LINE: A hand lead line is a 25-metre length of rope with a 3-kilo piece of lead bar tied at one end. It is marked off at every metre or every second or third metre and is used for measuring depth of water. Three different materials, serge, duck and bunting, are used for markings so that the soundings can be identified in the dark. There is a recess in the base of the lead, which can be filled with tallow (wax) in order to recover samples of the seabed - sand, shingles, etc.

MARKINGS OF A HAND LEAD LINE
2 metres   -   two strips of leather
3 metres   -   three strips of leather

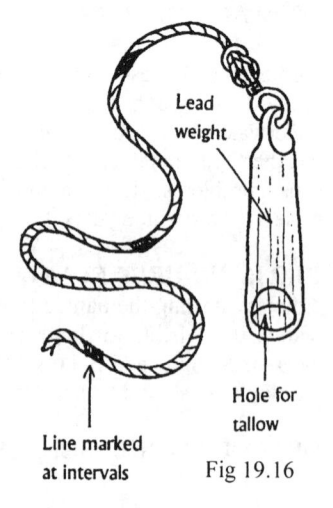

Fig 19.16

Chapter 19.3: OH&S - Safety Equipment, Sea Survival & Rescue

5 & 15 metres	-	a piece of white duck
7 & 17 metres	-	a piece of red bunting
10 metres	-	a piece of leather with a hole in it
13 metres	-	a piece of blue serge
20 metres	-	two knots
30 metres	-	three knots

HELIOGRAPH: Included in liferaft stores, a heliograph is a daylight signalling mirror made of metal. A hole in the mirror is aligned with a sight vane to reflect sunlight onto a ship or aircraft at a distance and attract its attention. The attached figure on the use of a heliograph appeared in the Australian Maritime Safety Authority's Safety Education Article No. 4.

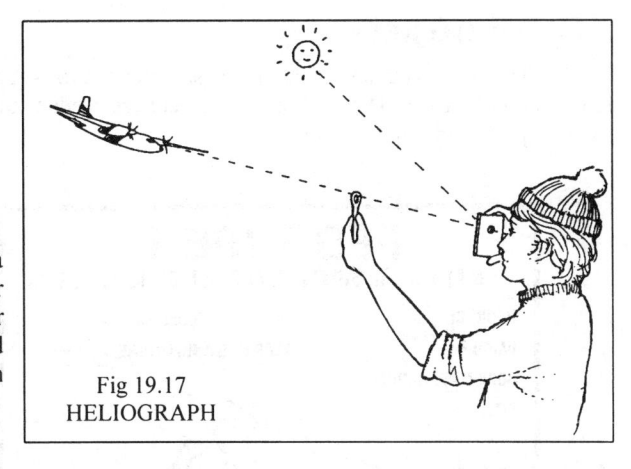

Fig 19.17 HELIOGRAPH

LIFEBUOY: This is a buoyant ring (or a horse-shoe shape) made of fibreglass and filled with kapok, fibreglass or plastic foam, capable of supporting at least 13 kilograms concentrated mass in water. The ring is fitted with four handholds or beckets, and four strips of retro-reflective tape wrapped around it.

Lifebuoys are designed to be quickly thrown overboard for a person overboard or when abandoning ship. They must be marked with the name of the vessel and stowed on float-free brackets. Lifebuoys and their attachments should not be lashed to the vessel in any way. They must remain free for immediate use. One lifebuoy (or two for larger vessels) should be stowed near the steering position. Others distributed about the vessel in accessible positions.

At least half the lifebuoys should have self-igniting lights of 2-candela intensity for 2 hours. At least one lifebuoy on each side near the stern of the vessel should be attached with a buoyant line, usually 27.5 metres in length, to haul in a person overboard. The exact requirement for commercial vessels is specified in their survey records.

Inspect lifebuoys regularly for their correct positioning, access, wear and tear of the casing, fraying of becket anchor points and other lines, the condition of lights and reflective tapes as well as the stowage brackets.

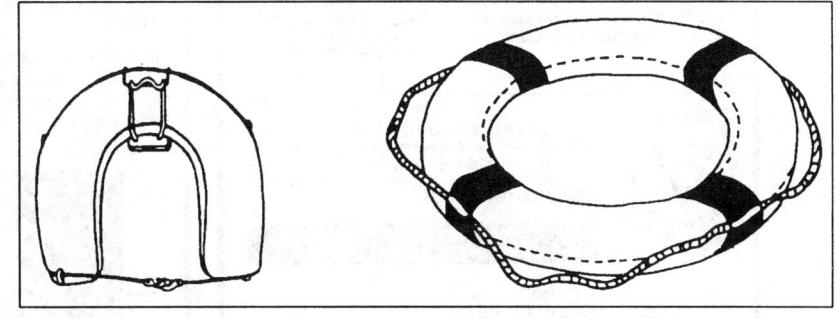

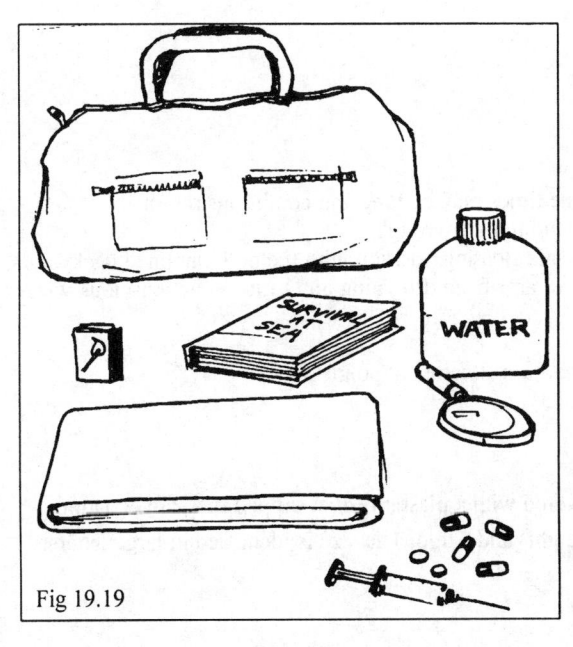

Fig 19.18: HORSE-SHOE & ROUND LIFEBUOYS

GRAB BAG: Also known as a **PANIC BAG**, it is designed to supplement the limited contents of a liferaft. Always keep it ready and take it with you if you need to abandon ship. Its contents may include:

- A spare pair of spectacles (if applicable)
- Personal medication (if applicable)
- Extra water and/or solar still
- Tins of food
- Waterproof matches
- Magnifying glass
- Space blankets
- Survival at Sea - instruction manual
- Personal EPIRB/GPIRB

Some sailing vessels carry, lashed on deck, plastic jars of 12 litres or of larger capacity, partly filled with drinking water. Secured to them is 2 metres of 6 mm clear plastic tube which can be used as a suction line to drink water whilst both the jar and the drinker float together in the sea. Contamination of the water in a wide range of sea conditions is thus minimised.

Fig 19.19

PFDs & LIFEJACKETS

The lifejackets and buoyancy vests required to be carried by non-commercial vessels are known as the Personal Floatation Devices (PFDs). Various manufacturers produce these in a variety of designs. The Australian Standards classify and stamp them in three categories:

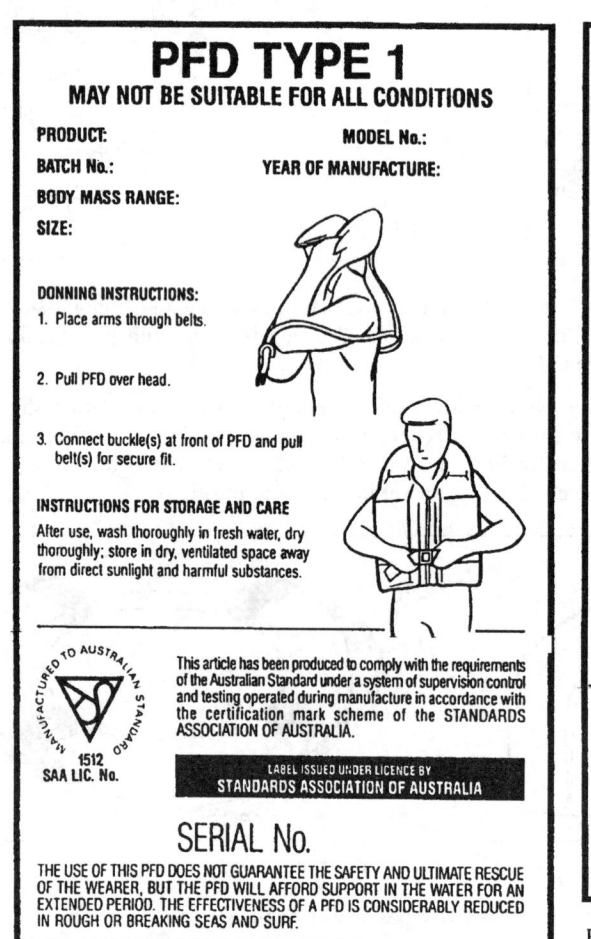

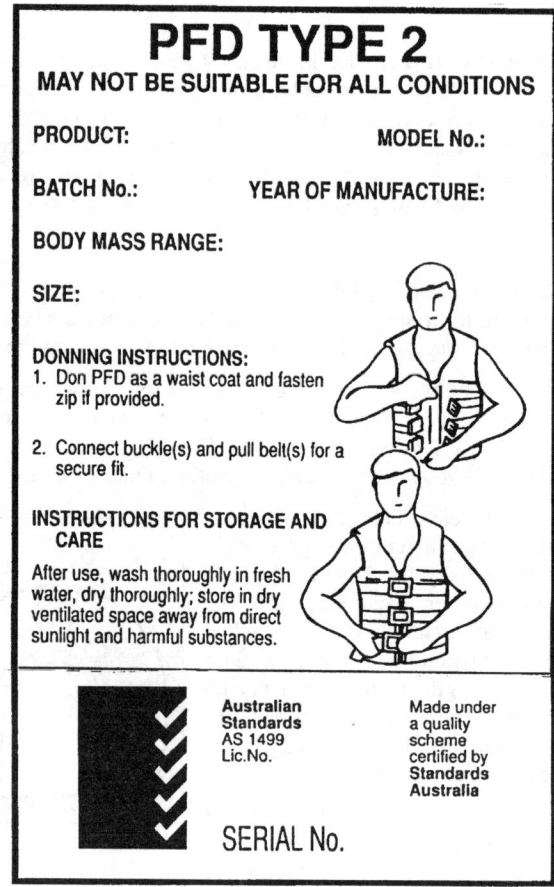

Fig 19.20 (A & B)

PFD1:
- Two types:
  - Fixed buoyancy type
  - Self or manually inflatable type
- Compulsory for all persons on board vessels on the open seas.
- Recommended for remote inland waters where search and rescue times may be long and conditions rough.
- Recommended for all persons aboard small vessels whenever conditions are rough.
- Fitted with a buoyancy collar, which maintains the wearer in a safe floating position with the body inclined backwards, and nose and mouth clear of the water. The collar protects the wearer from drowning and turns an unconscious wearer into face-up position
- Fitted with strips of retro-reflective tape.
- Many incorporate battery operated lights, whistles, and safety harness attachment points.
- The commonly used fixed flotation material is:
  - Pads of kapok, encased in sealed plastic film
  - Fibrous glass material, encased in sealed plastic film, or
  - Sections of specifically shaped unicellular plastic foam covered with a plastic film or dipped in a vinyl coating.
- Available in two sizes: Adult (for persons 40 kgs or more in weight) and Child. The size is identified in large letters.

PFD2:

- Smaller and less bulkier than PFD 1 - providing less buoyancy and safety.
- Known as buoyancy vest.
- Suitable for short-term flotation in sheltered waters during daylight hours, such as for aquatic sports.
- Personal Watercraft (PWC) users can also use a PFD 2.
- Available in numerous styles.
- Most types are fitted with strips of retro-reflective tape.
- Contain the same flotation materials as PFD1s.

PFD3:

- Also known as buoyancy vest.
- Buoyancy characteristics similar to PFD 2.
- Available in wide range of colours.
- Preferred by water-skiers & PWC riders.
- Not recommended for general boating because the colours are less visible in search and rescue.

CHILDREN'S PFDs & LIFEJACKETS:

As advised in AMSA's Safety Education Article No. 59, a child is difficult to float in a face up position because of the distribution of body mass and because the child tends to panic in an unfamiliar environment. The violent movement of the arms and legs in an attempt to "climb out" of the water tends to nullify the stability of the lifejacket.

Small children tend to have big heads for their bodies. They thus have a higher centre of gravity. However, a properly designed lifejacket of the correct size will keep a child's mouth and nose clear of the water. A child should be taught how to put on the device and, preferably, should be allowed to try it out in the water. It is important that the child feels comfortable and knows what a lifejacket is for and how it functions.

PFD 4 & 5:

In some countries, there also exist PFD 4s & 5s. Most commonly, a PFD4 is a buoyancy cushion made of kapok, fibrous glass material or unicellular plastic foam in a cover of fabric or coated cloth fitted with grab straps. As in PFD1s, kapok or fibrous glass material pads are encased in sealed plastic film covers. These cushions, although containing more buoyancy than PFD2s, do not provide safety for an exhausted or unconscious person. A PFD5 is any device designed for a specific and restricted use. One example is the work vests worn by workmen when employed over or near water under favourable conditions and properly supervised.

BUYING PFDs

- Buy your own PFD/lifejacket. One size does not fit all. Try it on. It should fit snugly without confining or riding up. With straps and buckles secured, the jacket should not ride up over your face. Such a PFD can plunge your face beneath the water surface.

- If buying for children, find PFDs appropriate to each child's weight. A wrong size PFD may ride up dangerously high over the child or the child may slip out of it in the water. With the child wearing it, try lifting him or her by the top of the PFD. It shouldn't ride up over the chin. Look for one with a crotch strap.

- Look at the label for size and weight.

- Make sure it complies with the Standards Australia or is approved by a government body.

- Simulate being overboard. Go into HELP position (discussed in this chapter) while wearing your PFD/lifejacket over some heavy clothing. It should not inhibit your breathing or ride up to your chin.

- Following the 1998 Sydney-Hobart disaster, the NSW Coroner recommended that future competing yacht crews should not use the "bib" ("Mae West") type of PFDs. The Cruising Yacht Club of Australia went further by banning kapok-filled PFDs. This type of PFDs and lifejackets, although the norm on commercial vessels, can be quite uncomfortable and a hindrance to the helicopter rescue strop.

LIFEJACKETS: For commercial vessels there are two categories of lifejackets.

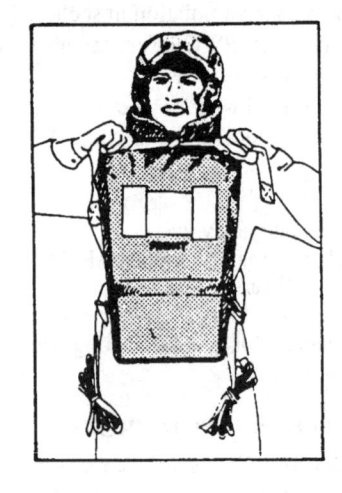

Fig 19.21: COASTAL (Left)
& SOLAS (Right) LIFEJACKETS

COASTAL LIFEJACKET:
- Reversible bib type construction.
- Provides full flotation support as described for PFD 1 above.
- Fitted with six strips of retro-reflective tape on each side,
- Fitted with a whistle and a salt-water activated light.
- The instructions for wearing are printed on it.
- Tested and stamped by the State marine authority or the A.A.P.M.A. (Australian Association of Ports and Marine Authorities).

S.O.L.A.S. LIFEJACKET:
- Compulsory for seagoing commercial vessels - generally those operating more than 30 miles offshore.
- Manufactured and tested under more stringent international standard of the SOLAS (Safety of Life at Sea) Convention. Australia is a signatory to the SOLAS Convention.
- Carries the stamp of the Australian Maritime Safety Authority.
- Fitted with six retro-reflective strips on each side.
- Fitted with a light and a whistle. The light is designed to burn for at least 8 hours with a power of 0.75 candelas.
- Fitted with an anti-hypothermia hood, a water deflecting foam collar and many other features.
- Designed to lift the mouth of an exhausted or unconscious person not less than 120 mm clear of the water with the body inclined backwards at an angle of no less than 20° and no more than 50°.
- Some recent developments in lifejackets include the use of closed cell foam plastics, synthetic outer covers, snap locks and quick release buckle fastenings and visors to protect head and face.

INFLATABLE LIFEJACKETS & PFDs
The inflatable or the so-called "double bubble" SOLAS and Coastal lifejackets incorporate at least two separate compartments so that if one springs a leak, the other can do the work. These lifejackets inflate automatically on immersion and must also be capable of being blown up by mouth.

LIFEJACKETS & PFDs BUOYANCY TESTS
A PFD assigned with 10-kilogram buoyancy rating will have been tested by hanging from it a 10-kilogram weight of lead or iron in the water. It would easily supports a person weighing 100 kilograms, but not an anchor of the same weight. This is because approximately 80% of human body is water, which has no weight when immersed in water. A further 15% of human body is fat which is lighter than water. Therefore, a 100-kilogram person weighs only about 5 kilograms in water, making the 10 kilograms buoyancy in a PFD quite adequate.

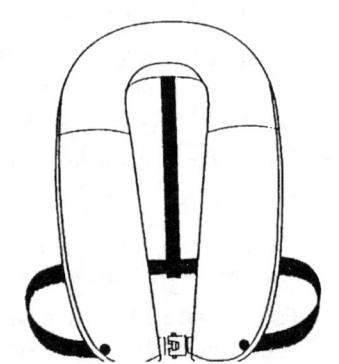

Fig 19.22: INFLATABLE PFD

STOWAGE, CARE & USE OF PFDs & LIFEJACKETS

- Stow in a dry place where they are available for immediate use, in separately designed lockers or beneath bunks or seats. Display suitable signs indicating their location and donning procedure.
- Remove plastic wrapping on lifejackets and PFDs to prevent condensation.
- Regularly inspect for mildew, condition of the fabric, straps and stitching; readability of donning instructions; attachments such as whistle, light, battery & retro-reflective tape. Check inflatable jackets for air leak. Faded fabric indicates deterioration by UV. Such lifejackets should be replaced.
- Unusually hard or heavy buoyancy cushions call for professional inspection for damage.
- Replace and account for all the old lifejackets and PFDs.
- Check PFDs annually for flotation and fit.
- Inflatable PFDs/lifejackets must be serviced as recommended by the manufacturer.
- Regularly muster all crew and practice wearing PFDs.
- Make suitable log book entries of musters, drills and inspections. You may need such records when your competency is challenged following an accident or death.
- Never alter a jacket. It could lose its effectiveness.

IS WEARING PFD OR LIFEJACKET COMPULSORY?
It is compulsory when:
- Riding a personal watercraft such as a Jet Ski
- On board a commercial vessel while crossing a coastal bar.

It should also be worn when:
- Sea conditions deteriorate
- Fishing near shallow reef outcrops ('bombies')
- Crossing a coastal bar in a recreational vessel.

It should be worn at all times when afloat by:
- Children
- Poor swimmers

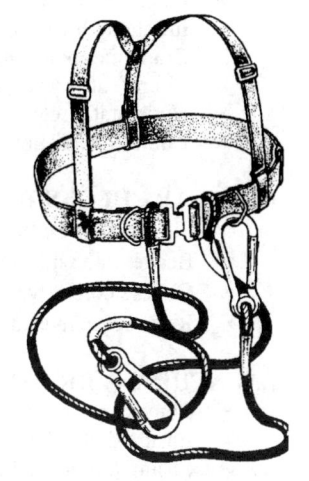

Fig 19.23
SAFETY HARNESS

SAFETY HARNESSES
Safety harnesses prevent sailors from being washed overboard from the decks of small vessels. Everyone on deck should wear them at all times, especially in rough weather. They are made of webbing material, in the form of a belt and two shoulder straps. They have a webbing tether (cord) with a hook on each end. (Single-ended harnesses dropped out of favour after a crewman on the yacht 'Waikikamukau' was dragged down with the sinking boat because he was unable to get to the end of his harness to unclip). The double-ended harnesses are also easier to clip and unclip on board as required, thus encouraging people to wear them.

Strong attachment points for harnesses are as important as the harnesses themselves. The attachments should also be sufficient in number, and suitably positioned on the vessel. Some craft are rigged with wire jackstays on deck to permit crewmembers to clip to and be able to move about the deck along the jackstay without the risk of unclipping from one point to another.

LIFERAFTS
Liferafts are of either inflatable or rigid construction. Both types must be orange in colour, equipped with hand beckets or holds and capable of supporting the number of persons for which they are designed.

Inflatable liferafts are the most compact and complete survival craft. In a seagoing vessel, if inflatable liferafts are the only survival craft on board (i.e., there are no lifeboats) they must be of sufficient capacity to carry everyone on board. Vessels operating in smooth and partially smooth waters may carry only the rigid liferafts, CARLEY FLOATS or dinghies. Carley floats are designed to float free from the sinking vessel and support people in the water.

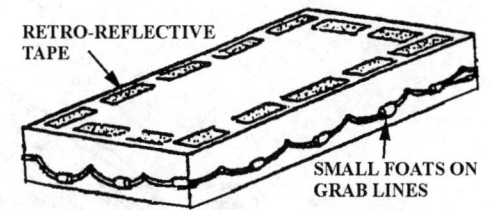

Fig 19.24: CARLEY FLOAT

The location of liferafts and Carley floats on board depends on the size of the vessel and her survey requirements. In general, they should be instantly accessible, which is usually on an upper deck, and clear of all obstructions. They should be able to float free from the vessel even when she is listed by 10° to either side and trimmed by 15° to either end.

Liferaft containers for commercial vessels are marked with the name of the vessel, capacity, instructions and the date of last inspection. An EPIRB/GPIRB, if contained, is indicated. Liferaft certificates are valid for one year. This means they are required to be inspected and serviced annually by an approved service centre.

RAFTS ARE FITTED WITH
- strips of reflective tape on canopy & underside of raft
- water (stability) pockets underneath
- a righting strap underneath
- a light outside
- a white light inside
- a lifeline all round
- boarding ladder

Fig 19.25: INFLATABLE LIFERAFT

REGULAR CHECKS TO LIFERAFTS
➢ Painter fast to railing.
➢ Inflatable liferaft canister & seal in good condition.
➢ Cradle in good condition.
➢ Liferaft always clear for launching.
➢ Hydrostatic release mechanism in date and in place. The manual release should also be operating.
➢ Liferaft inspection label current.

REGULAR CHECKS TO CARLEY FLOATS
- Float should be free of cracks and in good condition
- Becket line with small floats in good condition
- Chocks correctly positioned and clean
- Retro-reflective tape not deteriorated

LAUNCHING A RIGID LIFERAFT
➢ Remove straps and cover, if any.
➢ Ensure that painter is made fast on board.
➢ Ensure that it is all clear below.
➢ Launch raft.
➢ Pull on painter, and tug hard to fully inflate both liferaft canopies. There is a canopy on each side of the raft. There is no capsized condition. The submerged canopy acts as a drogue.
➢ Remove shoes & sharp objects. Pull raft alongside with painter and board. Try to avoid injury to persons in the raft if it is necessary to jump.
➢ When all persons are on board, cut the painter as far away from the raft as possible. (Knife stowed near painter).
➢ Paddle clear of vessel to join other craft.
➢ Stream the sea anchor or drogue.

INFLATABLE LIFERAFTS – PARTS & LAYOUT

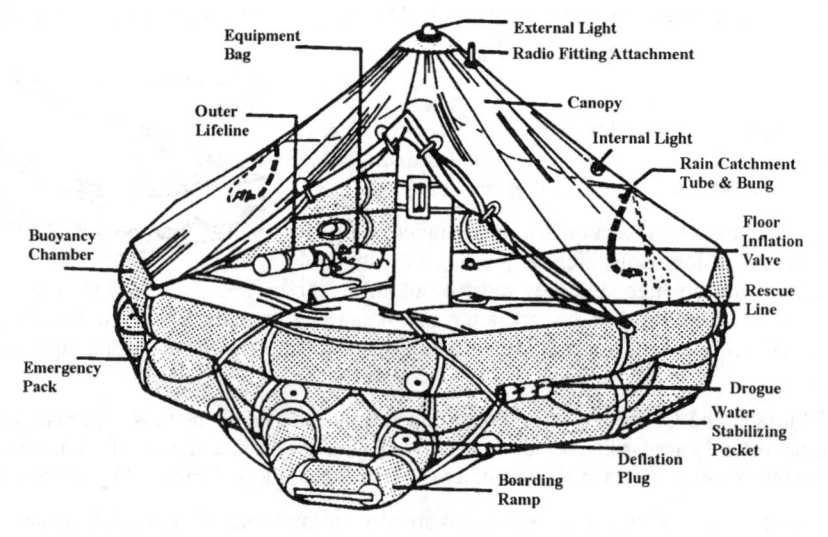

Fig 19.26: DUNLOP/RFD

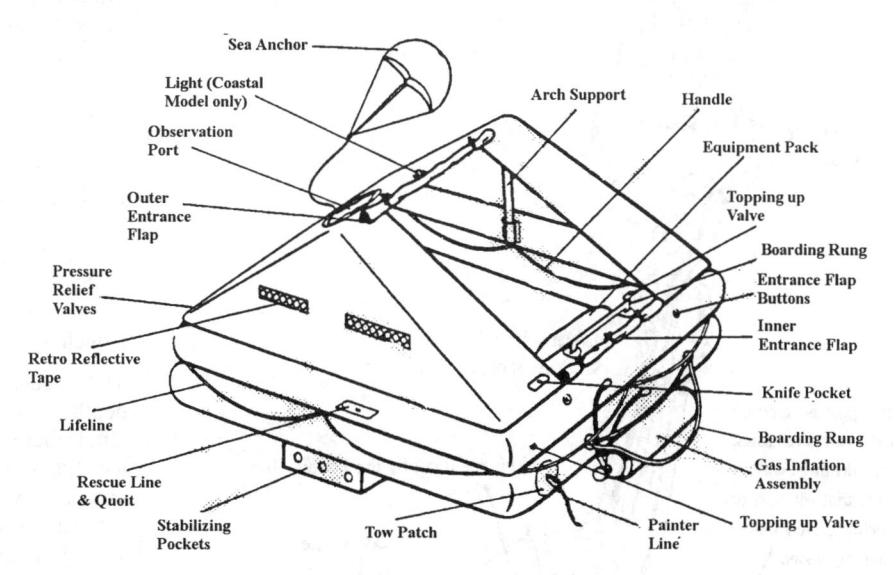

Chapter 19.3: OH&S - Safety Equipment, Sea Survival & Rescue

Fig 19.27: BEAUFORT "DOLPHIN"

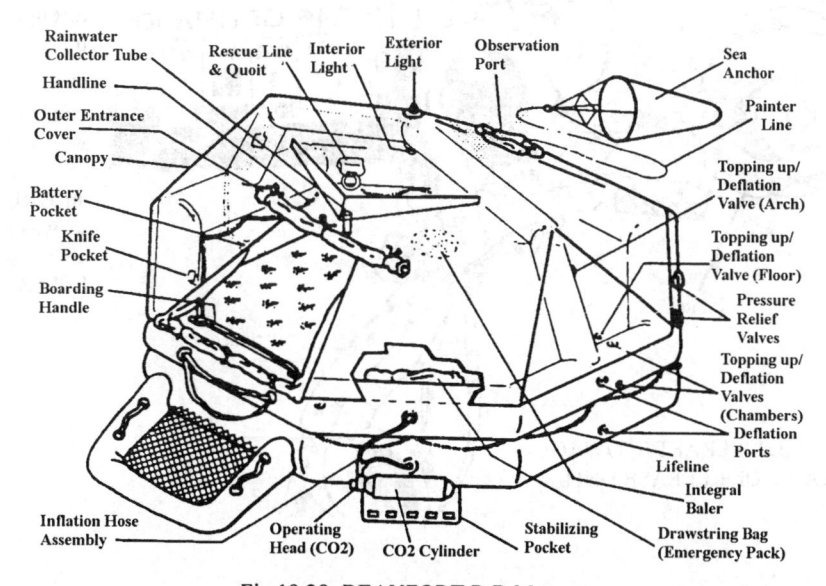

Fig 19.28: BEAUFORT R.B.M

LIFERAFT STOWAGE & RELEASE UNIT

Inflatable liferafts are packed either in fibreglass canisters or a soft valise. They are stowed in cradles or chocks near the ship's side. The canister or packing is automatically discarded when the raft inflates on launching.

Liferafts are fitted with a painter (a rope attached to the raft inside the casing), which must be secured to a strong point on the vessel as soon as the raft is stowed in position. The painter serves two functions: First, it holds the raft alongside after it is thrown overboard for launching. Second, a sharp tug on the painter releases the gas from the cylinder attached to the raft and inflates it.

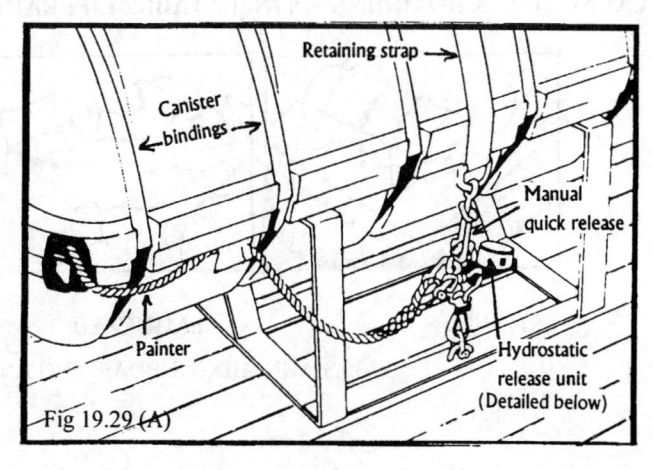

Fig 19.29 (A)

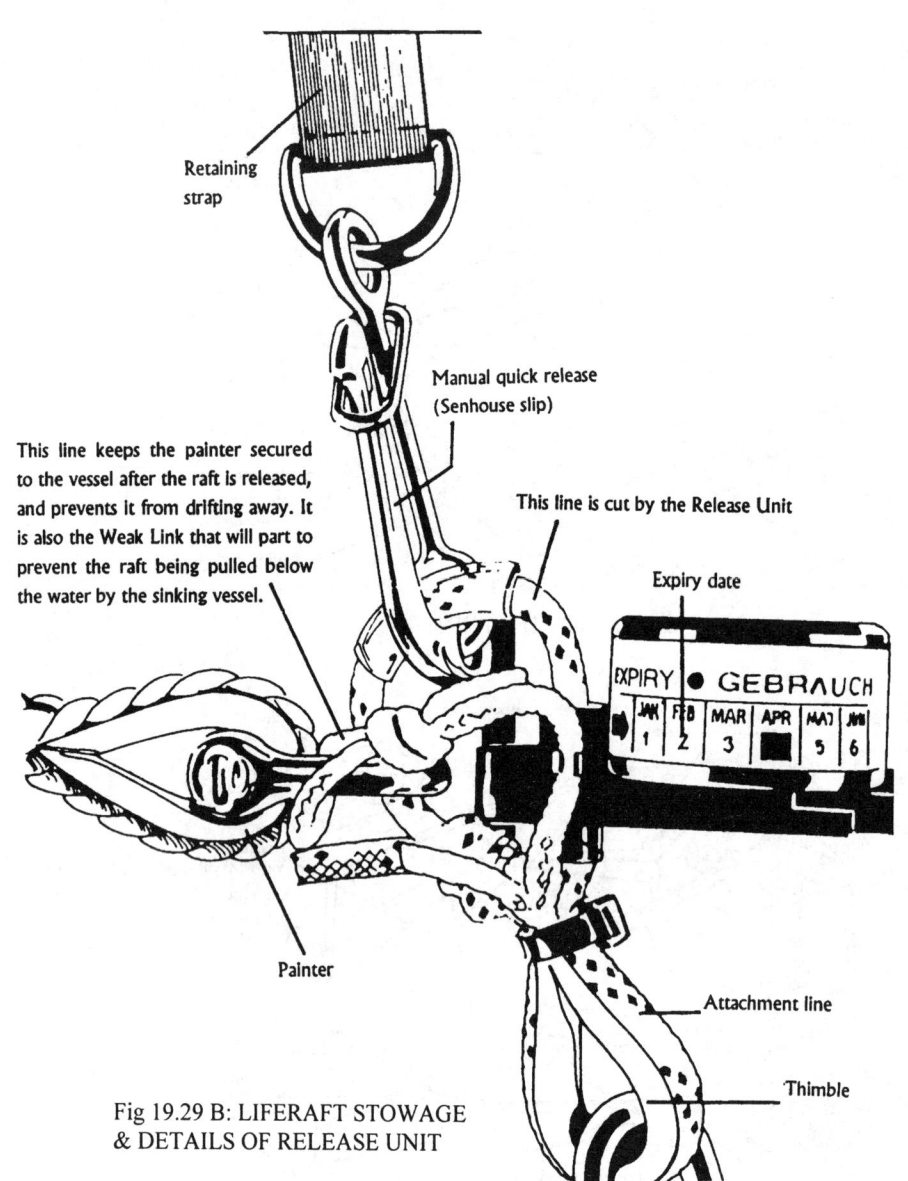

Do not secure the liferaft in a manner that would make its release difficult in an emergency. The lashings securing it should incorporate a manually operated quick release device and a **HYDROSTATIC RELEASE UNIT (HRU)**. The latter allows the liferaft to float free if the vessel sinks before the crew had the opportunity to launch it.

The mechanism works at a depth of about 3 metres (1.5 atmosphere). Modern HRUs are fitted with a pyrotechnic charge. When the unit is submerged to a depth between 1 and 4 metres, a pressure diaphragm activates the charge that drives the cutting blade through the rope or plastic bolt holding the liferaft.

However, under exceptional circumstances, a large wave washing on board in rough seas can accidentally operate an HRU and cause the liferaft to wash overboard. HRUs are also sometimes used for EPIRBS/GPIRBS.

Fig 19.29 B: LIFERAFT STOWAGE & DETAILS OF RELEASE UNIT

LAUNCHING & BOARDING AN INFLATABLE LIFERAFT

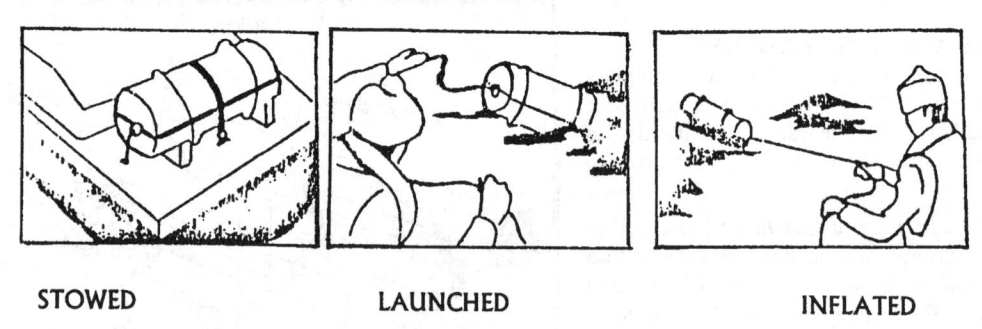

STOWED   LAUNCHED   INFLATED

Fig 19.30: LAUNCHING & BOARDING AN INFLATABLE LIFERAFT

Chapter 19.3: OH&S - Safety Equipment, Sea Survival & Rescue

- Ensure that the painter is made fast on board.
- Create a lee for the launch. (However, when abandoning ship, a liferaft launched in the lee of a vessel may find it difficult to shove off due to the wind blowing the vessel towards the raft.)
- Remove lashings.
- Ensure that it is all clear below.
- Throw liferaft overboard.
- Pull out the painter till the end, and give it a hard tug to fire the gas bottle. The liferaft will inflate in 20 to 30 seconds.
- A loud whistling noise is heard after the liferaft is inflated and excess CO_2 is released from the bottle. Excess gas is also vented through relief valves on warming. Topping up with air is necessary on cooling. The liferaft tubes in operation must be kept rigid for better stability. A hand pump or bellows is supplied in the raft.
- Remove shoes & sharp objects. Pull raft alongside with painter and board. Try to avoid injury to persons in the raft if it is necessary to jump.
- When all persons are on board, cut the painter as far away from the raft as possible. (Knife stowed near painter).
- Paddle clear of vessel.
- Stream the sea anchor or drogue.

DAVIT LAUNCHING & BOARDING AN INFLATABLE LIFERAFT (Fig 19.31, *Dunlop Marine*)

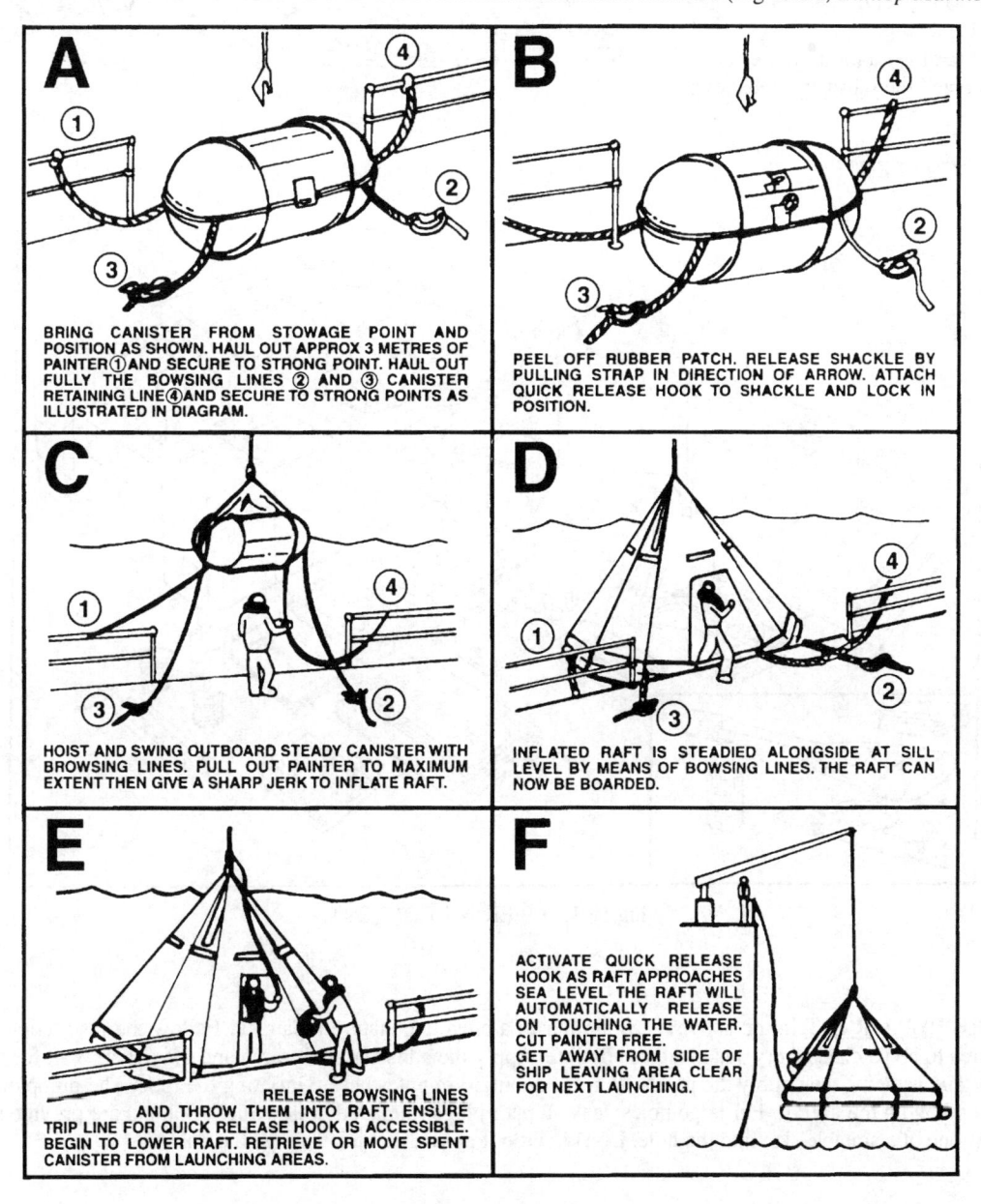

A — BRING CANISTER FROM STOWAGE POINT AND POSITION AS SHOWN. HAUL OUT APPROX 3 METRES OF PAINTER ① AND SECURE TO STRONG POINT. HAUL OUT FULLY THE BOWSING LINES ② AND ③ CANISTER RETAINING LINE ④ AND SECURE TO STRONG POINTS AS ILLUSTRATED IN DIAGRAM.

B — PEEL OFF RUBBER PATCH. RELEASE SHACKLE BY PULLING STRAP IN DIRECTION OF ARROW. ATTACH QUICK RELEASE HOOK TO SHACKLE AND LOCK IN POSITION.

C — HOIST AND SWING OUTBOARD STEADY CANISTER WITH BROWSING LINES. PULL OUT PAINTER TO MAXIMUM EXTENT THEN GIVE A SHARP JERK TO INFLATE RAFT.

D — INFLATED RAFT IS STEADIED ALONGSIDE AT SILL LEVEL BY MEANS OF BOWSING LINES. THE RAFT CAN NOW BE BOARDED.

E — RELEASE BOWSING LINES AND THROW THEM INTO RAFT. ENSURE TRIP LINE FOR QUICK RELEASE HOOK IS ACCESSIBLE. BEGIN TO LOWER RAFT. RETRIEVE OR MOVE SPENT CANISTER FROM LAUNCHING AREAS.

F — ACTIVATE QUICK RELEASE HOOK AS RAFT APPROACHES SEA LEVEL THE RAFT WILL AUTOMATICALLY RELEASE ON TOUCHING THE WATER. CUT PAINTER FREE. GET AWAY FROM SIDE OF SHIP LEAVING AREA CLEAR FOR NEXT LAUNCHING.

IF LIFERAFT IS INVERTED ON LAUNCHING

Turn around the inverted liferaft until the gas bottle is downwind. Climb onto the inverted liferaft. With feet on gas bottle, haul on the righting strap. Get the wind to help you. Remember to stand on the gas bottle to right it, or it may hit you over the head.

LAUNCHING A LIFEBOAT OR A DINGHY IN ROUGH WEATHER

- Turn the vessel to create a lee for the launch.
- Test the motor, if any.
- Rig a boat rope well forward.
- Wear lifejackets.
- Tend to fore and aft falls (lines) equally during launching.
- Do not start the motor until the boat is launched and the falls have been let go.

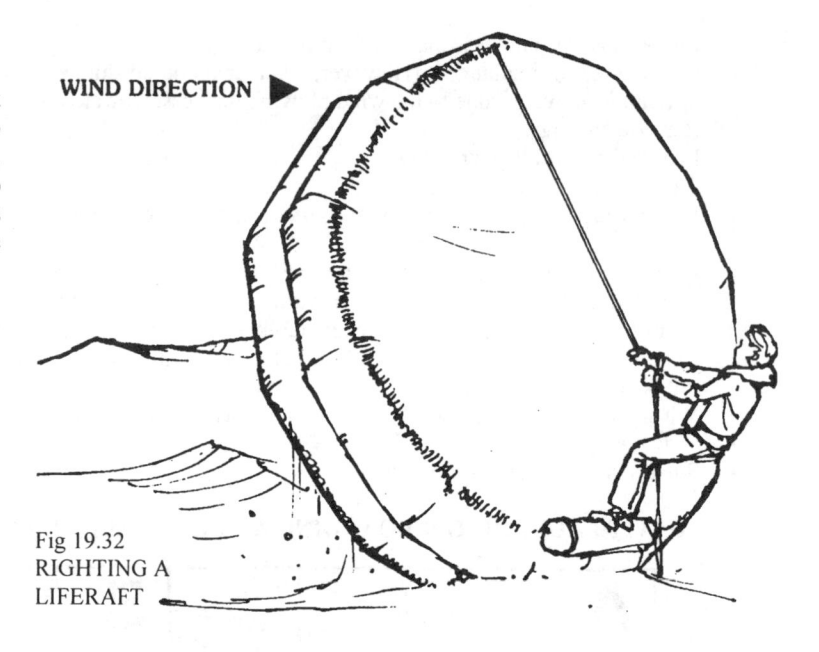

Fig 19.32 RIGHTING A LIFERAFT

LIFERAFT CONTENTS

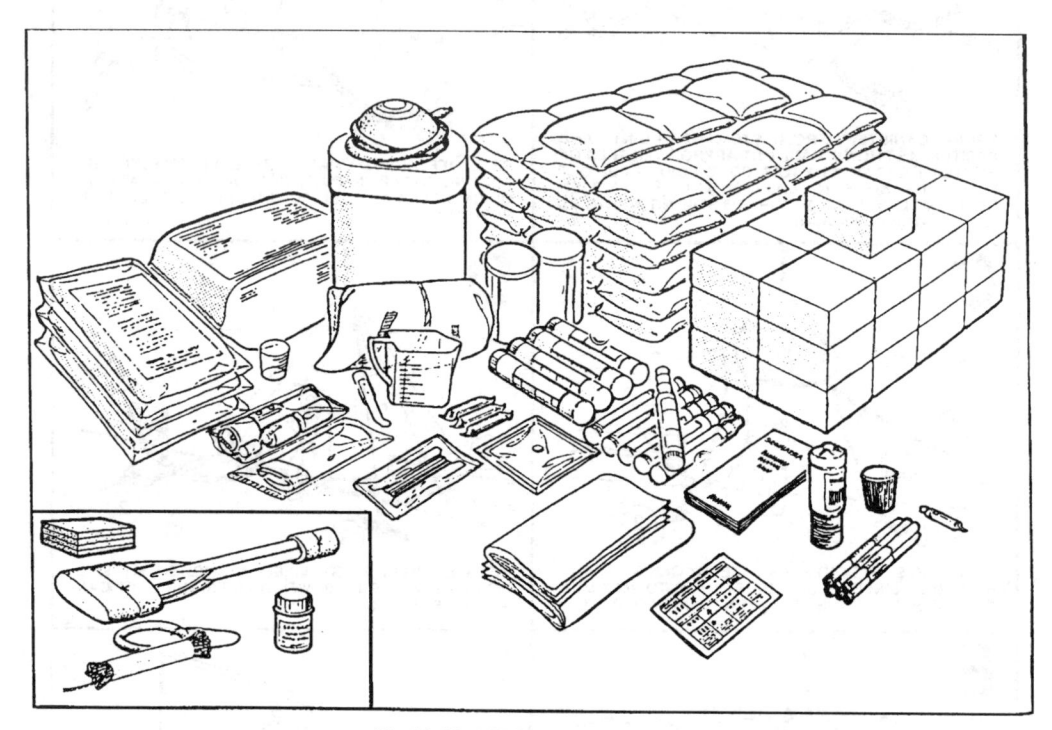

Fig 19.33: LIFERAFT CONTENTS

PUNCTURE REPAIR KIT in the above list is similar to a bicycle puncture repair kit. Follow the instructions provided. Make the area to be patched as <u>dry</u> and clean as possible. Apply the adhesive to the area and the patch. Wait for a minute or so until the glue is tacky, then apply the patch. Wait for about 30 minutes before inflating the tube. The pump or bellows is meant for topping up the liferaft. For large holes, leak stopper plugs (threaded hollow rubber plugs) are provided in various sizes. Screw one of a suitable size into the hole. Don't overdo it or you will just enlarge the hole.

Chapter 19.3: OH&S - Safety Equipment, Sea Survival & Rescue

LIFERAFT RATIONS

Packed inflatable liferafts do not have space to carry food as such. The idea behind any emergency rations is to provide the longest possible survival, maximum energy and minimum thirst. The only food in them is biscuits or barley sugar. This is rich in carbohydrates and free from protein. Consumption of protein (meat, fish, fish juice) causes thirst. Birds and fish should be eaten only when there is abundant supply of drinking water (rain water, etc.).

	COASTAL	SOLAS	AYF
Sea anchors	1 (attached)	2 (1 attached)	2
Bailers (buoyant)	1	1	1
	(2 in each type if above 12 persons)		
Pump or bellows	1	1	1
Puncture repair kit	1	1	1
Paddles (*buoyant*)	2	2	2
Torch (*waterproof*)	1	1	1
	(Suitable for Morse signalling)		
	(+1 set spare batteries & 1 bulb)		
Whistle	--	1	1
Rescue signal table	1	1	1
Survival manual	--	1	1
Smoke signals (*orange*)	1	2	2
Hand flares (*red*)	2	6	4
Parachute rockets	--	4	2
Dye marker	--	--	2
Thermal protective aids	--	2	--
		(3 for 25-person raft)	
Seasickness bags (*per person*)	--	1	5
Seasickness tablets (*per person*)	6	6	6
Can openers (*safety type*)	1	1	1
First Aid kit (*approved*)	1	1	1
Matches (*containers*)	1	1	--
	(Waterproof container with striker & 25 matches, not readily extinguishable by wind.)		
Heliograph	1	1	1
Rescue quoit (*with 30 m line*)	--	1	1
Sponges (*per person*)	1	1	1
Knives (*safety type*)	1	1	1
	(2 in each type if above 12 persons)		
Fishing kit (*1 line & 6 hooks*)	--	1	--
Drinking vessel (Graduated)	1	1	1
Chemi-luminescent lights	6	6	1
EPIRB/GPIRB	--	1	--
Radar reflectors	--	1	--
Sunburn cream	--	--	2
Leak stoppers	3	3	3
Relief valve plugs	2	2	2
Food per person	500g	667g	500g
	(Biscuits, barley sugar, etc.: free from protein & fat).		
Water per person	1 litre	1.5 litre	0.5 litre

(In SOLAS liferaft, 0.5 litres of water may be replaced by a suitable desalting apparatus)

Fig 19.34: CONTENTS OF AN INFLATABLE LIFERAFT

SHIP REPORTING AND SEARCH & RESCUE SYSTEM

(See contact details on the last page of this book)

Under an international agreement, Australia is responsible for search and rescue (SAR) in an extensive area of the oceans. In the West it extends half way across the Indian Ocean, in the South to Antarctica, and to the North and East half way into the waters shared with Indonesia and New Zealand. In order to meet its responsibility, the Australian Maritime Safety Authority (AMSA) operates the Australian Ship Reporting System (**AUSREP**) from its Rescue Coordination Centre (RCC Australia) situated in Canberra. It is also known as the Australian Search and Rescue (AusSAR) Centre, and operates 24 hours a day.

AMSA and AUSREP require that SOLAS vessels and vessels trading on inter-state or international voyages must be fitted with HF DSC or INMARSAT *(discussed in Chapter 18)*, and must report their position, course and speed direct to RCC via the AMSA Radio Network or a Land Earth Station on a 24-hour schedule. On departure from an Australian port or on entry into the Australian area of responsibility, vessels send their Sailing Plans (SP). Each vessel follows this with a daily Position Report (PR) on an agreed time at intervals not exceeding 24 hours. Should the vessel be out by more than two hours steaming time from the reported SP or PR, she sends out a Deviation Report (DR). Ultimately the vessel sends the Final Report (FR) on arrival at the next Australian port or on exiting the Australian area of responsibility.

When applicable, vessels also send the following special reports using the International Maritime Organization (IMO) message format: Dangerous Goods report (DG), Harmful Substances report (HS) and Marine Pollutants Report (MP).

The RCC computer is alerted if a vessel fails to make the scheduled call. If necessary, the Centre can identify and call other vessels in the vicinity to provide assistance to a vessel in distress. It is an active SAR watch for participating vessels. Trading vessels of other countries are actively encouraged to report on entering or departing the Australian area of responsibility, but reporting is compulsory for all Australian trading vessels and for foreign vessels on voyages between Australian ports.

There is also the Ship Reporting System (**SRS**), also known as **REEFREP,** which is designed specifically to enhance navigational safety in the Torres Strait and the inner route of the Great Barrier Reef. It is aimed at minimising risk of marine pollution in this highly sensitive area. The REEFREP centre (call sign "REEFCENTRE") is situated at Hay Point near Mackay. It is operated 24 hours a day jointly by AMSA and Queensland Transport. Just as with RCC in Canberra, vessels communicate with REEFCENTRE via coast radio stations or telephone. REEFCENTRE interacts with shipping in providing improved information on the presence, movements and patterns of shipping in the area. A data exchange interface between AUSREP and REEFREP systems avoids the need for duel reporting by ships participating in both systems.

The following ships are required to report their position and details to REEFCENTRE, which, in return, provides ship traffic information and other navigational safety related details:

- All ships of 50 metres or greater in overall length.
- All oil tankers, liquefied gas carriers, chemical tankers or ships coming within the INF (Irradiated Nuclear Fuels) Code, regardless of length.
- Ships engaged in towing or pushing where the length of the tow, measured from the stern of the towing ship to the after end of the tow, exceeds 150 metres.

Vessels should also advise the REEFCENTRE or RCC whenever their operation or seaworthiness is about to adversely affect the environment or when there is likelihood of pollution or loss of cargo from the vessel.

POSITION REPORTING BY NON-SOLAS (SMALL) VESSELS

As stated above, AUSREP is compulsory for SOLAS vessels, i.e., vessels trading vessels on inter-state or international voyages. However, any vessel may use the system free of charge so long as they meet the following criteria:

- have submitted a "Sea Safety - Small Craft Particulars Form" (shown below)
- are equipped with an HF DSC radio for regular communication with the AMSA Radio Network or equipped with the INMARSAT equipment that can maintain communication with a Land Earth Station (LES) (222 or 322) for the duration of the voyage.
- are equipped with an appropriate EPIRB/GPIRB.
- are proceeding on a passage of over 200 miles between ports.

VESSELS NOT FITTED with HF DSC OR INMARSAT are encouraged to transmit their Position Reports by radiotelephony to Voluntary Coast Stations, *as detailed in Chapter 18*, and not to the State-run Coast Radio Network (CRN). *Radio Equipment required for <u>NON-SOLAS VESSELS</u> is also discussed in Chapter 18*.

Chapter 19.3: OH&S - Safety Equipment, Sea Survival & Rescue

UNIVERSAL AUTOMATIC IDENTIFICATION SYSTEMS (UAIS or AIS): As discussed in Chapter 14, AIS units on board ships can automatically transmit ship's position, speed and heading to shore-based AIS Base Stations. This technology is currently being incorporated into the Ship Reporting System (SRS) in the Torres Strait and the inner route of the Great Barrier Reef.

AIR SEA RESCUE KITS (ASRKs): These kits consist of a liferaft, rations, bilge pump and/or marine radio. When necessary, the search and rescue authorities may drop such a kit from an aircraft close to the survivors with a buoyant line attached which is trailed onto or as near as possible to the survivors.

PADS (Precision Aerial Dropping System) is another method designed for dropping emergency equipment and supplies to survivors.

HELIBOX: This is a lightweight box used for dropping an emergency marine radio to survivors. The radio is quite simple to operate. It has only three push buttons for ON, OFF and PUSH TO SPEAK.

SEA SAFETY - SMALL CRAFT PARTICULARS FORM

Shown below is the double-sided form designed for small vessels to register their details with RCC Australia, if they wish. The form provides the RCC with a readily available record of essential particulars for the time-critical task of search planning, should this become necessary. The form should be updated every 3 years and on change of vessel ownership or contact details or if there is a change in vessel fitout, communication equipment or area of operation. These forms are available from RCC in Canberra, local marine authorities, police stations and marine education centres.

AMSA
Australian Maritime
Safety Authority

Reply Paid 1294
Manager Operations
AusSAR
Australian Maritime Safety Authority
GPO Box 2181
Canberra ACT 2601

SEA SAFETY REPORTING GUIDELINES

(A) Leave individual trip details with a responsible member of your family or friend on the understanding that they will institute follow-up action, i.e., inform the local marine search & rescue group, local police or the RCC on Toll Free 1800 641 792.
(B) Maintain Contact, via HF DSC or INMARSAT with RCC or via radiotelephone with the Volunteer Coast Radio Network.
(C) Send a Small Craft Particulars Form to RCC or the appropriate local search & rescue body. Its purpose is to make vessel details readily available for search and rescue, if required.

Fig 19.35: DOUBLE-SIDED SMALL CRAFT PARTICULARS FORM, SIDE 1
(Note: Guidelines are paraphrased by the author)
(SIDE 2 ON NEXT PAGE)

Sea Safety - Small Craft Particulars - PLEASE PRINT CLEARLY

Description of Vessel -
Please include photograph if possible

Name of vessel	
Licenced radio call sign	Sail No.
Reg. No.	

Type
eg yacht, motor sailer, runabout, etc.

Rig	Class
Length _____ Metres	
Hull material	
Engine type(s)	

Colour
- Hull
- Superstructure
- Sails
- Mast _____ Deck _____

Distinctive features or markings

Fuel
- Type
- Amount carried _____ Litres | Range on engine

Average cruising speed
- Power _____ Sail _____

Other pertinent information

General Operating Pattern and Area

At discretion of owner, this section records additional information of potential use to SAR authorities, for example: "fishing trips about 2 weekends per month during summer months from ramps between Port Stephens and Batemans Bay", or "most events on WA ocean racing calendar and one cruise to northern waters per year", or return to Sydney from Hobart third week in January with following crew ..."

Safety and Emergency Equipment

Flares/smoke

Type (Para., hand held, other)	Colour (W,R)	Number

Ship's tender
- Yes ☐ No ☐
- Type
- Colour
- Propulsion

Liferaft
- Yes ☐ No ☐
- Type
- Manufacturer
- Capacity

Lifejackets
- Yes ☐ No ☐
- Number
- Colour

Lifebuoys
- Yes ☐ No ☐
- Type
- Colour
- Distinguishing marks

Radio transceivers
- HF ☐ VHF ☐ 27MHz ☐ INMARSAT ☐
- Digital selective calling ☐ Other (specify) _____
- Transmitter frequencies
- Frequencies most often monitored

Distress beacon (EPIRB)

- Yes ☐ Operating frequency(ies) 406.0 MHz ☐ 121.5 / 243.0 MHz ☐ INMARSAT E ☐
- No ☐ Float free Yes ☐ No ☐

Signalling lights
- Yes ☐ Spotlight ☐ Strobe ☐ Torch ☐
- No ☐

Navigational aids
- Radar ☐ GPS/Satnav ☐ Other ☐

Food
- Number of person days _____

Water
- Amount _____ Litres

Owner and Contacts

Give names, addresses and phone numbers of owner and 24 hour contact, also other intermediate contacts with whom you would normally leave sailing plan details.

Owner
- Name
- Address _____ Phone _____

24 hour contact
- Name
- Address _____ Phone _____

Person/Authority holding a copy of this form
- Name
- Address _____ Phone _____

Form filled in by
Please give name, address and phone number if different from above
- Name
- Address _____ Phone _____

For sea safety's sake, leave a completed copy of this form with your sail plan contact as noted in Guideline A overleaf.

Chapter 19.3: OH&S - Safety Equipment, Sea Survival & Rescue

IF YOU FALL OVERBOARD

If your shouting and waving hasn't helped to attract your vessel's attention, do not swim or panic. Concentrate on staying afloat. If you are wearing a flotation device, you may have a whistle or a light to attract the vessel's attention. If not, try using your trousers (or shirt) as a flotation device. Take it off, tie its legs (or arms), force air into it and wrap it around your neck or under your arms.

If you are wearing a flotation device, concentrate on preventing HYPOTHERMIA *(discussed in more details below)*. Minimise your body contact with water and do not swim (see H.E.L.P. and HUDDLE Positions below). This is to reduce the circulation of the cold surface blood cooling the inner core of the body. Hypothermia is a complex reaction but a drop of inner core body temperature of 3 to 5 degrees can set off fatal hypothermia.

Fig 19.36: SURVIVAL TIME IN WATER
(Holding still, wearing lifejacket & light clothing)

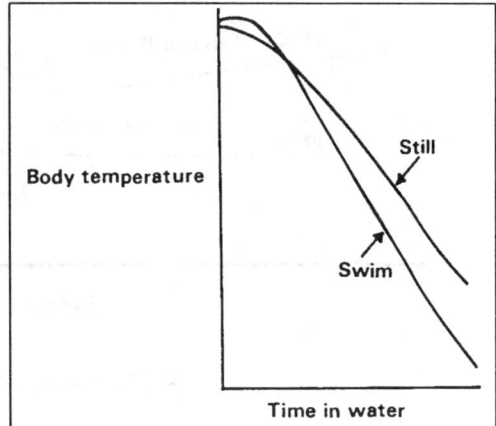

Fig 19.37: DROP IN BODY TEMPERATURE
(Swimming vs. Keeping still)

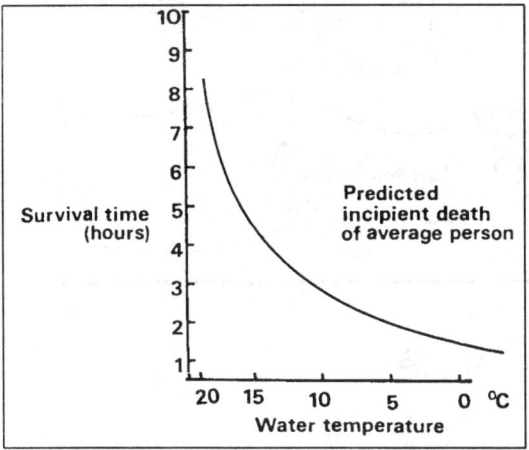

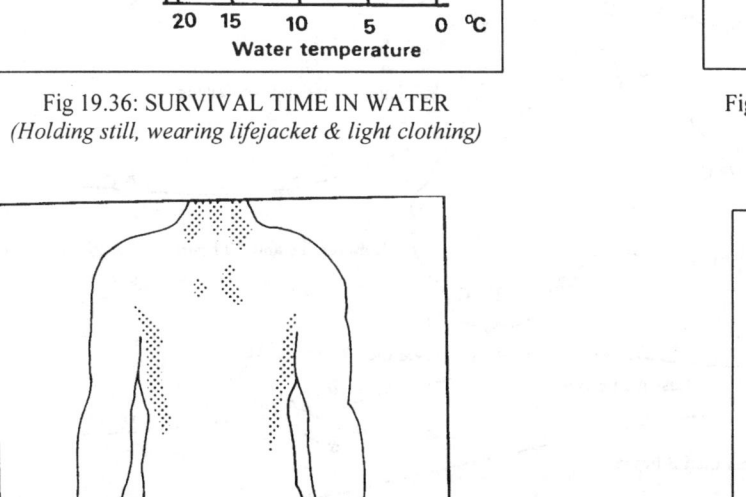

Fig 19.38: BODY REGIONS WITH HIGH RATE OF HEAT LOSS

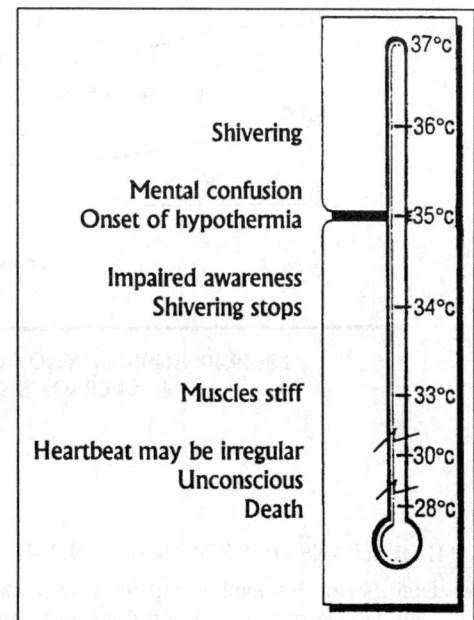

Fig 19.39: ONSET OF HYPOTHERMIA

The body regions with a high rate of body heat loss are the HEAD, SIDES OF THE CHEST *(due to little fat or muscle)*, and THE GROIN REGION *(due to large blood vessels near the surface)*. A lifejacket will reduce the amount of water coming in contact with some of these parts.

In the absence of a flotation device, *treading water* will help to keep your head out of the water. However, it will also increase your body-cooling rate by 34% as compared to holding still in a lifejacket.

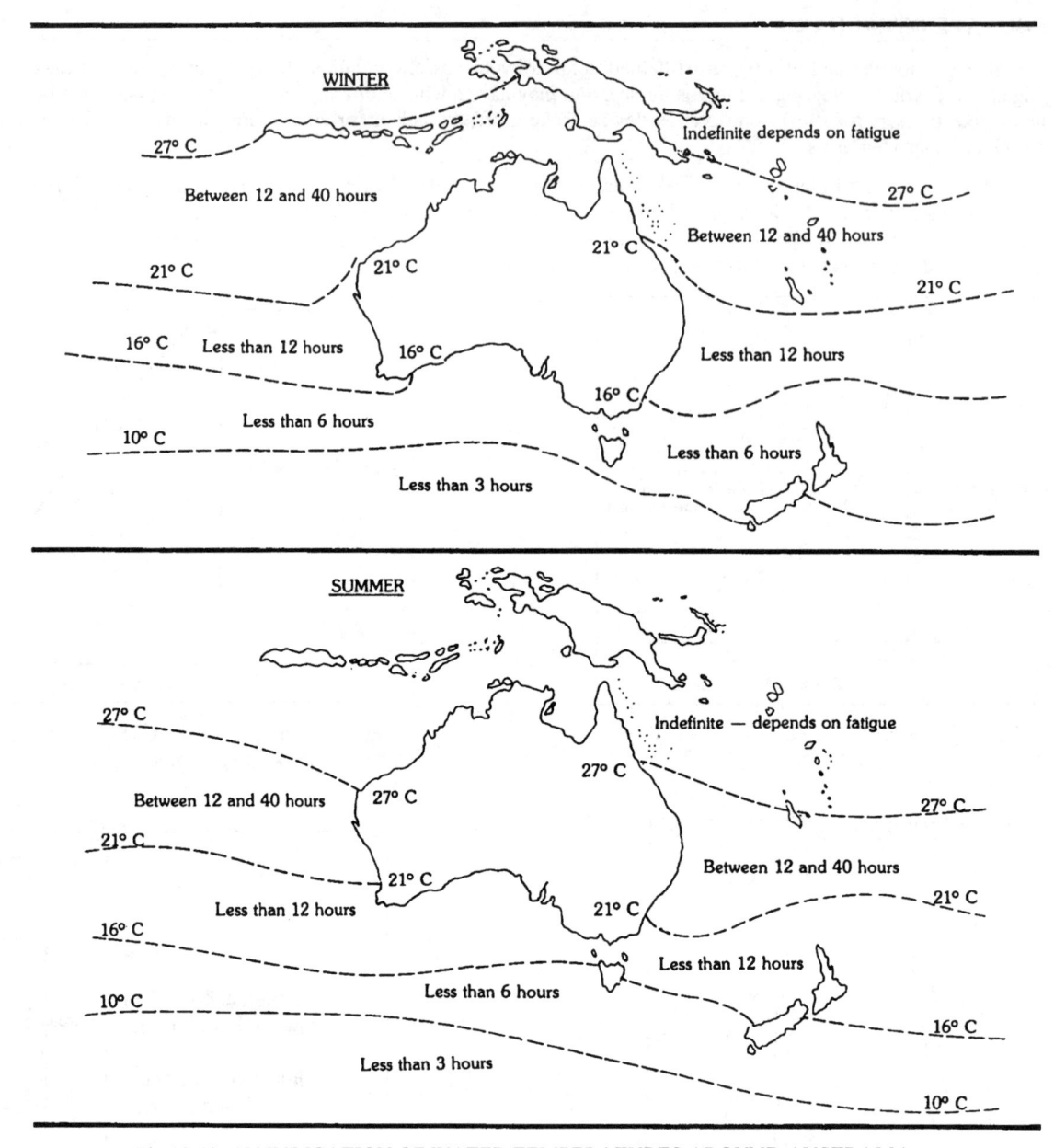

Fig 19.40: AN INDICATION OF WATER TEMPERATURES AROUND AUSTRALIA
& A PERSON'S SURVIVAL TIME IN WINTER & SUMMER

OTHER MEANS OF REDUCING BODY-HEAT LOSS:

- Don't swim. Swimming will increase blood flow to the cool surface of the body, resulting in rapid temperature reduction.
- Curl up into **H.E.L.P.** *(Heat Escape Lessening Position)* in order to reduce exposure of the above-mentioned parts of the body.
- **HUDDLE** up: Heat loss is reduced because the sides of the chest are held together. This method also increases survival time by being close together.

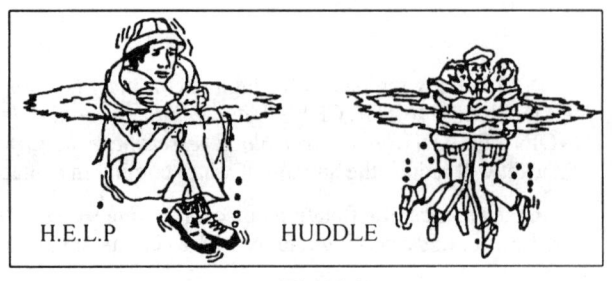

Fig 19.41

PERSON OVERBOARD

In the event of a person falling overboard from a power-driven vessel, the following actions should be taken simultaneously:

- Shout "Person Overboard" (or Man Overboard).
- Wheel hard over to the side he/she fell.
- Throw a lifebuoy (or a lifejacket, etc.) over the side, upwind and close to the person without hitting him/her.
- Mark position on GPS. Switch GPS to MOB Mode.
- Post one or more lookouts to keep the person in sight at all times (with a spotlight, if necessary) and to continue pointing at the person overboard while the skipper manoeuvres the vessel.
- Perform one of the turns listed below, whichever is suitable, to pick him/her up.
- Indicate to the person in the water that he/she has been seen and recovery action is being taken.

WILLIAMSON'S TURN

In rough seas, swell or waves, and in reduced visibility or darkness, it is extremely likely to lose the sight of a person fallen overboard. The purpose of the Williamson's turn is to return the vessel as close to its original track as possible in order to pick up the person. It is recommended for power driven vessels. The steps are:

➢ Note the course being steered.
➢ Put the wheel hard over to the side the person fell.
➢ Hold the wheel there until the vessel is about 70° off her original course.
➢ Then put the wheel hard over to the other side, until the vessel comes round to the reciprocal of the original course.

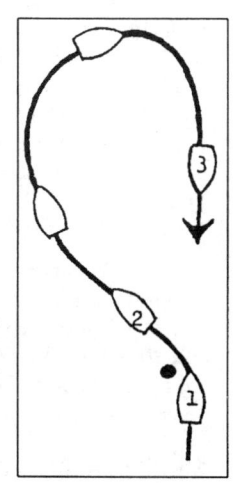

Fig 19.42 WILLIAMSON'S TURN

ELLIPTICAL OR DOUBLE TURN

This is another method for recovering a person overboard. It is less common because it requires that the person be kept in sight all the time. The steps are:

➢ Put the wheel hard over towards the person until on a reciprocal course.
➢ When the person is three points (say, 35°) abaft the beam, put the wheel over the same way until back on the original course.

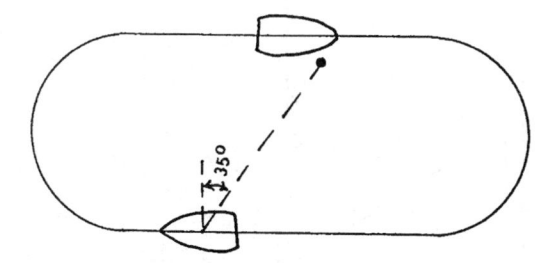

Fig 19.43: ELLIPTICAL TURN

"Y" TURN

This is quite a suitable and quick turn for small vessels. The steps are:

➢ Put the wheel hard over to the side the person fell.
➢ Keeping the wheel hard over, stop engine and go full astern.
➢ Still keeping the wheel hard over, go full ahead.
➢ Then slow down, straighten up and stop near the person

QUICK-STOP METHOD

For a sailing vessel, the quickest way to reduce speed is to head into the wind for the sails to go aback. Then, continue the turn until the boat is heading towards the person overboard. Sails can be reduced as required. To bring the vessel near the person, approach him/her from downwind, heading into the wind. You will thus be able to manoeuvre at slow speed.

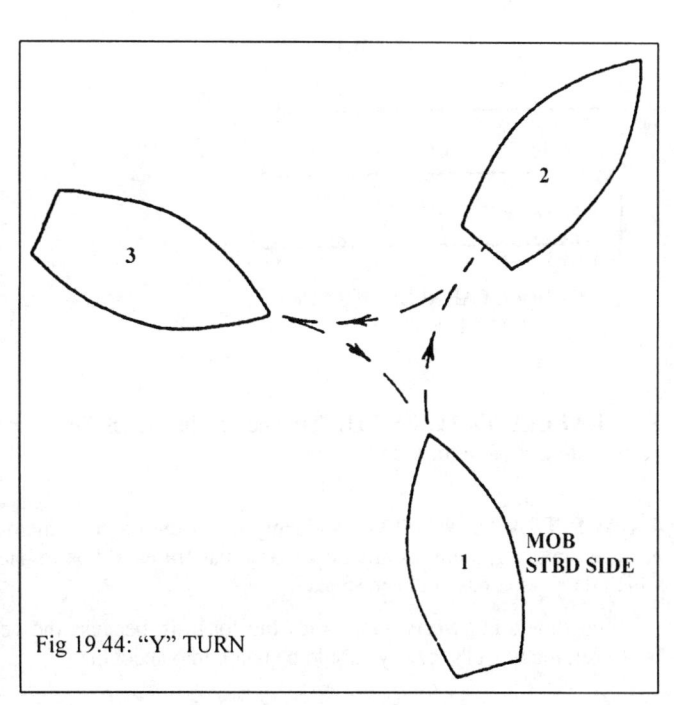

Fig 19.44: "Y" TURN — MOB STBD SIDE

PERSON MISSING OVERBOARD - SEARCH PATTERNS

If you are unable to immediately find the person fallen overboard, it is advisable to contact a Coast Radio Station (Maritime Coast Station) for assistance without delay. Meanwhile, if there are other vessels in the vicinity, you should hoist the international code flag "O" and start searching - using one of the following search patterns recommended in the International Aeronautical and Maritime Search and Rescue Manual **(IAMSAR)** Volume III. It is published jointly by the International Maritime Organisation (IMO) and International Civil Aviation Organisation (ICAO). It replaces the Merchant Ship Search and Rescue **(MERSAR)** Manual, which was published by IMO.

1. EXPANDING SQUARE SEARCH PATTERN

This is used when you have a probable position (datum) from which to search, but you wish to gradually expand the search area. In the attached figure, the space between the tracks is such that the vessel must be able to see the missing person up to half way between the lines. Naturally, the larger (higher) the vessel, and the smoother the sea and swell, the larger the spacing.

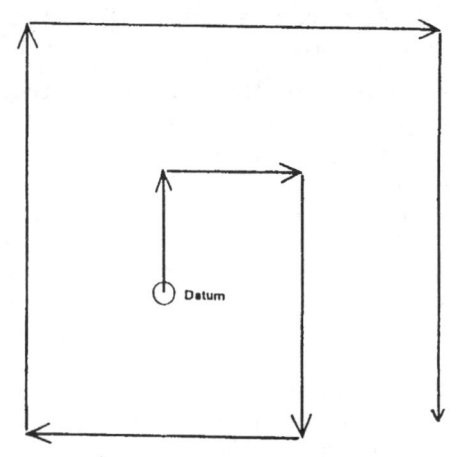

Fig 19.45: EXPANDING SQUARE SEARCH PATTERN

2. SECTOR SEARCH PATTERN

This pattern is used when you wish to continue searching a given area. The vessel continues to return to her starting position. All turns are 120 degrees to starboard. Upon completion of first search, change the pattern 30 degrees to the right and re-search as shown by the broken line. The length of each line will depend on the size of the vessel. Large ships may use 2 mile legs.

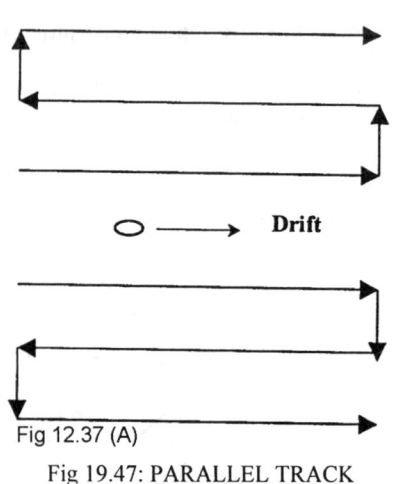

Fig 12.37 (A)

Fig 19.47: PARALLEL TRACK SEARCH PATTERN

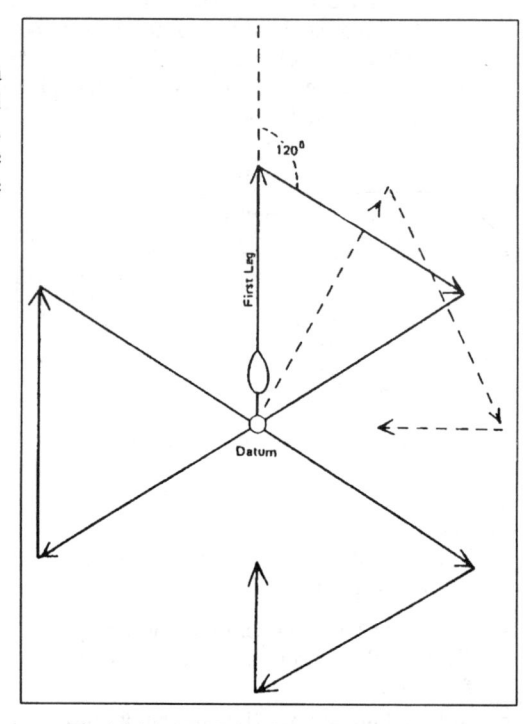

Fig 19.46: SECTOR SEARCH PATTERN

3. PARALLEL TRACK SEARCH: The Parallel Track Search pattern, shown below, is recommended when two or more vessels are available for search.

4. BACK TRACKING: If you suddenly discovered a crew member missing on passage and don't know how long he/she has been missing, the most practical search pattern would be to back track (after doing a Williamson's turn) for the time for which the person has not been seen.

You should not worry about allowing for drift, because the vessel and the person would have drifted at the same rate. However, the vessel's leeway should be taken into account.

RETRIEVING PERSON OVERBOARD

In small vessels, both power and sail, it is often sufficient to stop the vessel close to the overboard person to be able to swim to the craft. If the person is weak or unconscious you may need to lower a sail, net a tarpaulin in the shape of a cradle or hammock into the water, scooping the person head first so that a stretched arm or a leg does not cause an obstruction. Once in position, gently heave on the outer side of the cradle and the person will come on board.

The cradle method is particularly helpful in retrieving a person suffering from hypothermia. The horizontal lift uniformly transfers the person from the influence of **HYDROSTATIC SQUEEZE** into a state of gravity. The gravitational effect of a vertical lift, on the other hand, is to send the blood rushing to the legs with the possibility of the person collapsing.

The retrieval of a person by a vessel with a large freeboard may require lowering a dinghy or a rescue boat. The vessel would need to create a lee for the launch as well as for recovery.

DANGERS TO SURVIVORS & PRECAUTIONS

HYPOTHERMIA: *As shown in figures 19.36 to 19.40*, Hypothermia is a killer. It is the condition of low body-core temperature. All survivors are easy targets of hypothermia, due to immersion, insufficient or wet clothing, windy conditions, physical exhaustion, hunger, anxiety or low morale.

Hypothermia is not easily recognisable. A victim is exhausted, reluctant to do anything, difficult to reason with and has slowed mental and physical reactions. Sense of touch is poor, speech may be slurred, and lips, hands and feet may swell. Victims of hypothermia often do not realise their condition. They can die within one hour of its onset, without even complaining of the cold.

As advised in the AMSA's Survival at Sea Manual, protect yourself from the possibility of hypothermia by protecting against cold:
- Wear plenty of clothing - especially waterproof and windproof.
- Keep liferaft canopy closed.
- Keep liferaft floor inflated to insulate against cooling from the sea.
- Do not exhaust yourself by swimming unnecessarily.
- Survivors in the water should follow "person overboard" precautions discussed above.

TREATMENT OF HYPOTHERMIA OF A CASUALTY
- Prevent further heat loss due to evaporation or exposure.
- Place the survivor next to other people for warmth. Huddling together under covers will promote heat transfer to the victim.
- Avoid unnecessarily handling of the person.
- When conscious, give a warm sweet drink.
- Do not wrap in a blanket unless the air temperature is less than the water temperature or unless the blankets have been preheated. (Unheated blankets insulate the cold body surface from the source of external heat).
- Do not massage the body or limbs.
- Do not feed solids or liquids to an unconscious survivor.
- Do not give alcohol.

SHARKS
- Leave them alone.
- Don't trail legs in the water.
- Wear clothing.
- Don't fish if sharks are present.
- Don't throw food scraps overboard.
- Remain quiet.

DEHYDRATION
Conserve body fluids by minimising sweating and avoiding sea sickness. Sea sickness is extremely demoralising. Seasick people just want to die. Take SEA SICKNESS TABLETS before abandoning vessel or immediately on boarding liferaft, thereafter take them as follows:

Morning = one tablet
Midday = half tablet
Evening = half tablet
(Maximum 2 tablets per day)

- If you become adjusted/accustomed to the liferaft motion, cease taking seasickness tablets.
- Alcohol consumption causes dehydration of liver and kidneys. Don't consume it.
- Conserve the water supplied in the liferaft to make it last as long as possible. You do not need to drink water for the first 24 hours after abandoning ship. There is sufficient water in your body for 24 hours.
- Seawater or urine should not be drunk. The salt and chemicals in them will make you more thirsty.
- Do not consume any protein (meat, fish or fish juice) unless there is a liberal supply of drinking water. Protein dehydrates the body.

ABANDONING SHIP & THE RESCUE PHASE
(As advised in AMSA's Safety Education Article No. 62)

STAY WITH YOUR VESSEL
➢ Your vessel is your best liferaft. Do not abandon it until you absolutely have to. Try & keep the disabled vessel afloat & stay with it. The vessel will not only be more stable but also a better target for detection by rescuers.
➢ The order to abandon ship must be given only by the master or the person in charge.
➢ Use a sea anchor or drogue to reduce drift.

To prevent the drogue spinning or tipping the raft in steep seas, make sure the drogue line is payed out a distance equal to 1½ or 2½ wavelengths. The drogue and the raft should not be allowed to ride wave crests simultaneously.

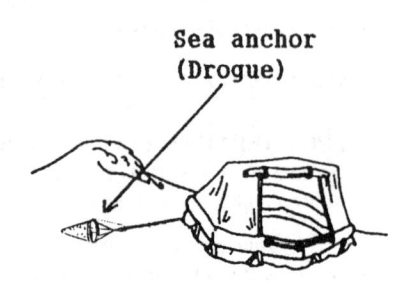

Fig 19.48: SEA ANCHOR

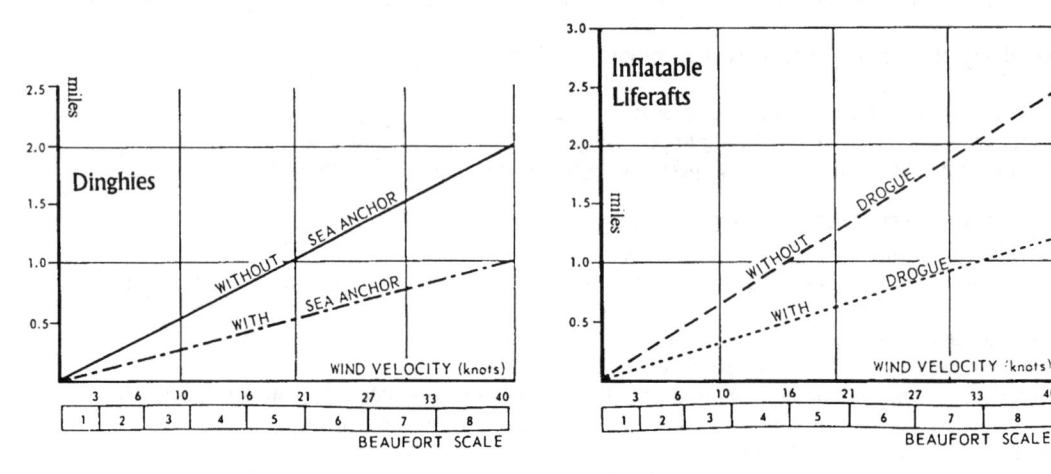

Fig 19.49: RATE OF DOWNWIND MOVEMENT (KNOTS)

BASIC RULES WHEN ABANDONING SHIP
➢ Send out a distress signal and activate EPIRB/GPIRB.
➢ Put on as many clothes as possible.
➢ Drink water to your fill.
➢ Plan orderly abandon.
➢ Take extra blankets and food.
➢ Stop propeller.
➢ Shut all watertight doors and hatches.
➢ Avoid abandoning in way of masts, booms, gallows and lifting gear.
➢ Stay close to the wreck.
➢ Board liferaft without getting wet.
➢ If forced to leave directly into water, pull lifejacket downwards in front and jump feet first, legs crossed.
➢ Swim away from a vessel sinking close by. There is a risk of suction.
➢ A makeshift raft can be made from empty sealed drums and timber lashed together, sorting trays, refrigeration box, etc.

Chapter 19.3: OH&S - Safety Equipment, Sea Survival & Rescue

LIFERAFT: LIVING IN IT
- Skipper or a competent person to take charge.
- Give seasick tablets to everyone for the first two days, whether they are needed or not. (Loss of body fluids through vomiting can quickly lead to dehydration).
- Search for survivors and rescue them.
- Post lookout.
- Treat injuries, small or large.
- Plan protection from hypothermia, exhaustion and dehydration.
- Maintain discipline.
- Maintain positive attitude.
- Consider physical safety from dangerous fish etc.
- Keep all survival craft together. Tie them to each other with the parts of the painters retrieved after launching the rafts. Don't tie too close together to avoid bumping and damage. Check the lines and rafts for damage due to pulling & rubbing.
- Make the elderly & the children comfortable first.
- Keep liferaft dry & inflated, using bailers, sponges & pump.
- Keep canopy erected & inflated. Entry flaps can be closed or opened depending on weather & temperature.
- To protect against cold, inflate the floor. Wear warm clothes. Secure against the weather.
- To protect against heat, do not inflate the liferaft floor. Wear light clothing. Increase raft ventilation.
- Stream sea anchor or drogue to reduce drift away from location of distress.
- Regularly inspect the drogue's line for breakage due to sudden strain or chafing.
- Space out survivors evenly on the floor of the raft with feet towards the centre and arms through hand lines secured inside the raft. This will prevent the liferaft from blowing over.
- In the evenings or in cold weather, top up air chambers as required. Plug the relief valves after inflation. In hot weather, frequently remove them to prevent pressure build up. Ventilate interior of raft after each "blow off".
- Keep the air chambers well topped up and rigid to improve raft stability and habitability.
- Prepare all distress equipment.
- Make proper use of distress signals and EPIRB/GPIRB.
- Maintain proper lookout.
- Eat birds and fish only when abundant (rain) water is available.
- Be visible - use radar reflector, distress sheet, etc.
- Stay close to the wreck.

Hot weather - Maximum ventilation & Floor deflated

Cold weather - Minimum ventilation & Floor inflated

Fig 19.50: PROTECTION FROM HEAT & COLD IN INFLATABLE LIFERAFTS

MORALE IN A LIFERAFT
- Early & confident action is needed.
- Keep survivors comfortable & busy with duties.
- Organise immediately a lookout roster
- Study & understand the safety equipment.
- Organisation & routine are important.

DIRECTION OF LAND
Birds' flight path in the evening indicates the direction of land. Muddy water flowing from rivers, smoke and floating wood are a good indication of being close to land.

RESCUING (RECOVERING) SURVIVORS FROM WATER

A buoyant quoit and a buoyant line are supplied in every inflatable liferaft. Survivors in the water can be rescued by throwing the quoit to them or swimming out to them with the quoit. Secure the line by linking an arm through the quoit. Make sure it is secured to the raft.

Unconscious or weak persons can be pulled by the becket on the back of their lifejacket, and lifted into the raft by their armpits. The becket is not designed for lifting persons out of the water.

DRINKING WATER - THREE WAYS OF MAKING AT SEA

1. **USING A REVERSE OSMOSIS WATER MAKER**:
   The Reverse Osmosis desalination watermakers are discussed in Chapter 29.

2. **USING A STOVE**:
   If you are on board the vessel and the stove is functioning, boil seawater in a kettle or a similar container. Pipe the resulting steam into another container. Lag the middle of the pipe with a rag and continue to keep it cool by pouring seawater over it. This will condense the steam inside the pipe into distilled water.

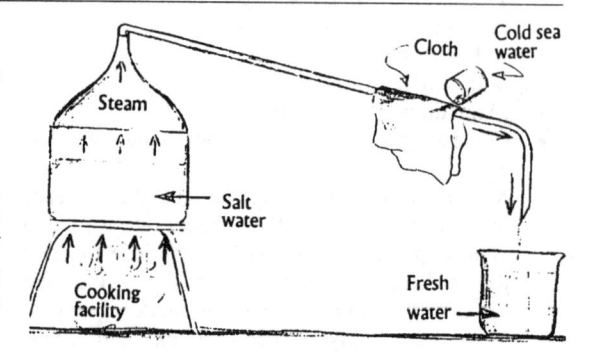

Fig 19.51: MAKING FRESH WATER BY USING SHIP'S STOVE

3. **MAKING A SOLAR STILL**
   Place a small shallow container inside a large shallow container. Place rags soaked in seawater in the outer container, cover it with a plastic sheet and tie it around it. Place a stone or a weight in the middle of the sheet so that it dips slightly towards the middle - over the inner container. Place it in the sun, if you can. Pure water evaporating from seawater in the outer container will accumulate on the inner surface of the plastic sheet and run into the smaller container. Commercial solar stills are also available. They are made entirely of plastic and can be folded for storage.

PRECAUTION WHEN COLLECTING RAIN WATER

Allow the liferaft canopy to be rinsed of salt deposit before collecting rainwater from it.

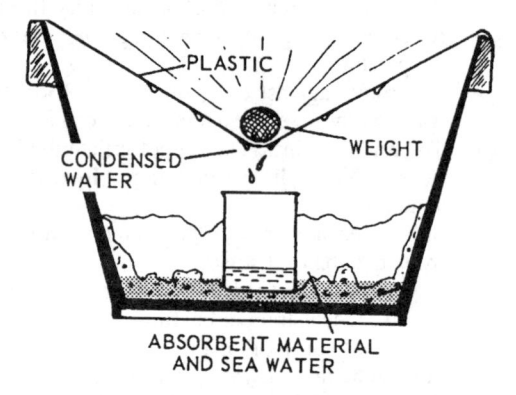

Fig 19.52: AN IMPROVISED SOLAR STILL

LIFERAFT RATIONS – CONTROL

No rations should be issued for the first 24 hours, except to the sick and injured. After that, serve rations at pre-determined times, preferably three times a day. Daily water ration of 500 ml is sufficient to sustain survivors providing they avoid dehydration and reduce sweating. Water should be drunk slowly. Holding it in the mouth and gargling before swallowing helps to get the most value from the water.

No other food is essential for survival over a short period of time. However, barley sugar or other non-thirst provoking food (fat and protein free biscuits) is included in the liferaft rations. It is to provide energy enabling work to be done, and to aid morale. In a coastal liferaft, 500 grams of food is supplied for each person, for consumption at the rate of 125 grams daily after the first 24 hours.

USE OF DISTRESS SIGNALS & PENALTY FOR MISUSE

- The Australian Rescue Co-ordination Centre (RCC) in Canberra will initiate search & rescue on receipt of distress alert.
- Conserve flares & smoke markers until there is a reasonably good chance of them being seen. For example, don't let them off if you see a vessel at a distance heading away from you.

Severe penalties including heavy fines and jail apply to the intentional misuse of distress signals of any type. Such activities could expose others to danger and the perpetrator to action to recover costs incurred in response to the false distress signal.

DANGEROUS FISH & SHELLFISH (Fig 19.53)
(As shown in the Survival at Sea Manual in liferafts)

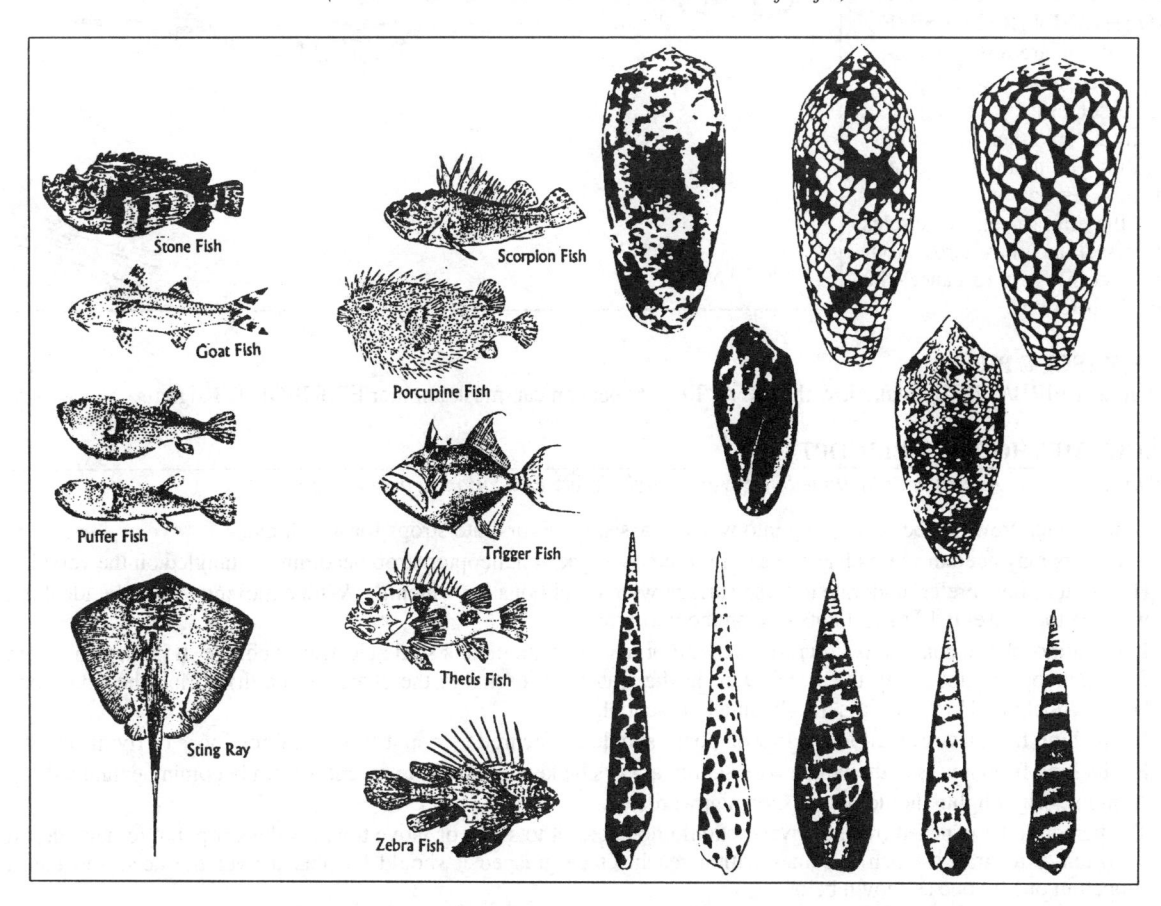

SEARCH AIRCRAFT SIGNALS & ACKNOWLEDGEMENTS

An aircraft searching at night for pyrotechnic equipped survivors will either fire a green flare or, in the case of non military aircraft, switch on landing lights at 3 to 5 minutes intervals and at each turning point in the search pattern. Survivors should acknowledge it by firing a distress flare, but not until after the aircraft's signal has ended. Wait for a full minute before firing a second flare. Fire an additional flare when the aircraft is about a mile away.

The sighting of the distress flare is acknowledged by the aircraft by firing a succession of green flares and/or switching on the aircraft's landing lights.

Fig 19.54

THE AIRCRAFT HAS RECEIVED & UNDERSTOOD YOUR MESSAGE (Fig 19.54): It will rock from side to side. On a dark night it will make GREEN FLASHES with signal lamp.

Fig 19.55

THE AIRCRAFT HAS RECEIVED BUT NOT UNDERSTOOD YOUR MESSAGE (Fig 19.55): It will make a complete right hand circle. On a dark night it will make RED FLASHES with signal lamp.

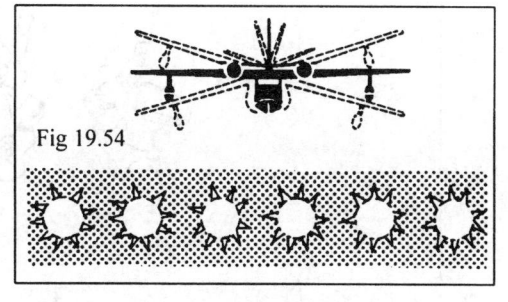

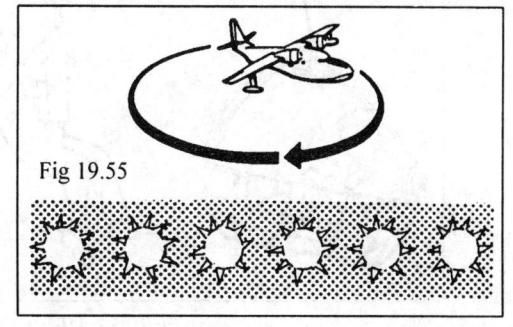

THE AIRCRAFT DIRECTING A VESSEL TO ANOTHER IN DISTRESS (Fig 19.56): If you are **not** in distress and you see an aircraft circle your vessel, cross your path at low altitude and then head off, it is guiding you to another vessel in distress. Follow it.

If the aircraft returns, circles your vessel and flies across your track (astern), it is to cancel the above instructions.

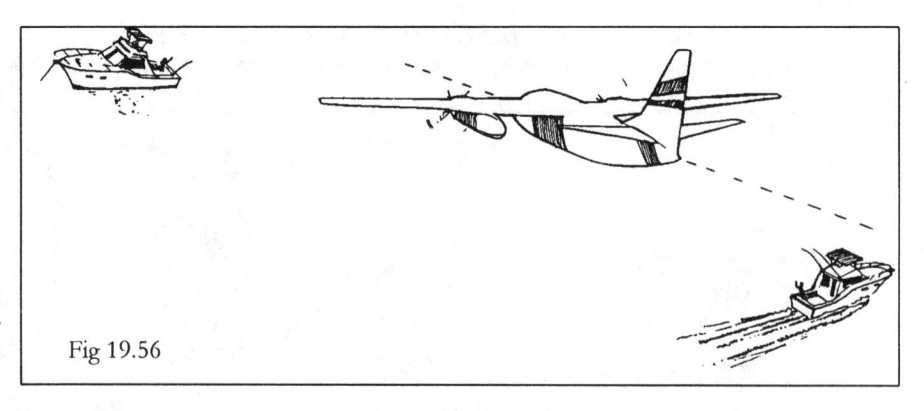

Fig 19.56

POST-RESCUE PHASE
Switch off EPIRB/GPIRB & take it with you. A "Rogue" beacon can mask another EPIRB/GPIRB signal.

RESCUE METHOD BY HELICOPTER
(An abstract of the Australian Maritime Safety Authority's Safety Education Article No. 55)

- A helicopter crewmember may jump into water to assist survivors into strops for winching.
- Survivors may need to swim clear of vessel if there is a risk of helicopter strop becoming entangled on the vessel.
- Helicopter pilots prefer moderate to fresh surface wind conditions for winching. With a moving vessel the ideal relative wind on the vessel is 10 to 15 knots fine on the port bow.
- In dry atmospheric conditions, aircraft when airborne can accumulate a static electricity charge. It is therefore common for a helicopter pilot to dip the winch wire in the water (to discharge the charge to earth) during the final approach before winching. However, it is not absolutely essential.
- If winching from a vessel, all loose gear on board should be secured to stop it from breaking loose or flying about.
- If winching from a vessel, the winch wire should always be kept in hand to prevent it from becoming entangled
- At night, bright lights should not be shown upwards.
- The person to be winched from the water should not attempt to swim or move towards the strop, but remain alert to the opportunity to grab it when it comes within reach. Once grasped it should be placed over the head and then arms slipped in one by one as shown below.
- Once in the strop, the survivor should steady it in front with one hand and extend the other arm outwards with a "thumb up" signal indicating readiness for hoisting. (Vertical lift and hydrostatic squeeze are discussed above)

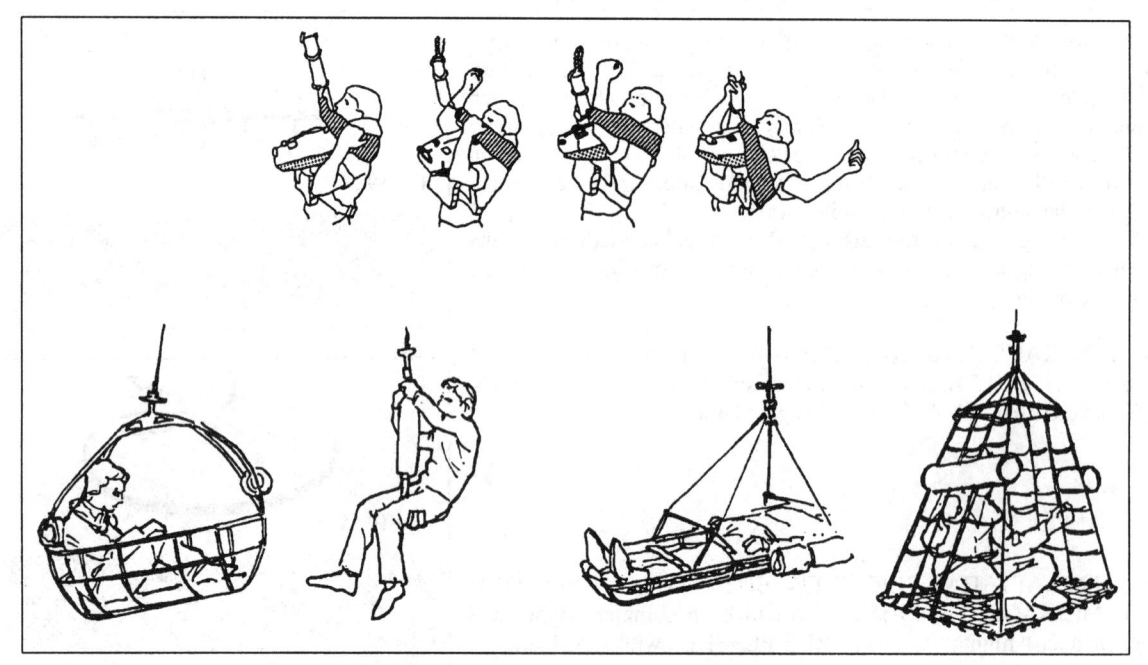

Fig 19.57: HELICOPTER RESCUE -VARIOUS METHODS

HELICOPTER WINCHING FROM SMALL VESSELS
Helicopter engines can fail. Therefore only the twin-engine winch equipped helicopters are used, especially for rescues more than 10 miles offshore.

Approximate range: Military helicopters 200 miles
                            Civilian: Squirrel 80 miles
                            Bell 412 100 miles

ASSISTING OTHER VESSELS & AIRCRAFT IN DISTRESS

DISTRESS SIGNAL: IF YOU SIGHT ONE
Report its range and bearing and your position by radio to the nearest Coast Station. Proceed to the vessel's assistance and search if necessary. Follow the instructions from the Coast Station. If you cannot contact a Coast Station, try calling the vessel in distress.

SUBMARINE SIGNALS
> *White or yellow smoke Or green pyrotechnic signal:*
This is not a distress signal. It indicates a submarine operating below periscope depth.

> *White or yellow smoke candles released 3 minutes apart:*
This is not a distress signal. It indicates a submarine preparing to surface. Clear the immediate vicinity. Do not stop propellers.

> *Red pyrotechnic or smoke signal:*
Submarine is in distress - emergency surfacing. Clear the area immediately. Do not stop propellers. Stand by to render assistance. The sound of your propeller may be heard by the sub. It will reassure survivors that help is at hand. If the sub does not surface within 5 minutes of the red pyro, assume she has sunk.

> *Sub Sunk Indicator Buoy:*
The submarine releases an indicator buoy when she is unable to surface (she is in distress).

SUBMARINE INDICATOR BUOY
It is a cylindrical or semi-spherical buoy made of aluminium or plastic, orange in colour, about 60 cms in diameter. It is fitted with cat's eyes reflectors, flashing light (2 secs), one or two whip aerials (about 1.5 metres) and red and white reflective tape. It is marked with a serial number and message "Finder inform navy, coast guard or police. Do not secure or touch." The buoy is attached to submarine by a 200-fathom thin wire rope.

> *Other Submarine Distress Signals:*
Red smoke candles
Blowing air
Discharging oil
Yellow/green dye marker. (The dye marker container may contain a message).

ACTION ON SIGHTING A SUBMARINE DISTRESS SIGNAL OR BUOY
- Call a Coast Station, Police, Navy or the Australian Rescue Co-ordination Centre. (Naval vessels in the area listen on VHF distress and calling channels).
- Remain in the area and search for survivors.
- Try to establish if the buoy is still secured or adrift.
- If secured: It will have no leeway
  It may have a bow wave caused by current.
  It may submerge in big swells.

If it is in water deeper than 100 fathoms, it is likely to be adrift. If it is in water deeper than 200 fathoms, it is almost certain to be adrift. Do not approach it at close range. Do not touch.

NOTE: SUBMARINE NAVIGATION LIGHTS: Because of possible poor recognition, the submarines may show, in addition to the navigation lights for a power driven vessel, a quick flashing amber/yellow light (90-105 flashes/minute). [The hovercraft signal is 60 flashes/minute.]

AIRCRAFT IN DISTRESS: An aircraft may indicate distress in one of the following manners:
o By firing one or more distress signals.
o By radio.

- By lowering and raising its landing gear.
- By jettisoning cargo.
- By flashing its landing lights.

Your possible responses and action:

- Alert a Coast Station
- Acknowledge the message by releasing black smoke, turning on deck lights and/or shining a searchlight upwards.
- Head into the wind.

SAFETY DRILLS (EMERGENCY DRILLS) ON SMALL VESSELS

The purpose of the emergency procedure is two-fold: to notify the crew of the existence of a problem and to account for everyone on board. Emergency Drills are designed to muster everyone to a muster point dressed in warm clothes and with a lifejacket. They are then allocated duties for emergencies such as fire, person overboard, etc. In some drills, the crew directly proceed to perform their nominated function, such as reporting with a nominated fire extinguisher, starting the fire pump, casting off the liferaft lashings, etc. If lifejackets are stowed in a locker on deck, a crewmember would assist passengers with dispensing and donning them.

The USL Code, Section 15, states that each crewmember shall be allocated an emergency station and special fire station, and shall be informed of the stations when joining the vessel. If there are more than 4 crewmembers, an **EMERGENCY STATION LIST (CREW MUSTER LIST)** shall be displayed. An example of such a list is shown below. Each berth shall be numbered. Identification on the Station List shall be by name, or designation, or by berth number.

> **CREW EMERGENCY PROCEDURES** shall be practised **at intervals of not more than 1 month**. In PASSENGER VESSELS the emergency station list must indicate the emergency stations of passengers and the survival craft allotted to each passenger.

> **SURVIVAL CRAFT DRILLS** shall be held **at intervals of not more than 2 months** on passenger vessels (Class 1) and **3 months on non-passenger vessels** (Classes 2 & 3).

> **FIRE DRILLS**: as for the Survival Craft drills above.

> **COLLISION DRILLS**: as for the Survival Craft drills above.

EMERGENCY SIGNALS

Emergency Station signal: At least 7 short & 1 long blast on whistle or ringing of the bell in a similar manner.
Fire Station signal: Continuous ringing of the bell.
Abandon Ship signal: One short & one long blast at least 3 times in succession.

SAFETY DRILLS - **RECORD KEEPING**

Entries of all musters and drills held shall be made in the vessel's Survey Record Book and the logbook. Failure to comply may jeopardise safety, Permits to operate, Certificates of Competency and/or insurance claims.

SAFETY DRILLS - **EXEMPT VESSELS** *(Ref: NSW Commercial Vessels Regulations 1987, Reg. 256, S.1, Para 2.3)*

In NSW, commercial vessels under 25 metres in length or with less than 4 crew members are exempt from conducting formal safety drills. If "exempt", the masters must:

* Familiarise the crew with the emergency procedures;
* Properly maintain the fire fighting and survival equipment;
* Familiarise the crew with the fire fighting procedures;
* Familiarise the crew with the after-collision procedures.

Fig 19.58: EMERGENCY STATION LIST FOR MV "DIANA"
(an example of a small vessel's Muster List)

STATIONS	FIRE	COLLISION/ GROUNDING	BOAT MUSTER (Prepare to abandon ship)	ABANDON SHIP	MAN OVERBOARD
SIGNALS	Continuous ringing of bell	At least 7 short + 1 long blasts on whistle or bell	At least 7 short + 1 long blasts on whistle or bell	1 short + 1 long blasts, at least 3 times in succession on whistle or bell	Shout "Man Overboard" & if necessary, sound 3 long blasts on whistle or bell
RANK					
Master-cum-Engineer	* Navigation & manoeuvres * Radio * Public Address * Appraise situation * Shut off fans & air conditioning * Close fire dampers * Attend fire detection panel * Take charge of fire party	* Navigation & manoeuvres * Radio * Public Address * Appraise situation * In charge of damage control	* Navigation & manoeuvres * Radio * Public Address * Supervise launching of liferafts	* Sound signal * Take charge of evacuation	* Navigation & manoeuvres * Radio * Supervise man-overboard pick up
CREW 1	* Start fire pump * Fight fire	* Damage control	* Prepare liferafts for launching	* Manoeuvre liferafts alongside & assist passengers to embark	* Prepare to pick up man overboard * Render First Aid
CREW 2	* Organise passenger muster & comfort them * Assist Crew 1	* Move passengers if necessary * Render First Aid * Assist Crew 1	* Prepare & ready passengers with life jackets on upper deck * Assist Crew 1	* As Crew 1	* Assist Crew 1

SYDNEY TO HOBART YACHT RACE – SOME SAFETY RECOMMENDATIONS

(Some safety recommendations made by the Coroner following the death of six men in the 1998 Sydney to Hobart yacht race in heavy weather.)

EPIRBS: Crewmembers wear a personal 406 MHz EPIRB when on deck in all weather conditions and be trained in their use. The competing yachts too should carry a 406 MHz Epirb. The 121.5 MHz Epirb is less accurate and is in the process of being superseded. (Epirbs are discussed in Chapter 18).

LIFERAFTS: The inflatable rafts carried on competing yachts should comply with regulation 15 of the SOLAS convention. This requires the raft floor to be insulated (with about 25 mm thick foam) to provide protection against the chill of the water underneath. Furthermore, the sealing capability of raft entrances against shipment of waves should be improved and their buoyancy tubes should be made of a highly visible colour, instead of black. (Liferafts are discussed in this chapter).

WEATHER FORECASTING: Weather forecasting specifically provided for yacht racing should contain not only the expected average winds, but also the maximum likely gusts of winds and the significant wave heights likely to be encountered. (Meteorology is discussed in Chapter 20).

BATTERIES: To prevent leakage of battery acid vapour in a yacht rolling through 360 degrees, their batteries should be of the closed or gel cell type. (Batteries are discussed in Chapter 30)

CLOTHING: The crew on deck during rough weather should wear clothing that would protect them from hypothermia.

PFDs: The crew should wear PFDs other than the "Mae West" or "Bib" type. The survivors had difficulty with the use of this type of PFDs as well as in placing the helicopter rescue sling over their heads when wearing them.

STROBES: Crewmembers should have with them a personal strobe light when on deck in all conditions. It allows them to be seen at greater distances if they are washed overboard.

HULL IDENTIFICATION: Each competing yacht should carry on her hull or deck some form of marking that can readily identify her to the air rescuers. This will prevent the wrong yacht being identified by rescuers, while the yacht being searched may already have sunk and her crew may have taken to liferafts.

CREW TRAINING: At least 50% of a competing yacht's crew should have completed a yacht safety and survival course every three years. The course should include assessment criteria that require all participants to: Board a liferaft; locate and use a rescue quoit; bring a simulated casualty on board a liferaft; escape from a capsized liferaft and don and be lifted in a helicopter rescue strop.

CHAPTER 19.3: QUESTIONS & ANSWERS

(Qns. 1-56 are in Ch. 19.1 & 57-145 in Ch. 19.2)

OH&S - SAFETY EQUIPMENT, SEA SURVIVAL & RESCUE

CHAPTER 19.3.1: PERSON OVERBOARD

146. A person falls overboard on the starboard side of a power driven vessel. You should:

    (a) turn the wheel to port side.
    (b) turn the wheel to starboard side.
    (c) slow down without changing course.
    (d) go full astern.

147. When helping a person overboard, the most effective throw of a lifebuoy is:

    (a) near the person upwind.
    (b) near the person downwind.
    (c) between the vessel and the person.
    (d) downwind of the vessel.

148. For a person overboard on port side, start the Williamson's turn by:

    (a) turning the wheel to port side.
    (b) turning the wheel to starboard side.
    (c) putting the engines astern.
    (d) stopping the vessel.

149. A person falls overboard on the port side at night. You should:

    (a) perform the Williamson's turn.
    (b) perform a U-turn.
    (c) turn the wheel to starboard side.
    (d) stop and engage stern power.

150. In order to retrieve a person fallen overboard from a large vessel in rough seas, the vessel will:

    (a) make a lee for the rescue boat to be launched and retrieved.
    (b) lower a ladder on the leeward side.
    (c) approach the survivor head-on.
    (d) approach the survivor stern-on.

151. In order to retrieve a person fallen overboard from a large vessel in calm waters, the vessel will:

    (a) lower a ladder on each side.
    (b) approach the survivor head-on.
    (c) not need to make a lee for the rescue boat to be launched and retrieved.
    (d) approach the survivor stern-on.

CHAPTER 19.3.2: SAFETY EQUIPMENT & SURVIVAL

(IMPORTANT: Also study questions on SARTS & Radar Reflective Distress Signals in Chapter 16, and on EPIRBS/GPIRBS in Chapter 18)

152. Compressed air cylinders or bottles:

    (a) should be stored below decks.
    (b) should be stored above decks.
    (c) contain liquefied air under pressure.
    (d) should be stored in warm places.

153. It is recommended by AMSA that a Sea Safety Small Craft Particulars Form should be lodged with RCC or other appropriate body:

    (a) prior to proceeding interstate.
    (b) monthly.
    (c) annually.
    (d) once every three years.

154. Telling someone where your vessel is going and when she'll return will be most helpful in case of:

    (a) poor stability.
    (b) person overboard.
    (c) radio breakdown.
    (d) vessel breakdown.

155. Safety equipment in a vessel should be installed as follows:

    (i) marine radio near an entrance.
    (ii) portable fire extinguishers near the entrances
    (iii) distress signals well inside a cabin.
    (iv) life jackets in lockers.

    The correct answer is:
    (a) all except (i).
    (b) all except (ii).
    (c) all except (iii).
    (d) all except (iv).

156. For a vessel operating within 15 miles of a coast station, the following safety item is of little use:

    (a) EPIRB/GPIRB.
    (b) VHF marine radio.
    (c) pyrotechnic distress signals.
    (d) MF/HF marine radio.

157. When passing close to a much larger vessel making way, a small vessel faces the risk of:

    (i)   interaction.
    (ii) suction from the larger vessel.
    (iii) being pushed away by the larger vessel.
    (iv) the larger vessel drifting on to her.

    The correct answer is:
    (a) all except (i).
    (b) all except (ii).
    (c) all except (iii).
    (d) all except (iv).

158. A radar reflector is useful in small vessels because it helps them to:

    (a) scan vessels operating radar.
    (b) become more visible on other vessels' radar.
    (c) keep a better lookout.
    (d) estimate other vessels' distance.

159. Which of the following devices is the most unlikely installation in a vessel's bilge?

    (a) bilge pump suction.
    (b) bilge water alarm.
    (c) gas sniffer.
    (d) smoke alarm.

160. The purpose of a line throwing apparatus is to:

    (a) assist rigging halyards on the mast.
    (b) control sails in a sailing vessels.
    (c) rig sails in a sailing vessel.
    (d) send a rescue line to another vessel or shore.

161. A vessel is foundering slowly but inevitably, the master and crew should:

    (a) abandon her as quickly as possible.
    (b) not abandon her unless they have to.
    (c) swim ashore as fast as they can.
    (d) get into a life raft and set drift.

162. In the following list, the quickest killer of an "abandon ship" victim is:

    (a) hypothermia.
    (b) hunger.
    (c) dehydration.
    (d) drinking sea water.

163. The best way to abandon vessel, when sea surface is on fire, is to:

    (a) jump downwind, swim using breast stroke.
    (b) jump upwind, swim using breast stroke.
    (c) jump downwind, swim freestyle.
    (d) jump upwind, swim freestyle.

164. The safety equipment required to be carried by a 10-metre commercial passenger vessel on enclosed waters includes:

    (a) Life Jackets.
    (b) Inflatable Liferafts.
    (c) EPIRB/GPIRB.
    (d) Personal Flotation Devices.

165. The following vessels are not required to carry any pyrotechnic distress signals:

    (a) non-commercial vessels.
    (b) non-commercial vessels on enclosed waters.
    (c) all vessels on enclosed waters.
    (d) sailing vessels.

166. A vessel has capsized. Three survivors are in the water, wearing life jackets, about two miles offshore. They should not exercise the following option for survival:

    (a) try to right the vessel.
    (b) try to swim ashore.
    (c) climb on top of the upturned vessel.
    (d) adapt a HUDDLE or HELP position.

167. A life jacket is designed to:

    (a) be worn loose around the body.
    (b) be tied at the back of the wearer.
    (c) turn an unconscious person face up.
    (d) assist swimming.

168. Before abandoning the vessel, the survivors should:

    (a) reduce the amount of clothing on their body.
    (b) fill extra water containers to the brim.
    (c) leave vessel's doors and hatches open.
    (d) drink water to their fill.

169. The least important item in a vessel's grab bag is:

    (a) water.
    (b) food.
    (c) passports.
    (d) blankets.

170. Fishing gear is supplied in a liferaft so that the survivors may:

    (a) supplement meals with fish.
    (b) supplement water with fish juice.
    (c) eat fish when water supply is ample.
    (d) use fish as bait for birds.

171. On abandoning a vessel into a liferaft, the first action among the group should be to:

    (a) equally distribute the rations.
    (b) take seasickness pills.
    (c) organise duties and a routine.
    (d) study and understand the safety equipment.

Chapter 19.3: OH&S - Safety Equipment, Sea Survival & Rescue

172. In case of an emergency on board a vessel it is advisable to:

    (a) abandon her as soon as possible.
    (b) delay abandoning her as long as possible.
    (c) launch the liferaft as soon as possible.
    (d) get everyone into a liferaft but stay close by.

173. An upturned rigid liferaft:

    (a) does not need to be righted.
    (b) should be righted prior to boarding.
    (c) will lose the use of its canopy.
    (d) will lose the use of its stores.

174. If forced to leave directly into water, the survivors should jump:

    (a) pulling lifejacket downwards in front.
    (b) with arms on the side.
    (c) with arms crossed in front.
    (d) in a "huddle" position.

175. A sinking vessel creates:

    (a) interaction.
    (b) suction.
    (c) interaction and suction.
    (d) none of the choices stated here.

176. On abandoning a vessel in hot summer months:

    (a) double up drinking water rations.
    (b) do not inflate the life raft's bottom.
    (c) remove most of your clothing.
    (d) do not inflate the liferaft canopy.

177. The hydrostatic release mechanism on liferafts:

    (a) should be allowed to work automatically.
    (b) prevents the raft from sinking with the vessel.
    (c) is independent of the manual release.
    (d) works as soon as it is underwater.

178. When an inflatable liferaft is stowed on deck in its container:

    (a) fasten the painter to a railing or deck.
    (b) by-pass the hydrostatic release mechanism.
    (c) take extra lashings.
    (d) remove the container seal.

179. If a loud whistling noise is heard after a liferaft has inflated:

    (a) do not board it.
    (b) board it and find the leak.
    (c) ignore it, it is the expansion valve.
    (d) find the leak and then board it.

180. In a liferaft with an uninsulated floor, the protection against cold weather requires:

    (a) inflating the liferaft floor.
    (b) increasing the consumption of food rations.
    (c) reducing drinking water rations intake.
    (d) consuming alcohol if available.

181. In a liferaft with an uninsulated floor, the protection against hot weather requires:

    (a) deflating the liferaft floor.
    (b) increasing drinking water rations intake.
    (c) decreasing raft ventilation.
    (d) deflating the raft canopy.

182. When in a liferaft:

    (i) take seasickness pills for the first 2 days.
    (ii) secure to other rafts.
    (iii) rig a drogue.
    (iv) attempt to contact an aircraft on 2182 kHz.

    The correct answer is:
    (a) all except (i).
    (b) all except (ii).
    (c) all except (iii).
    (d) all except (iv).

183. In order to haul in other survivors from the water, inflatable liferafts are supplied with:

    (a) a buoyant rescue quoit.
    (b) a buoyant rescue line.
    (c) a buoyant rescue quoit and a buoyant line.
    (d) a life buoy and a line.

184. A liferaft is stabilised by:

    (a) the water pockets fitted underneath.
    (b) rigging a drogue.
    (c) the aerodynamic shape of the canopy.
    (d) the CO2 cylinder underneath.

185. A capsized inflatable liferaft is best righted by:

    (a) standing on the gas bottle.
    (b) facing downwind.
    (c) standing on the raft ladder.
    (d) lifting it from underneath.

186. In a life raft, seasickness tablets should be taken as follows:

    (a) Once daily.
    (b) Three times daily.
    (c) one tablet per day.
    (d) as necessary.

187. Sea-sickness:
    (a) cannot be prevented.
    (b) leads to dehydration.
    (c) affects only those with full stomachs.
    (d) does not affect survival time.

188. On boarding a liferaft, the first ration of drinking water should be issued:
    (a) immediately.
    (b) after twelve hours.
    (c) after twenty-four hours.
    (d) after forty-eight hours.

189. Among ship wrecked survivors, the consumption of alcohol:
    (a) provides warmth.
    (b) causes dehydration.
    (c) can supplement shortage of fresh water.
    (d) prevents dehydration.

190. The drinking of sea water or urine by survivors:
    (a) can supplement shortage of fresh water.
    (b) leads to increased thirst & kidney failure.
    (c) is not harmful when diluted in fresh water.
    (d) is not harmful when consumed with fish.

191. The drift of a survival craft can be reduced by:
    (a) rowing gently upwind.
    (b) rowing gently up current.
    (c) using a drogue.
    (d) pouring oil on water surface.

192. If you were in a liferaft at night and a lookout reported hearing the engines of an aircraft, you would attract its attention by a:
    (a) smoke signal.
    (b) rocket parachute flare.
    (c) hand flare.
    (d) dye marker.

193. The survivors in a liferaft on a dark night can extend the sighting range of a search aircraft by the use of a:
    (a) heliograph.
    (b) smoke signal.
    (c) hand flare.
    (d) die marker.

194. A 6-person inflatable coastal liferaft contains the following quantity of water:
    (a) 1 litre.
    (b) 6 litres.
    (c) 9 litres.
    (d) none of the choices stated here.

195. A vessel's liferafts are required to be internally inspected and repacked:
    (a) once a year.
    (b) once every two years.
    (c) once every five years.
    (d) only if opened.

196. In a vessel lifebuoys should not be stowed:
    (a) near the steering position.
    (b) spread about the vessel.
    (c) in float-free brackets.
    (d) secured to the brackets.

197. A personal floatation device or a lifejacket helps the wearer to:
    (a) reduce loss of body heat.
    (b) swim more easily.
    (c) keep shoulders above water.
    (d) swim a longer distance.

198. For a survivor in the water, the act of swimming:
    (a) increases loss of body heat.
    (b) decreases loss of body heat.
    (c) has no effect on body heat.
    (d) is necessary for blood circulation.

199. A drogue attached to a life raft reduces its:
    (a) drift.
    (b) rolling.
    (c) drift and rolling.
    (d) capsizing effect.

200. The following actions by a vessel will assist a helicopter in its rescue operation from the vessel:
    (i) Fly a flag.
    (ii) Steer an upwind course.
    (iii) Motor at slow speed.
    (iv) Motor slowly in a circle.

    The correct answer is:
    (a) all except (i).
    (b) all except (ii).
    (c) all except (iii).
    (d) all except (iv).

201. The following signal with an outstretched arm indicates to the rescue helicopter that you are ready to be hoisted in the rescue strap:
    (a) thumb up.
    (b) rotating motion with a fore finger
    (c) rotating motion with both fore fingers.
    (d) rotating motion with one hand.

Chapter 19.3: OH&S - Safety Equipment, Sea Survival & Rescue

202. In the emergency provisions of a life raft, the following element is not catered for:

(a) protein.
(b) vitamins.
(c) carbohydrates.
(d) energy.

203. To a survivor at sea, the evening flight path of birds indicates the:

(a) direction for a supply of fish.
(b) direction of land.
(c) change in weather.
(d) direction for a supply of fresh water.

204. The normal loss of body fluids is one litre per day. It can be reduced by minimising perspiration and preventing sea sickness. In addition, it can also be reduced by:

(a) sucking on a lolly or a button.
(b) sun bathing.
(c) minimising physical exertion.
(d) not emptying the bladder.

205. In a liferaft, in addition to making an exception for the sick and injured, emergency rations should be distributed as follows:

(a) Allow everyone to control their own share.
(b) Issue only the first day's rations on boarding.
(c) Don't issue any rations for the first 24 hours.
(d) Don't issue any rations for the first 48 hours.

206. A coastal liferaft is supplied with the following ration of drinking water:

(a) four litres.
(b) one litre per person per day.
(c) one litre per person.
(d) one and a half litre per person.

207. After the first day, the minimum advisable issue of drinking water ration per person per day in a life raft is:

(a) one litre.
(b) 500 ml.
(c) one and a half litre.
(d) 250 ml.

208. In a survival craft, fish juice squeezed or sucked from fish:

(a) may be consumed to supplement water rations.
(b) is not a substitute for water rations.
(c) may be consumed when mixed with water rations.
(d) is a good supplement for food & water rations.

209. If you had to jump into water from a vessel listing heavily to port, the most suitable side to jump will be:

(a) port side.
(b) starboard bow.
(c) bow or stern.
(d) starboard quarter.

210. A grab bag is a:

(a) bag containing the life raft rations.
(b) panic bag.
(c) bag of unnecessary supplies.
(d) bag of vessel's supplies.

211. A SOLAS lifejacket is designed to allow the wearer to jump, without injury or dislodgement, from a vertical distance of:

(a) 3 metres
(b) 6 metres
(c) 12 metres
(d) 9 metres

212. The Standard Survival Pack in a liferaft does not include:

(a) Sponges
(b) Pyrotechnic Distress Signals
(c) Seasickness pills
(d) Hypothermia Suits

213. The advantage of a davit-launched liferaft over an ordinary liferaft is that:

(a) It can be inflated & boarded before being lowered into the water.
(b) It does not require the painter made fast to the vessel.
(c) It does not require a check on overboard obstructions and discharges prior to launching.
(d) It can be launched on the higher side of the listed vessel.

214. Which of the following is the preferred method of boarding an inflated liferaft?

(a) Jump into the water and climb into it.
(b) Step into it, climbing down a ladder if necessary.
(c) Jump from the deck directly into it.
(d) Enter the water, inflate the liferaft from there and enter it.

215. When in the water, the best way to deter a shark is to:

(a) vigorously splash water.
(b) swim away from it.
(c) remain still and quiet.
(d) distract it by taking off lifejacket and hurling it as far as possible.

216. A liferaft should inflate in:

    (a) 20 to 30 seconds
    (b) 50 to 60 seconds
    (c) 4 to 5 minutes
    (d) 1 to 2 minutes

217. With sufficient number of survivors in a liferaft, visual lookout should be maintained as follows:

    (a) 24 hours a day at both openings of the liferaft.
    (b) 24 hours a day at one opening.
    (c) Daylight hours at both openings.
    (d) Nighttime at both openings.

218. Locally operating commercially vessels are required to carry buoyant lifesaving appliances for:

    (a) 150% of the complement.
    (b) 100% of the complement.
    (c) 50 % of the complement.
    (d) 60% of the complement.

219. In relation to rigid liferafts, which of the following statements is incorrect?

    (a) They must be orange in colour.
    (b) The must be fitted with hand beckets.
    (c) They must be carried by vessels operating offshore.
    (d) They are designed to float free from the sinking vessel.

220. The number of persons a lifebuoy is designed to support is:

    (a) One.
    (b) Two
    (c) Three
    (d) Four

221. In relation to rigid & inflatable liferafts, which of the following statements is incorrect?

    (a) They should be instantly accessible.
    (b) They should be able to float free from the vessel even when she is trimmed by 10° to either end.
    (c) They should be able to float free from the vessel even when she is listed by 10° to either side.
    (d) Vessels operating in smooth waters may carry only the rigid liferafts.

222. In relation to small vessels being able to use the AMSA AUSREP System, which of the following statements is incorrect?

    (a) They should be equipped with a VHF or HF radiotelephone.
    (b) They should be equipped with a HF DSC transceiver or INMARSAT terminal.
    (c) They should be equipped with an EPIRB/GPIRB.
    (d) They should have submitted a Small Craft Particulars Form.

223. In relation to REEFREP, the following ships are required to report their positions to the REEF CENTRE:

    (a) Only the Oil Tankers 50 metres or greater in length.
    (b) Vessels of 50 metres or greater in length.
    (c) Vessels of 60 metres or greater in length.
    (d) Vessels engaged in towing.

224. The group Huddle in the water reduces:

    (a) the body heat loss on all sides.
    (b) the risk of shark attacks.
    (c) the drift
    (d) the body heat loss on one side of the body.

225. To reduce the risk of hypothermia when in the water:

    (a) Adopt HUDDLE position when alone.
    (b) Adopt HELP position when in a group.
    (c) Lie on your back with hands behind your head.
    (d) Reduce contact of chest & groin regions with the water.

226. To prevent dehydration:

    (a) Maximise sweating.
    (b) Drink fish juice, if available.
    (c) Eat protein-rich foods.
    (d) Take seasickness tablets.

227. On abandoning ship:

    (a) Row the liferaft as far away from the abandoned vessel as possible.
    (b) Do not activate EPIRB/GPIRB until after boarding the liferaft.
    (c) If there are two or more liferafts, secure them together.
    (d) Do not take seasickness tablets for the first 24 hours.

228. The operating cord on an inflatable liferaft is known as the:

   (a) Painter
   (b) Quoit
   (c) Heaving line
   (d) Drogue

229. On a sinking vessel, the hydrostatic release mechanism releases the liferaft container by:

   (a) exerting a pull on the painter.
   (b) releasing the container's tie down strap.
   (c) activating the gas bottle.
   (d) busting open the liferaft sealing tape.

230. In addition to being received by satellites, an Epirb signal may also be heard by:

   (a) Coast radio stations
   (b) Water police
   (c) Large commercial ships
   (d) Aircraft

231. A Carley Float is a:

   (a) Quoit
   (b) Buoyant rescue line.
   (c) Lifejacket.
   (d) Rigid raft or similar structure

232. A life jacket or PFD must have:

   (a) the vessel's name written on it.
   (b) most of its buoyancy around the chest and neck.
   (c) a hood.
   (d) most of its buoyancy in its lower section.

233. When planning to carry out an emergency drill:

   (a) do not advise the crew that it is a drill.
   (b) advise the crew that it is a drill.
   (c) catch the crew off guard by raising the alarm in the middle of the night.
   (d) conduct it only when the vessel is alongside a wharf.

234. On hearing the "Abandon Ship Signal" below decks:

   (a) run to your emergency station carrying a lifejacket.
   (b) jump overboard wearing a lifejacket.
   (c) wear as much warm clothing as possible followed by a lifejacket.
   (d) take a few gulps of alcohol to keep warm, wear warm clothes followed by a lifejacket.

235. The lifebuoy self-igniting lights are capable of burning for at least:

   (a) 2 hours
   (b) 45 minutes
   (c) 6 hours
   (d) 15 minutes

236. A vessel's rescue boat should not be carried:

   (a) in the bow of the vessel.
   (b) in the after part of the vessel.
   (c) near the engine room.
   (d) On the side of the vessel.

237. A PFD of 10-kilogram buoyancy rating:

   (a) is an adult PFD, because only about 5% of an overboard person's weight is not water & fat, & needs supporting.
   (b) is a child PFD.
   (c) is an adult PFD, because only the unsubmerged head of a person overboard needs supporting.
   (d) does not exist.

238. Liferafts and Carley floats should be able to float free from the vessel even when the vessel is in the following condition:

   (a) Listed or trimmed by 10° to either side or end.
   (b) Listed by 10° to either side & trimmed by 15° to either end.
   (c) Listed by 15° to either side.
   (d) Trimmed by 10° to either end

CHAPTER 19.3.3: DISTRESS & OTHER SIGNALS

239. A distress flare sighted at sea may be from:

   (a) a naval vessel engaged in practice.
   (b) a yacht requesting a ship to alter course.
   (c) an aircraft searching for a vessel in distress.
   (d) none of the choices stated here.

240. A signal consisting of a series of alternative short and long blasts on the whistle or rings on the bell in the same order is the:

   (a) abandon ship signal.
   (b) continuous ringing of the bell signal.
   (c) emergency signal.
   (d) muster signal.

241. The ranges of visibility of distress signals in good visibility are as follows:

    (i) Hand flares 5 miles during the day.
    (ii) Rocket parachutes 25 miles during the night.
    (iii) Hand flares 5 miles during the night.
    (iv) Rocket parachutes 8 miles during the day.

    The correct answer is:
    (a) all except (i).
    (b) all except (ii).
    (c) all except (iii).
    (d) all except (iv).

242. To attract the attention of a ship on the horizon the following distress signal should be used:

    (a) hand flare.
    (b) smoke signal.
    (c) dye marker.
    (d) rocket parachute flare.

243. Radar reflective rocket distress signals:

    (a) eject a cloud of radar reflective material.
    (b) are radar reflector fitted rockets.
    (c) produce an echo on radar for two minutes.
    (d) can only be used during the day.

244. A make-shift visual distress signal can be contrived from two shirts by hanging one rolled up (resembling a ball) and the other from its sleeves (resembling a square) as follows:

    (a) they must be painted black before hanging.
    (b) ball above the square.
    (c) ball below the square.
    (d) ball above or below the square.

245. The least effective method of attracting the attention of an aircraft during daylight is to use a:

    (a) smoke signal.
    (b) torch light.
    (c) hand flare.
    (d) dye marker.

246. A "V" sheet is:

    (a) an Australian distress signal.
    (b) an international distress signal.
    (c) not a distress signal.
    (d) a signal for medical assistance required.

247. When firing hand-held flares and smoke signals, you should:

    (a) keep them downwind.
    (b) keep them upwind.
    (c) not remove the top and bottom lids.
    (d) not remove the top lid.

248. When wet, the ignition capability of the pyrotechnic distress signals is affected as follows:

    (a) they can still be ignited.
    (b) they cannot be ignited.
    (c) they lose the self-igniting capability.
    (d) only the hand flare can be ignited.

249. The average life of pyrotechnic distress signals is:

    (a) one year.
    (b) three years.
    (c) five years.
    (d) an unspecified period.

250. An aircraft rolling from side to side in the vicinity of a vessel in distress is an indication of its:

    (a) inability to assist.
    (b) acknowledgement of distress signal.
    (c) distress signal.
    (d) inquiry of the nature of distress.

251. A signal consisting of seven short and one prolonged blast on a ship's whistle, bell or siren may be used for:

    (i) abandon ship.
    (ii) fire.
    (iii) general emergency.
    (iv) muster.

    The correct answer is:
    (a) all except (i).
    (b) all except (ii).
    (c) all except (iii).
    (d) all except (iv).

252. Green pyrotechnic flares fired by a search aircraft means:

    (i) it is searching for the vessel in distress.
    (ii) it acknowledges a distress signal.
    (iii) the vessel in distress should fire a signal.
    (iv) any vessel in the area should respond.

    The correct answer is:
    (a) all except (i).
    (b) all except (ii).
    (c) all except (iii).
    (d) all except (iv).

253. A signal consisting of a white pyrotechnic flare means:

    (a) distress.
    (b) searching for the vessel in distress.
    (c) cancellation of distress.
    (d) a collision warning give way.

Chapter 19.3: OH&S - Safety Equipment, Sea Survival & Rescue

254. A floating yellow or white pyrotechnic signal means a:

    (a) submarine surfacing/ engaged in naval exercises
    (b) submarine in distress and surfacing.
    (c) cancellation of a distress.
    (d) searching for the vessel in distress.

255. The time-expired pyrotechnic distress signals should be:

    (a) retained on board as additional spares.
    (b) disposed overboard beyond 12 miles offshore.
    (c) used for fireworks.
    (d) handed over to a fire station or water police.

256. The sighting radius of an orange smoke distress signal is approximately:

    (a) 1 mile (2 km)
    (b) 2 miles (4 km)
    (c) 5 miles (10 km)
    (d) 10 miles (20 km)

257. The visible range of a red parachute flare at night is approximately:

    (a) 10 miles (18 km)
    (b) 15 miles (27 km)
    (c) 25 miles (40 km)
    (d) 40 miles (72 km)

258. Commercial aircraft are able to receive distress signals on the following frequency:

    (a) VHF Channel 16
    (b) 2182 MHz
    (c) EPIRB frequency
    (d) 4215 MHz

259. The Dye Marker distress signal is designed to attract the attention of:

    (a) Aircraft by day
    (b) Ships by day
    (c) Aircraft by day or night
    (d) Ships by day or night

260. A signal consisting of continuous ringing of a bell indicates:

    (a) abandon ship.
    (b) general emergency.
    (c) muster station.
    (d) there is a fire on board.

261. You have attracted the attention of an aircraft by use of a heliograph. The aircraft has altered course. To assist it to locate your liferaft, you should:

    (a) provide your position by VHF.
    (b) light a hand flare.
    (c) fire a rocket parachute flare.
    (d) take no further action.

262. The Red Hand & Parachute Flares & the Orange Smoke Signal remain visible for the following <u>approximate</u> period:

    (a) 2 minutes
    (b) 30 seconds
    (c) 60 seconds
    (d) 5 minutes

CHAPTER 19.3.4: SEARCH & RESCUE

(IMPORTANT: YOU MUST ALSO STUDY CHAPTER 18 FOR DISTRESS & URGENCY COMMUNICATION BY RADIO)

263. A commercial airliner flying overhead can be contacted on the following radio frequency:

    (a) 2182 kHz.
    (b) VHF channel 16.
    (c) 27 MHz radio channel 88.
    (d) none of the choices stated here.

264. There are 50 people in the water, seeking rescue. Your 10-metre vessel is the first to arrive at the scene. You may:

    (i) load your vessel beyond the certified limit.
    (ii) use your vessel's survival gear.
    (iii) allow survivors to hang from your vessel.
    (iv) risk the safety of your own vessel.

    The correct answer is:
    (a) all except (i).
    (b) all except (ii).
    (c) all except (iii).
    (d) all except (iv).

265. When assisting a vessel in distress, your legal obligation ceases when:

    (a) the vessel is safe.
    (b) the lives are safe.
    (c) the lives and the vessel are safe.
    (d) the tow line is cast off.

266. If you came across an empty liferaft at sea:

    (a) you should not use PANPAN signal to report it.
    (b) you may use PANPAN signal to report it.
    (c) you should sink it.
    (d) you should fire a distress signal.

267. Your vessel's collision with another vessel's midships section has shattered her hull. She has lost her radio contact and is fast taking water. You are not certain if your vessel is capable of taking all the survivors on board. Your first and foremost action should be:

   (a) secure her alongside to prevent her sinking.
   (b) take off all the survivors.
   (c) transmit MAYDAY RELAY.
   (d) tow her as fast as you can.

268. You are in a vessel by yourself and in a position to rescue people from another vessel on fire. You should:

   (a) not take charge.
   (b) stay downwind.
   (c) advise the authorities prior to rescue attempt.
   (d) not rig lifeline around your vessel.

269. A vessel is loaded to her permitted capacity. Rescuing survivors from another vessel is:

   (a) in breach of the law.
   (b) within the law.
   (c) required by the law.
   (d) none of the choices stated here.

270. A vessel is unable to take on board all the survivors from a vessel in distress without sacrificing her own safety. Her master should:

   (a) hoist the international code flag Zulu.
   (b) tow the other vessel.
   (c) help the few.
   (d) seek alternative means to help everyone.

271. In relation to the Elliptical Turn to retrieve a person overboard:

   (a) For the second alteration of course, the wheel is put hard over in the opposite direction to that of the first alteration.
   (b) The first alteration of course is commenced by putting the wheel hard over away from the person.
   (c) The manoeuvre requires only one alteration of course.
   (d) The second alteration of course is commenced when the person is about 3 points abaft the beam.

272. In relation to the Expanding Square Search Pattern:

   (a) The space between the tracks is 0.5 miles when the sea is smooth.
   (b) This pattern is designed for gradually expanding the search area.
   (c) The space between the tracks is 1 mile when the sea is rough.
   (d) All turns are 120° to starboard.

273. In relation to the Sector Search Pattern:

   (a) The space between the tracks is 0.5 miles when the sea is smooth.
   (b) The turn is designed for gradually expanding the search area.
   (c) The space between the tracks is 1 mile when the sea is rough.
   (d) All turns are 120° to starboard.

274. When retrieving a person fallen overboard who may be suffering from hypothermia:

   (a) A vertical lift will minimise the influence of hydrostatic squeeze.
   (b) The gravitational effect is minimised by the hydrostatic squeeze.
   (c) A horizontal lift will minimise the influence of hydrostatic squeeze.
   (d) The hydrostatic squeeze is minimised by the gravitational effect.

275. The symptoms of hypothermia include:

   (a) Hyperactivity and lack of reason.
   (b) Shouting and incoherent speech.
   (c) Slurred speech and lack of reason.
   (d) Hyperactivity & incoherent speech

276. When treating a person for hypothermia:

   (a) Prevent further loss of person's heat due to evaporation.
   (b) Massage the person's body & limbs.
   (c) Give the person a drink of brandy.
   (d) Immediately wrap the person in blankets.

277. To transfer an unconscious survivor from the water into the liferaft:

   (a) Lift by the becket on the back of the person's lifejacket.
   (b) Lift by the person's armpits.
   (c) Lift the person on his/her side..
   (d) Use the buoyant quoit.

278. When being rescued by helicopter:

   (a) At night, a bright light should be pointed upwards from the vessel.
   (b) The survivor in the water should swim towards the strop lowered by the helicopter.
   (c) The survivor in the strop should indicate readiness for winching with a "thumb up" signal on an extended arm.
   (d) The helicopter winch wire should never be touched.

279. The most effective way to attract the attention of an aircraft flying overhead at night is to:

    (a) light a red hand flare.
    (b) fire a red parachute flare.
    (c) use a heliograph.
    (d) Use the orange smoke signal.

280. The following signal with an outstretched arm indicates to the rescue helicopter that you are not ready to be hoisted in the rescue strap:

    (a) thumb pointing down.
    (b) rotating motion with a fore finger.
    (c) downward rotating motion with both fore fingers.
    (d) downward rotating motion with one hand.

281. To retrieve a conscious and able person from overboard in heavy seas, the safest approach is:

    (a) Manoeuvre the vessel astern so that the person can climb up from astern.
    (b) Stop the vessel and jump overboard to help the person.
    (c) Stop a short distance away & throw a rope to the person.
    (d) Manoeuvre the vessel upwind so that the person can climb up from the bow.

CHAPTER 19.3.5: ADDITIONAL FOR COMMERCIAL VESSELS

282. Some commercial vessels are exempt from conducting regular safety drills. The masters of such vessels need to:

    (a) familiarise the crew with emergency procedures.
    (b) practice operating all safety equipment.
    (c) conduct drills only after crew changes.
    (d) regularly update the muster list.

283. When changing ownership of a vessel in survey, the following action should be taken:

    (a) advise the local authority at the next survey.
    (b) submit a change of ownership form.
    (c) call in the authority's surveyor.
    (d) seek authority's permission.

284. In commercial vessels, "PFDs":

    (a) are carried in lieu of lifejackets.
    (b) are the life jackets.
    (c) are inflatable flotation devices.
    (d) have not replaced life jackets

285. The number of life jackets required to be carried on board a commercial vessel are as follows:

    (a) at least one per person.
    (b) depends on other flotation equipment on board.
    (c) none if sufficient life rafts are carried.
    (d) none if PFDs are carried.

286. The minimum safety equipment requirements for your particular commercial vessel is listed in the:

    (a) USL Code.
    (b) vessel's station list.
    (c) NSCV.
    (d) vessel's survey documents.

287. In commercial vessels (unless exempt), crew emergency procedures are required to be practised at intervals of not more than:

    (a) one month.
    (b) two months.
    (c) three months.
    (d) whenever necessary.

288. In commercial passenger vessels (unless exempt), survival craft drills are required to be held at intervals of not more than:

    (a) one month.
    (b) two months.
    (c) three months.
    (d) whenever necessary.

289. In commercial passenger vessels (unless exempt), fire drills are required to be held at intervals of not more than:

    (a) one month.
    (b) two months.
    (c) three months.
    (d) whenever necessary.

290. In commercial non-passenger vessels (unless exempt), survival craft drills are required to be held at intervals of not more than:

    (a) one month.
    (b) two months.
    (c) three months.
    (d) whenever necessary.

291. In commercial non-passenger vessels (unless exempt), fire drills are required to be held at intervals of not more than:

    (a) one month.
    (b) two months.
    (c) three months.
    (d) whenever necessary.

292. In commercial vessels, an Emergency Station List is required to be displayed, when the crew complement is:

    (a) 4 or more
    (b) 10 or more
    (c) 2 or more
    (d) 6 or more

293. A vessel's Emergency Station List includes:

    (a) list of the safety equipment carried.
    (b) passenger list.
    (c) location of the safety equipment carried.
    (d) duties of each crew member in case of an emergency.

294. On hearing an Emergency Station signal, the following must muster at the appointed Emergency Station(s):

    (a) All passengers and crew.
    (b) All passengers & crew, except those on essential duties.
    (c) All passengers & crew, except those on duty.
    (d) All crew.

295. Where there are two lifeboats on each side of a vessel, the second lifeboat on the starboard side may be numbered as follows:

    (a) S2
    (b) 2
    (c) S2 or 3
    (d) 1

296. The following vessels should send their Position Reports to the State-run CRN:

    (a) None of the choices stated here
    (b) DSC/INMARSAT equipped vessels
    (c) SOLAS vessels
    (d) Non-SOLAS vessels

297. Under SOLAS regulations, the radio equipment requirement for non-SOLAS vessels includes:

    (a) Two independent means of transmitting distress
    (b) Two independent means of receiving MSI
    (c) One DSc capable HF radio
    (d) One VHF radio

CHAPTER 19.3: ANSWERS

(Answers 1-56 are in Ch. 19.1 & 57-145 in Ch. 19.2)

146 (b), 147 (a), 148 (a), 149 (a), 150 (a),
151 (c), 152 (b), 153 (d), 154 (c), 155 (c),
156 (d), 157 (d), 158 (b), 159 (d), 160 (d),
161 (b), 162 (a), 163 (b), 164 (a), 165 (b),
166 (b), 167 (c), 168 (d), 169 (c), 170 (c),
171 (b), 172 (b), 173 (a), 174 (a), 175 (b),
176 (b), 177 (b), 178 (a), 179 (c), 180 (a),
181 (a), 182 (d), 183 (c), 184 (a), 185 (a),
186 (b), 187 (b), 188 (c), 189 (b), 190 (b),
191 (c), 192 (c), 193 (c), 194 (b), 195 (a),
196 (d), 197 (a), 198 (a), 199 (a), 200 (d),
201 (a), 202 (a), 203 (b), 204 (c), 205 (c),
206 (c), 207 (b), 208 (b), 209 (c), 210 (b),
211 (b), 212 (d), 213 (a), 214 (b), 215 (c),
216 (a), 217 (a), 218 (b), 219 (c), 220 (b),
221 (b), 222 (a), 223 (b), 224 (d), 225 (d),
226 (d), 227 (c), 228 (a), 229 (b), 230 (d),
231 (d), 232 (b), 233 (b), 234 (c), 235 (b),
236 (a), 237 (a), 238 (b), 239 (d), 240 (a),
241 (a), 242 (d), 243 (a), 244 (d), 245 (b),
246 (a), 247 (a), 248 (a), 249 (b), 250 (b),
251 (a), 252 (d), 253 (d), 254 (a), 255 (d),
256 (b), 257 (c), 258 (c), 259 (a), 260 (d),
261 (b), 262 (c), 263 (d), 264 (d), 265 (b),
266 (b), 267 (c), 268 (c), 269 (c), 270 (d),
271 (d), 272 (b), 273 (d), 274 (c), 275 (c),
276 (a), 277 (b), 278 (c), 279 (a), 280 (a),
281 (c), 282 (a), 283 (b), 284 (d), 285 (a),
286 (d), 287 (a), 288 (b), 289 (b), 290 (c),
291 (c), 292 (a), 293 (d), 294 (b), 295 (c),
296 (a), 297 (a)

Chapter 19.4

OCCUPATIONAL HEALTH & SAFETY -
WORK PRACTICES

OCCUPATIONAL HEALTH & SAFETY (OH&S) AT SEA

During a charity yacht race in 1997, the yacht Condor's genoa block dislodged itself from the track and injured an <u>experienced</u> member of its crew. The injured crew member sued the yacht, its owner and skipper for damages (Bollen v Condor of Bermuda, Federal Court of Australia 1998/99). The case provides a good summation of the duties and responsibilities of vessel owners, operators as well as crewmembers on both recreational and commercial vessels:

Court's Summation Of The Vessel Owner's Duty

The vessel owner owed a duty of care to all on board the vessel "to take reasonable care to ensure that they were not injured as a result of the yacht having inadequate equipment, including inadequately maintained or defective equipment". (The owner of Condor was found to have discharged this duty).

Court's Summation Of The Crew Member's Duty

"As an experienced sailor, (the crew member) failed to take reasonable care for his own safety in standing (where he apparently did)".

OH&S LEGISLATION

- The Australian OH&S legislation is based on the concept of DUTY OF CARE TO YOUR NEIGHBOUR. In the marine environment, every person carries the responsibility as well as the expectation that their vessel (work place) is safe for him or her as well as for everyone else on board - safe from physical as well as psychological harm.

- Owners and operators have a responsibility for the safety of employees under OH&S legislation to ensure that persons on board are trained for the jobs that they are required to perform. They are also responsible for the supervision of employees and providing a safe workplace. Employees have a responsibility to contribute to the health and safety of the workplace and to work with employers in providing this.

- OH&S legislation requires all workplaces to effectively manage health and safety. This is to be done in a systematic way by identifying hazards, assessing risks and taking appropriate action to control these risks. The risk assessment is an important element in determining the action required. In some cases this might be as simple as a visual inspection of the workplace. In other cases more detailed assessment may be required. The legislation also requires employees to be consulted on matters that may affect their health and safety. For more information on the requirements of the OH&S legislation contact WorkCover in your State or Territory. Other organisations for guidance are Worksafe Australia, Standards Australia and the Australian Maritime Safety Authority (AMSA). The websites of all these bodies are listed on the last page of this book.

DUTY OF CARE

Vessel owners and operators have a general duty of care for the safety of people on or in their property. The duty of care is independent of marine legislation. Attempted exclusion of duty of care written on placards and on the back of tickets and dockets usually has little or no legal value. The duty of care and OH&S requirements expose an owner or operator of any vessel to considerable risk of damages should an accident occur. By providing appropriate safety briefings and training, a vessel owner or operator can greatly improve safety and reduce exposure to risk if damages.

WHY THE NEED FOR EXTRA CARE ON VESSELS?

- Essential services such as medical assistance, police rescue and fire brigades are not easily accessible.
- You live and work on a rolling and pitching platform.
- You are involved in handling dangerous equipments and perform dangerous tasks.
- You are exposed to heavy seas, gale force winds and the risk of hypothermia.
- You are vulnerable to falling overboard
- Your chances of dealing with life threatening situations are high.
- You may have to survive without food and water.
- You may have to survive suspended in water or enclosed in a tiny rubber raft that is wet inside, smells of rubber and bobs up and down, making you seasick. Seasickness makes you want to die.

COMPONENTS OF OH&S ON VESSELS:
- Knowledge of your rights and responsibilities on vessels
- Knowledge of your working environment
- A Senior First Aid Certificate (See Chapter 19.1 for practical first aid)
- Knowledge of basic fire prevention and control
- Knowledge of survival at sea.

RISKS & SAFETY

RISK is the chance of something happening that will have a detrimental impact upon safety. It is measured in terms of the consequences and likelihood of injury, illness or environmental damage.

RISK ASSESSMENT is the process of evaluating the probability and consequences of injury, illness or environmental damage arising from exposure to identified hazards associated with a vessel.

DEVELOPING A RISK MANAGEMENT PLAN
- Assess the risk to crew arising from vessel's normal activities.
- Carry out risk assessment for all operations, such as anchoring and berthing, loading and discharging, refuelling and galley operation, engine and winch operation, watchkeeping, rough weather, emergencies, maintenance and inspections.
- Identify those crew who may be particularly at risk in performing their tasks, and then take measures to reduce or avoid such risks.
- Make practical and effective plan to manage risk for each task and each area in a vessel.
- Devise procedural risk management plans. For example, what is the procedure to follow if the master falls overboard with the engine keys in his/her pocket. Or, in a crew of two, what is the procedure if one falls overboard and other is unable to get him/her back on board.
- Set up a training strategy.
- Promote, maintain and improve strategies.
- Take an appropriate insurance cover.

THE VESSEL OWNER OR OPERATOR MUST PROVIDE THE CREW WITH
- Safe system of work (regular information, instructions and procedural training, etc.)
- Safe working environment
- Safe entry and exit to the vessel and all spaces in it
- Safe machinery and equipment.

THE CREW MEMBER MUST
- Inform the owner or operator of the discovery of any risks to their own safety or the safety of the vessel that may result from performing a task.
- Take reasonable care for the safety of themselves and others on board
- Cooperate with the owner and operator in matters of safety
- Not misuse or interfere with provisions of safety and heath.

STRESS
- Stress causes physical, emotional and chemical changes in the body. It even contributes to seasickness.
- Both physical and psychological factors - real and imagined - need to be addressed.
  Physical factors:
  - Diet, fitness and lifestyle
  - Nautical/dive knowledge, skills and experience
  - Correct gear, suitable working conditions and exertion level

  Psychological factors:
  - Task loading, potential activity hazards
  - Perceived degree of crew support, peer pressure
  - Perceptual narrowing risks (overlooking hazards when preoccupied with performance or other worries)
- Perceptual narrowing (a psychological term) is said to occur when a person, rightly or wrongly, perceives that he/she is under pressure to perform. They may thus become prone to accidents or overlook occupational health and safety issues. A person concerned with lack of job security or financial hardship may not, for example, notice a leaking fuel line.

Chapter 19.4: OH&S - Work Practices

- Stress management involves both the individual (crew or passenger) and those in position of authority around them.
- Stress (above the water and below the water, if diving) needs to be recognised and managed. Introducing crew to basic stress management principles will assist all persons on board.
- Stress management should start in the pre-embarkation phase when planning a trip. Prevention is the key to stress reduction. Do not leave it until on the high seas.
- Establish communication with relevant others who will be on board. Discuss potential stressors and manage each factor or anticipated factor before they arise.
- The individual (whether a novice crew member or an experienced hand) needs to know that they can call on superiors or peers at any time for assistance, without pressure or guilt, and are not alone in how they may feel.
- Diet and lifestyle influence one's ability to enjoying one's time on and in the water. It is important to eat healthy reduced-fat high-carbohydrate meals, and drink adequate non-alcoholic fluids such as water or juice. Eat light but adequate breakfast. Caffeine should be limited to regular strength tea or coffee (1-2 cups maximum). Try not to leave port on an empty stomach.
- A good sleep the night before the vessel leaves port is essential. It may be helpful to acclimatise yourself against motion sickness by sleeping on board the night before departure. Alternatively, on boarding, a brief nap in your berth is helpful in getting your 'sea legs'.
- Taking drugs (tobacco, cannabis, etc) normally alters (reduces) the cognitive ability of the participant, often with a reduction in the availability of oxygen in the body. It may increase anxiety and cramps, and cause withdrawal symptoms. Prescription medication and other common 'cough-and-cold' preparations may cause similar adverse symptoms. Discuss your condition and treatment options with a doctor or pharmacist. Inform them of your intending voyage.
- Anti-motion seasickness medication ideally needs to be taken before going embarking on a vessel, or well prior to facing the open water swell.

OTHER COMMON RISKS ON BOARD

- Damaging behaviour and accidents resulting from alcohol or drug consumption.
- Health risks from chemicals, paints and solvents used in cleaning and maintenance of vessels.
- Injury from unprotected welding and cutting activities.
- Hepatitis infection passed on from one person to another through poor hygiene.
- Damage to spine through careless lifting.
- Health risk from passive smoking.
- Risk of skin cancer from working in the sun without wearing sun protection.
- Risk of injury or of falling overboard from not wearing protective clothing or equipment.
- Risk of fire or injury from transportation or storage of dangerous goods or accepting improperly labelled containers.
- Injury and damage caused by exceeding safe working loads of lifting equipment.
- Injury and damage resulting from horseplay or practical jokes.

SOME DO'S AND DON'TS FOR THE CREW

- You have a duty of care on board the vessel. Do not allow known hazards to exist. As the saying goes, you are not only not to throw a banana skin on deck, but also pick it up if you see one.
- Remain alert, remove obstructions and change any process or equipment that is likely to hurt you or anyone else.
- Your habits and practices at work must not be harmful to you or to others.
- Make others aware of safe practices. Do not allow anyone to proceed unknowingly into danger.
- Very hot or cold conditions can lead to exhaustion and injury. Wear suitable clothing at all times.
- Carry out and participate in appropriate safety drills.
- Keep an eye on the fire extinguishers pressure gauges. Regularly look inside their nozzles for chewing gum blockage or insect infestation. Also keep an eye on the fire hose boxes, if any. Fix or report any irregularities.
- Fix or report any fire hazards: overloaded electrical sockets, unattended oily rags, uncovered or leaking flammable liquids or gases.
- Report all accidents, even those resulting in minor injuries, which the master or person in charge must record.
- The master or person in charge must send an accident report for serious injuries to the relevant Workcover Authority within seven days.

A SHORT SAFETY CHECKLIST

- Are you dressed for the job with loose clothing tucked in and loose hair tied up?
- Are you equipped for the job with safety harness, safety boots, goggles, etc?
- Do you have the right tools?
- Is your work environment clean and uncluttered?
- Are all items properly secured?
- Are flammable liquids and gases safely stored?
- Are there any fire hazards?
- Is the work area ventilated? Will it continue to remain free of smoke, dust and fumes?
- Are electrical tools and wiring free of damage?
- Have you informed others of your whereabouts? Is someone keeping an eye on you when working in a confined space or aloft?
- Is there an electrical or radiation risk from working near a radio or radar antenna?
- Are you "keeping one hand for yourself and one for the ship"? (use one hand to hold onto something to prevent falling)
- Are you "cutting towards your mate"? (when using a knife, cut away from yourself)
- Are there "fools and first trippers to sit on ship's rail"? (encourage newcomers into safe practices so that they get to make the next rip).

GALLEY SAFETY

- Cooking burners should be fitted with flame failure devices and fiddle rails with adjustable potholders. Better still if the burners are gimballed against heeling, rolling and pitching.

- Wear overalls when deep-frying, in case hot oil spills on your body. Don't fill cooking pots to the top. Keep long hair and clothing away from food and equipment. Do not smoke in galleys or toilets. Don't overload electrical outlets. Galley equipment is often heavy and made of metal. Stow it safe and firmly. Safely secure knives and all sharp implements. Unsecured cleaning equipment can be dangerous. Stow flammable liquids, matches and lighters separately. Do not stow bleaches with disinfectants. Oily rags can self-combust; dispose them in proper receptacles.

- In commercial galleys: keep food preparation area clean. Don't sit on food preparation areas or tables. Wash hands before handling food and after using toilet. Remove hand and wrist jewellery when preparing food.

SAFETY IN ACCOMMODATION SPACES

- Stow all gear in lockers and drawers. Have faulty wiring, safety equipment, emergency lighting, etc. fixed immediately. Properly stow life saving gear – to be available for immediate use. Keep heads (toilets) clean and clear.

CONFINED SPACES (ENCLOSED SPACES): PRECAUTIONS WHEN ENTERING

- Where access to enclosed spaces in a vessel is required as part of normal operation, the persons may become entrapped, exposing them to increased risk due to heat, cold or lack of oxygen, then the following should be provided:
    - Emergency lighting
    - Means of opening the door from both sides
    - Alarm systems

- There are spaces on board vessels, which, due to the nature of their contents or by simply having been closed for a period of time, are contaminated or lack oxygen to sustain human life. All spaces, left closed for a time, become deficient in oxygen because of the oxidation occurring in most materials. Processes such as rotting food, petroleum products and rust increase the speed at which the oxygen level is depleted..

- A space may also lack oxygen due to the presence of a dangerous gas, such as LPG (the onboard cooking gas), hydrogen (from batteries), exhaust fumes, refrigeration gas, methane (from sullage tanks or pipes) or solvents (for or in paints). List all the possible gasses on board and learn how to identify them.

- There is also the physical danger to consider when entering enclosed spaces: you may hurt yourself by falling off a damaged ladder, slipping or tripping over.

- THE RULES FOR ENTRY INTO CONFINED, CLOSED AND VOID SPACES are governed by the Occupational Health & Safety Act in each State and Commonwealth.

- The regulations define a confined space at a place of work as a space of any volume which a person may at any time enter or be allowed to enter and in which the atmosphere is liable to be contaminated at any time by dust, fumes, mist, vapour, gas or other harmful substances; or the atmosphere is liable to be oxygen deficient. The definition includes any shipboard spaces entered through a small hatchway or manhole, cargo tanks, cellular double bottom tanks, duct keel, cofferdams, ballast, sewage and oil tanks and void spaces (but not cargo holds).

Chapter 19.4: OH&S - Work Practices

- Prior to entry, the atmosphere of the confined space must be tested, and should include testing for flammables, oxygen deficiency and toxic gases. Liquids should be pumped out of the spaces before testing for toxic gases. The toxic gases that may be present can include hydrogen sulphide, carbon monoxide, carbon dioxide and others specific to individual workplaces.
- It is not sufficient that the atmosphere is free from toxic gases before entry. Everybody associated with the entry must have confidence, gained from local knowledge and experience, that the atmosphere will not change while work is carried out in the space.
- CONTINUOUS MONITORING: If there is any reason to suspect that the person may be accidentally injured or the atmosphere may alter then the space should be continuously monitored while it is occupied by a worker. This can be done by a watchkeeper standing by at the entry to the space and the worker entering the space wearing a continuous monitoring device. The device should be capable of sounding an alarm if dangerous gas levels or oxygen deficiency develop. Continuous monitoring devices are readily available. If there is strong indication that the atmosphere may change suddenly, all people entering the space must wear an air-supplied respirator.
- If the initial testing shows that the space is contaminated, the atmosphere must be purged by blowing air through the space or extracting toxic gases with a suitable exhaust system; or a combination of blowing and exhausting. Care should be taken to prevent people outside the confined space being exposed to gas while the atmosphere is being decontaminated. Alternatively, people can enter the space if they wear an approved air-supplied respirator, provided there are no flammable gases present.
- TRAINING, RESCUE AND FIRST AID: Where applicable, training of selected people is essential in accordance with the Australian Standard AS 2865. Rescue and first aid procedures should be planned, established and regularly rehearsed. Untrained, unequipped and unprepared people must not enter a confined space to rescue a person because multiple fatalities have occurred in this way.
- The AMSA Marine Notice on entry into confined spaces reads as follows: The failure to adopt a systematic and careful approach to confined space entry can result in injury and sometimes death. Multiple fatalities can occur in confined spaces especially when rescue teams fail to assess the risk... Identification, assessment and controlling the risks associated with confined space are important steps in minimising risks that can occur during confined space entry. Procedures for confined space entry should be reviewed periodically and crewmembers should be appropriately trained.

DOCUMENTING HAZARDS, RISKS & OPTIONS FOR CONTROLLING RISK

The following tables provide a convenient format for identifying hazards and analysing risks and the options for controlling risks. When using these tables, decisions need to be made whether a risk is acceptable or unacceptable. Guidance on what constitutes an acceptable or unacceptable risk is provided in the NSCV Part B Annex C.

REF	THE HAZARD	THE CONSEQUENCES OF A HAZARD HAPPENING WITHOUT CONTROLS		EXISTING CONTROLS	THE CONSEQUENCES OF A HAZARD HAPPENING WITH EXISTING CONTROLS		LEVEL OF RISK	RISK PRIORITY
	What can happen and how can it happen	Likelihood Rating	Consequence Rating		Likelihood Rating	Consequence Rating		

Fig 19.59 Risk Register (*NSCV Part B*)

REF	RISK	PRIORITY	POSSIBLE TREATMENT OPTIONS	PREFERRED OPTION	RISK RATING AFTER TREATMENT	COST/BENEFIT ANALYSIS RESULT
			1. 2. 3.			

Fig 19.60: RISK TREATMENT OPTIONS *(NSCV PART B)*

ON BOARD COMMUNICATION

- To communicate is to make known; to reveal thoughts or feelings.
- The joy of boating is in being involved and in having the ability to make correct decisions. This is possible only if the crew understand the responsibility weighing on the skipper's shoulders.
- Before heading off to do a job find out what to do.
- The problems of lack of nautical terminology can be overcome by using hand signals.
- To hear each other above the wind on deck, speak directly to each other. To hear each other in a cabin, speak quietly.
- Know the marine radio procedures and practice using the radio. Know the international phonetic alphabets. Do not use the calling channel for chitchat.
- Listen to weather broadcasts carefully.
- Learn to laugh at yourself and end each day on a happy note.

INTERNATIONAL SAFETY MANAGEMENT (ISM) CODE

(See IMO & ISM website addresses on the last page of this book)

In the late 1980s, a number of very serious maritime accidents were identified as the cause of human errors and management failure. Lord Justice Sheen in his inquiry into the loss of the **Herald of Free Enterprise** famously described the management failures as "the disease of sloppiness".

In 1993, the International Maritime Organisation (IMO) adopted the International Management Code for the Safe Operation of Ships and for Pollution Prevention (the ISM Code). Its purpose is to provide those responsible for the operation of ships with a framework for the proper development, implementation and assessment of safety and pollution prevention management within the industry as a whole. Its objective is to ensure safety, to prevent human injury or loss of life, and to avoid damage to the environment.

In 1998, the ISM Code became mandatory under SOLAS *(Safety of Life at Sea Convention)* for vessels and mobile offshore drilling units of 500 gross tonnes and above.

The Code establishes safety-management objectives and requires a SAFETY MANAGEMENT SYSTEM (SMS) to be established by "the Company", which is defined as the shipowner or any person, such as the manager or bareboat charterer, who has assumed responsibility for operating the ship. The Company is then required to establish and implement a policy for achieving these objectives. This includes providing the necessary resources and shore-based support.

Every company is expected "to designate a person or persons ashore having direct access to the highest level of management".

The procedures required by the Code should be documented and compiled in a Safety Management Manual, a copy of which should be kept on board.

Chapter 19.4: OH&S - Work Practices

INTERNATIONAL MARITIME SAFETY SYMBOLS
(Including symbols for Dangerous/Hazardous Goods)

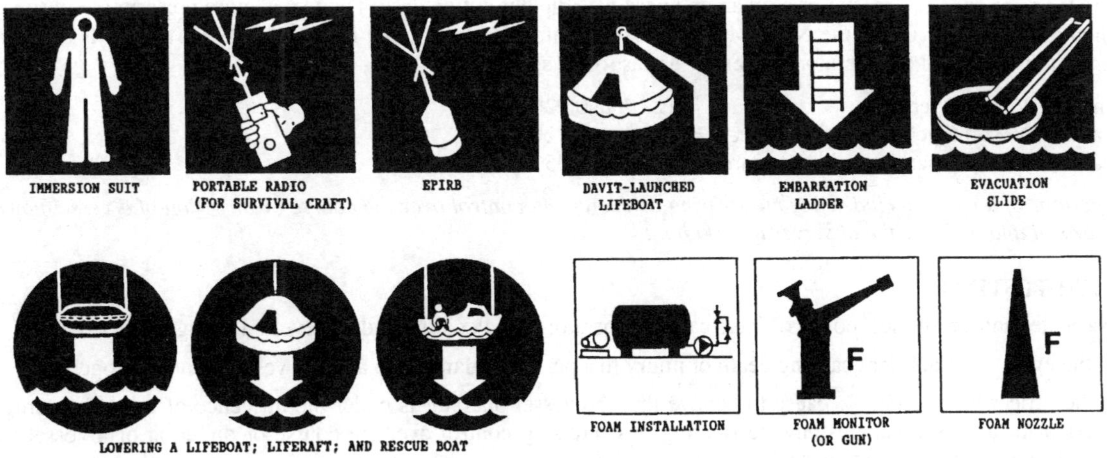

IMMERSION SUIT   PORTABLE RADIO (FOR SURVIVAL CRAFT)   EPIRB   DAVIT-LAUNCHED LIFEBOAT   EMBARKATION LADDER   EVACUATION SLIDE

LOWERING A LIFEBOAT; LIFERAFT; AND RESCUE BOAT   FOAM INSTALLATION   FOAM MONITOR (OR GUN)   FOAM NOZZLE

I.M.O. SAFETY SYMBOLS
(IMO: International Maritime Organisation)

EXPLOSIVE 1 (orange)   FLAMMABLE GAS 2 (red)   FLAMMABLE LIQUID 3 (red)   FLAMMABLE SOLID 4 (red/white)   OXIDIZING AGENT 5.1 (yellow)

POISON 6 (white)   RADIOACTIVE 7 (yellow/white/red)   CORROSIVE 8 (white)   9 (white/black/white)

I.M.D.G. CODE LABELS
(IMDG: International Maritime Dangerous Goods)
Nos. 1 to 9 indicate nine classes of dangerous or hazardous substances.
Class 9 is the Miscellaneous Dangerous Substances.

Symbol for Harmful Substances   MARINE POLLUTANT

Fig 19.61

ALCOHOL CONSUMPTION – THE LAW

It is prohibited to operate a vessel under the influence of alcohol or drugs, including prescription drugs. This law applies to all vessels in the Australian territorial seas and to all Australian vessels worldwide. Under the Commonwealth Navigation Act 1912, the specified limit of blood alcohol content is 0.04% in the case of a master or seaman while on duty, and 0.08% in the case of a master or seaman on board the ship but not on duty. The law is more prescriptive in some States and should be checked. Under the NSW Marine (Boating Safety - Alcohol and Drugs) Act 1991, the limits of permissible concentration of alcohol for operators of vessels are as follows:

Commercial Vessel Operators = less than 0.02%
Recreational Vessel Operators under 18 years = less than 0.02%
Recreational Vessel Operators over 18 years = less than 0.05%

The operator of a vessel includes anyone steering or exercising control over the course or direction of a vessel powered by motor or sail and includes the observer in a ski boat.

BREATH TESTING:

In most States and Territories, police officers can test operators for alcohol and drug levels. Testing is carried out:

- In the event of an accident causing death or injury to a person or damage to another vessel or other property.
- If the officer has reasonable cause to believe that the vessel operator is under the influence of alcohol or drugs. [The operator of a vessel includes anyone steering or exercising control over the course or direction of a vessel and may include a waterskier or aquaplaner].

DEALING WITH A DRUNK & DISORDERLY PASSENGER ON BOARD

The following is an extract of the Commonwealth Navigation Act, 1912:

- It is the duty of the owner, operator and master to ensure that the passengers are in a safe and seaworthy vessel.
- The master of a vessel may refuse to receive on board any person who, by reason of drunkenness or otherwise, is in such a state, or misconducts himself/herself in such a manner, as to cause annoyance or injury to passengers on board. If such a person is already on board, the master may, after tendering to the person the amount of his/her fare (if paid) less a proper deduction in respect of the distance conveyed (if any), put him/her ashore at a suitable place.
- The master of a vessel may refuse to receive on board any person who appears to be suffering from any disease likely to endanger the health or safety of those on board.
- If it is necessary to restrain a drunk, dangerous or disorderly person on board, only the absolute necessary (minimum) force must be used.
- Any complaints from passengers regarding safety or unseaworthiness of the vessel must be investigated, logged and acted upon.

THE PRIVACY ACT (Privacy And Personal Information Protection Act, 1998)

Members of public supply personal information to boat operators and other maritime organisations in a variety of circumstances. These include boat hire, vessel registration, boat licence and mooring application.

Under the Privacy Act, organisations holding such information must maintain a Privacy Management Plan. It is to ensure that:

- The personal information is only collected for a lawful purpose that is directly related to the function or activity of the organisation.
- The personal information is securely stored.
- The individual concerned has access to their personal information, and to correct and update any details.
- The personal information is only disclosed to authorised bodies.

Unless permitted otherwise by the individual, the (maritime) organisation may only disclose such information to the agencies or organisations with statutory rights. These include Police, Taxation Office and Immigration authorities.

CHAPTER 19.4: QUESTIONS & ANSWERS

(Qns. 1-56 are in Ch 19.1, 57-145 in Ch. 19.2 & 146-297 in Ch. 19.3)

OH&S – WORK PRACTICES

298. The factors that cause the air in confined spaces to lose its oxygen content include:

    (i) chemical changes in materials.
    (ii) rotting food.
    (iii) stagnation of air itself.
    (iv) rust.

    The correct answer is:
    (a) all except (i).
    (b) all except (ii).
    (c) all except (iii).
    (d) all except (iv).

299. You are the standby person for someone who has gone inside a confined space. The person inside has fainted. You should:

    (a) go in and help the person out.
    (b) go in and give the person CPR.
    (c) raise an alarm.
    (d) wear a breathing apparatus and go in to help

300. The Australian OH&S legislation is based on the concept of:

    (a) duty of care to the neighbour.
    (b) safety from physical harm.
    (c) management's responsibilities.
    (d) employees' responsibilities.

301. The OH&S legislation requires matters of health and safety at workplaces to be managed by:

    (a) managers.
    (b) employees.
    (c) Workplaces.
    (d) Workcover inspectors.

302. The management of health & safety at workplace does not require:

    (a) identification of hazards.
    (b) assessment of risks.
    (c) actions to control risks.
    (d) monthly meetings.

303. On a charter vessel, prominently positioned placards read: "We are not responsible for any loss or injury." As a result of a passenger sustaining injury after slipping on a banana skin on deck, the above signage:

    (a) protects the vessel from legal liability.
    (b) has little or no legal value on its own.
    (c) fulfils the vessel's duty of care obligation.
    (d) protects the vessel from legal liability if the disclaimer is also printed on the passengers' tickets.

304. You are the ship's cook. On your way home from the vessel, you notice that a ship's painter has left a tin of paint in the middle of an alleyway and gone home for the night. You should:

    (a) not move it in case the painter has left it there for a reason.
    (b) ignore it because it is not your problem.
    (c) move it to where no one is likely to trip over it.
    (d) assume that others will step over it just you did.

305. There is a need to take extra safety and health precautions on vessels. But the following could make us over-confident and ignore our responsibility:

    (a) Vessels are rolling & pitching platforms to live & work on.
    (b) The chances of dealing with life threatening situations are high at sea.
    (c) Medical and rescue services are not easily available at sea.
    (d) Every crewmember has a First Aid Certificate.

306. Which of the following factors is a not part of developing an OH&S Risk Management Plan?

    (a) Assessment of risk to crew.
    (b) Carrying out risk assessment for every shipboard operation.
    (c) Identifying crewmembers who may be particularly at risk in performing their tasks.
    (d) Discharging any crewmember with a physical disability.

307. In relation to OH&S, crewmembers must take reasonable care for the safety of:

    (a) themselves.
    (b) others on board
    (c) themselves and others on board.
    (d) passengers.

308. Which of the following actions is essential to ensure safety in the galley?

    (a) Secure knives and sharp implements against rolling & pitching.
    (b) Wear only shorts when deep-frying.
    (c) Do not use burners on gimbals.
    (d) Stow flammable liquids, bleaches and disinfectants in one common cupboard.

309. Emergency lighting, the means of opening doors from both sides and alarm systems are required in vessels in:

    (a) all spaces.
    (b) enclosed spaces.
    (c) spaces where a person may become entrapped and exposed to increased risk due to lack of oxygen.
    (d) spaces where a person may become entrapped and exposed to increased risk due to heat, cold or lack of oxygen.

310. In relation to confined spaces, which of the following statements is incorrect?

    (a) The atmosphere of the space must be tested prior to entry of a worker.
    (b) The test for toxic gases must be carried out prior to pumping out any liquid in the space.
    (c) The atmosphere of the space must be continuously monitored while it is occupied by a worker.
    (d) Everybody associated with the entry must have confidence gained from local knowledge & experience.

311. The OH&S requirements are set out in the:

    (a) USL Code.
    (b) NSCV Code.
    (c) State Marine Acts.
    (d) Workplace Health & Safety legislation and the applicable Commonwealth or State Marine Act.

312. The OH&S regulations require employers to:

    (a) provide safe systems of work.
    (b) appoint one safety officer per site.
    (c) hold safety meetings once a month.
    (d) appoint a Safety Officer for every 10 employees.

313. The OH&S regulations require employers to ensure that:

    (a) all employees have a First Aid Certificate.
    (b) all employees have a Senior First Aid Certificate.
    (c) arrangements are in place for the safe handling of equipment and substances.
    (d) weekly safety meetings are held.

314. The OH&S regulations require all employees to:

    (a) provide their own safety clothing.
    (b) cooperate with the employer in matters of OH&S.
    (c) maintain the currency of their First Aid Certificates.
    (d) Attend at least 50% of the safety meetings.

315. Under OH&S legislation, entering an empty fuel tank, freezer compartment, store room or a cargo tank is to enter a:

    (a) restricted space.
    (b) void space.
    (c) dangerous space.
    (d) confined space.

316. For occupations requiring safety shoes, such shoes should be worn:

    (a) only when working on a task.
    (b) to & from workplace & at workplace.
    (c) only in the machinery spaces.
    (d) at all times.

317. If feeling cold in your uniform, the most obvious protection against cold is to:

    (a) wear more clothes only if they are part of the uniform.
    (b) refuse to work.
    (c) work only below decks.
    (d) wear more clothes.

318. When asked to work in a noisy area of the vessel, the most obvious protection is:

    (a) allow your ears to acclimatise to the noise.
    (b) request permission to work elsewhere.
    (c) wear ear protection.
    (d) take regular breaks.

319. When asked to work in the sun, the most obvious protection is to:

    (a) wear the necessary sun protection.
    (b) use the opportunity to get a tan.
    (c) ask for safety clothing to be supplied.
    (d) Stay close to the water to reduce sunburn.

320. Prior to commencing work in the vicinity of radar, Inmarsat and radio antennas:

    (a) take the antennas down.
    (b) switch off the generator.
    (c) switch off the power supply.
    (d) advise the master and post notices on the switches to ensure no one will switch on the equipment.

Chapter 19.4: OH&S - Work Practices

321. Being close to an operating radar or Inmarsat dish antenna exposes the person to:

    (a) fire hazard.
    (b) radiation hazard.
    (c) blockage of signals.
    (d) creation of blind sectors.

322. In the absence of a crane or derrick, transferring heavy weights from one deck to another is safely done by:

    (a) carrying them on the shoulders.
    (b) balancing them in two hands.
    (c) Rigging a tackle.
    (d) Carrying them in front of the body.

323. Which of the following statements is incorrect? To communicate is to:

    (a) get the job done.
    (b) make known.
    (c) reveal thoughts
    (d) reveal feelings

324. Which of the following statements is incorrect? The joy or boating or the job satisfaction of working on a vessel lies in:

    (a) being involved.
    (b) having little or no responsibility.
    (c) having the ability to make correct decisions.
    (d) understanding master's responsibility.

325. In relation to on board communication, which of the following statements is incorrect?

    (a) The problem of lack of nautical terminology can be overcome by using hand signals.
    (b) Learn to laugh at yourself.
    (c) Speak as loud as you can whether or not it is necessary to do so.
    (d) Know the international phonetic alphabets.

326. Under OH&S legislation, an employee's obligation towards safety is:

    (a) continuous.
    (b) limited to working under supervision.
    (c) limited to job at hand
    (d) limited to watchkeeping hours.

327. Under OH&S legislation, injuries at work are to be dealt with as follows:

    (a) Report only the serious injuries.
    (b) Report & log all injuries.
    (c) Log only the serious injuries.
    (d) Report & log only the injuries attended by a doctor.

328. Wearing loose gear on board may be a hazard under the following circumstances:

    (a) only in rough weather.
    (b) only below decks.
    (c) only above decks.
    (d) at any time.

329. Under OH&S legislation, a crewmember must not use a tool until:

    (a) directed by the owner.
    (b) directed by the master
    (c) properly instructed in its use.
    (d) provided with written instructions

330. The hood above the cooking range in the galley is a fire hazard because it:

    (a) collects grease in the filter.
    (b) collects grease in the ducting
    (c) collects grease in the filter & ducting.
    (d) does not allow the heat to escape.

331. On discovering a leak in a fuel line, you first action would be to:

    (a) shut off fuel supply to the line.
    (b) pump out the bilge.
    (c) disconnect the line.
    (d) wrap a sealing tape around the leak.

332. The OH&S legislation:

    (a) replaces the Commonwealth & State Marine Acts.
    (b) complements the Commonwealth & State Marine Acts.
    (c) replaces only the State Marine Acts
    (d) replaces only the Commonwealth Marine Act.

333. The chain of command for reporting cases of personal injuries ends with the:

    (a) Master.
    (b) Owner
    (c) Workcover Authority
    (d) Duty officer

334. Onboard welding activity in port does not requires:

    (a) readiness of fire fighting appliances.
    (b) port authority's permission.
    (c) vessel to be off the wharf.
    (d) gas freeing certificate.

335. In relation to the Privacy Act, which of the following statements is incorrect?

   (a) An organisation may disclose the personal information supplied by an individual to the Australian Taxation Office.
   (b) An individual is entitled to alter details of his/her personal information held on an organisation's computer.
   (c) An organisation holding personal information of individuals must have a Privacy Management Plan.
   (d) Personal information supplied to a charter boat operator in relation to boat hire may be disclosed to the Boating Industry Association.

336. The main purpose of emergency drills is to ensure that:

   (a) the crew understand the use of equipment & procedures.
   (b) the passengers can be evacuated safely & quickly.
   (c) survey requirements are met.
   (d) the master becomes aware of any deficiencies.

337. Stress causes the following change(s) in the human body:

   (a) Physical
   (b) Chemical
   (c) Emotional
   (d) All three changes listed here.

338. Which of the following factors is beyond the control of master or vessel owner in reducing the level of stress among the crew?

   (a) Suitability of working gear
   (b) A person's family circumstances
   (c) Working conditions
   (d) Exertion levels of the crew

339. The following factor is beyond the influence of a master or vessel owner in reducing the stress level of a crewmember?

   (a) Crewmember's nautical knowledge
   (b) Crewmember's length of experience
   (c) Crewmember's fitness
   (d) Crewmember's diet

340. Which of the following statements is incorrect? A crewmember's stress level is influenced by his/her:

   (a) gender.
   (b) perceived degree of crew support.
   (c) peer pressure.
   (d) perceptual narrowing.

341. In relation to stress management, which of the following statements is correct?

   (a) Stress reduction is entirely in the hands of the individual concerned.
   (b) Stress management should commence immediately after embarkation.
   (c) Stress is reducible, if not preventable.
   (d) Discussing potential stressors before they arise causes unnecessary worries.

342. Which of the following statements is correct?

   (a) What we eat or drink has no bearing on the level of stress we experience.
   (b) Drinking coffee reduces stress.
   (c) Drinking alcohol reduces stress.
   (d) Keeping stomach empty is the best remedy for seasickness.

343. Which of the following statements is correct?

   (a) Smoking tobacco or cannabis reduces anxiety.
   (b) Cough-and-cold medicines do not alter a person's level of stress.
   (c) Drug usage increases crewmembers' ability to participate.
   (d) Drug usage reduces the oxygen available to the body.

344. The hazardous goods are listed in the International Dangerous Goods Code in the following number of classes:

   (a) six
   (b) twelve
   (c) nine
   (d) eighteen

CHAPTER 19.4: ANSWERS

(Answers 1-56 are in Ch. 19.1, 57-145 in Ch. 19.2 & 146-297 in Ch. 19.3)

298 (c), 299 (c), 300 (a), 301 (c), 302 (d),
303 (b), 304 (c), 305 (d), 306 (d), 307 (c),
308 (a), 309 (d), 310 (b), 311 (d), 312 (a),
313 (c), 314 (b), 315 (d), 316 (b), 317 (d),
318 (c), 319 (a), 320 (d), 321 (b), 322 (c),
323 (a), 324 (b), 325 (c), 326 (a), 327 (b),
328 (d), 329 (c), 330 (c), 331 (a), 332 (b),
333 (c), 334 (c), 335 (d), 336 (a), 337 (d),
338 (b), 339 (b), 340 (a), 341 (c), 342 (d),
343 (d), 344 (c)

Chapter 20

METEOROLOGY
(Weather)

BEAUFORT WIND SCALE	
	Mean wave height in metres
BEAUFORT NUMBER....0	CALM 0-1 KTS Sea like a mirror.
....1	LIGHT AIR 1-3 KTS. 0.1 WAVES Ripples with appearance of scales; no foam crests
....2	LIGHT BREEZE 4-6 KTS. 0.2 WAVES Small wavelets; crests of glassy appearance, not breaking.
....3	GENTLE BREEZE 7-10 KTS. 0.6 WAVES Large wavelets; crests begin to break; scattered whitecaps.
....4	MODERATE BREEZE 11-16 KTS. 1.0 WAVES Small waves, becoming longer; numerous whitecaps
....5	FRESH BREEZE 17-21 KTS. 2.0 WAVES Moderate waves, taking longer form, many whitecaps; some spray.

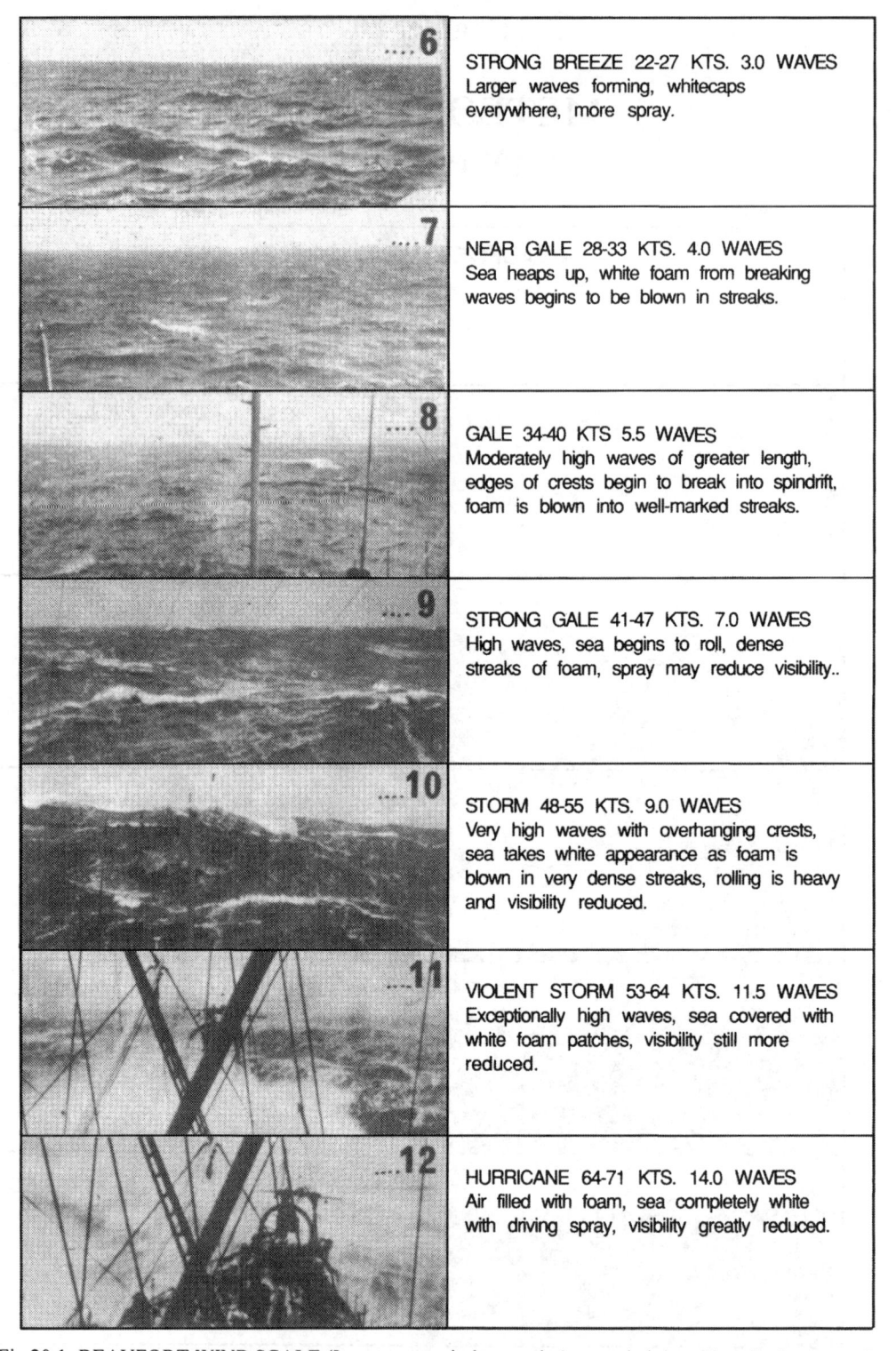

....6	**STRONG BREEZE 22-27 KTS. 3.0 WAVES** Larger waves forming, whitecaps everywhere, more spray.
....7	**NEAR GALE 28-33 KTS. 4.0 WAVES** Sea heaps up, white foam from breaking waves begins to be blown in streaks.
....8	**GALE 34-40 KTS 5.5 WAVES** Moderately high waves of greater length, edges of crests begin to break into spindrift, foam is blown into well-marked streaks.
....9	**STRONG GALE 41-47 KTS. 7.0 WAVES** High waves, sea begins to roll, dense streaks of foam, spray may reduce visibility..
....10	**STORM 48-55 KTS. 9.0 WAVES** Very high waves with overhanging crests, sea takes white appearance as foam is blown in very dense streaks, rolling is heavy and visibility reduced.
....11	**VIOLENT STORM 53-64 KTS. 11.5 WAVES** Exceptionally high waves, sea covered with white foam patches, visibility still more reduced.
....12	**HURRICANE 64-71 KTS. 14.0 WAVES** Air filled with foam, sea completely white with driving spray, visibility greatly reduced.

Fig 20.1: BEAUFORT WIND SCALE (It expresses wind strength & wave height with a single numeral)

You don't have to memorise each Beaufort wind scale. Get a general idea. Looking at the scale, you will notice that:
- Zero is for calm
- Scale 2-6 is for breezes
- 7-9 is for gales
- 10-12 is storms and hurricane.

Chapter 20: Meteorology

Furthermore, the Beaufort scale can be converted into average wind speed in knots as follows:
- Up to Force 8, multiply the scale by 4
- Above force 8 multiply it by 5

For example, Beaufort Scale 6 = a wind speed of 6 x 4 = 24 knots. Also, don't confuse wind speed in knots (nautical miles per hour) with kilometres per hour. 1 knot = approx. 1.8 km/h. (It is roughly double) Therefore, Force 8 = gale force winds = wind speed 40 knots or 80 km/h.

AIR MASSES, ATMOSPHERE, CLIMATE ZONES, PRESSURE BELTS & TRADE WINDS

The earth's atmosphere is an ocean of air. Technically the atmosphere is said to extend out to 100 kilometres from the earth's surface. However, on average, the ocean of air is only about 8 kilometres deep over the poles and 16 kilometres deep over the equator. It is known as the troposphere, and its upper boundary is called the tropopause. Above it is the stratosphere. We live at the bottom of the atmosphere. The polar air is cold and dense. Its sinking character exerts high atmospheric pressure. The tropical air is warm, light and rising. The atmospheric pressure in the tropics is therefore low. The atmospheric pressure also falls with height, from about 1000 hPa at the surface of the earth to about 200 hPa at the tropopause.

Warm tropical air rises all the way to the top of the troposphere and then moves off towards the poles. Some of it sinks along the way to create different pressure zones. Cold air at the poles sinks and moves down towards the equator. These cycles of rising and falling air and, due to earth's rotation, the swirling movement of air masses of various densities, temperatures and humidity give shape to a variety of climates on earth.

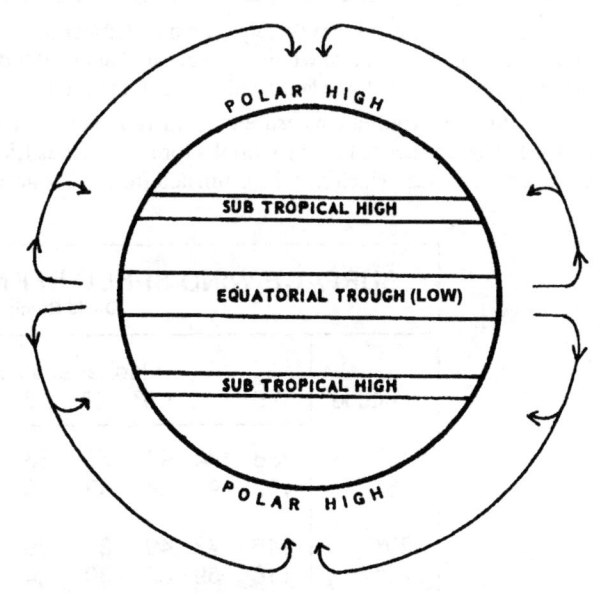

Fig 20.2: ATMOSPHERIC WIND CIRCULATION

The sun does not warm the earth uniformly. The wind too is obstructed from blowing uniformly by the mountains. The atmosphere thus remains in constant turbulent motion. Air masses (bodies of air) measuring thousands of kilometres across are heated or cooled as they pass over land or oceans. The air masses that form over the oceans are called "maritime", and those that form over the land are called "continental". Thus an air mass can be polar maritime, tropical maritime, polar continental or tropical continental.

The prevailing wind directions, called Trade Winds since the days of the old sailing ships, can best be understood by looking at the globe in bands of alternating high and low pressure belts or climate zones. The winds blow from high-pressure belts to low-pressure belts. If the earth did not spin, the trade winds would blow in the north or south direction. However, the spin of the earth makes the wind appear deflected by up to 45 degrees from the north south axis. Imagine firing a canon from the South Pole towards Sydney. By the time the canon reaches where Sydney was supposed to have been, the earth has spun around by a few degrees and Sydney has moved to the right. The canon lands somewhere in Western New South Wales, appearing to have deflected to the left. Had we fired it somewhere in the Northern hemisphere it would have appeared to deflect to the right. This is because if a globe were seen to be spinning anti-clockwise when viewed from the top, it would appear to spin in the opposite direction (clockwise) when viewed from the

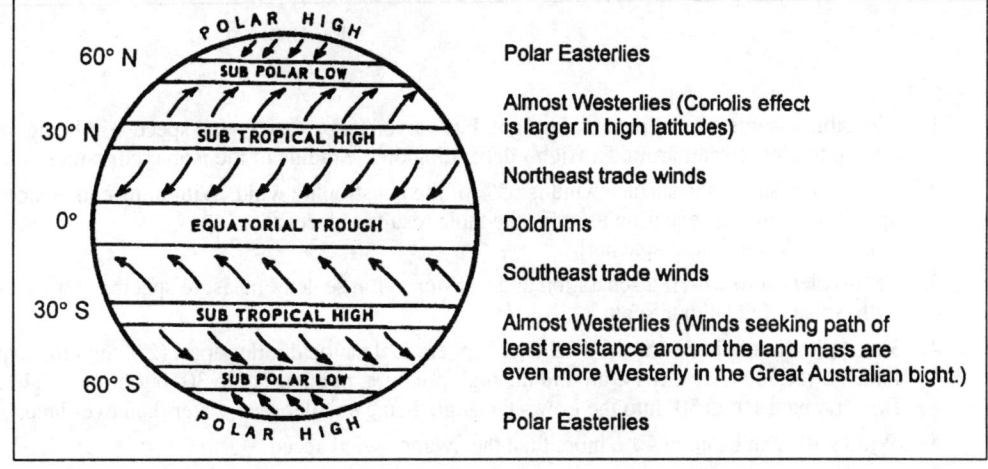

- Polar Easterlies
- Almost Westerlies (Coriolis effect is larger in high latitudes)
- Northeast trade winds
- Doldrums
- Southeast trade winds
- Almost Westerlies (Winds seeking path of least resistance around the land mass are even more Westerly in the Great Australian bight.)
- Polar Easterlies

Fig 20.3: TRADE WINDS

bottom. This deflection is known as the **CORIOLIS EFFECT**, which is zero at the equator and maximum at the poles. It is to the right in the northern hemisphere and to the left in the southern hemisphere. The wind is also slowed down by the earth's friction. The resultant spiralling movement of wind in the southern hemisphere is clockwise into low-pressure centres and anti-clockwise out of high-pressure centres. It is in reverse in the northern hemisphere.

In the short flight of a football or a cricket ball, the Coriolis effect is insignificant. However, if aircraft pilots did not make navigational corrections for the Coriolis effect, they would land at wrong airports; and uncorrected artillery shells would destroy wrong targets.

GEOSTROPHIC & SURFACE WINDS

Geostrophic winds are largely driven by temperature differences, i.e., pressure differences. Using weather balloons and such implements, these winds are measured at an altitude of about 1 kilometer above the ground, i.e., outside the effects of the Coriolis force and friction due to the ground.

Surface winds are measured within 100 metres above the ground (10 metres above the sea surface). Their direction and speed are influenced by the Coriolis force as well as friction due to the earth's surface. Mariners are mostly concerned with surface winds, which may be estimated from the following table when used with a weather map.

SURFACE WIND SPEED IN KNOTS FROM 4-HPa ISOBAR SPACING
(David Burch, STARPATH, USA)

Latitude	0.5°	1°	1.5°	2°	2.5°	3°	3.5°	4°	4.5°	5°	6°	8°	10°	12°
10°	288	144	96	72	58	48	41	36	32	29	24	18	14	12
15°	193	97	64	48	39	32	28	24	21	19	16	12	10	8
20°	146	73	49	37	29	24	21	18	16	15	12	9	7	6
25°	118	59	39	30	24	20	17	15	13	12	10	7	6	5
30°	100	50	33	25	20	17	14	12	11	10	8	6	5	4
35°	87	44	29	22	17	15	12	11	10	9	7	5	4	4
40°	78	39	26	19	16	13	11	10	9	8	6	5	4	3
45°	71	35	24	18	14	12	10	9	8	7	6	4	4	3
50°	65	33	22	16	13	11	9	8	7	7	5	4	3	3
55°	61	31	20	15	12	10	9	8	7	6	5	4	3	3
60°	58	29	19	14	12	10	8	7	6	6	5	4	3	2
65°	55	28	18	14	11	9	8	7	6	6	5	3	3	2
70°	53	27	18	13	11	9	8	7	6	5	4	3	3	2
75°	52	26	17	13	10	9	7	6	6	5	4	3	3	2

Isobar spacing in degrees of latitude

Fig 20.4

NOTES:
1. The table assumes the isobars are straight. For curved isobars, the wind speed will be up to 30% less around a Low and up to 30% greater around a High - depending on the radius of the isobar curvature.
2. The table assumes the surface wind is 65% of the geostrophic wind. With units conversions and numerical approximations, the equation used for the table results reduces to:
   Winds = 25/[sin (Lat) x spacing].
3. Set dividers across 4-HPa separation and transfer to Lat scale to measure spacing from a weather map, showing both isobars and latitude scale.
4. **Example**: If at latitude 45° the 4-HPa isobars are 120 nautical miles apart (2°), then the expected surface wind is 18 knots, directed anticlockwise around the high pressure, pointed 10° to 30° out of the High, or clockwise around a Low, pointed 10° to 30° into the Low - the angle being smaller over water than over land.
5. Wind gusts can be up to 40% more than the average wind speed.

THE AUSTRALIAN CLIMATE

The southeasterly trade winds over Australia are the result of air moving from the "calm sinking air" sub-tropical high-pressure belt to the "thundery rising air" equatorial low-pressure belt.

Countries in the calm sinking-air high-pressure belt (such as Australia) generally enjoy better climate than those in the changeable rising-air low-pressure belt (such as northern Europe). The sinking air generally creates dry and settled weather in the high-pressure belts. It tends to compress and evaporate any clouds. In the low-pressure belts, on the other hand, the rising air cools in the upper atmosphere and condenses into clouds, resulting in disturbed weather conditions. The bad weather is further fuelled by air masses of varying characteristics moving in from the adjacent high-pressure belts to replace the rising air.

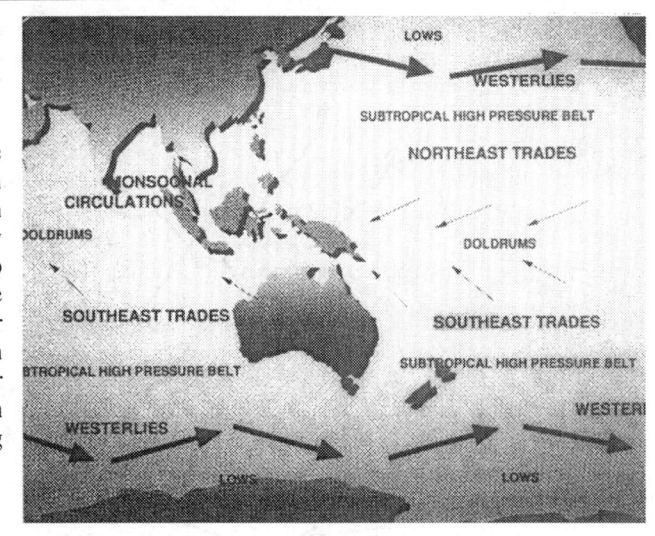

Fig 20.5

The equatorial trough or the Inter-tropical Confluence (ITC), also known as the Inter-tropical Convergence Zone (ITCZ) - which is like the weather equator - moves northwards or southwards with the sun during the year. Thus the summer months in Australia (particularly the tropical part) experience a lower atmospheric pressure than the winter months.

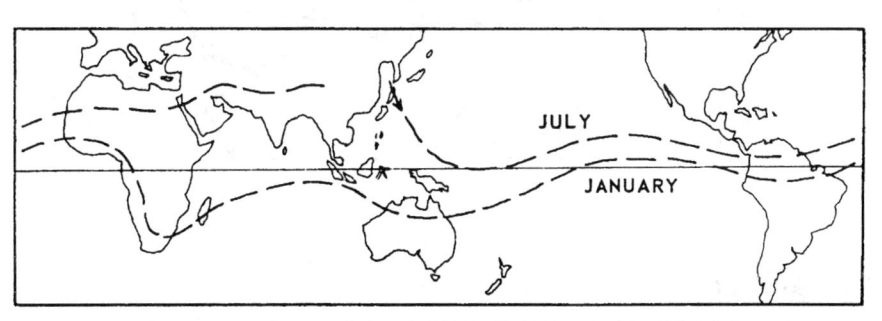

Fig 20.6: THE LOCATION OF ITC IN WINTER AND SUMMER

The movement of ITC causes the global pressure belts to move with it, which results in the difference in the winter and summer winds. In our summer the equatorial low pressure belt moves southwards over Australia, which moves the sub-tropical high pressure belt from overland Australia to the south of the continent. The Southeast Trade winds too move about 10° southwards and can be experienced as low as in 30°S latitude. The northern Australia experiences cyclone and monsoon season

In our winter the opposite happens. The sub-tropical high-pressure belt moves back over the continent, where it is reinforced by the cooler land mass. The changeable "rising air" low-pressure belt also moves up to the south of the continent, where it gives rise to a series of lows and strong westerlies.

The above seasonal changes must not be confused with the local day-to-day weather patterns such as sea and land breezes and thunderstorms.

Fig 20.7 AUSTRALIAN SUMMER (OCTOBER – MARCH)

Fig 20.8
AUSTRALIAN WINTER
(ARIL – SEPTEMBER)

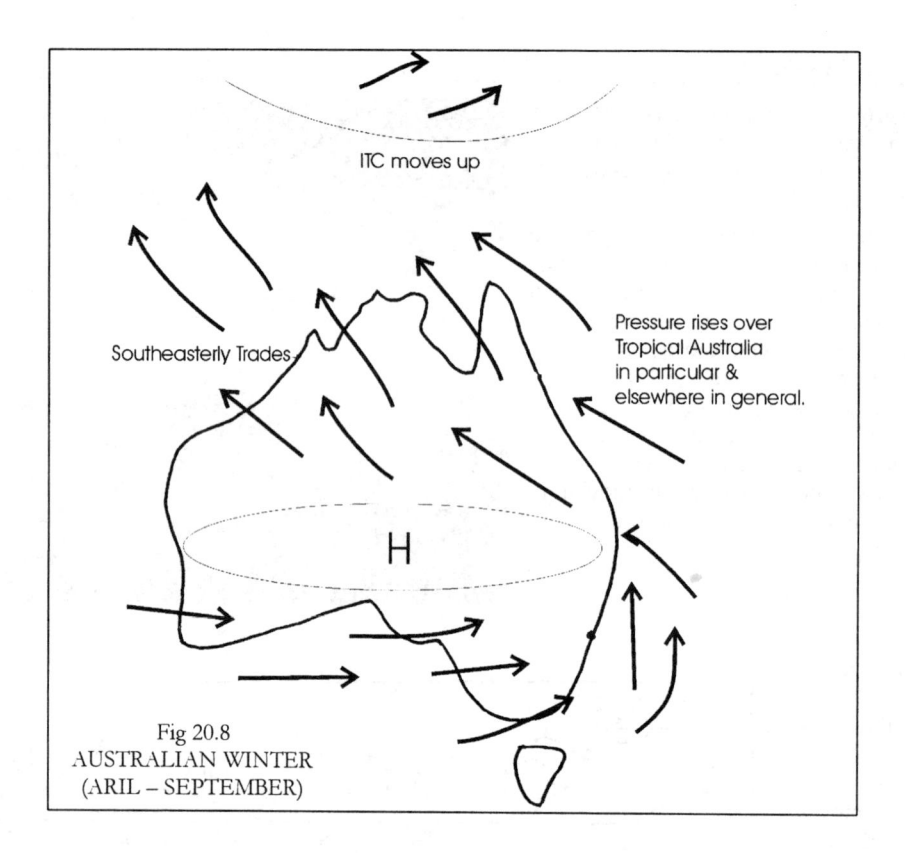

BAROMETER

Barometer measures atmospheric pressure, which is the force exerted by the earth's atmosphere. It can be thought of as the weight of air.

By indicating the atmospheric pressure, the rise or fall in it and the amount and speed (rate) of such a change, a barometer helps us to forecast weather.

There are two types of mechanical barometers: mercury-filled and aneroid. The former is found only in the meteorological offices and laboratories. It is not as convenient an instrument as the aneroid barometer. In addition, there are electronic barometers, either in the form of wristwatches or 'stand alone' instruments. A pressure sensing electronic component inside these instruments measures the pressure. They are relatively cheap and usually more accurate than the aneroid barometers.

Inside an aneroid barometer is a closed sealed capsule with flexible sides (marked "A"). Any change in pressure alters the thickness of the capsule. A set of levers is fitted to magnify these changes, causing a pointer to move on a dial or numbers to change on a digital readout device.

The capsule must be kept exposed to the air pressure that it is designed to measure. To prevent the air passages on the back of the barometer becoming sealed with paint, you must remove the instrument before painting the bulkhead.

The barometric pressure is measured in hectoPascals (hPa). The term "millibar" is the old name for hectoPascal.

A barometer should be installed in a well-ventilated dry position away from the weather and direct sunlight. It should be securely fastened to a vertical bulkhead where it is safe from physical damage and easily readable.

An aneroid barometric reading is subject to two errors: index error (instrument error) and the height above the sea. The barometer correction for the index error, if any, can be checked by calling a Coast Radio Station (Maritime Coast Station). It doesn't matter if you are 20 or 30 miles away from the station.

A barometer can be corrected for the index error by turning a screw that is usually fitted at the back of the instrument. Small vessels usually don't need to worry about the correction for the height above the sea, unless the instrument is fitted on a high flying bridge. **Pressure falls 1 hPa for every 10 metres of height**. The mercury barometers also need to be corrected for latitude and temperature.

A barometer, which is uncorrected, is good enough to indicate a change in weather. The direction of change in pressure (its fall or rise) and the rate of change (amount of fall or rise in a period) are all indications of the expected weather.

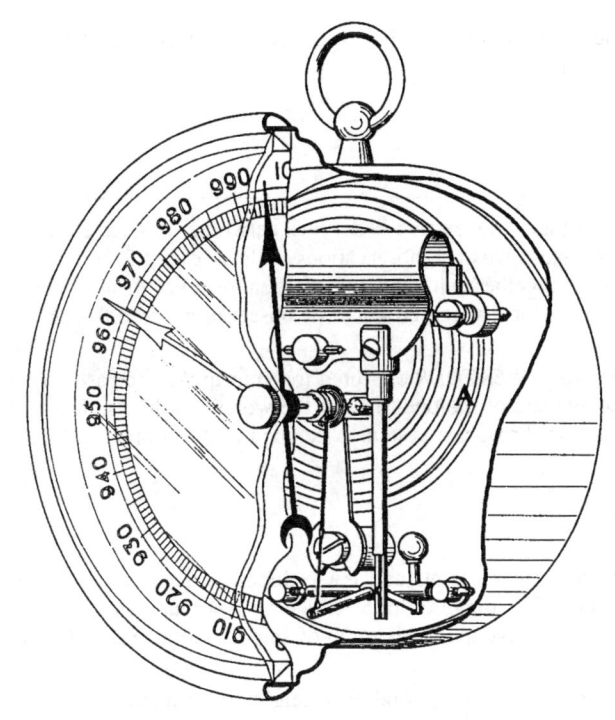

Fig 20.9: (Left): THE ANEROID BAROMETER
("A" is the closed sealed capsule)
(Handbook of Meteorological Instruments, Meteorological Office, London)

Fig 20.10 (Above): ELECTRONIC BAROMETER–CUM–HYGROMETER *(Dick Smith)*
(Battery-operated display of temperature, humidity, barometric pressure, barographic record & time)

A ROUGH GUIDE TO FALL IN PRESSURE & WIND STRENGTH
(Corrected for diurnal variation, mentioned below)

3 hPa fall in 3 hours	Strong to gale-force winds within 6 to 12 hours
6 hPa fall in 3 hours	Gale force winds within next 6 hours
9 hPa fall in 3 hours	Gale to storm-force winds within 3 hours

Before reading a barometer, tap the glass lightly but firmly to make sure that linkages are not sticky. After reading it, turn the set hand by the knob at the centre of the glass so that it covers the reading hand. This will help you to see at your next reading whether the pressure has risen or fallen and by how much.

A good mariner enters the barometer readings in the logbook every 3 hours in normal weather and more frequently on indication of poorer weather. Some vessels carry a BAROGRAPH. In a mechanical barograph, a pen attached to a barometer traces changes in atmospheric pressure on a 7-day graph paper wrapped around a clockwork driven drum.

Many of the electronic barometers are also barographs. The atmospheric pressure history can be accessed at any time.

DIURNAL CHANGES IN BAROMETRIC PRESSURE

The word "diurnal" means twice a day. Unaffected by any change in weather, the barometer rises from 0400 to 1000 then falls until 1600. It then rises till 2200 when it once more falls until 0400. These times are approximate and are in Local Mean Time of the place.

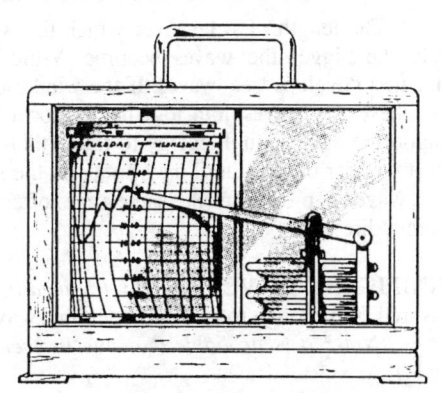

Fig 20.11: BAROGRAPH

These changes result from atmospheric pressure waves that sweep regularly around the earth from east to west. The effect is maximum at the equator where the variation is about 3 hPa. At latitude 51° the range is about 0.8 hPa. In higher latitudes it is inappreciable. To avoid mistaking the diurnal change for a change in weather, observations should be made over 24 hour periods.

Atmospheric pressures vary throughout the world and between seasons. It is therefore incorrect to print "fair", "rain" and "change" on a barometer's face based on one set of pressures.

WEATHER TERMINOLOGY & DEFINITIONS

WEATHER FORECASTS: Most national weather services and their cooperative observing networks of selected ships, aircraft, airports, weather balloons, drifting buoys and farmers make weather observations at least every 3 hours. The weather forecasts are complied from the 6-hourly observations, together with the information received from weather satellites and the historical data stored in the computer. The internationally agreed hours of observations are 0000, 0600, 1200 & 1800 UTC (GMT).

The reports are then produced to describe the *SIGNIFICANT WEATHER* expected to occur in the forecast period, which is not description of the worst weather condition but the average of the one-third of the worst conditions. The high seas weather reports are given in UTC. The worldwide radio weather broadcast system is detailed in the Admiralty List of Radio Signals, Volume 3.

SYNOPTIC CHART: Weather map. (Figures 20.33 & 20.34)

WIND DIRECTION & SPEED: Wind direction is given in 16 compass points and is the direction the wind is coming from. The direction of waves (not swell) is the true wind direction. The direction of a telltale on a shroud or of a flag or funnel smoke indicates the relative or apparent wind direction. It is relative to the direction of ship's motion and of the true wind.

Wind speed (in knots) refers to the average speed over a 10-minute period at a height of 10 metres above the surface. One knot = 1.85 km/h (roughly 1 knot = 2 km/h).

It is important to note that weather observations and forecasts provide only the average wind direction and speed. The wind can gust up to 40% more than the average wind speed.

GUST: Gust is a sudden increase in wind strength of very short duration, usually a few seconds. It can be up to 40% stronger than the average speed.

SQUALL: A squall is a sudden and often violent increase in wind speed that lasts for at least several minutes. It may include many gusts and a change in wind direction, usually blowing at twice the speed of the wind that will follow. It can come with rain or snow. A heavy rainsquall can flatten the sea, at times.

WAVES, FETCH & SWELL: Waves are created by wind passing over the water surface. Wave height is the vertical distance between the top of the crest and bottom of trough. The significant wave (or swell) height stated in a weather report or forecast is the average of the highest one-third of the waves (See below). Some waves will be higher and some lower than the significant height. Typically one in 2000 waves will be twice this height *(See NSCV definition below).*

Sometimes, wind waves and/or swell waves join to produce a very high wave. This is known as the KING, FREAK or ROGUE WAVE. Such a wave is much higher than the significant wave height.

The length of water over which the wind blows is known as the FETCH. The stronger the wind and longer the Fetch, the bigger the waves become. Wind blowing over water first develops 'sea' waves. If the wind blows for a long time, these sea waves join together to form 'swell' waves, which often have enough energy to travel thousands of miles over a number of days until they break on the shore. Sea and Swell waves can occur at the same time, often coming from different directions.

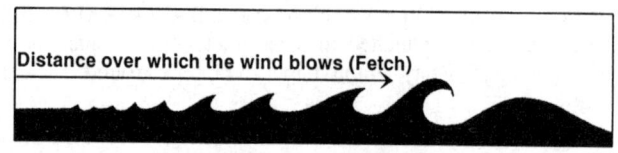

Figure 20.12: FETCH

SIGNIFICANT WAVE HEIGHT *(defined in NSCV Part B)*: It is the mean value of the highest one third of the heights measured from trough to crest recorded in a wave time history.

(Note: It is probable that one in every 1000 waves will have a height at least 1.86 times the significant wave height.)

LINE SQUALL (SQUALL LINE): Thunderstorms in a cold front are forced into forming a long narrow band. The forward edge of this formation is known as the Line Squall or squall line. It can be 10-15 miles wide and up to a hundred miles long.
   See Thunderstorms.

VISIBILITY: The only correct way to measure visibility at sea is by measuring the radar distance of an echo that has just become visible. Fog and Mist are defined below.

CLOUDS: Cloud cover is expressed in the eighths of the sky, such as $1/8^{th}$, $2/8^{th}$, etc. If the clouds cover 50% of the sky, it is a $4/8^{th}$ cover. An overcast sky is an 8/8 cover.

   According to the Howard system of 1803 still in use, clouds are divided into three categories based on the height of the bottom (base) of the cloud, even though these heights vary considerably between latitudes. Forecasting weather from clouds is far from accurate.

   Low Clouds: their bases are less than 2 km above the horizon. The typical puffy looking Cumulus clouds (Cumulus meaning heap or puffy in Latin) can be quite tall, but their bases remain low. A tall cumulus cloud can, at times, become a cumulonimbus (Nimbus is Latin for rain) or thunderstorm cloud at times. A cumulonimbus is tall with a cauliflower shape and distinctive anvil top.

   Medium Clouds: their bases are between 2 and 6 km above the horizon, such as Alto Cumulus and Alto Stratus clouds (Stratus is Latin for layer). They are associated with unstable weather ahead of a cold front.

   High Clouds: their bases are over 6 km above the horizon. Cirrus (meaning curls of hair in Latin) are the most common high clouds. They are white thin fibrous ice clouds, thin enough to allow the sun or moon to shine through. No rain falls from them.

   (NOTE: For more detailed cloud types, a cloud chart should be consulted.)

SHOWERS: Meteorologically speaking, only the rain from cumulus clouds is referred to as showers, whether it is periods of light intermittent rain or torrential downpours of heaviest rains on record.

PRESSURE: As discussed above, atmospheric pressure is the force exerted by the earth's atmosphere. It can be thought of as the weight of air. A high pressure is usually associated with good weather and a low pressure with poor weather. In the southern hemisphere, wind circulation around the pressure systems is as follows:

   Low pressure: clockwise
   High pressure: anti-clockwise

Reverse for the northern hemisphere. In the Southern Hemisphere, wind rotates clockwise around a cyclone (low) and anti-clockwise around an anticyclone (high). It is the reverse in the Northern Hemisphere.

ISOBARS & PRESSURE GRADIENT: An isobar is a line on a weather map that joins places having the same atmospheric pressure. By examining the spacing between isobars (known as pressure gradient) we can get an idea of the direction and strength of wind. Isobars closer together means steeper pressure gradient, which means stronger wind in the direction parallel or nearly parallel to the isobars. (Figures 20.13)

GRADIENT WIND: Wind resulting the pressure gradient (assuming no friction). It blows parallel to isobars, clockwise around the Lows and anticlockwise around the Highs.

DEPRESSION: Another name for a low-pressure area, which is an area of low atmospheric pressure within a closed system of isobars. Low pressure means the air is light so it will rise from its centre into the upper atmosphere, allowing more surface air to flow in towards its centre. (Figure 20.13(A)). Lows quite readily disperse moisture and pollution from the earth's surface, giving rise to clouds and strong winds.

TROPICAL DEPRESSION / CYCLONE / TROPICAL CYCLONE: It is a Low originating in tropics. These are the Lows that sometimes deepen and become cyclones as they travel away from the equator. The signs of an

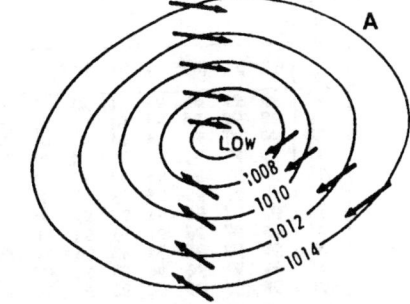

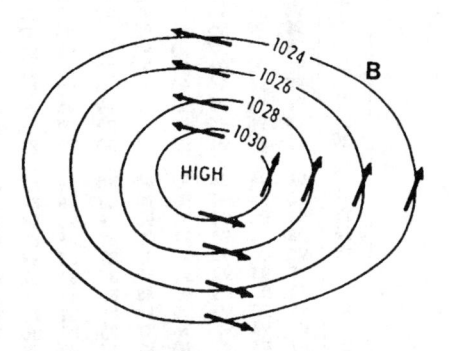

Fig 20.13
(A) A LOW, DEPRESSION OR CYCLONE
(B) A HIGH OR ANTICYCLONE

advancing depression or a cyclone are the same - only the severity may differ. Cyclones are discussed in more details below.

ANTICYCLONE: Another name for a high-pressure area, which is an area of high atmospheric pressure within a closed system of isobars. High pressure means the air is heavy so it will be sinking down from the upper atmosphere to its centre, and then flowing outward on the earth's surface. (Figure 20.13(B)). Descending air does not form clouds, and the weak pressure gradient does not generate strong winds. Anticyclones do not allow water vapour and pollution from earth's surface to disperse. Fog and smog are therefore a common phenomenon in the outer regions of anticyclones.

RELATIVE HUMIDITY: Humidity is the amount of moisture in the air. Relative humidity is the comparison of how much moisture there is to how much the air could hold. A 95% humidity means the air is loaded with moisture to almost its capacity. Relative humidity is measured by a pair of "wet and dry bulb thermometers", known as the **HYGROMETER** or an electronic hygrometer (see "electronic barometer-cum-hygrometer" above), and an associated table.

HOW WET & DRY BULB THERMOMETERS & THE TABLE WORK

Water evaporating from the surface of the wet bulb cools the bulb, just as water evaporating from the surface of a human body cools the body. When the air is dry, there is greater evaporation, causing the wet bulb thermometer to record a lower temperature. The difference between the wet and dry bulb thermometers is then greater. When the air is moist, there is less evaporation, and the difference between the wet and dry bulb thermometers is smaller. Reading the table - an example: For a dry bulb reading of 30°C and the difference between the wet and dry bulb thermometers of 2°C, the Relative Humidity is 85%.

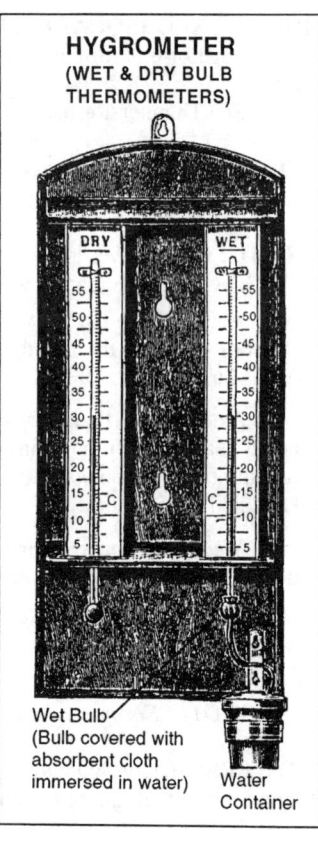

HYGROMETER
(WET & DRY BULB THERMOMETERS)

Wet Bulb
(Bulb covered with absorbent cloth immersed in water)
Water Container

TABLE FOR FINDING THE RELATIVE HUMIDITY (%)
(For use with Wet & Dry Bulb Thermometers)

Dry Bulb °C	0.2°	0.4°	0.6°	0.8°	1.0°	1.2°	1.4°	1.6°	1.8°	2.0°	2.5°	3.0°	3.5°	4.0°	4.5°	5.0°	5.5°	6.0°	6.5°	7.0°	7.5°	8.0°	8.5°	9.0°
40	99	97	96	95	94	92	91	90	89	88	85	82	79	76	73	71	68	66	63	61	58	56	53	51
39	99	97	96	95	94	92	91	90	89	87	84	82	79	76	73	70	68	65	63	60	58	55	53	50
38	99	97	96	95	94	92	91	90	89	87	84	81	78	75	73	70	67	65	62	59	57	54	52	50
37	99	97	96	95	93	92	91	90	88	87	84	81	78	75	72	69	67	64	61	59	56	54	51	49
36	99	97	96	95	93	92	91	90	88	87	84	81	78	75	72	69	66	63	61	58	55	53	50	48
35	99	97	96	95	93	92	91	89	88	87	83	80	77	74	71	68	65	63	60	57	55	52	49	47
34	99	97	96	95	93	92	91	89	88	86	83	80	77	74	71	68	65	62	59	56	54	51	49	46
33	99	97	96	94	93	91	90	89	87	86	83	80	76	73	70	67	64	61	58	56	53	50	48	45
32	99	97	96	94	93	91	90	89	87	86	83	79	76	73	70	67	64	61	58	55	52	49	47	44
31	99	96	96	94	93	91	90	88	87	86	82	79	75	72	69	66	63	60	57	54	51	48	46	43
30	98	97	95	94	93	91	90	88	87	85	82	78	75	72	68	65	62	59	56	53	50	47	44	42
29	98	97	95	94	92	91	89	88	86	85	81	78	74	71	68	65	61	58	55	52	49	46	43	40
28	98	97	95	94	92	91	89	88	86	85	81	77	74	70	67	64	60	57	54	51	48	45	42	39
27	98	97	95	94	92	90	89	87	86	84	81	77	73	70	66	63	60	56	53	50	47	44	41	38
26	98	97	95	93	92	90	89	87	86	84	80	76	73	69	66	62	59	55	52	49	46	42	39	36
25	98	97	95	93	92	90	88	87	85	84	80	76	72	68	65	61	58	54	51	47	44	41	38	35
24	98	97	95	93	91	90	88	86	85	83	79	75	71	68	64	60	57	53	50	46	43	39	36	33
23	98	96	95	93	91	90	88	86	84	83	79	75	71	67	63	59	56	52	48	45	41	38	35	31
22	98	96	95	93	91	89	88	86	84	82	78	74	70	66	62	58	54	51	47	43	40	36	33	29
21	98	96	94	93	91	89	87	85	84	82	78	73	69	65	61	57	53	49	45	42	38	34	31	27
20	98	96	94	92	91	89	87	85	83	81	77	73	68	64	60	56	52	48	44	40	36	33	29	25
19	98	96	94	92	90	88	86	85	83	81	76	72	67	63	59	55	50	46	42	38	34	31	27	23
18	98	96	94	92	90	88	86	84	82	80	76	71	66	62	58	53	49	45	41	36	32	28	25	21
17	98	96	94	92	90	88	86	84	82	80	75	70	65	61	56	52	47	43	39	34	30	26	22	18
16	98	96	94	91	89	87	85	83	81	79	74	69	64	60	55	50	46	41	37	32	28	24	20	16
15	98	96	93	91	89	87	85	83	81	78	73	68	63	58	53	49	44	39	35	30	26	21	17	13
14	98	95	93	91	89	86	84	82	80	78	72	67	62	57	52	47	42	37	32	28	23	18	14	10
13	98	95	93	91	88	86	84	81	79	77	71	66	61	55	50	45	40	35	30	25	20	16	11	6
12	98	95	93	90	88	86	83	81	78	76	70	65	59	54	48	43	38	32	27	22	17	12	8	3
11	97	95	92	90	87	85	83	80	78	75	69	63	58	52	46	41	35	30	25	19	14	9	4	
10	97	95	92	90	87	84	82	79	77	74	68	62	56	50	44	38	33	27	22	16	11	5		

Fig 20.14 (A & B): HYGROMETER & RELATIVE HUMIDITY TABLE

Chapter 20: Meteorology

FRONT: Two air masses of different characteristics (i.e., of different temperature & moisture, etc.) do not want to mix when brought together. They thus form a front between them where they clash, generally causing bad weather. Fronts are discussed below. (Figure 20.20)

COLD FRONT: When a mass of cold air meets a mass of warm air, the cold air being heavier, pushes its way under the warm air and forces it to rise almost vertically. This can often result in violent weather for a short period of time, as discussed below. (Figures 20.20 to 20. 22)

WARM FRONT: When a mass of warm air meets a mass of cold air, the warm air being lighter, rises over it in a slope extending 300-400 miles horizontally. Its moisture content causes overcast sky and possibly rain, drizzle, fog, etc. (Figures 20.20 & 20.23)

OCCLUDED FRONT: An occluded front is a combination of a warm and a cold front. Occlusion occurs when a faster moving cold front catches up with a warm front. It pushes the warm air mass off the ground, which becomes wedged between the two cold air masses.

The rising warm air is likely to produce cloud and rain. But, as the depression becomes surrounded by cold air, it decreases in intensity. The resultant weather is often similar to that of a warm front. Occluded fronts are not common in Australia.

COL: A region between diagonally opposed two Highs and two Lows. It is a no-person's land. The wind may blow from any direction and any type of weather can be expected. The centre of a Col is associated with very light and variable winds, and fog in colder months.

STORM SURGE: Read the effect of weather on tides in Chapter 17.3.

FOG: Suspension of very small water droplets in the air, reducing visibility - a form of low clouds. In a cloud, when the horizontal visibility is less than 1000 metres then technically we are in fog conditions.

MIST: Less dense fog -Horizontal visibility greater than 1000 metres.

RIDGE: Elongated area of high pressure extending out from a High.

TROUGH: Elongated area of low pressure, extending south from a Low.

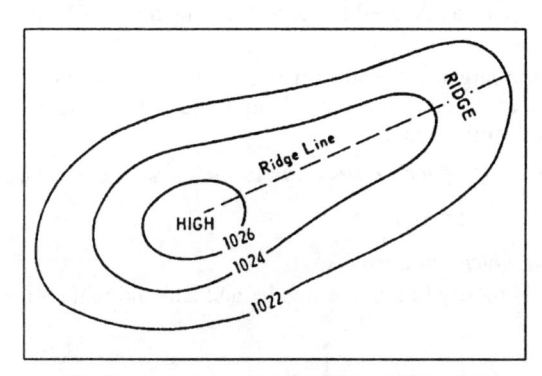

Fig 20.15: A RIDGE OF HIGH PRESSURE

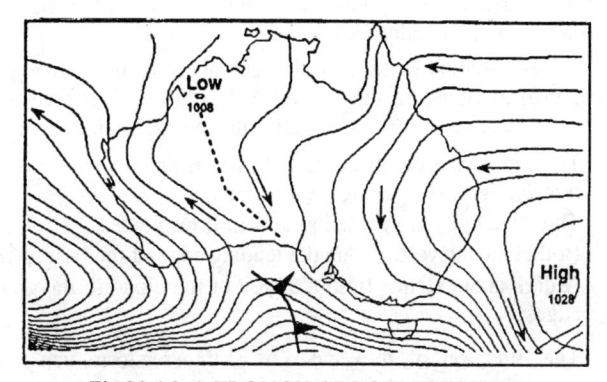

Fig 20.16: A TROUGH OF LOW PRESSURE
(The dotted line)

WEATHER FORECASTS & WARNINGS

(Bureau of Meteorology's weather service by phone, fax and radio and the Website address are listed on the last two pages of this book.)

- Boating weather information on the Bureau of Meteorology's State-wise phone numbers – see back page of the book.
- Routine coastal waters forecasts and observation reports for areas within 60 nautical miles of the coast are issued several times daily via marine radio and Inmarsat.
- Coastal warnings are issued via marine radio and Inmarsat whenever strong winds, gales, storm-force or greater winds are expected, and are renewed every 6 hours.
- Routine high seas forecasts are issued twice daily via marine radio and Inmarsat for the areas beyond the coastal waters

surrounding Australia.
- High seas warnings are issued via marine radio and Inmarsat whenever gale, storm-force or greater winds are expected, and are renewed every 6 hours. (See NAVAREA X warnings in Chapter 12 & AUSCOAST warnings in Chapter 18).

RADIO WIND WARNINGS PREFIXED BY SECURITÉ (SAFETY MESSAGE)
- Strong Wind Warning (for coastal waters only): Winds 25 – 33 knots (Beaufort Force 6 - 7)
- Gale Warning: Winds 34 - 47 knots (Beaufort Force 8 - 9)
- Storm Warning: Winds exceeding 47 knots (Beaufort Force 10 - 11)
- Hurricane Warning (for tropical high seas only): Winds exceeding 63 knots (Beaufort Force 12)

WHEN TO LISTEN TO WEATHER FORECASTS
You should listen to weather forecasts on radio:
- While in port and before departure
- At sea: to the regular schedule weather reports & forecasts
- When there are signs of approaching bad weather
- During bad weather

THUNDERSTORMS

Thunderstorms can be a serious hazard for small vessels. They usually produce strong gusty winds that blow out from the front of the storm. In general, they are local storms produced by Cumulo-nimbus clouds. (Cumulus means 'heaps', i.e., clouds piled up on each other. Nimbus means rain bearing.) These are the grey towering clouds that you sometimes see on the horizon. The clouds often move in different directions to the wind at the surface. If you observe that clouds will pass over or within a few miles of your position, you should head for shore.

The lightning is a discharge of electricity between cloud and ground (or sea) or between clouds. The thunder is due to the rapid expansion of atmospheric gases in the path of a lightning strike. The lightning and thunder occur at the same time, but we see the flash before hearing the thunder. It is because the light travels faster than sound. A 5 second interval between the flash and sound indicates that the lightning struck at a distance of approximate 1 mile. This relationship is constant and can be used for general calculations, as long as the flash and the thunder are from the same source.

Fig 20.17 CUMULO-NIMBUS CLOUD

A cumulus (cotton wool appearance) cloud growing larger and larger is the main danger signal of a developing thunderstorm. Static crashes on the AM radio receiver are also a good early indication. The cloud will eventually develop four distinct features. (Some of these may not be visible, being blocked by other clouds):

- The top of the cloud is shaped like an anvil. *[The "anvil" is made up of ice crystals due to the cloud rising above freezing level (becoming cirrus clouds)].*
- The main body is very tall with cauliflower sides.
- Roll clouds develop along the leading edge of the base. *[Caused by violent air currents]*
- A dark area extends from the base of the cloud to the ground. [It is mostly rain in the middle and hail and rain on the edges].

The direction of movement of a thunderstorm can sometimes be estimated from the direction of the 'anvil' on top of the thunderstorm cloud. The storm usually moves in the direction in which the 'anvil' is pointing.

The thunderstorm weather is usually showers, possibly with hail, thunder and lightning; and occasionally strong squally winds and rough seas. It consists of abrupt fluctuations of pressure, temperature and wind.

PROTECTION AGAINST ELECTRICAL STORMS

An electrical storm can upset the vessel's compass; burn out equipment such as radio and radar. These should be switched off. It can also cause fire on board due to lightning. You don't often hear of boats being

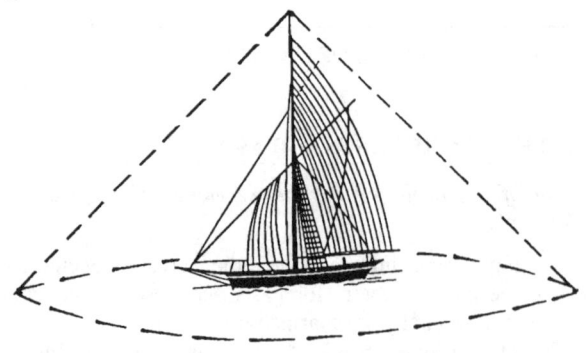

Fig 20.18: LIGHTNING CONDUCTOR
(The higher the conductor, the greater the area of protection)

struck by lightning, but it can happen. It is advisable to rig a lightning conductor on the vessel. A properly rigged conductor will provide 99.9% protection. The greater the height of the conductor (or the lightning protective mast), the greater the area of protection around it. It protects a conical area of a radius equal to its own height above the water.

The following information is based on the Safety Standard E-4 of the American Boat & Yacht Council:

➢ The conducting wire or the copper strip should be of a suitable gauge (#8 gauge for wire and #20 gauge for copper strip). The path followed by the grounding conductor should be as straight as possible with no sharp bends. All metal objects close to the lightning conductor and large metal objects in other parts of the vessel should be connected to it. This is to safeguard against sparks and side flashes jumping from one metal to another or dangerous levels of voltage rising in them.

➢ Sailing vessels with metal standing rigging are adequately protected if all rigging is bonded together and grounded. A metal chain, shackled to a stay, can be hung into the water as a temporary measure. The modern fibreglass whip aerial with an internal wire conductor will not provide lightning protection. However, the older metal rod aerials, when connected to earth, can do the job.

➢ Whether or not a lightning conductor is fitted in a vessel, metal objects, such as spotlights, projecting through cabin tops should be solidly grounded. The submerged ground connection in non-metal hulls can be achieved via the dyna plate used for grounding the radio, the metal propeller or the metal rudder.

➢ For personal protection from lightning, stay inside a closed vessel. Avoid touching any item connected to the lightning conductor. Do not touch two separately grounded objects at the same time. If you do, you may become a bridge between their grounding systems. Do not immerse any part of your body in the water.

TORNADOS & WATERSPOUTS

These are offshoots of tall cumulus clouds. There is a large difference of temperature between the top and bottom ends of the cloud. The air in the bottom of the cloud, being humid and warm, starts to rise. In the top end it is cold and heavy and starts to fall. This sets off a strong vertical upward air current within the cloud.

The lifting effect thus caused at the base of the cloud sometimes sucks air from underneath the cloud. This can result in a funnel-shaped protuberance appearing at the base of the cloud growing downward toward the ground or the sea. Over the ground this whirlpool of air is called a tornado; and over the sea, in the form of sea spray, it is a waterspout. The average diameter of a tornado is about 250 metres and of a waterspout about 30 metres. At the core of the waterspout there may appear a mound of water, about 30 cm high. This is because the pressure inside the funnel can be up to 40 HPa less than its surroundings.

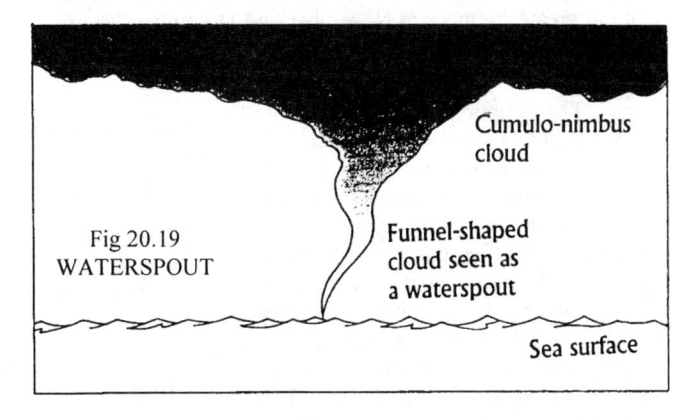

Fig 20.19 WATERSPOUT

The waterspouts, being heavy with water, break and fall apart within about half an hour. But the tornados, once formed, may travel for long distances causing wind speeds of up to 150 knots. Their destructive effect can extend up to half a mile from the centre. They do occur in Australia as "Willie - Willies", but those in the southern states of the USA are known to cause greater destruction.

FRONTS

Two air masses of different characteristics (i.e., of different temperature & moisture, etc.) do not want to mix when brought together. They thus form a front between them where they clash, causing bad weather.

As shown in the figures 20.13 & 20.20, wind rotates clockwise in a low-pressure system (in the southern hemisphere). It brings in air from adjacent different regions. For example, on its eastern side, the wind is coming from a warm sea. It is thus warm and humid. On the southern side, it is coming from the cold Antarctic region. On the western side it is again warm and humid. On the northern side it is blowing over the desert. It is thus hot and dry.

This pushing and shoving by different air masses into one system causes one or more fronts to build up. Typically, in the right hand section, the warm and highly moist tropical air is trying to push its way into the colder and less moist air. The front formed between them is known as the warm front. [It's easy to remember: in a warm front, the warm air pushes its way in]. In the left hand section, on the other hand, the cold polar air is pushing its way into warm air. The front formed between them is known as the cold front. [Remember: In a cold front, cold air pushes its way in].

A COLD FRONT

As discussed above, when a mass of cold air meets a mass of warm air, the cold air being heavier, pushes its way under the warm air and forces it to rise almost vertically. A steadily falling barometer heralds the approach of a cold front. Immediately on or after the passage of the front the barometric reading increases rapidly, and then flattens out as the high winds accompanying it, ease down.

The early signs of a Cold Front are a significant drop in barometer and large cumulous clouds aligned along the leading edge of the Front. As a result of the warm moist air rising so steeply, you may also see the towering cumulonimbus clouds with the distinctive anvils on top. However, occasionally, and particularly in summer months, a southerly change (Cold Front) can be cloud-free and treacherous. This is especially true for the more serious cold fronts, known as Southerly Busters in NSW.

The NSW coast experiences about 60 cold fronts a year, mainly in the warmer months. About 10 of these are strong enough to be called Busters, which are almost exclusively confined to the period of September to March, with the greater number in November and December. They are less frequent on the north NSW coast. They most often arrive in the afternoon and evening.

The important thing to note about the Cold Fronts is that because of their steep slope, the change is usually violent but short lived. An average Cold Front stretches only about 30 miles over the ground, but the wind speed can at times reach 40 knots or more, accompanied by thunderstorm and heavy rain showers. As shown in figure 20.20, the air behind the Cold Front is colder and blows from southwest to south. This is the change in temperature and wind direction experienced after a Cold Front.

Fig 20.20 FORMATION OF FRONTS
(Cold Front is the line or arc with barbs, and Warm Front has semi-circles; pointing in the direction of movement)

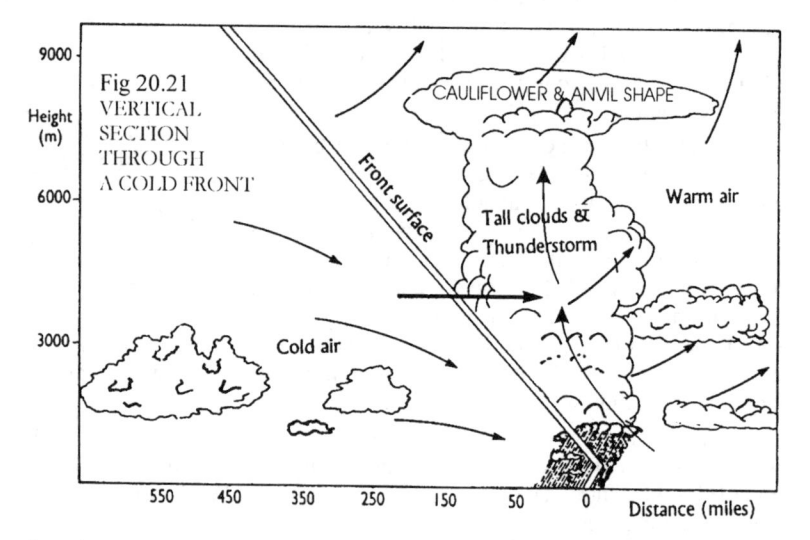

Fig 20.21 VERTICAL SECTION THROUGH A COLD FRONT

EAST COAST LOWS AND BOMBS

An East coast low is a tightly closed coastal cold front (within 5° of latitude or longitude from the coast). On average, we experience about two of them in a year, one of which is strong enough to be referred to as a Bomb or <u>Explosive Cyclogenesis</u> The Bomb is more likely to occur in the winter or either side of winter when there is a large difference in the sea and land temperatures and the conditions in the upper atmosphere are generally stable.

The weather before an east coast low can be deceptively good – so good that you don't bother listening to weather reports. But, severe weather is usually around the corner. The low can develop in a space of 12 hours. Unlike most major weather systems, it does not show up in weather maps until about 12 hours before the event. Then, suddenly the barometer drops sharply and winds can rapidly build up to a squally 50 knots with gusts to 70 knots. Seas can build up to 6 or more metres high, and the rain so heavy that you can barely see. It is like being in a tropical cyclone. Indeed, many of these lows tend to form a distinct eye and look like a tropical cyclone in the satellite picture, including spiral bands, though they may be very small. Forecasting the movement of east coast lows is also difficult because they can speed up or stall.

Be warned if the wind speed drops. If this happens, look to the south for signs of an approaching low. Don't rely on its position being indicated by a roll cloud or long line of cloud. The most common indicator is a line of haze or low clouds approaching from the south, along with a change in the colour of the sea surface. In Sydney area, you may also see a line of small cumulous clouds over land and along the leading edge of the front in the southwest to northeast line. Of course, get into a port if there is a significant drop in barometer.

Fig 20.22: COLD FRONT – A PHOTOGRAPH

A WARM FRONT

When a mass of warm air meets a mass of cold air, the warm air being lighter, rises over it in a slope extending 300-400 miles horizontally. Its moisture content causes overcast sky and possibly rain, drizzle, fog, etc.

The temperature, barometer, wind direction and wind speed change gradually as the Front takes a few days to pass. Because of their gentle slope, Warm Fronts are usually not violent, but they stay much longer. Wind speeds usually do not reach beyond Force 5 (17 - 21 knots).

THE WEATHER EXPECTED IN A COLD FRONT

➢ It is usually a sudden and violent change
➢ Steady drop in barometer before the change
➢ Towering high grey (Cumulo-nimbus) clouds
➢ Shift in wind direction. The synoptic or gradient wind of a Front backs to southward during the Front's passage. Due to the Coriolis effect, the gradient wind always backs in the southern hemisphere. (See Cyclone terminology for the meaning of Wind Veering or Backing).
➢ Increase in wind speed up to force 8 (34 - 40 knots)
➢ Choppy seas
➢ Possible thunderstorms
➢ Poor visibility
➢ Drop in temperature

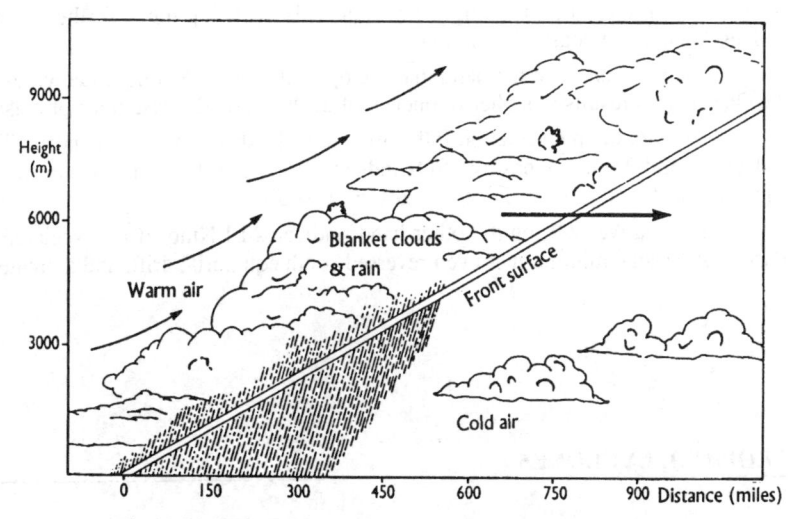

Fig 20.23 VERTICAL SECTION THROUGH A WARM FRONT

SAFETY PRECAUTIONS WHEN A COLD FRONT APPROACHES

- Good seamanship and safe practice.
- If in port: wait for it to pass.
- If close to shore: return immediately.
- If far from shore: keep in contact with a Maritime Coast Station (Coast Radio Station), batten down, heave-to if necessary and turn your vessel into a watertight cocoon, switch off all fires and shut off gas.

OCCLUDED FRONT

As mentioned earlier, an occluded front is a combination of a warm and a cold front. Therefore, on weather maps it is shown as a line of alternative barbs and semi-circles. Occlusion occurs when a faster moving cold front catches up with a warm front. It pushes the warm air mass off the ground, which becomes wedged between the two cold air masses.

The rising warm air is likely to produce cloud and rain. But, as the depression becomes surrounded by cold air, it decreases in intensity. The resultant weather is often similar to that of a warm front. Occluded fronts are not common in Australia.

CONCLUSION ABOUT THE FRONTS

Both the Cold and Warm Fronts bring a change. From a mariner's point of view, there is no such thing as a nice Front. The difference is that the Cold Fronts are quick and violent, whereas Warm Fronts are not so violent and tend to set in for a few days. A good example of the Warm Front weather is the weather in Britain and Tasmania. [Both places also experience Cold Fronts].

Warm Fronts are not common on mainland Australia. It is because the deep Lows that develop south of the mainland and rotate clockwise, as shown in figure 20.20, tend to swing the Cold Fronts northwards (towards the southern mainland) and the Warm Fronts southwards (towards Tasmania and beyond).

EL NINO, LA NINA & SOUTHERN OSCILLATION INDEX

The El Nino Southern Oscillation (**ENSO**) causes the southeast trade winds to weaken. Therefore, the warm surface water of the Southern Pacific stays much farther to the east than usual. This causes draught in Eastern Australia and floods in Peru. Cyclones, needing warm water, also develop much further to the east, in places as far as French Polynesia and less on the east coast of Australia.

La Nina, on the other hand, has the opposite effect. Strong trade winds bring the warm water to the western part of the Ocean. This results in higher frequency of cyclones on the east coast of Australia.

The Southern Oscillation Index measures the difference from "normal" of the mean atmospheric pressure recorded in Darwin and Tahiti. A positive **SOI** indicates La Nina. It means strong SE trade winds, strong westerly flowing south equatorial drift and more cyclones in Eastern Australia.

A negative SOI, on the other hand, indicates El Nino. It is associated with weak SE trade winds, a drought over Eastern Australia, minimum or even reversed south equatorial drift, and a greater likelihood of cyclones in central Pacific.

TROPICAL CYCLONES

CYCLONE WARNING SYSTEM

- Tropical Cyclone Threat Maps and Warning Messages are available from the Bureau of Meteorology's weather service by radio, phone, fax and website. Details are listed on the last two pages of this book.

    Further information on cyclone preparedness can be obtained from the State/Territory Emergency Services phone numbers, also listed on the last two pages of the book.

HIGH SEAS WARNING

> Warnings for cyclones and other dangers existing or likely to move into the region are broadcast every hour via AMSA Radio Network commencing at 0000 hours EST & WST. The broadcast radio frequencies and phone and fax numbers are listed on the last two pages of this book. (See also NAVAREA X warning system in Chapter 12)

Chapter 20: Meteorology

CYCLONE WARNINGS ISSUED BY THE METEOROLOGICAL OFFICE

- *CYCLONE WATCH OR CYCLONE ALERT:* Winds above gale force from a cyclone will affect coastal or island communities within 24 to 48 hours. These messages are renewed every 6 hours.
- *CYCLONE WARNING:* Gales or stronger winds are expected to affect coastal or island communities within 24 hours. These warnings are issued every 3 hours, and every hour when the threat becomes severe.
- *FLASH CYCLONE WARNING:* An area not previously affected has now come under threat. Or, there is an urgent amendment to an earlier warning, e.g., the cyclone has changed its course.
- *SEVERE WEATHER WARNING:* It may be issued if the system is no longer a cyclone but there is still a threat of damaging winds, flooding rain or pounding seas.

THE INFORMATION BROADCAST WITH A CYCLONE WARNING:

- The cyclone's name.
- The coastal area under threat.
- The location of the cyclone's centre, its central pressure and the CYCLONE CATEGORY on a scale of 1 to 5 with 5 being the most severe. The strongest gust in Category 1 is of less than 70 knots. In category 5, it is of more than 150 knots. Cyclone Winifred was of category 3, Tracy of category 4 and Orson of category 5.
- The direction of movement and speed at which the storm is travelling and intensity (including maximum wind gusts).
- A forecast of heavy rain, flooding, abnormal tides and storm surges.

WHAT ARE CYCLONES:

The word Cyclone is derived from the Greek word 'kukloma', meaning 'coiled snake'. Tropical cyclones are extremely violent. Even a small one has enough power to theoretically provide for Australia's electrical needs for several years. Cyclones generate extreme winds and flood rains. The sea conditions become dangerous both for vessels out at sea and those moored in harbours. Storm surges cause havoc on low-lying coastal areas. Storm surges are a raised dome of water up to 80 metres across and up to 5 metres higher than the normal tide level. The damage is worst when a storm surge coincides with a high tide. The cyclone centre or "eye" would produce a temporary lull in the wind but this would soon be replaced by extreme winds from another direction.

From November to April the coastline of the northern half of Australia is vulnerable to up to 10 tropical cyclones a year. A quarter of these are classified as severe. Australia experiences the world's highest percentage of slow-moving, stationary and erratic cyclones.

Tropical cyclones are also known as:
- Tropical Revolving Storms (TRS)
- Cyclones
- Tropical Storms
- Hurricanes
- Typhoons

DEVELOPMENT OF A CYCLONE

The tropical region, being hot, is a low-pressure belt around the globe. It constantly gives rise to lows, resulting in rain showers and sunshine almost every day many times a day. At times one of these tropical lows deepens and becomes a cyclone as it travels away from the equator. This happens in summer months when the sea temperature and humidity are at their highest and the atmospheric pressure is at its lowest, resulting in maximum converge of warm and moist air over the sea – an essential ingredient for cyclones.

A cyclone often begins its life as a tropical wave, which is a non-circulatory trough of low pressure. The Coriolis force causes a part of the wave to start revolving, making it a circulatory tropical disturbance.

Cyclones do not form within 5° of the equator where there is not enough Coriolis effect. The wind cannot deflect to form a circular motion without the Coriolis effect. Even outside the 5° limits, very few lows develop into cyclones. A whole range of conditions has to be met for a cyclone to form, but the following four events must take place for a cyclone to form:

- Sea surface temperature must reach at least 27°C.
- There must exist a surface low-pressure area over the ocean.
- Surface winds must converge around an eye (not in the eye, as in an ordinary low). This would require the existence of a divergent mechanism (jet stream) in the upper atmosphere.
- The Coriolis wind shift must be assisted by an activity such as the tropical airwave disturbance.

Cyclone develop in three stages, taking anywhere from hours to several days:
- **Tropical depression** – swirling clouds and rain with wind speeds of less than 33 knots.
- **Tropical storm** – wind speeds of 34 to 63 knots.
- **Tropical cyclone** – wind speeds greater than 64 knots.

They are formally designated Tropical Cyclones and named when winds of at least gale force have developed. The name given to a cyclone at this time is used throughout its life. The Bureau of Meteorology's public warning system is activated if a cyclone threatens coastal and island communities within 48 hours. Other cyclones and developing depressions are mentioned in the Bureau's weather notes and advice to shipping and aviation.

The normal path of a cyclone is parabolic (Fig. 20.25). In the southern hemisphere they usually start moving to west-southwest due to the Coriolis effect and then curve towards southeast. On average their speed is under 10 knots before recurving and about 15 knots after recurving. However the behaviour of cyclones is erratic and variable. Almost everything about them is unpredictable.

CYCLONE SEASON (in the southern hemisphere): About November to April

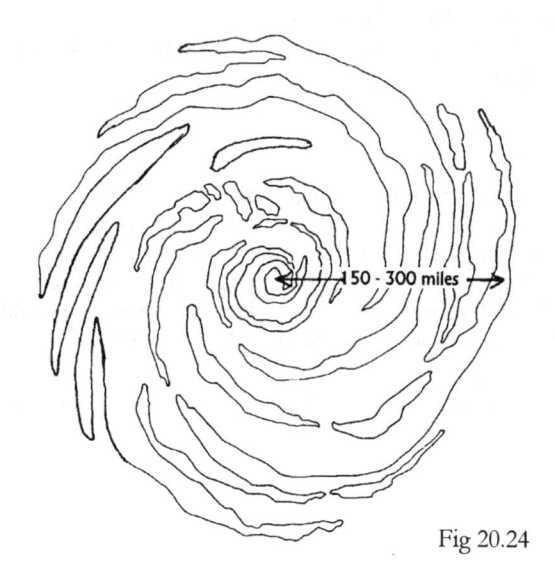

Fig 20.24

A TROPICAL CYCLONE IN THE SOUTHERN HEMISPHERE
(Left: Satellite View
Below: Section View)

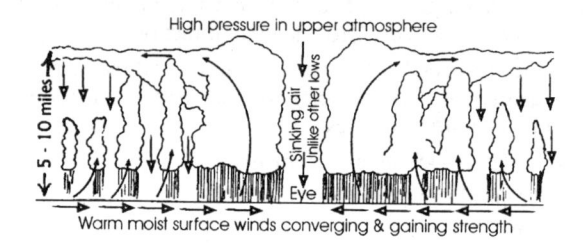

CYCLONE MOVEMENT

For some reason, paths taken by tropical cyclones in the Australian region are more erratic than in any other part of the world. Cyclones can last for up to two or three weeks. They may take sharp turns and even form loops.

However, as shown in Figures 20.25 & 20.27, cyclones in the southern hemisphere usually curve eastward in their travel. Therefore, a vessel on the eastern side of the cyclone (i.e., on the left side of its path) is likely to remain in it for a longer period and has a greater possibility to get trapped in its centre than a vessel on the western side. The winds are also typically stronger on the eastern side, because they are moving in the same direction as the eye. On the western side the winds are going to the opposite side of the movement of the cyclone. Some of their speed is thus taken off. For this reason, the eastern semicircle of the cyclone is known as the NON-NAVIGABLE (OR MORE DANGEROUS) SEMICIRCLE and the western semicircle as the NAVIGABLE SEMICIRCLE.

The front half of the Non-navigable semi-circle is known as the MOST DANGEROUS QUADRANT. It is so for three reasons: First, the wind and waves are running almost head-on into the path of the storm. They are thus strongest and biggest. Second, the wind will try to blow a vessel in this quadrant into the path of the storm. Third, any recurving of the cyclone is most likely to be towards this quadrant. A vessel in this quadrant is therefore more likely to run into the path of the storm.

Some books use the term "Dangerous" for the "Non-navigable" semi-circle. Do not interpret it to mean that the Navigable semi-circle is safe. It is dangerous to be anywhere in a cyclone, but one half of it is more risky than the other.

Chapter 20: Meteorology

CYCLONE TERMINOLOGY
(Figures 20.25 & 20.26)

TROUGH LINE: It is the line of least barometric pressure, drawn at right angle to the path. The barometer starts to rise when the vessel crosses the trough line.

PATH: The predicted course of a cyclone.

TRACK: The trail left by the passage of a cyclone.

VORTEX: The cyclone's eye

VERTEX: The most westerly point of a cyclone's track.

ANGLE OF INDRAFT: It is the angle at which the wind cuts isobars. On the outskirts of a cyclone the wind may cut the isobars at 45° (i.e., 135° to the path), whereas near the centre it would be almost parallel to the isobars (an angle of 0° to the isobars or 90° to the path). Wind tends to eddy round the centre. See Buys Ballot Law below.

Fig 20.25: AVERAGE TRACK OF CYCLONES (in the southern hemisphere) *[The sketch indicates that the southern States usually do not experience cyclones]*

BUYS BALLOT'S LAW: The Buys Ballot's Law is a way of determining the approximate direction of the centre of a Low of a High pressure system - in the absence of a reliable weather forecast.

In the Southern hemisphere, if you face the wind, the centre of the low pressure lies between 90 and 135 degrees (8-12 points) on your left hand side, depending on your distance from the centre of the storm. When the pressure begins to fall, the centre is at about 12 points (135°). When it has fallen 10 hPa the centre is at about 10 points (112.5°). And when it has fallen 20 hPa the centre is at about 8 points (90°). See "Angle of Indraft" above.

Reverse this rule for the high pressure [and for the northern hemisphere].

WIND BACKING & VEERING: In the absence of a reliable weather forecast, a vessel in a cyclone (i.e., a low pressure) can also assess whether she is in the navigable or the non-navigable semicircle. In a cyclone the wind direction changes constantly, except on its path. A vessel can assess the direction of

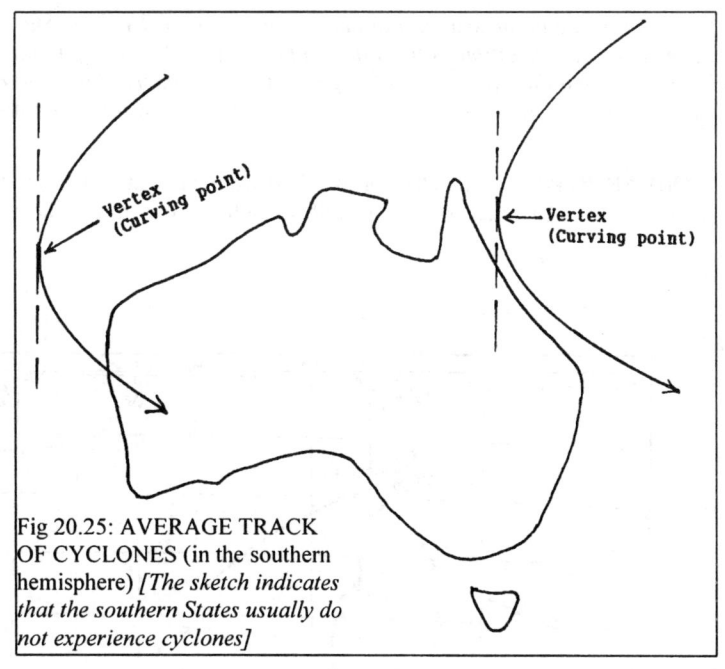

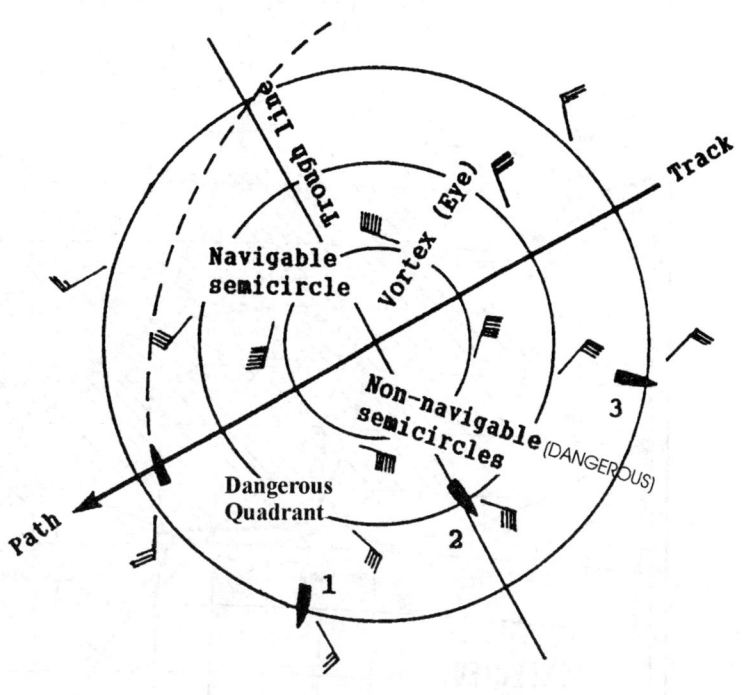

Fig 20.26: PARTS OF A CYCLONE

the shift of wind by recording it over a period of time. In the southern hemisphere, if the shift is clockwise (i.e., the wind VEERS), the vessel is in the navigable semi-circle. If the shift is anti-clockwise (i.e., it BACKS), then she is in the non-navigable semicircle. If the wind direction remains steady, then the vessel is on the cyclone's path. It is best to make this observation from a heaved-to vessel. Observations of wind shift made from a vessel making way through a storm may be inaccurate.

STORM SURGES & TIDE SURGES: Storm Surge is the name given to flooding in low-lying coastal areas caused by onshore winds and the low atmospheric pressure in a storm. They are raised domes of water of some 80 kilometres across and up to 5 metres higher than the normal tide level. The damage is worst when a storm surge coincides with a high tide. Slow moving storms of large diameter create higher storm surges.

Some authorities split the above description into two: Storm Surges and Storm Tides. In which case, the Storm

Surge is the rise in the sea level of about a metre or so. The Tide Surge is the combined effect of a storm surge, high tide, wave set-up (the addition to the water column by broken waves), and wave run-up (the rushing of broken waves up the beach and sand dunes). This is what generates the 5 metres high and 80 kilometres across mound of seawater. Also note that a negative storm surge can drop the sea level to below that of predicted tide.

WIND ARROWS: Each feather on wind arrows represents 10 knots (5k for half feathers). Wind is stronger closer to the centre of a low. There are thus more feathers on the arrows nearer the centre.

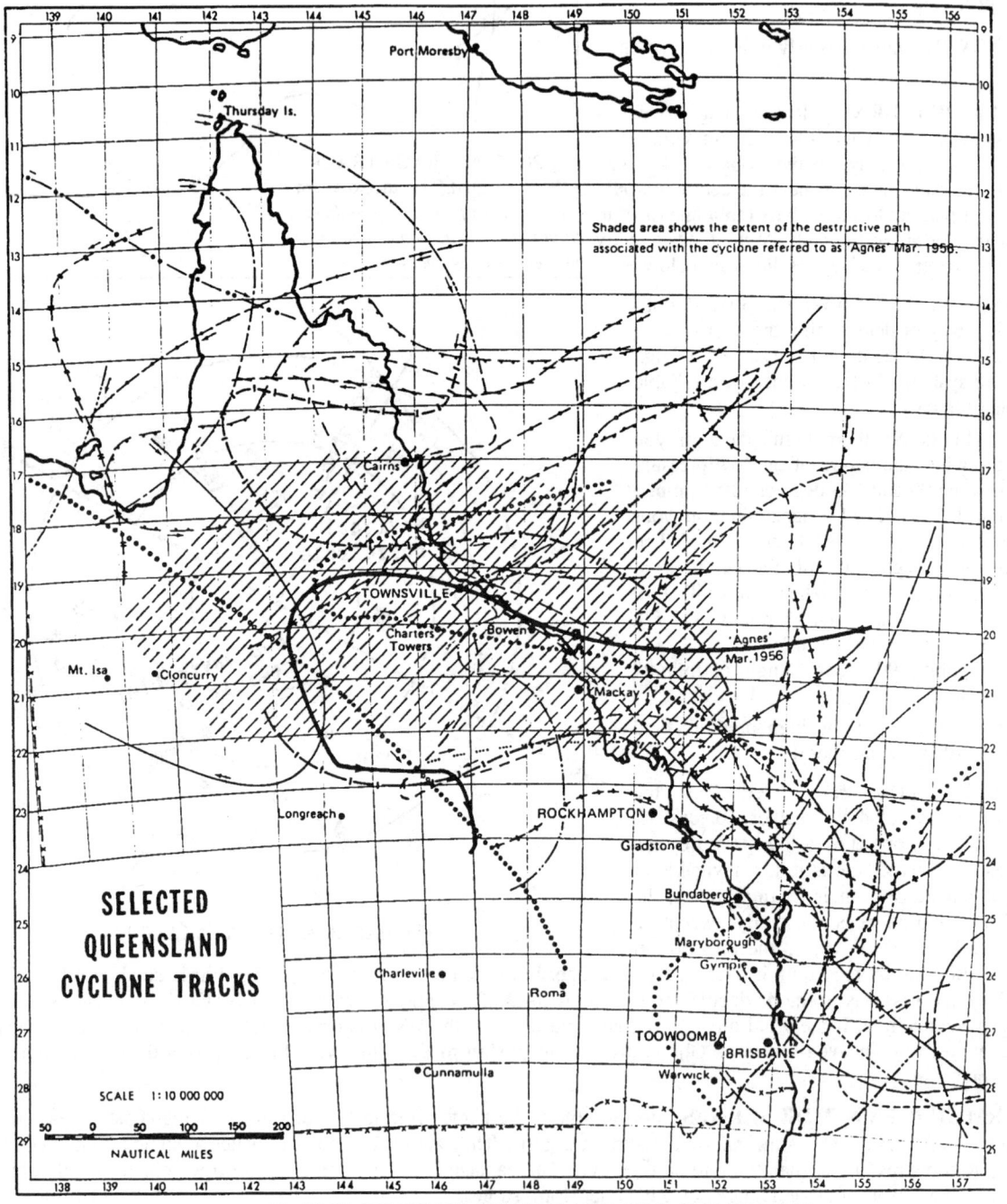

Fig 20.27: SELECTED QUEENSLAND CYCLONE TRACKS

Chapter 20: Meteorology

CYCLONE CONTINGENCY PLANS & SAFE HAVENS

- In cyclonic regions, contingency plans need to be made during calm weather.
- Visit and check out suitable safe havens, obtain extra large-scale charts and notes from pilot books for these places.
- Note down relevant names and telephone numbers.
- Listen to weather forecasts at schedule times.
- Start plotting cyclone's position as soon as broadcast.
- Brief the crew, batten down the vessel and prepare a supply of sandwiches.
- Change course as necessary and radio in your deviation report.
- Allow at least 24 hours barrier to be in safe haven.

If a cyclone is expected, travel up an inlet and check the depths and clearances. You should know your boat's draft and check that you have plenty of lines to secure it. A trailer boat could be left with a friend further inland where the risk of cyclonic winds is lower than on the coast. If going out of town during the cyclone season, get a responsible friend to look after your boat. The local Coast Guard or Coastal Patrol may look after your boat if you leave written instructions and the access keys with them.

Australian cyclone-affected ports have cyclone contingency plans in place. Mariners operating in a cyclone season in these areas should find out the plan from the local maritime centre. The contingency plans are aimed at advising vessels to move from the risky areas of ports into designated shelters in creeks and waterways at least 6 hours before the destructive winds commence. Vessels with shallower drafts are moved to the upper reaches of the creeks.

CAIRNS (QLD) EXAMPLE

The port of CAIRNS has been divided into alphabetically marked sections. The maps are available at any time from Cairns Maritime Office (Ph. 07 4052-7400). At the possibility of a cyclone being in the vicinity the Harbour Master's Office becomes the Maritime Control Centre and the Pier Master's Office becomes the initial Communication Centre. Both the centres operate under the radio Callsign of "Port Emergency Control". All vessels must contact the Communication Centre before moving to their area of shelter.

Vessels can telephone on 4051-2558 or call on VHF Channel 16 or on 27 MHz Channel 88. The Control Centre telephone numbers are 4035-1025 and 4052-7412. See also Bureau of Meteorology's broadcast schedule and contact details on the last two pages of this book.

COLOUR CODED CYCLONE ALERT SYSTEM

Some States use a colour coding system to describe the closeness of a cyclone. The code may vary between States and is not used by the Bureau of Meteorology. The following colour code is used in Cairns.

YELLOW ALERT means destructive winds are expected within 20 hours. Vessels are to proceed to allocated safety moorings and anchorages.
- Vessels in port area A move to ... ... Creek
- Vessels in port area B move to ... ... Creek
- Etc.

BLUE ALERT means destructive winds are expected within 16 hours. Vessels should be in their allocated safety moorings and anchorages. They should prepare to batten down. This is the last opportunity to evacuate to shore. Large ships are ordered to sea and wharves cleared.

RED ALERT means destructive winds are expected within 6 hours. Vessels should be well battened down. All necessary should be steps taken to secure people and vessels. Port closed. No vessel movement.

EXMOUTH BOAT HARBOUR (WA) EXAMPLE:

A Cyclone Contingency Plan is made available to the boating public. The Plan can also be downloaded from the WA Department of Planning & Infrastructure website listed on the last page of this book. It recommends minimum mooring arrangements for various sizes of vessels. Boat pens for trawlers, charter boats and for general purpose are shown on the plan. Mooring priority is given to vessels covered by an existing mooring agreement.

Names and phone numbers of Harbour Coordinators for different interest groups (Transport, Marina, Fisheries, Customs, etc.) are written on the Plan. In cases of any difficulty, contact can be established through Exmouth Police (08 9949 2444), Exmouth SES (08 9949 1488) or Exmouth Sea Rescue (on VHF, HF or 27 MHz radio).

CYCLONE ALERT SYSTEM AT EXMOUTH

CYCLONE WATCH ISSUED (Threat of gale force winds within 48 hours): Vessels to maintain contact with the harbour Coordinator, and plan to be in the harbour at least 24 hours before the gale force winds.

CYCLONE WARNING ISSUED (Gale force winds forecast within 24 hours): Vessels have to ensure that their vessel and the area of responsibility have been secured.

SES STAGE RED (High winds imminent): Seek appropriate shelter until the State Emergency Service declares the "All Clear".

WARNING SIGNS OF AN APPROACHING CYCLONE

- Radio broadcasts.
- Long heavy swell from the direction of the storm centre is experienced up to a distance of 1000 miles from the centre. This is a good long-range indicator.
- Tides become higher due to drop in the atmospheric pressure.
- Unusually good visibility.
- Initially falling or unsteady barometric pressure. This is followed by a definite steep fall (If 3 hPa fall = beware; 5 hPa fall = cyclone probably within 200 miles. If the wind is above force 8, the centre is probably within 100 miles)
- Lurid (odd coloured) sky caused by ice crystals in the upper atmosphere giving the effect of a horizontal rainbow. [The high fibrous (cirrus) clouds make the sunny sky appear to glow through a haze, as flames enveloped by smoke. Fiery copper-coloured sunrises and sunsets are common, especially in tropics. The direction of the streaky cirrus clouds can sometimes indicate the direction of the storm centre.] As you get closer to the cyclone, there is a significant increase in cloud formation.
- Changes in the strength and direction of wind, as it shift due to the Coriolis effect and becomes stronger closer to the cyclone.
- Sultry weather - very humid. Heavy rain within 150 miles of the centre.
- Unusual behaviour of sea birds. They will either roost ashore all day or disappear all together.

Fig 20.28: TYPICAL BAROGRAPH TRACE FOR A TRS

SAFETY PRECAUTIONS NEAR CYCLONES

As discussed above, when a cyclone is forecasted to hit a port, small vessels are better out of water or in a sheltered creek or anchorage. Large vessels usually handle storms better at sea. So, they secure for sea and sail out.

Large vessels can usually handle cyclonic bad weather as long as they are not within four or five hundred miles of the centre. However, it is advisable for all vessels to avoid this area due to the damage it can cause. Stay as far away from the storm centre or the eye as possible, which is about 20 miles across. The eye wall has the strongest winds and heaviest rainfall. The wind blows around the eye in the form of an inescapable eddy. Inside the eye there is little wind and the sky may be blue, but the sea is confused and swell heaves like a sleeping giant. The barometer will read very low. The eye can be a deceptively calm trap, from which escape can be difficult. A vessel trapped in the eye goes where the storm takes it.

If a cyclone warning is issued, monitor the storm and avoid it at all cost. Small vessels should head inshore away from violent winds to a safe anchorage or shallow mangrove creeks. In a creek, allow the vessel settle in the mud with anchors leading well ahead secured ashore without letting the vessel get too close to the shoreline.

Stow all gear as low in the vessel as possible and secure her watertight and wind tight. Secure deadlights and tape down glass portholes. Make sure the bilge pump is in working order. Take the motor off the tender and submerge the tender so it can't be tossed around. If it has positive buoyancy it will not sink. Filling any small light boat is a good way to secure it ashore.

If shelter is out of reach and you are near a reef, head for deeper water to ride out the storm. The major factors that would influence a vessel's decision in the proximity of a cyclone are:

- Size of vessel
- Sea-keeping capabilities of vessel
- Proximity of a suitable shelter
- Radio contact
- Position of the storm
- Path of the storm
- Position of vessel relative to storm - whether in navigable or non-navigable (more dangerous) semicircle.

Wind speeds must average at least 65 km/h (35 knots) for a low-pressure system to be labelled a tropical cyclone. It is therefore possible to inadvertently overtake a developing cyclone and be misled by the rising barometer while on its path. So, be careful.

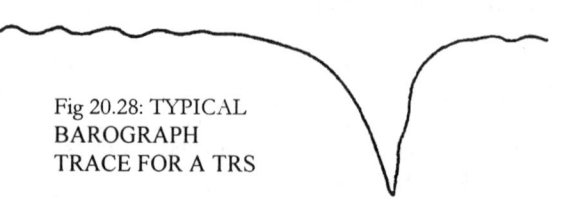

Chapter 20: Meteorology

Cyclone plotting maps are available from maritime centres in relevant ports as well as in the local telephone directories. The successive positions of the cyclone centre are continuously broadcast. The positions are broadcast in degrees and decimals of degrees (not degrees and minutes) of latitude and longitude, and in bearing and distance from a town or coastal landmark. The cyclone advice also defines the zones of dangerous winds that may be expected. These dangerous zones can be experienced on the coast several hours ahead of the arrival of the cyclone. The forecast speed and path of the cyclone are also broadcast.

CYCLONE PLOTTING MAP (& KEEPING CLEAR OF DANGER)

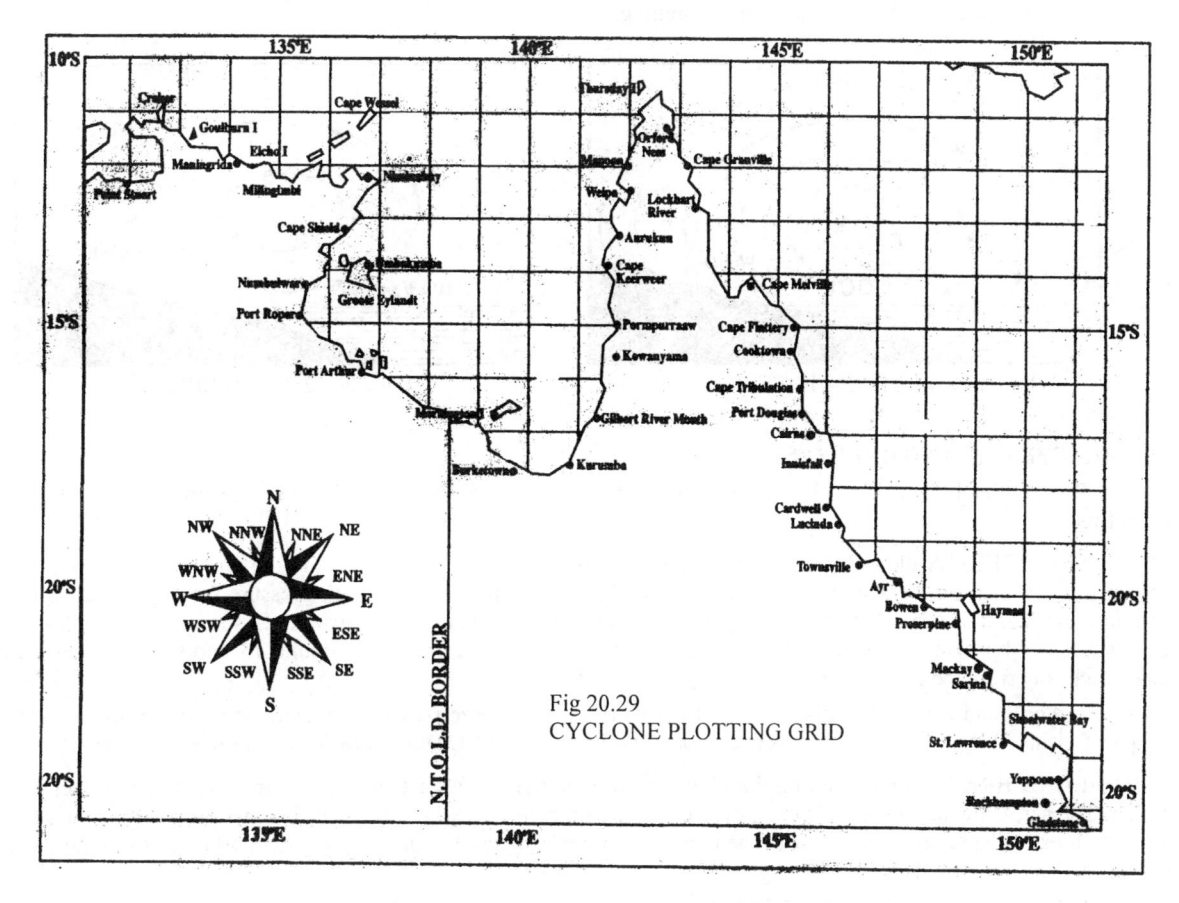

Fig 20.29 CYCLONE PLOTTING GRID

If you are unable to seek shelter from a cyclone, you must give it as wide a berth as possible as follows: Maintain a continuous plot of the cyclone's track by plotting the positions of its centre. From each successive position of its centre draw an angle of 40° on each side of its forecast path. This is to allow for the possibility of the storm deviating from this path. Then draw a curve between the above lines equal to 48 hours of its speed. For example, if the cyclone is travelling at 10 knots, draw the curve at 48 x 10 = 480 miles. Plan your courses outside of this area, updating it as and when new information comes to hand.

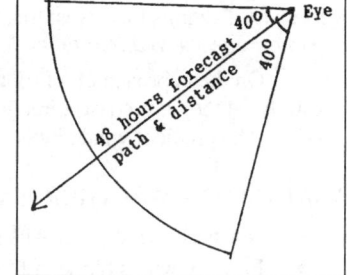

Fig 20.30: AVOIDING (RUNNING CLEAR) OF A CYCLONE (When unable to seek shelter)

HOW TO GET OUT OF A CYCLONE *(Figure 20.26)*

If you are already in a cyclone, you should take the following course in order to avoid its eye.

- If you are in the non-navigable semi-circle: steam to windward with wind on port bow. You will thus be steering a course away from and at right angles to the path of the storm.
- If you are in the path of the storm or in the navigable semi-circle: run with wind on the port quarter. Similarly, you will be steering a course away from and at right angles to the path of the storm.

(For the <u>Northern Hemisphere</u>, see the back of figures 20.24 and 20.26 in front of a light.)

YOUR LOCAL WEATHER

It is essential for all boat skippers to become familiar with their local weather patterns. Nationally speaking, the Northern half of Australia is prone to tropical cyclones, and the southern half to cold fronts. Warm fronts are restricted mainly to Tasmania.

More specifically, on the New South Wales Coast, winds are predominantly NE'ly in summer and SE'ly in winter. Sometimes hot westerlies blow during the summer months. The temperature on the coast is also influenced by sea breezes during the day and land breezes at night. Cold fronts and thunderstorms are common. Weather can suddenly become violent with gale force winds, rain, hale, thunder and lightening.

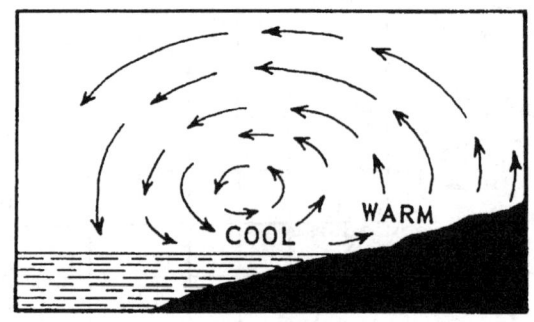

Fig 20.31: SEA BREEZE 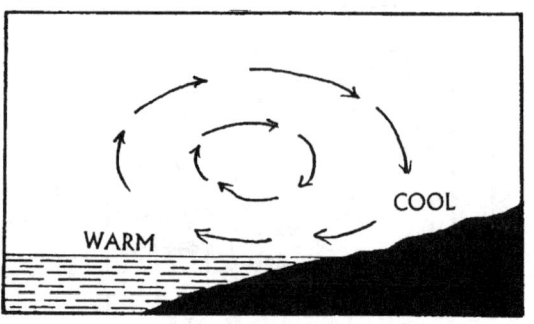   Fig 20.32: LAND BREEZE

SEA BREEZES & LAND BREEZES

Breezes that blow in coastal areas due to uneven heating and cooling of land and sea are known as Sea Breezes and Land Breezes.

SEA BREEZES blow during the day when land heats faster than the sea, and the atmospheric pressure on land becomes lower than on the sea. The surface wind thus blows from an area of high pressure (the sea) to an area of low pressure (the land). Cause = Effect. Being temperature driven, sea breezes attain maximum strength about mid-afternoon. Unless affected by trade winds or a pressure gradient, the speed of sea breezes is around 15 knots. They extend up to 20 miles each way from the coast.

In Perth and some other places around the world, a sea breeze may cool a scorching hot day by as much as 10 degrees Celsius. People thus refer to it as a 'doctor'. The **FREMANTLE DOCTOR** is well known for its strength.

LAND BREEZES blow during the night because land cools faster than the sea, making its atmospheric pressure higher. The surface wind direction is thus reversed. The land breeze flowing off the land is not as strong as the sea breeze, nor does it extend as far inland or to sea. Their speed is around 5 knots and they extend up to 10 miles from the coast. Land breezes become most evident after midnight.

On an east coast in the southern hemisphere, a sea breeze starts from the easterly direction in the morning. Then, on picking up speed as the day warms up, it swings to the north-easterly direction due to the Coriolis effect. Similarly, on a west coast it swings from west to southwest. As discussed earlier, the *Coriolis effect is the apparent shift in wind direction due to the earth's rotation: to the left in the southern hemisphere and to the right in the northern hemisphere.*

On the approach of a Front, the sea breeze, which is a local wind, weakens and swings back to the easterly (westerly on the west coast) direction. The synoptic or gradient wind of a Front backs to southward during the Front's passage. The gradient wind always backs in the southern hemisphere.

ANALYSING A WEATHER MAP: SOME POSSIBLE FORECASTS

- Northerly winds over Australia will carry hot, dry air from inland towards coastal areas.
- Easterly winds blowing from the sea over eastern Australia will create humid and sultry weather - often leading to "unstable" weather such as showers and thunderstorms. Westerly winds will do the same on the west coast.
- A cold front on the south Coast will replace hot dry north-westerlies with southerlies carrying cooler, often relatively humid air.
- Southerly wind blowing over northern Australia will carry cool to mild, relatively more moist air, as it travels from southern to northern parts.
- A high pressure normally results in stable weather.
- A low pressure normally causes unstable weather.
- Closer the isobars stronger the wind.

Chapter 20: Meteorology

WEATHER INFORMATION SOURCES FOR VESSELS

- Coastal: Newspapers, radio, TV, Telstra recorded weather service, Coast Radio Networks.
- Marine: Maritime Communication Stations (Coast Stations). See Chapter 18 and the last two pages of the book.
- Boating weather information on the Bureau of Meteorology's State-wise phone numbers listed on the last page of the book.
- Bureau's National Weather fax service, listed on the last two pages of the book.

Fig 20.33: SYNOPTIC CHART (WEATHER MAP) No. 1

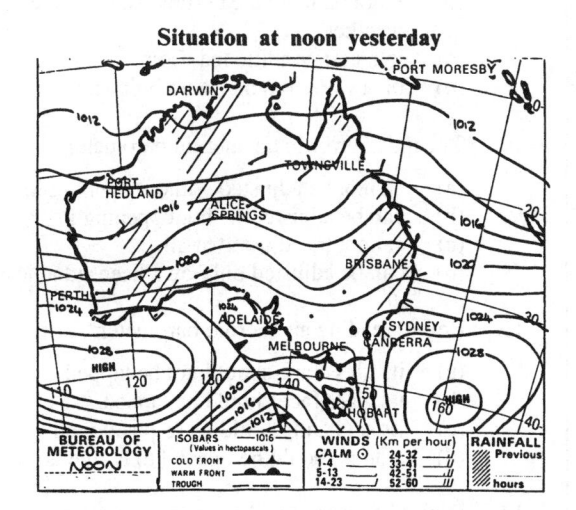

Satellite picture — noon yesterday.

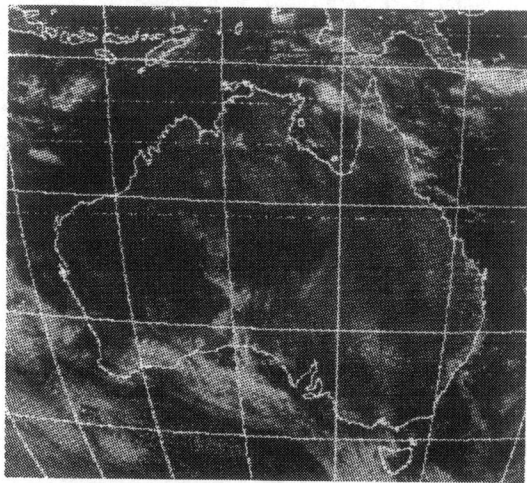

Cloud approaching WA is associated with an upper level trough. Cloud over the Bight is associated with a cold front. Cloud about the E coast is associated with a moist SE airstream.

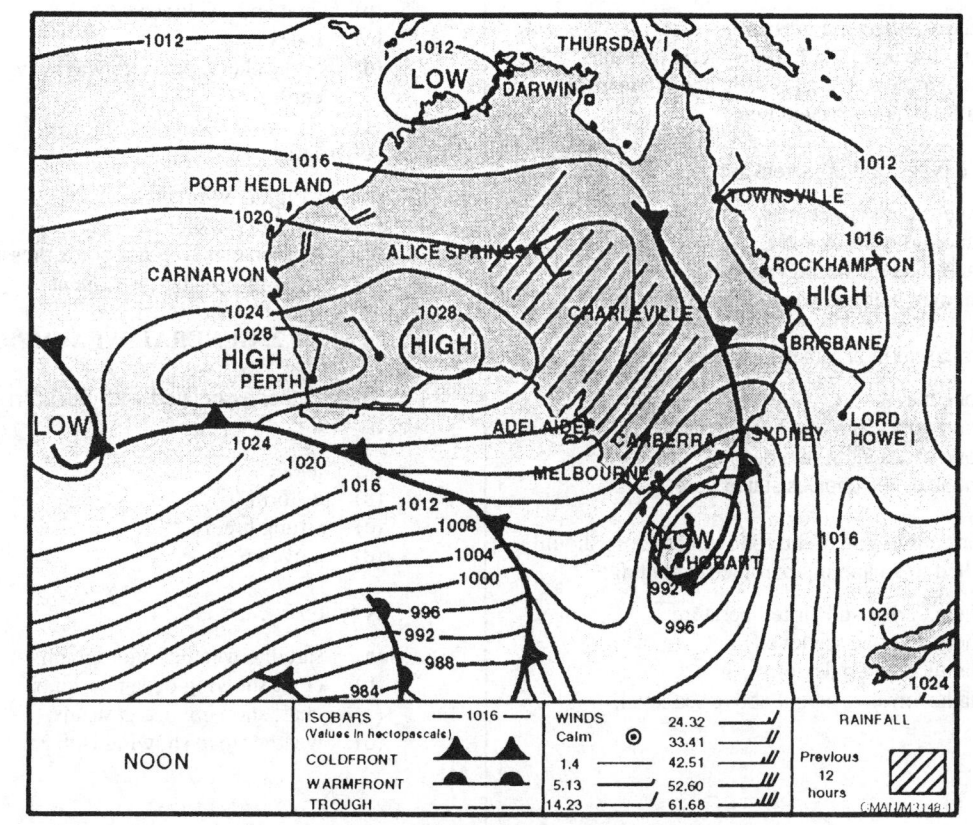

Fig 20.34: SYNOPTIC CHART (WEATHER MAP) No. 2

CHAPTER 20: QUESTIONS & ANSWERS

CHAPTER 20.1: BAROMETER

1. For observing changes in weather, an uncorrected aneroid barometer:

    (a) is of no use.
    (b) is good enough.
    (c) may be used if its error is known.
    (d) must be printed with "fair", rain" & "change".

2. The words "fair", "rain", and "change" written on the face of an aneroid barometer:

    (a) are only decorative.
    (b) apply only in the northern hemisphere.
    (c) apply only in the southern hemisphere.
    (d) are accurate weather information.

3. When painting a bulkhead on which an aneroid barometer is mounted, the following precaution must be observed:

    (a) don't take it down.
    (b) don't paint surface under the barometer.
    (c) don't alter its position.
    (d) take it down before painting.

4. A barometer or a barograph assists weather forecasting by indicating:

    (a) the atmospheric pressure.
    (b) the rise and fall of atmospheric pressure.
    (c) the amount and rate of change in pressure.
    (d) all the factors stated here.

5. The two types of barometers are:

    (a) aneroid and mercurial.
    (b) aneroid types A and B.
    (c) aneroid and partial vacuum types.
    (d) human hair and vacuum chamber types.

6. A hectoPascal in barometric pressure is:

    (a) one bar.
    (b) one-hundredth of a bar.
    (c) one-thousandth of a bar.
    (d) none of the choices stated here.

7. The corrections due to the following factors should be applied to an aneroid barometric reading:

    (a) index error and air temperature.
    (b) height above sea level.
    (c) latitude and index error.
    (d) index error & height above sea level.

8. A barometer can be checked for accuracy:

    (a) by calling a maritime coast station.
    (b) by comparing its reading with weather map.
    (c) by comparing its rise and fall over a week.
    (d) only by a licensed adjuster.

9. An aneroid barometer on board a vessel should be installed:

    (a) outdoors.
    (b) indoors, in an area protected from the weather.
    (c) in an air conditioned room.
    (d) on a warm engine room bulkhead.

10. The index error in an aneroid barometer:

    (a) cannot be adjusted without opening it.
    (b) can be adjusted without opening it.
    (c) cannot be adjusted at all.
    (d) can be adjusted only by a licensed adjuster.

11. Before reading an aneroid barometer:

    (a) its glass face should not be tapped.
    (b) its glass face should be tapped.
    (c) its set hand should be re-positioned.
    (d) a door should be opened to let the air in.

12. Diurnal changes in a barometer are caused by the:

    (a) sudden changes in atmospheric pressure.
    (b) land and sea breezes.
    (c) index error.
    (d) atmospheric pressure waves sweeping the earth.

13. A barograph is:

    (a) a digital barometer.
    (b) not a barometer.
    (c) a barometer which records pressure changes.
    (d) wind measuring instrument.

CHAPTER 20.2: GENERAL WEATHER

14. The weather associated with Beaufort wind force 8 is:

    (a) a gale.
    (b) a storm.
    (c) strong breeze.
    (d) a cyclone.

15. A synoptic chart is a:

    (a) satellite weather photograph.
    (b) chart showing ocean currents.
    (c) chart showing ocean winds.
    (d) weather map showing isobars.

Chapter 20: Meteorology

16. A heavy rain squall:
    (a) whips up the sea surface.
    (b) has no effect on the sea surface.
    (c) flattens the sea surface.
    (d) confuses the sea surface.

17. Coastal winds direction and speed are largely influenced by:
    (a) land and sea breezes.
    (b) cold fronts.
    (c) the mountain ranges.
    (d) warm fronts.

18. According to Buys Ballots Law for the southern hemisphere, if you face the wind the centre of a low pressure lies:
    (a) approximately 45° to your right.
    (b) approximately 45° to your left.
    (c) approximately 90° to your right.
    (d) approximately 90° to your left.

19. In meteorology, a Col is:
    (a) an abbreviation for Cold Front.
    (b) the centre of a low pressure.
    (c) the region between two highs and two lows.
    (d) none of the choices stated here.

20. In meteorology, a squall is a:
    (a) sudden increase in wind strength.
    (b) brief windstorm.
    (c) ridge of high pressure.
    (d) trough of low pressure.

21. In meteorology, a depression is:
    (a) any low pressure.
    (b) a tropical cyclone.
    (c) a cold front.
    (d) a ridge.

22. In meteorology, a trough is:
    (a) an elongated area of high pressure.
    (b) a line squall.
    (c) any low pressure.
    (d) an elongated area of low pressure.

23. In meteorology, a line squall is:
    (a) the forward edge of a band of thunderstorms.
    (b) a line of squalls along a coast.
    (c) an elongated squall.
    (d) none of the choices stated here.

24. In meteorology, a ridge is:
    (a) an elongated area of high pressure.
    (b) a line squall.
    (c) any high pressure.
    (d) an elongated area of low pressure.

25. The signs of an advancing depression (low pressure) are similar to that of a:
    (a) tropical cyclone.
    (b) cold front.
    (c) line squall.
    (d) squall.

26. Sea breezes are caused by atmospheric pressure being:
    (a) lower on land than on water during the day.
    (b) higher on land than on water during the day.
    (c) lower on land than on water during the night.
    (d) higher on land than on water during the night.

27. On the east coast of Australia, there is a high-pressure cell, and north of it there is a low-pressure cell. The coastal community between the two cells should experience the wind blowing from the:
    (a) north.
    (b) south.
    (c) east.
    (d) west.

28. In a low-pressure system in either hemisphere, the angle of the wind to the isobars (the angle of indraft) is as follows:
    (a) the angle is larger nearer to the centre.
    (b) the angle is smaller nearer to the centre.
    (c) the angle does not change.
    (d) there is no angle.

29. In the southern hemisphere, the wind around a low pressure blows as follows:
    (a) clockwise, inwards.
    (b) clockwise, outwards.
    (c) anti-clockwise, inwards.
    (d) anti-clockwise, outwards.

30. Around an anti-cyclone in the southern hemisphere, the wind blows:
    (a) clockwise, inwards.
    (b) clockwise, outwards.
    (c) anti-clockwise, inwards.
    (d) anti-clockwise, outwards.

31. Thunderstorms generate lightning and thunder in the following order:
    (a) lightning is generated before thunder.
    (b) thunder is generated before lightning.
    (c) both are generated at the same time.
    (d) there is no fixed order.

32. It is important to regularly log changes in barometric readings in order to establish:

    (a) weather changes.
    (b) diurnal changes.
    (c) weather and diurnal changes.
    (d) a safe location in a storm.

33. The symbol on a weather map for 20 km/h (not 20 knots) wind is:

    (a) an arrow with two feathers.
    (b) an arrow with four feathers.
    (c) an arrow with one feather.
    (d) the letters "MR".

34. For personal protection from lightning during electrical storms, don't:

    (a) stay on open decks.
    (b) touch two separately grounded objects.
    (c) immerse your limbs or body in water.
    (d) ground lights fitted on cabin tops.

35. The atmospheric pressure falling more than 5 hPa below normal indicates:

    (a) the barometer is faulty.
    (b) a severe storm is in the vicinity.
    (c) a diurnal variation.
    (d) the weather would improve.

36. The Coriolis effect:

    (a) is zero at the poles.
    (b) is maximum at the equator.
    (c) is to the left in the southern hemisphere.
    (d) causes clockwise spiralling of wind into high pressure centres in the southern hemisphere.

37. The Coriolis effect:

    (a) is zero at the poles.
    (b) is maximum at the equator.
    (c) is to the right in the southern hemisphere.
    (d) causes clockwise spiralling of wind into low pressure centres in the southern hemisphere.

38. The Maritime air masses:

    (a) are called trade winds.
    (b) blow from high to low pressure.
    (c) form over the land.
    (d) form over the oceans.

39. The sinking air generally generates:

    (a) dry and settled weather.
    (b) low pressure belts.
    (c) unsettled weather.
    (d) clouds.

40. The ITC:

    (a) is another name for tropical cyclones.
    (b) is like the weather equator.
    (c) causes the atmospheric pressure to rise during summer months in Australia.
    (d) keeps the equatorial low pressure belt stationary.

41. A Gentle Breeze has the following range of wind speed:

    (a) 15 to 20 knots.
    (b) 1 to 5 knots.
    (c) 7 to 10 knots
    (d) 5 to 10 knots.

42. A Strong Breeze has the following range of wind speed:

    (a) 22-27 knots.
    (b) 10 to 15 knots.
    (c) 30-40 knots.
    (d) 5 to 10 knots.

43. A 9 hPa fall in barometric pressure in 3 hours is most likely an indication of:

    (a) diurnal variation.
    (b) strong winds within the next 12 hours.
    (c) a gale or storm within the next 3 hours.
    (d) a faulty barometer.

44. A 6 hPa fall in barometric pressure in 3 hours is most likely an indication of gale force winds within the next:

    (a) 12 hours.
    (b) 6 hours.
    (c) 3 hours.
    (d) 24 hours.

45. National weather observations are made at least every:

    (a) 3 hours.
    (b) 6 hours.
    (c) 1 hour.
    (d) 4 hours.

46. Routine weather forecasts are compiled from the:

    (a) 4-hourly weather observations.
    (b) 3-hourly weather observations.
    (c) hourly weather observations.
    (d) 6-hourly weather observations.

47. The strength of gusts of wind:

    (a) is the forecast wind speed.
    (b) can be up to 40% greater than the forecast wind speed.
    (c) depends on the swell height.
    (d) is measured over a 10-minute period.

48. The true wind direction is indicated by the:
    (a) funnel smoke.
    (b) telltale on a shroud.
    (c) direction of the swell.
    (d) direction of the waves.

49. Wave height is the:
    (a) distance between two wave crests.
    (b) vertical distance between the top of the wave crest and the bottom of the trough.
    (c) vertical distance between the top of the wave crest and the flat sea.
    (d) average height of ten waves.

50. The "significant wave height" stated in weather forecasts is the:
    (a) average of the highest one-third of the waves.
    (b) maximum wave height measured.
    (c) maximum wave height forecast.
    (d) average wave height.

51. A rogue wave:
    (a) is typically one in every 100 waves.
    (b) is typically one in every 200 waves.
    (c) is usually produced by the joint forces of wind waves and swell waves.
    (d) is usually produced by the wind blowing over the sea.

52. Swell waves:
    (a) is another name for sea waves.
    (b) are caused by wind blowing over the sea for a long period.
    (c) are short lived.
    (d) are localised waves.

53. Which of the following statements is incorrect?
    (a) The length of water over which the wind blows is known as the fetch.
    (b) The stronger the wind and longer the fetch, the bigger the waves become.
    (c) Wind blowing over water first develops swell waves, then wind waves.
    (d) Swell waves travel thousands of miles over a number of days.

54. Sea and swell waves:
    (a) do not occur at the same time.
    (b) always come from the same direction.
    (c) always occur at the same time.
    (d) can occur at the same time and can come from different directions.

55. Which of the following statements is incorrect?
    (a) It is probable that 1 in every 1000 waves has a height at least 1.86 times the significant wave height.
    (b) It is probable that 1 in every 2000 waves is twice the height of the significant wave height.
    (c) In meteorological observations, the cloud cover is expressed in tenths of the sky.
    (d) The only correct way to measure visibility at sea is by measuring the radar distance of an echo that has just become visible.

56. Which of the following statements is incorrect?
    (a) Low clouds are those with bases less than 2 km above the horizon.
    (b) Medium clouds are those with bases between 2 and 6 km above the horizon.
    (c) High clouds are those with bases over 6 km above the horizon.
    (d) In the Howard system, the clouds are categorised based on the height of the top of the cloud.

57. In relation to Relative Humidity, which of the following statements is incorrect?
    (a) It is the amount of moisture in the air.
    (b) A 100% humidity means the air is loaded with moisture to its capacity.
    (c) It is measured with a hydrometer.
    (d) A pair of wet & dry bulb thermometers is known as the hygrometer.

58. Which of the following statements is incorrect?
    (a) Fog is a form of low clouds.
    (b) Fog is caused by suspension of small water droplets in the air.
    (c) In a cloud, when the horizontal visibility is less than 100 meters then technically it is fog.
    (d) Mist is a less dense form of fog.

59. Which of the following statements is incorrect? In Wind Warnings issued by radio, Strong Wind Warnings:
    (a) are issued for the high seas.
    (b) are for winds of 25-33 knots.
    (c) are for the Beaufort Force 6-7.
    (d) are prefixed by SECURITÉ.

60. In relation to thunderstorms, which of the following statements is incorrect?

    (a) They can be a serous hazard for small vessels.
    (b) They are local storms.
    (c) Clouds often move in different directions to the wind at the surface.
    (d) They usually move in the direction opposite to in which the 'anvil' is pointing.

61. Which of the following statements is incorrect?

    (a) La Nina causes southeast trade winds to strengthen.
    (b) El Nino causes draught in Eastern Australia.
    (c) La Nina causes higher frequency of cyclones on the east coast of Australia.
    (d) A positive Southern Oscillation Index indicates El Nina.

62. The worldwide radio weather broadcast system is detailed in the following publication:

    (a) Admiralty List of Radio Signals, Volume 3
    (b) Annual Notices to Mariners
    (c) Weekly Notices to Mariners
    (d) The Mariners Handbook

CHAPTER 20.3: FRONTS

63. In the absence of a weather forecast, the most advanced and reliable warning of an approaching cold front or a Southerly Buster is the:

    (a) sharp drop in barometer.
    (b) shift in wind direction.
    (c) towering high grey clouds.
    (d) thunderstorms.

64. In a cold front, it is common for wind speed to reach up to force:

    (a) two.
    (b) eight.
    (c) ten.
    (d) twelve.

65. The difference between a cold front and a warm front is:

    (a) the difference between good and bad weather.
    (b) that one is more violent than the other.
    (c) that one is a low and the other a high.
    (d) none.

66. Some parts of Australia experience Warm Fronts. The weather associated with them is:

    (a) good weather.
    (b) the bad weather that sets in for a few days.
    (c) more unpredictable than with the Cold Fronts.
    (d) more violent than with the Cold Fronts.

67. Warm Fronts are not common on the Australian mainland, because:

    (a) Australia is a large island.
    (b) deep lows develop mainly south of the mainland.
    (c) Australia has a desert in the middle.
    (d) they are not as strong as Cold Fronts.

68. With the knowledge of a severe cold front approaching, the following steps are advisable for the safety of your vessel at sea:

    (i) Seek shelter if close to shore.
    (ii) Advise your passage details to Coast Station.
    (iii) Batten down.
    (iv) Rig sea anchor from astern.

    The correct answer is:
    (a) all except (i).
    (b) all except (ii).
    (c) all except (iii).
    (d) all except (iv).

69. A cold front causes more wind velocity than a warm front because:

    (a) it doesn't last as long.
    (b) it has a steeper pressure gradient.
    (c) it originates from a low pressure.
    (d) it is cyclonic.

70. Immediately after the cold front has passed, the barometric pressure...

    (a) rises and winds are gusty.
    (b) falls and winds are gusty.
    (c) rises and there is no wind.
    (d) falls and there is no wind.

71. An occluded front:

    (a) is the combination of a warm front and a cold front.
    (b) occurs when a faster moving warm front catches up with a cold front.
    (c) produces weather similar to a cold front.
    (d) is commonly seen on Australian weather maps.

72. In weather maps, an Occluded Front is shown as a line of:
    (a) alternative barbs and semi-circles.
    (b) barbs.
    (c) semicircles.
    (d) none of the choices stated here.

73. Which of the following statements is incorrect?
    (a) East Coast Lows develop with 5° of latitude or Longitude from the coast.
    (b) A Bomb is a severe East Coast Low.
    (c) East Coast Lows average one a month.
    (d) The weather before an East Coast Low can be deceptively good.

74. Which of the following is not in the usual combination of signs of an approaching East Coast Low?
    (a) Poor visibility.
    (b) Drop in wind speed.
    (c) A line of haze or low clouds approaching from the south.
    (d) A change in the colour of the sea.

CHAPTER 20.4: CYCLONES

75. The warning signs of an approaching tropical cyclone are:
    (i) long heavy swell.
    (ii) poor visibility.
    (iii) lurid (gloomy) sky.
    (iv) falling barometer.

    The correct answer is:
    (a) all except (i).
    (b) all except (ii).
    (c) all except (iii).
    (d) all except (iv).

76. A "Flash Cyclone Warning" is issued by the meteorological office to notify the coastal communities of:
    (a) a cyclone within 400 miles of the coast.
    (b) an urgent amendment to an earlier warning.
    (c) a cyclone expected on the coast in 24 hours.
    (d) a cyclone within 800 miles of the coast.

77. The warning signs of an approaching tropical cyclone are:
    (i) falling barometric pressure.
    (ii) increasing wind speed.
    (iii) shifting wind direction.
    (iv) Wind blowing onshore.

    The correct answer is:
    (a) all except (i).
    (b) all except (ii).
    (c) all except (iii).
    (d) all except (iv).

78. A "Cyclone Watch" is issued by the meteorological office to notify the coastal communities of:
    (a) a cyclone within 400 miles of the coast.
    (b) an urgent amendment to an earlier warning.
    (c) a cyclone expected on the coast in 24 hours.
    (d) a cyclone within 800 miles of the coast.

79. A "Cyclone Warning" is issued by the meteorological office to notify the coastal communities of:
    (a) a cyclone within 400 miles of the coast.
    (b) an urgent amendment to an earlier warning.
    (c) a cyclone expected on the coast in 24 hours.
    (d) a cyclone within 800 miles of the coast.

80. The information broadcasted to the coastal communities of an approaching tropical cyclone includes:
    (i) rate of rise of barometric pressure.
    (ii) cyclone's name.
    (iii) forecast of storm surges.
    (iv) location of the cyclone's centre.

    The correct answer is:
    (a) all except (i).
    (b) all except (ii).
    (c) all except (iii).
    (d) all except (iv).

81. A "Flash Cyclone Warning" is issued under the following circumstances:
    (i) A new coastal area is under threat.
    (ii) An urgent amendment to an earlier warning.
    (iii) Cyclone has changed course.
    (iv) Cyclone on predicted course and speed.

    The correct answer is:
    (a) all except (i).
    (b) all except (ii).
    (c) all except (iii).
    (d) all except (iv).

82. Faced with a cyclone, the following steps are advisable for the safety of your vessel at sea:

    (i)   Seek shelter in its relatively calm eye.
    (ii) Advise your passage details to Coast Station
    (iii) Batten down.
    (iv) Stream warps.

    The correct answer is:
    (a) all except (i).
    (b) all except (ii).
    (c) all except (iii).
    (d) all except (iv).

83. The recommended rule for avoiding a tropical cyclone in the southern hemisphere is to:

    (a) stay out of a given arc ahead from its centre
    (b) sail with wind on port bow.
    (c) sail wind on port quarter.
    (d) heave to.

84. The recommended rule for getting out of navigable semicircle of a tropical cyclone in the southern hemisphere is to:

    (a) sail with wind on port bow.
    (b) sail with wind on port quarter.
    (c) heave to.
    (d) stay out of a given arc ahead from its centre

85. The recommended rule for getting out of non-navigable (dangerous) semicircle of a tropical cyclone in the southern hemisphere is to:

    (a) sail with wind on port bow.
    (b) sail with wind on port quarter.
    (c) heave to.
    (d) stay out of a given arc ahead from its centre

86. According to the Buy Ballot Law for the southern hemisphere, a person facing the wind has the centre of the low pressure:

    (a) right ahead.
    (b) about 90° on the right.
    (c) about 90° on the left.
    (d) directly behind.

87. In a cyclone, a trough line is:

    (a) the line of lowest pressure.
    (b) another name of the eye.
    (c) its curving point.
    (d) the southern limit of its travel.

88. A vessel on a cyclone's path will experience the wind:

    (a) backing.
    (b) veering.
    (c) from a steady direction.
    (d) gusting in strength.

89. A vessel in the eye of a cyclone will experience:

    (a) the lowest barometric pressure.
    (b) good weather.
    (c) steady wind direction.
    (d) no wind.

90. The following items belong in the list of parts of a tropical cyclone:

    (i)   trough line.
    (ii) vortex.
    (iii) vertex.
    (iv) ridge line.

    The correct answer is:
    (a) all except (iv).
    (b) all except (iii).
    (c) all of them.
    (d) all except (i).

91. After a cyclone's trough line has passed over a vessel, her barometer will:

    (a) start to rise.
    (b) start to fall.
    (c) remain steady.
    (d) show the highest reading.

92. A vessel in a cyclone experiencing a steady wind direction and rising barometric pressure is:

    (a) on cyclone's trough line.
    (b) on cyclone's path.
    (c) on cyclone's track.
    (d) in cyclone's dangerous quadrant.

93. A vessel in a cyclone experiencing a steady wind direction and falling barometric pressure. She is:

    (a) on cyclone's trough line.
    (b) on cyclone's path.
    (c) on cyclone's track.
    (d) in cyclone's dangerous quadrant.

94. A vessel in the left (non-navigable) semi-circle of a tropical cyclone in the southern hemisphere will experience wind shift as follows:

    (a) veering.
    (b) backing.
    (c) steady.
    (d) unpredictable.

95. The highest storm surges are caused by storms of the following description:

    (a) slow moving & large diameter.
    (b) fast moving & small diameter.
    (c) slow moving & small diameter.
    (d) fast moving & large diameter.

Chapter 20: Meteorology

96. The cyclone season in the southern hemisphere is:
    (a) April to November.
    (b) January to April.
    (c) November to April.
    (d) April to July.

97. The cyclone warning originating centres in Australia are:
    (i) Perth
    (ii) Sydney
    (iii) Darwin
    (iv) Brisbane

    The correct answer is:
    (a) all except (i).
    (b) all of them.
    (c) none of them.
    (d) all except (ii).

98. The significance of the knowledge of Trough Line in a Cyclone is that one should try & manoeuvre the vessel to a position:
    (a) behind the trough line.
    (b) on the trough line.
    (c) South of the trough line.
    (d) North of the trough line.

99. In relation to seeking a safe haven from a cyclone, which of the following is incorrect?
    (a) Consider the distance of the safe haven from the vessel's position.
    (b) Every vessel should head for the nearest port.
    (c) Mangroves provide a good windbreak without being too high.
    (d) Consider the proximity of the safe haven in relation to the cyclone's path.

100. The information broadcast with a tropical cyclone warning does not include:
    (a) cyclone categories above 10.
    (b) cyclone's name.
    (c) location of the cyclone's centre.
    (d) forecasts of storm surges.

101. Four events must take place for a tropical cyclone to form. Which of the following events is not included among them?
    (a) Sea surface temperature must be above 27°C.
    (b) A surface low-pressure area must exist over the ocean.
    (c) Surface winds must converge into an eye.
    (d) A tropical air wave disturbance must assist the Coriolis wind shift.

102. Tropical cyclones develop in three stages, in the following order:
    (a) Depression – Storm - Cyclone.
    (b) Storm – Depression - Cyclone.
    (c) Cyclone – Depression – Storm.
    (d) Depression – Cyclone - Storm.

103. In relation to tropical cyclones in the southern hemisphere, which of the following statements is incorrect?
    (a) The normal path of a cyclone is parabolic.
    (b) A vessel on the eastern side of a cyclone is likely to remain in it for a longer period.
    (c) A vessel on the eastern side of a cyclone has a greater possibility of getting trapped in its centre.
    (d) The winds are stronger on the western side of a cyclone because they are moving in the same direction as the eye.

104. In relation to the most dangerous quadrant of a tropical cyclone, which of the following statements is incorrect?
    (a) The winds and waves are running almost head-on into the path of the storm.
    (b) Any recurving of the cyclone is most likely to be away from this quadrant.
    (c) The winds will try to blow a vessel into the path of the storm.
    (d) The winds and waves are the strongest and the biggest.

105. In relation to the Angle of Indraft, which of the following statements is incorrect?
    (a) It is the angle at which the wind cuts isobars.
    (b) Wind tends to eddy round the centre of the cyclone.
    (c) On the outskirts of a cyclone, the wind is almost parallel to the isobars.
    (d) An angle of 0° to the isobars is 90° to the path of the cyclone.

106. In relation to contingency plans for Tropical Cyclones, which of the following statements is incorrect?
    (a) Vessels should make contingency plans during calm weather.
    (b) The Australian contingency plans are aimed at advising vessels to move into designated sheltered areas on commencement of destructive winds.
    (c) The cyclone-affected Australian ports have cyclone contingency plans in place.
    (d) Vessels with shallower drafts are moved to the upper reaches of the creeks.

107. In relation to the colour-coded cyclone alert system used in Australia, which of the following statements is correct?
    (a) Some cyclone affected ports use colour-coding system to describe the closeness of a cyclone.
    (b) The colour-coding is also used by the Bureau of Meteorology.
    (c) The colour-coding is uniform in all cyclone affected ports.
    (d) In Cairns, Yellow Alert means destructive winds are expected within 6 hours.

108. In relation to the warning signs of an approaching cyclone, which of the following statements is incorrect?
    (a) If the wind is above force 8, the centre is probably within 100 miles.
    (b) Tides become higher.
    (c) There is a marked absence of swell.
    (d) If the barometer falls 5 hPa, the centre is probably within 200 miles.

109. In relation to Tropical Cyclones in the southern hemisphere, which of the following statements is incorrect?
    (a) They originate between 5° and 15° degrees south latitude.
    (b) They generally start their journey by moving in the SE'ly direction.
    (c) A vessel trapped in the eye will find it very difficult to get out.
    (d) A vessel in the eye will generally experience light winds but heavy confused seas.

110. In relation to the Tropical Cyclone Watch Messages and Warnings, which of the following statements is incorrect?
    (a) The Watch Messages are reviewed every 3 hours & issued every 6 hours.
    (b) Warnings are issued at least every 3 hours, and possibly every hour when the cyclone is close to the coast.
    (c) The Watch Messages are reviewed & issued every 3 hours.
    (d) Warnings are issued when gale force or stronger winds are expected to develop in a coastal area within 24 hours.

111. There is a relationship between the fall of barometric pressure and the approximate direction of the centre of the cyclone. In this regard, which of the following statements is incorrect?
    (a) 10 hPa fall – the centre is 10 points to the left of the wind direction.
    (b) 20 hPa fall – the centre is 8 points to the left of the wind direction.
    (c) Barometer begins to fall – the centre is 12 points to the left of the wind direction.
    (d) Barometer begins to fall - the centre is in the same direction as the wind.

112. There is a tropical revolving storm in the vicinity. The wind strength is increasing and the direction is veering. While proceeding at your best speed, which of the following courses would you steer in relation to the wind direction?
    (a) Keep wind on the port bow.
    (b) Keep wind on the starboard bow.
    (c) Keep wind on the port quarter.
    (d) Keep wind on the starboard quarter.

113. In the vicinity of a tropical revolving storm, a vessel is experiencing winds, which are slowly backing and increasing in speed. The vessel is:
    (a) in the navigable semi-circle
    (b) in the dangerous semi-circle
    (c) on cyclone's path
    (d) on cyclone's track

CHAPTER 20: ANSWERS

1 (b), 2 (a), 3 (d), 4 (d), 5 (a),
6 (c), 7 (d), 8 (a), 9 (b), 10 (b),
11 (b), 12 (d), 13 (c), 14 (a), 15 (d),
16 (c), 17 (a), 18 (d), 19 (c), 20 (b),
21 (a), 22 (d), 23 (a), 24 (a), 25 (a),
26 (a), 27 (c), 28 (b), 29 (a), 30 (d),
31 (c), 32 (c), 33 (c), 34 (a), 35 (b),
36 (c), 37 (d), 38 (d), 39 (a), 40 (b),
41 (c), 42 (a), 43 (c), 44 (b), 45 (a),
46 (d), 47 (b), 48 (d), 49 (b), 50 (a),
51 (c), 52 (b), 53 (c), 54 (d), 55 (c),
56 (d), 57 (c), 58 (c), 59 (a), 60 (d),
61 (d), 62 (a), 63 (a), 64 (b), 65 (b),
66 (b), 67 (b), 68 (d), 69 (b), 70 (a),
71 (a), 72 (a), 73 (c), 74 (a), 75 (b),
76 (b), 77 (d), 78 (a), 79 (c), 80 (a),
81 (d), 82 (a), 83 (a), 84 (b), 85 (a),
86 (c), 87 (a), 88 (c), 89 (a), 90 (a),
91 (a), 92 (c), 93 (b), 94 (b), 95 (a),
96 (c), 97 (d), 98 (a), 99 (b), 100 (a),
101 (c), 102 (a), 103 (d), 104 (b), 105 (c),
106 (b), 107 (a), 108 (c), 109 (b), 110 (c),
111 (d), 112 (c), 113 (b)

Chapter 21

HOW ENGINES WORK & THEIR PARTS

MOTOR CAR PETROL ENGINE   - THE CONCEPT

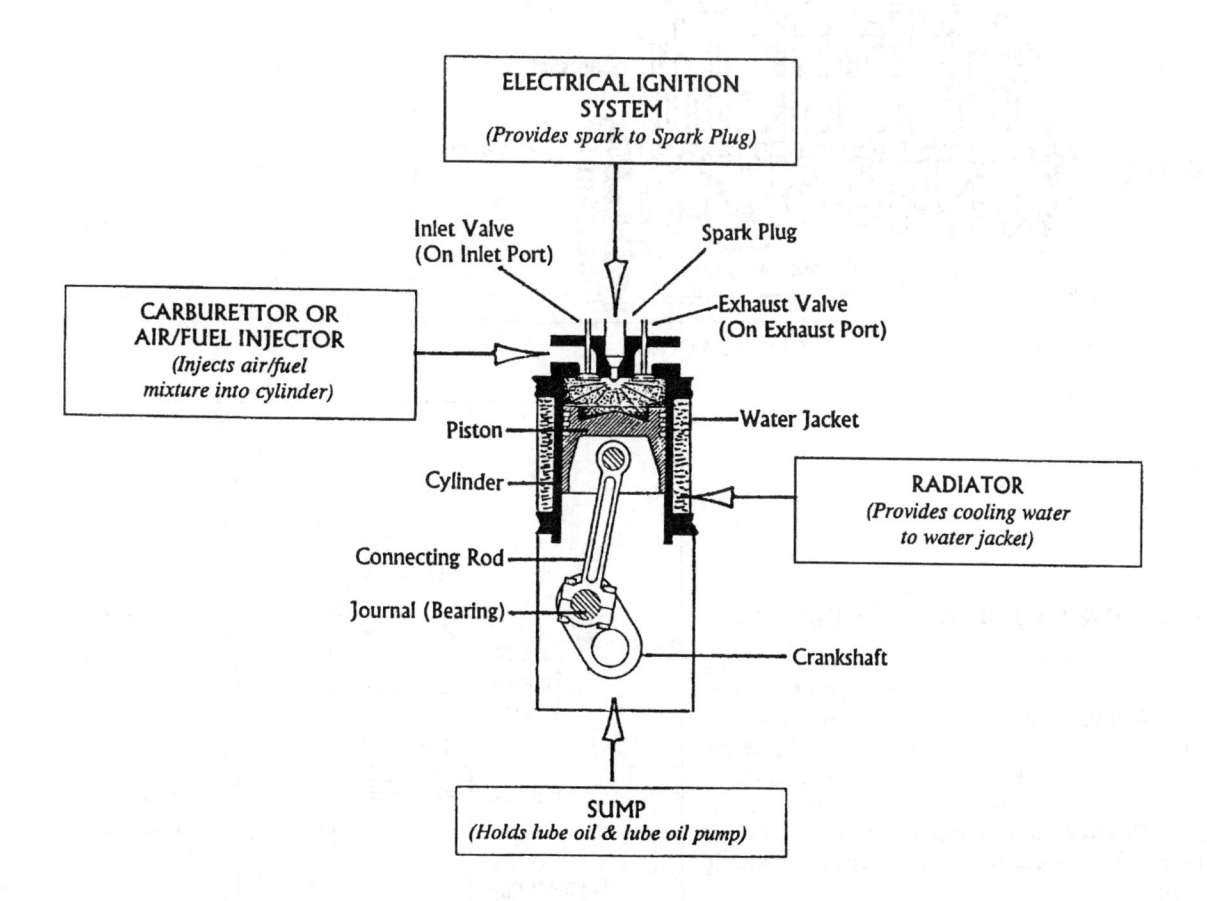

FIG 21.1:
BASIC ARRANGEMENT OF A FOUR-CYCLE (or FOUR-STROKE)
INTERNAL COMBUSTION ENGINE

INTERNAL COMBUSTION ENGINES

These are petrol, LPG or diesel driven engines. Fuel and air mixture is burnt inside one or more cylinders. The generated heat increases gas pressure, which moves the piston in the cylinder. A crankshaft, which is connected to the pistons by connecting rods, thus rotates and turns the wheels in a motorcar or the propeller in a boat. An engine may consist of any number of cylinders - four or six being most common.

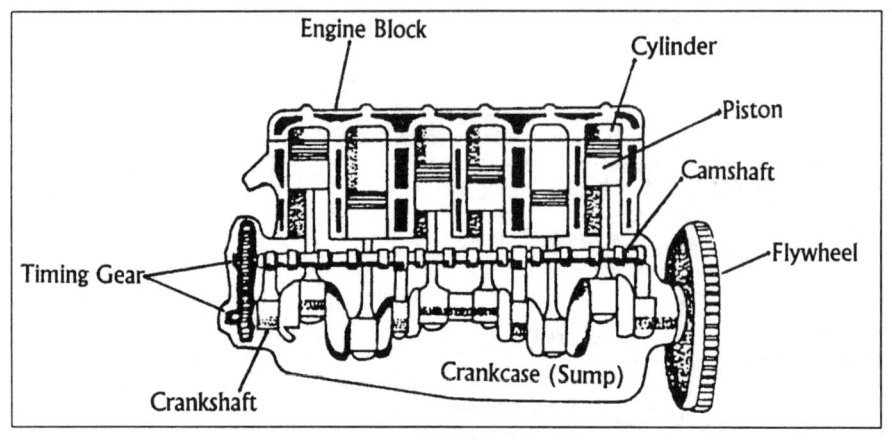

Fig 21.2: A 6-CYLINDER ENGINE LAYOUT

FOUR-STROKE & TWO-STROKE ENGINES

Engines come in two operating varieties: 4-stroke and 2-stroke (also referred to as **four-cycle and two-cycle engines**). A 4-stroke engine uses four piston strokes to complete one power cycle. The engine fires once for every two revolutions of the crankshaft. A 2-stroke engine, on the other hand, uses two piston strokes to complete one power cycle. Therefore, it fires once for each revolution of the crankshaft. The stroke cycles are discussed below in more details.

Figure 21.3 illustrates the operation of inlet and exhaust valves on a cylinder in a 4-stroke engine. The crankshaft (through a gear and chain drive) operates the valves. The firing is timed so that each cylinder completes one power cycle per two revolutions of the crankshaft.

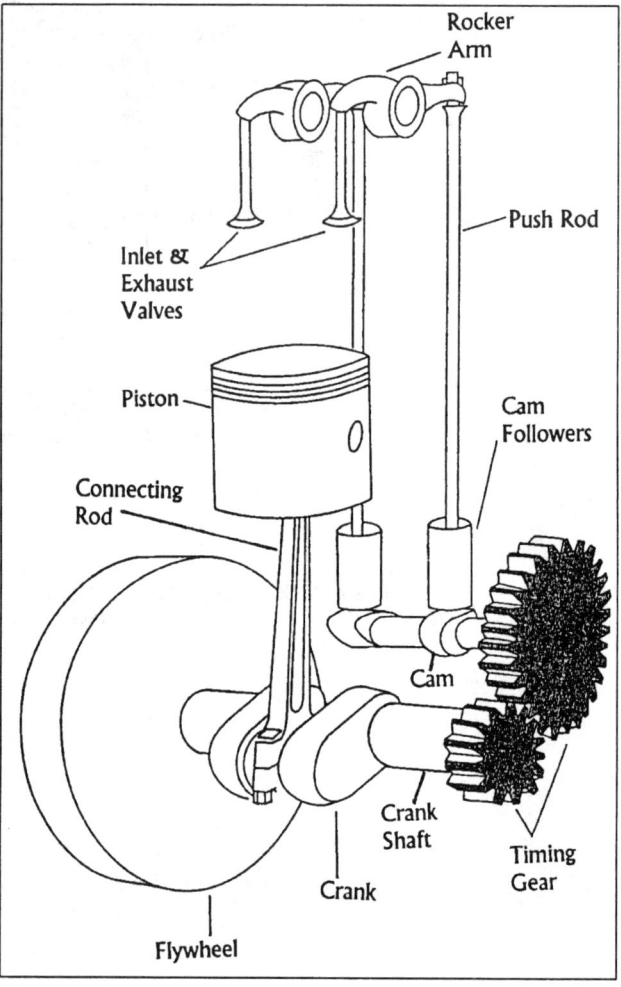

Fig 21.3: OPERATION OF INLET & EXHAUST VALVES ON A CYLINDER IN A 4-STROKE ENGINE

Chapter 21: How Engines Work & Their Parts

PETROL & DIESEL ENGINES

PETROL ENGINES

In petrol engines, power is produced by spark plugs igniting fuel-air mixture received from an Electronic Fuel Injection (EFI) system (or a Carburettor in older engines) and compressed by pistons.

Compared with diesel engines of similar weight, petrol engines generate greater power at higher engines revolutions. Therefore, they are more popular in motorcars as well as in stern drives and outboards. The weight of a heavy diesel stern drive or outboard would push the vessel out of trim.

However, petrol engines are less reliable than diesel engines in the cold and damp marine environment, coupled with the irregular use of boats. They also present a greater fire hazard due to a lower flash point of petrol than that of diesel fuel.

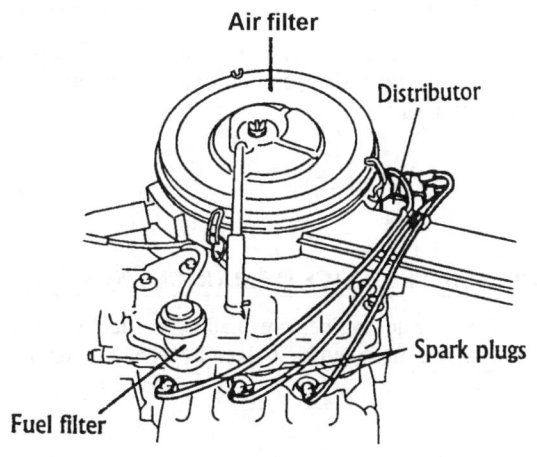

Fig 21.4: A 3-CYLINDER PETROL ENGINE

DIESEL ENGINES

In a conventional diesel engine, power is produced by hot compressed air igniting fuel sprayed under very high pressure into the cylinder head. A diesel engine does not employ a carburettor to mix fuel and air, nor an electrical system (spark plugs, etc.) to ignite the mixture. Instead, it uses the pistons to compress the air to 3000 kPa (which causes it to become very hot) and the fuel is ignited as soon as it is injected into the cylinder. Some are also fitted with a *Heater Plug* in the inlet manifold or a *Glow Plug* in the pre-combustion chamber of each cylinder to provide additional heat to the combustion air during starting. They help to break the chill in the air before it enters the cylinders.

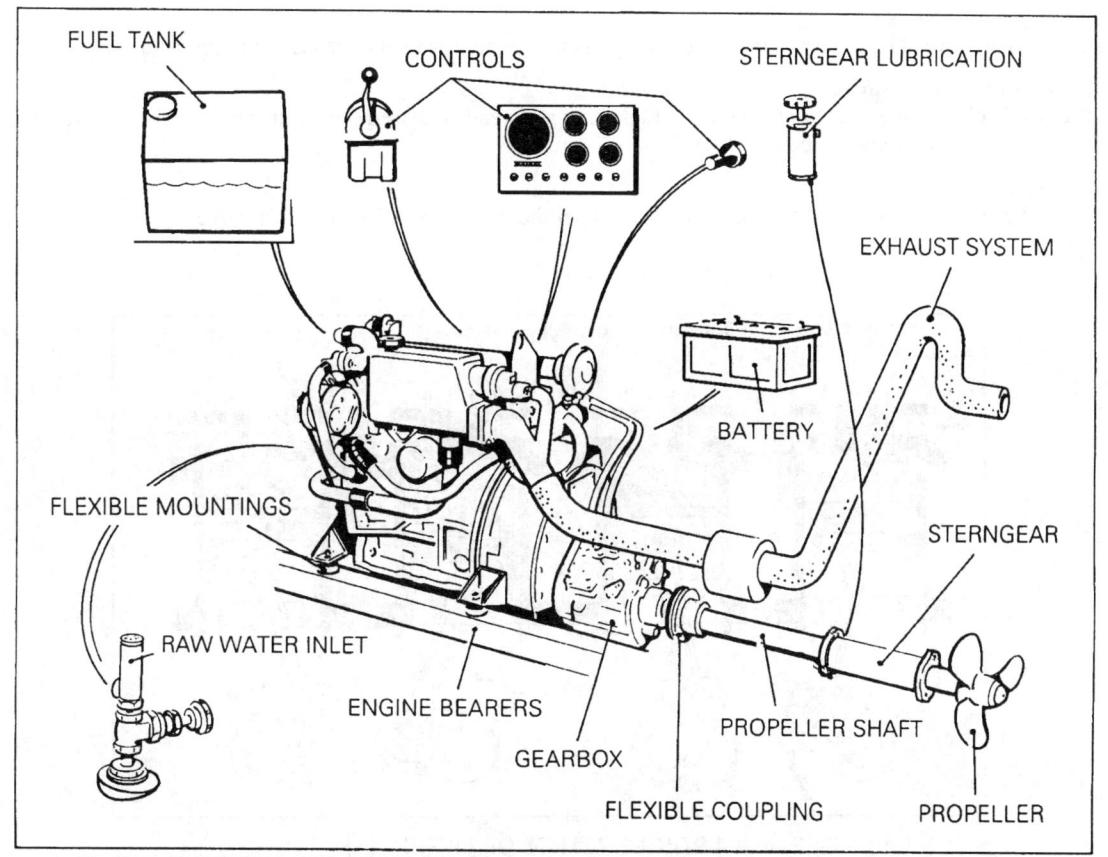

Fig 21.5: PRINCIPAL COMPONENTS OF AN INBOARD MARINE ENGINE (A Lister Engine)

Modern engines employ the ***Electronic Fuel Injection System*** in which fuel and air are mixed more thoroughly in a small pre-combustion chamber before entering a cylinder. This system burns the fuel much better to maximise fuel economy and power. Such engines are also less polluting. In order to meet the IMO 2000 emission regulations (which have not yet become universal), most new engines are being fitted with EFI, turbocharger and aftercooler.

As stated earlier, diesel engines are heavier and slower revving. But they are more reliable than petrol engines, because they do not depend on external carburetion or an electrical spark for ignition. Most inboard marine engines are diesel engines.

CHARACTERISTICS OF MARINE ENGINES

Marine engines are made specifically for use in confined spaces and to cope with the rolling and pitching of vessels. The sump can be drained by a hand pump or, as in motorcars, by unscrewing a drain plug in the base of the sump.

FOUR-STROKE ENGINES – THE STROKES

A 4-stroke engine uses four piston strokes to complete one power cycle. During the four strokes, the inlet and exhaust valves of each cylinder open and close once and the crankshaft turns twice. The engine fires once for every two revolutions of the crankshaft.

THE FOUR STROKE CYCLE OF A **PETROL OR LPG** ENGINE

1st Stroke: *Induction* or Inlet Stroke:
   The piston moves down and the inlet valve opens to admit a mixture of fuel and air.

2nd Stroke: *Compression* Stroke:
   As the inlet valve closes, the piston rises to compress the mixture, which increases its temperature.

3rd Stroke: *Power* or Ignition Stroke:
   The spark plug ignites the compressed air/fuel mixture, creating extreme heat and pressure - forcing the piston down. Both inlet and exhaust valves remain closed.

4th Stroke: *Exhaust* Stroke:
   The exhaust valve opens and the piston rises to expel the burnt gases. The exhaust valve closes at the end of the stroke.

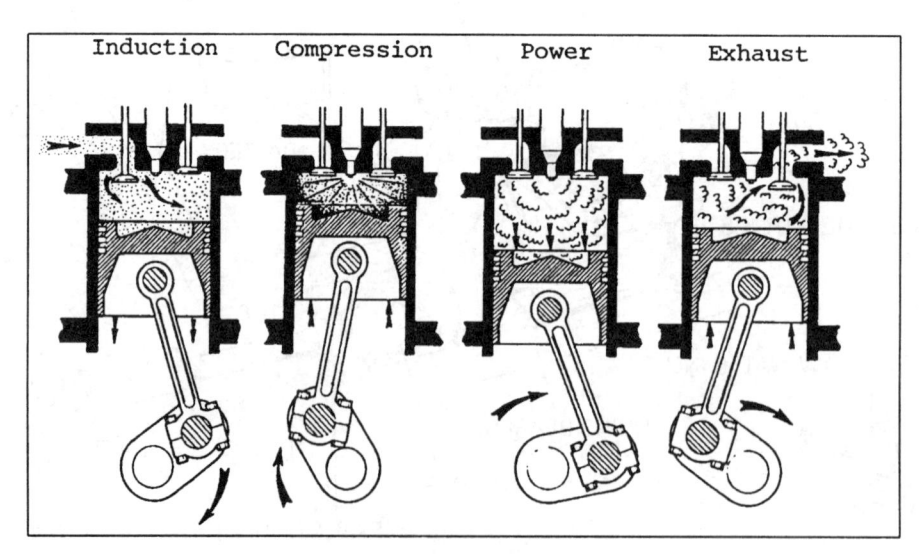

Fig 21.6: A 4-STROKE PETROL OR LPG ENGINE

Chapter 21: How Engines Work & Their Parts

THE FOUR STROKE CYCLE OF A **DIESEL** ENGINE

1st Stroke: *Induction* or inlet Stroke:
The piston moves down and the inlet valve opens to admit a charge of air.

2nd Stroke: *Compression* Stroke:
The inlet valve closes and piston rises, compressing the air to 3000 kPa. This causes the compressed air to become very hot. The fuel is injected just before the end of the stroke, and continues slightly into the third stroke.

3rd Stroke: *Power* or Firing Stroke:
The fuel ignites, creating extreme heat and pressure - forcing the piston down. Both inlet and exhaust valves remain closed.

4th Stroke: *Exhaust* Stroke:
The exhaust valve opens and the piston rises to expel the burnt gases. The exhaust valve closes at the end of the stroke.

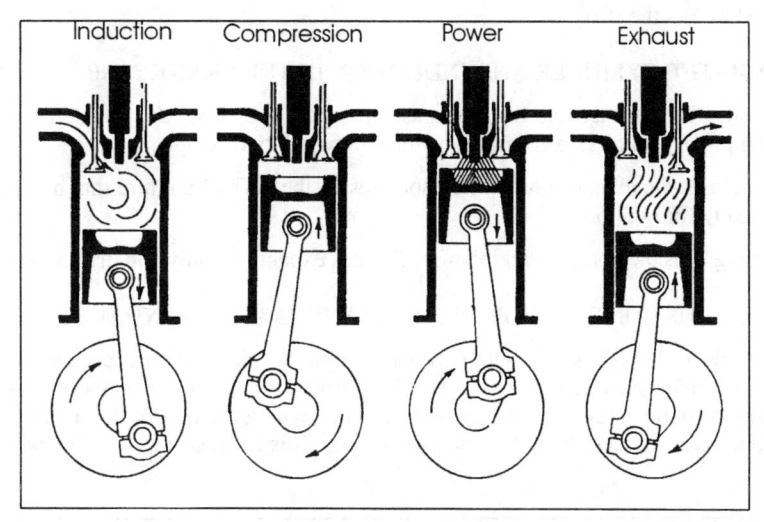

FIG. 21.7: A 4-STROKE DIESEL ENGINE (CAV/Lucas)

TDC & BDC

When a piston is at the extreme top of its travel, its position is known as TDC (Top Dead Centre). When it is at the extreme bottom, it is known as BDC (Bottom Dead Centre). The movement of the piston between these extremes either up or down is referred to as one stroke that causes half a turn of the crankshaft.

TWO-STROKE ENGINES – THE STROKES

Two stroke engines use two piston strokes to complete one power cycle. They fire twice as often as a four-stroke engine. They are thus smaller, simpler, have fewer moving parts and are more responsive. This means that on size for size basis two stroke engines have the potential to produce twice as much power as four stroke engines.

However, extra fittings such as governors and blowers on two stroke diesel engines make them more expensive to produce. Two stroke engines also tend to consume more fuel than their equivalent four stroke engines, making their exhaust equally more polluting. The two-stroke petrol engine exhaust, in particular, is more polluting for two reasons. Firstly, it burns lubricating oil that has to be mixed in its fuel. Secondly, in compressing the petrol-air mixture, some of the unburnt fuel leaks out each time the cylinder is recharged. The two stroke diesel engines do not face either of these disadvantages. Firstly, lubricating oil is not mixed in the fuel. Secondly, the fuel is injected into the cylinder after the air is compressed.

Until recently almost all outboards and many inboards were two stroke engines. There is now a shift towards four stroke petrol as well as diesel engines, which are getting smaller and more efficient.

TWO STROKE **PETROL** ENGINE – the strokes

STROKE 1: COMPRESSION IN THE CYLINDER AND INDUCTION IN THE CRANKCASE.

As the piston rises from BDC, it uncovers the crankcase inlet port and closes the transfer ports. This generates a partial vacuum under the rising piston and fresh petrol/air mixture is drawn into the crankcase.

The rising piston also closes the exhaust port, which compresses the air/fuel mixture that had passed earlier up the transfer ports. in the cylinder ready for ignition.

Near TDC the spark plug fires the compressed mixture. The gas expands rapidly and drives the piston down.

STROKE 2: POWER STROKE, PLUS COMPRESSION IN THE CRANKCASE AND EXHAUST IN THE CYLINDER.

The descending piston closes the crankcase inlet port and compresses the air/fuel charge in the crankcase. Before BDC, the piston crown uncovers the exhaust port and the expanding burnt gas in the cylinder escapes. It then uncovers the transfer port to transfer the compressed charge from the crankcase to the cylinder ready for the next cycle. The transferred charge also forces out of the open exhaust port the last of the exhaust gas. This is called SCAVENGING.

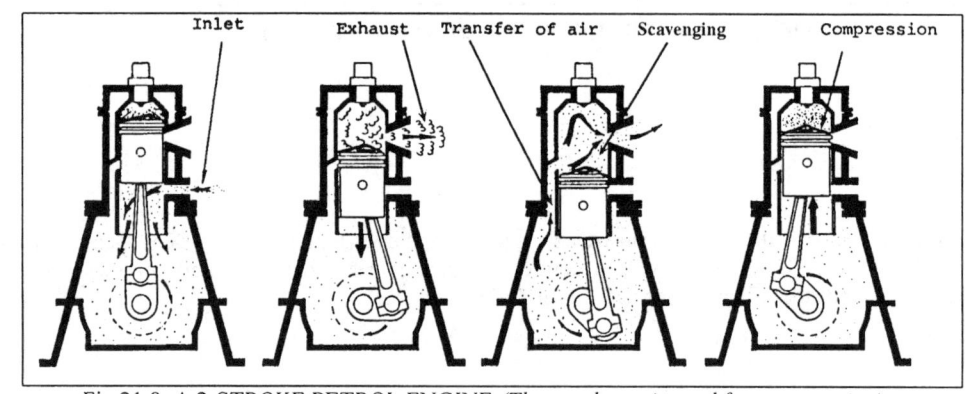

Fig 21.8: A 2-STROKE PETROL ENGINE *(The crankcase is used for compression)*

TWO STROKE **DIESEL** ENGINE – the strokes

The strokes are the same as for a 2-stroke petrol engine. The basic differences in the engines are:

- Two stroke diesels are fitted with a supercharger (blower) to compress the incoming air. The crankcase is thus free to be used as a sump just like a four stroke engine. There are inlet ports for the induction of pressurised air that are fully open at BDC.
- There are no spark plugs in diesel engines.
- The last of the exhaust gases are expelled (scavenged) not by the transfer charge, as in a two-stroke petrol engine, but by the incoming pressurised air.
- There are exhaust valves and timing gear, instead of just exhaust ports.

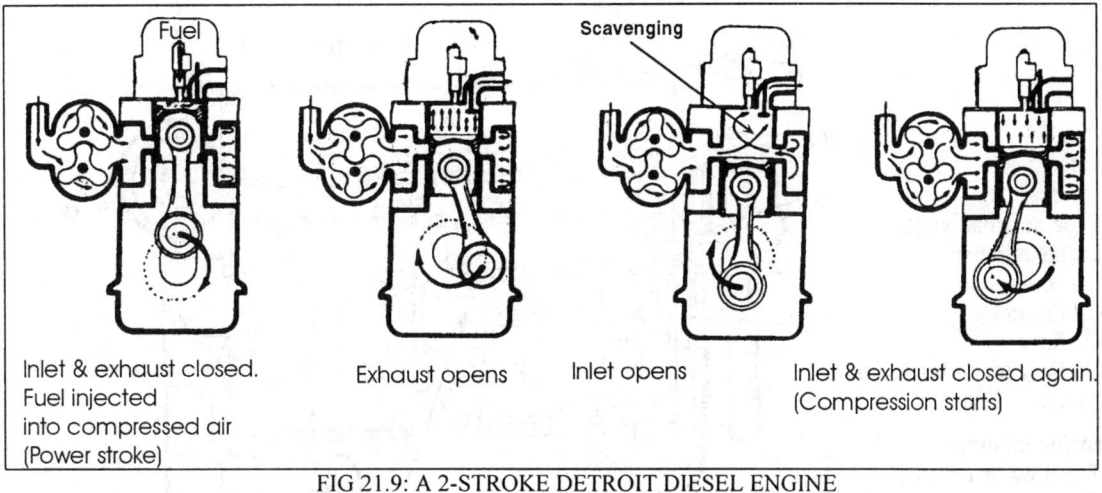

Chapter 21: How Engines Work & Their Parts

FIG 21.9: A 2-STROKE DETROIT DIESEL ENGINE
(Fitted with a mechanically driven blower to compress the incoming air)

Inlet & exhaust closed. Fuel injected into compressed air (Power stroke) — Exhaust opens — Inlet opens — Inlet & exhaust closed again. (Compression starts)

TYPICAL MARINE DIESEL ENGINES

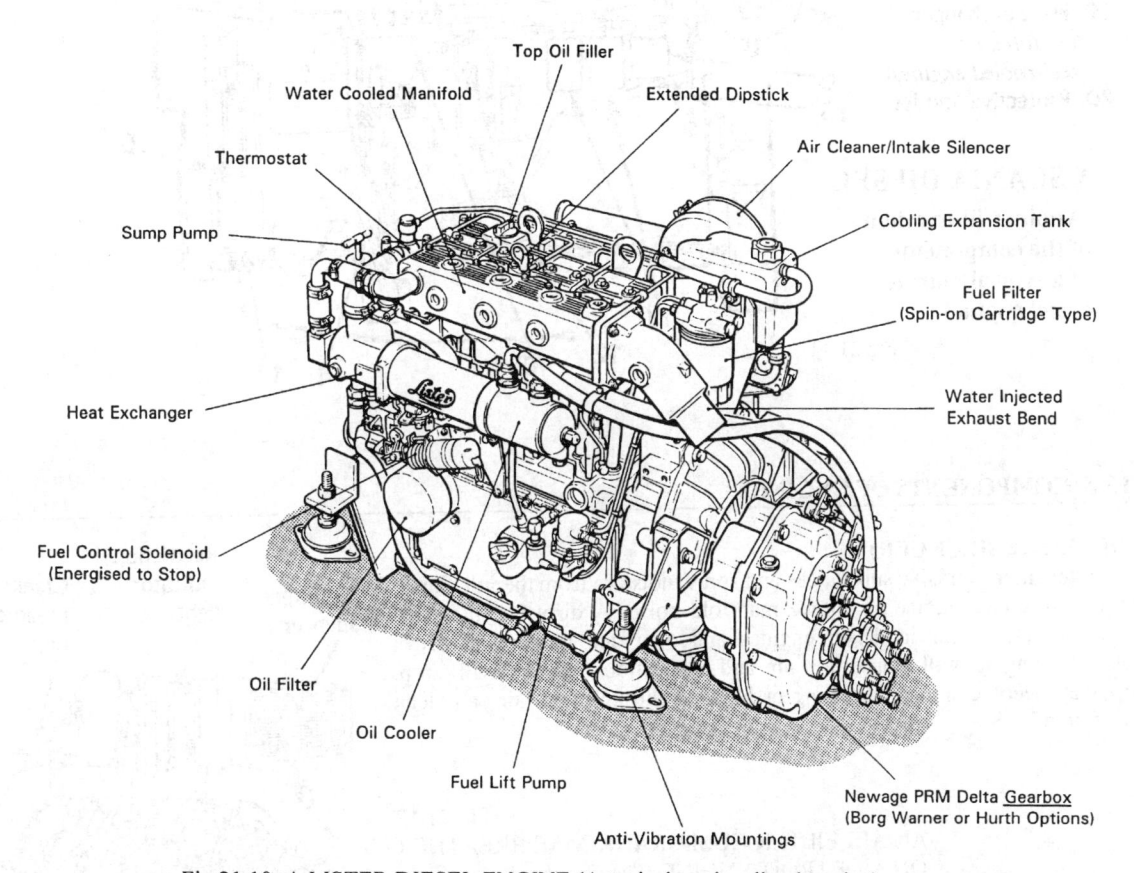

Fig 21.10: A LISTER DIESEL ENGINE (A typical marine diesel engine)

PORT VIEW

1. Type designation plate
2. Engine number
3. Oil cooler
4. Oil dipstick
5. Oil filter
6. Oil filter for turbocharger *(if fitted)*
7. Engine oil drain
8. Fresh water pump
9. Turbocharger *(if fitted)*
10. Fuel injection pump
11. Fuel filters
12. Stop solenoid
13. Starter
14. Alternator
15. Fan belt, adjustment
16. Drain cock for coolant
17. Engine oil filler
18. Sea water pump
19. Heat exchanger *(not fitted on keel-cooled engines)*
20. Protective anodes

STARBOARD VIEW

A SCANIA DIESEL
(Another illustration of the components of a typical marine diesel engine)

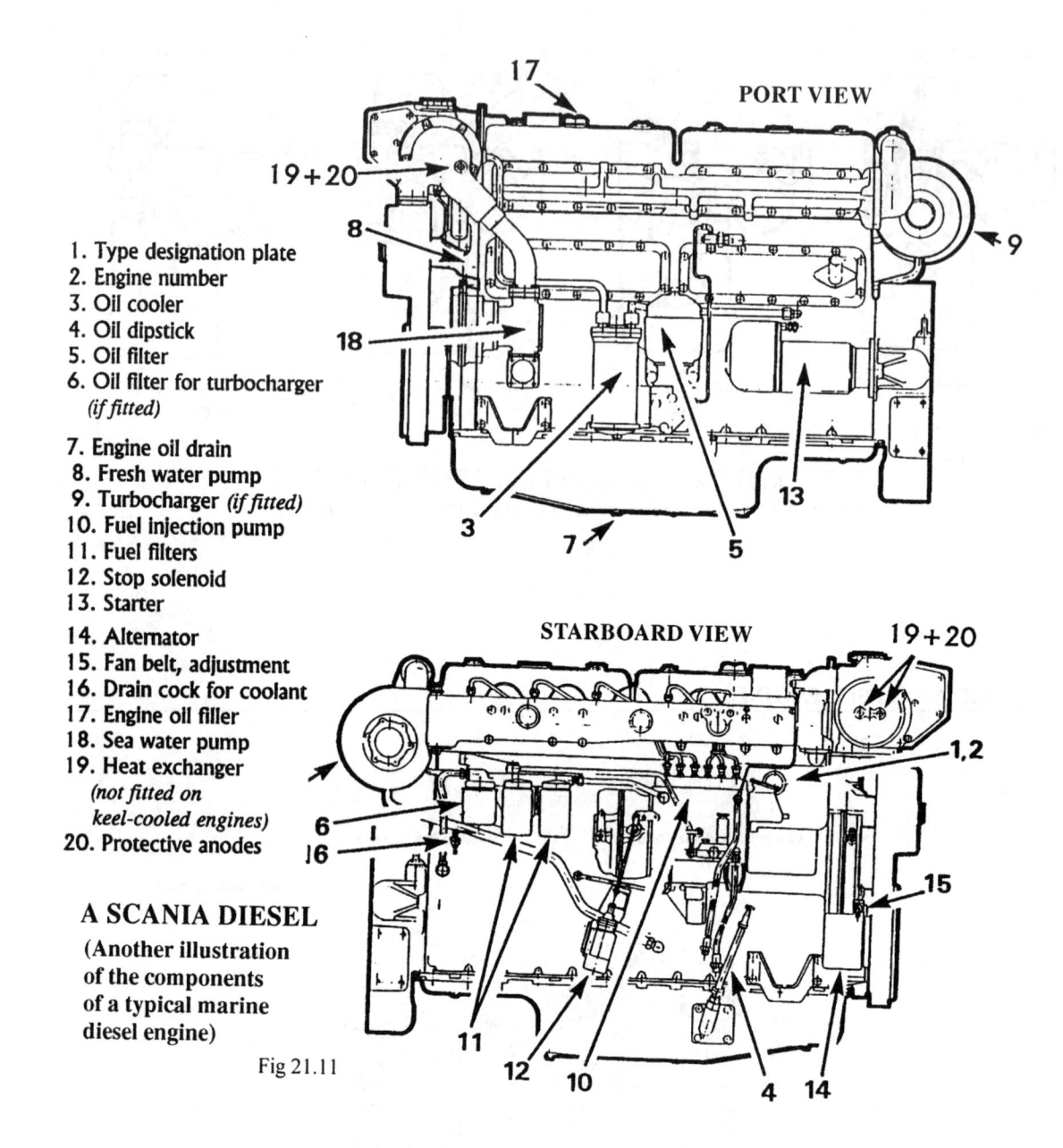

Fig 21.11

PARTS & COMPONENTS OF ENGINES

AIR FILTER (& SILENCER)

The Air Filter filters abrasive substances and some moisture from the inlet air. It also serves to reduce engine air intake manifold noise. A dirty air filter will result in shortage of air in diesel engines and over-rich mixture in petrol engines. The engine will starve for air and revolutions/ power will drop. Exhaust may become smoky. Washing or blowing through with air can clean most air filters.

Fig 21.12
AN AIR FILTER-CUM-CRANKCASE BREATHER ON A DETROIT ENGINE *(See Crankcase Breather)*

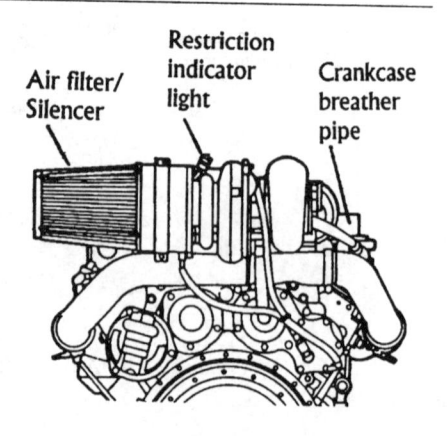

Chapter 21: How Engines Work & Their Parts

AIR SHUTDOWN VALVE

Diesel engines have been known to overspeed for no apparent reason, running faster and faster out of control. Shutting off fuel supply may not shut the engine. It will first burn the lubricating oil and then blow up, unless stopped by shutting off the air supply. Blocking the air intake with pillows and blankets has worked at times. A better system is an electric or manual (pull cable-actuated) air shutdown valve, which is fitted to some engines. When activated, it cuts off air flow to the engine and shuts it down during a mechanical emergency. This valve must be tested and reset at specified intervals.

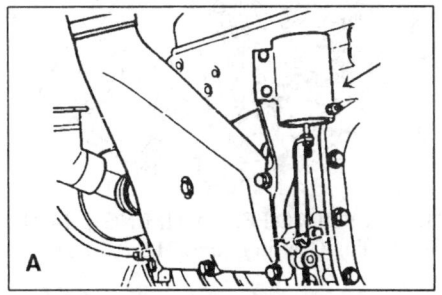

Fig 21.13: A DETROIT AIR SHUTDOWN ASSEMBLY (ELECTRIC)

Fig 21.14: A CATERPILLAR SHUT OFF SWITCH

ALTERNATOR & AMMETER

The Alternator is not a component of the engine, but one of its essential companions. It is a dynamo, belt-driven from the engine, which produces AC current. A rectifier is used to change the AC into DC current to charge one or more batteries.

ALTERNATOR DRIVE BELT (Fig 21.15)

The Alternator belt tension should be such as to deflect 10 mm with moderate thumb pressure. To adjust tension, loosen the pivot fasteners (1) on the alternator and the adjustment link fasteners (2). Change the position of the alternator to give the correct tension, and tighten fasteners (1) and (2).
A Perkins Alternator Drive Belt is shown here.

The **AMMETER** (shown here) indicates the rate of battery charge or discharge. (Fig 21.16)

ANODES
See Sacrificial Anodes

ATOMISERS
See Electronic Fuel Injection.

BREATHER VALVE
See Crankcase Breather.

CAMSHAFT

Driven from crankshaft with a timing gear, timing chain or timing belt, it controls opening and closing of inlet and exhaust valves. Modern lightweight engines of greater performance and reliability are fitted with a **double overhead camshaft (DOHC)**. The belt tension needs to be adjusted as per manufacturer's instructions (usually if it can be stretched by 10 mm by pushing it). It should be replaced when:

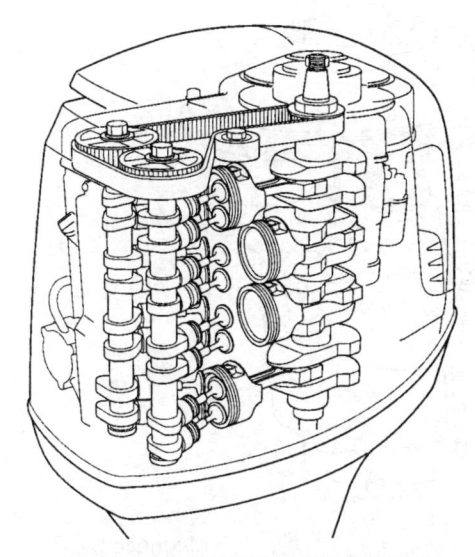

Fig 21.17: A YAMAHA OUTBOARD 16-VALVE DOHC

- Cracks appear on the back of the belt or in the base of the belt teeth.
- Excessively worn at the roots of the cogs.
- The rubber portion is swollen by oil.
- Belt surface is roughened.
- There are signs of wear on edges or outer surface of the belt.

CARBURETTOR
The unit in the inlet system of older style petrol and LPG engines in which fuel is mixed with air to provide the air-fuel mixture to the engine. One of its jets is provided with a screw head to permit manual adjustment of air/fuel mixture if necessary. Modern petrol engines use Electronic Fuel Injection (EFI), which eliminates the need for a carburettor.

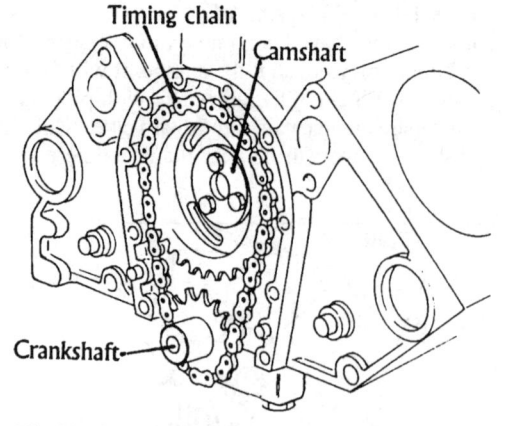

Fig 21.18: A VOLVO PENTA TIMING CHAIN
(See Timing Gear in figures 21.2 & 21.3)

CHOKE CONTROL
It is an air valve located in the inlet area of a carburettor used for restricting intake air when starting an engine. The engines with electronic fuel injection are fitted with an electronic spark-advancing device.

CONNECTING RODS
Shown in figure 21.3, they connect pistons to Crank Journal.

CRANKCASE
See Sump

CRANKCASE BREATHER OR VACUUM LIMITER
A sump must breathe in order to prevent build up of vacuum due to movement of pistons or pressure created by heat. In conventional petrol engines, it usually breathes into the carburettor. In diesel engines and fuel injected petrol engines, it breathes into the air filter (figure 21.12) or the engine is fitted with a breather valve and a vapour-collecting filter (figure 22.8 in chapter 22). The filter may consist of steel wool, charcoal or cardboard-foam material. It should be cleaned or replaced as specified by the manufacturer.

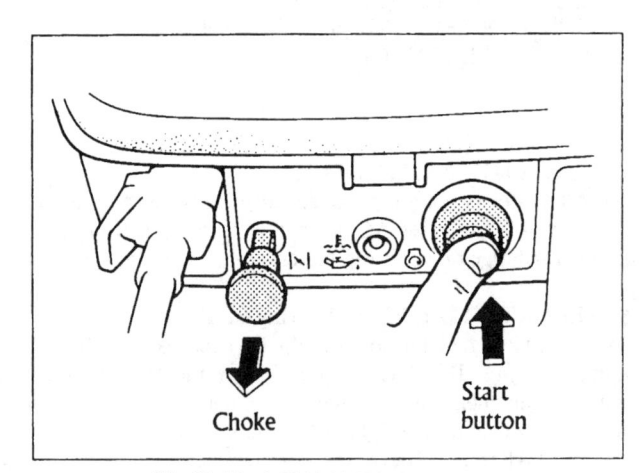

Fig 21.19: A YAMAHA OUTBOARD "PULLOUT" CHOKE

CRANKING
To crank is to start an engine.

Fig 21.20: YAMAHA OUTBOARD CRANKING MECHANISM

Chapter 21: How Engines Work & Their Parts

CRANKSHAFT

Crankshaft is built with cranks (arms) at right angles to the shaft in order to convert reciprocating (up-and-down) motion into rotary motion. Connecting rods transmit the reciprocating motion of pistons to crankshaft, which transmits it in a rotary motion to the propeller. (Figures 21.2 & 21.3)

Forces generated by pistons and connecting rods during the power stroke can deflect the crankshaft and cause its angular displacement. The resulting **TORSIONAL VIBRATIONS** may fracture the crankshaft and/or the flywheel bolts. They may also overheat the vibration damper. Where necessary, the engine manufacturers perform a **Torsional Vibration Calculation (TVC)** in order to establish the safe range of operating speeds of a vessel.

DISTRIBUTOR

Fitted in the ignition system of petrol and LPG engines, a distributor directs high voltage current via a rotor to each spark plug in turn. (See figure 21.4 & 21.24). Older distributors contain breaker points to break the current to the coil and control the timing of the spark. Spark timing in modern engines is much more critical. Instead of using points, they use a sensor that tells the engine control unit (ECU) the exact position of the pistons. The engine computer then controls a transistor that opens and closes the current to the coil. See also Ignition Coil.

Fig 21.21: TORSIONAL VIBRATIONS IN CRANKSHAFT *(Volvo Penta)*

You may come across a distributorless ignition system on a rare motor car. Instead of being equipped with one common coil, such an engine is fitted with a coil for each spark plug. This type of engine could run for 100,000 miles before needing a tune-up. It would have no spark plug leads and would offer even more precise control of spark timing, thus improving efficiency, emissions and the overall engine power.

ELECTRONIC FUEL INJECTION (EFI)

See Fuel System

EMERGENCY SHUTDOWN VALVE

See Air Shutdown Valve

ENGINE (OR CYLINDER) BLOCK

It is a cast iron metal block containing cylinders and coolant passages (Figures 21.2 & 21.22). For the purpose of cooling the cylinder walls are thin and the engine block quite hollow.

Fig 21.22 CYLINDER BLOCK

ENGINE MOUNTINGS:
Fittings on the engine block securing the engine to the hull through flexible mounts to the engine bearers. The engine manufacturers recommend a periodic inspection of flexible mounts. Engine beds are discussed below.

ENGINE SENSORS
See Fuel System

EXHAUST PIPE
It should be as straight and with as few bends as possible in order to reduce the possibility of backpressure in the combustion chamber. Caution must also be exercised when installing a water-cooled exhaust pipe. It must be designed to prevent the cooling water from accidentally flooding back into the engine.

EXHAUST MANIFOLD
See Manifold.

EXHAUST VALVES
They allow burnt gases to be exhausted from the cylinders. (Figure 21.1)

FLYWHEEL
It is a metal wheel fitted to the crankshaft of reciprocating engines in order to absorb and smooth out the intermittent power strokes and carry the crankshaft over the non-working strokes of a four-cycle engine. The flywheel is heavier for smaller engines. Multi-cylinder engines have light flywheels. (See figures 21.2 & 21.3.)

FUEL FILTER
A unit designed to trap water and solid contaminants in the fuel and provide the means to remove such material (Figures 21.4 & 21.10). Fuel filters are discussed in more detail in the Fuel System.

FUEL INJECTOR
See Injectors.

FUEL LIFT PUMP
Unless a gravity-fed day-tank feeds fuel to the engine, a fuel lift pump is required to get fuel from the tank to the fuel pump. It can be a gear, diaphragm or plunger type pump.

FUEL PUMP (FUEL INJECTION PUMP)
The fuel pump feeds fuel to the engine under pressure. Fuel is drawn from the tank - through the shut off valve - through the filter(s) - to the fuel pump suction - to the carburettor or the injectors. See inboard and outboard fuel system.

Located on the side of modern fuel-injected engines, the fuel injection pump is connected directly to fuel injectors by an injection pipe. With the help of the Governor, it accurately meters the fuel and delivers it under pressure at the required precise moment to the fuel injectors.

The **Rotary Type** pump incorporates a single pumping element with a port for every cylinder in the engine. Its element does not need to be balanced, as is the case with the multi-element pumps.

The **Jerk Type** pump can be of the single-element or multi-element type. The former is a separate unit for each cylinder, with its own cam operating off the camshaft. Being of single-element, it too does not need to be balanced.

The multi-element jerk type pump is made up of a number of single-element jerk-type pumps, one element for each cylinder. It has its own camshaft in the main engine drive, which is lubricated either from the engine's lubrication system or its own oil reservoir. The elements in this type of pump may periodically need to be adjusted by a mechanic to ensure correct metering of fuel to each injector.

GOVERNOR
Fitted only to diesel engines, it helps in keeping the revs constant by matching the fuel received by the engine with its load-bearing. It gives more fuel for greater loads and less fuel for smaller loads. For example, if the propeller comes out of

Fig 21.23: GOVERNOR (IN A HOUSING) ON A CATERPILLAR ENGINE

Chapter 21: How Engines Work & Their Parts

water, the governor will prevent the engine from speeding up by reducing the fuel supply. It works by controlling the rotation of the fuel rack on the fuel injection pump. Attempts to override the governor or to obtain more speed by tinkering with it will result in frequent engine breakdowns, higher maintenance costs and potential failure.

HARMONIC BALANCER
It is a small flywheel sometimes fitted on long crankshafts at the opposite end to the flywheel to harmonise or balance their rotation.

HEAT EXCHANGER
Part of the engine cooling system in which the fresh water is cooled by the sea water (Figure 21.10). The Cooling System is discussed below in more detail.

IGNITION SYSTEM (ELECTRICAL SYSTEM)
In petrol and LPG engines, it consists of battery, ignition coil, distributor (either electronic or old fashioned), spark plugs and starter motor. All these components are briefly described individually.

10. Alternator. 11. Starter motor. 12. Automatic choke. 13. Fuse. 14. Main switch. 15. Battery. 16. Temperature sender. 17. Oil pressure sender. 18. Distributor. 19. Ignition coil. 20. Relay. 21. Resistor

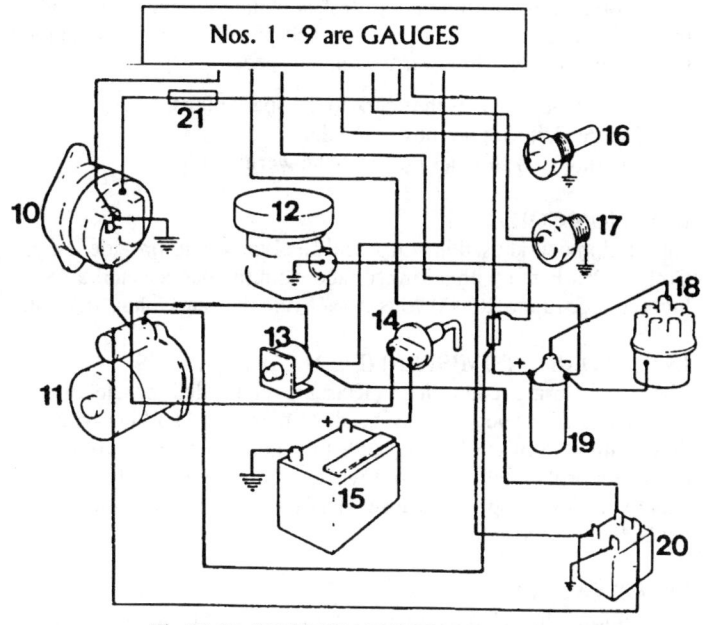

Fig 21.24: IGNITION SYSTEM – AN EXAMPLE

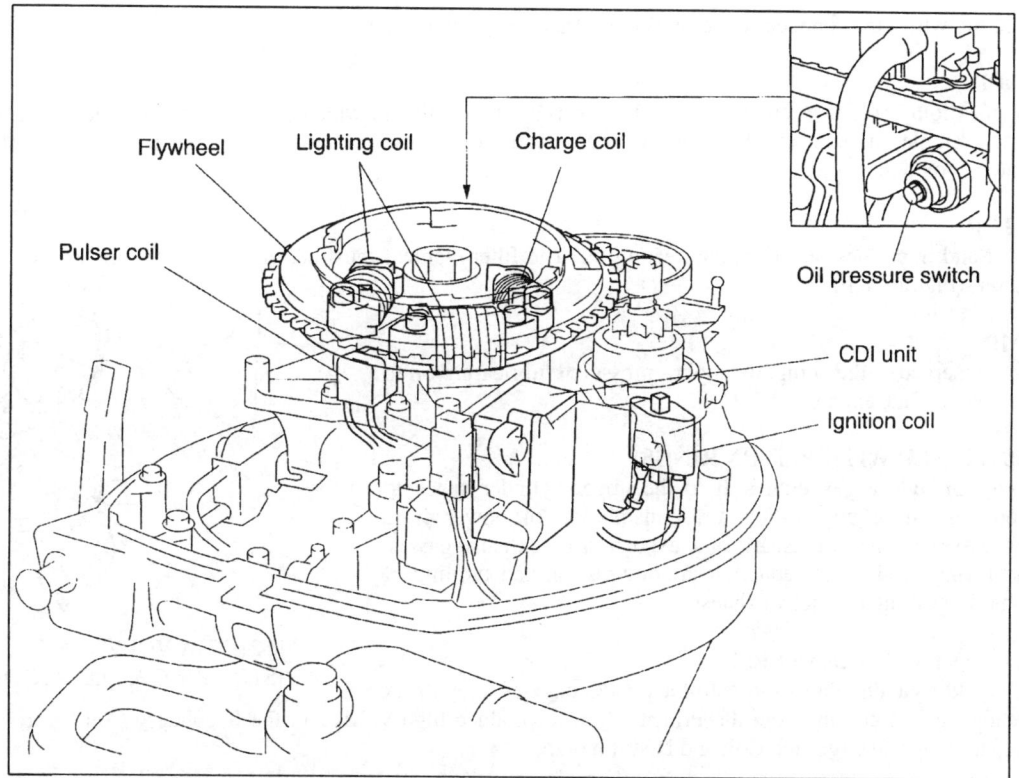

Fig 21.25: COMPUTERISED IGNITION SYSTEM (Yamaha petrol outboard)

PETROL AND LPG IGNITION SYSTEM
Figure 21.25 illustrates a computer controlled digital ignition system on 4-stroke 2-cylinder Yamaha petrol outboard. Being a pull-start engine, it is not fitted with components such as battery, alternator and starter motor. (These can be seen in figures 21.20 in this chapter and 27.3 in chapter 27).

The **CDI (Capacitor Discharge Ignition) Unit** controls the ignition timing at various speeds, including when the oil pressure drops too low or when the engine over-revs. If the lubricating oil pressure drops too low, the oil pressure switch is turned on and the warning lamp flashes while the ignition of both cylinders is cut to reduce the engine speed. Similarly, if the engine over-revs (due to mismatched propeller or on sudden loss of propeller load), the ignition of the spark plugs for both chambers is cut to lower the engine speed.

- Pulser Coil provides the spark to the spark plugs.
- Charge Coil supplies power to CDI
- Lighting Coil supplies power to the warning lights.

IGNITION COIL
An ignition coil is used in petrol and LPG motors to provide a high voltage charge to spark plugs. It consists of a "Primary Coil" which is fed with a low voltage from the battery, and a "Secondary Coil" that steps it up for an instant to extremely high voltage (up to 20,000 volts) to activate the spark plugs in rotation. See also distributor. (Figure 21.24)

INJECTORS (ATOMISERS) (Fig 21.26)
These are spring controlled, electronic controlled or electronic-hydraulic controlled valves located in each cylinder head of an engine. Working with the fuel-injection pump, they squirt (atomise) fuel through tiny nozzles directly into the engine's combustion chambers or its intake manifold where it mixes with air before entering the combustion chambers. A leaking valve causes misfiring and irregular speed, particularly under heavy loads. See Fuel System.

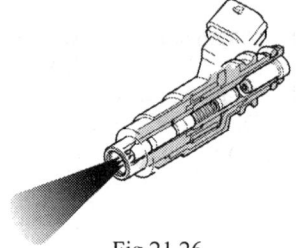

Fig 21.26

INLET VALVES
They allow air/fuel mixture into cylinders - only air in the case of a diesel engine. (Figure 21.1)

JOURNAL
It is the bearing where the connecting rod joins the crankshaft. (Figure 21.1)

MANIFOLD
The term "manifold" originates from "many folds". It is a pipe or tube in which one or more inlets join with one or more outlets. For example, the manifold between the engine and the exhaust is known as the exhaust manifold. (Figure 21.10 shows Water Cooled manifold)

OIL FILTER
It filters dirt and impurities out of the lubricating oil. The filter element must be replaced as recommended by the engine manufacturer. (Figure 21.10)

OIL PUMP
Normally submerged in the sump, it supplies pressured lubricating oil to the moving parts of the engine.

PISTONS (FITTED WITH PISTON RINGS)
Pistons rings provide a gas tight seal to the upper cylinder space to permit compression of combustion air, transmission of power from expanding gasses to the crankshaft, and expulsion of exhaust gasses. Worn piston rings will cause reduction in compression and burning of lubrication oil, resulting in smoky exhaust.

POINTS (CONTACT BREAKERS)
Fitted inside old style distributors in petrol and LPG engines, they break

Fig 21.27: A VOLVO PENTA PISTON, PISTON RING & CONNECTING ROD

up the steady flow of current from the Primary Coil to produce high voltage in the Secondary Coil to produce ignition sparks in spark plugs. See Ignition Coil and Distributor.

Chapter 21: How Engines Work & Their Parts

SOLENOID
It is an electromagnetic switch designed to ensure a positive electrical contact in high current, low voltage circuits. It is most commonly found in engine starter circuits to provide battery current to the starter motor.

SACRIFICIAL ANODES
These are small `pencils' of zinc fitted in the engine cooling water system and held in place by easily recognizable small plugs or cups with square heads. They should be regularly checked and replaced as indicated by their wasting.

The illustration shows a Caterpillar engine with seven zinc plugs located as follows: (1) three on the heat exchanger; (2) two on the gear oil cooler; and (3) two on the exhaust risers or exhaust elbows. The zinc plugs are painted red for identification.

Fig 21.28: SACRIFICIAL ANODES

SENSORS (ENGINE SENSORS)
See Fuel System

SPARK PLUGS
Fitted one per cylinder in petrol and LPG engines, they ignite the compressed fuel-air mixture inside cylinders.

SUMP & SUMP BREATHER
The metal casing enclosing the crankshaft is called crankcase or sump. (Figure 21.2) It is the lubricating oil reservoir below the engine (Figure 21.2). However, in two-stroke petrol and LPG engines the sump is needed for compression. Their lubricating oil is therefore mixed in the fuel. See Crankcase Breather for Sump Breather.

Fig 21.29 SPARK PLUG   0.9~1.0 mm (0.035~0.039 in)

SUMP PUMP
It is a hand pump for pumping out the sump when changing oil. (Figure 21.10) Pumps are discussed below.

TIMING CHAIN, TIMING BELT & TIMING GEAR
See Camshaft

TEMPERATURE INDICATORS
Temperature gauges and warning lights are used to monitor temperatures of engine coolant, engine oil and transmission oil. They may use mercury, gas or bimetal connections to transmit temperature changes or operate a switch at a critical temperature. In one mechanical type gauge, a coiled tube ('Bourdon' tube) is fitted. As the tube heats, its unrolling motion turns the gauge pointer. The electrical type gauges incorporate a 'sender' to operate a warning light or to vary the current, which would operate an electrical gauge.

THERMOSTAT
It is a heat sensitive valve that opens and closes to control the flow of coolant in a cooling system. It thus regulates the engine temperature. There are various types of thermostats, made of bi-metal springs, bellows or wax elements. Only the right type must be fitted to prevent temperature problems. An engine taking too long to warm up or overheating may be due to a stuck thermostat. To get back home, a sticking thermostat may be temporarily removed (after the engine has cooled down) and then replaced with a new one. (Figure 21.10)

TROLLING VALVE
In order to run an auxiliary (such as a pump during fishing) the main engine may need to be run at high rpm while idling or when the vessel is at slow speed. Trolling valve is an accessory that can be fitted to the gearbox to reduce oil pressure on the disc pack in order to achieve this objective. The valve is engaged or disengaged by operating a dedicated control lever. The trolling valve control cable is installed the same way as the gearshift control cable, discussed in Propulsion Control Systems.

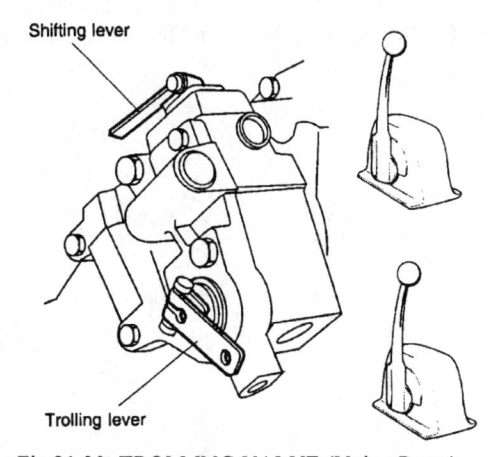

Shifting lever
Trolling lever

Fig 21.30: TROLLING VALVE (Volvo Penta)

TURBOCHARGER

An engine breathes in natural air in order to burn fuel. In some engines, slightly compressed (additional) air is blown into the cylinders, allowing more fuel to burn without altering the fuel-to-air mixture. It can increase an engine's output by up to 50 per cent. Such engines are known as turbocharged or "*Blown*" engines. The term SUPERCHARGED is also sometimes used.

The turbo unit consists of an exhaust turbine and a compressor fitted on the same shaft. In other words, the exhaust gas turbine blades are fitted on one end of its shaft and the rotary air compressor blades on the other, both running at some 100,000-rpm. The shaft bearings are lubricated either from the engine-driven oil pump or they may have their own reservoir within the turbocharger unit, with an oil-level indicator.

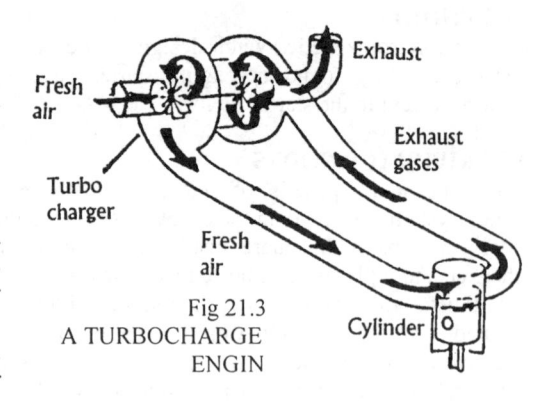

Fig 21.3
A TURBOCHARGE ENGIN

There are two methods of running the turbine rotor (which in turn rotates the compressor to force air into cylinders): (1) mechanically by a belt or a gear train, which is used in <u>all</u> two stroke diesels, and (2) by the engine's exhaust gasses, which can theoretically be added to all types but are mostly found in four stroke diesels. Planing hulls need engines of maximum power but minimum weight. They tend to use turbocharged engines.

Turbocharging can be further improved by *Aftercooling*. It means that the compressed air blown by the turbocharger is cooled by a heat exchanger before entering the combustion chamber. Cooler air is denser and allows more fuel to burn.

Fig 21.32
(*Left*) TURBOCHARGER ON CATERPILLAR ENGINE

(*Right*) TURBO FILTER ON SCANIA ENGINE

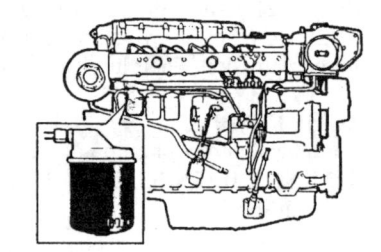

WATER CIRCULATING PUMPS

As shown in figures 21.11 in this chapter and 22.24 in Chapter 22, an engine may drive one or more water pumps to operate the Cooling System. Pumps and the cooling systems are discussed in more detail below.

CHAPTER 21: QUESTIONS & ANSWERS

HOW ENGINES WORK & THEIR PARTS

1. In a petrol engine fuel is ignited by:

    (a) spark plugs.
    (b) hot compressed air.
    (c) spark plugs after mixing with air.
    (d) compression after mixing with air.

2. A "four stroke" engine completes one firing cycle in:

    (a) each piston stroke.
    (b) every second piston stroke.
    (c) four piston strokes.
    (d) none of the choices stated here.

3. The power stroke in a petrol engine is also known as:

    (a) intake.
    (b) compression.
    (c) ignition.
    (d) exhaust.

4. In engines, the term TDC means the piston is:

    (a) on its way up.
    (b) at the extreme top of its travel.
    (c) at the extreme bottom of its travel.
    (d) on its way down.

5. In an engine, a camshaft:

    (a) controls opening of valves.
    (b) connects pistons to crankshaft.
    (c) allows burnt gases to escape.
    (d) allows air/fuel mixture into cylinder.

6. A magneto is a dynamo sometimes connected with and run by internal combustion engines to:

    (a) break up the steady current flow.
    (b) produce current in the primary coil.
    (c) generate current providing a spark for ignition
    (d) to magnetise the components.

7. In the following list, a diesel engine is fitted with only the:

    (a) distributor.
    (b) carburettor.
    (c) spark plugs.
    (d) fuel injection pump

8. In a two-stroke engine, Stroke 2 produces:

    (i) compression in the cylinder.
    (ii) power.
    (iii) compression in the crankcase.
    (iv) exhaust in the cylinder.

    The correct answer is:
    (a) all except (i).
    (b) all except (ii).
    (c) all except (iii).
    (d) all except (iv).

9. In a two stroke diesel engine:

    (a) diesel fuel is also the lubricant.
    (b) the lubricant is in the sump.
    (c) the lubricant is mixed with fuel.
    (d) air-fuel mixture is compressed in the sump.

10. In a diesel engine fuel is ignited by:

    (a) spark plugs.
    (b) hot compressed air.
    (c) spark plugs after mixing with air.
    (d) compression after mixing with air.

11. Turbocharging:

    (a) is improved by aftercooling.
    (b) must use exhaust gases.
    (c) alters the fuel-to-air ratio.
    (d) burns less fuel.

12. The most common and efficient method of turbocharging is to make use of:

    (a) hot air.
    (b) fresh air.
    (c) cold air.
    (d) engine exhaust.

13. A choke control in a petrol engine provides:

    (a) more fuel.
    (b) more air.
    (c) richer petrol/air mixture.
    (d) greater oil pressure.

14. The purpose of a solenoid in the engine starting system is to:

    (a) start the engine.
    (b) switch battery current to starter motor.
    (c) create a spark.
    (d) prevent a spark.

15. A carburettor:

    (a) is fitted in EFI petrol engines.
    (b) controls the fuel supply.
    (c) mixes air and fuel in correct proportion.
    (d) controls the air supply.

16. In a two-stroke diesel engine, scavenging is done by the:

    (a) incoming pressurised air.
    (b) exhaust blower.
    (c) compression stroke.
    (d) transfer charge.

17. In a two-stroke petrol engine, scavenging is done by the:

    (a) incoming scavenging air.
    (b) incoming pressurised air.
    (c) compression stroke.
    (d) transfer charge.

18. A distributor:

    (a) is fitted on all engines.
    (b) directs high voltage current.
    (c) provides high voltage current.
    (d) is an electrically operated switch.

19. An ignition coil:

    (a) is fitted on all engines.
    (b) directs high voltage current.
    (c) provides high voltage current.
    (d) is an electrically operated switch.

20. Spark plugs:

    (a) are fitted on all engines.
    (b) direct high voltage current.
    (c) provide high voltage current.
    (d) ignite the fuel-air mixture.

21. A vacuum limiter:

    (a) pushes air into exhaust.
    (b) limits air in a carburettor.
    (c) is a crankcase breather.
    (d) does not require cleaned.

22. In a vessel the following compartment has the greatest risk of fire:

    (a) engine room.
    (b) battery compartment.
    (c) fuel compartment.
    (d) all the compartments stated here.

CHAPTER 21: ANSWERS

1 (c), 2 (c), 3 (c), 4 (b), 5 (a),
6 (c), 7 (d), 8 (a), 9 (b), 10 (b),
11 (a), 12 (d), 13 (c), 14 (b), 15 (c),
16 (a), 17 (d), 18 (b), 19 (c), 20 (d),
21 (c), 22 (d)

Chapter 22

ENGINE LUBRICATION & COOLING SYSTEMS

ENGINE LUBRICATION SYSTEMS
(Excluding 2-stroke petrol outboards)

Fig 22.1: **OLD STYLE** LUBRICATION SYSTEMS – EXTERNAL (DRY-SUMP)
1. Suction strainer.
2. Lube oil pump
3. Pressure relief valve (Crankcase breather)
4. Lube oil cooler
5. Lube oil filter
6. Engine block
7. Dipstick

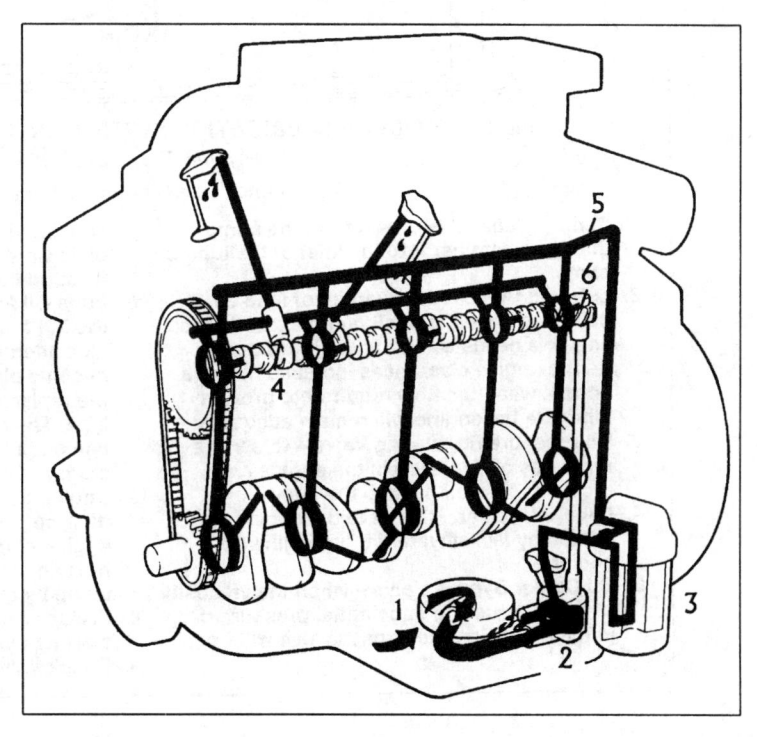

Fig 22.2: **MODERN** LUBRICATION SYSTEM – INTERNAL (WET-SUMP)
(Volvo Petrol Engine)

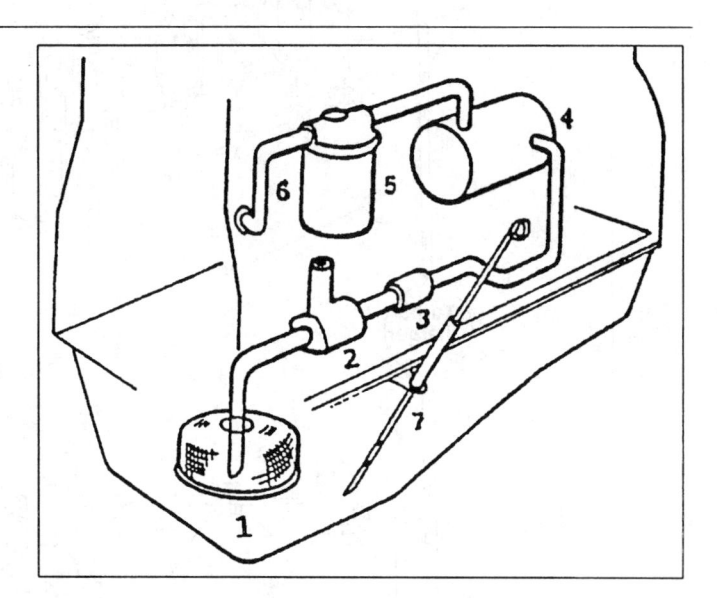

1. Lube oil strainer
2. Lube oil pump & Relief valve (Crankcase breather)
3. Lube oil filter
4. Hydraulic valve lifters (Not always fitted)
5. Lube oil gallery in engine
6. Drive for lube oil pump

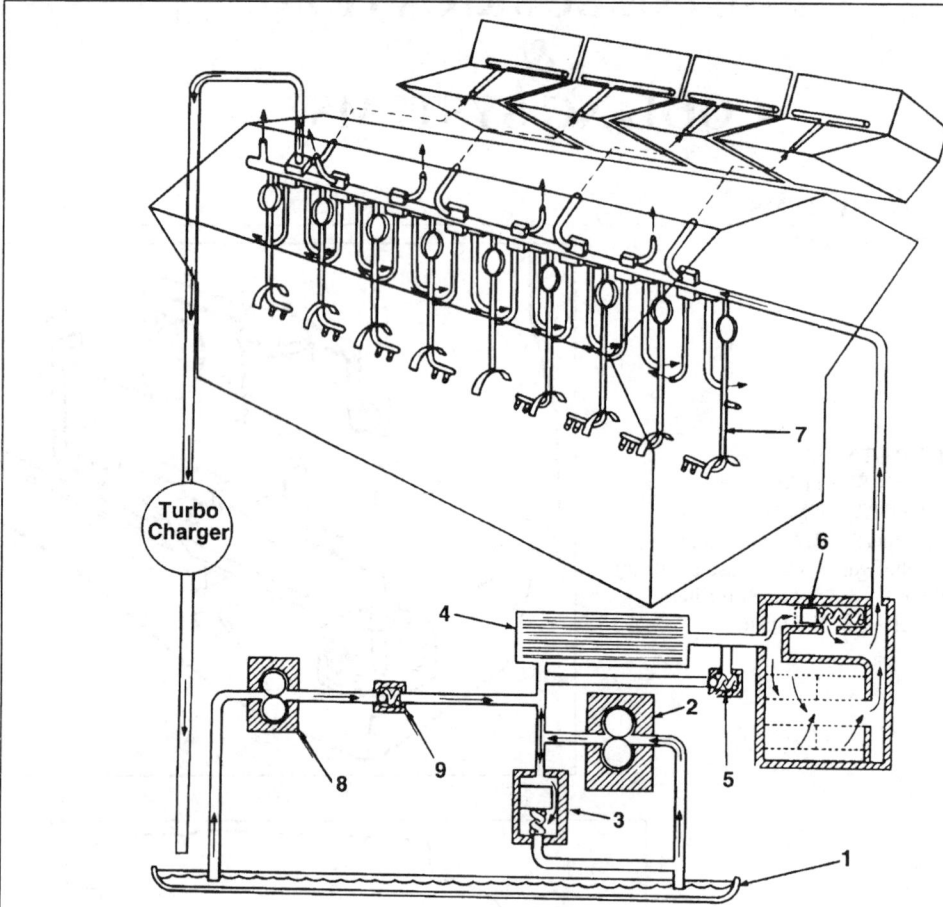

Fig 22.3: MODERN LUBRICATION SYSTEM - INTERNAL (WET-SUMP)
(Caterpillar Diesel Engine)
Engine Lube Oil Flow Schematic

1. Sump – lube oil is drawn from the sump through a strainer into the inlet of the lube oil pump.
2. Lube Oil Pump – the quantity of lube oil delivered by the lube oil pump exceeds the engine's needs when the engine is new. As the engine clearances increase through normal wear, the flow required to properly lubricate the engine will remain adequate.
3. Oil Pressure Regulating Valve – this valve regulates oil pressure in the engine and routes excess oil back to the sump.
4. Lube Oil Cooler – the oil to the engine is cooled by jacket water in the engine oil cooler.
5. Oil Cooler Bypass Valve – when the viscosity of the oil causes a substantial pressure drop in the oil cooler, the bypass valve will open, causing the oil to bypass the cooler until the oil is warm enough to require full oil flow through the cooler.
6. Lube Oil Filter – Caterpillar lube oil filters are the full-flow type with a bypass valve to provide adequate lubrication should the filter become plugged. The filter system may have the replaceable element type or the spin-on type. The oil filter bypass valve is a protection against lube oil starvation if the oil filter clogs.
7. Engine Oil Passages – the main oil flow is distributed through passages to internal engine components. The oil flow carries away heat and wear particles and returns to the sump by gravity.
8. Prelubrication Pump – used only during starting cycle on largest engines.
9. Check Valve.

ENGINE LUBRICATION IS NECESSARY FOR THE FOLLOWING REASONS:
- It reduces wear and prevents metal-to-metal contact between moving parts. This minimises friction and allows an engine to run smoothly without becoming too hot.
- It acts as a secondary engine cooling system with excess heat being extracted via the oil cooler.

Chapter 22: Engine Lubrication & Cooling Systems

- It removes grit and other contaminants from machine surfaces, accumulating them in the oil filter.
- It cushions the engine's bearings from shock of cylinder firing.
- It neutralises the corrosive combustion products.
- It seals the engine's metal surfaces from rust.

THE CLASSIC SYMPTOMS OF OIL RELATED ENGINE FAILURE ARE:
- Bearing failure
- Piston ring sticking
- Excessive oil consumption

HOW LUBRICATION OPERATES

In today's engines (other than in 2-stroke petrol outboards), lubricating oil is circulated from the sump. It is known as the *internal* or *wet-sump* lubrication system. The oil is forced through pipe lines, drilled passages and splash feeds or is pressure fed through galleries in the crankcase, hence the term *force-feed* lubrication system. In some older engines, the oil reservoir is an *external* tank. They are said to have the *dry-sump* lubrication system.

Lube oil pumps are provided with a *relief valve* to cope with build up of high pressure. And, the filtration system incorporates a *bypass valve*. This allows unfiltered oil to reach the engine should the filter become blocked. However, when the bypass valve remains open, the debris from the filter may also be flushed into it. Filter plugging can also lead to distortion and cracks in the filter element due to increase in the pressure difference between outside and inside the element. This too would allow debris to flow into the engine.

As discussed in Parts & Components of an Engine, the sump (crankcase) is vented to the atmosphere in order to prevent vacuum build up due to movement of pistons or pressure created by heat.

The oil thickness (its resistance to flow) is known as VISCOSITY. The more viscous (thicker) oil is, the stronger the film it will provide. The thicker the oil film, the more resistant it will be to being wiped or rubbed from lubricated surfaces. On the other hand, too thick oil will have excess resistance to flow quickly to those parts in need of lubrication. Oil supply to all parts must be adequate regardless of the temperature.

For the maximum protection of an engine, a multigrade marine oil of the maker's specifications should be used. Commonly, 20-grade oil is used in cold regions, 40-grade in the tropics, and 30-grade being the best all-rounder. It must be maintained to the correct sump level, and changed, along with its filters, as per the maker's recommendation, to prevent build up of acids and contaminants in the oil.

PRELUBRICATION (AFTER LONG PERIODS OF ENGINE IDLENESS)

The Caterpillar prelubrication system provides the capability to prelubricate all critical bearing and journals before starting the engine. It is designed to minimise the sometimes-severe engine wear associated with starting an engine after periods of idleness.

The *automatic* system utilizes a small pump that fills the engine oil galleries from the sump until the presence of oil is sensed at the upper portion of the lubrication system. The starter motors are automatically energised only after the engine has been prelubricated.

The *manual* system uses the engine's manually operated sump pump and allows the engine operator to fill all engine oil passages after oil changes, filter changes, periods of idleness, and before activating the starter motors.

LUBE OIL CHECKS

Check the oil level daily with the engine stopped. If necessary, add sufficient oil to raise the level to the proper mark on the dipstick. Make sure the dipstick marking has been adjusted if the engine is installed in a tilted position. All diesel engines are designed to use some oil, so the periodic addition of oil is normal. Never overfill the sump. The pressure created by too much oil can force it to seep past the main bearing seals.

If the oil level is constantly above normal, and excess lube oil has not been added to the crankcase, seek expert advice for the cause. Fuel or coolant dilution of lube oil can result in serious engine damage.

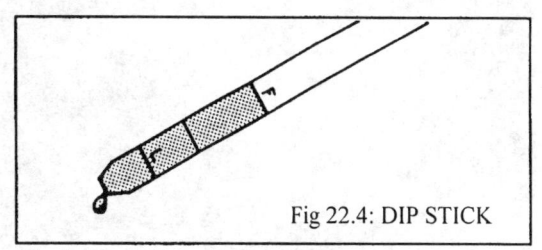

Fig 22.4: DIP STICK

Check oil level with engine stopped

Check for normal operating pressure and temperature. Too much oil pressure will scour the bearings and too little seize them. Generally speaking, an average diesel engine will start cold at 450 kPa (65 psi) then drop back to around 240 kPa (35 psi) at a cruising speed producing optimum temperature.

Proper gauges are superior to Low Pressure Alarms. They usually do not switch on until the pressure is down to 48 kPa (7 psi). A combination of the two is an ideal solution.

On starting the engine, the Low Oil Pressure Alarm may sound for a few seconds. If the oil pressure does not rise within a few seconds, stop the engine and check the oil level, filter and breather.

Fig 22.5: OIL PRESSURE GAUGE
(Indicating Normal Pressure)

Fig 22.6: OIL PRESSURE SWITCH

Figure 22.6 shows an Oil Pressure Sender on a Caterpillar engine, which triggers the alarm when the oil pressure is below the rated pressure. It may be fitted with an Override Button, which is pushed to permit the engine to start without sounding the alarm. As the oil pressure increases, the button automatically comes out to RUN position.

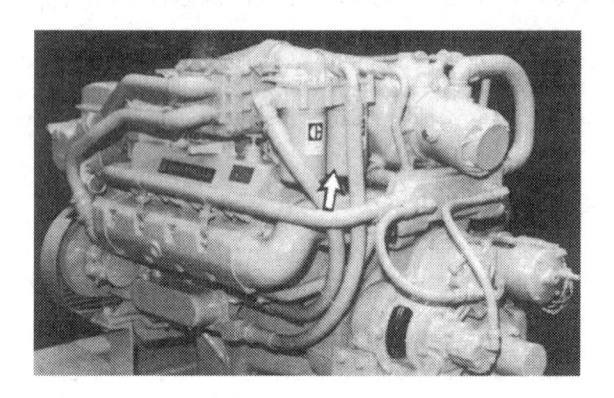

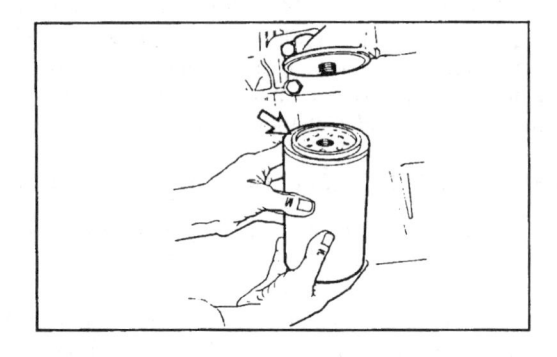

Fig 22.7: OIL FILTER & REPLACING THE FILTER ELEMENT (*A Caterpillar Engine*)

Fig 22.8
CRANKCASE
BREATHER
WITH VAPOUR
COLLECTOR
(Remove, clean, dry & reinstall)

Left: A crankcase breather valve with a vapour collector (Caterpillar)
Right: A vapour collector on a crankcase breather connected to air cleaner
(*Perkins showing securing clips*)

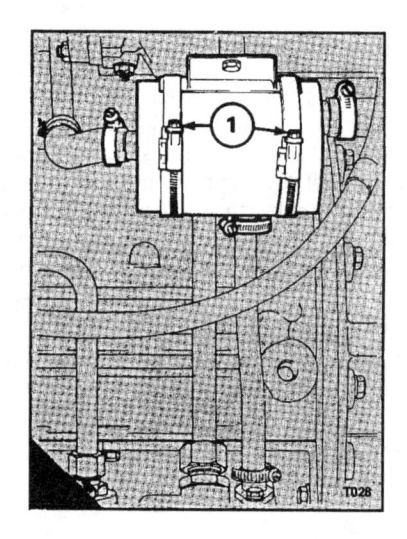

Chapter 22: Engine Lubrication & Cooling Systems

COOLING SYSTEMS – INBOARD ENGINES

The marine internal combustion engines must be operated at a specific working temperature. Overheating causes excessive expansion, which can cause an engine to seize. Most marine engines are water-cooled. Their cylinders are surrounded by cooling water spaces known as a *Water-jacket*. With a few exceptions, the transmission oil too is cooled by either the engine's water-jacket or a water jacket of its own.

On rare occasions, you may come across an AIR-COOLED marine engine or an OPEN CIRCUIT seawater cooling system. The latter is a direct seawater (raw water) cooling system.

Most marine engines are cooled by a closed-circuit fresh water system. The fresh water carrying the engine heat is circulated through an outboard mounted *Keel Cooler* or an inboard mounted *Heat Exchanger (tube-nest cooler)*, where it is cooled by seawater. Seawater does not enter the *Keel or Skin Cooler*. It makes contact on the outside. In the case of the heat exchanger, the seawater is pumped in and circulated through it. It is then discharged overboard either through the exhaust or directly over the side. If discharged through the exhaust, it is known as the "*Wet Exhaust*" system. The keel cooling system has only a "*Dry Exhaust*".

HEAT EXCHANGER COOLING SYSTEM (OPEN CIRCUIT/CLOSED CIRCUIT)

Part A — Sea-water system
Part B — Fresh-water system

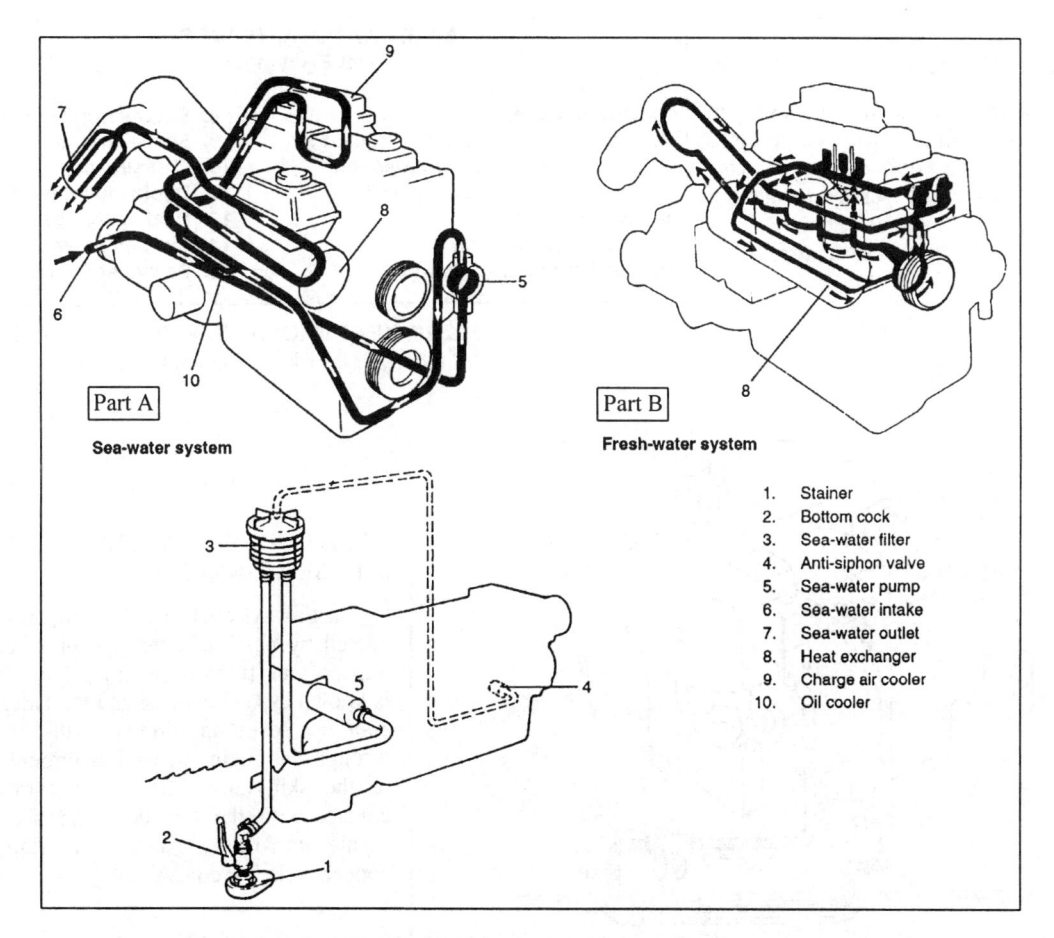

1. Stainer
2. Bottom cock
3. Sea-water filter
4. Anti-siphon valve
5. Sea-water pump
6. Sea-water intake
7. Sea-water outlet
8. Heat exchanger
9. Charge air cooler
10. Oil cooler

Fig 22.9: HEAT EXCHANGER COOLING SYSTEM *(Volvo Penta)*
(Anti-siphon/vacuum valve is fitted where engine is fitted below the waterline. See also wet-exhaust system)

The heat-exchanger system is the most common form of cooling system employed in marine engines. The heat exchanger can be engine-mounted or remote from the engine. It has two water circuits: the closed circuit, which circulates through the engine block, and the open (or exchange) circuit. The latter uses raw water which, having been once used, is rejected. Two pumps are needed; one to circulate the fresh water through the engine and the heat exchanger, and the other to feed raw water to the heat exchanger.

The advantages of the Heat-Exchanger System are:

- The raw water only comes into contact with parts intended to withstand it. These parts are resistant to corrosion.
- A sufficient flow of water is available for cooling the exhaust pipe and exhaust gas.
- The amount of water in the sealed circuit is relatively small. This means that the engine will warm up fairly rapidly. The flow of water is easy to control with a thermostat.
- Addition of anti-freeze to protect the block in the winter.

AFTERCOOLING

The term *Aftercooling* is used in turbocharged engines where compressed air blown by the turbocharger is also cooled by a heat exchanger before entering the combustion chamber. Cooler air is denser and allows more fuel to burn.

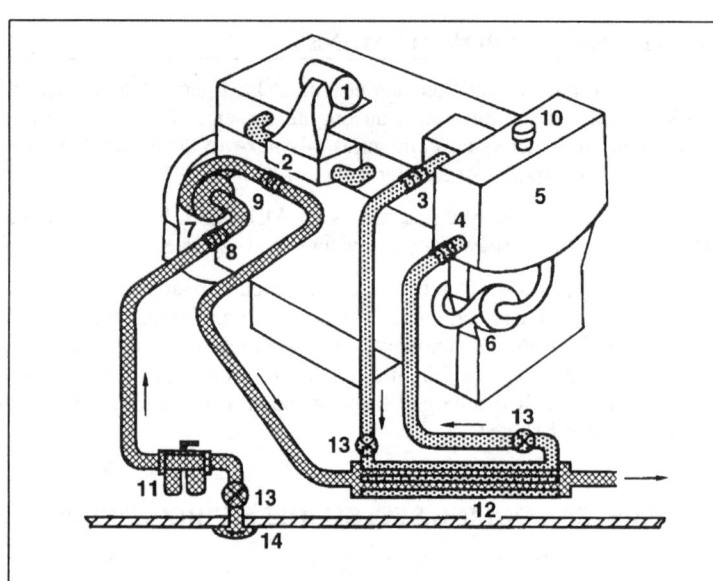

JACKET WATER AFTERCOOLED
Heat Exchanger

1. Turbocharger
2. Aftercooler, jacket water cooled
3. Jacket water outlet connection
4. Jacket water inlet connection
5. Expansion tank
6. Jacket water pump
7. Auxiliary pump, seawater
8. Seawater inlet connection
9. Seawater outlet connection
10. Pressure cap
11. Duplex full-flow strainer
12. Heat exchanger
13. Shut-off valve
14. Seawater intake

Fig 22.10: HEAT EXCHANGER COOLING SYSTEM
(Aftercooler is jacket-water cooled) (*Caterpillar*)

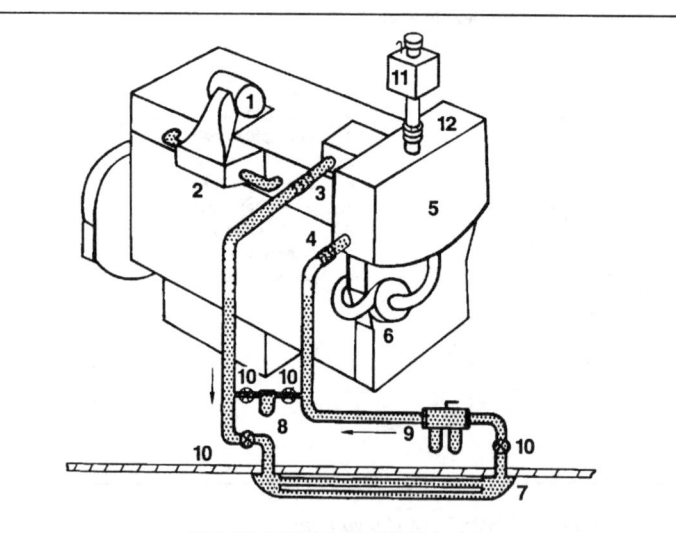

JACKET WATER AFTERCOOLED
Keel Cooler

1. Turbocharger
2. Aftercooler, jacket water cooled
3. Jacket water outlet connection
4. Jacket water inlet connection
5. Expansion tank
6. Jacket water pump
7. Keel cooler
8. Bypass filter
9. Duplex full-flow strainer
10. Shut-off valve
11. Auxiliary expansion tank
12. Flexible connection

KEEL COOLING SYSTEM (CLOSED CIRCUIT)

In this system, closed circuit fresh water is cooled by passing it through pipes or, in steel hulls, a tank fitted beneath the hull. The former is known as *keel cooling* and the latter *skin tank cooling*. A pump circulates the fresh water through the engine as well as through the pipes or the skin tank. The cooler components in contact with the outside water should be of highly corrosion resistant material, such as copper-nickel or equivalent.

Fig 22.11: KEEL COOLING
(Aftercooler is jacket-water cooled) (*Caterpillar*)

Chapter 22: Engine Lubrication & Cooling Systems

REGULATING THE TEMPERATURE

A *thermostat* and a bypass (*at "ABC" in the illustration*) regulate the *jacket-water* temperature. Depending on the temperature of the water discharged by the *engine jacket*, the thermostat closes and opens as necessary in order to direct all or a part of the discharged water to the cooler (*through "Outlet" in the illustration*). The remainder is bypassed to the expansion tank for mixing with water from the cooler. The following illustration shows the thermostat directing jacket water into the expansion tank and the cooler's inlet. It is known as the *Controlled Inlet Thermostat*. In some engines, thermostats are configured to direct jacket water into the cooler's outlet. They are known as *Controlled Outlet Thermostats*. Regardless of whether it is the inlet or the outlet control system, the thermostat is always placed at the jacket water outlet.

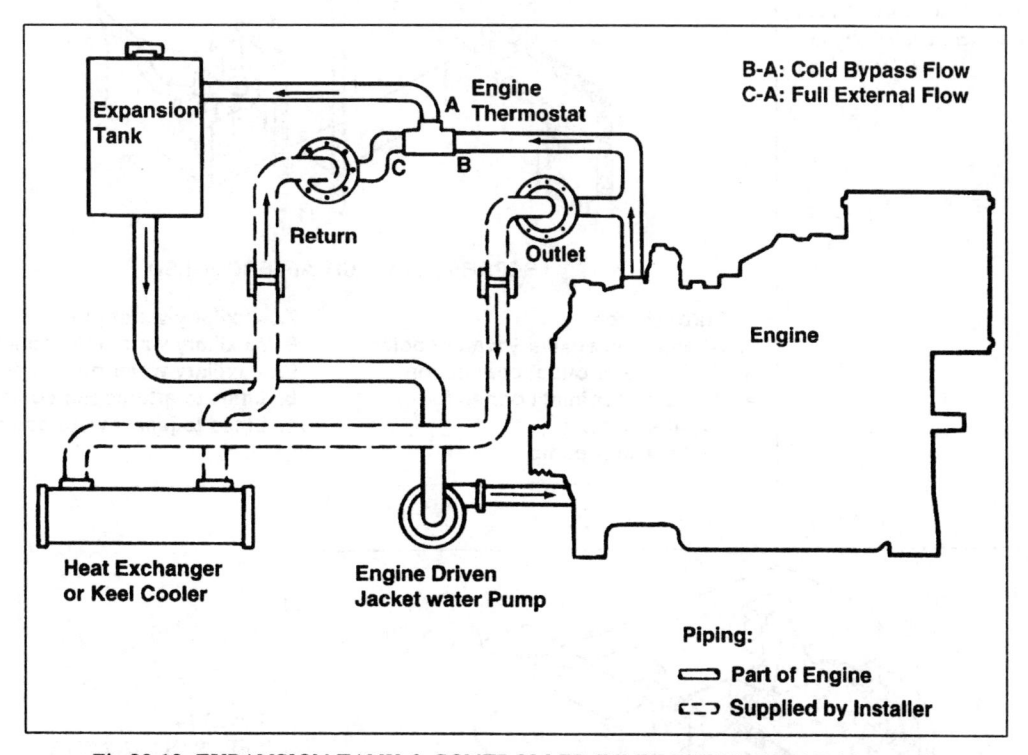

Fig 22.12: EXPANSION TANK & CONTROLLED INLET THERMOSTAT (*Caterpillar*)

As illustrated, the inlet control system uses the expansion tank for mixing water from the engine with water from the cooler. The water returning to the engine is therefore less subject to temperature variations. In the outlet control system, water mixing is done at the water pump inlet. This can cause serious temperature variations in the engine when the sea is very cold.

EXAMPLES OF AFTERCOOLER ARRANGEMENTS

Fig 22.13
SEPARATE CIRCUIT AFTERCOOLED
(Auxiliary seawater-cooled aftercooler on keel-cooled engine)
(*Caterpillar*)

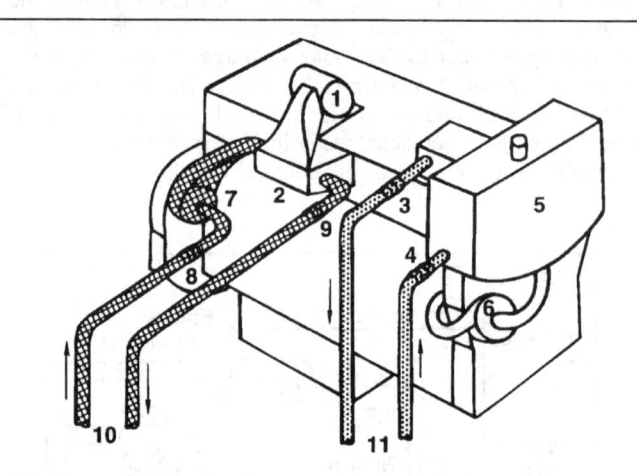

SEPARATE CIRCUIT AFTERCOOLED

1. Turbocharger
2. Aftercooler, auxiliary water cooled
3. Jacket water outlet connection
4. Jacket water inlet connection
5. Expansion tank
6. Jacket water pump
7. Auxiliary water pump
8. Auxiliary water inlet connection
9. Auxiliary water outlet connection
10. Lines to aftercooler cooler
11. Lines to jacket watercooler

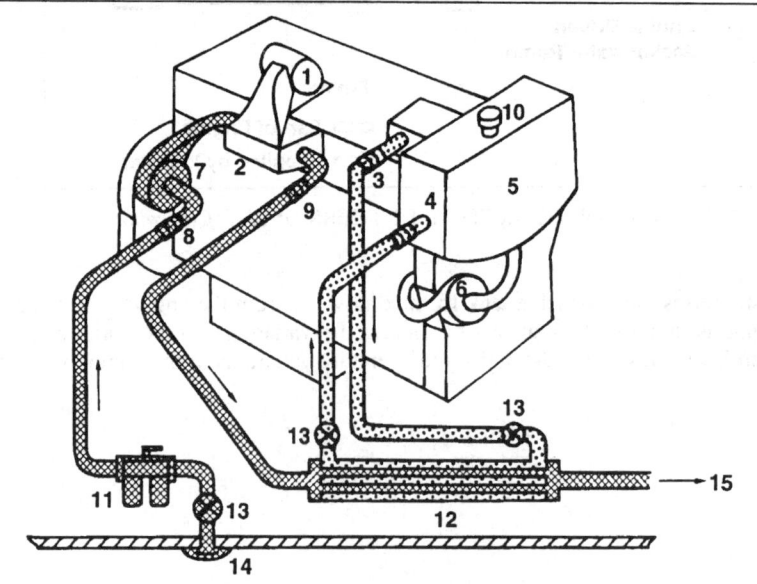

Fig 22.14
SEPARATE CIRCUIT AFTERCOOLED
(Auxiliary seawater-cooled aftercooler on heat-exchanger cooled engine)
(*Caterpillar*)

SEPARATE CIRCUIT AFTERCOOLED
Seawater Aftercooled

1. Turbocharger
2. Aftercooler, seawater cooled
3. Jacket water outlet connection
4. Jacket water inlet connection
5. Expansion tank
6. Jacket water pump
7. Auxiliary seawater pump
8. Auxiliary seawater inlet connection
9. Aftercooler outlet connection
10. Pressure cap
11. Duplex full-flow strainer
12. Heat Exchanger
13. Shut-off valves
14. Seawater intake
15. Seawater discharge

Chapter 22: Engine Lubrication & Cooling Systems

Fig 22.15
SEPARATE CIRCUIT AFTERCOOLED
(Two keel coolers)
(*Caterpillar*)

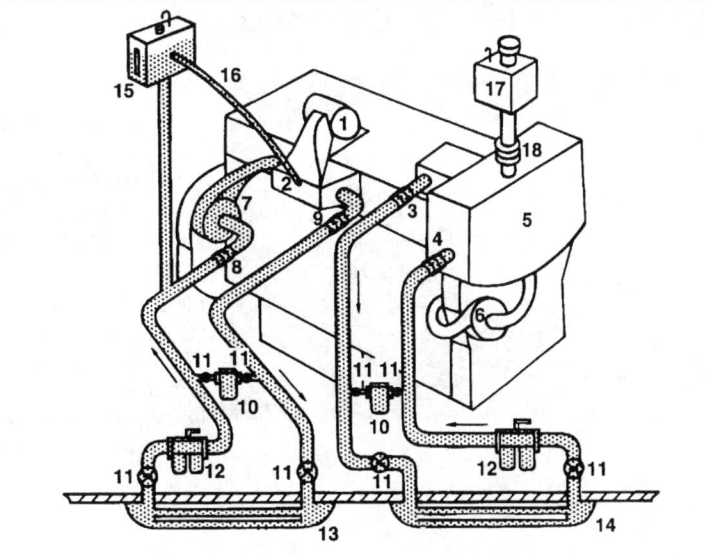

SEPARATE CIRCUIT AFTERCOOLED
Keel Coolers

1. Turbocharger
2. Aftercooler, keel cooled
3. Jacket water outlet connection
4. Jacket water inlet connection
5. Expansion tank
6. Jacket water pump
7. Auxiliary fresh water pump
8. Auxiliary fresh water inlet connection
9. Aftercooler outlet connection
10. Bypass filter
11. Shut-off valve
12. Duplex full-flow strainer
13. Keel cooler for aftercooler
14. Keel cooler for jacket water
15. Expansion tank for aftercooler circuit
16. Vent line for aftercooler circuit
17. Auxiliary expansion tank
18. Flexible connection

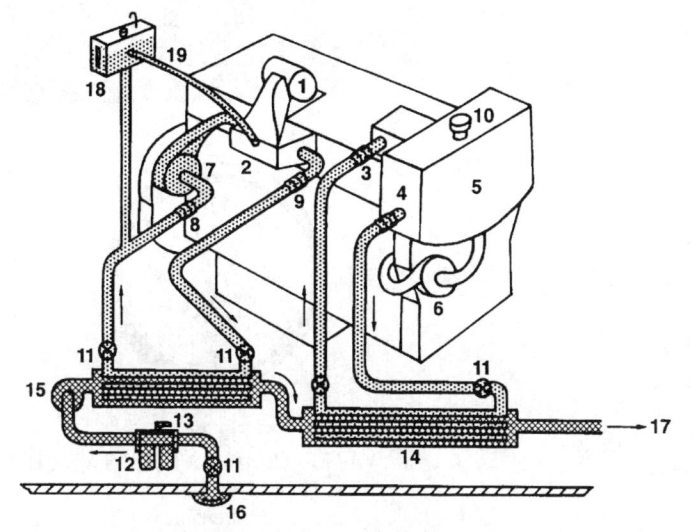

Fig 22.16
SEPARATE CIRCUIT AFTERCOOLED
(Two heat exchangers)
(*Caterpillar*)

SEPARATE CIRCUIT AFTERCOOLER
Heat Exchangers

1. Turbocharger
2. Aftercooler, heat exchanger cooler
3. Jacket water outlet connection
4. Jacket water inlet connection
5. Expansion tank
6. Jacket water pump
7. Auxiliary fresh water pump
8. Auxiliary fresh water inlet connection
9. Aftercooler outlet connection
10. Pressure cap
11. Shut-off valve
12. Duplex full-flow strainer
13. Heat exchanger for aftercooler
14. Heat exchanger for jacket water
15. Customer provided seawater pump
16. Seawater intake
17. Seawater discharge
18. Expansion tank for aftercooler circuit
19. Vent line for aftercooler circuit

HEAT EXCHANGER & TEMPERATURE GAUGE

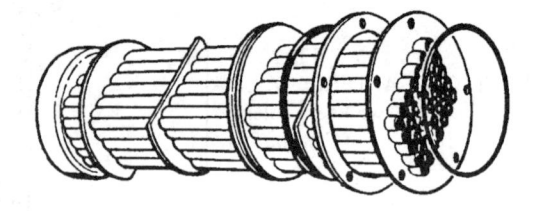

Fig 22.17 (A), (B)
HEAT EXCHANGER AND ITS CORE (*Caterpillar*)
The core (tube-nest) must be inspected and cleaned for scale and debris inside and outside the tubes. Use a 3 mm diameter brazing-rod to clean the tubes. A mechanic can pressure test the Heat Exchanger for leaks.

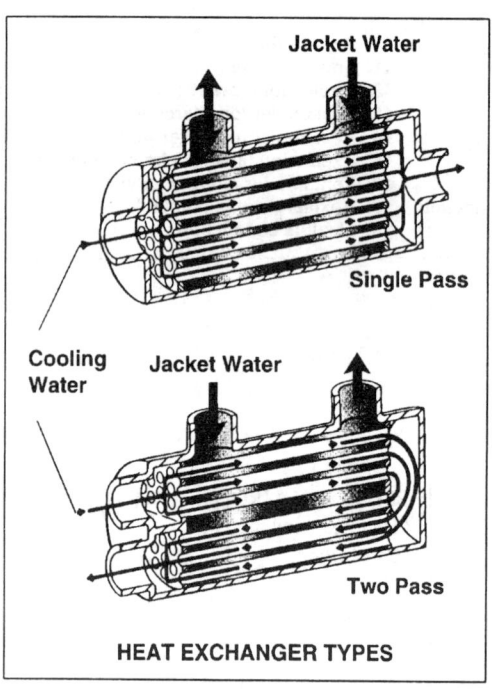

Fig 22.18
SINGLE-PASS AND DOUBLE-PASS
HEAT EXCHANGER (*Caterpillar*)

Fig 22.19
WATER TEMPERATURE SWITCH
& TEMPERATURE GAUGE
(*Caterpillar*)

Chapter 22: Engine Lubrication & Cooling Systems

SEAWATER SUCTION SYSTEM

- For unrestricted waterflow, the piping should be at least of the same size as the sea suction opening, and a size larger if covering a large distance or containing many elbows or bends.

- As shown in figure 22.20, the suction inlet should be a scoop strainer with a double-bottom. This would eliminate the possibility of obstruction by plastic sheeting or similar. The scoop strainer should be so located that it remains underwater even when vessel heels over or rolls. The scoop must face the direction of the boat travel except when sailing, when its direction should be reversed.

- The seawater pressure at the pump suction should never be less than 24 kPa or 3.5-psi vacuum.

- A seacock is fitted to allow the suction to be shut for cleaning the strainer or the plumbing,

- As shown in figures 22.20 and 22.21, the strainer could be remote type or an integral part of the inlet system.

- As much of the inlet piping as possible should be below the vessel's light waterline without air traps.

- A non-return check valve installed downstream of the strainer and as close to it as possible will prevent water from draining out of the pump inlet while the pump is not operating or during cleaning of the strainer.

- Any air trapped after cleaning the strainer and opening the seacock can be vented through a vent valve installed between the strainer and the check valve.

- Air forced under the hull during manoeuvring is purged through vent connections in the sea chest, so that it does not reach the centrifugal pump.

- If the pump is above the vessel's light waterline, then the pump and its priming chamber are kept filled by installing a piping loop above the inlet elbow of the pump.

Typical sea cock and integral strainer

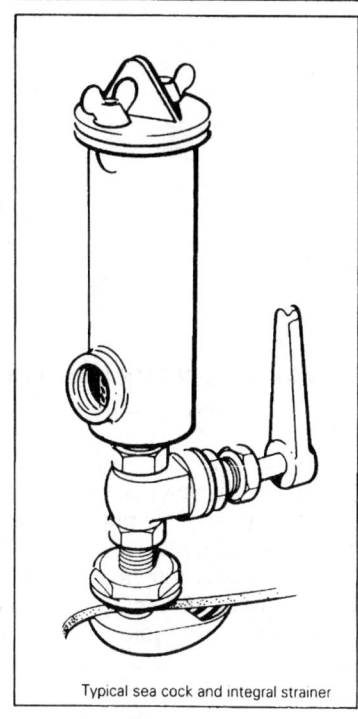

Fig. 22.20
SEACOCK & INTEGRAL STRAINER
(*Lister Petter Marine*)

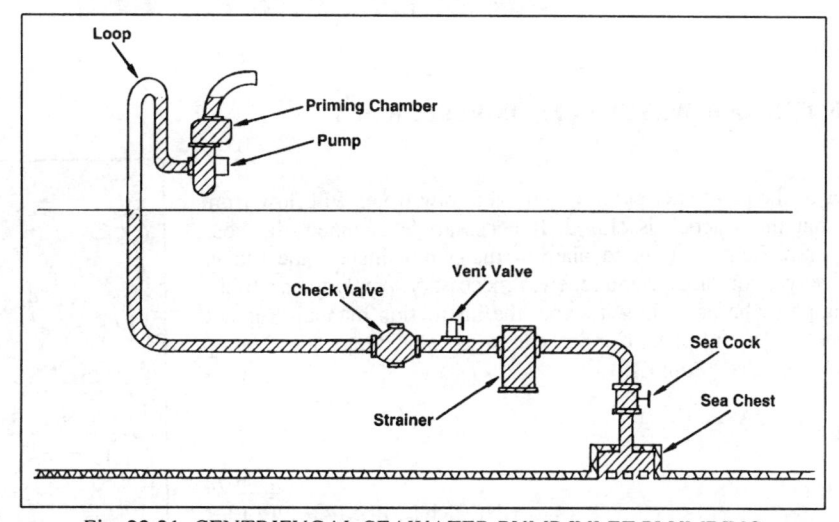

Fig. 22.21: CENTRIFUGAL SEAWATER PUMP INLET PLUMBING

HOW TO DRAIN THE COOLANT CIRCUIT
(A Perkins engine, type TW)

Wait till the engine has cooled down and the coolant is no longer under pressure. Remove the coolant filler cap (1). Remove the drain plugs from the heat exchanger/ manifold assembly (2) and the cylinder block (3). Ensure that the drain holes are not restricted.

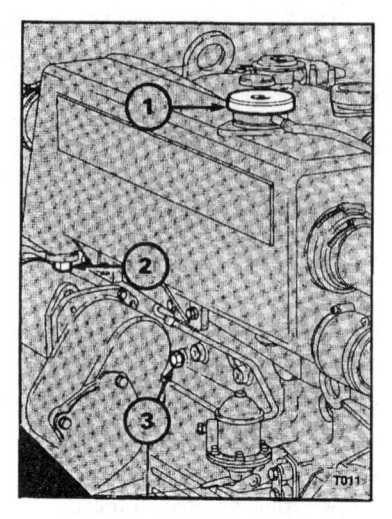

Fig 22.22
DRAINING THE COOLANT

HOW TO DRAIN THE RAW WATER CIRCUIT
(A Perkins engine, type LD)

Ensure that the seacock is closed. Release the setscrews and remove the inlet connection (A1) of the raw water pump and disconnect the outlet hose (A2). Remove the drain plugs from the heat exchanger (B1) and from the engine oil cooler (B2). Ensure that the drain holes are not restricted.

Fig 22.23
DRAINING RAW WATER CIRCUIT

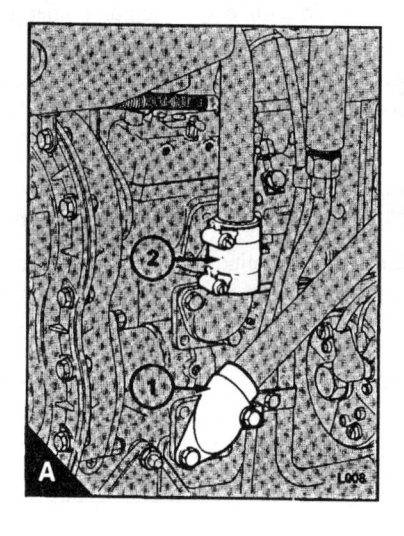

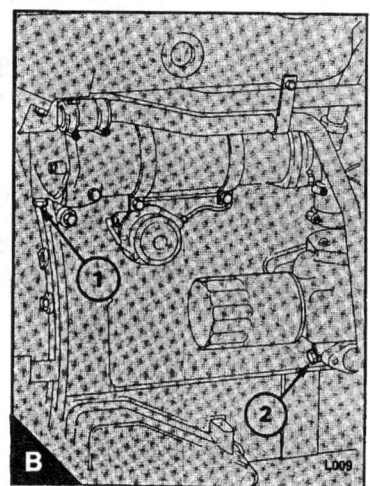

HOW TO CHECK THE RAW WATER PUMP IMPELLER
(Perkins engine)

When the cover of the raw water pump is removed, raw water will flow from the pump. Ensure that the seacock is closed. If necessary, disconnect the hose connection at the pump. Remove the end plate of the pump. Inspect the rubber impeller for excessive wear or damage and renew, if necessary. Apply grease to the blades of the new impeller before fitting it as per the instruction manual. Replace the gasket and re-secure the end plate. Connect the hose connections at the pump and open the seacock. See also "Pumps".

Fig 22.24
CHECKING PUMP IMPELLER

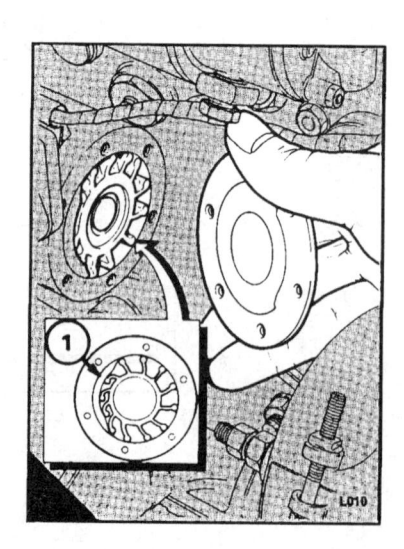

POSSIBLE PROBLEMS WITH WATER COOLING SYSTEMS

- Blockage of seawater inlet.
- An obstruction blocking flow of water around underwater pipes.
- Salt deposits in sea water pipes, reducing their internal diameter.
- Damage due to corrosion.
- Air trapped inside the cooling system.
- Pump drive belt slipping.
- Pump impeller failure.
- Marine growth on the keel cooling system.

As discussed in the fuel system, plastic or other non-metallic hoses should also not be used in the cooling system. They are liable to fail under pressure and collapse under suction. They can be easily damaged and restrict flow, become soft and detached from clips when heated. They also melt easily in a fire. All flexible hoses on an engine must be made of woven wire casings with screwed hose fittings. They should have as few joints as possible, and routed to avoid sharp bends and edges.

Even a slight fall in the amount of cooling water discharge could indicate a problem. The pump impeller could be slowly failing or the raw water intake could be slightly blocked. If it is not a problem at low revs, it will become one when you increase speed.

CHECKING THE WATER COOLING SYSTEM

A) ITEMS TO CHECK BEFORE STARTING

- Fresh water strainer and filter clean
- Zinc anode protection in place
- Trapped air bled through vent plugs
- (Vents are fitted in the higher parts of the cooling system)
- Raw water grill (underwater) clear of obstructions
- Pumps operational
- Pressure cap fitted and sealing

B) ITEMS TO CHECK WHEN RUNNING

- Seawater overboard discharge evident
- Water level in expansion tank correct
- Operating temperature normal.

CHAPTER 22: QUESTIONS & ANSWERS

ENGINE LUBRICATING & COOLING SYSTEMS

1. The engine lubrication system:

    (i) reduces engine wear.
    (ii) transfers heat to moving parts.
    (iii) reduces fuel consumption.
    (iv) prevents metal to metal contact.

    The correct answer is:
    (a) all except (i).
    (b) all except (ii).
    (c) all except (iii).
    (d) all except (iv).

2. Lubrication to Camshaft is supplied by:

    (a) force through pipe lines.
    (b) splash oil.
    (c) pressure feed.
    (d) pressure feed or splash.

3. In most cases, the sump of an internal combustion engine:

    (a) is vented through the fuel filter.
    (b) is vented through the air filter.
    (c) is vented through a separate ventilator.
    (d) does not need to be vented.

4. The following temperature should be maintained in order to reduce salt deposits in the inboard engine cooling system:

    (a) Below 100° C.
    (b) Between 50° & 100° C.
    (c) Below 55° C.
    (d) Below 20° C.

5. The raw-water cooling system in marine engines offers the following advantages:

    (i) installation cost.
    (ii) weight of the installation.
    (iii) fuel consumption.
    (iv) availability of water.

    The correct answer is:
    (a) all except (i).
    (b) all except (ii).
    (c) all except (iii).
    (d) all except (iv).

6. Closed circuit engine cooling is another name for:

    (a) keel cooling system.
    (b) skin tank cooling system.
    (d) heat exchange cooling system.
    (d) any of the cooling systems stated here.

7. In running a vessel's keel-cooling system, you should:

    (i) keep the expansion tank dry.
    (ii) bleed off trapped air.
    (iii) treat water with rust inhibitor.
    (iv) keep fresh water strainer clean.

    The correct answer is:
    (a) all except (i).
    (b) all except (ii).
    (c) all except (iii).
    (d) all except (iv).

8. The heat exchanger engine cooling system should be operated as follows:

    (i) Sea cock open.
    (ii) Pressure cap fitted and sealing.
    (iii) Raw water grill above waterline.
    (iv) Zinc block protection in place.

    The correct answer is:
    (a) all except (i).
    (b) all except (ii).
    (c) all except (iii).
    (d) all except (iv).

9. The reasons for a vessel's cooling system failure include:

    (i) faulty thermostat.
    (ii) faulty pressure cap.
    (iii) no air in the system.
    (iv) blocked sea water inlet.

    The correct answer is:
    (a) all except (i).
    (b) all except (ii).
    (c) all except (iii).
    (d) all except (iv).

10. In a marine engine, a water jacket:

    (a) is fitted under the hull.
    (b) surrounds the engine pistons.
    (c) is not part of an engine cooling system.
    (d) cannot be checked for smooth water flow.

11. The parts of an engine lubrication system include:

    (i) distributor.
    (ii) suction strainer.
    (iii) pressure relief valve.
    (iv) dip stick.

    The correct answer is:
    (a) all except (i).
    (b) all except (ii).
    (c) all except (iii).
    (d) all except (iv).

Chapter 22: Engine Lubrication & Cooling Systems

12. The lubricating oil in the engine is contaminated by:
    - (a) water/carbon/fuel/dust/engine scrapings.
    - (b) carbon only.
    - (c) fuel & carbon only.
    - (d) water & engine scrapings only.

13. A sudden loss of lube oil pressure may be due to:
    - (i) an oil leak.
    - (ii) choked oil strainer.
    - (iii) choked oil cooler.
    - (iv) worn piston rings.

    The correct answer is:
    - (a) all except (i).
    - (b) all except (ii).
    - (c) all except (iii).
    - (d) all except (iv).

14. If the seawater cooling pump stops working, it wouldn't be due to:
    - (a) drive belt slippage.
    - (b) impeller damage.
    - (c) radiator leakage.
    - (d) pump blockage.

15. If the seawater cooling pump gradually starts to lose suction, it wouldn't be due to:
    - (a) blocked suction strainer.
    - (b) sheared impeller drive.
    - (c) vessel rolling.
    - (d) broken drive belt.

16. The cooling water pump should be replaced:
    - (a) regularly.
    - (b) before operating in the shallows.
    - (c) prior to wintering.
    - (d) at the time of cleaning filters.

17. During the warming-up period of the engine, a thermostat in a heat exchanger:
    - (a) bypasses the heat exchanger.
    - (b) stops water circulation through the engine.
    - (c) switches off the raw water pump.
    - (d) restricts the flow of raw water.

18. In keel-cooled engines in small vessels, the keel-cooler cools the:
    - (a) circulating seawater.
    - (b) water from the heat exchanger.
    - (c) engine's seawater.
    - (d) engine's fresh water.

19. The heat exchanger engine cooling system:
    - (a) is a closed circuit system.
    - (b) does not permit addition of antifreeze.
    - (c) is an open circuit system.
    - (d) is an open circuit/closed circuit system.

20. Running a marine diesel engine in an overheated state can:
    - (a) damage the exhaust manifold.
    - (b) seize the engine.
    - (c) burn out the inlet valves.
    - (d) damage the wet exhaust system.

21. The wet sump lubrication system in a four-stroke marine diesel engine means the oil reserve container is:
    - (a) an external tank.
    - (b) an enclosed sump below the engine.
    - (c) also the fuel container.
    - (d) not required.

22. Zinc plugs fitted in the engine cooling water system should:
    - (a) never be replaced.
    - (b) be replaced when wasted.
    - (c) be replaced even if not wasted.
    - (d) be prevented from wasting away.

CHAPTER 22: ANSWERS

1 (b), 2 (d), 3 (b), 4 (c), 5 (c),
6 (d), 7 (a), 8 (c), 9 (c), 10 (b),
11 (a), 12 (a), 13 (d), 14 (c), 15 (d),
16 (a), 17 (a), 18 (d), 19 (d), 20 (b),
21 (b), 22 (b)

Chapter 23

INBOARD DIESEL FUEL SYSTEM

ARE DIESEL & PETROL FUEL DELIVERY SYSTEMS DIFFERENT?

It can be argued that the diesel and petrol fuel delivery systems are entirely different. It can be equally argued that in modern engines the difference lies not in the delivery of fuel, but in the engines they deliver to.

In diesel engines the fuel is injected directly into its combustion chambers (cylinders) where it mixes with hot air and ignites by compression. In petrol engines it is delivered to one or more pre-combustion chambers where it mixes with cold air and then enters the cylinders to be ignited by spark plugs.

The pre-combustion chamber in traditional petrol engines is the carburettor. In the modern EFI engines it is the intake manifold, which is located just before the intake valve of each cylinder. Petrol was pumped into the carburettor, but in the EFI engines it is injected by a fuel injection system similar to diesel engines.

DIESEL FUEL DELIVERY - TANK TO ENGINE

In its simplest form, the fuel flows from the tank – through a primary filter - to fuel lift pump – through a secondary filter – to fuel-injection pump - to fuel injectors, which squirt it directly into the combustion chambers (cylinders).

The fuel injectors are usually delivered more fuel than is required for combustion. The excess is returned to the fuel tank, taking with it small amounts of air entrained in the supplied fuel.

NOTE: Should large amount of air leak into the fuel system, it will prevent the engine from starting or it will result in power loss. It is bled out as illustrated later in this section.

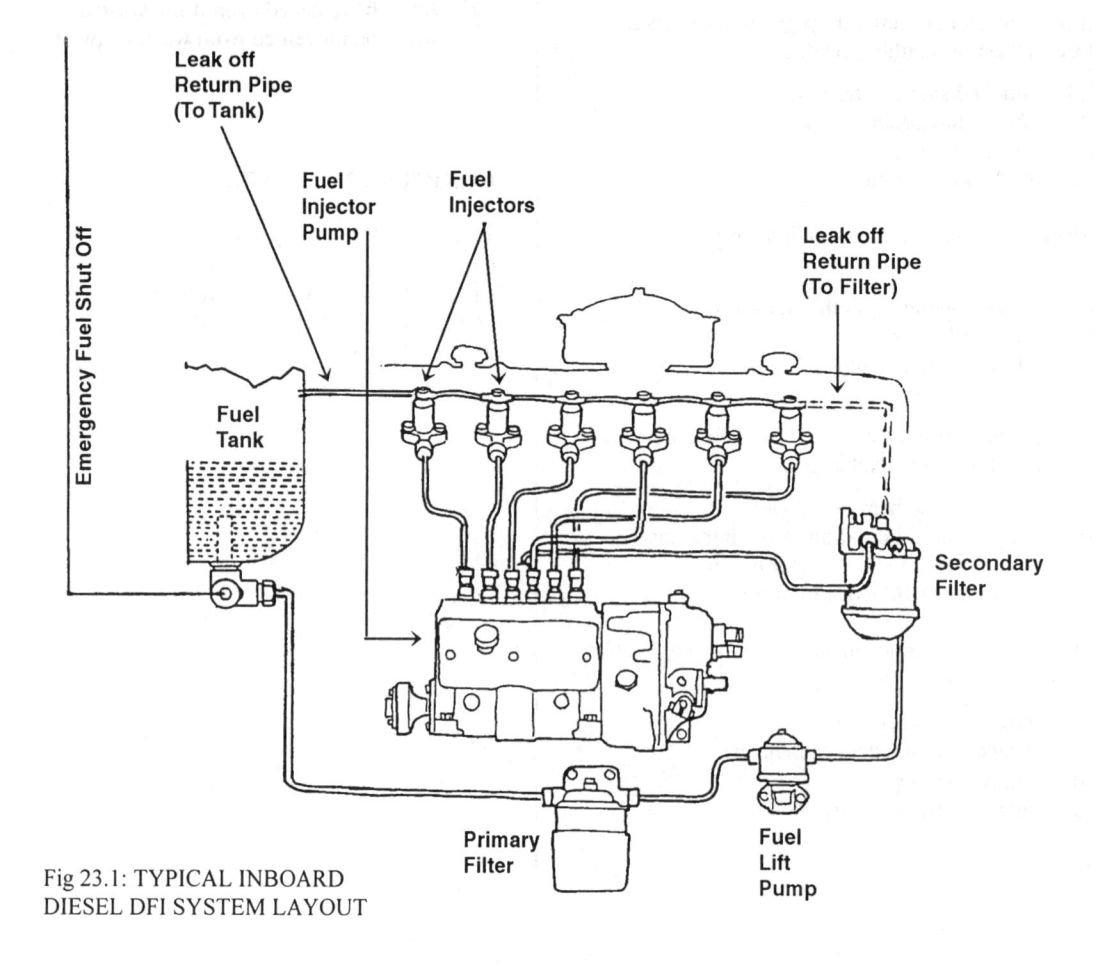

Fig 23.1: TYPICAL INBOARD DIESEL DFI SYSTEM LAYOUT

Chapter 23: Inboard Diesel Fuel System

Illustrated above and below are two diesel fuel delivery systems. One is the traditional or direct fuel injection (DFI) and the other an electronic fuel injection (EFI) system. In both types the fuel is injected directly into combustion chambers. In the latter, however, the mechanics of delivery may vary between manufacturers.

ELECTRONIC FUEL INJECTION (EFI)

Fuel injection is fuel injection, whether direct or electronic. The difference is that the latter is controlled by a computer, known as the electronic control unit or module (ECU or ECM), which also controls various electronic components and sensors. In DFI engines, the injectors are supplied with fuel through individual high-pressure fuel lines. In EFI engines, they are supplied with fuel from a common fuel rail at the cylinder head.

On acceleration, the throttle valve opens to permit more air to enter the engine. The ECM responds by opening more fuel. Sensors monitor the amount of air entering the engine and the amount of oxygen in the exhaust. They allow ECM to fine tune the delivery so that the air-to-fuel ratio is just right.

The Caterpillar HEUI (Hydraulic Electronic Unit Injection) system supposedly goes one step further in the delivery of 'just right' fuel to the engine. The fuel injectors are operated 'electronic-hydraulically' instead of 'electronic-mechanically'. The engine lubricating oil (sump oil) is utilised to generate the hydraulic pressure by boosting it with a high-pressure oil pump. It is delivered to the injectors through a manifold between the cylinder head and valve covers. As illustrated, the ECM regulates the oil pressure, which determines the pressure of the fuel to be injected.

Being more fuel efficient and lower in emissions, EFI has become almost universal in modern engines.

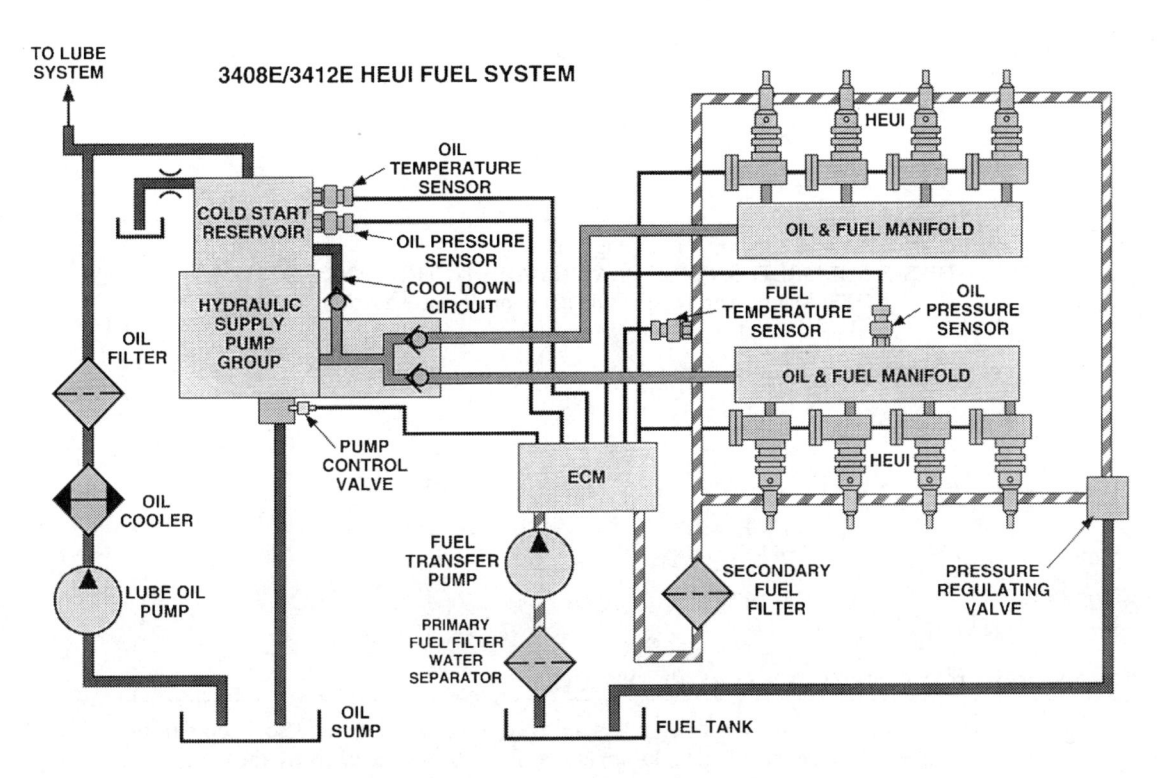

Fig. 23.2: CATERPILLAR 'HEUI' HYDRAULIC EFI
(4-STROKE INBOARD DIESEL)

ELIMINATING AIR IN FUEL SUPPLY – DESIGN FEATURES

Inboard fuel injector systems are usually designed to deliver more fuel to the engine than is required for combustion. The excess is returned to the fuel tank, taking with it small amounts of air entrained in the supplied fuel. The return line runs to the bottom of the tank. This avoids air from entering the fuel system when engine is stopped.

Another simple method of eliminating the air problem is to install a standpipe between the fuel tank and the engine as illustrated below by Caterpillar. The fuel will flow from the tank to the bottom of the standby by gravity. This is the point where the engine picks up fuel. The fuel return line must enter the standpipe at a point a few centimetres above the higher of either the supply or delivery point. The top of the standpipe can be vented into the top of a fuel tank or to atmosphere. This system works satisfactorily with any number of fuel tanks. There must be no upward loops in piping between the fuel tank and the standpipe, as entrapped air may block fuel

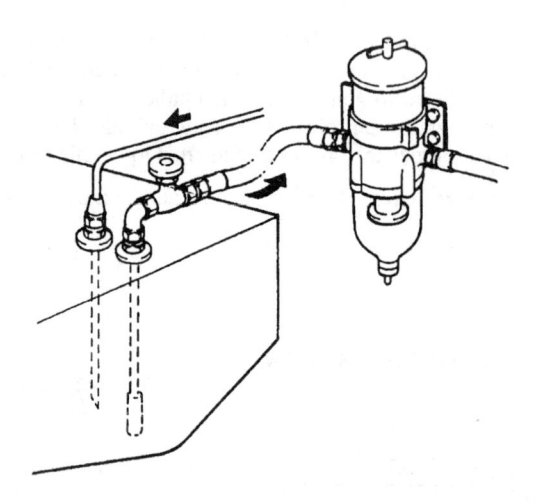

Fig 23.3
DIESEL FUEL-RETURN LINE DRAWN BACK TO THE TANK BOTTOM
(NOTE: Marine petrol engines do not normally have a return line)
(*Volvo Penta*)

DAY TANK (SERVICE TANK, AUXILIARY FUEL TANK)

Day tanks are useful in providing a settling reservoir to separate air, water and sediments from the fuel. Usually, they are required to be fitted where the main fuel tanks are too far from the engine or above the engine.

The day tank should be close enough to the engine so that the suction lift is not more than 3.65 metres. The smaller this figure, the easier the engine will start. It should be located so that the level of the fuel is not higher than the fuel injection valves on the engine. This is to safeguard against the fuel leaking into combustion chambers when the engine is not running. Liquid fuel in combustion chambers at the time of starting the engine is likely to cause engine failure.

Chapter 23: Inboard Diesel Fuel System

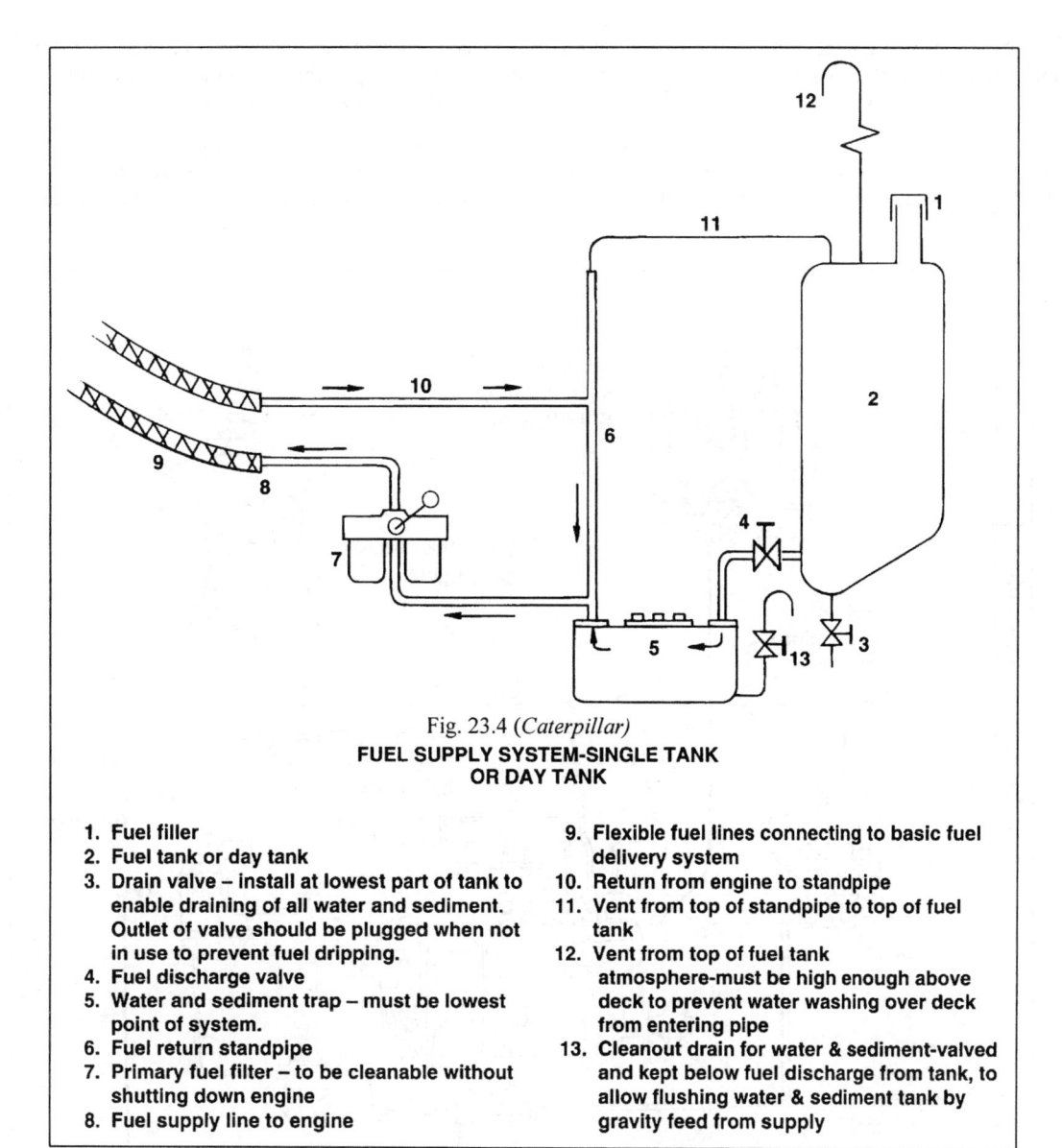

Fig. 23.4 (*Caterpillar*)
FUEL SUPPLY SYSTEM-SINGLE TANK OR DAY TANK

1. Fuel filler
2. Fuel tank or day tank
3. Drain valve – install at lowest part of tank to enable draining of all water and sediment. Outlet of valve should be plugged when not in use to prevent fuel dripping.
4. Fuel discharge valve
5. Water and sediment trap – must be lowest point of system.
6. Fuel return standpipe
7. Primary fuel filter – to be cleanable without shutting down engine
8. Fuel supply line to engine
9. Flexible fuel lines connecting to basic fuel delivery system
10. Return from engine to standpipe
11. Vent from top of standpipe to top of fuel tank
12. Vent from top of fuel tank atmosphere-must be high enough above deck to prevent water washing over deck from entering pipe
13. Cleanout drain for water & sediment-valved and kept below fuel discharge from tank, to allow flushing water & sediment tank by gravity feed from supply

BLEEDING (PURGING) THE FUEL SYSTEM

A large quantity of air can enter the fuel system when the fuel tank is drained, a fuel pipe is disconnected, the fuel system leaks or the fuel pump or filter is cleaned. It must be vented to prevent engine power loss.

Many engines are fitted with the leak-off pipe (excess fuel line) to the fuel filter. In such cases, getting rid of air from the fuel system requires only to open the bleeder screw on the filter and manually pumping at the fuel pump until bubble-free fuel flows from the bleeder. It will remove the air right through the injectors. Rarely do modern engines need bleeding at the injector pump or the injectors. However, if necessary, figure 23.5 illustrates the method of bleeding a DFI fuel system.

After bleeding, if the engine starts correctly for a short period and then stops or runs roughly, check again for air in the fuel system. If air has returned, there is probably a leakage in the low-pressure system.

For bleeding the injectors, the fuel injection pump has to be made to pulsate without fuel being supplied by the fuel delivery pump. This is achieved by turning over the engine – small ones by hand and larger ones by use of the starter motor. Mechanics sometimes bleed the injectors while the engine is running. This is not advisable for the average boating person.

Air in an EFI engine can also be bled through an air bleed valve at the fuel filter followed by priming with the priming pump. After bleeding at the fuel filter, prime the pump as much as possible, until the pressure builds up. Now try starting the engine. It could take 15 to 20 seconds of cranking before the fuel system is fully purged and the engine starts.

Bleeding of injectors in EFI engines is not possible because, unlike the DFI injectors, they are not supplied with fuel through individual high-pressure fuel lines. They are supplied with fuel from a common fuel rail at the cylinder head. It is not advisable to disassemble EFI fuel injectors or the high-pressure fuel pump unit or to stick a pin into the injector hole to remove debris.

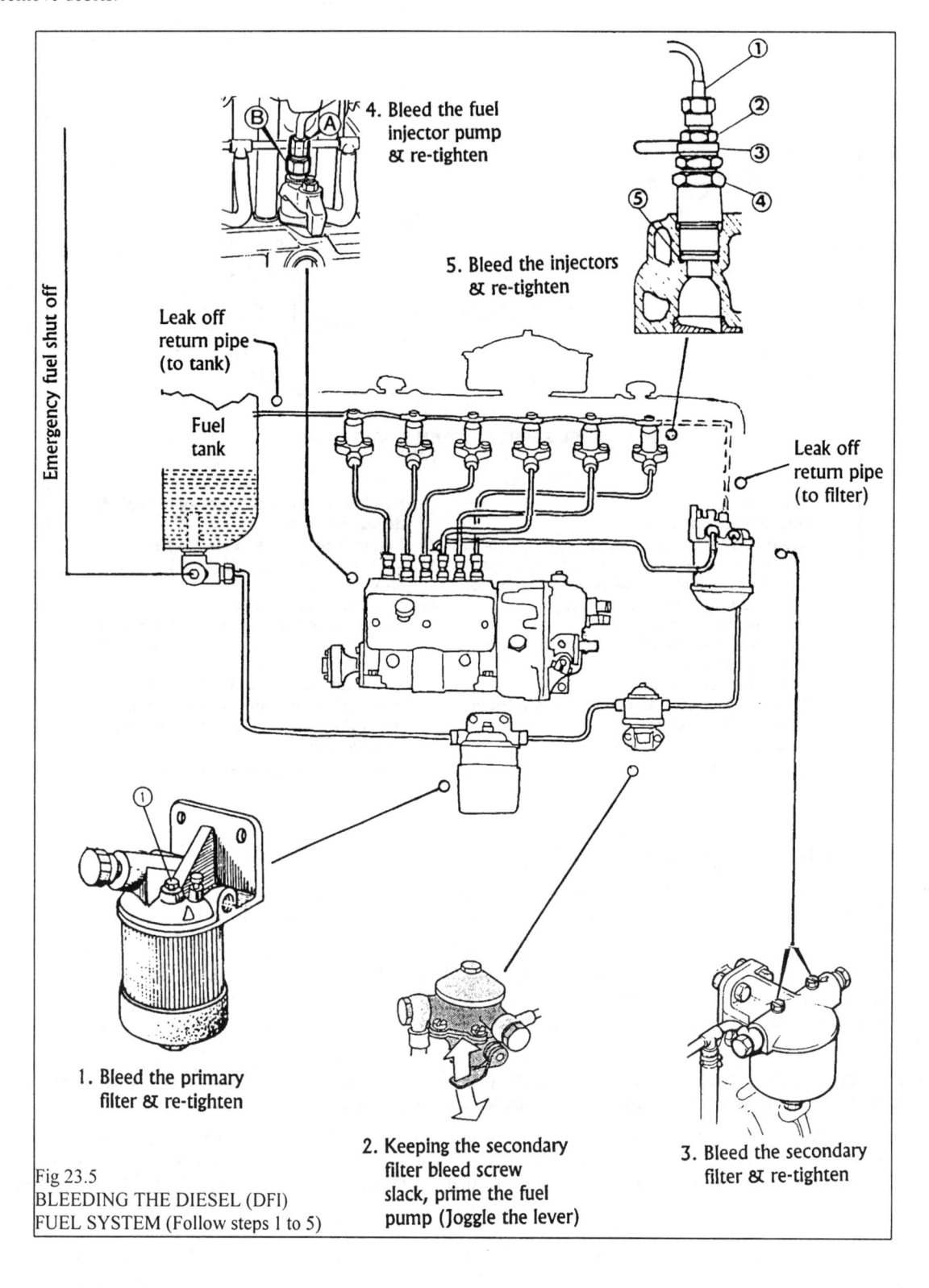

Fig 23.5
BLEEDING THE DIESEL (DFI) FUEL SYSTEM (Follow steps 1 to 5)

Chapter 23: Inboard Diesel Fuel System

FUEL TANKS, PIPES & CONNECTIONS

(Bonding and grounding of refuelling system is also discussed in Chapter 12)

Fuel tanks are usually constructed of stainless steel or aluminium sheet metal. No part of the fuel tank must depend on soft solder for tightness. Zinc is not used with diesel fuels because it is unstable in the presence of sulphur.

Fuel tanks are designed without any openings in the bottom, sides or ends. Filling and venting connections are on the top. However, diesel tanks may have a lower pocket to test for the presence of water in fuel. There should be no locations for moisture to accumulate on the outside of the tanks.

A fuel tank has connections for filling, venting, suction line, sender for tank gauge, ground connection and a manhole with cover. The suction line and the return line should be separated as illustrated. The filler fitting should be installed such that overfilling and fuel entering air intakes is avoided. A shut-off valve is installed in the suction line as close to the tank as possible. The valve can be closed from a position outside the compartment in which the tank is situated. In case of a fire in the engine room, the fuel can be cut off without entering the compartment.

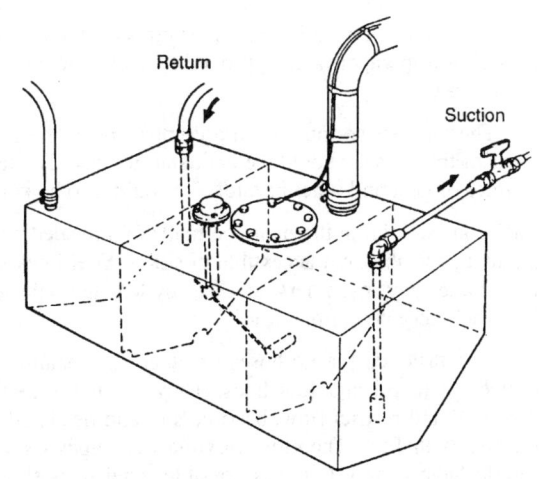

Fig 23.6: FUEL TANK FITTINGS
(Volvo Penta)

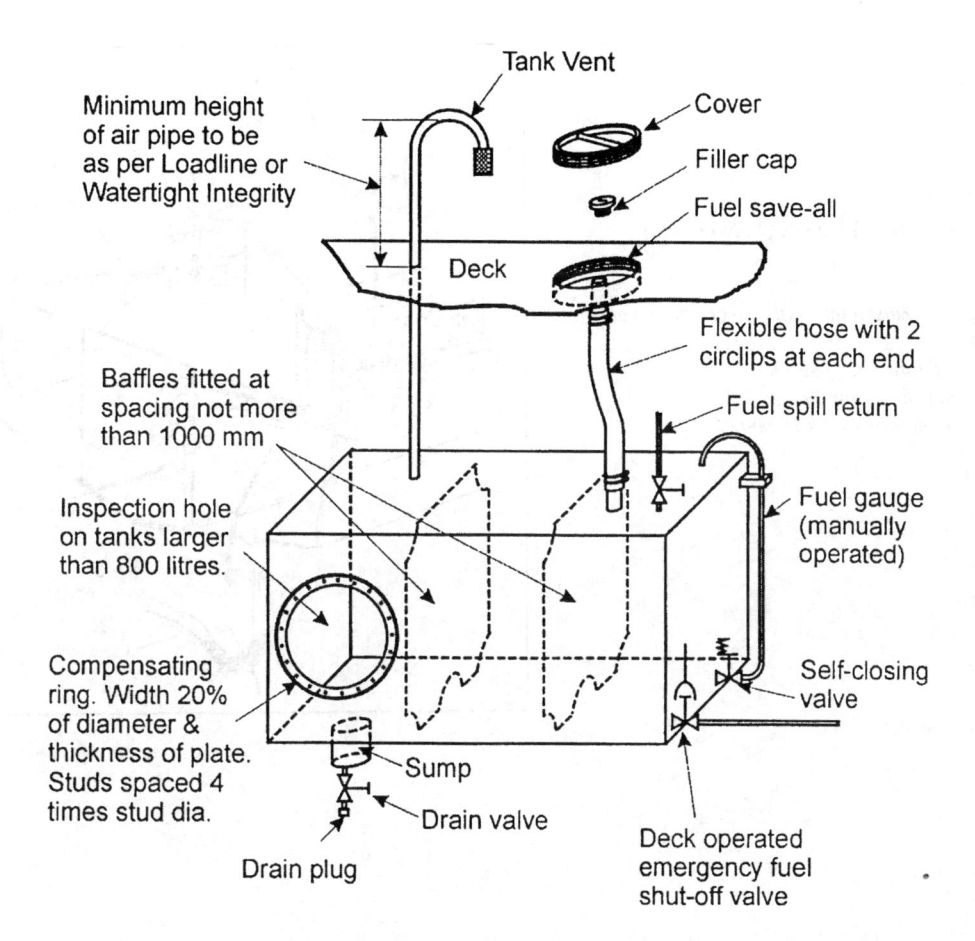

Fig 23.7: TYPICAL FREESTANDING NON-PORTABLE FUEL TANK INSTALLATION
(For fuel over 60° C flash point)
(NSCV PART C Section 5 Subsection 5A)

Fittings supporting and holding the tanks should neither reduce the strength of the tanks, nor cause them to become distorted. Do not position tanks on wooden blocks or on any other uneven bedding. This might cause abnormal stresses with subsequent risk of developing cracks in the tank.

At least one baffle for each 200 litres of volume should be fitted. However, baffles in tanks should not restrict the flow of fuel or trap vapour. An approved type electric gauge mounted on the dashboard usually indicates the fuel level of under-floor tanks.

The tank space must form a compartment of its own in order to prevent fuel and fuel fumes from leaking into other areas of the vessel, should the tank start to leak. The tanks should be installed in well-ventilated places, away from possible sources of heat and other hazards. It should also be kept away from accommodation space.

Fuel lines should, in the main, be made of annealed copper of marine standard. They should be firmly secured to the hull in an always visible and accessible position. Fuel lines should be regularly checked for weeps. As far as possible, route fuel lines under any machinery, so that any leakage will be confined to the bilges. Leaks from overhead fuel pipes falling onto hot machinery are a fire hazard.

Clip fastened plastic hoses or other non-metallic hoses are a threat to vessel's safety. They do not provide sufficient protection to oil and fuel lines. They are liable to fail under pressure and collapse under suction. They can be easily damaged and restrict flow, become soft and detached from clips when heated. They also melt easily in a fire, thus feeding the fire with fuel. Therefore flexible fuel pipes must be made of woven wire casings with screwed hose fittings. They should have as few joints as possible. Fuel lines should be routed avoiding sharp bends and edges. They should be kept clear of decks to prevent fuel starvation due to crimping. Fittings on flexible fuel hoses must be grounded to the common ground system.

Fig 23.8: FUEL TANK FITTINGS (*Volvo Penta*)

Key:
1. *Tank venting line, raised internally to create a water lock*
2. *Filler fitting*
3. *Filler fitting and tank are grounded*
4. *Fuel filling line is free from "traps"*
5. *Compartment containing tank is vented*

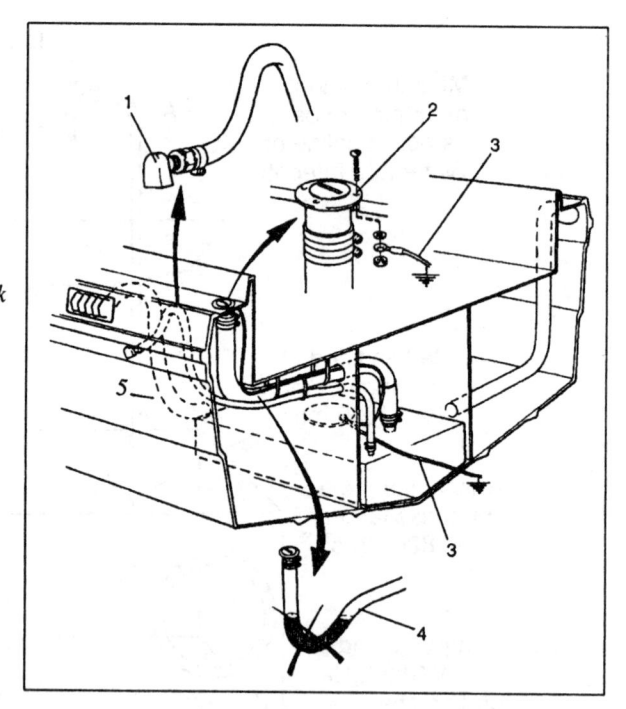

Chapter 23: Inboard Diesel Fuel System

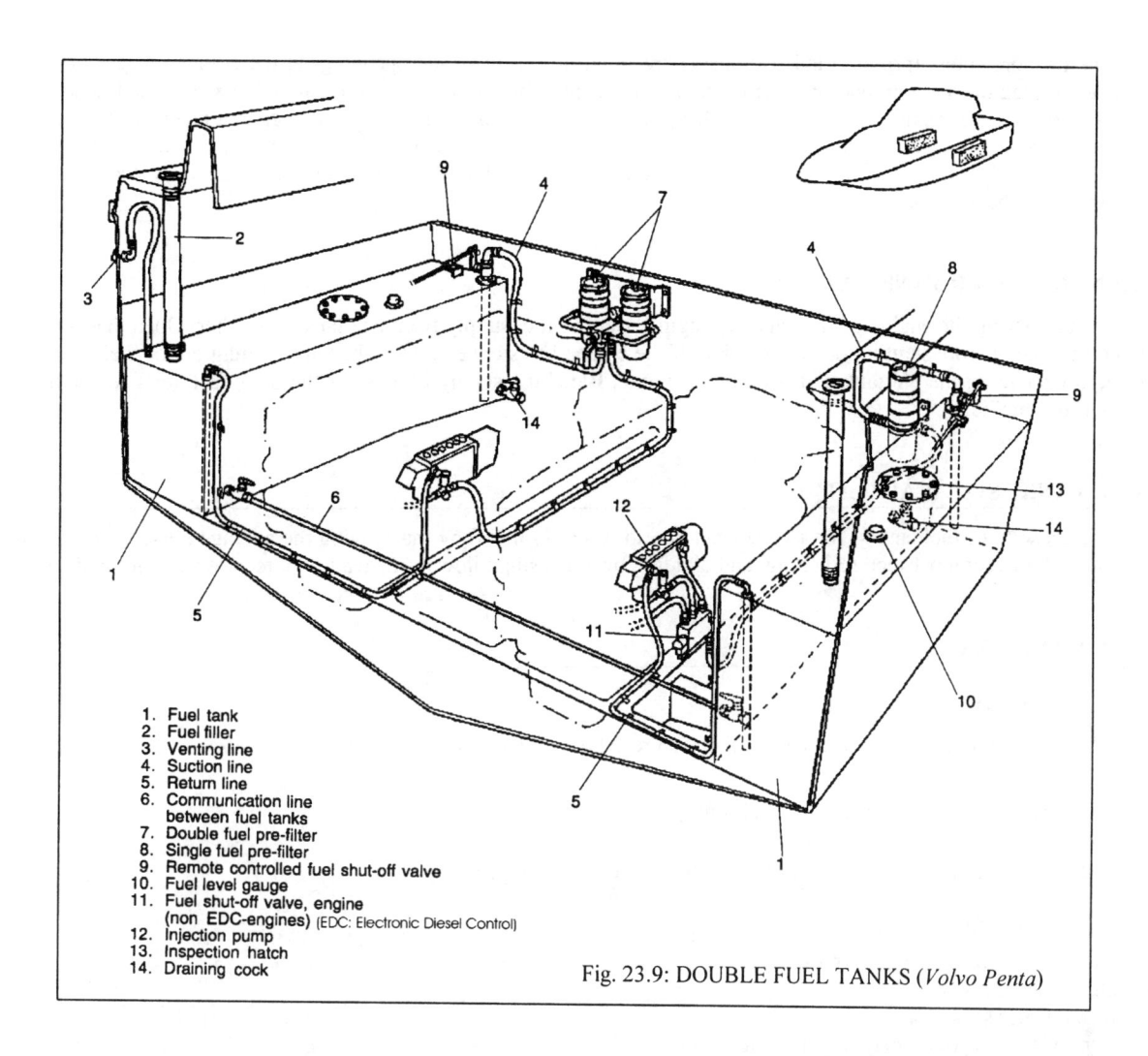

1. Fuel tank
2. Fuel filler
3. Venting line
4. Suction line
5. Return line
6. Communication line between fuel tanks
7. Double fuel pre-filter
8. Single fuel pre-filter
9. Remote controlled fuel shut-off valve
10. Fuel level gauge
11. Fuel shut-off valve, engine (non EDC-engines) (EDC: Electronic Diesel Control)
12. Injection pump
13. Inspection hatch
14. Draining cock

Fig. 23.9: DOUBLE FUEL TANKS (*Volvo Penta*)

Tanks should be vented by pipes to the outside, fitted well away from and high enough to keep fumes away from vessel's interior and cockpit. Wire gauze diaphragms (mesh screens) are fitted to the open mouths of these pipes to act as flame arresters, and prevent flames from travelling into tanks. A gooseneck or cowl fitting on the pipe's mouth will prevent water entering the fuel tank. A blocked vent can hold back the fuel, causing the engine to fade and possibly stop. This can be easily confirmed by opening the filler cap. If the engine picks up, the vent is blocked. Fuel tanks should be fitted with an anti-siphoning device to prevent fuel from siphoning back to the tank. This eliminates the need for lengthy priming prior to starting.

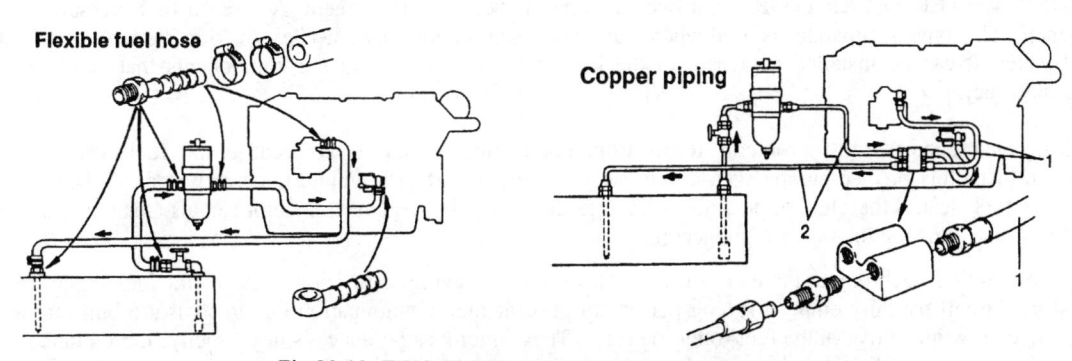

Fig 23.10: TWO TYPES OF FUEL PIPE CONNECTIONS
(Illustration K-2 shows transition from flexible fuel hoses (1) to copper pipes (2)) (*Volvo Penta*)

One or more filters (primary and secondary) are fitted in the fuel line depending on the engine design. Filters allow water to be bled off and removal of other contaminants. If possible, do not fill the main and auxiliary fuel tanks from the same pump - in case the fuel is contaminated. If a filter becomes blocked or damaged and cannot be cleaned or replaced in an emergency, it can be by-passed. The fuel filter must not contain parts made of glass that can crack. The most common type of copper pipe joint is shown. However, should a flexible fuel hose be used between the tank and the fuel filter, the filter must also be grounded.

CHANGING OVER TANKS AT SEA

Avoid starving the engine of fuel when changing over the fuel supply from one tank to another. Open the supply valve on the new tank before shutting off the old tank. If the engine has stopped due to the tank running out of fuel, then shut off the supply valve on the old tank and open the new one. Bleed the supply line as per the manufacturer's recommendation and start the engine.

FUEL TANKS SIZE (CAPACITY)

The capacity of fuel tank(s) for a vessel can be estimated (in litres) by multiplying the average horsepower demand by the hours of operation between refuelling, and dividing by 4. If using kilowatts instead of horsepower, then divide by 3.

FUEL MAINTENANCE

WATER IN FUEL

Water can enter into the fuel during shipment or as a result of condensation during storage. It causes:

- Excessive sludge
- Wear and tear in the engine and its fuel supply system
- Power loss due to fuel starvation
- Swelling up of the filter media, cutting off the engine's fuel supply altogether

To minimise the effect of water:

- Obtain fuel from reliable sources.
- Drain fuel tanks daily.
- Install water separators
- Removal of salt water may require centrifuges.

WATER SEPARATORS

There are two types of water separators:

SEDIMENT TYPE WATER SEPARATOR does not have a filtering media in the element. Therefore, it does not normally need scheduled element replacement. This type of filter is installed ahead of the engine's fuel transfer pump, and as close to the tank as possible so that a minimum length of the fuel line is subjected to water and sediment contamination.

COALESCING WATER SEPARATOR has a two-stage paper media in its element. A drop in fuel pressure indicates blocked element. This type of separator is used when water is broken up into tiny particles, making the fuel cloudy. It will separate all water. It can be installed anywhere in the fuel line, such as next to the components that needs the most protection from water.

In fuel injected engines most problems result from contaminated fuel. It is because the fuel-injection system incorporates an incredibly precise pump. Even a microscopic piece of dirt or a minute trace of water can disable it. The need to keep the fuel clean is therefore of paramount importance. Water droplets in an injector can also turn to steam during a compression stroke, blowing the top off the injector.

Top up the fuel tank each day at the end of the day's operation, leaving some room (ullage) for fuel expansion. Fuel tanks should not be left partially empty for long periods to prevent fuel contamination due to moisture build up and the resultant microbe growth that live in the fuel/water interface. These microbes produce a slimy, smelly film on tank surfaces that plugs injectors, filters and pumps. Use a fuel treatment to absorb moisture and fight microbe growth, and to keep the injectors clean. Drain water and sediment from the fuel service tank at the start of a shift, as well as, 5 to 10 minutes after filling it. Drain a cupful at the start of every shift for inspection. Drain storage tanks weekly and before refilling them. This

Chapter 23: Inboard Diesel Fuel System

will help prevent water and sediment being pumped from storage tanks to the service tank. Install and maintain a water separator before the primary fuel filter. Maintain fuel filters as discussed below.

FUEL FILTERS

Fuel cleanliness is very important. Two filters should be fitted. The first must have a water trap whilst the second can be a plain element type. Both should be regularly drained, cleaned and replaced according to the manufacturer's instructions. Check the fuel pressure differential, which can indicate a dirty fuel filter.

Some engines are fitted with a separate water trap, as shown below for a Caterpillar engine. No (1) is the vent and (2) the drain valve. It is recommended to drain it daily before starting the engine. The fuel supply line must be closed before draining. The water trap can be opened for cleaning as shown.

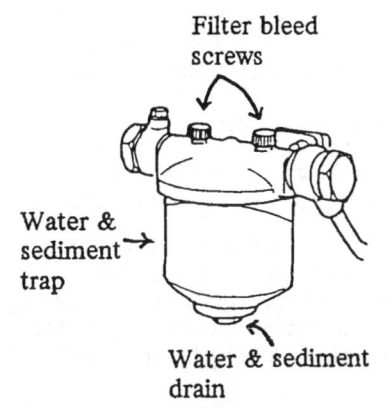

Fig 23.11: A FUEL FILTER

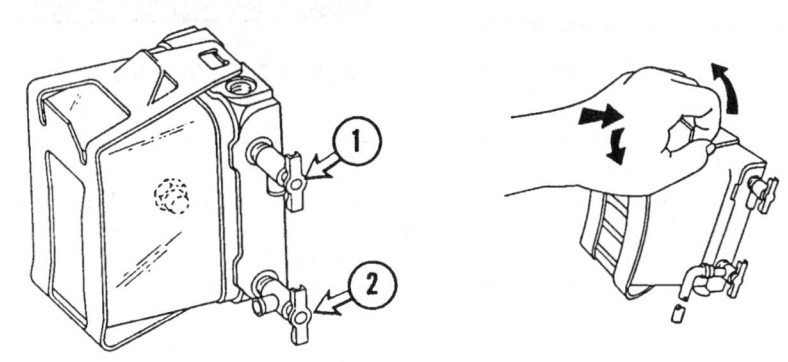

Fig 23.12: DRAINING & CLEANING WATER SEPARATOR
(*Caterpillar*)

Some engine control consoles are fitted with a fuel pressure gauge to indicate a dirty fuel filter. In the attached illustration, when the filter element becomes clogged the indicator moves to OUT. A struggling and eventually dying engine is another sign of dirty filters, unless the tank vent is blocked, as discussed earlier. Before cleaning or changing a filter, the engine should be stopped, fuel shut off, surrounding area cleaned and rags kept ready to clean up any escaping fuel.

Fig 23.13: FUEL PRESSURE GAUGE

The Perkins fuel filter is renewed as shown. The steps are:

- Clean around the filter housing.
- Remove filter without introducing dirt into the housing.
- Inspect the new filter for debris or metal filings. Especially check the thread of spin-on filters. Any filings already in the filter will go straight to the fuel pump and injectors. (New filters should be properly stored to remain dust free).
- Do not pour fuel into the new filter element before installing it. Contaminated fuel will cause problems and damage.
- Clean the inside surface of the filter head
- Renew joints, lightly lubricating with clean fuel.
- Re-assemble filter.
- Prime the fuel pump to bleed air out of fuel system. If further necessary, bleed the fuel system as shown below.
- Cut apart the used filter to inspect for contaminants and to compare for quality and filtering effectiveness.

Fig 23.14: RENEWING FUEL FILTER ELEMENT
(*Perkins*)

FUEL LIFT PUMP

Fuel lift pumps are more often diaphragm pumps. Rotary (centrifugal) pumps are also used. Both types of pumps are discussed under "Pumps". To clean a Perkins rotary fuel lift pump: Remove cover and joint (1) from the top of the pump (3). Remove gauze strainer (2). Wash all sediment from the pump body. Clean the gauze strainer, joint and cover. Assemble the pump, making it air tight. Bleed air from the fuel system as discussed below.

FUEL COOLERS

The excess fuel returned from the engine can be quite hot. In most cases it does not present a problem because of one the following factors:

- The operating periods are not long enough to build up heat in the tank.
- The time between operating periods is quite long to dissipate the heat.
- The tank is quite large (Tanks larger than 11,000 litres accept a great deal of heat before it becoming a problem).
- The tank being in contact with the hull dissipates its heat to seawater.

Only in some cases a fuel cooler is necessary for proper engine performance.

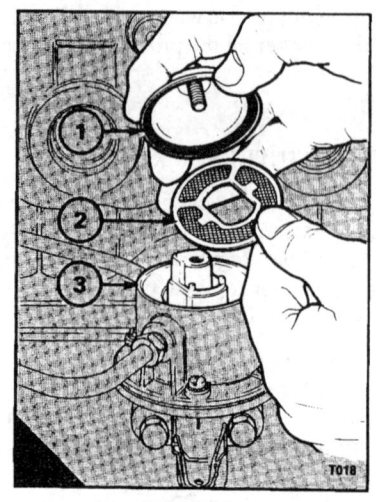

Fig 23.15
A PERKINS FUEL LIFT PUMP

FUEL SAFETY

GAS FREEING A FIXED FUEL TANK

Before entering a fuel tank or carrying out any repairs in its vicinity, the tank must be made gas free. There are test meters available and industrial chemists can be hired to test a compartment's environment. In most ports, large vessels are required to advise the authorities before carrying out hot works, such as oxy-cutting and welding.

After draining all the fuel from the tank, it can either be filled with water or ventilated from bottom up.

WELDING - SAFETY PRECAUTIONS
- Follow port regulations regarding "hot" works
- Gas-free the vessel
- Keep necessary fire-fighting gear in readiness and close by
- Keep a bucket of water close by to sprinkle on welding spatter, if any
- Make note of contents of the adjoining compartments
- Make note of what's under deck and above deck
- Wear safety gear
- Assess risk of hazardous fumes from burning

FUEL TERMINOLOGY

FLASH POINT

Flash point of a fuel is the lowest temperature at which its vapour would ignite with a flash. It is determined by the type of fuel and the air-fuel ratio. If a fuel were heated to its flash point, it would give off vapour that would ignite with a flash.

Both petrol and LPG have a flash point less than 60° C. However, technically speaking, LPG does not have a flash point because it is not a liquid unless stored under pressure. Petrol starts to vaporise at -40°C, therefore its flash point is often stated as -40°C. It can catch fire well below freezing temperature.

The minimum flash point for diesel fuels is about 38°C. However, by regulation, the closed flash point for fuel used at sea must be at least 60°C. It is typically between 72° to 80°C. Diesel fuel poses fire danger in cases where high-pressure fuel line between fuel injection pump and injectors sprays atomised fuel onto a hot surface such as an unlagged exhaust pipe. To avoid this situation, engines are normally designed with fuel system on the opposite side of the exhaust system.

Overall, diesel is a safer fuel to handle than petrol. Still, given the right conditions, it is also prone to ignition due to static and prone to explosion. It should be treated with the same care as petrol.

For safety, maintain fuel tanks at least 10°C below the flash point of the fuel. Know the flash point of your fuel, especially when working with heavy fuels that need heating to a higher temperature to flow readily.

FIRE POINT

A fuel heated to its fire point would give off vapour which would burn continuously when ignited by a spark or flame. There is usually not much difference in a fuel's Fire Point and Flash Point.

Chapter 23: Inboard Diesel Fuel System

AUTO-IGNITION TEMPERATURE (A.I.T.) (SPONTANEOUS COMBUSTION POINT)

Spontaneous Combustion or Self Ignition Point is the minimum temperature at which a fuel would burst into flames without any external source of ignition. Diesel would self-combust at a lower temperature (350°) than petrol (390°). However, keep all fuels away from the exhaust, which can get much hotter than the A.I.T. of fuels.

FLAMMABILITY RANGE

Flammable Range is the percentage amount of fuel in the air that supports ignition. The flammable range of petrol is 1.4% to 7.6%, and is similar for diesel and LPG. Outside this range the fuel/air mixture is either too poor or too rich to ignite. Hydrogen and acetylene, on the other hand, burn over a very wide fuel/air range (about 3% to 80%).

DENSITY RELATIVE TO AIR

The density of air is one. Gases of density less than that of air rise because they are lighter than air, and vice versa. Petrol, diesel and LPG are all heavier than air with density of about 1.5. However, hydrogen is very light (Density 0.1) and Acetylene's density relative to air is 0.9.

CLOUD POINT

The cloud point of a fuel is the temperature at which a cloud or haze appears in the fuel. This appearance is caused by the temperature falling below the melting point of waxes or paraffins that occur naturally in petroleum products. The importance of the cloud point is that it is at this temperature that fuel filter plugging starts to occur, hampering the fuel flow to the engine. The refiner determines the fuel's cloud point. It is kept at least 6°C below the lowest outside (ambient) temperature.

On rare occasions, and in consultation with the fuel refiner, wax in the fuel can be maintained in a melted state by one of three ways:
- Employ a fuel heater.
- Dilute the high cloud-point fuel with a low cloud-point fuel like kerosene.
- Add a wax crystal modifier to the fuel. The modifiers do not change the cloud point of the fuel but keep the wax crystals small enough to pass through the fuel filter.

POUR POINT

The pour point of a fuel is that temperature which is 3°C above the temperature at which it just fails to flow or turns solid. It too is determined by wax or paraffin content of the fuel, and improved by adding a flow improver, kerosene or by employing a fuel heater.

VISCOSITY

Viscosity is a measure of liquid's resistance to flow. Fuel with the wrong viscosity can cause engine damage. A low viscosity fuel may not provide adequate lubrication to plungers, barrels and injectors. A high viscosity fuel, on the other hand, is thicker and does not flow easily. By causing higher injection pressure, it will increase wear and tear on gear train, cam and follower on the fuel pump assembly. It will atomise less efficiently and the engine will be more difficult to start. The fuel viscosity problem can be resolved by knowledgeable heating, cooling or blending.

SPECIFIC GRAVITY

Specific gravity is also referred to as *Density* or *Relative Density*. The easiest way to understand it is as follows:
- One cubic metre of fresh water weighs one tonne. Therefore the density of freshwater is 1 (i.e., 1 tonnes/cubic metre or 1000 kg/cubic metre).
- One cubic metre of steel weighs 7.8 tonnes. Therefore, the density of steel is 7.8.
- One cubic metre of seawater weighs 1.025 tonnes. Therefore the density of seawater is 1.025 (or 1025 kilograms/cubic metre)
- One cubic metre of a diesel fuel weighs 0.8 tonnes. Therefore, the density of that fuel is 0.8.

In summary, the density or specific gravity of a substance is the weight of its one cubic metre volume. Properly defined, however, the specific gravity is the weight of one cubic metre of the substance compared to the weight of one cubic metre of freshwater (at the same temperature), the latter of which is 1 tonnes.

Heavier substances (and heavier fuels) have higher specific gravity. Heavier fuels provide more energy or power (per volume) to the engine. But they tend to create more deposits in the combustion chamber, resulting in abnormal liner and ring wear. Lighter fuels like kerosene, on the other hand, do not produce the rated power.

Adjusting the engine fuel settings by an inexperienced person may seriously reduce the life of the engine. Inexperienced blending of fuels to alter the specific gravity of the final product is also not advisable. For example, blending diesel fuel with alcohol (ethanol or methanol) or petrol will create an explosive atmosphere in the fuel tank. Water condensation in the tank can also cause the alcohol to separate and stratify the tank.

CHAPTER 23: QUESTIONS & ANSWERS

INBOARD DIESEL FUEL SYSTEM

1. The presence of water in fuel can be detected by observing:

    (i) a sample of fuel in a glass container.
    (ii) the bottom content of the fuel tank.
    (iii) the bottom content of the fuel filter.
    (iv) the bottom content of the sump oil.

    The correct answer is:
    (a) all except (i).
    (b) all except (ii).
    (c) all except (iii).
    (d) all except (iv).

2. Fuel filters on engines perform the following functions:

    (i) trap water.
    (ii) trap contaminants.
    (iii) filter air.
    (iv) provide means to remove contaminants.

    The correct answer is:
    (a) all except (i).
    (b) all except (ii).
    (c) all except (iii).
    (d) all except (iv).

3. The filters of a vessel's engines should be cleaned at the following intervals:

    (a) daily.
    (b) 400 hours.
    (c) 4800 hours.
    (d) 10,000 hours.

4. The baffles in fuel tanks are fitted in order to:

    (a) restrict the fuel supply.
    (b) trap vapour.
    (c) reduce free surface effect.
    (d) trap moisture.

5. Wire gauze diaphragms are fitted to fuel oil tank vents in order to prevent:

    (a) flames from travelling into tanks.
    (b) contamination of fuel.
    (c) vapour loss.
    (d) flames from travelling out of tanks.

6. In relation to fuel tanks and pipe lines in vessels, the following precautions must be observed:

    (i) Pipes fitted out of sight.
    (ii) Shut-off valve fitted directly to tank.
    (iii) Wire-braided connection to engine.
    (iv) pipes made of annealed copper.

    The correct answer is:
    (a) all except (i).
    (b) all except (ii).
    (c) all except (iii).
    (d) all except (iv).

7. To bleed air out of the complete fuel system:

    (i) purge tanks through their drain valves.
    (ii) purge pipes via bleed screws at filters.
    (iii) purge the pump through its bleed screw.
    (iv) purge injectors.

    The correct answer is:
    (a) all except (i).
    (b) all except (ii).
    (c) all except (iii).
    (d) all of them.

8. Before entering a supposedly gas-freed fuel tank or undertaking any cutting, welding (hot works) or repairs in its vicinity, the following means of gas-free testing may be employed:

    (a) a gas test meter.
    (b) an industrial chemist.
    (c) all of the choices stated here.
    (d) port authority's recommendations.

9. On observing a leak in a vessel's fuel line while under way, the following precautions should be taken:

    (i) switch off motor.
    (ii) isolate battery.
    (iii) remove the ignition key.
    (iv) shut off ventilation.

    The correct answer is:
    (a) all except (i).
    (b) all except (ii).
    (c) all except (iii).
    (d) all except (iv).

10. The parts of a vessel's fuel system include:

    (a) pistons and journals.
    (b) radiator and sump.
    (c) inlet and outlet ports.
    (d) injectors and filters.

Chapter 23: Inboard Diesel Fuel System

11. If a diesel engine runs out of fuel and stops, the system should:
    (a) not be bled.
    (b) be bled before replenishing fuel tank.
    (c) be bled after replenishing fuel tank.
    (d) be bled immediately.

12. The fuel pump is:
    (a) a displacement pump (D.P.)
    (b) a centrifugal pump (C.P.)
    (c) either a D.P. or a C.P.
    (d) neither a D.P. nor a C.P.

13. Partly filled fuel tanks:
    (a) should be topped up at every opportunity.
    (b) should not be topped up too frequently.
    (c) do not accumulate moisture.
    (d) are not subject to algae growth.

14. When fuelling a vessel, the build up of static electricity charge in the tank can be prevented by:
    (a) earthing the hose.
    (b) keeping fuel nozzle in contact with filler pipe
    (c) discharging static prior to fuelling.
    (d) none of the choices stated here.

15. In EFI engines, the correct air-to-fuel ratio is determined by the:
    (a) atomisers.
    (b) carburettor.
    (c) computer.
    (d) fuel injectors.

16. The EFI system is designed to inject into the engine:
    (a) fuel vapour.
    (b) vapour-free fuel.
    (c) mixture of fuel and vapour.
    (d) petrol only.

17. The diesel fuel injection system:
    (a) injects fuel into the combustion chamber.
    (b) is another name for EFI.
    (c) injects fuel into the intake manifold.
    (d) provides better fuel efficiency than EFI.

18. In an EFI system, a mechanical fuel pump:
    (a) is not required.
    (b) supplies fuel to the fuel injection pump.
    (c) is the fuel injection pump
    (d) is an emergency fuel pump.

19. To get rid of live algae in fuel:
    (a) let the fuel sit for a few days.
    (b) ventilate the tank.
    (c) Top up (fill up) the fuel tank.
    (d) run fuel through an electronic filter.

20. The output of a diesel engine is controlled by the:
    (a) pressure from the fuel lift pump.
    (b) setting on the fuel injection pump rack.
    (c) injector settings.
    (d) air intake valve setting.

21. The sloping or hopper-shaped bottom of a fuel tank is to:
    (a) reduce free surface effect.
    (b) match the hull shape.
    (c) collect sediment and water contaminants.
    (d) improve fuel suction.

22. Fuel tanks should be kept as full as possible in order to:
    (a) reduce free surface effect.
    (b) reduce condensation of moisture.
    (c) increase stability of the vessel.
    (d) improve fuel suction.

23. Air cannot enter the fuel system as a result of:
    (a) draining a fuel tank.
    (b) excess fuel returning to the fuel tank.
    (c) disconnecting a fuel pipe.
    (d) cleaning a fuel filter.

24. Engines fitted with the leak-off pipe to the fuel filter can get rid of air from the fuel system through the:
    (a) return line to the fuel tank.
    (b) bottom of the tank.
    (c) fuel filter.
    (d) bleeder screw.

25. Air in an EFI fuel system can be removed:
    (a) through the fuel return line.
    (b) through fuel injectors.
    (c) through the bleed valve at the fuel filter following by priming with the priming pump.
    (d) through fuel filters, injection pump and injectors, followed by priming with the priming pump.

26. A fuel shut off valve is installed in the suction line of a fuel tank:

    (a) as close to the tank as possible.
    (b) as far from the tank as possible.
    (c) such that it cannot be closed without entering the engine room.
    (d) such that air in the fuel line cannot return to the tank.

27. Fuel lines should not:

    (a) be made of annealed copper.
    (b) be routed over the engine.
    (c) secured to the hull.
    (d) remain visible.

28. Fittings on flexible fuel lines must be:

    (a) secured to the decks.
    (b) grounded individually.
    (c) grounded to the common ground system.
    (d) clip fastened.

29. In relation to fuel tank vent pipes, which statement is incorrect?

    (a) Mesh screens act as flame arresters.
    (b) Goosenecks prevent water entry into tanks.
    (c) A blocked vent can cause an engine to fade away.
    (d) Mesh screens do not prevent flames from travelling into tanks.

30. In relation to fuel maintenance, which of the following is unnecessary & impracticable?

    (a) Drain water and sediment from the fuel service tank at the start of a shift.
    (b) Drain water and sediment from the fuel service tank 5 to 10 minutes after filling it.
    (c) Top up the fuel tank each day at the start of the day's operation.
    (d) Install a water separator ahead of the primary fuel filter.

31. In relation to fuel water separators, which statement is incorrect?

    (a) The sediment type separator is installed ahead of the fuel transfer pump and as close to the tank as possible.
    (b) The sediment type separator does not have a filtering media in its element.
    (c) The coalescing type separator has a two-stage paper media in its element.
    (d) The coalescing type separator is installed as far from the tank as possible.

32. A fuel rack on a diesel engine works in a similar manner to:

    (a) an automatic choke on a petrol engine.
    (b) an electric heater plug.
    (c) a glow plug.
    (d) a jacket water heater.

CHAPTER 23: ANSWERS

1 (d), 2 (c), 3 (b), 4 (c), 5 (a),
6 (a), 7 (a), 8 (c), 9 (d), 10 (d),
11 (c), 12 (c), 13 (a), 14 (b), 15 (c),
16 (b), 17 (a), 18 (b), 19 (d), 20 (b),
21 (c), 22 (b), 23 (b), 24 (d), 25 (c),
26 (a), 27 (b), 28 (c), 29 (d), 30 (c),
31 (d), 32 (a)

Chapter 24

PROPULSION, GEARBOXES & PROPELLERS

TYPES OF PROPULSION (DRIVELINES)

There are two basic means of propulsion:

- Jet Drives: The thrust is applied to the hull through a pump housing into which engine driven pumps accelerate a large flow of water.
- Screw Propellers: The propeller converts the engine power into a thrust force.

JET DRIVE

A Jet Drive or a Waterjet engine works by drawing water with a pump into a housing in the vessel and then ejecting it out through a steering nozzle in the stern. This provides high-speed propulsion to the vessel without the torque (paddle wheel effect) of a propeller. The Waterjet is normally driven by a high-speed marine engine.

The vessel is steered with a steering deflector, which deflects the direction of the water jet to port or starboard. A thrust reversing deflector is employed to go astern or in neutral. This "CLAM SHELL" deflector cuts the jet stream aft of the steering deflector, providing a range of ahead, neutral and astern manoeuvring speeds. There is no gearbox. The vessel can shift instantly from full ahead to full astern by using the "clam shell" without causing any strain on the engine. The water jet is simply redirected from aft to forward. However, some vessels are fitted with a small rudder (fin) to supplement the steering deflector at slow speeds.

Advantages of Jet Drives:

- They respond instantly to acceleration, stopping and turning. They are thus well suited to water skiing. Their ability to spin around in their own length is sometimes referred to as the Hamilton Turn.
- There is no propeller to get fouled or to pose a danger to skiers in the water.
- With nothing protruding under the vessel, she can operate in shallow water.
- There is no need for a reverse gear.
- Less prone to damage from floating debris.

Fig 24.1
A HAMILTON JET DRIVE

However the following precautions should be kept in mind:

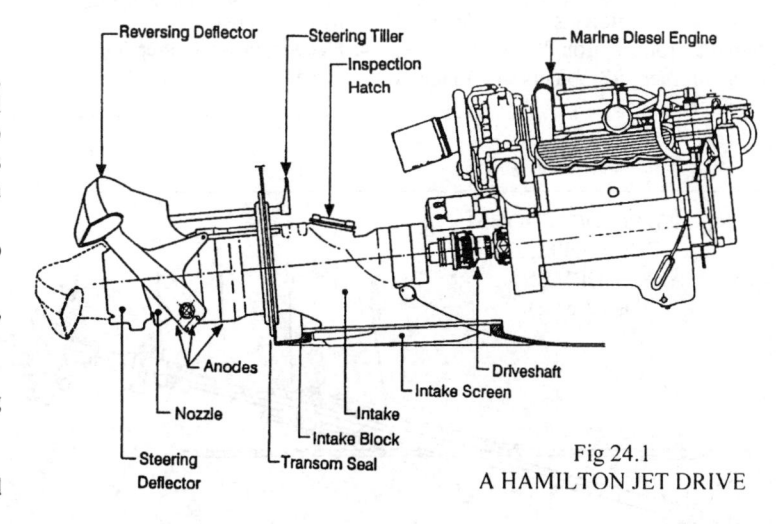

➢ Do not use their ability to stop suddenly (within their own length) without first warning the passengers.
➢ Do not accelerate too quickly with a skier in tow. You may strain the skier's arms or pitch him or her head first into the water with a jackrabbit start.

Disadvantages of Jet Drives:

- The pump suction grill tends to get plugged with debris, unless fitted with self-cleaning mechanism.
- The engine is susceptible is *BLOCK LOADING* if the vessel comes off the water ingesting air and loss of load. Block loading can also occur when the vessel comes back in the water if the engine speed is not matched to the load.

SCREW PROPELLERS

There are several ways of connecting the screw propeller to the engine:

(I) Conventional in-line propeller arrangement
(II) Surface Drive
(III) Stern (or Sail) Drive
(IV) Vee Drive
(V) Z Drive

(I) CONVENTIONAL IN-LINE PROPELLER ARRANGEMENT

This is the most common arrangement. It employs a straight and rigid propeller shaft between the marine gear output flange and the propeller. The engine is located low near the longitudinal centreline of the hull and the gearbox generally accepts the full propeller thrust. The shaft bearings are placed close enough to prevent shaft whip, but far enough apart not to subject the shaft to hull flexing. To prevent inducing unwanted forces on the transmission thrust bearing, the first shaft bearing is located at least 12 shaft diameters (preferably 20) from the transmission output flange. The tail shaft, being more susceptible to damage from a propeller's contact with submerged objects, is made stronger.

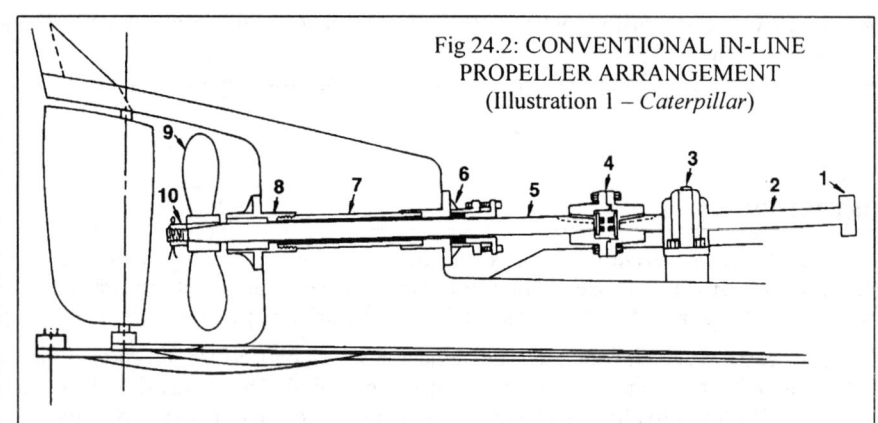

Fig 24.2: CONVENTIONAL IN-LINE PROPELLER ARRANGEMENT
(Illustration 1 – *Caterpillar*)

DRIVELINE COMPONENTS – CENTERLINE MOUNTED THROUGH STERN POST

1. Shaft companion flange
2. Intermediate shaft
3. Shaft bearing – pillow block, expansion type
4. Flange type shaft coupling
5. Tail shaft
6. Stuffing box – may or may not contain bearing
7. Stern tube – one end threaded, the other slip fit
8. Stern bearing
9. Propeller
10. Retaining nut

Fig 24.3: CONVENTIONAL IN-LINE PROPELLER ARRANGEMENT
(Illustration 2)

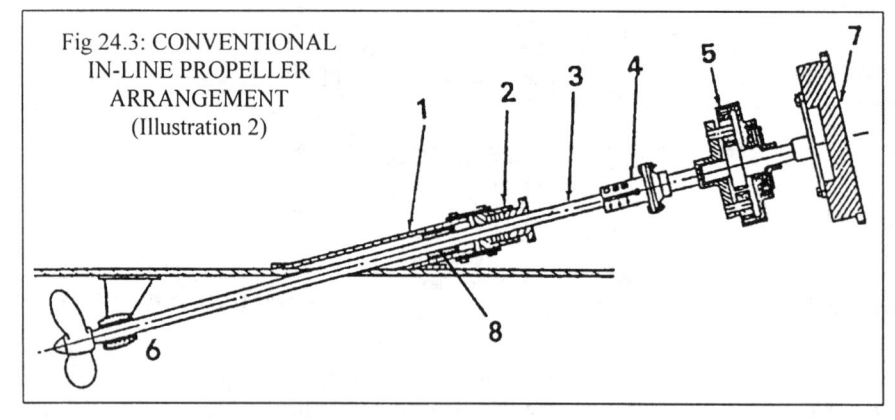

1. Stern tube
2. Stern gland
3. Propeller shaft
4. Coupling
5. Reversing gear
6. Skeg bearing
7. Engine flywheel
8. Stern tube bearing

STERN TUBE

The stern tube is a metal tube designed to support the tail end of the propeller shaft (known as the tail shaft) on bearings, and, to provide a watertight seal (stern gland) between the sea and machinery space. There are many types of stern tubes. They can either be water or grease/oil lubricated.

In the water lubricated type, water is allowed to enter the outer end of the tube to circulate and keep the bearings cool. The bearings are made of hard plastic - the Cutlass brand being common.

The inboard end of the tube houses the stern gland. It is made watertight by inserting in it a packing of greasy hemp and covering it with a tightening gland. A grease nipple is fitted over the packing so that grease can be injected into it from

Chapter 24: Propulsion, Gearboxes & Propellers

time to time.

The gland must be adjusted to allow low-friction on the drive shaft while also keeping out the water. It is allowed to drip slowly to provide a check in case of a blockage of circulation of seawater inside the tube. If the gland is over-tightened to stop leaking, it may increase friction and strain the shaft. This can create grooves in the shaft, making the adjustment more difficult.

The oil lubricated stern tubes are sealed at both ends. *Oil-filled stern tubes* provide better lubrication and reduce corrosion by keeping out the seawater. The oil header tank for such a system must be checked regularly and kept topped up.

A mechanical shaft seal is used in some vessels in lieu of the stern gland. It is similar to the seals found in pumps of all types. It is made of nylon-reinforced rubber. This shaft seal works by the water pressure pushing a high-density carbon face against a polished stainless steel face, rotating with the shaft. This system is claimed not only to eliminates the need to have sea water dripping into the engine room, but also to reduce vibration because of the rubber assembly.

Improper alignment of propeller shaft with the gearbox shaft will cause noise, vibrations and overheating of bearings.

In *outboard engines*, a sealed oil filled 'foot' houses gearbox, bearings, shafts and seals. Their oil change is illustrated in Chapter 27.

The packing gland(s) must be readily accessible to permit repacking and adjustment. For their lubrication, large capacity greasers are available and can be mounted in a convenient position remote from the gland. They must allow free rotation of the shaft to prevent it from overheating.

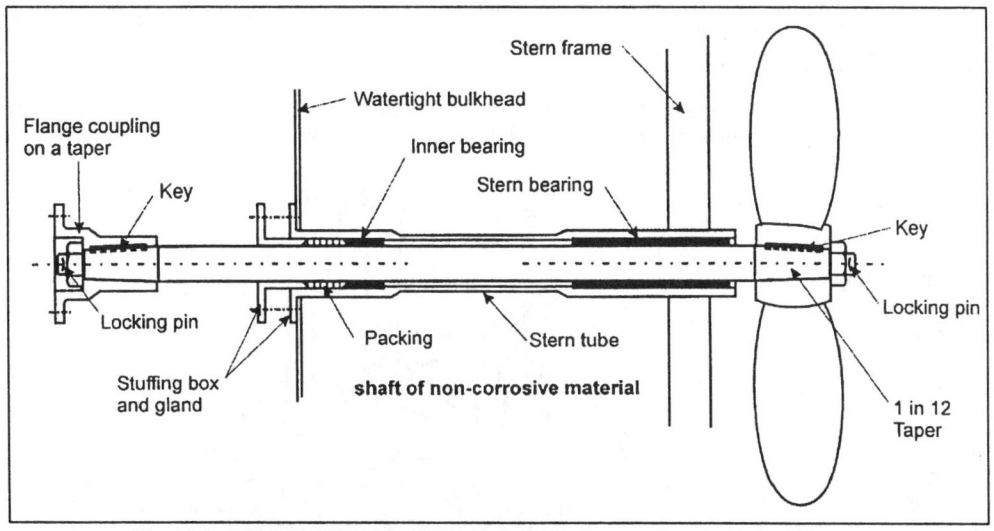

Fig 24.4: CONVENTIONAL IN-LINE PROPELLER ARRANGEMENT
(Illustration 3 - *NSCV Part C Section 5 Subsection 5A*)

PROPELLER SHAFT

The propeller shaft should be of sufficient size and strength to withstand the load put on it by the engine and propeller. The bending tolerance of the shaft should be 0.3 mm per metre or less. As discussed under Engine Beds, the shaft angle should be as shallow as possible to the water level. The efficiency of the propeller in the water will significantly decrease as the inclination angle of the propeller shaft increases. Angles greater than 15° cause loss of engine efficiency and abnormal wear. If, for some reason, the engine bed and propeller shaft angles do not correspond, universal joints and an intermediate shaft are incorporated.

MEASURING "WEAR-DOWN" IN STERN TUBE BEARING & TAIL SHAFT

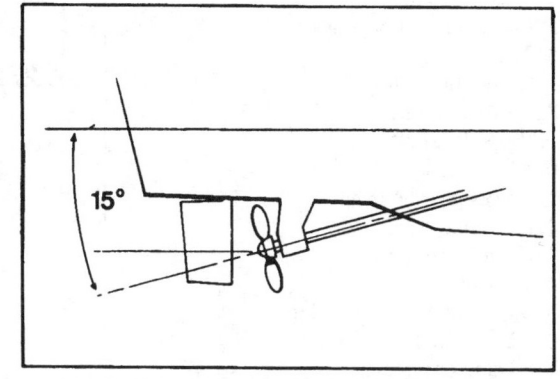

Fig. 24.5: PROPELLER SHAFT INCLINATION
(*Lister Petter Marine*)

The forward bearing takes only a part of the shaft's weight. The aft bearing, on the other hand, not only shares the shaft's weight but also carries and the whole of the propeller's weight. It is, therefore, more susceptible to wear.

Tail shafts should be withdrawn for inspection of the shaft as well as the bearings at regular intervals when the vessel is on a slip. If the shaft deflection exceeds the specified amount, then the bearing (usually only the aft bearing) is replaced. The stress in the shaft due weight of the propeller is thus kept to a minimum.

II) SURFACE DRIVE

This is a surface piercing in-line propeller system for high-speed applications. It is available with rudder arrangement or as a steerable drive unit. The propeller is designed to operate half submerged at planing speeds and fully submerged at low speeds.

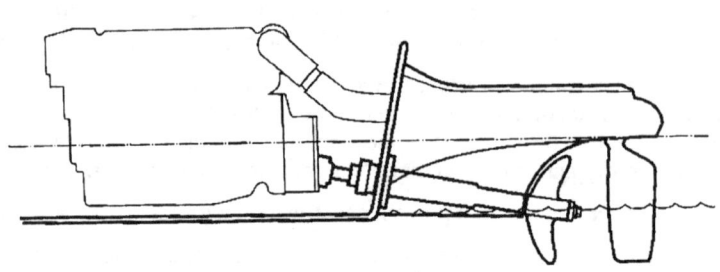

Fig 24.6: SURFACE DRIVE (*Volvo Penta*)

III) STERNDRIVE & SAILDRIVE

Also known as the INBOARD - OUTBOARD or the OUTDRIVE engines. As the name suggests, they are a hybrid of an inboard and outboard system. An inboard motor is mounted inside the hull at the stern, but its driving unit is external and is similar to the lower unit of an outboard motor. A double universal joining shaft is employed to drive two right-angled gear sets, and the engine flywheel faces aft.

Sail drives are the auxiliary engines found in sailing vessels. As discussed in Propellers below, they are usually fitted with a folding propeller. By folding it when not in use the vessel minimises propeller drag.

1. Steering arm
2. Electro-mechanical tilting device
3. Steering yoke
4. Rubber cushion
5. Universal joint
6. Steering helmet
7. Input gear
8. Forward/reverse gear
9. Forward/reverse cone
10. Upper gear housing
11. Coupling box
12. Shift mechanism
13. Shift plate
14. Vertical shaft
15. Intermediate housing
16. Trim tab
17. Circulation pump
18. Propeller shaft
19. Propeller shaft gear
20. Water intake
21. Lower gear housing
22. Retaining pawl
23. Exhaust bellows
24. Transom shield
25. Supporting rubber cushion
26. Drive shaft
27. Damper plate
28. Flywheel

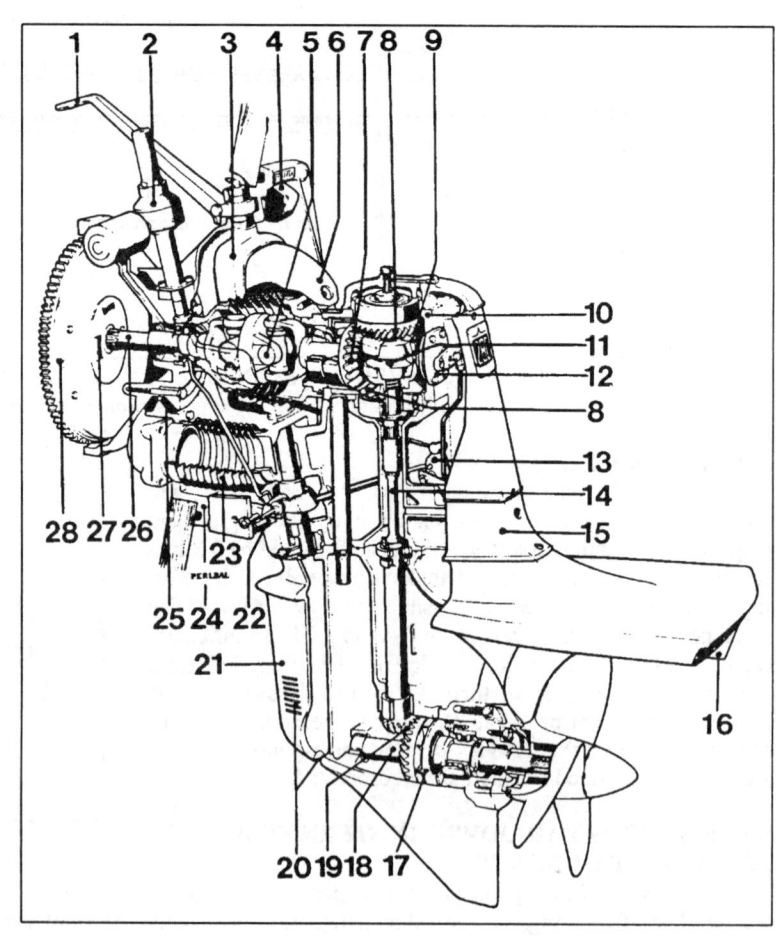

Fig 24.7: A VOLVO PENTA STERN DRIVE

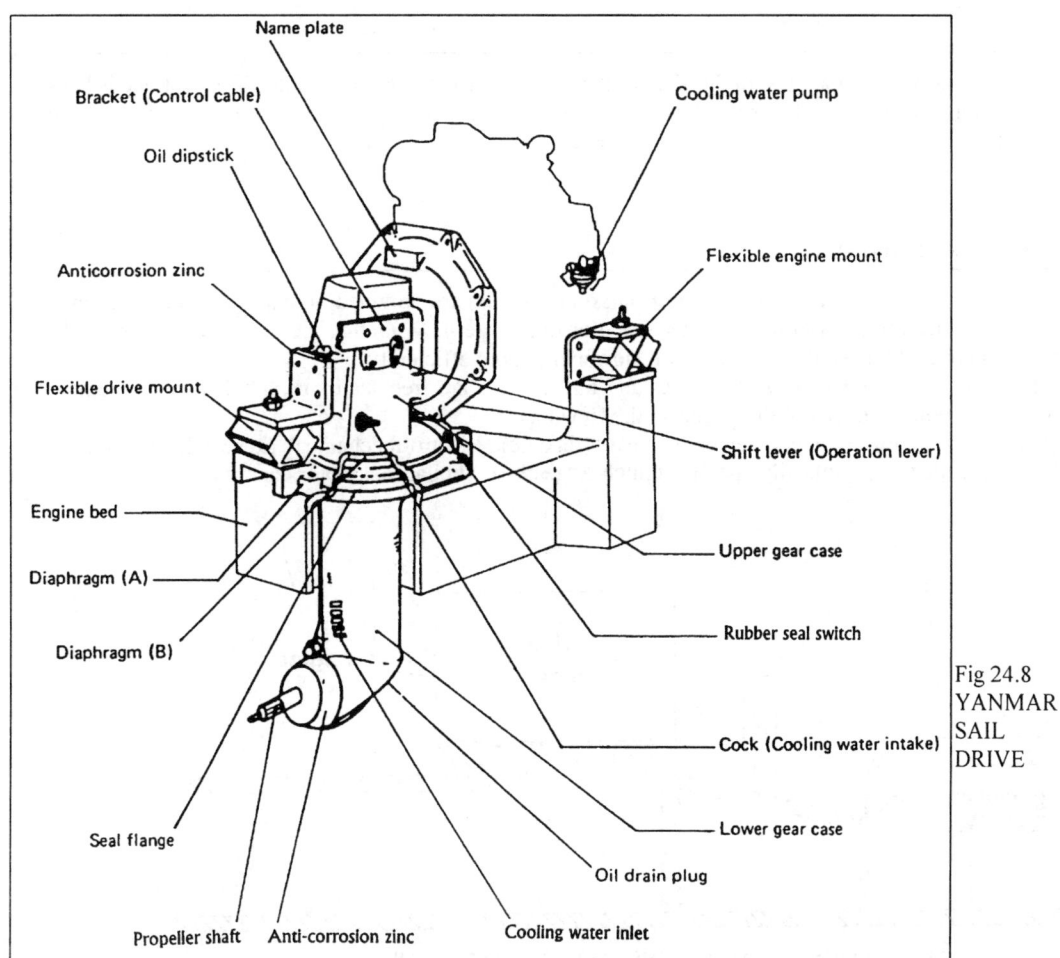

Chapter 24: Propulsion, Gearboxes & Propellers

Fig 24.8 YANMAR SAIL DRIVE

IV) VEE-DRIVE

Inboard engines are normally mounted in the middle of the vessel, and fitted with a gearbox and a drive shaft aft of the engine. But in a Vee-Drive arrangement, the engine is mounted aft and the drive shaft extends forward to a gearbox. The direction of the shaft is then reversed inside the gearbox, and made to come out under the hull in a normal manner, as shown below.

There are two advantages of Vee drive engines:

The engine is located at the extreme aft end of the boat, taking minimal usable space inside the hull.

The shaft between the engine and the Vee drive unit is not loaded with the propeller thrust forces. It can therefore have universal joints or other soft couplings. This would permit fitting of soft engine mounts, resulting in a much quieter installation.

The disadvantage of a Vee drive lies in the relatively high centre of gravity of the engine and a larger stern trim in the vessel.

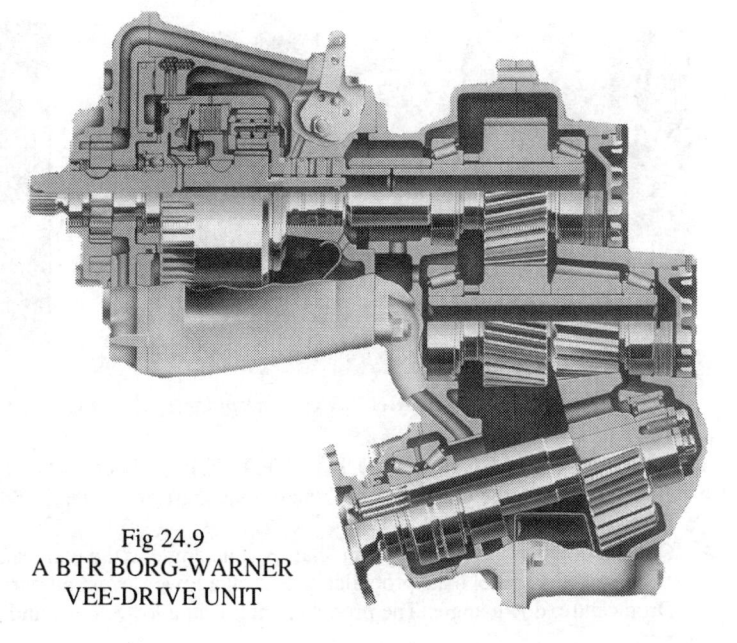

Fig 24.9
A BTR BORG-WARNER
VEE-DRIVE UNIT

V) Z-DRIVE

Imagine the letter 'Z" with two right angles. In the Z drive there are two right-angle gear units. The first one is between the engine and a shaft running vertically down through the hull. The second, submerged, gear unit is between this vertical shaft and another short horizontal drive shaft leading to the propeller.

MARINE REVERSING GEARBOXES

Reverse gears *(reversing gears, marine gearboxes, transmission systems)* are provided in marine engines to change the direction of rotation of propeller for a vessel to break her speed or to go astern. *Reduction gears* are often combined with the reversing mechanism in order to make the propeller turn slower and more efficiently.

Gearboxes are designed to match a variety of engines and hulls. Some are free standing (also known as 'remote' or 'island mounted'), others are bolted directly to the engine at the flywheel.

The gearshift may be Mechanical (a gear lever, cable or linkage), Electro-mechanical (a switch operating gear-changing solenoid) or Hydraulic (a hydraulically operated clutch or gearbox, similar to a motor car).

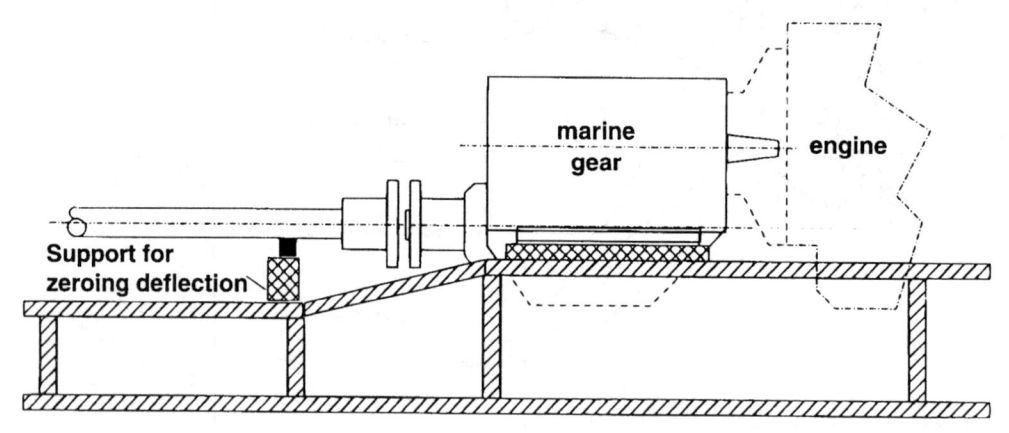

Fig. 24.10: MARINE GEARBOX INSTALLATION (*Caterpillar*)

As discussed in Engine Lubrication in Chapter 22, the lubricating oil used in gearboxes should be maintained to the correct level, and changed as per the maker's recommendation. It must not be overfilled as the increased pressure may damage the seals.

Fig 24.11: GEAR OIL DRAIN PLUG (*Caterpillar*)   Fig 24.12: GEAR OIL FILLER & DIP STICK (*Caterpillar*)

A GEARBOX MAY BE BOLTED TO THE ENGINE IN ONE OF FOUR WAYS:
➤ Coaxial (the engine crankshaft and the output shaft of the gearbox are at the same level. Thus, the propeller shaft is in line with the crankshaft)
➤ Coaxial down angle (the propeller shaft is at an angle to the crankshaft)
➤ Drop centre, parallel (The propeller shaft is at a lower level and parallel to the crankshaft)
➤ Drop centre, down angle (The propeller shaft is at a lower level and at an angle to the crankshaft)

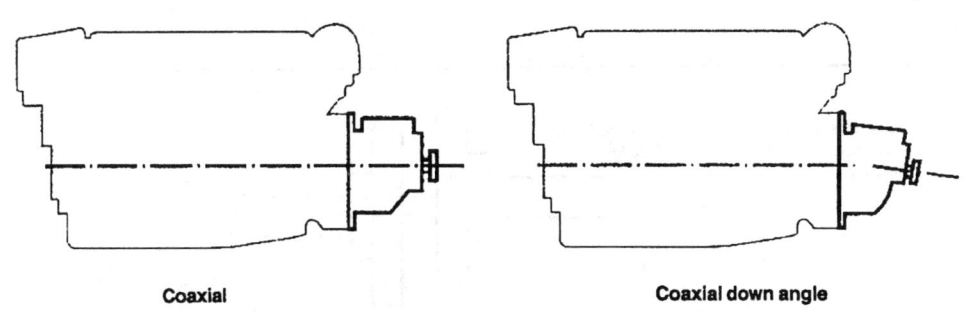

Chapter 24: Propulsion, Gearboxes & Propellers

Coaxial **Coaxial down angle**

Fig. 24.13: TYPES OF GEARBOX ARRANGEMENTS (1) (*Volvo Penta*)

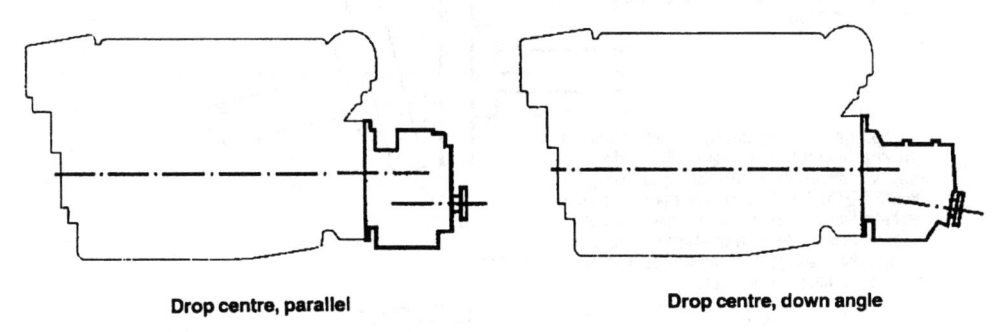

Drop centre, parallel **Drop centre, down angle**

Fig. 24.14: TYPES OF GEARBOX ARRANGEMENTS (2) (*Volvo Penta*)

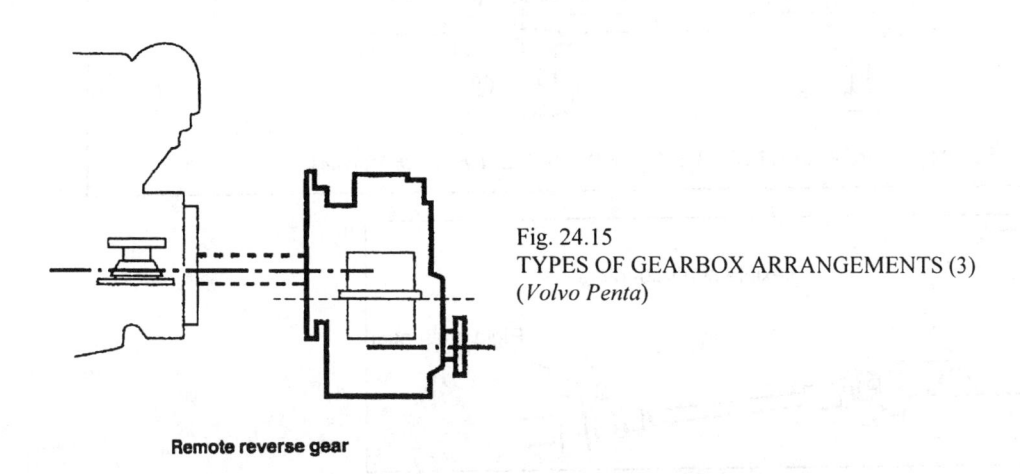

Fig. 24.15
TYPES OF GEARBOX ARRANGEMENTS (3)
(*Volvo Penta*)

Remote reverse gear

ALIGNMENT OF PROPELLER SHAFT (DRIVE TRAIN) WITH GEARBOX SHAFT

After taking into account *propeller shaft droop or deflection due to unsupported shaft,* the gearboxes must be aligned to the propeller shaft within the specified tolerances. Misalignment will cause noise, vibrations and overheating of bearings. Misalignment can occur even with a flexible coupling fitted between them. Two principal types of misalignment are:

- Parallel Misalignment
- Conical (or angular) misalignment

By rotating the shaft, each type of misalignment can be checked with a digital or clock gauge. Readings taken should not vary by more than the specification (usually, 0.127 mm or 0.005" per revolution).

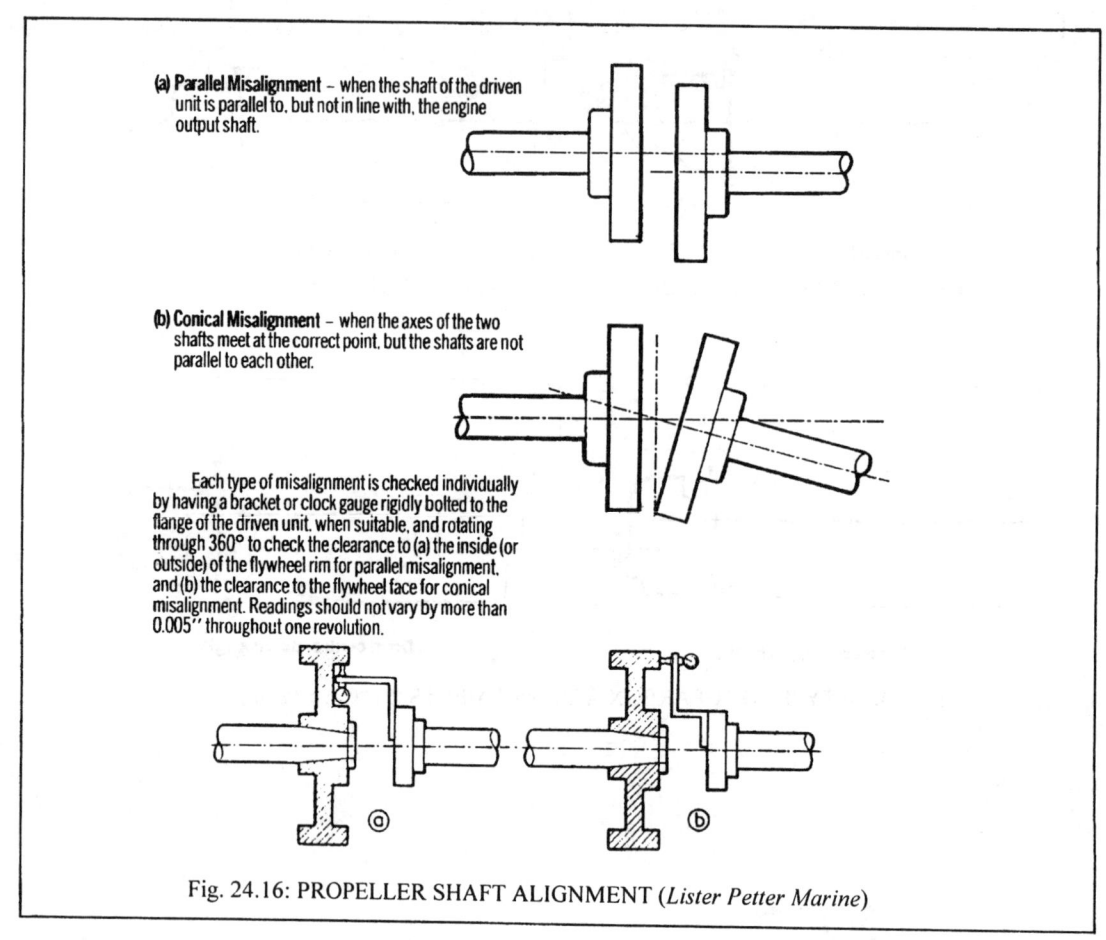

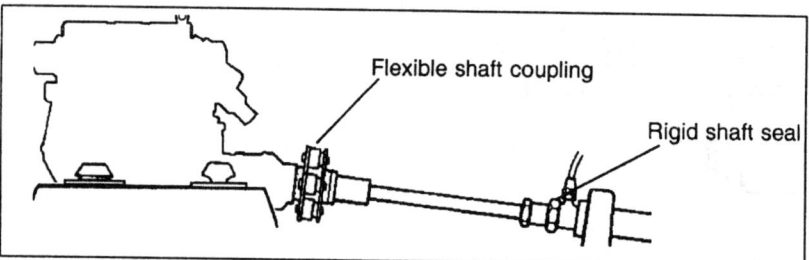

(a) Parallel Misalignment – when the shaft of the driven unit is parallel to, but not in line with, the engine output shaft.

(b) Conical Misalignment – when the axes of the two shafts meet at the correct point, but the shafts are not parallel to each other.

Each type of misalignment is checked individually by having a bracket or clock gauge rigidly bolted to the flange of the driven unit, when suitable, and rotating through 360° to check the clearance to (a) the inside (or outside) of the flywheel rim for parallel misalignment, and (b) the clearance to the flywheel face for conical misalignment. Readings should not vary by more than 0.005″ throughout one revolution.

Fig. 24.16: PROPELLER SHAFT ALIGNMENT (*Lister Petter Marine*)

Fig 24.17: FLEXIBLE COUPLING, FLEXIBLE ENGINE MOUNTS, RIGID SHAFT SEAL (*Volvo Penta*)

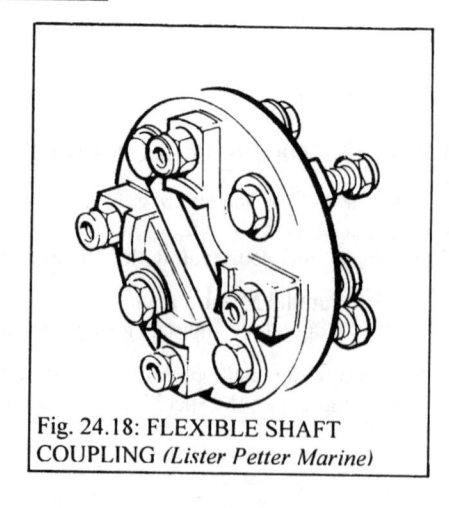

Fig. 24.18: FLEXIBLE SHAFT COUPLING (*Lister Petter Marine*)

Chapter 24: Propulsion, Gearboxes & Propellers

SHAFT INSTALLATION: TWO ALTERNATIVES

- Alternative 1 is usually found in vessels up to 12 metres in length. It does not incorporate a flexible shaft coupling.

- Alternative 2 shows a flexible shaft coupling and a rigid mounted shaft seal (stuffing box).

NOTES:

- *Flexible coupling must not be fitted together with a flexible mounted shaft seal (stuffing box). It can cause vibration problems.*

- *An installation of rigid mounted shaft seal and flexible engine suspension must be fitted with a flexible coupling.*

ALTERNATIVE 1: FLEXIBLE STUFFING BOX (No. 2)
1. Engine mounts
3. Stern bearing (water lubricated)

ALTERNATIVE 2: FLEXIBLE COUPLING (No. 2)
1. Engine mounts
3. Rigid mounted stuffing box
4. Stern bearing (water lubricated)

Fig 24.19 (A) & (B)
SHAFT INSTALLATION
TWO ALTERNATIVES
(*Volvo Penta*)

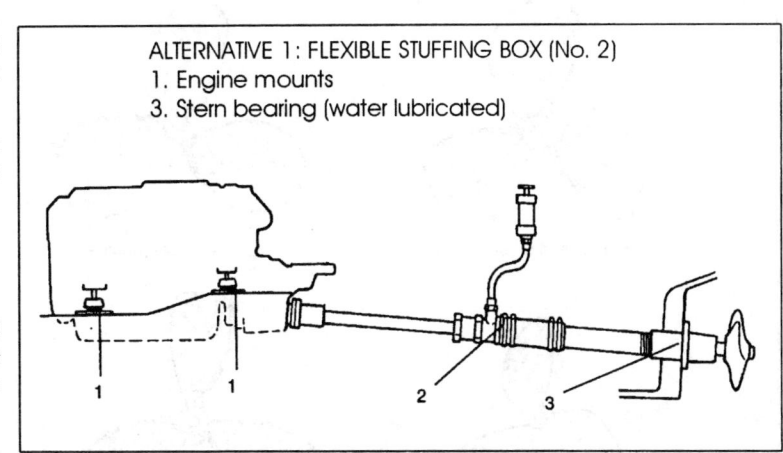

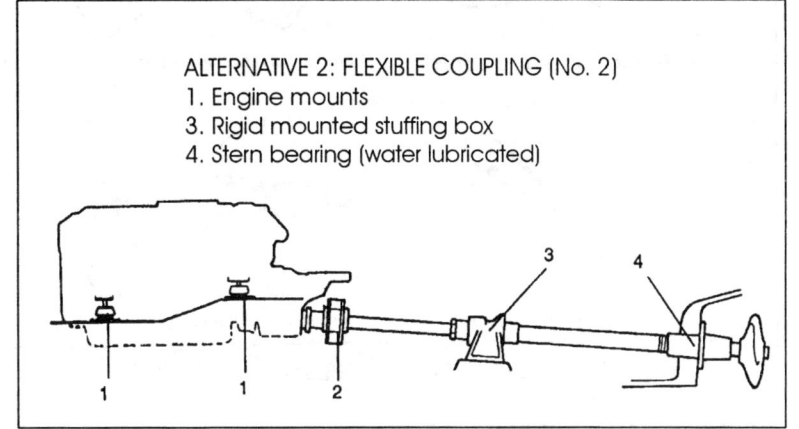

PROPELLERS

A household electric fan pulls air in from behind it and blows it out in front. A marine propeller acts in much the same way in water. The pressure difference thus created provides the vessel a forward or stern thrust.

A single-blade propeller is the most efficient, but it will cause excessive vibration. The more blades you add to a propeller, less efficient but also less vibrating it becomes. Furthermore, more the blades, greater the transverse thrust.

Therefore, as a compromise, most propellers are three-bladed. However, a single screw vessel, operating in restricted waters, may consider fitting a four or five bladed propeller. It will improve her turning circle by providing greater transverse thrust.

The fast (racing type) vessels usually have two bladed propellers. This is to minimise the interference of water flow between the trailing edge of one blade and the leading edge of the other. But they are subject to excessive vibration when both blades lose traction at the same time. This is overcome by making the aperture (the space in which the propeller rotates) as large as possible. The extreme example of a large aperture would be to fit the propeller at the end of a long shaft as far behind the vessel as possible.

(Transverse Thrust or Paddle Wheel Effect of propellers is discussed in Chapter 8)

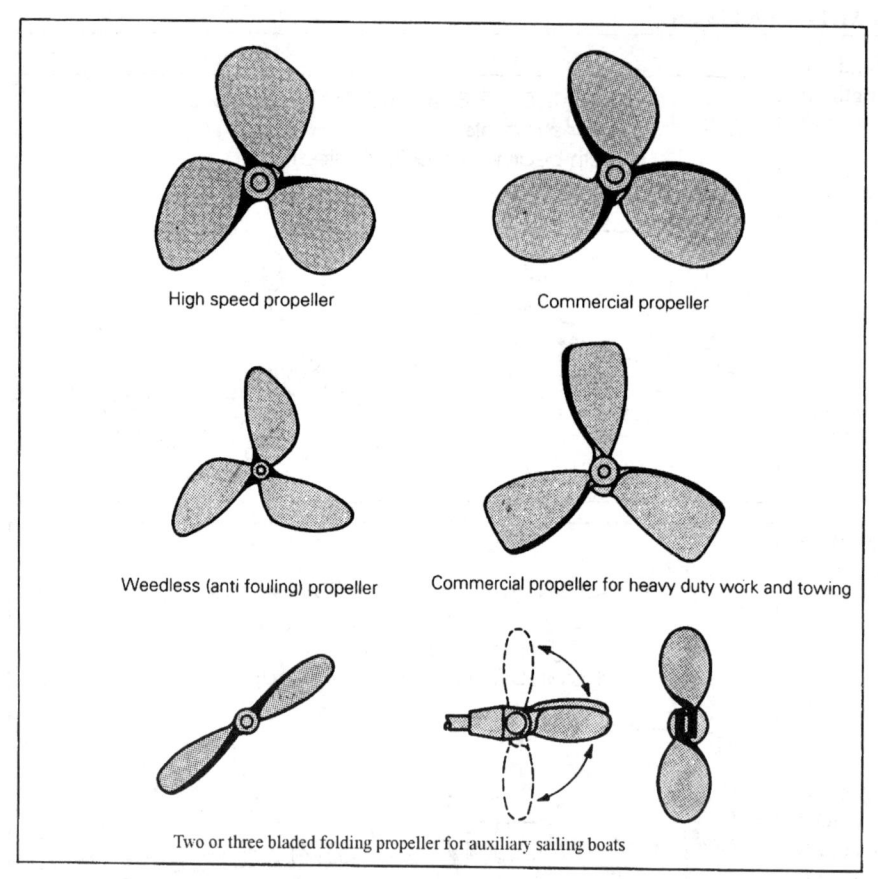

Fig. 24.20: PROPELLER TYPES (*Lister Petter Marine*)

PROPELLER MEASUREMENTS

Most manufacturers mark propellers with a part number and the diameter and pitch. Simplistically speaking, the blade diameter (blade area) gets the boat going from her stationary position. Thereafter, the pitch gives her the forward push. As shown in the attached Volvo illustrations, No. 1 is the part number, and No. 2 the diameter and the pitch.

Propellers are measured in *Diameter* and *Pitch*. For example, a 12 x 13 propeller is of 12 inches diameter and 13 inches pitch. A 300 x 250 propeller is of 300 mm diameter and 250 mm pitch. Diameter means twice the distance from the centre of the hub to the tip of a blade. The pitch is the theoretical distance the propeller will travel through the water in one revolution (not allowing for slippage, i.e., if it was 100% efficient). Therefore, a propeller of 13-inch pitch turning at 500 RPM will cover a distance of 13 x 500 = 6500 inches per minute. This is equal to 541 feet or 162.5 metres per minute, or 9750 metres per hour or 5.26 knots (1 nautical or sea mile = 1852 metres). In actual operation, however, the propeller will move approximately 70% to 90% of this distance; the remaining 10% to 30% is called "slip", which is discussed below.

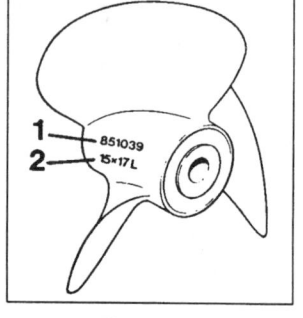

Fig 24.21
PROPELLER MARKINGS

Propeller pitch is determined by both the diameter of the propeller and the angle of its blades. Pitch cannot be substituted for diameter, nor can diameter be substituted for pitch. Experts in the field match it for various engines and vessels. For example, water skiing runabouts with outboard motors are fitted with a 2 inches smaller pitch than standard. This results in better acceleration and load carrying with skiers.

PROGRESSIVE PITCH: Most propellers are manufactured with the true or flat pitch, which is the same at all points of the blade. High performance stainless steel propellers, on the other hand, are sometimes manufactured with a 'progressive' pitch. The pitch at the leading edge is lower than that at the trailing edge of the blades. In such cases, the progressive pitch is the average pitch.

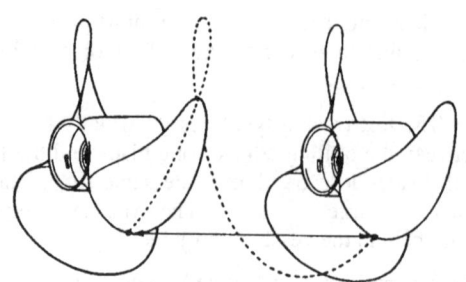

Fig 24.22 PROPELLER PITCH

Chapter 24: Propulsion, Gearboxes & Propellers

RAKE & CUPPING: Rake and cupping too are associated with high performance propellers.

Rake is the blade's slant (at the tip) from the perpendicular of the hub. It is also defined as the blade's angle of attack. The rake of a standard propeller is around 15°, and that of a performance propeller anything up to 30°.

Cupping is to increase the propeller pitch, thereby increasing a racing boat's top speed. Sometimes, it also works in reducing the revs of an over-stressed engine.

SKEW: Propeller blades are sometimes skewed (swept back in shape) for lower horsepower boats working in areas where there are excessive weeds or other obstacles. This shape of blades sheds weeds easier.

Higher skew is also considered beneficial in surface-piercing or high performance propellers in reducing shock sensation as the blades repeatedly enter and exit water.

PROPELLER SLIP

Slip has been defined in more ways than any other propeller terms, probably because it has different meanings for different purposes. The Quicksilver manufacturer's book on propellers defines it as the difference between the actual and the theoretical travel resulting from a necessary propeller blade angle of attack. The Webster's dictionary defines it as the difference between the actual speed of a vessel and the speed at which it would move if the propeller were acting against (i.e., moving through) a solid.

This book deals with its practical meaning: The slip is the effort that is wasted by the propeller because it is not uniformly submerged while rotating. Even the best-designed propellers must suffer from slip. It is maximum when the propeller is partially submerged due to the vessel being light, badly trimmed or pitching or rolling. It ranges between 10 to 40%. On conventionally fitted outboard hulls the slip is about 12%, runabouts 20%, planing cruisers 25%, and displacement cruises 35%. Slip can be calculated by comparing engine speed with the actual speed of the vessel. The illustration shows a 20% slip on a 16 x 20 propeller.

MATERIALS USED FOR CONSTRUCTION OF PROPELLERS

Propellers are made of stainless steel, bronze, aluminium or plastic. Bronze propellers are most common with the inboard engines. For the outboards, the stainless steel propellers are the strongest, and most suited for better engine performance. They are resistant to the typical accumulation of small nicks and bends found on aluminium propellers.

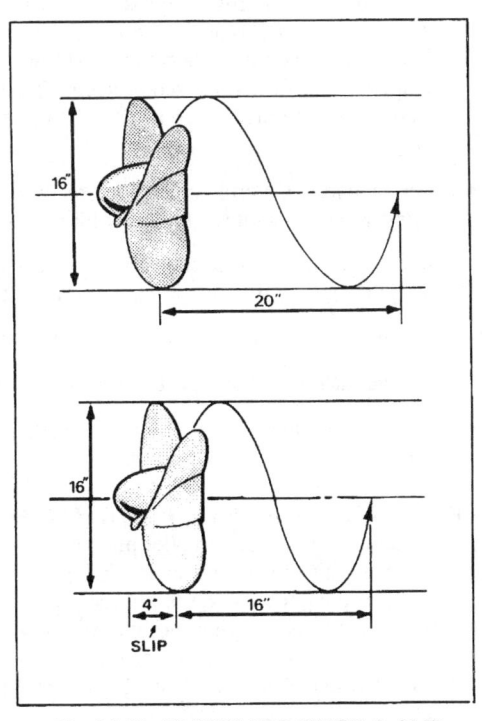

Fig. 24.23: PROPELLER PITCH & SLIP
(Lister Petter Marine)

Aluminium is by far the most popular material for the outboards and stern drives. It is relatively low in cost, has good strength, good corrosion resistance and is easily repaired.

Plastic propellers are used on electric outboard motors (trolling motors) and small petrol outboards. Plastic is more resistant to impact damage than aluminium, comparatively half the weight of aluminium, corrosion free, resistant to grease, oil, fuel and battery acid.

SELECTING THE RIGHT PROPELLER

- To get the best performance out of your boat, you need to select the propeller and gearing that will suit your particular boat, engine and speed range. The selection of propeller is a very complicated procedure requiring specialist knowledge. A propeller of wrong dimension or design can adversely affect vessel speed and put additional unnecessary stresses on the engine components.

- The best people to consult are usually the engine and propeller manufacturers, not necessarily the dealers. But, borrowing a 'recommended' type of second-hand propeller from a dealer to try out may also be a suitable option.

- Fast vessels, such a high-speed patrol boats and yachts, need more precise propellers than slow work boats. They must be of exact specified pitch. A propeller pitch error of even an inch or two could cost them 2 or 3 knots of their top speed, but it would be insignificant on a 10-knot river tug.

- A high performance propeller must therefore be precisely manufactured and carefully maintained. Repairing or repitching does not easily restore its precision. Even if the machinist is adequately skilled, most propeller pitch measurement machines cannot resolve or detect small errors. Even the slightest defect in its leading or trailing edge would cause an invisible error in its profile, which would set up cavitation. In more severe cases it could result in blade failure or loss after as little as a day or two of high speed running.

- The angle of propeller shaft should be as small as possible. Shaft angles of more than 12 degrees should be avoided. For shaft angles greater than 12 degrees, the use of smaller propellers should be considered.

- The profile of the keel or the propeller shaft brackets in front of the propeller should be such that it creates a minimum of drag and turbulence. The problem can be minimised by reducing the shaft diameter and the surface area of rudders and propeller supports by use of stronger materials. The shape of the tunnel is also important. A poor tunnel design can create a lot of turbulence in the propeller and reduce the boat's buoyancy at the stern.

- For outboards (2-stroke and 4-stroke), Yamaha Motors recommend that at full throttle and under a maximum boat load, the engine rpm should be within the manufacturer's specified upper half of the full throttle operating range. For example, it should be between 5500 and 6000 if the specified full throttle rpm range is 5000 – 6000. Select a propeller that fulfils this requirement. If operating under conditions that allow the engine to rise above the maximum recommended range (such as light boat loads), reduce the throttle setting to maintain the rpm in the proper operating range.

- For inboards, Caterpillar Engines recommend that the propeller must be of the size to allow the engine to operate slightly above its rated rpm under the maximum boat load and adverse sea conditions.

- In practice, if your propeller gives the best speed under full revolutions of the engine, your vessel is fitted with the right propeller. When rotating, it must also clear the vessel's bottom by about 20% of its diameter.

- Engine may over-rev if propeller is too small.

- Engine may overload if propeller is too large.

- In petrol engines, always remove spark plugs before removing a propeller. This is to prevent the engine starting suddenly.

PROPELLERS FOR WATER SKIING & PLANING HULLS:

> Generally speaking, a smaller pitch propeller is needed to power a heavily loaded boat, while a larger pitch is needed to achieve high speeds. A boat towing water skiers comes into the former category. She needs a smaller pitch propeller to give her the punch to get out of the 'hole' – just like using a low gear in a motorcar to get it out of a 'hole'. Even after take off, she is carrying the load of the skiers. Therefore, she is normally fitted with a propeller of 2 inches smaller pitch than standard. A boat engaged in high-speed work, on the other hand, is fitted with at least a 2 inches higher pitch propeller than standard - like a high gear in a fast motorcar.

> Boats wanting flexibility of acceleration and top-end speed, would need to have two propellers or a removable-blade propeller. The removable blade-propeller would allow you to change the pitch very quickly. It is also a convenient way of dealing with a damaged blade at sea. Having two propellers, on the other hand, gives you the choice of purchasing high-performance stainless steel models for high performance applications. Fuel savings easily offsets the higher cost of such propellers in 'sports' application. However, as mentioned earlier, the best choice is usually the one recommended by the engine and propeller manufacturers, not necessarily the dealers.

> Solid propellers (without a rubber bush) are predominantly for competitions. They do not offer any shock absorption and cause considerable vibrations.

> Varying the number of blades may be necessary for an upper end of the performance scale. In most cases, however, the standard propeller that comes with the engine is usually the best choice, whether it is stainless steel, aluminium or composite glass-reinforced plastic.

> In planing hulls over 20 knots, the size of the propeller depends on the engine power. Volvo Penta recommend about 7-8 sq. cm. propeller blade surface per kW of shaft power. If the shaft is at an angle to the flow of the water, the surface area of 8-15 sq. cm./kW would be more reasonable, depending on the angle. For example, for a shaft power of 400 kW, the propeller blade surface may need to be 400 x 9 sq. cm. = 3600 sq. cm. This surface may be divided over three, four or five blades.

> The efficiency of a propeller blade diminishes when it becomes far too wide in relation to its length. Since the propeller diameter is usually limited in size, it is better to select a propeller with a greater number (four or five) of narrower blades than a smaller number (three) of wide ones.

> At boat speeds of over 25 knots, the resistance of shafts, rudders and propeller supports starts to reduce propeller efficiency. As mentioned earlier, the resistance can be minimised by reducing the shaft diameter and the surface area of rudders and propeller supports by use of stronger materials.

Chapter 24: Propulsion, Gearboxes & Propellers

> Generally, a larger propeller with narrow blades turning at low speed is more efficient than a small propeller turning at high speed. A suitable pitch ratio (Pitch ÷ Diameter) for planing hulls at 20 knots is 0.9-1.15; at 30 knots 1.0-1.3; and at 35 knots 1.05-1.35.

PROPELLERS FOR DISPLACEMENT & SEMI-PLANING HULLS

- Displacement hulls need larger slow revving propellers than planing hulls. For example, by increasing the propeller diameter by 50% and reducing the propeller speed by 40%, a trawler can save 20-30% fuel or gain 20% greater thrust when trawling.

- Although three-blade propellers are often more efficient, the four-blade propellers usually produce less vibration and tend to be the preferred choice.

- A suitable pitch ratio (Pitch ÷ Diameter) at 10 knots is 0.7-0.9, and at 15 knots 0.8-1.05. Since the pitch ratio varies with speed, a working boat, such as a trawler, needs to decide whether the propeller should be at its best when trawling (with a pitch ratio of, say, 0.7) or when not trawling (with a pitch ratio of, say, 1.0). *Adjustable pitch propellers* may be the ideal solution for boats such as trawlers and tugs.

PROPELLERS FOR SAILING VESSELS

The auxiliary-powered sailing vessels are usually fitted with either a Folding or a Feathering propeller.

FOLDING PROPELLER: As discussed in Sail Drive engines, by folding the propeller when not in use the vessel minimises propeller drag. When the engine is needed, the propeller opens automatically due to the centrifugal force of the rotating propeller shaft. Some expensive propellers employ a gearing mechanism to ensure that the blades open smoothly and evenly. Some sailing vessels also have their propeller shaft set to one side of the keel. This eliminates the need for an aperture, making their rudder more effective.

The **FEATHERING (OR SELF-FEATHERING) PROPELLER**, instead of folding, rotates the blades into feathering position when the yacht's engine is switched off. The pitch is automatically adjusted by an internal mechanism, which rotates the blades to feather, forward and reverse positions.

Fig 24.24: FOLDING PROPELLER ON A SAIL DRIVE

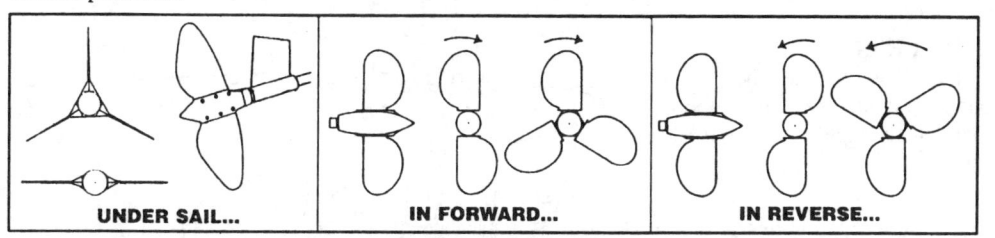

Fig 24.25: TWO & THREE-BLADE FEATHERING PROPELLER (*Max-Prop*)

Fig 24.26
INTERNAL ADJUSTABLE PITCH IN A TWO-BLADE FEATHERING PROPELLER
(*Max-Prop*)

CONTROLLABLE (VARIABLE) PITCH PROPELLERS

Seagoing tugs and fishing trawlers are often fitted with propellers whose blade-angle can be altered to suit the speed or power requirement of the vessel. While proceeding to their destination and not pulling a tow or nets in the water, they set the blades at the maximum pitch in order to gain the maximum speed of, say, 12 knots. But when pulling a tow or nets in the water, they reduce the pitch to get maximum power at, say, 4 knots.

Instead of reversing the engine rotation, the vessel can be put into forward or reverse by simply changing the pitch of the propeller. A sliding control rod inside a hollow propeller shaft operates the propeller blades.

Fig 24.27 VARIABLE-PITCH PROPELLER

KORT NOZZLE (or DUCTED) PROPELLERS

The Kort Nozzle is an airfoil shaped duct wrapped around a generally square-tipped propeller. The nozzle, made of mild steel, is made corrosion resistant by providing it with a stainless steel liner. It can be fitted to a timber, steel or fibreglass hull.

The Kort Nozzle can be fixed or it can be steered like a rudder to direct water flow in the required direction. It thus propels and steers the vessel without a rudder. However, a rudder is usually fitted to the unit. The nozzle eliminates transverse thrust, but in the case of fixed nozzles the steerage is usually poor until power is applied.

Kort propellers provide greater bollard pull than open-water propellers. They are used where more thrust and power are required, e.g., commercial towing, pushing and trawling. High drag of the nozzle at high speeds makes them unsuitable for fast vessels.

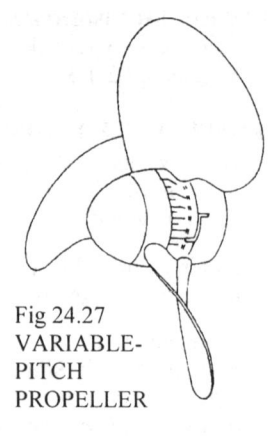

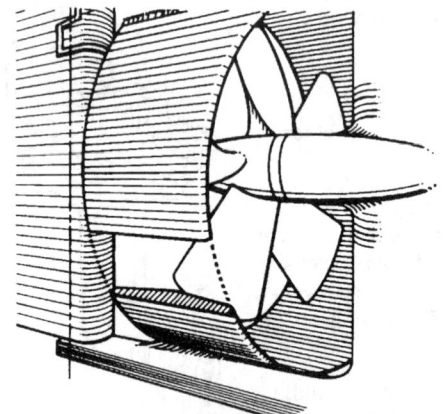

Fig 24.28: KORT NOZZLE (*Caterpillar*)

AZIMUTH OR "Z" PROPELLERS

Similar in construction to Kort Nozzle, these twin-screw units rotate through 360 degrees and operate independent of each other. Such highly manoeuvrable configuration is useful in tugs,

DUAL PROPELLERS

Volvo Penta and some other manufacturers also make counter rotating duel propellers driven by a single coaxial shaft. These are designed to minimise transverse thrust.

TWIN-SCREW VESSELS

A vessel with two engines driving two independent propellers is known as the twin-screw vessel. Twin outboards are similar in characteristics to twin screws.

Fig 24.29: VOLVO PENTA DUOPROP FITTED TO A STERN DRIVE

Advantages:
- Two screws are safer than one in case of an engine breakdown.
- They make the vessel more manoeuvrable. A twin-screw vessel can be steered without a rudder whether going ahead or astern. She can also be turned on the spot. See Chapter 8.
- They cancel each other's transverse thrust (paddle wheel effect). They are contra-rotating, and usually each turns outwards at the top when going ahead and inwards when going astern.

However:
> A vessel fitted with two screws is not necessarily faster than a single screw vessel. Furthermore, two engines are more expensive to install, maintain and run.

Chapter 24: Propulsion, Gearboxes & Propellers

PROPELLER DAMAGE

- Propeller shafts and propellers are damaged by hazards such as objects in water, cavitation, vibration and corrosion.
- Even slight propeller damage can cause around 15% drop in top speed, 40% reduction in acceleration and 20% reduction in cruise economy.
- Uneven damage to blades can set up imbalance vibrations, resulting in fatigue to other parts of engine and vessel.
- If a propeller hits a submerged object, the vessel's stern area can suddenly start to vibrate. (Speed should be reduced immediately).
- A badly positioned outboard engine (too high on the transom or the leg trimmed out too far) can create uneven or irregular water flow to the propeller.
- Badly designed or damaged stern area or brackets (struts) holding the shaft can create uneven or irregular water flow past the propeller.
- Alterations to the hull such as a new water intake or a transducer can interrupt the flow of water to the propeller.

All the above factors can create air bubbles around the blades - a process known as **Cavitation**. These bubbles have the same effect on the blades as corrosion. They become pitted and cause vibrations in the stern area. If the pits are left unfilled, vibration may damage the bearing and break the shaft. Hundreds of little pinholes often seen along the leading edge of the propeller is known as CAVITATION BURNING.

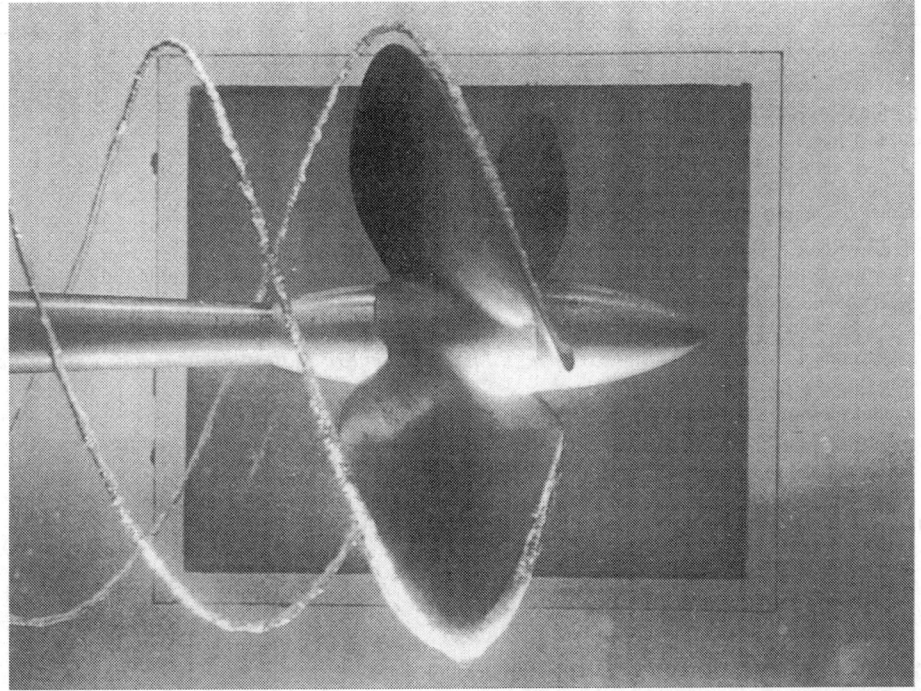

Photo courtesy Admiralty Research Establishment

Fig 24.30
THE PHOTO SHOWS A MODEL PROPELLER UNDER TEST IN A LABORATORY.
OVERLOADING IS CAUSING CAVITATION, WHICH IS SWEPT AWAY DOWNSTREAM.
IT WOULD CAUSE NOISE IN THE VESSEL AND POSSIBLY EROSION OF THE RUDDER.

Propellers on outboard motors are fitted with a shear pin, which is designed to break if the propeller hits a solid object, thus protecting the drive shaft from damage. The pin is easily replaced. Larger motors are fitted with a Slip Clutch instead of a shear pin.

Zinc anodes are suitably placed to minimize corrosion in the propeller and the stern area of the vessel. Propeller efficiency can be improved by keeping the blades clean and polished.

PROPELLER REPLACEMENT

CAUTION: *In petrol engines, if the propeller shaft is rotated while the engine is in gear, there is a possibility that the engine will crank over and start. To prevent a possible serious injury caused by a rotating propeller due to engine starting, always shift the engine into neutral position and remove spark plug leads when servicing a propeller. A diesel engine, on the other hand, would normally not crank when in gear. It is therefore a common practice to engage it in gear when servicing a propeller. It stops the shaft from turning when tightening propeller nuts. Still, it is wise to always test to make certain that the engine will not crank in forward or reverse gear positions.*

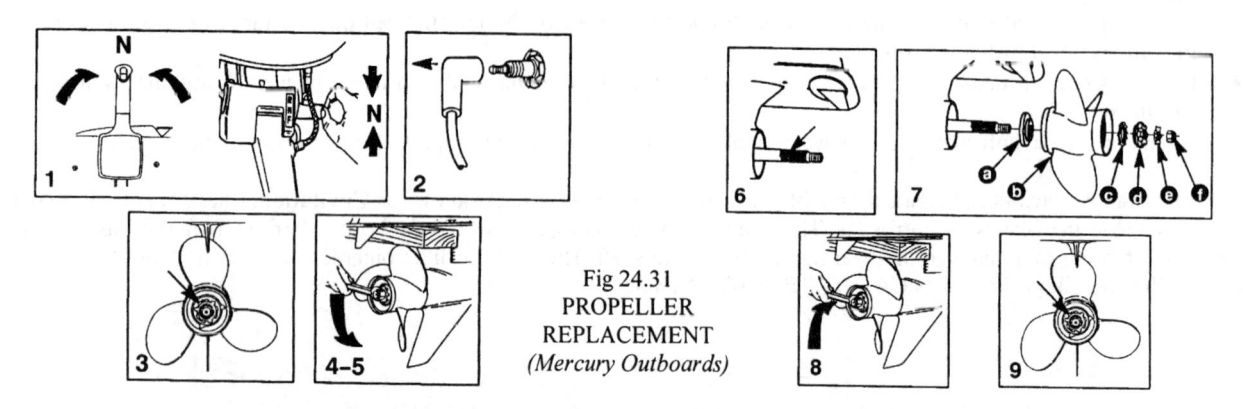

Fig 24.31
PROPELLER REPLACEMENT
(Mercury Outboards)

PROPELLER REPLACEMENT STEPS FOR PETROL OUTBOARDS
(Mercury Marine)

- Shift the engine into neutral position.
- Remove spark plug leads to prevent engine from starting.
- Straighten the bent tabs on the propeller nut retainer.
- Place a block of wood between gear case and propeller to hold the propeller, and remove the nut.
- Pull propeller straight off the shaft. If it is seized to the shaft and cannot be removed, have it removed by an authorised dealer.
  (To prevent this from recurring, apply anti-corrosive grease to the entire shaft at recommended maintenance intervals and each time the propeller is removed.)
- Coat the shaft with recommended anti-corrosive grease.
- Install Thrust Washer (a), Propeller (b), Continuity Washer (c), Thrust Hub (d), Propeller Nut Retainer (e), and Propeller Nut (f) onto the shaft.
- Place a block of wood between gear case and propeller, and torque the propeller nut to 55 lb. ft. (75 N-m).
- Secure propeller nut by bending three of the tabs into the thrust hub grooves.

PROPELLER FITTING STEPS FOR DIESEL STERNDRIVE
(Volvo Penta)

- Set the controls in "Forward" position.
- Grease the propeller shaft.
- Fit the fishing line protection (1) and the propeller (2).
- Tighten the nut (3), using the tool supplied, with a torque of 70-80 N-m (7-8 kpm).
- Set the controls in "Reverse" position.
- Install the other fishing line protection (4) and the propeller (5).
- Place a plastic washer (6) on the propeller cone (7).
- Place a washer (8) on the screw (9) and tighten the screw in the propeller shaft to a torque of 70-80 N-m (7-8 kpm).
- Set the control in "Neutral" position prior to starting the engine.

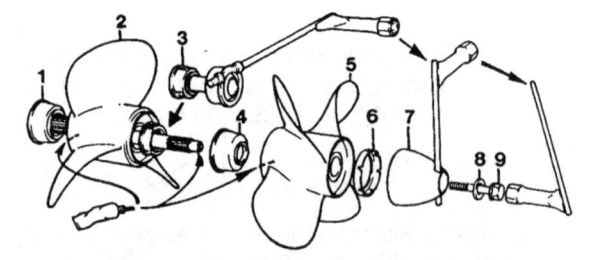

Fig. 24.32: FITTING PROPELLERS ON VOLVO PENTA STERNDRIVE
(The propellers are supplied in pairs and must not be mixed)

Chapter 24: Propulsion, Gearboxes & Propellers

THRUSTER UNITS

Thrusters, fitted on larger vessels, such as ferries, facilitate their manoeuvring when berthing and unberthing. Cross winds, tides and lack of space are often difficult for them to overcome with conventional propulsion and steering gear. Thrusters, particularly bow thrusters, provide additional control, usually at the touch of a button or the move of a joystick.

Thrusters are driven by Pump Drive Systems fitted to the main engine, generator or gearbox. Engines with a higher torque are more suitable. The Variable Displacement Pump Drives Systems offer fingertip control of power selection. They are quite energy efficient as only the power demanded is generated. They are also less noisy.

Fig 24.33: TYPES OF THRUSTERS (*Lewmar*)

The pump can be disconnected when not required by fitting an electromagnetic clutch. This saves energy, heat and noise. Fitting a speed trip (Overspeed/Underspeed Sensor) ensures that the engine is not damaged at high RPM.

If the vessel's generator can provide sufficient usable power, this can be an excellent way of achieving thrust without disturbing the main engine. The advantage of this system is that the generator runs at a constant high RPM, so the pump size can be smaller for the same horsepower. The RPM being constant, a speed trip is also not required. The generator manufacturers can install a safety device to shed electrical load, when the thrusters is engaged. There is wide range of thrusters available from various manufacturers. Examples of Lewmar thrusters are illustrated.

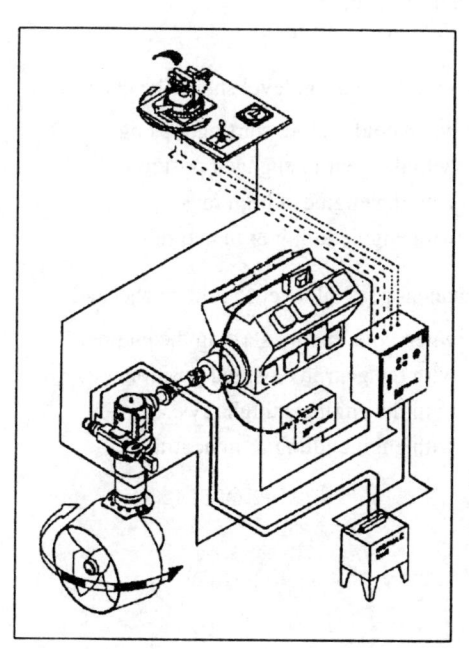

Fig 24.34
PUMP DRIVE FOR
THRUSTER UNIT
FITTED TO MAIN ENGINE
(*Lewmar*)

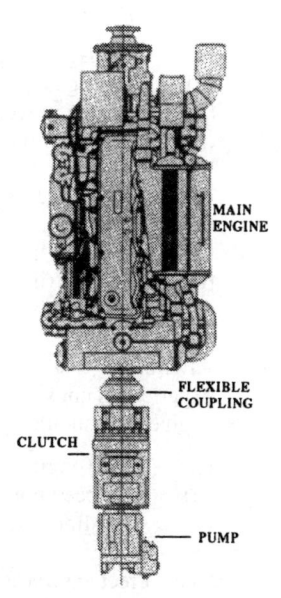

Fig 24.34 (A)
THRUSTER STEERING
& SPEED CONTROL *(HRP)*

CHAPTER 24: QUESTIONS & ANSWERS

CHAPTER 24.1: PROPULSION SYSTEMS

1. The parts of a vessel's propulsion gear include:

    (i) Coupling.
    (ii) Filter.
    (iii) Skeg bearing.
    (iv) Flywheel.

    The correct answer is:
    (a) all except (i).
    (b) all except (ii).
    (c) all except (iii).
    (d) all except (iv).

2. A stern gland can become very hot in the following circumstances:

    (a) Stuffing box full of grease.
    (b) Gland not tight enough.
    (c) Bent shaft.
    (d) Vessel on slow speed for a long period.

3. After hitting a submerged object, a vessel experiences unusual vibrations while running her engine. The reasons include:

    (i) damaged propeller.
    (ii) damaged propeller shaft.
    (iii) engine moved out of alignment.
    (iv) damaged rudder.

    The correct answer is:
    (a) all except (i).
    (b) all except (ii).
    (c) all except (iii).
    (d) all except (iv).

4. After hitting a submerged object, a vessel experiences unusual vibrations while running her engine. She should:

    (i) stop engine.
    (ii) proceed home at reduced speed.
    (iii) be slipped to assess damage.

    The correct answer is:
    (a) all except (i).
    (b) all except (ii).
    (c) all except (iii).
    (d) all of them.

5. In relation to jet drives, which statement is incorrect?

    (a) Less prone to damage from floating debris.
    (b) Not susceptible to block loading.
    (c) Can operate in shallow water.
    (d) Well suited to water skiing.

6. In relation to jet drives, which statement is incorrect?

    (a) Warn passengers before crash stopping.
    (b) Do not accelerate too quickly with a skier in tow.
    (c) Jet drives respond quickly to acceleration, stopping & turning.
    (d) Do not operate in shallow water.

7. The "wear-down" measured at the stern tube bearing with the shaft in place represents the wear on the:

    (a) aft bearing.
    (b) forward bearing.
    (c) shaft.
    (d) shaft and bearing.

8. With regard to propeller shafts, which statement is incorrect?

    (a) Misalignment of propeller shaft will cause noise, vibrations & overheating of bearings.
    (b) Misalignment of propeller shaft cannot occur when fitted with a flexible coupling.
    (c) Flexible shaft coupling is not fitted together with a flexible mounted shaft seal.
    (d) A rigid mounted shaft seal and flexible engine suspension is fitted with a flexible coupling.

CHAPTER 24.2: GEARBOXES

9. A marine reversing gearbox is usually located:

    (a) inside the engine.
    (b) parallel the engine.
    (c) aft of the engine.
    (d) forward of the engine.

10. The hydraulic gear oil level should be checked:

    (a) when cold, before starting the engine.
    (b) with the gear in engaged position.
    (c) with the engine at high revs.
    (d) with engine idling & in neutral.

11. The manual gear oil level should be checked:

    (a) when cold, before starting the engine.
    (b) with the gear in engaged position.
    (c) with the engine at high revs.
    (d) with engine idling & in neutral.

12. The reduction gear oil pressure drops slowly during the first half hour of running the vessel. The most likely reason is that the oil:
    (a) is now at normal working temperature.
    (b) has overheated.
    (c) is leaking.
    (d) is not heating.

CHAPTER 24.3: PROPELLERS

13. With a mis-matched propeller, the engine:
    (i) over-revs if propeller is too small.
    (ii) overloads if propeller is too large.

    The correct answer is:
    (a) (i) only.
    (b) (ii) only.
    (c) both (i) & (ii).
    (d) neither (i) or (ii).

14. A 12 x 13 propeller:
    (a) is of 12-inch diameter & 13-inch pitch.
    (b) is of 12-inch pitch & 13-inch width.
    (c) will cover 156 inches in one revolution.
    (d) is of 13-inch pitch & 12-inch diameter.

15. A propeller of 13-inch pitch turning at 500 rpm will cover a distance of:
    (a) 6500 feet per minute.
    (b) 6500 inches per minute.
    (c) 13 inches per minute.
    (d) 13 feet per minute.

16. A 300 x 250 propeller:
    (a) is of 300 mm pitch & 250 mm diameter.
    (b) will cover 75000 mm in one revolution.
    (c) will cover 550 mm in one revolution.
    (d) is of 300 mm diameter & 250 mm pitch.

17. A propeller of 250 mm pitch turning at 500 rpm will cover a distance of:
    (a) 550 metres per minute.
    (b) 750 mm per minute.
    (c) 125 metres per minute.
    (d) 250 metres per minute.

18. Propeller pitch is the:
    (a) theoretical distance the propeller will travel through water in one revolution.
    (b) difference between the actual and theoretical travel of propeller.
    (c) speed at which the propeller will move through solid.
    (d) effort that is wasted by the propeller for not being uniformly submerged.

19. Propeller pitch is:
    (a) about 10% to 30% of slip.
    (b) the difference between the actual and theoretical travel of propeller.
    (c) determined by both the diameter of the propeller & the angle of its blades.
    (d) determined by the angle of the propeller blades.

20. Propeller slip is:
    (a) the effort wasted by the propeller because it is not uniformly submerged while rotating.
    (b) determined from its pitch.
    (c) determined by both the diameter of the propeller & the angle of its blades.
    (d) determined by the angle of the propeller blades.

21. Propellers are made from:
    (a) bronze, aluminium or plastic.
    (b) stainless steel, bronze or plastic.
    (c) stainless steel, bronze or aluminium.
    (d) stainless steel, bronze, aluminium or plastic.

22. In relation to propeller selection for a vessel, which statement is <u>incorrect</u>?
    (a) Displacement hulls need larger slow revving propellers than planing hulls.
    (b) In petrol engines, remove spark plugs before removing a propeller.
    (c) Adjustable pitch propellers are suited to boats such as trawlers & tugs.
    (d) Slow workboats need more precise propellers than fast patrol boats & yachts.

23. In relation to a feathering propeller, which statement is <u>incorrect</u>?

    (a) The blades rotate into feathering position when the yacht's engine is switched off.
    (b) The blades assume the folded position when not needed.
    (c) The blades are rotated into feather, forward or reverse position by an internal mechanism.
    (d) The pitch is automatically adjusted by an internal mechanism.

24. In relation to a folding propeller, which statement is <u>incorrect</u>?

    (a) The blades rotate into feathering position when the yacht's engine is switched off.
    (b) When the engine is needed, the propeller opens automatically due to the centrifugal force of the rotating shaft or through a gearing mechanism.
    (c) By folding the propeller, a sailing vessel minimises propeller drag.
    (d) Some sailing vessels have their propeller shaft set to one side of the keel.

25. In relation to variable pitch propellers, which statement is correct?

    (a) When pulling a tow or fishing nets, the vessel can reduce the pitch to get maximum power.
    (b) When pulling a tow or fishing nets, the vessel can set the blades at maximum pitch.
    (c) They are fitted on yachts.
    (d) The engine can be put into forward or reverse by changing the propeller pitch.

26. In relation to a kort nozzle propeller, which statement is <u>incorrect</u>?

    (a) It provides greater bollard pull than an open-water propeller.
    (b) The Kort Nozzle can be fixed or it can be steered like a rudder.
    (c) The nozzle provides excellent steerage even at slow speeds.
    (d) The nozzle helps to eliminate transverse thrust.

27. In relation to azimuth propellers, which statement is <u>incorrect</u>?

    (a) The twin-screw units operate independent of each other.
    (b) They rotate through 360 degrees.
    (c) They are a highly manoeuvrable configuration.
    (d) The engine is put into forward or reverse by changing the propeller pitch.

28. In relation to twin-screw vessels, which statement is <u>incorrect</u>?

    (a) Twin-screws make the vessel more manoeuvrable.
    (b) A twin-screw vessel is always faster than a single-screw vessel.
    (c) The counter-rotation of twin-screws helps to cancel each other's transverse thrust.
    (d) A twin-screw vessel can be steered without a rudder.

29. In relation to propeller damage, which statement is <u>incorrect</u>?

    (a) A slight propeller damage has little effect on vessel's speed, acceleration or fuel economy.
    (b) A shear pin fitted to propellers on some outboard motors is designed to break if the propeller hits a solid object.
    (c) Cavitation has the same effect on the blades as corrosion.
    (d) Fitting a new transducer on the hull can affect propeller performance.

30. Which of the following statement is <u>incorrect</u>? When servicing a propeller:

    (a) Shift petrol engine into neutral and remove spark plugs.
    (b) Engage diesel engine in gear.
    (c) Do not grease the propeller shaft.
    (d) Always test a diesel engine to make certain that it will not crank in gear.

31. In relation to thrusters fitted to vessels, which statement is <u>incorrect</u>?

    (a) They are fitted to ferries to facilitate berthing & unberthing.
    (b) They help overcome crosswinds, tides and lack of space during berthing or unberthing.
    (c) They provide additional control.
    (d) They are fitted only in the bow.

CHAPTER 24: ANSWERS

1 (b), 2 (c), 3 (d), 4 (d), 5 (b),
6 (d), 7 (d), 8 (b), 9 (c), 10 (d),
11 (a), 12 (a), 13 (c), 14 (a), 15 (b),
16 (d), 17 (c), 18 (a), 19 (c), 20 (a),
21 (d), 22 (d), 23 (b), 24 (a), 25 (d),
26 (c), 27 (d), 28 (b), 29 (a), 30 (c),
31 (d)

Chapter 25

ENGINE BEDS, EXHAUST, VENTILATION & SOUNDPROOFING

ENGINE BEDS

Engine bearers and mountings are shown in figure 21.5 in Chapter 21. Proper mounting of the engine, gearbox and generators etc. on firm foundations is critical for maintaining good alignment and smooth, quiet operation. The engine bed generally consists of two longitudinal rails or bearers with liberal transverse bracing. It must be strong enough not only to carry the weight of the engine and gearbox, but also to withstand the forces due to thrust, torque, inertia, rolling, pitching and occasional grounding. Making the bearers as long as possible helps to distribute the load more widely, thus limiting hull deflection. Ideally they should be at least twice the length of the engine. The bed must also be more rigid than the propulsion system so that the latter is least affected by the flexing of the hull.

Careful consideration must be made to the actual position of the engine on the bed. It must allow easy access to the engine for maintenance, and no part of the engine or gearbox should touch the bed. The inclination of the bed must not exceed the maximum list and trim angles given in the engine specifications.

As discussed under Sterngear, the propeller shaft angle should be as shallow as possible to the water level. If, for some reason, the engine bed and propeller shaft angles do not correspond, universal joints and an intermediate shaft may have to be incorporated.

The engine bed structure may be of steel, aluminium, timber or fibreglass to suit the vessel's own construction. In fibreglass hulls the bed is generally the *foam filled* type or timber. In the case of the **FOAM FILLED** foundations, a metal raft is placed between the machinery and the fibreglass foundation to distribute the load more evenly. The **WOOD** foundation does not need special means of load distribution.

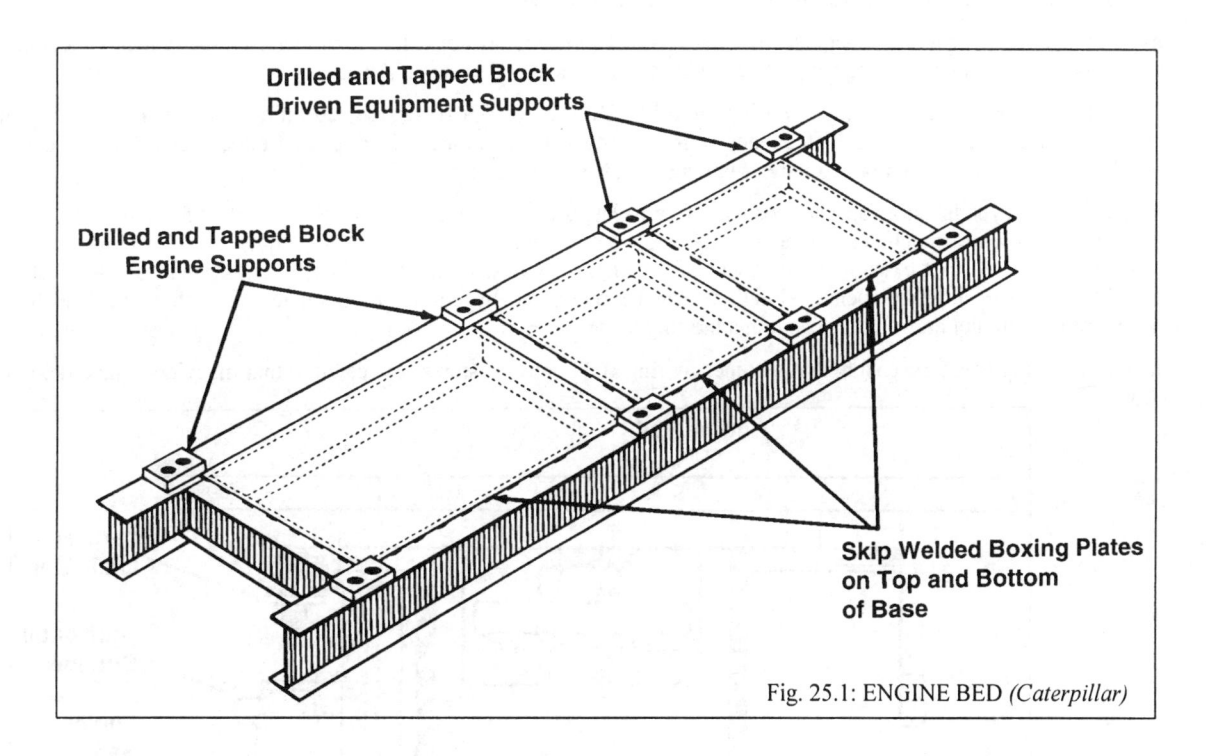

Fig. 25.1: ENGINE BED *(Caterpillar)*

ENGINE MOUNTS OR SUSPENSION

Vibration is not particularly noticeable in workboats with heavy hulls. Besides, the torsion forces and propeller axial force caused by their high gear ratios can be excessive for the rubber mounts. Their engines are therefore usually bolted directly to the engine bearers. However, in most small vessels with low gear ratios and where personal comfort is a high priority, the engine is mounted on flexible mounts to eliminate vibration and reduce unnecessary noise. When these are used, the engine's ancillary equipment such as ducting, piping, etc. must also be flexibly mounted to accommodate this extra movement. The propeller shaft too must have a flexible stuffing box or a flexible shaft coupling. However, a flexible shaft coupling must never be fitted together with a flexible mounted stuffing box. This can cause vibration problems.

The flexible mounts are fully adjustable to enable correct alignment to be maintained, which should be professionally checked and adjusted annually to allow for any setting of the mounting rubbers. The elasticity of the rubber mounts must never be utilised to compensate for an inclined bed. In installations with a down angle propeller shaft there will be a lifting force transmitted from the propeller shaft to the mounts fitted on the gearbox-end. Therefore, engines with a close coupled V-drive must be equipped with mounts designed for this type of installation at the rear end.

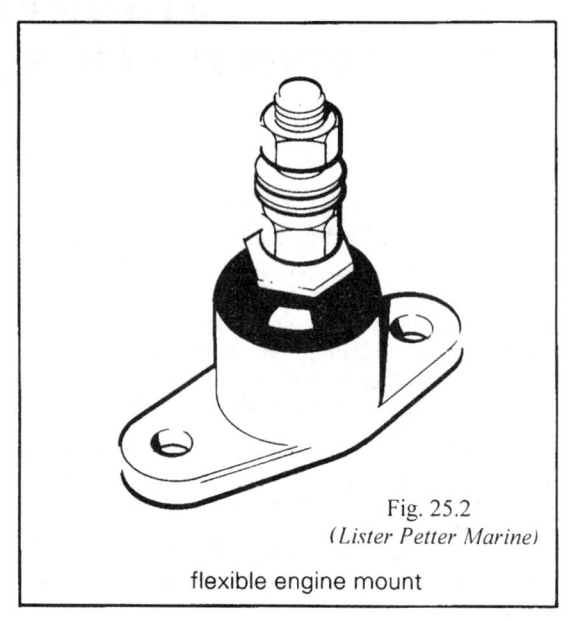

Fig. 25.2 (*Lister Petter Marine*)
flexible engine mount

VIBRATION ISOLATORS

Vibration Isolators are often used when mounting generators and other auxiliary power packages. Care must be taken in selecting the isolator. Generally speaking, softer isolators provide greater deflection, thus more effective isolation. However, the loading limit of the isolator must also be taken into account.

Rubber isolators are generally adequate in most cases. However, do not select a rubber isolator that has the same natural frequency as the engine-exiting frequencies in both the vertical and horizontal planes.

Isolators made of Fibreglass, Felt, Composition and Flat Rubber (of Waffle design) do little to isolate major vibration forces, but do isolate much of the high frequency noise. Pad-type isolators are effective for frequencies above 2000 Hz. However, their fabric materials tend to compress with age and become ineffective.

The steel Spring Type Isolators are the most effective isolators of low frequency vibration range of 5 to 1000 Hz. They are suitable for all types of machinery other than the propulsion machinery. They are also equipped with all-directional limit-stops designed to restrict excessive movement of the engine against forces of rolling, pitching and pounding. Their lack of ability to isolate high frequency vibrations can be overcome by adding rubber pads beneath them. The high frequency vibrations are not harmful but result in annoying noise.

When using spring isolators, ensure that all pipes, wiring and control systems connected to that machinery have flexible connections.

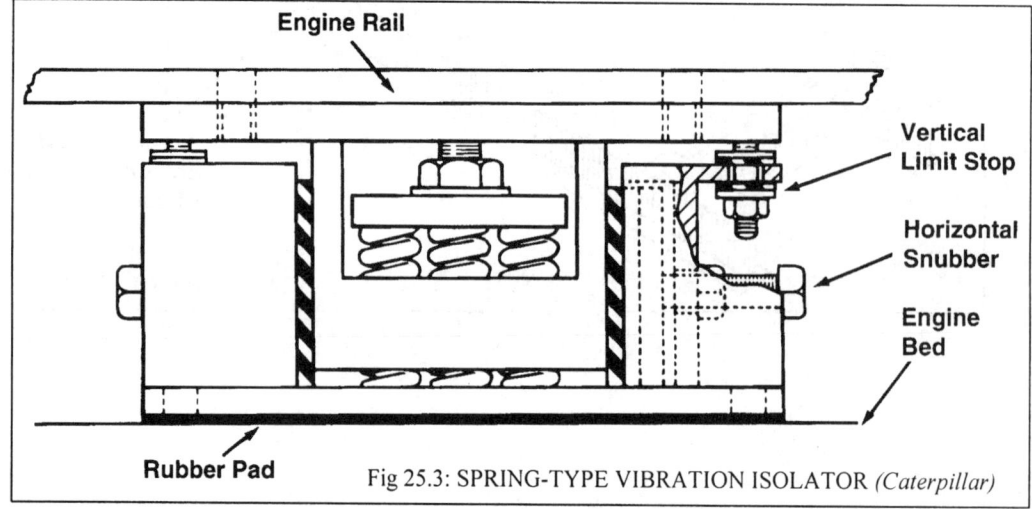

Fig 25.3: SPRING-TYPE VIBRATION ISOLATOR (*Caterpillar*)

ENGINE INCLINATION

Each type of engine has a maximum permitted engine inclination while the boat is underway. Exceeding this angle may affect the engine lubrication and cooling systems. In the illustration: "A" is the engine's static inclination; "B" is the boat's trim when underway; "C" is the total inclination of the engine underway (maximum permissible inclination = A + B).

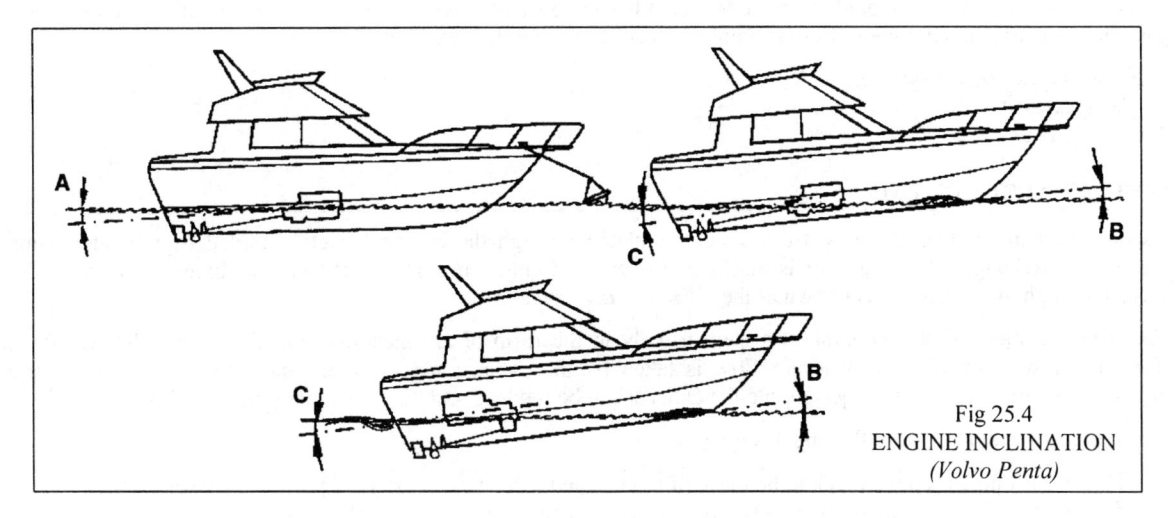

Fig 25.4
ENGINE INCLINATION
(Volvo Penta)

ENGINE ROOM WEIGHT DISTRIBUTION

As discussed in Chapter 13, a vessel's centre of gravity has a major influence on her static and dynamic stability. For planing and semi-planing hulls in particular it is important that heavy equipment such as engines, fuel tanks and batteries be positioned to obtain the best possible trim of the boat in the water. The boat's longitudinal centre of gravity (LCG) should not move too far aft of her centreline even with different levels of fuel and water in her tanks. It is also advisable to install the fuel tanks away from the warm engine room. The batteries should be placed in a separate, well-ventilated area if possible. In the illustration, Figure "A" represents a good longitudinal weight distribution. Figure "B" represents a wrong installation, with a subsequent bad running attitude.

DISTANCE APART BETWEEN TWIN ENGINES

There must be sufficient distance between the engines to allow accessibility for service work. Larger distances also give better manoeuvring capability. When designing the engine room always pay attention to the accessibility needed to allow proper service and repair to the engine. Ensure that the complete engine can be removed without damage to the boat structure. There must be sufficient space for the soundproofing material. Study the dimensional drawings of the relevant engine carefully.

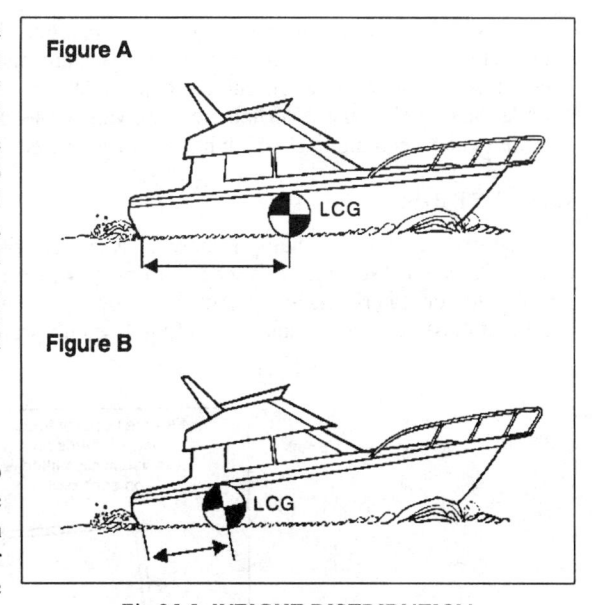

Fig 25.5: WEIGHT DISTRIBUTION
(Volvo Penta)

EXHAUST SYSTEMS

The exhaust system carries the engine's exhaust gas out of the engine room, through piping, to the atmosphere. It should have the following features:

- The pipe runs should be as straight as possible without sharp bends. Sharp bends will increase backpressure, which will reduce the engine performance, overheat parts adjacent to it and shorten the life of exhaust valves and turbocharger.
- The engine movement should be accommodated by the use of flexible exhaust bellows at the manifold flange (dry exhaust system only).
- The exhaust system should be adequately lagged to minimise burns and fire hazard.

- The exhaust discharge opening should be located far enough from the hull to minimise discolouration of the hull and deck caused by smoke when the engine gets older.
- The noise output level should be within an acceptable limit.
- The water should not be able to force its way back through the exhaust into the engine when the engine is switched off.
- Exhaust pipes (wet or dry) should not be constructed from copper or brass due to the corrosive quality of diesel exhaust gases in the marine environment. Iron or stainless steel fittings are the most suitable.

There are two types exhaust systems:
1. Wet Exhaust
2. Dry Exhaust

1. WET EXHAUST SYSTEM

As discussed in Cooling Systems, fresh water circulated through the engine jacket is cooled by seawater pumped through a heat exchanger. The seawater is discharged overboard either through the exhaust or directly over the side. If discharged through the exhaust, it is known as the "*Wet Exhaust*" system.

The water is injected into the exhaust system usually in the form of an injection bend fitted immediately after the manifold. In most cases all the raw water flow is taken by the exhaust but in cases where this results in excessive backpressure, the amount of water injected into the exhaust can be reduced and the remainder piped overboard.

Wet Exhaust system has the following advantages:

➢ The exhaust pipe is cool enough to be made of uninsulated, fibreglass reinforced plastic (FRP) or rubber.
➢ The use of a flexible exhaust pipe more easily accommodates the movement of the engine when running.
➢ The wet exhaust is quieter than the dry exhaust due to the silencing effect of the water.
➢ The wet exhaust line is particularly suitable with a flexible mounted engine because it can mostly be made of oil- and-heat resistant rubber exhaust hose.

Many types of mixers, silencers and mufflers are available, which should be fitted only in consultation with the engine manufacturer. The design of the wet exhaust system must be such that water cannot run back or siphon into the engine when it is at rest (even when subjected to violent motions, such as at an exposed anchorage). Engines installed with cylinder heads below the waterline are more vulnerable. Therefore, an anti-siphon valve (vacuum breaking device) is fitted on their exhaust cooling line, which helps prevent water siphoning back into the engine (Figures 25.6 to 25.13)

EXHAUST RISER

The exhaust riser is a hump (elbow or swan neck) in the wet exhaust pipe that prevents backflow of water into the engine. Seawater discharge is connected to its downward sloping portion. The upward sloping portion is very hot. It must be insulated or water-jacketed to protect persons in the engine room from burns. If the hose passes through bulkheads or similar, it must be protected against chafing. Use only stainless hose clips.

Fig. 25.6
Typical Water-cooled Exhaust
(Incorporating a water cooling jacket, riser and no valve at the discharge end)
(Slope "a" must be adequate to avoid water entering the engine)
(NSCV PART C Section 5 Subsection 5A)

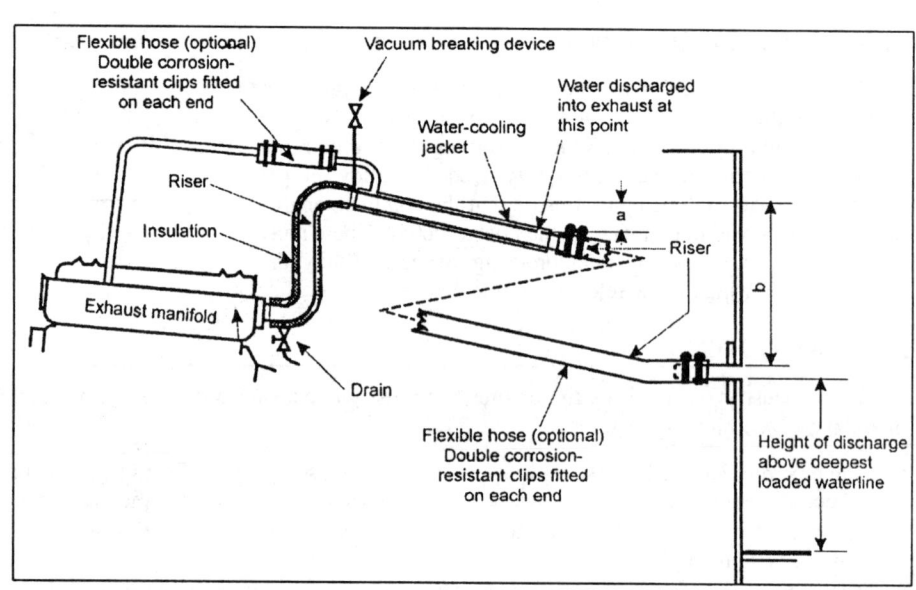

Chapter 25: Engine Beds, Exhaust, Ventilation & Soundproofing

PREVENTING WAVES FORCING WATER INTO WET EXHAUST

In a badly designed wet exhaust system and in rough seas, waves can force water to reach the engine. The result can be early turbocharger failure or piston seizure. This can be prevented in a number of ways, such as:

- Position the engine far enough above the water line (minimum 560 mm) so that breaking waves do not reach the highest point (exhaust elbow) in the exhaust pipe.
- Position the exhaust outlet opening so that it remains above the waterline even when the vessel is loaded. Waves can be kept out of the exhaust pipe by placing in it a one-way hinge valve where it enters the hull.
- Employ a waterlift muffler (see below).

Fig 25.7
TYPICAL WATER-COOLED
EXHAUST
(Incorporating a water-jacketed riser loop & shut-off valve at the discharge)
(Height "a" prevents the injected water entering the engine)
(Source: NSCV PART C Section 5 Subsection 5A)

WATERLIFT MUFFLER

A waterlift muffler minimises the possibility of water entering the engine from backflow in the wet exhaust system. It is a small, sealed tank, mounted to the deck in the engine compartment. It has an inlet and an outlet connection. An additional small drain connection in the bottom is often provided. The inlet tube enters the tank through the top or side, and does not extend past the tank wall. The outlet tube enters the tank wall through the top, and extends to the bottom of the tank, where it terminates on an angle.

Fig 25.8
TYPICAL WATER-COOLED
EXHAUST
(Incorporating a waterlift muffler, riser & flapper valve at the discharge)
(Slope "a" prevents the injected water entering the engine)
(NSCV PART C Section 5 Subsection 5A)

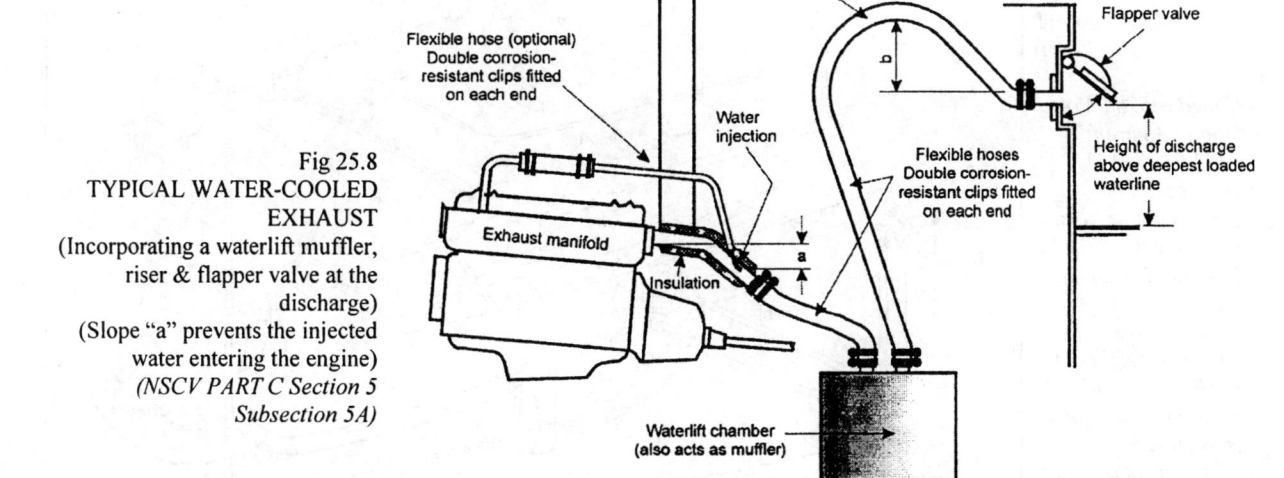

The mixture of seawater and exhaust gas enters the tank from the inlet hose. The rising water level in the tank reduces the gas flow area entering the discharge pipe. This causes the gas flow to speed up, resulting in water dividing into fine mist. This flow of mist, instead of water, prevents the backflow of water into the engine.

However, if due to poor design the speed of gas flow is not maintained, the mist-forming water droplets will not remain in suspension. The water will then be forced out of the waterlift muffler as a solid slug of water. This will cause the exhaust backpressure.

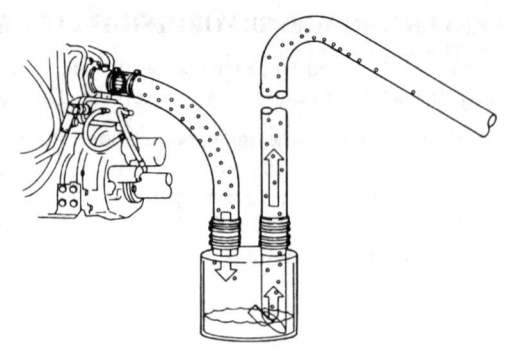

Fig 25.9
WATERLIFT MUFFLER *(Caterpillar)*

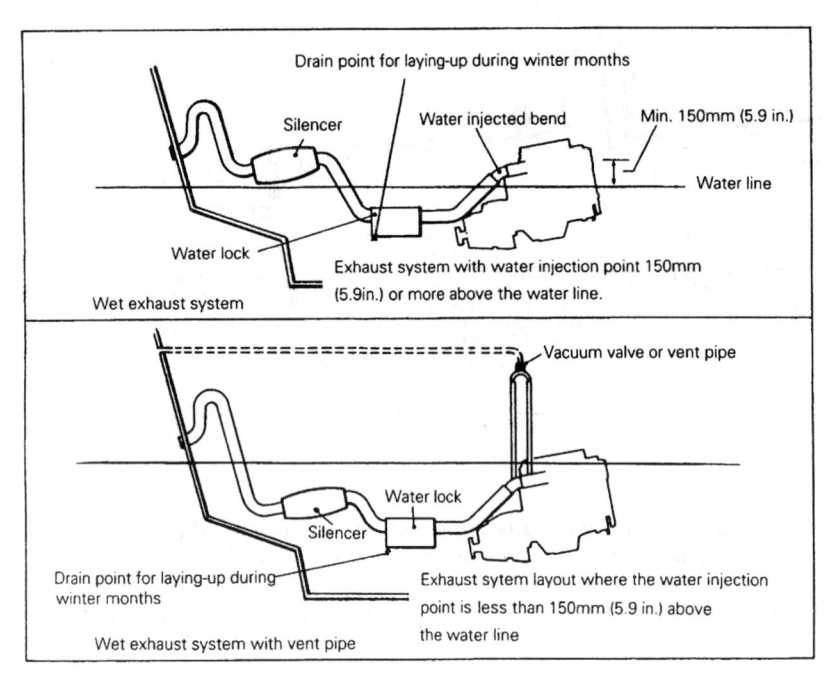

Fig. 25.10
WET EXHAUST SYSTEM
(Bottom illustration shows an anti-siphon valve for engine fitted below waterline)
(Lister Petter Marine)

Fig 25.11
WET EXHAUST ON A SAILING YACHT
(Volvo Penta)

(Fitted with the Volvo Penta "Aqua-lift" muffler.
C = height of exhaust elbow from waterline. If "C" is less than 200 mm, fit an anti-siphon/vacuum valve at a minimum height "D" of 500 mm. A suitable position for the valve is as close to the centreline of the boat as possible.)

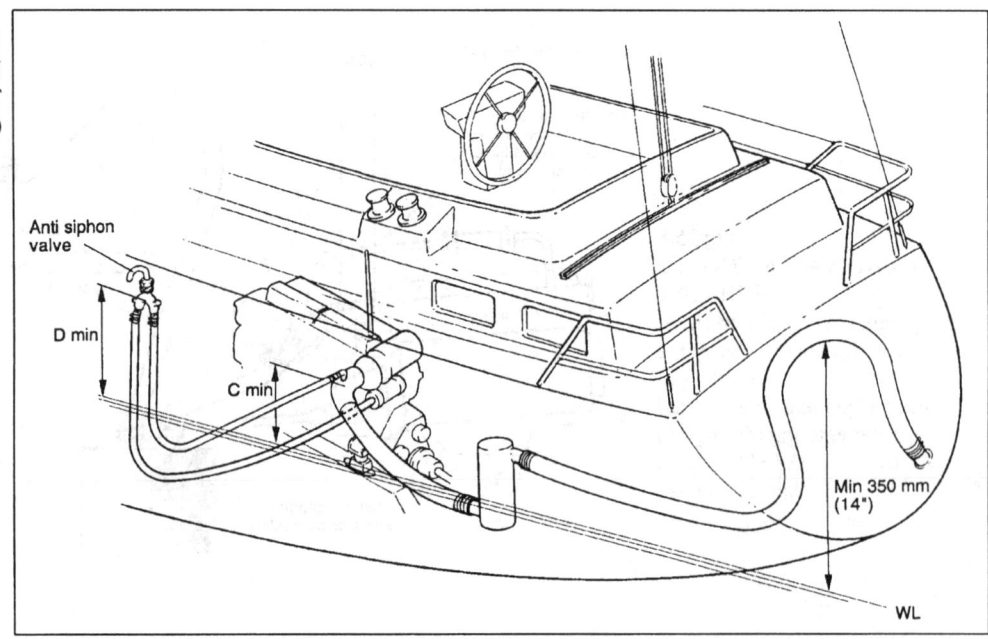

Chapter 25: Engine Beds, Exhaust, Ventilation & Soundproofing

VENTED LOOP (LOOP WITH AN ANTI-SIPHON VALVE)

The anti-siphon valve is a one-way valve fitted on top of a loop in plumbing. It allows air to enter the line when not in use, but seals itself when liquid passes through the line. Such non-corroding GRP loops with built-in valves are commonly used in plumbing wherever there is a risk of seawater being siphoned back into the vessel; such as in marine toilets and wet exhaust systems.

SILENCER (MUFFLER)

The illustration shows mufflers fitted fore and aft and athwartship. They should be fitted on a bulkhead or similar as close to the engine as possible, and always lower than the exhaust elbow. The angle of inclination (A) of the fore and aft muffler should be 5°–75° with the inlet facing upwards. The height of the riser (B) above the waterline should be at least 350 mm. A silencer installed athwartship should be inclined (angle C) between 25°-45° with the silencer inlet facing upwards. This inclination is important to prevent water ingress in the engine when the vessel heels over (particularly with sailing vessels).

On a new installation, after first starting the engine, stop the engine and check the water level in the silencer. The level should be well below the lower edge of the silencer intake pipe. This will eliminate the risk of water entering into the exhaust system.

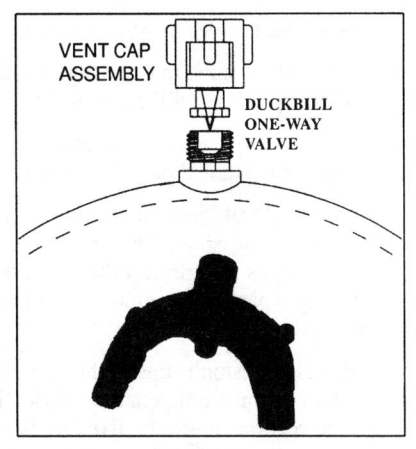

Fig 25.12
VENTED LOOP & ANTI-SIPHON VALVE *(Forespar)*

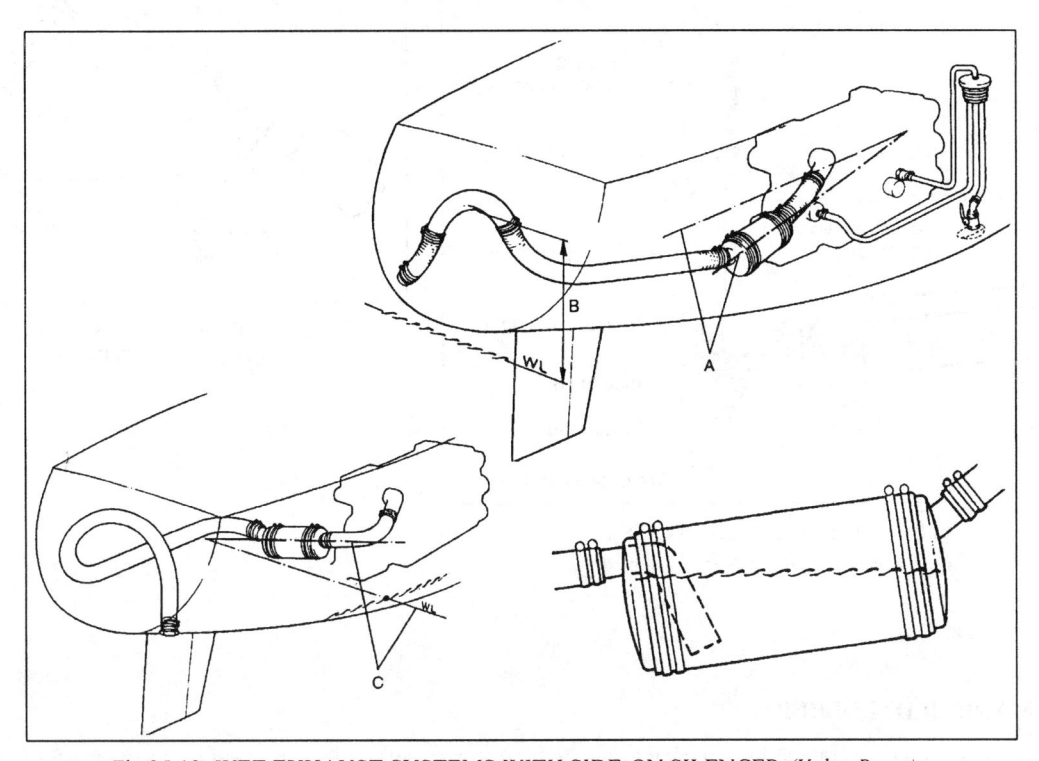

Fig 25.13: WET EXHAUST SYSTEMS WITH SIDE-ON SILENCER *(Volvo Penta)*

2. DRY EXHAUST SYSTEM

Dry exhaust system is cheaper and simpler than the wet exhaust system. However, being noisy and exposed, it is found only on workboats or where noise is not a problem. It may also be found on keel-cooled engines. The dry exhaust system is necessary in cold climates with temperatures below zero. Its installation requires the following precautions:

➢ Exposed parts of the dry exhaust piping can exceed 650°C. Combustible parts of the vessel and the personnel must be protected from its heat.

➢ The engine must be protected from rain and sea spray entering the dry exhaust pipe. This is achieved by fitting an elbow-type discharge fitting and drain slots in the exhaust pipe. Flapper type rain caps are discouraged because they are likely to deflect the exhaust gases downward towards the inlet air duct. Long pipes are fitted with moisture traps where the exhaust outlet slopes to its lowest point.

- The exhaust stack must be high enough and clear of the air turbulence created by the vessel's superstructure. Uncleared exhaust products will cause engine failures.

- In order to prevent overloading the engine room ventilation system, mufflers and other large exhaust components should be installed outside the engine room.

- The weight of the exhaust pipe must be isolated from the engine by way of a flexible connection fitted as close to the engine as possible. The engine should support no more than 28 kg of exhaust piping weight. Additional flexible connections will protect the exhaust components from engine vibrations, and shifting relative to the vessel's structure due to expansion, contraction and creeping. It would also keep the vessel much quieter and more comfortable for the occupants.

- Metal expansion is one of the main problems that must be overcome. Steel pipe expands 1.21 mm/metre for each 100°C rise in exhaust temperature. Divide long runs of exhaust pipe into sections having expansion joints between sections. Each section should be fixed at one end and be allowed to expand at the other. They should be able to freely expand and contract even when insulated. This generally requires either a soft material or an insulated sleeve to encase the flexible pipe connection.

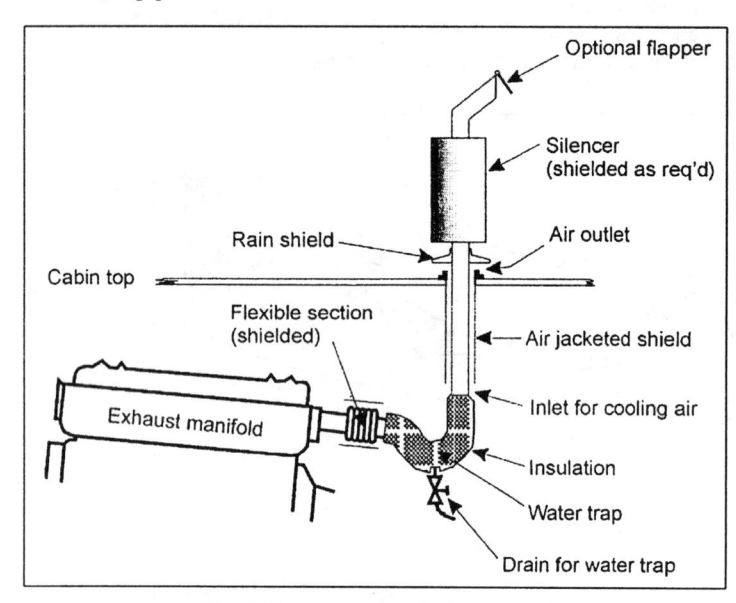

Fig. 25.14: DRY EXHAUST SYSTEM
(NSCV Part C Section 5 Sub-section 5A)

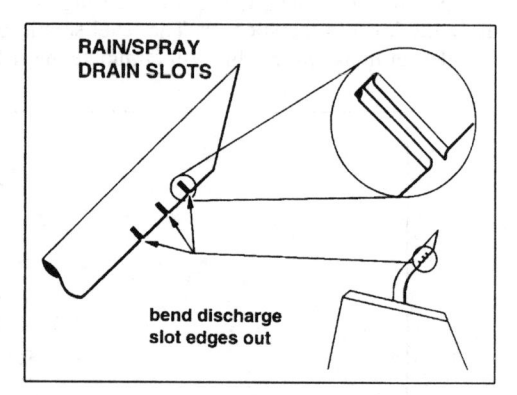

Fig. 25.15: DRAIN SLOTS IN DRY EXHAUST SYSTEM

EXHAUST MANIFOLD SLOBBER

Engines are designed to operate at loaded conditions. When operated with little or no load (at less than 15% load), the sealing capability of some of the engine components is adversely affected; and black oily fluid, known as slobber, starts to leak from the exhaust joints. Slobber is produced when soot in the exhaust manifold mixes with fuel and/or oil from the engine. The presence of slobber is not necessarily an indication of engine problem, nor does it usually harm the engine. But it is always unsightly.

Normally, an engine should be able to run at light loads for at least an hour without significant slobber. Some engines may run upto four or five hours without slobbering. But, eventually, any engine running at light loads will slobber. If extended idle or light load periods are unavoidable, slobbering can be minimised by loading the engine to at least 30% load for approximately 10 minutes every four hours. This will remove any fluids that have accumulated in the exhaust manifold.

It must be noted, however, that worn valve guides, worn piston rings or worn turbocharger seals also cause oil slobber. Similarly, fuel slobber can be an indication of combustion problems. The best way of minimising slobber is to correctly size the engine for the application.

Chapter 25: Engine Beds, Exhaust, Ventilation & Soundproofing

WAGON-BACK EFFECT

Due to a difference in air pressure the air flow behind a forward moving car or vessel has a tendency to draw dirt, water spray and exhaust emissions back into the vehicle. This phenomenon is known as the "Wagon-back effect". A vessel with a sheer, broad transom and high superstructure is most susceptible to the Wagon-back effect. Opening a ventilator or porthole under such conditions merely draws more fumes into the vessel.

Exhaust from a diesel engine does not contain as much carbon monoxide (CO) as from a petrol engine. However, a dangerous level of concentrated gases can build up from any fuel in the hull and occupied spaces A suitably arranged exhaust outlet is therefore essential.

The SLIPSTREAM METHOD utilizes the water flow from the propeller to transport the exhaust fumes as far away from the vessel as possible. The fumes are guided into the water via a BOOT positioned directly over the mid-point of the propeller. They are thus carried into the wake behind the propeller.

Fig 25.16: WAGON BACK EFFECT
(Volvo Penta)

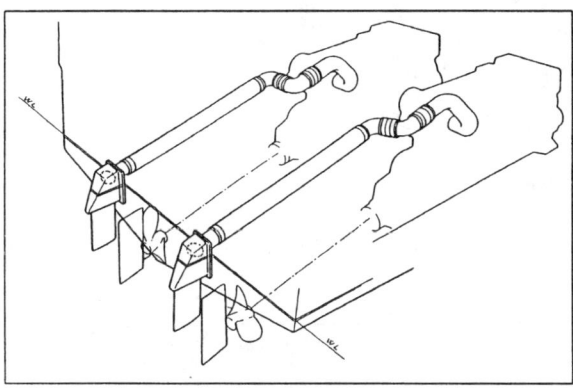

Fig 25.17: SLIPSTREAM METHOD OF DISPERSING EXHAUST EMISSIONS
(Volvo Penta)

SAFEGUARDS AGAINST ACCUMULATION OF CO

- Permanent ventilation by a continuous flow of air from forward to aft.
- A properly tuned engine will release less CO.
- Fit a CO gas detector. These are of two types:
  - The electronic type sound an alarm and flash a light. They are similar to bilge alarms.
  - The simpler disk-type warns of CO build up by changing colour. They can be restored to original colour and re-used after a few hours in fresh air. However, the life of a disk, once opened, is only about a month.

A TWIN-ENGINE INSTALLATION & GENERAL ARRANGEMENT

(NOTE: The ghost-view illustration makes the starboard engine appear on the centerline)

The illustration shows an example of a twin-engine installation with wet-exhaust systems. The muffler on the starboard exhaust is the upright type. On the port exhaust it is fitted on its side.

The starboard propeller shaft is mounted with a water-lubricated stuffing box and a rubber seal. The port propeller shaft has a grease-lubricated stuffing box as the seal. On both shafts, there is an "vane", just outside the through hull, which reinforces the water stream into the stuffing box.

The engines are equipped with the Volvo Penta EDC system (Electronic Diesel Control) and the boat has a hydraulic steering system with a Volvo Penta steering pump, hydraulic cylinder and tie bar.

The Australian Boating Manual

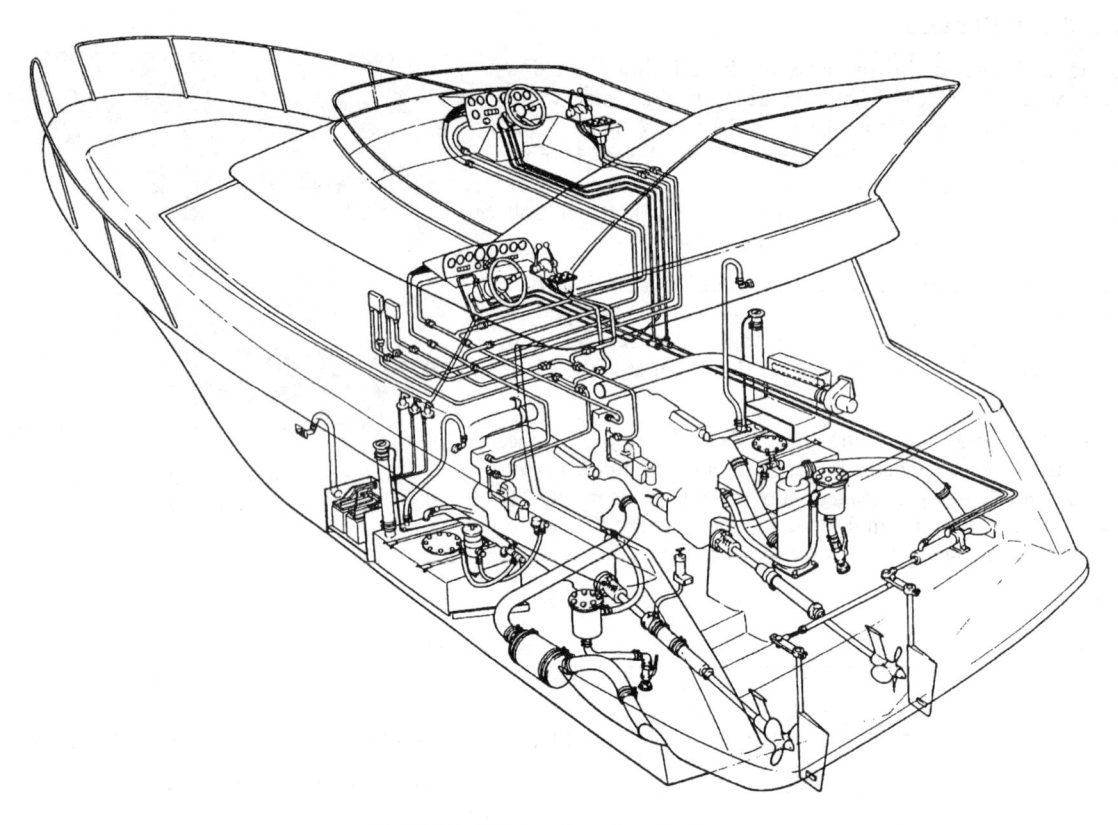

Fig. 25.18: A Volvo Penta Installation

ENGINE ROOM VENTILATION

VENTILATION
Diesel engines use a large amount of air for combustion. Their performance is affected by air pressure, air temperature and exhaust backpressure. Deviations from the normal values show up first with an increase in black smoke, and then with the loss of power. Good engine room ventilation is also necessary for the disposal of crankcase fumes. The crankcase fumes must either be ingested by the engine or piped out of the engine room. The engine's electrical equipment and fuel system should also be kept at a low temperature. Hence, both the inlet and outlet air ducts to the engine room must be of adequate dimensions.

VENTILATION AIR
Drawbacks of a poorly ventilated engine room:

- The engine room will be hot for personnel.
- The engine cooling system would need to work harder.
- The engine will have insufficient combustion air, affecting its power output.

The engine room ventilation should meet the following requirements:

➢ The engine room temperature should be within 9°C above the outside temperature.
➢ At a distance of 20 mm from electrical equipment, such as generators, alternators and solenoids the air temperature should not exceed 60°C.
➢ The intake air temperature should not exceed 45°C

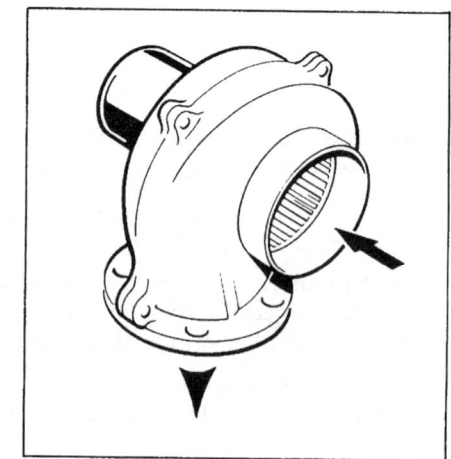

Fig. 25.19
A DRAFT FAN (DISCHARGE FAN)

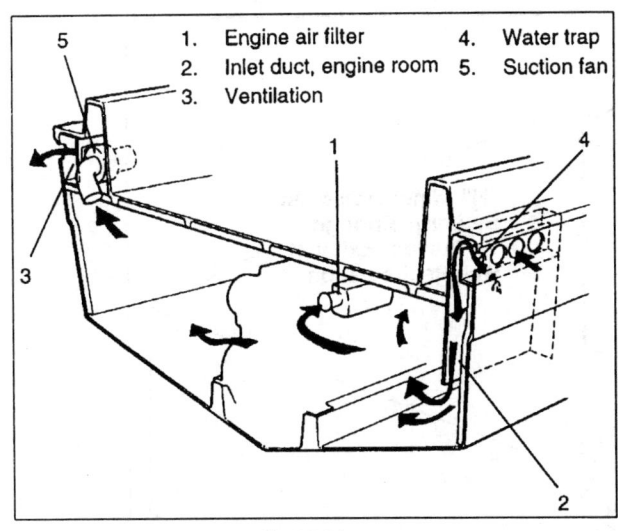

Chapter 25: Engine Beds, Exhaust, Ventilation & Soundproofing

1. Engine air filter
2. Inlet duct, engine room
3. Ventilation
4. Water trap
5. Suction fan

Fig 25.20
ENGINE ROOM VENTILATION *(Volvo Penta)*

The simplest form of ventilation system is the use of an intake and exhaust ducting arrangement. The duct openings are protected from entry of seawater and rain by mushroom caps, cowls or drains. The use of sharp bends should be avoided, as any restriction will impair airflow.

The ventilation system should be designed to make fresh air enter the engine room from as far from the source of heat as possible and from a level as low as possible, but not so low that bilge water, if any, can block the air supply. It should discharge hot air from the highest point in the engine room, preferably directly over the engine. A two-speed discharge fan would make it easier to vary the airflow in summer and winter. Fans must never be installed in the inlet air ducts, as this could lead to overpressure in the engine room with the risk of gas or air leaking out into other parts of the vessel. The incoming ventilation duct should not blow cool air onto the hot engine components. Such mixing of the hottest air with the incoming cool air would raise temperature throughout the engine room.

Test the ventilation system with the vessel battened down for bad weather. It must provide a safe working temperature and adequate airflow with hatches and doors secured for bad weather. The air should be agitated, but not violently, throughout the engine room. High velocity air only around the engine or other heat sources is not good ventilation. It will spread the heat throughout the engine room, raising its average temperature.

Exhaust pipes and other hot surfaces in the engine room, if left uninsulated, can dissipate more heat into the engine room than all the machinery surfaces combined. They must therefore be insulated.

If the engine room temperature cannot be brought down within reasonable limits, use larger or more ducts.

OTHER DESIGN CONSIDERATIONS FOR AIR DUCTS

- Cross winds and following winds make it impossible to avoid some recirculation of ventilation air. However, recirculation can be minimised by locating the intake opening forward or abreast of and at a lower elevation than the discharge opening.
- Discharge fans are more effective than inlet fans. Fit only discharge fans.
- Centrifugal fans (squirrel cage blowers) are better than the axial flow fans (propeller fans)
- Fans should be powered-type.
- Discharge fan motors should be located outside the direct flow of hot ventilating air.
- The *nameplate* rating of fans does not necessarily reflect its actual air flow when installed in an opening.
- As a rule of thumb, use 10 sq cm (1.25 sq in.) of duct cross-section area per engine kW, and no more than 3 right angle bends. If more right angle bends are required, increase the pipe diameter by one pipe size.
- Prevent refrigerant leakage into the engine's intake system Freon or ammonia drawn into the engine's combustion chamber would cause severe engine damage. The chemicals in refrigerants become highly corrosive acids in the engine's combustion chamber.
- For a gas engine, in particular, the air inlet must not be located at the same side as the fuel filler fitting. The filler should be also be located at least 1 metre away from the air outlet.

CRANKCASE FUMES DISPOSAL

The pressure inside an internal combustion engine causes some fumes to BLOWBY (escape) from around the pistons into the crankcase. Two methods are employed to prevent pressure building up in the crankcase. Smaller engines consume their crankcase fumes by drawing them into the engine's air intake system. Larger engines are provided with a crankcase vent tube to pipe the gas out of the engine room, so that their high-efficiency paper air filter elements do not become clogged. If there is more than one engine, a separate vent line for each engine is required.

The blowby of engine fumes into crankcase increases as the time for engine overhaul gets nearer, thus increasing the crankcase pressure. Increase in crankcase pressure is also an indication of deterioration of valve guides and piston rings.

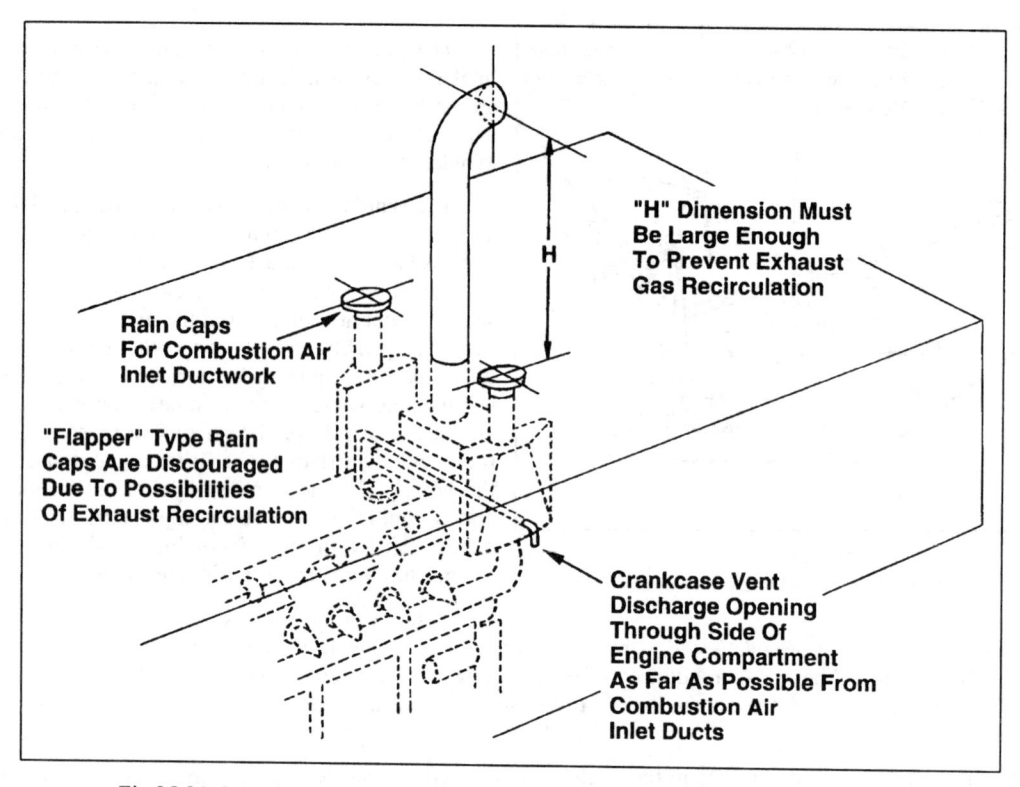

Fig 25.21: LARGE VESSEL ENGINE ROOM VENTILATION *(Caterpillar)*

Generally the size the crankcase vent pipe should be equal to the size of the crankcase fumes vent. If the pipe run is longer than approximately 3 metres or if there are more than three 90° elbows, the inside diameter of the pipe should be increased by one pipe size. Any horizontal runs in the pipe should have a minimum slope of 40 mm per metre. To prevent air ventilation ducts from becoming coated with oil deposits, crankcase fumes should not be discharged through them. The crankcase vent pipe should either be vented into the exhaust flow at the terminal of the exhaust duct, or preferably, directly into the atmosphere through the top or the side of the engine room.

EXHAUST EJECTED VENTILATION SYSTEM

As illustrated, the hot air from the engine room can also be ejected by utilizing the kinetic energy of the exhaust gases. This system can draw out a quantity of ventilating air approximately equal to the flow of the exhaust gases. The ejector is placed in the exhaust stack just prior to the exhaust discharge to atmosphere. This arrangement minimises the risk of backpressure from the mixture of exhaust gases and ventilation air in the exhaust stack. The stack also remains cooler and cleaner by keeping the exhaust gases inside the exhaust pipe during its run through the stack.

The exhaust ejectors are most effective on vessels with only one engine. With twin engines, if one engine is operated at reduced load, its ejected airflow can reverse and pull exhaust gas from the other heavily loaded engine into the engine room.

Chapter 25: Engine Beds, Exhaust, Ventilation & Soundproofing

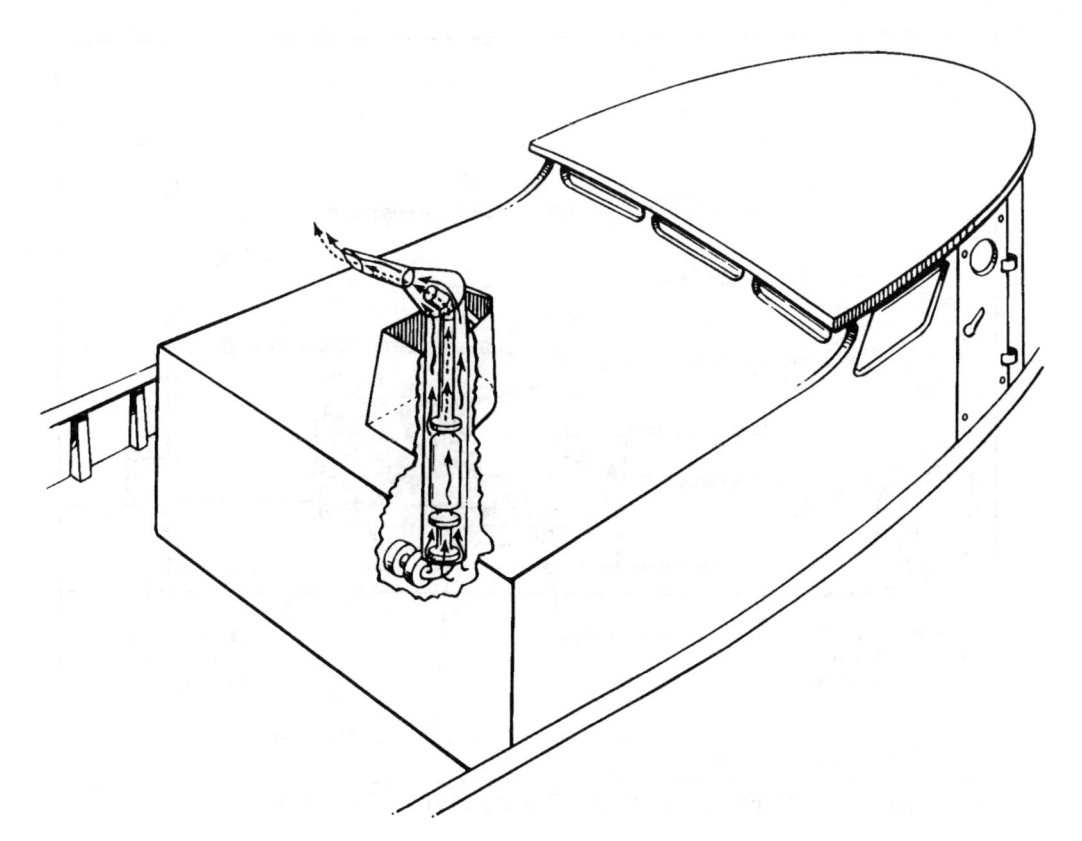

Fig. 25.22 EXHAUST EJECTED VENTILATION SYSTEM

SOUNDPROOFING ENGINE & MACHINERY NOISE

The machinery generates airborne as well as structural noise. The latter are in the form of vibrations and pressure pulses. Vibrations from the engine are transmitted in the hull, propulsion systems, exhaust pipes, coolant pipes, fuel pipes, electric cables and control cables. Water pressure pulses from the propeller are transmitted in the hull via support blocks, bearings and seals.

The airborne noise is minimised through proper sealing of the engine room, sound insulation and sound traps in the air inlets. Make sure the sound insulation material does not become a trap for bilge water. Also make allowances for engine inspections, repairs and service.

Shift cables, throttle cables and electrical wires passing through bulkheads should be drawn through a tube or grommet. Fuel hoses passing through bulkheads should also be drawn through grommets. Other cables, such as electrical wires and battery leads, can be drawn through a rubber hose or PVC electrical tube built into a GRP bulkhead. Clearances between the tubing and wires can be sealed off with insulation material or sealing compound. Such devices not only provide a seal against noise but also protect hoses and cables from sharp edges.

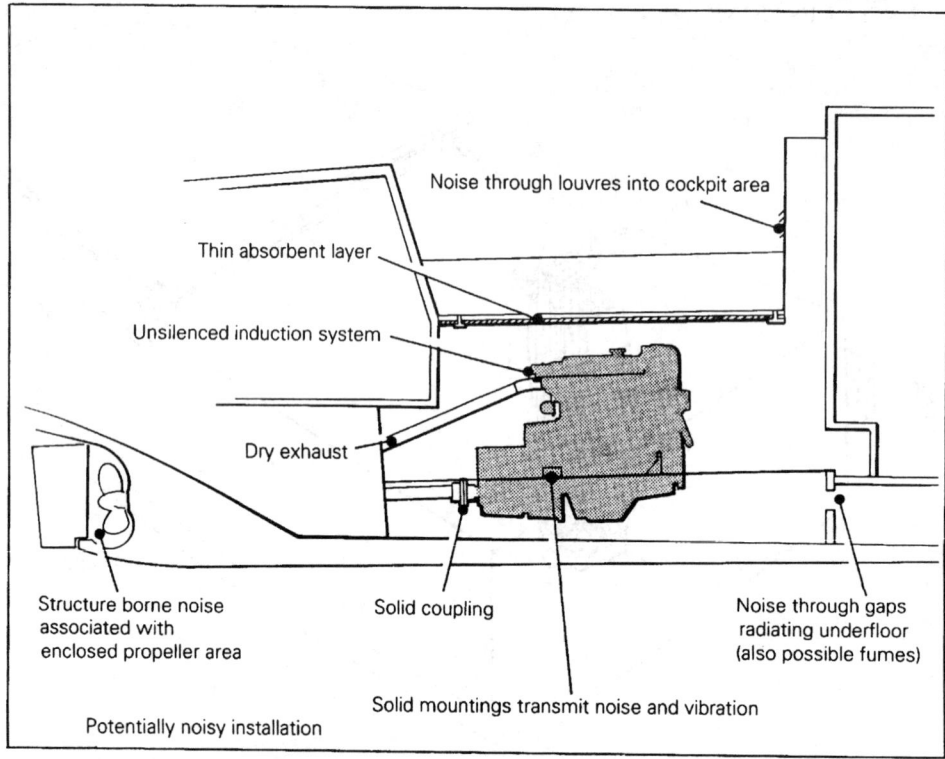

Fig, 25.23: POTENTIALLY NOISY INSTALLATION *(Lister Petter Marine)*

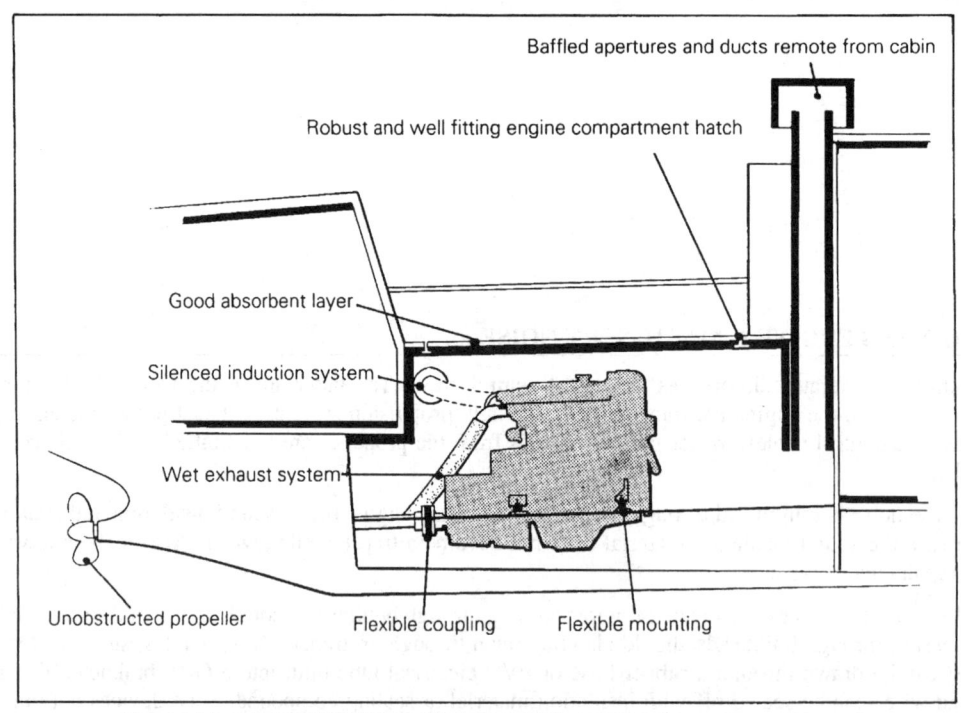

Fig, 25.24: RECOMMENDED SOUND INSULATION *(Lister Petter Marine)*

CHAPTER 25: QUESTIONS & ANSWERS

EXHAUST & VENTILATION

1. In relation to a vessel's exhaust system, which statement is incorrect?
    (a) Sharp bends in exhaust pipes reduce engine performance.
    (b) The exhaust pipe should be constructed from copper or brass.
    (c) The exhaust discharge opening should be located far from the superstructure.
    (d) It should be adequately lagged to minimise burns & fire hazard.

2. In relation to wet exhaust system, which statement is incorrect?
    (a) It is not suitable with a flexible mounted engine.
    (b) The exhaust pipe is cool enough to be made of uninsulated FRP or rubber.
    (c) Seawater discharge is connected to the downward sloping portion of the exhaust riser.
    (d) The upward sloping portion of the exhaust riser is very hot.

3. Waves forcing water into wet exhaust:
    (a) cannot be prevented by positioning the engine above the vessel's waterline.
    (b) cannot be prevented by positioning the exhaust opening above the vessel's waterline.
    (c) cannot be prevented by employing a waterline muffler.
    (d) can cause piston seizure.

4. In relation a vented loop, which statement is incorrect?
    (a) It is an anti-siphon valve.
    (b) It allows air to enter the line when liquid passes through it.
    (c) It is used in plumbing such as wet exhausts and marine toilets.
    (d) It is a one-way valve.

5. In relation to dry exhaust system, which statement is incorrect?
    (a) Exposed parts of a dry exhaust system can exceed 650°C.
    (b) A flexible connection must be fitted between the exhaust pipe and the engine.
    (c) It is noisier than the wet exhaust system.
    (d) The flapper type rain caps should be fitted to prevent rain entering the exhaust pipe.

6. Which of the following statements is incorrect for safeguarding against accumulation of carbon monoxide in the vessel?
    (a) Maintain permanent ventilation by a continuous flow of air from forward to aft.
    (b) Open only the aft portholes.
    (c) Install a CO gas detector.
    (d) Keep the engine properly tuned.

7. There is a greater risk of build up of exhaust fumes in vessels with:
    (a) forward to aft ventilation system.
    (b) blunt sterns.
    (c) petrol engines.
    (d) single engines.

CHAPTER 25: ANSWERS

1 (b),   2 (a),   3 (d),   4 (b),   5 (d),
6 (b),   7 (b)

Chapter 26

ENGINE OPERATION, ALARMS, CONTROLS & MAINTENANCE

SAFETY IN THE ENGINE ROOM

- Cleanliness is the greatest safeguard against fire or explosion.
- The engine, fuel and battery compartments are at the greatest fire risk.

FIRE RISK IS INCREASED BY
- Poor exhaust installations - uninsulated and unguarded.
- Accumulation of fuel in bilges.
- Oil soaked timberwork.
- Incorrect materials used for fuel pipes.
- Fitting batteries without protective covers or adequate ventilation. [Hydrogen gas accumulation].
- Faulty electrical equipment.
- Lack of maintenance.
- Volatile gases in confined spaces.
- LPG allowed to drain into bilges.
- Build up of flammable material.
- Smoking or naked flames.

OTHER SAFETY HAZARDS
- Loose clothing.
- Long loose hair.
- Tools lying about.
- Electrical extension cords.
- Oily rags and other waste.
- Flammable liquids on board.
- Spilt oil and fuel.

FIRE RISK CAN BE REDUCED BY
- Adequate sealing of engine room and the fuel and battery compartments from the rest of the vessel.
- Constructing and furnishing vessels with non-combustible materials.
- Keeping bilges thoroughly clean.
- Not tolerating any oil leakages.
- Keeping all tools and equipment secure.
- Being aware of fire hazard around the engine exhaust system.
- Ensuring adequate ventilation.
- Fitting the battery switches inside the engine bay:
- This way, the skipper will have to open the engine room hatch or door to turn on the batteries, and smell or see any fuel leak before starting the engine.

ENGINE STARTING SYSTEMS

Based on the methods of storing and recharging, there are three types of engine starting systems:

1. **ELECTRIC**
   Batteries – recharged by an engine-driven alternator or an external source

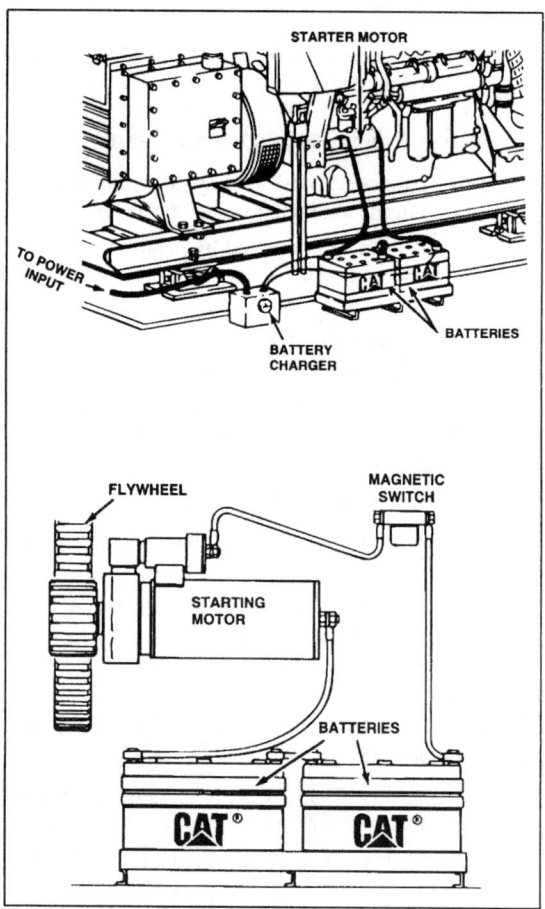

Fig 26.1: ELECTRIC STARTING SYSTEM
(Caterpillar)

Chapter 26: Engine Operation, Alarms, Controls & Maintenance

2. PNEUMATIC
Compressed air tanks – recharged by an electric motor-driven air compressor

3. HYDRAULIC
Hydraulic oil tanks – recharged by a small engine-driven hydraulic pump

1. ELECTRIC STARTING SYSTEM

Battery-powered electric motors use low voltage DC current. They provide a fast, convenient and reasonably reliable push button start. Their associated components are also lightweight, compact and engine mounted.

Precautions:

- Do not use a battery-isolating switch that causes spikes (voltage transients).
- Cranking an engine continuously for more than 30 seconds can overheat the starter motor. Allow it to cool for two minutes before resuming cranking.
- Use only the recommended cables.

2. PNEUMATIC STARTING SYSTEM

Air starting, whether manual or automatic, is highly reliable. Greater torque provided by an air motor can accelerate the engine to twice the cranking speed in about half the time required by an electric starter.

The system consists of an electric motor-driven air compressor, an air receiver and tanks to store compressed air at about 759 kPa. From the tanks, air is piped to the air starting motor at the time of starting the engine. The tanks are fitted with pressure valves and pressure gauges. A dryer at the compressor outlet or a small quantity of alcohol in the starter tank helps prevent water vapour in the sub-zero air supply from freezing. Manual or automatic oil and water traps are installed at the lowest parts of the piping and in the air receiver, and all piping is made to slope towards these traps.

Precautions:

- The air pipe to the starting motor should be a short, direct iron pipe with a flexible connection at the starter.
- To prevent damage to the starting motor, the accumulation of oil and water contaminants must be removed daily.

3. HYDRAULIC STARTING SYSTEM

Although the hydraulic starting system has many qualities, it also has many disadvantages. It is used mainly where the use of electrical connections is unsafe.

Qualities:

- High cranking speed
- Relatively compact
- The engine-driven recharging pump is quite small.
- Hydraulic recharging time is not long

Disadvantages:

- Requires high pressure pipes and fittings and extremely tight connections
- Hydraulic accumulators, if used, contain large amounts of stored mechanical energy. They must be protected from perforation or breakage.
- Recharging oil lost through leakages requires special equipment
- Repairs require special tools.

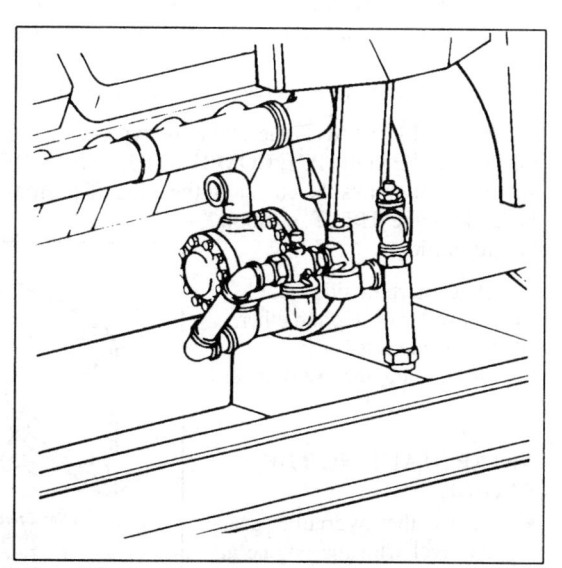

Fig 26.2
PNEUMATIC STARTING SYSTEM
(Caterpillar)

STARTING AN INBOARD ENGINE

The starting ability of a diesel engine is affected by its:

- Ambient (surrounding) temperature
- Jacket water temperature
- Lubricating oil viscosity
- Parasitic loads

Diesel engines need heat of compression of air in their cylinders to ignite fuel. A cold engine would need longer cranking periods or higher cranking speed to develop adequate ignition temperature. Similarly, the drag due to cold lube oil would impose greater load on the cranking motor. The type of oil and its temperature determine its viscosity. For example, the SAE 30 oil below 0° C approaches the consistency of grease.

PRE-START CHECKS
- Make a thorough inspection of the engine compartment. Look for items such as fuel, oil or coolant leaks. Correct the cause of any leak and clean the compartment. Also look for loose bolts, worn fan belts and rubbish build-up. Remove rubbish and carry out repairs as needed.
- Check crankcase oil level. The dipstick should read HIGH.
- Check the coolant level.
- Observe the air cleaner service indicator, if fitted.
- Drain water from the water separator, if fitted.
- Disconnect the battery charger, if it is attached.
- If the engine has not been run for several weeks, fuel may have drained and allowed air into the filter housing. Also, when fuel filters have been changed, some air space will be left in the housing. In these instances, prime (bleed) the fuel system.

STARTING THE ENGINE
- Open raw water valve, if equipped.
- Open fuel supply valve.
- Open fuel return valve to prevent engine damage.
- Disengage marine gear. Do not start engine under load.
- Start engine by turning the starter switch to START. The Low Oil Pressure Alarm may sound for a few seconds. If oil pressure does not rise as recommended, stop engine and investigate.

NOTE: The Oil Pressure Switch may have an override button to push, to permit the engine to start without sounding the alarm. As the oil pressure increases, the button automatically comes out to RUN position.

The illustration shows two types of Lister engine starting devices.

For COLD START, diesel engine type (A) is fitted with a HEATER PLUG in the inlet manifold and type (B) with a GLOW PLUG in the PRE-COMBUSTION CHAMBER of each cylinder. In type (A), the heater plug is energised when the starter motor cranks the engine in the START position. In type (B), the glow plug circuit is energised when the ignition key is held in the HEAT position before starting.

Other starting aids include a jacket water heater and ether. As discussed under troubleshooting, the use of ether is not encouraged.

AFTER STARTING THE ENGINE
- Check the hydraulic gear oil level with the engine at idle speed, and the gear in neutral. The dipstick should read FULL. (The manual gear oil level is usually checked prior to starting the engine, when the oil is cold).

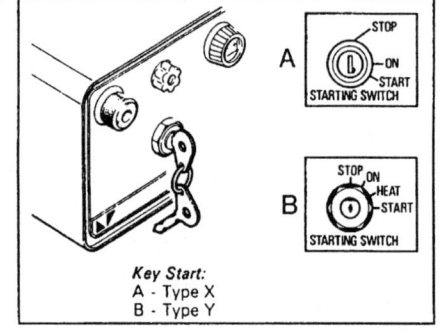

Key Start:
A - Type X
B - Type Y

STARTING A LISTER DIESEL ENGINE
Fig 26.3

Heater Plugs:
A - Manifold Heater Plug
B - Glow plug

- Do not apply load to the engine or increase engine speed until oil pressure gauge indicates normal and all temperature gauges start to move. Check all gauges during the warm-up period to ensure they are working. Do not fully load the engine until temperature gauges indicate normal.
- Check for any fluid or air leaks at idle and at half full RPM with no load on the engine.
- Operate the engine at low load until all systems reach operating temperatures. Check all gauges during the warm up period.
- Do not run an engine on light load or at a low jacket temperature for prolonged periods. Some sailing vessel skippers think that to keep their engines in good order it is necessary to run them at the moorings now and again or to let them tick over when motor-sailing. Such a practice will result in premature carbon build up (Glazing) in the cylinders. The

Chapter 26: Engine Operation, Alarms, Controls & Maintenance

engines should be either left alone or run under load, or, at least, at the optimum battery charging speed.
- Do not reduce idling speed below the manufacturer's setting. It can cause damage to crankshaft and connecting rods, especially in high-speed diesels whose cylinder diameter and stroke tends to be the same. Such engines are known as "SQUARE" ENGINES. It is not so critical in some of the older, long stroke, slow speed engines.

TURBOCHARGER CARE & MAINTENANCE

➢ When first starting a turbocharged engine, the engine speeds should be kept to a minimum for a short period to allow the turbocharger shaft bearings to be pressure lubricated before being subjected to high speeds and temperatures.
➢ If a turbocharged engine has not been started for a long period, the engine lubrication system should be primed before starting.
➢ The turbocharger would keep running for some time after the engine is stopped, which can damage its turbine blades from excessive heat and deprive the turbine shaft of pressure lubrication. Therefore, engine should be run at idle speed for 3-4 minutes before shutting down, to permit cooling of the turbine wheel and housing. Some engines employ a *Turbo Timer* to achieve this automatically.
➢ Do not run at idling speed for too long. The temperature in the combustion chambers then drops too low for effective combustion, and the engine produces unpleasant exhaust emissions.
➢ Do not charge batteries by running the engine at idling speed. As a matter of fact, the generator does not charge at idling.
➢ The Caterpillar engine manual recommends a regular inspection of the turbocharger, as follows: Remove the inlet and exhaust piping from the turbocharger. Make sure the turbine and compressor wheels freely turn by hand without contact with the compressor or turbocharger housing. Look for evident oil leaks.
➢ Some manufacturers recommend changing the turbo filter with every oil change.

DASHBOARD DISPLAYS

Engines in boats are usually fitted with an ignition key at the steering position. The ignition switch has various positions to switch on the engine accessories, and to start or stop the engine.

1. Master warning light
2. Instrument lighting
3. Stop switch & warning light
4. Tachometer
5. Temperature gauge
6. Oil pressure gauge
7. Interlock switch (Push to override oil pressure monitor during starting)
8. Ignition switch

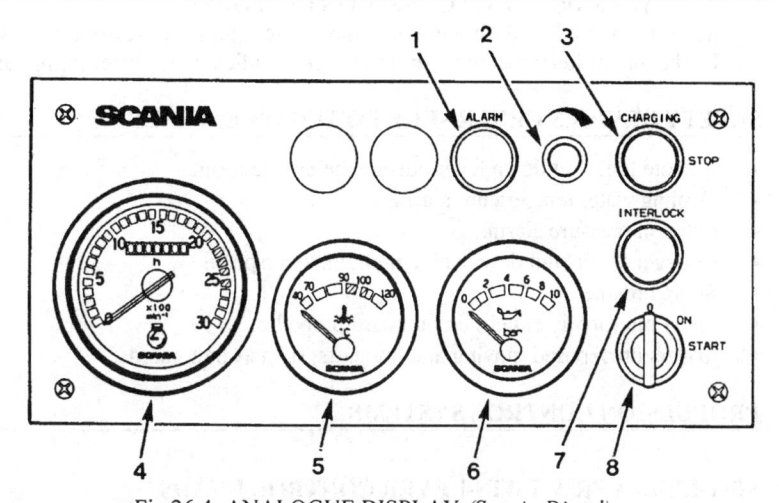

Fig 26.4: ANALOGUE DISPLAY *(Scania Diesel)*

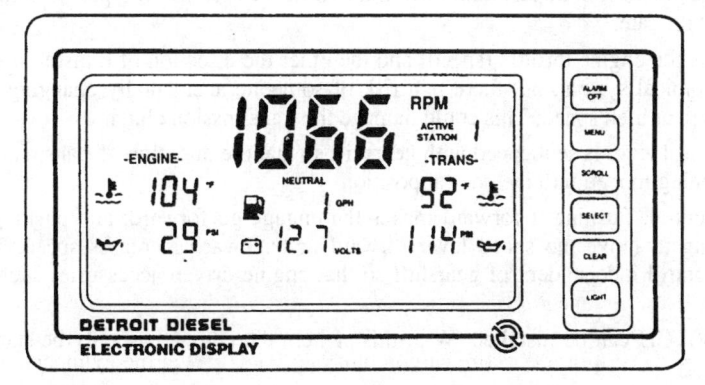

Fig 26.5
DIGITAL DISPLAY
(Detroit Diesel)

ENGINE INSTRUMENTS AND ALARMS

(For Non-Engine Alarms, See Chapter 15)

The weak link in any instrumentation system is the sensor unit (transducer). A high quality sensor or its annual replacement would ensure that the system is not sabotaged by false alarms.

MUST HAVE INSTRUMENTS
- Engine lubrication oil pressure
- Jacket water temperature
- Engine speed (rpm) (Relationship between engine speed, vessel load and throttle position allows informed judgement about engine load and need for maintenance)

HIGHLY DESIRABLE INSTRUMENTS
➢ Transmission oil pressure
➢ Voltmeter

USEFUL INSTRUMENTS
- Transmission oil temperature (indicates clutch slippage, insufficient clutch pressure, bearing wear, cooler blockage)
- Exhaust stack temperature (indicates air filter restriction, injector condition, valve problems and engine load)

MUST HAVE ALARMS & SWITCHES (CONTACTORS)
➢ Low lube oil pressure
➢ High coolant temperature (Caution: the sensing bulb must be submerged in the coolant)
➢ Overspeed (indicates high fuel flow while some part of engine has failed)

HIGHLY DESIRABLE ALARMS & SWITCHES (CONTACTORS)
- Water level alarm (indicates loss of coolant)

USEFUL ALARMS & SWITCHES (CONTACTORS)
➢ Seawater pump low differential pressure switch (indicates restriction in seawater flow to heat exchanger)
➢ Intake manifold temperature alarm switches (indicates seawater pump failure or sea strainer plugging)

SAFETY DEVICES GENERALLY FITTED ON ENGINES

- Remote fuel shut down lever outside the engine room.
- Cooling water temperature alarm.
- Lube oil pressure alarm.
- Emergency air shut down valve if engine overspeeds.
- Sump breather.
- Shear pin (or slip clutch) on outboard propeller.
- Covers on exposed moving and hot parts, such as belt guards.

PROPULSION CONTROL SYSTEMS

SINGLE-LEVER & TWIN-LEVER CONTROL HEADS

In its most basic form a vessel's propulsion control head, fitted at the steering position, allows the operator to adjust the engine speed and change gears. There are two types of control heads: one-lever and twin-lever type. Both types need to be reliable, precise and dependable, as well as easy to operate.

On a twin-lever control head, one lever adjusts the engine throttle (speed) and the other the direction of transmission – ahead, astern or neutral. It is a simpler and economical system, but there is a risk of someone accidentally changing the direction of transmission while the engine is running at high speed. This could damage the transmission clutch.

On a single-lever control head, the throttle control lever is sequenced with gearshift so that the direction of transmission cannot be changed without the throttle lever first being moved into the neutral position.

In the upright position the gearshift lever is in neutral. Pushing it forward makes the engage go forward; and pushing it backward makes the engine go astern. Continuing to move the speed lever forward or backward increases speed. The neutral position is built with a degree of speed control independent of gearshift so that engine-driven accessories such as generators, pumps and winches can be run.

In most controls, a NEUTRAL SAFETY SWITCH can be installed. With this switch, the engine can only be started when the control is in the neutral position.

Chapter 26: Engine Operation, Alarms, Controls & Maintenance

CAUTION: An inappropriate engine and transmission control system can cause the engine to stall or reverse itself.

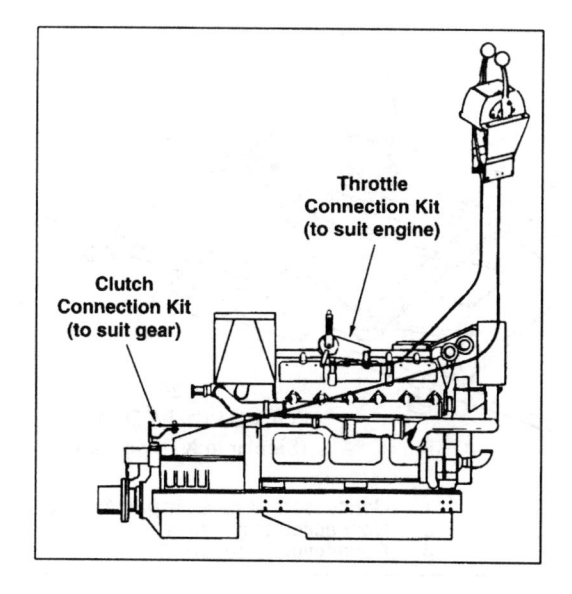

Fig. 26.6 (A) & (B)
CABLE ARRANGEMENT – TWIN & SINGLE LEVER CONTROLS
(*Twin S Control & MT Control*)

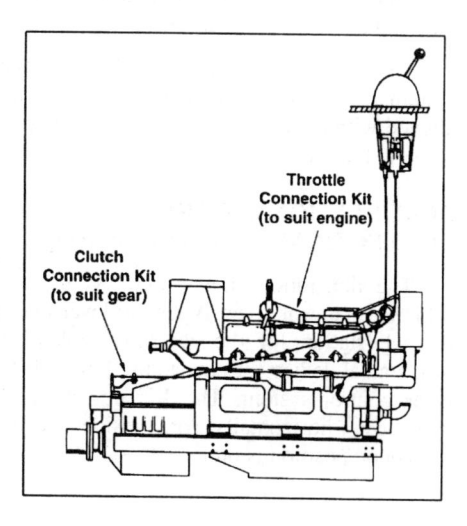

Fig. 26.7
SINGLE LEVER CONTROL POSITIONS
(*Caterpillar*)

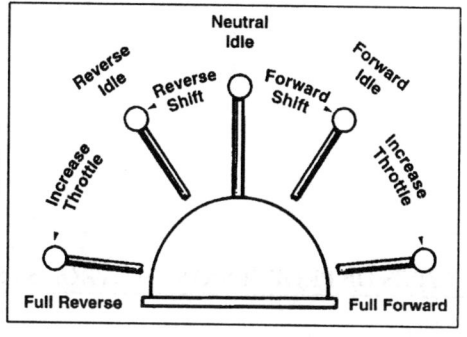

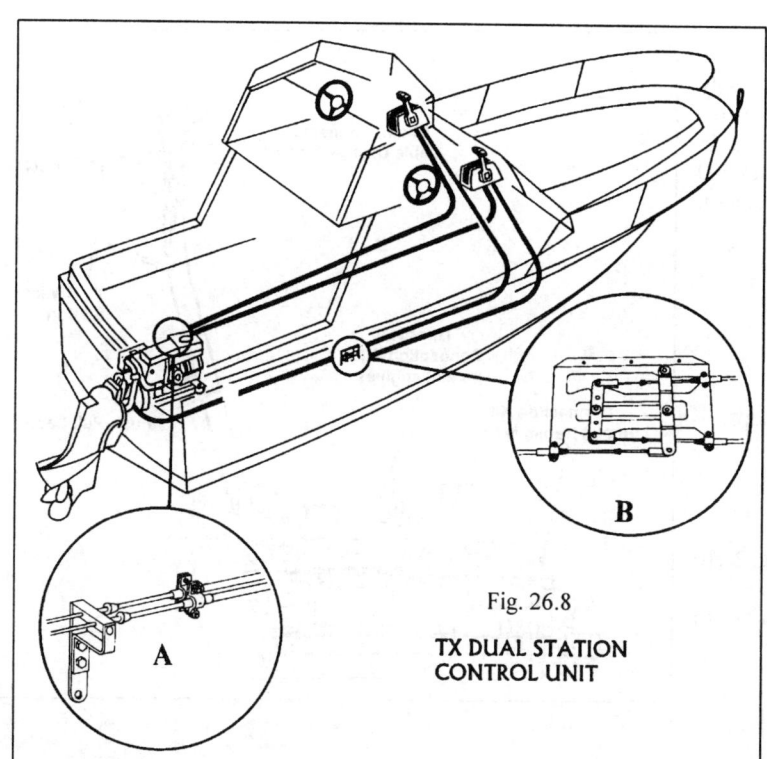

Fig. 26.8
TX DUAL STATION CONTROL UNIT

MULTIPLE PROPULSION CONTROL STATIONS – PARALLEL & SERIES

Many vessels have only a single propulsion control station in the wheelhouse for steering as well as engine controls. However, it is useful to have an outside control station for better view during activities such as docking and fishing.

There are two types of multiple (dual) station systems: PARALLEL AND SERIES. The parallel system is simpler and generally more efficient. Cables from each lever-control run directly to the engine clutch and throttle levers where they are connected with a parallel dual station kit.

In the series system, cables from the upper control station run to the lower station, where they are connected with a cable attachment kit. From there cables are run to the engine clutch and throttle levers, where they are connected with an engine connection kit. The series system is not as precise as the parallel system. It should be used only when a parallel installation is impracticable due to long cable runs and excessive or sharp bends.

The support brackets installed throughout control systems must be rigid. The cable-ends must be well aligned with the engine's throttle and clutch control levers to avoid binding.

DUAL STATION CONTROL UNIT - CABLE CONNECTIONS (TX Control System)

The illustration shows unit "A" connecting throttle cable to a control lever. A second lever must be provided at each station for the gearshift cable. Unit "B" is a mechanical device for connecting both cables to a single-lever control system. The XT device can be mounted vertically, horizontally or upside down. However, horizontal position is preferred.

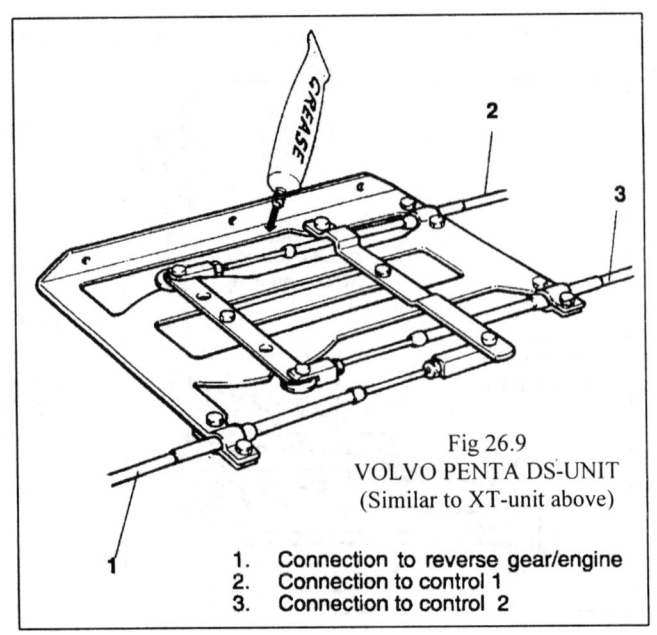

Fig 26.9
VOLVO PENTA DS-UNIT
(Similar to XT-unit above)

1. Connection to reverse gear/engine
2. Connection to control 1
3. Connection to control 2

TYPES OF PROPULSION CONTROL SYSTEMS

1. PUSH-PULL CABLE PROPULSION CONTROL SYSTEM

This is an economical, reliable and easily installed system for cable lengths of up to 10 metres in length. There should not be too many bends in the cable, and the bends should not be less than 200 mm in radius. Stiffness or binding in its operation could be caused by:

- Excessive bends
- Sharp bends
- Tight engine linkage
- Misaligned engine linkage
- Cable supports compressing cable
- Engine clutch lever hitting its limit stops.

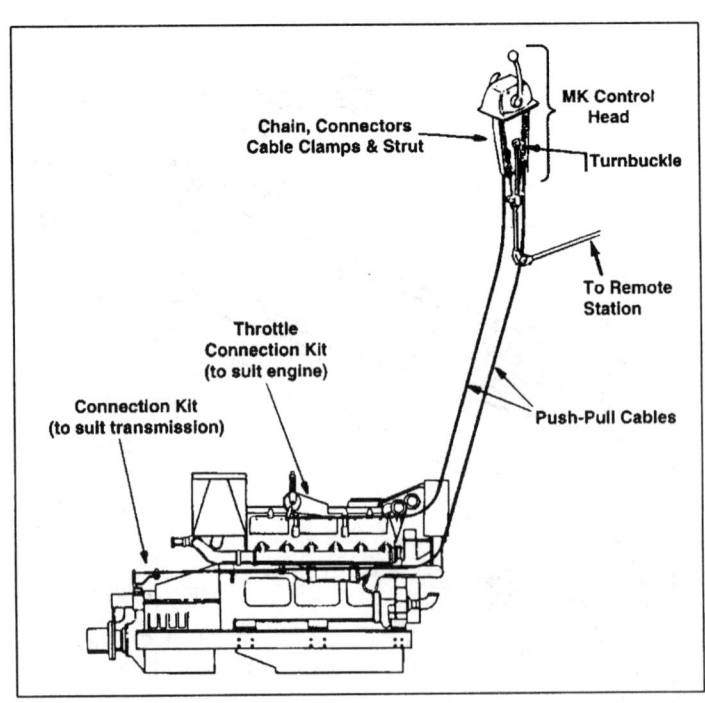

Fig. 26.10
PUSH-PULL CABLE SYSTEM
(*MK Control System*)

Chapter 26: Engine Operation, Alarms, Controls & Maintenance

2. CABLE-IN-TENSION PROPULSION CONTROL SYSTEM

This propulsion control system overcomes some of the problems of long cable runs and too many bends. It employs two cables in tension that run over pulleys mounted on anti-friction bearings. They are the "neutral throttle" and "shut off" cables. In order to keep the number of cables to a minimum it employs a single lever control system. A control gearbox to the governor and reverse gear is installed on the engine.

3. HYDRAULIC PROPULSION CONTROL SYSTEM

This propulsion control system offers smoother and more precise control. It can incorporate more than two control stations, and it works over a greater distances than the mechanical cable systems.

4. PNEUMATIC PROPULSION CONTROL SYSTEM

The pneumatic control system can be employed over long distances. 90 metre long air lines are not unrealistic. The only real limitation is the speed of response over very long air lines.

Other advantages of the pneumatic control system include:

- Ability to incorporate an unlimited number of control stations
- Ability to incorporate electronic logic into the system. It protects against the abuse of the propulsion system. (The logic circuitry is combined with a propulsion monitoring system in a cabinet in the engine room.).
- Ability to add a sequencing and timing device (See Engine Stall and Reversal Problem).

The disadvantages of the pneumatic control system are:

- A relatively heavy and expensive compressor and an air storage tank are required.
- The tank and lines require regular maintenance, such as draining of condensation.

5. ELECTRONIC PROPULSION CONTROL SYSTEM

An electronically controlled propulsion system is simpler and easier to install than the equivalent mechanical or hydraulic system. The forces needed to drive rheostats and switches are much less than those needed to operate cables or hydraulic cylinders.

The electronic control system for a conventional engine involves fitting an electric-to-mechanical converter box in the engine room. It converts electrical signals received from the control station into mechanical forces - generally via push-pull cables – which operate the engine

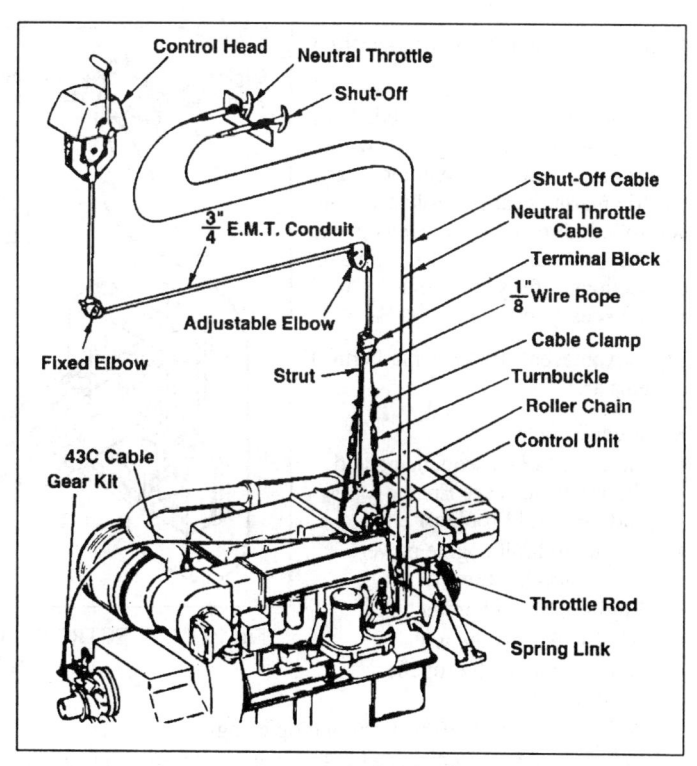

Fig. 26.11
CABLE-IN-TENSION SYSTEM (*MD 24 Control System*)

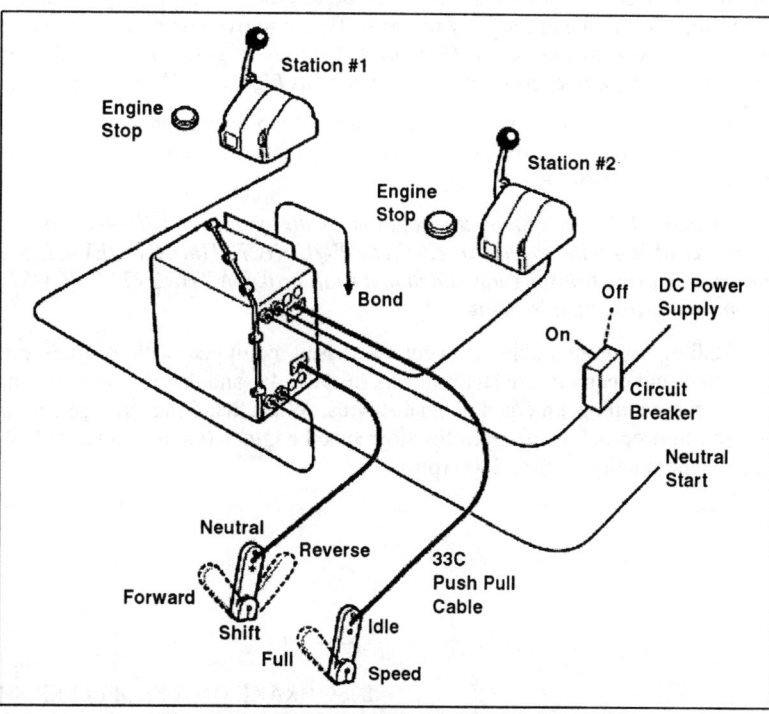

Fig. 26.12: ELECTRONIC CONTROL SYSTEM
FOR CONVENTIONAL ENGINE (*ZF Marine Micro Commander*)

throttle and the marine gear-shifting valve.

The converter box is not needed for controlling modern engines, whose electronic engine governors and electric marine gear control valves are controlled directly by the electronic control systems. With plug-in connections a fully electronic control system is easy to install.

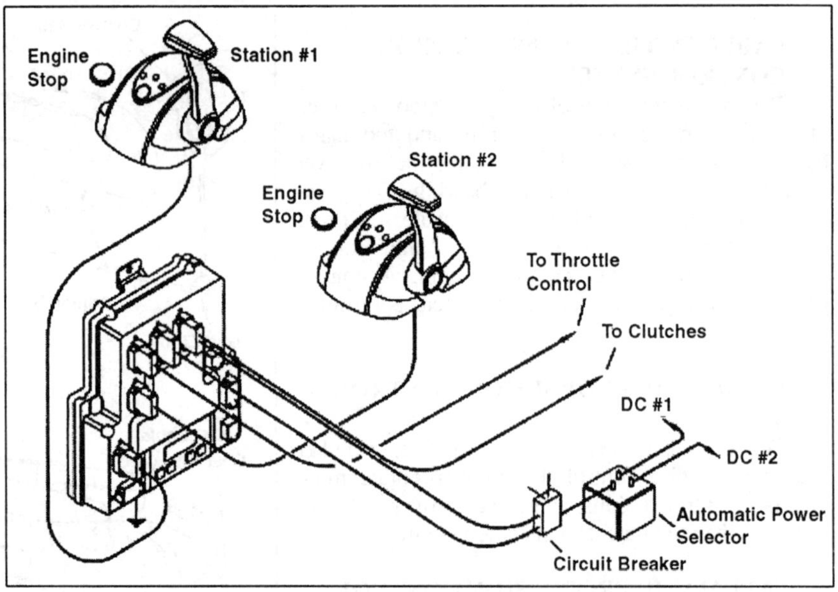

The advantages of electronic control systems are:

- Precise control
- Ability to add a sequencing and timing device (See Engine Stall and Reversal Problem)
- Ability to limit engine power during acceleration
- Engine overload protection
- Ability to integrate with controllable pitch propeller control systems
- Ability to share loads between multiple engines
- Effective over long distances between control station and engine
- Ability to add addition control systems after vessel completion

Fig. 26.13: ELECTRONIC CONTROL SYSTEM FOR ELECTRONIC ENGINE
(ZF Marine Cruise Command)

ENGINE STALL AND REVERSAL PROBLEM

The engine torque is usually sufficient to overcome the inertia of a slowly turning drive shaft and propeller. Therefore, during low speed manoeuvres vessels respond quite appropriately. However, during high speed manoeuvres, such as a crash stop, the engine torque may not be sufficient to overcome propeller and drive shaft inertia, marine transmission inertia and the slip stream torque. *(SLIP STREAM torque is generated in a freewheeling propeller, being turned by the water flowing past the hull. Slipstream torque can be as high as 75% of the engine's rated torque.)*

This may cause the engine to stall during high speed manoeuvres, unless the controls are prudently handled or the driveline is fitted with auxiliary manoeuvring devices, such as a shaft brake, an event sequencing timing device and a throttle boost system.

[A SHAFT BRAKE stops the rotation of the propeller whenever the transmission clutches are disengaged and the engine is at low idle speed. An EVENT SEQUENCE TIMING DEVICE sequences and times the actions of the engine governor, marine transmission clutch and shaft brake. A THROTTLE BOOST system increases the engine speed just before the marine gear clutch is engaged.]

Stalling is more likely to occur in vessels equipped with engines producing over 500 hp (375 kw), fixed pitch propellers and deep ratio reduction gears (usually 4:1 and deeper). Raising an engine's idle speed setting would increase its torque for handling emergency manoeuvres. When handling emergency manoeuvres, the engine speed should not be allowed to drop below 300 rpm for slow speed engines (rated at nominally 1200 rpm) or 400 rpm for high-speed engines (rated at nominally 1800 to 2300 rpm).

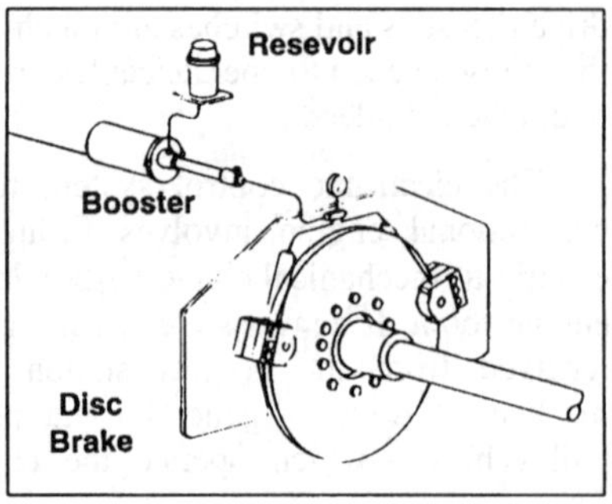

Fig 26.14
DISC BRAKE ON A PROPELLER SHAFT
(ZF Marine – Mathers Shaft Brake)

TOOLS, SPARES & LEAKS

Every vessel should have on board: bilge pump impeller, raw water pump impeller, spare filters, torch, screw driver, pliers, emery paper, electric tape, stainless steel wire, temporary hose and pipe repair tape, de-watering spray can, adjustable spanners, hacksaw, trouble lamp, hammer, filters, fuses, electrical cable, sea valve packing, stern gland packing, drive belts, hoses, fuses, bulbs, hydrometer, grease for stern tube, penetrating oil, distilled water, gasket sealer, spare spark plugs and plug spanner for petrol engines, etc. Other items - as appropriate to the motor and its use.

LEAKY PIPE –TEMPORARY REPAIRS
Wrap a sheet of rubber or soft gasket around the leak and secure with one or more pipe clips. Commercial repair tapes are also available for such jobs.

LEAKY JOINT OR FLANGE
Drain or isolate the line. Open and clean the joint, renew the gasket and retighten.

TROUBLESHOOTING

CAUTION:
- *When working with fuel system, switch off ignition and keep a fire extinguisher near-by.*
- *All checks and repairs must be done with the engine stopped.*
- *Run the engine room blower for 2 to 3 minutes before starting work.*

BEFORE REPAIRING ENGINE OR ELECTRICAL SYSTEM
➢ Secure long loose hair or clothing.
➢ Prevent tools falling down.
➢ Take precautions against generation of heat or sparks.
➢ Shut off all electricity and machinery.
➢ Take precautions against electricity or machinery being accidentally turned on.
➢ Make fire-fighting equipment ready and available.
➢ Gas free the vessel.
➢ Advise port authority regarding hot works.
➢ Secure vessel properly.
➢ Provide good lighting.
➢ Take weather into account.
➢ Supervise all work being carried out.

AFTER REPAIRING ENGINE
- Engine or engine room/space should be checked to ensure that all connections have been made, parts are replaced and tools removed.
- Motor should be turned over by hand to ensure that all moving parts move as intended.

BEFORE REMOVING A FUEL FILTER
➢ Turn off the fuel or disconnect the fuel line.
➢ Observe fire safety precautions.
➢ Provide for safe disposal of waste fuel.
➢ Clean the area around the filter before removing it - to avoid the risk of fuel system becoming contaminated.

AFTER REASSEMBLING A FUEL FILTER
- Test pipe connections and fuel bowl joint for possible fuel leakage.

ELECTRONIC DIAGNOSTIC SYSTEMS
In the event of a malfunction, some engines provide a diagnostic code and a brief description of the fault on an electronic data reader or a personal computer.

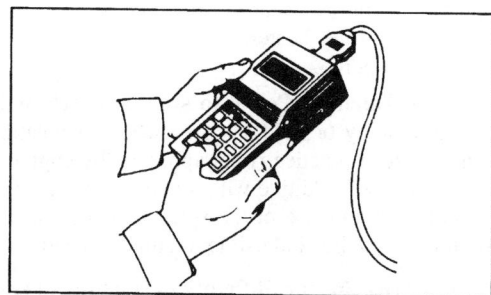

Diagnostic Codes — Marine Engines

TO READ CODES: Use diagnostic data reader or press and hold the diagnostic switch on the BBIM. The latter method will flash codes at the CEL.

Error Code #	Description	Error Code #	Description
11	Hand Throttle Speed Adj Lo Volt	43	Low Coolant Level
12	Hand Throttle Speed Adj Hi Volt	44	Engine Overtemperature
13	Coolant Sensor Lo Volt	45	Low Oil Pressure
14	Eng Temp Sensor Hi Volt	46	Low Battery Voltage
15	Eng Temp Sensor Lo Volt	47	Hi Fuel Pressure
16	Coolant Sensor Hi Volt	48	Lo Fuel Pressure
23	Fuel Temp Sensor Hi Volt	51	ECM Calibration Memory Failure
24	Fuel Temp Sensor Lo Volt	52	ECM A/D Fail
25	No Codes	53	ECM Nonvolatile Memory Failure
26	External Warning Switch Enabled	54	Tach Sync Circuit Fault
31	Fault on Auxiliary Output	55	Proprietary Data Link Circuit Fault
32	ECM Backup Fail	56	ECM A/D Fail
33	Turbo Bst Sensor Hi Volt	57	To Be Determined
34	Turbo Bst Sensor Lo Volt	58	Auxiliary Switch Circuit Fail
35	Oil Prs Sensor Hi Volt	61-68	Inj Response Time Long
36	Oil Prs Sensor Lo Volt	71-78	Inj Response Time Short
37	Fuel Prs Sensor Hi Volt	81	Crankcase Monitor - Hi Volt
38	Fuel Prs Sensor Lo Volt	82	Crankcase Monitor - Lo Volt
41	Timing Reference Sensor	84	Crankcase Pressure - Hi
42	Synchronous Ref Sensor	85	Engine Overspeed

Fig 26.15 (A) & (B)
DIAGNOSTIC DATA READER & DIAGNOSTIC CODES
(Detroit Diesel)

DIESEL ENGINE & GENERATOR PROBLEMS
(SOME ALSO APPLICABLE TO PETROL & LPG ENGINES)

It should be clear to the reader by now that diesel engines are internal combustion engines just like petrol and LPG engines, with similar fuel filters, pumps and cooling systems. They differ only in the method of fuel ignition. Diesel engines do not employ carburettors and spark plugs. As discussed earlier, the same principles apply to the working and maintenance of diesel-powered generators.

NO PROBLEM: DIESEL ENGINE STARTS EASILY & RUNS WELL

Leave it alone. Do not service the injectors or the pump in the hope of improving performance. Do not tinker with it as you might with a petrol engine in a motorcar.

ENGINE WON'T START
- Engine in gear
- Flat or disconnected battery
- No fuel (Bleeding usually necessary after refuelling)
- Fuel cock turned off
- Faulty fuel line
- Propeller is fouled with a rope or wire

DIESEL ENGINE DIFFICULT TO START

Cold engine: In older diesel engines, you may need to lift the overload (overspeed) trip to give extra fuel. In modern engines, a "FUEL RACK" fitted with a solenoid supplies a regulated amount of fuel depending on the engine temperature. Fitted on the fuel injection pump, it works in a similar manner to an automatic choke in petrol engines. As discussed earlier, some engines are fitted with an electric Heater Plug to the inlet manifold or a Glow Plug to each cylinder to provide additional heat to the combustion air during starting. Some are fitted with a jacket water heater. Still others are fitted with a cold start control on the fuel injection pump.

Fig 26.16: COLD START CONTROL ON INJECTION PUMP OF A SCANIA ENGINE

Use of Ether or Petrol to start a diesel engine: For a diesel engine that is really difficult to start, it is a common practice to spray or hold a rag soaked in a volatile liquid, such as ether or petrol, over the air intake. Unless absolutely necessary, such practice is not approved by engine manufacturers. It is a fire hazard, and a prolonged and excessive use of an alcohol-based additive will contaminate the oil and increase engine wear. There is also a risk of the rag getting sucked in, and a generous use of the spray or liquid may cause the engine to start with a scream. When an engine is frequently difficult to start, it is indicating a fault or a need for servicing.

Other reasons for difficulty in starting an engine include:
- Fouled propeller
- No fuel
- Dirty fuel filter
- Blocked fuel injectors
- Low battery charge or terminals corroded
- Air lock in the fuel line (Solution: Bleed it)
- Unsuitable engine oil (Solution: Drain & refill)
- Incorrect grade of fuel (Solution: Drain & refill)

ENGINE STARTS & THEN STOPS ALMOST IMMEDIATELY
➢ Shortage of fuel due to fuel being turned off
➢ Air in fuel line

BLACK SMOKY EXHAUST
- Engine overloaded due to dirty propeller or bottom. Solution: Reduce speed
- Unsuitable fuel
- Water in fuel. Solution: Change tank in use
- Incomplete combustion from too little air or too much fuel. There is a restriction in air supply due to choked air filter, blocked air intake or poor engine room ventilation. Solution: Change ducting to draw from cooler source. Clean filter, check & clear air intake & engine room ventilation.

NOTE: Engines overload when there is insufficient or poor combustion; or if there is a faulty fuel injector in one of the cylinders. The other cylinders thus become overloaded.

Chapter 26: Engine Operation, Alarms, Controls & Maintenance

BLUE SMOKY EXHAUST
- ➢ Burning oil. It is usually a sign of worn piston rings, cylinders or engine valves. Solution: Seek expert assistance.

WHITE SMOKY EXHAUST (BUT ENGINE RUNNING SMOOTH)
- Moisture in exhaust due to condensation in cold weather. Solution: No action necessary.

WHITE SMOKY EXHAUST (ENGINE RUNNING ERRATIC & WATER NEEDS REGULAR TOPPING UP)
- ➢ Steam or water in fuel or in combustion chamber due to a blown head gasket or cracked cylinder head. Solution: Seek expert assistance.

ENGINE MISFIRES OR SLOWS DOWN DURING OR AFTER A ROUGH TRIP
- This is usually the first sign of algae sediment in the fuel tank. A short-term solution is to change over the filter, and the tank if possible. But, the sediment would have to be removed by draining the fuel tank and flushing it. To get rid of the live algae, the fuel will have to be run through an electronic filter.

ENGINE MISFIRES
- ➢ Leaky inlet valve. Solution: Slow down to reduce speed, but don't stop engine until vessel in a safe place.
- ➢ A faulty atomiser (figures 21.26 in Chapter 21 and 23.5 in Chapter 23). Solution: In order to find the defective injector, operate the engine at fast idle speed. Loosen and tighten the union nut of the high pressure fuel line at each atomiser. When the union nut of the defective atomiser is loosened, it has little or no effect on the engine speed.

LOW OIL PRESSURE
- Low oil level. Solution: Check level with dip stick & top up.
- Oil leak. Solution: The leak may be into bilges or engine. Check bilge level.
- Choked oil strainer.
- Choked oil cooler.
- Suction bell uncovered due to engine tilt
- Worn piston rings

ENGINE OVER-REVS
- ➢ Propeller too small for the engine

ENGINE OVERLOADS
- ❖ Propeller too large for the engine. Also see "Black Smoky Exhaust"

ENGINE OVERHEATING

SUDDEN OVERHEATING
- ➢ Water inlet blocked
- ➢ Engine overloaded
- ➢ Water leakage
- ➢ Oil leakage
- ➢ Faulty thermostat. Solution: Change it
  (The faulty thermostat can be temporarily removed (after the engine cools down) to get back home)
- ➢ Air trapped in the system. Solution: Purge it

GRADUAL OVERHEATING
- Constriction in cooling circuit due to corrosion or dirt
- Worn pump impeller. Solution: Replace impeller
- Engine overloaded
- Oil cooler not working
- Incorrect fuel/oil mixture
- Propeller too large
- Shortage of lube oil, or wrong type
- Loose water pump belt. Solution: Tighten it
- Poor valve timing, ignition timing or injector timing. Solution: Seek expert assistance

  Eventually an overheated engine will seize up. If safe, disengage propeller until the system is cleared. If unable to clear the system and unable to stop due to unsafe sea conditions, run the motor at minimum revs and add more oil. This will help to control the temperature until the motor can be switched off and the problem solved.

ENGINE FAILS TO REACH NORMAL OPERATING TEMPERATURE
➢ Thermostat stuck open or incorrect type
➢ Temperature sender unit defective
➢ Temperature gauge defective

LOSS OF COOLANT
- Radiator leaking
- Radiator cap defective or incorrect type
- Loose or damaged hose connections
- Water pump leaking
- Water jacket plugs leaking
- Cylinder head gasket leaking:
    o Faulty sleeve sealing
    o Incorrect head tightening
    o Warped head or block faces

LOSS OF POWER
➢ Dirty air filter
➢ Dirty fuel filter
➢ Choked exhaust
➢ Faulty fuel injector pump
➢ Propeller cavitation: Propeller is turning in aerated water. If it is too close to the surface, it will vibrate and the engine will race

UNABLE TO ACHIEVE NORMAL SPEED
- Engine started on overload.
- Fuel system not primed.
- Insufficient fuel supply. Solution: Clean fuel filter. Fuel pump may need adjustment - seek expert assistance.
- Propeller is fouled with a rope or wire.

ENGINE STOPS
➢ Fuel tank empty.
➢ Air in fuel.
➢ Water in fuel.
➢ Blocked or choked fuel filter.
➢ Blocked nozzle.
➢ Overheated. Solution: Prevent hot air re-circulating.
➢ Overloaded. Solution: Allow engine to cool slowly. Turn it over by hand to ensure it is not seized. Restart & check loading.

WATER IN FUEL *will gather in the bottom of fuel tanks or filters. Take a sample in a clear glass container. If water is present, there will be a clear line of demarcation. If vibration has caused fuel and water to emulsify, a water sensing paste is available to detect it. The same paste can be used to detect presence of water in lube oil.*

FUEL SHORTAGE ON VOYAGE
- Reduce speed to maintain steady progress. Operating range increases at slower speed. Switch off excess auxiliaries.

SUDDEN INCREASE IN LEVEL OF LUBE OIL
➢ Fuel leaking into sump. Solution: Seek expert assistance.
➢ Cooling water leaking into sump. Solution: Seek expert assistance.

NOISY ENGINE
- Worn bearings. Solution: Seek expert assistance.

VESSEL EXPERIENCING UNUSUAL VIBRATIONS AFTER HITTING A SUBMERGED OBJECT
➢ Propeller or shaft damaged. Engine may also be out of alignment. Solution: Reduce engine revs to acceptable vibration level. Return home for inspection & repairs.

Chapter 26: Engine Operation, Alarms, Controls & Maintenance

SEA WATER PUMP STOPS WORKING
- Drive belt out of place, loose or broken. Solution: Adjust it.
- Pump choked with foreign matter.
- Impeller damaged - possibly due to pump running dry.
- Pump drive failure. Solution: Seek expert assistance.

LOSS OF PUMP SUCTION
- Blocked suction strainer.
- Sheared impeller drive.
- Vessel rolling. Solution: Alter course to reduce rolling.
- Air leak in valve gland or pump gland. Solution: Change gland.
- Damaged suction pipe. Solution: Change pipe.
- Drive belt slipping.
- Excessive clearance in pump impeller. Solution: Adjust clearance.

OPERATING RANGE & FUEL CALCULATIONS
- The "standard" fuel consumption for a vessel does not take into account unusual circumstances that may be experienced at sea, such as rough weather and poor engine performance.
- Always carry more fuel than required for the intended voyage. 20% reserve is recommended.
- Generally speaking, slowing down can increase the operating range. In case of fuel shortage when at sea, reduce speed to maintain steady progress (unless otherwise recommended by the manufacturer) and switch off excess auxiliaries.

SEE CHAPTER 17.2 FOR SPEED, TIME & DISTANCE CALCULATIONS

EXERCISE 1
A vessel's fuel consumption is 24 litres per hour at 16 knots. Her fuel tanks capacity is 250 litres. Calculate her range at 16 knots.

SOLUTION:

Fuel on board will last for $= \dfrac{\text{FuelCapacity}}{\text{FuelConsumption}} = \dfrac{250}{24} = 10.4 \text{ hours}$

Distance (Range) = Speed x Time = 16 x 10.4 = 166.4 nautical miles.

EXERCISE 2
A vessel consumes 5 litres per hour at 8 knots. Calculate the minimum amount of fuel required for a voyage of 10.5 hours at that speed.

SOLUTION:
Total fuel required = hourly consumption x the number of hours to travel
= 5 x 10.5 = 52.5 litres

EXERCISE 3
The external measurements of a rectangular fuel tank are length 3 metres, width 1.5 metres and depth 1.0 metres. The thickness of the tank material is 25 mm. Calculate the tank capacity in litres.

SOLUTION
25 mm = 2.5 cm, which divided by 100 = 0.025 metres

Tank's internal measurements are as follows:
L = 3.0 – 0.025 = 2.975 metres
W = 1.5 – 0.025 = 1.475 metres
D = 1.0 – 0.025 = 0.975 metres
Volume of the tank = L x W x D = 2.975 x 1.475 x 0.975 = 4.28 cubic metres

1 cubic metre = 1000 litres
Therefore, the tank capacity = 4.28 x 1000 = 4280 litres

EXERCISE 4

Your vessel's fuel consumption is 20 litres per hour at 14 knots and your genset consumes 3 litres per hour. Calculate the amount of fuel required for a round voyage of 400 nautical miles at that speed.

Total steaming time at 14 knots: $\text{Time} = \dfrac{\text{Distance}}{\text{Speed}} = \dfrac{400}{14} = 28.57 \text{ hours}$

Hourly consumption = 20 + 3 = 23 litres
Total fuel required = hourly consumption x Total number of hours
= 23 x 28.57 = 657.11 litres

EXERCISE 5

A vessel has two identical fuel tanks of rectangular cross-section. However, a bottom rectangular section along the entire length of one tank has been removed. The measurements of the latter tank are as follows: Length 2.2 metres; Width on top 1.0 metre, bottom 0.3 metres; Depth inside 1.6 metres, outside 1.2 metres. Calculate the vessel's total fuel capacity in litres.

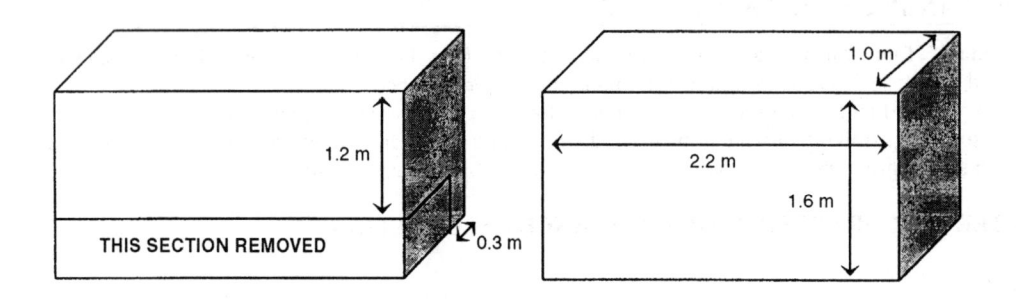

Volume of one complete tank = L x W x H = 2.2 x 1.0 x 1.6 = 3.52 cubic metres.
Volume of two complete tanks = 3.52 x 2 = 7.04 cubic metres.
Measurements of the section removed:
L = 2.2 m
W = 1.0 – 0.3 = 0.7 m
H = 1.6 – 1.2 = 0.4 m.
Volume of the section removed = 2.2 x 0.7 x 0.4 = 0.62 cubic metres
Therefore, volume of the tanks = 7.04 – 0.62 = 6.42 cubic metres.
Therefore, fuel capacity = 6.42 x 1000 = 6420 litres.

ENGINE MAINTENANCE

BE ABLE TO DO....

Every boat operator should at least be able to change the filter and primer bowl; Clean and change spark plugs (if fitted); Check for spark; Check and replace fuses; Change the propeller; Clean battery terminals.

The basic areas needing regular attention are:

- **WATER PUMP**: Replace regularly specially after operating in the shallows and stirring sand. The water pump impeller may deteriorate if not used for lengthy periods.
- **FUEL FILTERS & LINES**: Filters become clogged and lines hardened with age and exposure.
- **PROPELLERS**: The propeller bush can fail specially if it has been dredging considerable sand or trying to cut rocks.
- **SPARK PLUGS** (in petrol engines): Old plugs may cause difficulty in starting the engine or they break down completely.
- **GEARBOX OIL**: Gearbox seal can leak due to being damaged by a fishing line. Water in the gearbox will eventually cause it to fail. Regular oil change will keep it in good condition.

Chapter 26: Engine Operation, Alarms, Controls & Maintenance

DIESEL ENGINE MAINTENANCE & INSPECTION SCHEDULE
(As recommended by Caterpillar for diesel engines)

DAILY
Engine Crankcase - Check oil level
Cooling System - Check coolant level
Marine Gears - Check oil level
Water Separator (If Equipped) - Drain
Fuel Tank - Drain water
Leaks & Loose Connections - Inspect
Air Cleaner Indicator - Check
Clutch - Lubricate fittings

EVERY 50 SERVICE HOURS OR TWO WEEKS
Zinc Rods - Inspect
Clutch - Adjust/Lubricate

EVERY 200 OR 250 SERVICE HOURS*
Scheduled Oil Sampling - Test oil
Engine Oil & Filter - Change oil

EVERY 250 SERVICE HOURS*
Cooling System - Add conditioner
Marine Gear - Change oil & wash screen
Radiator Fins (If Equipped) - Inspect
Hoses - Inspect/Replace
Fuel Filters - Replace/Clean
PVC Valve - Check diaphragm & hoses
Air Cleaner - Clean/Replace elements
Batteries - Check
Alternator Belts - Inspect/Adjust
Oil Fumes Separator Canister (If Equipped) - Clean/Replace
Crankcase Breather - Clean

EVERY 1200 SERVICE HOURS OR ONE YEAR*
Engine Valve Lash - Adjust
Water Separator - Inspect/Replace
Engine Protective Devices - Check
Thermostat - Replace
Marine Gear - Change oil & clean filters
Marine Gear Output Shaft Seal - Lubricate fitting
Sea Water Strainer & Pump - Inspect

EVERY 2400 SERVICE HOURS OR TWO YEARS*
Fuel injection Nozzle - Test
Governor - Check
Fuel Pump/Governor Housing - Drain
Fuel Ratio Control - Check/Adjust
Marine Engine Analysis Report
Cooling System - Change Coolant

EVERY 3600 SERVICE HOURS*
Engine Mounts - Inspect/Check
Air Compressor - Inspect/Rebuild or Exchange
Electric Starting Motor - Inspect/Rebuild or Replace
Turbocharger - Inspect/Check
Vibration Damper - Inspect/Check

EVERY 6000 SERVICE HOURS OR 113700 L
Complete Overhaul – as recommended.

AFTER FAILURE OVERHAUL
As recommended by manufacturer

* First Perform Previous Service Hour Items.

NOTE: The Indirect Fuel Injected Engines require more frequent oil changes - roughly at half the interval of direct injection engines.

AIR COMPRESSOR INSPECTION

Some vessels are fitted with an air compressor. Before carrying out an inspection, service or replacement, its air content must be released until the air pressure is zero. The compressor service usually involves removing excess carbon and water deposits from its discharge line.

ENGINE LOG BOOK ENTRIES

REPAIRS & SERVICE
- What was repaired?
- When and who did it?
- Any new components?
- Details of normal & acceptable running conditions; i.e., oil pressure, cooling water temperature.

OPERATIONS
➢ Pre-sailing checks.
➢ Regular record of all performance readings. For example, oil pressure gauges and temperatures.
➢ Record of expendable items. For example, fuel, cooling water, lubricating oil, grease, etc.
➢ Observations of anything unusual. For example, oil leaks, hot spots, temperatures, noises, vibration, etc.
➢ Arrival entries of expendable items and need for service, if any.

CHAPTER 26: QUESTIONS & ANSWERS

CHAPTER 26.1: ENGINE OPERATION & TROUBLESHOOTING

1. Before starting an inboard engine at the start of a journey, it is possible to check the correct:

    (i) oil level.
    (ii) water level.
    (iii) fuel level.
    (iv) temperature.

    The correct answer is:
    (a) all except (i).
    (b) all except (ii).
    (c) all except (iii).
    (d) all except (iv).

2. Running a diesel engine on light load or at low jacket temperature will:

    (a) prolong its life.
    (b) extend maintenance periods.
    (c) result in premature carbon build up.
    (d) not charge the battery.

3. The safety devices provided for inboard marine engines include:

    (i) Remote emergency stop control.
    (ii) Cooling water temperature alarm.
    (iii) Lube oil pressure alarm.
    (iv) Underspeed cut-out.

    The correct answer is:
    (a) all except (i).
    (b) all except (ii).
    (c) all except (iii).
    (d) all except (iv).

4. The good housekeeping rules for the engine compartment are:

    (i) Keep bilges clean and dry.
    (ii) Minimise ventilation.
    (iii) Keep fire watch around the exhaust system.
    (iv) Store oily rags in a container.

    The correct answer is:
    (a) all except (i).
    (b) all except (ii).
    (c) all except (iii).
    (d) all except (iv).

5. In an internal combustion engine, the following is an indication of worn piston rings:

    (a) blue smoke.
    (b) engine misfiring.
    (c) black smoke.
    (d) engine back firing.

6. Lack of compression in an internal combustion engine may exist due to:

    (a) bad fuel mix.
    (b) low battery.
    (c) low engine oil.
    (d) worn piston rings.

7. A diesel engine may fail to start due to:

    (i) dirty spark plugs.
    (ii) blocked injectors.
    (iii) fuel cock turned off.
    (iv) Flat battery.

    The correct answer is:
    (a) all except (i).
    (b) all except (ii).
    (c) all except (iii).
    (d) all except (iv).

8. While underway offshore, the vessel's engine starts to stutter or misfire. The vessel should:

    (a) slow down.
    (b) stop immediately.
    (c) proceed to a safe place at full speed.
    (d) go astern for about five minutes.

9. In case of a shortage of fuel when at sea, a vessel's cruising range can be increased as follows:

    (i) increase speed.
    (ii) stop one engine & lift it out of water.
    (iii) switch off auxiliaries.
    (iv) exercise good seamanship.

    The correct answer is:
    (a) all except (i).
    (b) all except (ii).
    (c) all except (iii).
    (d) all except (iv).

10. A vessel's engine spins but won't start. It may be due to:

    (i) lack of fuel.
    (ii) flat battery.
    (iii) water in the fuel.
    (iv) air in the fuel line.

    The correct answer is:
    (a) all except (i).
    (b) all except (ii).
    (c) all except (iii).
    (d) all except (iv).

11. Blue smoke from the exhaust of an engine indicates:

    (a) insufficient combustion.
    (b) excessive cylinder lubrication.
    (c) condensation in the exhaust pipe.
    (d) overloaded engine.

Chapter 26: Engine Operation, Alarms, Controls & Maintenance

12. In trying to start a marine diesel engine, it turns over but does not fire. The reasons include:

    (i) blocked water intake.
    (ii) engine stop in the stopped position.
    (iii) lack of compression.
    (iv) poor cranking speed.

    The correct answer is:
    (a) all except (i).
    (b) all except (ii).
    (c) all except (iii).
    (d) all except (iv).

13. In a four-stroke engine, lack of compression may exist due to:

    (a) intake or exhaust valves not sealing properly.
    (b) high engine oil temperature.
    (c) low engine oil.
    (d) impurity in fuel.

14. After starting, a diesel engine may fail to develop adequate power due to:

    (i) leaking piston rings.
    (ii) dirty spark plugs.
    (iii) faulty fuel injectors.
    (iv) intake or exhaust valves not sealing properly.

    The correct answer is:
    (a) all except (i).
    (b) all except (ii).
    (c) all except (iii).
    (d) all except (iv).

15. A diesel engine starts but then stops almost immediately due to:

    (a) air in fuel line.
    (b) worn piston rings.
    (c) shortage of lubrication oil.
    (d) restricted water supply.

16. A diesel engine may give off black smoke due to:

    (i) engine overload.
    (ii) restricted inlet ports.
    (iii) air in fuel line.
    (iv) restricted air filter.

    The correct answer is:
    (a) all except (i).
    (b) all except (ii).
    (c) all except (iii).
    (d) all except (iv).

17. A diesel engine may overheat due to:

    (i) faulty thermostat.
    (ii) shortage of lubrication oil.
    (iii) damaged pump impeller.
    (iv) incomplete burning of fuel.

    The correct answer is:
    (a) all except (i).
    (b) all except (ii).
    (c) all except (iii).
    (d) all except (iv).

18. A diesel engine may fail to develop full power due to:

    (i) restricted air supply to cylinders.
    (ii) engine overloaded.
    (iii) faulty spark plugs.
    (iv) blocked air intake.

    The correct answer is:
    (a) all except (i).
    (b) all except (ii).
    (c) all except (iii).
    (d) all except (iv).

19. To start a reluctant diesel engine when cold or with low batteries, it is recommended to:

    (a) use ether spray over the air intake.
    (b) use petrol soaked rag over the air intake.
    (c) bleed the fuel system.
    (d) use the glow plugs.

20. A dirty air filter:

    (a) causes fuel starvation in diesel engine.
    (b) causes fuel starvation in petrol engine.
    (c) can be washed or blown with air.
    (d) increases engine revs.

21. An air shut down valve:

    (a) is not fitted on diesel engines.
    (b) shuts off exhaust valves.
    (c) must be manually operated.
    (d) needs periodic testing & resetting.

22. The camshaft belt should be replaced when:

    (a) it starts to deflect.
    (b) the belt surface is roughened.
    (c) it deflects about 10 mm.
    (d) none of the choices stated here.

23. The torsional vibration calculations are performed on engines in order to establish:

    (a) the type of vibrations in engine beds.
    (b) the safe range of propeller vibrations.
    (c) torsion in a propeller shaft.
    (d) the safe range of operating speeds of engines.

24. Overfilled engine lube oil:

    (a) ensures better lubrication.
    (b) creates unduly excessive pressure.
    (c) compensates for any loss of oil.
    (d) keeps the engine cleaner.

25. If an inboard diesel engine has not been run for several weeks:

    (a) air may have entered the filter housing.
    (b) fuel filter would be nice & clean.
    (c) it would be easier to start.
    (d) it may have flooded with fuel.

26. On starting an inboard diesel engine, the Low Oil Pressure Alarm:

    (a) will not sound.
    (b) may sound for a few seconds.
    (c) will sound only if the pressure is low.
    (d) will sound only if it is sticky.

27. The weakest link in any instrumentation system is the:

    (a) switch.
    (b) transducer.
    (c) wiring.
    (d) joints.

28. To facilitate cold starting, diesel engines are fitted with:

    (a) a glow plug on the inlet manifold.
    (b) an electric heater plug on each cylinder.
    (c) a fuel rack button.
    (d) the facility to lift the overload trip to give the engine extra fuel.

29. The use of ether or petrol to start a diesel engine:

    (a) is generally approved by manufacturers.
    (b) is a fire hazard.
    (c) does not increase engine wear.
    (d) starts the engine without screaming.

30. The steps before starting a diesel engine do not include:

    (a) open raw water valve, if fitted.
    (b) open fuel supply valve.
    (c) engage marine gear.
    (d) open fuel return valve.

31. After starting a diesel engine, you do not need to inspect:

    (a) the condition of the fuel filter.
    (b) the overside cooling water discharge.
    (c) for any fluid leaks.
    (d) for any air leaks.

32. While at sea, a DFI diesel engine misfires. You checked and found all fuel injectors were in working order. The reason must therefore be a leaky inlet valve. You should:

    (a) proceed at full speed to a safe place.
    (b) reduce the air intake.
    (c) stop engine without further delay.
    (d) proceed at reduced speed.

33. The engine fails to reach normal operating temperature. The possible reason is:

    (a) thermostat stuck open.
    (b) radiator leaking.
    (c) water pump leaking.
    (d) water jacket plug leaking.

34. Which of the following will not cause loss of engine power?

    (a) Dirty air filter.
    (b) Dirty fuel filter.
    (c) Leaking water pump.
    (d) Choked exhaust.

35. The reason for an engine to become noisy is likely to be:

    (a) worn bearings.
    (b) fuel leaking into sump.
    (c) cooling water leaking into sump.
    (d) blocked suction strainer.

36. A DFI diesel engine misfires due to a faulty atomiser. To find the defective injector, the engine is operated at fast idle speed, and the union nut of the high-pressure fuel line at each atomiser is loosened and tightened. When the union nut of the faulty atomiser is loosened, it will have the following effect on the engine:

    (a) Speed up the engine.
    (b) Little or no effect on the engine speed.
    (c) Slow down the engine.
    (d) None of the choices stated here.

37. To solve the problem of blue smoky exhaust in an engine, you would:

    (a) change the fuel tank in use.
    (b) check & clean the air intake.
    (c) check propeller for obstruction.
    (d) check the piston rings.

38. In case of temporary white smoky exhaust after starting an inboard engine, you should:

    (a) change the fuel tank in use.
    (b) seek expert assistance.
    (c) take no action.
    (d) check & clean air intake.

39. The white smoky exhaust in an inboard engine running erratic and needing regular topping up of water is an indication of:

    (a) water in fuel.
    (b) blown head gasket.
    (c) faulty water pump.
    (d) blocked air intake.

Chapter 26: Engine Operation, Alarms, Controls & Maintenance

40. An engine misfires or slows down during or after a rough trip. It is a sign of:

    (a) algae sediment in the fuel tank.
    (b) water in fuel.
    (c) dirty air filter.
    (d) faulty cooling system.

41. Which of the following may not cause black smoky exhaust in an engine?

    (a) Burning oil.
    (b) Choked air filter.
    (c) overloaded engine.
    (d) poor engine room ventilation.

42. Which of the following statements is incorrect with regard to slobber leaking from the exhaust joints?

    (a) Fuel slobber is an indication of combustion problems
    (b) The presence of slobber is not always an indication of engine problem.
    (c) Slobbering can be minimised by operating with little or no load.
    (d) Worn valve guides, piston rings or turbocharger seals can cause oil slobber.

43. In order to get home, a sticky thermostat on a marine diesel engine:

    (a) may be temporarily removed.
    (b) should be greased.
    (c) should never be removed.
    (d) should be reversed in direction.

44. After starting a marine diesel engine the oil pressure gauge reads very high. It could be due to the:

    (a) oil level in the sump being too high.
    (b) fuel leaking into the sump .
    (c) oil pressure relief valve being stuck.
    (d) oil filter bypass valve not working.

45. The water flow at the engine overboard discharge has been slowly reducing in volume for some time. It could be due to:

    (a) the change in seawater temperature.
    (b) the pump impeller becoming worn.
    (c) an air lock in the seawater pick up.
    (d) raw water leaking into the cooling jacket.

46. After starting an inboard engine, an alarm continues to sound. The most likely cause is:

    (a) dirty fuel supply.
    (b) engine overheating.
    (c) engine overloading.
    (d) low oil pressure.

47. With its throttle stuck in the open position, the quickest way to stop a small diesel engine is to:

    (a) close the fuel supply.
    (b) smother the air intake.
    (c) turn off the ignition switch.
    (d) apply the shaft brake.

48. Running a diesel engine at light loads on a regular basis in a yacht:

    (a) keeps it in good order.
    (b) promotes glazing in cylinders
    (c) prevents carbon build up in cylinders.
    (d) allows it to breathe

49. Idling speed of diesel engines can be reduced below the manufacturer's settings:

    (a) in all engines.
    (b) only in high-speed engines.
    (c) only in Square engines.
    (d) only in old, long stroke, slow speed engines.

CHAPTER 26.2: TURBOCHARGED ENGINES

50. Shutting down a turbo charged engine requires:

    (a) no special precaution.
    (b) some speed consideration.
    (c) priming the lubrication system.
    (d) aftercooling.

51. For a short period after first starting a turbocharged engine, the engine speed should be:

    (a) kept to a minimum.
    (b) increased to maximum.
    (c) kept constant.
    (d) matched with the turbocharger speed.

52. A turbocharged engine should be:

    (a) run at idle speed for a while before shut down.
    (b) shut down before shutting of the turbocharger.
    (c) run at high speed for a while before shut down.
    (d) starved of air for shut down.

53. If a turbocharged engine has not been started for a long time:

    (a) run it at full throttle for a while.
    (b) switch off its turbocharger.
    (c) start the turbocharger first.
    (d) prime its lubrication system before starting.

CHAPTER 26.3: ENGINE STARTING SYSTEMS

54. In relation to electric engine starting systems, which statement is <u>incorrect</u>?

    (a) Use as large a cable as possible.
    (b) Do not use a battery-isolating switch that causes spikes.
    (c) Cranking an engine continuously for more than 30 seconds can overheat the starter motor.
    (d) Battery-powered starter motors provide engine mounted push button start.

55. In relation to pneumatic engine starting systems, which statement is <u>incorrect</u>?

    (a) A small quantity of alcohol can be added to the starter tank.
    (b) Oil and water traps are fitted at the lowest parts of the piping.
    (c) The oil and water traps should be drained daily.
    (d) The air pipe should be rigidly connected to the starter motor.

56. In relation to hydraulic engine starting systems, which statement is <u>incorrect</u>?

    (a) It is more common than electric or pneumatic system.
    (b) It has a high cranking speed.
    (c) Its recharging pump is quite small.
    (d) Hydraulic charging time is not long.

CHAPTER 26.4: FUEL CALCULATIONS

57. A vessel's fuel consumption is 10 litres per hour at 8 knots. Her fuel tanks can hold 230 litres. How far can the vessel travel at that speed?

    (a) 287.5 miles.
    (b) 184 miles.
    (c) 2300 miles.
    (d) 1840 miles.

58. A vessel's fuel consumption is 12 litres per hour at 14 knots. Her fuel tanks can hold 340 litres. How far can the vessel travel at that speed?

    (a) 340 miles.
    (b) 168 miles.
    (c) 291.4 miles.
    (d) 396.7 miles.

59. A vessel's voyage time is 20 hours. She consumes 10 litres per hour at 14 knots. Allowing for 20% reserve, she would need to carry the following amount of fuel.

    (a) 280 litres.
    (b) 200 litres.
    (c) 240 litres.
    (d) 160 litres.

60. A vessel's voyage time is one and a half days. She consumes 8 litres per hour at 11 knots. Without allowing for 20% reserve, she would need to carry the following amount of fuel.

    (a) 288 litres.
    (b) 88 litres.
    (c) 240 litres.
    (d) 345,6 litres.

61. A cubical fuel tank's external side measures 1.5 metres, and metal thickness 20 mm. Its fuel capacity in litres is:

    (a) 2197 litres.
    (b) 4440 litres.
    (c) 324 litres.
    (d) 3240 litres.

62. A vessel has two identical fuel tanks of rectangular cross-section. However, a bottom section of each tank has been removed. The rectangular measurements of each tank are 2.0 x 1.5 x 0.9 metre. The measurements of each of the two sections removed are 0.5 x 0.9 x 0.4 metres. Calculate the vessel's total fuel capacity in litres:

    (a) 3600 litres.
    (b) 3240 litres.
    (c) 1980 litres.
    (d) 1620 litres.

63. A vessel's fuel tank is of rectangular cross-section of 1.8 m x 1.2 m x 0.8 m., and of metal thickness of 0.025 m. However, a bottom rectangular section of internal measurements of 0.35 cubic metres has been removed. Calculate its fuel capacity in litres:

    (a) 1616 litres.
    (b) 1966 litres.
    (c) 1378 litres.
    (d) 1266 litres.

CHAPTER 26: ANSWERS

1 (d), 2 (c), 3 (d), 4 (b), 5 (a),
6 (d), 7 (a), 8 (a), 9 (a), 10 (b),
11 (b), 12 (a), 13 (a), 14 (b), 15 (a),
16 (c), 17 (d), 18 (c), 19 (d), 20 (c),
21 (d), 22 (b), 23 (d), 24 (b), 25 (a),
26 (b), 27 (b), 28 (d), 29 (b), 30 (c),
31 (a), 32 (d), 33 (a), 34 (c), 35 (a),
36 (b), 37 (d), 38 (c), 39 (b), 40 (a),
41 (a), 42 (c), 43 (a), 44 (c), 45 (b),
46 (d), 47 (b), 48 (b), 49 (d), 50 (b),
51 (a), 52 (a), 53 (d), 54 (a), 55 (d),
56 (a), 57 (b), 58 (d), 59 (c), 60 (a),
61 (d), 62 (b), 63 (d)

Chapter 27

OUTBOARD ENGINES

Outboards come in various sizes, ranging from one to eight cylinders. Outboard engines steer well ahead and astern when in gear, but not when in neutral. Being very responsive to throttle movements, they are extremely manoeuvrable.

TWO vs. FOUR STROKE OUTBOARDS

The two stroke outboards are gradually being replaced by FOUR STROKE outboards, which offer all the advantages of a modern engine:
- Electronic fuel injection
- Quicker starting. (The crankcase doesn't have to primed with fuel/air mixture)
- Oil not required to be mixed in fuel (thus, less fouling of spark plugs and more reliability)
- No carburettor (thus no gummy residue left behind by the evaporated fuel).
- 50% saving on fuel (therefore, less fuel to be carried) (However, the higher capital cost of the motor is offset by the fuel saving only for those clocking up at least 80 hours a year. Most trailer-sailors do about 20 hours.)
- Less air pollution (because they don't burn oil mixed in petrol)
- Softer and quieter exhaust (because they develop maximum torque at low revs)

However, regulations permitting, trailer-sailors may still find two-stroke outboards more economical, as do the yachties for yacht tenders. TWO-STROKE outboards are:

- Cheaper to purchase
- Cheaper to insure
- Possibly cheaper to service (No oil and filter changes, and no valve adjustment)
- Simpler
- Lighter (higher power-to-weight ratios) in most cases (Unlike a 4-stroke, a 2-stroke motor does not require mechanisms to operate inlet and exhaust valves through the cylinder head)

SINGLE vs. TWIN CYLINDER MOTORS

Most people are familiar with the one-cylinder petrol lawn mowers. Such engines are simpler to build than twin cylinders. They are also cheaper, smaller and probably marginally lighter, despite needing a heavier flywheel.

The twin cylinder engines, on the other hand, suffer less vibration and require less force to start. Being able to tolerate misfiring by one cylinder, they are also more reliable.

Two-stroke engines under 4 hp usually have only one cylinder, and between 4 and 8 hp they may have either one or two cylinders. One-cylinder 4-stroke engines are usually available only in 2 to 5 hp models. Above 8 hp, both the 2-stroke and 4-stroke engines usually have two cylinders.

OTHER CONSIDERATIONS

LEG LENGTH
Outboards are manufactured with various shaft lengths to suit different transom heights. When choosing, consider the variety of transoms and drafts where the outboard may be mounted during its life. It is better to err on the deeper side than to have the propeller out of water.

POWER GENERATORS IN OUTBOARDS
An outboard of about 5 hp and above can be set up to produce un-regulated power to light up navigation lights. Regulated 12-volt outputs are available only on larger or longer shaft motors.

GEAR SHIFT
Very small engines do not usually have a clutch or reverse gear. They always start in gear, and the reverse propulsion is attained by rotating the motor 180°. Provided the motor is not accidentally started or manoeuvred at high throttle setting, their idling thrust is not large enough to cause a problem.

Larger motors incorporate a dog clutch and gearing in the leg. They can be engaged into neutral, forward or reverse.

Most modern engines also incorporate a 'no-start-in-gear' safety mechanism.

CONTROLS

The main outboard controls are throttle, stop and gearshift. The choke is usually automatic. The controls are located either on or beside the powerhead cowling or in the tiller. The cowling location requires two hands to simultaneously steer and adjust a control. The controls in the tiller are more desirable, but they have to be able to pivot vertically when the motor is tilted.

Engines upwards of about 10 hp, particularly those with an electric start, can be fitted with permanently installed remote control via cables and electric wires. Outboards up to about 40 hp are also fitted with a recoiling rope start.

EXHAUST GAS RE-BURNING SYSTEM

Modern engines achieve cleaner exhaust by higher combustion efficiency and by re-burning unburnt fuel in exhaust gases.

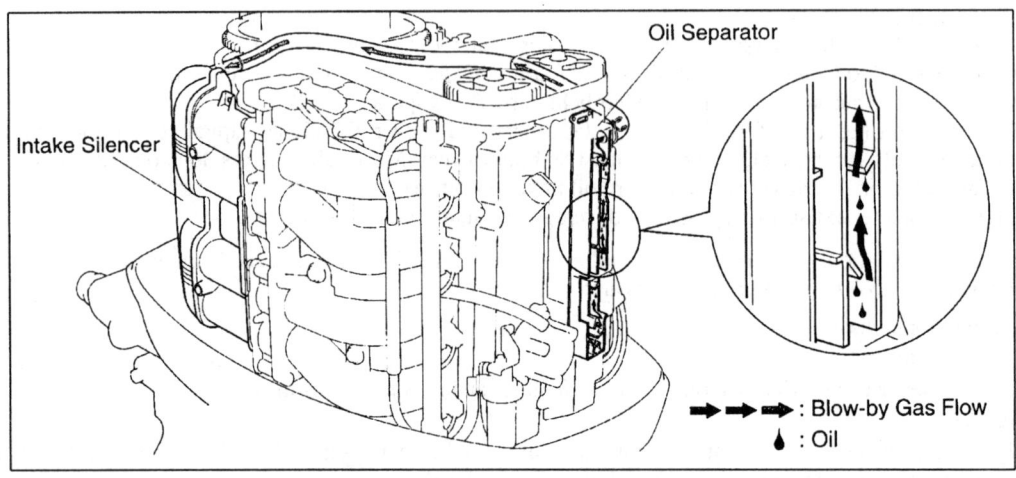

Fig 27.1: YAMAHA 4-STROKE OUTBOARD 'BLOW-BY' GAS RE-BURNING SYSTEM

TWO-STROKE OUTBOARDS

1. Manual start recoil handle
2. Choke knob
3. Overheat warning lamp
4. Electric start button
5. Gear-shift lever (Usually absent in small motors. They use a 180° pivot for reverse drive)
6. Throttle control & Steering handle
7. Engine stop (lock plate) lanyard switch
8. Engine securing clamp
9. Tilt lock for preventing accidental tilting
10. Anti splash plate
11. Cooling water inlet
12. Propeller
13. Trim tab for steering adjustment
14. Anti-cavitation plate
15. Trim-angle adjusting rod
16. Rope attachment
17. Shallow water lever for raising motor
18. Battery lead
19. Wiring harness
20. Remote control attachment

Fig 27.2: A 2-CYLINDER 2-STROKE 25HP YAMAHA OUTBOARD

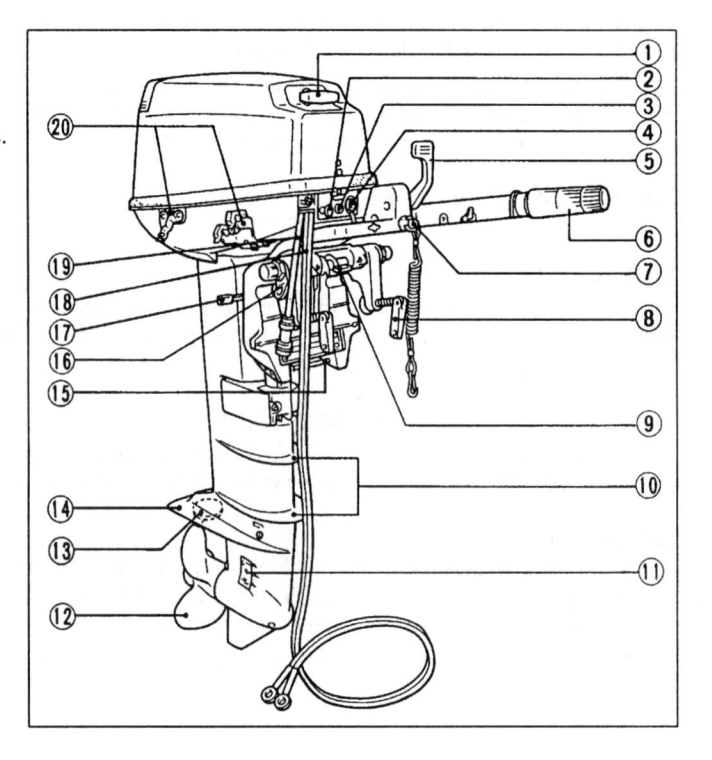

Chapter 27: Outboard Engines

FOUR-STROKE OUTBOARDS

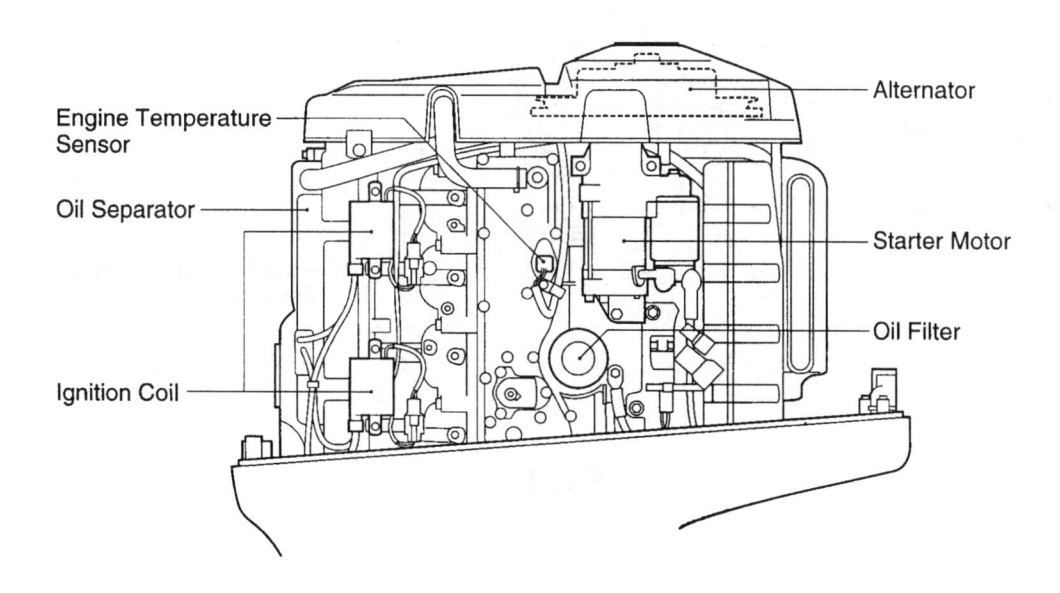

Starboard Side

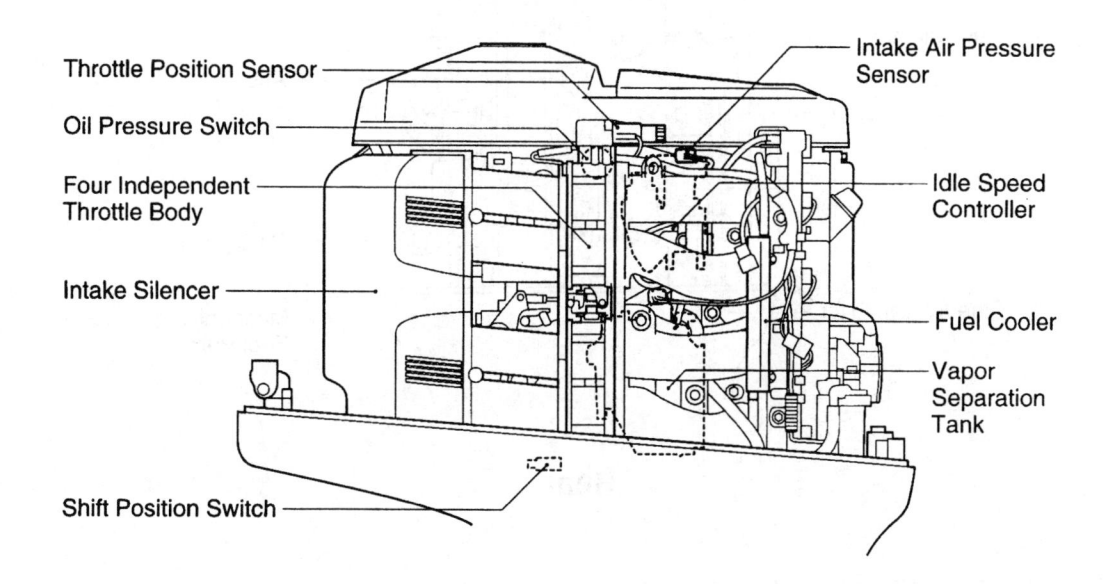

- Engine Temperature Sensor
- Oil Separator
- Ignition Coil
- Alternator
- Starter Motor
- Oil Filter

Port Side

- Throttle Position Sensor
- Oil Pressure Switch
- Four Independent Throttle Body
- Intake Silencer
- Shift Position Switch
- Intake Air Pressure Sensor
- Idle Speed Controller
- Fuel Cooler
- Vapor Separation Tank

Fig 27.3: A 4-CYLINDER 4-STROKE FUEL-INJECTED 115HP YAMAHA OUTBOARD

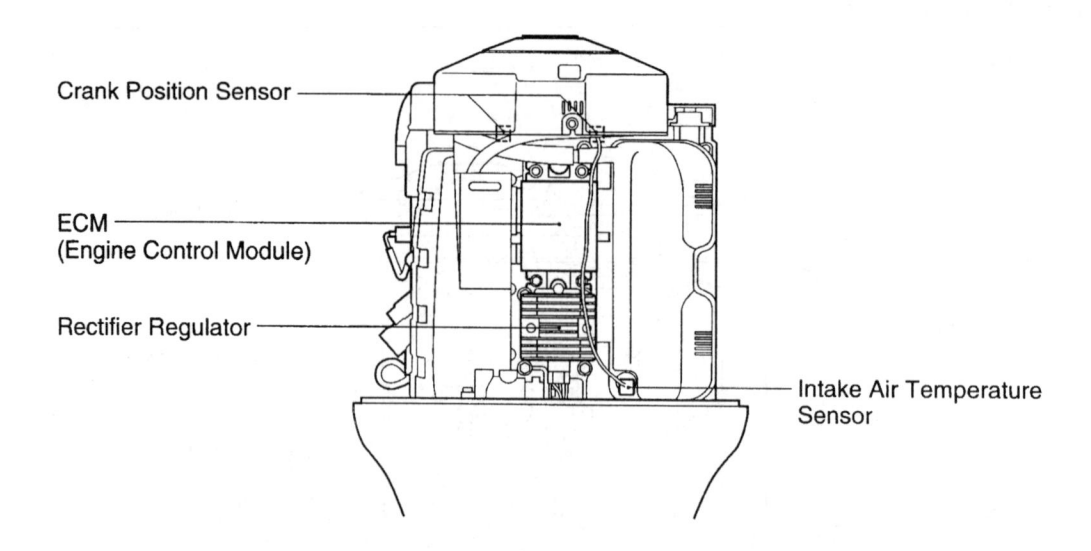

Front

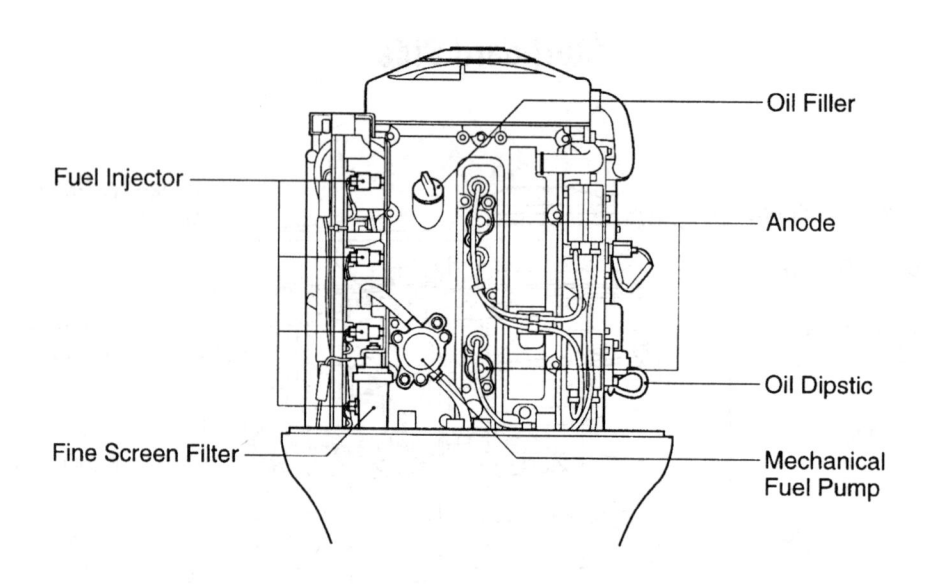

- Crank Position Sensor
- ECM (Engine Control Module)
- Rectifier Regulator
- Intake Air Temperature Sensor

Rear

- Fuel Injector
- Fine Screen Filter
- Oil Filler
- Anode
- Oil Dipstic
- Mechanical Fuel Pump

ELECTRONIC FUEL INJECTION (EFI) -PETROL

As discussed for diesel engines, electronic fuel injection provides better fuel efficiency and lower emissions in engines.

The engine is fitted with various electronic components and sensors that are controlled by a microcomputer known as the Electronic Control Unit or Module (ECU or ECM). A precision-built injector pump delivers the correct quantity of pressurised fuel to electronically controlled valves, known as injectors or atomisers. They squirt (atomise) fuel through tiny nozzles into the intake manifold where it mixes with air before entering the combustion chamber.

On acceleration, the throttle valve opens to permit more air to enter the engine. The ECM responds by opening more fuel. Sensors monitor the amount of air entering the engine and the amount of oxygen in the exhaust. They allow ECM to fine tune the delivery so that the air-to-fuel ratio is just right.

Chapter 27: Outboard Engines

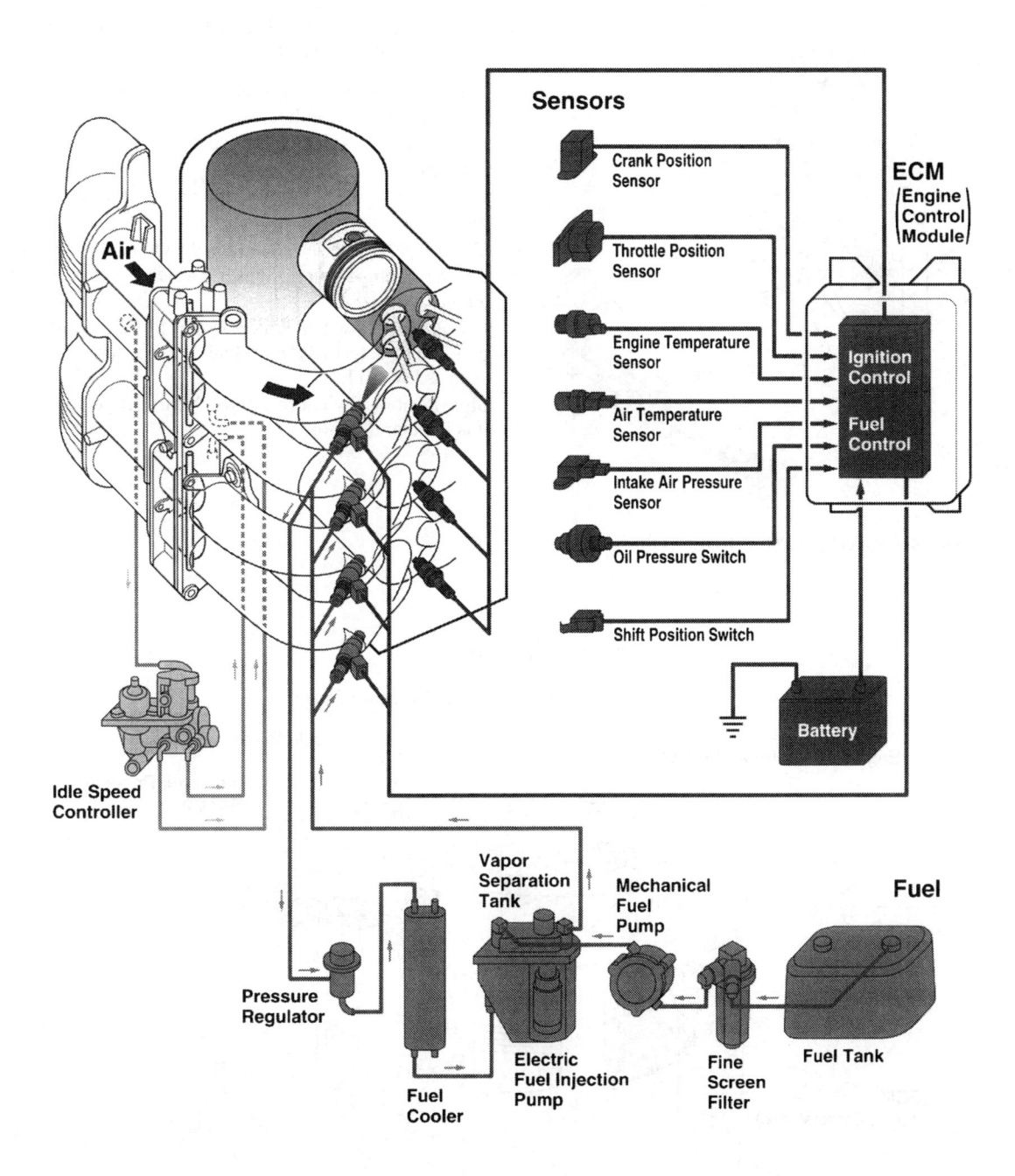

NOTES
- *Vapour separation tank separates vapour from fuel, sending only fuel to the fuel injectors.*
- *Fuel Cooler prevents vapour generation in fuel.*
- *Idle speed controller assists stable idling.*

Fig 27.4: YAMAHA EFI (4-STROKE PETROL OUTBOARD)

The Australian Boating Manual

The number of sensors and their names vary between engines. The Yamaha sensors are as follows:

- Throttle Position Sensor detects the opening position of the throttle valve to calculate the intake of air.
- Intake Air Pressure Sensor detects the pressure of the induced air.
- Intake Air Temperature Sensor detects the temperature of the induced air.
- Crank Position Sensor identifies the engine speed (rpm).
- Engine Temperature Sensor watches out for engine temperature.
- Oil Pressure Switch watches out for the lubrication oil pressure to detect any abnormality in the lubrication system such as low oil level.
- Shift Position Switch recognizes gearshift position; whether it is in neutral or not.

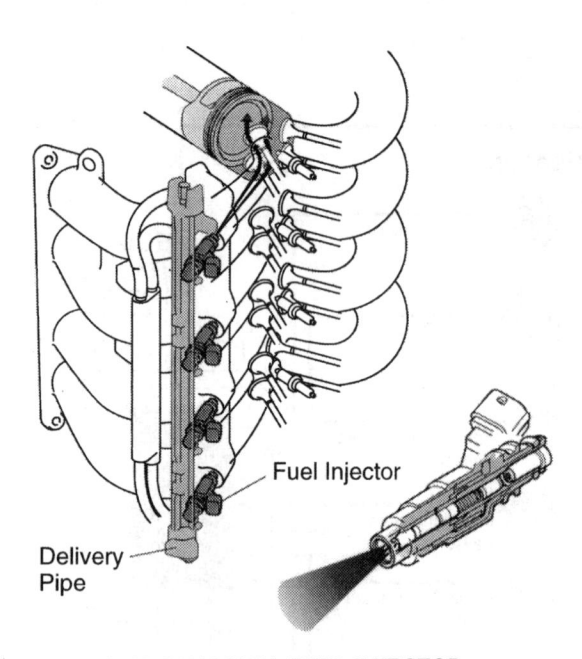

Fig 27.5 YAMAHA FUEL INJECTOR

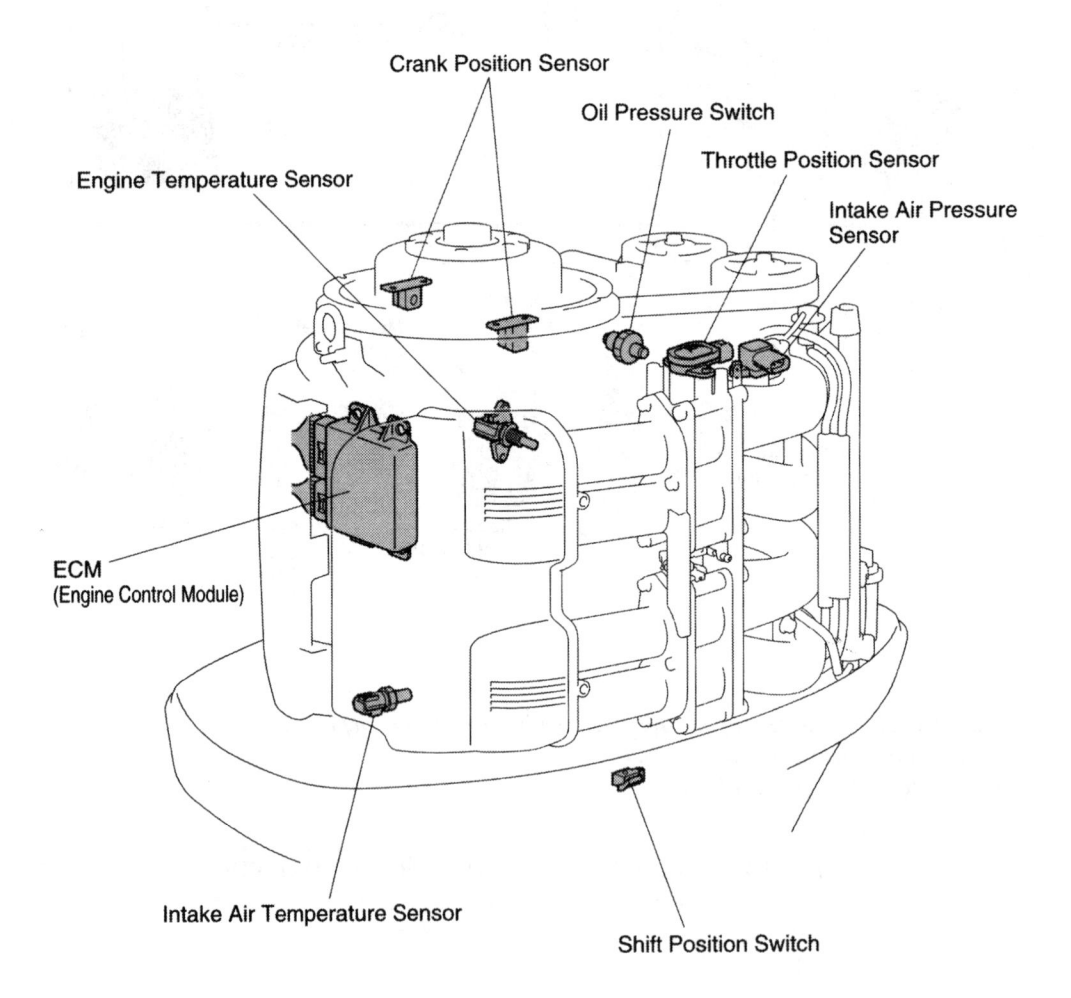

Fig 27.6: YAMAHA EFI CONTROL SYSTEM (EMC & SENSORS)

Chapter 27: Outboard Engines

ELECTRIC TROLLERS (ELECTRIC OUTBOARDS)

The DC electric outboards are small units sometimes used for trolling (fishing) at low speeds on lakes and for propelling canoes. Although some models are rated for saltwater, by and large they are designed for freshwater use. Therefore, only in some cases they are an attractive option for yacht tenders as a substitute for rowing, One 12 volt battery charge would propel a tender for about an hour.

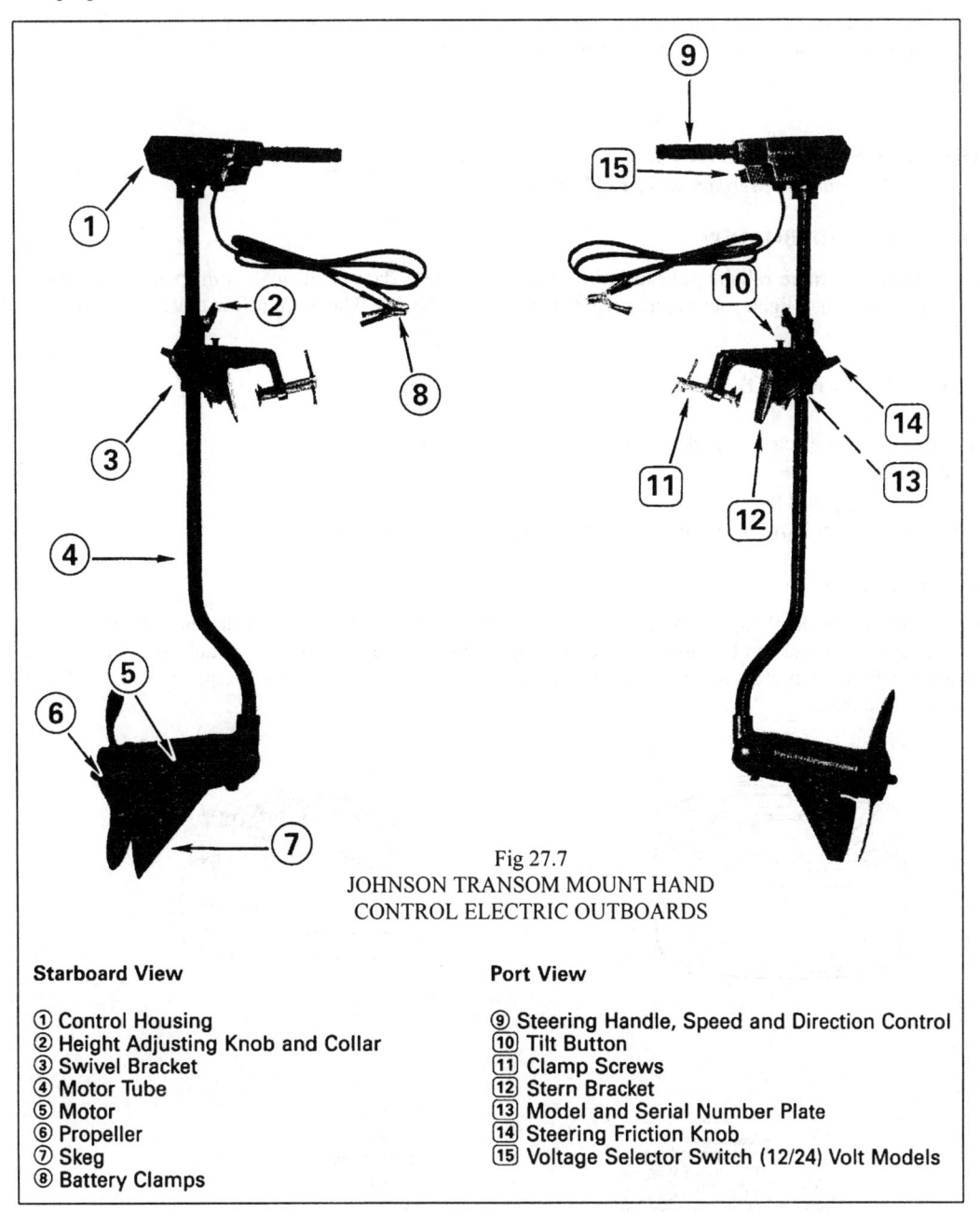

Fig 27.7
JOHNSON TRANSOM MOUNT HAND
CONTROL ELECTRIC OUTBOARDS

Starboard View

① Control Housing
② Height Adjusting Knob and Collar
③ Swivel Bracket
④ Motor Tube
⑤ Motor
⑥ Propeller
⑦ Skeg
⑧ Battery Clamps

Port View

⑨ Steering Handle, Speed and Direction Control
⑩ Tilt Button
⑪ Clamp Screws
⑫ Stern Bracket
⑬ Model and Serial Number Plate
⑭ Steering Friction Knob
⑮ Voltage Selector Switch (12/24) Volt Models

The motor in these units is inside the propeller hub, which keeps the winding cool and the weight low down. They generally have several speed settings in forward and reverse. They are extremely quiet, easy to start, very lightweight (ignoring the battery) and don't take up much room on the transom. Depending on the model, they can be mounted on transom or bow, and controlled with hand or foot. A conventional 12-volt lead acid type battery is connected via crocodile clips.

There are also some larger electric outboards with the motor at the top. For example, the Swiss made 6 hp, 36 volts AeroTech Sol-Z drives the propeller via belts.

Separate batteries should be provided for the electric outboard. If an electric and a gasoline outboard must share two batteries, the gasoline outboard must be connected to the No. 2 battery to avoid severe damage caused by galvanic corrosion. Be sure the speed control is turned off before connecting an electric outboard to a battery.

The electric outboards should be installed and adjusted so that the motor tube is perpendicular to the water surface under normal operating conditions.

The motor height should be adjusted just low enough to eliminate propeller noise and ventilation. The top edge of the propeller rotation should be approximately 20 cm below the water surface.

OUTBOARD FUEL SYSTEM - PETROL

(Illustrations Yamaha, Johnson, Mercury & Suzuki Outboards)

PETROL vs. DIESEL OUTBOARDS

Diesel outboards are more reliable, cheaper to run and use a less flammable fuel. Still, petrol outboards remain the norm because they are much lighter and more responsive than the diesels. There are a few diesels in commercial use and some in military.

PRECAUTIONS WITH PETROL
- Highly flammable.
- Highly explosive near electrical sparks, naked flames or static charge.
- Heavier than air.
- Vapour in an empty petrol tank can explode.
- Do not pour water over burning petrol. It may force it into crevices and drains.

PETROL OUTBOARDS

Small portable outboards always have a built-in fuel tank, which gravity feeds to the engine via an on/off tap. For longer journeys additional fuel can be carried in a petrol can. Larger motors incorporate a fuel lift pump, which draws fuel from an external tank. Some motors (often called "Compacts") incorporate both a built-in tank as well as a fuel lift pump to draw from an external tank.

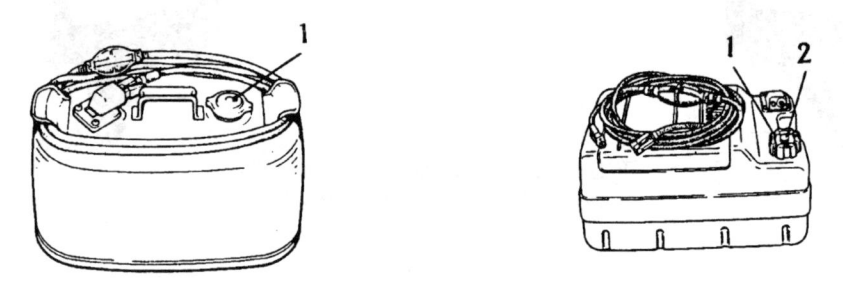

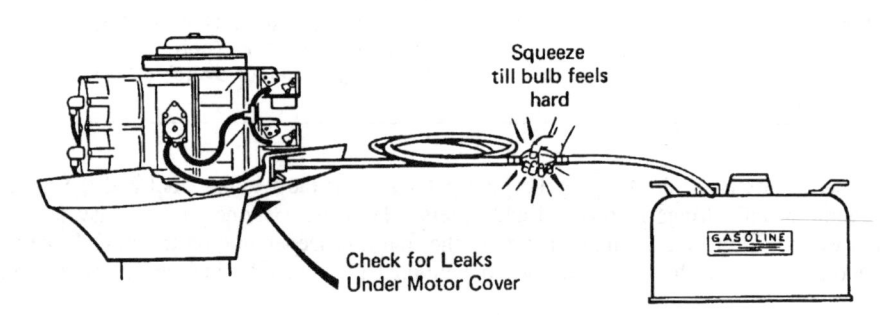

Fig 27.8: PORTABLE EXTERNAL FUEL SYSTEM
(1. Filler cap; 2. Breather screw)

Chapter 27: Outboard Engines

Portable fuel tanks for outboard motors are fitted with a manual priming-bulb. It is squeezed a few times to draw fuel when first starting the engine or to continue to provide fuel to the motor in case of fuel pump failure. Care must be exercised not to continue squeezing the bulb after it has become firm and full of fuel. Over-squeezing may cause leaks in the system or flooding of the engine with fuel, making it difficult to start.

Plastic fuel tanks are cheaper, lighter and available in various shapes and sizes. However, the mechanical connection into plastic tanks is difficult to seal. The escaping fuel vapour is likely to cause an explosion. Insufficient sealing may also allow the rainwater to get into the tank. The tern-plated steel tanks are the most common type, but the stainless steel tanks are more durable.

Petrol vapour is heavier than air. Therefore, outboard fuel tanks should not be stored below decks or in areas where the vapour can accumulate.

The fuel line connection to the tank should be self-locking, and the connection to the motor should be of either the quick-release type or automatic shut-off type when the fuel line is disconnected.

Portable fuel tanks should be filled ashore to avoid spillage on board. They should be stowed in clamps to avoid movement; and connected with the SA (Standards Australia) lines and plugs - suitably positioned to avoid damage.

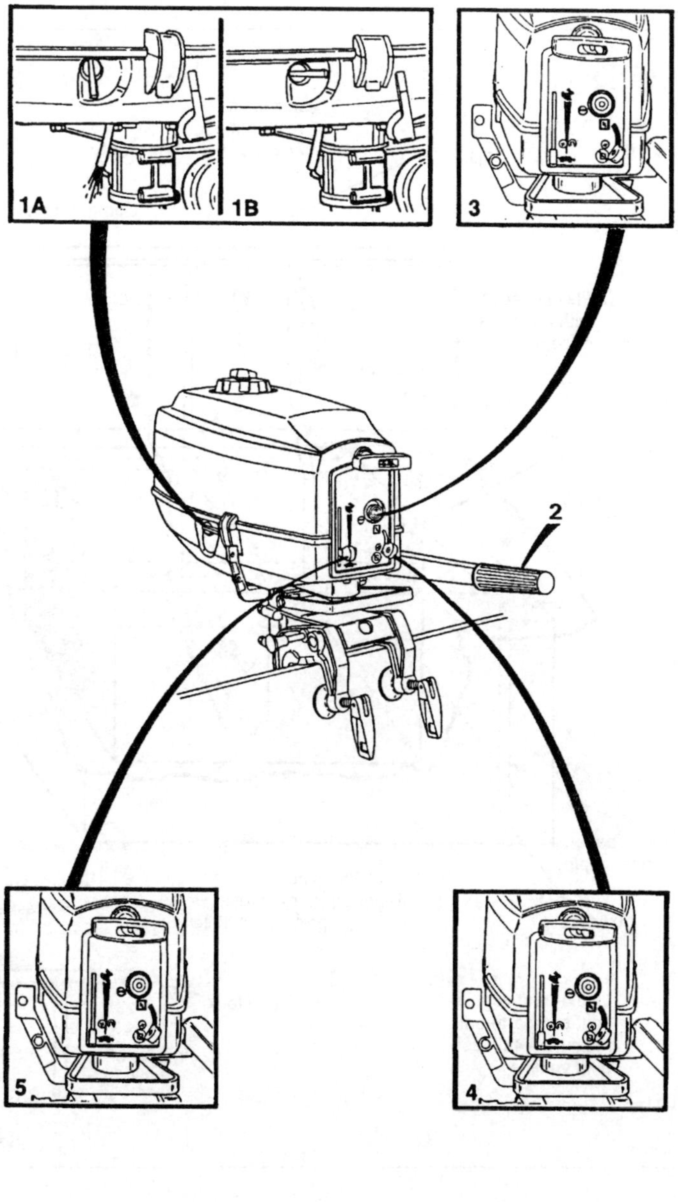

Fig 27.9
INTERNAL FUEL SYSTEM
(Mercury Outboards)

1A. Fuel cock in the open position
1B. Fuel cock in the closed position
2. Tiller
3. Stop button
4. Choke lever (move up for fast)
5. Throttle lever (move up for speed)

The Australian Boating Manual

Fig 27.10
FIXED FUEL TANK

a. Fuel filling point/cap
b. Anti-siphon fitting
c. Fuel tank breather
d. Fuel level indicator (float)
e. Manual priming bulb
f. Fuel filter
g. Flexible fuel line

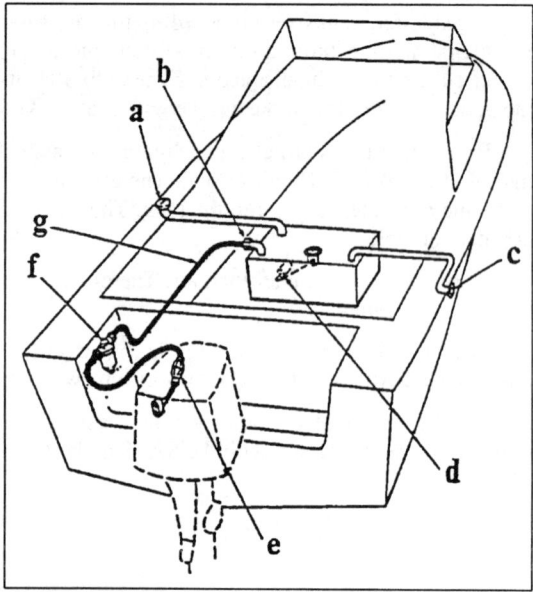

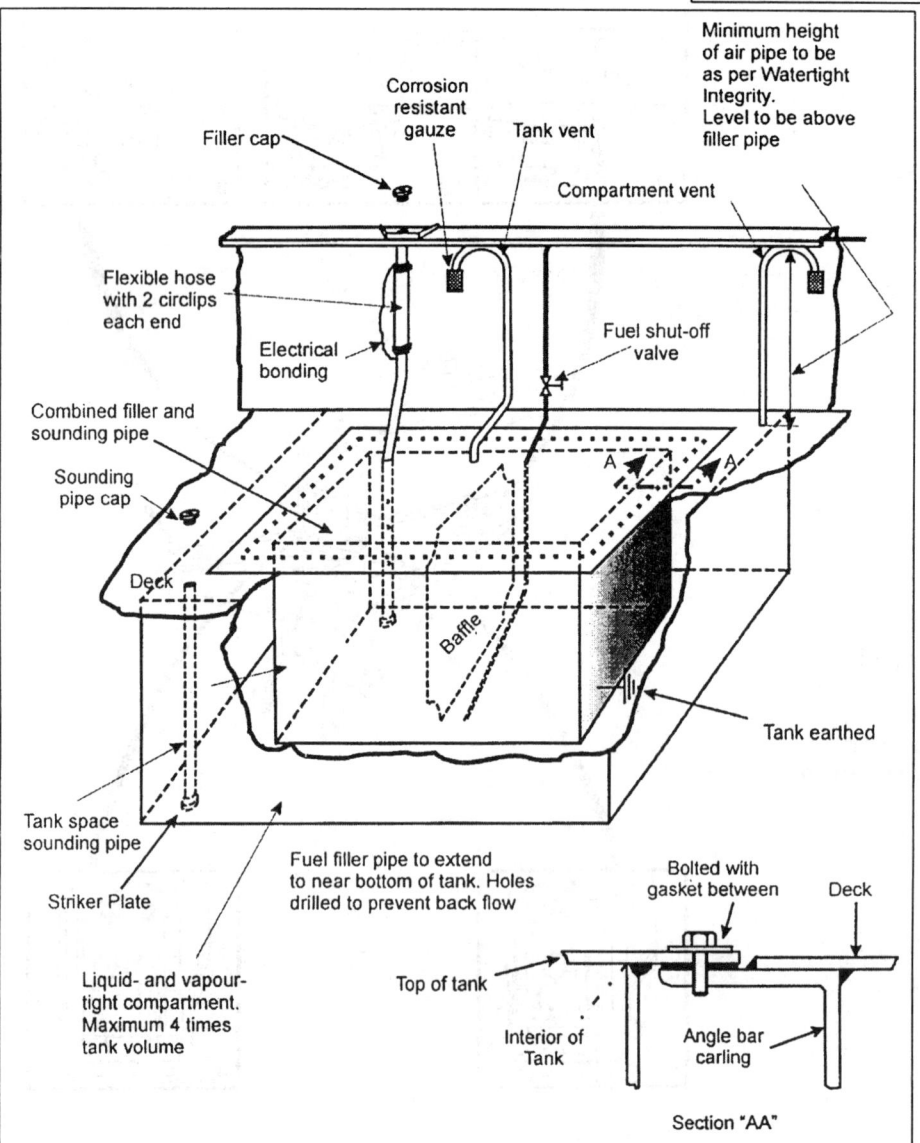

Fig 27.11
TYPICAL UNDER-FLOOR FUEL TANK INSTALLATION
(For fuel less than 60° C flash point, e.g., petrol and LPG)
(NSCV PART C Section 5 Subsection 5A)

Chapter 27: Outboard Engines

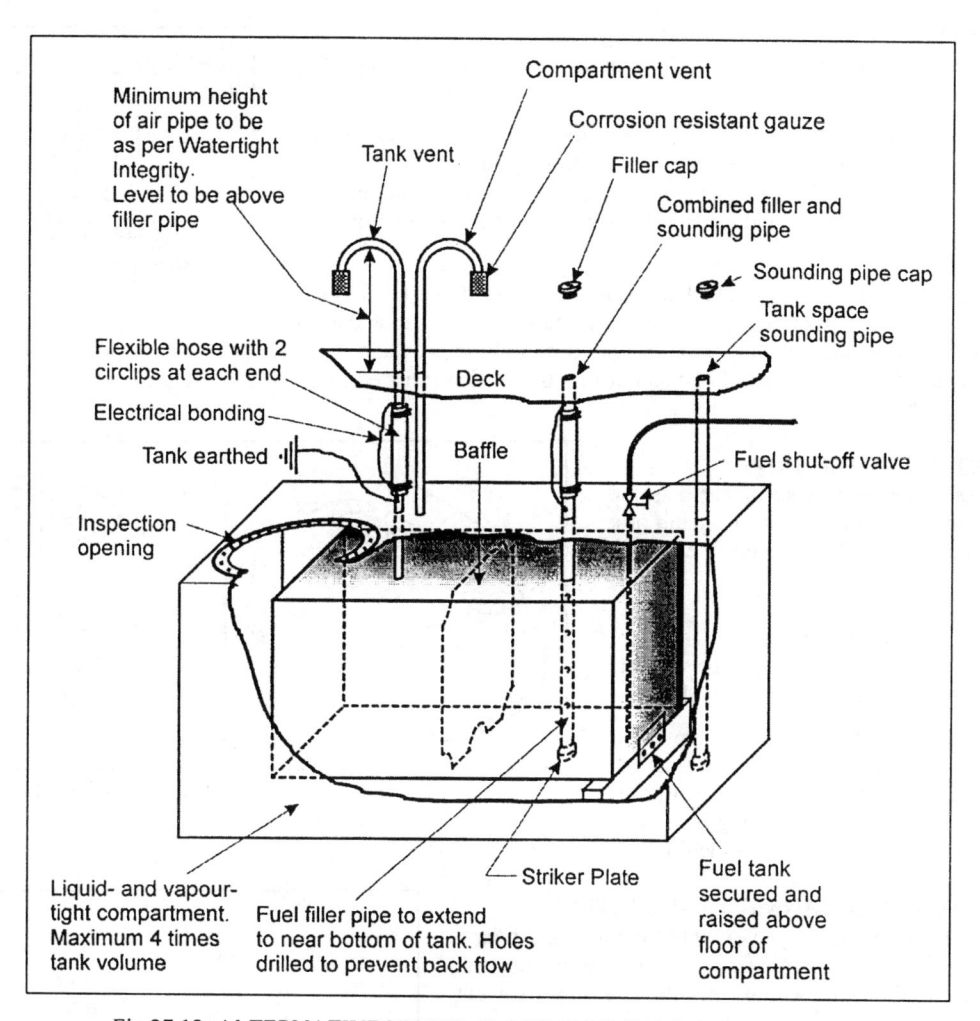

Fig 27.12: ALTERNATIVE UNDER-FLOOR FUEL TANK INSTALLATION
(For fuel less than 60° C flash point, e.g., petrol and LPG)
(NSCV PART C Section 5 Subsection 5A)

Petrol tank fittings can also be manufactured in one unit as shown here. This unit contains filling, venting, feeding line, possibly a return line (marine petrol engines do not normally have a return line), sender for tank gauge (sounding pipe) as well as the ground connection. It also functions as a manhole cover.

Fig 27.13 MULTI-PURPOSE TANK FITTING
(*Volvo Penta*)

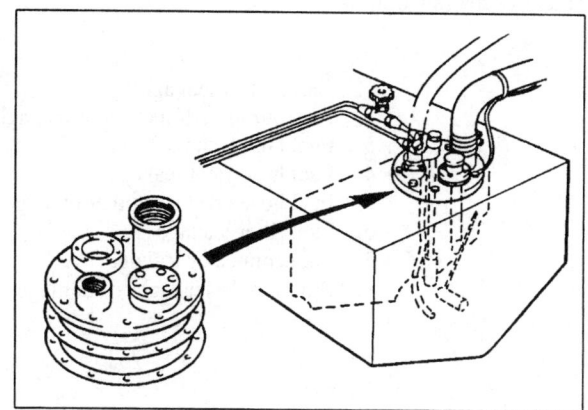

FUEL MAINTENANCE

Do not use fuel mixed with oil in EFI outboards. Oil can damage fuel injectors.

In two-stroke petrol engines use only a fresh mixture of fuel and two-stroke outboard oil, in the ratio recommended by manufacturer (Usually 50:1). Do not use fuel contaminated with dirt or water. Make sure the fuel pipe is not blocked with dirt. If blocked, remove it and clear it by blowing through it. Remove and clean fuel filter.

As discussed for inboard engines, fuel tanks should not be left partially empty for long periods to prevent fuel contamination due to moisture build up and algae growth on the inner surface of the tanks. For long periods of stowage, tanks should be drained of fuel. They should be stored in a dry, well-ventilated place, not in direct sunlight.

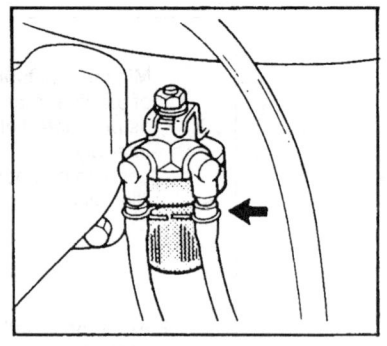

Fig 27.14 FUEL FILTER

TRANSPORTING OUTBOARDS

Always keep the powerhead higher than the leg. Water from the leg or oil from the sump of a 4-stroke engine must not be allowed to enter the cylinders.

CLEANING THE FUEL FILTER

The outboard fuel filter should be cleaned every 100 hours of operation or six monthly.

Stop the engine before removing the filter. Keep away from sparks, cigarettes, flames or other sources of ignition. Remove the fuel hoses and wash the filter in suitable cleaning solvent. If compressed air is available, blow-dry the filter.

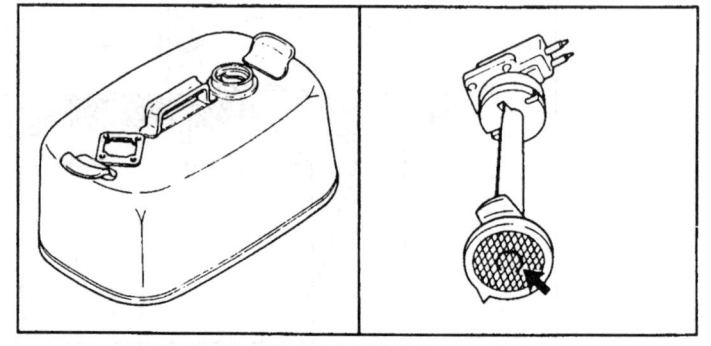

Fig 27.15: FUEL TANK & TANK FILTER

CLEANING THE FUEL TANK & TANK FILTER

Every 200 hours or once a year, empty the fuel tank, pour a small quantity of suitable cleaning solvent, and clean the tank thoroughly by shaking it. Drain off the cleaning solvent completely.

Thoroughly clean the tank filter (located at the end of the suction pipe) in a suitable cleaning solvent. If compressed air is available, blow-dry the filter.

FUEL SYSTEM INSPECTION (CARBURETTOR TYPE)

The fuel system should be regularly inspected for leaks, cracks or malfunction.

1. Carburettor leakage
2. Fuel pump malfunction or leakage
3. Fuel tank leakage
4. Fuel hose joint leakage
5. Fuel hose cracks or other damage
6. Fuel filter leakage
7. Fuel connector leakage
8. Primer bulb leakages or damage

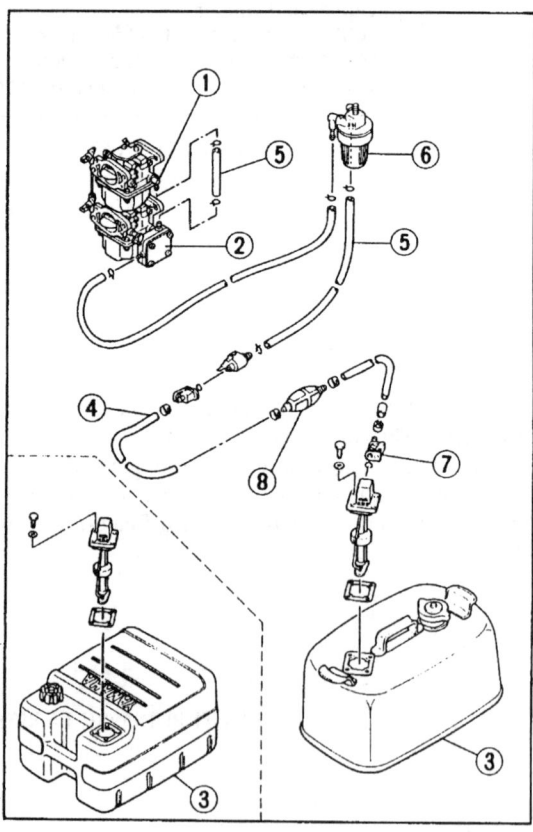

Fig 27.16: FUEL SYSTEM INSPECTION

Chapter 27: Outboard Engines

FUEL SYSTEM INSPECTION (EFI TYPE)

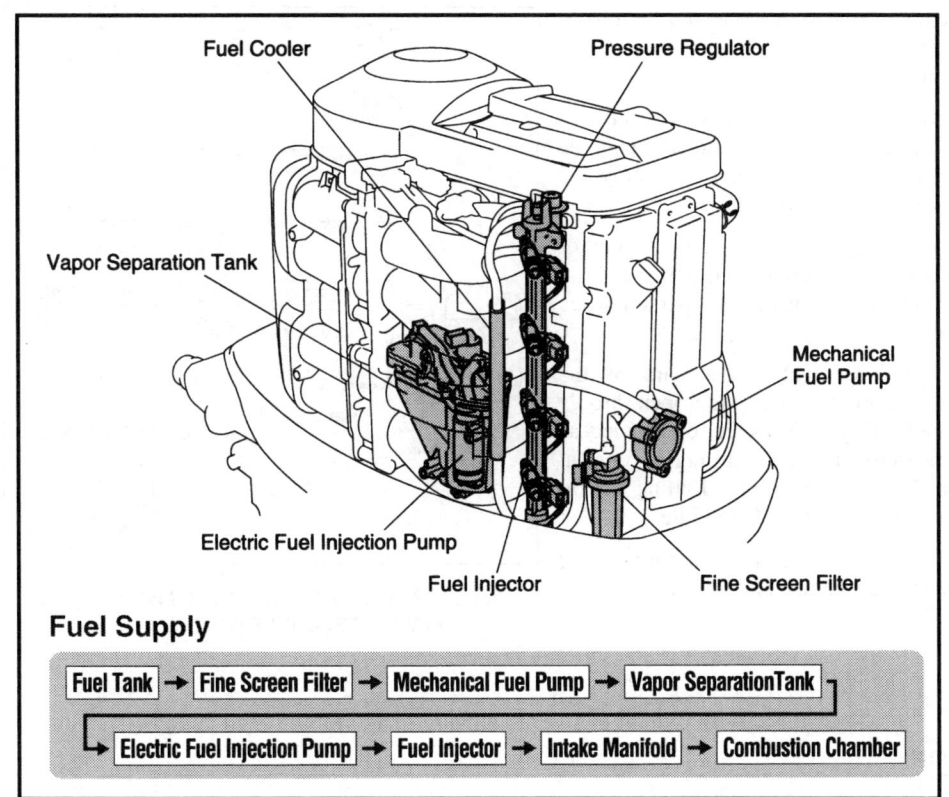

Fig 27.17 YAMAHA EFI FUEL SYSTEM

Fuel Supply

Fuel Tank → Fine Screen Filter → Mechanical Fuel Pump → Vapor Separation Tank → Electric Fuel Injection Pump → Fuel Injector → Intake Manifold → Combustion Chamber

TWO-STROKE OUTBOARD INTERNAL LUBRICATION SYSTEM

Two-Stroke petrol and LPG engines need their crankcase for compression. Therefore, they cannot have a wet-sump for a circulating oil system. Their lubrication is achieved by mixing oil with petrol or by injecting specially formulated oil into the dry fuel (LPG) engines. Oil is either poured into the fuel tank (and mixed by shaking) or is mixed as required by an oil injection pump known as the Variable Ratio Oiling **(VRO) from an attached lube oil tank**. Modern two-stroke engines are usually fitted with such precision blend system.

NOTE: Diesel engines (including most two-stroke) and four-stroke petrol engines have a conventional wet-sump, discussed in Chapter 22.

Fig 27.18
YAMAHA PRECISION
BLEND LUBRICATION
SYSTEM
(TWO STROKE PETROL)

1. Lube oil tank
2. Oil injection pump
3. Carburettor
4. Fuel filter
5. Fuel pump

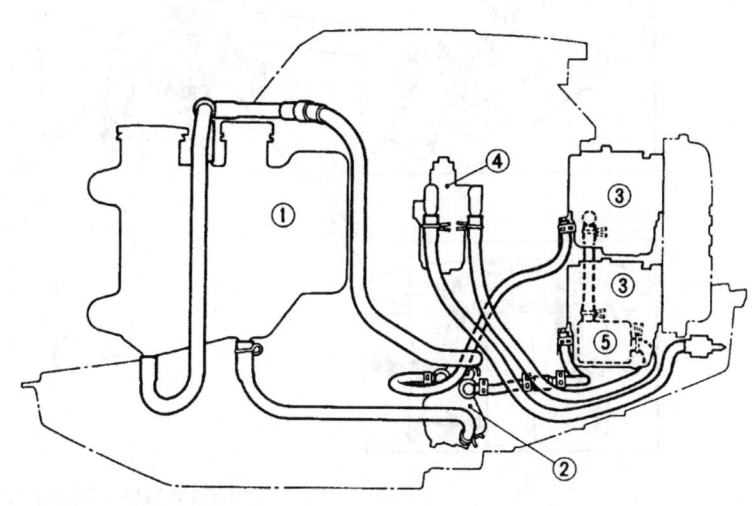

PETROL-OIL RATIO IN 2-STROKE PETROL ENGINES

Insufficient oil will cause the engine to overheat and eventually seize. Excessive oil will foul spark plugs, cause smoky exhaust and heavy carbon deposits. Both will cause engine failure.

The required amount of oil per litre of petrol varies between 1:50 and 1:100. The manufacturer's recommendations should be followed. Paint the correct ratio on the fuel tank. Use only the recommended outboard petrol motor two-stroke oil.

The VRO pump automatically adjusts and mixes the amount of oil required by the engine under different conditions and speeds. It may even be fitted with a no-oil alarm to warn the operator when engine is not receiving any oil.

Oil is usually injected into the fuel downstream of the carburettor. Petrol dilutes it at a ratio of 50:1 at wide-open throttle (W.O.T.) and deposits it on the bearing surfaces. Petrol also acts as a solvent to the oil on the bearing surfaces requiring a constant replenishment of the oil film. To reduce spark plug fouling and oil consumption, the petrol/oil ratio increases to 150:1 at idle.

If the reservoir of the V.R.O. pump runs dry during engine operation, the operator should stop the engine and refill it.

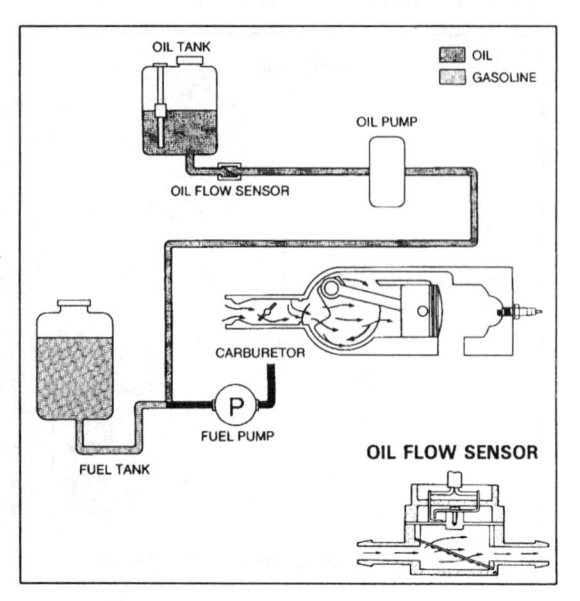

Fig 27.19: SUZUKI'S OIL INJECTION SYSTEM (TWO STROKE PETROL ENGINE)

EXTERNAL LUBRICATION

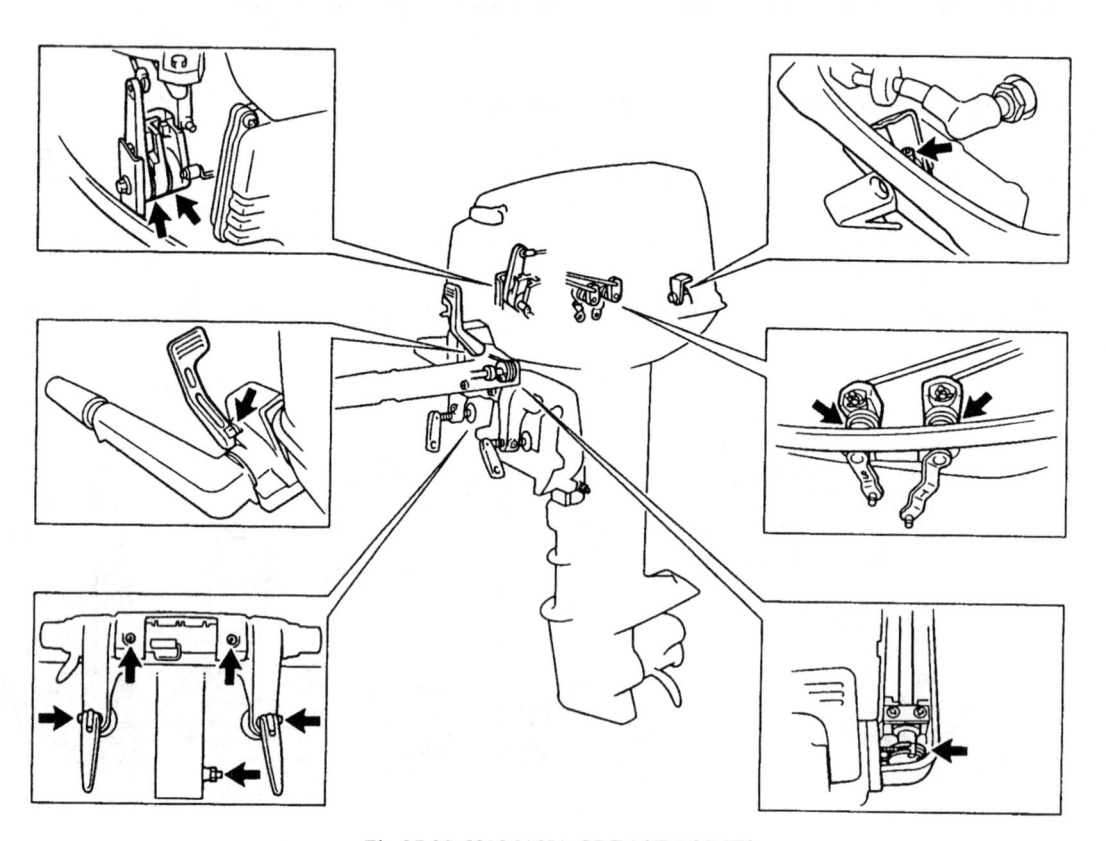

Fig 27.20: YAMAHA GREASE POINTS

Chapter 27: Outboard Engines

GEARBOX OIL CHANGE

Gearbox oil should be changed every 100 hours of operation or six monthly.

Drain the gearbox in a container by opening the oil-drain plug (2) and the oil-level plug (1). With the outboard motor in the upright position, inject the recommended gearbox oil into the oil-drain plughole until it starts to flow out of the oil-level plughole. Insert and tighten both the drain plugs.

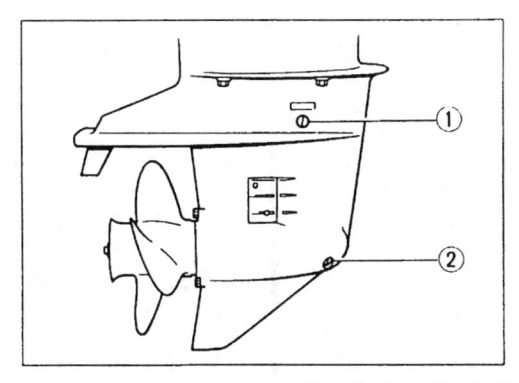

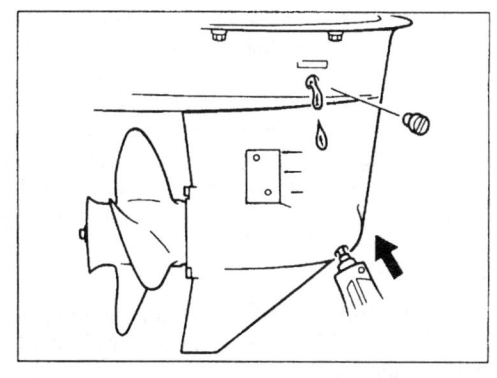

Fig 27.21: GEARBOX OIL CHANGE

THE OUTBOARD MOTOR COOLING SYSTEM

The outboard cooling system is the direct, raw water type. An impeller pump made of plastic or rubber, which is located in the lower leg, draws up seawater, which then passes through the galleries in the engine and out through the exhaust.

A small stream of water is also bled off somewhere in the system as a telltale sign, indicating to the operator that it is circulating. A thermostat maintains a minimum operating temperature. An audio alarm and a "*hot light*" are also sometimes fitted.

Fig 27.22
THE OUTBOARD COOLING SYSTEM

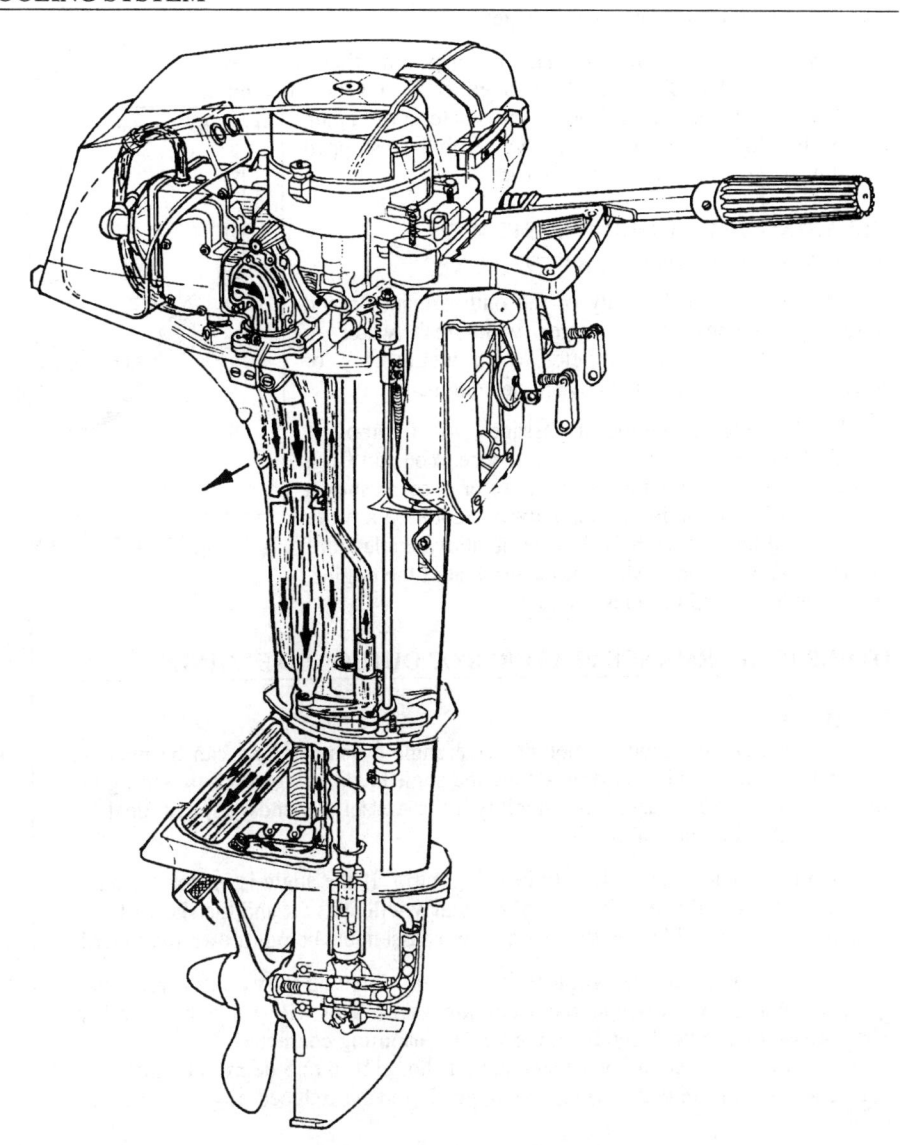

SHORTCOMINGS & PRECAUTIONS

The outboard cooling system is perhaps the most vulnerable part of the engine. Water pumps wear out due to constant abrasive action of sand, dirt and salt. For this reason, they should be checked every 200 hours.

Over a period of time the pump impellers (made of rubber or plastic) become set in the direction of travel. They also become brittle. Therefore if for any reason the motor is turned in an anti-clockwise direction, the impeller tips can not only break off, but also get sucked into the cooling system - blocking the system almost completely.

The impeller blades are designed to run in the cooling medium of water. Testing of motors out of water can heat up, melt and break their tips, sucking them into the cooling system. The heat generated can also distort the pump housing, not to mention the engine itself seizing up. Outboard Motor "Ears" are available for testing the engine using a garden tap.

When testing an outboard motor before launching a boat at a ramp, the trailer should be backed into the water until the water inlets (pick ups) are submerged.

The operator should be constantly aware of proper functioning of the cooling system. The biggest enemy of an operating engine is plastic bags floating just below the water surface. If there is a drop in the telltale stream of water or an alarm or overheating of the motor indicates a problem in the water flow, the motor should be taken out of "ahead" gear and inspected for blockages.

Further checks can also be made by inspecting that all the water pick ups are clear. They are located either in a scoop behind the impeller or in the gear case. The motor is safe to run again if the "hot light" has gone out, the alarm has stopped sounding and the telltale stream is back.

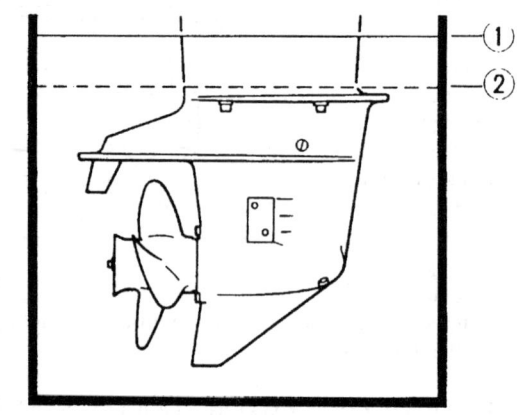

Fig 27.23
CLEANING THE OUTBOARD MOTOR
(1) Water surface. (2) Lowest water level.

CLEANING THE COOLING SYSTEM
As recommended by Yamaha motors)

After use, wash the body of the outboard motor and flush the cooling-water passages with fresh water to remove mud, salt, seaweed, etc. which would clog or corrode the passages and thereby shorten engine life.

To clean the cooling-water passages, some outboards are fitted with a fresh water flushing-device. You can flush down the water jacket with a running water hose, as shown in figure 27.24. If not, install the outboard motor on a water tank, and fill the tank with fresh water to above the level of the anti-cavitation plate. Shift into neutral, start the engine, and run at low speed for a few minutes.

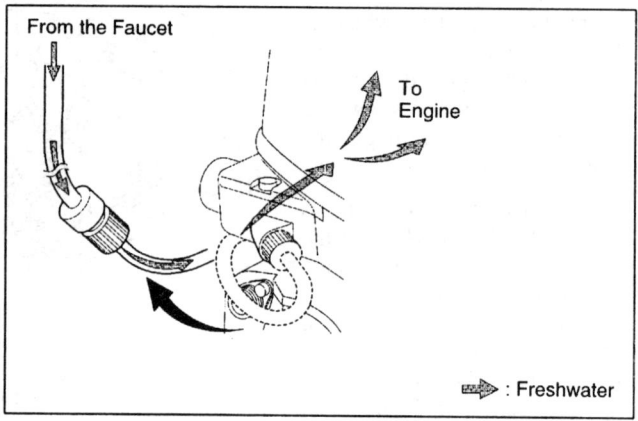

Fig 27.24: FRESH WATER FLUSHING DEVICE

OTHER PERFORMANCE FEATURES OF OUTBOARD ENGINES

1. TRIM ANGLE

By changing an outboard motor's drive angle a vessel's bow can be made to rise or drop. The bow is raised by setting the angle back and lowered by setting the angle in. The performance and stability of a vessel depends a great deal on correctly trimming the outboard.

Incorrect angle can also lead to loss of control. If the angle is set too far back, the vessel's bow will rise, the centre of buoyancy will shift aft and the she will tend to "porpoise". It could lead to the operator and passengers being thrown overboard.

If it is set too far in, she will tend to "plough". She will throw extra spray, lose speed and may bury her bow in the oncoming waves. This will reduce her stability. The correct trim angle depends on the vessel's handling characteristics, the size of the outboard and sea conditions. Generally, it should be 3 to 5 degrees by the stern. For better steering into the wind, the angle should be reduced slightly, and visa versa.

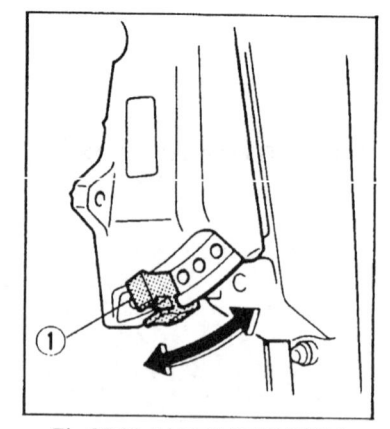

Fig 27.25: YAMAHA'S TRIM ANGLE ADJUSTING PIN

Chapter 27: Outboard Engines

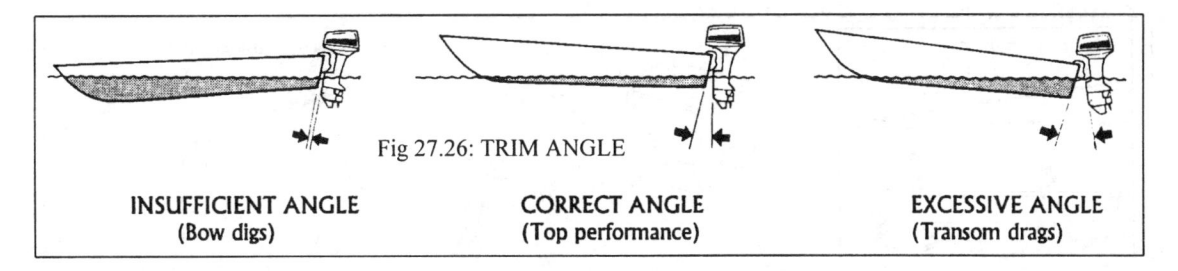

Fig 27.26: TRIM ANGLE

INSUFFICIENT ANGLE (Bow digs) **CORRECT ANGLE** (Top performance) **EXCESSIVE ANGLE** (Transom drags)

On smaller outboards the trim angle is adjusted manually by moving a pin or an adjusting rod to different holes in the mounting bracket. On remote controlled outboards, a *Tilt/Trim Switch* or a *Power Trim Switch* is fitted on the remote control panel or lever. The term "*Trim*" applies to small changes in angle under normal operations; and "*Tilt*" to larger changes in angle when cruising in shallow water or to raise the propeller out of water when stopped. See Tilt-Up Mechanism.

The motor should not be operated with the trim-adjusting pin removed. The pin must also not be used to support the weight of the motor.

2. TILT-UP MECHANISM

The tilt-up mechanism is provided on outboards for tilting the engine 10-15 degrees upwards when cruising in shallow water, or tilting it higher to raise the propeller out of water. The Yamaha mechanism consists of a tilt-lock lever, a tilt-up lever and a tilt-support bar.

Precaution: Keep the power unit higher than the propeller at all times to prevent water running into the cylinder and damaging it. To prevent fuel leaking out, disconnect the fuel-line if the engine is to be tilted up for more than a few minutes.

As discussed above for Trim Angle, on remote control outboards a combined Tilt/Trim Switch is usually fitted on the remote control panel or lever.

Unlock   Pull up   Disconnect fuel line

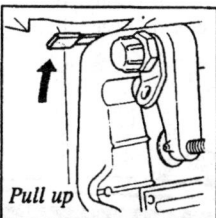

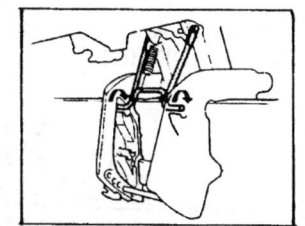

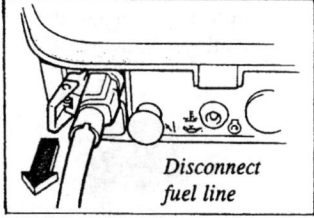

Fig 27.27: YAMAHA TILT-UP MECHANISM
(Unlock, Pull up & tilt the engine. The tilt-support bar will lock automatically)

3. TRIM TAB

When an outboard motor is trimmed at an angle, its propeller shaft is not parallel to the water surface. The propeller steering torque then tends to pull the boat in one direction. The outboards are usually fitted with a trim tab to compensate for this steering torque. The trim tab should be adjusted so that the boat can be steered to either side by applying the same amount of force.

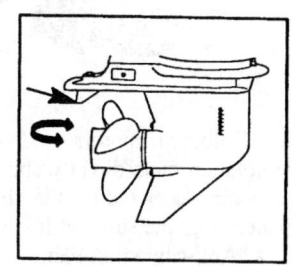

Fig 27.28: TRIM TAB

Be sure to tighten the bolt after adjusting the trim tab, and test run the boat to be sure the steering is correct. It should be noted that trim tab adjustment has little effect on reducing the steering torque on outboards fitted with an anti-ventilation plate, discussed below.

4. AFTER PLANES (TRIM PLANES)
The tilt/trim control of an outboard motor must not be confused with the After Planes fitted on the back of some vessels. These too are individually adjustable up or down (manually or remote controlled) and provide another method of trimming the vessel.

Fig 27.29: AFTER PLANES (TRIM PLANES)

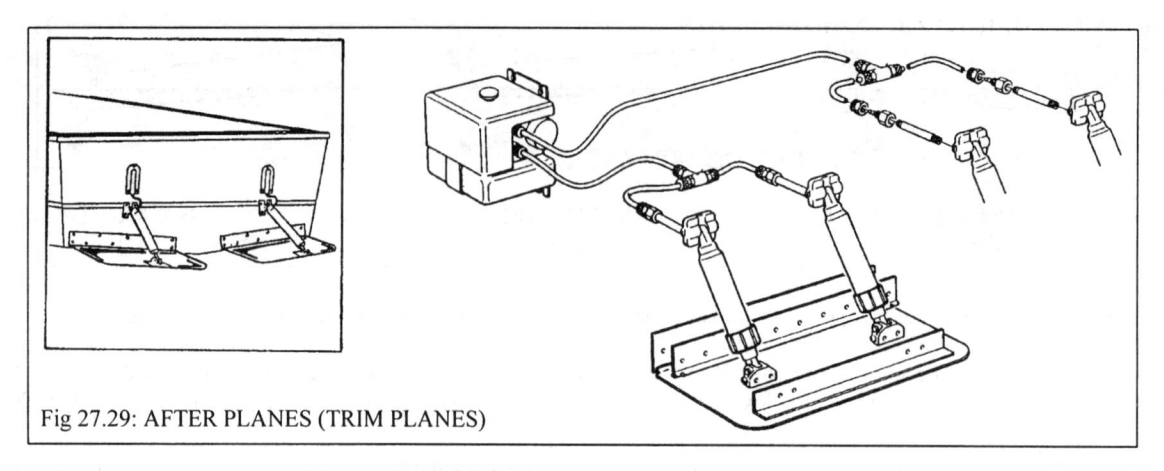

5. ANTI-VENTILATION (ANTI-CAVITATION) PLATE

VENTILATION OR 'BLOW-OUT' is the loss of grip or bite by the propeller. The aeration from the water surface or from the exhaust are drawn into the propeller blades. This interruption to the flow of 'good' water results in reduction in water pressure on the propeller, causing it to over-rev, and eventually resulting in massive cavitation.

Ventilation can be caused by:

- The sucking of surface air due to the outboard engine set up too high on the transom or the leg trimmed too far out
- Dirty, scratched or chipped hull
- Badly designed or damaged stern area or brackets (struts) holding the shaft
- Alterations to the hull such as a new water intake or a transducer

Ventilation invariably progresses on to be come CAVITATION, i.e., formation of air bubbles along the leading edges of the blades. As discussed in 'Propeller Damage' and illustrated in Figure 24.30 (Chapter 24), these bubbles have the same effect on the blades as corrosion. The blades become pitted and cause vibrations in the stern area. If the pits are left unfilled, vibration may damage the bearing and break the shaft. Hundreds of little pinholes often seen along the leading edge of the propeller is known as CAVITATION BURNING.

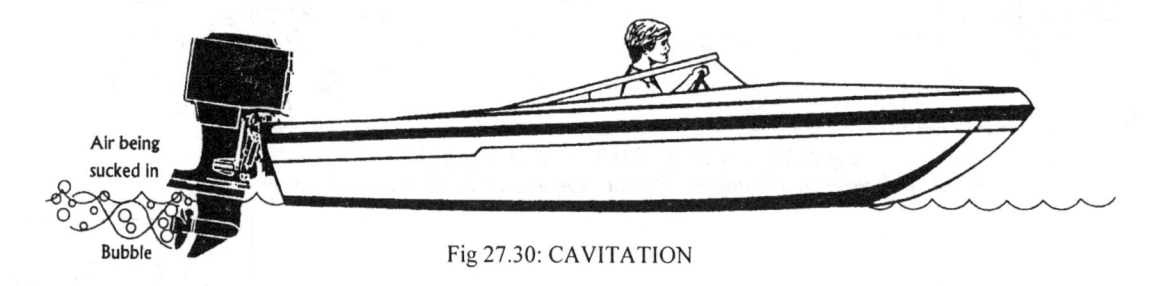

Fig 27.30: CAVITATION

Outboard motors and sterndrives are fitted with a large anti-ventilation plate directly above the propeller. The purpose of the plate is to eliminate or reduce the air being sucked into the tips of the propeller plates from the surface. It is frequently referred to as a "Cavitation Plate" or a "Anti-cavitation Plate".

Many propellers are designed with a hub and a flared trailing edge or a diffuser ring to assist proper exhaust gas flow, preventing it from feeding back into the propeller.

Fig 27.31
A QUICKSILVER ANTI-VENTILATION PLATE

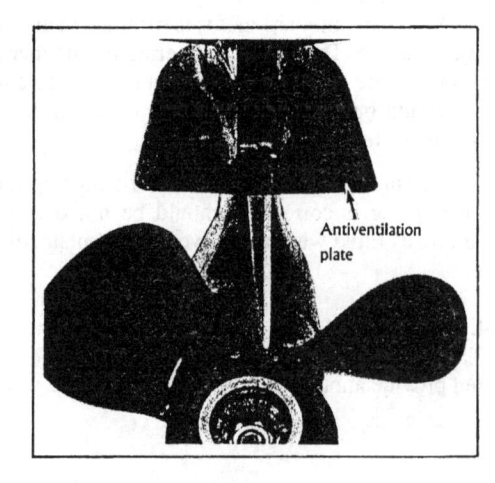

Chapter 27: Outboard Engines

6. MOUNTED HEIGHT OF AN OUTBOARD

With the motor in upright position, its mounted height should be such that the anti-ventilation plate is in line with the vessel's bottom or up to 25 mm below it.

Setting the engine deeper in water will tend to cause excessive spray, increased drag and reduction in underwater clearance. Raising it may reduce the flow of cooling water in the engine, increase cavitation and vibration, and reduce propulsion. If propeller tips cut the air, the engine will speed abnormally and become hot. In dual-engine installations one engine may lift out of water when turning the vessel.

However, the engine height is sometimes varied to suit the circumstances. For example, raising it 5 cm will reduce drag and improve handling for high speed applications. Lowering it a similar amount will improve its load carrying capacity at slow speed.

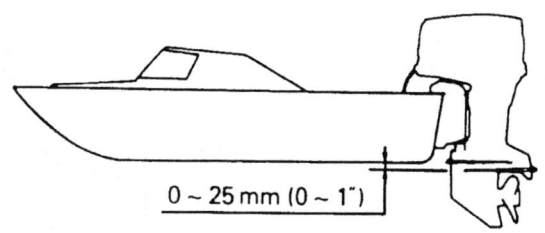

Fig 27.32
MOUNTED HEIGHT OF A YAMAHA OUTBOARD

7. TRIMMING BY WEIGHT DISTRIBUTION

Every vessel - outboard or inboard - must also be suitably trimmed by proper weight distribution, as shown.

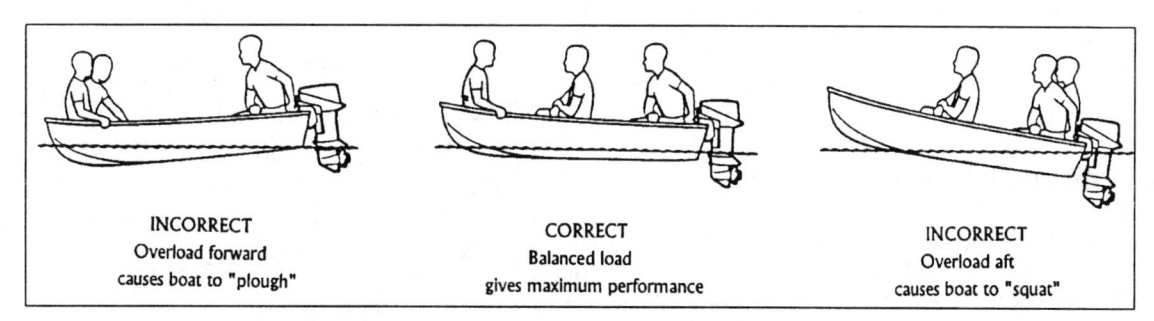

INCORRECT
Overload forward
causes boat to "plough"

CORRECT
Balanced load
gives maximum performance

INCORRECT
Overload aft
causes boat to "squat"

Fig 27.33: WEIGHT DISTRIBUTION

8. HYDROFOIL STABILIZERS OR FINS

These fins reduce bow lift at low speed. They also get the boat on a plane more quickly by increasing the lift generated by the engine. This can be beneficial for activities such as water skiing. The acceleration is improved and the skier's pull line is pulled out of water more quickly. The young and the beginner skiers may stay up more easily at lower speeds. The rooster tail is reduced and the wake is smoother, making the boat more stable when turning.

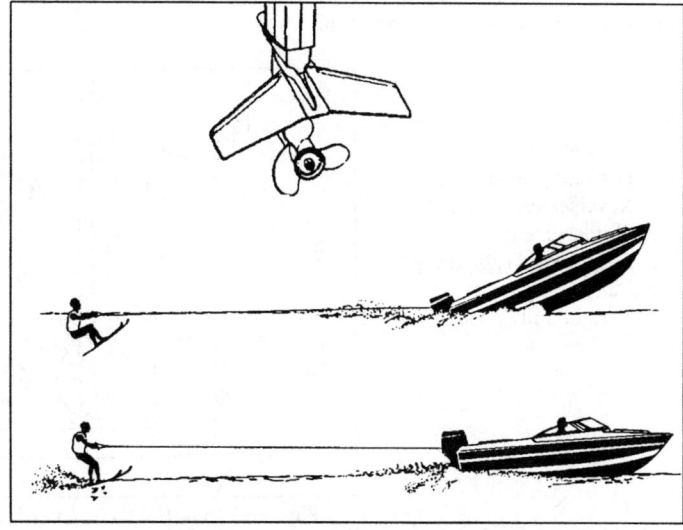

Fig 27.34: HYDROFOIL STABILIZERS (OR FINS) PULL A SKIER OUT OF WATER MORE QUICKLY

REMOTE PROPULSION CONTROL FOR OUTBOARDS

(Illustrations Yamaha, Suzuki & Johnson Outboards)

Fig 27.35 (A) & (B)

1. Switch panel
2. Remote control lever
3. Remote control cable

The *kill switch* on the Switch Panel usually consists of a lock plate inserted under the engine stop switch. On pulling the emergency lanyard, the lock plate is pulled out and the Stop Switch stops the engine by opening the ignition circuit. A kill switch is also fitted on the outboard motor itself for operation without a remote control.

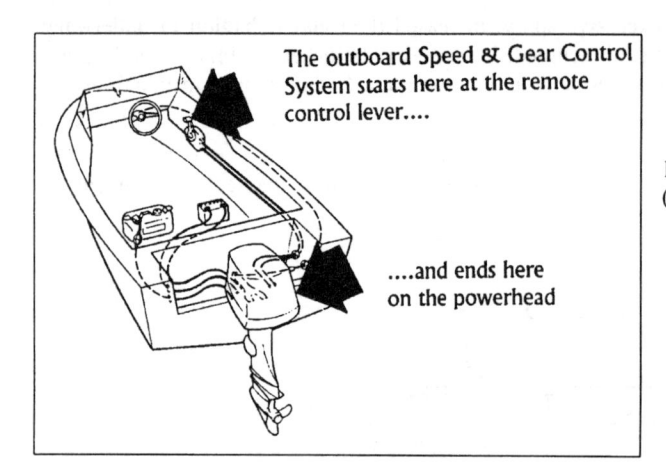

Fig 27.35 (C) SWITCH PANEL

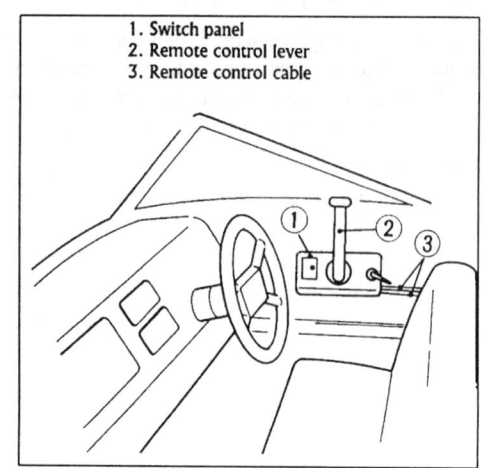

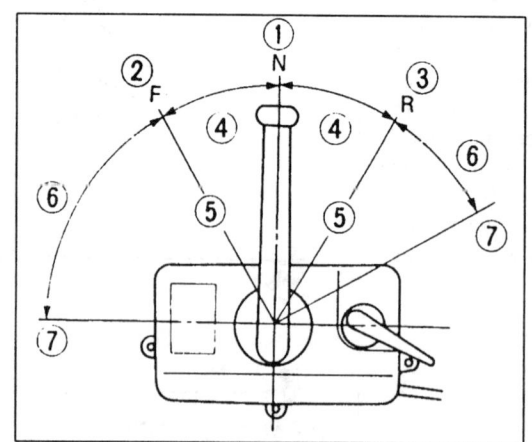

REMOTE CONTROL LEVER
(ONE TYPE)

To engage gear, turn the control lever forwards or backwards through approximately 35° (The detent position), whereupon the vessel will commence to move forwards or backwards at idle revs. Turning the lever further downwards opens the throttle.

To open the throttle when idling, turn the Fast Idle Lever (Neutral Throttle Lever) up. It will warm up a cold engine. For safety, this lever can only be operated when the remote control lever is in the neutral position.

1. Neutral
2. Forward gear engaged
3. Reverse gear engaged
4. Shift segment
5. Fully-closed (idle revs)
6. Throttle segment
7. Throttle fully open

Fig 27.36 (A): REMOTE CONTROL LEVER (ONE TYPE)

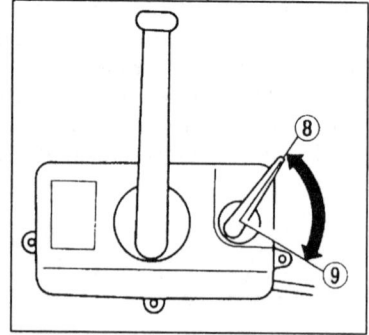

Fig 27.36 (B): FAST IDLE LEVER
8. Fully-open. 9. Fully-closed

Chapter 27: Outboard Engines

REMOTE CONTROL LEVER (ANOTHER TYPE)

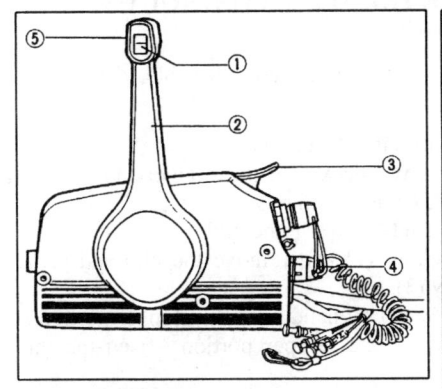

1. Trim switch
2. Remote control lever
3. Warm-up (Fast idle) lever
4. Emergency stop switch (Pull out)
5. Hand lever
6. Buzzer
7. Ignition switch
8. Plate assembly
9. Cable guide
10. Shift gear assembly
11. Neutral switch
12. Throttle arm
13. Neutral lock button

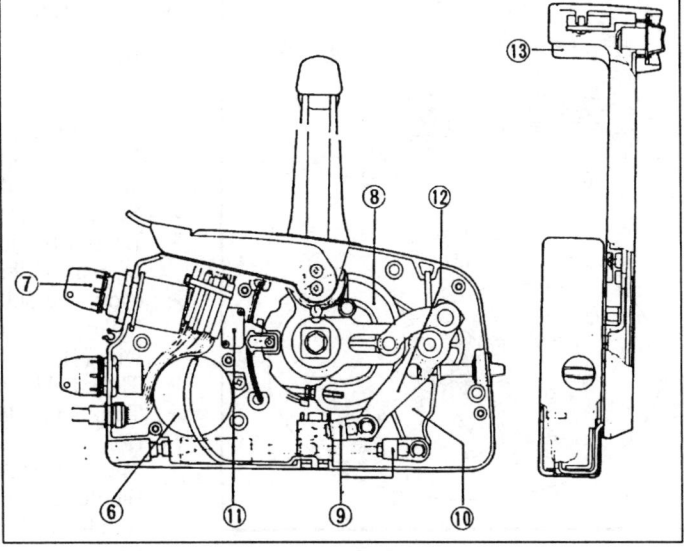

Fig 27.37 (A) & (B)
REMOTE CONTROL LEVER (ANOTHER TYPE)

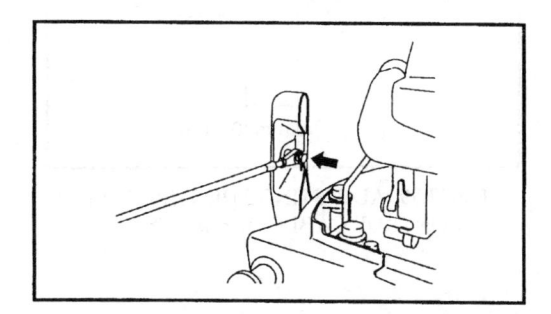

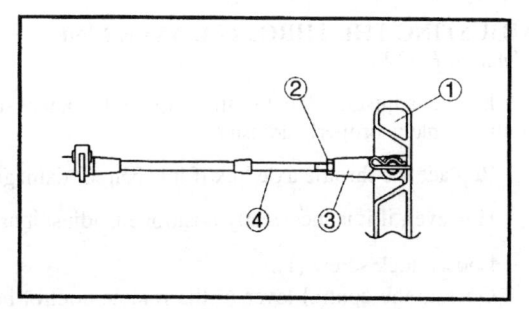

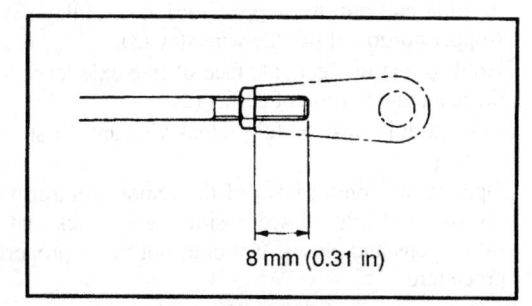

ADJUSTING THE SHIFT CONTROL CABLE
(Yamaha F8CW)

A damaged cable should be replaced. However, in case of poor shifting or incorrect shift handle positions, the shift control cable can be adjusted as follows:

- Set the shift handle (1) to neutral position.
- Loosen locknut (2) and remove the clip and the cable end (3) from the shift control cable (4).
- Set the remote control lever to neutral position.
- Turn the cable end (3) so that it aligns with the pin on the shift handle (1).
- Reinstall the cable end (3) onto the shift handle (1). The cable end should be screwed in more than 8 mm.
- Install the clip and tighten locknut (2).
- Check that the remote control lever functions properly. If not, repeat the above procedure.
- If the shift lever cannot be set to the reverse position, rotate the propeller slightly and try again.
- If the shift control cable cannot be properly adjusted at the engine end, make the adjustment at the remote control lever end.

8 mm (0.31 in)

Fig. 27.38
ADJUSTING THE SHIFT CONTROL CABLE
(Yamaha F8CW)

ADJUSTING THE THROTTLE CONTROL CABLE
(Yamaha F8CW)

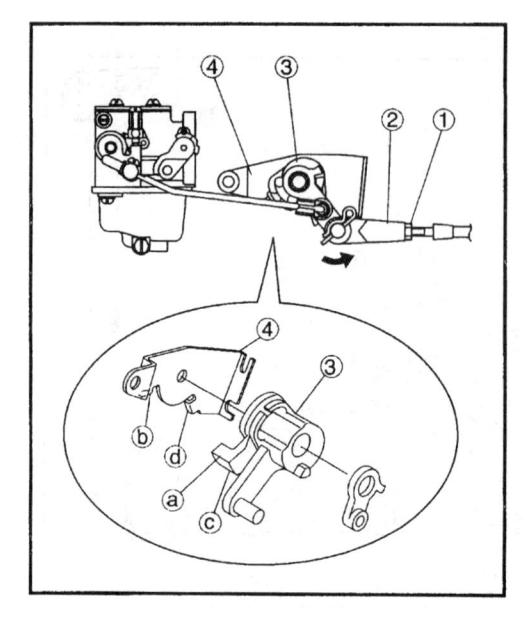

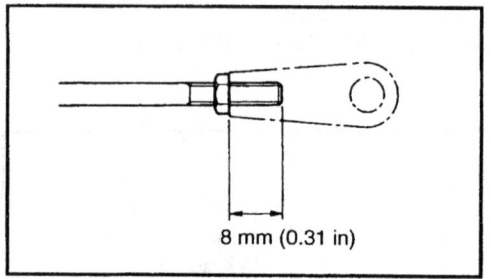

Before adjusting the throttle control cable, make sure the shift control cable is properly adjusted.

- A damaged cable should be replaced.
- In case of excessive slack or if the stoppers do not contact each other, adjust the cable as follows:
- Set the remote control lever to neutral.
- Loosen the locknut (1) and remove the clip and cable end (2) from the pulley (3).
- Move the pulley (3) by hand until the stopper portion (c) on the pulley (3) contacts the stopper portion (d) on the throttle stay wire (4).
- Then, hold it in position.
- Turn the cable end (2) so that it aligns with the pin on the pulley (3).
- Then, reinstall the cable end (2) onto the pulley. The cable end should be screwed in more than 8 mm.
- Install the clip and tighten locknut (1).
- Fully open and close the remote control lever and make sure the stopper portion (a) contacts the stopper portion (b) (at the fully opened position), and the stopper portion (c) contacts the stopper portion (d) (at the fully closed position).
- If proper contact is not made, repeat the above procedure.
- If the throttle control cable cannot be properly adjusted at the engine end, make adjustment at the remote control lever end.

Fig 27.39: ADJUSTING THE THROTTLE CONTROL CABLE *(Yamaha F8CW)*

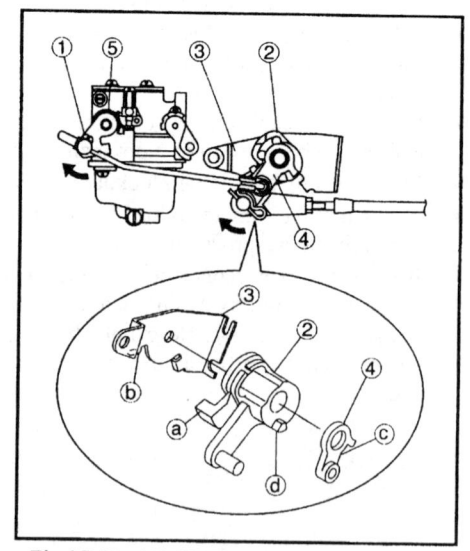

ADJUSTING THE THROTTLE AXLE LINK
(Yamaha F8CW)

Before adjusting the throttle axle link, make sure that the throttle control cable is properly adjusted.

Replace the throttle axle link if it is bent or damaged.

However, if it is incorrectly positioned, adjust it as follows:

- Loosen lock screw (1).
- Operate the control lever of the remote control box to the fully open position and put the stopper portion of pulley (2) in contact with the stopper portion of throttle wire stay (3).
- Hold so that the rear side face of free axle lever (4) is in contact with the actuating portion of pulley (2).
- Turn throttle link (5) fully clockwise until it stops, then tighten lock screw (1).
- Operate the control lever of the remote control box to the fully open position and fully closed position, and check that the throttle link (5) fully opens and closes. If it does not move properly, repeat the above procedure.

Fig 27.40: ADJUSTING THE THROTTLE AXLE LINK *(Yamaha F8CW)*

Chapter 27: Outboard Engines

ADJUSTING THE ENGINE IDLE SPEED
(Yamaha F8CW)

Before adjusting the engine idle speed, make sure the throttle axle link is properly adjusted. Warm up the engine for several minutes. Measure the engine idle speed by attaching the tachometer (1) to the spark plug lead (2) of the cylinder No. 1. It should read 1050 ± 50 rpm (for Yamaha F8CW)

If needed, adjust as follows:

- Turn in the pilot screw (3) in direction (a) until it is lightly seated.
- Turn out the pilot screw (3) in direction (b) to specified number of turns (between 1 & $\frac{3}{8}$).
- Turn the pilot screw (3) in direction (a) or (b), no more than ± $\frac{3}{8}$ of a turn in either direction, until the engine idling speed is at its highest.
- Turn the throttle stop screw (4) in direction (c) or (d) until the specified engine idling speed is obtained. [Direction (c) increases and direction (d) decreases idling speed]

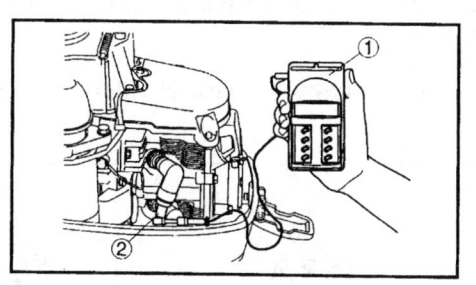

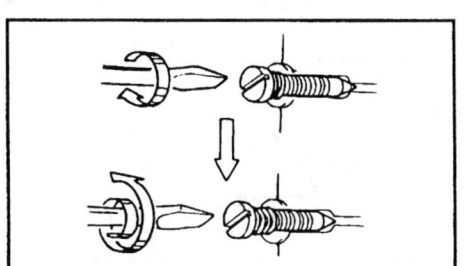

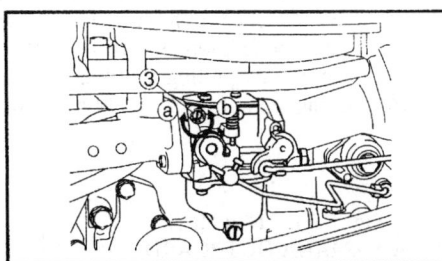

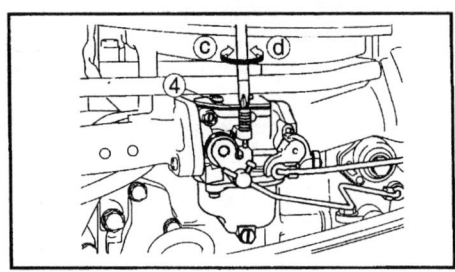

Fig 27.41: ADJUSTING THE ENGINE IDLING SPEED *(Yamaha F8CW)*

ADJUSTING THE START-IN-GEAR PROTECTION DEVICE
(Yamaha F8CW)

Check the start-in-gear protection device operation. If incorrect, adjust the start-in-gear protection cable as follows:

- Set the shift lever into neutral position.
- Loosen locknut (1)
- Adjust the adjusting nut (2) so that the point (a) on the wire connecter (3) aligns with the mark (b) on the starter cover.
- Tighten locknut (1).
- Shift the shift lever into neutral and make sure the starter can be pulled. If it cannot be pulled, repeat the above procedure.

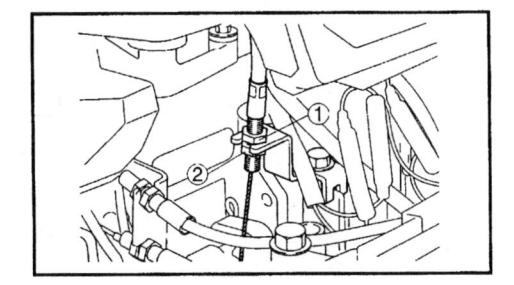

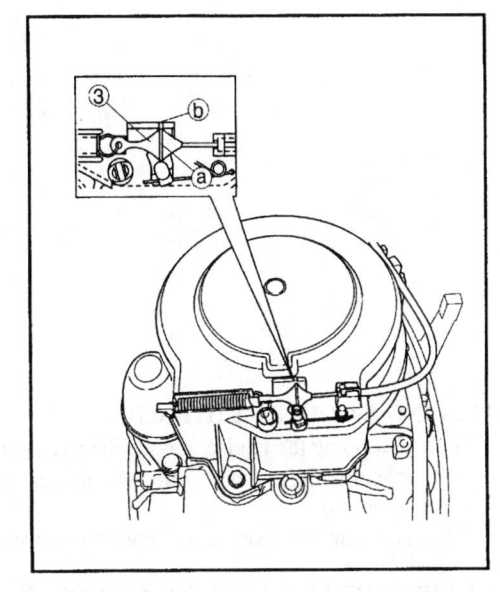

Fig. 27.42
ADJUSTING THE START-IN-GEAR PROTECTION DEVICE
(Yamaha F8CW)

STARTING A PETROL DRIVEN OUTBOARD

1. Carry out pre-start checks as discussed for inboard engines.
2. Loosen the air-vent screw on the fuel-tank cap by two or three turns (if fitted).
3. Firmly connect the fuel joint between the fuel tank and the motor.
4. Squeeze the primer bulb with the outlet end up until it becomes firm.
5. Ensure that the engine is in neutral, and the throttle grip on the steering handle in the START position.
6. Tie the line attached to the stop (kill) switch to your wrist.
7. Pull out the choke knob, if starting a cold engine.
8. Start the motor by pulling the (manual) starter handle or by pushing the (electric) starter switch.
9. Push the choke knob back into home position.

Fig 27.43: STARTING A YAMAHA OUTBOARD

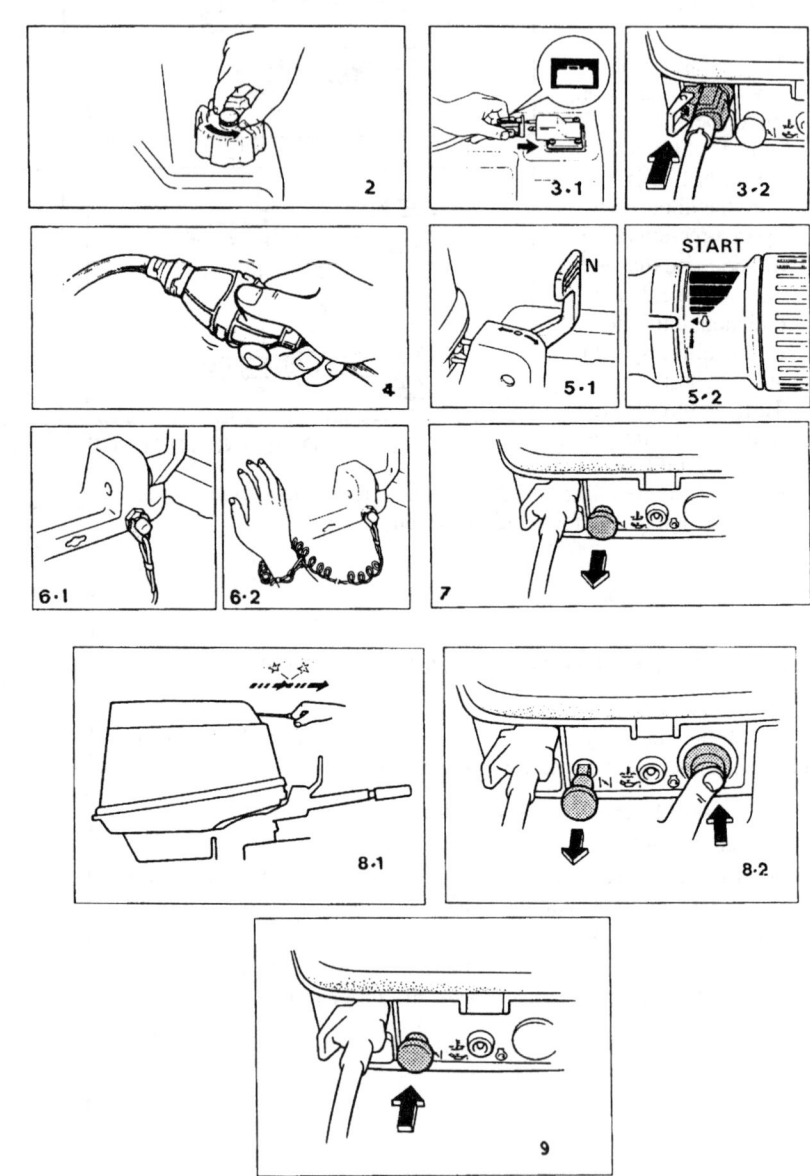

STOPPING AN OUTBOARD ENGINE

- Before stopping the engine, reduce the engine temperature by running it at idling speed or low speed for 2 to 3 minutes.
- After stopping the engine, tighten the air-vent screw (if fitted) on the fuel tank, and remove the fuel line connection from the motor.
- After stopping, remove the fuel connection from the motor.

ENGINE TOOLS & SPARES TO CARRY FOR OUTBOARDS

Bilge pump impeller, spare spark plugs, filters, plug spanner, torch, screw driver, pliers, emery paper, electric tape, spare

Chapter 27: Outboard Engines

hose (tank-motor connector), spare fuel line, stainless steel wire, plastic piping, hose clips, temporary hose and pipe repair tape, de-watering spray can. Other items as appropriate to motor and use.

PETROL AND LPG ENGINE - TROUBLESHOOTING

THE STARTER WILL NOT OPERATE (See Emergency Starting below)
- Battery capacity weak or low
- Battery connections loose or corroded
- Fuse for electric start circuit is blown
- Starter components faulty
- Engine stop switch lanyard not attached
- Shift lever in gear

THE STARTER OPERATES BUT ENGINE WILL NOT START
- Fuel tank empty
- Fuel contaminated or stale
- Fuel filter clogged
- Starting procedure incorrect. Solution: Read manual
- Fuel pump malfunctions
- Spark plug(s) fouled or incorrect type
- Spark plug cap(s) fitted incorrectly
- Poor connections or damaged ignition wire
- Ignition parts faulty
- Engine stop switch lanyard not attached
  Shift lever in gear
- Engine inner are parts damaged
- Bad settings or flooded carburettor (on conventional engine). See Carburettor below

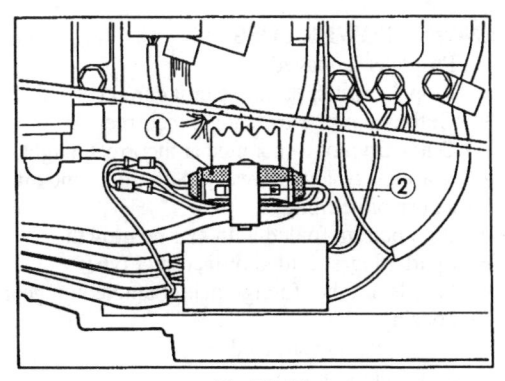

Fig 27.44
YAMAHA FUSE BOX (1) & FUSE (2)
(Electric Start Model)

ENGINE IDLES IRREGULARLY OR STALLS
- Spark plug(s) fouled or incorrect type
- Fuel system obstructed
- Fuel contaminated or stale
- Fuel filter clogged
- Failed ignition parts
- Warning system activated. Solution: Find and correct cause of warning
- Spark plug gap incorrect
- Poor connections or wet or damaged ignition wiring
- Specified engine oil not used
- Thermostat faulty or clogged
- Fuel pump damaged
- Air vent screw on the fuel tank closed
- Outboard motor angle too high
- Fuel joint connection incorrect
- Battery lead disconnected
- Bad settings or not enough fuel in the carburettor (on conventional engine). See Carburettor below
- Damaged pistons or valves
- Poor compression/induction. (Remove spark plugs and squirt some light oil into each cylinder. Rotate the motor a few times to lubricate rings and improve compression. Replace plugs and start the motor again)

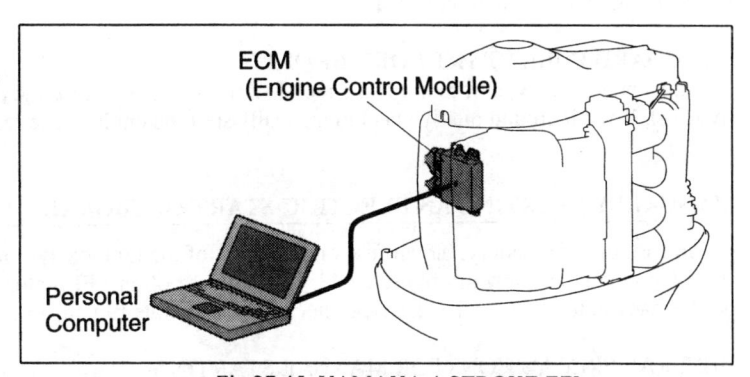

Fig 27.45: YAMAHA 4-STROKE EFI OUTBOARD DIAGNOSIS SYSTEM

WARNING BUZZER SOUNDS OR INDICATOR LAMP LIGHTS
- Cooling system clogged
- Engine oil level low
- Heat range of spark plugs incorrect. Solution: Replace spark plugs

- Specified engine oil not used
- Engine oil contaminated or deteriorated
- Oil filter clogged
- Oil feed/injection pump malfunctions
- Load on boat improperly distributed
- Water pump/thermostat faulty

ENGINE POWER LOSS

- Propeller damaged
- Propeller pitch or diameter incorrect
- Outboard motor trim angle incorrect
- Outboard motor mounted at incorrect height
- Warning system activated. Solution: Find and correct ca use of warning
- Boat bottom fouled with marine growth
- Spark plug(s) fouled or incorrect type
- Weeds or other foreign matter tangled on gear housing
- Fuel system obstructed
- Fuel filter clogged
- Fuel contaminated or stale

- Spark plug gap incorrect
- Poor connections or damaged ignition wiring
- Failed ignition parts
- Specified engine oil not used
- Thermostat faulty or clogged
- Air vent screw closed
- Fuel pump damaged
- Fuel joint connection incorrect
- Heat range of spark plug incorrect. Solution: Replace spark plugs
- Engine not responding properly to shift lever position

ENGINE VIBRATES EXCESSIVELY

- Propeller damaged
- Propeller shaft damaged
- Weeds or other foreign matter tangled on propeller
- Motor mounting bolt loose
- Steering pivot loose or damaged

OUTBOARD PULL-START CORD BROKEN

In a "pull start" type of motor if the recoil start cord breaks, take the cord off the motor, cutting its full length from the flywheel. Wrap it around the flywheel to manually start the engine. The recoil starter can be fixed later.

MANUALLY STARTING AN ELECTRIC-START OUTBOARD

In case of a flat battery, blown fuse or dampness of the ignition system, or in an emergency, an outboard engine can be started manually as shown in figure 27.43. Only in case of an EFI motor, the manual starting would not be possible if the battery has run too low for the electrical fuel pump to operate (9 volts or less on a Yamaha motor).

GENERAL PRECAUTIONS FOR MANUAL STARTING

- If battery power is insufficient, use the manual-lowering device to tilt the motor back into water in order to gain access to cooling water. Unscrewing the lowering device slowly tilts the engine down. It must be re-tightened after lowering.
- Turn off the unnecessary electrical system.
- Make sure the remote control lever is in neutral, the fast idle lever is raised and the kill switch is positioned correctly. This would ensure that the vessel does not unexpectedly start to move.
- Keep loose clothing and other objects away from the engine.
- Turn the ignition to "on" position (not to "off" or "start" position).
- No one should be standing behind you when pulling the starter rope.
- Follow the manufacturers instructions for using the starter rope (See instructions for the Yamaha EFI motor below). Otherwise, generally, wrap the emergency starter rope around the flywheel and pull as hard as you can. A few pulls should start an engine in good condition. All outboard motors are fitted with a grooved flywheel to fit the rope. Two turns of rope are sufficient.
- If the cowling had to be removed to access the flywheel, then do not touch it or other moving parts when the engine is running. An unguarded rotating flywheel is very dangerous.
- Do not install the starter mechanism or top cowling after the engine is running.
- Do not touch the ignition coil or any other electrical component when starting or operating the motor. It could give you a nasty electrical shock.

Chapter 27: Outboard Engines

MANUALLY STARTING A YAMAHA EFI OUTBOARD

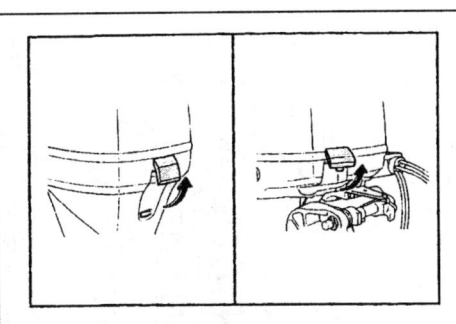

Emergency Starting Engine
1) Remove the top cowling.
2) Remove the two bolts holding the flywheel cover. 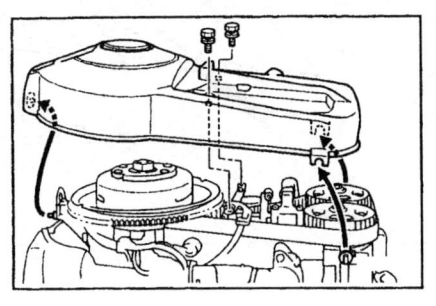
3) Lift up the rear of flywheel cover and pull it out forward. 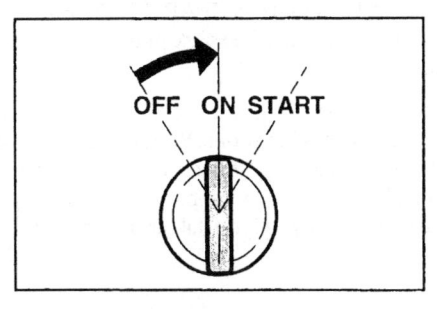
4) Prepare the engine for starting. See "STARTING ENGINE" for procedures. Be sure the engine is in Neutral and that the lock plate is attached to the engine stop lanyard switch. The main switch must be "ON".
5) Insert the knotted end of the emergency starter rope into the notch in the flywheel rotor and wind the rope clockwise.
6) Pull the rope slowly until resistance is felt. 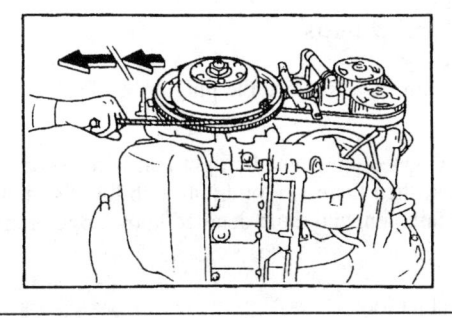
7) Remove the rope from the flywheel temporarily.
8) Rewind the rope approximately 3/4 of a turn clockwise.
9) Give a strong pull straight out to crank and start the engine. Repeat if necessary.

Fig. 27.46: EMERGENCY STARTING (YAMAHA EFI MODEL F115A)

ENGINE FLOODED WITH FUEL DUE TO REPEATED STARTING EFFORT
Let it sit for a while. Wait for a few minutes for the fuel to evaporate before trying again. This problem may occur with a 4-stroke engine. 2-stroke engines like extra fuel; it is hard to flood them.

CARBURETTOR PROBLEMS (IN NON-EFI CONVENTIONAL ENGINES)

Carburettors are usually calibrated and pre-set at the factory. Further adjustments should be carried out only in extreme changes in weather or elevation. If you don't have the knowledge to make carburettor adjustment, seek expert assistance.

Fuel spitting out through the carburettor is a serious fire hazard. The LEAF VALVES inside the carburettor are faulty. Leaf Valves time the injection of air/fuel mixture into engine, and prevent engine pressure passing back through the carburettor. Seek expert assistance.

Fig 27.47
IDLE SPEED ADJUSTMENT
ON A MERCURY CARBURETTOR

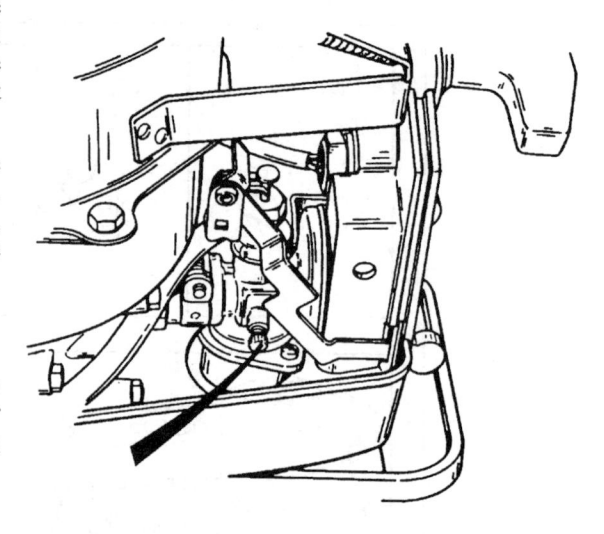

SPARK PLUG PROBLEMS

CHECKING SPARK PLUGS FOR SPARK

The safest and easiest way to check the ignition system for sparks is to use a SPARK TESTER. More than one type is available. Remove the end of the spark plug lead from the spark plug, leaving the spark plug in its hole, Connect the lead to the tester, turn on the ignition, and look for sparks inside the tester.

In the absence of a tester, it would be necessary to remove the spark plug to look for sparks travelling across its gap. Ensure there is good ventilation and no presence of fuel vapour in the engine room. Remove a spark plug from the engine and connect it back to its high-tension lead. In a fuel-injected engine, there is a greater risk of fire from fuel being injected into the uncovered spark plug hole during testing. It would therefore be safer to temporarily insert a spare spark plug in it. Keeping it away from the open hole, earth the body of the spark plug that you have taken out against the engine block, and turn the engine. Spark should be seen across the gap.

- IF SPARK PRESENT: Remove and check the spark plugs
    - PLUGS ARE DRY: Check fuel supply, fuel lines & filters
    - PLUGS ARE DIRTY: Clean or replace the plugs
    - PLUGS ARE WET: The engine is flooded with fuel. Replace the plug & HT lead. Open throttle & crank engine about 10 times to clear the cylinders.

- IF NO SPARK: Loose or wet wiring. Check all wiring. Connections must be tight and wires should be clean and dry. Spray dirty wiring with a water repellent. If still no spark, then there is a problem with the distributor, coil, switch or the capacitor (condenser). Tracing it requires some test equipment and more detailed knowledge of the ignition system.

SPARK PLUGS OILED
Excessive oil in the fuel or engine running rich for a prolonged period.

SPARK PLUG CLEANING & ADJUSTMENT

The spark plugs (in petrol and LPG engines) should be periodically removed and inspected. The heat and deposits will slowly cause them to break down and erode. They should be replaced when badly eroded or excessively covered with carbon and other deposits. The condition of a spark plug can also indicate the condition of the engine. For example, if the centre electrode is very white, there could be an intake air leak or carburetion problem in that cylinder, which requires expert assistance.

The spark plug gap should be set as recommended by the manufacturer (usually 0.9 - 1.0 mm). When fitting the plug, clean the gasket surface and use a new gasket. Wipe off any dirt from the threads and screw in the spark plug with a torque wrench or ¼ to ½ turn past finger tight. Be careful not to damage the spark plug insulator. If damaged, it could allow external sparks, leading to explosion or fire.

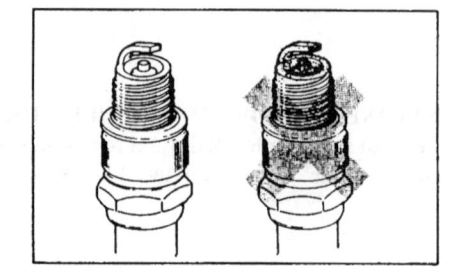

Fig 27.48: SPARK PLUG CONDITION

Chapter 27: Outboard Engines

OTHER OUTBOARD PROBLEMS

REMOTE CONTROL DAMAGED
Disconnect the remote control. Drive the motor manually.

TWO STROKE PETROL VRO PUMP FAILURE
Add two-stroke oil directly to the fuel tank in the ratio of 50:1.

OUTBOARD IMPACT DAMAGE
If the outboard motor hits an object in the water, stop engine immediately. Inspect the control system, all components as well as the vessel for damage. Even if the damage is not found, return to the nearest harbour slowly and carefully for a thorough specialist inspection.

RUNNING ON ONE OUTBOARD
When running a duel outboard vessel on one engine, operate it at low speed while keeping the unused motor tilted up. This will prevent engine trouble arising from water running into the exhaust pipe of the unused motor.

POWER TRIM AND TILT NOT OPERATING
In case of the failure of battery or of the power trim and tilt unit, the engine can be tilted manually. In a Yamaha motor, loosen the manual valve screw, put the engine in the required position and retighten the valve screw.

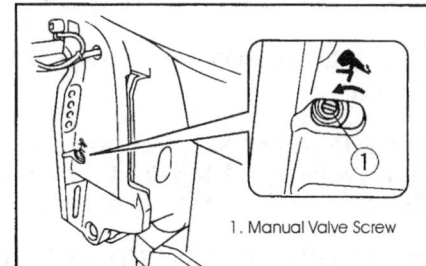

Fig 27.49
YAMAHA OUTBOARD
MANUAL TILT VALVE SCREW

OUTBOARD ENGINE DROPPED OVERBOARD (NOT RUNNING)
Outboard motors can sometimes shake loose and fall overboard. Securing it with a chain or lanyard to the transom will prevent it from being lost altogether. The securing clamps should be tightened before starting, and then again after about 15 minutes of running.

If, however, it does fall overboard, recover it and service it within three hours. If seawater submersion is not treated quickly enough, even if the engine can be restarted, it must be disassembled and cleaned and electrical components cleaned or replaced where necessary. Salt water can cause excessive corrosion of electrical components and internal parts.

If engine cannot be started or serviced immediately, it should be resubmerged in fresh water to avoid exposure to the atmosphere. If it is exposed to the air, both fresh and salt water will start etching the highly polished bearing surfaces of the crankshaft, connecting rods and bearings.

SERVICING AFTER SUBMERSION
(As recommended by Johnson Evinrude)
- Remove engine cover and rinse powerhead with freshwater.
- Disconnect spark plug leads and remove spark plugs.
- Disconnect fuel lines from engine. Drain and clean all fuel lines and fuel tank.
- Place engine in horizontal position (spark plug openings down) and work out all of the water by slowly rotating flywheel approximately 20 times or until there is no sign of water.
- Drain carburettor (in a conventional engine) - place engine in upright position and remove carburettor for disassembly and draining.
- Fill the sump (of a 4-stroke engine) with fresh engine oil.
- Disassemble, clean, and completely flush the starter, electrical connectors, and all electrical equipment with fresh water. Then treat them with a water displacing electrical spray, and thoroughly dry them before assembly.
- Inject engine fogging oil or engine oil through the carburettor (if fitted) and into spark plug holes while cranking the engine with the manual starter or emergency starter rope.
- If engine shows evidence that sand or silt may have entered it (sand or silt under the engine cover or a slight grinding or scraping when flywheel is rotated), do not attempt to start then engine. It must be disassembled and cleaned.
- Reassemble the parts. Start engine and run for 30 minutes in water.
- If engine fails to start, remove spark plugs again to see if water is present on electrodes. Blow out any water from electrodes and reinstall or replace with new plugs. Repeat starting procedure.

OUTBOARD ENGINE DROPPED OVERBOARD (RUNNING)
Follow the above procedure. However, if there is any binding when flywheel is rotated it may indicate a bent connecting rod. Do not attempt to start the engine. Powerhead must be disassembled and serviced immediately.

PROPELLER REPLACEMENT: See Propulsion Systems & Propellers

OUTBOARD MAINTENANCE & INSPECTION SCHEDULE

(The following intervals are for salt-water runs. Double each interval for fresh water runs. Follow manufacturer's recommendations.)

EVERY 30 DAYS: Lubricate all parts including: Clamp Screws, Steering system, Reverse Locking System, Starter Motor Pinion Gear, Carrying handle, Fuel Shut-Off Linkage, Choke & Throttle Linkage, Swivel Brackets, Shift lever Fitting, Trail Lock Linkage, Tilt Linkage, Carburettor Linkage (if fitted), Propeller Shaft & Drive Shaft.

GEARCASE OIL: Check level every 50 operating hours. Change every 100 operating hours or once each season, whichever comes first.

EVERY 60 DAYS: Check level of lubricant in Power Trim Pump. Lubricate Starter Motor Pinion Gear.

TWICE IN SEASON: Inspect batteries & terminals. Inspect fuel lines & connections. Inspect, clean & touch up paint on entire unit.

ONCE IN SEASON: Inspect spark plug leads & all electrical connections. Inspect condition of spark plugs. Clean fuel filter(s) & fuel tank Filter. Check for loose, damaged or missing parts in the entire unit. Inspect distributor points (if fitted). Inspect anodes. Inspect or replace water pump impeller. Change 4-Stroke engine oil and oil filter. Inspect propeller.

OUTBOARDS - OFF-SEASON STORAGE
- In order to prevent sump oil entering the cylinders, transport and store a 4-stroke outboard in the upright position. If it must be placed in a horizontal position, place it on a cushion after draining the engine oil.
- Remove the fuel line connection from the motor or shut off the fuel valve, if fitted.
- Drain any fuel from the vapour separator of an EFI engine. Fuel left in the vapour separator for a prolonged period would break down and possibly damage the fuel line.
- Wash the motor body with fresh water. Flush the cooling water passages with fresh water. Run the engine at idling speed while supplying fresh water to the cooling water passages until the fuel system becomes empty and the engine stops. Drain out all the water. Do not place the engine on its side until the cooling water has been drained out completely. The water can enter the cylinders through the exhaust ports and cause problems.
- Remove the spark plugs, pour a teaspoonful of clean engine oil into the cylinders, crank several times manually and replace the spark plugs.
- Remove batteries; inspect condition & store in a cool dry place. Charge periodically during storage.
- Service the fuel filter.
- Remove propeller and check for damage. Clean and lubricate propeller shaft.
- Drain & refill gearcase. Lubricate engine.
- Coat all outside surfaces with automotive wax.

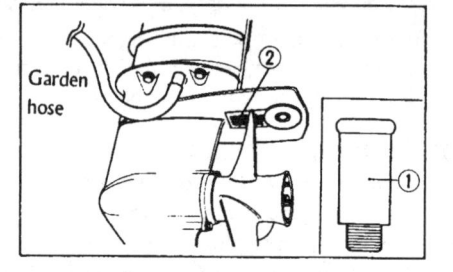

1. Shift the motor into "NEUTRAL".
2. Remove the flush plug and install the flushing attachment ①. Plug the water intake hole ② with a piece of duct tape.
3. Connect a garden hose to the flushing attachment and turn on the water to obtain a good water flow.

Fig 27.50: FLUSHING A SUZUKI

PRE-SEASON SERVICE
- Check lubrication in gearcase, power trim/tilt reservoir.
- Apply a light coating of marine grease to the ribbed portion of spark plug ceramics and the opening of the spark plug covers. Connect spark plug leads.
- Check battery for water level and charge.
- Check fuel system for leaks.
- Check sump oil (in 4-stroke motors).
- Before starting the engine, supply water to its cooling water passage. Running it dry would damage the water pump and/or the engine.

FUEL TANK SERVICE
- Drain and flush fuel tank at least once a year and at every tune-up or major repair.

Chapter 27: Outboard Engines

JOHNSON EVINRUDE

OUTBOARD SAFETY GUIDE

DO NOT SUBSTITUTE PARTS

"They look the same, but . . . . . are they the same?

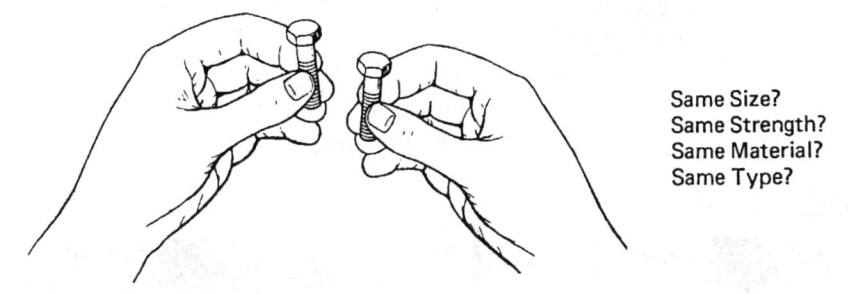

Same Size?
Same Strength?
Same Material?
Same Type?

Don't Substitute unless <u>You Know</u> they are the <u>Same</u> in <u>All</u> characteristics.

- Special Locking bolts and nuts are often used to hold steering, shift, and throttle remote control cables to the motor.

- When you take any motor off a boat, Keep Track of Special Nuts and Bolts. Don't mix with other parts.

- When motor is returned to boat, Use Only the Special Nuts and Bolts to hold remote Steering, Shift, and Throttle cables to the motor.

The Australian Boating Manual

Outboard Shift System and Persons Safety

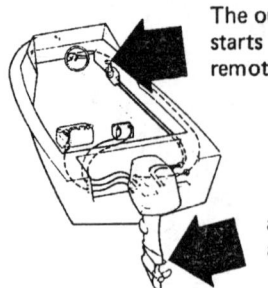

The outboard Shift System starts here at the remote control lever.....

and ends here..... at the propeller.

What's Most Important?

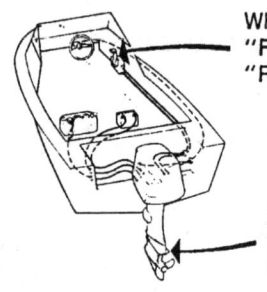

When control lever is in "Forward," "Neutral" or "Reverse" position,

Shift linkage must match control lever position.

What Could Happen?

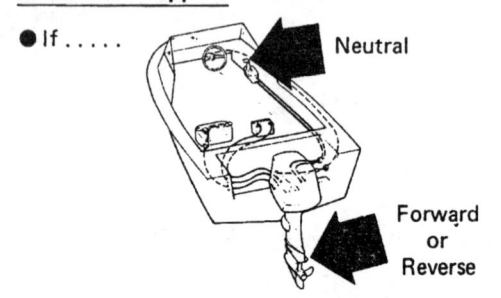

- If..... Neutral — Forward or Reverse

..... propeller still powered (turning) unknown to operator, or motor will Start in Gear, boat will move suddenly.

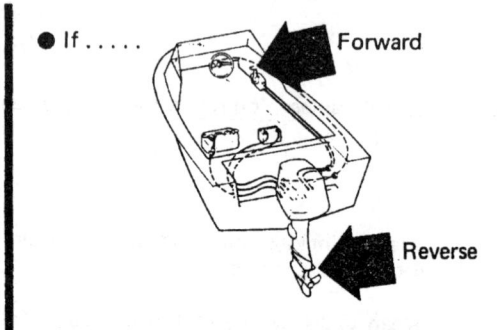

- If..... Forward — Reverse

..... boat will move Opposite to direction wanted by operator.

when Rigging or after Servicing

How Can Loss of Shift Control be Minimized?

- <u>Read</u>, <u>Understand</u>, and <u>Follow</u> manufacturers <u>Instructions</u>.
- Follow warnings marked " ⚠ " closely.
- Assemble Parts Carefully.
- Make Adjustments Carefully.
- Test Your Work. Don't Guess. Make Sure Propeller Does Just What the Operator Wants and Nothing Else.
- Do Not Shift Gears When Motor is Stopped. Adjustments Can be Lost and Parts Weakened.

Chapter 27: Outboard Engines

Outboard Speed Control System and Persons Safety

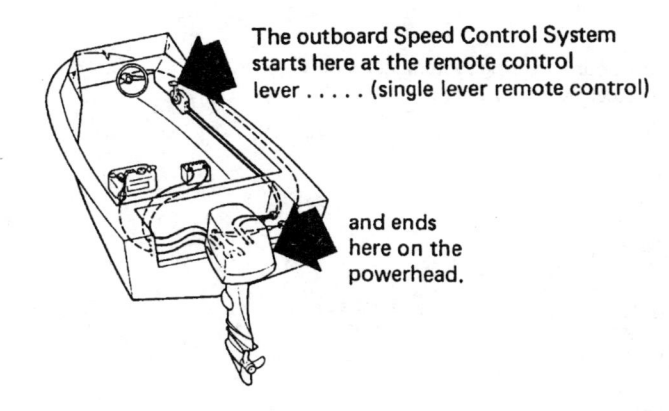

The outboard Speed Control System starts here at the remote control lever . . . . . (single lever remote control)

and ends here on the powerhead.

What's Most Important?

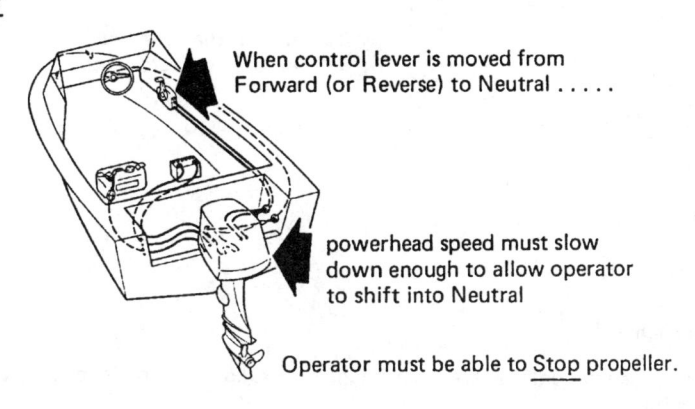

When control lever is moved from Forward (or Reverse) to Neutral . . . . .

powerhead speed must slow down enough to allow operator to shift into Neutral

Operator must be able to Stop propeller.

What Could Happen?

If Operator can't slow down the motor Or shift into Neutral gear (stop propeller), Operator could panic and lose control of boat.

How Can Loss of Speed Control be Minimized?

- Read, Understand, and Follow manufacturers Instructions.
- Follow warnings marked " ⚠ " closely.

when Rigging or after Servicing

- Assemble Parts Carefully
- Make Adjustment Carefully
- Test Your Work. Don't Guess. Make Sure Motor Changes Speed Smoothly, Quickly.
- Make Sure Full Throttle Can be Obtained so Operator Won't Overload Parts.

Outboard Steering System and Persons Safety

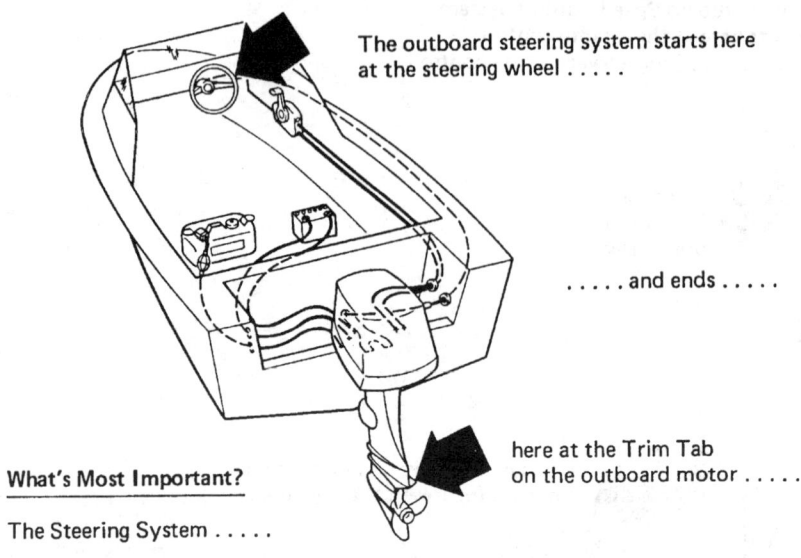

The outboard steering system starts here at the steering wheel . . . . .

. . . . . and ends . . . . .

here at the Trim Tab on the outboard motor . . . . .

What's Most Important?

The Steering System . . . . .

- must not come apart
- must not jam
- must not be sloppy or loose

What Could Happen?

- If steering system comes apart, boat might turn suddenly and circle . . . . . persons thrown into water could be hit.

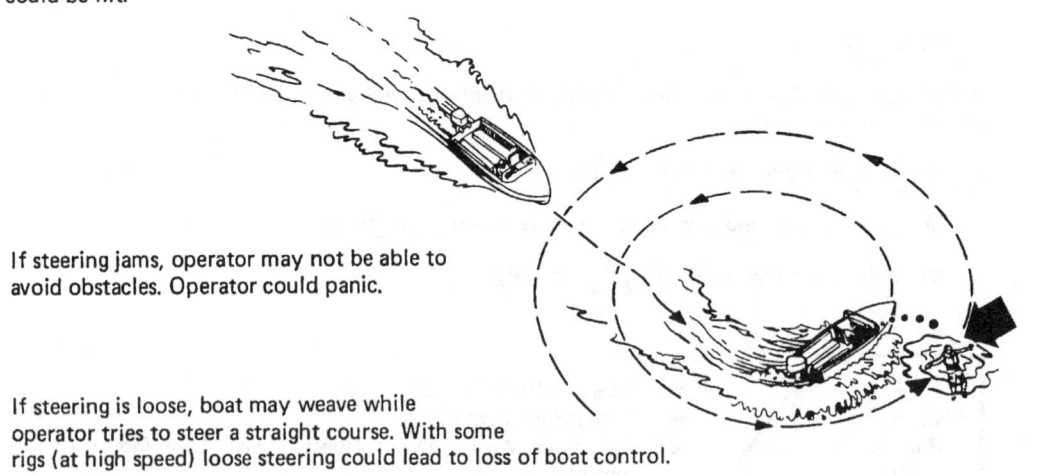

- If steering jams, operator may not be able to avoid obstacles. Operator could panic.

- If steering is loose, boat may weave while operator tries to steer a straight course. With some rigs (at high speed) loose steering could lead to loss of boat control.

Chapter 27: Outboard Engines

How Can Loss of Steering Control be Minimized?

When Rigging or after Servicing

- Use a steering system recommended by the motor manufacturer which meets Marine Industry Safety Standards (ABYC).
- <u>Read</u>, <u>Understand</u>, and <u>Follow</u> manufacturers <u>Instructions</u>.
- Follow warnings marked " ⚠ " closely.
- Assemble Parts Carefully
- Make Adjustments Carefully
- Keep Parts Moving Freely . . . . . Lubricate Parts as Shown in Manuals
- Use the bolts, nuts, and washers supplied with steering attachment kits . . . . they're a Special Locking Type That Won't Loosen, Rust, or Weaken.

- When Transom Mounted steering systems (see picture) are used, check to uncover possible Trouble!

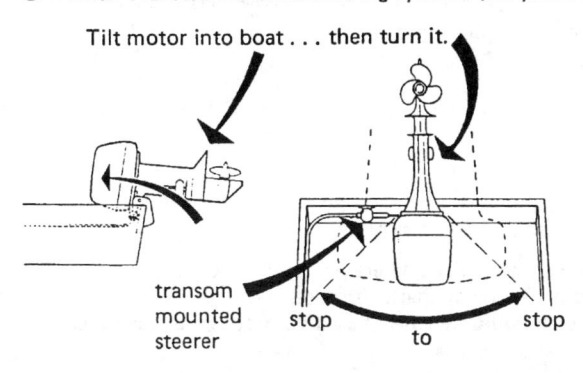

Tilt motor into boat . . . then turn it.

transom mounted steerer — stop to stop

During this procedure, steering parts

- Must Not Bind
- Must Not <u>Touch</u> Other boat, motor or accessory Parts in Transom Area.

<u>Why?</u> A hard blow to the motor's gearcase can result in damage to steering parts here

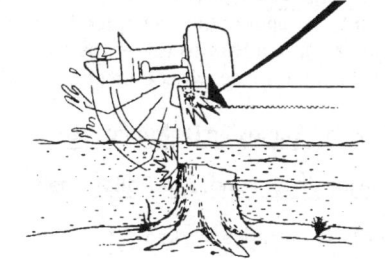

- Be aware that Raising or Lowering motor on transom can change a set up which was O.K. earlier. If moved up or down even 1/2 inch, run test again to make sure steering parts are free and clear.

- Check for Damaged parts . . . . . Blows to the Motor like this

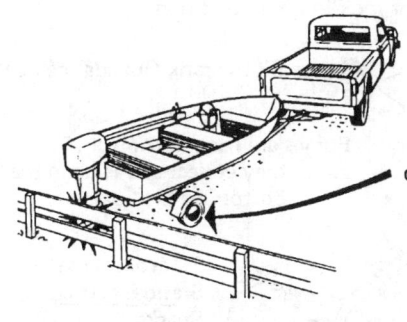

or this can put Heavy Loads on steering parts. Look for . . . . .

- Cracked parts, including steering parts, swivel brackets and transom brackets
- Bent parts
- Loose nuts and bolts

- Replace damaged parts. If weakened, parts could fail later . . . . on the water . . . . when least expected.

Outboard Fuel, Electrical System and Persons Safety

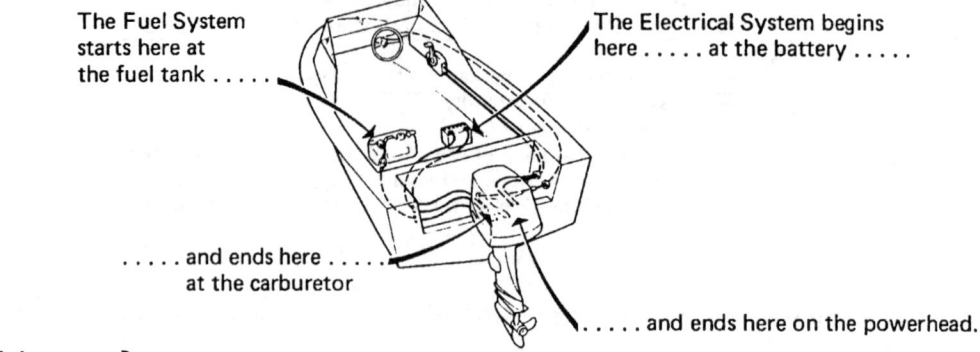

The Fuel System starts here at the fuel tank . . . . .

The Electrical System begins here . . . . . at the battery . . . . .

. . . . . and ends here . . . . . at the carburetor

. . . . . and ends here on the powerhead.

What's Important?

- Fuel Leakage must be eliminated.
- Stray electric sparks must be avoided.

What Could Happen?

Gasoline can ignite and burn easily. Gasoline vapors can ignite and explode.

- When not boating, fuel leaking in car trunk or van, or a place where portable tank is stored (basement, cottage) could be ignited by any open flame or spark (furnace pilot light, etc.).
- When boating, fuel leaking under the motor cover could be ignited by a damaged or deteriorated electrical part or loose wire connection making stray sparks.

How Can Fire and Explosion Be Minimized?

- Read, Understand, and Follow manufacturers Instructions.
- Follow warnings marked "⚠" closely.
- Do not substitute fuel or electrical system parts with other parts which may look the same. Some electrical parts, like starter motors, are of special design to prevent stray sparks outside their cases.
- Replace wires, sleeves, boots which are cracked or torn or look in poor condition.
- When Mixing and Refueling . . . . . Always Mix gas and oil Outside . . . . .

And remember

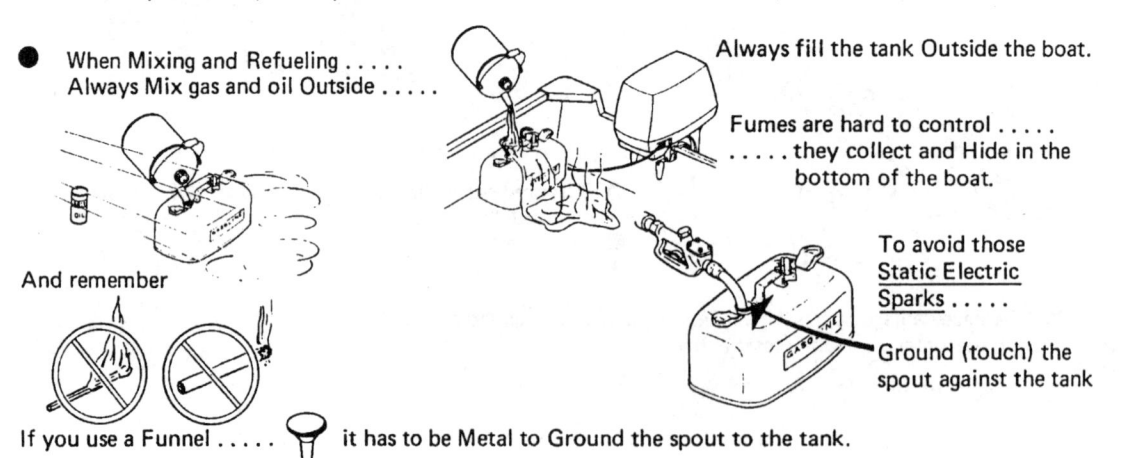

If you use a Funnel . . . . . it has to be Metal to Ground the spout to the tank.

Always fill the tank Outside the boat.

Fumes are hard to control . . . . they collect and Hide in the bottom of the boat.

To avoid those Static Electric Sparks . . . . .

Ground (touch) the spout against the tank

Chapter 27: Outboard Engines

- After Repair on any part of the fuel system, pressure test engine portion of fuel system as shown: (See Section 2 for testing the fuel tank portion.)

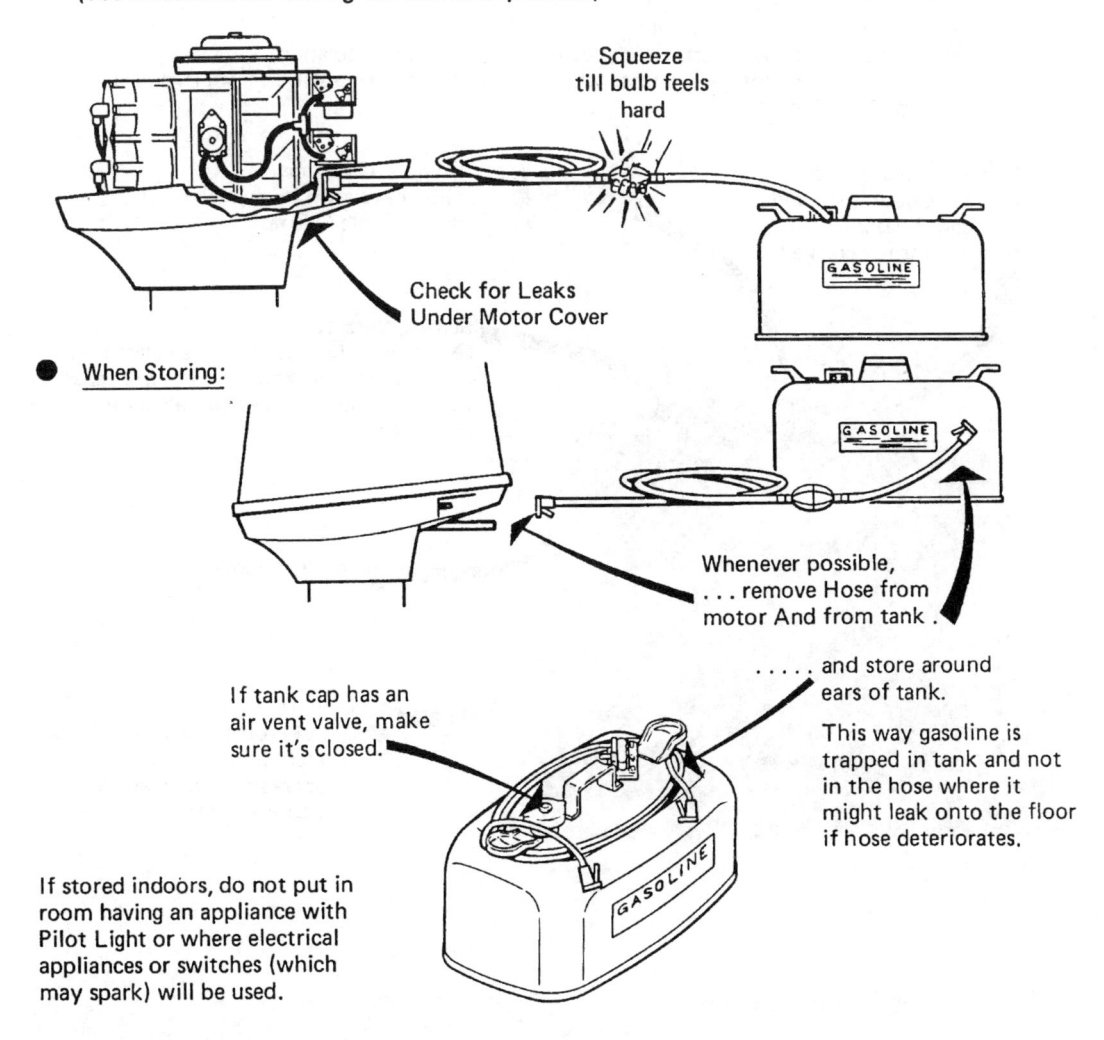

Squeeze till bulb feels hard

Check for Leaks Under Motor Cover

- When Storing:

Whenever possible, ...remove Hose from motor And from tank.

If tank cap has an air vent valve, make sure it's closed.

..... and store around ears of tank.

This way gasoline is trapped in tank and not in the hose where it might leak onto the floor if hose deteriorates.

If stored indoors, do not put in room having an appliance with Pilot Light or where electrical appliances or switches (which may spark) will be used.

- When Running:

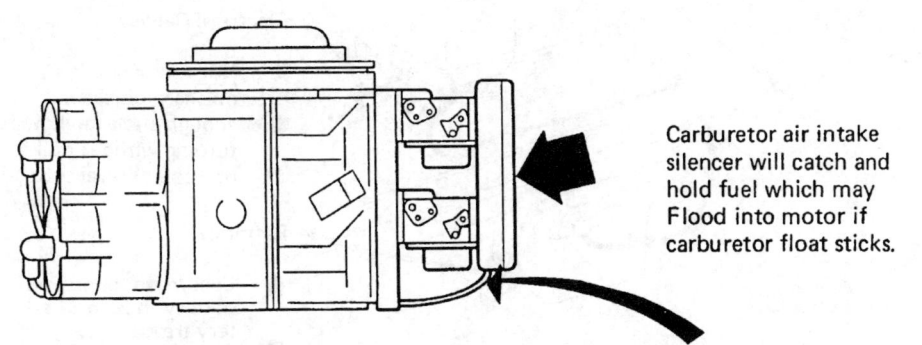

Carburetor air intake silencer will catch and hold fuel which may Flood into motor if carburetor float sticks.

So make sure Silencer and all its Gaskets are on motor And Drain Hose is in place.

Air silencer mounting Screws are special lock screws. Use only the special Screws.

- If electrical parts are replaced or even removed from the motor, check the following:

    Wire and High Voltage Lead Routing
    - as shown in service manual
    - away from moving parts which could cut wires or wire insulation
    - away from motor cover latches which can catch and cut insulation from High Voltage spark plug leads.

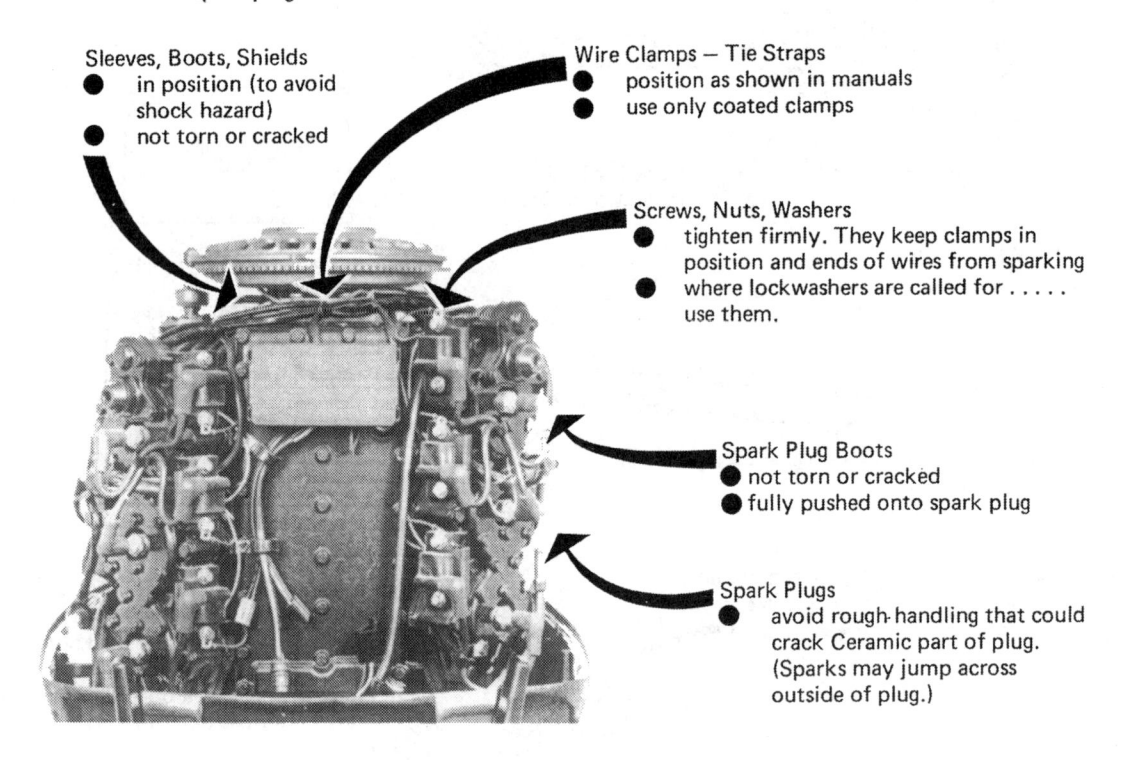

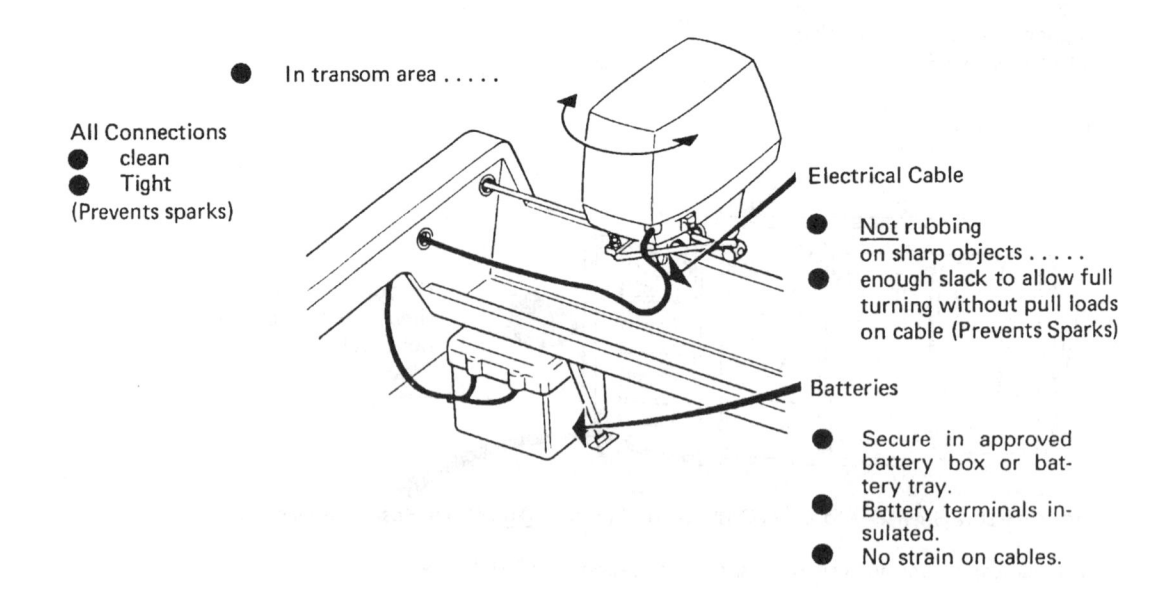

Sleeves, Boots, Shields
- in position (to avoid shock hazard)
- not torn or cracked

Wire Clamps — Tie Straps
- position as shown in manuals
- use only coated clamps

Screws, Nuts, Washers
- tighten firmly. They keep clamps in position and ends of wires from sparking
- where lockwashers are called for . . . . . use them.

Spark Plug Boots
- not torn or cracked
- fully pushed onto spark plug

Spark Plugs
- avoid rough handling that could crack Ceramic part of plug. (Sparks may jump across outside of plug.)

- In transom area . . . . .

All Connections
- clean
- Tight

(Prevents sparks)

Electrical Cable
- Not rubbing on sharp objects . . . . . enough slack to allow full turning without pull loads on cable (Prevents Sparks)

Batteries
- Secure in approved battery box or battery tray.
- Battery terminals insulated.
- No strain on cables.

Chapter 27: Outboard Engines

Outboard Mounting System and Persons Safety

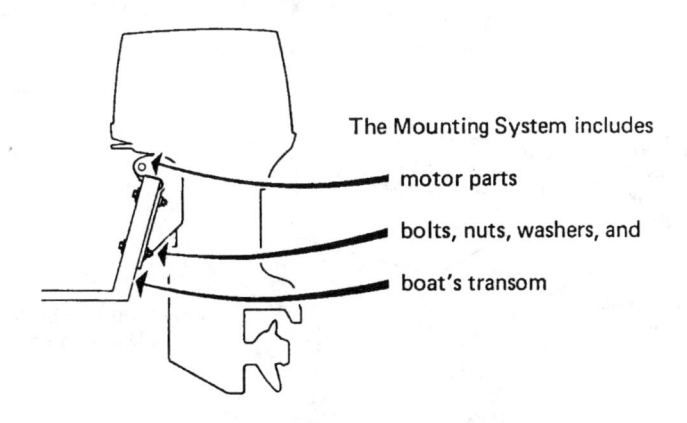

The Mounting System includes
- motor parts
- bolts, nuts, washers, and
- boat's transom

What's Important?

Motor must <u>stay</u> in position on boat's transom

What Could Happen?

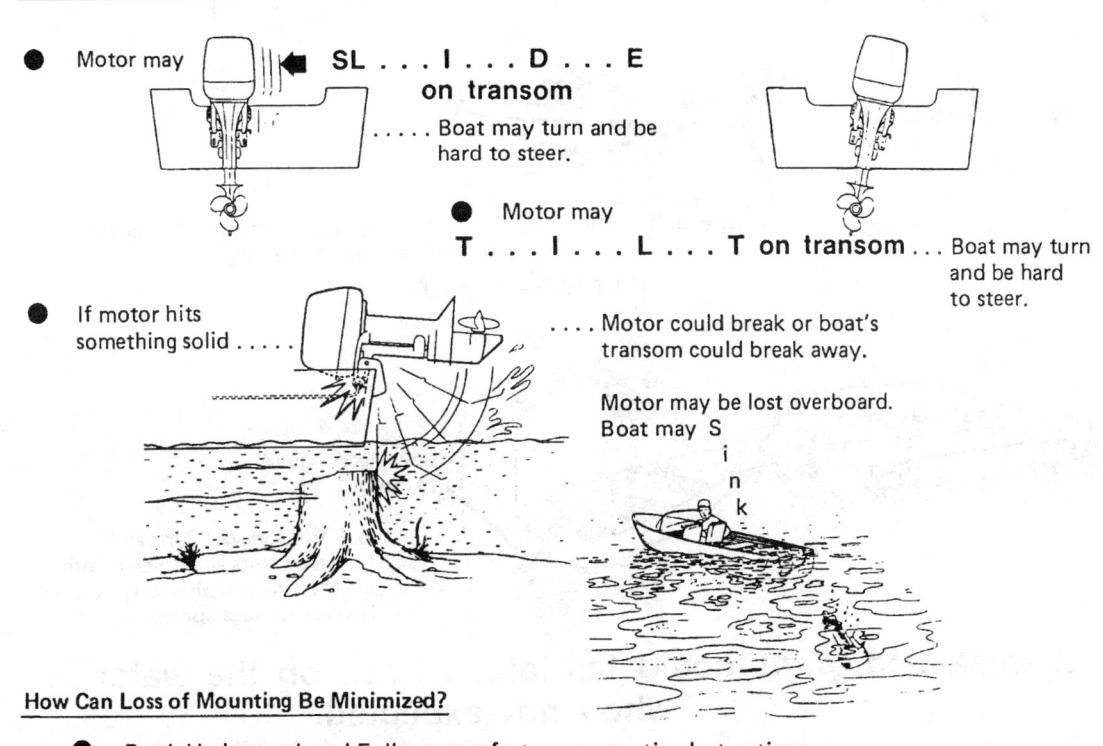

- Motor may **SL...I...D...E on transom**
  ..... Boat may turn and be hard to steer.

- Motor may **T...I...L...T on transom** ... Boat may turn and be hard to steer.

- If motor hits something solid ..... Motor could break or boat's transom could break away.

  Motor may be lost overboard. Boat may S i n k

How Can Loss of Mounting Be Minimized?

- <u>Read</u>, <u>Understand</u>, and <u>Follow</u> manufacturers mounting <u>Instructions</u>.
- Follow warnings marked " ⚠ " closely .....

- If Boat Plate shows.....

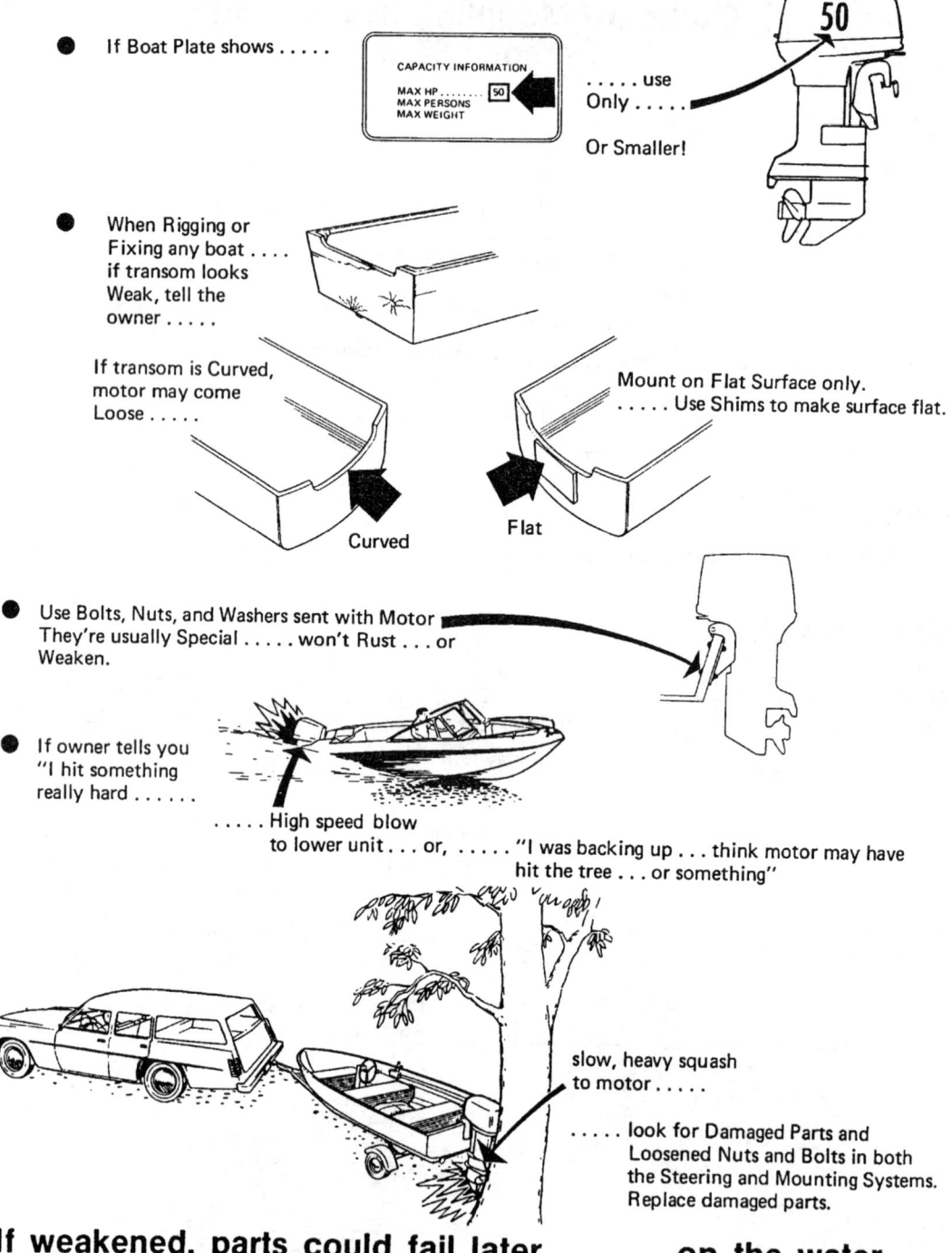

  CAPACITY INFORMATION
  MAX HP........ 50
  MAX PERSONS
  MAX WEIGHT

  .....use Only.....

  Or Smaller!

- When Rigging or Fixing any boat.... if transom looks Weak, tell the owner.....

  If transom is Curved, motor may come Loose.....

  Curved

  Mount on Flat Surface only.
  ..... Use Shims to make surface flat.

  Flat

- Use Bolts, Nuts, and Washers sent with Motor
  They're usually Special..... won't Rust... or Weaken.

- If owner tells you "I hit something really hard ......

  ..... High speed blow to lower unit... or, ..."I was backing up... think motor may have hit the tree... or something"

  slow, heavy squash to motor.....

  ..... look for Damaged Parts and Loosened Nuts and Bolts in both the Steering and Mounting Systems. Replace damaged parts.

If weakened, parts could fail later ..... on the water .. .. when not expected.

Chapter 27: Outboard Engines

Outboard Hydraulic Tilt/Trim Shock Absorption System and Persons Safety

What's Important?

- Shock Absorption system must always be ready to Absorb some Blows to the lower parts of the motor.

- Motor must not Trim "In" too far Suddenly

What Can Happen?

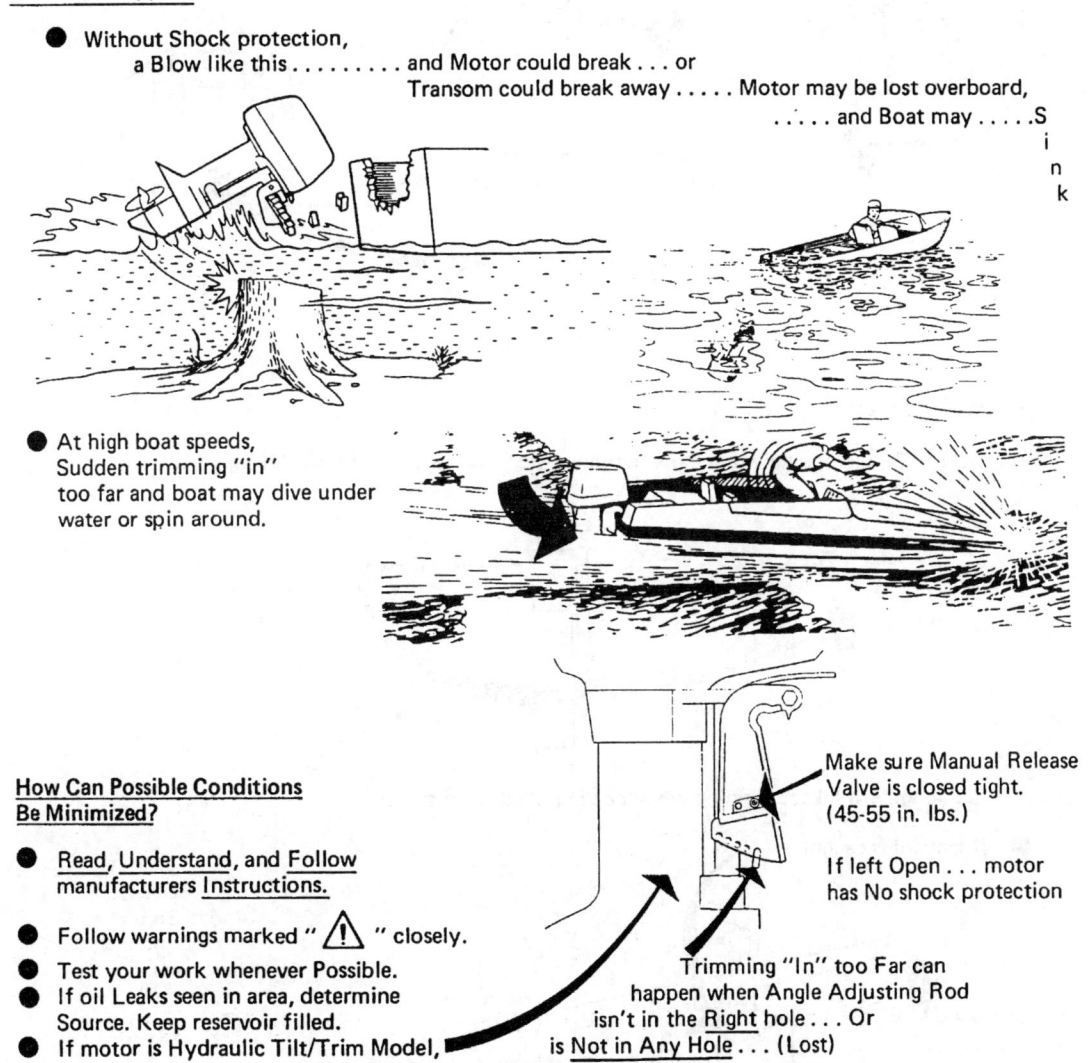

- Without Shock protection, a Blow like this......... and Motor could break... or Transom could break away..... Motor may be lost overboard, ... and Boat may.....S
  i
  n
  k

- At high boat speeds, Sudden trimming "in" too far and boat may dive under water or spin around.

How Can Possible Conditions Be Minimized?

- Read, Understand, and Follow manufacturers Instructions.
- Follow warnings marked " ⚠ " closely.
- Test your work whenever Possible.
- If oil Leaks seen in area, determine Source. Keep reservoir filled.
- If motor is Hydraulic Tilt/Trim Model,

Make sure Manual Release Valve is closed tight. (45-55 in. lbs.)

If left Open... motor has No shock protection

Trimming "In" too Far can happen when Angle Adjusting Rod isn't in the Right hole... Or is Not in Any Hole... (Lost)

Always return Rod to Hole Position determined earlier by Boat Operator... and make sure Angle Adjusting Rod Retainer is in locked position.

Outboard Emergency Stop System and Persons Safety

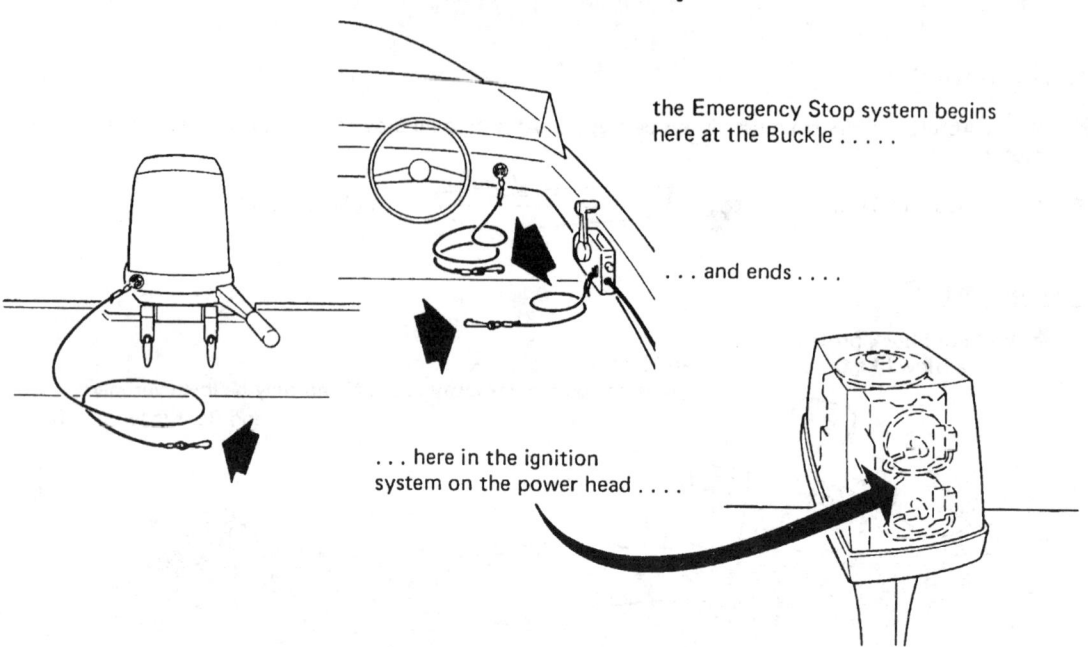

the Emergency Stop system begins here at the Buckle . . . . .

. . . and ends . . . .

. . . here in the ignition system on the power head . . . .

The Emergency System . . . . .

- must STOP the engine when the emergency stop switch's cap is pulled from switch.

- If switch fails . . . . .

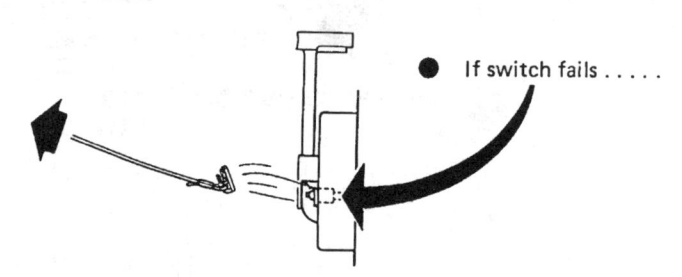

. . . . engine will keep running when cap is pulled from switch.

- If lanyard is caught . . . .

. . . engine will keep running.

Chapter 27: Outboard Engines

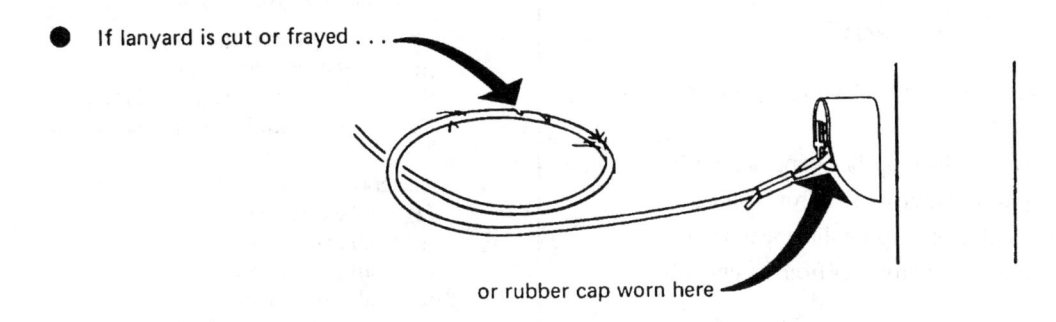

- If lanyard is cut or frayed . . .

or rubber cap worn here

. . . lanyard or cap may break when pulled . . . .

and engine will continue running.

- If engine does NOT stop when lanyard is pulled, an operator thrown from boat could be hit as boat circles area . . . . or boat may not turn but leave area as a runaway . . . . operator may drown and boat WILL run into something.

How can failure of the Emergency Stop system be minimized?

- **Read, Understand,** and **Follow** manufacturers **Instructions.**
- Follow warnings marked ⚠ closely.

When Rigging or after Servicing

- Assemble parts carefully
- Inspect lanyard for cuts or fraying, rubber cap for wear. Replace with original parts. Do not substitute.
- Locate control box and other items in area to keep lanyard from being caught.
- ALWAYS TEST EMERGENCY STOP SYSTEM. PULL LANYARD. ENGINE MUST STOP. IF IT DOESN'T, REPAIR BEFORE NEXT USE.

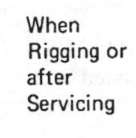

CHAPTER 27: QUESTIONS & ANSWERS

OUTBOARDS

1. The shear pin on an outboard propeller is designed to:

    (a) protect the propeller from damage.
    (b) protect the gear drive from damage.
    (c) cut lines or nets fouling the propeller.
    (d) protect the propeller from coming off.

2. An outboard motor's cruise tank and the auxiliary tank should be filled:

    (a) simultaneously.
    (b) at different times.
    (c) from different pumps.
    (d) from the same pump.

3. Portable fuel tanks for outboard engines should be:

    (i) refilled on board.
    (ii) stored in the after part of vessel.
    (iii) clamped in stowage.
    (iv) fitted with ASA lines and plugs.

    The correct answer is:
    (a) all except (i).
    (b) all except (ii).
    (c) all except (iii).
    (d) all except (iv).

4. Petrol fuel tanks may be constructed of the following materials:

    (i) galvanised steel.
    (ii) Lead-coated steel.
    (iii) PVC
    (iv) nickel-copper alloy.

    The correct answer is:
    (a) all except (i).
    (b) all except (ii).
    (c) all except (iii).
    (d) all except (iv).

5. On board vessels, flexible fuel pipes:

    (a) must not be used.
    (b) must be of woven wire casing type.
    (c) must be made of plastic.
    (d) must be fastened with clips.

6. Before removing a fuel filter of an outboard engine for inspection, you should:

    (i) disconnect the fuel line.
    (ii) pull out the choke.
    (iii) clean the area around the filter.
    (iv) take additional fire precautions.

    The correct answer is:
    (a) all except (i).
    (b) all except (ii).
    (c) all except (iii).
    (d) all except (iv).

7. After assembling the fuel pump filter of an outboard engine, it should be checked for leakages in the following areas:

    (i) the "in" pipe connection.
    (ii) fuel bowl joint.
    (iii) pivot joint.
    (iv) the "out" pipe connection.

    The correct answer is:
    (a) all except (i).
    (b) all except (ii).
    (c) all except (iii).
    (d) all except (iv).

8. Outboard engines are lubricated by:

    (a) circulating oil from a sump.
    (b) mixing oil in petrol.
    (c) both the methods stated here.
    (d) either one of the methods stated here.

9. In 2-stroke outboard engines lubricating oil is mixed in petrol because:

    (a) the crankcase is not used for compression.
    (b) a circulating oil system is not possible.
    (c) they have a "wet sump" system.
    (d) they are four cycle engines.

10. The commonly used oil-petrol mixture ratio for 2-stroke outboard engines is:

    (a) 1:50
    (b) 1:100
    (c) between 1:50 and 1:100
    (d) none of the choices stated here.

Chapter 27: Outboard Engines

11. A 2-stroke outboard engine running with too much oil in petrol will show the following symptoms:

    (i) overheating.
    (ii) blue smoke.
    (iii) fouled plugs.
    (iv) engine missing.

    The correct answer is:
    (a) all except (i).
    (b) all except (ii).
    (c) all except (iii).
    (d) all except (iv).

12. It is important to ensure correct petrol/oil mix in a 2-stroke outboard engine because:

    (a) spark plugs foul without sufficient oil.
    (b) engine overheats due to excessive oil.
    (c) oil in petrol lubricates engine.
    (d) oil stabilises petrol.

13. In the periodic lubrication of an outboard engine, the following areas should be included:

    (i) pivot joints and linkage.
    (ii) drive shaft bearing.
    (iii) gearbox.
    (iv) piston rings.

    The correct answer is:
    (a) all except (i).
    (b) all except (ii).
    (c) all except (iii).
    (d) all except (iv).

14. In a 2-stroke outboard engine, VRO is an abbreviation for:

    (a) Variable Ratio Oiling.
    (b) Variable Rotation Oscillation.
    (c) Variable Revolution Odometer.
    (d) Variable Revolving Oiler.

15. The parts of an outboard engine control system include:

    (i) lanyard for emergency cut-off switch.
    (ii) ignition/starter/priming switch.
    (iii) trim/tilt switch.
    (iv) lube oil pump switch.

    The correct answer is:
    (a) all except (i).
    (b) all except (ii).
    (c) all except (iii).
    (d) all except (iv).

16. An outboard motor may stop with little or no warning due to:

    (i) Low resistance on battery connections.
    (ii) Safety cut off switch disengaged.
    (iii) Water intake blocked.
    (iv) Spark plug lead disconnected.

    The correct answer is:
    (a) all except (i).
    (b) all except (ii).
    (c) all except (iii).
    (d) all except (iv).

17. An engine may overheat for the following reasons:

    (i) Lack of lubricating oil.
    (ii) Blockage in water inlet.
    (iii) Drop in oil pressure.
    (iv) Lack of fuel.

    The correct answer is:
    (a) all except (i).
    (b) all except (ii).
    (c) all except (iii).
    (d) all except (iv).

18. Generally speaking, in big seas an outboard motor's drive angle should be set at equivalent of the:

    (a) first pin.
    (b) second pin.
    (c) third pin.
    (d) fourth pin.

19. Where possible, all outboard motors, but in particular those over 10hp, should be secured to a vessel's transom:

    (a) by through bolts.
    (b) without piercing the transom.
    (c) by a single bolt.
    (d) additional chain.

20. Any of the following deficiencies will make it difficult to start a small outboard with no gearbox:

    (i) lube oil level too low.
    (ii) contaminated fuel.
    (iii) engine too cold.
    (iv) fouled propeller.

    The correct answer is:
    (a) all except (i).
    (b) all except (ii).
    (c) all except (iii).
    (d) all except (iv).

21. An outboard engine may not start for the following reasons:

    (i)   It is in gear.
    (ii) The fuel tank is not vented.
    (iii) Improper idling speed.
    (iv) Carburettor is restricted.

    The correct answer is:
    (a) all except (i).
    (b) all except (ii).
    (c) all except (iii).
    (d) all except (iv).

22. The following conditions will overheat an outboard engine:

    (i)   Blocked water intake.
    (ii) Impeller damage.
    (iii) Engine overloading.
    (iv) Slow cranking speed.

    The correct answer is:
    (a) all except (i).
    (b) all except (ii).
    (c) all except (iii).
    (d) all except (iv).

23. Spark plugs in an outboard engine can oil up due to:
    (a) engine running rich for long periods.
    (b) insufficient oil in petrol.
    (c) restriction in the fuel line.
    (d) air in the fuel system.

24. Four-stroke outboard engines are lubricated by:
    (a) oil mixed in fuel.
    (b) oil circulated from a sump.
    (c) oil injected into combustion chambers.
    (d) oil injected into intake manifold.

25. The level of lubricating oil in a 4-stroke outboard engine:
    (a) can only be checked when adding it to fuel.
    (b) cannot be checked.
    (c) can be checked at any time with a dip stick.
    (d) is fixed because it is in a sealed container.

26. The computerised ignition system on a pull-start petrol outboard does not include:
    (a) a ignition coil.
    (b) a timing device.
    (c) spark plugs.
    (d) a starter motor.

27. The petrol EFI system:
    (a) injects fuel into the combustion chamber.
    (b) is superior to petrol direct fuel injection (DFI) system.
    (c) injects fuel into the intake manifold.
    (d) injects fuel into the carburettor.

28. An excessively vibrating outboard engine is an indication of:
    (a) damaged propeller.
    (b) dirty fuel.
    (c) lack of lubrication.
    (d) faulty alternator.

29. If an EFI outboard fails to start with the electrical starter, it can be manually pull-started:
    (a) even if the battery is flat.
    (b) if the battery charge is sufficient to operate the fuel pump.
    (c) if the main switch is in the 'start' position.
    (d) if it is fitted with two flywheels.

30. The engine is flooded with fuel due to repeated starting effort. You should:
    (a) let it sit for a while.
    (b) try starting it with fuel shut off.
    (c) let it sit for a while if it is a 2-stroke engine.
    (d) let it sit for a while if it is a 4-stroke engine.

31. Excessive whiteness of the centre electrode porcelain of a spark plug in a cylinder indicates that the:
    (a) cylinder is clean.
    (b) spark plug gap is too large.
    (c) intake air is leaking.
    (d) spark plug gap is too small.

32. On recovering an outboard engine that had fallen overboard, if it cannot be started or serviced immediately, it should be:
    (a) submerged in fresh water.
    (b) wiped clean.
    (c) submerged in turpentine.
    (d) left in a vertical position.

33. The spark plug gap should be:
    (a) left alone.
    (b) nil.
    (c) about 1 mm.
    (d) about 10 mm.

34. Before stopping an outboard engine:
    (a) tighten the air-vent screw.
    (b) cut off the fuel supply.
    (c) run it at low speed for 2-3 minutes.
    (d) drain the fuel back to the tank.

Chapter 27: Outboard Engines

35. The manual starting of an EFI outboard would not be possible:

    (a) under any circumstances.
    (b) if the battery has run too low.
    (c) the propeller is damaged.
    (d) if the steering pivot is loose.

36. A damaged spark plug insulator could lead to:

    (a) explosion or fire.
    (b) engine flooding with fuel.
    (c) excessive carbon deposit.
    (d) spark plug not sparking.

37. When running a duel outboard vessel on one engine, with the unused motor still down in the water:

    (a) water entering its exhaust pipe causes no harm.
    (b) water entering its exhaust pipe can cause engine trouble.
    (c) you will need to operate at higher speed to maintain steerage.
    (d) the vessel will be impossible to steer.

38. You have recovered your outboard engine that had dropped overboard while still running. Rotating the flywheel appears difficult. You should:

    (a) not disassemble the powerhead.
    (b) leave it to dry out and drain.
    (c) immediately start the engine to clear the obstruction.
    (d) not attempt to start the engine following the after-submersion cleaning procedure because the connecting rod appears bent.

39. With regard to an outboard engine flooding with fuel due to repeated starting effort, which of the following statements is _incorrect_:

    (a) it's usually a 4-stroke engine problem.
    (b) it's usually a 2-stroke engine problem.
    (c) let it sit for a while before re-starting.
    (d) let the flooded fuel evaporate.

40. The likely reason for fuel spitting out through a carburettor is:

    (a) adjustable screw too loose.
    (b) adjustable screw too tight.
    (c) faulty leaf valves.
    (d) dirty filter.

41. In case of the VRO pump failure on a two-stroke petrol outboard:

    (a) do not add oil directly to fuel tank.
    (b) add oil directly to fuel tank.
    (c) operate at reduced speed.
    (d) pour oil into carburettor.

42. When transporting an outboard motor, always keep the:

    (a) powerhead in level with the leg.
    (b) leg higher than the powerhead.
    (c) powerhead higher than the leg.
    (d) leg separated from the powerhead.

43. An outboard engine starter operates but the engine will not start. It may be due to:

    (a) shift lever in gear.
    (b) blown fuse.
    (c) fuel pump malfunction.
    (d) the stop switch lanyard not attached.

44. While operating normally, an outboard motor has suddenly dropped its revs from 5000 to 2500, and it is sounding of an alarm. The most likely cause is:

    (a) overheated motor.
    (b) broken shear pin.
    (c) overloaded motor.
    (d) fouled propeller.

45. The presence of water in a carburettor-fitted fuel system would:

    (a) block the carburettor needle valve.
    (b) have no effect.
    (c) block the carburettor jets.
    (d) block the carburettor butterfly valve.

46. The adjustable screw head on a carburettor:

    (a) is to adjust air-fuel mixture.
    (b) is fitted only on EFI engines.
    (c) should always be kept loose.
    (d) should always be kept tight.

47. Which of the following you would _not do_ when manually starting an outboard?

    (a) Loosen the motor lowering device.
    (b) Raise the fast idle lever.
    (c) Switch ignition to "on" position.
    (d) Shift remote control lever to neutral.

CHAPTER 27: ANSWERS

1 (b), 2 (c), 3 (a), 4 (c), 5 (b),
6 (b), 7 (c), 8 (d), 9 (b), 10 (c),
11 (a), 12 (c), 13 (d), 14 (a), 15 (d),
16 (a), 17 (d), 18 (b), 19 (a), 20 (a),
21 (c), 22 (d), 23 (a), 24 (b), 25 (c),
26 (d), 27 (c), 28 (a), 29 (b), 30 (d),
31 (c), 32 (a), 33 (c), 34 (c), 35 (b),
36 (a), 37 (b), 38 (d), 39 (b), 40 (c),
41 (b), 42 (c), 43 (c), 44 (a), 45 (c),
46 (a), 47 (a)

Chapter 28

LPG & LPG ENGINES

PRINCIPLE CHARACTERISTICS OF LPG

Liquefied Petroleum Gas (LPG) is used on board as fuel for cooking and in some vessels for propulsion. It is composed of Butane or Propane or a mixture of the two, and is a hydrocarbon like petrol. It is more volatile and remains in a vapour state under normal atmospheric pressure. LPG is stored under pressure, compressed 270 times its normal vapour pressure to turn it into liquid. Mixed with air, it can be highly explosive.

LPG is heavier than air. Leaked gas will flow downwards, collecting in low spaces in the vessel. Pockets of LPG collected in this way create a serious fire hazard. LPG is non-toxic and non-corrosive, but if inhaled in the absence of oxygen, it will cause suffocation.

BASIC SAFETY RULES

- Use licensed gas fitter and approved fittings for installation and service. Use a marine-endorsed automotive gas fitter for installation and maintenance of an LPG engine. Do not tamper with the safety relief valve or fittings. The cylinder compartment must be outside the vessel's interior and vented outside.
- Maintain all LPG appliances and fittings in first class condition. Regularly check for performance, corrosion, rust and minor leaks. The gas cylinder should be kept in a secure, upper deck area so that any escaping gas drains directly overboard. Disconnected cylinders should be fitted with a thread protection cap. If stored in a closed locker, the locker must be vented downwards and outboard.
- Ensure that all electrical equipment close to the LPG storage and dispensing points complies with the Standards Association of Australia (SAA) Wiring Rules.
- If gas burner fails to ignite promptly, turn it off. Ventilate for 3 minutes to disperse the gas before attempting to re-ignite.
- Gas appliances need ventilation. Don't use them in enclosed cabins.
- Turn off the complete system when not in use for an appreciable period.
- Display "No Smoking" signs below deck & "Turn Off The Gas At The Bottle" signs near appliances.
- Don't open cylinder valve with the appliances cocks in the open position.
- Turn off every LPG appliance before petrol refuelling.
- Familiarise all crew with the LPG system and emergency procedures.

CARE OF LPG APPLIANCES

- The "POL" compression fitting on the LPG cylinders has a left hand thread. The hexagonal nuts have a circular groove cut in them for identification of the left hand thread. POL (Presto Light) is a brand name that has become synonymous with the left handed "gas thread".

- LPG and all other gas cylinders must be pressure tested by an authorised Test Station at least once every 10 years. Inspection dates should be stamped on the cylinder collar, neck or foot ring. (This rule applies under Standards Australia as well as the USL Code)

- The automatic safety relief valve, fitted on the cylinder's gas space, should be positioned so that any escaping gas is directed overboard.

- In its natural state, LPG gas is odourless. To ensure that escaping gas can be detected, a distinctive odorant (usually Ethyl Mercaptan) is added. It makes the gas smell similar to rotten onions. In case of a gas leak, shut off all appliances, close cylinder valves, stop engines and ventilate the craft. If the Safety Relief Valve leaks, move the cylinder away from any possible source of ignition. Slowly release its contents. Return the empty cylinder to the Test Station. Where a leaking cylinder is part of a fixed installation, shut off the main valve and notify a gas supplier at the next port of call.

- Always store, use and transport LPG cylinders in the designed (upright or horizontal) position. This ensures that the Safety Relief Valve remains in vapour space and not subjected to undue pressure by being immersed in liquid gas.

- Automotive grade LPG must not be used for domestic appliances.

- Vessels using LPG, when out of water, should be stored outdoors or under cover which has at least two open sides. When this is impracticable, provide mechanical ventilation. Eliminate all possible sources of ignition such as open flame heaters and unrestrained electrical cables. Enforce strict No Smoking rule.

Chapter 28: LPG & LPG Engines

- The appliance LPG cylinder valve must be either "fully closed" or "fully open". In the latter condition the gland on the valve spindle is sealed against possible leaks.

LPG ENGINE

LPG fuelled engines work just like petrol engines. In theory, a litre of LPG generates 33% less heat energy than a litre of petrol. Therefore, an LPG fuelled engine should consume 33% more fuel than a petrol engine. However, the power difference between the two types of engines is claimed to be negligible. The LPG fuel in an engine is said to improve efficiency for the following reasons:

> LPG has a higher octane rating or antiknock properties. It can tolerate more advance engine tuning without detonation.
> LPG fuel-air mixture, being of better composition, does not require a device for enriching the mixture, such as a choke or an acceleration pump.
> The engine does not have to deal with occasional drops of liquid fuel entering the cylinders. This results in better fuel combustion.
> For all of the above reasons, it is claimed that the need for engine maintenance is reduced in the areas of oil changes, carburettor maintenance and decarbonisation and spark plug maintenance. There is also less carbon monoxide in the exhaust fumes.
> LPG in liquid weighs almost half that of equal volume of petrol (Density or Specific Gravity of LPG = 0.52, Petrol = 0.89). Therefore, in spite of its container being heavier, LPG is a smaller deadweight on a vessel.

LPG CONVERSION KITS

Factory equipped LPG marine engines are not yet common. Therefore, a conversion kit is required to convert a petrol engine into a petrol-LPG (duel-fuel) system. Such a kit consists of the following specially designed components:

- Fuel tank
- Filling valve & connections
- Filter
- Safety switch
- Converter (Vaporiser & Pressure Regulator)
- Lock off valve
- Carburettor/Mixer.
- Hoses

To modify a diesel engine into an LPG engine the diesel injection equipment is removed, the compression ratio is reduced and spark ignition equipment is fitted.

For most applications, LPG is withdrawn from a fuel tank as liquid. Such a cylinder is known as the *Liquid Withdrawal (or Liquid Take-off) Cylinder*. It is passed then through a filter, into a converter. The converter vaporises it and regulates the vapour delivery pressure to the carburettor. Most converters are set to give a zero or slightly negative pressure at the converter outlet.

The LPG vapour is induced into the carburettor by the engine vacuum where it is mixed with the incoming air. The air-fuel mixture is then passed into the engine for combustion in the same manner as a petrol engine.

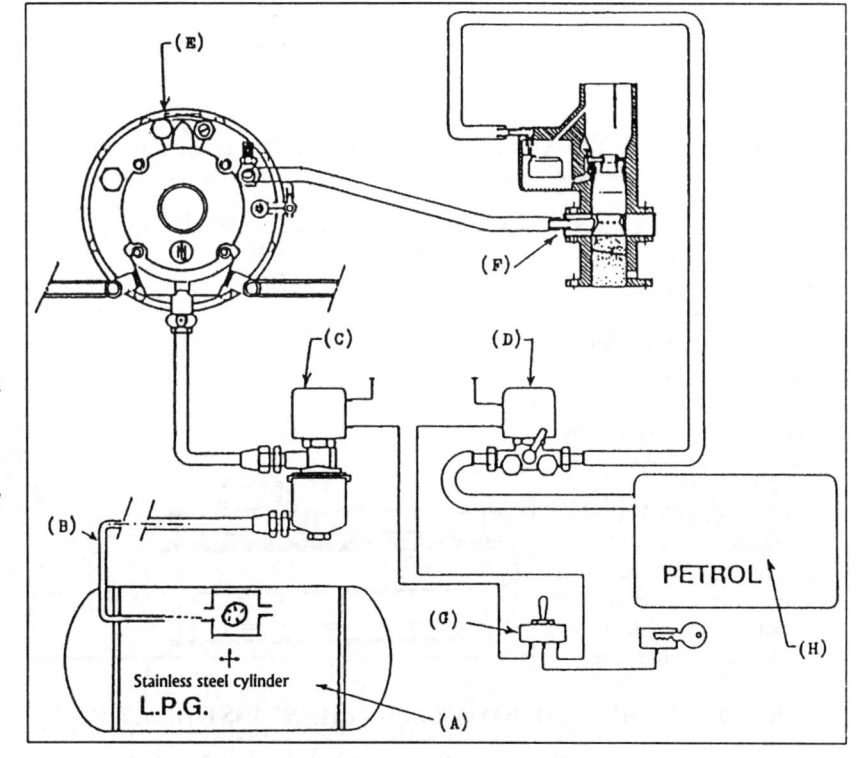

Fig 28.1: A DUEL FUEL (LPG/PETROL) ARRANGEMENT
(A) LPG container, (B) Service line, (C) LPG lock-off & filter, (D) Petrol lock-off, (E) Converter, (F) Mixer (Gas carburettor), (G) Fuel selector switch, (H) Petrol tank

IN THE CASE OF SMALL ENGINES (below 10 hp or 7.5 kw), the converter may be eliminated if the small amount of LPG vapour consumed can be adequately vaporised in a BBQ-type cylinder, known as the *Vapour Withdrawal (or Vapour Take-off) Cylinder*. This arrangement requires only a pressure regulator in place of a converter that is a vaporiser and a regulator.

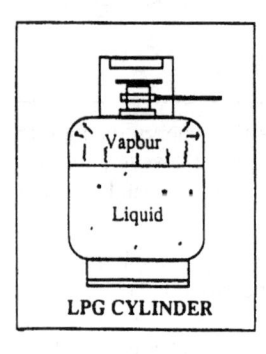

The Australian Boating Manual

OTHER CHANGES TO ENGINE WHEN CONVERTING TO LPG FUEL
- Due to a slightly higher octane rating of LPG, the advance curves in the distributor may need to be altered to maintain optimum performance. In the case of a fuel-injected engine, the complete module may need to be reprogrammed.
- An LPG engine runs slightly hotter in the combustion chamber. Therefore, the valves and valve seats in some engines may need to be replaced with those with hardened inserts. The cooling system must also be in good working order.

VALVES & FITTINGS ON AN LPG CYLINDER

A liquid withdrawal LPG cylinder can be a single-port multi-valve system or a 4-Valve system. The stainless steel cylinders designed for marine use are generally the multi-valve type. In both cases their valves must perform the following functions:

Fig 28.2
VAPOUR
WITHDRAWAL
LPG CYLINDER

1. SAFETY RELIEF VALVE (SRV) OR PRESSURE RELIEF VALVE (PRV): It protects the cylinder against rupture due to excessive pressure. The valve is fully recessed to guard against physical damage and breaking off.

2. FILLER VALVE & 80% AUTOMATIC FILL LIMITER (AFL): LPG cylinders are designed to fill only up to 80% of their liquid capacity. On reaching this level, the Filler Valve closes automatically and the filling stops. The operator can thus disconnect the refuelling hose with minimum fuel loss. The Filler Valve is a non-return valve, which prevents any fuel loss in the event of hose failure or disconnection of the filling nozzle. Any remote filler valve is also required to be non-return.

3. LIQUID LEVEL (OR CONTENTS) GAUGE: It indicates the percentage of liquid contents in the cylinder. For it to provide accurate reading, the cylinder must be mounted in the correct plane and at correct valve angle (usually 30° above the horizontal).

4. SERVICE VALVE: It performs four functions: (1) Opens and shuts the gas supply. (2) Isolates the tank when the service line is disconnected (3) It is fitted with an electric Lock-Off Solenoid which opens only when there is power supply to the engine and the engine is turning. A safety switch is fitted in the ignition switch to transmit an electrical pulse to open the valve during cranking. (4) It is fitted with an EXCESS FLOW VALVE that shuts off the gas if its flow rate exceeds the designed value. This may happen, for example, in the event of a hose bursting.

Fig 28.3: LPG CYLINDER & IT MULTI-VALVE SYSTEM
(STM Auto Gas Tanks, Melbourne)

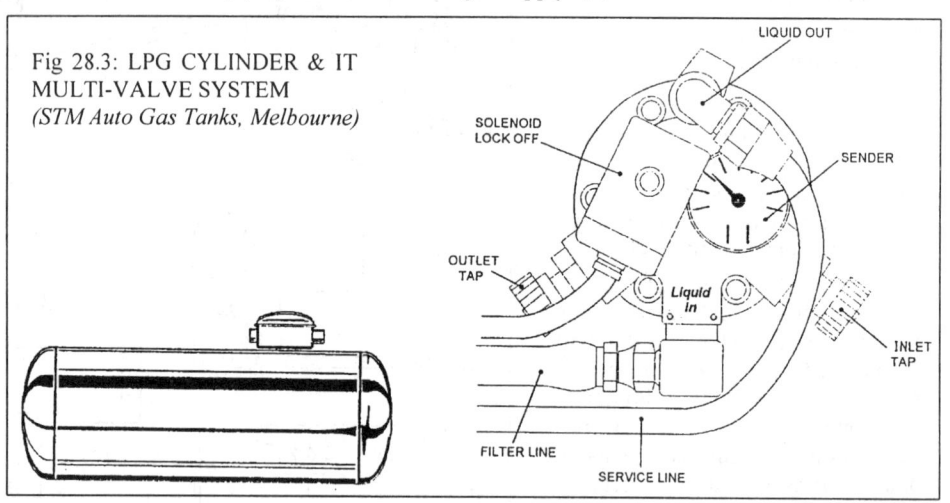

GUIDELINES FOR AUTOMOTIVE LPG MARINE INSTALLATIONS

- The installation should be carried out by a licensed installer with marine endorsement, and inspected annually by an approved person.
- The cylinder compartment should be constructed of an approved material, and substantially sealed. The authorities may approve aluminium only or may allow plastic or fibreglass. It should be clearly marked to indicate LPG. Its vents should be baffled to prevent water entry into the compartment. Electrolytic corrosion in the compartment should be minimised - perhaps with the aluminium stand acting as the sacrificial anode.
- The cylinders should be constructed of stainless steel, and approved for automotive use. They should be capable of being isolated from outside the cylinder compartment. Each cylinder should be positioned so that its longitudinal axis is parallel to the water surface when the vessel is stationary and parallel to the ground when on a trailer. This ensures that

Chapter 28: LPG & LPG Engines

the PRV remains in vapour space to prevent the liquid gas escaping should the valve open.

- A separate fill point should be installed for each cylinder. The cylinder relief valves should be vented separately at a termination point outside the vessel. If the relief valve is not located centrally along the length of the cylinder, the cylinder should be mounted so that the relief valve is forward of the centre towards the bow. It will thus not be in the liquid space when the vessel (and the cylinder) is trimmed by the stern.
- The electrical equipment associated with the cylinders should be non-sparking. The flexible part of the service line should be an approved gas hose with the reinforcing consisting of a textile component (not steel).

OPERATING CHARACTERISTICS OF LPG-PETROL (DUEL-FUEL) OUTBOARD

- To start a cold engine, a prime switch is activated to provide LPG for ignition. The idle lever should be raised to draw gas into the engine. Once the engine has fired, the idle lever is lowered to fast idle, then to idle when warm.
- When low on LPG, the engine lets you know that you have ignored the gauge by dropping revs and picking up again. At this time, you have sufficient gas for only two or three nautical miles.
- Changing to petrol involves purging the engine of gas, switching to petrol on the console, pumping the fuel bowl and starting on petrol.
- Changing to LPG requires switching the petrol off, running the engine out of petrol, switching to gas and refiring the engine. If the engine is flooded, switch off petrol and gas, and turn the engine over with throttle wide open. If flooding is severe, don't flatten the battery by continually trying to start, remove plugs and clear engine.

LPG EMERGENCY PROCEDURE

1. LEAK SUSPECTED - LOCATION UNCERTAIN: There is little likelihood of a large LPG leak. An excess flow valve fitted in the service line automatically shuts off in case of a drop in back pressure.

Small leaks in the service line or around the fittings can sometimes be detected by engine performance, smell and frosting of the area where the liquid loses pressure. LPG being a refrigerant often causes a frost to form on the area surrounding the leak. The system can also be checked by applying soapy water or a specially formulated foam with a brush to look for bubbles made by the escaping gas. Never use a match or flame to test. Electronic LPG sniffers are also available. After fixing a leak, rejoin and test again with soapy water. Ventilate before using the appliance.

2. LEAK DETECTED - NOT ON FIRE: Stopping the motor should automatically shut off the Service Valve on the gas cylinder. In the case of a cooking gas cylinder, shut the cylinder valve. Ventilate the area. Keep hands and face clear of any escaping liquid. The automotive gas cylinders are installed on a pod outside the vessel. In the case of an appliance cylinder stored on deck, if unable to stop the leak, tow it well clear astern until the gas is exhausted. If the cylinder cannot be removed, turn off electrical motors and other sources of ignition. If possible disperse gas with fine water spray and provide maximum ventilation.

3. LEAKING CYLINDER OR APPLIANCE IS ON FIRE: As discussed earlier, the automotive LPG Service Valve lock-off solenoid shuts off as soon as the engine is turned off.

On a cooking gas cylinder, close the valve if safe to do so and let the fire go out. Do not use again until inspected. If the valve cannot be closed, quickly evaluate the situation. Keep appliance cool. Control the spread of fire. Do not extinguish flame if fire is in an enclosed space. Unburnt gasses can explode when re-ignited. Send out a distress signal and prepare to abandon vessel. Don't forget the Grab Bag.

4. CYLINDER EXPOSED TO EXCESSIVE HEAT FROM A FIRE: Keep the cylinder cool with a water hose, sprayed from the maximum practicable distance. Remove it from the heat source if possible and when safe to do so. Cool it by water hose or by immersion in water.

LPG ENGINES TROUBLESHOOTING

See Outboards, for petrol and LPG engine problems.

CHAPTER 28: QUESTIONS & ANSWERS

LPG FUEL SYSTEM & ITS RISKS

1. Gas cylinders should be pressure tested every:

    (a) year.
    (b) two years.
    (c) five years.
    (d) ten years.

2. LPG is:

    (a) Heavier than petrol.
    (b) Lighter than air.
    (c) Volatile.
    (d) unable to flow when leaked.

3. An LPG fuelled engine:

    (a) is not fitted with a carburettor.
    (b) requires more maintenance.
    (c) runs hotter than a petrol engine.
    (d) requires petrol as a standby fuel.

4. The marine automotive LPG fuel cylinders are constructed of:

    (a) stainless steel only.
    (b) aluminium or stainless steel.
    (c) aluminium only.
    (d) PVC

5. The following is not fitted on a modern multi-port LPG cylinder:

    (a) Ullage valve
    (b) PRV
    (c) AFL
    (d) Service valve

6. The service valve on a LPG cylinder for an outboard motor automatically shut off when:

    (i) the fuel line is disconnected.
    (ii) the engine is switched off.
    (iii) there is a major gas leak.
    (iv) there is any gas leak.

    The correct answer is:
    (a) all except (i).
    (b) all except (ii).
    (c) all except (iii).
    (d) all except (iv).

7. The LPG cylinder for an outboard should be positioned parallel to the:

    (a) vessel's stern.
    (b) deck at all times.
    (c) water surface when vessel is underway.
    (d) water surface when vessel is stationary.

8. On a duel-fuel outboard engine, changing the fuel from LPG to petrol requires:

    (i) purging the engine of gas.
    (ii) switching to petrol on the console.
    (iii) pumping fuel into the fuel bowl.
    (iv) keeping the engine turning.

    The correct answer is:
    (a) all except (i).
    (b) all except (ii).
    (c) all except (iii).
    (d) all except (iv).

9. On a duel-fuel outboard engine, changing the fuel from petrol to LPG requires:

    (v) switching off petrol.
    (vi) running the engine out of petrol.
    (vii) switching to gas on the console.
    (viii) opening service valve on the LPG cylinder.

    The correct answer is:
    (a) all except (i).
    (b) all except (ii).
    (c) all except (iii).
    (d) all except (iv).

10. It is true that LPG is:

    (a) liquid stored under pressure.
    (b) lighter than air.
    (c) not odourless in its natural state.
    (d) toxic.

11. LPG gas cylinders should be:

    (a) always stored under deck.
    (b) stored under deck only when empty.
    (c) stored under deck only when full.
    (d) always stored above deck.

12. A suspected LPG leak can be confirmed by:

    (a) a flame test.
    (b) applying soapy water to pipes.
    (c) sniffing for rotten eggs smell in higher areas.
    (d) looking for emitting white liquid.

CHAPTER 28: ANSWERS

1 (d), 2 (c), 3 (c), 4 (a), 5 (a),
6 (d), 7 (d), 8 (d), 9 (d), 10 (a),
11 (d), 12 (b)

Chapter 29

BILGES, PUMPS, AUXILIARY POWER TAKE-OFFS, REFRIGERATION & PLUMBING

THE BILGE SYSTEM

The purpose of a bilge system is to pump out the vessel's bilges. It allows pumping out of one or more compartments.

Bilge systems should have the following characteristics:

- A separate suction line from each compartment
- All lines fitted with non-return valves to prevent cross flooding or back flooding.
- All suctions installed in the lowest points in the bilges
- Overboard discharge fitted with a non-return valve to prevent back flooding from the sea
- An Isolating valve fitted on each line.
- Plastic or other non-metallic hoses should not be used (For reasons discussed in the "fuel system"). Flexible hoses must be made of woven wire casings with screwed hose fittings.
- A STRUM BOX, FILTER, STRAINER or MUD BOX must be fitted at each suction. Strainers and Mud Boxes in large vessels may be fitted with a perforated steel plate instead of a screen filter, and the bonnet may be bolted instead of being screwed in. The mud box has a recess to collect the dirt, preventing it from falling back into the bilge.

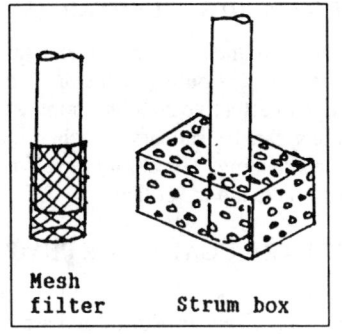

Fig 29.1 SUCTION FILTERS

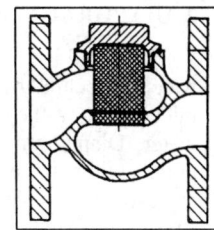

Fig 29.2 STRAINER WITH DIRT-COLLECTOR (MUD BOX) *(Ernest Gopfert)*

Bilge pumps can be manual, electric or engine-driven. Most are *positive-displacement pumps. (NOTE: Pumps, valves and seacocks are discussed below).* They are mounted above the bilge where they are easily accessible and, if manual, wouldn't be tiring to operate. The removable handle of a manual pump should be secured where it will stay until needed. However, submersible rotary (centrifugal) bilge pumps are also around, which are of non-positive-displacement type.

Bilge suction is positioned low in the bilge, and the discharge high above the waterline so that it remains out of water at all angles of heel. A non-return valve must be fitted on the discharge line, particularly in vessels with low freeboard where the discharge is likely to go underwater during heeling or rolling and cause back flooding. Fitting a very long discharge line on a small-volume bilge may return an excessive column of residual water to the bilge each time the pump stops. This would repeatedly activate the *level switch* (shown below), causing the pump to continue to cycle "on" and "Off". Overcoming this problem by fitting a *check valve* in the line is not desirable. These valves tend to stick and are inaccessible to maintain.

On shutting down the bilge pump there is always some back flushing from the water column in the discharge line. To allow the heavier sediments to fall well below the filter, the *strum box* or the submersible bilge pump should be mounted on a small pedestal.

BILGE PUMPING SYSTEMS

1. DIAPHRAGM BILGE PUMPS - MANUAL

The old *rotary and semi-rotary manual bilge pumps* were tiring to operate. The *plunger-type manual pumps* have also now disappeared. Their leather sealing-flap suffered constantly from shrinking, cracking and loss of shape.

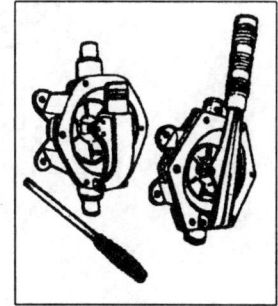

Fig 29.3 ECONOMY SINGLE ACTION MANUAL BILGE PUMP (Through-deck or above-deck bulkhead-mounted)

Manual bilge pumps today are mostly the *diaphragm pumps*. They can be mounted to bulkhead or deck. They are operated by moving a handle back and forth. A diaphragm attached to the handle moves with it to draw fluid into the pump chamber and then expel it out. One-way flap valves on the inlet and outlet allow the fluid to move in the correct direction. However, the flow direction can be revered by exchanging the inlet and outlet ports.

A double-diaphragm pump has a diaphragm and pump chamber on both sides of the lever. As one diaphragm moves in to draw water into one chamber, the other moves out to expel it from the other. These are double-action larger capacity pumps.

Manual pumps are the traditional mainstay of the small craft bilge pumping systems and can offer surprisingly good flow rates, as well as a variety of installation options. Yacht racing rules state that at least one pump must be capable of being pumped from the cockpit with all hatches shut. The best way to accomplish this is with a through deck kit which allows the pump to be installed safely below decks while the operator pumps from above. A well-designed manual bilge pump is very hard to block due to its large boreholes and valves and one-piece diaphragm. The larger manual pumps are also suitable for ballast and waste transfer.

2. DIAPHRAGM BILGE PUMPS - ELECTRIC (See Figure 29.22)

Mounted in a dry location, these pumps are excellent for shallow bilge boats where the water left behind by other types of pumps will slop from side to side. With the small size of inlet pipework they are very effective as bilge 'hoovers', removing all but the last drops of water. Diaphragm pumps can be run dry which removes the need to watch overboard outlets whilst the pump is running. The ability to self-prime means they can be mounted in an easily accessible position to aid in servicing.

3. SUBMERSIBLE CENTRIFUGAL BILGE PUMPS

By far the most popular type of electric bilge pumps, submersibles give very high outputs, are cheap to buy, have low amp draw and are easy to install. They are designed to fit in the lowest part of the bilge and only require discharge pipework and electricity supply. However, flow rates decrease more rapidly than other types of pumps as the discharge head increases, so check the manufacturer's maximum discharge heads and if in doubt, always increase the size of the pump. Although modern submersibles can be run dry for an hour or two, pump life can be extended by avoiding dry running whenever possible. All submersibles can be operated using a float or hydro-air switch and remotely controlled outside the bilge compartment from a switch panel.

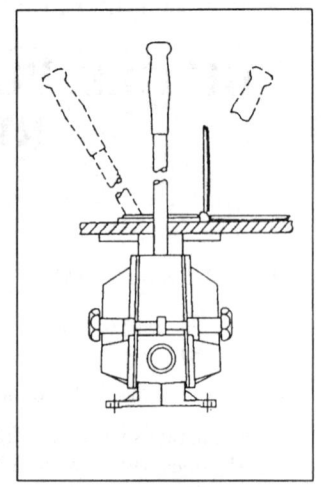

Fig 29.4
JABSCO MANUAL DOUBLE ACTION TWIN-DIAPHRAGM PUMP
(Through-deck or on-deck mounted)

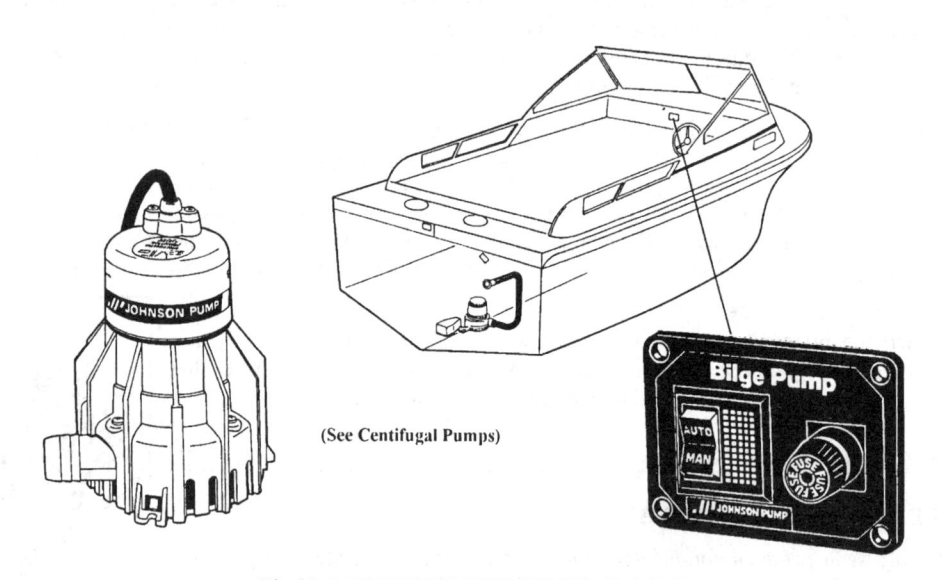

(See Centifugal Pumps)

Fig 29.5: JOHNSON SUBMERSIBLE BILGE PUMP

Chapter 29: Bilges, Pumps, Auxiliary Power Take-offs, Refrigeration & Plumbing

4. FLEXIBLE IMPELLER BILGE PUMPS

Flexible impeller pumps make excellent bilge pumps, principally because of their ability to handle bilge debris without damage. They also give the benefits of good flow, low cost, low size and weight, and are easily serviced and maintained. Flexible impeller pumps must not be run dry, as this will damage the impeller, which will require replacing. However, some pumps allow a few minutes of dry running after initial priming. As with submersible pumps, flexible impeller pumps can be operated using a float or hydro-air switch, and remotely controlled from outside the bilge compartment.

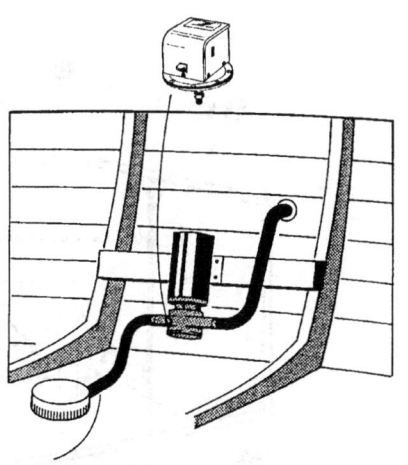

Fig 29.6: JABSCO FLEXIBLE IMPELLER BILGE PUMP WITH A VACUUM SWITCH
(In pumps not designed to run dry, a vacuum switch should be fitted to cut off current to the pump when the bilge is dry.)

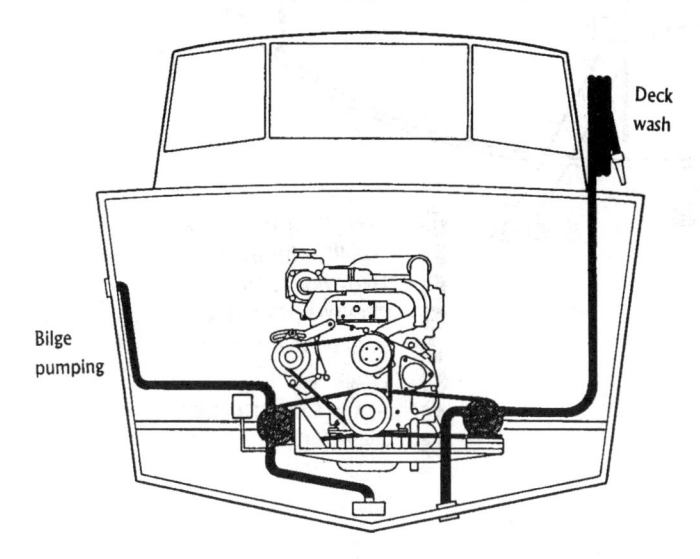

Fig 29.7: JABSCO ENGINE DRIVEN BILGE PUMP & DECK WASH SYSTEM

5. ENGINE DRIVEN BILGE PUMPS

Engine driven pumps are probably the strongest and most reliable pumps. Combined with other benefits such as self-priming and very high flow rates, these pumps are the firm favourites on commercial vessels. They come in a wide range of performances and sizes and can be direct driven by a clutch, pulley or hydraulic drive; or can be close coupled to a suitable power take off on the engine.

The illustration shows the bilge pumping system separated from the deck wash system. One is for pumping water out of the vessel and the other for pumping seawater on deck. However, in most small vessels, the two systems use a common pump with an "L-port Cock" (a two-way cock) fitted in the line to make sure that the pump suction is either on the sea suction or the bilge suction. When one is in use the other is automatically isolated. This prevents any possibility of bilges being flooded from the sea. Priming is necessary prior to sucking a distant bilge with a pump not designed to run dry. As mentioned earlier, running an impeller pump dry for any length of time would burn out the impeller vanes.

The outlet from the bilge pump should not be joined into any other outlet, such as the sink outlet of the galley. With the outlet below sea level due to vessel being heeled or otherwise, there is a risk of water siphoning back into the vessel after the bilge pump has been used. A further precaution is to fit non-return valves at both the suction and the outlet. It is unsafe to leave a water hose hanging over the side in water in an unattended vessel. Water may siphon back into the vessel.

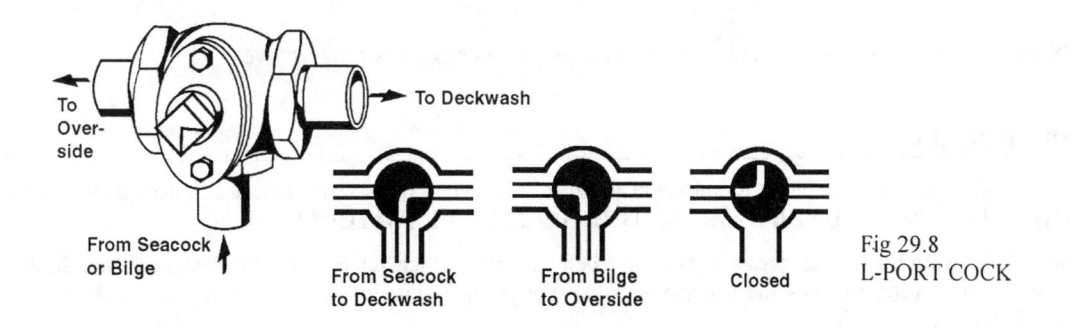

Fig 29.8 L-PORT COCK

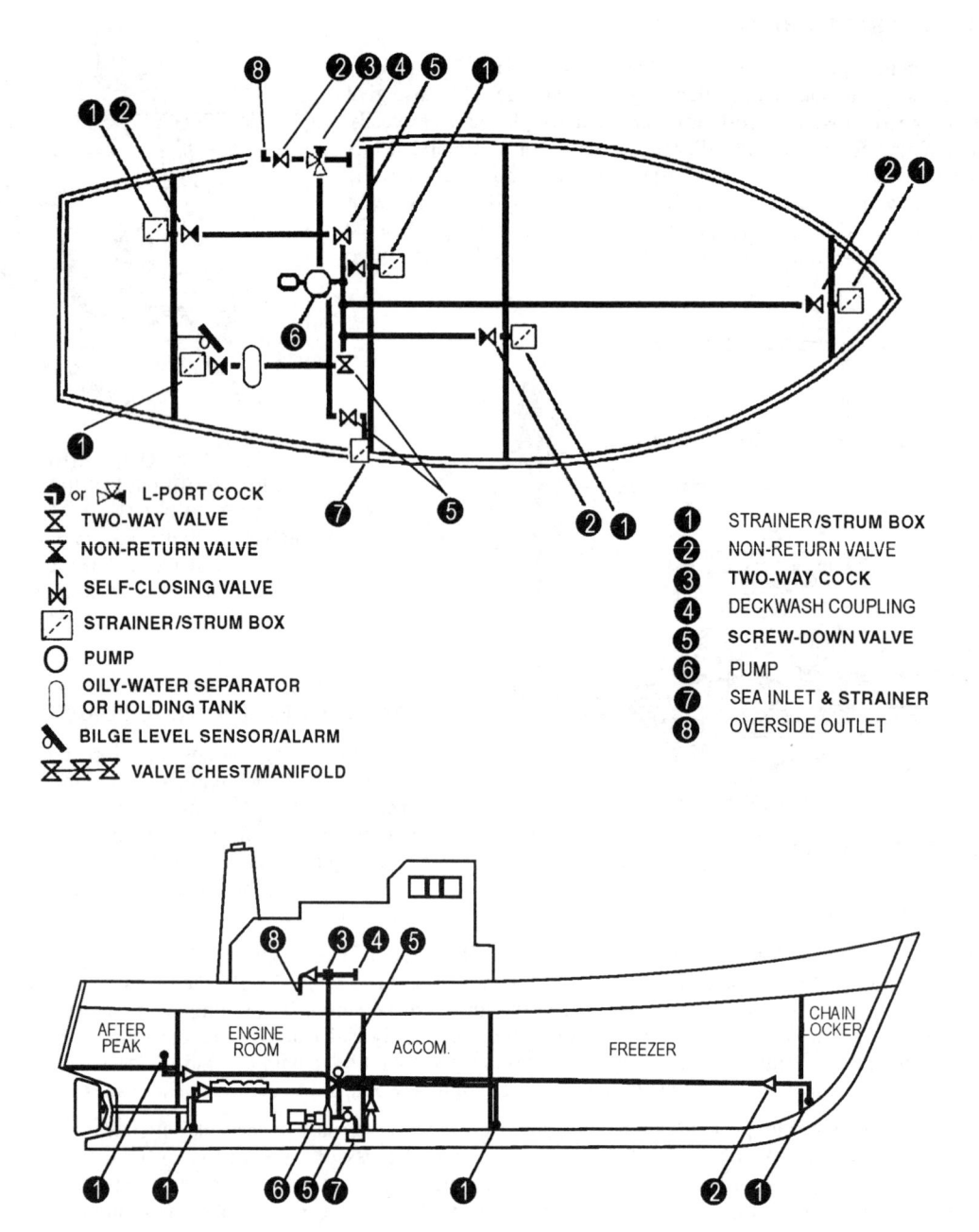

Fig. 29.9: TYPICAL BILGE PUMPING SYSTEM FOR MULTI-COMPARTMENT VESSELS

BILGE ALARMS must be fitted in bilges to sound an alarm when bilge water exceeds a preset level.

AUTOMATIC BILGE PUMPING

A switch can be attached to the bilge pump to automatically turn it on when bilge water reaches an acceptable level. The two common types of switches are: **HYDRO AIR SWITCHES** and **FLOAT SWITCHES**.

In a Hydro Air Switch, rising bilge water pressurises air in a bell and tube, which activates the diaphragm switch. In the case of a Float Switch, when water exceeds the acceptable level, the pump is turned on by a rolling steel ball, which changes the pressure on the micro switch.

A switch panel (shown above) provides a control for manual or automatic bilge pumping. Only the heavily insulated

Chapter 29: Bilges, Pumps, Auxiliary Power Take-offs, Refrigeration & Plumbing

wiring should be used to avoid electrical leaks and the pump grounded to minimise electrolysis.

A word of caution about the float switches: They should be changed at least every 2 years, whether they appear to need it or not. Located in the grimy salt-water bilge, unused and unattended most of the time, the micro-switch may fail to operate, an electric cable may corrode or come undone, or the ball may fail to roll.

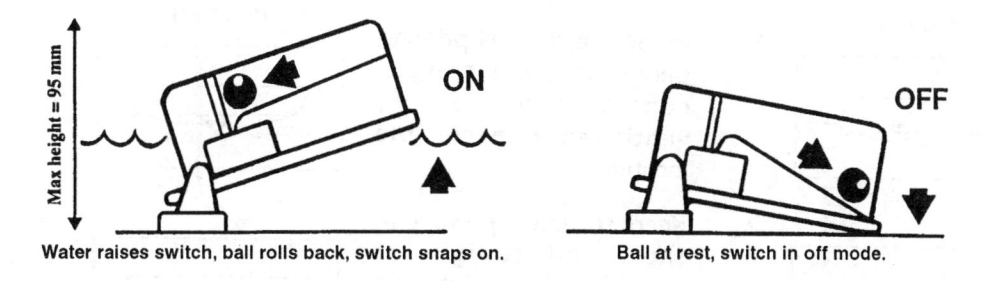

Fig 29.10: JABSCO AUTOMATIC BILGE PUMP SWITCH

BILGE MAINTENANCE

Bilges must be inspected on every watch and l*imber holes* kept clean so that water from all sections of the bilge moves freely towards the bilge suction. Sediments settled below the suction filter must not be allowed to accumulate.

Sinking of boats on their moorings is not uncommon. A large amount of water can accumulate even from a slowly dripping stern gland. The bilge pump would operate a few times until the battery goes flat, the level switch fails or the fuse blows due to corrosion of the fuse holder. After that the water is free to fill the boat. *Wet-type batteries* are particularly prone to over-discharging in hot weather to the point of being useless within a few months.

Therefore, it is unwise to rely on bilge pumps on unattended vessels. It is better to make sure that water does not enter the vessel. Keep batteries charged and electrical circuits in good condition. A secondary high-level alarm system or a pump-timer may be a worthwhile investment.

BILGE PUMP REQUIREMENTS FOR COMMERCIAL VESSELS (Fig 29.11)
(NSCV PART C Section 5 Subsection 5A)

Measured length of vessel (m)	Manual pumps Qty.	Manual pumps Capacity per pump, as installed, in kL/hr	Powered pumps Qty.	Powered pumps Capacity per pump, as installed, in kL/hr
Less than 7.5	1	4.0		
7.5 and over but less than 10	2	4.0		
10 and over but less than 12.5	1	5.5	1	5.5
12.5 and over but less than 17.5	1	5.5	1	11.0
17.5 and over but less than 20	1	8.0	1	11.0
20 and over but less than 25			2	11.0
25 and over but less than 35			2	15.0

NOTE: The installed capacity of a bilge pump is normally less than the nominal figure specified by the manufacturer due to the head of the discharge above the suction and losses through valves and piping.

GOOD PUMPING PRACTICE — BILGE PUMPS

Correct / *Incorrect*

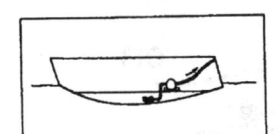

1. Install electric self-priming pumps as low as possible, consistent with a dry, ventilated & accessible location.

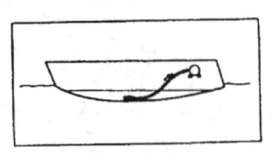

2. Keep suction pipework as short and straight as possible. Use reinforced hose that will not deform or collapse under suction conditions. Ensure that all connections are airtight.

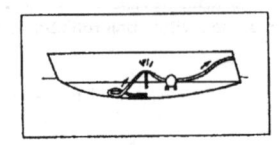

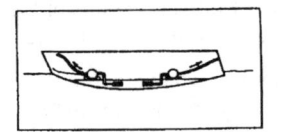

3. Keep pipework simple. Complex valving in the suction system increases the risk of air leaks and loss of priming ability.

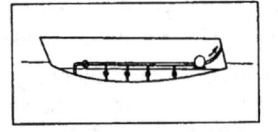

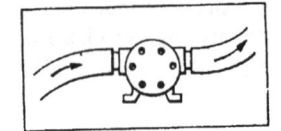

4. Use pipes of internal diameter at least as large as the nominal bore of the pump ports.

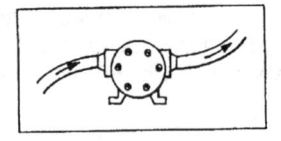

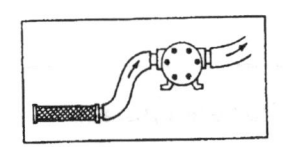

5. Always fit an adequately sized suction strainer to protect the pump from debris. Make sure that it is accessible, and remember to inspect it periodically.

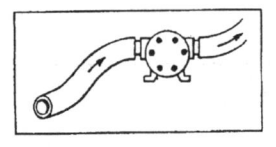

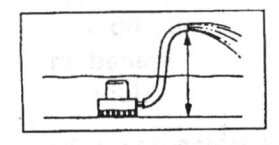

6. Remember that small electric submersible pumps are rarely useful at more than 1.2m (4ft) vertical discharge head. Medium / large submersibles are similarly ineffective above 2m (7ft).

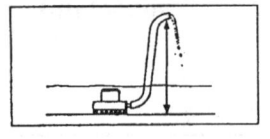

7. Always fit a manually operated bilge pump as a back-up to electric or engine-driven pumps.

GOOD BILGE PUMPING PRACTICE (Fig. 29.12)
(Recommended by Cleghorn Waring Pumps)

Chapter 29: Bilges, Pumps, Auxiliary Power Take-offs, Refrigeration & Plumbing

BILGES - TROUBLESHOOTING

SUDDEN APPEARANCE OF OIL OR OILY WATER IN BILGE

- Damage to the engine sump.
- Oil pipes damaged or disconnected.
- Spillage of stored oil.
- Damaged tank(s).

ACTION FOLLOWING SUDDEN APPEARANCE OF OIL IN BILGE

- Find and rectify cause.
- Check oil pressure gauge.
- Slow down, if necessary.
- Dip or check oil in engine and storage.
- Remove oil into a container for disposal - to comply with pollution regulations.

LOSS OF PUMP SUCTION

- Blocked suction strainer.
- Sheared impeller drive.
- Air leak in a valve gland or a pump gland or due to vessel rolling or damaged suction pipe.
- Pump not turning fast enough - drive belt slipping.
- Excessive clearance in pump impeller.

BILGE PUMP NOT DRAWING WATER FROM A FULL BILGE

- See "loss of pump suction".
- Incorrect or unnecessary valves open.
- Non-return valve seized in seat.
- Pump priming device not working - air lock in the system.

PUMPS

POSITIVE DISPLACEMENT PUMPS: Any pump that alternately increases and decreases its volume is referred to as a positive-displacement pump. For example, the flexible impeller pump discussed below.

SELF-PRIMING PUMPS: Self-priming pumps have the capability to draw fluid to themselves. They do not have to be installed below the liquid level they are to pump. In the following list all pumps are self-priming, except for the centrifugal pump. However, flexible impeller pumps, although self-priming, need to be primed before drawing water through a long pipeline. This can be achieved by installing a one-way check valve (non-return valve) in the suction line. However, a long pipeline is not recommended for the engine cooling line. Keep it as short and straight as possible, and fit an in-line strainer instead of a non-return valve. Risking running the pump dry for any period of time would not only damage the impeller vanes, but also the engine.

Remember that the self-priming ability of pumps usually deteriorates with age. Therefore, where self-priming ability is important, choose a pump with a slightly greater self-priming ability than needed.

The pump capacity is expressed in litres per minute.

TYPES OF PUMPS

1. FLEXIBLE IMPELLER PUMP

Almost all engine raw water pumps are of this type, as are the engine driven deckwash pumps. Carrying spares is therefore essential.

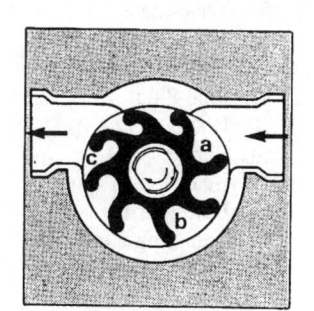

Fig. 29.13: FLEXIBLE IMPELLER PUMP (*Jabsco*)

How it Works:
- Flexible impeller blades create a nearly perfect vacuum for instant self-priming.
- As the impeller rotates, each successive blade draws in liquid and carries it from intake to outlet port.
- As the flexible impeller blades come into contact with the offset cam they bend with a squeezing action, thus providing a continuous, uniform flow. It is therefore a positive displacement pump.

Features:

Versatile: The flexible impeller pump combines the priming features of positive displacement type pumps with the general transfer ability of centrifugals. It will pump either thin or viscous liquids and can handle more solids in suspension than other types of rotary pumps. The pump can be mounted at any angle and will pump in either direction with equal efficiency.

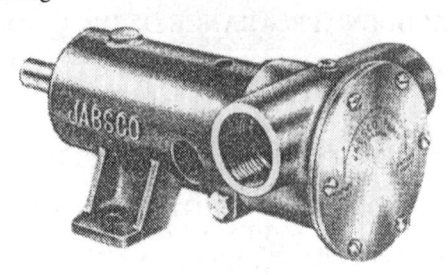

Fig. 29.14: FLEXIBLE IMPELLER PUMP HOUSING (*Jabsco*)

Self-Priming: Pumps instantly with dry suction lifts up to 3m (10 ft), up to 8m (25 ft) when wetted.

Simplicity: One moving part - a tough, long-life, wear-resistant flexible impeller.

Flexibility: The flexible impeller pump offers both high flow and high pressure according to motor and impeller design.

Good pumping practice: Pipe runs should always be kept as short and straight as possible, avoiding rising and dipping over obstructions, as this can lead to air-locks. Pipework should always be reinforced, non-collapsible hose of the recommended size. Electric pumps should always be installed in a dry, well-ventilated position as close to the liquid to be pumped as possible. Flexible impeller pumps must not be run dry.

A VACUUM SWITCH can be fitted to the pump to cut off the power to the motor when there is no fluid in the inlet pipe. Bilge pumps and cooling water pumps in particular must be fitted with a filter or strainer on the inlet pipework to protect them from debris. These should be fastened to the boat structure to ensure their permanent location. Mount flexible impeller pumps so that some water is left in the pump body when the pump is shut off. This will prolong impeller life and speed priming. Pumps must be installed with the overboard discharge well above the waterline (both static and heeled) to avoid water siphoning back into the vessel.

IMPELLER USAGE - IMPORTANT

- ENGINE COOLING: Use only neoprene compound rubber impellers.
- BILGES, ETC.: Use nitrile compound rubber impellers. Nitrile impellers are excellent for oil/water/diesel mixture. However, they are not suitable for engine cooling. Their flow capacity is 10 to 15% less than that of neoprene impellers, which could cause engine overheating, especially in larger engines.

FLEXILE IMPELLER PUMP DETAILS
1. Screw, End Cover
2. End Cover
3. Gasket
4. Impeller (neoprene or nitrile rubber)
5. "O" Ring
6. Wear plate
7. Screw
8. Cam
9. Pipe Plug
10. Body
11. Slinger
12. Bearing Seal (inner)
13. Ball Bearing
14. Retaining Ring
15. Retaining Ring
16. Bearing Seal (outer)
17. Seal Assembly
18. Shaft, Stainless Steel
19. Key

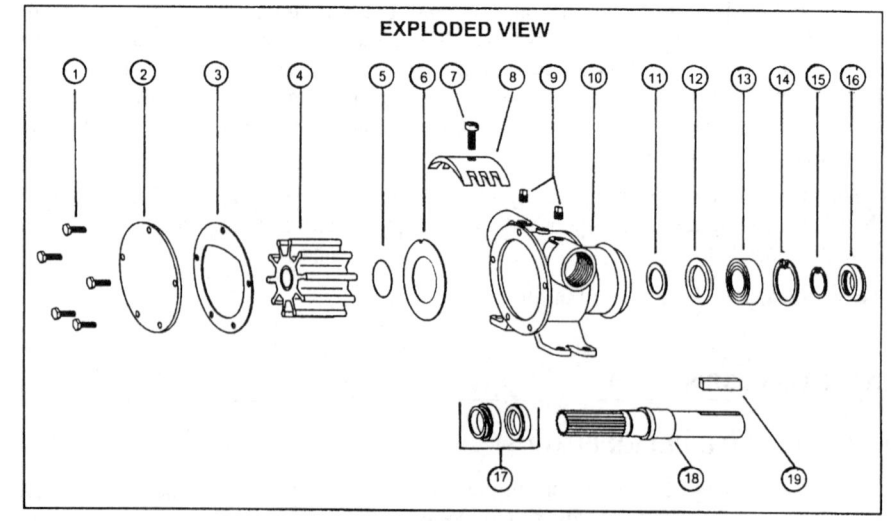

Fig 29.15: FLEXIBLE IMPELLER DETAILS (*Jabsco*)

Chapter 29: Bilges, Pumps, Auxiliary Power Take-offs, Refrigeration & Plumbing

Impeller Trouble Shooting Guide

This guide is designed to help spot typical application problems that show up in flexible impellers during normal use.

Problem 1

Pieces missing from blades tips especially in center of impeller.

Edges look hollowedout or eaten away.

Pitting on ends of impeller.

Causes:
- Cavitation, i.e. too much vacuum at pump inlet, fluid boils locally.

Remedies:
- Reduce pump speed.
- Increase inlet pipe diameter.
- Reduce inlet pipe length and restrictions.

Problem 2

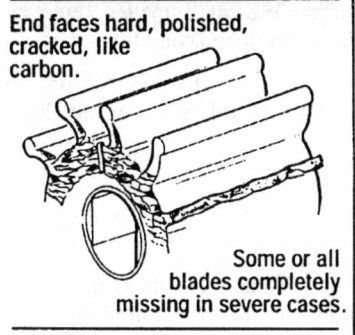

End faces hard, polished, cracked, like carbon.

Some or all blades completely missing in severe cases.

Causes:
- Dry running.

Remedies:
- Do not run more than 30 seconds without liquid in pump.
- Stop pump as soon as liquid is exhausted.
- Arrange pipe work to trap liquid in pump on discharge side. Prevents dry running for several minutes.

Problem 3

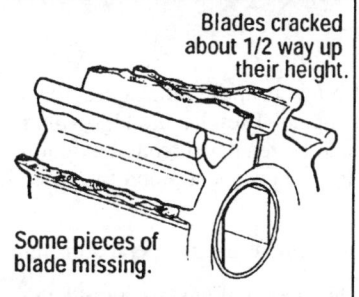

Blades cracked about 1/2 way up their height.

Some pieces of blade missing.

Causes:
- Normal end of useful life.
- Excessive outlet pressure reduces impeller life.
- A crease on trailing side of each blade can also indicate excessive pressure.

Remedies:
- Reduce pressure and/or pump speed.
- Increase outlet pipe diameter.
- Reduce outlet pipe length and restrictions.
- Can also be due to dry running.

Problem 4

Blades permanently and excessively curved.

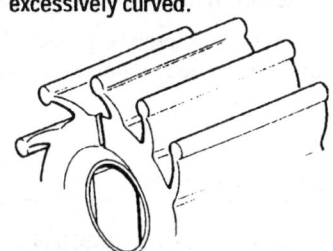

Causes:
- Long term storage in pump.
- Normal end of useful life (especially nitrile impellers).

Remedies:
- Remove impeller for long term storage.
- Refit impeller to rotate in opposite direction.

Problem 5

Worn blade tips and faces.

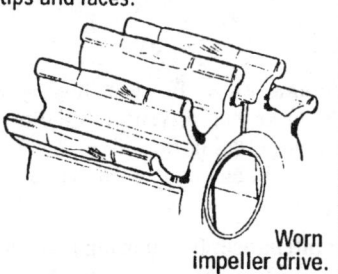

Worn impeller drive.

Causes:
- Abrasive wear from pump or fluid.
- Worn impeller drive can also be due to excessive pressure or dry running.

Remedies:
- Pump should continue to operate satisfactorily in worn condition.
- Replace severely worn pump parts.

Fig 29.16: FLEXIBLE IMPELLER TROUBLESHOOTING GUIDE (*Jabsco*)

FLEXIBLE IMPELLER REPLACEMENT

The flexible impellers should be replaced at least once every year. Spare impellers (of the correct part number) should be carried on board in the event of emergency. The impeller is one of the most vital components of the engine cooling system, and should always be treated as such. Replacing the impeller is easy. By removing the end cover screws, it is possible to remove the impeller either using a dedicated impeller removal tool, or a plumber's wrench that grips the impeller hub. It is not advisable to use two screwdrivers, as these will damage the face of the pump body causing leaks and can be dangerous in confined spaces. The Jabsco Flexible Impeller Removal Tool is the easiest way to remove impellers, especially when the pump is mounted in tight and cramped conditions.

The new impeller should be greased for two reasons: firstly it makes it easier to install the impeller into the pump bore, and secondly it gives added protection to the impeller whilst under initial prime. After replacing the gasket and end cover the pump is ready to use.

2. CENTRIFUGAL PUMP

This is not a not positive displacement pump. The example of this type of pump is the submersible bilge pump. It is also common to find such a pump plumbed with the suction fitted with an L-port cock so that it can draw from the bilge or the sea, as required.

How it Works:

- Centrifugal pumps are not self-priming. Their inlet must be flooded before they can start pumping. They are thus usually submersible pumps. The pump can run dry periodically without damage. However, for maximum seal life, the dry run periods should be kept to a minimum. They can suck from a small height if initially primed and fitted with a non-return valve at the bottom of the inlet pipe.

- The rotating impeller gives velocity energy to the liquid. The liquid drawn into the centre of the impeller is forced to its periphery by centrifugal force and towards the discharge port.

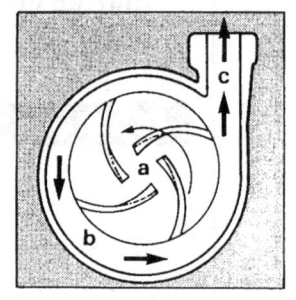

- The momentum generated in the liquid keeps it moving and the pump continues to work.

Fig. 29.17 CENTRIFUGAL PUMP (*Jabsco*)

Features:

High Volume Flow: Being of high-speed rotation, these pumps are always electric motor driven or belt driven from the main engine. They handle high volumes with a smooth, non-pulsating flow. The flow rate can be regulated from maximum output to no flow without any damage to the pump. It is an excellent pump for general transfer applications.

Low Maintenance: Few moving parts mean that wear due to operation is minimal.

Easy Installation: Compact size for flow rate. Option of port positions simplifies pipe runs.

Versatility: Centrifugal pumps can be built in submersible form making excellent bilge pumps. They can also handle dirty water. They do not generally stall unless physically jammed with debris. A coarse bilge strainer is therefore sufficient.

Low Power Consumption: Electric centrifugal pumps consume less power than most other bilge pumps.

Fig. 29.18: CENTRIFUGAL PUMP HOUSING (*Johnson*)

INSTALLATION

Always use hose of the recommended size. Pipe runs should be kept as short and straight as possible, avoiding rising and dipping over obstructions as this can cause air-locks. However, modern pumps may incorporate an anti-lock design so that a dip or water-lock in the hose is cleared automatically when the pump is started. In fact, in some instances it is necessary to include a water-trap in the discharge hose to prevent exhaust fumes from blowing into the vessel through the bilge discharge when it is not in use.

Submersible bilge pumps should be located in the lowest part of the bilge. They must also be plumbed to a thru-hull overboard discharge that remains above the waterline at all angles of heel. This is to avoid water siphoning back into the vessel. Sailing vessels generally discharge through or below the transom.

Chapter 29: Bilges, Pumps, Auxiliary Power Take-offs, Refrigeration & Plumbing

MAINTENANCE

Submersible centrifugal pumps generally require no periodic maintenance other than occasionally checking and possibly cleaning the pump inlet port and the strainer base. This can usually be done by depressing the base release tabs and lifting the pump assembly from the strainer base. At the same time inspect the hose connections to ensure they are tight.

3. SLIDING VANE PUMP

Although now uncommon in small vessels, you may still come across them in applications such as manual fuel pumping into a header tank or the manual bilge pump.

How it works:

- The vanes create a partial vacuum for priming.
- As the rotor rotates, each successive vane draws and carries liquid from the intake to the discharge port.
- When the vanes contact the eccentric portion of the pump body, they force liquid out the discharge port.

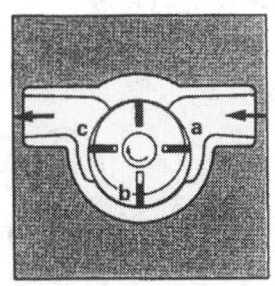

Fig 29.19: SLIDING VANE PUMP (*Jabsco*)

Fig 29.20: SLIDING VANE PUMP HOUSING (*Jabsco*)

Features:

Durable: Heavy-duty construction gives long life.

Self-Priming: Self-priming up to 4m (13 ft).

Simplicity: There are few moving parts to replace. The liquid being pumped lubricates the rotor, vanes and seal. Full access to the pump head can be gained by loosening about 3 screws.

Versatile: An excellent compact unit for general utility or transfer applications which will pump thin or somewhat viscous liquids, and can be mounted at any angle and run in either direction.

Good pumping practice: Pipe runs should be kept as short and straight as possible, avoiding rising and dipping over obstructions as this can cause air-locks. Always use hose of the recommended size and of a rigid or reinforced type that will not collapse under suction conditions. Electric pumps should always be installed in a dry, well ventilated, position as close as possible to the liquid to be pumped.

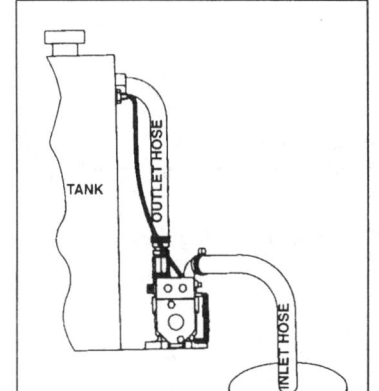

Fig 29.21: SLIDING VANE PUMP APPLICATION (*Jabsco*)

4. DIAPHRAGM PUMP

Examples of these self-priming pumps are the non-submersible manual and electric bilge pumps; pressurised water supply pumps and engine fuel lift pumps.

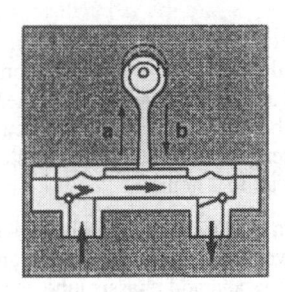

Fig. 29.22 (A) & (B) DIAPHRAGM PUMP & HOUSING (Jabsco)

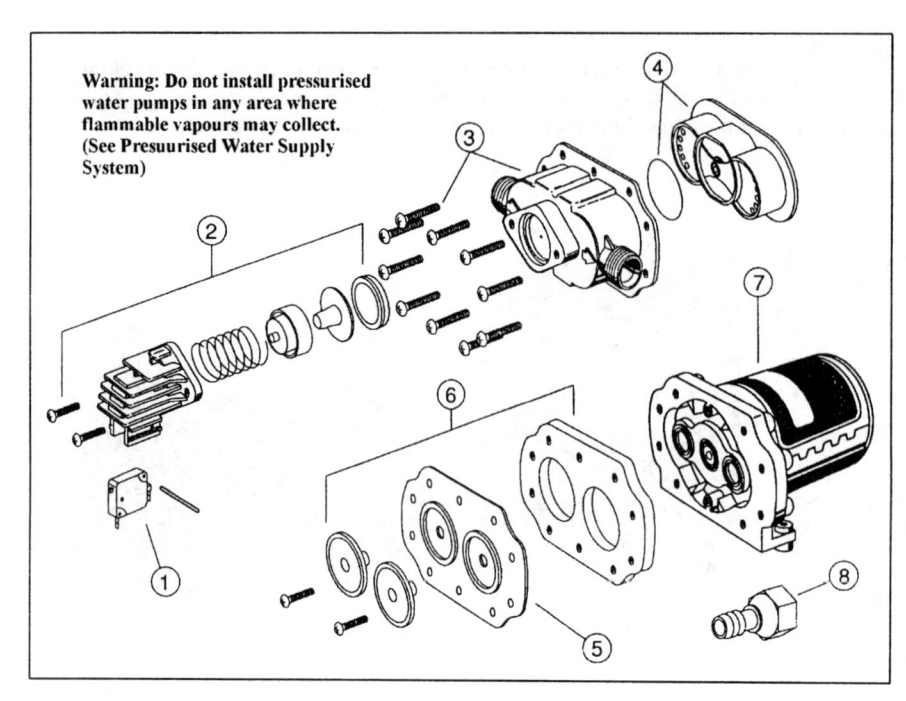

Fig 29.23
EXPLODED VIEW OF A
TWIN-DIAPHRAGM
PRESSURISED
WATER SUPPLY PUMP
(*Jabsco*)

1. Micro Switch
2. Pressure Switch
3. Body Kit
4. Valve Kit
5. Diaphragm
6. Plate/Piston Kit
7. Motor
8. Hose Adaptor

Warning: Do not install pressurised water pumps in any area where flammable vapours may collect. (See Presurrised Water Supply System)

How it works:

- The pump may be fitted with one or more diaphragms, which are constructed of several layers of fabric. The diaphragm(s), pulled upwards by the movements of a piston or a handle causes a partial vacuum, opening the inlet port and closing the outlet, drawing in liquid.
- Downward movement of the diaphragm(s) pressurizes the liquid, closing the inlet valve and opening the outlet valve through which liquid is expelled by pressure.

Features:

Self Priming: Vertical lift up to 5m (16 ft) for manual pumps and 1.5m (5 ft) for electric pumps.

Dry Running: Diaphragm pumps can be run dry for extended periods with no damage, therefore requiring less attention in use.

Versatility: Self-priming and dry running capabilities mean few limitations on use. However, they are best suited for low-pressure situations, such as bilges. Pumps with plastic bodies and synthetic diagrams are ideal for corrosive liquids such as saltwater and sewage. Larger manual diaphragm pumps have the ability to handle some solids in suspension. Electric diaphragm pumps are most widely used in pressurized freshwater systems. They also make excellent bilge pumps for boats with shallow bilges, virtually hoovering the bilges dry.

Quiet Running: Electric pumps feature pulsation dampeners, which smooth flow and reduce noise levels, resulting in less interference when in operation.

Good pumping practice: Electric pumps should always be installed in a dry, well-ventilated position as close as possible to the liquid to be pumped. Pipework should always be reinforced, non-collapsible hose of the recommended size. Bilge pumps and water pumps in particular should always be fitted with a filter and strainer on the inlet pipework to protect them from debris. These should be fastened to the boat structure to ensure their permanent location. Pumps must be installed with the overboard discharge well above the waterline (both static and heeled) to avoid water siphoning back into the vessel. Electric diaphragm pumps can be run dry for up to 2 hours.

Maintenance: To maintenance a diaphragm pump, a regular check would indicate when to get a screwdriver out and replace the diaphragm or any of the flap valves. Make sure the glands and clamps are tight, and flap valves pliant and free moving. Check connecting rod bearing annually and add chassis lube as needed.

Winter Storage: Where possible, it is preferred that the complete pump be removed and stored in a warm dry place. If this is not possible, the pump must be completely drained, hoses removed and pump run until all water is expelled.

Chapter 29: Bilges, Pumps, Auxiliary Power Take-offs, Refrigeration & Plumbing

Fig 29.24
EXPLODED VIEW OF A
FOUR-DIAPHRAGM
PRESSURISED WATER
SUPPLY PUMP (*Jabsco*)

1. Micro Switch
2. Pressure Switch
3. Body Kit
4. Valve Kit
5. Diaphragm Kit
6. Plate/Piston Kit
7. Motor

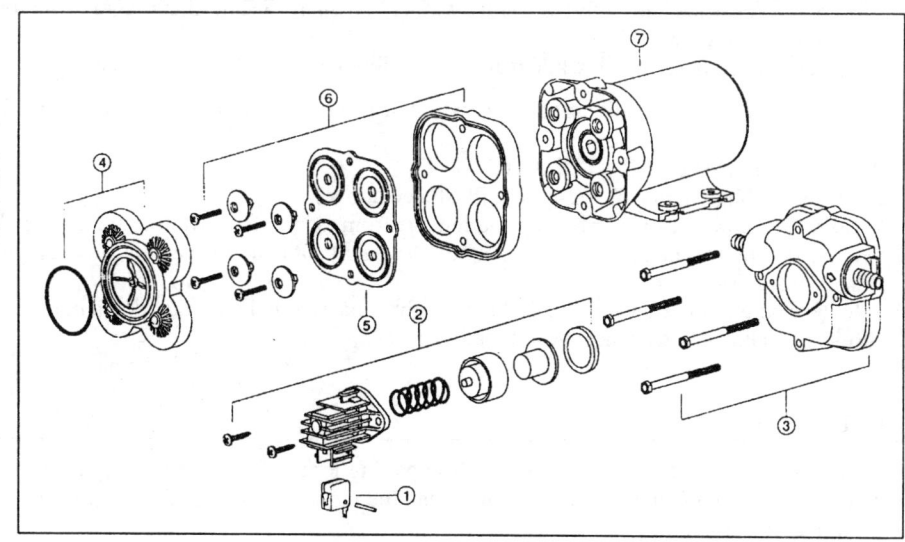

DIAPHRAGM PUMP – TROUBLESHOOTING

If the pump fails to prime, it could be due to:
- Air leak in the suction line.
- Bilge pickup not submerged.
- Intake hose kinked or plugged.
- Dirt preventing a valve from proper seating. (Fouled intake or discharge valve.)
- Rubber flaps swelled up due to chemicals or calcium build up.
- Flaps torn or becoming brittle with age.
- Diaphragm ruptured, delaminating or becoming brittle with age.
- Galvanic corrosion (due to stainless steel hinges and pins being used in cast aluminium pump housing and levers).

Rough or noisy operation can be due to:
- Intake or discharge hose kinked or plugged.
- Pump not mounted firmly.
- Loosened eccentric setscrew or worn connecting rod bearing.
- Ruptured or collapsed pulsation dampener.

VALVE REPLACEMENT
(Jabsco electric bilge pump, 37202-Series, illustrated above)
- Turn off power to pump. Remove four tie-down bolts.
- Expose valves by lifting motor mount and the attached diaphragm assembly from pump base.
- Remove and clean or replace valves.
- Install valves, making sure rubber flapper is UP on intake and DOWN on discharge. Replace valve-retaining plate.
- Replace motor-mount-diaphragm assembly and fasten evenly to base with the four tie down bolts.

DIAPHRAGM & CONNECTING ROD REPLACEMENT
(Jabsco electric bilge pump, 37202-Series, illustrated above)
- Turn off power to pump. Remove four tie down bolts.
- Lift motor mount and the attached diaphragm assembly from pump base.
- Remove two diaphragm retainer screws and the diaphragm retainer.
- Pull connecting rod and diaphragm from eccentric and remove from motor mount; then unscrew bolt to separate diaphragm and plates.
- Check diaphragm for cuts and cracks. Check rod assembly bearing and eccentric for breaks, cracks or excessive wear. Replace, if badly worn.
- During reassembly, ensure that eccentric is firmly seated on motor shaft and set screw is tightened firmly against flat side of shaft.
- Loosely reassemble diaphragm, diaphragm plates, washer and diaphragm bolt onto connecting rod. Slide connecting

rod onto eccentric shaft with two thrust washers on shaft. Secure diaphragm to motor mount with the diaphragm retainer and two screws.
- Tighten connecting nod bolt while maintaining alignment of rod bearing and eccentric.

CAUTION: Avoid misalignment or twisting of rod on eccentric shaft or excessive bearing wear will result.

PULSATION DAMPENER REPLACEMENT
(Jabsco electric bilge pump, 37202-Series, illustrated above)
- Disconnect power leads from pump and remove from mount. Remove four tie down bolts.
- Remove bottom plate screw and the bottom plate. Pull out and replace pulsation dampener; position the 3-ribbed cavity on discharge side.
- Replace bottom plate and screw. Tighten evenly to base with the four tie down bolts.
- Reinstall pump and reconnect power leads.

5. LOBE PUMP

It too is a *Positive Displacement Pump*. The parting lobes of counter-rotating rotors draw in the liquid at the inlet. They carry it around the pump and then discharge it when they mesh at the discharge port.

You are unlikely to find this type of pump on board a vessel. Because of their hygienic construction (in 316 stainless steel), lobe pumps are more commonly used in the food and pharmaceutical industries. They are also known to be used in submarine ballast systems.

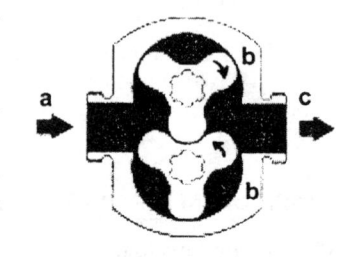

Fig 29.25
LOBE PUMP (*Jabsco*)

VALVES & SEACOCKS

It is not uncommon to find valves and seacocks seized in the open position or corroded to the point of being non-repairable. Such negligence can only lead to flooding of the vessel. Ideally, the valves should be of corrosion-resistant materials, which would make brass valves unsuitable. Even in metal hulls, reinforced plastic valves can be installed. If fitting brass valves and through-hull fittings, they must be insulated from the metal hull and piping to prevent galvanic corrosion. Valves in the ship's hull (overside valves) should be dismantled and overhauled whenever the vessel is on the slip.

1. SEACOCK & BALL VALVE

Traditionally all seacocks and skin fittings (through-hull fittings) in small vessels were made of bronze. The seacock was a tapered bronze plug with a hole through it. Turned one way, the hole in the plug lined up with the pipeline. Turned the other way it closed the pipeline. A quarter of a turn of its handle opened or closed the line. It was open when the handle was in line with the pipe, and closed when at right angle to the pipe. Such a seacock was susceptible to excessive friction due to the lubrication grease leaking out, over-tightening and the plug becoming wasp-waisted.

The traditional seacock has now been replaced by fittings such as ball valves, gate valves and globe valves, and they are made of materials such as stainless steels, brass (bronze), rubber and plastic.

The ball valve is the direct descendant of the seacock. It consists of a ball with a hole through it. Just like a seacock, it is opened or closed by turning the ball to line up the hole with or at right angle to the pipeline. Balls can be designed to direct flow in more than two directions. Ball valves are much more efficient and durable than the old seacocks. They are generally made of bronze or reinforced plastic, both with Teflon seals. Both types should be greased annually.

Fig. 29.26
BALL VALVE (*Dayco*)

Chapter 29: Bilges, Pumps, Auxiliary Power Take-offs, Refrigeration & Plumbing

2. NON-RETURN VALVES (NRV) AND SCREW-DOWN NON-RETURN VALVES (SDNRV)

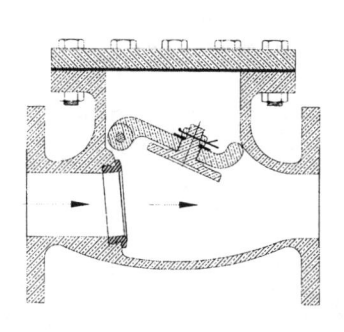

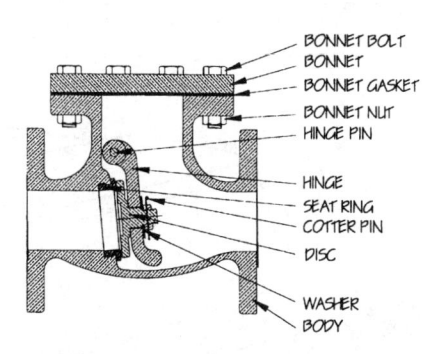

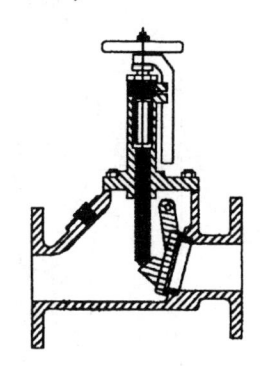

Fig. 29.27 (A): FLOW THROUGH A HINGED NON-RETURN CHECK VALVE (*Pima*)
Fig. 29.27 (B): PARTS OF A HINGED NON-RETURN CHECK VALVE (*Pima*)
Fig. 29.27 (C): A SCREW-DOWN HINGED NON-RETURN VALVE (*Pima*)

NRVs (or non-return check valves) permit only one-way flow of a fluid. They are employed to prevent back flooding into bilge compartments. They come in a variety of designs and are made of various materials. A NRV may consist of a hinged flap, as shown here, a spring loaded plunger or a valve sitting on a seat such that liquid flowing from the bottom inlet lifts it open while a back flow from the side outlet forces it against the seat. The hinged flaps are better in low-pressure applications, where the liquid pressure may be insufficient to lift the plunger or the 'sit-down' type of valve, especially if it is sticking to the seat. However, flaps must be mounted such that gravity will close them when the flow stops. Backpressure holds them closed.

SDNRVs are fitted on multi-bilge suction valve chests. They permit shutting off empty bilge suctions to prevent air being sucked from them when other bilges are pumped out. Screw-down non-return globe valves are discussed in Globe Valves below.

3. GATE VALVES (SCREW-LIFT VALVES)

A gate valve is a screw-lift valve. It has a tapered gate (metal disc) that slides firmly in a slot in the body of the valve. The gate is lowered and raised by turning a threaded spindle on which it rides. A small tightening nut around the stem allows the leaky gland to be tightened. The valve should be disassembled and inspected annually. It should be corrosion free, the spindle should be greased, and the gland repacked if necessary. Some gate valves are fitted with position indicators and limit switches.

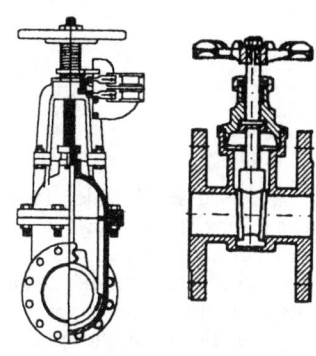

Fig 29.28 (A): GATE VALVE WITH DOUBLE LIMIT-SWITCHES (*Pima*)
Fig 29.28 (B): GATE VALVE (*Ernest Gopfert*)

4. GLOBE VALVES (SCREW-LIFT & SDNR VALVES)

Globe valves are similar to gate valves. They too are screw-lift valves. However, instead of a gate, there is a disc at the end of the spindle, which is screwed down against a seat to shut the valve. The valve maintenance is also similar to that of gate valves. In screw-down non-return globe valves the disc is not attached to the screw-down spindle. The disc can only be screwed down against the seat by the spindle; it cannot be lifted by it. Only the pressure of the liquid from the inlet can lift it.

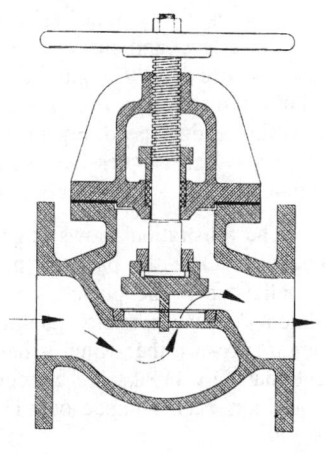

Fig. 29.29
GLOBE VALVE (*Pima*)

AUXILIARY POWER TAKE-OFF SYSTEMS

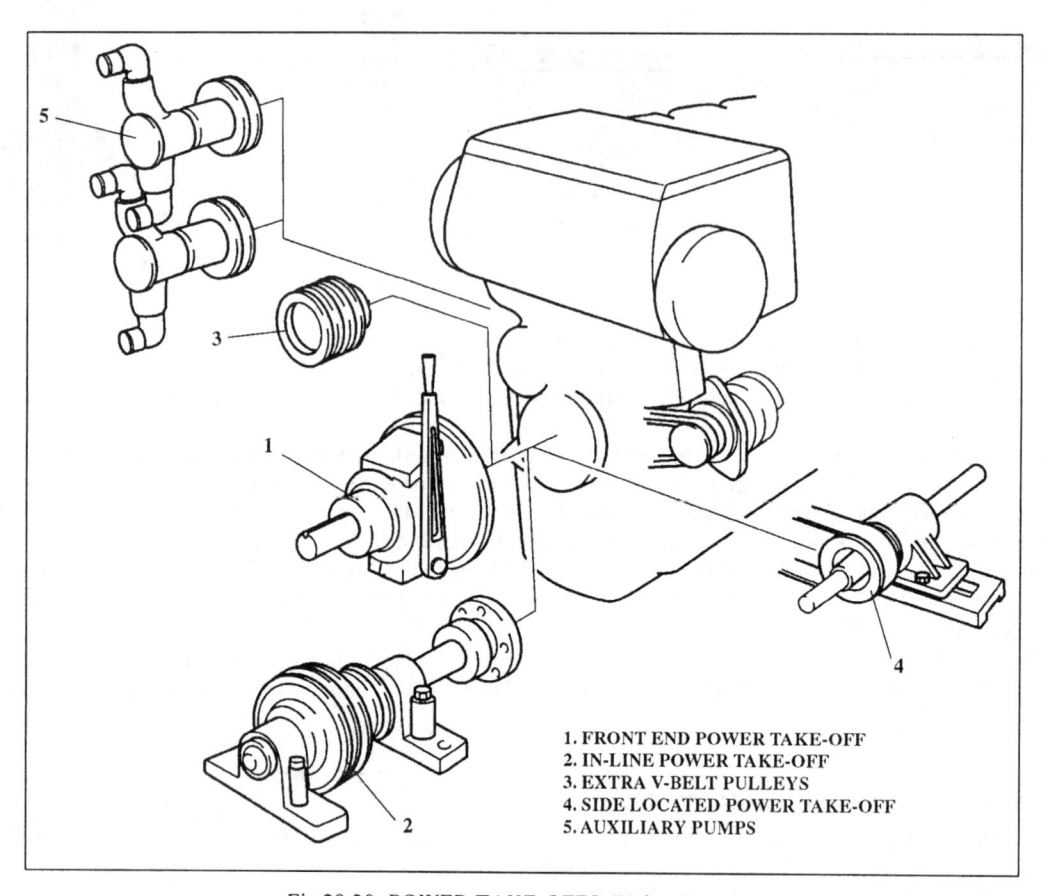

Fig 29.30: POWER TAKE-OFFS (*Volvo Penta*)

1. FRONT END POWER TAKE-OFF
2. IN-LINE POWER TAKE-OFF
3. EXTRA V-BELT PULLEYS
4. SIDE LOCATED POWER TAKE-OFF
5. AUXILIARY PUMPS

As discussed in Bilge Pumps, accessories such as water pumps, bilge pumps and steering pumps are often run from a power take-off groove on the main engine crankshaft pulley. They may also be run from a power take-off (PTO) auxiliary drive fitted to the main engine at its front end, in-line or at its side. A disconnectable power take-off (No. 1 in the illustration) is required if it needs to be engaged or disengaged while the engine is running.

V-belt types of transmissions are quite flexible. By using different pulley sizes, they can be easily adapted for different ratios. They have low noise level and are relatively free of maintenance, as long as they have been carefully aligned to allow the V-belt tension to be easily adjusted. Insufficient installation tension can reduce belt life and cause it to slip at high speed. Equally, excessive belt tension on the auxiliary drive may cause excessive side loads, which may result in crankshaft failure.

The illustration shows bilge and flushing pumps mounted on the engine PTO at the back of the timing gear casing. The pumps are impeller-type. The power is transferred through an electromagnetic clutch. A vacuum switch monitors the connection timing of the bilge pump. At start, the switch is held down for about 20 seconds. It then automatically breaks the current to the magnetic coupling when the liquid has been pumped out. The flushing pump is used for services

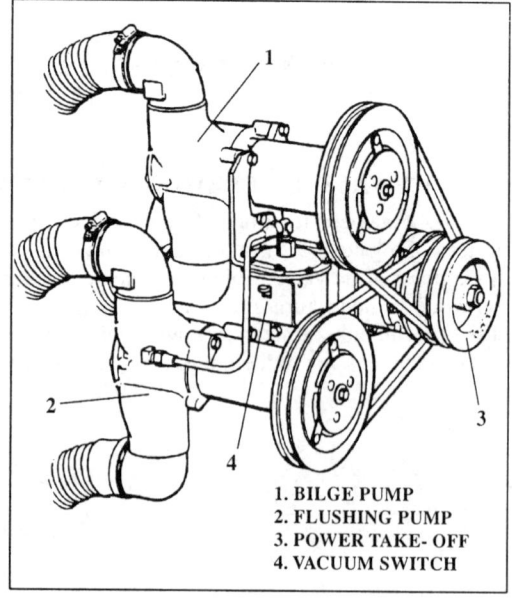

1. BILGE PUMP
2. FLUSHING PUMP
3. POWER TAKE-OFF
4. VACUUM SWITCH

Fig 29.31
V-BELTS DRIVEN PUMPS (*Volvo Penta*)

Chapter 29: Bilges, Pumps, Auxiliary Power Take-offs, Refrigeration & Plumbing

such as deck washing and fish washing.

The belt tension can be estimated by applying pressure to it midway between the pulleys, and adjusted until the belt deflects by the amount shown in the table. IDLER PULLEYS used for tensioning the V-belts should be fitted on the slack side of the belt. They should not be smaller than the recommended diameter. A spring-loaded idler pulley is preferable to one that is adjusted and clamped, especially with larger PTO values or where there could be movement between a flexibly mounted engine and the drive PTO drive mounted on a separate chassis.

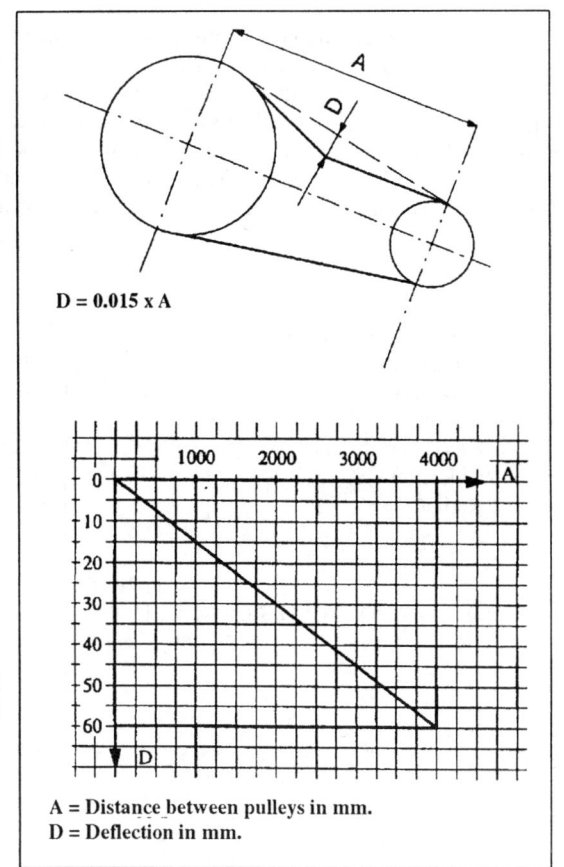

$D = 0.015 \times A$

A = Distance between pulleys in mm.
D = Deflection in mm.

Fig 29.32
BELT TENSION (*Volvo Penta*)

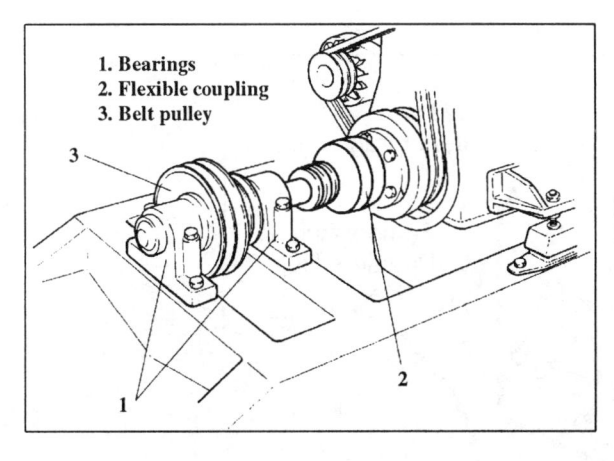

1. Bearings
2. Flexible coupling
3. Belt pulley

Fig 29.33
IN-LINE POWER TAKE-OFF (*Volvo Penta*)

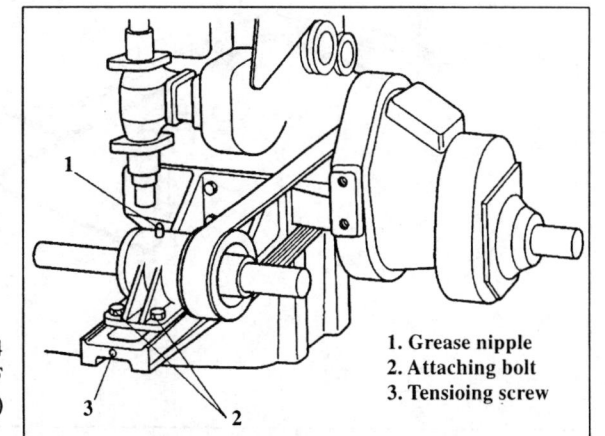

Fig 29.34
SIDE-MOUNTED POWER TAKE-OFF
(*Volvo Penta*)

1. Grease nipple
2. Attaching bolt
3. Tensioing screw

HOT & COLD WATER SUPPLY & CABIN HEATER

The volume of an engine's cooling water (freshwater) can usually be expanded by fitting a larger expansion tank (in consultation with the manufacturer). This allows additional circuits for *cabin heaters* and *calorifiers*, to be fitted. (Calorifier is a heat exchanger for heating potable water). When a cabin heater is installed, it must always have a manual-venting nipple (4) at its highest point. The system is vented once pressurised. The heater or the calorifier must not be higher than the specified maximum height above the expansion tank. In large heating systems, a hose thermostat (5) should be mounted in the line of the hot water circuit. This ensures that the engine quickly reaches its operating temperature. Incorporating shut off valves (2, 3) is useful when it is time to repair or service the system.

Components:

1. Cabin heater with defroster unit
2. Outlet valve
3. Inlet valve
4. Venting nipple
5. Hose thermostat
6. Calorifier
7. Radiator

Fig 29.35
HOT WATER CONNECTIONS
(*Volvo Penta*)

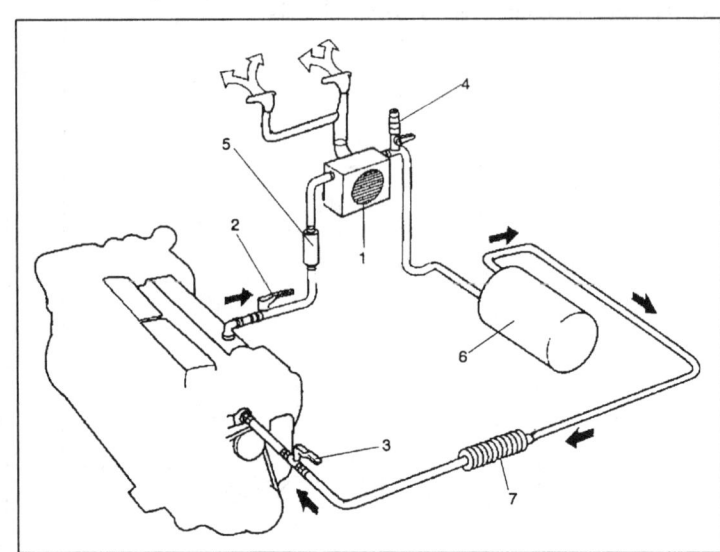

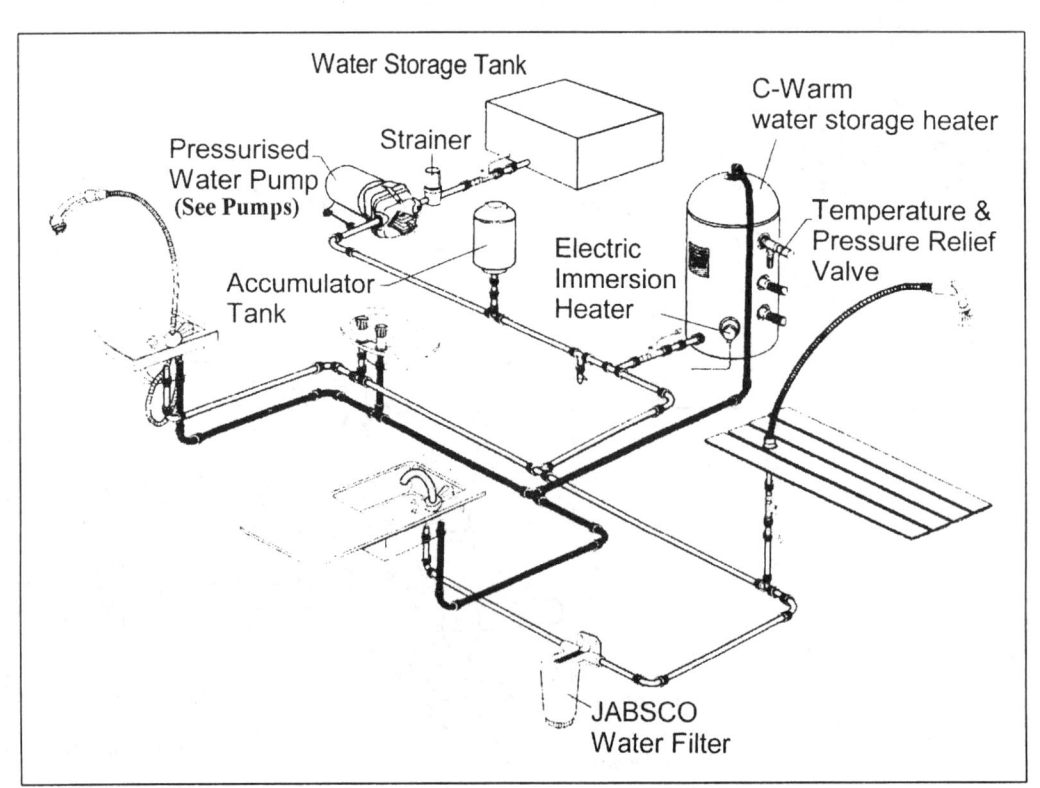

Fig 29.36
JABSCO
PRESSURISED
WATER SUPPLY
SYSTEM

Chapter 29: Bilges, Pumps, Auxiliary Power Take-offs, Refrigeration & Plumbing

REVERSE OSMOSIS WATER MAKERS

Reverse Osmosis desalination watermakers convert seawater to fresh water. They consist of a pump which forces seawater at very high pressure (800 psi or 5516 KPa) through a semi-permeable membrane, which excludes the passage of, dissolved salts in seawater. The product is fresh water suitable for drinking but not as tasty as normal fresh water.

Only about 10% of water passes through as fresh water. The remaining waste brine, still under high pressure, is either discharged overboard through a pressure-reducing valve or redirected to assist in pressurising the pump for better efficiency. Because of the high pressure involved, manual operation requires much energy and is very exhausting. Cruising vessels are more likely to carry a battery-operated unit. A hand operated Reverse Osmosis pump, weighing about 1 kilogram, can produce about half a litre of fresh water in 30 minutes. A 12-Volt battery operated pump drawing 4 amps can produce about 6 litres in an hour. The hand-operated units are included in some sophisticated liferafts.

The unit's pre-filter cartridge needs replacing after a specified period of use. The membrane in the pump, once used, needs regular cleaning to prevent biological growth.

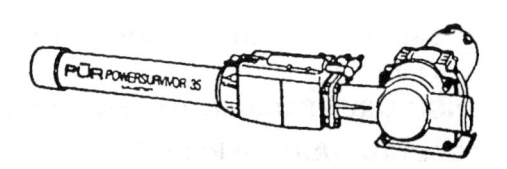

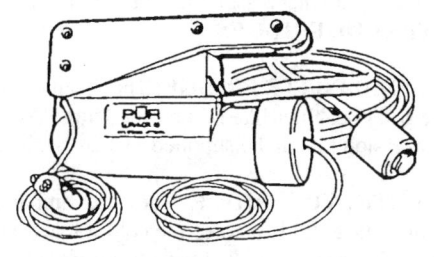

Fig 29.37: "PUR" WATER MAKERS (HAND & BATTERY OPERATED MODELS)

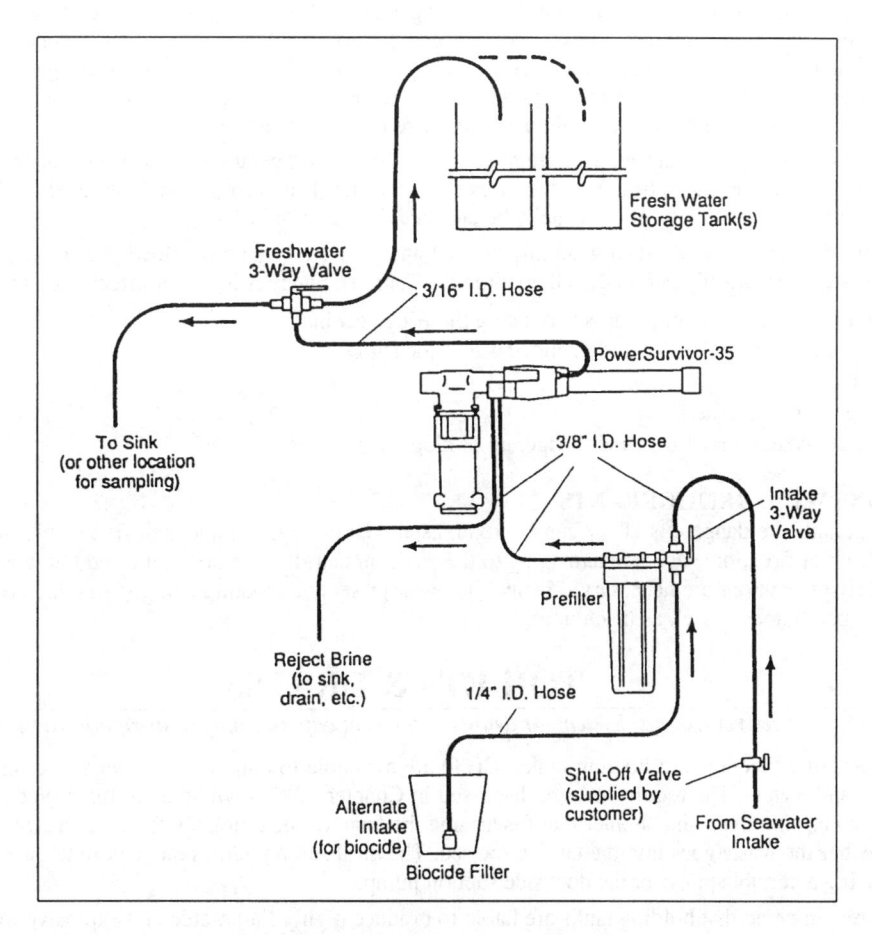

Fig 29.38: THE WATER MAKER PLUMBING DIAGRAM

REFRIGERATION

Installation of an AC/DC fridge on board is a luxury that requires considerable thought and expense. There are two basic choices:

ENGINE-RUN FRIDGE/FREEZER: It requires someone to be on board every day to run the engine or generator.

ELECTRIC FRIDGE/FREEZER: It requires deep cycle batteries of large capacity (min. 120 amp-hours) and a charging system, such as solar panels, wind generator or towed generator/propeller shaft alternator. There will always be some days when the engine would need to be run to charge the batteries.

A towed generator, towing impeller and propeller shaft alternator are essentially the same thing. An impeller is towed behind a moving vessel, the rotation of which is converted into electrical energy by a low-speed alternator or generator.

TYPES OF FRIDGES

1. COLD PLATE FRIDGE: This is an on-demand refrigeration system. It incorporates a rapidly cooling cold plate inside the fridge cabinet like a domestic fridge. Standard models run on 12/24 volts battery system, but those fitted with a power conversion kit can be supplied 110/240 volts from the ship's engine or a generator.

2. EUTECTIC FRIDGE: They are similar to cold plate fridges. The eutectic tank contains a fluid of very low freezing temperature. It also stores the cold longer. This type of fridge is more suitable if the boat's engine or generator can be run on daily basis to pull down its initial temperature, after which it may be run off the batteries, if fitted with a power conversion kit.

COMMERCIAL REFRIGERATION

Vessels engaged in commercial fishing and such activities may carry large refrigeration units.

The principle of all refrigeration is to use a liquid refrigerant to absorb heat from inside the plant and release it to the outside. Driven by a motor, a compressor pushes the refrigerant into a cooling unit (an evaporator) fitted in the walls of the fridge cabinet, where the heat from the plant is used to evaporate it. This warm high-pressure gas is then pumped into an outside condenser for cooling and becoming liquid again - releasing its heat to the surrounding outside air. The cycle continues until the refrigeration plant is brought down to the required temperature.

In small refrigeration plants, the motor is regulated only by the temperature control setting. In larger units, it is also regulated by the pressure control setting that monitors the pressure of the refrigerant. The level of the refrigerant and the presence of any moisture in the system can usually be checked through a sight glass.

Equipped with greater capacity than generally needed and protected by an overload protector, refrigeration plants are designed to operate automatically and under all conditions. Their efficiency is however affected by the following factors:

- Air temperature and the amount of air surrounding the fridge cabinet;
- Size of the evaporator for the size and intended use of the fridge;
- Leaking refrigerant;
- Incorrect type of refrigerant;
- Moisture in the system may freeze the refrigerant or clog the controls.

PRECAUTIONS WITH REFRIGERANTS

Some refrigerants are dangerous. They can be toxic, asphyxiating, eye-irritants and flammable. Mishandling them can cause injury, death or fire. Some are also damaging to the environment. In case of a leak, find out what type of refrigerant it is and what safety precautions are necessary. Meanwhile, wear protective clothing and goggles and keep the area ventilated. Switch off all naked flames as well as machinery.

TOILETS & DRAINS

(See Wet Exhaust System for details of the vent cap assembly (anti-siphon valve))

Various types of collection and holding tanks (CHT) are available to comply the toilet waste discharge regulations in harbours and inland waters. The regulations are discussed in Chapter 12. Shown here are three types of holding tanks, one of which is the collapsible type and another that fits around the front of the toilet. On flushing the toilet, a "Y" valve setting determines whether the waste goes into the tank or the sea. The third setting allows the tank to be pumped overboard or to a shore facility using a portable pump or the dockside suction pump.

It must be remembered that holding tanks are liable to produce highly flammable and explosive methane gas, as well as the asphyxiating hydrogen sulphide gas. The material of some holding tanks is also likely to absorb the smell of its contents over time.

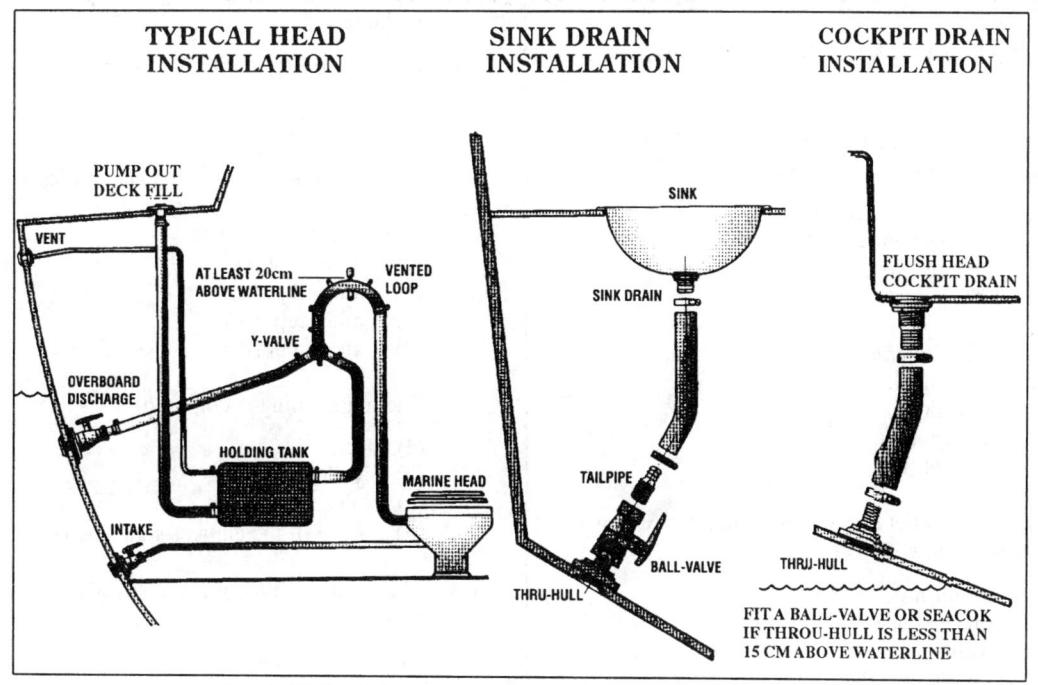

Chapter 29: Bilges, Pumps, Auxiliary Power Take-offs, Refrigeration & Plumbing

Fig 29.39: TOILETS & DRAINS PLUMBING (*Forespar*)

COLOUR CODING OF PIPELINES

The Australian Standard (AS 1345 of 1995 + Appendix) colour coding system is a combination of primary and secondary colours. For example, green is the primary colour for water; and the colour dark blue indicates that the contents are suitable for human consumption. Therefore, a drinking water pipe could be painted green with dark blue bands OR green with pipe markers in text OR it could simply be painted dark blue to indicate that its contents are safe for human consumption. The secondary colour for a dedicated fire fighting system (water, foam, etc.) is red. Hot water above 60°C or below -10°C must be additionally marked.

Pollutant discharge (Bilge/Sewage, etc)	= Black
Communication (Low voltage lines)	= White
Electric	= Orange
Fire main	= Red or Green with red bands
Drinking Water	= Dark Blue or Green with dark blue bands
Fuel	= Brown
Gas (dangerous gas)	= Yellow
Sea suction	= Green
Steam	= Silver
Compressed air	= Light Blue

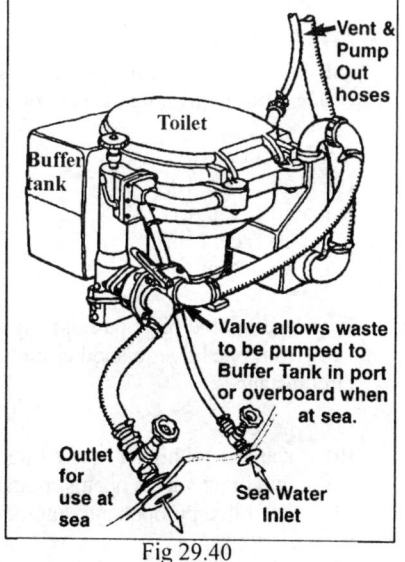

Fig 29.40 A "*RM*" TOILET BUFFER TANK

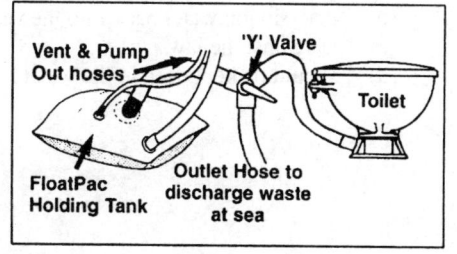

Fig 29.41 A "*FLOATPAC*" COLLAPSIBLE TOILET HOLDING TANK

CHAPTER 29: QUESTIONS & ANSWERS

BILGES, PUMPS, VALVES, COLOUR CODING, ETC.

1. Plastic hoses when used in sea water and bilge pumping systems in vessels are likely to:

    (i) burst under pressure.
    (ii) collapse under suction.
    (iii) melt due to heat.
    (iv) resist mechanical damage.

    The correct answer is:
    (a) all except (i).
    (b) all except (ii).
    (c) all except (iii).
    (d) all except (iv).

2. The overside outlet of a vessel's pumping system should be fitted with a:

    (a) non-return valve.
    (b) L-shaped cock.
    (c) no valve.
    (d) screw-down valve.

3. Your bilge pump fails to draw water from a full bilge. It may be due to the:

    (i) incorrect valve open.
    (ii) overboard discharge valve open.
    (iii) non-return valve seized in seat.
    (iv) pump priming device not working.

    The correct answer is:
    (a) all except (i).
    (b) all except (ii).
    (c) all except (iii).
    (d) all except (iv).

4. In an operational bilge pumping system, water can flow from one bilge to another under the following circumstances:

    (a) never.
    (b) water level higher in one bilge.
    (c) two bilge valves open simultaneously.
    (d) one bilge pumped into another accidentally.

5. A water hose hanging in the water over the side of a vessel in port:

    (a) provides readiness for fire fighting.
    (b) may siphon water back into the vessel.
    (c) is against the law.
    (d) should be kept deep in the water.

6. The parts of a vessel's bilge pumping system include:

    (i) an L-port cock.
    (ii) a common suction line to all compartments.
    (iii) strum boxes.
    (iv) a non-return valve in overboard discharge.

    The correct answer is:
    (a) all except (i).
    (b) all except (ii).
    (c) all except (iii).
    (d) all except (iv).

7. The bilge piping is coloured:

    (a) red
    (b) yellow
    (c) blue
    (d) none of the choices stated here

8. The colour coding of a vessel's piping system is as follows:

    (i) Fuel: red
    (ii) Fresh water: blue
    (iii) Sea suction: green
    (iv) Bilge: black

    The correct answer is:
    (a) all except (i).
    (b) all except (ii).
    (c) all except (iii).
    (d) all except (iv).

9. The colour coding of a vessel's piping system is as follows:

    (i) Fuel: brown
    (ii) Fresh water: blue
    (iii) Steam: silver
    (iv) Dangerous gas: red

    The correct answer is:
    (a) all except (i).
    (b) all except (ii).
    (c) all except (iii).
    (d) all except (iv).

10. Engine driven bilge pumps and fire pumps are usually run by:

    (a) a direct drive off the engine.
    (b) a chain drive from the engine.
    (c) a Vee belt and remotely controlled clutch.
    (d) a dog clutch off the engine.

Chapter 29: Bilges, Pumps, Auxiliary Power Take-offs, Refrigeration & Plumbing

11. When pumping out bilges with the combined bilge/deckwash line, the L-port cock would be in the following position:

    (a) Closed.
    (b) Seacock to deckwash.
    (c) Bilge intake to overside outlet.
    (d) Bilge intake to seacock.

12. A leaky bilge suction line makes pumping-out difficult because the leak:

    (a) reduces the flow.
    (b) prevents a vacuum building up.
    (c) creates air pockets.
    (d) prevents the pump from starting.

13. The test-pumping of a dry bilge involves:

    (a) sucking in air through the line.
    (b) back-flooding the bilge.
    (c) checking that it is clean.
    (d) pouring some water in it & pumping it out.

14. The symbol for one-way valve is two arrowheads pointing towards each other, one blank and one filled-in. The permitted direction of flow of liquid is indicated by the direction of:

    (a) the filled-in arrowhead.
    (b) the blank arrowhead.
    (c) neither of the arrowheads.
    (d) either of the arrowheads.

15. The symbol of a valve on a pipeline shows two arrowheads pointing towards each other with a dog-leg emerging from the middle. It is a:

    (a) one-way valve.
    (b) two-way valve.
    (c) self-closing valve.
    (d) L-port cock.

16. A series of small triangles in pairs with apexes pointing towards one another is the symbol for:

    (a) a self-closing valve.
    (b) a manifold.
    (c) valves with strainers.
    (d) a pump.

17. A symbol consists of three arrowheads, pointing inwards. Two of them are blank and the third one is filled in. It is the symbol for:

    (a) a non-return valve.
    (b) a two-way valve.
    (c) none of the choices stated here.
    (d) an L-port cock.

18. A small leak on the outlet side of a bilge pump would cause:

    (a) the pump to slow down.
    (b) the pump to speed up.
    (c) the flow to decrease due to sucking of air.
    (d) some water to escape during pumping.

19. A pump that alternately increases & decreases its volume is known as the:

    (a) self-priming pump.
    (b) flexible impeller pump.
    (c) positive displacement pump.
    (d) centrifugal pump.

20. Self-priming pumps:

    (a) cannot be run dry.
    (b) are of centrifugal type.
    (c) must be installed below the liquid level.
    (d) draw fluid to themselves.

21. The following type of pump is not a self-priming pump:

    (a) Flexible impeller.
    (b) Centrifugal.
    (c) Diaphragm.
    (d) Sliding vane.

22. A flexible impeller pump:

    (a) needs to be primed before drawing water through a long pipeline.
    (b) is not self-priming.
    (c) is not a positive displacement pump.
    (d) can be of a smaller size than the required self-priming ability.

23. Almost all engine raw water pumps and engine driven deckwash pumps are of the following type:

    (a) Flexible impeller.
    (b) Centrifugal.
    (c) Diaphragm.
    (d) Sliding vane.

24. If the blades of a flexible impeller pump become permanently and excessively curved, it indicates:

    (a) cavitation.
    (b) too much vacuum at the pump inlet.
    (c) dry running.
    (d) long term storage or the end of useful life.

25. In a flexible impeller pump, pieces are missing from the blade tips, especially in the middle of the impeller. It is due to:

    (a) cavitation.
    (b) excessive pressure.
    (c) abrasive wear from the fluid.
    (d) long term storage.

26. When replacing the impeller in a flexible impeller pump:

    (a) seek expert assistance.
    (b) grease the new impeller.
    (c) cut the blades of the new impeller to size.
    (d) bring the pump out on deck.

27. A good example of a centrifugal pump is the:

    (a) raw water cooling pump.
    (b) engine driven deckwash pump.
    (c) submersible bilge pump.
    (d) flexible impeller pump.

28. Which of the following precautions does not apply to submersible bilge pumps?

    (a) replace the flexible impeller when worn.
    (b) locate in the lowest part of the bilge.
    (c) the overboard discharge should remain above waterline at all angles of heel.
    (d) a water trap in the discharge line prevents exhaust fumes blowing into vessel.

29. A good example of a diaphragm pump the:

    (a) raw water cooling pump.
    (b) engine driven deckwash pump.
    (c) submersible bilge pump.
    (d) non-submersible hand-operated bilge pump.

30. Diaphragm pumps:

    (a) are submersible pumps.
    (b) do not suffer from galvanic corrosion.
    (c) can be run dry.
    (d) are flexible impeller pumps.

31. Valves and seacocks that would corrode the least are made of:

    (a) reinforced plastic.
    (b) brass.
    (c) steel.
    (d) stainless steel.

32. The following type of non-return valve is most suited in low-pressure applications:

    (a) spring-loaded plunger.
    (b) valve sitting on a seat.
    (c) hinged flap.
    (d) spring-loaded sitting type.

33. Which of the following statements is incorrect:

    (a) A gate valve is a screw-lift valve.
    (b) A globe valve is a screw lift valve.
    (c) A globe valve can be a SDNRV.
    (d) Only the pressure of liquid from the inlet can open a gate valve.

34. Which of the following characteristics does not belong to the bilge system:

    (a) A non-return valve is fitted common to all bilge suctions.
    (b) Overboard discharge is fitted with a non-return valve.
    (c) Plastic hoses are not used.
    (d) There is a separate suction line from each compartment.

35. Fitting a very long discharge line on a small-volume bilge pump:

    (a) is acceptable when fitted with a level switch.
    (b) is not advisable.
    (c) is acceptable when fitted with a check valve.
    (d) is acceptable when fitted with a non-return valve.

36. On shutting down the bilge pump there is always some back flushing from the discharge line. To allow heavier sediments to fall well below the filter:

    (a) a non-return valve should be fitted.
    (b) submersible pumps should not be employed.
    (c) the bilge suction should be positioned high in the bilge.
    (d) the strum box or the submersible pump should be mounted on a small pedestal.

37. The float switches in bilges:

    (a) are fitted with fail-proof micro-switches.
    (b) need little attention.
    (c) should be changed periodically whether they appear to need it or not.
    (d) should never be used.

38. On sudden appearance of oil in a bilge, which of the following actions is not advisable:

    (a) Pump it out immediately.
    (b) Check the oil pressure gauge.
    (c) Find and rectify the cause.
    (d) Check oil levels in the engine & storage.

39. Which of the following statements is correct?

    (a) Impeller pumps must not be run dry.
    (b) Bellows pumps must not be run dry.
    (c) Displacement pumps must be submerged.
    (d) Self priming pumps never need priming.

Chapter 29: Bilges, Pumps, Auxiliary Power Take-offs, Refrigeration & Plumbing

40. The possible reasons for a pump to lose suction are:

    (i) airtight valve gland.
    (ii) excessive clearance in pump impeller.
    (iii) slipping drive belt.
    (iv) pump choked with foreign matter.

    The correct answer is:
    (a) all except (i).
    (b) all except (ii).
    (c) all except (iii).
    (d) all except (iv).

41. In refrigeration plants, which of the following components is located in the body of the cold room?

    (a) Compressor.
    (b) Evaporator.
    (c) Liquid reservoir.
    (d) Condenser.

CHAPTER 29 = ANSWERS

1 (d), 2 (a), 3 (b), 4 (a), 5 (b),
6 (b), 7 (d), 8 (a), 9 (d), 10 (c),
11 (c), 12 (b), 13 (d), 14 (b), 15 (c),
16 (b), 17 (d), 18 (d), 19 (c), 20 (d),
21 (b), 22 (a), 23 (a), 24 (d), 25 (a),
26 (b), 27 (c), 28 (a), 29 (d), 30 (c),
31 (a), 32 (c), 33 (d), 34 (a), 35 (b),
36 (d), 37 (c), 38 (a), 39 (a), 40 (a)
41 (b)

Chapter 30

SOLAR & OTHER ELECTRICAL SYSTEMS

ELECTRICAL INSTALLATION

The electrical installation requires careful planning and care, keeping the design simple.

STARTING BATTERY CONNECTION:
The starting battery may be connected to the starter motor in one of two ways:
- In the ONE-POLE connecting system, the positive battery terminal is connected to the starter motor, while the negative terminal is connected to earth (the flywheel housing).
- In the TWO-POLE system, both the terminals are connected to the starter motor.

LIGHT SWITCH CONNECTION:
In AC light switches (with active, neutral and earth cables), it is common to connect only the active cable to the switch. Such switches are referred to as one-pole or single-pole switches. The main battery isolator switch, on the other hand, is usually a two-pole or double-pole switch, where both the active and neutral cables are connected to the switch. The added safety feature of the latter is that it not only cuts off the power supply, but also protects the neutral side from any feedback of power.

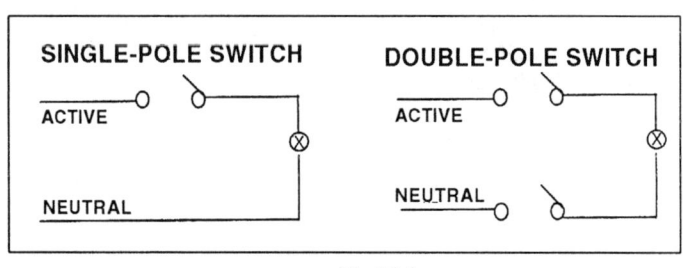

Fig 30.1

PRECAUTIONS
- In twin-engine installations or with separate battery banks, the engines must share the same battery ground for the synchronizing function to operate.
- All wires and connectors must be of types approved for marine use. The cable areas (sizes) and lengths should be as recommended in the charging distribution installation kit.
- Wires should be routed in a protective sheath and properly clamped. The leads should not be installed too close to heated parts of the engine or other sources of heat. The leads must not be subject to mechanical wear. If necessary, draw the leads through protective tubing.
- Keep the joints to a minimum, making sure the cables and joints are accessible for inspection and repair. None of the joints in the engine room should end up deep down. All cable joints should be higher than the alternator.
- Keeping a wiring diagram of the complete electrical installation will simplify fault tracing and installation of additional equipment.
- Spray all electrical equipment with a moisture repellent spray.
- Install the MAIN BATTERY SWITCH on the plus side. Provide the positive and negative cable leads with grommets as needed. Position the main switch outside but as close to the engine compartment as possible to keep the cable length to a minimum. Install it in a locked compartment to prevent theft of the boat.
- To avoid voltage drop in the charging circuits, the cables should be of correct dimensions and the terminals correctly finished. The cables between the charging distributor and the two batteries should be as short as possible, and should have the same length. However, if necessary, the cable between the charging distributor and the start battery (C) can be longer than the cable between the charging distributor and the accessory battery (B).

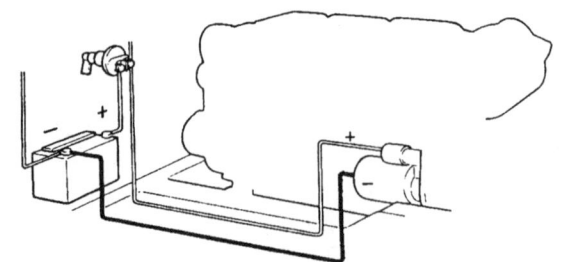

Fig 30.2: MAIN BATTERY SWITCH

Chapter 30: Solar & other Electrical Systems

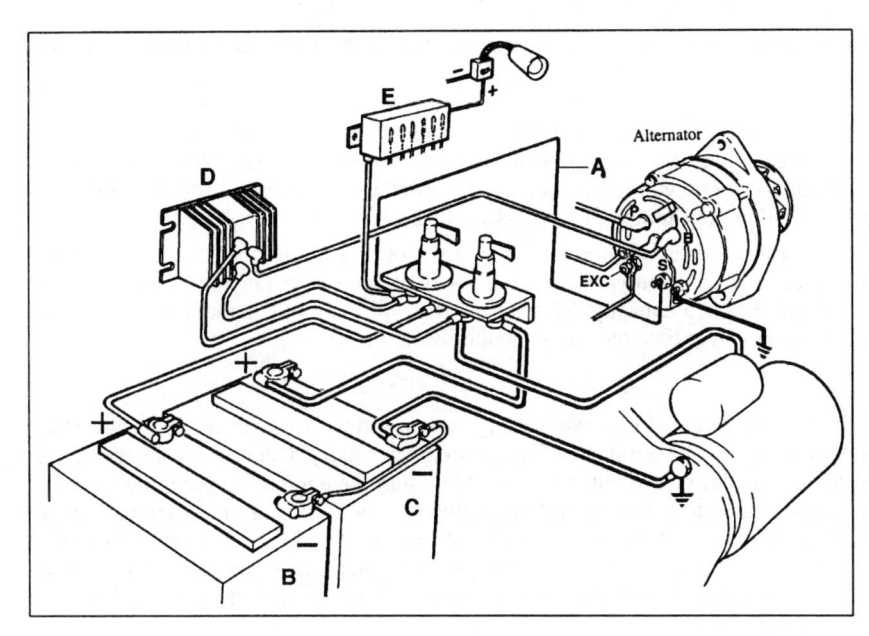

Fig 30.3
BATTERY CHARGING SYSTEM
(*Volvo Penta*)

Key
A. Sensor cable
B. Accessories battery
C. Starting battery
D. Charging distributor
E. Fuse box, other loads

COLOUR CODING OF ELECTRICAL WIRING
- Red/Brown – Active
- Black/Blue – Neutral
- Green - Earth

SHORT CIRCUIT
The connection between the positive and negative sides of an electrical circuit without going though the equipment itself is known as short-circuiting. The excessive flow of current thus caused generates heat. Unless protected by a fuse or circuit breaker, it often results in a fire, usually at the smallest wire in the circuit. Dropping a metal tool such that it lands on the two poles of a battery is a good example of a short circuit.

FUSES AND CIRCUIT BREAKERS
Fuses and circuit breakers perform a similar function. Both protect electrical wiring and equipment from failure or fire due to electrical overloads. The difference between the two is that fuses are designed to melt instantaneously when overheated due to excessive current flowing through an electrical circuit. The circuit breakers, on the other hand, are designed to trip (or open) to break the circuit. [There is a time delay]. Circuit breakers are more convenient because they can be re-set by simply re-switching them to the original position.

Only the correct size electrical fuses and circuit breakers must be used. In the event of repeated failure in either system, check the system for overloading, short-circuiting, major earth leakages and faulty equipment.

AUTOMATIC BATTERY CHARGING SYSTEM
Just as in motorcars, marine engines are fitted with a battery-charging unit, which automatically charges batteries when the engine is running. Sailing vessels need to run their motors periodically in order to recharge the batteries. The battery-charging unit consists of the following parts:

- **ALTERNATOR**: It is a dynamo producing AC current from the rotation of the engine. It is essentially a rotor, composed of an electromagnetic coil wrapped around an iron core. Six diodes convert the AC into DC current. Diodes are is a one-way valve for electricity. They allow the current to flow in one direction only, conducts during half of the AC cycle, and not conducting during the other half. Six diodes working together make a smooth flow of the DC current.

- **VOLTAGE REGULATOR** (electronic or transistorised) (internal or external): It controls the output current of the alternator by controlling the supply of the field current to its magnetic coil. When the engine is running slowly or when the electrical demand is high, the regulator allows longer periods of the field current flow. As the engine speeds up, the regulator interrupts the field circuit as necessary to keep voltage within the required range.

- **CHARGING DISTRIBUTOR**: It is a double diode installed in vessels with high demand for electrical power. It independently and automatically charges two batteries. One circuit is used for starting the engine and the other for the electrical equipment. Under this arrangement, if the accessories battery has been emptied, you can still start the engine.

The charging distributor should be installed on a bulkhead or equivalent, as close to the batteries as possible and well protected from splashing water.

The alternator can generate quite a high voltage even with the engine just idling. The unit is really three alternators in one body. Each of the three sections generates its voltage out of phase with the other two sections. Therefore, one complete cycle (revolution) of the alternator puts out three separate voltages – each phase shifted by 120 degrees from the next. Each of the three phases has its own winding of the coil in the alternator and each winding has its own pair of diodes.

The failure of an alternator can thus be in stages. Each of its windings and/or diodes can fail, one set at a time. If this happens, the alternator can still charge the batteries, but only with two-thirds of its capacity. On failure of two of its systems, the alternator would put out one-third its rated capacity. This means that you may not immediately know of a failing alternator. It also means that you can go for a long time on a limping alternator.

TESTING AN ALTERNATOR

The best way to test an alternator is with a professional battery charging system tester. However, you can also test it with a simple voltmeter by checking the voltage across the terminals of the accessories battery. Connect a voltmeter across the battery terminals with the engine and all accessories off. Note the reading. Now start the engine and see if the voltmeter reading starts to rise. You may have to rev the engine to get some voltmeter relays to kick in. With the engine idling and with all accessories switched off, the voltage across the terminals of a 12-volt battery should read around 14 volts. A reading of 12 volts may indicate a failed alternator. Now test it under load by switching on as many accessories as possible. It should still read around 14 volts. If it reads lower than 13 volts the chances are the alternator is on its way out. If voltage continues to rise above the upper limit, the regulator is faulty.

ELECTRICAL ACCESSORIES

Before installing any additional accessory, such as a navigation instrument or a lamp, make sure that the additional power consumption can be met by the system's charging capacity. Clamp the equipment leads at close intervals and mark them at the junction box (1 and 2 in the illustration) with their purpose, such as radio, refrigerator, navigation lights, etc. Position the switches control panel (switch box) in a position free from moisture with easy access and close to the instrument panel. If a 220 V system is installed, clearly mark and identify its area in the control panel.

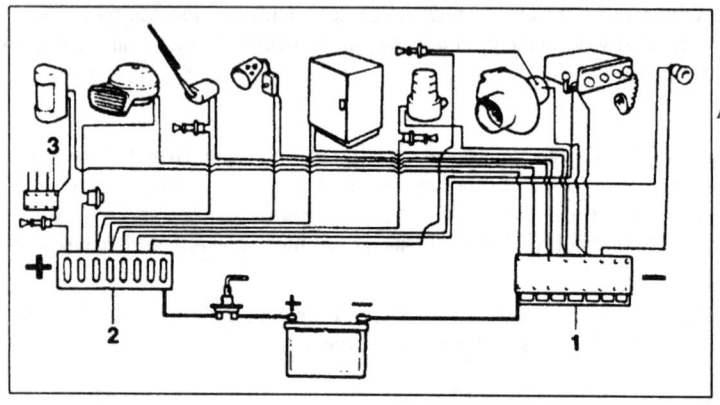

Fig 30.4
ELECTRICAL ACCESSORIES
(Volvo Penta)

1. Junction box for ground leads (-)
2. Fuse box (+)
3. Junction box for navigation lights

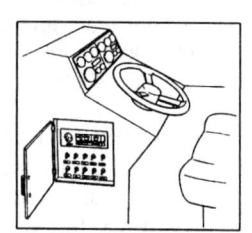

Fig 30.5
SWITCH BOX
(Volvo Penta)

CALCULATING THE FEEDER CABLE AREA

The length and area of the feeder cable (A) is dependent on how many accessories are to be connected to it. The more the power demanded by accessories, the bigger the area of the feeder cable. A feeder lead of 10 AWG (6 sq. mm) may be used for maximum 50 amps, while a lead of 8 AWG (10 sq. mm) may be used for maximum 70 amps alternator capacity. *(AWG = American Wire Gauge).*

CALCULATING CABLE AREA FOR AN ACCESSORY

Measure and double the distance of the accessory from the terminal box. Then enter the appropriate table shown below. The table illustrates the example for a 12V refrigerator of 45W consumption and the distance between the terminal block and the refrigerator of 3 metres. The straight line drawn between the refrigerator's load of 45W and its cable length of 6 metres (i.e. 3 x 2) indicates the required size of cable being 1.5 sq. mm. *(This calculation is based on the*

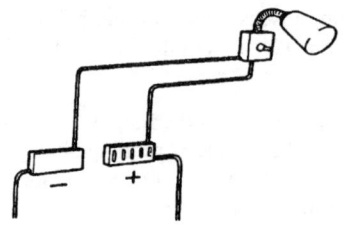

Fig 30.7: CABLE FOR AN ACCESSORY *(Volvo Penta)*

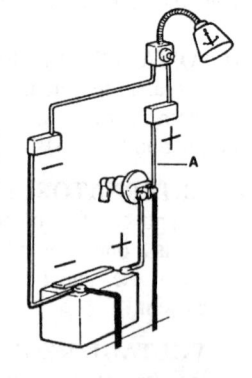

Fig 30.6
FEEDER CABLE
(Volvo Penta)

maximum permitted total voltage drop in all cables between the positive terminal to the accessories and back to the negative terminal. The total voltage drop should not exceed 0.4V.)

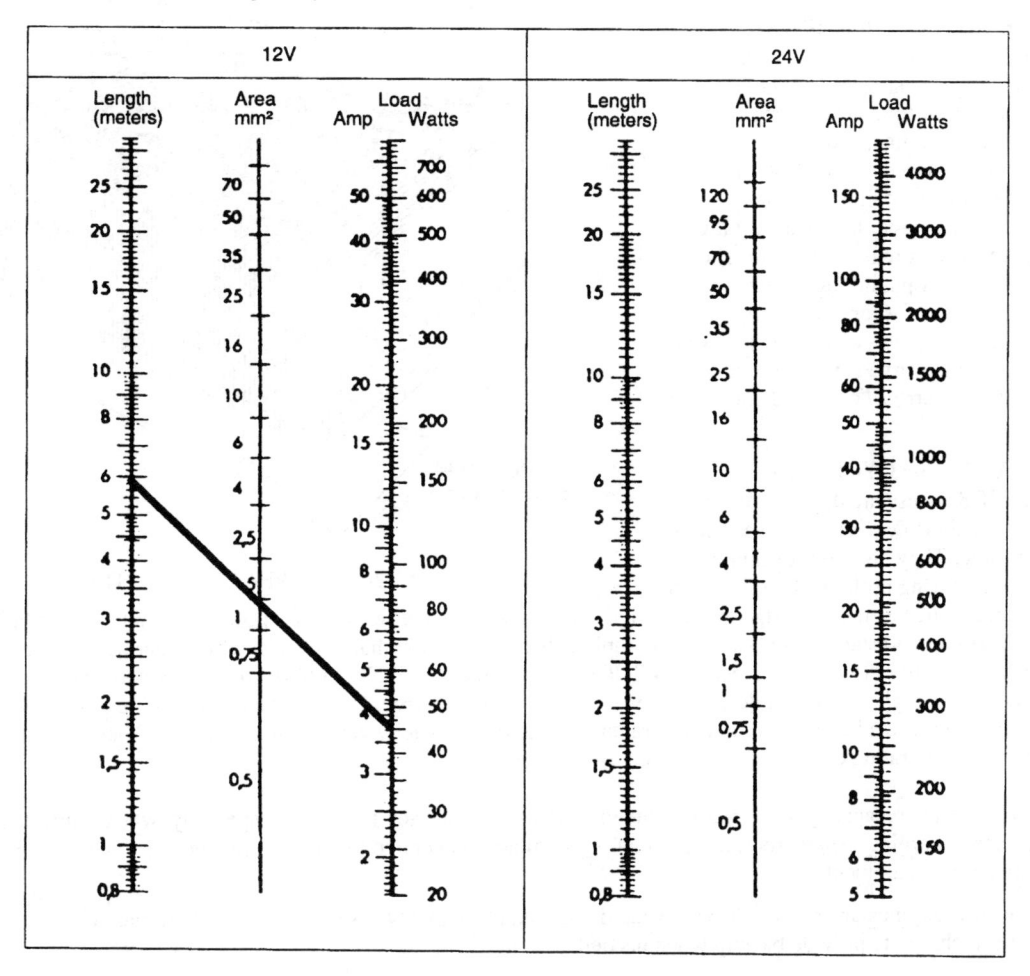

Fig 30.8: TABLE INDICATING CABLE AREAS FOR ELECTRICAL ACCESSORIES *(Volvo Penta)*

BATTERIES

Two dissimilar metals when immersed in a weak acid (known as *electrolyte*) form a 2 Volt battery cell. Six cells joined together form a 12 Volt battery, and 12 cells form a 24 Volt battery. Due to chemical reaction between the metals and electrolyte, a current flows from the less noble or *ignoble metal* (*anode*) to the one that is more noble (*cathode*). The circuit is completed by the current flowing through a circuit outside the battery from cathode (negative terminal) to anode (positive terminal). The process gradually corrodes (breaks down) the anode.

Battery capacity is measured in ampere-hours (Ah). If a battery can produce 3 amps current for 20 hours, its capacity is 60 Ah. This capacity is stated at 20°C. Decrease in temperature reduces battery capacity. For every degree drop in temperature, the battery capacity decreases approximately 1%. At -18°C, the battery capacity is reduced to 55%.

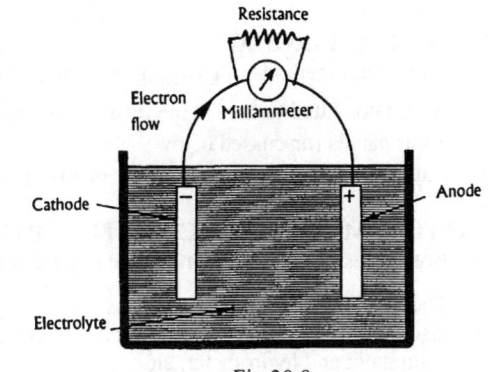

Fig 30.9
A SIMPLE ELECTROLYTIC CELL

MARINE BATTERIES differ from automotive-type batteries. They are constructed with additional plates to suit the variations in the rated performance levels. They are made vibration resistant by lock-bonding and fibreglass separators.

IN RESPECT OF THEIR APPLICATION,
THERE ARE TWO TYPES OF BATTERIES:

- **SHALLOW CYCLE (or CRANKING) BATTERIES**: These are engine-starting batteries. They discharge a very large amount of current for a very short period of time to start the engine, and then immediately recharge as you drive. They use a large number of thin plates to maximise surface area, so that a very high current can flow from the battery for short periods of time. If put through slow discharge, such as for running a refrigerator all night, the thin plates of such a battery are likely to become warped and distorted. These batteries carry a "cc" (cold cranking) rating.

- **DEEP CYCLE (or CYCLING) BATTERIES**: These are designed to provide a steady flow of current over a long period of time. They are used for running equipment requiring prolonged discharge of low amperage, such as radio, refrigeration and lighting. (Also see batteries in 'solar power' below). Their tubular or thicker plates can withstand several hundred complete discharge-and-recharge cycles, while a car battery is not designed to be totally discharged. Electrical WINCHES and other LIFTING EQUIPMENT also require deep cycle batteries, but of a higher rating. A deep cycle battery can provide a surge when needed, but nothing like the surge a car battery can. Deep cycle batteries are not "cc" rated. Instead their capacity is rated in amp-hours. They will also be heavier for a given rating due to the extra lead in their plates.

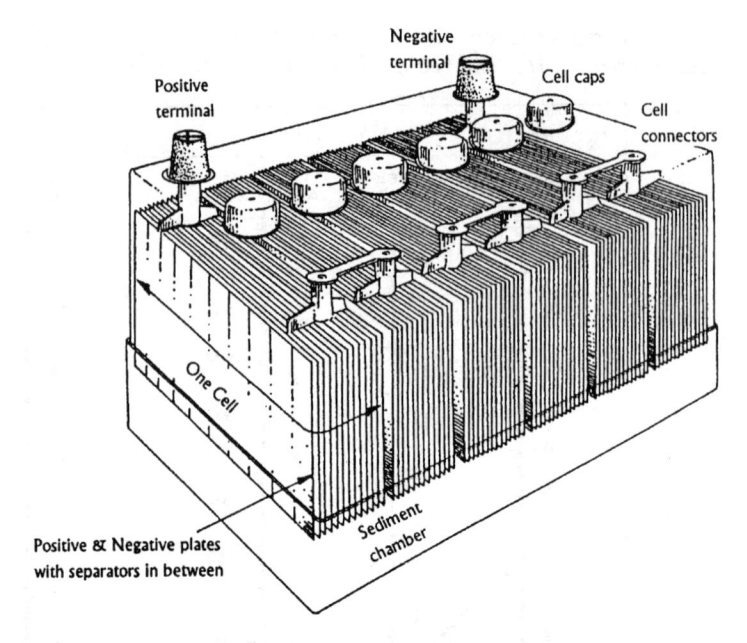

Fig 30.10: A 12-VOLT WET CELL BATTERY

Batteries of smaller capacity than recommended will have a shorter life span, especially when subjected to non-regulated repeated charging. Apart from the resultant high voltages generated in a non-regulated system, damage can occur to sensitive electrical equipment.

Vessels whose engines are not fitted with a hand start should carry two sets of starting batteries. One set should be maintained at full charge to provide backup when needed.

TWO TYPES OF WET CELL BATTERIES
- Lead-acid batteries [the electrolyte is dilute sulphuric acid]
- Alkaline batteries [the electrolyte is potassium hydroxide]

BATTERY CHARGING
The three main methods of charging the batteries are:

- Alternator fitted to the engine (discussed above)
- Solar panels (discussed below)
- Battery charger powered by either mains or generator

BATTERY MAINTENANCE & SAFETY PRECAUTIONS:
- Both types of batteries demand the same safety precautions.

- The alkaline battery cells can be damaged by the slightest contact with acid. Therefore, if both types of batteries are installed, they should be kept apart - in separate compartments - each with its own distilled water bottle, topping-up equipment and hydrometer, etc.

- The level of Electrolyte should be maintained 10 mm above the plates. The cells should be topped up with distilled water as necessary. The use of distilled water prevents the build-up of organic and iron particles during chemical changes in the normal charging and discharging activity. If distilled water is not available, use drinking water or other cleanest water you can find - rainwater would be better than drinking water. Don't use seawater. Get the battery checked at the first opportunity. If a battery requires constant topping up, the charging method is likely to be faulty.

- The batteries should be secured in brackets. Their terminals, cables, casing, and the sealing compound should be kept

Chapter 30: Solar & other Electrical Systems

clean, free from dirt, moisture and grease. Dirt or moisture on top of the battery can form a conducting path, which takes energy from the battery; and if allowed to become bad enough, it may eventually damage the battery. The terminals should be lightly coated with Vaseline to safeguard against corrosion. Corrosion and crystalline matter on terminals can be removed by dipping them in a solution of baking soda and hot water. Then rinse them clean, dry, and smear lightly with petroleum jelly. Build up of copper sulphate beneath the plastic shroud of battery leads will case the leads to appear swollen or distorted near the terminals. It will impair the efficiency of the system. The leads should be replaced immediately.

➢ A battery connected with incorrect polarity can damage the regulator, diodes and the alternator due to overheating. The positive terminal of a battery is coloured Red. It is also larger in size, and marked with a positive (+) symbol.

➢ Salt is a good conductor of electricity and it absorbs moisture. it can destroy and cause leakage and short-circuiting in electrical equipment. Therefore the insulation on electrical equipment must not be damaged. Once a battery is fully immersed in salt water, it should be replaced, as salt water will contaminate the electrolyte and can produce toxic gas. Replacing the electrolyte will not eliminate the problem.

➢ Batteries should be charged at a suitable **CHARGING RATE**, and they should not be overcharged. The charging system should charge the battery from fully discharged "flat" to charged within 16 hours (preferably less), but not at such a rate as to cause the battery to become hot or the electrolyte to boil. Fast charging should be particularly avoided for deep cycle batteries. The deep cycle batteries should never be discharged beyond 80% of their rated capacity. Overcharging causes the plates to corrode and fracture. The build up of sediment on the bottom may also bridge the plates and cause internal short-circuiting.

➢ Batteries give off explosive hydrogen gas, which is lighter than air. When charging, they should be protected from sparks, naked flames and heat. The gas hangs above the electrolyte at all times. This is due to a continuous weak chemical reaction in the battery, even when disconnected. The most dangerous time is just after charging. To prevent a spark igniting it, caps should be on cells when connecting or disconnecting a battery. The battery charger or ignition should be turned off before disconnecting a battery.

➢ Wash any battery fluid, accidentally splashed on skin or clothing, with running water. It can cause burns. Seek medical advice if serious. Usually it is not serious on healthy skin. If splashed in the eyes, flush eyes and under eyelids with cold flowing water. Cover the eye. Seek medical advice. It can be serious.

➢ Don't leave a battery in a discharged condition for too long, particularly a lead-acid battery. The capacity of such a battery will be reduced due to the "sulphating" of its plates, i.e., a hard white layer of lead-sulphate will form on the plates.

➢ An over-discharged battery may require bench charging, i.e. charged independently of the operating system. Should two or more batteries require charging it is essential that all batteries be of similar size. If not, the smaller battery will govern the rate of charge and absorb more of the available energy, whereby the larger battery may not fully respond.

➢ Likewise, multiple parallel charging in *Solar Systems* is not recommended because minor internal differences are difficult to balance out, and uneven discharging and recharging will occur. Should one cell in any battery become defective, all other cells will discharge into the faulty battery.

➢ The *Density* (used to be referred to as *Relative Density or Specific Gravity*) of batteries should be checked regularly with a *Hydrometer*. A good quality float hydrometer is essential. Cheap hydrometers with red and green markings tend to be misleading. A fully charged battery will lift the glass float to around 1.26 Density (1260 SG) in each cell. An even reading in all cells below 1.26 indicates a need for charging. An uneven reading, e.g., 5 cells at 1.26 and 1 cell at 1.20 indicates the battery failure. The figures may vary slightly due to variations in manufacturing techniques. The density should not be measured immediately after topping-up with water. Charging for thirty minutes or more after topping-up will mix the electrolyte to give accurate readings. Batteries should not be tested immediately after coming off charge as the temperature has increased and immediate readings will not be accurate.

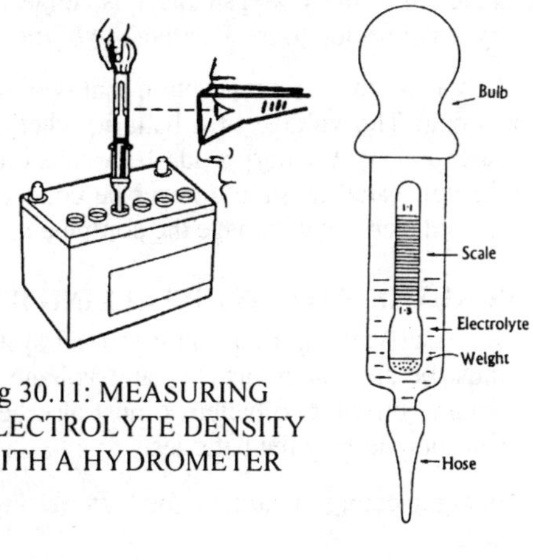

DENSITY READINGS FOR VARIOUS STATES OF CHARGE
    1.260   Fully charged
    1.230   75% charged
    1.200   50% Charged
    1.170   25% Charged
    1.100   Completely discharged

Fig 30.11: MEASURING ELECTROLYTE DENSITY WITH A HYDROMETER

Consistent readings below 1.200 would indicate the battery is being over discharged and could be under capacity for the particular function or the charging pattern is irregular.

In the case of a sealed battery, use a voltmeter to measure **VOLTAGE** across the battery terminals. However, this measurement should be carried out only when the battery is on-load. For a 24 volt battery, the terminal voltage should not fall below approximately 23.2 volts when on-load. It is considered discharged if it measures 21 V with a light load applied to it. Measuring the off-load voltage across the terminals of a battery is not appropriate. It overlooks the health of individual cells.

A 12 V battery on charge and nearing full charge will measure about 14 V across its terminals. When charging is stopped, the voltage will drop to between 12 and 12.5 V almost straight away. Battery is considered discharged if it measures 10.5 V with a light load applied to it.

AN ALKALINE (NICKEL-CADMIUM OR "NICAD") BATTERY has many times longer commercial life than a lead-acid battery. However, its condition of charge cannot be measured with a hydrometer. The density of its electrolyte does not vary during charge/discharge cycle. It should normally be in the range of 1.190 to 1.250, and the electrolyte should be replaced when it falls below 1.160.

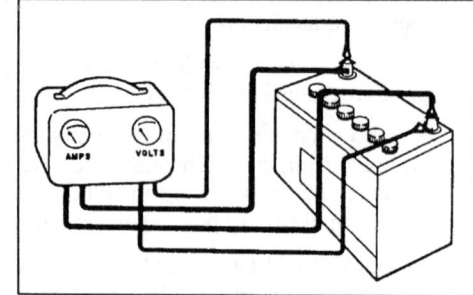

Fig 30.12
MEASURING BATTERY VOLTAGE

BATTERIES IN SERIES & PARALLEL
Two or more batteries can be connected together in two different ways for two different functions.

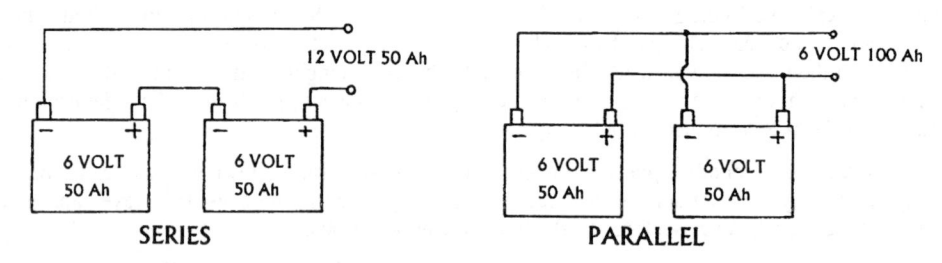

Fig 30.13: BATTERIES CONNECTED IN SERIES & PARALLEL

Batteries connected in series (i.e., Positive (+) terminal of one battery connected to the Negative (−) terminal of the next) add up their voltages. But their capacities (i.e. current in ampere-hours) do not add up. Therefore, a vessel with a 12-volt electrical system may install a single 12-volt battery or hook up two 6-volt batteries in series.

The opposite is the case for batteries connected in parallel (i.e., all Positive (+) terminals connected together and all Negatives (−) connected together). They add up their capacities (i.e., current in ampere-hours), but their combined voltage supply remains the same as of a single battery. This arrangement is suitable when there is a need to run electrical equipment requiring high amperage, or when a vessel wants to start an engine with two half discharged batteries. It will combine their current in ampere-hours, while maintaining the designed voltage.

Batteries are also connected in parallel when using jumper cables to jump start an engine from an external battery: Negative (−) of the vessel's battery is connected to the Negative (−) of the external battery; and Positive (+) of the vessel's battery is connected to the Positive (+) of the external battery.

It is a common misconception that two half discharged 12-volt batteries can be hooked up in series to provide one 12-volt circuit. The voltage of a battery, whether charged or discharged, remains close to its designed value. Even a fully discharged 12-volt battery reads 10.5 volts on a voltmeter. What a discharged battery loses is its current supply. Therefore, two 12-volt batteries should never be connected in series in a vessel with a 12-volt electrical system. The resulting high voltage will seriously damage the equipment.

PRECAUTIONS WHEN INSTALLING BATTERIES
- A vessel seeking a 24-volt electrical system would usually connect two 12-volt batteries in series. The two batteries must be similar in capacity and voltage. They must be of the same age (The charging rate changes with age). The loading on the two batteries must also be equal. Even a small load such as a radio connected across only one battery can soon destroy the batteries.

- For connecting in parallel, the batteries must have the same nominal voltage, but they may have different capacities and ages.

Chapter 30: Solar & other Electrical Systems

- Copper nails and fastenings in timber boats are particularly vulnerable to corrosion if electrolyte leaks out of batteries. Metal hulls have the same problem.
- Install the batteries in a tight-fitting, acid proof battery box, protected by a cover to prevent tools dropping on them. Vent the box with 25 mm hoses into the outside atmosphere. If batteries are installed in the engine compartment, its ambient temperature must be kept down to avoid battery boiling during charging.
- The ventilation hose must end up outside the vessel to allow highly volatile battery gas to escape.
- Secure the batteries so that the do not move more than 10 mm
- Locate them above the level of the bilge and away from ignition sources.
- Locate them as near as practicable to starter motor.
- Do not install alkaline and lead-acid batteries in the same compartment - as discussed above.
- Don't top-up whilst on charge. Wear safety goggles when topping up or handling electrolyte.
- Fit only non-sparking ventilation fans.
- Install away from motors and other sources of sparks or heat.
- Do not install batteries in living areas.
- Install battery-monitoring gear: ammeter, voltmeter, etc.
- Fit fuses or circuit breakers on the accessory wiring. Do not fit them to the main battery leads to avoid the risk of a spark igniting the hydrogen gas.

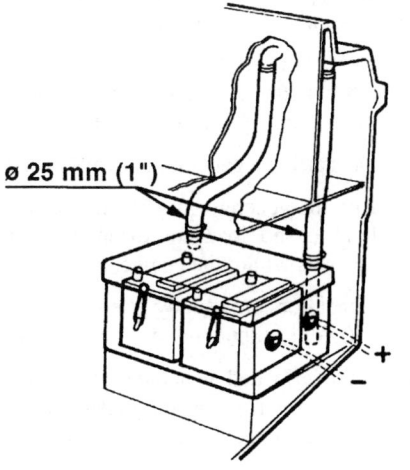

Fig 30.14: A VENTILATED BATTERY BOX (Volvo Penta)

PRECAUTIONS WHEN REMOVING OR REPLACING BATTERIES
- Switch off all sources of power drain (switch off isolating switch).
- Mark the terminals and the battery leads [+tive terminal is red and the –tive is black].
- Disconnect the black return terminal first and reconnect it last to prevent the risk of shorting.
- Use non-metallic carry strap or suitably designed handles to lift the battery.
- The replacement battery must be of correct voltage.
- Secure it against movement.

STORAGE OF BATTERIES WHEN NOT IN USE
Disconnect and remove the battery from service, check electrolyte levels, add distilled water as necessary to the required levels. Give the battery a full bench charge. Clean and dry the battery top and terminal posts. Store in a cool, dry place (not directly on a concrete floor). Recharge fully at least every two months and immediately before putting into service. Ensure vent plugs are free of dirt or grease.

BATTERIES - TROUBLESHOOTING

BATTERY NEEDS TOPPING UP TOO OFTEN
- Overcharging. Solution: Seek expert assistance.
- Cracked casing. Solution: Replace battery.
- High surrounding temperature. Solution: Find better location.

BATTERY NOT FULLY CHARGING
- Drive belt slipping. Solution: Adjust it.

THE AUXILIARY GENERATOR

In large sailing vessels and motor cruisers, it is usually uneconomical to run a vessel's main engine just to charge the batteries. Some diesel engines dislike running with a light load, and this can cause glazing of cylinder walls, leading to a major overhaul. Some vessels are therefore fitted with a separate diesel powered auxiliary generator producing 240-volt AC power supply. It is suitable to run most appliances on board, including refrigerators and microwave ovens. A transformer is fitted to convert the 240-volt AC to 12 volt DC for charging batteries.

➢ Generators are like main engines in operation and maintenance. They need cooling, lubrication, fuel and filters like any other 4-stroke engine. A generator should be supplied with a separate starting battery, so that it can be started if the main engine is out of service. Such a battery can also provide emergency back-up power for the main engine.

➢ On starting a generator, always visually check the seawater cooling overboard discharge.

➢ Run the generator for five minutes to reach its normal operating temperature before putting on load. But don't run on light or no load for extended periods. It causes cylinder glazing and deposits within the engine.

➢ Generators are usually fitted with a voltmeter and an amp meter to indicate 240 volts output and the amperage being drawn.

➢ An overloaded generator will run at reduced RPM, tripping the circuit breaker or tripping the machinery. Switching off some of the equipment will reduce the load.

➢ Use a voltmeter to ensure the charge voltage is approximately between 13.5 and 14.5 volts.

➢ Check rubber mountings for fatigue and cracks.

➢ Keep the battery and starter connections tight and clean.

➢ Check and tighten alternator, water pump and fuel pump drive belts once a month.

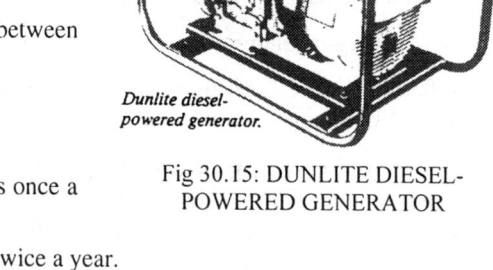

➢ Check the condition of the sacrificial anodes in the cooling system twice a year.

Dunlite diesel-powered generator.

Fig 30.15: DUNLITE DIESEL-POWERED GENERATOR

Some vessels are fitted with a generator that is coupled to the main engine. One such brand is AUTO-GEN. It has the advantage of a lower installation cost because it does not require separate cooling, exhaust or fuel tank.

GENERATOR PROBLEMS:

See diesel engines.

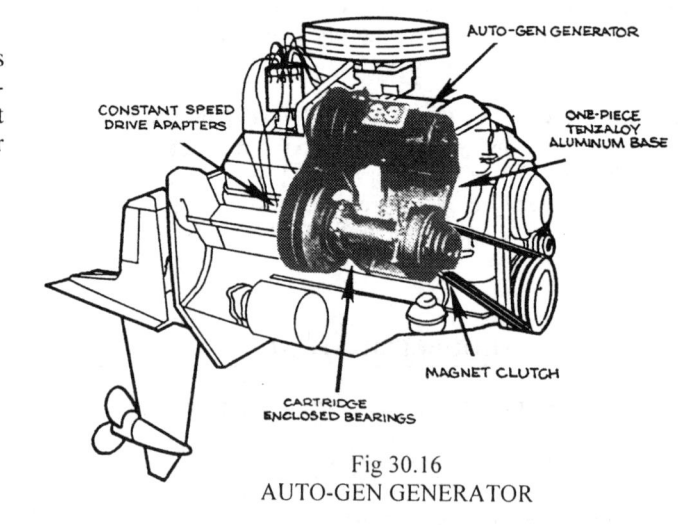

Fig 30.16 AUTO-GEN GENERATOR

SHORE POWER SUPPLY

Vessels equipped with a 240-volt electrical installation are usually fitted to take a power supply from shore outlets with the necessary circuit protection and monitoring devices for safety. Care must be taken when connecting to shore power supply. The polarity of the shore connection must match with that of the vessel's switchboard. This can be done either with a polarity changeover switch (if fitted) on the vessel or by changing the polarity in the shore cable, in order to connect "active with active". The shore power must not be grounded (earthed) to the engine or any part of the vessel. Instead, draw the connection cabinet's ground terminal to the shore.

Only an electrician qualified to work on high voltage installations may carry out installation and work using shore-connected equipment. Incorrect installation can result in danger to life.

SOLAR POWER

Although the initial investment in solar energy equipment is high, it is an excellent and silent way of maintaining the battery charge, thus minimising the need for running the generator, and when vessel is left unattended.

Solar panels are made up of small PV (photovoltaics) cells, typically measuring 5 or 10 cm across and 0.25 mm thick. They are solid-state semiconductors that convert light directly into electricity. A cell measuring 5 x 1.25 cm (6.45 sq. cm or 1 sq. inch) can produce 70 milliwatts of power depending on the season and location.

PV cells are usually made of silicon with traces of other elements in them. The bulk of the cell is doped with a small quantity of boron to make it a POSITIVE or *p-type* layer. But a thin layer on the face of the cell is doped with phosphorous, which makes it a NEGATIVE or *n-type* layer. The interface between the two layers contains an electric field and is called a JUNCTION.

A PV MODULE is made by connecting several cells (usually 36) in series and parallel to achieve useful levels of voltage and current, and putting them in a sturdy frame complete with a glass cover and positive and negative terminals on the back. The photovoltaic process of the module of converting particles of light (photons) into electricity (volts) is completely solid-state and self-contained. There are no moving parts and no materials are consumed or emitted.

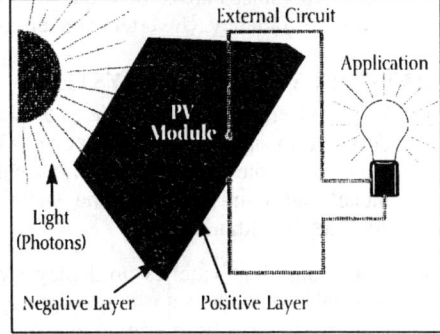

Fig 30.17
PV MODULE CONVERTING LIGHT INTO ELECTRICITY (*BP Solarex*)

WHY ARE SOLAR CELLS INEFFICIENT?

There are two types of solar cells: Single-junction and Multi-junction. Both types are inefficient. The efficiency of *single junction cells* is only about 13%, and that of the *multi-junction cells* a little higher.

As indicated by the colours of rainbows, light is made up of different wavelengths or energy levels. Not all the photons in the light spectrum are of a suitable energy level (measured in electron-volts or eV) to become either the n-type electrons or the p-type holes of the opposite charge. Photons with too much or too little energy do not suit the *band-gap energy* of the cell. Using a material of a really low band-gap, so that more photons can be used, is not the answer. The band-gap determines the strength (voltage) of the electric field of the cell. Lowering the band-gap will certainly provide extra current, but at a low voltage. The optimal band-gap, balancing voltage and current, is around 1.4 eV for a cell made from a single material.

Multi-junction cells use two or more (usually three) layers of materials of different band-gaps. This allows the higher band-gap material on the upper surface of the cell to absorb the high energy 'blue' photons while allowing the middle energy 'green' photons and the lower energy 'red' photons to pass through so that they can be absorbed by the middle and lower band-gap materials. Such cells can have more than one electric field. The ability of multi-junction solar panels to split the light spectrum certainly improves their efficiency, but only to a degree. 60 - 70% of the radiation energy is still lost to the atmosphere.

There are other losses as well. The electrons have to flow from one side of the cell to the other through an external circuit. The bottom of the cell can be made a good conductor by covering it with metal, but if we did the same with the top surface, the photons wouldn't be able to get through the opaque conductor. The silicon cover is only a semi-conductor. It is not as good as a metal for transporting current. Improvements are being made in this area by using transparent conductors in the top surface of some of the cells, but not in all. Silicon, being a very shiny material, is also very reflective. It thus loses many of the photons. An antireflective coating is applied to the top of the cell to reduce loss due to reflection to less than 5%.

CALCULATING SOLAR PANEL REQUIREMENT

Solar panels on the market range from 2 watts to 83 watts. In considering a solar panel's output, the following should be taken into account:

➤ P (watts) = V (volts) x A (amps)

➤ Solar panels are generally rated at 17 volts. If they produced only 12 volts they would not be able to push the charging current into a 12 volt battery. Therefore, a 60-watt panel produces approximately 3.5 amps charging current (60 = 17 x 3.53).

➤ Solar panels provide their full power when placed in full sun. Shade, cloud cover and the angle of the sun relative to the panel influence their power output. On overcast days, a panel's charging current drops down to 25 to 50% of its capacity.

- If you need a system only to maintain a charge in a battery when not in use, solar panels in the range of 2 watts to 30 watts would be sufficient. These panels are either self regulating or not sufficiently powerful to require a regulator. For power production for continuous use, on the other hand, you would need panels in the range of 40 to 83 watts. Larger panels are more economical to purchase because their price per watt is lower than that of smaller panels. The higher power rated panels require a regulator between the panels and the bank of batteries to prevent overcharging. Regulators range from 5 amps upwards.
- If AC appliances are to be used, the solar panel's power rating would have to be increased by 20% to reflect an 85% inverter efficiency. (Inverters convert DC power into AC, as discussed below).

BATTERIES FOR SOLAR SYSTEM

- A PV system can be as simple as a solar panel and a load (such as a direct driven fan). However, most PV systems are designed to supply power on demand. Therefore, deep cycle batteries (discussed above) are employed to charge during the day and then supply a small current for many hours during the night. These can be lead-acid (sealed or vented) or nickel cadmium batteries. The latter are more expensive, but they also last longer and can be discharged more completely without harm.
- Not recharging batteries until they are fully discharged shortens their life. It is better to recharge them at regular intervals when they are only slightly discharged. Therefore, a good quality charge-controller (discussed below) is employed in a solar system to ensure that the batteries are not discharged to more than 40-50% of their total charge. In a well-controlled system, the daily load should not discharge the batteries more than 20%. In the event of extended cloudy periods, the batteries should never be discharged more than 80% of their capacity.
- The amount of energy that a battery provides depends on its rate of discharge. For example, a typical 12-volt deep cycle battery may have a rating of 100 amp-hours when discharged over 20 hours (i.e., at a discharge rate of 5 amps). If the discharge rate is reduced to 1 amp, it may be possible to receive 120 amp-hours from the same 100 amp-hour battery (i.e., a discharge period of 120 hours).
- Battery power (amp-hours) can be converted to photovoltaic power (watt-hours) by simply multiplying the battery voltage by the amp-hour rating. Thus, a 12-volt, 100 amp-hour, battery is capable of providing 1200 watt-hours of energy.
- The term BATTERY AUTONOMY refers to the length of time a bank of fully charged batteries could operate its electrical load without input from the solar panel. Applications requiring 2 or 3 days of battery autonomy need a smaller battery bank than those requiring 7 or 8 days of autonomy.

INVERTERS

- Solar panels and batteries are inherently DC devices. Therefore, larger solar systems usually employ a power inverter to convert 12 or 24 volt DC power into 230-240 volt AC. This allows the boat owner to make use of standard household appliances. Without an inverter, only DC appliances may be used. Some PV modules, called AC modules, have an inverter built into each module. This eliminates the need for a large central inverter and simplifies the wiring requirements.
- Inverters of the past had a bad reputation for eating up batteries, producing poor quality power and for being generally unreliable. However, today's electronic inverters are both efficient (85% to 95%) and reliable. You would need to consider the following points when purchasing an inverter:
- Inverters are rated by their wattage output, varying from 150 watts to 1500 watts and more. They will always have a surge rating at least double their continuous rating to allow them to handle loads requiring large start-up currents.
- SINE WAVE INVERTERS produce negligible RFI (radio frequency interference) and have quite a pure waveform. They run all appliances, including television sets, computers, refrigerators, microwave ovens, power tools and other kitchen appliances, at their maximum efficiency. They generally have many features and if they are within your budget they are definitely worth the extra investment.
- The MODIFIED SINE WAVES INVERTERS are also capable of running most of the appliances. However, they do create some RFI that can affect TV reception and create some noise in audio equipment. Cheaper, less known, brands usually give more problems.
- Some of the larger inverters are bi-directional. They also act as battery chargers. They are a better option in larger systems as they reduce complexity and save on components cost.
- Loads requiring continuous high power such as air conditioners, heaters and some cooking appliances are not practical to use with inverters as they pull too much power from the batteries.

Chapter 30: Solar & other Electrical Systems

OTHER COMPONENTS

- As mentioned earlier, solar systems with batteries need to install a CHARGE CONTROLLER to control the charging and limit the discharging of batteries. Even the nickel cadmium batteries last longer if they aren't overcharged or drained too much. Sometimes, charge controllers are referred to as CHARGE REGULATORS. However, make sure they perform both the above functions.

- The installation of a solar system would require mounting hardware, wiring, junction boxes, grounding equipment, overcurrent protection and DC and AC disconnects. Ensure that the equipment is manufactured and fitted in compliance with the national PV electrical safety codes.

INSTALLATION

➢ Seek advice on the solar panel size and secure installation. The sighting of the panels on a boat must be considered carefully. Shading due to booms and dodgers etc. should be avoided, if possible. Laying them flat on the deckhouse may be the most suitable alignment for unattended operation at anchorages and moorings. But make sure the panels are in a well-ventilated place, because although they thrive on light, they deteriorate in heat. For their operation underway, mounting them on the aft guardrail would allow you to trim them several times during the day. But, they might need to be taken in whenever going alongside.

➢ It is recommended that a licensed electrician with experience in PV systems should carry out the installation.

➢ Safe installation of battery bank including DC fusing of all loads is essential.

➢ Pay particular attention to power flow from the alternator, including correct wire sizes and cable routes.

➢ The inverter installation should include high capacity DC fusing and correct 240-volt connections.

RATING OF PV MODULES

PV modules are typically rated by their peak power output when exposed to the following Standard Test Conditions (STC):

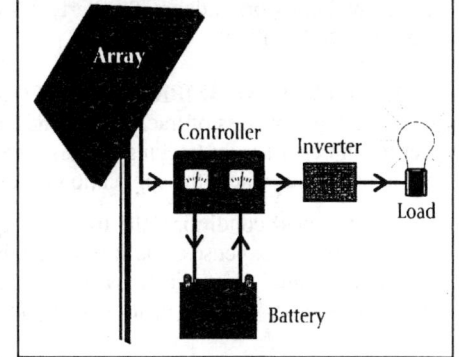

Fig 30.18
A SOLAR SYSTEM (*BP Solarex*)

- Temperature of PV cell = 25° C
- Air temperature = 0° C
- Intensity of solar radiation = 1000 watts/sq. metre.
- Spectral distribution of light through air mass (AM) = 1.5. (This is the spectrum of light that has passed through 1.5 thickness of the earth's atmosphere.)

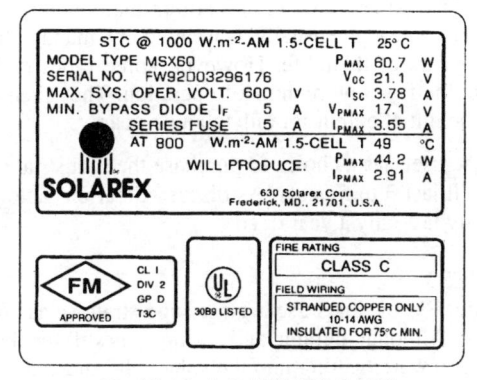

Fig 30.19: A RATINGS CARD
(*BP Solarex*)

These conditions correspond to noon on a clear sunny day with sun about 60° above the horizon, the PV module directly facing the sun, and an air temperature of 0° C. In production, PV modules are tested in a chamber known as a flash simulator. This device contains a flash bulb and filter designed to mimic sunlight as closely as possible. Because the flash takes place in only 50 milliseconds, the cells do not heat up appreciably. This allows the electrical characteristics of the module to be measured in the ambient temperature of the factory and the cell, which is 25° C. Solar cells become less efficient as the air temperature rises. In summer, cells can easily reach 45° C, reducing power output to 92% of STC. The BP Solarex rating card illustrated below provides the additional rating information at operating condition of 80% sun and a cell temperature of 49° C, which represents conditions more common in actual operation.

DOES PV WORK IN THE COLD?

Yes, very well in fact. PVs are electronic devices that generate electricity from light, not heat. Like all electronics they work better at lower temperatures. They generate more power at low temperatures. The fact that they generate less electricity during winter months than in the summer is due to the shorter days, lower sun angle and greater cloud cover in some areas.

DOES IT WORK IN CLOUDY WEATHER?

In cloudy weather the output of PVs diminishes linearly down to about 10% of the normal full intensity. Since flat plate PVs correspond to a 180° window, they do not need direct sun and can generate 50-70% of their rated output under a bright overcast. A dark overcast might produce only 5-10% of the full sun intensity. So, the output could be diminished proportionally.

DOES IT WORK INDOORS?

PVs designed for the outdoor light intensity, which is several hundred times more than that produced by the indoor lights, will not work indoors. Similarly, PVs designed for lower light intensity, like those found on calculators, will perform poorly in full sunlight.

DOES TRACKING IMPROVE PERFORMANCE?

The effectiveness of tracking depends a lot on the climate and the application. Areas with a lot of haze or clouds won't get much benefit from tracking because the light is scattered. Tracking also can't improve the performance during the poor weather periods where the application needs the same load every month.

Under ideal conditions, the tracking equipment can improve PV output per day up to 40%, but it adds to the system complexity and expenses, and it is generally not as strong as the fixed mounting system. The use of tracking equipment is generally limited to applications where the increased output matches the output demand (such as livestock watering) in areas that are dry (such as central Australia).

DO REFLECTORS INCREASE OUTPUT?

Reflectors can increase the output of PV arrays. However, the improvement is not linear because the increased light intensity also increases the module temperature, which reduces its efficiency. More importantly, the increased module temperatures and light intensities can lead to premature failure of the module. The use of artificial reflectors is therefore not recommended and will in fact void the module's warranty.

HOW LONG DOES THE PV SYSTEM LAST?

Top quality polycrystalline PV modules last over 30 years. They can withstand arctic cold, desert heat, tropical humidity, winds in excess of 125 mph and 25 mm hail at terminal velocity. The power output of most modules falls off a little during their operating life. They are usually warranted by manufacturers to produce only 80% of their original minimum rating for 20 or so years of their life. However, some types of PV modules, using thin film silicon, have a predictable fall-off in output in the first few months of operation. They then become stabilized. However, they are usually warranted to produce 80% of the initial output for only about 5 years.

With regard to batteries, the user has a choice to purchase the industrial strength units, which will last up to 7 years, or the smaller sealed units that will last 3 to 5 years. As discussed earlier, automotive batteries are a poor substitute for deep cycle batteries and will generally last only a year or so.

WHAT ABOUT BREAKAGES?

The most reliable, long life, PV modules use a glass superstrate (glass front). It is usually a low-iron tempered glass, laminated with layers of plastic. Although durable and strong, it is still breakable. If the glass is shattered or punctured, the module will eventually fail due to water getting into the solar cells and causing corrosion. It may take years for the module to completely fail (produce no power). On the other hand, if the electrical connections between any given pair of cells are severed, the module's output will cease immediately.

Some lightweight modules, designed for lightweight and rugged activities such as camping, use an aluminium substrate and a plastic superstrate. They are shatterproof, but do not last as long as the modules with glass front. This is because the plastic covering is not as inert as glass, and the aluminium is not a good thermal expansion match for solar cells made of silicon.

Chapter 30: Solar & other Electrical Systems

WHAT TO LOOK FOR WHEN PURCHASING PV MODULES

- The equipment should be approved and verified for long-tem reliability by the appropriate testing and safety authorities.

- The manufacturer should regularly qualify production units (rather than laboratory samples) to international standards.

- Check out the module. It should have a solid feel. The frame should not twist easily.

- The junction box should be suitable for marine use. It should be solidly attached (glue the conduit with silicone). It should be capable of taking heavy gauge wire and accommodate standard electrical fittings. It should also be capable of accommodating diodes and regulators, if needed.

- The solar cells should not be perilously close to the module frame, which can lead to electrical breakdown and premature failure.

- The module bus bars should be open and well isolated. They should not be folded behind the cells where they can cause electrical shorts and delamination.

- Study the label. It should not be a generic label. The actual tested power of that individual module should be printed on the back. The manufacturer's tolerance on power should be stated, making it clear how far below the nominal can the power be and the module still be considered within specifications.

- The module should have enough voltage to charge batteries under all conditions. It should be at least 16.5 volts at maximum power.

- The warranty document should not be vague. It should guarantee a specific level of performance.

- Check to see if the manufacturer and distributor are reputable. Would they still be in business in, say, 10 years?

MAINTENANCE

Although arrays of PV modules can be wired for higher voltages, they are generally low voltage DC devices with no moving or wearing parts. They thus require no maintenance other than occasional cleaning to help them work more efficiently. The maintenance and care of batteries has been discussed under 'batteries'.

The PV systems are generally subject to the same safety codes that govern the installation of electrical wiring and equipment in vessels. Properly installed PV systems with safety approved components are covered by most insurance policies in the same way as any other electrical equipment installed in the vessel, but individual policies should be consulted to determine the limit of the coverage.

CHAPTER 30: QUESTIONS & ANSWERS

CHAPTER 30.1: BATTERIES

1. When connecting or disconnecting a wet-cell battery:

    (a) all cell caps should be on.
    (b) all cell caps should be off.
    (c) alternative cell caps should be on.
    (d) none of the choices stated here apply.

2. The following is one of the reasons for a battery cell needing constant topping-up with distilled water:

    (i) battery too small for the job.
    (ii) overcharging.
    (iii) cracked casing.
    (iv) high surrounding temperature.

    The correct answer is:
    (a) all except (i).
    (b) all except (ii).
    (c) all except (iii).
    (d) all except (iv).

3. Wrong polarity connections during a battery installation will damage the:

    (i) battery.
    (ii) regulator.
    (iii) alternator.

    The correct answer is:
    (a) all except (i).
    (b) all except (ii).
    (c) all except (iii).
    (d) all of them.

4. When charging a lead-acid battery:

    (a) ventilation should be restricted.
    (b) there is no risk of overcharging.
    (c) engine should be running.
    (d) hydrogen gas is emitted.

5. Build up of dirt and grease on the body of a battery:

    (a) does not affect its performance.
    (b) may reduce its performance.
    (c) may increase its performance.
    (d) may increase its life.

6. The following hydrometer reading indicates a fully charged battery:

    (a) 1.200
    (b) 1.270
    (c) 1.150
    (d) 1.000

7. The voltage across the terminals of a 24-volt battery will read about 23.5 volts when it is:

    (a) fully charged and on load.
    (b) fully charged and off load.
    (c) fully discharged.
    (d) partially discharged.

8. The following may cause damage to a lead-acid battery:

    (a) maintaining water level above the plates.
    (b) slow charging.
    (c) leaving it discharged for long periods.
    (d) leaving it charged for a prolonged period.

9. Partially discharged batteries connected in parallel:

    (a) supply their combined voltage.
    (b) supply their combined capacity.
    (c) turn an engine more easily.
    (d) supply their combined capacity & voltage.

10. The acid level in your batteries is well below the plate level. You have no distilled water on board. You are in a remote area where battery failure could lead you into serious trouble. You should preferably use:

    (a) sea water.
    (b) rain water.
    (c) tap water.
    (d) no water.

11. The electrolyte in lead-acid batteries is:

    (a) potassium hydroxide.
    (b) distilled water.
    (c) dilute sulphuric acid.
    (d) dilute hydrochloric acid.

12. With regard to installing batteries in a vessel, the following statement is correct:

    (a) Install fuses in the battery compartment.
    (b) Keep all types of batteries together.
    (c) Do not ventilate battery compartment.
    (d) Insulate bottom and sides of batteries.

13. Overcharging a battery causes:

    (a) the plates to erode, corrode & fracture.
    (b) excessive voltage.
    (c) sulphating of the plates.
    (d) excessive current.

14. The following statement is correct in relation to an alkaline battery:

    (a) Lasts shorter than lead-acid battery.
    (b) Lasts longer than nickel-cadmium battery.
    (c) Specific gravity is always above 1.250
    (d) Replace electrolyte when SG below 1.160

Chapter 30: Solar & other Electrical Systems

15. A battery's negative lead is coloured:
    (a) red.
    (b) black.
    (c) blue.
    (d) red or black

16. The circumstances under which a battery is most likely to explode are when the battery is:
    (a) in use.
    (b) being charged.
    (c) fully charge.
    (d) left unused for a prolonged period.

17. A charged battery has an explosive gas on top of the electrolyte:
    (a) only when connected.
    (b) when being charged.
    (c) when in use.
    (d) at all times.

18. The lead-acid and alkaline (nickel-cadmium) batteries differ as follows:
    (a) lead-acid battery charge is easier to measure.
    (b) Nickel cadmium battery needs less care.
    (c) only a lead-acid battery is chargeable.
    (d) Nickel cadmium battery is a dry cell.

19. The most common cause of a low battery charge rate from generators and alternators is:
    (a) a slipping drive belt.
    (b) incompatible battery.
    (c) incompatible generator or alternator.
    (d) all the factors stated here.

20. Batteries in a vessel should be located:
    (a) in an unventilated place.
    (b) far away from starter motor.
    (c) below the level of bilge.
    (d) in an insulated container.

21. The terminals of batteries in series are connected as follows:
    (a) Positive to negative.
    (b) Positive to positive.
    (c) Negative to negative.
    (d) None of the choices stated here.

22. The terminals of batteries in parallel are connected as follows:
    (a) Negative to positive.
    (b) Positive to negative.
    (c) Negative to negative.
    (d) None of the choices stated here.

23. Which of the following statements is incorrect with regard to batteries?
    (a) Marine batteries are constructed with more plates than the automotive-type batteries.
    (b) If a battery produces 3 amps current in 20 hours, its capacity is 60 Ah.
    (c) Decrease in temperature decreases battery capacity.
    (d) The engine starting batteries are deep cycle batteries.

24. Which of the following statements is correct with regard to batteries?
    (a) Shallow cycle batteries provide a steady flow of current over a long period of time.
    (b) The engine starting battery should be a shallow cycle battery.
    (c) Deep cycle batteries discharge a very large amount of current for a very short period of time.
    (d) The battery for operating marine radio should be a shallow cycle battery.

25. A vessel seeking to start an engine with two half charged batteries will need to connect them in:
    (a) series.
    (b) parallel.
    (c) parallel and series.
    (d) parallel or series.

26. The gas produces by charging lead acid batteries is:
    (a) Chlorine (poisonous)
    (b) Sulphur dioxide (offensive)
    (c) Hydrogen (explosive when mixed with air)
    (d) Nitrogen (odourless)

27. Two 12-volt batteries connected in series will produce:
    (a) 24 volts, with increased current supply.
    (b) 12 volts, with increased current supply.
    (c) 24 volts.
    (d) 12 volts.

28. Two 12-volt batteries connected in parallel will produce:
    (a) 24 volts, with increased current supply.
    (b) 24 volts.
    (c) 12 volts.
    (d) 12 volts, with increased current supply.

29. The white-green powder (corrosion) on the terminals of a lead acid battery used for marine radio may cause:

    (a) battery damage.
    (b) drop in radio transmission.
    (c) radio damage.
    (d) the fuse to blow up frequently.

30. The following battery connection arrangement is most suitable to run electrical equipment requiring high amperage:

    (a) parallel.
    (b) series.
    (c) parallel and series.
    (d) parallel or series.

31. An electric storage battery can cause an acid burn to a sensitive part of a person's body, by coming in contact with the:

    (a) electrodes.
    (b) anodes.
    (c) electrolyte.
    (d) cathodes.

32. The density (specific gravity) of a fully charged lead acid battery is approximately:

    (a) 1100
    (b) 1150
    (c) 1180
    (d) 1250

33. The following would provide the most complete check of overall battery condition, including the state of its charge:

    (a) Hydrometer
    (b) Voltmeter
    (c) Easy engine starting
    (d) Shorting with a wire to see if it heats up.

34. If the battery plates are not covered with electrolyte, the following should be added:

    (a) Nothing, as the liquid will restore during the charging cycle.
    (b) A weak mixture of Sulphuric acid & water
    (c) Distilled water.
    (d) Sea water.

35. The following can damage a battery:

    (a) Greasing the terminals with Vaseline.
    (b) Trickle charging for long periods.
    (c) Charging heavily for long periods.
    (d) tightening the terminal clamps.

36. The following will keep a battery in good condition for the longest period:

    (a) keep fully charged.
    (b) Regular periods of charge & discharge.
    (c) Keeping in a low state of charge.
    (d) Never quite fully charging.

37. The following would be the preferred location for a radio battery:

    (a) Near the motor.
    (b) In the bottom of the vessel & under cover.
    (c) Close to the radio & under cover.
    (d) As far from the radio as possible.

38. Submerging a battery in the sea will affect its state of charge as follows:

    (a) No effect
    (b) Discharge quite quickly
    (c) Discharge much slower
    (d) Discharge as normal

39. The gas given off by a charging battery can be dangerous in an enclosed space because it is:

    (a) flammable.
    (b) corrosive.
    (c) poisonous.
    (d) flammable, poisonous & corrosive.

40. A battery should be filled with electrolyte to the following approximate level:

    (a) The top of the filler holes.
    (b) Just level with the top of the plates.
    (c) 10 mm above the level of the plates.
    (d) Just below the level of the plates.

41. The following condition is the most likely cause of one cell of a battery needing to be topped up more often than the others:

    (a) Some batteries have a spare cell.
    (b) Overcharging.
    (c) Cell stronger than the others.
    (d) Cell breaking down.

42. A hydrometer may be used for measuring:

    (i) dock water density.
    (ii) an electrolyte's density.
    (iii) humidity.
    (iv) specific gravity of a liquid.

    The correct answer is:
    (a) all except (i).
    (b) all except (ii).
    (c) all except (iii).
    (d) all except (iv).

CHAPTER 30.2: ELECTRICAL SYSTEMS

43. A larger than necessary fuse element in an electrical circuit may:

    (i) provide greater protection to wiring.
    (ii) overload the wiring.
    (iii) cause a fire.
    (iv) increase the circuit rating.

    The correct answer is:
    (a) all except (i).
    (b) all except (ii).
    (c) all except (iii).
    (d) all except (iv).

44. In a charging system if an alternator was connected to a battery with the incorrect polarity, the following component would be damaged:

    (a) alternator.
    (b) battery.
    (c) alternator & battery.
    (d) none

45. Build up of salt on the exterior surfaces of electrical equipment:

    (a) improves their performance.
    (b) provides greater insulation.
    (c) destroys their insulation.
    (d) prevents their short-circuiting.

46. While at sea, you replaced a blown fuse. But it blew as soon as you switched on the power. You should:

    (a) use larger fuse element.
    (b) use a higher amperage fuse.
    (c) replace it without switching off power.
    (d) follow none of the choices stated here.

47. When excessive current flows through an electrical system, an electrical fuse is designed to:

    (a) trip open to break the circuit.
    (b) re-switch to its original position.
    (c) melt instantaneously.
    (d) none of the choices stated here..

48. An electrical circuit breaker is re-set by:

    (a) replacing its fuse wire.
    (b) re-switching to its original position.
    (c) replacing it with a new circuit breaker.
    (d) any of the choices stated here.

49. A glow plug:

    (a) provides a gentle light in engine room.
    (b) assists with ignition on initial start up.
    (c) is another name for a spark plug.
    (d) is fitted in petrol engines only.

50. When overloaded, an electrical fuse:

    (a) melts.
    (b) trips off.
    (c) prevents rise in voltage.
    (d) prevents rise in current.

51. With regard to circuit breakers, which of the following statements is incorrect?

    (a) Circuit breakers & fuses perform a similar function.
    (b) They melt instantly when overloaded.
    (c) They are designed to trip an overloaded circuit.
    (d) They prevent electrical wiring & equipment from failure or fire due to overloading.

52. A double-pole switch:

    (a) is less safe than a single-pole switch.
    (b) is not permitted for marine use.
    (c) disconnects active and earth cables.
    (d) disconnects active and neutral cables.

53. The one-pole connection between a battery and a starter motor is as follows:

    (a) Negative to starter motor, positive to earth.
    (b) Both terminals to starter motor.
    (c) Positive to starter motor, negative to earth.
    (d) None of the choices stated here.

54. When connecting to AC power supply from a shore outlet:

    (a) ground shore supply to the vessel's engine.
    (b) connect active to neutral.
    (c) the polarity must be changed over.
    (d) do not earth shore supply to the vessel.

55. With regard to the main battery switch, which of the following statements is incorrect?

    (a) Install it on the plus end of the battery.
    (b) Position it as far from the engine compartment as possible.
    (c) Position it outside the engine compartment.
    (d) Keep the cable length to a minimum.

56. A battery-charging distributor:

    (a) independently & automatically charges two batteries.
    (b) converts AC into DC current.
    (c) is a voltage regulator.
    (d) can generate high voltage even when the engine is idling.

57. The advantage of a battery-charging distributor is that:
    (a) it controls the output current of the alternator.
    (b) the engine can be started even if the accessories battery has been emptied.
    (c) the batteries can be charged even when the engine is idling.
    (d) it can be installed far from the batteries.

58. An alternator:
    (a) is a voltage regulator.
    (b) always fails all at once.
    (c) can fail in stages.
    (d) should be fitted just outside the machinery space.

59. In relation to an auxiliary generator, which statement is incorrect?
    (a) It is like a main engine in operation & maintenance.
    (b) It should be supplied with a separate starting battery.
    (c) Run the generator for five minutes to reach its normal operating temperature before putting o load.
    (d) Use a voltmeter to ensure the charge voltage is 12 volts.

60. The alternator drive belt tension should be such that with moderate thumb pressure it:
    (a) deflects 10 mm.
    (b) deflects 5 cm.
    (c) does not deflect at all.
    (d) is free to move in and out.

CHAPTER 30.3: SOLAR SYSTEMS

61. In relation to solar panels, which statement is incorrect?
    (a) They are made up of small PV cells.
    (b) They convert light directly into electricity.
    (c) A PV module is made by connecting several cells either in parallel or series.
    (d) They can be made up of single-junction or multi-junction cells.

62. In relation to solar systems, which statement is incorrect?
    (a) Small solar systems usually employ a power inverter.
    (b) Deep cycle batteries are employed in solar systems.
    (c) A charge-controller is employed in solar systems.
    (d) Some large inverters are bi-directional. They also act as battery chargers.

63. In relation to solar systems, which statement is incorrect?
    (a) PVs generate electricity from light, not heat.
    (b) In cloudy weather the output of PVs diminishes linearly down to about 10%
    (c) PVs generate more power at high temperatures.
    (d) In winter months PVs generate less electricity due to shorter days and lower angle of the sun.

64. In relation to solar systems, which statement is incorrect?
    (a) The PV module will continue to function for quite some time in spite of damage to its glass superstrate (front).
    (b) The PV module will continue to function for quite some time in spite of the electrical connections between one pair of cells being severed.
    (c) The power output of most modules falls off a little during their operating life.
    (d) PVs require no maintenance other than occasional cleaning for better performance.

CHAPTER 30: ANSWERS

1 (a), 2 (a), 3 (a), 4 (d), 5 (b),
6 (b), 7 (a), 8 (c), 9 (b), 10 (b),
11 (c), 12 (d), 13 (a), 14 (d), 15 (b),
16 (b), 17 (d), 18 (a), 19 (a), 20 (d),
21 (a), 22 (c), 23 (d), 24 (b), 25 (b),
26 (c), 27 (c), 28 (d), 29 (b), 30 (a),
31 (c), 32 (d), 33 (a), 34 (c), 35 (c),
36 (b), 37 (c), 38 (b), 39 (a), 40 (c),
41 (d), 42 (c), 43 (a), 44 (c), 45 (c),
46 (d), 47 (c), 48 (b), 49 (b), 50 (a),
51 (b), 52 (d), 53 (c), 54 (d), 55 (b),
56 (a), 57 (b), 58 (c), 59 (d), 60 (a),
61 (c), 62 (a), 63 (c), 64 (b)

Chapter 31

RUDDERS & STEERING SYSTEMS

RUDDERS

A rudder operates by deflecting the water flowing past it. Its effectiveness is governed by the size of the rudder, the helm angle applied and the speed of the water flowing past it. Slow moving large displacement vessels require larger rudders. The most effective helm angle is 35 to 40 degrees. It is at this angle that the stops are fitted for a full helm to port or starboard.

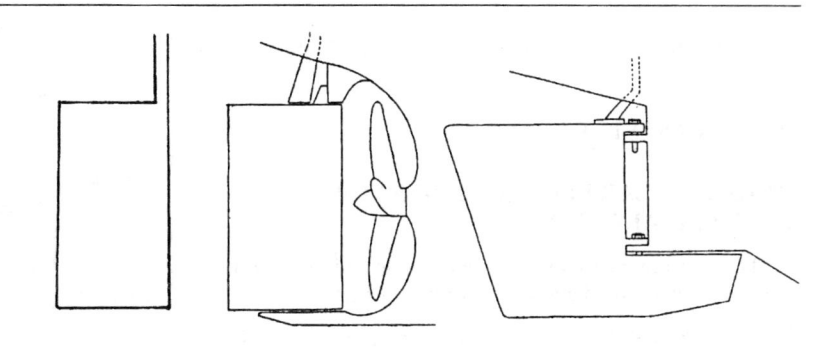

A rudder can be of one of three types:
- Unbalanced
- Semi-balanced
- Balanced.

UNBALANCED   BALANCED   SEMI-BALANCED

Fig 31.1: TYPES OF RUDDERS

The shape of a rudder can be just a flat plate or streamlined into a hydrofoil design.

Fig 31.2: HYDROFOIL SHAPE

Fig 31.3: BALANCED SPADE RUDDER
OF FRP CONSTRUCTION
(NSCV Part C Section 5 Subsection 5A)

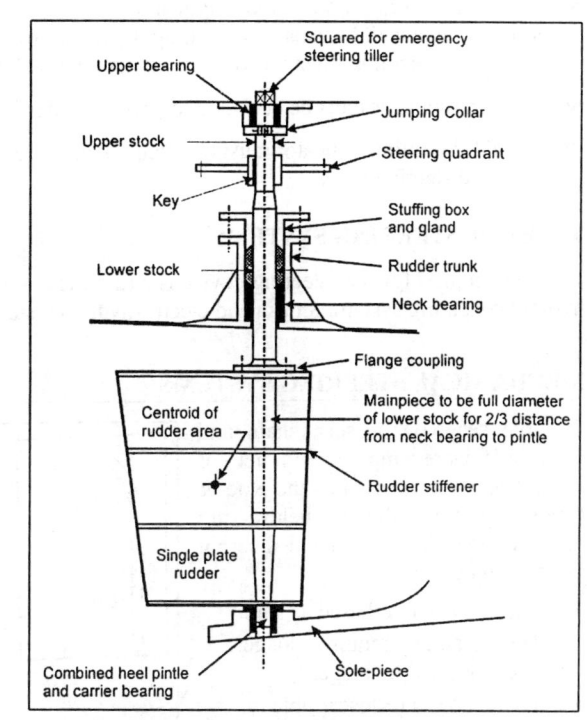

Fig 31.4: BALANCED RUDDER
OF SINGLE PLATE CONSTRUCTION
(NSCV Part C Section 5 Subsection 5A)

The principle of a balanced rudder is that a leading edge (forward of the rudder) assists to turn the blade. The pressure applied to a balanced rudder through the rudderstock is nearer to its turning axis or the centre of effort. It requires less turning torque. It is therefore easier to turn than an unbalanced rudder. However, a fully balanced rudder does not offer significant advantage at small angles.

The streamlined hydrofoil-shaped balanced rudders offer less resistance to water flow and a better hydrodynamic turning effect.

STEERING SYSTEMS

STEERING GEAR REQUIREMENTS FOR COMMERCIAL VESSELS

- The steering gear must be able to withstand maximum helm at maximum ahead and astern speed.
- The person at the helm must have a clear view ahead while at the normal steering position.
- All vessels, except twin screw vessels, shall be fitted with two independent means of steering unless steering is normally achieved via a hand tiller, in which case a second means of steering need not be provided.
- The secondary or emergency means of steering must be capable of being brought speedily into action.
- Rudder movement must be no less than 35° to port to 35° to starboard.
- In vessels of 12.5 m measured length and over, the steering gear shall be capable of putting the rudder over from 35° on one side to 30° *(no, it's not a misprint!)* on the other in 30 seconds when the vessel is at maximum ahead service speed with the rudder totally submerged.
- The steering gear must be designed to prevent violent recoil of the steering wheel.
- A rudder position indicator must be fitted on all vessels of 15 m measured length and over which are fitted with power-operated steering gear.

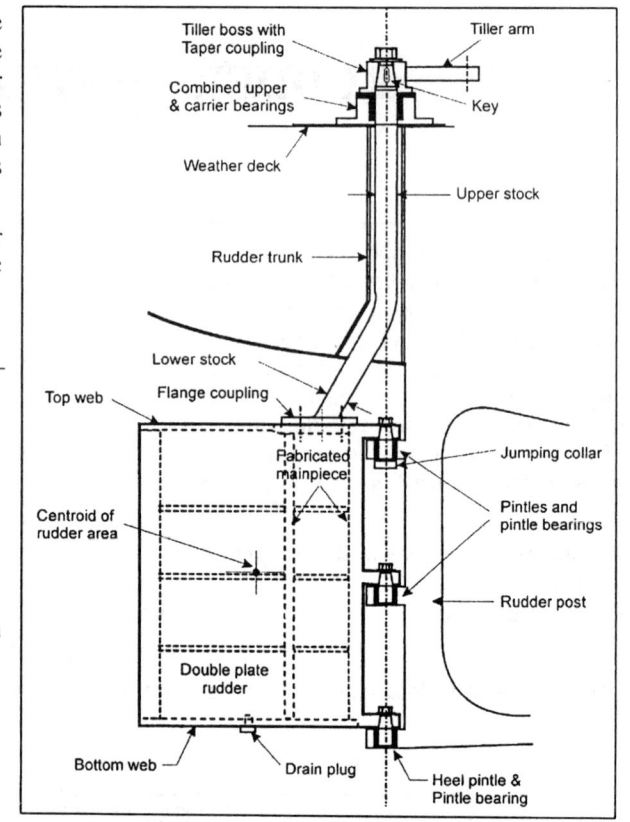

Fig 31.5: UNBALANCED RUDDER OF DOUBLE PLATE CONSTRUCTION
(NSCV Part C Section 5 Subsection 5A)

TYPES OF STEERING SYSTEMS

The simplest form of steering device is a tiller. However, most vessels are steered with some form of a mechanical or a hydraulic steering system. Ships need electro-hydraulic steering to turn their large rudders.

MECHANICAL STEERING SYSTEMS

In large seagoing vessels, chain and steel rods were employed in lieu of wire. They were known as the Rod & Chain system. Shown below are various types of mechanical steering systems in use today.

- They can be purchased as a unit.
- They generally require a longer tiller for good leverage.
- Their cables are susceptible to become slack, requiring periodic tightening.

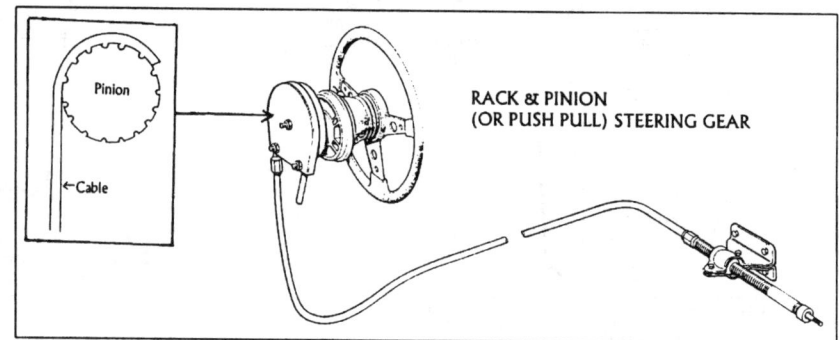

Fig 31.6: RACK & PINION (PUSH-PULL) STEERING SYSTEM

Chapter 31: Rudders & Steering Systems

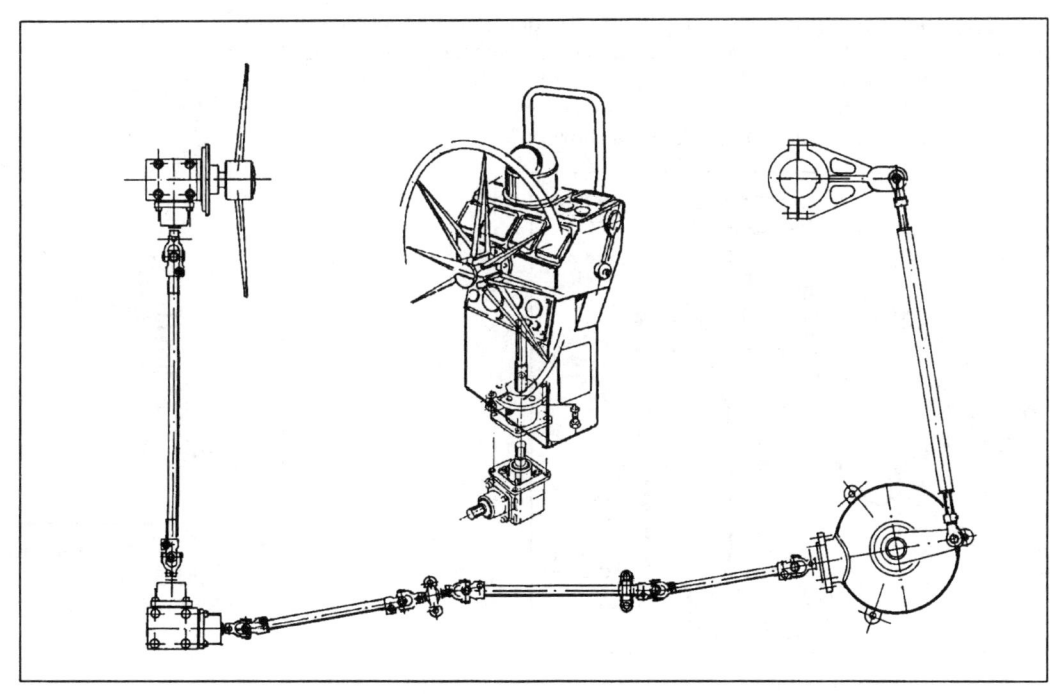

Fig 31.7: MECHANICAL STEERING SYSTEM FOR LARGE BOATS *(Solimar)*

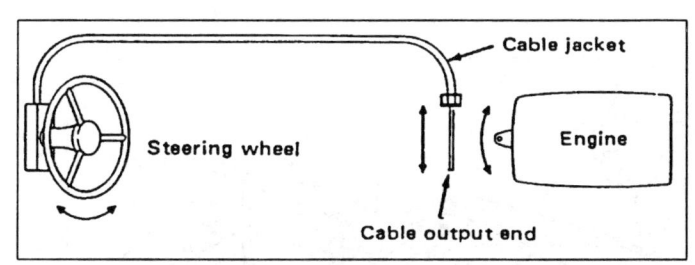

Fig 31.8: PUSH-PULL CABLE SYSTEM FOR OUTBOARDS

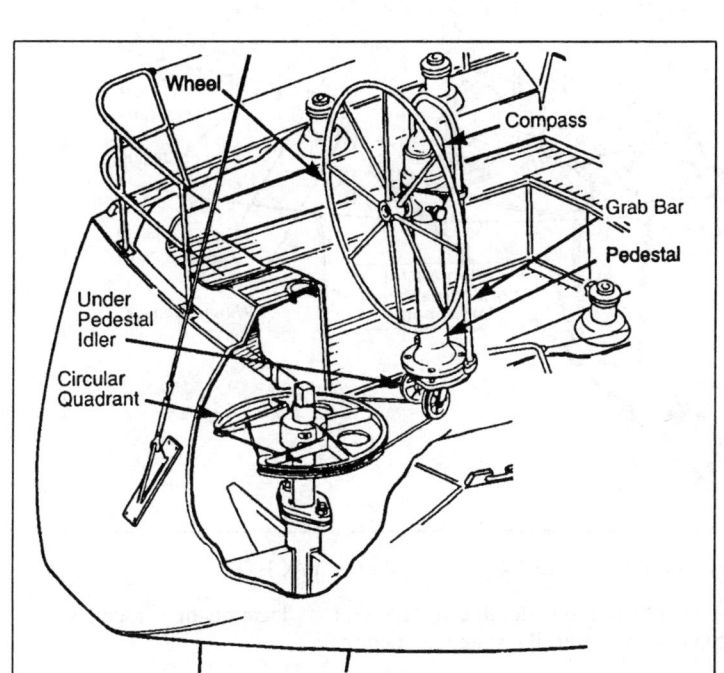

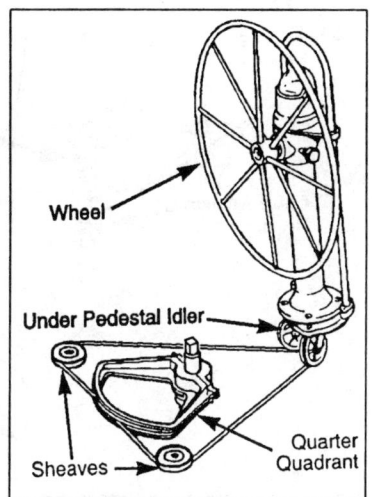

Fig 31.9 (A & B): A *WHITLOCK* CABLE & SHEAVE STEERING SYSTEM FOR YACHTS

A typical aft cockpit steering system, using a circular quadrant and an under pedestal idler, is illustrated. Correct vertical alignment is achieved by adjusting the height of the quadrant on the rudder shaft or by packing the idler under the cockpit floor. Horizontal sheave angle is also adjustable. In some cases, a quarter quadrant is used instead of the circular quadrant, as shown.

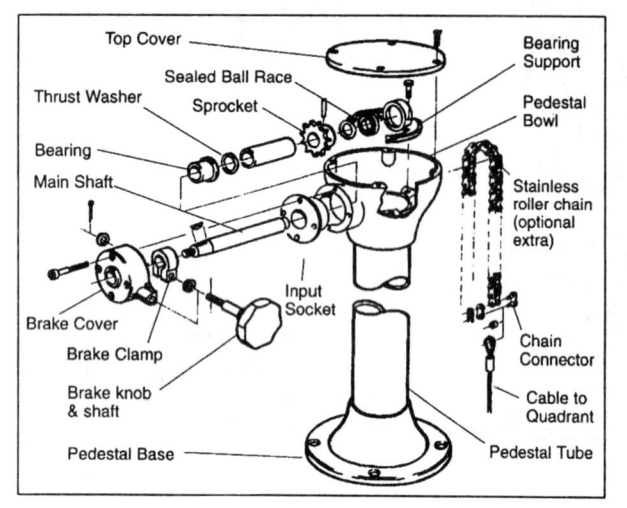

Fig 31.11
PARTS OF A *WHITLOCK* STEERING PEDESTAL

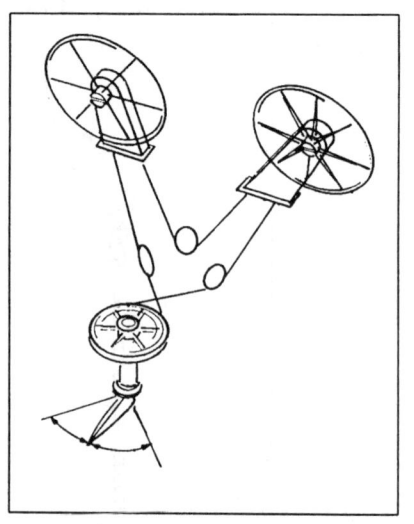

Fig 31.10: DOUBLE PEDESTAL CABLE & SHEAVE STEERING SYSTEM FOR LARGE RACING YACHTS *(Solimar)*

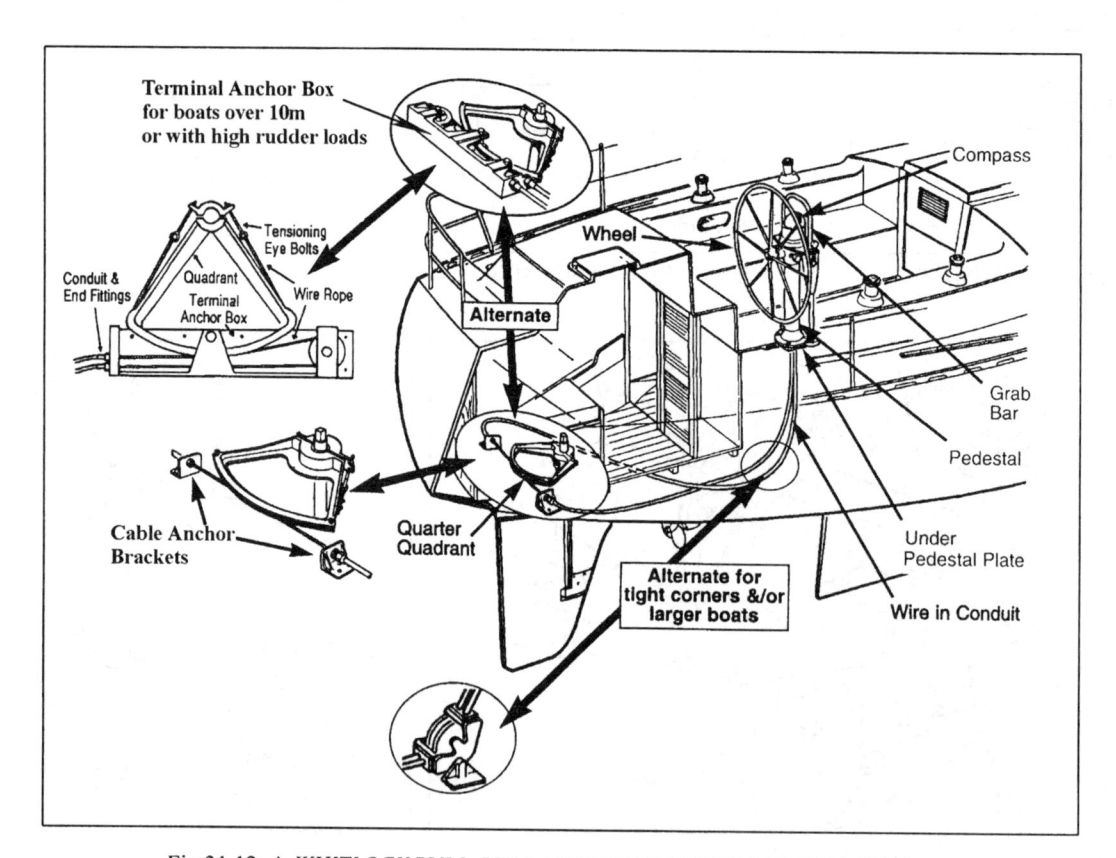

Fig 31.12: A *WHITLOCK* PULL-PULL STEERING SYSTEM FOR YACHTS

The Pull-Pull 'wire in conduit' steering system is more flexible in terms of its location in a yacht. However, to retain the correct feel it is important to carefully route the conduit.

Chapter 31: Rudders & Steering Systems

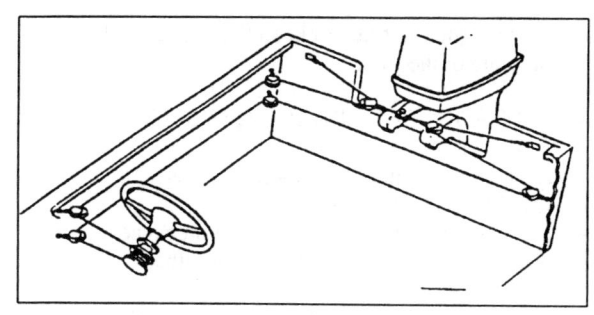

Fig 31.13
ROPE (WIRE) & PULLEY STEERING SYSTEM
FOR OUTBOARDS

HYDRAULIC STEERING SYSTEMS

In the direct (hand-operated) hydraulic system, the turn of a wheel causes a pump to force hydraulic oil into a cylinder containing a double-acting ram in the after part of the vessel. The ram is attached to the tiller and rudder, which moves with it. The force, leverage and movement in this system can be altered by changing the sizes of rams, gears, ratios and linkages. It can be set up as a single or a dual station installation.

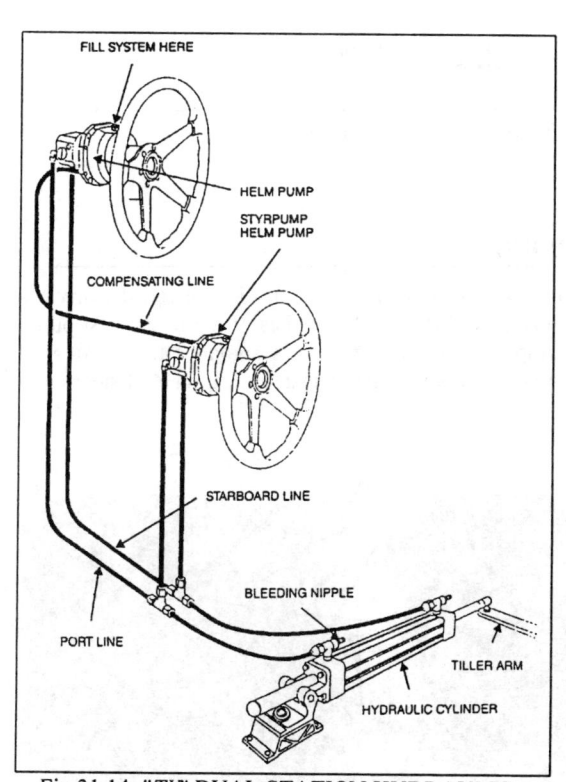

Fig 31.14: "*TX*" DUAL STATION HYDRAULIC STEERING SYSTEM

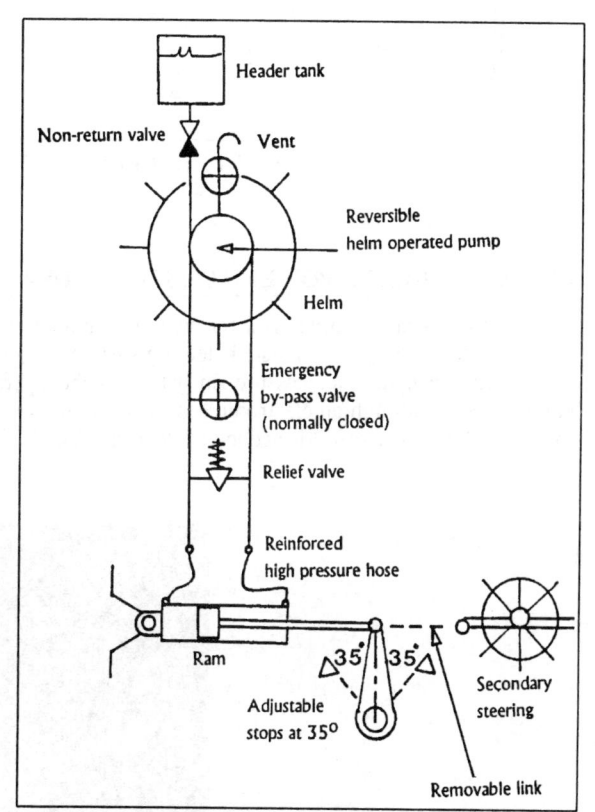

Fig 31.15: HYDRAULIC STEERING SYSTEM DETAILS FOR BOATS IN SURVEY
(The figures show a Hand-hydraulic Steering System. It does not incorporate electric or telemotor steering.)

COMPONENTS OF HYDRAULIC SYSTEM

- An oil reservoir or a header tank holds a reserve of oil. It also helps to keep the oil cool and allows it to settle and filter out contaminants.
- A bleed screw is provided to bleed air out of the system after repairs. For normal operation it is usually self-bleeding.
- For vessels in Survey, a by-pass relief valve (or regulator) is required to be fitted to relieve pressure in the hydraulic system should the rudder hit an obstruction or be subjected to excessive load in a heavy seaway.

- A mesh filter is fitted to keep the oil free of larger contaminants. A "filter condition indicator" can be fitted to indicate the state of the filter.

ADVANTAGES OF HYDRAULIC SYSTEM
➢ Easy to attach to an auto-pilot
➢ Greater mechanical advantage
➢ Can be installed in any position
➢ Few wearing parts - needs less maintenance
➢ No backlash on the wheel in bad weather

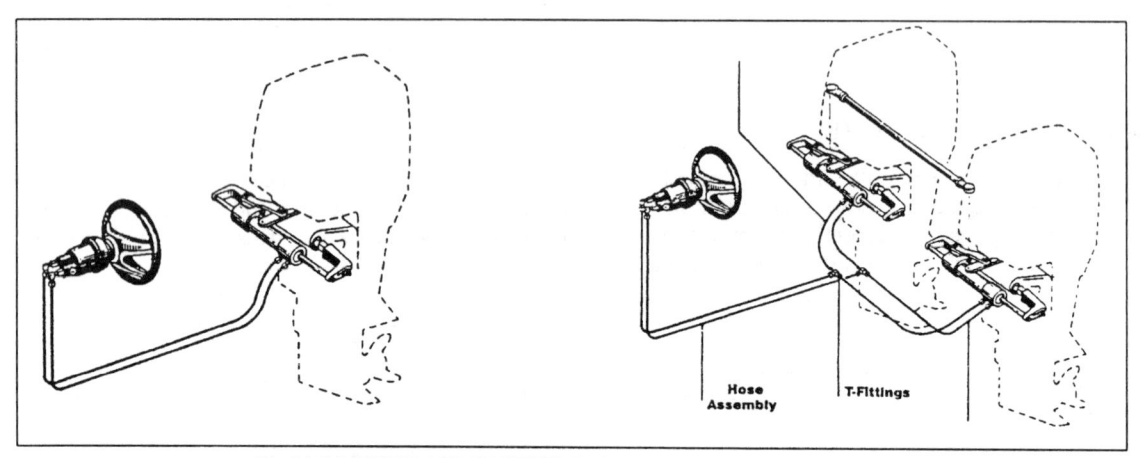

Fig 31.16: HYDRAULIC STEERING SYSTEM FOR OUTBOARDS

POWER STEERING (POWER-ASSISTED HYDRAULIC STEERING)

Most of us are familiar with the hydraulic power-assisted steering system used in motorcars. It reduced the driver's steering effort. The power-assisted steering system used in small vessels is no different. It consists of a power-assist pump that exerts the hydraulic pressure on behalf of the operator. The pump receives the operator's signals either through a control line or through an electronic sensor. As in motorcars, should a vessel's power-assist system fail, manual steering is maintained but extra effort is needed to steer the vessel.

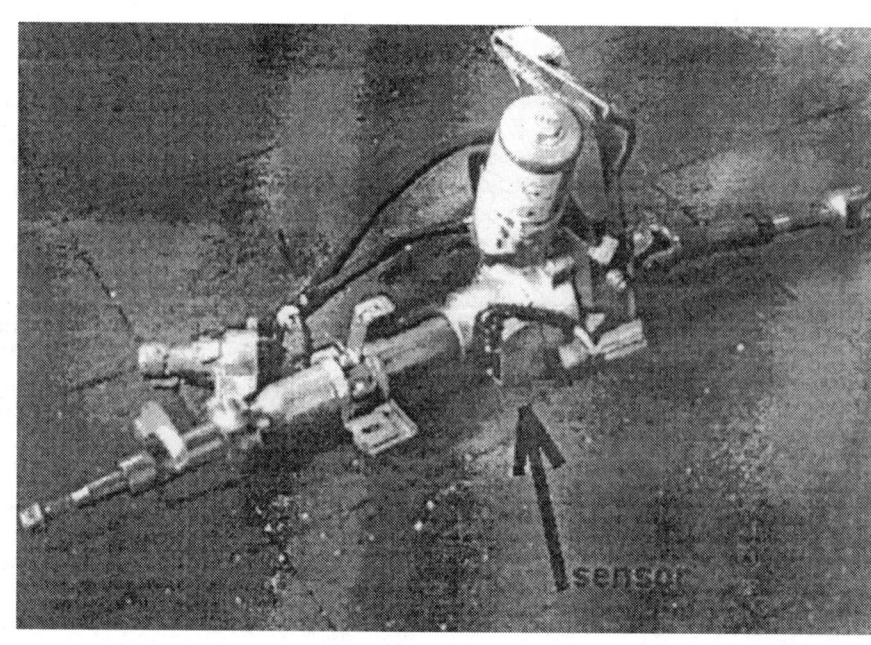

Fig 31.17
POWER-ASSIST STEERING PUMP
& ELECTRONIC SENSOR
IN A MOTOR CAR

Chapter 31: Rudders & Steering Systems

RAM ARRANGEMENTS

Larger boats employ two cylinders with double acting rams in order to increase the thrust (steering torque) on the rudder. This is done by either placing one cylinder on each side of the tiller or both on the same side of a double acting tiller. The latter arrangement has the additional advantage of reducing the side-force on the rudderpost.

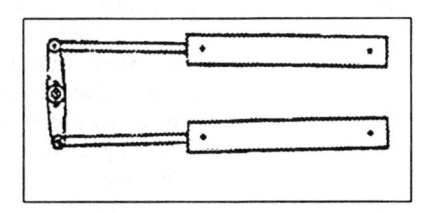

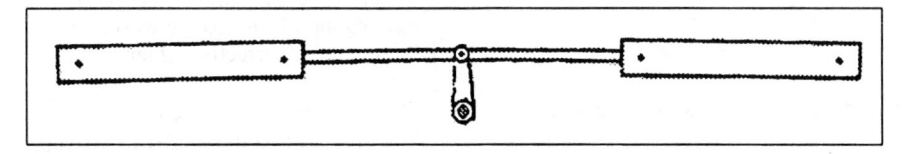

Fig 31.18 (A) & (B)

MAINTENANCE OF ALL TYPES OF STEERING GEAR

- Maintain it according to the manufacturers instructions.
- Regularly inspect, clean and lubricate all moving parts, including the bearings in the wheel assembly.
- Regularly inspect the hydraulic lines for any leaks.
- Keep the header tank (oil reservoir) topped up.
- Regularly inspect the tiller head fittings for tightness.
- Maintain the emergency tiller assembly in good condition & stow in a readily available storage.
- Test the steering gear before every departure.

STEERING SYSTEM (MECHANICAL) – TROUBLESHOOTING

SYMPTOM	POSSIBLE CAUSES
The cables don't move when the steering wheel is turned.	Slipping wheel, slipping or stripped sprockets teeth; disengaged clutch; or a jumped chain (whichever applicable).
The rudder quadrant doesn't turn when the steering wheel is turned.	Broken or jumped cable between the pedestal and quadrant.
The rudder quadrant turns, but not the rudder.	The quadrant clamp on the rudderstock is slipping.

STEERING SYSTEM (HYDRAULIC) – TROUBLESHOOTING

SYMPTOM	POSSIBLE CAUSES
Steering is stiff or the rudder doesn't turn full travel.	Incorrect type, incorrect level or dirty hydraulic oil or air in the system. Wrong oil (too viscous) can cause the steering to become stiff. Something is binding some part of the steering gear or the wheel brake is engaged. The rudder assembly, and not the steering gear, is stiff. This can be confirmed by disconnecting the steering cylinder from the tiller and trying out the steering Dirt trapped in the system is causing blockage. Solution: Check filter. Water has leaked into hollow rudder. Rudder shaft bearings are worn. Ram gland incorrectly tightened.
Rudder does not respond to the steering wheel.	The cylinder valve is open. The tiller arm on the rudderpost is slipping. The hydraulic oil has leaked out. The check valves are defective (In a multi-helm station system, this would cause the second wheel or the autopilot to turn). The piston seals in the steering cylinder at the tiller have perished (This can be confirmed by disconnecting the cylinder and trying moving the cylinder rod in and out. You should not be able to move it. Helm pump is defective, although this is uncommon.
Steering feels spongy.	Air leakage into the system. Solution: Check tightness and purge the system. Use of wrong grade oil. Solution: Seek expert assistance.
Leaking oil.	Faulty joints. (Excessive vibrations and inadequate clamping can also cause damage to pipes and joints. However, flexible hoses, where fitted, must be allowed freedom to move as required by the system.)
Header tank needs frequent topping.	Oil leakage.

CHAPTER 31: QUESTIONS & ANSWERS

RUDDERS & STEERING SYSTEMS

1. In relation to rudders, which statement is incorrect?
   (a) A full helm is an angle of about 35°.
   (b) The effectiveness of a rudder is governed by its size, helm angle applied & the speed of the water flowing past it.
   (c) A hydrofoil-shaped rudder is susceptible to excessive resistance to water flow.
   (d) A balanced rudder requires less turning torque.

2. In relation to rudders, which statement is correct?
   (a) All rudders are supported by the sole-piece.
   (b) Rudder stock is another name for rudderpost.
   (c) Stuffing box is another name for the carrier bearing.
   (d) Pintles are not found on all rudders.

3. In relation to rudders, which statement is correct?
   (a) All rudders should be fitted with a jumping collar.
   (b) All rudders should be fitted with a drain plug.
   (c) Neck bearing is another name for carrier bearing.
   (d) Rudder trunk is the body of the rudder.

4. In relation to steering gear on commercial vessels, which statement is incorrect?
   (a) It must be capable of putting rudder over from 35° on one side to 30° on the other in 30 seconds.
   (b) All vessels must be fitted with two independent means of steering.
   (c) Rudder movement must be no less than 35° on each side.
   (d) A rudder position indicator is not required in all vessels.

5. The cable in a mechanical steering system doesn't move when the steering wheel is turned. It could be due to:
   (a) broken cable.
   (b) slipping steering wheel.
   (c) slipping quadrant clamp on the rudder stock.
   (d) jumped cable between the pedestal & quadrant.

6. In a mechanical steering system, the cable moves when the steering wheel is turned but the rudder quadrant doesn't turn. It could be due to:
   (a) jumped cable between the pedestal and quadrant.
   (b) slipping steering wheel.
   (c) slipping quadrant clamp on the rudder stock.
   (d) disengaged clutch.

7. In a mechanical steering system, when the steering wheel is turned the rudder quadrant turns but not the rudder. It could be due to:
   (a) jumped cable between the pedestal and quadrant.
   (b) slipping steering wheel.
   (c) broken cable.
   (d) the slipping of the quadrant clamp on the rudderstock.

8. In a hydraulic steering system, the steering wheel turns, but the rudder does not respond to the steering wheel. It could be due to:
   (a) the cylinder valve left open.
   (b) air in the system.
   (c) dirty hydraulic oil.
   (d) the ram gland being incorrectly tightened.

9. A hydraulic steering system feels spongy. It could be due to:
   (a) the cylinder valve left open.
   (b) worn rudder shaft bearings.
   (c) air leakage into the system.
   (d) the ram gland being incorrectly tightened.

10. The best method of protecting hydraulic hoses on deck is to:
    (a) paint them.
    (b) coat them with an oil repellent.
    (c) wrap them in waterproof tape.
    (d) enclose them.

11. To bleed air out of the starboard side of a hydraulic steering gear, one of the steps is:
    (a) turn the helm to port.
    (b) open both the bleed screws.
    (c) turn the helm to starboard.
    (d) keep the helm amidships.

12. In terms of rams, the hydraulic steering system requires at least:
    (a) one single acting ram.
    (b) one double acting ram.
    (c) two double acting rams.
    (d) two single or double acting rams.

13. While underway, you suddenly lose response from the electro-hydraulic steering and the vessel starts turning in a tight circle. The possible reasons are:

    (i) Electrical failure.
    (ii) Slipped V-belt.
    (iii) Rudder jammed to one side.
    (iv) steering overload.

    The correct answer is:
    (a) all except (i).
    (b) all except (iv).
    (c) all of them.
    (d) none of them.

14. The function of a by-pass valve in a hydraulic steering system is to:

    (a) equalize pressure.
    (b) bleed trapped air after repairs.
    (c) keep the header tank topped up.
    (d) eliminate "spongy" steering.

15. A blockage in a hydraulic steering system:

    (i) may prevent rudder's full travel.
    (ii) may be eliminated by cleaning the filter.
    (iii) may be bled.
    (iv) may cause "spongy" steering.

    The correct answer is:
    (a) all except (i).
    (b) all except (ii).
    (c) all except (iii).
    (d) all except (iv).

CHAPTER 31: ANSWERS

1 (c), 2 (d), 3 (a), 4 (b), 5 (b),
6 (a), 7 (d), 8 (a), 9 (c), 10 (d),
11 (c), 12 (b), 13 (b), 14 (a), 15 (d)

INDEX

Abaft, 33
Abandon Ship Signal, 564
Abandoning & Rescue Phase, 558
Abandoning (Capsized), 186
Abandonment (for Salvage), 182
Abbreviations on Charts, 100
Abeam (Chartwork), 391
Abeam (Symbol), 393
Abeam, 33
Abroad, Arrival, 249
AC Current, 633
ACA Website, 852
Accidents (ISM Code), 584
Accidents (Nautical Emergencies), 181
Accommodation Safety, 582
Accounting Authority (AAIC), 248
A-Cockbill, 33
Acts of Parliament (in bookshops), 251
Adult, Weight, 255
Advance (Turning Circle), 144
Aft, 33, 40
After Peak, 33
Aftercooling, 640, 648
Aground, 182, 183
Air Bubble in Compass, 361
Air Cleaner, 631
Air Compressor Inspection, 723
Air Cooled Marine Engine, 647
Air Ducts (Engine Room), 703
Air Filter (Engine), 632
Air in Fuel Supply, 660
Air Sea Rescue Kits, 551
Air Shutdown Valve (Engines), 633
Aircraft Circling Vessel, 562
Aircraft in Distress, 563
Aircraft Signals, 561
AIS Transponder, 298
AIT, 669
Alarm Signal, Radiotelephony, 467
Alarms (Engine), 712
Alarms, Detectors & Indicators, 320
Alcohol Consumption – the Law, 586
Alcohol Consumption, Dehydration, 558
Algal Blooms, 261
Alphabet Codes, 480
Alternator, 633, 807
Aluminium & Stainless Fittings, 80
Aluminium Boats, 12, 80
American Bureau of Shipping, 277
Amidships, 33, 36
Ammeter, 633
AMSA Notices, 418
AMSA Radio Network, 462
AMSA Website, 852
AMVER, 248
Anchor Aweigh, 33
Anchor Bearings, 123
Anchor Buoy, 122
Anchor Cable & its Survey, 117
Anchor Cable Marking, 117
Anchor Dragging, 123
Anchor Dropped under Foot, 124
Anchor Light in Mooring Areas, 240

Anchor Line Size, 62
Anchor under Submarine Cable, 123
Anchor Watch, 123
Anchor Windlass, 36, 117
Anchor, Forces acting on it, 123
Anchor, Recreational Vessel Requirement, 536
Anchorages - Chart Symbol, 103
Anchoring & Fishing Prohibited - Chart Symbol, 103
Anchoring Procedure, 122
Anchoring Prohibited Areas, 122
Anchoring Terminology, 114
Anchors -Types, 115
Angle of Indraft, 609
Angle of Loll, 282
Anodes, 77, 639, 809
Anti-Collision Radar Transponders, 348
Anticyclone, 600
Anti-Fouling Pollution, 263
Anti-Fouling, 84
Anti-Roll Tanks, 284
Anti-Siphon Valve, 699
ANTT, 432
Aqua Scooters, 243
Aquaplaning, Distances Off, 243
Archimedes Principle, 274
Areas A1, A2, A3, 459
Areas of Operation, Seaward Limits, 255
ARPA (Radar), 345
Arrival Overseas, 249
Aspect Ratio, 51
ASRKs, 551
Assisting Others in Distress, 563
Astern, 33
Asymmetrical Wind Forces in Sails, 284
Athwartship, 33, 40
Atmosphere (Meteorology), 593
Atmospheric Pressure Fall due to Height, 596
Atmospheric Pressure, 593
Atomisers (Fuel), 638, 732
Attenuation, 336
ATU, 477
AUSCOAST Areas, 464
AUSREP, 550
AusSAR Website, 852
AusSAR, 550, 852
Australian & Middle Waters, 256
Australian Boating Regulations, 241
Australian Climate, 595
Australian Communications Authority, Website, 852
Australian Customs Website, 852
Australian Defence Qualifications, 22
Australian Environmental Protection (Sea Dumping) Act, 262
Australian Immigration Website, 852
Australian Maritime Group, 251
Australian Maritime Safety Authority Website, 852
Australian National Tide Tables, 432
Australian Quarantine Website, 852
Australian Registrar of Ships, 248
Australian Search & Rescue, 550

Australian Ship Reporting System, 550, 852
Australian Standards, Equipment, 241
Australian Yachting Federation Courses, 20
Australian Yachting Federation Website, 852
Auto-Ignition Temperature, 669
Automated Mutual-Assistance Vessel Rescue System (AMVER), 248
Autopilot, 317
Autopilot, Troubleshooting, 319
Auxiliary Power Take-Off Systems, 796
Aweigh (Anchor), 33, 114
AYF Certification & Courses, 20
Azimuth Mirror, 366
Azimuth or "Z" Propellers, 686
Back Splice, 68
Back Tracking, S&R, 556
Backing & Veering (Wind), 609
Backing a Sail, 33
Backing the Jib, 133
Backing, 33
Backstay, 41
Baffles, 281
Bahamian Moor, 125
Bailer & Lanyard, 537
Ballasted Keel, 284
Baltic Moor for Careening, 88
Baltic Moor, 143
Bar, 145
Barge, 137
Barograph, 597
Barometer & its Corrections, 596
Barometer, Diurnal Changes, 597
Barometer, Index Error, 596
Barometric Fall & Wind Speed, 597
Barometric Pressure Fall due to Height, 596
Barrier Reef, 261
Basic Flotation, Definition, 247
Battens (of a Sail), 44
Batteries - Cranking, 810
Batteries - Cycling Type, 810
Batteries - Installation Precautions, 812
Batteries - Lead Acid & Alkaline, 810
Batteries - Marine, 809
Batteries - Precautions When Removing or Replacing, 813
Batteries - Recommendation, 566
Batteries - Series & Parallel, 812
Batteries - Shallow Cycle, 810
Batteries - Storage When not in Use, 813
Batteries - Troubleshooting, 813
Batteries (See Also Electrical), 809
Batteries for Solar System, 816
Battery - Alkaline (NICAD), 812
Battery - Charging Rate, 811
Battery - Charging System, 807
Battery - Hydrometer, 811
Battery - Isolator Switch, 806
Battery - Main Switch, 806
Battery - Maintenance & Safety, 810
Battery - not Fully Charging, 813
Battery - One Pole Connection, 806
Battery - Specific Gravity, 811

Battery - Topping up too often, 813
Battery - Two-Pole Connection, 806
BDC, 629
Beaches, Regulations, 242
Beaching a vessel, 87, 88, 182, 183
Beam Reaching, 131
Beam, 33, 40, 42
Bearing (Symbol), 393
Bearing Correction Box, 390
Bearing Mirror, 366
Bearings & Courses (Navigation), 358
Bearings to Avoid (Navigation), 394
Beating, 131
Belay (Secure), 134
Belaying a Rope, 66
Belt Tension in V-Drives, 796
Bend (knot), 33, 64
Bermuda Rig, 51
Bernoulli's Principle, 130
Berthing a Twin Screw Vessel, 140
Berthing in Bad Weather, 141
Berthing Ropes Sizes, 62
Berthing Ropes, Terminology, 134
Berthing with a Berthing Line, 140
Berthing, 142, 143
BIA Websites, 852
Bilge Alarms, 784
Bilge Float Switch, 784
Bilge Hydro Air Switch, 784
Bilge Keels, 284
Bilge L-Port Cock, 783
Bilge Maintenance, 785
Bilge Pump Requirement for Recreational Vessels, 537
Bilge Pump Requirements for Commercial Vessels, 785
Bilge Pumping - Automatic, 784
Bilge Pumping - Good Practice, 786
Bilge Pumps - Engine Driven, 783
Bilge System, 781
Bilge Trouble Shooting, 787
Bilge Two-Way Cock, 783
Bird Droppings, 263
Birds' Flight Path, 559
Bites & Stings, 510
Bitter End, 33, 119
Bitts, 33, 38
Blake Slips, 120
Block (Pulley), 33, 165
Block Coefficient, 274
Block Loading (Jet Drives), 673
Block Loading, Angle of the Rope, 167
Block Systems on Yachts, 165
Blooper, 44
Blowby of Engine Fumes, 703
Blown Engines, 640
Blue Alert (Cyclone), 611
Blue Bottle (First Aid), 511
Blue Ringed Octopus (First Aid), 510
Boarding Ladders, 71
Boat (see also Vessel)
Boat Capacity Limits, 246
Boat Code, 246
Boat Handling in Heavy Weather, 146
Boat Handling, Ducted Propeller, 140
Boat Licence, Cancellation, 245

Boat Licence, Interstate Recognition, 245
Boat Licences, National Standard, 245
Boat Maintenance, 79
Boat Operator Competencies, 253
Boat Operator's Permit, 245
Boat Plate, National Standard, 247
Boat Ramps, 134
Boat Stability, 278
Boat Trailers, 133
Boat Training Schemes, 20
Boat, Minimum Length of Seagoing Passenger Vessels, 256
Boating Equipment, Standards, 241
Boating Industry Association Websites, 852
Boating Licence, 245
Boating Restrictions, Limitations & Prohibitions, 242
Boatswain's Chair, 71
Body Heat Loss in Water, 553
Body Plan, 42
Body Regions of High Body Heat Loss, 553
Body Sections (Vessel), 42
Bollard Hitch, 121
Bollard Pull - Tow Ropes, 149
Bollard, 33, 38, 66
Bolt Rope, 33
Bombora, 33
Bombs (East Coast Lows), 604
Bonding System (Boat Maintenance), 77
Boom, 41
Boot Positioned over Propeller, 701
Bosun's Chair, 65, 71
Bottle Screws, 120
Bottom Dead Centre, 629
Bottom Lock (Doppler Log), 316
Bow Chock, 33
Bow Thrusters, 689
Bow, 33, 40
Bower Anchor, 33
Bowline on the Bight, 64
Bowline, 64
Bowsprit, 33
Box Jellyfish (First Aid), 511
Braided Rope, 68
Brake on Propeller Shaft, 716
Breaking out the Anchor, 33, 114
Breaking Stress, Ropes & Chains, 161
Breast Rope, 134
Breath Testing, 586
Breather Valve (Engines), 633
Bridge Clearance - Chart Symbols, 101
Bridge Logbook, 417
Bridges, Clearances under them, 385
Bridle, 150
Broached, Broaching, 33, 148
Broad Reach, 131
Broome, Tidal Streams, 438
Brought up (Anchor), 33, 114, 122
Bubble in Compass Bowl, 361
Bulkhead, 33, 46
Bulkheads (floodability), 284
Bulldog Grips, 60
Buoyage System, 104

Buoyancy Standards for Recreational Vessels, 247
Buoyancy, 13, 284
Buoyancy, Foam, 284
Buoys Direction on Charts, 108
Buoys on Victoria's Inland Waters, 97
Bureau of Meteorology Telephone Marine Weather Service, 851
Bureau of Meteorology Weather Fax, 851
Bureau Veritas, 277
Burns (First Aid), 509
Buttock Lines (Vessel), 42
Buying a Boat, 10
Buying or Selling a Registrable or In-Survey Vessel, 247
Buys Ballot's Law, 609
Cabin Heater, 798
Cable (Anchor Cable), 33, 117
Cable Lifter, 117
Cable Lockers, 119
Cable or Chain Pipe, 120
Cable-in-Tension Propulsion Control System, 715
Cables - Chart Symbol, 103
Calorifier, 798
Camber, 33
Camshaft, 633
Cap Shroud, 41
Capacity Limits, 246
Capsize, 186
Capstan & Windlass, 33, 117
Captain & Master Mariner, 257
Carbon build-up in Engine, 710
Carbon Monoxide Accumulation, 701
Carburettor, 625, 634
Carburettor, Leaf Valves, 756
Cardinal Marks, 106
Cardinal Points, 33
Careening, 33, 87
Carley Floats, 543
Catamaran Stability, 280
Catamaran, 45
Catamarans - Handling, 145
Catenary in Tow Line, 150
Catenary, 121
Caterpillar HEUI, 659
Cathead, 33
Cathode, 809
Cathodic Protection, 77
Cavitation & Burning, 687, 746
Cb, 274
CDI Unit, 638
Cement Box, 184
Centerboard, 33
Centre of Buoyancy, 279
Centre of Effort, 133
Centre of Gravity, 278
Centre of Lateral Resistance, 133
Certificate of Survey, 254
Certificates of Competency, 21
Chain Blocks, 168
Chain Hoists, 168
Chain Lockers, 119
Chain Locks & Compressors, 120
Chain Pennant, 120, 121

Chain Pipe, 120
Chain Safe Working Load, 161
Chain Stopper, 33, 70
Chain Types, 116
Chain, 116
Channel (Boating Regulations), 242
Channel Closed Signal, 212
Channels - Tidal Streams, 142
Charge Controller, Solar Power, 817
Charging Distributor, 807
Chart Agents, Class A & B, 381
Chart Catalogue, 418
Chart Correction, 381
Chart Datum, 385
Chart Datums (GPS), 296
Chart Index, 418
Chart Plotter, 291
Chart Plotting Instruments & Aids, 379
Chart Plotting Terminology, 393
Chart Projections, 386
Chart Symbols, 100, 380
Chart Work (see also Navigation, Position Fixing, Plotting & individual topic entries)
Chartered Data, 385
Chartless Track Plotter (GPS), 290
Charts (Digitised), 291
Charts (Scanned), 291
Charts Indicating Colours of Lights, 105
Charts, & How to Read, 380, 381
Charts, Position Plotting, 396
Charts, Reliability, 381
Charts, Types, 386
Charts: How Buoys are numbered, 107
Chartwork, 379
Cheesed Down Rope, 67
Child, Definition, 246
Children's PFDs & Lifejackets, 541
Chine, 33, 38, 42
Choke Control, 634
Christmas Island, travel to, 249
Circuit Breakers, 807
Clam Shell Deflector, Jet Drives, 673
Clamp, 42
Classed Vessel, 276
Classification Certificate, 276
Classification of Fires, 520
Classification of Ships, 276
Classification Societies, 276
Classification, Commercial Vessels, 255
Clear Anchor, 34, 114
Clearance over Shoal Calculation, 432
Clearances under Bridges, 385
Clearing Lines - Chart Symbol, 102
Clearing or Danger Bearing, 395
Cleat, 38, 66
Clew, 51
Climate Zones (Meteorology), 593
Close Hauled, 131
Close Reaching, 131
Closed & Void Spaces, 582
Closed Circuit Cooling System, 648
Clothing, Recommendation, 566
Cloud Cover Description, 599
Cloud Point (Fuel), 669
Clouds Types, 602

Clove Hitch, 65
Club, 51
CMG & SMG by applying Leeway & Current, 399
CMG by Ratios, 411
CMG, 393
Coalescing Water Separator, 666
Coaming, 34, 284
Coast Radio Network, 463
Coastal Lifejacket, 542
Coastlines - Chart Symbols, 102
Cock-A-Bill, 34, 114, 171
Cocked Hat (Chartwork), 394
Cockpit, 41
Cocos (Keeling) Island, travel to, 249
Code Flags & Pennants, 97
Coefficient of Fineness, 274
Coiling a Rope, 57
Col, 601
Cold Front, 601, 604
Cold Start Control, 718
Cold Start, 710
Collapsed Person (First Aid(, 504
Collection & Holding Tanks, 800
Collision Avoidance by Radar, 344
Collision Drills, 564
Collision Hull Damage, 184
Collision Mat, 184
Collision Prevention Regulations, 193
Colour Coding, Electrical Wiring, 807
Colour Coding, Pipelines, 801
Colour Illustrations (*Fire Extinguishers, Buoys, Flags, Pyrotechnics, Chart Symbols*), 96
Commercial Certificates of Competency, 21
Commercial Vessels Act, 251
Commercial Vessels, 254
Commonwealth Bodies, Websites, 852
Communication among the Crew, 584
Communications Authority, Website, 852
Communication Equipment Requirement for non-SOLAS vessels, 465
Compacts (Outboards), 736
Compass Alignment, 360
Compass Behaves Erratic, 361
Compass Construction, 359
Compass Course to Steer, 389
Compass Dampened, 359
Compass Deviation, 360
Compass Error, 367
Compass Internal Magnets, 361
Compass Jewel & Pivot, 359
Compass Points, 37
Compass Points, 37
Compass Requirement for Recreational Vessels, 537
Compass, Dead Beat, 359
Compass, Fluxgate, 364
Compass, Fluxgate, Error, 370
Compass, Gyro, 362
Compass, Gyro, Error, 367
Compass, Magnetic Bearings & Courses, 369
Compass, Magnetic Influences, 361

Compass, Magnetic, 359
Compass, Magnetic, Error, 367
Compass, Steering Compass, 359
Compasses, 358
Compasses, Dumb Compass, 366
Compasses, Hand Bearing, 364, 369
Competency Based Training, 28
Competent Person, Definition, 254
Complaints from Passengers, Safety, 586
Compliant Equipment, 252
Components of Engines, 632
Con, 34
Cone Shell (First Aid), 511
Confined Spaces, 582
Connecting Rods (Engines), 625, 634
Contact Breakers, 638
Contactors (Switches) for Engines, 712
Control Systems for Propulsion, 712
Controllable Pitch Propellers, 686
Controls to Test, 135
Convention for Pollution Prevention, 260
Conventional in-Line Propeller, 674
Coolant & Oil Drainage System, 263
Cooling System & Cleaning (Outboards), 743, 744
Cooling Systems – Inboards, 647
Copper Oxide Paints, 77
Coral Spawns, 261
Coriolis Effect, 594
Correction Box (Chartwork), 390
Corrections, Charts & Publications, 381
Corrosion Prevention in Propellers, 687
Corrosion, 76
COSPAS, 481
Cospas-Sarsat Satellite System, 481
Councils Have Power, 242
Counter Rotating Duel Propellers, 686
Counter Stern, 34
Coupling, Flexible, 681
Course & Bearing Correction Box, 390
Course & Speed Made Good by applying Leeway & Current, 399
Course Made Good by Ratios, 411
Course Made Good, 393
Course over the Ground, 393
Course to Make Good, 393
Course to Steer to Intercept a Vessel, 413
Course to Steer, 389, 393, 397
Course to Steer, Finding, 397
Courses & Bearings, 358
Coxswain Qualification, 21
Cradle, 86
Crane or Travel Lift, 86
Cranes, 168
Crank, 625
Crankcase & Breather, 632, 634
Crankcase Fumes Disposal, 703
Cranking an Engine, 634
Cranking Batteries, 810
Crankshaft, 625, 635
Crash Stop, 145
Crevice & Deposit Corrosion, 76
Crew Emergency Procedures, 564
Crew List, 257
Crew Member Must, 580
Crew Muster List, 564

The Australian Boating Manual

Crew Training, Recommendation, 566
Cringle, 52
Cropping, 60
Cross Index Ranging, 344
Cross River Ferries, Lights, 212
Cross Trees, 52
Crossing Bars, 145
Crosstree, 41
Crowd (Sail), 52
Crown Knot, 68
Cruising Boats, Round Hull, 38
Cruising Permit, 248
CTS to Intercept a Vessel, 413
CTS, 393
Cumulo-Nimbus Clouds, 602
Cunningham, 52
Cupping (Propellers), 683
Current, 633
Current, Counteracting it, 400
Currents Chart Symbols, 101
Currents Seasonal Maps, 441
Customs Service, Website, 852
Customs, Quarantine, Immigration, 249
Cutlass Bearing, 46
Cutter with Topsail, 45
Cutter-Rigged, 44
Cycle of an Engine, 628
Cycling Batteries, 810
Cyclone Alert System, 611
Cyclone Contingency Plans & Safe Havens, 611
Cyclone How to Get out of, 613
Cyclone Movement, 608
Cyclone Path, 609
Cyclone Plotting Map, 613
Cyclone Preparedness Information, 851
Cyclone Season, 608
Cyclone Terminology, 609
Cyclone Track, 609
Cyclone Warning Signs, 612
Cyclone Warnings, 607
Cyclone Watch, 612
Cyclone, 599
Cylinder Block, 635
Cylinder, 625
Cylinders, Glazing/Carbon Build up, 710
Damage Below the Waterline, 184
Dan Buoy, 34
Danger Angle, Plotting, 406
Danger Bearing, 395
Dangers to Survivors & Precautions, 557
Dashboard Displays, 711
Davit, 34
DC Current, 633
Dead Reckoning, 393
Deadweight Capacity, 276
Deadweight, 276
Deadwood, 46
Deck Beam, 42
Deck Fittings, 38
Deck Loading, 276
Deck Wash System, 783
Deckhand Qualification, 21
Deckhead, 46
Deckline Mark, 277
Decks Non-Slippery, 84

Defence Force Marine Qualifications, 22
Dehydration Prevention, 557
Density Relative to Air (Fuels), 669
Density, 274, 669
Density, LPG (0.52), Petrol (0.89), 777
Departure from Australia, 248
Departure from Overseas, 249
Deposit Corrosion, 76
Depression (Weather), 599
Depth Sounder, 304
Depths Printed on Charts, 385
Depths, Position by Line of Soundings, 410
Desalination (for Survival), 560
Desalination Watermakers, 799
Designed Waterline, 42
Detectors, Indicators & Alarms, 320
Detectors: Smoke, Heat & Radiation, 524
Deviation & Variation, Combined, 389
Deviation by Transit Bearing, 372
Deviation Card, 369
Deviation, 369
Devil's Claw, 34, 120
DFI, 659
DGPS, 297
Diesel Engine Starting using Ether/Petrol, 718
Diesel Engine, 627
Diesel Engine, Generator Problems, 718
Diesel Fuel System, 658
Differential GPS (DGPS), 297
Digital Compasses, 364
Digital Selective Calling, 469
Dinghy Docks, 145
Dipping of the Eye, 134
Dipstick, 631
Displacement Hull, 38
Displacement Table, 275
Displacement, 275
Displays on Dashboard, 711
Distance & Bearing by Radar, 337
Distance Apart, Twin Engines, 695
Distance, Time, Speed Calculations, 388
Distress (Contact Numbers), 852
Distress & Priority Communication by Inmarsat-C, 488
Distress (See also Radio & individual topic entries)
Distress Acknowledging, 468
Distress Alert Procedure by DSC, 472
Distress Alert Relay Via DSC, 475
Distress Call from Liferaft, Lifeboat, 468
Distress Flares, 537
Distress or Urgency?, 466
Distress Signal, Sighting One, 563
Distress Signals - Annex IV, 222
Distress Signals & Misuse Penalty, 560
Distress Signals, 99, 212
Distress Signals, Radar Reflective, 348
Distress Transmission, 466
Distress, Master's Legal Duty, 467
Distributor (Ignition System), 635
Diurnal Changes in Barometer, 597
Diurnal Tides, 432
Diving Restrictions, 243
Docking Ropes Sizes, 62

Docking Ropes Terminology, 134
Docking Strip, 47
Docking, Dredging, Surveying by RTK, 298
Doctor Service, 469
Documenting Hazards & Risks, 583
Documents on Commercial Vessels, 257
Dog's-Leg Course (Sailing), 131
Doppler Effect, 315
Double Luff, 163
Doubling Angle on the Bow, 409
Downflooding, 280
Downstream, 34
Downwind, 131
DR (Chartwork), 393
Draft & Displacement Table, 275
Draft (Draught), 34, 40, 274
Drain Slots in Exhaust Pipe, 699
Drainage System for Oil & Coolant, 263
Drains (Scuppers), 800
Dredger, Speed Limit when passing, 242
Dredging Down, Dredging Anchor, 144
Drift, 34
Drinking Water, 3 Ways of Making, 560
Drive Train Alignment with Gearbox Shaft, 679
Drivelines (Types), 673
Drogue, 34, 150
Drunk & Disorderly Person, 586
Dry Docking Stress, 147
Dry Rot, 92
Drying Height Calculation, 432
Drying Heights, 385
Dry-Sump, 645
DSC Call Options, 472
DSC Distress Acknowledgement, 472
DSC Distress Alert Cancellation, 475
DSC Distress Alert Procedure, 472
DSC Distress Alert Relay Procedure, 475
DSC Epirbs, 484
DSC Identification (MMSI), 471
DSC Modernisation (DSC2, DSC3), 476
DSC Safety Alert, 475
DSC Test, 476
DSC Urgency Alert, 475
Ducting (Radar/Radio signals), 341, 462
Dumping at Sea, 262
Duplex & Simplex Radio, 464, 478
Duty of Care, 579
DWT, 276
Dye Marker, 538
Dynamo, 807
E.P by Applying Leeway & Current, 399
E.P, 393
E.P, Finding, 397
East Coast Lows & Bombs, 604
Ebb Tide, 34, 429
ECDIS, 291, 292
Echo Sounder, 304
Echo Sounder, Buying One, 314
ECS, 291
Eddy, 34
Effect of using Engine at a Wharf, 137
EFI Outboards, 732
EFI, 659
EGC Receivers (Inmarsat), 489

El Nino, 606
Electric Starting System, 709
Electric Trollers (Outboards), 735
Electrical - Accessories, 808
Electrical - Alternator, 807
Electrical - Batteries (See Batteries), 809
Electrical - Battery Charging, 807
Electrical - Battery Isolator Switch, 806
Electrical - Battery Main Switch, 806
Electrical - Calculating Cable Area, 808
Electrical - Charging Distributor, 807
Electrical - Double Pole Switch, 806
Electrical - Fires, 520
Electrical - Fuses, Circuit Breakers, 807
Electrical - Generator, 814
Electrical - Installation, 806
Electrical - Interference, 465
Electrical - Light Switch Connection, 806
Electrical - One Pole Switch, 806
Electrical - Shore Power Supply, 814
Electrical - Short Circuit, 807
Electrical - Solar Power, 815
Electrical - Starting Battery Connection, 806
Electrical - Storms, 602
Electrical - System (Engines), 637
Electrical - Transformer, 814
Electrical - Voltage Regulator, 807
Electrical - Wiring Colour Coding, 807
Electrolyte, 809
Electrolytic Corrosion, 76
Electronic Chart & Display Information System, 292
Electronic Chart System, 291
Electronic Diagnostic (Engines), 717
Electronic Fuel Injection - Petrol, 732
Electronic Fuel Injection, 635, 659
Electronic Navigation Charts, 291
Electronic Propulsion Control, 715
Elliptical or Double Turn, 555
Emergencies, 182
Emergencies (Contact Numbers), 852
Emergency Anchor, 115
Emergency Fire Pump, 526
Emergency Hull Repair Kits, 91
Emergency Radio Transmissions, 466
Emergency Services (SESs), 851
Emergency Shutdown Valve (Engines), 635
Emergency Station List & Signals, 564
Employee Must, 580
Enamel, 81
ENC, 291
Encounter with Waves, 147
End-for-Ending, 60
Engine (Cylinder) Block, 635
Engine (Diesel) Starting, using Ether/Petrol, 718
Engine (Inboard) Starting, 709
Engine (see also Fuel, Outboards & individual topic entries)
Engine Alarms, 712
Engine Bearers & Mountings, 693
Engine Beds, 693
Engine Control Systems, 712

Engine Cooling Systems (Inboard), 647
Engine Driven Bilge Pumps, 783
Engine Driven Pumps, 796
Engine Foundations, 693
Engine Idling Speed, 711
Engine Inboard-Outboard, 676
Engine Inclination, 695
Engine Installation, 701
Engine Instruments & Alarms, 712
Engine Journal, 625, 638
Engine Log Book Entries, 723
Engine Lubrication Systems, 643
Engine Maintenance & Inspection Schedule (Outboards), 758
Engine Maintenance, 722
Engine Mountings, 631, 636
Engine Mounts or Suspension, 694
Engine on Light Load, 710
Engine Operation, 708
Engine Problems, 718
Engine Reversal Problem, 716
Engine Room Safety, 708
Engine Room Temperature, 702
Engine Room Ventilation, 702
Engine Room Weight Distribution, 695
Engine Safety Devices, 712
Engine Shut Off Switch, 633
Engine Size, 11
Engine Soundproofing, 705
Engine Stall & Reversal Problem, 716
Engine Starting Systems, 708
Engine Throttle Boost System, 716
Engine Tools & Spares to Carry, 717
Engine Tools & Spares, Outboards, 752
Engine Troubleshooting, Petrol & LPG, 753
Engine Vibration Isolators, 694
Engine, 4-Cycle & 2-Cycle, 626
Engine, 4-Stroke & 2-Stroke, 626
Engine, Flywheel, 636
Engine, Glazing/Carbon Build up, 710
Engine, Glow Plug, 627, 710
Engine, Governor, 636
Engine, Heat Exchanger & Temperature Gauge, 637, 647, 652, 696
Engine, Heater Plug, 627, 710
Engine, Running at Slow Speed, 711
Engine, Stroke Cycle, 628
Engine, Supercharged, 640
Engine, using alongside a Wharf, 137
Engines Troubleshooting, 717
Engines, Harmonic Balancer, 637
Engines, High Speed Diesels, 711
Engines, How They Work, 625
Engines, Twin (Distance Apart), 695
Enhanced Group Calling Receivers, 489
Environmental Protection Act, 262
EPIRB & Gpirbs, 481
EPIRB, L-Band, 490
EPIRB, personal, 485
EPIRBS (VHF DSC), 484
EPIRBS, Recommendation, 566
EPIRBS/GPIRBS, Design Features, 485
EPIRBS/GPIRBS, Testing, 485
Epoxy, 81

Estimated Position by Applying Leeway & Current, 399
Estimated Position, 393
Estimated Position, Finding, 397
Event Sequence Timing Device, 716
Exemptions from Recreational Vessel Registration, 247
Exemptions Register of Generic Vessel
Exemptions, 252
Exhaust Backpressure, 698
Exhaust Bend, 631
Exhaust Ducting (Engine Room), 703
Exhaust Ejected Ventilation System, 704
Exhaust Fans (Engine Room), 703
Exhaust Gas Re-Burning System, 730
Exhaust Manifold Slobber, 700
Exhaust Manifold, 636
Exhaust Pipe, 636
Exhaust Pipe, Slots, 699
Exhaust Port, 625
Exhaust Riser, 696
Exhaust Systems, 695
Exhaust Valve, 625, 636
Exhaust, Preventing Waves Entry, 697
Expanding Square Search Pattern, 556
Expansion Tank, 631, 649
Expeditions Vessel Classification, 255
Explosive Cyclogenesis, 604
External Cardiac Compression, 508
Fairlead, 34, 38
Fairway, 34, 242
Faked Rope, 66
Fall in Pressure & Wind Strength, 597
Fans (Engine Room Exhaust), 703
Fast Craft (F1 & F2) & Risks, 255, 256
Fathom, 69, 385
Fender, 34, 134
Ferries in Chains - Lights, 212
Ferro Hull Repairs, 80
Ferro-Cement Hulls, 12
Fetch, 598
Fibreglass (GRP) Hull, 92
Fibreglass Boats, 80
Fibreglass Laminate Repairs, 79
Figure Codes, 480
Figure Eight Knot, 64
Filter, Air Filter on Engine, 632
Filters (Fuel) Maintenance, 667
Fire (see also individual topic entries)
Fire Bucket with Lanyard, 537
Fire Detectors, 524
Fire Detectors, 524
Fire Drills, 564
Fire Extinguisher (Recreational Vessel Requirement), 537
Fire Extinguishers Maintenance, 522
Fire Extinguishers Safety, 521
Fire Extinguishers, 98, 521
Fire Extinguishers, How to Use, 521
Fire Extinguishing Installation, 522
Fire Fighting Hoses (Foam), 263
Fire Fighting, Accumulated Water, 527
Fire Hoses, 526
Fire Point (Fuel), 668
Fire Prevention & Control, 520, 535
Fire Pump, 526

Fire Risk Is Increased by, 708
Fire Safety on Board, 526
Fire Station Signal, 564
Fire Triangle & Tetrahedron, 520
Fire-Fighting Appliances, 536
Fires Classification, 520
Fires, How They Spread, 525
First Aid - OH& S Standard, 504
First Aid Kit, 513
First Aid, 504
First Aid, Bites & Stings, 510
First Aid, Bleeding First Aid, 510
First Aid, C.P.R., 508
First Aid, Cardiac Arrest, 508
First Aid, Carotid Pulse, 508
First Aid, Dehydration, 514
First Aid, Drowning, Near Drowning, 506
First Aid, Drug Overdose, 506
First Aid, E.A.R., 506
First Aid, E.C.C., 508
First Aid, Haemorrhage, 510
First Aid, Heart Attack When Alone, 509
First Aid, Scalds, 510
First Aid, Sea Snake Bite, 511
First Aid, Shake & Shout Test, 504
First Aid, Shock, Air Hunger, 505
Firwst Aid, Unconscious Person, 505
Fish & Shellfish, Dangerous Type, 561
Fish Finder, 304
Fish or Fish Juice for Survival, 558
Fish Tailing, 145
Fisherman's Bend, 66
Fishing Means Commercial Fishing, 21
Fishing or Oyster Vessel, 255
Fishing Prohibited - Chart Symbol, 103
Fishing Skipper Qualification, 21
Fishing Vessels, 21
Fittings, Aluminium & Stainless, 80
Fix (on Chart), 393
Fixed Fire Extinguishing, 522
Fixing Positions, 398
Flags & Pennants, 97
Flaked Rope, 66
Flammability Range, 669
Flare (Hull), 34, 40
Flares that are not Red or Orange, 240
Flares, Distress Signals, 537, 538
Flash Cyclone Warning, 607
Flash Point, 668
Flat Aft, 133
Flat-Bottom Barge, 137
FleetNET, 490
Flemish Coil, 67
Flexible Mounts (Engines), 694
Flexible Shaft Coupling, 681
Flexible Steel Wire Ropes - FSWR, 59
Flinder's Bar, 361
Float - Why Ships Float, 274
Float Switch, 784
Floating Dock, 87
Floating Sea Scum, 261
Flood Tide, 34, 429
Floor, 42, 47
Fluke (Anchor), 34
Fly up into the Wind, 143

Flying Doctor Service, 469
Flywheel (Engine), 636
Foam Buoyancy, 284
Foam Fire Fighting Hoses, 263
Focsle, 34
Fog, 601
Foot (of a Sail), 44
Fore & Aft Rig, 52
Fore & Aft, 34
Fore Sheet (Rope), 52
Forecast, 598
Forecastle, 34
Forecasts & Warnings (Weather), 601
Forelock, 171
Forepeak, 34
Foresail, 44, 52
Foreshore - Chart Symbols, 102
Forestay, 41, 52
Forward, 34, 40
Foul Anchor, 34, 114
Foul Ground, 100
Foul Hawse, 34, 114
Foundered, 34
Four-Point Bearing (Plotting), 409
Fractional Rig, 52
Fraction-Rigged Sloop, 45
Frames, 34, 42
Freak or Rogue Wave, 598
Freak Propagation, 341
Free Surface Effect, 281, 284
Freeboard, 40, 47, 274
Freeing Ports, 34, 47
Fremantle Doctor, 614
Fresh Water Allowance, 277
Freshen the Nip, 124, 150
Freshwater Loadline, 277
Friction in a Pulley, 167
Fridge/Freezer, 800
Fronts, 601, 604
FSWR, 59
Fuel - Maintenance, 666
Fuel (see also Engines, & individual topic entries)
Fuel Amount Calculations, 721
Fuel Atomisers, 638
Fuel Capacity, 666
Fuel Connections, 663
Fuel Consumption - Rule of Thumb, 144
Fuel Expansion in Hot Weather, 262
Fuel Filter Cleaning (Outboards), 740
Fuel Filter, 636
Fuel Filter, After Reassembling, 717
Fuel Filter, Before Removing, 717
Fuel Filters Maintenance, 667
Fuel Injection System, 628, 635, 659
Fuel Injection, Direct, 659
Fuel Injection, Electronic - Petrol, 732
Fuel Injection, Electronic, 635, 659
Fuel Injection, Indirect, 628
Fuel Injectors, 638
Fuel Maintenance (Outboards), 740
Fuel Operating Range Calculation, 721
Fuel Pipes, 663
Fuel Pump/ Fuel Lift Pump, 636
Fuel Pumps, Types & Maintenance, 668
Fuel Rack, 718

Fuel Safety, 668
Fuel Shortage on Voyage, 720
Fuel Spitting through Carburettor, 756
Fuel Supply, Eliminating Air, 660
Fuel System (Inboard Diesel), 658
Fuel System Inspection (Outboards), 740
Fuel System, Bleeding/Purging, 661
Fuel System, Outboards, 736
Fuel Tank - Day Tank, 660
Fuel Tank Calculations, 721
Fuel Tank Cleaning (Outboards), 740
Fuel Tank Installation for Outboards, 738
Fuel Tank Leaks, 526
Fuel Tank Service (Outboards), 758
Fuel Tank, Auxiliary, 660
Fuel Tank, Changing over at Sea, 666
Fuel Tanks Size (Capacity), 666
Fuel Tanks, Pipes & Connections, 663
Fuel Terminology, 668
Fuel Volume Calculations, 721
Fumes Disposal, 703
Fuses & Circuit Breakers, 807
Gaff, 52
Gaff-Rigged Cutter with Topsail, 45
Gale at Anchor, 124
Gale Warning, 602
GALILEO Positioning System, 287
Galley Safety, 582
Galvanic Corrosion, 76
Galvanic Series, 76
Gantline, 34, 71
Garbage Disposal Regulations, 260
Garbage Record Book, 260
Garboard Strake, 47
Gas Detector, 701
Gas Freeing a Fixed Fuel Tank, 668
Gauges are Superior, 646
Gearbox Oil Change (Outboards), 743
Gearboxes, 678
Gelcoat Repairs, 79
General Purpose Hand Qualification, 21
Generator Problems, 718
Generator, 814
Generic Vessel Exemptions Register, 252
Genoa, 44, 52
Geographical Range Table, 414
GEOSAR, 482
Geostationary S&R System, 482
Geostrophic & Surface Winds, 594
Germanischer Lloyd, 277
Girting (of Tugs), 151
Give Way, 34
Glazing in Engine, 710
Global Maritime Distress & Safety System, 469
Global Navigational Satellite Systems, 287
Global Positioning System, 287
GLONASS Positioning System, 287
Glow Plug (Engine), 627, 710
GMDSS & DSC, 470
GMDSS Ships, VHF Aural Watchkeeping, 464
Gnomonic Projections, 386
GNSS Positioning Systems, 287
Gob Rope, 151

Going About, 131
Gooseneck, 48
Goosewinged, 132
Governor (Engine), 636
Gpirbs, 481, 483
GPS - Accuracy, 296
GPS - AMSA Marine Notice, 296
GPS - Buying One, 299
GPS - Calculate Boat Speed, 316
GPS – Dithering or Selective Availability, 293
GPS - Ephemeris, 293
GPS - HDOP, 294
GPS - Ionospheric Delay, 290
GPS - Multiplexing Channel, 293
GPS - Parallel-Channel, 292
GPS - RAIM, 294
GPS - Sequencing Channel, 292
GPS - UAIS or AIS, 298
GPS - UERE, 294
GPS II - Accuracy Improvement Initiative (AII), 298
GPS II - Modernization of GPS, 298
GPS, 287
Grab Bag, 539
Gradient Wind, 599
Graving Dock, 87
Gravitational Effect of a Vertical Lift of Man Overboard, 557
Great Barrier Reef, 261
Great Circle Navigation, 386
Green & Red Relative Bearings, 358
Green Flares, 240, 561
Grey Line Function (Sounders), 310
Grey Water, 259
Groove & Rope Size, 168
Gross & Net Registered Tonnes, 276
Grounding, 182
Growing, 34, 114
GRP Hulls, 12, 92
GRT, 276
Gun Tackle, 163
Gunwale (Gunwhale), 34, 47
Gusset, 42
Gust, 598
Guy, 48
Gybing, 132
Gyn Tackle, 163
Gypsy, 34, 117
H.E.L.P., 554
Halyard, 48
Hammond Rock, Tidal Streams, 437
Hand Lead Line, 538
Handing or Taking over a Watch, 181
Handy Billy, 163
Harbour Control, VHF Channels, 460
Hard Chine, 38, 42
Hard Stand, 87
Hard-A-Port, 34, 36
Harmonic Balancer (Engine), 637
Harnesses, 543
Hatch, 41
Hawse Pipe, 34, 114, 124
Hazards to Safety, 708
Head (of a Sail), 44
Head Reach, 145

Head Rope, 134
Head, 52
Headings, 358
Heads (Toilet), 800
Headsail, 44
Headsail, 52
Headstay, 52
Head-up, 330
Heart Attack When Alone, 509
Heat Loss from Body, 553
Heater for Cabin, 798
Heave to, 34, 149
Heaving Line, 135
Heavy Weather Boat Handling, 146
Heavy Weather Jib, 44
HectoPascals (HPa), 596
Heel, List & Loll, 282
Height & Fall in Pressure, 596
Height of Tide by Form AH130, 430
Height of Tide by Rule of Twelfth, 430
HELIBOX, 551
Helicopter Rescue, 562
Helicopter Winching from Vessel, 563
Heliograph, 539
Helm, 34, 133
Helm, Rudder or Steering Orders, 36
HEUI, 659
Herald of Free Enterprise, 584
High Seas Warning, 606
High Speed Diesels, 711
High Water Mark, 242
Higher (No Higher - Sailing), 132
High-Speed Boats Fitted with Chine, 38
High-Speed Craft (HSC) Code, 255
HIN, 246
Hire & Drive Vessels, 259
Hitch, 34
Hitches - Types, 64
Hog, 34, 42
Hogging, 146
Holding Tanks, 800
Holed Below the Waterline, 184
Hollow Keel, 281
Hooks, 171
Horizon Distance Calculation, 343
Horizontal Angles, Plotting, 404
Horizontal Lift of Man Overboard, 557
Horizontal Sextant Angles, Plotting, 404
Hose Hanging Overside, 783
Hoses, Foam Fire Fighting, 263
Hot & Cold Water Supply, 798
Hove in Sight, 34, 114
Hovercraft Signal, 563
Hove-to, 132
HSA, Plotting, 404
HSC Code, 255
Huddle, 554
Hugging, 133
Hull Damage, 184
Hull ID, Recommendation, 566
Hull Identification Number (HIN), 246
Hull Materials, 12
Hull Repair Kits, 91
Hull Speed & Fuel Consumption, 144
Hull Stresses in Heavy Weather, 146
Hull Stresses While on a Slip, 88

Hulls –Types, 38
Humidity Measurement, 600
Hurricanes – see Cyclones
Hydraulic Electronic Unit Injection, 659
Hydraulic Propulsion Control, 715
Hydraulic Starting System, 709
Hydraulic Steering Systems, 829
Hydro Air Switches, 784
Hydrometer, 811
Hydroplanes, 284
Hydrostatic Release Unit, 546
Hydrostatic Squeeze, 557
Hygrometer, 600
Hypothermia, 553, 557
IALA Buoyage System "A", 104
IAMSAR, 556
Idler (Steering Component), 828
Idler Pulley (Auxiliary Power Take-Off), 797
Idling Speed (Engine), 711
Ignition Coil Engines), 638
Ignition System (Engines), 637
Ignoble Metal, 809
Immigration Department Website, 852
Immigration, Quarantine, Customs, 249
IMO High-Speed Craft Code, 255
IMO, 251
In Irons or In Stays, 133
Inboard Engine Starting, 709
Inboard-Outboard Engines, 676
Indicators, Detectors & Alarms, 320
Inertia Stop (Boat Handling), 145
Inflatable Lifejackets & PFDs, 542
Inflatable Liferafts, 543
Initial Survey, 254
Injectors (Fuel), 638, 732
Inland Waters, 97
Inlet Port, 625
Inlet Valve, 625, 638
In-Line Propeller Arrangement, 674
Inmarsat A, B or C Licence, 461
Inmarsat D+, 489
Inmarsat Endorsement Certificate, 486
Inmarsat Enhanced Group Calling, 489
Inmarsat Epirbs, 484
Inmarsat Epirbs, Description, 490
Inmarsat Satellite Communication, 486
Inmarsat Systems, Non-GMDSS, 489
Inmarsat-C Antenna Siting, 489
Inmarsat-C Features, 488
Inmarsat-C Performance Test, 489
Inmarsat-C, Distress & Priority, 488
Inmarsat-E, Description, 490
Inmarsat-M, 489
Inshore Operations, 256
Inshore Waters, 256
Inshore, 256
Inspection, Definition, 254
Inspection, Portable Extinguishers, 522
Instruments & Alarms (Engine), 712
Insuring a Boat, 19
Intake & Exhaust Ducting, 703
Intake Silencer, 631
Interaction in Narrow Channels, Shallow Water & with Larger Ships, 135
Intercept Another Vessel (C.T.S), 413

Interlock Switch, 711
Internal Combustion Engines, 626
International & Local Collision Prevention Regulations, 193
International & Local Distress Signals, 212
International Aeronautical & Maritime Search & Rescue Manual, 556
International Chamber of Commerce, 248
International Code Flags & Pennants, 97
International Collision Regulations, 193
International Convention for Prevention of Pollution from Ships, 260
International Maritime Organization, 251
International Mobile Satellite Organization, 486
International Safety Management Code, 584
International Safety Symbols, 585
International Sea Areas (Radio), 459
International Telecommunication Union, 248
International VHF Setting, 464
International Voyages, 248
Internet Addresses, 852
Interstate Visitors, 247
Interstate Voyages - the Law, 257
Inter-Tropical Confluence, 595
Inter-Tropical Convergence Zone, 595
Inversion, 341
Inverters (Solar Power), 816
Ionospheric Delay (GPS), 290
ISM Code, 584
Isobar, 599
Isolated Danger Marks, 106
Isometric View, 43
ITC or ITCZ, 595
ITU, 248
JANUS Configuration (Doppler Log), 315
Jet Boat Regulations, 243, 259
Jet Drive, 673
Jib, 44, 52
Jibing, 132
Journal (Engine), 625, 638
Jump Start & Jumper cables, 812
Jury Rig, 34
Jury Rudder, 185
Kedge & Kedging, 34, 183
Keel Cooler, 647
Keel, 41
Keeling, Islands travel to, 249
Keels – Types, 39
Keelson, 47
Kenter, 171
Ketch, 45, 52
King, Freak or Rogue Wave, 598
Kitchen & Toilet Waste, 259
Knee, 47
Knots - Types, 64
Knots (Speed), 34
Kort Nozzle – Boat Handling, 140
Kort's Nozzle Propellers, 686
La Nina, 606
Ladders, 71

Laminate Repairs, 79
Land Breezes, 614
Land Earth Stations, 487
Land, Indication of being close, 559
Landing Barges - Handling, 145
Langs Lay, 34
Lanyard, 34
Large Vessels Less Manoeuvrable & Slow to Respond, 135
Lateen, 53
Latitude, Lat (Chartwork), 382, 385
Lazarette, 34
Lead Block, 163
Lead Lights & Lighthouses, 385
Lead Line, 538
Lead or Copper Oxide Paints, 77
Leading Lights - Chart Symbol, 102
Leads, 34
Leaf Valves (Carburettor), 756
Leaky Joint or Flange, 717
Leaky Pipe –Temporary Repairs, 717
Lee Board, 35
Lee Helm, 133
Lee Shore, 35, 114
Lee Tide, 35, 114
Leech (of a Sail), 44, 53
Leeward, 35
Leeway (Chartwork), 398
Leeway Angle, 143
Leeway, 35
Length & Approx Tonnage, 256
Length Measurement, 241
Length Overall, 35, 40
Length, Seagoing Passenger Vessels, 256
LEOLUTs, 482
LEOSAR, 482
LES, 487
Let Go, 35, 114, 118
Letting Go & Picking up Anchor, 122
Level Flotation, Definition, 247
Lies to, 132
Life Support Flow Chart, 509
Lifebuoy, 539
Lifejacket & PFD Buoyancy Rating, 542
Lifejackets, 540, 542
Lifejackets, Double Bubble, 542
Lifelines, 41
Liferaft Rations Control, 560
Liferaft Rations, 549
Liferaft, 543
Liferaft, Living in it, 559
Liferafts, Recommendation, 566
Life-Saving Appliances, 536
Lifting Equipment, 162
Lifting Gear Inspection & Log, 63, 174
Light Characteristics, 105
Light Displacement, 275
Light Planing Hull at High Speeds, 144
Lighterman's Hitch, 121
Lighthouses, 105, 385
Lightning, 603
Lights, List, 418
Limited Coast Stations, 462
Line of Soundings, Position Fixing, 410
Line Squall, 599
Lines Plan, 42

List & Loll, 282
List of Lights, 418
List of Radio Signals, 248
Living on Board, 247
Lizard, 35, 70
Lloyd's Register of Shipping, 276
Lloyds Length, 35
Lloyds Open Form of Salvage, 181
LOA, 35, 40
Load Displacement, 275
Loading of Vessels, 274
Loadlines Certificate, 277
Loadlines Exempt, 277
Local Boating Regulations, 241
Local Distress Signals, 212
Local Knowledge Examination, 244
Local Signals on Vessels, 212
Local User Terminal, 481
Local Weather, 614
Log (Terminology), 35
Log Book (Navigational), 417
Log Book (Radio) 466
Log Book Entries (Engines), 723
Log Book Requirements, 257
Log Keeping of Lifting Gear, 174
Logs (Measuring Speed), 315
Loll, 282
Long Stay (at Anchor), 35, 114
Long Stroke, Slow Speed, Engines, 711
Longitude, 382
Loose-Footed, 132
Loss of Propeller at Sea, 185
Loss of Steering at Sea, 185
Low Clouds, 599
Low Earth Orbit S&R System, 482
Low Pressure, 599
Lowering Hitch, 70, 71
Lowest Astronomical Tide (LAT), 385
LPG Density or Specific Gravity, 777
LPG Emergency Procedure, 779
LPG Engine Troubleshooting, 753
LPG Engine, 777
LPG Engine, Liquid Withdrawal & Vapour Withdrawal Cylinder, 777
LPG Safety Rules, 776
LPG Valves & Fittings, 778
LPG, 776
L-Shaped Graph, Tide Calculation, 430
Lubber's Line, 360
Lubrication System (Outboards), 741
Lubrication Systems (Engines), 643
Luff on Luff, 164
Luff or Luff-Up, 133
Luff or Watch Tackle, 163
Luff, 53, 131
Lugless Joining Shackle, 171
Luminous Range Diagram, 415
LUT, 481
LWL, 35, 40
Machinery Guidance System, 298
Machinery Noise Soundproofing, 705
Made Fast, 35
Magnetic Bearings & Courses, 369
Magnetic Deviation, 369
Magnetic Variation, 367
Main Sheet, 53

Main Sheets, 41
Main Stay, 53
Mainsail, 44
Maintenance & Inspection Schedule (Outboards), 758
Maintenance (Hull), 79
Maintenance of Engines, 722
Making Way, 35
Man Overboard Button, 290
Man Overboard, 555
Manifold (Engines), 631, 638
Manoeuvring Difficulties with Large Vessels, 135
Manoeuvring in Wind, Currents & Tides, 142, 143
Marconi Rig, 53
Marine Act, 251
Marine Animal Bites & Stings, 510
Marine Authorities, Websites, 852
Marine Batteries, 809
Marine Casualties & Incidents, 181
Marine Driver's Licence, 245
Marine Gearboxes, 678
Marine Guidance Manuals, 252
Marine Notices (State), 381
Marine Notices by AMSA, 418
Marine Pollution Regulations, 259
Marine Radio, (See Radio)
Marine Railway, 86
Marine Reversing Gearboxes, 678
Marine Safety Act, 251
Marine Satellite Communication Certificate, 461
Maritime Communication Stations, 462
Maritime Rescue Co-Ordination, 261
Maritime Safety Authority Website, 852
Maritime Safety Info by Radio, 463, 851
Maritime Safety Info by Satellite, 489
Maritime Safety Symbols, 585
Maritime Search & Rescue Manual, 556
Marker Buoys on Inland Waters, 97
MARPOL, 260
Mast Supports, Failure of, 186
Master Class V & IV Qualifications, 21
Master Mariner & Captain, 257
Master, Captain & Master Mariner, 257
Master's Instructions to Watchkeepers, 181
Masthead Rig, 53
Masthead Sloop, 45
Masthead, 35
Mayday Call from Liferaft, 467
Mayday or PanPan?, 467
Mayday Relay, 468
Mayday Transmission, 466
Measured Length, 35
Measuring "Wear-Down" in Stern Tube Bearing & Tail Shaft, 675
Meat, Fish or Fish Juice, for Survival 558
Mechanical Advantage (Purchase), 162
Medical Emergencies Transmission, 469
Mediterranean Moor, 143
Mercator Projection, 386
Merchant Ship S&R Manual, 556
MERSAR, 556
Messenger, 135

Metal Primers, 84
Meteorological Visibility, 416
Meteorology (see also Weather & individual topic entries)
Meteorology, 591
Meteorology, Earth's Atmosphere, 593
Meteorology, East Coast Lows & Bombs, 604
Meteorology, Terminology & Definitions, 598
Methods of Berthing, 143
MHHW & MHWS, 385
Middle Water Operation, 256
Midships, 35
Mile = 1852 Metres, 682
Mile, 382 (measuring on Chart)
Millibar, 596
Mini-M, 489
MINISAT, 489
Mirage Effect, 341
Mist, 601
Mizzen, 44
MMSI, 470
Mob Button on GPS, 290
Mobile Phone not sufficient, 459
Monkey's Fist, 35, 135
Monohull or Multihull, 15
Mooring Areas - No Anchor Lights, 240
Mooring Between Two Buoys, 143
Mooring Buoy, 140
Mooring Lines Sizes, 62
Mooring Lines, Terminology, 35, 134
Mooring with Two Anchors, 125
Morse Code, 96
Motion Sickness, 512, 515
Motor Car Petrol Engine, Concept, 625
Moused, 171
MRCC, 261
MSI by Satellite, 489
MSI HF Broadcasts, 463, 851
MSL, 385
Mud Box, 781
Muffler, Waterlift Type, 697
Multihulls, 15
Muster List, 564
Mute Control, 478
Narrow Channel Effect, 137
National Marine Electronics Assoc, 288
National Marine Guidance Manuals, 252
National Marine Safety Committee, 251
National Maritime Safety Committee Website, 852
National Recreational Boat Plate, 247
National Standard for Boat Licences, 245
National Standard for Commercial Vessels (NSCV), 251
Nature of Seabed - Chart Symbols, 101
Nautical Mile = 1852 Metres, 682
Nautical Mile, 35, 382
Nautical Publications, Corrections, 381
Nautical Terms, 33
Naval Vessels, distance off, 243
NAVAREA X, 250
Navigable Semicircle (Cyclone), 608

Navigation (see also Chartwork, Position Fixing, Plotting & individual topic entries)
Navigation Act, 251
Navigation Lights Alarm, 320
Navigation, 379
Navigation, Bearings Observing & Plotting, 379
Navigational by Radar, 344
Navigational Charts, 380
Navigational Charts, How to Read, 381
Navigational Charts, Reliability, 381
Navigational Rulers, 379
Navigational Satellite Systems, 287
NCS, 487
Net Registered Tonnes, 276
Network Coordination Stations, 487
Neutral Safety Switch in Propulsion Control Heads, 712
New Danger Marks (Buoys), 107
Night Orders, 181
Nipped Cable, 35, 114
Nippon Kaiji Kyokai, 277
NMEA Cable, 288
NMSC Website, 852
NMSC, 251
No Cure No Pay Salvage, 181
No Go Zone, 131
No Higher (Sailing), 132
Noise Soundproofing, 705
Nominal Ranges, 416
Non-GMDSS Inmarsat Systems, 489
Non-SOLAS Vessel Position Report, 550
Non-Standard Refraction (Radar), 341
Norfolk Island, travel to, 249
Northern Territory Marine Dept Website, 852
North-up Display (Radar), 330
Nothing Off (Sailing), 132
Notices to Mariners, Annual, 250
Notices to Mariners, Weekly, 381
Novel (Nov) Vessel Classification, 255
NRT, 276
NSCV, 251
NSW Waterways Authority Website, 852
Occluded Front, 601, 606
Ocean Currents, Seasonal Maps, 441
Ocean Passages of the World, 418
Ocean Rip, 242
Octahedral Cluster, 346
Off-Course Alarm, 320
Official Log Book, 257
Offshore Operations, 255
Offshore Waters, 256
Occupational Health & Safety, 504
OH & S - Court Case, 579
OH & S - First Aid, 504
OH & S - Responsibility, 579
OH & S - Work Practices, 579
OH & S - Yacht Race, 579
OH &S - Need Extra Care onboard, 579
OH&S - Accommodation Safety, 582
OH&S - at Sea, 579
OH&S - Closed & Void Spaces, 582
OH&S - Confined Spaces, 582
OH&S - Crew Member Must, 580

OH&S - Do's & Don'ts, 581
OH&S - Documenting Hazards, 583
OH&S - Duty of Care, 579
OH&S - Employee Must, 580
OH&S - Enclosed Spaces, 582
OH&S - Galley Safety, 582
OH&S – ISM Code, 584
OH&S - Legislation, 579
OH&S - Operators Responsibility, 579
OH&S - Owner's Duty, 579
OH&S - Risk Management Plan, 580
OH&S - Risks & Safety, 580
OH&S - Risks on Board, 581
OH&S - Safety Checklist, 582
OH&S - Standard for First Aid, 504
OH&S - Stress, 580
OH&S - Training, Rescue, First Aid, 583
OH&S - Vessel Owner's Duty, 579
OH&S - Void Spaces, 582
Oil & Coolant Drainage System, 263
Oil Change (Outboards), 743
Oil Changes, 723
Oil Cooler, 631
Oil Filter, 631, 638
Oil Lubricated Stern Tubes, 675
Oil Pressure Gauge, 710
Oil Pressure, 645
Oil Pump (Engines), 638
Oil Spill in Harbour, 262
OLB (Official Log Book), 257
On Board Communication, 584
On Board Training, 28
One-Pole Battery Connection, 806
Open Beaches Regulations, 242
Open Chock, 38
Open Circuit Seawater Cooling, 647
Open Circuit/Closed Circuit Engine Cooling System, 647
Operating Range & Fuel Calculations, 721
Operational Sea Areas, 255
Orange Diamond, Sydney Harbour, 212
Orange Smoke Signals, 537
Osmosis, 79, 560, 799
Others in Distress, 563
Outages (GPS), 294
Outboard, Controls, 730
Outboard, Cooling System, 743
Outboard, Engines, 729
Outboard, Fuel System, 736
Outboard, Fuel Tank Service, 758
Outboard, Internal Lubrication, 741
Outboard, Maintenance & Inspection Schedule, 758
Outboard, Manually Starting an Electric-Start, 754
Outboard, Mounted Height, 747
Outboard, Starting & Stopping, 752
Outboard, VRO Pump, 742
Outboards, (see also Engines, Fuel & individual topic entries)
Outboards, 2-Stroke & 4-Stroke, 730
Outboards, Adjusting Idle Speed, 751
Outboards, Adjusting Shift Control Cable, 749

Outboards, Adjusting Start-in-Gear Protection Device, 751
Outboards, Adjusting Throttle Axle Link, 750
Outboards, Adjusting Throttle Control Cable, 750
Outboards, After Planes, 745
Outboards, Anti Splash Plate, 730
Outboards, Anti-Cavitation Plate, 730
Outboards, Anti-Cavitation Plate, 746
Outboards, Anti-Ventilation Plate, 746
Outboards, Carburettor Leaf Valves, 756
Outboards, Carburettor Problems, 756
Outboards, Carburettor, 625, 634
Outboards, Cavitation Plate, 730
Outboards, Cavitation Plate, 746
Outboards, Compacts, 736
Outboards, Cooling System Cleaning, 744
Outboards, Drive Angle, 744
Outboards, Electric, 735
Outboards, Electronic Fuel Injection, 732
Outboards, Exhaust Gas Re-Burning, 730
Outboards, External Lubrication, 742
Outboards, Fins, 747
Outboards, Fuel Filter Cleaning, 740
Outboards, Fuel Maintenance, 740
Outboards, Fuel System Inspection, 740
Outboards, Fuel Tank Cleaning, 740
Outboards, Fuel Tank Installation, 738
Outboards, Gear Shift, 729
Outboards, Gearbox Oil Change, 743
Outboards, Hydrofoil Stabilizers, 747
Outboards, Kill Switch, 748
Outboards, Leg Length, 729
Outboards, LPG-Petrol (Duel-Fuel), 779
Outboards, Performance Features, 744
Outboards, Petrol-Oil Ratio, 742
Outboards, Power Generators, 729
Outboards, Power Trim Switch, 745
Outboards, Propeller Replacement, 757
Outboards, Remote Propulsion Control, 748
Outboards, Safety Guide, 759
Outboards, Single & Twin Cylinder, 729
Outboards, Spark Plug Problems, 756
Outboards, Splash Plate, 730
Outboards, Stabilizers, 747
Outboards, Things to Consider, 729
Outboards, Tilt/Trim Mechanism, 745
Outboards, Tools & Spares to Carry, 752
Outboards, Transporting, 740
Outboards, Trim Angle, 744
Outboards, Trim Planes, 745
Outboards, Trim Tab Adjustment, 730
Outboards, Trim Tab, 745
Outboards, Trim-Adjusting Pin, 745
Outboards, Trimming by Weight Distribution, 747
Outboards, Troubleshooting, 753
Outboards, Two &. Four-Stroke, 729
Outboards, Ventilation or Blowout, 746
Outdrive Engines, 676
Overboard (Man Overboard), 553
Overfalls, 35
Overhand Knot, 64

Overseas Voyages, 248
Overspeed/Underspeed Sensor, 689
Owner's Duty, 579
Ownership Change, 247
Paddle Wheel Effect, 138
Pads, 551
Painter, 35
Paints & Painting, 81
Paintwork, Testing, 82
Panic Bag, 539
PanPan or Mayday?, 467
PanPan Transmission, 466
Panting, 35, 47
Parachute Flares, 538
Paraflying, Distances Off, 243
Parallax Error (Radar), 342
Parallel & Series Propulsion Control, 713
Parallel Indexing (Radar), 344
Parallel Ruler, 379
Parallel Track Search, 556
Paravanes, 284
Partially Smooth Water Operations, 256
Parts & Components of Engines, 632
Parts of Boat & Hull Terminology, 40
Parts of Boat, Rigs, Rigging, 33
Passage Planning, 418
Passenger Vessel, Definition, 255
Passenger Vessels, Minimum Length for Seagoing, 256
Patch Microstrip (GPS), 287
Patch Scanners (Radar), 326
Patent Slip, 86
Pawl, a Pivoted Lever, 169
Paying out, 118
Pelorus, 366
Pennants, 97
Period of Encounter, 148
Period of Pitch, 147
Period of Roll, 147
Periodic Inspection, Extinguishers, 522
Permit Label, 254
Permit or Certificate of Survey, 254
Person Collapsed (First Aid), 504
Person in Shock (First Aid), 505
Person Overboard Button on GPS, 290
Person Overboard, 555
Personal Epirb, 485
Personal Floatation Devices, 540
Personal Information - Privacy Act, 586
Personal Locator Beacons (PLBs), 484
Personal Water Craft Regulations, 243, 259
Person's Weight, 255
Petrol & LPG Engine - Troubleshooting, 753
Petrol Density or Specific Gravity, 777
Petrol Engine, 627
Petrol to Start a Diesel Engine, 718
PFD Buoyancy Rating, 542
PFD, Recreational Requirement, 537
PFDs & Lifejackets, 540
PFDs, 540
PFDs, Buying Them, 541
PFDs, Recommendation, 566
Phonetic Alphabet & Figure Codes, 480
Picking up Anchor, 122

Pilot Books, 418
Pilot Flag, 212
Pilot on Board Flag, 242
Pilotage Exemption Certificate, 244
Pilotage Ports, 244
Pinching, 133
Pipelines - Chart Symbol, 103
Pipelines, Colour Coding, 801
Pipes (Fuel), 663
Piston (Engine), 625, 638
Piston Rings, 638
Pitch of propeller, 682
Pitch Period, 147
Pitching, 147
Pitchpoling, 149
Pivoting Point of Vessel, 143
Plan Position Indicator (Radar), 326
Planing Hull, 38
Planing Hulls Heel at High Speeds, 144
Planing Hulls, Propeller Selection, 684
Plastic (Polyethylene), 12
PLBs, 484
Plimsoll Mark, 277
Plotter (GPS), 290
Plotting Instruments & Aids, 379
Plotting Position & Course on Chart, 390
Plotting Position by Radar, 391
Plotting Positions on Charts, 396
Plumbing - Toilets & Drains, 800
Plumbing - Water, 798
Plywood, 12
Pneumatic Propulsion Control, 715
Pneumatic Starting System, 709
Point Loading, 276
Points (Contact Breakers), 638
Points of Compass, 37
Pollution from Anti-Fouling, 263
Pollution Prevention from Boats, 262
Pollution Regulations, 259
Polyethylene, 12
Polymer, 81
Polyurethane, 81
Poop Deck, 35
Pooping, 148
Port Closed Signal, 212
Port Control, VHF Channels, 460
Port Side, 35, 40
Port Tack, 131
Port Ten, Command, 35
Portable Fire Extinguishers, 521
Portuguese Man-of-War (First Aid), 511
Position by Radar, 391
Position Circle (Symbol), 393
Position Circles, 391
Position Fixing by Radar, 338
Position Fixing, 398
Position Line (Symbol), 393
Position Lines, 391
Position Plotting on Chart, 382
Position Reports (Radio), 464, 550, 852
Positions on Charts (Plotting), 396
Potable Water, 798
Pounding, 147
Pour Point (Fuel), 669
Power Boat Licence, 245
Power Skis, 243

Power Supply (See also Electrical, Solar Power, Batteries & Generators), 806
Power Supply, 814
Power Take-Off Systems, 796
Precision Aerial Dropping System, 551
Precision Docking of Ships, 298
Pre-Combustion Chamber, 628, 710
Preformed Ropes, 60
Prelubrication (Engines), 645
Preparation for Sea, 135
Pre-Sea Safety Qualification, 21
Pressure & Wind, 597
Pressure (Meteorology), 593, 599
Pressure Belts (Meteorology), 593
Pressure fall due to Height, 596
Pressure Testing Extinguishers, 522
Pressurised Water Supply System, 798
Pre-Start Engine Checks, 710
Preventing Collisions at Sea, 193
Preventing Waves Entering Exhaust, 697
Prevention of Pollution Act, 260
Primary Coil, 638
Primer (Paint), 81
Priority-over-Sail Signal, 212
Privacy & Personal Information Protection Act, 586
Profile (or Sheer Plan), 42
Profile View, 43
Prohibited Anchorages - Chart Symbol, 103
Prop Walk, 138 (Boat Handling)
Propeller Damage, 687
Propeller Fitting Steps, 688
Propeller Fouled at Sea, 185
Propeller May Have Become Fouled, 185
Propeller Measurements, 682
Propeller Replacement, 688
Propeller Selection, 683
Propeller Shaft - Size, Strength & Angle, 675
Propeller Shaft (Drive Train) Alignment with Gearbox Shaft, 679
Propeller Shaft Droop or Deflection, 679
Propeller Shaft Installation, 681
Propeller Shaft Misalignment, 679
Propeller Shaft, 674
Propeller Shaft, Wear-Down Measurement, 675
Propeller Slip, 683
Propeller Vibrations, 687
Propeller, Loss of, 185
Propellers on Outboard Motors, 687
Propellers, Corrosion Prevention, 687
Propellers, Types & Description, 674, 681
Propellers, Zinc Anodes, 687
Propulsion (Types), 673
Propulsion Control (Outboards), 748
Propulsion Control Systems, 712
Protecting Sea from Pollution Act, 260
Protein Dehydrates the Body, 558
Public Jetty, 242
Pulley Systems on Yachts, 165
Pulley Systems, 163
Pulpit, 35, 41
Pump - Diaphragm Type, 781

Pump - Plunger-Type Manual, 781
Pump Drive Systems, 689
Pump not Drawing Water, 787
Pump, Raw Water, Impeller, 654
Pump, Rotary & Semi-Rotary, 781
Pumps - Centrifugal, 790
Pumps - Diaphragm Troubleshooting, 793
Pumps - Diaphragm Type, 791
Pumps - Engine Driven, 796
Pumps - Flexible Impeller Replacement, 790
Pumps - Flexible Impeller Types, 783
Pumps - Flexible Impeller, 787
Pumps - Impeller Usage, 788
Pumps - Lobe Type, 794
Pumps - Positive Displacement, 787
Pumps - Self-Priming, 787
Pumps - Sliding Vane, 791
Pumps - Submersible Centrifugal, 782
Pumps - Vacuum Switch, 788
Pumps, 787
Punts, Lights, 212
Purchases & Tackles, 162
Pushpit, 35, 41
Push-Pull Cable Propulsion Control, 714
PWCs, 243
Pyrotechnic Distress Signals, 99, 537
Quadrant (a Steering Component), 828
Quadrantal Correctors (Compass), 361
Quadrifilar (GPS), 287
Quarantine Service Website, 852
Quarantine, Customs, Immigration, 249
Quarter, 35
Queensland Registration of Commercial/Fishing Vessels, 255
Queensland Transport Department, 852
Quickstop Method, 555
Racking, 35, 47, 147
Racon (Radar Reflector), 347
Radar Anomalous Propagation, 341
Radar Automatic Plotting (ARPA), 345
Radar Bearing Measurement, 337
Radar Blind & Shadow Sectors, 338
Radar Buying One, 346
Radar Dead Angles, 338
Radar Dead Range, 344
Radar Distances, plotting on Chart, 391
Radar Errors, 342
Radar False Echoes, 339
Radar Interference, 341
Radar Navigational & Collision Avoidance, 344
Radar Parallax Error, 342
Radar Parallel Indexing, 344
Radar Performance Monitor, 329
Radar Plotting Position on Chart, 391
Radar Plotting, 345
Radar PPI, 326
Radar Pulse Length & PRF, 336
Radar Radio Waves Bend, 341
Radar Range Calculation, 343
Radar Reflective Distress Signals, 348
Radar Reflectors, 346
Radar T/R Switch, 325

Radar Transponders (Anti-Collision), 348
Radar, 324
Radiation Detectors (Fire), 524
Radiator (Car Engine), 625
Radio (See also Distress & individual topic entries)
Radio Antenna & Earth, 476
Radio Antenna Tuning Unit (ATU), 477
Radio Callsign, 462
Radio Check, 464, 852
Radio Communication Rules & Procedures, 465
Radio Controls, 477
Radio Distress Acknowledging, 468
Radio Doctor, 469
Radio Documents Required, 466
Radio Duplex & Simplex, 464, 478
Radio Duplex & Simplex, 464, 478
Radio Electrical Interference, 465
Radio Equipment Licence, 462
Radio Equipment, 476
Radio Equipment Small Vessel, 465, 550
Radio Frequencies, Re-Designation, 461
Radio GMDSS, 470
Radio DSC, 470
Radio Interference, Electrical, 465
Radio Interference, Human, 468
Radio Log Book, 466
Radio Master's Authority & Duty, 467
Radio Mayday or PanPan?, 467
Radio Medical, 469
Radio MF/HF, HF, 460
Radio Operator's Certificate, 461
Radio Position Reports, 464, 550, 852
Radio Propagation, 462
Radio Repeater Towers, 464
Radio Report Boat Positions, 464, 550, 852
Radio Required, non-SOLAS, 465, 550
Radio Satellite, 459
Radio Sea Areas, 459
Radio Signals, List of, 248
Radio Silence Periods, 465
Radio Spares Kit, 476
Radio Trouble Shooting, 478
Radio VHF International Setting, 464
Radio VHF Repeaters, 464
Radio VHF Service - Summary, 464
Radio Watchkeeping System, 465
Radio Weather Service, 851
Radio Wind Warnings, 602
Radiotelephony Alarm Signal, 467
Radphone, 465
RAIM (GPS), 294
Rain Caps on Dry Exhaust, 699
Rain Water (Survival), 560
Rake & Cupping (Propellers), 683
Rake, 35, 40
Ram Arrangements (Steering), 831
Ramark (Radar Reflector), 347
Range (Distance) & Bearing Measurement by Radar, 337
Range of Tide, 429
Range Table, 414
Raster Navigation Charts, 291

Raster Scan Display (RSD), 328
Raw Water Circuit, 654
Raw Water Pump Impeller, 654
RCC Australia, 550
Real Time Kinematics (GPS), 298
Receiver Autonomous Integrity Monitoring (GPS), 294
Recognition of Australian Defence Force Marine Qualifications, 22
Recognition of Visiting Vessel's Registration, 247
Recognition of Visitors' Speed Boat, RSM (Qld), & PWC Licences, 245
Recommended Shipping Routes - Chart Symbol, 102
Record Books (Log Books) - Types, 257
Recovering Survivors from Water, 560
Recreation Boat Training Schemes, 20
Recreational Boat Operator Competencies, 253
Rectifier (AC into DC), 633
Red Alert (Cyclone), 611
Red Hand-Held Flares, 537
Red Relative Bearings, 358
Re-Designation of HF Frequencies, 461
Reduction Gears, 678
Reef & Bar Openings, 146
Reef Knot, 64
Reef Points (Sails), 35
Reefing (Sails), 132
REEFREP, 550, 852
Reeve of a Sling, 173
Refrigerant Leakage Into Engine's Intake System, 703
Refrigerant Plant - Precautions, 527
Refrigeration Space Alarm, 320
Refrigeration, 800
Refuelling a Vessel, Precautions, 262
Register for Compliant Equipment, 252
Registered Tonnes, 276
Registration Number, 254
Registration of Commercial/Fishing Vessels, 255
Registration of Vessels, 246
Regulations, Collision Prevention, 193
Relative Bearings, 358, 366
Relative Density, 274, 669
Relative Humidity, 600
Relative Motion Unstabilised Display (Radar), 330
Reliability of Charts, 381
Rendezvous Another Vessel (CTS), 413
Repeater (VHF) Towers, 464
Reporting Casualties & Incidents, 181
Reporting Boat Positions, 464, 550, 852
Rescue Coordination Centre, 550
Rescue Method by Helicopter, 562
Rescuing Survivors from Water, 560
Research Vessel, Classification, 255
Restricted Offshore Operations, 255
Restrictions, Limitations & Prohibitions for boating, 242
Reversal Problem (Engine), 716
Reverse Osmosis (desalination), 560, 799
Reversing Gear & gearboxes, 674, 678
Rhumb (or Rhomb) Line, 386

Ridge (Meteorology), 601
Riding out Gale at Anchor, 124
Rigged to Advantage (Purchase), 163
Rigged to Disadvantage, 163
Rigging & Parts, 41
Rigging a Stage, 70
Righting Moment, 279
Rigid Liferaft, 544
Rigid Mounted Shaft Seal, 681
Rip Current, 242
Rip Tide, 242
Risk Management Plan, 580
Risks & Safety on Board, 580, 581
Rivers & Channels - Tidal Streams, 142
RNC, 291
Roach (of a Sail), 44
Rode, 120
Rogue Wave, 598
Roll Damping by Sails, 284
Roll Period, 147, 283
Roller Chock, 38
Rolling & Pitching affected by Load Distribution, 149
Rolling Reduction, 284
Rolling Rule, 379
Rolling, 147
Rooster Tail, 149
Rope Construction, 55
Rope Size & Sheave Groove Matching, 168
Rope Sizes - Guide, 61
Rope Sizes, 56
Rope Stopper, 69
Rope Uncoiling, 57
Rope under Tension, 69
Ropes Life, 62
Ropes Safe Working Load, 161
Ropes to Condemn/Retire, 63
Ropes with Knots & Splices, Reduction in Strength, 57
Ropes, 55
Rot (Boat Maintenance), 92
Round Turn & Two Half Hitches, 66
Round Turn, 66
Routes - Chart Symbol, 102
Routing Charts, 418
Royal Flying Doctor Service, 469
RTK, 298
Rudder or Steering Orders, 36
Rudder, Losing it, 185
Rudders, Types, 825
Rule of Twelfths (Tides), 430
Runner, 41
Running Engine on Light Load, 710
Running Fix (Chartwork), 401
Running Moor (Anchoring), 125
Running or Sailing Downwind, 131
Running, 131
S&R Radar Transponders, 347
SA Transport Department Website, 852
Sacrificial Anodes, 77, 639
Saddle Grips, 60
Safe Water Marks (Buoys), 106
Safe Working Load, Ropes/Chains, 161
Safety Checklist, 582
Safety Devices Fitted on Engines, 712

Safety Drills, Exempt Vessels, 564
Safety Drills, Record Keeping, 564
Safety Equipment, 535
Safety for High-Speed Craft, 255
Safety Guide (Outboards), 759
Safety Harnesses, 543
Safety Hazards, 708
Safety in Accommodation Spaces, 582
Safety in Engine Room, 708
Safety of Life at Sea, 258
Safety Management System (ISM), 584
Safety Recommendations - Sydney to Hobart Yacht Races, 566
Safety Symbols, 585
Safety Transmission (Radio), 466
SafetyNET, 490
Sag, 35
Sagging, 146
Sail Area, 142
Sail Training Certificates, 21
Sail with High Aspect Ratio, 51
Sail, Failure, 186
Sailboarding, 243
Saildrive, 676
Sailing Boats & Layout, 14, 41
Sailing Directions, 418
Sailing Dog's-Leg Course, 131
Sailing Downwind, 131
Sailing Rigs - Types, 45
Sailing Theory, 130
Sailing Vessel Stability, 279
Sailing Vessels, Propeller Selection, 685
Sails – Taking Care, 90
Sails - Types & Parts, 44
Salvage Claim, 181
Samson Post, 35, 47
SARSAT, 481
SARTS, 347
Satcom, 486
Satellite Communication Certificate, 461
Satellite Communications Endorsement Certificate, 486
Satellite Systems (Navigation), 287
Satellite-Derived Positions, 295
Scantlings, 35
Scend, 145
Schooner, 45, 53
Scope to Use (Anchoring), 121
Scope, 35
Scupper, 35, 48
Sea Anchor, 35, 115
Sea Area Classification, SOLAS, 459
Sea Breezes, 614
Sea Cock, 35
Sea Dumping Act, 262
Sea Levels Reference Chart, 385
Sea Mile = 1852 Metres, 682
Sea Safety - Small Craft Particulars Form, 551
Sea Scum, 261
Sea Sickness Tablets, 557
Sea Survival & Rescue, 535
Sea Water or Urine Drinking, 558
Seabed - Chart Symbols, 101
Seacock & Ball Valve, 794
Seaphone, 465

Search & Rescue Centre Website, 852
Search & Rescue Manual, 556
Search & Rescue System, 550
Search Aircraft Signals, 561
Seasickness, 512, 515
Seaward Limits - Definitions, 256
Seawater Suction System, 653
Seaworthy Boat - the Law, 257
Secondary Coil, 638
Secondary Ports, 432
Sector Lights, 394
Sector Search Pattern for MOB, 556
Securing a Rope, 66
Securing the Cable on Board, 121
SÉCURITÉ Transmission, 466
Sediment Type Water Separator, 666
Seelonce Distress, 468
Seelonce Mayday, 468
Self-Locking Worm Reduction Gearing, (Anchor Windlass), 118
Self-Steering, 317
Self-Tailing of the Rope Rode, 119
Self-Tailing Winches, 169
Selling Registrable or In-Survey Vessel, 247
Senhouse Slip, 119
Series & Parallel Propulsion Control, 713
SESs Stage Red, 612
SESs, 487
Set & Drift (Chart Plotting Symbol), 393
Set & Drift, Finding in Chartwork, 400
Set, 35
Set, Rate & Drift, Plotting, 397
Severe Weather Warning, 607
Sewage Management Plan, 259
Sewage Tanks, 800
Shackle, Cant, 171
Shackles - Types, 171
Shackles or Shots of Cable, 117
Shadow Sectors (Radar), 338
Shaft Alignment, 679
Shaft Brake, 716
Shaft Coupling, Flexible, 681
Shaft Droop or Deflection, 679
Shaft Installation, 681
Shaft Misalignment, 679
Shaft Sea, Rigid Mounted, 681
Shaft, Wear-Down Measurement, 675
Shallow Water Effect, 136
Shallow Waters Soundings, 313
Sharks (Liferaft Survival), 557
Shear Pin, 687
Sheave & Rope Size, 168
Sheave Diameter, 168
Sheave's Groove, 168
Sheepshank, 65
Sheer Clamp, 42
Sheer Plan, 42
Sheer Strake, 48
Sheer, 35, 40, 48
Sheet Bend, 65
Sheets, 41, 131
Shelf, 48
Shellfish, Dangerous Type, 561
Sheltered Waters, 256
Ship Earth Stations, 487

Ship Reporting & S&R System, 550
Ship Reporting System (REEFREP), 550, 852
Ship Station Licence, 462
Ship's Register, 248
Shipboard Waste Management Plan, 260
Shipping Routes - Chart Symbol, 102
Shock (First Aid for Shock), 505
Shore Power Supply, 814
Shoring Bulkheads, Damaged Boat, 184
Short Circuit, 807
Short Splice, 67
Short Stay (Anchoring), 35, 114
Shots of Cable, 117
Showers, Definition, 599
Shroud, 41
Shut Off Switch (Engine), 633
Significant Wave Height (NSCV), 598
Significant Wave Height, 256
Significant Weather, 598
Silencer (Engines), 632
Silencer (Muffler), 699
Simplex & Duplex, Radio, 464, 478
Sine Wave Inverters, 816
Single Side Band (SSB), 461
Single Whip (Lead Block), 163
Skeg Bearing, 674
Skeg, 48
Skewed Propeller, 683
Skin Cooler, 647
Skipper Qualification, 21
Skipping (Radio Signal), 478
Slack Water, 35, 430
Slings Lifting Capacity Calculations, 174
Slings, 172
Slip – Propeller Slip, 683
Slip Clutch, 687
Slip Ropes, 141
Slip Stream Torque, 716
Slipping, 86
Slipstream Method of Preventing Exhaust Entering Vessel, 701
Slobber, 700
Sloop, 45, 53
Small Craft Particulars Form, 551
Small Craft Position Reporting, 550
SMG, 393
Smoke Inhalation, 506
Smoke Signals, 537
Smoke, Heat & Radiation Detectors, 524
SMS (ISM Code), 584
Snatch Block, 165
Snub Cable (Anchoring), 36, 114
Snubbing Round, 144
Soft Chine, 38
Soft Iron Correctors (Compass), 361
SOI, 606
Solar - Batteries, 816
Solar - Systems, 811
Solar Cells - Why Inefficient, 815
Solar Power - Installation, 817
Solar Power - Inverters, 816
Solar Power - Maintenance, 819
Solar Power - Panel Requirement, 815
Solar Power – PV Module Rating, 817
Solar Power - PV Module, 815

Solar Power - What to Look for, 819
Solar Power, 815
Solar Still, 560
SOLAS AUSREP Equipment, 550
SOLAS Communication Equipment, 550
SOLAS –ISM Code, 584
SOLAS Lifejacket, 542
SOLAS Radio Requirements, 465, 550
SOLAS, Sea Areas Classification, 459
SOLAS, 258
Soldier's Wind, 131
Solenoid, 631, 639, 778
Sonar, 304
Sounder, Buying One, 314
Sounding, 36
Soundings (Chartwork), 385
Soundings, Line of Soundings, 410
Soundproofing Machinery Noise, 705
South Australia Marine Department Website, 852
Southern Oscillation Index, 606
SP Vessel, 255
Spares - to Carry on Board, 717
Spark Plug, 625, 639, 756
Special Marks (Buoys), 106
Special Purpose (SP) Vessel, 255
Special Service Notation, 256
Specific Gravity, 669
Specific Gravity, LPG & Petrol, 777
Speed Boat Licence, 245
Speed Boat Licence, Inter-State Recognition, 245
Speed Calculation from RPM, 315
Speed Limits, Regulations, 242
Speed Logs, 315
Speed Made Good, 393
Speed over the Ground (Chartwork), 393
Speed Trip (Overspeed/Underspeed Sensor on Thrusters), 689
Speed, 34
Speed, Time, Distance, Calculations, 388
Spine-Bearing (First Aid), 511
Spinnaker, 44, 53
Splices, 64
Splicing a Braided Rope, 68
Spoking, 341
Sponson, 36
Spontaneous Combustion Point, 669
Spontaneous Combustion, 525
Spring Tide, 36, 429
Spring, 134
Spurling Pipe, 36, 120
Squall Line, 599
Squall, 598
Square Engines, 711
Squelch (Mute) Control, 478
SRS (Ship Reporting System), 550
SSB (Single Side Band Radio), 461
Stabilisers, 284
Stability, 278
Stability, Lack of, Danger Signs, 283
Stabilized Display (Radar), 330
Stage, 70
Staghorn, 38, 66
Staining Timber, 84
Stainless Steel Fittings, 80

Stall & Reversal Problem, 716
Stand on, 36
Standard Meteorological Visibility, 416
Standards of Training & Watchkeeping Convention, 251
Standards of Training & Watchkeeping, 256
Standing Moor, 125
Standing Orders, 181
Standing Rigging & Parts, 41
Starboard Side, 36, 40
Starboard Tack, 131
Starboard Ten (Helm Order), 36
State Emergency Services, 851
State Marine Authorities, Websites, 852
State Yachting Assoc, Websites, 852
State/Territory Emergency Services, 851
Static Electrical Charge (Danger when Refuelling), 263
Station Identification (Radio), 462
Stay, 36
Staysail, 44
STCW Certification Standard, 251, 256
STCW-95 Endorsement, 21
Steady, 36
Steaming Light, 36
Steel Hulls, 13, 80
Steer 010 (Helm Order), 36
Steer, 36
Steering Gear Maintenance, 831
Steering Gear Requirements for Commercial Vessels, 826
Steering Orders, 36
Steering Ram Arrangements, 831
Steering Rod Parted, 184
Steering System Troubleshooting, 831
Steering Systems, 826
Steering, Loss of, 185
Stem, 36, 48
Stern & Bow, 14
Stern Gland, 674
Stern Rope, 134
Stern Thrusters, 689
Stern Tube Bearing & Tail Shaft, Wear-Down Measurement, 675
Stern Tube, 674
Stern, 36, 40
Sterndrive, 676
Stiff, 278
Stings (First Aid), 510
Stopper, Stopper Hitch, 36, 69
Stopping Distances, Large vessels, 144
Storm Jib, 44
Storm Sails, 132
Storm Surge, 429, 609
Storm Warning, 602
Stranding, 182
Stray Current Corrosion, 78
Stress (OH&S), 580
Stress Line (Towing Telltale), 150
Stresses on Hull in Heavy Weather, 146
Stresses While on a Slip, 88
Stringer, 36, 42, 48
Strobes, Recommendation, 566
Strong Wind Warning, 602
Strop - Load Slung from It., 173

Strum Box, 781
Stuffing Box, Rigid Mounted, 681
Sub Refraction (Radar), 341
Submarine Cable & Pipelines, 123
Submarine Cables - Anchoring Prohibited Signs, 242
Submarine Cables - Chart Symbol, 103
Submarine Indicator Buoy, 563
Submarine Navigation Lights, 563
Submarine Pipelines - Chart Symbol, 103
Submarine Signals, 563
Summer Zone, Loadlines, 277
Sump Level, 645
Sump Must Breathe, 634
Sump Pump, 631, 639
Sump, 625, 639
Sump, Dry Sump, 645
Sump, Internal or Wet, 645
Super High Frequency (satellites), 487
Super Refraction (radar), 341
Supercharged Engine, 640
Surf Zone Regulations, 243
Surface Currents, Seasonal Maps, 441
Surface Drive, 676
Surface Winds, 594
Surge Cable, 36, 114
Survey Exemption Certificate, 254
Survey Number, 254
Survey Vessels Classification, 255
Survey, Definition, 254
Survival & Rescue, 535
Survival Craft Drills, 564
Survival Craft Radar Transponders, 347
Survivors - Transferring from a Stricken Vessel, 186
Survivors, Recovering from Water, 560
Swell, 598
Swimming Zone Regulations, 243
Switches (Contactors) for Engines, 712
Sydney & Broome, Tidal Streams, 438
Sydney to Hobart Yacht Race Safety, 566
Symbols & Abbreviations, Charts, 100
Synchronization, 148
Synoptic Chart, 598
Tachometer, 711
Tack (of a Sail), 44
Tack, Tacking 36, 131
Tackle on Tackle, 164
Tackles, 162
Tactical Diameter (Turning Circle), 144
Tail Shaft, 674
Tail Shaft, Wear-Down Measurement, 675
Take in Anchor, 122
Taken-Aback, 133
Taking over a Watch, 181
Tanks - Collection & Holding, 800
Tanks Sizes (Fuel Capacity), 666
Tasmania Navigation & Survey Authority Website, 852
TDC & BDC, 629
Telephone Marine Weather Service, 851
Telltale in Tow Rope, 150
Temperature Inversion, 341

Temporarily Holding a Rope under Tension, 69
Temporary Exemptions from Registration Requirements for Recreational Vessels, 247
Temporary Visitor Use of Private Moorings, 248
Tender (Stability), 278
Tensile Strengths of Chains, 161
Teredo Worm, 93
Terminology (Fuel), 668
Terminology (Weather), 598
Tetrahedron, 520
Thermostat, 631, 639, 649
Three Point Turn, 139
Three-Bearings Method, find CMG, 411
Throat, 53
Throttle Boost System (Engines), 716
Thruster Units, 689
Thrusters Pump Drive Systems, 689
Thunderstorms, 602
Tidal Current, 429
Tidal Levels & Chartered Data, 385
Tidal Prediction, 429
Tidal Range Large, Tying a Boat, 142
Tidal Stream, 429
Tidal Streams - Chart Symbols, 101
Tidal Streams in Rivers & Channels, 142
Tidal Streams, in Tide Tables, 437
Tide Arrows, 397
Tide Calculation (by Form AH130), 430
Tide Calculation, Rule of Twelfth, 430
Tide Calculations, LAT Values, 436
Tide Chart, 385
Tide Rip, 242
Tide Rode, 142
Tide Surges, 609
Tide Tables Extract, 445
Tide Tables, 432
Tides - Ebb & Flood, 34, 429
Tides - Mean High Water Springs (MHWS), 385
Tides - Mean Higher High Water (MHHW), 385
Tides - Mean Sea Level (MSL), 385
Tides - Mixed, 432
Tides - Neap, 35, 429
Tides - Semi-Diurnal, 432
Tides - Standard & Secondary Ports, 432
Tides & Currents, 429
Tides (see also individual item entries)
Timber & Plywood, 12
Timber Hitch, 65
Timber Hull Maintenance, 84
Time, Speed, Distance Calculations, 388
Timing Chain, 633
Timing Device (Event Sequence), 716
Timing Gear, 633
Toilet Waste, 259
Toilets, 800
Tools & Spares - to Carry on Board, 717
Tools & Spares for Outboards, 752
Top Dead Centre, 629
Top Sail, 53
Topping Lift, 48
Topsail Schooner, 45

Torchlight Requirement, 537
Tornados, 603
Torque, Slip Stream, 716
Torres Strait, Tide Predictions, 437
Total Compass Error, 370
Tow Ropes, 149
Towing Hitch, 121
Towing Telltale, 150
Towing, 149
TPL (Symbol), 393
TPL, Plotting, 401
Track Plotter (GPS), 290
Traction Winches, 169
Trade Winds, 593, 595
Trading Vessels, 21
Trailer Maintenance, 89
Trailers, 133
Training Schemes, 20
Training Vessels Classification, 255
Training, 28
Training, Sydney to Hobart Recommendation, 566
Transceiver (Radio) Controls, 477
Transducer Installation, 311
Transducer, 304
Transfer (Turning Circle), 144
Transfer Position Circle (Symbol), 393
Transferred Position Circle, Plotting, 401
Transferred Position Line, Plotting, 401
Transferring Survivors from Stricken Vessel, 186
Transformer, 814
Transit (Symbol), 393
Transit Bearing, 372
Transit, 36
Transits - Chart Symbol, 102
Transits & Sector Lights, 394
Transmission Systems (Gearboxes), 678
Transom Stern, 36, 41
Transverse Section, Hard Chine Hull, 42
Transverse Thrust & Boat Handling, 138
Travel Lift, 86
Treading Water (Survival), 553
Triatic Stay, 54
Trichodesmium, 261
Trim, 36
Trimaran, 45
Trimmed, 131
Trollers, Electric, 735
Trolling Valve, 639
Tropical Cyclone Preparedness Info, 851
Tropical Cyclone/ Depression/ Low/ Storm /Revolving Storm, 599, 606
Troubleshooting Engines, 717
Troubleshooting Petrol & LPG Engines, 753
Trough Line, 609
Trough, 601
True Bearings, 389
True Course, 389
True Motion Stabilized Display, 331
Tube-Nest Cooler, 647
Tumblehome, 36
Turbo Timer, 711
Turbocharger Care & Maintenance, 711
Turbocharger, 640

Turn Differently to Motorcars, 136
Turn, 66
Turning Circle, 139, 144
Turning Point of Vessel, 143
Turning Short Around, 139
Twelfths Rule for Tides, 430
Twenty-Seven (27) MHz Radio, 459
Twin-Engine Installation, 701
Twin-Engines (Distance Apart), 695
Twin-Screw Vessel Handling, 139
Twin-Screw Vessels, 686
Two-Cycle & Four-Cycle Engines, 626
Two-Pole Battery Connection, 806
Two-Stroke & Four-Stroke Engines, 626
Typhoons (See Cyclones), 607
UAIS or AIS (Universal Automatic Identification Systems), 298
Unberthing, 141
Uncoiling a Rope, 57
Under Foot (Anchoring), 114
Under Keel Clearance Calculation, 432
Undercoat & Undercoating, 81, 83
Undertow (Tide Rip), 242
Underwater Operation Regulations, 243
Underway, 36
Uniform Shipping Laws Code, 250
Unlimited Domestic Operations, 255
Unstabilised Displays (Radar), 330
Up-&-Down (Anchor Chain), 36, 114
Upstream, 36
Urgency or Distress?, 467
Urgency Transmission, 466
Urine not to drink, 558
Using Engine alongside a Wharf, 137
USL Code, 250
Vacuum Limiter (Engines), 634
Vacuum Switch, 788, 796
Valves & Seacocks - Types, 794
Vapour-Collecting Filter (Engines), 634
Variable Displacement Pump Drives, 689
Variable Pitch Propellers, 686
Variable Ratio Oiling, 741
Variation & Deviation, Combined, 389
Variation, 367
Varnishing, 84
V-Belt Driven Pumps, 796
Vector Charts, 291
Vee Shaped Hull for Planing, 38
Vee-Drive Propulsion, 677
Veer Cable, 36, 114, 124
Veering (Wind), 609
Ventilation (Engine Room), 702
Ventilation or Blow-out (Outboards), 746
Ventilation System (Exhaust Ejected Type), 704
Vertex (Cyclones), 609
Vertical Lift of Man Overboard, 557
Vertical Sextant Angles, Plotting, 406
Vessel (See also Boat)
Vessel Exemptions Register, 252
Vessel Handling, 137
Vessel Length Approx. Tonnage, 256
Vessel Monitoring System (VMS) (Inmarsat-C), 461, 486
Vessel Owner's Duty, 579
Vessel Record Books - Types, 257

Vessel Types, 255
Vessel Working in Chains - Lights, 212
Vessel, Minimum Length of Seagoing Passenger Vessels, 256
Vessels & Aircraft in Distress, 563
VHF Aural Watchkeeping by GMDSS Ships, 464
VHF DSC Epirbs, 484
VHF DSC, 470
VHF Duplex & Simplex, 464, 478
VHF International Setting, 464
VHF Phone Calls, 464
VHF Position Reports, 464, 852
VHF Radio, 460
VHF Repeater Towers, 464
VHF Service - Summary, 464
VHF Simplex & Duplex, 464, 478
Vibration Isolators (Engines), 694
Vibrations After Hitting a Submerged Object, 720
Victoria's Inland Waters, 97
Victorian Marine Board Website, 852
Video Chart Plotter, 291
Viscosity (Oils & Fuel), 669
Viscosity, 645
Visibility (Standard Meteorological), 416
Visibility, Measuring It, 599
Visiting Vessel's Registration, 247
Visitor Use of Private Moorings, 248
VMR, 462, 852
VMS, 486
Void Spaces, 582
Voltage Regulator, 807
Voltmeter, 812
Volunteer Coast Stations, 462
Volunteer Marine Rescue, 462
Vortex (Cyclones), 609
VSA, Plotting, 406
V-Sheet Requirement for Recreational Vessels, 537
W.O.T. (Wide Open Throttle), 742
WA Transport Department Website, 852
Wagon-Back Effect, 701
Walk Back (Rope or Cable), 36, 114
Warm Front, 601, 605
Warp a Boat along a Jetty, 134
Warp, 36
Warping, 117
Warship, 242
Wash from Speeding Boat, 241
Watch Tackle, 163
Watchkeeping Alarm, 320
Watchkeeping in Port, 180
Watchkeeping Log, 417
Watchkeeping When Underway, 180
Watchkeeping, 180
Water - Three Ways of Making, 560
Water Hose Hanging Overside, 783
Water Jacket, 625
Water Lock (Doppler Log), 316
Water Lubricated Stern Tubes, 674
Water Makers, 799
Water Potable, 798
Water Sensing Paste, 720
Water Separators (Fuel), 666
Water Skiing, Distances Off, 243

Water Skiing, Propeller Selection, 684
Water Supply System, 798
Water-Cooled Exhaust, 696
Water-Jacket, 647
Waterjet Engine, 673
Waterlift Muffler, 697
Waterline & Waterline Length, 40
Watertight Bulkheads, 284
Wave Encounter (Boat Handling), 147
Wave Height (NSCV), 598
Wave Height Criteria, 256
Wave Jammers Regulations, 243
Wave Period, 148
Wave Runners Regulation, 243
Waveguide (Radar), 325
Waves (Meteorology), 598
Waypoint Navigation, 289
Wear-Down Measurement of Stern Tube Bearing & Tail Shaft, 675
Weather – by Phone, Fax & Radio, 851
Weather (see also Meteorology & individual topic entries)
Weather Anchor, 144
Weather Broadcast, AMSA Radio Network, 851
Weather Fax, 851
Weather Forecasts & Warnings, 598, 601
Weather Forecasts Radio Broadcast, 462, 851
Weather Helm, 133
Weather Information Sources, 615
Weather Map Reading, 614
Weather Service, 851
Weather Terminology & Definitions, 598
Weather Tide, 36, 114
Weather, 591
Web Frame, 42
Website Addresses, 852
Weigh Anchor, 36, 114, 122
Weight Distribution (Engine Room), 695
Weight of a Person, 255
Welding Safety Precautions, 668
Western Australia Transport Department Website, 852
Wet & Dry Bulb Thermometers, 600
Wet Exhaust, Prevent Waves Entry, 697
Wet-Sump, 645
Wetted Surface Area, 137
WGS84, 295
Whales, Distance to Keep, 243
Wharf with a Large Tidal Range, 142
Whip, 163
Whipping a Rope, 67
White Flag, 242
White Flare, 240
White/Grey Line Function (Sounder), 310
Why Do Vessels Float?, 274
Wildcat, 117
Williamson's Turn, 555
Winch Drum, Wire Rope Spooling, 170
Winch Failure, 186
Winches, 169
Wind Arrows (Meteorology), 610
Wind Backing & Veering, 609
Wind Direction & Speed Forecast, 598

Wind Rode, 142
Wind Speed & Pressure Fall, 597
Wind Speed Forecast, 598
Wind Speed Measurement, 594
Wind Warnings, 602
Windage - Manoeuvring, 142
Windage, 36
Windlass Failure, 186
Windlass Maintenance, 119
Windlass, 36, 117
Windvanes, 317
Windward, 36
Wire Rope Grips, 60
Wire Rope Uncoiling, 57
Wire Ropes to Condemn, 63
Wire Ropes, 59
Wiring - Colour Coding, 807
WLL of Slings, 172
Workcover Websites, 852
Work Practices, 579
Working Jib, 44
Working Load Limit -Slings, 172
Working to Windward, 131
World Geodetic System, 295
Worm Reduction Gearing (Anchor Windlass), 118
Wrecks (Chart Symbol), 100
Y- Turn, 555
Yacht Winch Maintenance, 170
Yachting Associations, Websites, 852
Yachting Federation Website, 852
Yard, 48
Yawing, 147
Yawl, 45, 52
Yellow Alert (Cyclone), 611
Z-Drive Propulsion, 678
Zinc Anodes on Propellers, 687
Zinc Anodes, 77
Zone of Confidence Diagram on Navigational Charts, 381
Z-Propellers, 686

WEATHER - Bureau of Meteorology, via Phone, Fax & Radio

TELEPHONE (Forecasts are broadcast on 1900 numbers, and Warnings on 1300 numbers)

NATIONAL	1900 955 370				
NSW		**VIC**		**WA**	
Sydney Waters	1900 969 955	Coastal Waters	1900 969 966	Marine Service	1900 926 150
NSW Coastal Waters	1900 926 101	Port Phillip & W'nport	1900 920 557	Perth Local Waters	1900 955 350
NSW/ACT Warnings	1300 659 218	Bass Strait	1900 969 930	N. Coastal Waters	1900 969 901
TAS		N. Bass Strait	1900 969 931	W. Coastal Waters	1900 969 902
Boating Weather	1900 969 940	S. Bass Strait	1900 969 932	S. Coastal Waters	1900 969 903
Warnings	1300 659 216	E. Bass Strait	1900 969 933	Warnings	1300 659 223
QLD		W. Bass Strait	1900 969 934	Tropical Cyclone Info	1300 659 210
Coastal Waters	1900 969 923	Warnings	1300 659 217		
SE Qld Boating Weather	1900 926 115	**SA**		**NT**	
Warnings	1300 360 427	Coastal Waters	1900 969 975	NT Service	1900 955 367
Tropical Cyclone Info	1300 659 212	Warnings	1300 659 215	Tropical Cyclone Info	1300 659 211

FAX (NATIONAL): Schedule: 1902 935 046. Directory: 1902 935 200, 1800 630 100 or (03) 9273 8200.

HF RADIO WEATHER BROADCAST via AMSA NETWORK
(See Chapter 18 for AMSA, Coast & Volunteer Radio Networks.)

- ❖ The Bureau of Meteorology weather service callsigns are:
  - VMC Australia Weather East (for services from Charleville)
  - VMW Australia Weather West (for services from Wiluna)

  The stations are programmed to simultaneously broadcast voice and radio fax services 24 hours a day.

- ❖ WEATHER BROADCASTS (VOICE)
  - VMC Broadcast Frequencies are 2201, 4426, 6507, 8176, 12365 & 16546 kHz
  - VMW Broadcast Frequencies are 2056, 4149, 6230, 8113, 12356 & 16528 kHz

  Transmissions are 4-hourly on 4 frequencies simultaneously. Lower frequencies are used at night and higher frequencies by day.

- ❖ WEATHER BROADCASTS (FAX)
  - VMC Broadcast Frequencies are 2628, 5100, 11030, 13920 & 20469 kHz
  - VMW Broadcast Frequencies are 5755, 7535, 10555, 15615 & 18060 kHz

COASTAL WATERS BROADCAST SCHEDULE (VOICE) *in Eastern, Western & Central Standard Times*

WARNINGS: Every hour commencing 000 EST & WST & 0030 CST (via VMC & VMW)

SPECIAL ANNOUNCEMENTS: 5 minutes to every hour EST & WST (via VMC & VMW)

FORECASTS: 4-HOURLY COMMENCING -

0330 EST (Qld via VMC)
0030 EST (Qld Gulf via VMW)
0130 EST (NSW & VIC via VMC)
0230 EST (TAS, via VMC)
0200 & 0300 CST (SA, via VMC & VMW resp.)
0300 CST (NT via VMC & VMW)
0030, 0430, 0830, 1230, 1630, 2030 WST (WA via VMW)

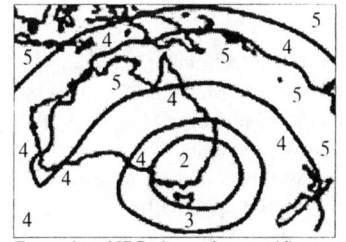

Example of IPS charts for specific dates & times on BOM website
(IPS = Ionospheric Prediction Service)
1 = 2201 kHz 2 = 4426
3 = 6507 4 = 8176kHz 5 = 12365

CYCLONE PREPAREDNESS INFORMATION
(State/ Territory Emergency Services Phone Nos.)

- ➢ Western Australia (Perth)   (08) 9277 5333
- ➢ Northern Territory (Darwin)   (08) 8922 3630
- ➢ Queensland (Brisbane)   (07) 3247 4172

EMERGENCIES, POSITION REPORTS & RADIO CHECK – via radiotelephone

- **EMERGENCIES**: VHF Distress Channel 16 or MF/HF Distress Frequencies 2182, 4125, 6215, 8291 kHz, etc. AusSAR Rescue Coordination Centre Australia Toll Free Phone 1800 641 792 (24 hours)
- **POSITION REPORT & RADIO CHECK**: Call on a VMR/Voluntary Coast Radio VHF Repeater Channel for the area (Channel 21, 22, 80, 81 or 82). HF Frequency 8176 kHz or any other Supplementary Frequency (2201, 4426 kHz etc.) may also be used. Avoid using distress channels & frequencies for this purpose. *Radio Communication is discussed in Chapter 18.*
- **AUSREP & REEFREP**: *See Chapter 19.3 for applicability.* RCC 1800 641 792. REEFREP (07) 4956 3581, VHF 5, 18, 19

USEFUL INTERNET ADDRESSES

NATIONAL BODIES
- Australian Communications Authority — www.aca.gov.au
- Australian Hydrographic Service — www.hydro.navy.gov.au
- Australian Maritime Safety Authority/AusSAR — www.amsa.gov.au/aussar
- Australian Yachting Federation — www.yachting.org.au
- Bureau of Meteorology — www.bom.gov.au/marine
- Customs Service — www.customs.gov.au
- Department of Immigration — www.immi.gov.au
- National Maritime Safety Committee — www.nmsc.gov.au
- Quarantine & Inspection Service — www.affa.gov.au
- Royal Australian Navy — www.navy.gov.au
- Standards Australia — www.standards.com.au
- WorkSafe Australia — www.nohsc.gov.au

STATE	MARINE AUTHORITY	YACHTING ASSOC.	BOATING INDUSTRY ASSOC.
ACT		www.act.yachting.org.au	
NSW	www.waterways.nsw.gov.au	www.nsw.yachting.org.au	www.bia.org.au
NT	www.nt.gov.au	www.nt.yachting.org.au	
QLD	www.transport.qld.gov.au	www.qld.yachting.org.au	www.biaq.com
SA	www.marine.transport.sa.gov.au	www.yachtingsa.org.au	
TAS	www.mast.tas.gov.au	www.tas.yachting.org.au	
VIC	www.marinesafety.vic.gov.au	www.vic.yachting.org.au	www.biavic.com.au
WA	www.dpi.wa.gov.au	www.wa.yachting.org.au	

OTHERS
- Australian Institute of Navigation — www.gmat.unsw.edu.au/aion/aion.html
- International Maritime Organisation — www.imo.org
- International Safety Management Code — www.ismcode.net/content
- Nautical Institute — www.nautinst.org
- NZ Maritime Safety Authority — www.msa.govt.nz
- UK Hydrographic Office — www.ukho.gov.au
- US Coast Guard — www.uscg.mil
- Royal Institute of Navigation — www.rin.org.uk
- Workcover (NSW, ACT, QLD, VIC, WA) — www.workcover.nsw.gov.au (Substitute nsw with act, qld, vic, wa)
- Workcover NT — www.deet.nt.gov.au
- Workcover TAS — www.wsa.tas.gov.au